Cooperative and Graph Signal Processing

Cooperative and Graph Signal Processing

Principles and Applications

Edited by

Petar M. Djurić

Cédric Richard

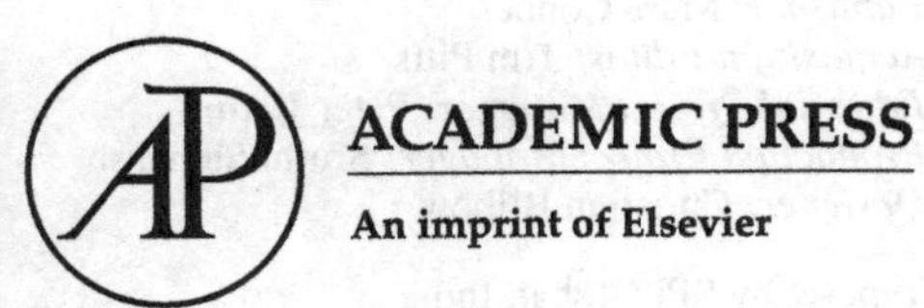

Academic Press is an imprint of Elsevier
125 London Wall, London EC2Y 5AS, United Kingdom
525 B Street, Suite 1650, San Diego, CA 92101, United States
50 Hampshire Street, 5th Floor, Cambridge, MA 02139, United States
The Boulevard, Langford Lane, Kidlington, Oxford OX5 1GB, United Kingdom

Library of Congress Cataloging-in-Publication Data
A catalog record for this book is available from the Library of Congress

British Library Cataloguing-in-Publication Data
A catalogue record for this book is available from the British Library

ISBN 978-0-12-813677-5

For information on all Academic Press publications visit our website at https://www.elsevier.com/books-and-journals

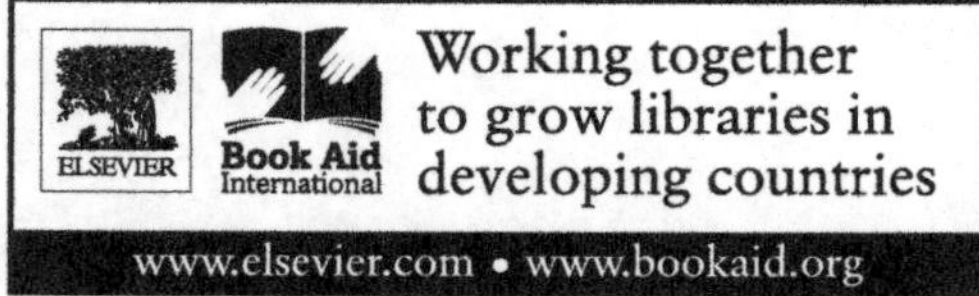

Publisher: Mara Conner
Acquisition Editor: Tim Pitts
Editorial Project Manager: Peter Jardim
Production Project Manager: Sruthi Satheesh
Designer: Christian Bilbow

Typeset by SPi Global, India

Contents

PART 2 SIGNAL PROCESSING ON GRAPHS

Contributors

Selin Aviyente
Department of Electrical and Computer Engineering, Michigan State University, East Lansing, MI, United States

Paolo Banelli
Department of Engineering, University of Perugia, Perugia, Italy

Sergio Barbarossa
Department of Information Engineering, Electronics, and Telecommunications, Sapienza University of Rome, Rome, Italy

Pierre Borgnat
ENS de Lyon, Univ Lyon 1, CNRS, Laboratoire de Physique & IXXI, Univ Lyon, Lyon, France

Elena Ceci
Department of Information Engineering, Electronics, and Telecommunications, Sapienza University of Rome, Rome, Italy

Shih-Fu Chang
Department of Electrical Engineering and Department of Computer Science, Columbia University, New York, NY, United States

Chao Cheng
Dartmouth-Hitchcock Medical Center, Lebanon, NH, United States

Jie Chen
Center of Intelligent Acoustics and Immersive Communications, School of Marine Science and Technology, Northwestern Polytechnical University, Xi'an, China

Sundeep Prabhakar Chepuri
Delft University of Technology, Delft, The Netherlands

Luca Corinzia
CMLA, ENS Cachan, CNRS, University of Paris-Saclay, Cachan, France; ETH Zurich, Zurich, Switzerland

Kamil Dedecius
Department of Adaptive Systems, Institute of Information Theory and Automation, The Czech Academy of Sciences, Prague, Czech Republic

Petar M. Djurić
Department of Electrical and Computer Engineering, Stony Brook University, Stony Brook, NY, United States

Ceyhun Eksin
Department of Industrial and Systems Engineering, Texas A&M University, College Station, TX, United States

Andrey Garnaev
Wireless Information Network Laboratory (WINLAB), Rutgers, The State University of New Jersey, North Brunswick, NJ, United States

David Gesbert
Communication Systems Department, EURECOM, Biot, France

Georgios B. Giannakis
Department of Electrical and Computer Engineering, University of Minnesota, Minneapolis, MN, United States

Paulo Gonçalves
ENS de Lyon, Univ Lyon 1, CNRS, Inria, LIP & IXXI, Univ Lyon, Lyon, France

Alfred Hero
EECS Department, University of Michigan, Ann Arbor, MI, United States

Franz Hlawatsch
Institute of Telecommunications, TU Wien, Vienna, Austria

William Hoiles
Biosymetrics, New York, NY, United States

Hazer Inaltekin
Department of Electrical and Electronic Engineering, The University of Melbourne, Parkville, VIC, Australia

Kwang-Sung Jun
Wisconsin Institute for Discovery, Madison, WI, United States

Bhavya Kailkhura
Center for Applied Scientific Computing, Lawrence Livermore National Laboratory, Livermore, CA, United States

Argyris Kalogeratos
CMLA, ENS Cachan, CNRS, University of Paris-Saclay, Cachan, France

Sithan Kanna
Department of Electrical and Electronic Engineering, Imperial College London, London, United Kingdom

Soummya Kar
Department of Electrical and Computer Engineering, Carnegie Mellon University, Pittsburgh, PA, United States

Paul de Kerret
Communication Systems Department, EURECOM, Biot, France

Vikram Krishnamurthy
School of Electrical and Computer Engineering, Cornell Tech, Cornell University, New York, NY, United States

Amir Leshem
Faculty of Engineering, Bar-Ilan University, Ramat Gan, Israel

Geert Leus
Delft University of Technology, Delft, The Netherlands

Zhenguo Li
Huawei Noah's Ark Lab, Shatin, Hong Kong Special Administrative Region

Sijia Liu
EECS Department, University of Michigan, Ann Arbor, MI, United States

Ying Liu
Wireless Information Network Laboratory (WINLAB), Rutgers, The State University of New Jersey, North Brunswick, NJ, United States

Paolo Di Lorenzo
Department of Information Engineering, Electronics, and Telecommunications, Sapienza University of Rome, Rome, Italy

Danilo P. Mandic
Department of Electrical and Electronic Engineering, Imperial College London, London, United Kingdom

Morteza Mardani
Department of Electrical Engineering, Stanford University, Stanford, CA, United States

Antonio G. Marques
Department of Signal Theory and Communications, King Juan Carlos University, Madrid, Spain

Gonzalo Mateos
Department of Electrical and Computer Engineering, University of Rochester, Rochester, NY, United States

Vincenzo Matta
Department of Information Engineering, Electrical Engineering and Applied Mathematics, DIEM, University of Salerno, Fisciano, Italy

Mattia Merluzzi
Department of Information Engineering, Electronics, and Telecommunications, Sapienza University of Rome, Rome, Italy

José M.F. Moura
Department of Electrical and Computer Engineering, Carnegie Mellon University, Pittsburgh, PA, United States

Eric Moulines
CMAP, Ecole Polytechnique, Palaiseau, France

Robert Nowak
Department of Electrical and Computer Engineering, University of Wisconsin-Madison, Madison, WI, United States

Brandon Oselio
EECS Department, University of Michigan, Ann Arbor, MI, United States

Konstantinos Plataniotis
Electrical and Computer Engineering, University of Toronto, Toronto, ON, Canada

H. Vincent Poor
Department of Electrical Engineering, Princeton University, Princeton, NJ, United States

Michael G. Rabbat
McGill University and Facebook AI Research, Montreal, QC, Canada

Alejandro Ribeiro
Department of Electrical and Systems Engineering, University of Pennsylvania, Philadelphia, PA, United States

Cédric Richard
Laboratoire Lagrange, Université Côte d'Azur, CNRS, OCA, Nice, France

Stefania Sardellitti
Department of Information Engineering, Electronics, and Telecommunications, Sapienza University of Rome, Rome, Italy

Ali H. Sayed
School of Engineering, École Polytechnique Fédérale de Lausanne (EPFL), Lausanne, Switzerland

Anna Scaglione
School of Electrical, Computer and Energy Engineering, Arizona State University, Tempe, AZ, United States

Kevin Scaman
CMLA, ENS Cachan, CNRS, University of Paris-Saclay, Cachan, France; MSR, Inria Joint Center, Palaiseau, France

Mihaela van der Schaar
Electrical Engineering Department, University of California, Los Angeles, Los Angeles, CA, United States

Santiago Segarra
Institute for Data, Systems, and Society; Massachusetts Institute of Technology, Cambridge, MA, United States

Petros Spachos
School of Engineering, University of Guelph, Guelph, ON, Canada

Alex Sprintson
Department of Electrical and Computer Engineering, Texas A&M University, College Station, TX, United States

Brian Swenson
Department of Electrical and Computer Engineering, Carnegie Mellon University, Pittsburgh, PA, United States

Cem Tekin
Electrical and Electronics Engineering Department, Bilkent University, Ankara, Turkey

Shang Kee Ting
DSO National Laboratories, Singapore, Singapore

Wade Trappe
Wireless Information Network Laboratory (WINLAB), Rutgers, The State University of New Jersey, North Brunswick, NJ, United States

Nicolas Tremblay
CNRS, GIPSA-lab, Univ. Grenoble Alpes, Grenoble, France

Pramod K. Varshney
Department of EECS, Syracuse University, Syracuse, NY, United States

Nicolas Vayatis
CMLA, ENS Cachan, CNRS, University of Paris-Saclay, Cachan, France

Aditya Vempaty
IBM Thomas J. Watson Research Center, Yorktown Heights, NY, United States

Marisel Villafañe-Delgado
Department of Electrical and Computer Engineering, Michigan State University, East Lansing, MI, United States

Hoi-To Wai
School of Electrical, Computer and Energy Engineering, Arizona State University, Tempe, AZ, United States

Daifeng Wang
Department of Biomedical Informatics, Stony Brook University, Stony Brook, NY, United States

Xiao-Ming Wu
Department of Computing, The Hong Kong Polytechnic University, Kowloon, Hong Kong Special Administrative Region

Yili Xia
School of Information Science and Engineering, Southeast University, Nanjing, PR China

Jie Xu
Electrical and Computer Engineering Department, University of Miami, Coral Gables, FL, United States

Simpson Zhang
Economics Department, University of California, Los Angeles, Los Angeles, CA, United States

Xiaochuan Zhao
University of California at Los Angeles, UCLA, Los Angeles, CA, United States

Preface

Cooperative and graph signal processing are areas that have seen significant growth in recent years, a trend that without a doubt will continue for many years to come. These areas are well within the scope of network science, which deals with complex systems described by interconnected elements and/or agents. The networks can be of many types including physical, engineered, information, social, biological, and economic networks. Within the framework of cooperative and graph signal processing one aims at discovering principles of various tasks including distributed detection and estimation, adaptation and learning over networks, distributed decision making, optimization and control over networks, and modeling and identification. The range of possible applications is vast and includes robotics, smart grids, communications, economic networks, life sciences, ecology, social networks, wireless health, and transportation networks.

With this book we want to provide in a single volume the basics of cooperative and graph signal processing and to address a number of areas where they are applied. The chapters are grouped in five parts. In Part 1, the fundamentals of inference over networks are addressed, Part 2 is on graph signal processing, Part 3 focuses on communications, networking, and sensing, Part 4 studies social networks, and Part 5 describes applications that range from genomics and system biology to big data and brain networks. All the chapters are written by experts and leaders in the field.

The contents of Part 1 include the fundamentals of inference, learning, and optimization over networks in synchronous and asynchronous settings (Chapter 1); estimation and detection where the emphasis is on studying agents that track drifts in models and examining performance limits under both estimation and detection (Chapter 2); learning in networks where the agents make inference about parameters of both, common and local interest (Chapter 3), collaborative inference of agents within the Bayesian framework (Chapter 4); multiagent distributed optimization, where the aim of all agents is to agree on the minimizer (Chapter 5); sequential filtering using state-space models by Kalman and particle filtering (Chapter 6); and game-theoretic learning where the objective of the decision-makers in the network depends on both the actions of other agents and the state of the environment (Chapter 7).

Part 2 starts with the basics of graph signal processing (Chapter 8); then it continues with a review of recent advances in sampling and recovery of signals defined over graphs (Chapter 9); active learning in networks where one has the freedom of choosing nodes for querying and wants to select intelligently informative nodes to minimize prediction errors (Chapter 10); design of graph filters and filter banks (Chapter 11); extension of the notion of stationarity to random graph signals (Chapter 12); estimating a graph topology from graph signals (Chapter 13); and a unifying framework for learning on graphs and based on partially absorbing random walks (Chapter 14).

Part 3, with its theme on communications, networking, and sensing, addresses projection-based algorithms and projection-free decentralized optimization algorithms for performing big data analytics (Chapter 15); optimal joint allocation of communication, computation, and caching resources in 5G communication networks (Chapter 16); analysis of a network's structural vulnerability to attacks (Chapter 17); team methods for cooperation in device-centric setups (Chapter 18); cooperative data exchange so that clients can decode missing messages while minimizing the total number of transmissions (Chapter 19); and Byzantine attacks and defense for collaborative spectrum-sensing in cognitive radio networks (Chapter 20).

Part 4, on social networks, provides material on diffusion models for information in social networks, Bayesian social learning, and revealed preferences (Chapter 21); network identification for social networks (Chapter 22); search and data routing in dynamic social networks (Chapter 23); information diffusion in social networks and control of diffusion of rumors and fake news (Chapter 24); methods for dealing with multilayer social networks, including multilayer community detection and multilayer interaction graph estimation (Chapter 25); and investigation of how cooperative and strategic learning lead to different types of networks and how efficient these networks are in making its agents achieve their goals (Chapter 26).

Part 5 brings various applications of cooperative and graph signal processing including construction, modeling, and analysis of gene regulatory networks (Chapter 27); cooperative frequency estimation in power networks (Chapter 28); the use of Bluetooth low energy beacons in smart city applications (Chapter 29); scalable decentralized and streaming analytics that facilitate machine learning from big data (Chapter 30); and capturing the dynamics of brain activity by analyzing functional brain networks and multivariate signals on the networks with graph signal processing methods (Chapter 31).

This book is a collective effort of the chapter authors. We are thankful for their enthusiastic support in writing their parts and for their timely responses. We are grateful to the reviewers for taking the time to read the chapters and for their valuable feedbacks. Finally, we thank Elsevier for all its support throughout the duration of this project.

Petar M. Djurić
Cédric Richard

PART

BASICS OF INFERENCE OVER NETWORKS

1

CHAPTER 1

ASYNCHRONOUS ADAPTIVE NETWORKS[a]

Ali H. Sayed*, Xiaochuan Zhao†

School of Engineering, École Polytechnique Fédérale de Lausanne (EPFL), Lausanne, Switzerland University of California at Los Angeles, UCLA, Los Angeles, CA, United States†*

1.1 INTRODUCTION

Adaptive networks consist of a collection of agents with learning abilities. The agents interact with each other on a local level and diffuse information across the network to solve inference and optimization tasks in a decentralized manner. Such networks are scalable, robust to node and link failures, and are particularly suitable for learning from big datasets by tapping into the power of collaboration among distributed agents. The networks are also endowed with cognitive abilities due to the sensing abilities of their agents, their interactions with their neighbors, and the embedded feedback mechanisms for acquiring and refining information. Each agent is not only capable of sensing data and experiencing the environment directly, but it also receives information through interactions with its neighbors and processes and analyzes this information to drive its learning process.

As already indicated in [1,2], there are many good reasons for the peaked interest in networked solutions, especially in this day and age when the word "network" has become commonplace whether one is referring to social networks, power networks, transportation networks, biological networks, or other networks. Some of these reasons have to do with the benefits of cooperation over networks in terms of improved performance and improved robustness and resilience to failure. Other reasons deal with privacy and secrecy considerations where agents may not be comfortable sharing their data with remote fusion centers. In other situations, the data may already be available in dispersed locations, as happens with cloud computing. One may also be interested in learning and extracting information through data mining from large datasets. Decentralized learning procedures offer an attractive approach to dealing with such datasets. Decentralized mechanisms can also serve as important enablers for the design of robotic swarms, which can assist in the exploration of disaster areas.

[a]This work was supported in part by NSF grants ECCS-1407712 and CCF-1524250. Coauthor X. Zhao was a PhD student in electrical engineering at UCLA. The authors are grateful to IEEE for allowing the reproduction of substantial material from [1,3–5] in this book chapter.

Cooperative and Graph Signal Processing. https://doi.org/10.1016/B978-0-12-813677-5.00001-8

1.1.1 ASYNCHRONOUS BEHAVIOR

The survey article [1] and monograph [2] focused on the case of synchronous networks where data arrive at all agents in a synchronous manner and updates by the agents are also performed in a synchronous manner. The network topology was assumed to remain largely static during the adaptation process. Under these conditions, the limits of performance and stability of these networks were identified in some detail for two main classes of distributed strategies: consensus and diffusion constructions. In this chapter, we extend the overview from [1] to cover *asynchronous* environments. In such environments, the operation of the network can suffer from the occurrence of various random events, including randomly changing topologies, random link failures, random data arrival times, and agents turning on and off randomly. Agents may also stop updating their solutions or may stop sending or receiving information in a random manner and without coordination with other agents. Results in [3–5] examined the implications of such asynchronous events on network performance in some detail and under a fairly general model for the random events. The purpose of this chapter is to summarize the key conclusions from these works in a manner that complements the presentation from [1] for the benefit of the reader. While the works [3–5] consider a broader formulation involving complex-valued variables, we limit the discussion here to real-valued variables in order not to overload the notation and to convey the key insights more directly. Proofs and derivations are often omitted and can be found in the above references; the emphasis is on presenting the results in a motivated manner and on commenting on the insights they provide into the operation of asynchronous networks.

We indicated in [3–5] that there already exist many useful studies in the literature on the performance of consensus strategies in the presence of asynchronous events (see, e.g., [6–16]). There are also some studies in the context of diffusion strategies [17,18]. However, with the exception of the latter two works, the earlier references assumed conditions that are not generally favorable for applications involving continuous *adaptation* and learning. For example, some of the works assumed a decaying step-size, which turns off adaptation after sufficient iterations have passed. Some other works assumed noise-free data, which is a hindrance when learning from data perturbed by interferences and distortions. A third class of works focused on studying pure averaging algorithms, which are not required to respond to continuous data streaming. In the works [3–5], we adopted a more general asynchronous model that removes these limitations by allowing for various sources of random events and, moreover, the events are allowed to occur simultaneously. We also examined learning algorithms that respond to streaming data to enable adaptation. The main conclusion from the analysis in these works, and which will be summarized in future sections, is that asynchronous networks can still behave in a mean-square-error (MSE) stable manner for sufficiently small step-sizes and, interestingly, their steady-state performance level is only slightly affected in comparison to synchronous behavior. The iterates computed by the various agents are still able to converge and hover around an agreement state with a small MSE. These are reassuring results that support the intrinsic robustness and resilience of network-based cooperative solutions.

1.1.2 ORGANIZATION OF THE CHAPTER

Readers will benefit more from this chapter if they first review the earlier article [1]. We continue to follow a similar structure here as well as a similar notation because the material in both this chapter and the earlier reference [1] are meant to complement each other. We organize the presentation into three main components. The first part (Section 1.2) reviews fundamental results on adaptation and learning by *single* stand-alone agents. The second part (Section 1.3) covers asynchronous centralized solutions. The objective is to explain the gain in performance that results from aggregating the data from the agents and processing it centrally at a fusion center. The centralized performance is used as a frame of reference for assessing various implementations. While centralized solutions can be powerful, they nevertheless suffer from a number of limitations. First, in real-time applications where agents collect data continuously, the repeated exchange of information back and forth between agents and the fusion center can be costly, especially when these exchanges occur over wireless links or require nontrivial routing resources. Second, in some sensitive applications, agents may be reluctant to share their data with remote centers for various reasons including privacy and secrecy considerations. More importantly perhaps, centralized solutions have a critical point of failure: if the central processor fails, then this solution method collapses altogether.

For these reasons, we cover in the remaining sections of the chapter (Sections 1.4 and 1.5) distributed asynchronous strategies of the consensus and diffusion types, and examine their dynamics, stability, and performance metrics. In the distributed mode of operation, agents are connected by a topology and are permitted to share information only with their immediate neighbors. The study of the behavior of such networked agents is more challenging than in the single-agent and centralized modes of operation due to the coupling among interacting agents and the fact that the networks are generally sparsely connected.

1.2 SINGLE-AGENT ADAPTATION AND LEARNING

We begin our treatment by reviewing stochastic gradient algorithms, with emphasis on their application to the problems of adaptation and learning by stand-alone agents.

1.2.1 RISK AND LOSS FUNCTIONS

Thus, let $J(w) : \mathbb{R}^{M\times 1} \mapsto \mathbb{R}$ denote a twice-differentiable real-valued (cost or utility or risk) function of a real-valued vector argument, $w \in \mathbb{R}^{M\times 1}$. When the variable w is complex-valued, some important technical differences arise that are beyond the scope of this chapter; they are addressed in [2–5]. Likewise, some adjustments to the arguments are needed when the risk function is nonsmooth (nondifferentiable), as explained in the works [19,20]. It is sufficient for our purposes in this chapter to limit the presentation to real arguments and smooth risk functions without much loss in generality.

We denote the gradient vectors of $J(w)$ relative to w and w^{T} by the following row and column vectors, respectively:

$$\nabla_w J(w) \triangleq \left[\frac{\partial J(w)}{\partial w_1}, \frac{\partial J(w)}{\partial w_2}, \ldots, \frac{\partial J(w)}{\partial w_M}\right], \tag{1.1a}$$

$$\nabla_{w^{\mathsf{T}}} J(w) \triangleq [\nabla_w J(w)]^{\mathsf{T}}. \tag{1.1b}$$

These definitions are in terms of the partial derivatives of $J(w)$ relative to the individual entries of $w = \text{col}\{w_1, w_2, \ldots, w_M\}$, where the notation col$\{\cdot\}$ refers to a column vector that is formed by stacking its arguments on top of each other. Likewise, the Hessian matrix of $J(w)$ with respect to w is defined as the following $M \times M$ symmetric matrix:

$$\nabla_w^2 J(w) \triangleq \nabla_{w^{\mathsf{T}}}[\nabla_w J(w)]=\nabla_w[\nabla_{w^{\mathsf{T}}} J(w)], \tag{1.1c}$$

which is constructed from two successive gradient operations. It is common in adaptation and learning applications for the risk function $J(w)$ to be constructed as the expectation of some loss function, $Q(w;\boldsymbol{x})$, where the **boldface** variable $\boldsymbol{x}$ is used to denote some random data, say,

$$J(w)=\mathbb{E}\quad Q(w;\boldsymbol{x}) \tag{1.2}$$

and the expectation is evaluated over the distribution of $\boldsymbol{x}$.

Example 1.1 (Mean-square-error (MSE) costs). Let $\boldsymbol{d}$ denote a zero-mean scalar random variable with variance $\sigma_d^2 = \mathbb{E}\boldsymbol{d}^2$, and let $\boldsymbol{u}$ denote a zero-mean $1 \times M$ random vector with covariance matrix $R_u = \mathbb{E}\boldsymbol{u}^{\mathsf{T}}\boldsymbol{u} > 0$. The combined quantities $\{\boldsymbol{d},\boldsymbol{u}\}$ represent the random variable $\boldsymbol{x}$ referred to in Eq. (1.2). The cross-covariance vector is denoted by $r_{du} = \mathbb{E}\boldsymbol{d}\boldsymbol{u}^{\mathsf{T}}$. We formulate the problem of estimating $\boldsymbol{d}$ from $\boldsymbol{u}$ in the linear least-mean-squares-error sense or, equivalently, the problem of seeking the vector w^o that minimizes the quadratic cost function:

$$J(w) \triangleq \mathbb{E}(\boldsymbol{d}-\boldsymbol{u}w)^2 = \sigma_d^2 - r_{du}^{\mathsf{T}}w - w^{\mathsf{T}}r_{du} + w^{\mathsf{T}}R_u w. \tag{1.3a}$$

This cost corresponds to the following choice for the loss function:

$$Q(w;\boldsymbol{x})=(\boldsymbol{d}-\boldsymbol{u}w)^2. \tag{1.3b}$$

Such quadratic costs are widely used in estimation and adaptation problems [21–25]. They are also widely used as quadratic risk functions in machine learning applications [26,27]. The gradient vector and Hessian matrix of $J(w)$ are easily seen to be:

$$\nabla_w J(w)=2\,(R_u w - r_{du})^{\mathsf{T}}, \quad \nabla_w^2 J(w)=2R_u. \tag{1.3c}$$

■

Example 1.2 (Logistic or log-loss risks). Let $\boldsymbol{\gamma}$ denote a binary random variable that assumes the values ± 1, and let $\boldsymbol{h}$ denote an $M \times 1$ random (feature) vector with $R_h=\mathbb{E}\boldsymbol{h}\boldsymbol{h}^{\mathsf{T}}$. The combined quantities $\{\boldsymbol{\gamma},\boldsymbol{h}\}$ represent the random variable $\boldsymbol{x}$ referred to in Eq. (1.2). In the context of machine learning and pattern classification problems [26–28], the variable $\boldsymbol{\gamma}$ designates the class that feature vector $\boldsymbol{h}$ belongs to. In these problems, one seeks the vector w^o that minimizes the regularized logistic risk function:

$$J(w) \triangleq \frac{\rho}{2}\|w\|^2 + \mathbb{E}\left\{\ln\left[1+e^{-\boldsymbol{\gamma}\boldsymbol{h}^{\mathsf{T}}w}\right]\right\}, \tag{1.4a}$$

where $\rho > 0$ is some regularization parameter, $\ln(\cdot)$ is the natural logarithm function, and $\|w\|^2=w^{\mathsf{T}}w$. The risk (1.4a) corresponds to the following choice for the loss function:

$$Q(w;\boldsymbol{x}) \triangleq \frac{\rho}{2}\|w\|^2 + \ln\left[1+e^{-\boldsymbol{\gamma}\boldsymbol{h}^{\mathsf{T}}w}\right]. \tag{1.4b}$$

Once w^o is recovered, its value can be used to classify new feature vectors, say, $\{h_\ell\}$, into classes +1 or −1. This can be achieved by assigning feature vectors with $h_\ell^{\mathsf{T}} w^o \geq 0$ to one class and feature vectors with $h_\ell^{\mathsf{T}} w^o < 0$ to another class. It can be easily verified that:

$$\nabla_w J(w) = \rho w^{\mathsf{T}} - \mathbb{E}\left\{ \gamma h^{\mathsf{T}} \cdot \frac{e^{-\gamma h^{\mathsf{T}} w}}{1 + e^{-\gamma h^{\mathsf{T}} w}} \right\}, \tag{1.4c}$$

$$\nabla_w^2 J(w) = \rho I_M + \mathbb{E}\left\{ h h^{\mathsf{T}} \cdot \frac{e^{-\gamma h^{\mathsf{T}} w}}{\left(1 + e^{-\gamma h^{\mathsf{T}} w}\right)^2} \right\}, \tag{1.4d}$$

where I_M denotes the identity matrix of size $M \times M$. ■

1.2.2 CONDITIONS ON COST FUNCTION

Stochastic gradient algorithms are powerful iterative procedures for solving optimization problems of the form

$$\min_w \; J(w). \tag{1.5}$$

While the analysis that follows can be pursued under more relaxed conditions (see, e.g., the treatments in [29–32]), it is sufficient for our purposes to require $J(w)$ to be strongly convex and twice-differentiable with respect to w. The cost function $J(w)$ is said to be ν-strongly convex if, and only if, its Hessian matrix is sufficiently bounded away from zero [30,33–35]:

$$J(w) \text{ is } \nu\text{-strongly convex} \iff \nabla_w^2 J(w) \geq \nu I_M > 0 \tag{1.6}$$

for all w and for some scalar $\nu > 0$, where the notation $A > 0$ signifies that matrix A is positive-definite. Strong convexity is a useful condition in the context of adaptation and learning from streaming data because it helps guard against ill-conditioning in the algorithms; it also helps ensure that $J(w)$ has a *unique* global minimum, say, at location w^o; there will be no other minima, maxima, or saddle points. In addition, it is well known that strong convexity helps endow stochastic-gradient algorithms with geometric convergence rates in the order of $O(\alpha^i)$, for some $0 \leq \alpha < 1$ and where i is the iteration index [30,31].

In many problems of interest in adaptation and learning, the cost function $J(w)$ is either already strongly convex or can be made strongly convex by means of regularization. For example, it is common in machine learning problems [26,27] and in adaptation and estimation problems [23,25] to incorporate regularization factors into the cost functions; these factors help ensure strong convexity. For instance, the MSE cost (1.3a) is strongly convex whenever $R_u > 0$. If R_u happens to be singular, then the following regularized cost will be strongly convex:

$$J(w) \triangleq \frac{\rho}{2}\|w\|^2 + \mathbb{E}(d - uw)^2, \tag{1.7}$$

where $\rho > 0$ is a regularization parameter similar to Eq. (1.4a).

Besides strong convexity, we shall also assume that the gradient vector of $J(w)$ is δ-Lipschitz, namely, there exists $\delta > 0$ such that

$$\|\nabla_w J(w_2) - \nabla_w J(w_1)\| \leq \delta \, \|w_2 - w_1\| \tag{1.8}$$

for all w_1 and w_2. It can be verified that for twice-differentiable costs, conditions (1.6) and (1.8) combined are equivalent to

$$0 < \nu I_M \leq \nabla_w^2 J(w) \leq \delta I_M. \tag{1.9}$$

For example, it is clear that the Hessian matrices in Eqs. (1.3c) and (1.4d) satisfy this property because

$$2\lambda_{\min}(R_u) I_M \leq \nabla_w^2 J(w) \leq 2\lambda_{\max}(R_u) I_M \tag{1.10a}$$

in the first case and

$$\rho I_M \leq \nabla_w^2 J(w) \leq (\rho + \lambda_{\max}(R_h)) I_M \tag{1.10b}$$

in the second case, where the notation $\lambda_{\min}(R)$ and $\lambda_{\max}(R)$ refers to the smallest and largest eigenvalues of the symmetric matrix argument, R, respectively. In summary, we will be assuming the following conditions [2,3,36,37].

Assumption 1.1 (Conditions on cost function). *The cost function $J(w)$ is twice-differentiable and satisfies Eq. (1.9) for some positive parameters $\nu \leq \delta$. Condition (1.9) is equivalent to requiring $J(w)$ to be ν-strongly convex and for its gradient vector to be δ-Lipschitz as in Eqs. (1.6) and (1.8), respectively.* ■

1.2.3 STOCHASTIC-GRADIENT APPROXIMATION

The traditional gradient-descent algorithm for solving Eq. (1.5) takes the form:

$$w_i = w_{i-1} - \mu \, \nabla_{w^{\mathsf{T}}} J(w_{i-1}), \quad i \geq 0, \tag{1.11}$$

where $i \geq 0$ is an iteration index and $\mu > 0$ is a small step-size parameter. Starting from some initial condition, w_{-1}, the iterates $\{w_i\}$ correspond to successive estimates for the minimizer w^o. In order to run recursion (1.11), we need to have access to the true gradient vector. This information is generally unavailable in most instances involving learning from data. For example, when cost functions are defined as the expectations of certain loss functions as in Eq. (1.2), the statistical distribution of the data $\boldsymbol{x}$ may not be known beforehand. In that case, the exact form of $J(w)$ will not be known because the expectation of $Q(w;\boldsymbol{x})$ cannot be computed. In such situations, it is customary to replace the true gradient vector, $\nabla_{w^{\mathsf{T}}} J(w_{i-1})$, by an instantaneous approximation for it, and which we shall denote by $\widehat{\nabla_{w^{\mathsf{T}}} J}(\boldsymbol{w}_{i-1})$. Doing so leads to the following *stochastic-gradient* recursion in lieu of Eq. (1.11):

$$\boxed{\boldsymbol{w}_i = \boldsymbol{w}_{i-1} - \mu \, \widehat{\nabla_{w^{\mathsf{T}}} J}(\boldsymbol{w}_{i-1}), \quad i \geq 0.} \tag{1.12}$$

We use the **boldface** notation, $\boldsymbol{w}_i$, for the iterates in Eq. (1.12) to highlight the fact that these iterates are now randomly perturbed versions of the values $\{w_i\}$ generated by the original recursion (1.11). The random perturbations arise from the use of the approximate gradient vector. The boldface notation is therefore meant to emphasize the random nature of the iterates in Eq. (1.12). We refer to recursion

(1.12) as a *synchronous* implementation because updates occur continuously over the iteration index i. This terminology is meant to distinguish the above recursion from its *asynchronous* counterpart, which is introduced later in Section 1.2.5.

We illustrate construction (1.12) by considering a scenario from classical adaptive filter theory [21–23], where the gradient vector is approximated directly from data realizations. The construction will reveal why stochastic-gradient implementations of the form (1.12), using approximate rather than exact gradient information, become naturally endowed with the ability to respond to *streaming* data.

Example 1.3 (LMS adaptation). Let $\boldsymbol{d}(i)$ denote a streaming sequence of zero-mean random variables with variance $\sigma_d^2 = \mathbb{E}\boldsymbol{d}^2(i)$. Let $\boldsymbol{u}_i$ denote a streaming sequence of $1 \times M$ independent zero-mean random vectors with covariance matrix $R_u = \mathbb{E}\boldsymbol{u}_i^{\mathsf{T}}\boldsymbol{u}_i > 0$. Both processes $\{\boldsymbol{d}(i), \boldsymbol{u}_i\}$ are assumed to be jointly wide-sense stationary. The cross-covariance vector between $\boldsymbol{d}(i)$ and $\boldsymbol{u}_i$ is denoted by $r_{du} = \mathbb{E}\boldsymbol{d}(i)\boldsymbol{u}_i^{\mathsf{T}}$. The data $\{\boldsymbol{d}(i), \boldsymbol{u}_i\}$ are assumed to be related via a linear regression model of the form:

$$\boldsymbol{d}(i) = \boldsymbol{u}_i w^o + \boldsymbol{v}(i) \tag{1.13a}$$

for some unknown parameter vector w^o, and where $\boldsymbol{v}(i)$ is a zero-mean white-noise process with power $\sigma_v^2 = \mathbb{E}\boldsymbol{v}^2(i)$ and assumed independent of $\boldsymbol{u}_j$ for all i, j. Observe that we are using parentheses to represent the time dependency of a scalar variable, such as writing $\boldsymbol{d}(i)$, and subscripts to represent the time dependency of a vector variable, such as writing $\boldsymbol{u}_i$. This convention will be used throughout the chapter. In a manner similar to Example 1.1, we again pose the problem of estimating w^o by minimizing the MSE cost

$$J(w) = \mathbb{E}\left(\boldsymbol{d}(i) - \boldsymbol{u}_i w\right)^2 \equiv \mathbb{E}Q(w; \boldsymbol{x}_i), \tag{1.13b}$$

where now the quantities $\{\boldsymbol{d}(i), \boldsymbol{u}_i\}$ represent the random data $\boldsymbol{x}_i$ in the definition of the loss function, $Q(w; \boldsymbol{x}_i)$. Using Eq. (1.11), the gradient-descent recursion in this case will take the form:

$$w_i = w_{i-1} - 2\mu\left[R_u w_{i-1} - r_{du}\right], \quad i \geq 0. \tag{1.13c}$$

The main difficulty in running this recursion is that it requires knowledge of the moments $\{r_{du}, R_u\}$. This information is rarely available beforehand; the adaptive agent senses instead realizations $\{\boldsymbol{d}(i), \boldsymbol{u}_i\}$ whose statistical distributions have moments $\{r_{du}, R_u\}$. The agent can use these realizations to approximate the moments and the true gradient vector. There are many constructions that can be used for this purpose, with different constructions leading to different adaptive algorithms [21–24]. It is sufficient to focus on one of the most popular adaptive algorithms, which results from using the data $\{\boldsymbol{d}(i), \boldsymbol{u}_i\}$ to compute *instantaneous* approximations for the unavailable moments as follows:

$$r_{du} \approx \boldsymbol{d}(i)\boldsymbol{u}_i^{\mathsf{T}}, \qquad R_u \approx \boldsymbol{u}_i^{\mathsf{T}}\boldsymbol{u}_i. \tag{1.13d}$$

By doing so, the true gradient vector is approximated by:

$$\widehat{\nabla_{w^{\mathsf{T}}} J}(w) = 2\left[\boldsymbol{u}_i^{\mathsf{T}}\boldsymbol{u}_i w - \boldsymbol{u}_i^{\mathsf{T}}\boldsymbol{d}(i)\right] = \nabla_{w^{\mathsf{T}}} Q(w; \boldsymbol{x}_i). \tag{1.13e}$$

Observe that this construction amounts to replacing the true gradient vector, $\nabla_{w^{\mathsf{T}}} J(w)$, by the gradient vector of the loss function itself (which, equivalently, amounts to dropping the expectation operator).

Substituting Eq. (1.13e) into Eq. (1.13c) leads to the well-known (synchronous) least-mean-squares (LMS, for short) algorithm [21–23,38]:

$$\boxed{\boldsymbol{w}_i = \boldsymbol{w}_{i-1} + 2\mu \boldsymbol{u}_i^{\mathrm{T}} \left[\boldsymbol{d}(i) - \boldsymbol{u}_i \boldsymbol{w}_{i-1}\right], \quad i \geq 0.} \tag{1.13f}$$

The LMS algorithm is therefore a stochastic-gradient algorithm. By relying directly on the instantaneous data $\{\boldsymbol{d}(i), \boldsymbol{u}_i\}$, the algorithm is infused with useful tracking abilities. This is because drifts in the model w^o from Eq. (1.13a) will be reflected in the data $\{\boldsymbol{d}(i), \boldsymbol{u}_i\}$, which are used directly in the update (1.13f).

■

If desired, it is also possible to employ iteration-dependent step-size sequences, $\mu(i)$, in Eq. (1.12) instead of the constant step-size μ, and to require $\mu(i)$ to satisfy

$$\sum_{i=0}^{\infty} \mu^2(i) < \infty, \qquad \sum_{i=0}^{\infty} \mu(i) = \infty. \tag{1.14}$$

Under some technical conditions, it is well known that such step-size sequences ensure the convergence of $\boldsymbol{w}_i$ toward w^o almost surely as $i \to \infty$ [2,30–32]. However, conditions (1.14) force the step-size sequence to decay to zero, which is problematic for applications requiring continuous adaptation and learning from streaming data. This is because, in such applications, it is not unusual for the location of the minimizer, w^o, to drift with time. With $\mu(i)$ decaying toward zero, the stochastic-gradient algorithm (1.12) will stop updating and will not be able to track drifts in the solution. For this reason, we shall focus on constant step-sizes from this point onward because we are interested in solutions with tracking abilities.

Now, the use of an approximate gradient vector in Eq. (1.12) introduces perturbations relative to the operation of the original recursion (1.11). We refer to the perturbation as gradient noise and define it as the difference:

$$\boxed{\boldsymbol{s}_i(\boldsymbol{w}_{i-1}) \triangleq \widehat{\nabla_{w^{\mathrm{T}}} J}(\boldsymbol{w}_{i-1}) - \nabla_{w^{\mathrm{T}}} J(\boldsymbol{w}_{i-1}).} \tag{1.15}$$

The presence of this perturbation prevents the stochastic iterate, $\boldsymbol{w}_i$, from converging almost surely to the minimizer w^o when constant step-sizes are used. Some deterioration in performance will occur and the iterate $\boldsymbol{w}_i$ will instead fluctuate close to w^o. We will assess the size of these fluctuations by measuring their steady-state mean-square value (also called mean-square-deviation or MSD). It will turn out that the MSD is small and in the order of $O(\mu)$; see Eq. (1.28c) further ahead. It will also turn out that stochastic-gradient algorithms converge toward their MSD levels at a geometric rate. In this way, we will be able to conclude that adaptation with small constant step-sizes can still lead to reliable performance in the presence of gradient noise, which is a reassuring result. We will also be able to conclude that adaptation with constant step-sizes is useful even for stationary environments. This is because it is generally sufficient in practice to reach an iterate $\boldsymbol{w}_i$ within some fidelity level from w^o in a *finite* number of iterations. As long as the MSD level is satisfactory, a stochastic-gradient algorithm will be able to attain satisfactory fidelity within a reasonable time frame. In comparison, although diminishing step-sizes ensure almost-sure convergence of $\boldsymbol{w}_i$ to w^o, they nevertheless disable tracking and can only guarantee slower than geometric rates of convergence (see, e.g., [2,30,31]). The next example from [36] illustrates the nature of the gradient noise process (1.15) in the context of MSE adaptation.

Example 1.4 (Gradient noise). It is clear from the expressions in Example 1.3 that the corresponding gradient noise process is

$$s_i(\boldsymbol{w}_{i-1}) = 2\left(R_u - \boldsymbol{u}_i^{\mathsf{T}}\boldsymbol{u}_i\right)\tilde{\boldsymbol{w}}_{i-1} - 2\boldsymbol{u}_i^{\mathsf{T}}\boldsymbol{v}(i), \tag{1.16a}$$

where we introduced the error vector:

$$\tilde{\boldsymbol{w}}_i \triangleq w^o - \boldsymbol{w}_i. \tag{1.16b}$$

Let the symbol $\boldsymbol{\mathcal{F}}_{i-1}$ represent the collection of all possible random events generated by the past iterates $\{\boldsymbol{w}_j\}$ up to time $i-1$ (more formally, $\boldsymbol{\mathcal{F}}_{i-1}$ is the *filtration* generated by the random process $\boldsymbol{w}_j$ for $j \le i-1$):

$$\boldsymbol{\mathcal{F}}_{i-1} \triangleq \text{filtration}\left\{\boldsymbol{w}_{-1}, \boldsymbol{w}_0, \boldsymbol{w}_1, \ldots, \boldsymbol{w}_{i-1}\right\}. \tag{1.16c}$$

It follows from the conditions on the random processes $\{\boldsymbol{u}_i, \boldsymbol{v}(i)\}$ in Example 1.3 that

$$\mathbb{E}\left[s_i(\boldsymbol{w}_{i-1})|\boldsymbol{\mathcal{F}}_{i-1}\right] = 0, \tag{1.16d}$$

$$\mathbb{E}\left[\|s_i(\boldsymbol{w}_{i-1})\|^2|\boldsymbol{\mathcal{F}}_{i-1}\right] \le 4c\,\|\tilde{\boldsymbol{w}}_{i-1}\|^2 + 4\sigma_v^2\,\mathrm{Tr}(R_u) \tag{1.16e}$$

for some constant $c \ge 0$. If we take expectations of both sides of Eq. (1.16e), we further conclude that the variance of the gradient noise, $\mathbb{E}\|s_i(\boldsymbol{w}_{i-1})\|^2$, is bounded by the combination of two factors. The first factor depends on the quality of the iterate, $\mathbb{E}\|\tilde{\boldsymbol{w}}_{i-1}\|^2$ while the second factor depends on σ_v^2. Therefore, even if the adaptive agent is able to approach w^o with great fidelity so that $\mathbb{E}\|\tilde{\boldsymbol{w}}_{i-1}\|^2$ is small, the size of the gradient noise will still depend on σ_v^2. ■

1.2.4 CONDITIONS ON GRADIENT NOISE PROCESS

In order to examine the convergence and performance properties of the stochastic-gradient recursion (1.12), it is necessary to introduce some assumptions on the stochastic nature of the gradient noise process, $s_i(\cdot)$. The conditions that we introduce in the sequel are similar to conditions used earlier in the optimization literature, e.g., in [30, pp. 95–102] and [39, p. 635]; they are also motivated by the conditions we observed in the MSE case in Example 1.4. Following the developments in [3,36,37], we let

$$R_{s,i}(\boldsymbol{w}_{i-1}) \triangleq \mathbb{E}\left[s_i(\boldsymbol{w}_{i-1})s_i^{\mathsf{T}}(\boldsymbol{w}_{i-1})|\boldsymbol{\mathcal{F}}_{i-1}\right] \tag{1.17a}$$

denote the conditional second-order moment of the gradient noise process, which generally depends on i. We assume that, in the limit, the covariance matrix tends to a constant value when evaluated at w^o and is denoted by

$$\boxed{R_s \triangleq \lim_{i\to\infty} \mathbb{E}\left[s_i(w^o)s_i^{\mathsf{T}}(w^o)|\boldsymbol{\mathcal{F}}_{i-1}\right].} \tag{1.17b}$$

For example, comparing with expression (1.16a) for MSE costs, we have

$$s_i(w^o) = -\,2\boldsymbol{u}_i^{\mathsf T}\boldsymbol{v}(i), \tag{1.18a}$$

$$R_s = 4\sigma_v^2 R_u. \tag{1.18b}$$

Assumption 1.2 (Conditions on gradient noise). *It is assumed that the first- and second-order conditional moments of the gradient noise process satisfy Eq. (1.17b) and*

$$\mathbb{E}\left[\boldsymbol{s}_i(\boldsymbol{w}_{i-1})|\boldsymbol{\mathcal{F}}_{i-1}\right] = 0, \tag{1.19a}$$

$$\mathbb{E}\left[\|\boldsymbol{s}_i(\boldsymbol{w}_{i-1})\|^2|\boldsymbol{\mathcal{F}}_{i-1}\right] \le \beta^2\,\|\tilde{\boldsymbol{w}}_{i-1}\|^2 + \sigma_s^2 \tag{1.19b}$$

almost surely, for some nonnegative scalars β^2 and σ_s^2. ■

Condition (1.19a) ensures that the approximate gradient vector is unbiased. It follows from conditions (1.19a) and (1.19b) that the gradient noise process itself satisfies:

$$\mathbb{E}\boldsymbol{s}_i(\boldsymbol{w}_{i-1}) = 0, \tag{1.20a}$$

$$\mathbb{E}\|\boldsymbol{s}_i(\boldsymbol{w}_{i-1})\|^2 \le \beta^2\,\mathbb{E}\|\tilde{\boldsymbol{w}}_{i-1}\|^2 + \sigma_s^2. \tag{1.20b}$$

It is straightforward to verify that the gradient noise process (1.16a) in the MSE case satisfies conditions (1.19a) and (1.19b). Note in particular from Eq. (1.16e) that we can make the identifications $\sigma_s^2 \to 4\sigma_v^2\mathrm{Tr}(R_u)$ and $\beta^2 \to 4c$.

1.2.5 RANDOM UPDATES

We examined the performance of synchronous updates of the form (1.12) in some detail in [1,2]. As indicated earlier, the focus of the current chapter is on extending the treatment from [1] to asynchronous implementations. Accordingly, the first main digression in the exposition relative to [1] occurs at this stage.

Thus, note that the stochastic-gradient recursion (1.12) employs a constant step-size parameter, $\mu > 0$. This means that this implementation expects the approximate gradient vector, $\widehat{\nabla_{w^{\mathsf T}} J}(\boldsymbol{w}_{i-1})$, to be available at every iteration. Nevertheless, there are situations where data may arrive at the agent at random times, in which case the updates will also be occurring at random times. One way to capture this behavior is to model the step-size parameter as a *random* process, which we shall denote by the boldface notation $\boldsymbol{\mu}(i)$ (because boldface letters in our notation refer to random quantities). Doing so allows us to replace the synchronous implementation (1.12) by the following asynchronous recursion:

$$\boxed{\boldsymbol{w}_i = \boldsymbol{w}_{i-1} - \boldsymbol{\mu}(i)\widehat{\nabla_{w^{\mathsf T}} J}(\boldsymbol{w}_{i-1}), \quad i \ge 0.} \tag{1.21}$$

Observe that we are attaching a time index to the step-size parameter in order to highlight that its value will now be changing randomly from one iteration to another.

For example, one particular instance discussed further ahead in Example 1.5 is when $\boldsymbol{\mu}(i)$ is a Bernoulli random variable assuming one of two possible values, say, $\boldsymbol{\mu}(i) = \mu > 0$ with probability p_μ and $\boldsymbol{\mu}(i) = 0$ with probability $1 - p_\mu$. In this case, recursion (1.21) will be updating p_μ-fraction of the time. We will not limit our presentation to the Bernoulli model but will allow the random step-size to assume broader probability distributions; we denote its first- and second-order moments as follows.

Assumption 1.3 (Conditions on step-size process). *It is assumed that the stochastic process* $\{\boldsymbol{\mu}(i), i \geq 0\}$ *consists of a sequence of independent and bounded random variables,* $\boldsymbol{\mu}(i) \in [0, \mu_{ub}]$, *where* $\mu_{ub} > 0$ *is a constant upper bound. The mean and variance of* $\boldsymbol{\mu}(i)$ *are fixed over i, and they are denoted by:*

$$\bar{\mu} \triangleq \mathbb{E}\boldsymbol{\mu}(i), \tag{1.22a}$$

$$\sigma_{\mu}^2 \triangleq \mathbb{E}(\boldsymbol{\mu}(i) - \bar{\mu})^2 \tag{1.22b}$$

with $\bar{\mu} > 0$ *and* $\sigma_{\mu}^2 \geq 0$. *Moreover, it is assumed that, for any i, the random variable* $\boldsymbol{\mu}(i)$ *is independent of any other random variable in the learning algorithm.* ■

The following variable will play an important role in characterizing the MSE performance of asynchronous updates and, hence, we introduce a unique symbol for it:

$$\boxed{\mu_x \triangleq \bar{\mu} + \frac{\sigma_{\mu}^2}{\bar{\mu}}} \tag{1.23}$$

where the subscript "x" is meant to refer to the "asynchronous" mode of operation. This expression captures the first and second-order moments of the random variations in the step-size parameter into a single variable, μ_x. While the constant step-size μ determines the performance of the synchronous implementation (1.12), it turns out that the constant variable μ_x defined above will play an equivalent role for the asynchronous implementation (1.21). Note further that the synchronous stochastic-gradient iteration (1.12) can be viewed as a special case of recursion (1.21) when the variance of $\boldsymbol{\mu}(i)$ is set to zero, i.e., $\sigma_{\mu}^2 = 0$, and the mean value $\bar{\mu}$ is set to μ. Therefore, by using these substitutions, we will able to deduce performance metrics for Eq. (1.12) from the performance metrics that we shall present for Eq. (1.21). The following two examples illustrate situations involving random updates.

Example 1.5 (Random updates under a Bernoulli model). Assume that at every iteration i, the agent adopts a random "on-off" policy to reduce energy consumption. It tosses a coin to decide whether to enter an active learning mode or a sleeping mode. Let $0 < p_{\mu} < 1$ denote the probability of entering the active mode. During the active mode, the agent employs a step-size value μ. This model is useful for situations in which data arrives randomly at the agent: at every iteration i, new data is available with probability p_{μ}. The random step-size process is therefore of the following type:

$$\boldsymbol{\mu}(i) = \begin{cases} \mu, & \text{with probability} \quad p_{\mu}, \\ 0, & \text{with probability} \quad 1 - p_{\mu}. \end{cases} \tag{1.24a}$$

In this case, the mean and variance of $\boldsymbol{\mu}(i)$ are given by:

$$\bar{\mu} = p_{\mu}\mu, \quad \sigma_{\mu}^2 = p_{\mu}(1 - p_{\mu})\mu^2, \quad \mu_x = \mu. \tag{1.24b}$$

■

Example 1.6 (Random updates under a Beta model). Because the random step-size $\boldsymbol{\mu}(i)$ is limited to a finite-length interval $[0, \mu_{\text{ub}}]$, we may extend the Bernoulli model from the previous example by adopting a more general continuous Beta distribution for $\boldsymbol{\mu}(i)$. The Beta distribution is an extension of the Bernoulli distribution. While the Bernoulli distribution assumes two discrete possibilities for the random variable, say, $\{0, \mu\}$, the Beta distribution allows for any value in the continuum $[0, \mu]$.

Thus, let $\boldsymbol{x}$ denote a generic scalar random variable that assumes values in the interval $[0, 1]$ according to a Beta distribution. Then, according to this distribution, the pdf of $\boldsymbol{x}$, denoted by $f_{\boldsymbol{x}}(x; \xi, \zeta)$, is determined by two shape parameters $\{\xi, \zeta\}$ as follows [40,41]:

$$f_{\boldsymbol{x}}(x;\xi,\zeta)=\begin{cases}\dfrac{\Gamma(\xi+\zeta)}{\Gamma(\xi)\Gamma(\zeta)}x^{\xi-1}(1-x)^{\zeta-1}, & 0\le x\le 1,\\ 0, & \text{otherwise,}\end{cases} \tag{1.25a}$$

where $\Gamma(\cdot)$ denotes the Gamma function [42,43]. Fig. 1.1 plots $f_{\boldsymbol{x}}(x; \xi, \zeta)$ for two values of ζ. The mean and variance of the Beta distribution (1.25a) are given by:

$$\bar{x} = \frac{\xi}{\xi+\zeta}, \qquad \sigma_x^2 = \frac{\xi\zeta}{(\xi+\zeta)^2(\xi+\zeta+1)}. \tag{1.25b}$$

We note that the classical uniform distribution over the interval $[0, 1]$ is a special case of the Beta distribution for $\xi = \zeta = 1$; see Fig. 1.1. Likewise, the Bernoulli distribution with $p_\mu = 1/2$ is recovered from the Beta distribution by letting $\xi = \zeta \to 0$.

In the Beta model for asynchronous adaptation, we assume that the ratio $\boldsymbol{\mu}(i)/\mu_{\text{ub}}$ follows a Beta distribution with parameters $\{\xi, \zeta\}$. Under this model, the mean and variance of the random step-size become:

$$\bar{\mu}=\left(\frac{\xi}{\xi+\zeta}\right)\mu_{\text{ub}}, \qquad \sigma_\mu^2=\left(\frac{\xi\zeta}{(\xi+\zeta)^2(\xi+\zeta+1)}\right)\mu_{\text{ub}}^2. \tag{1.26}$$

■

1.2.6 MEAN-SQUARE-ERROR STABILITY

We now examine the convergence of the asynchronous stochastic-gradient recursion (1.21). In the statement below, the notation $a = O(\mu)$ means $a \le b\mu$ for some constant b that is independent of μ.

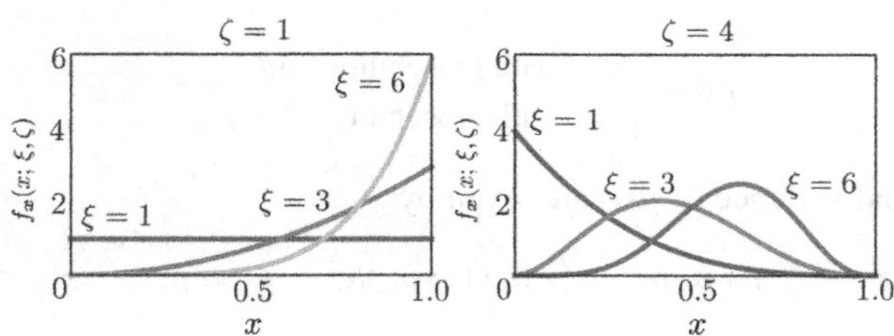

FIG. 1.1

Beta distribution. The pdf of the Beta distribution, $f_{\boldsymbol{x}}(x; \xi, \zeta)$, defined by Eq. (1.25a) for different values of the shape parameters ξ and ζ.

Figure reproduced with permission from Zhao X, Sayed AH. Asynchronous adaptation and learning over networks—Part I: Modeling and stability analysis. IEEE Trans Signal Process 2015;63(4):811–26.

Lemma 1.1 (Mean-square-error stability). *Assume the conditions under Assumptions 1.1–1.3 on the cost function, the gradient noise process, and the random step-size process hold. Let* $\mu_o = 2\nu/(\delta^2+\beta^2)$. *For any* μ_x *satisfying*

$$\mu_x < \mu_o \tag{1.27}$$

it holds that $\mathbb{E}\|\tilde{w}_i\|^2$ *converges exponentially (i.e., at a geometric rate) according to the recursion*

$$\mathbb{E}\|\tilde{w}_i\|^2 \le \alpha\,\mathbb{E}\|\tilde{w}_{i-1}\|^2 + (\bar{\mu}^2+\sigma_\mu^2)\sigma_s^2, \tag{1.28a}$$

where the scalar α *satisfies* $0 \le \alpha < 1$ *and is given by*

$$\begin{aligned} \alpha &\stackrel{\Delta}{=} 1 - 2\nu\bar{\mu} + (\delta^2+\beta^2)(\bar{\mu}^2+\sigma_\mu^2) \\ &= 1 - 2\nu\bar{\mu} + O(\mu_x^2). \end{aligned} \tag{1.28b}$$

It follows from Eq. (1.28a) that, for sufficiently small step-sizes:

$$\limsup_{i\to\infty}\ \mathbb{E}\|\tilde{w}_i\|^2 = O(\mu_x). \tag{1.28c}$$

Proof. We subtract w^o from both sides of Eq. (1.21) to get

$$\tilde{w}_i = \tilde{w}_{i-1} + \mu(i)\,\nabla_{w^\mathsf{T}} J(w_{i-1}) + \mu(i)\,s_i(w_{i-1}). \tag{1.29a}$$

We now appeal to the mean-value theorem [2,30,44] to write:

$$\nabla_{w^\mathsf{T}} J(w_{i-1}) = -\left[\int_0^1 \nabla_w^2 J(w_o - t\tilde{w}_{i-1})dt\right]\tilde{w}_{i-1} \ \stackrel{\Delta}{=}\ -H_{i-1}\tilde{w}_{i-1}, \tag{1.29b}$$

where we are introducing the *symmetric* and *random* time-variant matrix H_{i-1} to represent the integral expression. Substituting into Eq. (1.29a), we get

$$\tilde{w}_i = [I_M - \mu(i)H_{i-1}]\,\tilde{w}_{i-1} + \mu(i)\,s_i(w_{i-1}) \tag{1.29c}$$

so that from Assumption 1.2:

$$\mathbb{E}\left[\|\tilde{w}_i\|^2|\mathcal{F}_{i-1}\right] \le \left(\mathbb{E}\left[\|I_M - \mu(i)H_{i-1}\|^2|\mathcal{F}_{i-1}\right]\right)\|\tilde{w}_{i-1}\|^2 + \left(\mathbb{E}\mu^2(i)\right)\left(\mathbb{E}\left[\|s_i(w_{i-1})\|^2|\mathcal{F}_{i-1}\right]\right). \tag{1.30}$$

It follows from Eq. (1.9) that

$$\begin{aligned} \|I_M - \mu(i)H_{i-1}\|^2 &= \left[\rho\left(I_M - \mu(i)H_{i-1}\right)\right]^2 \\ &\le \max\left\{[1-\mu(i)\,\delta]^2,\ [1-\mu(i)\,\nu]^2\right\} \\ &\le 1 - 2\mu(i)\,\nu + \mu^2(i)\delta^2 \end{aligned} \tag{1.31a}$$

because $\nu \le \delta$. In the first line above, the notation $\rho(A)$ denotes the spectral radius of its matrix argument (i.e., $\rho(A) = \max_k|\lambda_k(A)|$ in terms of the largest magnitude eigenvalue of A). From Eq. (1.31a) we obtain

$$\mathbb{E}\left(\|I_M - \mu(i)H_{i-1}\|^2|\mathcal{F}_{i-1}\right) \le 1 - 2\bar{\mu}\nu + (\bar{\mu}^2+\sigma_\mu^2)\delta^2. \tag{1.31b}$$

Taking expectations of both sides of Eq. (1.30), we arrive at Eq. (1.28a) from Eqs. (1.20b) and (1.31b) with α given by Eq. (1.28b). The bound in Eq. (1.27) on the moments of the random step-size ensures that $0 \le \alpha < 1$. For the $O(\mu_x^2)$ approximation in expression (1.28b) note from Eq. (1.23) that

$$(\bar{\mu}^2 + \sigma_\mu^2) = \bar{\mu}\mu_x \le \mu_x^2. \tag{1.32}$$

Iterating recursion (1.28a) gives

$$\mathbb{E}\,\|\tilde{w}_i\|^2 \le \alpha^{i+1}\,\mathbb{E}\,\|\tilde{w}_{-1}\|^2 + \frac{(\bar{\mu}^2 + \sigma_\mu^2)\sigma_s^2}{1-\alpha}. \tag{1.33a}$$

Because $0 \le \alpha < 1$, there exists an iteration value I_o large enough such that

$$\alpha^{i+1}\,\mathbb{E}\|\tilde{w}_{-1}\|^2 \le \frac{(\bar{\mu}^2 + \sigma_\mu^2)\sigma_s^2}{1-\alpha}, \quad i > I_o. \tag{1.33b}$$

It follows that the variance $\mathbb{E}\,\|\tilde{w}_i\|^2$ converges exponentially to a region that is upper bounded by $2(\bar{\mu}^2 + \sigma_\mu^2)\sigma_s^2/(1-\alpha)$. It can be verified that this bound does not exceed $2\mu_x\sigma_s^2/\nu$, which is $O(\mu_x)$, for any $\mu_x < \mu_o/2$. ■

1.2.7 MEAN-SQUARE-ERROR PERFORMANCE

We conclude from Eq. (1.28c) that the MSE can be made as small as desired by using small step-sizes, μ_x. In this section we derive a closed-form expression for the asymptotic MSE, which is more frequently called the *mean-square-deviation* (MSD) and is defined as:

$$\boxed{\text{MSD} \triangleq \lim_{i\to\infty} \mathbb{E}\,\|\tilde{w}_i\|^2.} \tag{1.34a}$$

Strictly speaking, the limit on the right side of the above expression may not exist. A more accurate definition for the MSD appears in Eq. (4.86) of [2], namely,

$$\text{MSD} \triangleq \mu_x \cdot \left(\lim_{\mu_x\to 0}\ \limsup_{i\to\infty}\ \frac{1}{\mu_x}\mathbb{E}\|\tilde{w}_i\|^2\right). \tag{1.34b}$$

However, it was explained in [2, Sec. 4.5] that derivations that assume the validity of Eq. (1.34a) still lead to the same expression for the MSD to first-order in μ_x as derivations that rely on the more formal definition (1.34b). We therefore continue with Eq. (1.34a) for simplicity of presentation. We explain below how an expression for the MSD can be obtained by following the energy conservation technique of [2,23,24,45,46]. For that purpose, we need to introduce two smoothness conditions.

Assumption 1.4 (Smoothness conditions). *In addition to Assumptions 1.1 and 1.2, we assume that the Hessian matrix of the cost function and the noise covariance matrix defined by Eq. (1.17a) are locally Lipschitz continuous in a small neighborhood around $w = w^o$:*

$$\left\|\nabla_w^2 J(w^o + \delta w) - \nabla_w^2 J(w^o)\right\| \le \tau\,\|\delta w\|, \tag{1.35a}$$

$$\left\|R_{s,i}(w^o + \delta w) - R_{s,i}(w^o)\right\| \le \tau_2\,\|\delta w\|^\kappa \tag{1.35b}$$

for small perturbations $\|\delta w\| \le r$ and for some $\tau, \tau_2 \ge 0$ and $1 \le \kappa \le 2$. ■

The range of values for κ can be enlarged, e.g., to $\kappa \in (0,4]$. The only change in allowing a wider range for κ is that the exponent of the higher-order term, $O(\mu_x^{3/2})$, that will appear in several performance expressions, as is the case with Eqs. (1.41a), (1.41b), will need to be adjusted from $\frac{3}{2}$ to $\min\{\frac{3}{2}, 1+\frac{\kappa}{2}\}$, without affecting the first-order term that determines the MSD [2,4,47]. Therefore, it is sufficient to continue with $\kappa \in [1,2]$ to illustrate the key concepts though the MSD expressions will still be valid to first order in μ_x.

Using Eq. (1.9), it can be verified that condition (1.35a) translates into a global Lipschitz property relative to the minimizer w^o, i.e., it will also hold that [2,4]:

$$\|\nabla_w^2 J(w) - \nabla_w^2 J(w^o)\| \le \tau' \|w - w^o\| \tag{1.35c}$$

for all w and for some $\tau' \ge 0$. For example, both conditions (1.35a) and (1.35b) are readily satisfied by MSE costs. Using property (1.35c), we can now motivate a useful long-term model for the evolution of the error vector $\tilde{\boldsymbol{w}}_i$ after sufficient iterations, i.e., for $i \gg 1$. Indeed, let us reconsider recursion (1.29c) and introduce the deviation matrix:

$$\tilde{\boldsymbol{H}}_{i-1} \triangleq H - \boldsymbol{H}_{i-1}, \tag{1.36a}$$

where the constant (symmetric and positive-definite) matrix H is defined as:

$$\boxed{H \triangleq \nabla_w^2 J(w^o).} \tag{1.36b}$$

Substituting Eq. (1.36a) into Eq. (1.29c) gives

$$\tilde{\boldsymbol{w}}_i = (I_M - \boldsymbol{\mu}(i)H)\tilde{\boldsymbol{w}}_{i-1} + \boldsymbol{\mu}(i)\boldsymbol{s}_i(\boldsymbol{w}_{i-1}) + \boldsymbol{\mu}(i)\boldsymbol{c}_{i-1}, \tag{1.37a}$$

where

$$\boldsymbol{c}_{i-1} \triangleq \tilde{\boldsymbol{H}}_{i-1}\tilde{\boldsymbol{w}}_{i-1}. \tag{1.37b}$$

Using Eq. (1.35c) and the fact that $(\mathbb{E}\boldsymbol{a})^2 \le \mathbb{E}\boldsymbol{a}^2$ for any real-valued random variable, we can bound the conditional expectation of the norm of the perturbation term as follows:

$$\begin{aligned}
\mathbb{E}\left[\,\|\boldsymbol{\mu}(i)\boldsymbol{c}_{i-1}\|\,|\boldsymbol{\mathcal{F}}_{i-1}\right] &= (\mathbb{E}\boldsymbol{\mu}(i))\left(\mathbb{E}\left[\,\|\boldsymbol{c}_{i-1}\|\,|\boldsymbol{\mathcal{F}}_{i-1}\right]\right) \\
&\le \sqrt{(\mathbb{E}\boldsymbol{\mu}^2(i))}\,\mathbb{E}\left[\|\boldsymbol{c}_{i-1}\|\,|\boldsymbol{\mathcal{F}}_{i-1}\right] \\
&\overset{(1.35c)}{\le} \sqrt{\bar{\mu}^2+\sigma_\mu^2}\cdot\frac{\tau'}{2}\,\|\tilde{\boldsymbol{w}}_{i-1}\|^2 \\
&\le \frac{\bar{\mu}^2+\sigma_\mu^2}{\bar{\mu}}\cdot\frac{\tau'}{2}\,\|\tilde{\boldsymbol{w}}_{i-1}\|^2 \\
&= \mu_x\cdot\frac{\tau'}{2}\,\|\tilde{\boldsymbol{w}}_{i-1}\|^2
\end{aligned} \tag{1.38a}$$

so that using Eq. (1.28c), we conclude that:

$$\limsup_{i\to\infty}\ \mathbb{E}\,\|\boldsymbol{\mu}(i)\boldsymbol{c}_{i-1}\| = O(\mu_x^2). \tag{1.38b}$$

We can deduce from this result that $\|\boldsymbol{\mu}(i)\boldsymbol{c}_{i-1}\| = O(\mu_x^2)$ asymptotically with *high probability* [2,4]. To see this, let $r_c = m\mu_x^2$, for any constant integer $m \ge 1$. Now, calling upon Markov's inequality [48–50], we conclude from Eq. (1.38b) that for $i \gg 1$:

$$\begin{aligned}\Pr\left(\|\boldsymbol{\mu}(i)\boldsymbol{c}_{i-1}\| < r_c\right) &= 1 - \Pr\left(\|\boldsymbol{\mu}(i)\boldsymbol{c}_{i-1}\| \geq r_c\right) \\ &\geq 1 - \frac{\mathbb{E}\|\boldsymbol{\mu}(i)\boldsymbol{c}_{i-1}\|}{r_c} \\ &\overset{(1.38\text{b})}{\geq} 1 - O(1/m). \end{aligned} \tag{1.38c}$$

This result shows that the probability of having $\|\boldsymbol{\mu}(i)\boldsymbol{c}_{i-1}\|$ bounded by r_c can be made arbitrarily close to one by selecting a large enough value for m. Once the value for m has been fixed to meet a desired confidence level, then $r_c = O(\mu_x^2)$. This analysis, along with recursion (1.37a), motivates us to assess the mean-square performance of the error recursion (1.29c) by considering instead the following long-term model, which holds with high probability after sufficient iterations $i \gg 1$:

$$\boxed{\tilde{\boldsymbol{w}}_i = (I_M - \boldsymbol{\mu}(i)H)\tilde{\boldsymbol{w}}_{i-1} + \boldsymbol{\mu}(i)\boldsymbol{s}_i(\boldsymbol{w}_{i-1}) + O(\mu_x^2).} \tag{1.39}$$

Working with iteration (1.39) is helpful because its dynamics are driven by the constant matrix H as opposed to the random matrix $\boldsymbol{H}_{i-1}$ in the original error recursion (1.29c). If desired, it can be shown that, under some technical conditions on the fourth-order moment of the gradient noise process, the MSD expression that will result from using Eq. (1.39) is within $O(\mu_x^{3/2})$ of the actual MSD expression for the original recursion (1.29c); see [2,4,47] for a formal proof of this fact. Therefore, it is sufficient to rely on the long-term model (1.39) to obtain performance expressions that are accurate to first order in μ_x. Fig. 1.2 provides a block-diagram representation for Eq. (1.39).

Before explaining how model (1.39) can be used to assess the MSD, we remark that there is a second useful metric for evaluating the performance of stochastic gradient algorithms. This metric relates to the mean excess-cost, which is also called the *excess-risk* (ER) in the machine learning literature [26,27] and the excess-mean-square-error (EMSE) in the adaptive filtering literature [21–23]. We denote it by the letters ER and define it as the average fluctuation of the cost function around its minimum value:

$$\boxed{\text{ER} \triangleq \lim_{i\to\infty} \mathbb{E}\left\{J(\boldsymbol{w}_{i-1}) - J(w^o)\right\}.} \tag{1.40a}$$

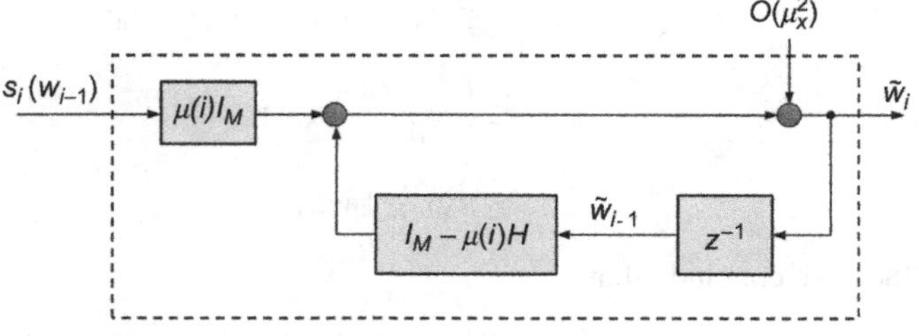

FIG. 1.2

Long-term dynamics. A block-diagram representation of the long-term recursion (1.39) for single-agent adaptation and learning.

Figure reproduced with permission from Sayed AH. Adaptive networks. Proc IEEE 2014;102(4):460–97.

Using the smoothness condition (1.35a), and the mean-value theorem [30,44] again, it can be verified that [2,4,47]:

$$\text{ER} \stackrel{\Delta}{=} \lim_{i\to\infty} \mathbb{E}\,\|\tilde{\boldsymbol{w}}_{i-1}\|^2_{\frac{1}{2}H} + O(\mu_x^{3/2}). \tag{1.40b}$$

Lemma 1.2 (Mean-square-error performance). *Assume the conditions under Assumptions 1.1–1.4 on the cost function, the gradient noise process, and the random step-size process hold. Assume further that the asynchronous step-size parameter μ_x is sufficiently small to ensure mean-square stability as required by Eq. (1.27). Then, the MSD and ER metrics for the asynchronous stochastic-gradient algorithm (1.21) are well approximated to first order in μ_x by the expressions:*

$$MSD^{asyn} = \frac{\mu_x}{2}\, Tr\left(H^{-1}R_s\right) + O(\mu_x^{3/2}), \tag{1.41a}$$

$$ER^{asyn} = \frac{\mu_x}{4}\, Tr(R_s) + O(\mu_x^{3/2}), \tag{1.41b}$$

where R_s and H are defined by Eqs. (1.17b) and (1.36b), and where we are adding the superscript "asyn" for clarity in order to distinguish these measures from the corresponding measures in the synchronous case (mentioned below in Eqs. (1.43a)–(1.43c)). Moreover, we derived earlier in Eq. (1.28b) the following expression for the convergence rate:

$$\alpha^{asyn} = 1 - 2\nu\bar{\mu} + O(\mu_x^2). \tag{1.41c}$$

Proof. We introduce the eigen-decomposition $H = U\Lambda U^{\mathsf{T}}$, where U is orthonormal and Λ is diagonal with positive entries, and rewrite Eq. (1.39) in terms of transformed quantities:

$$\overline{\boldsymbol{w}}_i = (I_M - \boldsymbol{\mu}(i)\,\Lambda)\,\overline{\boldsymbol{w}}_{i-1} + \boldsymbol{\mu}(i)\bar{\boldsymbol{s}}_i(\boldsymbol{w}_{i-1}) + O(\mu_x^2), \tag{1.42a}$$

where $\overline{\boldsymbol{w}}_i = U^{\mathsf{T}}\tilde{\boldsymbol{w}}_i$ and $\bar{\boldsymbol{s}}_i(\boldsymbol{w}_{i-1}) = U^{\mathsf{T}}\boldsymbol{s}_i(\boldsymbol{w}_{i-1})$. Let Σ denote an arbitrary $M \times M$ diagonal matrix with positive entries that we are free to choose. Then, equating the weighted squared norms of both sides of Eq. (1.42a) and taking expectations gives for $i \gg 1$:

$$\mathbb{E}\|\overline{\boldsymbol{w}}_i\|^2_\Sigma = \mathbb{E}\|\overline{\boldsymbol{w}}_{i-1}\|^2_{\Sigma'} + (\bar{\mu}^2 + \sigma_\mu^2)\,\mathbb{E}\|\bar{\boldsymbol{s}}_i(\boldsymbol{w}_{i-1})\|^2_\Sigma + O(\mu_x^{5/2}), \tag{1.42b}$$

where

$$\Sigma' \stackrel{\Delta}{=} \mathbb{E}(I_M - \boldsymbol{\mu}(i)\,\Lambda)\Sigma(I_M - \boldsymbol{\mu}(i)\,\Lambda) = \Sigma - 2\bar{\mu}\Lambda\Sigma + O(\mu_x^2). \tag{1.42c}$$

From Eqs. (1.17b), (1.19b), (1.28c), and (1.35b) we obtain:

$$\lim_{i\to\infty} \mathbb{E}\|\bar{\boldsymbol{s}}_i(\boldsymbol{w}_{i-1})\|^2_\Sigma = \mathrm{Tr}(U\Sigma U^{\mathsf{T}} R_s) + O(\mu_x^{\kappa/2}). \tag{1.42d}$$

Therefore, substituting into Eq. (1.42b) gives for $i \to \infty$:

$$\lim_{i\to\infty} \mathbb{E}\|\overline{\boldsymbol{w}}_i\|^2_{2\Lambda\Sigma} = \mu_x\,\mathrm{Tr}(U\Sigma U^{\mathsf{T}} R_s) + O(\mu_x^{3/2}). \tag{1.42e}$$

Because we are free to choose Σ, we let $\Sigma = \frac{1}{2}\Lambda^{-1}$ and arrive at Eq. (1.41a) because $\|\overline{\boldsymbol{w}}_i\|^2 = \|\tilde{\boldsymbol{w}}_i\|^2$ and $U\Sigma U^{\mathsf{T}} = \frac{1}{2}H^{-1}$. On the other hand, selecting $\Sigma = \frac{1}{4}I_M$ leads to Eq. (1.41b). ∎

We recall our earlier remark that the synchronous stochastic-gradient recursion (1.12) can be viewed as a special case of the asynchronous update (1.21) by setting $\sigma_\mu^2 = 0$ and $\mu_x = \bar{\mu} \equiv \mu$. Substituting these values into Eqs. (1.41a)–(1.41c), we obtain for the synchronous implementation (1.12):

$$\text{MSD}^{\text{sync}} = \frac{\mu}{2}\,\text{Tr}\left(H^{-1}R_s\right) + O(\mu^{3/2}), \tag{1.43a}$$

$$\text{ER}^{\text{sync}} = \frac{\mu}{2}\,\text{Tr}\,(R_s) + O(\mu^{3/2}), \tag{1.43b}$$

$$\alpha^{\text{sync}} = 1 - 2\nu\mu + O(\mu^2), \tag{1.43c}$$

which are the same expressions presented in [1] and which agree with classical results for LMS adaptation [51–56]. The matrices R_s that appear in these expressions, and in Eqs. (1.41a)–(1.41c), were defined earlier in Eq. (1.17b) and they correspond to the covariance matrices of the gradient noise processes in their respective (synchronous or asynchronous) implementations. The examples that follow show how expressions (1.41a) and (1.41b) can be used to recover performance metrics for MSE adaptation and learning under random updates.

Example 1.7 (Performance of asynchronous LMS adaptation). We reconsider the LMS recursion (1.13f) albeit in an asynchronous mode of operation, namely,

$$\boxed{\boldsymbol{w}_i = \boldsymbol{w}_{i-1} + 2\boldsymbol{\mu}(i)\boldsymbol{u}_i^{\mathsf{T}}\left[\boldsymbol{d}(i) - \boldsymbol{u}_i\boldsymbol{w}_{i-1}\right], \quad i \geq 0.} \tag{1.44}$$

We know from Example 1.3 and Eq. (1.18a) that this situation corresponds to $H = 2R_u$ and $R_s = 4\sigma_v^2 R_u$. Substituting into Eqs. (1.41a) and (1.41b) leads to the following expressions for the MSD and EMSE of the asynchronous LMS filter:

$$\text{MSD}_{\text{LMS}}^{\text{asyn}} \approx \mu_x M \sigma_v^2 = O(\mu_x), \tag{1.45a}$$

$$\text{EMSE}_{\text{LMS}}^{\text{asyn}} \approx \mu_x \sigma_v^2 \text{Tr}(R_u) = O(\mu_x), \tag{1.45b}$$

where here, and elsewhere, we will be using the notation $\approx$ to indicate that we are ignoring higher-order terms in μ_x. For example, let us assume a Bernoulli update model for $\boldsymbol{\mu}(i)$ where the filter updates with probability p_μ using a step-size value μ or stays inactive otherwise. In this case, we conclude from Eq. (1.24b) that $\mu_x = \mu$ so that the above performance expressions for the MSD and EMSE metrics will coincide with the values obtained in the synchronous case as well. In other words, the steady-state performance levels are not affected whether the algorithm learns in a synchronous or asynchronous manner. However, the convergence rate is affected because $\bar{\mu} = \mu p_\mu$ and, therefore,

$$\alpha_{\text{LMS}}^{\text{sync}} \approx 1 - 2\nu\mu, \tag{1.46a}$$

$$\alpha_{\text{LMS}}^{\text{asyn}} \approx 1 - 2\nu\mu p_\mu \; > \; \alpha_{\text{LMS}}^{\text{sync}}. \tag{1.46b}$$

It follows that asynchronous LMS adaptation attains the same performance levels as synchronous LMS adaptation, albeit at a slower convergence rate.

We may alternatively compare the performance of the synchronous and asynchronous implementations by fixing their convergence rates to the same value. Thus, consider now a second random update scheme with mean $\bar{\mu}$ and let us set $\mu = \bar{\mu}$. That is, the step-size used by the synchronous implementation is set equal to the mean step-size used by the asynchronous implementation. Then, in this case, we will get $\alpha_{\text{LMS}}^{\text{sync}} = \alpha_{\text{LMS}}^{\text{asyn}}$ so that the convergence rates coincide to first order. However, it now holds that $\text{MSD}_{\text{LMS}}^{\text{asyn}} > \text{MSD}_{\text{LMS}}^{\text{sync}}$ because $\bar{\mu}_x > \mu$ so that some deterioration in MSD performance occurs. ■

Example 1.8 (Performance of asynchronous online learners). Consider a stand-alone learner receiving a streaming sequence of independent data vectors $\{\boldsymbol{x}_i, i \geq 0\}$ that arise from some fixed probability distribution $\mathcal{X}$. The goal is to learn the vector w^o that optimizes some ν-strongly convex risk function $J(w)$ defined in terms of a loss function [57,58]:

$$w^o \stackrel{\Delta}{=} \text{argmin}_w\, J(w) = \text{argmin}_w\, \mathbb{E}Q(w; \boldsymbol{x}_i). \tag{1.47a}$$

In an asynchronous environment, the learner seeks w^o by running the stochastic-gradient algorithm with random step-sizes:

$$\boxed{\boldsymbol{w}_i = \boldsymbol{w}_{i-1} - \boldsymbol{\mu}(i)\, \nabla_{w^{\mathsf{T}}} Q(\boldsymbol{w}_{i-1}; \boldsymbol{x}_i), \quad i \geq 0.} \tag{1.47b}$$

The gradient noise vector is still given by

$$\boldsymbol{s}_i(\boldsymbol{w}_{i-1}) = \nabla_{w^{\mathsf{T}}} Q(\boldsymbol{w}_{i-1}; \boldsymbol{x}_i) - \nabla_{w^{\mathsf{T}}} J(\boldsymbol{w}_{i-1}). \tag{1.47c}$$

Because $\nabla_w J(w^o) = 0$, and because the distribution of $\boldsymbol{x}_i$ is stationary, it follows that the covariance matrix of $\boldsymbol{s}_i(w^o)$ is constant and given by

$$R_s = \mathbb{E}\nabla_{w^{\mathsf{T}}} Q(w^o; \boldsymbol{x}_i) \nabla_w Q(w^o; \boldsymbol{x}_i). \tag{1.47d}$$

The excess-risk measure that will result from this stochastic implementation is then given by Eq. (1.41b) so that $\text{ER} = O(\mu_x)$. ■

1.3 CENTRALIZED ADAPTATION AND LEARNING

The discussion in the previous section establishes the mean-square stability of *stand-alone* adaptive agents for small step-sizes (Lemma 1.1), and provides expressions for their MSD and ER metrics (Lemma 1.2). We now examine two situations involving a multitude of similar agents. In the first scenario, each agent senses data and analyzes it independently of the other agents. We refer to this mode of operation as noncooperative processing. In the second scenario, the agents transmit the collected data for processing at a fusion center. We refer to this mode of operation as centralized or batch processing. We motivate the discussion by considering first the case of MSE costs. Subsequently, we extend the results to more general costs.

1.3.1 NONCOOPERATIVE MSE PROCESSING

Thus, consider *separate* agents, labeled $k = 1, 2, \ldots, N$. Each agent, k, receives streaming data

$$\{\boldsymbol{d}_k(i), \boldsymbol{u}_{k,i}; i \geq 0\}, \tag{1.48a}$$

where we are using the subscript k to index the data at agent k. We assume that the data at each agent satisfies the same statistical properties as in Example 1.3, and the same linear regression model (1.13a) with a common w^o albeit with noise $\boldsymbol{v}_k(i)$. We denote the statistical moments of the data at agent k by

$$R_{u,k} = \mathbb{E}\boldsymbol{u}_{k,i}^{\mathsf{T}} \boldsymbol{u}_{k,i} > 0, \quad \sigma_{v,k}^2 = \mathbb{E}\boldsymbol{v}_k^2(i). \tag{1.48b}$$

We further assume in this motivating example that the $R_{u,k}$ are uniform across the agents so that

$$R_{u,k} \equiv R_u, \quad k = 1, 2, \ldots, N. \tag{1.48c}$$

In this way, the cost $J_k(w) = \mathbb{E}(\boldsymbol{d}_k(i) - \boldsymbol{u}_{k,i}w)^2$, which is associated with agent k, will satisfy a condition similar to Eq. (1.9) with the corresponding parameters $\{\nu, \delta\}$ given by (cf. Eq. 1.10a):

$$\nu = 2\lambda_{\min}(R_u), \quad \delta = 2\lambda_{\max}(R_u). \tag{1.49a}$$

Now, if each agent runs the asynchronous LMS learning rule (1.44) to estimate w^o on its own then, according to Eq. (1.45a), each agent k will attain an individual MSD level that is given by

$$\boxed{\text{MSD}^{\text{asyn}}_{\text{ncop},k} \approx \mu_x M \sigma^2_{v,k}, \quad k = 1, 2, \ldots, N} \tag{1.49b}$$

where we are further assuming that the parameter μ_x is uniform across the agents to enable a meaningful comparison. Moreover, according to Eq. (1.28b), agent k will converge toward this level at a rate dictated by:

$$\boxed{\alpha^{\text{asyn}}_{\text{ncop},k} \approx 1 - 4\mu_x \lambda_{\min}(R_u).} \tag{1.49c}$$

The subscript "ncop" is used in Eqs. (1.49b) and (1.49c) to indicate that these expressions are for the noncooperative mode of operation. It is seen from Eq. (1.49b) that agents with noisier data (i.e., larger $\sigma^2_{v,k}$) will perform worse and have larger MSD levels than agents with cleaner data. We are going to show in later sections that cooperation among the agents, whereby agents share information with their neighbors, can help enhance their individual performance levels.

1.3.2 CENTRALIZED MSE PROCESSING

Let us now contrast the above noncooperative solution with a centralized implementation whereby, at every iteration i, the N agents transmit their raw data $\{\boldsymbol{d}_k(i), \boldsymbol{u}_{k,i}\}$ to a fusion center for processing. In a *synchronous* environment, once the fusion center receives the raw data, it can run a standard stochastic-gradient update of the form:

$$\boldsymbol{w}_i = \boldsymbol{w}_{i-1} + \mu \left[\frac{1}{N} \sum_{k=1}^{N} 2\boldsymbol{u}^{\mathsf{T}}_{k,i}(\boldsymbol{d}_k(i) - \boldsymbol{u}_{k,i}\boldsymbol{w}_{i-1}) \right], \tag{1.50a}$$

where μ is the constant step-size and the term multiplying μ can be seen to correspond to a sample average of several approximate gradient vectors. The analysis in [1,2] showed that the MSD performance that results from this implementation is given by (using expression (1.64a) with $H_k = 2R_u$ and $R_{s,k} = 4\sigma^2_{v,k}R_u$):

$$\boxed{\text{MSD}^{\text{sync}}_{\text{cent}} \approx \frac{\mu M}{N} \left(\frac{1}{N} \sum_{k=1}^{N} \sigma^2_{v,k} \right).} \tag{1.50b}$$

Moreover, using expression (1.64b) given further ahead, this centralized solution will converge toward the above MSD level at the same rate as the noncooperative solution:

$$\boxed{\alpha_{\text{cent}}^{\text{sync}} \approx 1 - 4\mu\lambda_{\min}(R_u).} \tag{1.50c}$$

In an asynchronous environment, there are now *several* random events that can interfere with the operation of the fusion center. Let us consider initially one particular random event that corresponds to the situation in which the fusion center may or may not update at any particular iteration (e.g., due to some power-saving strategy). In a manner similar to Eq. (1.21), we may represent this scenario by writing:

$$\boldsymbol{w}_i = \boldsymbol{w}_{i-1} + \boldsymbol{\mu}(i)\left[\frac{1}{N}\sum_{k=1}^{N} 2\boldsymbol{u}_{k,i}^{\mathsf{T}}(\boldsymbol{d}_k(i) - \boldsymbol{u}_{k,i}\boldsymbol{w}_{i-1})\right] \tag{1.51a}$$

with a random step-size process, $\boldsymbol{\mu}(i)$; this process is again assumed to satisfy the conditions under Assumption 1.3. The analysis in the sequel will show that the MSD performance that results from this implementation is given by:

$$\text{MSD}_{\text{cent}}^{\text{asyn,1}} \approx \frac{\mu_x M}{N}\left(\frac{1}{N}\sum_{k=1}^{N}\sigma_{v,k}^2\right), \tag{1.51b}$$

where we are adding the superscript "1" to indicate that this is a preliminary result pertaining to the particular asynchronous implementation (1.51a). We will be generalizing this result soon, at which point we will drop the superscript "1". Likewise, using expression (1.64b) given further ahead, this version of the asynchronous centralized solution converges toward the above MSD level at the same rate as the noncooperative solution (1.49c) and the synchronous version (1.50c):

$$\alpha_{\text{cent}}^{\text{asyn,1}} \approx 1 - 4\mu_x\lambda_{\min}(R_u). \tag{1.51c}$$

Observe from Eqs. (1.50b) and (1.51b) that the MSD level attained by the centralized solution is proportional to $1/N$ times the *average* noise power across all agents. This scaled average noise power can be larger than some of the individual noise variances and smaller than the remaining noise variances. This example shows that it does not generally hold that centralized stochastic-gradient implementations outperform all individual noncooperative agents [59].

Now, more generally, observe that in the synchronous batch processing case (1.50a) as well as in the asynchronous implementation (1.51a), the data collected from the various agents are equally aggregated with a weighting factor equal to $1/N$. We can incorporate a second type of random events besides the random variations in the step-size parameter. This second source of uncertainty involves the possibility of failure (or weakening) in the links connecting the individual agents to the fusion center (e.g., due to fading or outage, or perhaps due to the agents themselves deciding to enter into a sleep mode according to some power-saving policy). We can capture these possibilities by extending formulation (1.51a) in the following manner:

$$\boxed{\boldsymbol{w}_i = \boldsymbol{w}_{i-1} + \boldsymbol{\mu}(i)\left[\sum_{k=1}^{N} 2\boldsymbol{\pi}_k(i)\boldsymbol{u}_{k,i}^{\mathsf{T}}(\boldsymbol{d}_k(i) - \boldsymbol{u}_{k,i}\boldsymbol{w}_{i-1})\right]} \tag{1.52a}$$

where $\boldsymbol{\mu}(i)$ continues to be a random step-size process, but now the coefficients $\{\boldsymbol{\pi}_k(i); k = 1, 2, \ldots, N\}$ are new random fusion coefficients that satisfy:

$$\boxed{\sum_{k=1}^{N} \boldsymbol{\pi}_k(i) = 1, \quad \boldsymbol{\pi}_k(i) \geq 0} \tag{1.52b}$$

for every $i \geq 0$.

Assumption 1.5 (Conditions on random fusion coefficients). *It is assumed that $\boldsymbol{\pi}_k(i)$ is independent of $\boldsymbol{\mu}(i)$ and of other random variables in the learning algorithm. The fusion coefficients are also independent over time, namely, $\boldsymbol{\pi}_k(i)$ and $\boldsymbol{\pi}_\ell(j)$ are independent for any $i \neq j$. For the same time i, the coefficients $\{\boldsymbol{\pi}_k(i)\}$ are correlated over space in view of the first requirement in Eq. (1.52b). The mean and covariance(s) of each $\boldsymbol{\pi}_k(i)$ are denoted by:*

$$\bar{\pi}_k \triangleq \mathbb{E}\boldsymbol{\pi}_k(i), \tag{1.53a}$$

$$c_{\pi,k\ell} \triangleq \mathbb{E}(\boldsymbol{\pi}_k(i) - \bar{\pi}_k)(\boldsymbol{\pi}_\ell(i) - \bar{\pi}_\ell) \tag{1.53b}$$

for all $k, \ell = 1, 2, \ldots, N$ and all $i \geq 0$. When $k = \ell$, the scalar $c_{\pi,kk}$ corresponds to the variance of $\boldsymbol{\pi}_k(i)$ and, therefore, we shall also use the alternative notation $\sigma_{\pi,k}^2$ for this case:

$$\sigma_{\pi,k}^2 \triangleq c_{\pi,kk} \geq 0. \tag{1.53c}$$

■

It is straightforward to verify that the first- and second-order moments of the coefficients $\{\boldsymbol{\pi}_k(i)\}$ satisfy:

$$\bar{\pi}_k \geq 0, \quad \sum_{k=1}^{N} \bar{\pi}_k = 1, \tag{1.54a}$$

$$\sum_{k=1}^{N} c_{\pi,k\ell} = 0, \quad \text{for any } \ell, \tag{1.54b}$$

$$\sum_{\ell=1}^{N} c_{\pi,k\ell} = 0, \quad \text{for any } k. \tag{1.54c}$$

Note that the earlier asynchronous implementation (1.51a) is a special case of the more general formulation (1.52a) when the mean of the random fusion coefficients is chosen as $\bar{\pi}_k = 1/N$ and the variances are set to zero, $\sigma_{\pi,k}^2 = 0$. In order to enable a fair and meaningful comparison among the three centralized implementations (1.50a), (1.51a), and (1.52a), we shall assume that the means of the fusion coefficients in the latest implementation are set to $\bar{\pi}_k = 1/N$ (results for arbitrary mean values are listed in Example 1.9 further ahead and appear in [5]). We will show in the sequel that for this choice of $\bar{\pi}_k$, the MSD performance of implementation (1.52a) is given by:

$$\boxed{\text{MSD}_{\text{cent}}^{\text{asyn}} \approx \frac{\mu_x M}{N}\left[\frac{1}{N}\sum_{k=1}^{N}\left(1+N^2\,\sigma_{\pi,k}^2\right)\sigma_{v,k}^2\right].} \tag{1.55a}$$

Moreover, using expression (1.62b) given further ahead, this asynchronous centralized solution will converge toward the above MSD level at the same rate as the noncooperative solution (1.49c) and the synchronous version (1.50c):

$$\boxed{\alpha_{\text{cent}}^{\text{asyn}} \approx 1 - 4\mu_x\lambda_{\min}(R_u).} \tag{1.55b}$$

1.3.3 STOCHASTIC-GRADIENT CENTRALIZED SOLUTION

The previous two sections focused on MSE costs. We now extend the conclusions to more general costs. Thus, consider a collection of N agents, each with an individual convex cost function, $J_k(w)$. The objective is to determine the unique minimizer w^o of the aggregate cost:

$$J^{\text{glob}}(w) \triangleq \sum_{k=1}^{N} J_k(w). \tag{1.56}$$

It is now the above aggregate cost, $J^{\text{glob}}(w)$, that will be required to satisfy the conditions of Assumptions 1.1 and 1.4 relative to some parameters $\{\nu_c, \delta_c, \tau_c\}$, with the subscript "c" used to indicate that these factors correspond to the centralized implementation. Under these conditions, the cost $J^{\text{glob}}(w)$ will have a unique minimizer, which we continue to denote by w^o. We will not be requiring each individual cost, $J_k(w)$, to be strongly convex. It is sufficient for at least one of these costs to be strongly convex while the remaining costs can be convex; this condition ensures the strong convexity of $J^{\text{glob}}(w)$. Moreover, although some individual costs may not have a unique minimizer, we require in this exposition that w^o is one of their minima so that all individual costs share a minimum at location w^o; the treatment in [1,2,60] considers the more general case in which the minimizers of the individual costs $\{J_k(w)\}$ can be different and need not contain a common location, w^o.

There are many centralized solutions that can be used to determine the unique minimizer w^o of Eq. (1.56), with some solution techniques being more powerful than other. Nevertheless, we shall focus on centralized implementations of the stochastic-gradient type. The reason we consider the *same* class of stochastic gradient algorithms for noncooperative, centralized, and distributed solutions in this chapter is to enable a *meaningful* comparison among the various implementations. Thus, we first consider a *synchronous* centralized strategy of the following form:

$$\boldsymbol{w}_i = \boldsymbol{w}_{i-1} - \mu\left[\frac{1}{N}\sum_{k=1}^{N}\widehat{\nabla_{w^{\mathsf{T}}} J_k}(\boldsymbol{w}_{i-1})\right], \quad i \geq 0 \tag{1.57a}$$

with a constant step-size, μ. When the fusion center employs random step-sizes, the above solution is replaced by:

$$\boldsymbol{w}_i = \boldsymbol{w}_{i-1} - \boldsymbol{\mu}(i) \left[\frac{1}{N} \sum_{k=1}^{N} \widehat{\nabla_{w^{\mathsf{T}}} J_k}(\boldsymbol{w}_{i-1}) \right], \quad i \geq 0, \tag{1.57b}$$

where the process $\boldsymbol{\mu}(i)$ now satisfies Assumption 1.3. More generally, the asynchronous implementation can employ random fusion coefficients as well such as:

$$\boxed{\boldsymbol{w}_i = \boldsymbol{w}_{i-1} - \boldsymbol{\mu}(i) \sum_{k=1}^{N} \boldsymbol{\pi}_k(i) \widehat{\nabla_{w^{\mathsf{T}}} J_k}(\boldsymbol{w}_{i-1}), \quad i \geq 0} \tag{1.57c}$$

where the coefficients $\{\boldsymbol{\pi}_k(i)\}$ satisfy Assumption 1.5 with means

$$\boxed{\bar{\pi}_k = 1/N.} \tag{1.57d}$$

1.3.4 PERFORMANCE OF CENTRALIZED SOLUTION

To examine the performance of the asynchronous implementation (1.57c) and (1.57d), we proceed in two steps. First, we identify the gradient noise that is present in the recursion; it is equal to the difference between the true gradient vector for the global cost, $J^{\text{glob}}(w)$, defined by Eq. (1.56) and its approximation. Second, we argue that Eq. (1.57c) has a form similar to the single-agent stochastic-gradient algorithm (1.21) and, therefore, invoke earlier results to write down performance metrics for the centralized solution (1.57c) and (1.57d).

We start by introducing the individual gradient noise processes:

$$\boxed{\boldsymbol{s}_{k,i}(\boldsymbol{w}_{i-1}) \triangleq \widehat{\nabla_{w^{\mathsf{T}}} J_k}(\boldsymbol{w}_{i-1}) - \nabla_{w^{\mathsf{T}}} J_k(\boldsymbol{w}_{i-1})} \tag{1.58a}$$

for $k = 1, 2, \ldots, N$. We assume that these noises satisfy conditions similar to Assumption 1.2 with parameters $\{\beta_k^2, \sigma_{s,k}^2, R_{s,k}\}$, i.e.,

$$R_{s,k} \triangleq \lim_{i \to \infty} \mathbb{E}\left[\boldsymbol{s}_{k,i}(w^o)\boldsymbol{s}_{k,i}^{\mathsf{T}}(w^o) | \boldsymbol{\mathcal{F}}_{i-1}\right] \tag{1.58b}$$

and

$$\mathbb{E}\left[\boldsymbol{s}_{k,i}(\boldsymbol{w}_{i-1}) | \boldsymbol{\mathcal{F}}_{i-1}\right] = 0, \tag{1.58c}$$

$$\mathbb{E}\left[\|\boldsymbol{s}_{k,i}(\boldsymbol{w}_{i-1})\|^2 | \boldsymbol{\mathcal{F}}_{i-1}\right] \leq \beta_k^2 \, \|\tilde{\boldsymbol{w}}_{i-1}\|^2 + \sigma_{s,k}^2. \tag{1.58d}$$

Additionally, we assume that the gradient noise components across the agents are uncorrelated with each other:

$$\mathbb{E}\left[\boldsymbol{s}_{k,i}(\boldsymbol{w}_{i-1})\boldsymbol{s}_{\ell,i}^{\mathsf{T}}(\boldsymbol{w}_{i-1}) \mid \boldsymbol{\mathcal{F}}_{i-1}\right] = 0, \quad \text{all } k \neq \ell. \tag{1.58e}$$

Using these gradient noise terms, it is straightforward to verify that recursion (1.57c) can be rewritten as:

$$\boldsymbol{w}_i = \boldsymbol{w}_{i-1} - \frac{\boldsymbol{\mu}(i)}{N}\left[\boldsymbol{s}_i(\boldsymbol{w}_{i-1}) + \sum_{k=1}^{N} \nabla_{w^{\mathsf T}} J_k(\boldsymbol{w}_{i-1})\right], \tag{1.59}$$

where $\boldsymbol{s}_i(\boldsymbol{w}_{i-1})$ denotes the overall gradient noise; its expression is given by

$$\boldsymbol{s}_i(\boldsymbol{w}_{i-1}) = \sum_{k=1}^{N}\left[N\boldsymbol{\pi}_k(i)\widehat{\nabla_{w^{\mathsf T}} J_k}(\boldsymbol{w}_{i-1}) - \nabla_{w^{\mathsf T}} J_k(\boldsymbol{w}_{i-1})\right]. \tag{1.60}$$

Because iteration (1.59) has the form of a stochastic gradient recursion with random update similar to Eq. (1.21), we can infer its MSE behavior from Lemmas 1.1 and 1.2 if the noise process $\boldsymbol{s}_i(\boldsymbol{w}_{i-1})$ can be shown to satisfy conditions similar to Assumption 1.2 with some parameters $\{\beta_c^2, \sigma_s^2\}$. Indeed, starting from Eq. (1.60), some algebra will show that

$$\mathbb{E}\left[\boldsymbol{s}_i(\boldsymbol{w}_{i-1})|\boldsymbol{\mathcal{F}}_{i-1}\right] = 0, \tag{1.61a}$$

$$\mathbb{E}\left[\|\boldsymbol{s}_i(\boldsymbol{w}_{i-1})\|^2|\boldsymbol{\mathcal{F}}_{i-1}\right] \le \beta_c^2\|\tilde{\boldsymbol{w}}_{i-1}\|^2 + \sigma_s^2, \tag{1.61b}$$

where

$$\beta_c^2 \triangleq \sum_{k=1}^{N}\left[\beta_k^2 + N^2\sigma_{\pi,k}^2(\beta_k^2 + \delta_c^2)\right], \tag{1.61c}$$

$$\sigma_s^2 \triangleq \sum_{k=1}^{N}(1 + N^2\sigma_{\pi,k}^2)\sigma_{s,k}^2. \tag{1.61d}$$

The following result now follows from Lemmas 1.1 and 1.2 [2,5].

Lemma 1.3 (Convergence of centralized solution). *Assume the aggregate cost (1.56) satisfies the conditions under Assumption 1.1 for some parameters $0 < \nu_c \le \delta_c$. Assume further that the individual gradient noise processes defined by Eq. (1.58a) satisfy the conditions under Assumption 1.2 for some parameters $\{\beta_k^2, \sigma_{s,k}^2, R_{s,k}\}$, in addition to the orthogonality condition (1.58e). Let $\mu_o = 2\nu_c/(\delta_c^2 + \beta_c^2)$. For any $\mu_x/N < \mu_o$, the iterates generated by the asynchronous centralized solution (1.57c) and (1.57d) satisfy:*

$$\mathbb{E}\|\tilde{\boldsymbol{w}}_i\|^2 \le \alpha\,\mathbb{E}\|\tilde{\boldsymbol{w}}_{i-1}\|^2 + \sigma_s^2(\bar{\mu}^2 + \sigma_\mu^2)/N^2, \tag{1.62a}$$

where the scalar α satisfies $0 \le \alpha < 1$ and is given by

$$\alpha_{cent}^{asyn} = 1 - 2\nu_c\,(\mu_x/N) + (\delta_c^2 + \beta_c^2)(\bar{\mu}^2 + \sigma_\mu^2)/N^2. \tag{1.62b}$$

It follows from Eq. (1.62a) that for sufficiently small step-size parameter $\mu_x \ll 1$:

$$\limsup_{i\to\infty}\mathbb{E}\|\tilde{\boldsymbol{w}}_i\|^2 = O(\mu_x). \tag{1.62c}$$

Moreover, under smoothness conditions similar to Eq. (1.35a) for $J^{glob}(w)$ for some parameter $\tau_c \geq 0$, and similar to Eq. (1.35b) for the individual gradient noise covariance matrices, it holds for small μ_x that:

$$MSD_{cent}^{asyn} = \frac{\mu_x}{2N} Tr\left[\left(\sum_{k=1}^{N} H_k\right)^{-1}\left(\sum_{k=1}^{N}(1+N^2\sigma_{\pi,k}^2)R_{s,k}\right)\right] + O\left(\mu_x^{3/2}\right), \tag{1.63}$$

where $H_k = \nabla_w^2 J_k(w^o)$. ■

We can recover from the expressions in the lemma, performance results for the particular asynchronous implementation described earlier by Eq. (1.57b) by setting $\sigma_{\pi,k}^2 = 0$ so that

$$\mathrm{MSD}_{\mathrm{cent}}^{\mathrm{asyn},1} \approx \frac{\mu_x}{2N} \mathrm{Tr}\left[\left(\sum_{k=1}^{N} H_k\right)^{-1}\left(\sum_{k=1}^{N} R_{s,k}\right)\right], \tag{1.64a}$$

$$\alpha_{\mathrm{cent}}^{\mathrm{asyn},1} \approx 1 - 2\nu_c\mu_x/N. \tag{1.64b}$$

It is seen from Eqs. (1.63) and (1.64a) that when the fusion center operates under the more general asynchronous policy (1.57c), the additional randomness in the fusion coefficients $\{\boldsymbol{\pi}_k(i)\}$ degrades the MSD performance relative to Eq. (1.64a) due to the presence of the factor $1 + N^2\sigma_{\pi,k}^2 > 1$, i.e., to first order in μ_x we have:

$$\mathrm{MSD}^{\mathrm{asyn},1} < \mathrm{MSD}^{\mathrm{asyn}}. \tag{1.65a}$$

In comparison, the added randomness due to the $\{\boldsymbol{\pi}_k(i)\}$ does not have a significant impact on the convergence rate because it is straightforward to see that to first order in the step-size parameter μ_x:

$$\alpha_{\mathrm{cent}}^{\mathrm{asyn},1} \approx \alpha_{\mathrm{cent}}^{\mathrm{asyn}}. \tag{1.65b}$$

Likewise, setting $\bar{\mu}_k = \mu$ and $\sigma_{\pi,k}^2 = 0$, the performance of the synchronous centralized solution (1.57a) is obtained as a special case of the results in the lemma:

$$\mathrm{MSD}_{\mathrm{cent}}^{\mathrm{sync}} \approx \frac{\mu}{2N} \mathrm{Tr}\left[\left(\sum_{k=1}^{N} H_k\right)^{-1}\left(\sum_{k=1}^{N} R_{s,k}\right)\right] \tag{1.66a}$$

$$\alpha_{\mathrm{cent}}^{\mathrm{sync}} \approx 1 - 2\nu_c\mu/N. \tag{1.66b}$$

These expressions agree with the performance results presented in [1,2].

Example 1.9 (Case of general mean-values). Although not used in this chapter, we remark in passing that if the mean values $\{\bar{\pi}_k\}$ are not fixed at $1/N$, as was required by Eq. (1.57d), then the MSD performance of the asynchronous centralized solution (1.57c) will instead be given by

$$\mathrm{MSD}_{\mathrm{cent}}^{\mathrm{asyn}} = \frac{\mu_x N}{2} \mathrm{Tr}\left[\left(\sum_{k=1}^{N} H_k\right)^{-1}\left(\sum_{k=1}^{N}(\bar{\pi}_k^2 + \sigma_{\pi,k}^2)R_{s,k}\right)\right] + O\left(\mu_x^{3/2}\right). \tag{1.67}$$

■

1.3.5 COMPARISON WITH NONCOOPERATIVE PROCESSING

We can now compare the performance of the asynchronous centralized solution (1.57c) against the performance of noncooperative processing when agents act independently of each other and run the recursion:

$$\boxed{\boldsymbol{w}_{k,i}=\boldsymbol{w}_{k,i-1} - \boldsymbol{\mu}(i)\,\widehat{\nabla_{w^{\mathsf{T}}} J}_k(\boldsymbol{w}_{k,i-1}), \quad i \geq 0.} \tag{1.68a}$$

This comparison is *meaningful* when all agents share the same unique minimizer so that we can compare how well the individual agents are able to recover the same w^o as the centralized solution. For this reason, we reintroduce the requirement that all individual costs $\{J_k(w)\}$ are ν-strongly convex with a uniform parameter ν. Because $J^{\text{glob}}(w)$ is the aggregate sum of the individual costs, then we can set the lower bound ν_c for the Hessian of $J^{\text{glob}}(w)$ at $\nu_c = N\nu$. From expressions (1.28b) and (1.64b) we then conclude that, for a sufficiently small μ_x, the convergence rates of the asynchronous noncooperative solution (1.68a) and the asynchronous centralized solution with random update (1.57b) will be similar:

$$\alpha_{\text{cent}}^{\text{asyn}} \approx 1 - 2\nu_c\,(\mu_x/N) = 1 - 2\nu\mu_x \approx \alpha_{\text{ncop},k}^{\text{asyn}}. \tag{1.68b}$$

Moreover, we observe from Eq. (1.41a) that the average MSD level across N noncooperative asynchronous agents is given by

$$\text{MSD}_{\text{ncop,av}}^{\text{asyn}} \approx \frac{\mu_x}{2N}\,\text{Tr}\left[\sum_{k=1}^{N} H_k^{-1} R_{s,k}\right] \tag{1.68c}$$

so that comparing with Eq. (1.64a), some simple algebra allows us to conclude that, for small step-sizes and to first order in μ_x:

$$\text{MSD}_{\text{cent}}^{\text{asyn},1} \leq \text{MSD}_{\text{ncop,av}}^{\text{asyn}}. \tag{1.68d}$$

That is, while the asynchronous centralized solution (1.57b) with random updates need not outperform every individual noncooperative agent in general, its performance outperforms the average performance across all noncooperative agents.

The next example illustrates this result by considering the scenario where all agents have the same Hessian matrices at $w = w^o$, namely, $H_k \equiv H$ for $k = 1, 2, \ldots, N$. This situation occurs, for example, when the individual costs are identical across the agents, say, $J_k(w) \equiv J(w)$, as is common in machine learning applications. This situation also occurs for the MSE costs we considered earlier in this section when the regression covariance matrices, $\{R_{u,k}\}$, are uniform across all agents, i.e., $R_{u,k} \equiv R_u$ for $k = 1, 2, \ldots, N$. In these cases with uniform Hessian matrices H_k, the example below establishes that the asynchronous centralized solution (1.57b) with random updates improves over the average MSD performance of the noncooperative solution (1.68a) by a factor of N.

Example 1.10 (N-fold improvement in MSD performance). Consider a collection of N agents whose individual cost functions, $J_k(w)$, are ν-strongly convex and are minimized at the same location $w = w^o$. The costs are also assumed to have identical Hessian matrices at $w = w^o$, i.e., $H_k \equiv H$. Then, using Eq. (1.64a), the MSD of the asynchronous centralized implementation (1.57b) with random updates is given by:

$$\text{MSD}_{\text{cent}}^{\text{asyn},1} \approx \frac{1}{N}\left(\frac{\mu_x}{2N}\sum_{k=1}^{N}\text{Tr}(H^{-1}R_{s,k})\right) \approx \frac{1}{N}\,\text{MSD}_{\text{ncop,av}}^{\text{asyn}}. \tag{1.69}$$

■

Example 1.11 (Random fusion can degrade MSD performance). Although the convergence rates of the asynchronous centralized solution (1.57c) and the noncooperative solution (1.68a) agree to first order in μ_x, the relation between their MSD values is indefinite (contrary to Eq. 1.68d), as illustrated by the following example.

Consider the same setting of Example 1.10 and assume further that the variances of the random fusion coefficients are uniform, i.e., $\sigma^2_{\pi,k} \equiv \sigma^2_\pi$. Then, using Eq. (1.63), the MSD of the asynchronous centralized implementation (1.57c) is given by

$$\text{MSD}^{\text{asyn}}_{\text{cent}} \approx \frac{1+N^2\sigma^2_\pi}{N}\left(\frac{\mu_x}{2N}\sum_{k=1}^{N}\text{Tr}(H^{-1}R_{s,k})\right) \approx \left(\frac{1+N^2\sigma^2_\pi}{N}\right)\text{MSD}^{\text{asyn}}_{\text{ncop,av}}. \tag{1.70}$$

Therefore, to first order in μ_x, we find that

$$\begin{cases} \text{MSD}^{\text{asyn}}_{\text{cent}} \le \text{MSD}^{\text{asyn}}_{\text{ncop,av}}, & \text{if } \sigma^2_\pi \le \frac{N-1}{N^2}, \\ \text{MSD}^{\text{asyn}}_{\text{cent}} > \text{MSD}^{\text{asyn}}_{\text{ncop,av}}, & \text{if } \sigma^2_\pi > \frac{N-1}{N^2}. \end{cases} \tag{1.71}$$

In other words, if the variance of the random fusion coefficients is large enough, then the centralized solution will generally have degraded performance relative to the noncooperative solution (which is an expected result). ■

Example 1.12 (Fully connected networks). In preparation for the discussion on networked agents, it is useful to describe one extreme situation where a collection of N agents are fully connected to each other; see Fig. 1.3. In this case, each agent is able to access the data from all other agents and, therefore, each individual agent can run a synchronous or asynchronous centralized implementation, say, one of the same form as Eq. (1.57b):

$$\boldsymbol{w}_{k,i} = \boldsymbol{w}_{k,i-1} - \boldsymbol{\mu}(i)\left[\frac{1}{N}\sum_{\ell=1}^{N}\widehat{\nabla_{w^{\mathsf T}} J_\ell}(\boldsymbol{w}_{k,i-1})\right], \quad i \ge 0. \tag{1.72}$$

When this happens, each agent will attain the same performance level as that of the asynchronous centralized solution (1.57b). Two observations are in place. First, note from Eq. (1.72) that the information that agent k is receiving from all other agents is their gradient vector approximations. Obviously, other pieces of information could be shared among the agents, such as their iterates $\{\boldsymbol{w}_{\ell,i-1}\}$. Second, note that the right most term multiplying $\boldsymbol{\mu}(i)$ in Eq. (1.72) corresponds to a convex combination of the approximate gradients from the various agents, with the combination coefficients being uniform and all equal to $1/N$. In general, there is no need for these combination weights to be identical. Even more importantly, agents do not need to have access to information from all other agents in the network. We are going to see in the sequel that interactions with a limited number of neighbors are sufficient for the agents to attain performance that is comparable to that of the centralized solution.

Fig. 1.4 shows a simple selection of connected topologies for five agents. The leftmost panel corresponds to the noncooperative case and the rightmost panel corresponds to the fully connected case. The panels in-between illustrate some other topologies. In the coming sections, we are going to present results that allow us to answer useful questions about such networked agents such as: (a) Which topology has the best performance in terms of MSE and convergence rate? (b) Given a connected topology, can it be made to approach the performance of the centralized solution? (c) Which aspects of the topology influence performance? (d) Which aspects of the combination weights (policy) influence

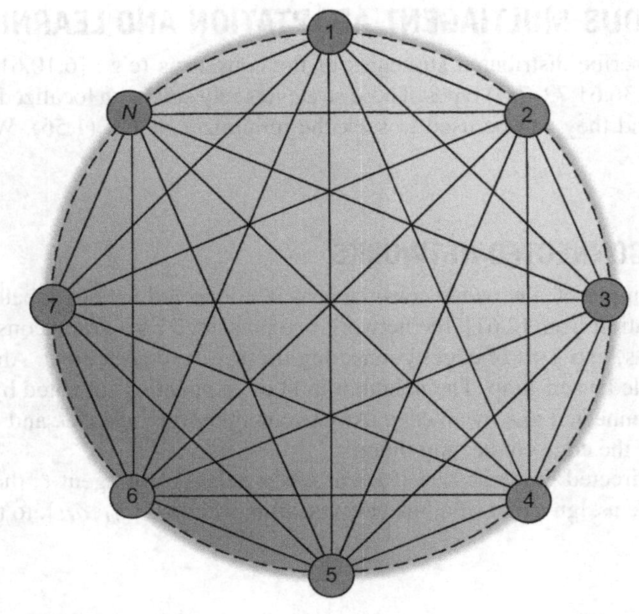

FIG. 1.3

Fully connected network. Example of a fully connected network where each agent can access information from all other agents.

Figure reproduced with permission from Sayed AH. Adaptive networks. Proc IEEE 2014;102(4):460–97.

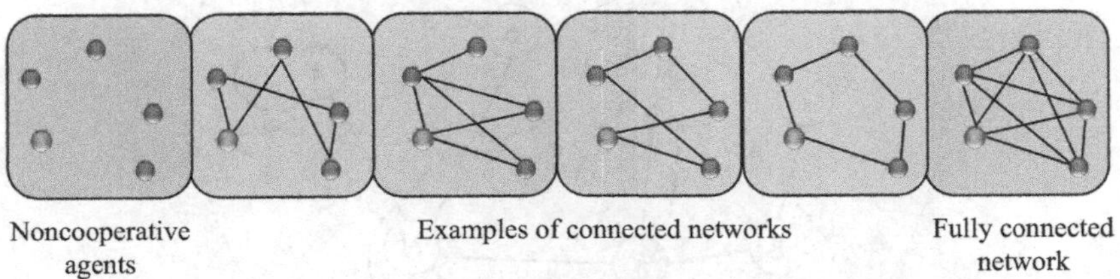

FIG. 1.4

Connected networks. Examples of connected networks, with the left most panel representing a collection of noncooperative agents.

Figure reproduced with permission from Sayed AH. Adaptive networks. Proc IEEE 2014;102(4):460–97.

performance? (e) Can different topologies deliver similar performance levels? (f) Is cooperation always beneficial? and (g) If the individual agents are able to solve the inference task individually in a stable manner, does it follow that the connected network will remain stable regardless of the topology and regardless of the cooperation strategy? ■

1.4 SYNCHRONOUS MULTIAGENT ADAPTATION AND LEARNING

In this section, we describe distributed strategies of the consensus (e.g., [6,10,61] and [62–71]) and diffusion (e.g., [1,2,22,36,61,72–74]) types. These strategies rely solely on localized interactions among neighboring agents, and they can be used to seek the minimizer of Eq. (1.56). We first describe the network model.

1.4.1 STRONGLY CONNECTED NETWORKS

Fig. 1.5 shows an example of a network consisting of N connected agents, labeled $k = 1, 2, \ldots, N$. Following the presentation from [2,61], the network is represented by a graph consisting of N vertices (representing the agents) and a set of edges connecting the agents to each other. An edge that connects an agent to itself is called a self-loop. The neighborhood of an agent k is denoted by $\mathcal{N}_k$ and it consists of all agents that are connected to k by an edge. Any two neighboring agents k and ℓ have the ability to share information over the edge connecting them.

We assume an undirected graph so that if agent k is a neighbor of agent ℓ, then agent ℓ is also a neighbor of agent k. We assign a pair of nonnegative scaling weights, $\{a_{k\ell}, a_{\ell k}\}$, to the edge connecting

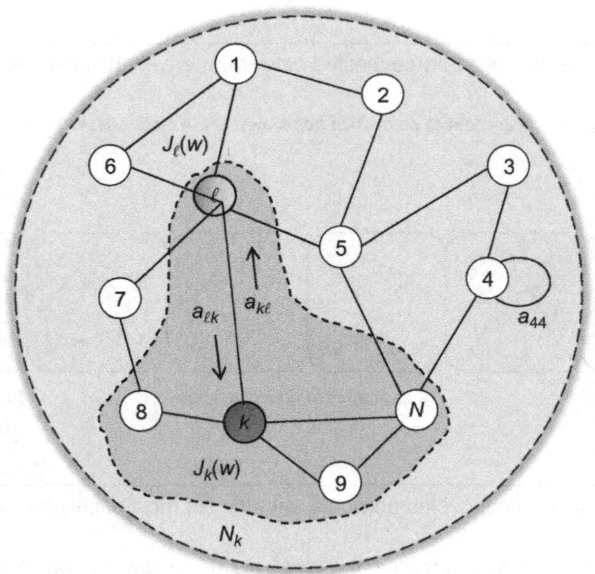

FIG. 1.5

Neighborhoods and interactions. Agents that are linked by edges can share information. The neighborhood of agent k is marked by the highlighted area.

Figure reproduced with permission from Sayed AH. Adaptive networks. Proc IEEE 2014;102(4):460–97.

k and ℓ. The scalar $a_{\ell k}$ is used by agent k to scale the data it receives from agent ℓ; this scaling can be interpreted as a measure of the confidence level that agent k assigns to its interaction with agent ℓ. Likewise, $a_{k\ell}$ is used by agent ℓ to scale the data it receives from agent k. The weights $\{a_{k\ell}, a_{\ell k}\}$ can be different so that the exchange of information between the neighboring agents $\{k, \ell\}$ need not be symmetrical. One or both weights can also be zero.

A network is said to be connected if paths with nonzero scaling weights can be found linking any two distinct agents in *both* directions, either directly when they are neighbors or by passing through intermediate agents when they are not neighbors. In this way, information can flow in both directions between any two agents in the network, although the forward path from an agent k to some other agent ℓ need not be the same as the backward path from ℓ to k. A strongly connected network is a connected network with at least one nontrivial self-loop, meaning that $a_{kk} > 0$ for some agent k.

The strong connectivity of a network translates into a useful property on the combination weights. Assume we collect the coefficients $\{a_{\ell k}\}$ into an $N \times N$ matrix $A = [a_{\ell k}]$, such that the entries on the kth column of A contain the coefficients used by agent k to scale data arriving from its neighbors $\ell \in \mathcal{N}_k$; we set $a_{\ell k} = 0$ if $\ell \notin \mathcal{N}_k$. We refer to A as the *combination* matrix or policy. It turns out that combination matrices that correspond to strongly connected networks are *primitive*—an $N \times N$ matrix A with nonnegative entries is said to be primitive if there exists some finite integer $n_o > 0$ such that all entries of A^{n_o} are strictly positive [2,61,75].

1.4.2 DISTRIBUTED OPTIMIZATION

Network cooperation can be exploited to solve adaptation, learning, and optimization problems in a decentralized manner in response to streaming data. To explain how cooperation can be achieved, we start by associating with each agent k a twice-differentiable cost function $J_k(w) : \mathbb{R}^{M\times 1} \mapsto \mathbb{R}$. The objective of the network of agents is to seek the unique minimizer of the aggregate cost function, $J^{\text{glob}}(w)$, defined by Eq. (1.56). Now, however, we seek a *distributed* (as opposed to a centralized) solution. In a distributed implementation, each agent k can only rely on its own data and on data from its neighbors.

We continue to assume that $J^{\text{glob}}(w)$ satisfies the conditions of Assumptions 1.1 and 1.4 with parameters $\{\nu_d, \delta_d, \tau_d\}$, with the subscript "d" now used to indicate that these parameters are related to the distributed implementation. Under these conditions, the cost $J^{\text{glob}}(w)$ will have a unique minimizer, which we continue to denote by w^o. For simplicity of presentation, we will also assume in the remainder of this chapter that the individual costs $J_k(w)$ are strongly convex as well. These costs can be distinct across the agents or they can all be identical, i.e., $J_k(w) \equiv J(w)$ for $k = 1, 2, \ldots, N$; in the latter situation, the problem of minimizing Eq. (1.56) would correspond to the case in which the agents work together to optimize the same cost function. If we let w_k^o denote the minimizer of $J_k(w)$, we continue to assume for this exposition that each $J_k(w)$ is also minimized at w^o:

$$w_k^o \equiv w^o, \quad k = 1, 2, \ldots, N. \tag{1.73}$$

The case where the individual costs are only convex and need not be strongly convex is discussed in [1,2,37]; most of the results and conclusions continue to hold but the derivations become more technical. Likewise, the case in which the individual costs need not share minimizers is discussed in these references. In that case, as was already shown in [60], the iterates $\boldsymbol{w}_{k,i}$ by the individual agents will not approach the minimizer w^o of Eq. (1.56), but rather the minimizer $w^\star$ of a weighted aggregate function with some positive scaling weights $\{q_k\}$. This function was defined in Eq. (57b) of [1]. It was also explained in [1] how this convergence property adds a useful degree of freedom to the operation of the network, and how it can be exploited advantageously to steer the network to converge to desirable limit points, including to w^o, through the selection of the combination policy A (which determines the scaling weights $\{q_k\}$ in the weighted aggregate cost).

It is, nevertheless, sufficient for the purposes of this chapter to continue with the case (1.73). There are many important situations in practice where the minimizers of individual costs coincide with each other. For instance, examples abound where agents need to work cooperatively to attain a common objective such as tracking a target, locating a food source, or evading a predator (see, e.g., [72,76,77]). This scenario is also common in machine learning problems [26,27,78–81] when data samples at the various agents are generated by a common distribution parameterized by some vector, w^o. One such situation is illustrated in the next example.

Example 1.13 (Mean-square-error (MSE) networks). Consider the same setting of Example 1.3 except that we now have N agents observing streaming data $\{\boldsymbol{d}_k(i), \boldsymbol{u}_{k,i}\}$ that satisfy the regression model (1.13a) with regression covariance matrices $R_{u,k} = \mathbb{E}\boldsymbol{u}_{k,i}^{\mathsf{T}}\boldsymbol{u}_{k,i} > 0$ and with the same unknown w^o, i.e.,

$$\boldsymbol{d}_k(i) = \boldsymbol{u}_{k,i} w^o + \boldsymbol{v}_k(i). \tag{1.74a}$$

The individual MSE costs are defined by

$$J_k(w) = \mathbb{E}(\boldsymbol{d}_k(i) - \boldsymbol{u}_{k,i} w)^2 \tag{1.74b}$$

and are strongly convex in this case, with the minimizer of each $J_k(w)$ occurring at

$$w_k^o \triangleq R_{u,k}^{-1} r_{du,k}, \quad k = 1, 2, \ldots, N. \tag{1.74c}$$

If we multiply both sides of Eq. (1.74a) by $\boldsymbol{u}_{k,i}^{\mathsf{T}}$ from the left, and take expectations, we find that w^o satisfies

$$r_{du,k} = R_{u,k} w^o. \tag{1.75}$$

This relation shows that the unknown w^o from Eq. (1.74a) satisfies the same expression as w_k^o in Eq. (1.74c), for any $k = 1, 2, \ldots, N$, so that we must have $w^o = w_k^o$. Therefore, this example amounts to a situation where all costs $\{J_k(w)\}$ attain their minima at the same location, w^o.

We shall use the network model of this example to illustrate other results in the chapter. For ease of reference, we shall refer to strongly connected networks with agents receiving data according to model (1.74a) and seeking to estimate w^o by adopting the MSE costs $J_k(w)$ defined above, as *MSE networks*. We assume for these networks that the measurement noise process $\boldsymbol{v}_k(i)$ is temporally white and independent over space so that

$$\mathbb{E}\boldsymbol{v}_k(i)\boldsymbol{v}_\ell(j) = \sigma_{v,k}^2 \delta_{k,\ell} \delta_{i,j} \tag{1.76}$$

in terms of the Kronecker delta $\delta_{k,\ell}$. Likewise, we assume that the regression data $\boldsymbol{u}_{k,i}$ is temporally white and independent over space so that

$$\mathbb{E}\boldsymbol{u}_{k,i}^{\mathsf{T}}\boldsymbol{u}_{\ell,j} = R_{u,k} \delta_{k,\ell} \delta_{i,j}. \tag{1.77}$$

Moreover, the measurement noise $\boldsymbol{v}_k(i)$ and the regression data $\boldsymbol{u}_{\ell,j}$ are independent of each other for all k, ℓ, i, j. These statistical conditions help facilitate the analysis of such networks. ■

In the next subsections we list *synchronous* distributed algorithms of the consensus and diffusion types for the optimization of $J^{\text{glob}}(w)$. We only list the algorithms here; for motivation and justifications, the reader may refer to the treatments in [1,2]. Moreover, Sec. V.D of [1] provides commentary on several other related works in the literature in addition to the history and evolution of the consensus and diffusion strategies.

1.4.3 SYNCHRONOUS CONSENSUS STRATEGY

Let $\boldsymbol{w}_{k,i}$ denote the iterate that is available at agent k at iteration i; this iterate serves as the estimate for w^o. The consensus iteration at each agent k is described by the following construction (see, e.g., [6,10,31,63,65,71]):

$$\boxed{\boldsymbol{w}_{k,i} = \sum_{\ell \in \mathcal{N}_k} a_{\ell k} \boldsymbol{w}_{\ell,i-1} - \mu_k \widehat{\nabla_{w^{\mathsf{T}}} J}_k(\boldsymbol{w}_{k,i-1})} \tag{1.78}$$

where the $\{\mu_k\}$ are individual step-size parameters, and where the combination coefficients $\{a_{\ell k}\}$ that appear in Eq. (1.78) are nonnegative scalars that are required to satisfy the following conditions for each agent $k = 1, 2, \ldots, N$:

$$\boxed{a_{\ell k} \geq 0, \quad \sum_{\ell=1}^{N} a_{\ell k} = 1, \quad a_{\ell k} = 0 \text{ if } \ell \notin \mathcal{N}_k.} \tag{1.79a}$$

Condition (1.79a) implies that the combination matrix $A = [a_{\ell k}]$ satisfies

$$A^{\mathsf{T}} \mathbb{1} = \mathbb{1}, \tag{1.79b}$$

where $\mathbb{1}$ denotes the vector with all entries equal to one. We say that A is left-stochastic. One useful property of left-stochastic matrices is that their spectral radius is equal to one [61,75,82–84]:

$$\rho(A) = 1. \tag{1.79c}$$

An equivalent representation that is useful for later analysis is to rewrite the consensus iteration (1.78) as shown in the following listing, where the intermediate iterate that results from the neighborhood combination is denoted by $\boldsymbol{\psi}_{k,i-1}$. Observe that the gradient vector in the consensus implementation (1.80) is evaluated at $\boldsymbol{w}_{k,i-1}$ and not $\boldsymbol{\psi}_{k,i-1}$.

Consensus strategy for distributed adaptation

for each time instant $i \geq 0$:

each agent $k = 1, 2, \ldots, N$ performs the update:

$$\begin{cases} \boldsymbol{\psi}_{k,i-1} = \sum_{\ell \in \mathcal{N}_k} a_{\ell k} \boldsymbol{w}_{\ell,i-1} \\ \boldsymbol{w}_{k,i} = \boldsymbol{\psi}_{k,i-1} - \mu_k \widehat{\nabla_{w^{\mathsf{T}}} J}_k(\boldsymbol{w}_{k,i-1}) \end{cases} \tag{1.80}$$

end

We remark that one way to motivate the consensus update (1.78) is to start from the noncooperative step (1.68a) and replace the first iterate $\boldsymbol{w}_{k,i-1}$ by the convex combination used in Eq. (1.78).

Example 1.14 (Consensus LMS networks). For the MSE network of Example 1.13, the consensus strategy reduces to:

$$\begin{cases} \boldsymbol{\psi}_{k,i-1} = \sum_{\ell \in \mathcal{N}_k} a_{\ell k}\, \boldsymbol{w}_{\ell,i-1}, \\ \boldsymbol{w}_{k,i} = \boldsymbol{\psi}_{k,i-1} + 2\mu_k \boldsymbol{u}_{k,i}^{\mathsf{T}} [\boldsymbol{d}_k(i) - \boldsymbol{u}_{k,i} \boldsymbol{w}_{k,i-1}]. \end{cases} \tag{1.81}$$

■

1.4.4 SYNCHRONOUS DIFFUSION STRATEGIES

There is an inherent asymmetry in the consensus construction. Observe from the computation of $\boldsymbol{w}_{k,i}$ in Eq. (1.80) that the update starts from $\boldsymbol{\psi}_{k,i-1}$ and corrects it by the approximate gradient vector evaluated at $\boldsymbol{w}_{k,i-1}$ (and not at $\boldsymbol{\psi}_{k,i-1}$). This *asymmetry* will be shown later, e.g., in Example 1.25, to be problematic when the consensus strategy is used for adaptation and learning over networks. This is because the asymmetry can cause an unstable growth in the state of the network [85]; see also the explanations in [1] and [2, Sec. 10.6]. Diffusion strategies remove the asymmetry problem.

Combine-then-Adapt (CTA) Diffusion. In the CTA formulation of the diffusion strategy, the *same* iterate $\boldsymbol{\psi}_{k,i-1}$ is used to compute $\boldsymbol{w}_{k,i}$, thus leading to description (1.82a) where the gradient vector is evaluated at $\boldsymbol{\psi}_{k,i-1}$ as well. The reason for the name "Combine-then-Adapt" is that the first step in Eq. (1.82a) involves a combination step while the second step involves an adaptation step. The reason for the qualification "diffusion" is that the use of $\boldsymbol{\psi}_{k,i-1}$ to evaluate the gradient vector allows information to diffuse more thoroughly through the network. This is because information is not only being diffused through the aggregation of the neighborhood iterates, but also through the evaluation of the gradient vector at the aggregate state value.

Diffusion strategy for distributed adaptation (CTA)

for each time instant $i \geq 0$:

each agent $k = 1, 2, \ldots, N$ performs the update:

$$\begin{cases} \boldsymbol{\psi}_{k,i-1} = \sum_{\ell \in \mathcal{N}_k} a_{\ell k}\, \boldsymbol{w}_{\ell,i-1} \\ \boldsymbol{w}_{k,i} = \boldsymbol{\psi}_{k,i-1} - \mu_k \widehat{\nabla_{w^{\mathsf{T}}} J_k}(\boldsymbol{\psi}_{k,i-1}) \end{cases} \tag{1.82a}$$

end

Adapt-then-Combine (ATC) Diffusion. A similar implementation can be obtained by switching the order of the combination and adaptation steps in Eq. (1.82a), as shown in the listing (1.82b). The structures of the CTA and ATC strategies are fundamentally identical: the difference lies in which variable we choose to correspond to the updated iterate $\boldsymbol{w}_{k,i}$. In ATC, we choose the result of the *combination* step to be $\boldsymbol{w}_{k,i}$, whereas in CTA we choose the result of the *adaptation* step to be $\boldsymbol{w}_{k,i}$.

Diffusion strategy for distributed adaptation (ATC)

foreach time instant $i \geq 0$:

each agent $k = 1, 2, \ldots, N$ performs the update:

$$\begin{cases} \boldsymbol{\psi}_{k,i} = \boldsymbol{w}_{k,i-1} - \mu_k \widehat{\nabla_{w^{\mathsf{T}}} J}_k(\boldsymbol{w}_{k,i-1}) \\ \boldsymbol{w}_{k,i} = \sum_{\ell \in \mathcal{N}_k} a_{\ell k} \, \boldsymbol{\psi}_{\ell,i} \end{cases} \tag{1.82b}$$

end

One main motivation for the introduction of the diffusion strategies (1.82a) and (1.82b) is the fact that they enable *single* time-scale distributed learning from *streaming* data under *constant* step-size adaptation and in a stable manner [3,60,73,74,86–90]; see also [2, Chs. 9–11]. The diffusion strategies further allow A to be left-stochastic, which permits larger modes of cooperation than doubly stochastic policies. The CTA diffusion strategy (1.83a) was first introduced for MSE estimation problems in [73,86–88]. The ATC diffusion structure (1.83b), with adaptation preceding combination, appeared in the work [91] on adaptive distributed least-squares schemes and also in the works [74,90,92,93] on distributed MSE and state-space estimation methods. The CTA structure (1.82a) with an iteration dependent step-size that decays to zero, $\mu(i) \to 0$, was employed in [8,94,95] to solve distributed optimization problems that require all agents to reach agreement. The ATC form (1.82b), also with an iteration dependent sequence $\mu(i)$ that decays to zero, was employed in [96,97] to ensure almost-sure convergence and agreement among agents.

Example 1.15 (Diffusion LMS networks). For the MSE network of Example 1.13, the ATC and CTA diffusion strategies reduce to:

$$\begin{cases} \boldsymbol{\psi}_{k,i-1} = \sum_{\ell \in \mathcal{N}_k} a_{\ell k} \, \boldsymbol{w}_{\ell,i-1} \qquad \text{(CTA diffusion)} \\ \boldsymbol{w}_{k,i} = \boldsymbol{\psi}_{k,i-1} + 2\mu_k \boldsymbol{u}_{k,i}^{\mathsf{T}} \left[\boldsymbol{d}_k(i) - \boldsymbol{u}_{k,i} \boldsymbol{\psi}_{k,i-1}\right] \end{cases} \tag{1.83a}$$

and

$$\begin{cases} \boldsymbol{\psi}_{k,i} = \boldsymbol{w}_{k,i-1} + 2\mu_k \boldsymbol{u}_{k,i}^{\mathsf{T}} \left[\boldsymbol{d}_k(i) - \boldsymbol{u}_{k,i} \boldsymbol{w}_{k,i-1}\right] \\ \boldsymbol{w}_{k,i} = \sum_{\ell \in \mathcal{N}_k} a_{\ell k} \, \boldsymbol{\psi}_{\ell,i} \qquad \text{(ATC diffusion)} \end{cases} \tag{1.83b}$$

■

Example 1.16 (Diffusion logistic regression). We revisit the pattern classification problem from Example 1.2, where we consider a collection of N networked agents cooperating with each other to solve the logistic regression problem. Each agent receives streaming data $\{\boldsymbol{\gamma}_k(i), \boldsymbol{h}_{k,i}\}$, where the variable $\boldsymbol{\gamma}_k(i)$ assumes the values ± 1 and designates the class that the feature vector $\boldsymbol{h}_{k,i}$ belongs to. The objective is to use the training data to determine the vector w^o that minimizes the cost

$$J_k(w) \triangleq \frac{\rho}{2}\|w\|^2 + \mathbb{E}\left\{\ln\left[1 + e^{-\boldsymbol{\gamma}_k(i)\boldsymbol{h}_{k,i}^{\mathsf{T}} w}\right]\right\} \tag{1.84}$$

under the assumption of joint wide-sense stationarity over the random data. It is straightforward to verify that the ATC diffusion strategy (1.82b) reduces to the following form in this case:

$$\begin{cases} \boldsymbol{\psi}_{k,i} = (1-\rho\mu_k)\boldsymbol{w}_{k,i-1} + \mu_k \left(\dfrac{\boldsymbol{\gamma}_k(i)}{1+e^{\boldsymbol{\gamma}_k(i)\boldsymbol{h}_{k,i}^{\mathsf{T}}\boldsymbol{w}_{k,i-1}}} \right) \boldsymbol{h}_{k,i}, \\ \boldsymbol{w}_{k,i} = \displaystyle\sum_{\ell \in \mathcal{N}_k} a_{\ell k}\, \boldsymbol{\psi}_{\ell,i}. \end{cases} \tag{1.85}$$

■

1.5 ASYNCHRONOUS MULTIAGENT ADAPTATION AND LEARNING

There are various ways by which asynchronous events can be introduced into the operation of a distributed strategy. Without loss in generality, we illustrate the model for asynchronous operation by describing it for the ATC diffusion strategy (1.82b); similar constructions apply to CTA diffusion (1.82a) and consensus (1.80).

1.5.1 ASYNCHRONOUS MODEL

In a first instance, we model the step-size parameters as random variables and replace Eq. (1.82b) by:

$$\begin{cases} \boldsymbol{\psi}_{k,i} = \boldsymbol{w}_{k,i-1} - \boldsymbol{\mu}_k(i)\, \widehat{\nabla_{w^{\mathsf{T}}} J}_k(\boldsymbol{w}_{k,i-1}), \\ \boldsymbol{w}_{k,i} = \displaystyle\sum_{\ell \in \mathcal{N}_k} a_{\ell k}\, \boldsymbol{\psi}_{\ell,i}. \end{cases} \tag{1.86}$$

In this model, the neighborhoods and the network topology remain fixed and only the $\boldsymbol{\mu}_k(i)$ assume random values. The step-sizes can vary across the agents and, therefore, their means and variances become agent-dependent. Moreover, the step-sizes across agents can be correlated with each other. We therefore denote the first- and second-order moments of the step-size parameters by:

$$\bar{\mu}_k \overset{\Delta}{=} \mathbb{E}\boldsymbol{\mu}_k(i), \tag{1.87a}$$

$$\sigma_{\mu,k}^2 \overset{\Delta}{=} \mathbb{E}(\boldsymbol{\mu}_k(i) - \bar{\mu}_k)^2, \tag{1.87b}$$

$$c_{\mu,k\ell} \overset{\Delta}{=} \mathbb{E}(\boldsymbol{\mu}_k(i) - \bar{\mu}_k)(\boldsymbol{\mu}_\ell(i) - \bar{\mu}_\ell). \tag{1.87c}$$

When $\ell = k$, the scalar $c_{\mu,kk}$ coincides with the variance of $\boldsymbol{\mu}_k(i)$, i.e., $c_{\mu,kk} = \sigma_{\mu,k}^2 \geq 0$. On the other hand, if the step-sizes across the agents happen to be uncorrelated, then $c_{\mu,k\ell} = 0$ for $k \neq \ell$.

More broadly, we can allow for random variations in the neighborhoods (and, hence, in the network structure), and random variations in the combination coefficients as well. We capture this more general asynchronous implementation by writing:

$$\boxed{\begin{cases} \boldsymbol{\psi}_{k,i} = \boldsymbol{w}_{k,i-1} - \boldsymbol{\mu}_k(i) \widehat{\nabla_{w^{\mathsf{T}}} J}_k(\boldsymbol{w}_{k,i-1}) \\ \boldsymbol{w}_{k,i} = \displaystyle\sum_{\ell \in \boldsymbol{\mathcal{N}}_{k,i}} \boldsymbol{a}_{\ell k}(i)\, \boldsymbol{\psi}_{\ell,i} \end{cases}} \tag{1.88}$$

where the combination coefficients $\{\boldsymbol{a}_{\ell k}(i)\}$ are now *random* and, moreover, the symbol $\boldsymbol{\mathcal{N}}_{k,i}$ denotes the *randomly* changing neighborhood of agent k at time i. These neighborhoods become random because the random variations in the combination coefficients can turn links on and off depending on the values of the $\{\boldsymbol{a}_{\ell,k}(i)\}$. We continue to require the combination coefficients $\{\boldsymbol{a}_{\ell k}(i)\}$ to satisfy the same structural constraint as given before by Eq. (1.79a), i.e.,

$$\boxed{\sum_{\ell\in\boldsymbol{\mathcal{N}}_{k,i}} \boldsymbol{a}_{\ell k}(i) = 1, \text{ and } \begin{cases}\boldsymbol{a}_{\ell k}(i) > 0, & \text{if } \ell \in \boldsymbol{\mathcal{N}}_{k,i},\\ \boldsymbol{a}_{\ell k}(i) = 0, & \text{otherwise.}\end{cases}} \tag{1.89}$$

Because these coefficients are now random, we denote their first- and second-order moments by:

$$\bar{a}_{\ell k} \triangleq \mathbb{E}\,\boldsymbol{a}_{\ell k}(i), \tag{1.90a}$$

$$\sigma^2_{a,\ell k} \triangleq \mathbb{E}(\boldsymbol{a}_{\ell k}(i) - \bar{a}_{\ell k})^2, \tag{1.90b}$$

$$c_{a,\ell k,nm} \triangleq \mathbb{E}(\boldsymbol{a}_{\ell k}(i) - \bar{a}_{\ell k})(\boldsymbol{a}_{nm}(i) - \bar{a}_{nm}). \tag{1.90c}$$

When $\ell = n$ and $k = m$, the scalar $c_{a,\ell k,nm}$ coincides with the variance of $\boldsymbol{a}_{\ell k}(i)$, i.e., $c_{a,\ell k,nm} = \sigma^2_{a,\ell k} \geq 0$.

Example 1.17 (Asynchronous diffusion LMS networks). For the MSE network of Example 1.13, the ATC diffusion strategy (1.86) with random update reduces to

$$\begin{cases}\boldsymbol{\psi}_{k,i} = \boldsymbol{w}_{k,i-1} + 2\boldsymbol{\mu}_k(i)\boldsymbol{u}_{k,i}^{\mathrm{T}}\left[\boldsymbol{d}_k(i) - \boldsymbol{u}_{k,i}\boldsymbol{w}_{k,i-1}\right],\\ \boldsymbol{w}_{k,i} = \sum\limits_{\ell\in\mathcal{N}_k} a_{\ell k}\,\boldsymbol{\psi}_{\ell,i} \quad \text{(diffusion with random updates)},\end{cases} \tag{1.91a}$$

whereas the *asynchronous* ATC diffusion strategy (1.88) reduces to:

$$\begin{cases}\boldsymbol{\psi}_{k,i} = \boldsymbol{w}_{k,i-1} + 2\boldsymbol{\mu}_k(i)\boldsymbol{u}_{k,i}^{\mathrm{T}}\left[\boldsymbol{d}_k(i) - \boldsymbol{u}_{k,i}\boldsymbol{w}_{k,i-1}\right],\\ \boldsymbol{w}_{k,i} = \sum\limits_{\ell\in\boldsymbol{\mathcal{N}}_{k,i}} \boldsymbol{a}_{\ell k}(i)\,\boldsymbol{\psi}_{\ell,i} \quad \text{(asynchronous diffusion)}.\end{cases} \tag{1.91b}$$

We can view implementation (1.91a) as a special case of the asynchronous update (1.91b) when the variances of the random combination coefficients $\{\boldsymbol{a}_{\ell k}(i)\}$ are set to zero. ■

Example 1.18 (Asynchronous consensus LMS networks). Similarly, for the same MSE network of Example 1.13, the asynchronous consensus strategy is given by

$$\begin{cases}\boldsymbol{\psi}_{k,i-1} = \sum\limits_{\ell\in\boldsymbol{\mathcal{N}}_{k,i}} \boldsymbol{a}_{\ell k}(i)\,\boldsymbol{w}_{\ell,i-1},\\ \boldsymbol{w}_{k,i} = \boldsymbol{\psi}_{k,i-1} + 2\boldsymbol{\mu}_k(i)\boldsymbol{u}_{k,i}^{\mathrm{T}}[\boldsymbol{d}_k(i) - \boldsymbol{u}_{k,i}\boldsymbol{w}_{k,i-1}].\end{cases} \tag{1.92}$$

■

1.5.2 MEAN GRAPH

We refer to the topology that corresponds to the average combination coefficients $\{\bar{a}_{\ell k}\}$ as the *mean graph*, which is fixed over time. For each agent k, the neighborhood defined by the mean graph is denoted by $\mathcal{N}_k$. It is straightforward to verify that the mean combination coefficients $\bar{a}_{\ell k}$ satisfy the following constraints over the mean graph (compare with Eq. 1.79a and Eq. 1.89):

$$\boxed{\sum_{\ell \in \mathcal{N}_k} \bar{a}_{\ell k} = 1, \quad \text{and} \quad \begin{cases} \bar{a}_{\ell k} > 0, & \text{if } \ell \in \mathcal{N}_k, \\ \bar{a}_{\ell k} = 0, & \text{otherwise.} \end{cases}} \tag{1.93}$$

One example of a random network with two equally probable realizations and its mean graph is shown in Fig. 1.6 [3]. The letter ω is used to index the sample space of the random matrix $\boldsymbol{A}_i$.

There is one useful result that relates the random neighborhoods $\{\boldsymbol{\mathcal{N}}_{k,i}\}$ from Eq. (1.88) to the neighborhoods $\{\mathcal{N}_k\}$ from the mean graph. It is not difficult to verify that $\mathcal{N}_k$ is equal to the *union* of all possible realizations for the random neighborhoods $\boldsymbol{\mathcal{N}}_{k,i}$; this property is already illustrated by the example of Fig. 1.6:

$$\mathcal{N}_k = \bigcup_{\omega \in \Omega} \boldsymbol{\mathcal{N}}_{k,i}(\omega) \tag{1.94}$$

for any k, where Ω denotes the sample space for $\boldsymbol{\mathcal{N}}_{k,i}$.

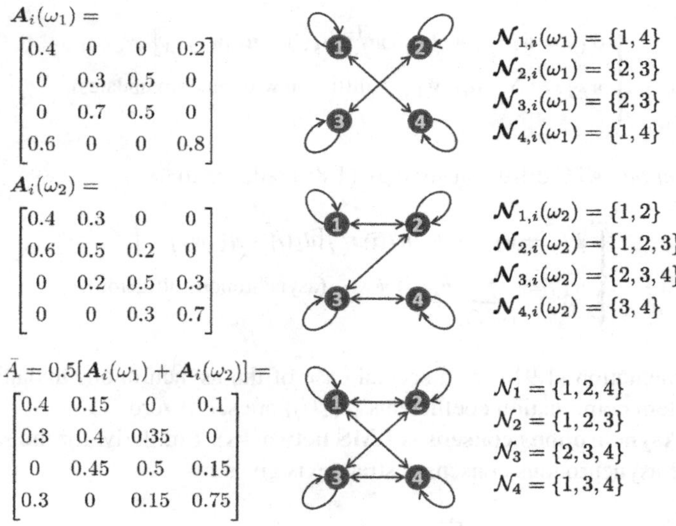

FIG. 1.6

Mean graph. The first two rows show two equally probable realizations with the respective neighborhoods. The last row shows the resulting mean graph.

This is a modified version of a Figure reproduced with permission from Zhao X, Sayed AH. Asynchronous adaptation and learning over networks—Part I: Modeling and stability analysis. IEEE Trans Signal Process 2015;63(4):811–26.

1.5.3 RANDOM COMBINATION POLICY

The first- and second-order moments of the combination coefficients will play an important role in characterizing the stability and MSE performance of the asynchronous network (1.88). We collect these moments in matrix form as follows. We first group the combination coefficients into a matrix:

$$\boldsymbol{A}_i \triangleq [\boldsymbol{a}_{\ell k}(i)]_{\ell,k=1}^{N} \quad (N \times N). \tag{1.95}$$

The sequence $\{\boldsymbol{A}_i, i \geq 0\}$ represents a stochastic process consisting of left-stochastic random matrices whose entries satisfy the conditions in Eq. (1.89) at every time i. We subsequently introduce the mean and Kronecker-covariance matrix of $\boldsymbol{A}_i$ and assume these quantities are constant over time; we denote them by the $N \times N$ matrix $\bar{A}$ and the $N^2 \times N^2$ matrix C_A, respectively:

$$\bar{A} \triangleq \mathbb{E}\boldsymbol{A}_i = [\bar{a}_{\ell k}]_{\ell,k=1}^{N}, \tag{1.96a}$$

$$C_A \triangleq \mathbb{E}[(\boldsymbol{A}_i - \bar{A}) \otimes (\boldsymbol{A}_i - \bar{A})]. \tag{1.96b}$$

The matrix C_A is *not* a conventional covariance matrix and is not necessarily Hermitian. The reason for its introduction is because it captures the correlations of each entry of $\boldsymbol{A}_i$ with all other entries in $\boldsymbol{A}_i$. For example, for a network with $N = 2$ agents, the entries of $\bar{A}$ and C_A will be given by:

$$\bar{A} = \left[\begin{array}{cc} \bar{a}_{11} & \bar{a}_{12}\bar{a}_{21} & \bar{a}_{22} \end{array}\right], \tag{1.97a}$$

$$C_A = \left[\begin{array}{cc|cc} c_{a,11,11} & c_{a,11,12} & c_{a,12,11} & c_{a,12,12} \\ c_{a,11,21} & c_{a,11,22} & c_{a,12,21} & c_{a,12,22} \\ \hline c_{a,21,11} & c_{a,21,12} & c_{a,22,11} & c_{a,22,12} \\ c_{a,21,21} & c_{a,21,22} & c_{a,22,21} & c_{a,22,22} \end{array}\right] \tag{1.97b}$$

We thus see that the (ℓ, k)th block of C_A contains the covariance coefficients of $\boldsymbol{a}_{\ell k}$ with all other entries of $\boldsymbol{A}_i$. One useful property of the matrices $\{\bar{A}, C_A\}$ so defined is that their elements are nonnegative and the following matrices are left-stochastic:

$$\boxed{(\bar{A})^{\mathsf{T}}\mathbb{1}_N = \mathbb{1}_N, \quad (\bar{A} \otimes \bar{A} + C_A)^{\mathsf{T}}\mathbb{1}_{N^2} = \mathbb{1}_{N^2}.} \tag{1.98}$$

Assumption 1.6 (Asynchronous network model). *It is assumed that the random processes* $\{\boldsymbol{\mu}_k(i), \boldsymbol{a}_{\ell m}(j)\}$ *are independent of each other for all* $k, \ell, m, i,$ *and* j. *They are also independent of any other random variable in the learning algorithm.* ■

The asynchronous network model described in this section covers many situations of practical interest. For example, we can choose the sample space for each step-size $\boldsymbol{\mu}_k(i)$ to be the binary choice $\{0, \mu\}$ to model random "on-off" behavior at each agent k for the purpose of saving power, waiting for data, or even due to random agent failures. Similarly, we can choose the sample space for each combination coefficient $\boldsymbol{a}_{\ell k}(i)$, $\ell \in \mathcal{N}_k \backslash \{k\}$, to be $\{0, a_{\ell k}\}$ to model a random "on-off" status for the link from agent ℓ to agent k at time i for the purpose of either saving communication cost or due to random link failures. Note that the convex constraint (1.89) can always be satisfied by adjusting the value of $\boldsymbol{a}_{kk}(i)$ according to the realizations of $\{\boldsymbol{a}_{\ell k}(i); \ell \in \boldsymbol{\mathcal{N}}_{k,i} \backslash \{k\}\}$.

Example 1.19 (The spatially uncorrelated model). A useful special case of the asynchronous network model of this section is the spatially uncorrelated model. In this case, at each iteration i, the random step-sizes $\{\boldsymbol{\mu}_k(i); k = 1, 2, \ldots, N\}$ are uncorrelated with each other across the network, and the random combination coefficients $\{\boldsymbol{a}_{\ell k}(i); \ell \neq k, k = 1, 2, \ldots, N\}$ are also uncorrelated with each other across the network. Then, it can be verified that the covariances $\{c_{\mu,k\ell}\}$ in Eq. (1.87c) and $\{c_{a,\ell k,nm}\}$ in Eq. (1.90c) will be fully determined by the variances $\{\sigma_{\mu}^2, \sigma_{a,\ell k}^2\}$:

$$c_{\mu,kk} = \sigma_{\mu,k}^2, \quad c_{\mu,k\ell} = 0, \; k \neq \ell \tag{1.99a}$$

and

$$c_{a,\ell k,nm} = \begin{cases} \sigma_{a,\ell k}^2, & \text{if } k = m, \ell = n, \ell \in \mathcal{N}_k \backslash \{k\}, \\ -\sigma_{a,\ell k}^2, & \text{if } k = m = n, \ell \in \mathcal{N}_k \backslash \{k\}, \\ -\sigma_{a,nk}^2, & \text{if } k = m = \ell, n \in \mathcal{N}_k \backslash \{k\}, \\ \sum\limits_{j \in \mathcal{N}_k \backslash \{k\}} \sigma_{a,jk}^2, & \text{if } k = m = \ell = n, \\ 0, & \text{otherwise.} \end{cases} \tag{1.99b}$$

■

1.5.4 PERRON VECTORS

Now that we introduced the network model, we can move on to examine the effect of network cooperation on performance. Some interesting patterns of behavior arise when agents cooperate to solve a global optimization problem in a distributed manner from streaming data [1,2]. For example, one interesting result established in [3,4,37,60,98] is that the effect of the network topology on performance is captured by the Perron vector of the combination policy. This vector turns out to summarize the influence of the topology on performance so much so that different topologies with similar Perron vectors will end up delivering similar performance. We explained the role of Perron vectors in the context of synchronous adaptation and learning in [1,2]. Here we focus on asynchronous networks. In this case, two Perron vectors will be needed because the randomness in the combination policy is now represented by two moment matrices, $\bar{A}$ and C_A. In the synchronous case, only one Perron vector was necessary because the combination policy was fixed and described by a matrix A. Although we are focusing on the asynchronous case in the sequel, we will be able to recover results for synchronous networks as special cases.

Let us first recall the definition of Perron vectors for synchronous networks, say, of the form described by Eq. (1.82b) with combination policy A. We assume the network is strongly connected. In this case, the left-stochastic matrix A will be primitive. For such primitive matrices, it follows from the Perron-Frobenius Theorem [75] that: (a) the matrix A will have a *single* eigenvalue at one; (b) all other eigenvalues of A will be strictly inside the unit circle so that $\rho(A) = 1$; and (c) with proper sign scaling, all entries of the right-eigenvector of A corresponding to the single eigenvalue at one will be *positive*. Let p denote this right-eigenvector with its entries $\{p_k\}$ normalized to add up to one, i.e.,

$$\boxed{Ap{=}p, \quad \mathbb{1}^{\mathsf{T}}p{=}1, \quad p_k > 0, \quad k = 1, 2, \ldots, N.} \tag{1.100}$$

We refer to p as the *Perron* eigenvector of A. It was explained in [1,2] how the entries of this vector determine the MSE performance and convergence rate of the network; these results will be revisited further ahead when we recover them as special cases of the asynchronous results.

On the other hand, for an asynchronous implementation, the individual realizations of the random combination matrix $\boldsymbol{A}_i$ in Eq. (1.88) need not be primitive. In this context, we will require a form of primitiveness to hold on average as follows.

Definition 1.1 (Strongly connected asynchronous model). *We say that an asynchronous model with random combination coefficients $\{\boldsymbol{a}_{\ell k}(i)\}$ is strongly connected if the Kronecker-covariance matrix given by $\bar{A} \otimes \bar{A} + C_A$ is primitive.* ■

Observe that if we set $\bar{A} = A$ and $C_A = 0$, then we recover the condition for strong-connectedness in the synchronous case, namely, that A should be primitive.

Definition 1.1 means that the directed graph (digraph) associated with the matrix $\bar{A} \otimes \bar{A} + C_A$ is strongly connected (e.g., [83, pp. 30,34] and [2]). It is straightforward to check from the definition of C_A in Eq. (1.96b) that

$$\bar{A} \otimes \bar{A} + C_A = \mathbb{E}(\boldsymbol{A}_i \otimes \boldsymbol{A}_i) \tag{1.101}$$

so that the digraph associated with $\bar{A} \otimes \bar{A} + C_A$ is the union of all possible digraphs associated with the realizations of $\boldsymbol{A}_i \otimes \boldsymbol{A}_i$ [99, p. 29]. Therefore, as explained in [3–5], Definition 1.1 amounts to an assumption that the *union* of all possible digraphs associated with the realizations of $\boldsymbol{A}_i \otimes \boldsymbol{A}_i$ is strongly connected. As illustrated in Fig. 1.7, this condition still allows individual digraphs associated with realizations of $\boldsymbol{A}_i$ to be weakly connected with or without self-loops or even to be disconnected [4].

It follows from property (1.98) and Definition 1.1, the matrix $\bar{A} \otimes \bar{A} + C_A$ is left-stochastic and primitive. It can be verified that mean matrix $\bar{A}$ is also primitive if $\bar{A} \otimes \bar{A} + C_A$ is primitive (although the converse is not true). Therefore, the matrix $\bar{A}$ is both left-stochastic and primitive. We denote the Perron eigenvector of $\bar{A} \otimes \bar{A} + C_A$ by $p_c \in \mathbb{R}^{N^2 \times 1}$, which satisfies:

$$\boxed{(\bar{A} \otimes \bar{A} + C_A) p_c = p_c, \qquad p_c^{\mathsf{T}} \mathbb{1} = 1.} \tag{1.102a}$$

Likewise, we denote the Perron eigenvector of $\bar{A}$ by $\bar{p} \in \mathbb{R}^{N \times 1}$, which satisfies:

$$\boxed{\bar{A}\bar{p} = \bar{p}, \qquad \bar{p}^{\mathsf{T}} \mathbb{1} = 1.} \tag{1.102b}$$

Observe that if we set $C_A = 0$ and $\bar{A} = A$, then we recover the synchronous case information, namely, $\bar{p} = p$ and $p_c = p \otimes p$.

The entries of the Perron vectors $\{p_c, \bar{p}\}$ are related to each other, as illustrated by the following explanation (proofs appear in [4, App. VIII]). Because the vector p_c is of dimension $N^2 \times 1$, we partition it into N subvectors of dimension $N \times 1$ each:

$$p_c \triangleq \mathrm{col}\{p_1, p_2, \ldots, p_N\}, \tag{1.103a}$$

where p_k denotes the kth subvector. We construct an $N \times N$ matrix P_c from these subvectors:

$$P_c \triangleq [p_1 p_2 \;\ldots\; p_N] = [p_{c,\ell k}]_{k,\ell=1}^{N}. \tag{1.103b}$$

We use $p_{\ell k}$ to denote the (ℓ, k)th element of matrix P_c, which is equal to the ℓth element of p_k. It can be verified that the matrix P_c in Eq. (1.103b) is symmetric positive semidefinite and it satisfies

$$\boxed{P_c \mathbb{1} = \bar{p}, \quad P_c^{\mathsf{T}} = P_c, \quad P_c \geq 0} \tag{1.103c}$$

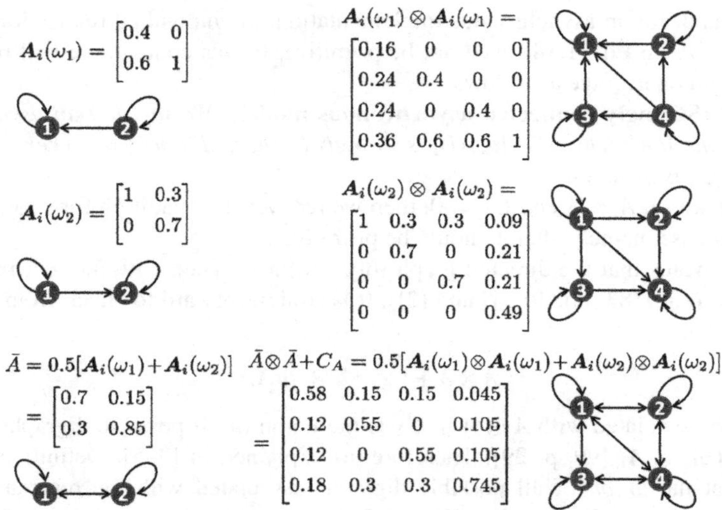

FIG. 1.7 Digraphs.

The combination policy $\boldsymbol{A}_i$ has two equally probable realizations in this example, denoted by $\{\boldsymbol{A}_i(\omega_1), \boldsymbol{A}_i(\omega_2)\}$. Observe that neither of the digraphs $\boldsymbol{A}_i(\omega_1) \otimes \boldsymbol{A}_i(\omega_1)$ or $\boldsymbol{A}_i(\omega_2) \otimes \boldsymbol{A}_i(\omega_2)$ is strongly connected due to the existence of the source and sink nodes. However, the digraph associated with $\mathbb{E}(\boldsymbol{A}_i \otimes \boldsymbol{A}_i)$, which is the union of the first two digraphs, is strongly connected, where information can flow in any direction through the network.

This is a modified version of a Figure reproduced with permission from Zhao X, Sayed AH. Asynchronous adaptation and learning over networks—Part II: Performance analysis. IEEE Trans Signal Process 2015;63(4):827–42.

where $\bar{p}$ is the Perron eigenvector in Eq. (1.102b). We can further establish the following useful relations:

$$p_{c,\ell k} = p_{c,k\ell}, \quad \sum_{k=1}^{N} p_{c,\ell k} = \bar{p}_\ell, \quad \sum_{\ell=1}^{N} p_{c,\ell k} = \bar{p}_k. \tag{1.103d}$$

We can also verify that the matrix difference

$$C_c \triangleq P_c - \bar{p}\bar{p}^{\mathsf{T}}, \quad C_c^{\mathsf{T}} = C_c, \quad C_c \geq 0 \tag{1.103e}$$

is symmetric, positive semidefinite, and satisfies $C_c \mathbb{1} = 0$. Moreover, it is straightforward to verify that

$$\boxed{c_{c,kk} = p_{c,kk} - \bar{p}_k^2 \geq 0} \tag{1.104}$$

where $c_{c,kk}$ denotes the (k, k)th entry in C_c.

1.6 ASYNCHRONOUS NETWORK PERFORMANCE

We now comment on the performance of asynchronous networks and compare their metrics against both noncooperative and centralized strategies.

1.6.1 MSD PERFORMANCE

We denote the MSD performance of the individual agents and the average MSD performance across the network by:

$$\text{MSD}_{\text{dist},k} \triangleq \lim_{i\to\infty} \mathbb{E}\,\|\tilde{w}_{k,i}\|^2, \tag{1.105a}$$

$$\text{MSD}_{\text{dist, av}} \triangleq \frac{1}{N}\sum_{k=1}^{N} \text{MSD}_{\text{dist},k}, \tag{1.105b}$$

where the error vectors are measured relative to the global optimizer, w^o. We further denote the gradient noise process at the individual agents by

$$s_{k,i}(w) \triangleq \widehat{\nabla_{w^{\mathsf{T}}} J_k}(w) \;-\; \nabla_{w^{\mathsf{T}}} J_k(w) \tag{1.106a}$$

and define

$$H_k \triangleq \nabla_w^2\, J_k(w^o), \tag{1.106b}$$

$$R_{s,k} \triangleq \lim_{i\to\infty} \mathbb{E}\left[s_{k,i}(w^o) s_{k,i}^{\mathsf{T}}(w^o) \mid \boldsymbol{\mathcal{F}}_{i-1}\right], \tag{1.106c}$$

where $\boldsymbol{\mathcal{F}}_{i-1}$ now represents the collection of all random events generated by the iterates from across all agents, $\{\boldsymbol{w}_{k,j},\ k = 1, 2, \ldots, N\}$, up to time $i-1$:

$$\boldsymbol{\mathcal{F}}_{i-1} \triangleq \text{filtration}\{\boldsymbol{w}_{k,-1}, \boldsymbol{w}_{k,0}, \boldsymbol{w}_{k,1}, \ldots, \boldsymbol{w}_{k,i-1},\ \text{all } k\}. \tag{1.106d}$$

It is stated later in Eq. (1.149a) that under some assumptions on the gradient noise processes (1.106a), that for strongly connected asynchronous diffusion networks of the form (1.86), and for sufficiently small step-sizes μ_x [2,4]:

$$\text{MSD}_{\text{dist},k}^{\text{asyn}} \approx \text{MSD}_{\text{dist,av}}^{\text{asyn}} \approx \frac{1}{2}\text{Tr}\left[\left(\sum_{k=1}^{N} \bar{\mu}_k \bar{p}_k H_k\right)^{-1} \left(\sum_{k=1}^{N} (\bar{\mu}_k^2 + \sigma_{\mu,k}^2) p_{c,kk} R_{s,k}\right)\right]. \tag{1.107}$$

The same result holds for asynchronous consensus and CTA diffusion strategies. Observe from Eq. (1.107) the interesting conclusion that the distributed strategy is able to *equalize* the MSD performance across all agents for sufficiently small step-sizes. It is also instructive to compare expression (1.149a) with Eqs. (1.63) and (1.67) in the centralized case. Observe how cooperation among agents leads to the appearance of the scaling coefficients $\{\bar{p}_k, p_{c,kk}\}$; these factors are determined by $\bar{A}$ and C_A.

Note further that if we set $\bar{\mu}_k = \mu$, $\sigma^2_{\mu,k} = 0$, $\bar{p}_k = p_k$, and $p_{c,kk} = p_k^2$, then we recover the MSD expression for synchronous distributed strategies:

$$\text{MSD}^{\text{sync}}_{\text{dist},k} \approx \text{MSD}^{\text{sync}}_{\text{dist,av}} \approx \frac{1}{2}\text{Tr}\left[\left(\sum_{k=1}^{N} \mu_k p_k H_k\right)^{-1} \left(\sum_{k=1}^{N} \mu_k^2 p_k^2 R_{s,k}\right)\right]. \tag{1.108}$$

This result agrees with expression (62) from [1].

Example 1.20 (MSE networks with random updates). We continue with the setting of Example 1.13, which deals with MSE networks. We assume the first- and second-order moments of the random step-sizes are uniform, i.e., $\bar{\mu}_k \equiv \bar{\mu}$ and $c_{\mu,kk} \equiv \sigma^2_{\mu}$, and also assume uniform regression covariance matrices, i.e., $R_{u,k} \equiv R_u$ for $k = 1, 2, \ldots, N$. It follows that $H_k = 2R_u \equiv H$ and $R_{s,k} = 4\sigma^2_{v,k} R_u$. Substituting into Eq. (1.149a), and assuming a fixed topology with fixed combination coefficients set to $a_{\ell k}$, we conclude that the MSD performance of the diffusion strategy (1.91a) with random updates is well approximated by:

$$\text{MSD}^{\text{asyn},1}_{\text{dist},k} \approx \text{MSD}^{\text{asyn},1}_{\text{dist,av}} \approx \mu_x M \left(\sum_{k=1}^{N} p_k^2 \sigma^2_{v,k}\right), \tag{1.109a}$$

where

$$\mu_x = \bar{\mu} + \frac{\sigma^2_{\mu}}{\bar{\mu}} \tag{1.109b}$$

and p_k is the kth entry of the Perron vector p defined by Eq. (1.100).

If the combination matrix A happens to be doubly stochastic, then its Perron eigenvector becomes $p = \mathbb{1}/N$. Substituting $p_k = 1/N$ into Eq. (1.109a) gives

$$\text{MSD}^{\text{asyn},1}_{\text{dist},k} \approx \text{MSD}^{\text{asyn},1}_{\text{dist,av}} \approx \frac{\mu_x M}{N} \left(\frac{1}{N} \sum_{k=1}^{N} \sigma^2_{v,k}\right), \tag{1.109c}$$

which agrees with the centralized performance (1.51b). In other words, the asynchronous diffusion strategy is able to match the performance of the centralized solution for doubly stochastic combination policies when both implementations employ random updates. Because the centralized solution can improve the average MSD performance over noncooperative networks, we further conclude that the diffusion strategy can also exceed the average performance of noncooperative networks. ■

Example 1.21 (Asynchronous MSE networks). We continue with the setting of Example 1.20 except that we now employ the asynchronous LMS diffusion network (1.91b). Its MSD performance is well approximated by:

$$\text{MSD}^{\text{asyn}}_{\text{dist},k} \approx \text{MSD}^{\text{asyn}}_{\text{dist,av}} \approx \mu_x M \left(\sum_{k=1}^{N} p_{c,kk} \sigma^2_{v,k}\right). \tag{1.110a}$$

If the mean combination matrix $\bar{A}$ happens to be doubly stochastic, then its Perron eigenvector becomes $\bar{p} = \mathbb{1}/N$. Substituting $\bar{p}_k = 1/N$ into Eq. (1.109a), and using $p_{c,kk} = \bar{p}_k^2 + c_{c,kk}$, where $c_{c,kk}$ is from Eq. (1.103e), gives

$$\text{MSD}_{\text{dist},k}^{\text{asyn}} \approx \text{MSD}_{\text{dist,av}}^{\text{asyn}} \approx \frac{\mu_x M}{N}\left[\frac{1}{N}\sum_{k=1}^{N}(1+N^2 c_{c,kk})\sigma_{v,k}^2\right]. \tag{1.110b}$$

It is clear that if $c_{c,kk} = \sigma_{\pi,k}^2$, then the MSD performance in Eq. (1.110b) will agree with the centralized performance (1.55a). In other words, the distributed diffusion strategy is able to match the performance of the centralized solution. ■

1.7 NETWORK STABILITY AND PERFORMANCE

In this section, we examine more closely the performance and stability results that were alluded to in the earlier sections. We first examine the consensus and diffusion strategies in a unified manner and subsequently focus on diffusion strategies due to their enhanced stability properties, as the ensuing discussion will reveal.

1.7.1 MSE NETWORKS

We motivate the discussion by presenting first some illustrative examples with MSE networks, which involve quadratic costs. Following the examples, we extend the framework to more general costs.

Example 1.22 (Error dynamics over MSE networks). We consider the MSE network of Example 1.13, which involves quadratic costs with a common minimizer, w^o. The update equations for the noncooperative, consensus, and diffusion strategies are given by Eqs. (1.68a), (1.81), (1.83a), and (1.83b). We can group these strategies into a single unifying description by considering the following structure in terms of three sets of combination coefficients $\{\boldsymbol{a}_{o,\ell k}(i), \boldsymbol{a}_{1,\ell k}(i), \boldsymbol{a}_{2,\ell k}(i)\}$:

$$\begin{cases} \boldsymbol{\phi}_{k,i-1} = \sum\limits_{\ell\in\mathcal{N}_{k,i}} \boldsymbol{a}_{1,\ell k}(i)\boldsymbol{w}_{\ell,i-1}, \\ \boldsymbol{\psi}_{k,i} = \sum\limits_{\ell\in\mathcal{N}_{k,i}} \boldsymbol{a}_{o,\ell k}(i)\boldsymbol{\phi}_{\ell,i-1} + 2\boldsymbol{\mu}_k(i)\boldsymbol{u}_{k,i}^{\mathsf{T}}\left[\boldsymbol{d}_k(i) - \boldsymbol{u}_{k,i}\boldsymbol{\phi}_{k,i-1}\right], \\ \boldsymbol{w}_{k,i} = \sum\limits_{\ell\in\mathcal{N}_{k,i}} \boldsymbol{a}_{2,\ell k}(i)\boldsymbol{\psi}_{\ell,i}. \end{cases} \tag{1.111}$$

In Eq. (1.111), the quantities $\{\boldsymbol{\phi}_{k,i-1}, \boldsymbol{\psi}_{k,i}\}$ denote $M \times 1$ intermediate variables while the nonnegative entries of the $N \times N$ matrices $\boldsymbol{A}_{o,i} = [\boldsymbol{a}_{o,\ell k}(i)]$, $\boldsymbol{A}_{1,i} = [\boldsymbol{a}_{1,\ell k}(i)]$, and $\boldsymbol{A}_{2,i} = [\boldsymbol{a}_{2,\ell k}(i)]$ are assumed to satisfy the same conditions (1.89). Any of the combination weights $\{\boldsymbol{a}_{o,\ell k}(i), \boldsymbol{a}_{1,\ell k}(i), \boldsymbol{a}_{2,\ell k}(i)\}$ is zero whenever $\ell \notin \mathcal{N}_{k,i}$. Different choices for $\{\boldsymbol{A}_{o,i}, \boldsymbol{A}_{1,i}, \boldsymbol{A}_{2,i}\}$, including random and deterministic choices, correspond to different strategies, as the following examples reveal:

$$\text{noncooperative:} \quad \boldsymbol{A}_{1,i} = \boldsymbol{A}_{o,i} = \boldsymbol{A}_{2,i} = I_N, \tag{1.112a}$$

$$\text{consensus:} \quad \boldsymbol{A}_{o,i} = \boldsymbol{A}_i,\ \boldsymbol{A}_{1,i} = I_N = \boldsymbol{A}_{2,i}, \tag{1.112b}$$

$$\text{CTA diffusion:} \quad \boldsymbol{A}_{1,i} = \boldsymbol{A}_i,\ \boldsymbol{A}_{2,i} = I_N = \boldsymbol{A}_{o,i}, \tag{1.112c}$$

$$\text{ATC diffusion:} \quad \boldsymbol{A}_{2,i} = \boldsymbol{A}_i,\ \boldsymbol{A}_{1,i} = I_N = \boldsymbol{A}_{o,i}, \tag{1.112d}$$

where $\boldsymbol{A}_i$ denotes some generic combination policy satisfying Eq. (1.89). We associate with each agent k the following three errors:

$$\tilde{\boldsymbol{w}}_{k,i} \triangleq w^o - \boldsymbol{w}_{k,i}, \tag{1.113a}$$

$$\tilde{\boldsymbol{\psi}}_{k,i} \triangleq w^o - \boldsymbol{\psi}_{k,i}, \tag{1.113b}$$

$$\tilde{\boldsymbol{\phi}}_{k,i-1} \triangleq w^o - \boldsymbol{\phi}_{k,i-1}, \tag{1.113c}$$

which measure the deviations from the global minimizer, w^o. Subtracting w^o from both sides of the equations in Eq. (1.111) we get

$$\begin{cases} \tilde{\boldsymbol{\phi}}_{k,i-1} = \sum\limits_{\ell \in \mathcal{N}_{k,i}} \boldsymbol{a}_{1,\ell k}(i)\, \tilde{\boldsymbol{w}}_{\ell,i-1}, \\ \tilde{\boldsymbol{\psi}}_{k,i} = \sum\limits_{\ell \in \mathcal{N}_{k,i}} \boldsymbol{a}_{o,\ell k}(i) \tilde{\boldsymbol{\phi}}_{\ell,i-1} - 2\boldsymbol{\mu}_k(i) \boldsymbol{u}_{k,i}^{\mathsf{T}} \boldsymbol{u}_{k,i} \tilde{\boldsymbol{\phi}}_{k,i-1} - 2\boldsymbol{\mu}_k(i) \boldsymbol{u}_{k,i}^{\mathsf{T}} \boldsymbol{v}_k(i), \\ \tilde{\boldsymbol{w}}_{k,i} = \sum\limits_{\ell \in \mathcal{N}_{k,i}} \boldsymbol{a}_{2,\ell k}(i)\, \tilde{\boldsymbol{\psi}}_{\ell,i}. \end{cases} \tag{1.114a}$$

In a manner similar to Eq. (1.16a), the gradient noise process at each agent k is given by

$$\boldsymbol{s}_{k,i}(\boldsymbol{\phi}_{k,i-1}) = 2\left(R_{u,k} - \boldsymbol{u}_{k,i}^{\mathsf{T}} \boldsymbol{u}_{k,i}\right) \tilde{\boldsymbol{\phi}}_{k,i-1} \; - \; 2\boldsymbol{u}_{k,i}^{\mathsf{T}} \boldsymbol{v}_k(i). \tag{1.114b}$$

In order to examine the evolution of the error dynamics across the network, we collect the error vectors from all agents into $N \times 1$ block error vectors (whose individual entries are of size $M \times 1$ each):

$$\tilde{\boldsymbol{w}}_i \triangleq \begin{bmatrix} \tilde{\boldsymbol{w}}_{1,i} \\ \tilde{\boldsymbol{w}}_{2,i} \\ \vdots \\ \tilde{\boldsymbol{w}}_{N,i} \end{bmatrix}, \tilde{\boldsymbol{\psi}}_i \triangleq \begin{bmatrix} \tilde{\boldsymbol{\psi}}_{1,i} \\ \tilde{\boldsymbol{\psi}}_{2,i} \\ \vdots \\ \tilde{\boldsymbol{\psi}}_{N,i} \end{bmatrix}, \tilde{\boldsymbol{\phi}}_{i-1} \triangleq \begin{bmatrix} \tilde{\boldsymbol{\phi}}_{1,i-1} \\ \tilde{\boldsymbol{\phi}}_{2,i-1} \\ \vdots \\ \tilde{\boldsymbol{\phi}}_{N,i-1} \end{bmatrix}. \tag{1.115a}$$

Motivated by the last term in the second equation in Eq. (1.114a), and by the gradient noise terms (1.114b), we also introduce the following $N \times 1$ column vectors whose entries are of size $M \times 1$ each:

$$\boldsymbol{z}_i \triangleq \begin{bmatrix} 2\boldsymbol{u}_{1,i}^{\mathsf{T}} \boldsymbol{v}_1(i) \\ 2\boldsymbol{u}_{2,i}^{\mathsf{T}} \boldsymbol{v}_2(i) \\ \vdots \\ 2\boldsymbol{u}_{N,i}^{\mathsf{T}} \boldsymbol{v}_N(i) \end{bmatrix}, \quad \boldsymbol{s}_i \triangleq \begin{bmatrix} \boldsymbol{s}_{1,i}(\boldsymbol{\phi}_{1,i-1}) \\ \boldsymbol{s}_{2,i}(\boldsymbol{\phi}_{2,i-1}) \\ \vdots \\ \boldsymbol{s}_{N,i}(\boldsymbol{\phi}_{N,i-1}) \end{bmatrix}. \tag{1.115b}$$

We further introduce the Kronecker products

$$\mathcal{A}_{o,i} \triangleq \boldsymbol{A}_{o,i} \otimes I_M, \quad \mathcal{A}_{1,i} \triangleq \boldsymbol{A}_{1,i} \otimes I_M, \quad \mathcal{A}_{2,i} \triangleq \boldsymbol{A}_{2,i} \otimes I_M \tag{1.116a}$$

and the following $N \times N$ *block* diagonal matrices, whose individual entries are of size $M \times M$ each:

$$\mathcal{M}_i \triangleq \text{diag}\{\boldsymbol{\mu}_1(i) I_M, \boldsymbol{\mu}_2(i) I_M, \ldots, \boldsymbol{\mu}_N(i) I_M\}, \tag{1.116b}$$

$$\mathcal{R}_i \triangleq \text{diag}\left\{ 2\boldsymbol{u}_{1,i}^{\mathsf{T}} \boldsymbol{u}_{1,i}, 2\boldsymbol{u}_{2,i}^{\mathsf{T}} \boldsymbol{u}_{2,i}, \ldots, 2\boldsymbol{u}_{N,i}^{\mathsf{T}} \boldsymbol{u}_{N,i} \right\}. \tag{1.116c}$$

From Eq. (1.114a) we can then easily conclude that the block network variables satisfy the recursions:

$$\begin{cases} \tilde{\boldsymbol{\phi}}_{i-1} &= \boldsymbol{\mathcal{A}}_{1,i}^{\mathrm{T}} \tilde{\boldsymbol{w}}_{i-1}, \\ \tilde{\boldsymbol{\psi}}_{i} &= \left(\boldsymbol{\mathcal{A}}_{o,i}^{\mathrm{T}} - \boldsymbol{\mathcal{M}}_i \boldsymbol{\mathcal{R}}_i\right) \tilde{\boldsymbol{\phi}}_{i-1} - \boldsymbol{\mathcal{M}}_i \boldsymbol{z}_i, \\ \tilde{\boldsymbol{w}}_{i} &= \boldsymbol{\mathcal{A}}_{2,i}^{\mathrm{T}} \tilde{\boldsymbol{\psi}}_{i} \end{cases} \tag{1.117a}$$

so that the network weight error vector, $\tilde{\boldsymbol{w}}_i$, evolves according to:

$$\tilde{\boldsymbol{w}}_i = \boldsymbol{\mathcal{A}}_{2,i}^{\mathrm{T}} \left(\boldsymbol{\mathcal{A}}_{o,i}^{\mathrm{T}} - \boldsymbol{\mathcal{M}}_i \boldsymbol{\mathcal{R}}_i\right) \boldsymbol{\mathcal{A}}_{1,i}^{\mathrm{T}} \tilde{\boldsymbol{w}}_{i-1} - \boldsymbol{\mathcal{A}}_{2,i}^{\mathrm{T}} \boldsymbol{\mathcal{M}}_i \boldsymbol{z}_i. \tag{1.117b}$$

For comparison purposes, if each agent operates individually and uses the noncooperative strategy (1.68a), then the weight error vector would instead evolve according to the following recursion:

$$\tilde{\boldsymbol{w}}_i = (I_{MN} - \boldsymbol{\mathcal{M}}_i \boldsymbol{\mathcal{R}}_i)\, \tilde{\boldsymbol{w}}_{i-1} - \boldsymbol{\mathcal{M}}_i \boldsymbol{z}_i, \quad i \geq 0, \tag{1.118}$$

where the matrices $\{\boldsymbol{\mathcal{A}}_{o,i}, \boldsymbol{\mathcal{A}}_{1,i}, \boldsymbol{\mathcal{A}}_{2,i}\}$ do not appear any longer, and with a block diagonal coefficient matrix $(I_{MN} - \boldsymbol{\mathcal{M}}_i \boldsymbol{\mathcal{R}}_i)$. It is also straightforward to verify that recursion (1.117b) can be equivalently rewritten in the following form in terms of the gradient noise vector, $\boldsymbol{s}_i$, defined by Eq. (1.115b):

$$\tilde{\boldsymbol{w}}_i = \boldsymbol{\mathcal{B}}_i \, \tilde{\boldsymbol{w}}_{i-1} + \boldsymbol{\mathcal{A}}_{2,i}^{\mathrm{T}} \boldsymbol{\mathcal{M}}_i \boldsymbol{s}_i, \tag{1.119a}$$

where

$$\boldsymbol{\mathcal{B}}_i \triangleq \boldsymbol{\mathcal{A}}_{2,i}^{\mathrm{T}} \left(\boldsymbol{\mathcal{A}}_{o,i}^{\mathrm{T}} - \boldsymbol{\mathcal{M}}_i \mathcal{R}\right) \boldsymbol{\mathcal{A}}_{1,i}^{\mathrm{T}}, \tag{1.119b}$$

$$\mathcal{R} \triangleq \mathbb{E} \boldsymbol{\mathcal{R}}_i = \mathrm{diag}\{2R_{u,1},\ 2R_{u,2}, \ldots, 2R_{u,N}\}. \tag{1.119c}$$

■

Example 1.23 (Mean-error behavior). We continue with the setting of Example 1.22. In MSE analysis, we are interested in examining how the quantities $\mathbb{E}\tilde{\boldsymbol{w}}_i$ and $\mathbb{E}\,\|\tilde{\boldsymbol{w}}_i\|^2$ evolve over time. If we refer back to the data model described in Example 1.13, where the regression data $\{\boldsymbol{u}_{k,i}\}$ were assumed to be temporally white and independent over space, then the stochastic matrix $\boldsymbol{\mathcal{R}}_i$ appearing in Eqs. (1.117b), (1.118) is seen to be statistically independent of $\tilde{\boldsymbol{w}}_{i-1}$. We further assume that, in the unified formulation, the entries of the combination policies $\{\boldsymbol{A}_{o,i}, \boldsymbol{A}_{1,i}, \boldsymbol{A}_{2,i}\}$ are independent of each other (as well as over time) and of any other variable in the learning algorithm. Therefore, taking expectations of both sides of these recursions, and invoking the fact that $\boldsymbol{u}_{k,i}$ and $\boldsymbol{v}_k(i)$ are also independent of each other and have zero means (so that $\mathbb{E}\boldsymbol{z}_i = 0$), we conclude that the mean-error vectors evolve according to the following recursions:

$$\mathbb{E}\tilde{\boldsymbol{w}}_i = \bar{\mathcal{B}}\left(\mathbb{E}\tilde{\boldsymbol{w}}_{i-1}\right) \qquad \text{(distributed)}, \tag{1.120a}$$

$$\mathbb{E}\tilde{\boldsymbol{w}}_i = (I_{MN} - \bar{\mathcal{M}}\mathcal{R})\left(\mathbb{E}\tilde{\boldsymbol{w}}_{i-1}\right) \quad \text{(noncooperative)}, \tag{1.120b}$$

where

$$\bar{\mathcal{B}} \triangleq \mathbb{E}\boldsymbol{\mathcal{B}}_i = \bar{\mathcal{A}}_2^{\mathrm{T}} \left(\bar{\mathcal{A}}_o^{\mathrm{T}} - \bar{\mathcal{M}}\mathcal{R}\right) \bar{\mathcal{A}}_1^{\mathrm{T}}, \tag{1.121a}$$

$$\bar{\mathcal{M}} = \mathbb{E}\boldsymbol{\mathcal{M}}_i = \mathrm{diag}\{\bar{\mu}_1 I_M,\ \ldots,\ \bar{\mu}_N I_M\}, \tag{1.121b}$$

$$\bar{A}_o = \mathbb{E}\,\boldsymbol{\mathcal{A}}_{o,i}, \tag{1.121c}$$

$$\bar{A}_1 = \mathbb{E}\,\boldsymbol{\mathcal{A}}_{1,i}, \tag{1.121d}$$

$$\bar{A}_2 = \mathbb{E}\,\boldsymbol{\mathcal{A}}_{2,i}. \tag{1.121e}$$

The matrix $\bar{\mathcal{B}}$ controls the dynamics of the mean weight-error vector for the distributed strategies. Observe, in particular, that $\bar{\mathcal{B}}$ reduces to the following forms for the various strategies (noncooperative (Eq. 1.68a), consensus (Eq. 1.81), and diffusion (Eqs. 1.83a–1.83b):

$$\bar{\mathcal{B}}_{\rm ncop} = I_{MN} - \bar{\mathcal{M}}\mathcal{R}, \tag{1.122a}$$

$$\bar{\mathcal{B}}_{\rm cons} = \bar{\mathcal{A}}^{\rm T} - \bar{\mathcal{M}}\mathcal{R}, \tag{1.122b}$$

$$\bar{\mathcal{B}}_{\rm atc} = \bar{\mathcal{A}}^{\rm T}\left(I_{MN} - \bar{\mathcal{M}}\mathcal{R}\right), \tag{1.122c}$$

$$\bar{\mathcal{B}}_{\rm cta} = \left(I_{MN} - \bar{\mathcal{M}}\mathcal{R}\right)\bar{\mathcal{A}}^{\rm T}, \tag{1.122d}$$

where $\bar{\mathcal{A}} = \bar{A} \otimes I_M$ and $\bar{A} = \mathbb{E}\boldsymbol{A}_i$. ■

Example 1.24 (MSE networks with uniform agents). The results of Example 1.23 simplify when all agents employ step-sizes with the same mean value, $\bar{\mu}_k \equiv \bar{\mu}$, and observe regression data with the same covariance matrix, $R_{u,k} \equiv R_u$ [61,85]. In this case, we can express $\bar{\mathcal{M}}$ and $\mathcal{R}$ from Eqs. (1.152b) and (1.119c) in Kronecker product form as follows:

$$\bar{\mathcal{M}} = \bar{\mu} I_N \otimes I_M, \quad \mathcal{R} = I_N \otimes 2R_u \tag{1.123}$$

so that expressions (1.122a)–(1.122d) reduce to

$$\bar{\mathcal{B}}_{\rm ncop} = I_N \otimes (I_M - 2\bar{\mu}R_u), \tag{1.124a}$$

$$\bar{\mathcal{B}}_{\rm cons} = \bar{A}^{\rm T} \otimes I_M - 2\bar{\mu}(I_M \otimes R_u), \tag{1.124b}$$

$$\bar{\mathcal{B}}_{\rm atc} = \bar{A}^{\rm T} \otimes (I_M - 2\bar{\mu}R_u), \tag{1.124c}$$

$$\bar{\mathcal{B}}_{\rm cta} = \bar{A}^{\rm T} \otimes (I_M - 2\bar{\mu}R_u). \tag{1.124d}$$

Observe that $\bar{\mathcal{B}}_{\rm atc} = \bar{\mathcal{B}}_{\rm cta}$, so we denote these matrices by $\bar{\mathcal{B}}_{\rm diff}$. Using properties of the eigenvalues of Kronecker products of matrices, it can be easily verified that the MN eigenvalues of the above $\bar{\mathcal{B}}$ matrices are given by the following expressions in terms of the eigenvalues of the component matrices $\{\bar{A}, R_u\}$ for $k = 1, 2, \ldots N$ and $m = 1, 2, \ldots, M$:

$$\lambda(\bar{\mathcal{B}}_{\rm ncop}) = 1 - 2\bar{\mu}\lambda_m(R_u), \tag{1.125a}$$

$$\lambda(\bar{\mathcal{B}}_{\rm cons}) = \lambda_k(\bar{A}) - 2\bar{\mu}\lambda_m(R_u), \tag{1.125b}$$

$$\lambda(\bar{\mathcal{B}}_{\rm diff}) = \lambda_k(\bar{A})\,[1 - 2\bar{\mu}\lambda_m(R_u)]. \tag{1.125c}$$

■

Example 1.25 (Potential instability in consensus networks). Consensus strategies can become unstable when used for adaptation purposes [2,85]. This undesirable effect is already reflected in expressions (1.125a)–(1.125c). In particular, observe that the eigenvalues of $\bar{A}$ appear multiplying $(1 - 2\mu\lambda_m(R_u))$ in expression (1.125c) for diffusion. As such, and because $\rho(\bar{A}) = 1$ for any left-stochastic matrix, we conclude for this case of uniform agents that $\rho(\bar{\mathcal{B}}_{\rm diff}) = \rho(\bar{\mathcal{B}}_{\rm ncop})$. It follows that, regardless of the choice of the mean combination policy $\bar{A}$, the diffusion strategies will be stable in the mean (i.e., $\mathbb{E}\tilde{w}_i$ will converge asymptotically to zero) whenever the individual noncooperative agents are stable in the mean:

$$\text{individual agents stable} \Longrightarrow \text{diffusion networks stable.} \tag{1.126a}$$

The same conclusion is not true for consensus networks; the individual agents can be stable and yet the consensus network can become unstable. This is because $\lambda_k(\bar{A})$ appears as an additive (rather than multiplicative) term in Eq. (1.125b) (see [2,72,85] for examples):

$$\text{individual agents stable} \nRightarrow \text{consensus networks stable.} \tag{1.126b}$$

The fact that the combination matrix $\bar{\mathcal{A}}^{\mathrm{T}}$ appears in an additive form in Eq. (1.122b) is the result of the asymmetry that was mentioned earlier in the update equation for the consensus strategy. In contrast, the update equations for the diffusion strategies lead to $\bar{\mathcal{A}}^{\mathrm{T}}$ appearing in a multiplicative form in Eqs. (1.122c), (1.122d). ■

Example 1.26 (Useful stability result). It is observed from expressions (1.122c) and (1.122d) in the asynchronous case as well as from the corresponding expressions (81c) and (81d) in the synchronous case studied in [1] that the mean stability of diffusion strategies usually involves examining the stability of a matrix product of the form:

$$\mathcal{B} \triangleq \mathcal{A}_2^{\mathrm{T}} \mathcal{D} \mathcal{A}_1^{\mathrm{T}}, \tag{1.127}$$

where $\mathcal{D}$ is a block diagonal symmetric matrix with blocks of size $M \times M$, while $\mathcal{A}_1$ and $\mathcal{A}_2$ are Kronecker product matrices defined in terms of $N \times N$ left-stochastic matrices A_1 and A_2 as $\mathcal{A}_1 = A_1 \otimes I_M$ and $\mathcal{A}_2 = A_2 \otimes I_M$. For example, in Eq. (1.122c) we have $A_1 = I_N$, $A_2 = \bar{A}$, $\mathcal{A}_1 = I_{MN}$, $\mathcal{A}_2 = \bar{\mathcal{A}}$, and $\mathcal{D} = I_{MN} - \mathcal{M}\mathcal{R}$.

Matrix products of the form (1.127) are induced by the cooperation mechanism that is inherent to diffusion learning. They have a useful property: it turns out that these matrix products are stable, regardless of A_1 and A_2, as long as $\mathcal{D}$ is stable (i.e., has all its eigenvalues strictly inside the unit disc). This useful result is easy to establish for *symmetric* left-stochastic matrices A_1 and A_2, as already noted in [73]. This is because for symmetric matrices, their spectral radii coincide with their 2-induced norms and, hence,

$$\rho(A_1) = \|A_1\|, \quad \rho(A_2) = \|A_2\|. \tag{1.128}$$

Consequently, because we already know that $\rho(A_1) = \rho(A_2) = 1$, it follows that

$$\begin{aligned}
\rho(\mathcal{B}) &\le \|\mathcal{B}\| \\
&\le \|\mathcal{A}_2\| \cdot \|\mathcal{D}\| \cdot \|\mathcal{A}_1\| \\
&= \rho(\mathcal{A}_2) \cdot \rho(\mathcal{D}) \cdot \rho(\mathcal{A}_1) \\
&= \rho(A_2) \cdot \rho(\mathcal{D}) \cdot \rho(A_1) \\
&= \rho(\mathcal{D}),
\end{aligned} \tag{1.129}$$

which confirms that a stable $\mathcal{D}$ guarantees a stable $\mathcal{B}$, regardless of A_1 and A_2.

The conclusion that $\mathcal{B}$ in Eq. (1.127) is stable whenever $\mathcal{D}$ is stable continues to hold even when the matrices A_1 and A_2 are *not* necessarily symmetric. However, the argument leading to Eq. (1.129) will need to be adjusted because property (1.128) need not hold anymore. This more general result was established in [61, App. D] and also in [100, App. A, pp. 3471–3473], where it was shown that multiplication of a symmetric block diagonal matrix $\mathcal{D}$ by any (not necessarily symmetric) left-stochastic Kronecker-product transformations from left and right generally reduces the spectral radius, i.e.,

$$\rho\left(\mathcal{A}_2^{\mathrm{T}}\mathcal{D}\mathcal{A}_1^{\mathrm{T}}\right) \leq \rho(\mathcal{D}). \tag{1.130}$$

Accordingly, a stable $\mathcal{D}$ again ensures a stable $\mathcal{B}$. This conclusion was established in the above references by replacing the 2-induced norm used to arrive at Eq. (1.129) by a more convenient block-maximum norm, denoted by $\|\cdot\|_{b,\infty}$ and defined as follows.

Let $x = \mathrm{col}\{x_1, x_2, \ldots, x_N\}$ denote an $N \times 1$ *block* column vector whose individual entries are themselves vectors of size $M \times 1$ each. Following [31,61,101], the block maximum norm of x is denoted by $\|x\|_{b,\infty}$ and is defined as

$$\|x\|_{b,\infty} \triangleq \max_{1\leq k\leq N} \|x_k\|. \tag{1.131}$$

That is, $\|x\|_{b,\infty}$ is equal to the largest Euclidean norm of its block components (this definition extends the regular notion of the ∞-norm of a vector to block vectors). The vector norm (1.131) induces a block-maximum matrix norm. Let $\mathcal{A}$ denote an arbitrary $N \times N$ block matrix with individual block entries of size $M \times M$ each. Then, the block-maximum norm of $\mathcal{A}$ is defined as

$$\|\mathcal{A}\|_{b,\infty} \triangleq \max_{x\neq 0} \frac{\|\mathcal{A}x\|_{b,\infty}}{\|x\|_{b,\infty}}. \tag{1.132}$$

The block-maximum norm has several useful properties; see [61]. In particular, when A is $N \times N$ left-stochastic and $\mathcal{A} = A \otimes I_M$, then it can be verified that $\|\mathcal{A}^{\mathrm{T}}\|_{b,\infty} = 1$. Likewise, when $\mathcal{D}$ is block diagonal and symmetric, then $\|\mathcal{D}\|_{b,\infty} = \rho(\mathcal{D})$. Consequently, repeating the argument leading to Eq. (1.129) and replacing the 2-induced norm used there by the block-maximum norm we have

$$\rho(\mathcal{B}) \leq \|\mathcal{B}\|_{b,\infty} \leq \|\mathcal{A}_2^{\mathrm{T}}\|_{b,\infty} \cdot \|\mathcal{D}\|_{b,\infty} \cdot \|\mathcal{A}_1^{\mathrm{T}}\|_{b,\infty} = \rho(D) \tag{1.133}$$

and we again conclude that a stable symmetric $\mathcal{D}$ guarantees a stable $\mathcal{B}$ for general left-stochastic matrices A_1 and A_2 because, for symmetric $\mathcal{D}$, it holds that $\rho(\mathcal{D}) = \|\mathcal{D}\|_{b,\infty}$.

Remark 1.1. The same argument (1.133) can be used to relax the requirement of symmetry and stability on the block diagonal matrix $\mathcal{D}$. Actually, as long as $\|\mathcal{D}\|_{b,\infty} < 1$, which is guaranteed by requiring the block diagonal entries of $\mathcal{D}$ to have their 2-induced norms bounded by one, we can again conclude that $\rho(\mathcal{B}) < 1$ so that $\mathcal{B}$ is stable.

Remark 1.2. The validity of property (1.130) for general left-stochastic matrices was already noted and exploited in earlier works, e.g., in Lemma 1 of [74] and in Lemma 2 of [102]. However, the statement of Lemma 1 in [74] left out the qualification "diagonal" for the center matrix, and the norm $\|\cdot\|_\rho$ that was used in the proof of the lemma in Appendix I of [74] should be replaced by the $\|\cdot\|_\infty$-norm. Although these corrections were already noted in the references [61,100], we restate below the correct form of Lemma 1 from [74] for accuracy and provide its adjusted proof using the current notation. That lemma deals with left-stochastic matrices A_1 and A_2 prior to extension by Kronecker products. Its correct statement should read as follows (the word "diagonal" is missing from the statement in [74]).

Restatement of Lemma 1 [74]: Let A_1, A_2, and D denote arbitrary $N \times N$ matrices, where A_1 and A_2 have real nonnegative entries, with columns adding up to one, i.e., $\mathbb{1}^{\mathsf{T}}A_1 = \mathbb{1}^{\mathsf{T}}$, $\mathbb{1}^{\mathsf{T}}A_2 = \mathbb{1}^{\mathsf{T}}$. Then, the matrix $B = A_2^{\mathsf{T}} D A_1^{\mathsf{T}}$ is stable for *any* choice of A_1 and A_2 if, and only if, the diagonal matrix D is stable.

Proof. One immediate derivation is to employ the ∞-norm and to note that

$$\|A_1^{\mathsf{T}}\|_\infty = 1 = \|A_2^{\mathsf{T}}\|_\infty, \quad \|D\|_\infty = \rho(D) \tag{1.134}$$

and, hence,

$$\rho(B) \le \|B\|_\infty \le \|A_2^{\mathsf{T}}\|_\infty \cdot \|D\|_\infty \cdot \|A_1^{\mathsf{T}}\|_\infty = \rho(D). \tag{1.135}$$

It follows that a stable D guarantees a stable B. A second derivation that fixes the argument from Appendix I of [74] relies on using the ∞-norm instead of the ρ-norm used there. Thus, note that we can alternatively argue that

$$\begin{aligned} \|B^i\|_\infty &\le \left(\|A_2^{\mathsf{T}}\|_\infty\right)^i \cdot (\|D\|_\infty)^i \cdot \left(\|A_1^{\mathsf{T}}\|_\infty\right)^i \\ &= (\rho(D))^i \longrightarrow 0, \text{ as } i \to \infty \end{aligned} \tag{1.136}$$

when D is stable. This completes the proof of Lemma 1 from [74].

Remark 1.3. These same arguments establish the validity of Lemma 2 from [102], which deals with a special case involving a matrix of the form $\mathcal{B} = \mathcal{A}_2^{\mathsf{T}}\mathcal{D}$ using the current notation (the matrix in [102] is denoted by $\mathcal{F} = \mathcal{C}^{\mathsf{T}}\mathcal{M}$ with $\mathcal{B} \leftarrow \mathcal{F}$, $A_2 \leftarrow \mathcal{C}$, and $\mathcal{D} \leftarrow \mathcal{M}$; moreover, the matrix $\mathcal{M}$ is now generally nonsymmetric but still block diagonal with stable entries of the form $\mathcal{M}_{kk} = (I - P_k S_k)F$ for some matrices $\{P_k, S_k, F\}$ defined in [102]. When $\|\mathcal{M}_{kk}\|_2 < 1$, these block diagonal entries will have 2-induced norms smaller than one so that $\|\mathcal{D}\|_{b,\infty} < 1$. Again, the ρ-norm used in the proof of Lemma 2 in [102] should be replaced by the $\|\cdot\|_{b,\infty}$ norm and the argument adjusted as in Eq. (1.133) or, alternatively, we note that

$$\|\mathcal{B}^i\|_{b,\infty} \le \left(\|\mathcal{A}_2^{\mathsf{T}}\|_{b,\infty}\right)^i \cdot (\|\mathcal{D}\|_{b,\infty})^i \longrightarrow 0, \text{ as } i \to \infty. \tag{1.137}$$

■

1.7.2 DIFFUSION NETWORKS

Given the superior stability properties of diffusion strategies for adaptation and learning over networks, we continue our presentation by focusing on this class of algorithms. The results in the previous section focus on MSE networks, which deal with MSE cost functions. We now consider networks with more general costs, $\{J_k(w)\}$, and apply diffusion strategies to seek the global minimizer, w^o, of the aggregate cost function, $J^{\text{glob}}(w)$, defined by Eq. (1.56). Without loss in generality, we consider ATC diffusion implementations of the form (1.88):

$$\boldsymbol{\psi}_{k,i} = \boldsymbol{w}_{k,i-1} - \boldsymbol{\mu}_k(i)\widehat{\nabla_{w^{\mathsf{T}}} J_k}(\boldsymbol{w}_{k,i-1}), \tag{1.138a}$$

$$\boldsymbol{w}_{k,i} = \sum_{\ell \in \mathcal{N}_{k,i}} \boldsymbol{a}_{\ell k}(i)\, \boldsymbol{\psi}_{\ell,i}. \tag{1.138b}$$

Similar conclusions will apply to CTA diffusion implementations.

Assumption 1.7 (Conditions on cost functions). *The aggregate cost $J^{glob}(w)$ in Eq. (1.56) is twice differentiable and satisfies a condition similar to Eq. (1.9) in Assumption 1.1 for some positive parameters $\nu_d \le \delta_d$. Moreover, all individual costs $\{J_k(w)\}$ are assumed to be strongly convex with their global minimizers located at w^o, as indicated earlier by Eq. (1.73).* ■

As explained before following Eq. (1.73), references [1,2,37] present results on the case in which the individual costs are only convex and need not be strongly convex. These references also discuss the case in which the individual costs need not share minimizers. With each agent k in Eq. (1.88), we again associate a gradient noise vector:

$$\boldsymbol{s}_{k,i}(\boldsymbol{w}_{k,i-1}) \triangleq \widehat{\nabla_{w^{\mathsf{T}}} J_k}(\boldsymbol{w}_{k,i-1}) - \nabla_{w^{\mathsf{T}}} J_k(\boldsymbol{w}_{k,i-1}). \tag{1.139}$$

Assumption 1.8 (Conditions on gradient noise). *It is assumed that the first- and second-order conditional moments of the gradient noise components satisfy:*

$$\mathbb{E}\left[\boldsymbol{s}_{k,i}(\boldsymbol{w}_{k,i-1}) \,|\, \boldsymbol{\mathcal{F}}_{i-1}\right] = 0, \tag{1.140a}$$

$$\mathbb{E}\left[\boldsymbol{s}_{k,i}(\boldsymbol{w}_{k,i-1})\boldsymbol{s}_{\ell,i}^{\mathsf{T}}(\boldsymbol{w}_{\ell,i-1}) \,|\, \boldsymbol{\mathcal{F}}_{i-1}\right] = 0, \quad \forall\, k \neq \ell, \tag{1.140b}$$

$$\mathbb{E}\left[\|\boldsymbol{s}_{k,i}(\boldsymbol{w}_{k,i-1})\|^2 \,|\, \boldsymbol{\mathcal{F}}_{i-1}\right] \le \beta_k^2\, \|\tilde{\boldsymbol{w}}_{k,i-1}\|^2 + \sigma_{s,k}^2 \tag{1.140c}$$

almost surely for some nonnegative scalars β_k^2 and $\sigma_{s,k}^2$, and where $\boldsymbol{\mathcal{F}}_{i-1}$ represents the collection of all random events generated by the iterates from across all agents, $\{\boldsymbol{w}_{\ell,j},\ \ell = 1, 2, \ldots, N\}$, up to time $i-1$. Moreover, it is assumed that the limiting covariance matrix of $\boldsymbol{s}_{k,i}(w^o)$ exists:

$$R_{s,k} \triangleq \lim_{i\to\infty} \mathbb{E}\left[\boldsymbol{s}_{k,i}(w^o)\boldsymbol{s}_{k,i}^{\mathsf{T}}(w^o) \,|\, \boldsymbol{\mathcal{F}}_{i-1}\right]. \tag{1.140d}$$

■

We collect the error vectors and gradient noises from across all agents into $N \times 1$ *block* vectors, whose individual entries are of size $M \times 1$ each:

$$\tilde{\boldsymbol{w}}_i \triangleq \begin{bmatrix} \tilde{\boldsymbol{w}}_{1,i} \\ \tilde{\boldsymbol{w}}_{2,i} \\ \vdots \\ \tilde{\boldsymbol{w}}_{N,i} \end{bmatrix}, \quad \boldsymbol{s}_i \triangleq \begin{bmatrix} \boldsymbol{s}_{1,i} \\ \boldsymbol{s}_{2,i} \\ \vdots \\ \boldsymbol{s}_{N,i} \end{bmatrix} \tag{1.141}$$

and where we are dropping the argument $\boldsymbol{w}_{k,i-1}$ from the $\boldsymbol{s}_{k,i}(\cdot)$ for compactness of notation. Likewise, we introduce the following $N \times N$ *block* diagonal matrices, whose individual entries are of size $M \times M$ each:

$$\boldsymbol{\mathcal{M}}_i = \text{diag}\{\, \boldsymbol{\mu}_1(i) I_M,\ \boldsymbol{\mu}_2(i) I_M,\ \ldots,\ \boldsymbol{\mu}_N(i) I_M \,\}, \tag{1.142a}$$

$$\boldsymbol{\mathcal{H}}_{i-1} = \text{diag}\left\{\, \boldsymbol{H}_{1,i-1},\ \boldsymbol{H}_{2,i-1},\ \ldots,\ \boldsymbol{H}_{N,i-1} \,\right\}, \tag{1.142b}$$

where

$$H_{k,i-1} \triangleq \int_0^1 \nabla_w^2 J_k(w^o - t\tilde{w}_{k,i-1})dt. \tag{1.142c}$$

Now, in a manner similar to Eq. (1.29b), we can appeal to the mean-value theorem [2,30,44] to note that

$$\nabla_{w^{\mathrm{T}}} J_k(w_{k,i-1}) = -H_{k,i-1}\tilde{w}_{k,i-1} \tag{1.143}$$

so that the approximate gradient vector can be expressed as:

$$\widehat{\nabla_{w^{\mathrm{T}}} J_k}(w_{k,i-1}) = -H_{k,i-1}\tilde{w}_{k,i-1} + s_{k,i}(w_{k,i-1}). \tag{1.144}$$

Subtracting w^o from both sides of Eqs. (1.138a), (1.138b), and using Eq. (1.144), we find that the network error vector evolves according to the following stochastic recursion:

$$\boxed{\tilde{w}_i = \mathcal{B}_{i-1}\tilde{w}_{i-1} + \mathcal{A}_i^{\mathrm{T}}\mathcal{M}_i s_i} \tag{1.145a}$$

where

$$\mathcal{B}_{i-1} \triangleq \mathcal{A}_i^{\mathrm{T}}\left(I_{NM} - \mathcal{M}_i\mathcal{H}_{i-1}\right). \tag{1.145b}$$

Recursion (1.145a) describes the evolution of the network error vector for general convex costs, $J_k(w)$, in a manner similar to recursion (1.119a) in the MSE case. However, recursion (1.145a) is more challenging to deal with because of the presence of the random matrix $\mathcal{H}_{i-1}$; this matrix is replaced by the constant term $\mathcal{R}$ in the earlier recursion (1.119a) because that example deals with MSE networks where the individual costs, $J_k(w)$, are quadratic in w and, therefore, their Hessian matrices are constant and independent of w. In that case, each matrix $H_{k,i-1}$ in Eq. (1.142c) will evaluate to $2R_{u,k}$ and the matrix $\mathcal{H}_{i-1}$ in Eq. (1.145b) will coincide with the matrix $\mathcal{R}$ defined by Eq. (1.119c).

The next statement ascertains that sufficiently small step-sizes exist that guarantee the MSE stability of the asynchronous diffusion strategy (1.138a) and (1.138b) [2,3].

Lemma 1.4 (MSE network stability). *Consider an asynchronous network of N interacting agents running the ATC diffusion strategy* (1.138a) *and (1.138b). Assume the conditions in Assumptions 1.6–1.8 hold. Let*

$$\mu_{x,\max} = \max_{1\le k\le N}\{\mu_{x,k}\} = \max_{1\le k\le N}\left(\bar{\mu}_k + \frac{\sigma_{\mu,k}^2}{\bar{\mu}_k}\right). \tag{1.146}$$

Then, there exists $\mu_o > 0$ such that for all $\mu_{x,\max} < \mu_o$:

$$\limsup_{i\to\infty} \mathbb{E}\|\tilde{w}_{k,i}\|^2 = O(\mu_{x,\max}). \tag{1.147}$$

Proof. See App. IV of [3]. ■

Result (1.147) shows that the MSD of the network is in the order of $\mu_{x,\max}$. Therefore, sufficiently small step-sizes lead to sufficiently small MSDs. As was the case with the discussion in Section 1.2.7, we can also seek a closed-form expression for the MSD performance of the asynchronous diffusion network and its agents. To do that, we first introduce the analog of Assumption 1.4 for the network case.

Assumption 1.9 (Smoothness Conditions). *The Hessian matrix of the individual cost functions* $\{J_k(w)\}$ *and the noise covariance matrices defined for each agent in a manner similar to Eq. (1.17a) and denoted by* $R_{s,k,i}(w)$*, are assumed to be locally Lipschitz continuous in a small neighborhood around* w^o*:*

$$\|\nabla_w^2 J_k(w^o + \delta w) - \nabla_w^2 J_k(w^o)\| \le \tau_{k,d}\|\delta w\|, \tag{1.148a}$$

$$\|R_{s,k,i}(w^o + \delta w) - R_{s,k,i}(w^o)\| \le \tau_{k,s}\|\delta w\|^{\kappa} \tag{1.148b}$$

for small perturbation $\|\delta w\| \le r_d$ *and for some* $\tau_{k,d}, \tau_{k,s} \ge 0$ *and* $1 \le \kappa \le 2$. ■

Lemma 1.5 (Asynchronous network MSD performance). *Consider an asynchronous network of* N *interacting agents running the asynchronous diffusion strategy* (1.138a) *and (1.138b). Assume the conditions under Assumptions 1.6, 1.7, 1.8, and 1.1 hold. Assume further that the step-size parameter* $\mu_{x,\max}$ *is sufficiently small to ensure mean-square stability, as already ascertained by Lemma 1.4. Then,*

$$MSD_{diff,k}^{asyn} \approx MSD_{diff,\,av}^{asyn} \approx \frac{1}{2}Tr\left[\left(\sum_{k=1}^{N}\bar{\mu}_k\bar{p}_k H_k\right)^{-1}\left(\sum_{k=1}^{N}(\bar{\mu}_k^2 + \sigma_{\mu,k}^2)p_{c,kk}R_{s,k}\right)\right], \tag{1.149a}$$

where $H_k = \nabla_w^2 J_k(w^o)$*. Moreover, for a large enough i, the convergence rate toward the above steady-state value is well approximated by the scalar:*

$$\alpha_{dist}^{asyn} = 1 - 2\lambda_{\min}\left(\sum_{k=1}^{N}\bar{\mu}_k\bar{p}_k H_k\right) + O\left(\mu_{x,\max}^{1+1/N^2}\right). \tag{1.149b}$$

Proof. See App. XII in [4]. ■

Example 1.27 (Gaussian regression data). MSE performance expressions of the form (1.149a) are accurate to first-order in the step-size parameters, i.e., they are on the order of $O(\bar{\mu}_k)$. The same is true of expression (112) from [1] in the synchronous case. There are situations, however, where exact expressions for the MSE performance can be derived for multiagent networks. We illustrate this possibility here for the case of MSE networks of the type described earlier in Example 1.13. We consider first synchronous networks and comment later on how the results should be adjusted to handle asynchronous behavior.

We refer to the same setting of Example 9 from [1] where we have N agents observing streaming data $\{\boldsymbol{d}_k(i), \boldsymbol{u}_{k,i}\}$ that satisfy the regression model:

$$\boldsymbol{d}_k(i) = \boldsymbol{u}_{k,i}w^o + \boldsymbol{v}_k(i). \tag{1.150a}$$

We assume the regression vectors are zero-mean Gaussian-distributed with diagonal covariance matrices denoted by Λ_k, say,

$$R_{u,k} = \mathbb{E}\boldsymbol{u}_{k,i}^{\mathsf{T}}\boldsymbol{u}_{k,i} \triangleq \Lambda_k > 0. \tag{1.150b}$$

We further assume that the regression data is temporally white and independent over space so that

$$\mathbb{E}\boldsymbol{u}_{k,i}^{\mathsf{T}}\boldsymbol{u}_{\ell,j} = R_{u,k}\delta_{k,\ell}\delta_{i,j}. \tag{1.150c}$$

We also assume that the measurement noise process $\boldsymbol{v}_k(i)$ is temporally white and independent over space so that $\mathbb{E}\boldsymbol{v}_k(i)\boldsymbol{v}_\ell(j) = \sigma_{v,k}^2\delta_{k,\ell}\delta_{i,j}$ in terms of the Kronecker delta sequence $\delta_{m,n}$. Likewise, the measurement noise $v_k(i)$ and the regression data $u_{\ell,j}$ are assumed to be independent of each other for all k, ℓ, i, j.

The Gaussian assumption on the regression data is useful because, for this case, fourth-order moments of the regression vectors can be evaluated in closed form (and such fourth-order moment calculations will arise in the process of determining closed-form expressions for the MSE). Specifically, for any matrix $Q \geq 0$ of size $M \times M$, it holds for independent *real-valued* Gaussian regressors that [23] (there is a typo in reproducing this expression in the second equation of App. II in [74]):

$$\mathbb{E}u_{k,i}^{\mathrm{T}}u_{k,i}Qu_{\ell,i}^{\mathrm{T}}u_{\ell,i} = R_{u,k}QR_{u,\ell} + \delta_{k,\ell}\left\{R_{u,k}\mathrm{Tr}\left(QR_{u,k}\right) + R_{u,k}QR_{u,k}\right\}. \tag{1.150d}$$

We assume the network is running a diffusion strategy of the form:

$$\begin{cases} \phi_{k,i-1} = \sum_{\ell\in\mathcal{N}_k} a_{1,\ell k}w_{\ell,i-1}, \\ \psi_{k,i} = \phi_{k,i-1} + 2\mu_k u_{k,i}^{\mathrm{T}}\left[d_k(i) - u_{k,i}\phi_{k,i-1}\right], \\ w_{k,i} = \sum_{\ell\in\mathcal{N}_k} a_{2,\ell k}\psi_{\ell,i}, \end{cases} \tag{1.151a}$$

which includes both the ATC and CTA LMS diffusion forms as special cases. We know from (77b) in [1] that the error network vector evolves according to the dynamics

$$\tilde{w}_i = \mathcal{A}_2^{\mathrm{T}}\left(I - \mathcal{M}\mathcal{R}_i\right)\mathcal{A}_1^{\mathrm{T}}\tilde{w}_{i-1} \; - \; \mathcal{A}_2^{\mathrm{T}}\mathcal{M}z_i, \quad i \geq 0, \tag{1.151b}$$

which is defined in terms of the following $N \times 1$ column vectors whose entries are of size $M \times 1$ each:

$$z_i \triangleq \begin{bmatrix} 2u_{1,i}^{\mathrm{T}}v_1(i) \\ 2u_{2,i}^{\mathrm{T}}v_2(i) \\ \vdots \\ 2u_{N,i}^{\mathrm{T}}v_N(i) \end{bmatrix}, \quad s_i \triangleq \begin{bmatrix} s_{1,i}(\phi_{1,i-1}) \\ s_{2,i}(\phi_{2,i-1}) \\ \vdots \\ s_{N,i}(\phi_{N,i-1}). \end{bmatrix} \tag{1.151c}$$

Moreover,

$$\mathcal{A}_1 \triangleq A_1 \otimes I_M, \quad \mathcal{A}_2 \triangleq A_2 \otimes I_M \tag{1.152a}$$

while the quantities

$$\mathcal{M} \triangleq \mathrm{diag}\{\; \mu_1 I_M, \; \mu_2 I_M, \; \ldots, \; \mu_N I_M \;\}, \tag{1.152b}$$

$$\mathcal{R}_i \triangleq \mathrm{diag}\left\{2u_{1,i}^{\mathrm{T}}u_{1,i}, 2u_{2,i}^{\mathrm{T}}u_{2,i}, \ldots, 2u_{N,i}^{\mathrm{T}}u_{N,i}\right\}, \tag{1.152c}$$

$$\mathcal{R} \triangleq \mathbb{E}\mathcal{R}_i = \mathrm{diag}\left\{2R_{u,1}, 2R_{u,2}, \ldots, 2R_{u,N}\right\}, \tag{1.152d}$$

$$\mathcal{S} \triangleq \mathbb{E}z_i z_i^{\mathrm{T}} = \mathrm{diag}\left\{4\sigma_{v,1}^2 R_{u,1}, \ldots, 4\sigma_{v,N}^2 R_{u,N}\right\} \tag{1.152e}$$

are $N \times N$ *block* diagonal matrices, whose individual entries are of size $M \times M$ each. Let Σ be any $N \times N$ nonnegative definite block matrix that we are free to choose, with blocks of size $M \times M$. Computing the Σ-weighted squared norm of the error vector in Eq. (1.151b) under expectation gives (see the derivation leading to (269) in [61] or (38)–(39) in [74]):

$$\mathbb{E}\|\tilde{w}_i\|_{\Sigma}^2 = \mathbb{E}\|\tilde{w}_{i-1}\|_{\Sigma'}^2 + \mathbb{E}\left(z_i^{\mathrm{T}}\mathcal{M}\mathcal{A}_2\Sigma\mathcal{A}_2^{\mathrm{T}}\mathcal{M}z_i\right), \tag{1.153a}$$

where the deterministic weighting matrix Σ' is given by:

$$\Sigma' = \mathcal{A}_1 \left\{ \mathcal{A}_2 \Sigma \mathcal{A}_2^{\mathrm{T}} - \mathcal{A}_2 \Sigma \mathcal{A}_2^{\mathrm{T}} \mathcal{M} \mathcal{R} - \mathcal{R} \mathcal{M} \mathcal{A}_2 \Sigma \mathcal{A}_2^{\mathrm{T}} + \mathbb{E}\left(\boldsymbol{\mathcal{R}}_i \mathcal{M} \mathcal{A}_2 \Sigma \mathcal{A}_2^{\mathrm{T}} \mathcal{M} \boldsymbol{\mathcal{R}}_i \right) \right\} \mathcal{A}_1^{\mathrm{T}}. \tag{1.153b}$$

We can evaluate the last expectations in Eqs. (1.153a), (1.153b) in closed form. But first we need to introduce a convenient block-vector notation, denoted by bvec(·). Thus, given an $N \times N$ block matrix, with blocks of size $M \times M$ each, say for $N = 3$,

$$X = \begin{bmatrix} X_{11} & X_{12} & X_{13} \\ X_{21} & X_{22} & X_{23} \\ X_{31} & X_{32} & X_{33} \end{bmatrix} \tag{1.154a}$$

its block vectorization is obtained as follows. We first vectorize each of the block entries and define the column vector $x_{k\ell} = \mathrm{vec}(X_{k\ell})$; this operation stacks the columns of $X_{k\ell}$ on top of each other. Subsequently, the quantity bvec(X) is obtained by stacking the vectors $\{x_{k\ell}\}$ on top of each other:

$$\mathrm{bvec}(X) \triangleq \mathrm{col}\{x_{11}, x_{21}, x_{31}, x_{12}, x_{22}, x_{32}, x_{13}, x_{23}, x_{33}\}. \tag{1.154b}$$

The following two useful properties can be easily verified for any block matrices $\{A, B, \Sigma\}$ of compatible dimensions [103,104]:

$$\mathrm{bvec}(A\Sigma B) = (B^{\mathrm{T}} \otimes_b A)\mathrm{bvec}(\Sigma), \tag{1.154c}$$

$$\mathrm{Tr}(A^{\mathrm{T}} B) = (\mathrm{bvec}(A))^{\mathrm{T}}\, \mathrm{bvec}(B), \tag{1.154d}$$

where the notation $A \otimes_b B$ denotes the block Kronecker product of two block matrices A and B (assumed here to be both of size $N \times N$ with $M \times M$ blocks); the $k\ell$th block of $A \otimes_b B$ has size $NM^2 \times NM^2$ and is given by [104]:

$$[A \otimes_b B]_{k\ell} = \begin{bmatrix} A_{k\ell} \otimes B_{11} & A_{k\ell} \otimes B_{12} & \dots & A_{k\ell} \otimes B_{1N} \\ A_{k\ell} \otimes B_{21} & A_{k\ell} \otimes B_{22} & \dots & A_{k\ell} \otimes B_{2N} \\ \vdots & \vdots & \ddots & \vdots \\ A_{k\ell} \otimes B_{N1} & A_{k\ell} \otimes B_{N2} & \dots & A_{k\ell} \otimes B_{NN} \end{bmatrix} \tag{1.154e}$$

in terms of the traditional Kronecker product operation. Using the block Kronecker properties (1.154c) and (1.154d) we now find that the last expectation in Eq. (1.153a) is given by:

$$\mathbb{E}\left(\boldsymbol{z}_i^{\mathrm{T}} \mathcal{M} \mathcal{A}_2 \Sigma \mathcal{A}_2^{\mathrm{T}} \mathcal{M} \boldsymbol{z}_i \right) = b^{\mathrm{T}} \sigma, \tag{1.155a}$$

$$\sigma \triangleq \mathrm{bvec}(\Sigma), \tag{1.155b}$$

$$b \triangleq (\mathcal{A}_2^{\mathrm{T}} \otimes_b \mathcal{A}_2^{\mathrm{T}})(\mathcal{M} \otimes_b \mathcal{M})\mathrm{bvec}(\mathcal{S}). \tag{1.155c}$$

This is the same expression (56) from [73] for the case of CTA diffusion, and the same expression (42) from [74]) for CTA and ATC diffusion (except that this latter reference used the traditional vec(.) and Kronecker notation $\otimes$ instead of bvec(.) and the block Kronecker notation $\otimes_b$).

Let us now evaluate the last expectation in Eq. (1.153a). Let $\mathcal{Q} = \mathcal{M}\mathcal{A}_2\Sigma\mathcal{A}_2^{\mathrm{T}}\mathcal{M}$ so that we can rewrite more compactly:

$$\mathcal{K} \triangleq \mathbb{E}\left(\boldsymbol{\mathcal{R}}_i\mathcal{M}\mathcal{A}_2\Sigma\mathcal{A}_2^{\mathrm{T}}\mathcal{M}\boldsymbol{\mathcal{R}}_i\right) = \mathbb{E}\boldsymbol{\mathcal{R}}_i\mathcal{Q}\boldsymbol{\mathcal{R}}_i. \tag{1.156a}$$

The $k\ell$th block entry in this matrix is given by

$$\mathcal{K}_{k\ell} = 4\mathbb{E}\boldsymbol{u}_{k,i}^{\mathrm{T}}\boldsymbol{u}_{k,i}\mathcal{Q}_{k\ell}\boldsymbol{u}_{\ell,i}^{\mathrm{T}}\boldsymbol{u}_{\ell,i} \tag{1.156b}$$

in terms of the $k\ell$th block of $\mathcal{Q}$. Using property (1.150d) for Gaussian regressors, we get

$$\mathcal{K}_{k\ell} = 4R_{u,k}\mathcal{Q}_{k\ell}R_{u,\ell} + 4\delta_{k\ell}\left\{R_{u,k}\mathrm{Tr}\left[\mathcal{Q}_{kk}R_{u,k}\right] + R_{u,k}\mathcal{Q}_{kk}R_{u,k}\right\}. \tag{1.156c}$$

It is clear from the above expression that the matrix $\mathcal{K}$ has the following general form involving two block diagonal matrices:

$$\frac{1}{4}\mathcal{K} = \mathcal{R}\mathcal{Q}\mathcal{R} + Z_1 + Z_2, \tag{1.156d}$$

$$Z_1 \triangleq \text{ blkdiag}\left\{R_{u,k}\mathcal{Q}_{kk}R_{u,k}\right\}, \tag{1.156e}$$

$$Z_2 \triangleq \text{ blkdiag}\left\{R_{u,k}\mathrm{Tr}\left(\mathcal{Q}_{kk}R_{u,k}\right)\right\}. \tag{1.156f}$$

Introduce the block diagonal matrices (written for $N = 3$):

$$\mathcal{L}_1 \triangleq \left[\begin{array}{ccccccccc} I_{M^2} & & & & & & & & \\ & 0 & & & & & & & \\ & & 0 & & & & & & \\ \hline & & & 0 & & & & & \\ & & & & I_{M^2} & & & & \\ & & & & & 0 & & & \\ \hline & & & & & & 0 & & \\ & & & & & & & 0 & \\ & & & & & & & & I_{M^2} \end{array}\right], \tag{1.157a}$$

$$\mathcal{L}_2 \triangleq \left[\begin{array}{ccccccccc} \lambda_1\lambda_1^{\mathrm{T}} & & & & & & & & \\ & 0 & & & & & & & \\ & & 0 & & & & & & \\ \hline & & & 0 & & & & & \\ & & & & \lambda_2\lambda_2^{\mathrm{T}} & & & & \\ & & & & & 0 & & & \\ \hline & & & & & & 0 & & \\ & & & & & & & 0 & \\ & & & & & & & & \lambda_3\lambda_3^{\mathrm{T}} \end{array}\right], \tag{1.157b}$$

where $\lambda_k = \mathrm{vec}(\Lambda_k)$. Then, it can be verified that

$$\mathrm{bvec}(Z_1) = \mathcal{L}_1(\mathcal{R} \otimes_b \mathcal{R})\mathrm{bvec}(\mathcal{Q}), \tag{1.158a}$$

$$\mathrm{bvec}(Z_2) = \mathcal{L}_2\mathrm{bvec}(\mathcal{Q}). \tag{1.158b}$$

Noting that

$$\text{bvec}(\mathcal{Q}) = (\mathcal{M} \otimes_b \mathcal{M})(\mathcal{A}_2 \otimes_b \mathcal{A}_2)\sigma \tag{1.159}$$

we conclude that

$$\text{bvec}(\mathcal{K}) = 4\mathcal{X}(\mathcal{M} \otimes_b \mathcal{M})(\mathcal{A}_2 \otimes_b \mathcal{A}_2)\sigma, \tag{1.160a}$$

$$\mathcal{X} \triangleq (I + \mathcal{L}_1)(\mathcal{R} \otimes_b \mathcal{R}) + \mathcal{L}_2. \tag{1.160b}$$

Note that the matrix $\mathcal{X}$ has the following block-diagonal structure:

$$\mathcal{X} \triangleq \text{diag}\{\mathcal{X}_1, \mathcal{X}_2, \ldots, \mathcal{X}_N\}, \tag{1.160c}$$

$$\mathcal{X}_k \triangleq \text{diag}\left\{\mathcal{X}_k^{(1)}, \mathcal{X}_k^{(2)}, \ldots, \mathcal{X}_k^{(N)}\right\}, \tag{1.160d}$$

$$\mathcal{X}_k^{(\ell)} = \begin{cases} \Lambda_k \otimes \Lambda_\ell, & \text{when } k \neq \ell, \\ \lambda_k \lambda_k^{\mathrm{T}} + 2\Lambda_k \otimes \Lambda_k, & \text{when } k = \ell. \end{cases} \tag{1.160e}$$

which is the same structure derived through equations (57)–(67) in [73]. Substituting Eq. (1.160a) into Eq. (1.153b) and using again Eqs. (1.154c), (1.154d), we find that the block vectorized forms of the weighting matrices $\{\Sigma, \Sigma'\}$ are related via (the expression for $\mathcal{F}$ below fixes the typo in Eq. (44) from [74]; in the derivation in this example we considered the special case in which $\mathcal{S}_m = I$ in (44) from [74]):

$$\sigma' = \mathcal{F}\sigma, \tag{1.161a}$$

$$\mathcal{F} \triangleq (\mathcal{A}_1 \otimes_b \mathcal{A}_1)\{I - (\mathcal{R}\mathcal{M} \otimes_b I) - (I \otimes_b \mathcal{R}\mathcal{M}) + 4(\mathcal{M} \otimes_b \mathcal{M})\mathcal{X}\}(\mathcal{A}_2 \otimes_b \mathcal{A}_2). \tag{1.161b}$$

This is the same expression for $\mathcal{F}$ in Eq. (69) from [73] for the case of CTA diffusion (where $A_2 = I_N$). This is also the same expression for $\mathcal{F}$ in Eq. (41) from [74] for CTA and ATC diffusion (except that this latter reference wrote $\otimes$ instead $\otimes_b$). Substituting Eq. (1.155a) and the above expression for σ' into Eq. (1.153a), and using the compact notation $\|x\|_\sigma^2$ for $\|x\|_\Sigma^2$, we rewrite Eq. (1.153a) in the form

$$\mathbb{E}\|\tilde{\boldsymbol{w}}_i\|_\sigma^2 = \mathbb{E}\|\tilde{\boldsymbol{w}}_{i-1}\|_{\mathcal{F}\sigma}^2 + b^{\mathrm{T}}\sigma. \tag{1.162}$$

In the steady state, as $i \to \infty$, the MSE approaches

$$\lim_{i\to\infty} \mathbb{E}\|\tilde{\boldsymbol{w}}_i\|_{(I-\mathcal{F})\sigma}^2 = b^{\mathrm{T}}\sigma. \tag{1.163}$$

As explained in Sec. 6.6 of [61], the network MSD can be assessed by selecting σ to satisfy

$$(I - \mathcal{F})\sigma = \frac{1}{N}\text{bvec}(I_{NM}), \tag{1.164}$$

which leads to the desired expression

$$\text{MSD}_{\text{diff, av}}^{\text{sync}} = \frac{1}{N} b^{\mathrm{T}}(I - \mathcal{F})^{-1}\text{bvec}(I_{NM}), \tag{1.165}$$

which is expression (105a) from [73].

To illustrate how these results are adjusted for asynchronous behavior, we consider the case in which the step-size parameters $\{\mu_k\}$ in Eq. (1.151a) are replaced by random values $\{\boldsymbol{\mu}_k(i)\}$. We denote the mean and variances of these random variables as follows:

$$\bar{\mu}_k \triangleq \mathbb{E}\boldsymbol{\mu}_k(i), \tag{1.166a}$$

$$\boldsymbol{\mathcal{M}}_i \triangleq \text{diag}\{\boldsymbol{\mu}_1(i)I_M, \boldsymbol{\mu}_2(i)I_M, \ldots, \boldsymbol{\mu}_N(i)I_M\}, \tag{1.166b}$$

$$\bar{\mathcal{M}} \triangleq \mathbb{E}\boldsymbol{\mathcal{M}}_i, \tag{1.166c}$$

$$\mathcal{C}_\mu \triangleq \mathbb{E}(\boldsymbol{\mathcal{M}}_i - \bar{\mathcal{M}}) \otimes_b (\boldsymbol{\mathcal{M}}_i - \bar{\mathcal{M}}). \tag{1.166d}$$

If we now repeat the same analysis, expressions (1.153a) and (1.153b) are replaced by

$$\mathbb{E}\|\tilde{\boldsymbol{w}}_i\|^2_\Sigma = \mathbb{E}\|\tilde{\boldsymbol{w}}_{i-1}\|^2_{\Sigma'} + \mathbb{E}\left(\boldsymbol{z}_i^{\mathsf{T}}\boldsymbol{\mathcal{M}}_i\mathcal{A}_2\Sigma\mathcal{A}_2^{\mathsf{T}}\boldsymbol{\mathcal{M}}_i\boldsymbol{z}_i\right), \tag{1.167a}$$

$$\Sigma' = \mathcal{A}_1\left\{\mathcal{A}_2\Sigma\mathcal{A}_2^{\mathsf{T}} - \mathcal{A}_2\Sigma\mathcal{A}_2^{\mathsf{T}}\bar{\mathcal{M}}\mathcal{R} - \mathcal{R}\bar{\mathcal{M}}\mathcal{A}_2\Sigma\mathcal{A}_2^{\mathsf{T}} + \mathbb{E}\left(\boldsymbol{\mathcal{R}}_i\boldsymbol{\mathcal{M}}_i\mathcal{A}_2\Sigma\mathcal{A}_2^{\mathsf{T}}\boldsymbol{\mathcal{M}}_i\boldsymbol{\mathcal{R}}_i\right)\right\}\mathcal{A}_1^{\mathsf{T}}. \tag{1.167b}$$

We can evaluate the last expectations in Eqs. (1.167a), (1.167b) as follows. First we have:

$$\mathbb{E}\left(\boldsymbol{z}_i^{\mathsf{T}}\boldsymbol{\mathcal{M}}_i\mathcal{A}_2\Sigma\mathcal{A}_2^{\mathsf{T}}\boldsymbol{\mathcal{M}}_i\boldsymbol{z}_i\right) = b^{\mathsf{T}}\sigma, \tag{1.168a}$$

where

$$\begin{aligned} b &\triangleq \mathbb{E}\text{bvec}\left(\mathcal{A}_2^{\mathsf{T}}\boldsymbol{\mathcal{M}}_i\mathcal{S}\boldsymbol{\mathcal{M}}_i\mathcal{A}_2\right) \\ &= (\mathcal{A}_2^{\mathsf{T}} \otimes_b \mathcal{A}_2^{\mathsf{T}})\mathbb{E}(\boldsymbol{\mathcal{M}}_i \otimes_b \boldsymbol{\mathcal{M}}_i)\text{bvec}(\mathcal{S}), \\ &= (\mathcal{A}_2^{\mathsf{T}} \otimes_b \mathcal{A}_2^{\mathsf{T}})(\mathcal{C}_\mu + \bar{\mathcal{M}} \otimes_b \bar{\mathcal{M}})\text{bvec}(\mathcal{S}). \end{aligned} \tag{1.168b}$$

Second, we have:

$$\begin{aligned} \text{bvec}\left\{\mathbb{E}\left(\boldsymbol{\mathcal{R}}_i\boldsymbol{\mathcal{M}}_i\mathcal{A}_2\Sigma\mathcal{A}_2^{\mathsf{T}}\boldsymbol{\mathcal{M}}_i\boldsymbol{\mathcal{R}}_i\right)\right\} &= \mathbb{E}(\boldsymbol{\mathcal{M}}_i \otimes_b \boldsymbol{\mathcal{M}}_i)\mathbb{E}(\boldsymbol{\mathcal{R}}_i \otimes_b \boldsymbol{\mathcal{R}}_i)(\mathcal{A}_2 \otimes_b \mathcal{A}_2)\sigma \\ &= (\mathcal{C}_\mu + \bar{\mathcal{M}} \otimes_b \bar{\mathcal{M}})\mathbb{E}(\boldsymbol{\mathcal{R}}_i \otimes_b \boldsymbol{\mathcal{R}}_i)(\mathcal{A}_2 \otimes_b \mathcal{A}_2)\sigma \end{aligned} \tag{1.169}$$

so that we replace the quantities $\mathcal{M}$ and $(\mathcal{M} \otimes_b \mathcal{M})$ in the expressions (1.155c) and (1.161b) for b and $\mathcal{F}$ by $\bar{\mathcal{M}}$ and $(\mathcal{C}_\mu + \bar{\mathcal{M}} \otimes_b \bar{\mathcal{M}})$, respectively. ■

1.8 CONCLUDING REMARKS

This chapter provides an overview of asynchronous strategies for adaptation, learning, and optimization over networks including noncooperative, centralized, consensus, and diffusion strategies. Particular attention is given to the constant step-size case in order to examine solutions that are able to adapt and learn continuously from streaming data. The presentation complements the results from [1,2]. We introduced a fairly general model for asynchronous behavior that allows for random step-sizes, link failures, random topology variations, and random combination coefficients. We examined the MSE performance and stability properties under asynchronous events and recovered results for synchronous operation as a special case. The results indicate that asynchronous networks are robust, resilient to failure, and remain mean-square stable for sufficiently small step-sizes.

There are of course several other aspects of distributed strategies that are not covered in this work. Comments on these aspects can be found in [1,2,61], including issues related to (a) the noisy exchange of information over links (e.g., [9,61,105–109]); (b) the use of gossip strategies (e.g., [7,10,15,17,65, 110,111]); (c) the exploitation of sparsity constraints (e.g., [112–115]); (d) the solution of constrained optimization problems (e.g., [8,95,116–118]); (e) the use of distributed solutions of the recursive least-squares type (e.g., [61,66,93]); (f) the development of distributed state-space solutions (e.g., [66,92,102, 119–124]); (g) the study of incremental-based strategies (e.g., [125–137]); (h) the study of distributed solutions under multitask environments [138–142]; (i) the case of nonsmooth risk functions in the context of subgradient learning [19,20,63,113]; and (j) the incorporation of proximal operators into the distributed setting [143–147].

REFERENCES

[1] Sayed AH. Adaptive networks. Proc IEEE 2014;102(4):460–97.
[2] Sayed AH. Adaptation, learning, and optimization over networks. In: Foundations and trends in machine learning, July 2014, vol. 7(4–5). Boston-Delft: NOW Publishers; 2014. p. 311–801.
[3] Zhao X, Sayed AH. Asynchronous adaptation and learning over networks—Part I: Modeling and stability analysis. IEEE Trans Signal Process 2015;63(4):811–26.
[4] Zhao X, Sayed AH. Asynchronous adaptation and learning over networks—Part II: Performance analysis. IEEE Trans Signal Process 2015;63(4):827–42.
[5] Zhao X, Sayed AH. Asynchronous adaptation and learning over networks—Part III: Comparison analysis. IEEE Trans Signal Process 2015;63(4):843–58.
[6] Tsitsiklis J, Bertsekas D, Athans M. Distributed asynchronous deterministic and stochastic gradient optimization algorithms. IEEE Trans Autom Control 1986;31(9):803–12.
[7] Boyd S, Ghosh A, Prabhakar B, Shah D. Randomized gossip algorithms. IEEE Trans Inf Theory 2006;52(6):2508–30.
[8] Srivastava K, Nedic A. Distributed asynchronous constrained stochastic optimization. IEEE J Sel Topics Signal Process 2011;5(4):772–90.
[9] Kar S, Moura JMF. Distributed consensus algorithms in sensor networks: link failures and channel noise. IEEE Trans Signal Process 2009;57(1):355–69.
[10] Kar S, Moura JMF. Convergence rate analysis of distributed gossip (linear parameter) estimation: fundamental limits and tradeoffs. IEEE J Sel Topics Signal Process 2011;5(4):674–90.
[11] Kar S, Moura JMF. Sensor networks with random links: topology design for distributed consensus. IEEE Trans Signal Process 2008;56(7):3315–26.
[12] Jakovetic D, Xavier J, Moura JMF. Weight optimization for consensus algorithms with correlated switching topology. IEEE Trans Signal Process 2010;58(7):3788–801.
[13] Jakovetic D, Xavier J, Moura JMF. Cooperative convex optimization in networked systems: augmented Lagrangian algorithms with directed gossip communication. IEEE Trans Signal Process 2011;59(8): 3889–902.
[14] Kar S, Moura JMF. Distributed consensus algorithms in sensor networks: quantized data and random link failures. IEEE Trans Signal Process 2010;58(3):1383–400.
[15] Aysal TC, Yildiz ME, Sarwate AD, Scaglione A. Broadcast gossip algorithms for consensus. IEEE Trans Signal Process 2009;57(7):2748–61.
[16] Aysal TC, Sarwate AD, Dimakis AG. Reaching consensus in wireless networks with probabilistic broadcast. In: Proceedings of the Allerton conference communication, control, computer, September and October 2009. IL: Allerton House; 2009, p. 732–9.

[17] Lopes C, Sayed AH. Diffusion adaptive networks with changing topologies. In: Proc. IEEE ICASSP, April 2008, Las Vegas; 2008. p. 3285–8.
[18] Takahashi N, Yamada I. Link probability control for probabilistic diffusion least-mean squares over resource-constrained networks. In: Proceedings of the IEEE international conference acoustics, speech, signal process, March 2010. Dallas, TX: ICASSP; 2010. p. 3518–21.
[19] Ying B, Sayed AH. Performance limits of stochastic sub-gradient learning, Part I: Single agent case: Signal Processing 2018;144:271–82.
[20] Ying B, Sayed AH. Performance limits of stochastic sub-gradient learning, Part II: Multi-agent case: Signal Processing 2018;144:253–64.
[21] Haykin S. Adaptive filter theory. Upper Saddle River, NJ: Prentice Hall; 2002.
[22] Widrow B, Stearns SD. Adaptive signal processing. Upper Saddle River, NJ: Prentice Hall; 1985.
[23] Sayed AH. Adaptive filters. NJ: Wiley; 2008.
[24] Sayed AH. Fundamentals of adaptive filtering. NJ: Wiley; 2003.
[25] Kailath T, Sayed AH, Hassibi B. Linear estimation. Upper Saddle River, NJ: Prentice Hall; 2000.
[26] Bishop CM. Pattern recognition and machine learning. Springer; 2007.
[27] Theodoridis S, Koutroumbas K. Pattern recognition. 4th ed. Academic Press; 2008.
[28] Hosmer DW, Lemeshow S. Applied logistic regression. 2nd ed. NJ: Wiley; 2000.
[29] Poljak BT, Tsypkin YZ. Pseudogradient adaptation and training algorithms. Autom Remote Control 1973;12:83–94.
[30] Poljak B. Introduction to optimization. NY: Optimization Software; 1987.
[31] Bertsekas DP, Tsitsiklis JN. Parallel and distributed computation: numerical methods. 1st ed. Singapore: Athena Scientific; 1997.
[32] Tsypkin YZ. Adaptation and learning in automatic systems. NY: Academic Press; 1971.
[33] Boyd S, Vandenberghe L. Convex optimization. Cambridge University Press; 2004.
[34] Bertsekas D. Convex analysis and optimization. Athena Scientific; 2003.
[35] Nesterov Y. Introductory lectures on convex optimization: a basic course. Kluwer Academic Publishers; 2004.
[36] Chen J, Sayed AH. Diffusion adaptation strategies for distributed optimization and learning over networks. IEEE Trans Signal Process 2012;60(8):4289–305.
[37] Chen J, Sayed AH. On the learning behavior of adaptive networks—Part I: Transient analysis. IEEE Trans Inf Theory 2015;61(6):3487–517.
[38] Widrow B, Hoff Jr ME. Adaptive switching circuits. In: IRE WESCON Conv. Rec., Pt. 4; 1960. p. 96–104.
[39] Bertsekas DP, Tsitsiklis JN. Gradient convergence in gradient methods with errors. SIAM J Optim 2000;10(3):627–42.
[40] Feller W. An introduction to probability theory and its applications, vol. 2. NY: Wiley; 1971.
[41] Hahn GJ, Shapiro S. Statistical models in engineering. NY: Wiley; 1994.
[42] Abramowitz M, Stegun I, editors. Handbook of mathematical functions with formulas, graphs, and mathematical tables. NY: Dover; 1972.
[43] Andrews GE, Askey R, Roy R. Special functions. Cambridge: Cambridge University Press; 1999.
[44] Rudin W. Principles of mathematical analysis. McGraw-Hill; 1976.
[45] Yousef NR, Sayed AH. A unified approach to the steady-state and tracking analysis of adaptive filters. IEEE Trans Signal Process 2001;49(2):314–24.
[46] Al-Naffouri TY, Sayed AH. Transient analysis of data-normalized adaptive filters. IEEE Trans Signal Process 2003;51(3):639–52.
[47] Chen J, Sayed AH. On the learning behavior of adaptive networks—Part II: Performance analysis. IEEE Trans Inf Theory 2015;61(6):3518–48.
[48] Papoulis A, Pilla SU. Probability, random variables, and stochastic processes. NY: McGraw-Hill; 2002.

[49] Durret R. Probability theory and examples. 2nd ed. Duxbury Press; 1996.
[50] Dudley RM. Real analysis and probability. 2nd ed. Cambridge University Press; 2003.
[51] Widrow B, McCool JM, Larimore MG, Johnson JCR. Stationary and nonstationary learning characteristics of the LMS adaptive filter. Proc IEEE 1976;64(8):1151–62.
[52] Horowitz L, Senne K. Performance advantage of complex LMS for controlling narrow-band adaptive arrays. IEEE Trans Acoust, Speech, Signal Process 1981;29(3):722–36.
[53] Jones S, Cavin III R, Reed W. Analysis of error-gradient adaptive linear estimators for a class of stationary dependent processes. IEEE Trans Inf Theory 1982;28(2):318–29.
[54] Gardner WA. Learning characteristics of stochastic-gradient-descent algorithms: a general study, analysis, and critique. Signal Process 1984;6(2):113–33.
[55] Feuer A, Weinstein E. Convergence analysis of LMS filters with uncorrelated Gaussian data. IEEE Trans Acoust, Speech, Signal Process 1985;33(1):222–30.
[56] Foley JB, Boland FM. A note on the convergence analysis of LMS adaptive filters with Gaussian data. IEEE Trans Acoust, Speech, Signal Process 1988;36(7):1087–9.
[57] Vapnik VN. The nature of statistical learning theory. NY: Springer; 2000.
[58] Towfic Z, Chen J, Sayed AH. On the generalization ability of distributed online learners. In: Proc. IEEE workshop on machine learning for signal processing (MLSP), Santander, Spain; 2012. p. 1–6.
[59] Zhao X, Sayed AH. Performance limits for distributed estimation over LMS adaptive networks. IEEE Trans Signal Process 2012;60(10):5107–24.
[60] Chen J, Sayed AH. Distributed pareto optimization via diffusion strategies. IEEE J Sel Topics Signal Process 2013;7(2):205–20.
[61] Sayed AH. Diffusion adaptation over networks. In: Chellapa R, Theodoridis S, editors. E-reference signal processing, vol. 3. Academic Press; 2014. p. 323–454. Also available as arXiv:1205.4220v1 [cs.MA], May 2012.
[62] Nedic A, Ozdaglar A. Cooperative distributed multi-agent optimization. In: Eldar Y, Palomar D, editors. Convex optimization in signal processing and communications. Cambridge University Press; 2010. p. 340–86.
[63] Nedic A, Ozdaglar A. Distributed subgradient methods for multi-agent optimization. IEEE Trans Autom Control 2009;54(1):48–61.
[64] Johansson B, Keviczky T, Johansson M, Johansson K. Subgradient methods and consensus algorithms for solving convex optimization problems. In: Proc. IEEE CDC, Cancun, Mexico; 2008. p. 4185–90.
[65] Dimakis AG, Kar S, Moura JMF, Rabbat MG, Scaglione A. Gossip algorithms for distributed signal processing. Proc IEEE 2010;98(11):1847–64.
[66] Xiao L, Boyd S, Lall S. A space-time diffusion scheme peer-to-peer least-squares-estimation. In: Proceedings of information processing in sensor networks (IPSN), Nashville, TN; 2006. p. 168–76.
[67] Ren W, Beard RW. Consensus seeking in multi-agent systems under dynamically changing interaction topologies. IEEE Trans Autom Control 2005;50:655–61.
[68] Olfati-Saber R, Shamma J. Consensus filters for sensor networks and distributed sensor fusion. In: Proceedings of 44th IEEE conference on decision and control (CDC), Seville, Spain; 2005. p. 6698–703.
[69] Barbarossa S, Scutari G. Bio-inspired sensor network design. IEEE Signal Process Mag 2007;24(3): 26–35.
[70] Sardellitti S, Giona M, Barbarossa S. Fast distributed average consensus algorithms based on advection-diffusion processes. IEEE Trans Signal Process 2010;58(2):826–42.
[71] Braca P, Marano S, Matta V. Running consensus in wireless sensor networks. In: Proc. 11th international conference on information fusion, Cologne, Germany; 2008. p. 1–6.
[72] Sayed AH, Tu SY, Chen J, Zhao X, Towfic Z. Diffusion strategies for adaptation and learning over networks. IEEE Signal Process Mag 2013;30(3):155–71.

[73] Lopes CG, Sayed AH. Diffusion least-mean squares over adaptive networks: formulation and performance analysis. IEEE Trans Signal Process 2008;56(7):3122–36.
[74] Cattivelli FS, Sayed AH. Diffusion LMS strategies for distributed estimation. IEEE Trans Signal Process 2010;58(3):1035–48.
[75] Horn RA, Johnson CR. Matrix analysis. Cambridge University Press; 2003.
[76] Tu SY, Sayed AH. Mobile adaptive networks. IEEE J Sel Topics Signal Process 2011;5(4):649–64.
[77] Cattivelli F, Sayed AH. Modeling bird flight formations using diffusion adaptation. IEEE Trans Signal Process 2011;59(5):2038–51.
[78] Dekel O, Gilad-Bachrach R, Shamir O, Xiao L. Optimal distributed online prediction. In: Proceedings of international conference on machine learning (ICML), Bellevue, WA; 2011. p. 713–20.
[79] Agarwal A, Duchi J. Distributed delayed stochastic optimization. In: Proceedings of neural information processing systems (NIPS), Granada, Spain; 2011. p. 873–81.
[80] Predd JB, Kulkarni SB, Poor HV. Distributed learning in wireless sensor networks. IEEE Signal Process Mag 2006;23(4):56–69.
[81] Towfic ZJ, Chen J, Sayed AH. Collaborative learning of mixture models using diffusion adaptation. In: Proceedings of the IEEE workshop machine learning signal processing (MLSP), Beijing, China; 2011. p. 1–6.
[82] Golub GH, Van Loan CF. Matrix computations. 3rd ed. Baltimore: The John Hopkins University Press; 1996.
[83] Berman A, Plemmons RJ. Nonnegative matrices in the mathematical sciences. PA: SIAM; 1994.
[84] Pillai SU, Suel T, Cha S. The Perron-Frobenius theorem: some of its applications. IEEE Signal Process Mag 2005;22(2):62–75.
[85] Tu SY, Sayed AH. Diffusion strategies outperform consensus strategies for distributed estimation over adaptive networks. IEEE Trans Signal Process 2012;60(12):6217–34.
[86] Lopes CG, Sayed AH. Distributed processing over adaptive networks. In: Proceedings adaptive sensor array processing workshop. MA: MIT Lincoln Laboratory; 2006. p. 1–5.
[87] Sayed AH, Lopes CG. Adaptive processing over distributed networks. IEICE Trans Fundam Electron Commun Comput Sci 2007;E90-A(8):1504–10.
[88] Lopes CG, Sayed AH. Diffusion least-mean-squares over adaptive networks. In: Proceedings IEEE ICASSP, Honolulu, Hawaii, vol. 3; 2007. p. 917–20.
[89] Lopes CG, Sayed AH. Steady-state performance of adaptive diffusion least-mean squares. In: Proceedings IEEE workshop on statistical signal processing (SSP), Madison, WI; 2007. p. 136–40.
[90] Cattivelli FS, Sayed AH. Diffusion LMS algorithms with information exchange. In: Proceedings Asilomar conference signals, system computer, Pacific Grove, CA; 2008. p. 251–5.
[91] Cattivelli FS, Lopes CG, Sayed AH. A diffusion RLS scheme for distributed estimation over adaptive networks. In: Proceedings IEEE workshop on signal process. Advances wireless communication (SPAWC), Helsinki, Finland; 2007. p. 1–5.
[92] Cattivelli FS, Sayed AH. Diffusion mechanisms for fixed-point distributed Kalman smoothing. In: Proceedings EUSIPCO, Lausanne, Switzerland; 2008. p. 1–4.
[93] Cattivelli FS, Lopes CG, Sayed AH. Diffusion recursive least-squares for distributed estimation over adaptive networks. IEEE Trans Signal Process 2008;56(5):1865–77.
[94] Ram SS, Nedic A, Veeravalli VV. Distributed stochastic subgradient projection algorithms for convex optimization. J Optim Theory Appl 2010;147(3):516–45.
[95] Lee S, Nedic A. Distributed random projection algorithm for convex optimization. IEEE J Sel Topics Signal Process 2013;7(2):221–9.
[96] Bianchi P, Fort G, Hachem W. Performance of a distributed stochastic approximation algorithm. IEEE Trans Inf Theory 2013;59(11):7405–18.

[97] Stankovic SS, Stankovic MS, Stipanovic DS. Decentralized parameter estimation by consensus based stochastic approximation. IEEE Trans Autom Control 2011;56(3):531–43.
[98] Chen J, Sayed AH. On the limiting behavior of distributed optimization strategies. In: Proceedings 50th annual Allerton conference on communication, control, and computing, Monticello, IL; 2012. p. 1535–42.
[99] Bondy JA, Murty USR. Graph theory. Springer; 2008.
[100] Zhao X, Tu SY, Sayed AH. Diffusion adaptation over networks under imperfect information exchange and non-stationary data. IEEE Trans Signal Process 2012;60(7):3460–75.
[101] Takahashi N, Yamada I, Sayed AH. Diffusion least-mean-squares with adaptive combiners: formulation and performance analysis. IEEE Trans Signal Process 2010;58(9):4795–810.
[102] Cattivelli F, Sayed AH. Diffusion strategies for distributed Kalman filtering and smoothing. IEEE Trans Autom Control 2010;55(9):2069–84.
[103] Tracy DS, Singh RP. A new matrix product and its applications in partitioned matrix differentiation. Statistica Neerlandica 1972;26(4):143–57.
[104] Koning RH, Neudecker H, Wansbeek T. Block Kronecker products and the vecb operator. Linear Algebra Appl 1991;149:165–84.
[105] Tu SY, Sayed AH. Adaptive networks with noisy links. In: Proceedings IEEE Globecom, Houston, TX; 2011. p. 1–5.
[106] Abdolee R, Champagne B. Diffusion LMS algorithms for sensor networks over non-ideal inter-sensor wireless channels. In: Proceedings IEEE international conference distribution computer sensor systems (DCOSS), Barcelona, Spain; 2011. p. 1–6.
[107] Khalili A, Tinati MA, Rastegarnia A, Chambers JA. Steady state analysis of diffusion LMS adaptive networks with noisy links. IEEE Trans Signal Process 2012;60(2):974–9.
[108] Zhao X, Sayed AH. Combination weights for diffusion strategies with imperfect information exchange. In: Proceedings IEEE ICC, Ottawa, Canada; 2012. p. 648–52.
[109] Mateos G, Schizas ID, Giannakis GB. Performance analysis of the consensus-based distributed LMS algorithm. EURASIP J Adv Signal Process 2009:1–19. https://doi.org/10.1155/2009/981030. Article ID 981030.
[110] Shah D. Gossip algorithms. In: Foundations and trends in networking, vol. 3; 2009. p. 1–125.
[111] Rortveit OL, Husoy JH, Sayed AH. Diffusion LMS with communications constraints. In: Proceedings of the 44th Asilomar conference on signals, systems and computers, Pacific Grove, CA; 2010. p. 1645–9.
[112] Di Lorenzo P, Sayed AH. Sparse distributed learning based on diffusion adaptation. IEEE Trans Signal Process 2013;61(6):1419–33.
[113] Chouvardas S, Slavakis K, Kopsinis Y, Theodoridis S. A sparsity-promoting adaptive algorithm for distributed learning. IEEE Trans Signal Process 2012;60(10):5412–25.
[114] Chouvardas S, Mileounis G, Kalouptsidis N, Theodoridis S. A greedy sparsity-promoting LMS for distributed adaptive learning in diffusion networks. In: Proceedings ICASSP, Vancouver, BC, Canada; 2013. p. 5415–9.
[115] Liu Y, Li C, Zhang Z. Diffusion sparse least-mean squares over networks. IEEE Trans Signal Process 2012;60:4480–5.
[116] Yan F, Sundaram S, Vishwanathan SVN, Qi Y. Distributed autonomous online learning: regrets and intrinsic privacy-preserving properties. IEEE Trans Knowl Data Eng 2013;25(11):2483–93.
[117] Theodoridis S, Slavakis K, Yamada I. Adaptive learning in a world of projections: a unifying framework for linear and nonlinear classification and regression tasks. IEEE Signal Process Mag 2011;28(1):97–123.
[118] Towfic Z, Sayed AH. Adaptive penalty-based distributed stochastic convex optimization. IEEE Trans Signal Process 2014;62(15):3924–38.
[119] Olfati-Saber R. Kalman-consensus filter: optimality, stability, and performance. In: Proceedings IEEE CDC, Shangai, China; 2009. p. 7036–42.

[120] Olfati-Saber R. Distributed Kalman filtering for sensor networks. In: Proceedings of the 46th IEEE conference decision control, New Orleans, LA; 2007. p. 5492–8.
[121] Cattivelli FS, Lopes CG, Sayed AH. Diffusion strategies for distributed Kalman filtering: formulation and performance analysis. In: Proceedings IAPR workshop on cognitive information process (CIP), Santorini, Greece; 2008. p. 36–41.
[122] Cattivelli F, Sayed AH. Diffusion distributed Kalman filtering with adaptive weights. In: Proceedings Asilomar conference on signals, systems and computers, Pacific Grove, CA; 2009. p. 908–12.
[123] Khan UA, Moura JMF. Distributing the Kalman filter for large-scale systems. IEEE Trans Signal Process 2008;56(10):4919–35.
[124] Alriksson P, Rantzer A. Distributed Kalman filtering using weighted averaging. In: Proceedings of the 17th international symposium on mathematical theory of networks and systems (MTNS), Kyoto, Japan; 2006. p. 1–6.
[125] Bertsekas DP. A new class of incremental gradient methods for least squares problems. SIAM J Optim 1997;7(4):913–26.
[126] Bertsekas DP. Nonlinear programming. 2nd ed. Belmont, MA: Athena Scientific; 1999.
[127] Nedic A, Bertsekas DP. Incremental subgradient methods for nondifferentiable optimization. SIAM J Optim 2001;12(1):109–38.
[128] Rabbat MG, Nowak RD. Quantized incremental algorithms for distributed optimization. IEEE J Sel Areas Commun 2005;23(4):798–808.
[129] Helou ES, De Pierro AR. Incremental subgradients for constrained convex optimization: a unified framework and new methods. SIAM J Optim 2009;20:1547–72.
[130] Johansson B, Rabi M, Johansson M. A randomized incremental subgradient method for distributed optimization in networked systems. SIAM J Optim 2009;20:1157–70.
[131] Blatt D, Hero AO, Gauchman H. A convergent incremental gradient method with a constant step size. SIAM J Optim 2008;18:29–51.
[132] Sayed AH, Lopes C. Distributed recursive least-squares strategies over adaptive networks. In: Proceedings of the 40th Asilomar conference on signals, systems and computers, Pacific Grove, CA; 2006. p. 233–7.
[133] Lopes CG, Sayed AH. Incremental adaptive strategies over distributed networks. IEEE Trans Signal Process 2007;55(8):4064–77.
[134] Sayed AH, Cattivelli F. Distributed adaptive learning mechanisms. In: Haykin S, Ray Liu KJ, editors. Handbook on array processing and sensor networks. NJ: Wiley; 2009. p. 695–722.
[135] Li L, Chambers J, Lopes CG, Sayed AH. Distributed estimation over an adaptive incremental network based on the affine projection algorithm. IEEE Trans Signal Process 2010;58(1):151–64.
[136] Cattivelli F, Sayed AH. Analysis of spatial and incremental LMS processing for distributed estimation. IEEE Trans Signal Process 2011;59(4):1465–80.
[137] Predd JB, Kulkarni SR, Poor HV. A collaborative training algorithm for distributed learning. IEEE Trans Inf Theory 2009;55(4):1856–71.
[138] Nassif R, Richard C, Ferrari A, Sayed AH. Multitask diffusion adaptation over asynchronous networks. IEEE Trans Signal Process 2016;64(11):2835–50.
[139] Bertrand A, Moonen M. Distributed adaptive node-specific signal estimation in fully connected sensor networks—Part I: Sequential node updating. IEEE Trans Signal Process 2010;58(10):5277–91.
[140] Bogdanovic N, Plata-Chaves J, Berberidis K. Distributed diffusion-based LMS for node-specific parameter estimation over adaptive networks. In: Proceedings IEEE ICASSP, Florence, Italy; 2014. p. 7223–7.
[141] Chen J, Richard C, Sayed AH. Multitask diffusion adaptation over networks. IEEE Trans Signal Process 2014;62(16):4129–44.
[142] Chen J, Richard C, Sayed AH. Diffusion LMS over multitask networks. IEEE Trans Signal Process 2015;63(11):2733–48.

[143] Chen AI, Ozdaglar A. A fast distributed proximal gradient method. In: Proceedings of annual Allerton conference on communication, control, and computing, Allerton, USA; 2012. p. 601–8.

[144] Wee W, Yamada I. A proximal splitting approach to regularized distributed adaptive estimation in diffusion network. In: Proceedings IEEE ICASSP, Vancouver, Canada; 2013. p. 5420–4.

[145] Vlaski S, Vandenberghe L, Sayed AH. Diffusion stochastic optimization with non-smooth regularizers. In: Proceedings IEEE ICASSP, Shanghai, China; 2016. p. 4149–53.

[146] Vlaski S, Sayed AH. Proximal diffusion for stochastic costs with non-differentiable regularizers. In: Proceedings IEEE ICASSP, Brisbane, Australia; 2015. p. 3352–6.

[147] Nassif R, Ferrari A, Richard C, Sayed AH. Proximal multitask learning over networks with sparsity-inducing coregularization. IEEE Trans Signal Process 2016;64(23):6329–44.

CHAPTER

ESTIMATION AND DETECTION OVER ADAPTIVE NETWORKS

2

Vincenzo Matta*, **Ali H. Sayed**†
*Department of Information Engineering, Electrical Engineering and Applied Mathematics, DIEM, University of Salerno, Fisciano, Italy** *School of Engineering, École Polytechnique Fédérale de Lausanne (EPFL), Lausanne, Switzerland*†

2.1 INTRODUCTION

As illustrated in Chapter 1, adaptive networks are well suited to model the complex behavior exhibited by several real-world systems as well as to perform decentralized information processing tasks. This chapter provides an overview of results pertaining to estimation and detection by networked agents, with a focus on adaptation to drifting conditions and learning from online streaming data. Specifically, we formulate two canonical problems for statistical inference: (i) the estimation problem, where the observer is tasked with approximating a vector parameter that models an unknown, continuous-valued system state; and (ii) the detection problem, where the observer is tasked with choosing among a finite number of hypotheses that model an unknown, discrete-valued system state. We consider distributed variations of these problems where a collection of interconnected agents is tasked with solving estimation or detection problems in a decentralized manner. We emphasize the need for adaptive distributed implementations in order to enable continuous tracking of system dynamics when drifts occur in the environmental conditions, which translate into drifts in the continuous-valued parameter for estimation problems or drifts in the discrete-valued state for detection problems.

We also explain how diffusion learning strategies with constant step-size updates form a powerful core for effective adaptive implementations. For these implementations, we review the basic formal representations and present the results available about their theoretical performance characterization. In particular, it will be seen that the theoretical characterization enables powerful understanding of the interplay between network properties (such as the connectivity of each agent and the combination policy used by these agents), the flow of information across the network, and the accuracy of the inferential process.

Special attention is paid to universal laws for network inference performance. We examine the resulting scaling laws with respect to the step-size parameter, and illustrate commonalities as well as distinctive features between the estimation and detection settings. These fundamental laws are also compared against known laws for estimation and detection in traditional (centralized or decentralized, nonadaptive) inferential systems.

Cooperative and Graph Signal Processing. https://doi.org/10.1016/B978-0-12-813677-5.00002-X

Table 2.1 List of Notation and Symbols Used in the Text and Appendices

Notation	Description
$\mathbb{R}$	field of real numbers
$\mathbb{1}$	column vector with all its entries equal to one
I_M	identity matrix of size $M \times M$
$\boldsymbol{x}$	boldface notation denotes random variables
x	normal font denotes realizations of random variables
$\mathbb{P}$	probability operator
$\mathbb{E}$	expectation operator
A	capital letters denote matrices
a	small letters denote vectors or scalars
$x(i)$	small letters with parenthesis denote scalars
x_i	small letters with subscripts denote vectors
$\top$	matrix transposition
$\text{col}\{a, b\}$	column vector with entries a and b
$\text{diag}\{a, b\}$	diagonal matrix with entries a and b
$\|x\|$	Euclidean norm of its vector argument
$\text{Tr}(A)$	trace of matrix A
$A \otimes B$	Kronecker product of A and B
$\alpha = o(\mu)$	signifies that $\alpha/\mu \to 0$ as $\mu \to 0$
$\alpha = O(\mu)$	signifies that $\|\alpha\| \leq c\|\mu\|$ for some constant $c > 0$
$\alpha(\mu) \doteq \beta(\mu)$	signifies that $\mu \ln[\alpha(\mu)/\beta(\mu)] \to 0$ as $\mu \to 0$

2.2 INFERENCE OVER NETWORKS

To start with, we introduce the two canonical inference problems of estimation and detection, with reference to the classic nondistributed setting.

2.2.1 CANONICAL INFERENCE PROBLEMS

Statistical inference [1–3] broadly refers to the problem of learning about a phenomenon of interest using a set of observations whose statistical distribution is governed by the *unknown* state of the phenomenon. The two fundamental problems of statistical inference are the estimation problem and the detection problem.

In the *estimation* problem, the phenomenon of interest is formally represented by a *continuous-valued* parameter vector that influences the statistical distribution of the observations. The parameter can be modeled as deterministic or random. In the latter case, a *prior* distribution is usually assigned to the unknown parameter, and the resulting setting is referred to as the Bayesian setting. There exist many criteria for optimizing the estimation performance, and many related estimation strategies. To mention a few, the classic choices include the maximum likelihood estimator (for the deterministic parameter setting), and the minimum mean-square-error estimator (for the Bayesian setting) [1,2,4–6]. Because in the following treatment we shall focus on the deterministic-parameter setting, the unknown parameter, $w^o \in \mathbb{R}^M$, is denoted by a normal font letter. The inference process amounts to computing an *estimator*, $\hat{\boldsymbol{w}}$, which is (except for trivial cases) random because it is a function of the available *random* data. We shall use boldface letters to refer to random quantities.

Example 2.1 (Classic nondistributed estimation). In wireless communications, it is often desirable to acquire a channel state information (CSI), namely, to get knowledge about the response of the channel connecting the source to the destination. Owing to channel-originated impairments, CSI estimation is usually based on *noisy* data measured at the output of the channel when the input is loaded with a sequence of training samples. This situation can be formally abstracted by saying that there is a vector parameter, w^o, which models the channel response, and two sequences of random variables, $\boldsymbol{x}_1, \boldsymbol{x}_2, \ldots, \boldsymbol{x}_N$ and $\boldsymbol{y}_1, \boldsymbol{y}_2, \ldots, \boldsymbol{y}_N$, which model the training samples as well as the measured output data. The statistical distribution of the received data depends on the actual value taken by w^o. Given a stream of input-output data, the inferential goal is to produce an estimator, $\hat{\boldsymbol{w}}$, that is able to approximate w^o with high fidelity, i.e., up to a small error. ■

The estimation error is evaluated through some distortion measure. A popular distortion measure is the mean-square-error, $\mathbb{E}\|w^o - \hat{\boldsymbol{w}}\|^2$, which, in particular, will play an important role in our treatment. When the chosen distortion measure is difficult to evaluate, it is often convenient to replace it with some *asymptotic* performance index. One widely adopted asymptotic descriptor is the Fisher information matrix, whose inverse represents a lower bound on the best performance attainable by an unbiased estimator (the Cramér-Rao lower bound) as well as the error-covariance matrix achieved (asymptotically) by a maximum-likelihood estimator [1,2,4]. These estimation methods and performance indices are well suited to those cases where exact knowledge of the data distributions (parametrized by the unknown w^o) is available. When such knowledge is limited or even absent, other methods must be employed. A primary role in the nonparametric setting is played by the *least-squares* criterion and by the theory of *linear* estimation [5]. Among other advantages, the estimators resulting from these two approaches exhibit high versatility under many distinct operational conditions, and are open to adaptive and sequential implementations, such as, for example, the famous least-mean-squares (LMS) and recursive-least-squares (RLS) algorithms [6]. As we shall see later in Section 2.4, these types of estimators play an important role in the development of distributed adaptive estimation (DAE).

Next, we describe the *detection* problem (aka hypothesis testing or decision problem) where the phenomenon of interest is modeled through a finite number of states (the hypotheses). Again, the unknown phenomenon can be modeled as deterministic or random. In the latter case, there is usually a probability assignment (the prior distribution) on the occurrence of the different hypotheses (Bayesian setting). The simplest and most typical setting is the *binary* detection setting, where one has a pair of hypotheses: $\mathcal{H}_0$ (the null hypothesis) and $\mathcal{H}_1$ (the alternative hypothesis). In a nontrivial binary detection problem, the dataset, which can be described through a sequence of random variables, $\boldsymbol{x}_1, \boldsymbol{x}_2, \ldots, \boldsymbol{x}_N$, has a distribution that is different under the different hypotheses. Such a difference is key to enabling successful discrimination between the hypotheses.

Example 2.2 (Classic nondistributed detection). There is a cat in a closed box. It is not known beforehand whether the cat is dead or alive. This model encompasses a binary state of nature, which can be formally represented by a pair of hypotheses: namely, hypothesis $\mathcal{H}_0$ (dead cat) and hypothesis $\mathcal{H}_1$ (cat alive). The observer (i.e., the agent who must perform detection) is not allowed to look inside the box. The observer is allowed to collect observations from the box, which can be related to the cat's state. Such observations might include, for instance, temperature of the box, noise or smell perceived in the proximity of the box, and so on. Multiple observations can be collected over time, and the stream of data is conveniently abstracted as a suite of random variables, $\boldsymbol{x}_1, \boldsymbol{x}_2, \ldots, \boldsymbol{x}_N$, featuring a different distribution under the different hypotheses. Based upon the gathered data, the observer must produce a binary decision as regards the cat's state. Due to the randomness of the data, the final decision will be affected by error. There will be a certain probability of declaring the cat as dead when it is alive, and of declaring the cat alive when it is dead. ■

The detection performance is typically measured in terms of the Type-I (false-alarm) and Type-II (miss-detection) error probabilities defined, respectively, as [3,7,8]:

$$\alpha \triangleq \mathbb{P}[\text{decide } \mathcal{H}_1 \text{ while } \mathcal{H}_0 \text{ is true}], \quad \beta \triangleq \mathbb{P}[\text{decide } \mathcal{H}_0 \text{ while } \mathcal{H}_1 \text{is true}]. \tag{2.1}$$

Simultaneous minimization of both errors is conflictual, and there exist different criteria to properly carry out the detection performance optimization, including the Neyman-Pearson criterion, where one error is constrained and the other is minimized; the Bayesian criterion, where the average (over the prior distribution) error probability is minimized; and the min-max criterion, where the maximum of the two errors is minimized. For all these criteria, the optimal test amounts to comparing the likelihood ratio (between the two distributions under the different hypotheses) against some threshold [3,7,8]. When the exact distribution of the observations is not known, the aforementioned optimization criteria cannot be pursued, and detection statistics different from the likelihood ratio must be employed. A common choice is the sum of some suitably chosen functions of the observations, which arise as the optimal choice in some specific frameworks, e.g., for locally optimum detection [9], for robust detection [10], and for universal/nonparametric hypothesis testing [11,12].

There exist also *asymptotic* criteria for optimizing and characterizing the performance of statistical tests. For example, in a *large deviations* framework [12,13], one studies the problem where the number of observations gets large and the error probabilities vanish exponentially fast with the number of observations. The related optimization criterion focuses on maximizing the decaying rate by maximizing the error exponents [12,13]. We shall see later in Section 2.5 that the large deviations tool plays an important role in the study of distributed adaptive detection (DAD).

2.2.2 DISTRIBUTED INFERENCE PROBLEM

One fundamental classification for distributed inference distinguishes network architectures with a fusion center from network architectures that are fully flat (and, therefore, fully decentralized).

Architectures with fusion center

Under this formulation, the network agents are geographically dispersed with respect to a central processing unit (the fusion center) that is tasked with producing the final inference, and the main concern is the necessity of communicating the data to and from the remote fusion center. The fundamental constraint that reflects the distributed nature of the problem is often determined by the data rate that each agent is allowed to sustain when communicating with the fusion center.

Example 2.3 (Distributed inference with fusion center). A firm's chief executive officer (CEO) is interested in solving an inferential (estimation and/or detection) task. The data sequence is not directly available to the CEO, perhaps because it represents tactical decisions by a firm competitor. Therefore, the CEO deploys a team of agents who gather noisy data about the underlying phenomenon of interest. Due to various limitations, the aggregate data rate at which agents may deliver information to the CEO is limited to, say, R bits per second. The fusion center must accomplish the inferential task based upon the received data. The inference performance will be affected by the bit-rate, namely, by the degree of compression employed by the remote agents. ■

The literature about decentralized inference under rate constraints is very abundant, and offers a number of elegant results and powerful solutions. An exhaustive overview of available results is not practical here. We limit ourselves to some useful entry-points and results. Let us consider decentralized estimation first. A careful management of the agents' data rates translates often into optimization of the agents' quantizers to attain the final estimation goal. Such optimization can be carried out under different settings, e.g., under the Bayesian setting [14–16], under the deterministic-parameter setting [17,18], or under the nonparametric setting [19–22]. It is also possible to take into account the role of the communication channel, as done, e.g., in [23,24]. Switching to the context of decentralized detection with the fusion center, a general overview can be found in [25,26]. Optimization of the remote quantizers for detection purposes is addressed in [27,28]. In many instances, decentralized detection often focuses on the limiting case where the agents are constrained to send just one bit (i.e., a local *decision*), and the main problem becomes that of establishing the best fusion rule [29–31]. Channel-aware strategies are also proposed in [32,33]. An alternative approach to manage the agents' data rates for detection purposes is the *censoring* approach, where the constraint is imposed in terms of number of channel accesses rather than in terms of bits [34,35].

The fundamental limits of distributed inference over networks employing a fusion center have also been investigated under an information-theoretic setting in the context of statistical inference under multiterminal data compression in [36], and multiterminal source coding in [37–39]. Some of these fundamental limits will be used later in Section 2.6, where they will be compared against the scaling laws derived for *adaptive* inference over *fully decentralized* networks.

Fully flat architectures

The *fully flat* (i.e., fully decentralized) architecture is the setting addressed in the present chapter. Under this setting, there is no fusion center, and each agent is demanded to produce its individual inference after (possibly iterated) consultation steps with the other agents. In order to keep the communication burden low, the agents are allowed to perform only *local* interactions, i.e., communication is permitted only among *neighboring* agents. Therefore, in contrast to the fusion center architecture, in the fully decentralized case the limitations are typically less constrained in terms of transmission rates. The limitations are more seriously dictated by the communication radius of each agent or by the network connection graph.

Great interest is devoted in the literature to distributed algorithms that are able to reproduce linear combinations of the *local* data (or some suitable function thereof) available at the individual agents. Besides the implementation convenience of linear combination rules, one important reason for the interest in these type of rules is that many optimal inference solutions require linear combinations of some local statistics. This is the case, for instance, when one has to compute likelihood functions in the presence of independent random variables. It is therefore not hard to explain why distributed averaging protocols (aka consensus protocols or gossip algorithms) are prevalent in the design of fully distributed inference algorithms [40–44]. In a nutshell, a consensus protocol prescribes that an agent sequentially updates its state by successively averaging it with that of its neighbors, and all agents perform simultaneously similar updates. Under proper conditions on the network connectivity, the asymptotic solution (with exponentially fast convergence) is driven to the desired linear combination of the initial states, i.e., of the desired local statistics.

Example 2.4 (Distributed inference with fully flat architecture). A wireless sensor network is tasked with estimating an underlying phenomenon of interest, which is formally represented in terms of a scalar parameter w^o. The kth sensor, for $k = 1, 2, \ldots, N$, possesses a noisy observation $\boldsymbol{x}_k$, which is a random variable with mean w^o. Each individual sensor must produce an estimate of w^o, after local consultation steps with its own neighbors. To this end, the sensors implement a consensus protocol by sequentially updating their state through averaging with the states of their neighbors. After a sufficient number of iterations, the state of each individual sensor becomes close enough to the arithmetic average of the observations, namely, sensor k owns an estimate of the parameter w^o given by $\hat{\boldsymbol{w}}_k \to \frac{1}{N}\sum_{\ell=1}^{N} \boldsymbol{x}_\ell$. ■

The aforementioned consensus construction has been applied in the context of distributed estimation and detection, e.g., in the framework known as *running consensus* [45–47] or *consensus + innovations* [48–51]. In this framework, the data acquisition and inferential process take place simultaneously, in the sense that the consensus algorithm runs when the agents are still gathering data and the updates are able to incorporate the new streaming data in the computed statistics.

These implementations belong to the class of stochastic approximation solutions with *decaying* step-sizes. For several useful formulations of distributed estimation and detection, the use of stochastic approximation solutions with *decaying* step-sizes leads to asymptotically optimal performance in the sense of asymptotic variance for the estimation framework [48], and in the sense of asymptotic relative efficiency [46] or of error exponents [49–51] for the detection framework. Optimality in these works is formulated with reference to the centralized solution, and the qualification "asymptotic" is used to refer either to a large number of observations or a large time window. The error performance (e.g., mean-square-error for estimation or error probabilities for detection) is shown in these works to decay with optimal rates as time elapses, provided that some conditions on the network structure are met. For these results to hold, it is critical for the statistical properties of the data to remain invariant and for the algorithms to rely on a recursive test statistic with a *decaying* step-size, which entails an inherent limitation as regards their applicability in *adaptive* contexts. Indeed, the asymptotic equivalence to centralized solutions goes hand in hand with a limited adaptation ability, namely, a limited capacity of following drifts in the phenomenon of interest and/or in the observation distributions and/or in the network topology. This is because, in order to converge to the same performance of the centralized solution, the running consensus strategies tend to give credit to observations coming from the remote past. The resulting "elephant's memory" implies that if the underlying state (e.g., the value of the parameter to be estimated) changes, the algorithm tends to remain trapped in the old value for a relatively long time before converging to the new value. This is one of the main reasons why, in adaptive network implementations, constant (as opposed to decaying) step-sizes are preferred. These constant step-sizes endow the distributed algorithms with continuous learning and adaptation abilities even in the face of drifting data and models.

2.2.3 INFERENCE OVER ADAPTIVE NETWORKS

In realistic distributed inference applications, the properties of the system should be allowed to vary over time. For example, the object of the inference problem (i.e., the unknown parameter in the estimation problem, or the unknown hypothesis in the detection problem) might drift over time.

Therefore, the adaptation aspect, i.e., the capability of tracking dynamic scenarios, becomes crucial. In such scenarios, the diffusion algorithms (with nondecaying, constant step-size) provide effective mechanisms for continuous adaptation and learning [52–54]. It is well known in the adaptation and learning literature that using *constant step-sizes* in the update relations automatically infuses the algorithms with a tracking mechanism that enables them to track variations in the underlying models. This effect does not happen for algorithms with decaying step-sizes (such as the aforementioned running-consensus algorithms) because the decaying step-size annihilates the gradient update term in the limit and ultimately stops adaptation. In contrast, when the system conditions change at a certain moment, an algorithm with a constant step-size will continue learning from that point onward and, given sufficient time to learn, will converge to a steady-state learning performance. However, the use of constant step-sizes allows gradient noise to seep into the operation of the algorithms. Therefore, a key challenge in the study of adaptive networks is to show, that despite the presence of the gradient noise, the algorithms are able to remain stable, learn, and deliver performance. We will find that the more an adaptive network algorithm learns, the more it reduces the size of the gradient noise and this feedback mechanism leads to effective learning.

Example 2.5 (Distributed adaptive estimation). A team of spy agents share the common goal of monitoring the activity of a dangerous criminal, who is known to be present in a certain extended area. Each agent collects, continually across time, information about the possible criminal's location within the area. Due to various forms of limitations (e.g., internal rules of the agencies, relative locations of the agents, etc.), a given agent can exchange information only with a given subset of the agents. The goal of each agent is that of estimating the current position of the criminal, based upon his own data, and the data exchanged with the agents he is "connected" to. After a sufficient amount of time, it is expected that the agents can converge to the same location. On the other hand, because the criminal can change his position across time, the distributed processing implemented by the agents must be open to adaptation, namely, the agents should be able to follow the variations in the criminal's state. As we shall see in the forthcoming treatment, such a continuous mechanism of adaptive estimation can be successfully implemented through a *diffusion* algorithm. ■

Example 2.6 (Distributed adaptive detection). A robotic swarm is deployed over a disaster area in order to perform exploration and/or rescuing operations. As the individual robots travel across the area, they must be aware of the environmental conditions in order to decide whether or not to enter a certain location. To this aim, the robots enhance their individual knowledge about the state of nature (danger/no-danger) by exchanging local information with neighbors. The knowledge of an individual agent is encoded into a certain state variable, which is continually updated, in order to keep alive the possibility of monitoring variations in the environmental conditions. At each time instant, each individual agent must make a binary decision (enter/do-not-enter), based upon the current individual state variable. As we shall see in the forthcoming treatment, such a continuous mechanism of adaptive detection can be successfully implemented through a *diffusion* algorithm. ■

Example 2.7 (Distributed adaptive detection + estimation). A school of fish must evade the attack of a predator. Information exchange across the network occurs through local interactions among adjacent members of the school. Using such an exchange of information, the fish school is able to carefully reconfigure its topology in the face of danger: when a predator is detected, the entire school of fish estimates the predator location and adapts its configuration to let the predator go through. When there is no predator, the fish coalesce and restart their schooling behavior. As time elapses, the individual agents must be continually alert in order to react to the presence/absence of a predator and to the variations in the predator's location. It is known that such a continuous mechanism of adaptive learning (simultaneous detection and estimation) can be emulated through a *diffusion* algorithm. ■

In the context of distributed inference, it is worth mentioning a third canonical problem, namely, Bayesian filtering. Useful methods (e.g., distributed Kalman filters, distributed particle filters) have been proposed for distributed (linear and nonlinear) Bayesian filtering. These methods will be treated in Chapter 6.

2.3 DIFFUSION IMPLEMENTATIONS

There exists an elegant and unifying formulation that captures the essence of many inference and learning problems over adaptive networks. Such a representation is based on three fundamental steps: (i) identification of a cost function that matches the learning objective of the network; (ii) implementation of a stochastic-gradient algorithm for minimizing the cost function; and (iii) implementation of a diffusion mechanism to enable information sharing across the network. Such a representation has been presented in Chapter 1 in greater detail as well as in references [52–54] for more general scenarios, and is briefly reported here for making the present chapter self-contained.

Let us consider a network with N agents. We assume that the learning objective of the network can be well described by the following optimization problem:

$$\boxed{\min_w J(w), \quad J(w) \triangleq \sum_{k=1}^{N} J_k(w), \quad J_k(w) \triangleq \mathbb{E} Q_k(w; \boldsymbol{x}_k).} \tag{2.2}$$

We see that agent k is associated with an individual cost function $J_k(w)$, which can be expressed as the expectation of a certain *loss function*, $Q_k(w; \boldsymbol{x}_k)$, with the expectation being evaluated with respect to certain random data, $\boldsymbol{x}_k$. We assume that the individual cost functions fulfill standard smoothness conditions. In particular, it is sufficient for the purposes of this chapter to assume that the individual costs are ν-strongly convex (see Chapter 1 and [53]), and that they attain their unique global minima at the same location, w^o, yielding

$$w^o \triangleq \arg\min_w J(w) = \arg\min_w J_k(w). \tag{2.3}$$

These smoothness conditions can be relaxed. For example, the individual costs do not need to attain their minima at the same location. It is also sufficient for them to be convex functions and to require only the aggregate cost $J(w)$ to be ν-strongly convex. These more general scenarios are discussed in detail in [53]. We continue with Eq. (2.3). Under the aforementioned smoothness conditions, one way to perform minimization relies on the evaluation of the gradient vector, $\nabla_w J(w) = \sum_{k=1}^{N} \nabla_w J_k(w)$, and to apply the steepest descent algorithm [6]. However, in our formulation we shall assume that the agents do not have information about the *true* gradient (as is the case in typical adaptive network applications). Therefore, agent k will replace the true gradient with an *instantaneous approximation*, chosen as

$$\boxed{\widehat{\nabla_w J_k}(w) = \widehat{\nabla_w J_k}(w; \boldsymbol{x}_k) \triangleq \nabla_w Q_k(w; \boldsymbol{x}_k), \quad [\textit{stochastic}\ \text{gradient}]} \tag{2.4}$$

which highlights how the gradient must be *learned from the data*. The task of each agent k consists in producing, at each time instant i, a value $\boldsymbol{w}_{k,i}$, which should approximate the desired value w^o. To this end, agents are allowed to exchange information with their neighbors. Randomness of the state $\boldsymbol{w}_{k,i}$ is originated from the fact that its determination is based on *random* data.

We are now ready to illustrate a *diffusion*-type algorithm that is able to guarantee distributed adaptation and learning over networks. This algorithm is traditionally referred to as the adapt-then-combine (ATC) algorithm [52,55]. While, as shown in Chapter 1 and [52–54], other forms and extensions are possible, in this chapter we focus on the ATC form to convey the main ideas. We choose the ATC form because it exhibits some inherent advantages in terms of a slightly improved mean-square-error performance relative to other forms [52].

In the ATC recursion, each node k updates its state from $\boldsymbol{w}_{k,i-1}$ to $\boldsymbol{w}_{k,i}$ through local cooperation with its neighbors as follows, for all $i \geq 0$:

$$\boxed{\begin{aligned} \boldsymbol{\psi}_{k,i} &= \boldsymbol{w}_{k,i-1} - \mu \nabla_w Q_k(\boldsymbol{w}_{k,i-1}; \boldsymbol{x}_{k,i}), && \text{[adaptation step]} \\ \boldsymbol{w}_{k,i} &= \sum_{\ell=1}^{N} a_{\ell k} \boldsymbol{\psi}_{\ell,i}, && \text{[combination step]} \end{aligned}} \tag{2.5}$$

where $0 < \mu \ll 1$ is a small step-size parameter. It is seen that node k first uses its locally available data $\boldsymbol{x}_{k,i}$ to update its state from $\boldsymbol{w}_{k,i-1}$ to an intermediate value $\boldsymbol{\psi}_{k,i}$. The other network agents simultaneously perform similar updates using their local data. Subsequently, node k aggregates the intermediate states of its neighbors using nonnegative convex combination weights $\{a_{\ell k}\}$ that add up to one. Again, all other network agents perform a similar calculation. Collecting the combination weights into a square matrix $A = [a_{\ell k}]$, then A is a *left-stochastic* matrix, namely, the entries on each column add up to one [53]. Formally:

$$a_{\ell k} \geq 0, \quad A^\top \mathbb{1} = \mathbb{1} \tag{2.6}$$

with $\mathbb{1}$ being a column-vector with all entries equal to one.

The network topology (or network graph) establishes whether two agents are connected, i.e., whether they are *in the position of* exchanging information; see Fig. 2.1 for an illustrative example. A relevant descriptive indicator of the network graph is the neighborhood of an agent. The neighborhood of the kth agent, denoted by $\mathcal{N}_k$, is the set containing the neighbors of agent k, including k itself. Clearly, when two agents are not connected (according to a given network graph), they *cannot* exchange information. If they are connected, the exchange of information can be bidirectional, unidirectional, or even absent, depending on the structure of the combination matrix. We shall say that the network is *strongly connected* [54] when there is always a path with nonzero weights between any pair of nodes, and at least one node in the network has a self-loop ($a_{kk} > 0$ for some agent k). Under this assumption, it holds true that there exists an eigenvector $p = [p_1, p_2, \ldots, p_N]^\top$ satisfying:

$$\boxed{Ap = p, \quad \mathbb{1}^\top p = 1, \quad p_k > 0, \quad k = 1, 2, \ldots, N} \tag{2.7}$$

where p is usually referred to as the Perron eigenvector of A; see, e.g., [54].

Example 2.8 (Combination policies). One of the simplest combination policies is the *uniform averaging rule*. The qualification "uniform" arises because each agent uses identical (nonzero) weights,

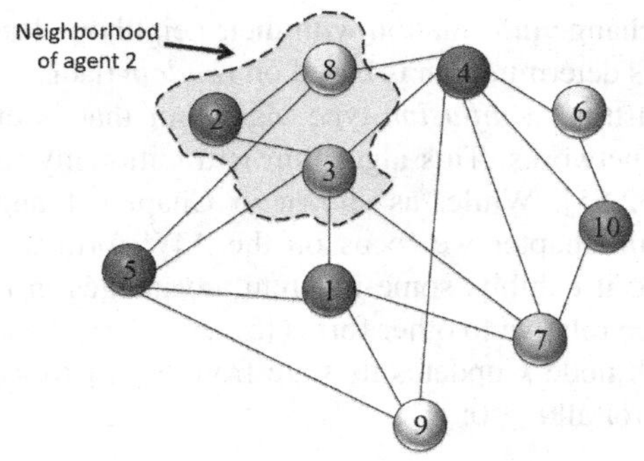

FIG. 2.1

Example of a connected network. A network with $N = 10$ agents is shown in the figure. The neighborhood of agent 2, $\mathcal{N}_2 = \{2, 3, 8\}$, is marked by the broken line. We note that agent 2 is the least connected (smallest neighborhood size). In contrast, agent 4, with neighborhood $\mathcal{N}_4 = \{3, 4, 6, 7, 8, 9, 10\}$, is the most connected. As we shall see later, the different agents' connectivity can play a role in the inference performance.

which implies that the nonzero weights of the kth agent are set as the inverse of the kth neighborhood size. Formally:

$$a_{\ell k} = \frac{1}{|\mathcal{N}_k|}, \quad \ell \in \mathcal{N}_k. \tag{2.8}$$

The resulting combination matrix is *left-stochastic*.

As a slight generalization of the uniform averaging rule, we present the *Metropolis rule*, which is defined as:

$$a_{\ell k} = \begin{cases} \dfrac{1}{\max\{|\mathcal{N}_k|, |\mathcal{N}_\ell|\}}, & \ell \in \mathcal{N}_k \setminus \{k\}, \\ 1 - \displaystyle\sum_{m \in \mathcal{N}_k \setminus \{k\}} a_{mk}, & \ell = k. \end{cases} \tag{2.9}$$

It is easy to verify that the Metropolis rule provides a *doubly-stochastic* combination matrix (i.e., the entries on each of its columns add up to one, and the entries on each of its rows also add up to one). ■

The formulation in Eq. (2.5) provides the fundamental equations of diffusion-based optimization, inference, and learning over adaptive networks. We shall now motivate the latter observation by examining in detail the structure of Eq. (2.5).

Where are inference and learning? The inference and learning aspects are encoded in the *stochastic gradient* terms employed during the adaptation step in Eq. (2.5). Indeed, the streaming data enter the update equation through these stochastic gradients.

Is learning effective? Effectiveness of learning resides in the fact that, as i gets large, the individual states, $\boldsymbol{w}_{k,i}$, will fluctuate within a small neighborhood of the desired solution, w^o. The radius of such a neighborhood is determined by the step-size: the smaller μ is, the smaller the neighborhood size is.

Where are the distributed features? The combination step in Eq. (2.5) encodes the distributed features because it shows how the agents aggregate information from their neighbors through the diffusion mechanism.

Where is adaptation? Adaptation is enabled by the small but *constant* step-size μ. When, for example, the value w^o changes or drifts over time, the constant step-size allows the system to react promptly and to start adapting the individual states to the new solution. Algorithms such as that described by Eq. (2.5) are naturally infused with adaptation and tracking capabilities. The larger the value of μ is, the faster the reaction to drifts (i.e., the convergence to a steady-state solution) will be.

We shall see in the forthcoming two sections how the general formulation in Eq. (2.5) can be specialized to the distinct settings of estimation and detection.

2.4 DISTRIBUTED ADAPTIVE ESTIMATION (DAE)

The diffusion paradigm illustrated in the previous section can be used to design estimators according to many different settings and optimization criteria. In order to illustrate the fundamental features of distributed estimation over adaptive networks, we focus here on the celebrated and widely used linear regression model. Specifically, we assume that each agent k senses data $\{\boldsymbol{d}_k(i), \boldsymbol{u}_{k,i}\}$ that satisfy a model of the form:

$$\boxed{\boldsymbol{d}_k(i) = \boldsymbol{u}_{k,i} w^o + \boldsymbol{v}_k(i). \quad \text{[linear regression model]}} \tag{2.10}$$

More general scenarios are considered in Chapter 1 and [53,54]. In the model (2.10):

- The unknown (deterministic) parameter to be estimated is w^o, an $M \times 1$ vector. All vectors in our treatment will be column vectors, with the exception of $\boldsymbol{u}_{k,i}$, which is a $1 \times M$ row vector.
- The *streaming* data available at time i to the kth agent are $\{\boldsymbol{x}_{k,i}\} = \{\boldsymbol{d}_k(i), \boldsymbol{u}_{k,i}\}$. In particular, the (scalar) data $\{\boldsymbol{d}_k(i)\}$ are usually referred to as *desired response* or *output data*. The data $\{\boldsymbol{u}_{k,i}\}$ are usually referred to as *regressors*, or *regression data*.
- The terms $\{\boldsymbol{v}_k(i)\}$ are usually referred to as *noise*.

We shall examine the performance of distributed adaptive algorithms with model (2.10) under the following technical assumptions:

1. The data and the noise are zero-mean jointly wide-sense stationary random processes.
2. The regression data are temporally white and independent over space, with covariance matrix $R_u \triangleq \mathbb{E}\boldsymbol{u}_{k,i}^\top \boldsymbol{u}_{k,i}$.
3. The cross-covariance vector between the output data and the regression data is defined as $r_{du} = \mathbb{E}\boldsymbol{d}_k(i)\boldsymbol{u}_{k,i}^\top$.
4. The noise data are temporally white and independent over space, with variances $\sigma_{v,k}^2 \triangleq \mathbb{E}\boldsymbol{v}_k^2(i)$.
5. The regression and noise processes $\{\boldsymbol{u}_{k,i}, \boldsymbol{v}_\ell(j)\}$ are independent of each other for all k, ℓ, i, j.

2.4.1 CONSTRUCTING THE DISTRIBUTED ADAPTIVE ESTIMATION ALGORITHM

In order to construct the distributed and adaptive estimator, we shall select the following cost function, for $k = 1, 2, \dots, N$:

$$\boxed{J_k(w) = \mathbb{E}(\boldsymbol{d}_k(i) - \boldsymbol{u}_{k,i}w)^2, \quad \text{[quadratic cost function]}} \tag{2.11}$$

which entails the following loss function and instantaneous approximation:

$$Q_k(w; \boldsymbol{x}_{k,i}) = (\boldsymbol{d}_k(i) - \boldsymbol{u}_{k,i}w)^2 \Leftrightarrow \widehat{\nabla_w J_k(w)} = 2\boldsymbol{u}_{k,i}^\top[\boldsymbol{u}_{k,i}w - \boldsymbol{d}_k(i)], \tag{2.12}$$

and, hence, the diffusion algorithm in Eq. (2.5) becomes, for all $i \geq 0$:

$$\boxed{\begin{aligned} \boldsymbol{\psi}_{k,i} &= \boldsymbol{w}_{k,i-1} + 2\mu\, \boldsymbol{u}_{k,i}^\top[\boldsymbol{d}_k(i) - \boldsymbol{u}_{k,i}\boldsymbol{w}_{k,i-1}], && \text{[adaptation step]} \\ \boldsymbol{w}_{k,i} &= \sum_{\ell=1}^{N} a_{\ell k} \boldsymbol{\psi}_{\ell,i}. && \text{[combination step]} \end{aligned}} \tag{2.13}$$

In the single-agent case, the particular stochastic-gradient implementation in Eq. (2.13) is traditionally known as LMS algorithm [6].

Before dwelling on the examination of the estimation performance, it is useful to draw some connections between the adaptive diffusion solution (2.13) and classic results on estimation theory. First, if we left-multiply by $\boldsymbol{u}_{k,i}^\top$ both sides of Eq. (2.10), and then take expectations, we see that the value of the true parameter, w^o, can be expressed as $w^o = R_u^{-1} r_{du}$. However, in a typical (and meaningful) estimation setting, both R_u and r_{du} are unknown. Therefore, one useful interpretation of the instantaneous approximation in Eq. (2.12) is that one tries to approximate the true gradient by replacing the unknown covariances, R_u and r_{du}, with their instantaneous (and local, i.e., corresponding to the individual agents) estimates, $\boldsymbol{u}_{k,i}^\top \boldsymbol{u}_{k,i}$ and $\boldsymbol{d}_k(i)\boldsymbol{u}_{k,i}^\top$. Then, in a steady-state situation, the instantaneous approximations feeding the algorithm combine with each other recursively so as to increase the precision in estimating the covariance matrix and the cross-covariance vector, and, hence, the parameter w^o. However, because the constant step-size leaves deliberately alive the gradient noise contribution, such an increase in precision does not proceed indefinitely: some deterioration in performance is accepted (i.e., the estimate will not become perfect as $i \to \infty$) in order to keep adaptation alive forever.

Another useful observation concerns the high versatility of the general diffusion implementation (2.5), which allows designing distributed-and-adaptive estimators even for statistical models other than the linear regression model. To illustrate this point, we consider an example based on the maximum-likelihood principle. Assume for simplicity identical distribution across time index i, and choose as loss function:

$$Q_k(w; \boldsymbol{x}_{k,i}) = -\ell(w; \boldsymbol{x}_{k,i}), \tag{2.14}$$

where $\ell(w; \boldsymbol{x}_{k,i})$ is the log-likelihood function corresponding to data $\boldsymbol{x}_{k,i}$. We immediately reach two conclusions. First, minimizing the expected loss corresponds to maximizing the expected log-likelihood, and it is well known that the maximizer of the *expected* log-likelihood is the *true* value of the parameter [1,3]. Second, in view of Eq. (2.14), the stochastic approximation of the true gradient corresponds to evaluating the (negative of the) *score vector*, namely,

$$\nabla_w Q(w; x_{k,i}) = -\nabla_w \ell(w; x_{k,i}). \tag{2.15}$$

Accordingly, provided that the statistical model guarantees the verification of standard conditions on the gradient noise associated with the chosen loss function, a diffusion implementation (2.5) that employs the score vectors in the adaptation step would enable each agent to estimate the desired parameter.

2.4.2 MEAN-SQUARE-ERROR PERFORMANCE

In the adaptation literature, the standard approach to get useful performance metrics consists of splitting the analysis into two complementary parts [6]:

(i) The *transient* analysis. Assuming that, from a certain point, the statistical conditions of the system remain stationary for a sufficiently long time, in the transient analysis the focus is on evaluating if and how convergence to a steady state takes place. Such an analysis is relevant because, when some variations in the statistical conditions occur at a certain time, the transient analysis allows quantifying the time to track such variations. Detailed studies about the transient phase of the diffusion learning algorithm have been carried out, which clarify the behavior of the learning curve during this stage of operation [56,57]. In particular, the convergence rate toward the steady-state regime is known to occur at an exponential rate in the order of $O(c^i)$ for some $c \in (0,1)$; this is a faster rate than $O(1/i)$ that is afforded, for example, by decaying step-sizes. Nevertheless, in the constant step-size case, the smaller the value of μ is, the closer the value of c gets to one. A compact qualitative index that is commonly employed to describe the effective duration of a transient is the time constant, which is the time needed to reduce by a certain amount the initial error state. For the problems considered in this chapter, it is known that the time constant is, for small values of μ, in the order of $1/\mu$ [56,57].

(ii) The *steady-state* analysis. This analysis provides the inference performance of the adaptive algorithm with reference to an infinitely long period of stationarity, which takes on the practical meaning of *giving the algorithm sufficient time to learn*. An exact analytical characterization of the inference performance is seldom affordable. However, closed-form results can be obtained by focusing on the regime of *small step-sizes*. As we shall see later, the inference performance is inversely related to the step-size: the smaller μ is, the smaller the steady-state error becomes.

Let us now focus on the performance analysis of the diffusion-based estimators in Eq. (2.13). A classic way to measure the estimation performance of the adaptive network algorithm (2.13) is to introduce the *error vector* corresponding to agent k at time i, and the related *time-dependent* mean-square-deviation,[1] for $k = 1, 2, \ldots, N$:

$$\boxed{\tilde{w}_{k,i} = w^o - w_{k,i}, \qquad \mathrm{MSD}_k(i) \triangleq \mathbb{E}\|\tilde{w}_{k,i}\|^2.} \tag{2.16}$$

We remark that the mean-square-deviations in Eq. (2.16) depend upon the statistical properties of the *entire* dataset used in the diffusion algorithm up to time i. In particular, the errors depend upon the different kinds of variations that may have occurred during the evolution of the algorithm.

[1] The terminology "mean-square-deviation" is used in lieu of "mean-square-error" in order to avoid confusion with another squared error that is typically measured in the linear regression context, namely, the cost function in Eq. (2.11).

Once a proper performance measure has been chosen, we can use it to obtain the estimation performance following the route traced at the beginning of this section: we assume that the data are possibly nonstationary up to a certain time instant, after which they are drawn from the same stationary distribution for a sufficiently long time. Technically, when performing this steady-state analysis, it suffices to assume that the data, for all $i \geq 0$, arise from the same distribution. The past history (including possible drifts that occurred in the statistical conditions) that influences the overall algorithm evolution is reflected in the initial state vectors, $\{\boldsymbol{w}_{k,-1}\}$. Accordingly, the *limiting* mean-square-deviation is defined as, for $k = 1, 2, \ldots, N$:

$$\boxed{\text{MSD}_k \triangleq \lim_{i\to\infty} \mathbb{E}\|\tilde{\boldsymbol{w}}_{k,i}\|^2.} \tag{2.17}$$

In order to evaluate MSD_k, it is convenient to introduce the *network* error vector at time i:

$$\tilde{\boldsymbol{w}}_i = \text{col}\{\tilde{\boldsymbol{w}}_{1,i}, \tilde{\boldsymbol{w}}_{2,i}, \ldots, \tilde{\boldsymbol{w}}_{N,i}\} \tag{2.18}$$

along with an *arbitrary* $N \times N$ block symmetric and nonnegative-definite matrix, defined as a block matrix with $M \times M$ block entries, and denoted by Σ. Different choices of Σ can be used to characterize different aspects of the inferential performance of the network. We further introduce the matrices $\mathcal{B}$ and $\mathcal{Y}$:

$$\mathcal{B} = A^\top \otimes (I_M - 2\mu R_u), \qquad \mathcal{Y} = 4\mu^2 \, (A^\top R_v A \otimes R_u), \tag{2.19}$$

where $R_v = \text{diag}\{\sigma_{v,1}^2, \sigma_{v,2}^2, \ldots, \sigma_{v,N}^2\}$. Elaborating on Eq. (2.13), by subtracting the true parameter vector w^o and then taking expectations, it is possible to derive the following recursions for the mean and mean-square evolution as time i progresses [52–54]:

$$\boxed{\mathbb{E}\tilde{\boldsymbol{w}}_i = \mathcal{B}\,\mathbb{E}\tilde{\boldsymbol{w}}_{i-1}, \qquad \mathbb{E}\tilde{\boldsymbol{w}}_i^\top \Sigma \tilde{\boldsymbol{w}}_i \approx \mathbb{E}\tilde{\boldsymbol{w}}_{i-1}^\top \mathcal{B}^\top \Sigma \mathcal{B} \tilde{\boldsymbol{w}}_{i-1} + \text{Tr}(\Sigma \mathcal{Y})} \tag{2.20}$$

where the second expression is an approximation for the mean-square-error evolution valid in the steady-state regime ($i \gg 1$) and for small step-sizes. The following important facts have been established in relation to the estimation performance of the network [52–54].

Steady-state unbiasedness. For sufficiently small step-sizes, the estimators $\{\boldsymbol{w}_{k,i}\}$ are asymptotically unbiased, namely, for $k = 1, 2, \ldots, N$:

$$\boxed{\lim_{i\to\infty} \mathbb{E}\tilde{\boldsymbol{w}}_{k,i} = 0.} \tag{2.21}$$

Let us now quantify the qualification "sufficiently small". The first equation in Eq. (2.19) implies the following relationship, which relates the MN eigenvalues of matrix $\mathcal{B}$ to the eigenvalues of matrices A and R_u, for $k = 1, 2, \ldots, N$ and $j = 1, 2, \ldots M$:

$$\lambda(\mathcal{B}) = \lambda_k(A)[1 - 2\mu\lambda_j(R_u)]. \tag{2.22}$$

Then, using the first relationship in Eqs. (2.20) and (2.22), we conclude that unbiasedness is guaranteed if the step-size is smaller than the quantity $1/\lambda_{\max}(R_u)$, where $\lambda_{\max}(R_u)$ is the largest eigenvalue of R_u. Notably, this condition is *independent of the combination matrix*, and, hence, coincides with the condition that would be required for mean stability of an *individual* LMS filter because the latter can be regarded as a special case of the truly distributed implementation (2.13). Accordingly, for the diffusion implementation in Eq. (2.13), it holds true that *individual* mean-stability implies *network* mean-stability, an implication that is not found, for instance, in other types of implementations, such as, for example, the consensus implementation [53].

Steady-state mean-square-error. Elaborating on the second relationship in Eq. (2.20), it can be shown that under the mean-stability condition one has [53]:

$$\lim_{i\to\infty} \mathbb{E}\tilde{w}_i^\top \Sigma \tilde{w}_i \approx \sum_{m=0}^{\infty} \mathrm{Tr}(\mathcal{B}^m \mathcal{Y} \mathcal{B}^{\top m} \Sigma). \tag{2.23}$$

With the particular choice $\Sigma = \mathcal{J}_k$, where $\mathcal{J}_k$ is the block diagonal matrix with I_M on the kth block and with zero blocks elsewhere, Eq. (2.23) yields:

$$\mathrm{MSD}_k \approx \sum_{m=0}^{\infty} \mathrm{Tr}(\mathcal{B}^m \mathcal{Y} \mathcal{B}^{\top m} \mathcal{J}_k). \tag{2.24}$$

Using the explicit expressions for $\mathcal{B}$ and $\mathcal{Y}$ in Eq. (2.19), it is finally possible to obtain the following revealing formula, for $k = 1, 2, \ldots, N$ [53,58]:

$$\boxed{\mathrm{MSD}_k = \mu M \sum_{\ell=1}^{N} p_\ell^2 \sigma_{v,\ell}^2 + O(\mu^2).} \tag{2.25}$$

Fundamental scaling law. Eq. (2.25) reveals the fundamental scaling law for distributed estimation with diffusion adaptation: as μ decreases, the mean-square performance of all agents scales proportionally to μ (i.e., inversely as a function of $1/\mu$).

Adaptation/learning trade-off. We see from Eq. (2.25) that the smaller the step-size, the finer the estimate of the parameter vector w^o becomes. Recalling that smaller values of μ entail less adaptation, we conclude that a better estimation accuracy costs in terms of adaptation speed. This result is a well-known trade-off in the adaptive filtering literature between estimation accuracy and adaptation [6].

Asymptotic equivalence among agents. By ignoring the $O(\mu^2)$ term appearing in Eq. (2.25), we see that MSD_k does not depend on the agent index k. As a result, for small step-sizes the estimation performance equalizes across agents. Actually, the differences across agents are contained in the $O(\mu^2)$ term appearing in Eq. (2.25). In summary, we can use the following formula to describe the performance of each individual agent:

$$\boxed{\mathcal{MSD} \triangleq \mu M \sum_{\ell=1}^{N} p_\ell^2 \sigma_{v,\ell}^2} \tag{2.26}$$

which does not depend on the particular agent, and which approximates the true mean-square-deviation, MSD_k, of each agent k, in the regime of small step-sizes.

2.4.3 USEFUL COMPARISONS

We now compare the distributed performance in Eq. (2.26) against the performance of some reference estimators.

Centralized (adaptive) processing. Let us consider first the case of a *centralized* stochastic-gradient (i.e., adaptive) processor that aims at minimizing the cost function $J(w) = \sum_{k=1}^{N} p_k J_k(w)$ for some convex and nonnegative combination weights $\{p_k\}$, and for the cost functions defined in Eq. (2.11). Because a centralized processor of this type can be derived by considering a *fully connected* network with a combination matrix with identical columns equal to $p = [p_1, p_2, \ldots, p_N]^\top$, the performance for the centralized case can be derived directly from Eq. (2.26), and is given by:

$$\boxed{\text{MSD}_{\text{cent}} \approx \mu M \sum_{\ell=1}^{N} p_\ell^2 \sigma_{v,\ell}^2.} \tag{2.27}$$

Accordingly, we see from Eq. (2.26) that if the chosen combination matrix has the Perron eigenvector p, then:

$$\boxed{\frac{\text{MSD}_k}{\text{MSD}_{\text{cent}}} \approx 1} \tag{2.28}$$

namely, we reach the important conclusion that the distributed diffusion solution *delivers, for sufficiently small step-sizes, the same performance of a centralized stochastic-gradient processor.*

Noncooperative processing. The case of noncooperative processing is derived from the general diffusion case by simply considering each agent separately (i.e., by considering a "combination matrix" $A = 1$, with $N = 1$), which yields, for $k = 1, 2, \ldots, N$:

$$\boxed{\text{MSD}_{\text{ncop},k} \approx \mu M \sigma_{v,k}^2.} \tag{2.29}$$

There are at least two meaningful ways to compare the performance of cooperative (i.e., diffusion) and noncooperative solutions. The first is in terms of an aggregate network performance measure, which we take here as the *network* mean-square-deviation, defined as:

$$\boxed{\text{MSD}^{(\text{net})} \triangleq \frac{1}{N} \sum_{k=1}^{N} \text{MSD}_k \approx \mathcal{MSD} = \mu M \sum_{\ell=1}^{N} p_\ell^2 \sigma_{v,\ell}^2.} \tag{2.30}$$

Because (for small step-sizes) diffusion equalizes performance across agents, *the network performance for the diffusion algorithm is equivalent (for small step-sizes) to the performance reached by each individual agent through cooperation.* The corresponding performance for the noncooperative case is:

$$\text{MSD}_{\text{ncop}}^{(\text{net})} \triangleq \frac{1}{N} \sum_{k=1}^{N} \text{MSD}_{\text{ncop},k} \approx \mu M \left(\frac{1}{N} \sum_{k=1}^{N} \sigma_{v,k}^2 \right). \tag{2.31}$$

If we now consider a doubly-stochastic combination matrix A (for which we know that the limiting Perron weights are $p_\ell = 1/N$ for all $\ell = 1, 2, \ldots, N$), from Eqs. (2.30) and (2.31) we get:

$$\mathrm{MSD}^{(\mathrm{net})} \approx \frac{1}{N}\mathrm{MSD}^{(\mathrm{net})}_{\mathrm{ncop}}, \tag{2.32}$$

namely, *diffusion adaptation yields an N-fold improvement in comparison to a noncooperative solution in terms of aggregate network performance.*

Let us now see what happens if the comparison with the noncooperative solution is made in terms of the *individual* agent's performance. Consider first the case that the agents feature the same reliability, namely, that the noise variances, $\sigma^2_{v,k}$, are equal across index k. Then, we immediately see from Eqs. (2.26) and (2.29) that the same N-fold performance improvement is observed. However, the situation might change drastically if the variances vary across agents. In the latter case, for doubly-stochastic combination policies, and even for general left-stochastic combination policies (i.e., for general limiting weights, $\{p_\ell\}$), *cooperation is not always beneficial for the individual agents*! For example, if one agent is particularly reliable (low noise variance), the sharing of information with unreliable neighbors might worsen its individual performance (with diffusion) with respect to the noncooperative performance (without diffusion). This is because the scaling corresponding to the limiting combination weights does not match the reliability of the individual agents: intuitively, an optimal solution must give less credit (i.e., lower Perron weights) to unreliable agents and more credit (i.e., higher Perron weights) to reliable agents. An optimal choice can be devised by seeking an optimal left-stochastic combination matrix that solves the following optimization problem [53,58]:

$$A^{(\mathrm{opt})} = \arg\min_{A\in\mathcal{A}} \sum_{\ell=1}^{N} p_\ell^2 \sigma^2_{v,\ell}, \qquad \text{subject to} \quad Ap = p, \quad \mathbb{1}^\top p = 1, \quad p_\ell > 0 \tag{2.33}$$

with $\mathcal{A}$ being the set of all primitive $N \times N$ left-stochastic matrices matching the underlying network connection graph. One solution to this problem is as follows:

$$a^{(\mathrm{opt})}_{\ell k} = \begin{cases} \dfrac{\sigma^2_{v,k}}{\max\left\{|\mathcal{N}_k|\sigma^2_{v,k}, \ |\mathcal{N}_\ell|\sigma^2_{v,\ell}\right\}}, & \ell \in \mathcal{N}_k \setminus \{k\}, \\ 1 - \displaystyle\sum_{m\in\mathcal{N}_k\setminus\{k\}} a^{(\mathrm{opt})}_{mk}, & \ell = k. \end{cases} \tag{2.34}$$

The combination rule in Eq. (2.34) is usually known as the Hastings combination rule. The corresponding mean-square-deviations are, for $k = 1, 2, \ldots, N$:

$$\boxed{\mathrm{MSD}_{\mathrm{opt},k} \approx \mathrm{MSD}^{(\mathrm{net})}_{\mathrm{opt}} \approx \frac{\mu M}{\sum_{\ell=1}^{N} \sigma^{-2}_{v,\ell}}.} \tag{2.35}$$

Comparing Eq. (2.35) with Eq. (2.31), we see that, for sufficiently small step-sizes:

$$\boxed{\mathrm{MSD}^{(\mathrm{net})}_{\mathrm{opt}} \leq \frac{1}{N}\mathrm{MSD}^{(\mathrm{net})}_{\mathrm{ncop}}, \qquad \mathrm{MSD}_{\mathrm{opt},k} \leq \mathrm{MSD}_{\mathrm{ncop},k}.} \tag{2.36}$$

The following important conclusions arise from Eq. (2.36): (i) the aggregate network performance for the diffusion algorithm with the optimized combination matrix is *at least N times better than the noncooperative solution*; and (ii) optimization of the combination matrix lets the *individual agent performance outperform the noncooperative solution even in the presence of different reliabilities across agents.*

2.4.4 DAE AT WORK

One illustrative example of adaptive estimation is offered in Fig. 2.2, leftmost panel. The network topology is that displayed in Fig. 2.1. The adaptive network aims at estimating a parameter vector $w^o \in \mathbb{R}^3$. For the sake of simplicity, the noise variables are generated as independent (both across time and space) standard Gaussian variables, and so are the regression vectors, with $R_u = I_3$. The plot shows the time evolution of the online estimates, $\boldsymbol{w}_{k,i}$, for two agents, namely, for $k = 2$ and $k = 4$. The three components of the estimated parameter vector and of the true parameter vector are displayed as functions of time index i. Starting from the time origin, we see that, after an initial transient, the online

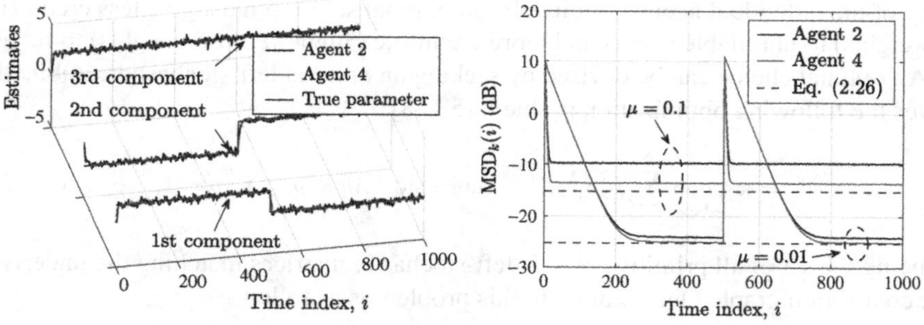

FIG. 2.2

Illustration of the distributed adaptive estimation problem. The estimation problem is formalized in Eq. (2.10). The noise variables are chosen as independent (both across time and space) standard Gaussian variables, and so are the regression vectors, with $R_u = I_3$. The network topology is shown in Fig. 2.1, and the combination rule is the Metropolis rule in Eq. (2.9). Leftmost panel: One realization of the diffusion algorithm in Eq. (2.13), namely, of the online estimators corresponding to agents 2 and 4 (*oscillating curves*). The three components of the *true* parameter are also displayed (piecewise constant curves). The first two components of the true parameter change when $i = 500$, and the *adaptive* algorithm is able to track such variation promptly. The step-size is set to $\mu = 0.1$. Rightmost panel: Empirical evaluation of the *time-dependent* mean-square-deviation, $\text{MSD}_k(i)$ [see Eq. (2.16)], obtained by averaging the squared error norm, $\|\tilde{\boldsymbol{w}}_{k,i}\|^2$, over statistical realizations similar to the one shown in the leftmost panel. Obviously, to encompass the variations of the true parameter vector, the definition of the error in Eq. (2.16) has been modified by letting the true parameter vector depend on time index i. The theoretical approximation of the steady-state mean-square-deviation, $\mathcal{MSD} \approx \text{MSD}_k$ [see Eq. (2.26)], is represented by the horizontal dashed lines. Two values of the step-size are considered.

estimates (oscillating curves) fluctuate in a close neighborhood of the true parameter vector components (piecewise constant curves). Then, at one-half of the observation window, the first two components of the *true* parameter vector drift to new values. Remarkably, the online estimates react promptly to such variations and *are able to adapt* their behavior to the new values.

Such a qualitative behavior is complemented by the results shown in the rightmost panel of Fig. 2.2, where the adaptation and learning performance are examined through a quantitative illustration of the mean-square performance of the network of estimators. This performance is evaluated by averaging over a number of statistical realizations (Monte Carlo trials) similar to the one shown in the leftmost panel. Specifically, the time-dependent mean-square-deviation, $\text{MSD}_k(i)$, is displayed as a function of time and is expressed in dB for a better reading. Obviously, to encompass the variations of the true parameter vector, the definition of the error in Eq. (2.16) has been modified by letting the true parameter vector depend on time index i. With respect to the example of the leftmost panel, where the step-size was set to a value $\mu = 0.1$, we add another example, for the case $\mu = 0.01$. Let us first refer to the case $\mu = 0.1$. After transient phases that correspond either to an initialization phase or to time instants where the true parameter drifts, we see that the mean-square-deviation reaches a steady-state behavior. Specifically, the curves corresponding to agents 2 and 4 become stable. For comparison purposes, the limiting theoretical mean-square-deviation, MSD_k, is displayed with a dashed line. Let us see what happens with a smaller step-size value, namely, with $\mu = 0.01$. Three basic features emerge that match perfectly the theoretical results illustrated in the previous section. First, the adaptation is slower for smaller step-sizes. Second, the accuracy of the steady-state estimates increases for smaller step-sizes. Third, as μ decreases, the curves pertaining to the different agents get closer to each other and are in turn closer to the theoretical performance: as a matter of fact, the estimation performance equalizes across agents and meets the approximation (2.26), which in fact holds in the small step-size regime. Moreover, we note that for both values of the step-size, the MSD curves pertaining to agent 2 are worse (i.e., higher) than those pertaining to agent 4. This could be explained through careful examination of Fig. 2.1, where we see that agent 2 is in a sense more peripheral than agent 4, as regards its connectivity properties. The discrepancies between the two curves (and between each of them and the theoretical dashed line) are contained in the higher order corrections, namely, in the $O(\mu^2)$ term appearing in Eq. (2.25) that is neglected by the approximation in Eq. (2.26).

2.5 DISTRIBUTED ADAPTIVE DETECTION (DAD)

The earlier sections provided an overview of the results on the DAE problem by focusing largely on the mean-square-error context, although most of the conclusions hold more generally [53,54]. We now turn to the DAD problem. This problem is abstracted as follows over adaptive networks. The network agents gather *streaming* data, $\{\boldsymbol{x}_{k,i}\}$ (with the agent index $k = 1, 2, \ldots, N$, and the time index $i = 0, 1, \ldots$) about a certain phenomenon of interest, which is modeled by a binary state of nature formally represented by the hypotheses $\mathcal{H}_0$ and $\mathcal{H}_1$. The statistical properties of the data depend upon the hypothesis that is in force. Under stationary conditions, the data $\{\boldsymbol{x}_{k,i}\}$ form a suite of independent and identically distributed (i.i.d.) variables, with the distribution depending on the underlying hypothesis that is in force. At each time instant, each individual agent is tasked with producing a decision about the state of nature, by exchanging information with its neighbors through a distributed processing algorithm.

As time elapses, the adaptive algorithm should allow each agent to learn about the true hypothesis. On the other hand, adaptation is required because variations might occur, e.g., in the statistical conditions, in the environmental conditions, and in the network topology, among other possibilities. For example, the hypothesis in force can drift over time and the agents must be able to promptly change their decisions in the presence of the aforementioned drifts.

2.5.1 CONSTRUCTING THE DISTRIBUTED ADAPTIVE DETECTION ALGORITHM

We now show how the construction of the distributed detection algorithm can be derived from Eq. (2.5). As it is well known for the i.i.d. data model, an optimal centralized (and nonadaptive) detection statistic is the sum of the log-likelihood ratios, which is compared against some threshold to make a decision. Therefore, it is meaningful to identify some individual cost functions, $J_k(w)$ in Eq. (2.2), that are able to drive the network agents toward distributed computation of the log-likelihood ratio. We now show that such computation is enabled, for example, by the following choice:

$$\boxed{Q_k(w; \boldsymbol{x}_{k,i}) = \frac{1}{2}\left(w - \ln\frac{f_1(\boldsymbol{x}_{k,i})}{f_0(\boldsymbol{x}_{k,i})}\right)^2} \tag{2.37}$$

where $\boldsymbol{x}_{k,i}$ describes the data collected by agent k at time i while $f_0(\cdot)$ and $f_1(\cdot)$ are the probability density functions (for continuous variables) or probability mass functions (for discrete variables) under $\mathcal{H}_0$ and $\mathcal{H}_1$, respectively. We note that, because the log-likelihood ratio is a scalar quantity, the argument of the functions $J_k(w)$ will be scalar as well. In view of Eq. (2.37), the exact (scalar) gradient corresponding to Eq. (2.37) becomes:

$$\nabla_w J_k(w) = w - \mathbb{E}_0 \ln\frac{f_1(\boldsymbol{x}_{k,i})}{f_0(\boldsymbol{x}_{k,i})} \quad \text{under } \mathcal{H}_0, \tag{2.38}$$

$$\nabla_w J_k(w) = w - \mathbb{E}_1 \ln\frac{f_1(\boldsymbol{x}_{k,i})}{f_0(\boldsymbol{x}_{k,i})} \quad \text{under } \mathcal{H}_1, \tag{2.39}$$

where $\mathbb{E}_0[\cdot]$ and $\mathbb{E}_1[\cdot]$ denote that the expectation operator acts under $\mathcal{H}_0$ and under $\mathcal{H}_1$, respectively. Now, the expectations of the log-likelihood ratio under the different hypotheses can be written as:

$$\mathbb{E}_0 \ln\frac{f_1(\boldsymbol{x}_{k,i})}{f_0(\boldsymbol{x}_{k,i})} = -\mathcal{D}_{01} < 0, \quad \mathbb{E}_1 \ln\frac{f_1(\boldsymbol{x}_{k,i})}{f_0(\boldsymbol{x}_{k,i})} = \mathcal{D}_{10} > 0, \tag{2.40}$$

where $\mathcal{D}_{hj}$ denotes the Kullback-Leibler divergence between hypothesis h and hypothesis j, for $h, j = 0, 1$, with $h \neq j$, and *strict* positivity of the divergences amounts to assume that the detection problem is nonsingular [1]. From Eqs. (2.38), (2.39) and (2.40), we see that the minimum of the true gradient is attained at value $-\mathcal{D}_{01}$ under $\mathcal{H}_0$, and value $\mathcal{D}_{10}$ under $\mathcal{H}_1$. Therefore, under the assumption of a strongly connected network, we know that each agent is able to approach (for μ sufficiently small) the negative value $-\mathcal{D}_{01}$ under $\mathcal{H}_0$, and the positive value $\mathcal{D}_{10}$ under $\mathcal{H}_1$. We conclude that the choice in Eq. (2.37) enables the network of detectors to achieve successful discrimination between the two hypotheses.

It is not mandatory to choose the log-likelihood ratio in Eq. (2.37) to let the algorithm work properly. Such observation is particularly relevant for the cases where the statistics of the data are not perfectly known. In these cases, it is useful to employ a detection statistic that is a linear combination of suitable local statistics, as is typical in some classical detection frameworks, e.g., in locally optimum detection [9], in robust detection [10], and in universal/nonparametric hypothesis testing [11,12]. Therefore, in the following treatment we introduce a scalar random variable, $\boldsymbol{\tau}_k(i)$, which will represent the local detection statistic computed at time i by agent k. This local statistic might be chosen as the local log-likelihood ratio but this requirement is not necessary. According to the adopted model, under stationary conditions, the local statistics $\boldsymbol{\tau}_k(i)$ will be spatially and temporally independent and identically distributed. To ensure detectability, we assume that $\mathbb{E}_0\boldsymbol{\tau}_k(i) \neq \mathbb{E}_1\boldsymbol{\tau}_k(i)$), and, without loss of generality, that $\mathbb{E}_0\boldsymbol{\tau}_k(i) < \mathbb{E}_1\boldsymbol{\tau}_k(i)$.

In order to allow the usage of a general local statistic, we now replace the log-likelihood ratio in Eq. (2.37) with the general local statistic, $\boldsymbol{\tau}_k(i)$, and, hence, the diffusion implementation in Eq. (2.5) takes on the following form, for all $i \geq 0$ [59]:

$$\boxed{\begin{aligned} \boldsymbol{\psi}_k(i) &= (1-\mu)\boldsymbol{w}_k(i-1) + \mu\,\boldsymbol{\tau}_k(i), && \text{[adaptation step]} \\ \boldsymbol{w}_k(i) &= \sum_{\ell=1}^{N} a_{\ell k}\boldsymbol{\psi}_\ell(i). && \text{[combination step]} \end{aligned}} \tag{2.41}$$

Moreover, at time i, the kth agent produces a decision based upon its current state value $\boldsymbol{w}_k(i)$. To this end, a decision rule must be designed, with a common choice being [59]:

$$\boxed{\boldsymbol{w}_k(i) \underset{\mathcal{H}_1}{\overset{\mathcal{H}_0}{\lesseqgtr}} \gamma} \tag{2.42}$$

for some threshold value γ.

Before ending this section, we would like to give another useful interpretation of the above adaptive detection algorithm. It is straightforward to verify that, in the *single-agent* case, the algorithm in Eq. (2.41) reproduces the so-called exponentially weighted (or geometric) moving average (EWMA) control chart [60]. The terminology EWMA stems from the fact that, through recursive application of the $(1-\mu)$ weighting, the most recent datum has higher weight (i.e., higher degree of importance) with respect to past data, and that the weights assigned to past data undergo an exponential decay as time elapses.

2.5.2 DETECTION PERFORMANCE

The general method presented at the beginning of Section 2.4.2 for evaluating the performance of adaptive estimation can be applied also in the context of adaptive detection. The first step is to introduce a proper performance measure for the detection framework. The classic choice is the error probability. The *time-dependent* detection error probability of agent k will be accordingly defined as:

$$\boxed{p_k^{(e)}(i) = \mathbb{P}\left[\text{agent } k \text{ chooses the } \textit{wrong} \text{ hypothesis at time } i\right].} \tag{2.43}$$

These error probabilities depend upon the statistical properties of the *entire* dataset used in the diffusion algorithm up to time i. In particular, the error probabilities depend upon the different variations that may have occurred during the evolution of the algorithm.

We shall focus on examining performance in the *steady state*. This means that the data are possibly nonstationary up to a certain time instant (which we conveniently set to $i = -1$), after which (i.e., for all $i \geq 0$) they are assumed to be drawn from the same stationary distribution for an infinitely long time. Since, under a stationary evolution, the hypothesis in force remains stable over time, the error probability in Eq. (2.43) can be split into two kinds of error probabilities, namely, the Type-I (false-alarm) and Type-II (miss-detection) error probabilities, defined, respectively, as:

$$\boxed{\begin{aligned} \alpha_k(i) &\triangleq \mathbb{P}_0\,[\text{agent } k \text{ decides } \mathcal{H}_1 \text{ at time } i], \quad [\text{Type-I}] \\ \beta_k(i) &\triangleq \mathbb{P}_1\,[\text{agent } k \text{ decides } \mathcal{H}_0 \text{ at time } i], \quad [\text{Type-II}] \end{aligned}} \tag{2.44}$$

where $\mathbb{P}_0[\cdot]$ and $\mathbb{P}_1[\cdot]$ denote that the probability operator acts under $\mathcal{H}_0$ and under $\mathcal{H}_1$, respectively. The limiting values of the aforementioned quantities give the steady-state error probabilities, namely,

$$\boxed{\alpha_k = \lim_{i\to\infty} \alpha_k(i), \qquad \beta_k = \lim_{i\to\infty} \beta_k(i).} \tag{2.45}$$

As a preliminary step toward the evaluation of Eq. (2.45), we show that the output of the diffusion algorithm, $\boldsymbol{w}_k(i)$, admits a steady-state distribution. First, we observe that, by simple recursion, Eq. (2.41) can be written in the form, for all $i \geq 0$:

$$\boldsymbol{w}_k(i) = \underbrace{(1-\mu)^{i+1} \sum_{\ell=1}^{N} b_{\ell k}(i+1)\boldsymbol{w}_\ell(-1)}_{\text{transient term}} + \underbrace{\mu \sum_{m=0}^{i} \sum_{\ell=1}^{N} (1-\mu)^m b_{\ell k}(m+1)\boldsymbol{\tau}_\ell(i-m)}_{\text{steady-state term}}, \tag{2.46}$$

where $b_{\ell k}(i)$ is the (ℓ, k)th entry of the matrix power $B_i \triangleq A^i$. The formulation in Eq. (2.46) emphasizes the presence of the transient term as opposed to the steady-state term. Disappearance of the transient term as $i \to \infty$, for any set of the initial agents' states, $\{\boldsymbol{w}_\ell(-1)\}$, is easy to see. Moreover, in Theorem 1 of [59], convergence of the diffusion output to a steady-state random variable (denoted by $\boldsymbol{w}_k^\star$) has been established. This property can be summarized by the following statement (the symbol $\rightsquigarrow$ means convergence in distribution, under the hypothesis that is in force):

$$\boxed{\boldsymbol{w}_k(i) \overset{i\to\infty}{\rightsquigarrow} \boldsymbol{w}_k^\star.} \tag{2.47}$$

In view of Eq. (2.47), the steady-state detection performance can now be expressed in terms of the steady-state random variable as follows[2]:

$$\alpha_k = \mathbb{P}_0[\boldsymbol{w}_k^\star > \gamma], \qquad \beta_k = \mathbb{P}_1[\boldsymbol{w}_k^\star \leq \gamma]. \tag{2.48}$$

The statistical characterization of $\boldsymbol{w}_k^\star$ is usually a formidable task. In order to gain useful insights, as shown for the DAE problem, we resort to an evaluation of the steady-state performance in the *small step-size* regime.

[2] According to the definition of weak convergence [1], the result in Eq. (2.47) implies Eq. (2.48) provided that $\mathbb{P}_0[\boldsymbol{w}_k^\star = \gamma] = \mathbb{P}_1[\boldsymbol{w}_k^\star = \gamma] = 0$, i.e., provided that the limiting random variable has no probability mass concentrated at γ. This is a mild condition that holds under most typical models; see the comments reported after Eq. (20) in [59] for more details.

2.5.3 WEAK LAW OF SMALL STEP-SIZES

It is useful to start with some results concerning the first two central moments of $\boldsymbol{w}_k^\star$, which have been established in [59]. Let us preliminarily make a remark about notation. When we deal with results holding for a generic distribution, it will not be necessary to specify the particular hypothesis that is in force. Accordingly, we will formulate our results without specifying the hypothesis. In contrast, when we need to distinguish the Type-I and Type-II error probabilities, a subscript $h \in \{0, 1\}$ will be appended to denote that the pertinent quantities are computed under the distribution corresponding to the particular hypothesis h.

The mean of $\boldsymbol{w}_k^\star$ equals, for all $k = 1, 2, \dots, N$, the mean of the local detection statistic, namely [59]:

$$\boxed{\mathbb{E}\boldsymbol{w}_k^\star = \mathbb{E}\boldsymbol{\tau}_k(i) \triangleq \mathbb{E}\boldsymbol{\tau}} \tag{2.49}$$

where, in the latter equality, the dependence on k and i is suppressed due to the i.i.d. assumption. The variance of the steady-state random variable is given by [59][3]:

$$\boxed{\sigma_k^2 \triangleq \mathrm{VAR}[\boldsymbol{w}_k^\star] = \sigma_\tau^2 \sum_{m=0}^{\infty} \sum_{\ell=1}^{N} \mu^2 (1-\mu)^{2m}\, b_{\ell k}^2(m+1).} \tag{2.50}$$

In contrast to the behavior of the mean, the variance of $\boldsymbol{w}_k^\star$ *does depend on the agent index k*. In addition, it depends on the variance of the local detection statistic, σ_τ^2, on the combination weights (through $\{b_{\ell k}(m)\}$), and on the step-size. It can be shown that the variances of the individual agents in Eq. (2.50) obey the following law [61]:

$$\boxed{\sigma_k^2 = \mu \frac{\sigma_\tau^2}{2} \sum_{\ell=1}^{N} p_\ell^2 + O(\mu^2).} \tag{2.51}$$

The knowledge of the first two moments of $\boldsymbol{w}_k^\star$ can be exploited to obtain an interesting limit law. Applying Chebyshev's inequality, we have, for any positive ϵ:

$$\boxed{\mathbb{P}[|\boldsymbol{w}_k^\star - \mathbb{E}\boldsymbol{\tau}| > \epsilon] \le \frac{\sigma_k^2}{\epsilon^2} \xrightarrow{\mu \to 0} 0} \tag{2.52}$$

which reveals that, for small step-sizes, *the steady-state output of the diffusion algorithm converges in probability, as $\mu \to 0$, to the expected value of the local statistic*, $\mathbb{E}\boldsymbol{\tau}$. Therefore, the limit law in Eq. (2.52) can be referred to as a *weak law of small step-sizes*.

The evident analogy with the weak law of large numbers can be further strengthened in light of the following observation. We see from the algorithm evolution in Eq. (2.46) that, in order to ensure *adaptation*, the algorithm uses an exponentially weighted moving window that gives progressively less credit (i.e., progressively smaller weight) to *old* data, following a geometric law $(1-\mu)^m$. Accordingly,

---

[3] In order to avoid confusion, we stress that, in the present treatment, we use a slightly different notation relative to the other works [59,61]. Such a choice is made to facilitate the illustration of the main ideas as well as to keep notation as uniform as possible across the estimation and detection settings.

given the current time instant, we can neglect all data having a weight lower than a certain value $0 < C < 1$. In this way, we approximate the *effective* number of data used by the algorithm as the current datum, plus all the past data having a weight greater or equal to C, which yields, for sufficiently small values of μ:

$$(1-\mu)^{N_{\rm eff}+1} = C \Rightarrow N_{\rm eff} \approx \frac{\ln C}{\ln(1-\mu)} \approx \frac{|\ln C|}{\mu}. \tag{2.53}$$

Accordingly, the quantity $1/\mu$ can be taken as a qualitative index of the effective amount of information used by the algorithm, and we now see why *the condition of small step-sizes takes on the meaning of a large number of samples.*

We now show a direct implication of the weak law of small step-sizes that is relevant for the distributed detection application. Assume that we set a detection threshold $\mathbb{E}_0\boldsymbol{\tau} < \gamma < \mathbb{E}_1\boldsymbol{\tau}$. Then we can write:

$$\alpha_k = \mathbb{P}_0[\boldsymbol{w}_k^\star - \mathbb{E}_0\boldsymbol{\tau} > \gamma - \mathbb{E}_0\boldsymbol{\tau}] \le \mathbb{P}_0[|\boldsymbol{w}_k^\star - \mathbb{E}_0\boldsymbol{\tau}| > \gamma - \mathbb{E}_0\boldsymbol{\tau}], \tag{2.54}$$

$$\beta_k = \mathbb{P}_1[\boldsymbol{w}_k^\star - \mathbb{E}_1\boldsymbol{\tau} \le \gamma - \mathbb{E}_1\boldsymbol{\tau}] \le \mathbb{P}_1[|\boldsymbol{w}_k^\star - \mathbb{E}_1\boldsymbol{\tau}| > \mathbb{E}_1\boldsymbol{\tau} - \gamma]. \tag{2.55}$$

In view of the choice $\mathbb{E}_0\boldsymbol{\tau} < \gamma < \mathbb{E}_1\boldsymbol{\tau}$, it is now legitimate to apply the weak law of small step-sizes in Eq. (2.52) to both Eqs. (2.54) and (2.55), with the choices $\epsilon = \gamma - \mathbb{E}_0\boldsymbol{\tau}$ and $\epsilon = \mathbb{E}_1\boldsymbol{\tau} - \gamma$, respectively, obtaining:

$$\boxed{\alpha_k \xrightarrow{\mu\to 0} 0, \qquad \beta_k \xrightarrow{\mu\to 0} 0.} \tag{2.56}$$

The results in Eq. (2.56) show that, by proper choice of the detection threshold, the Type-I and Type-II error probabilities can be made arbitrarily small as μ decreases. This is a remarkable conclusion as regards the performance of distributed detection over adaptive networks. It is now desirable to give a shape to the aforementioned vanishing behavior, i.e., to understand *how* α_k and β_k vanish as μ decreases.

2.5.4 ASYMPTOTIC NORMALITY

The weak law of small step-sizes can be refined to ascertain the following central limit theorem (CLT), which claims that, for μ small enough, the steady-state diffusion output is distributed as a Gaussian random variable. Formally, we have that [59,61]:

$$\boxed{\frac{\boldsymbol{w}_k^\star - \mathbb{E}\boldsymbol{\tau}}{\sigma_k} \overset{\mu\to 0}{\rightsquigarrow} \mathcal{N}(0,1)} \tag{2.57}$$

where $\mathcal{N}(0,1)$ denotes a standard normal distribution. There are several important implications of the asymptotic normality result as regards the DAD problem.

CLT for detection performance. First of all, Eq. (2.57) provides a simple closed-form solution that can be used to approximate the test performance of each individual agent, yielding:

$$\boxed{\alpha_k \approx \mathbb{Q}\left(\frac{\gamma - \mathbb{E}_0\boldsymbol{\tau}}{\sigma_{k,0}}\right), \qquad \beta_k \approx 1 - \mathbb{Q}\left(\frac{\gamma - \mathbb{E}_1\boldsymbol{\tau}}{\sigma_{k,1}}\right)} \tag{2.58}$$

where $\mathbb{Q}(\cdot)$ is the complementary cumulative distribution function of the standard normal distribution.

Asymptotic equivalence among agents? For small step-sizes, we can use Eq. (2.51) to replace the *actual* variances in Eq. (2.58) by their leading-order approximation, $\mu\,(\sigma_\tau^2/2)\sum_{\ell=1}^{N} p_\ell^2$, which turns out to be *identical across agents because it does not depend on the agent index k*. This additional simplification suggests that the diffusion process equalizes in some sense the detection performance across agents. On the other hand, neglecting the $O(\mu^2)$ correction appearing in Eq. (2.51) might ignore some useful information about the distinguishing attributes among the individual agents. Which perspective is correct? We shall answer this question in Section 2.6, where we shall see that both perspectives are correct: equivalence across agents holds in a precise sense, and the (possibly) different connectivity properties of the individual agents reflect in their detection performance.

CLT and small *deviations.* By applying the definition of convergence in distribution, the result in Eq. (2.57) can be written in a more explicit form as follows [1]:

$$\mathbb{P}[\boldsymbol{w}_k^\star > \mathbb{E}\boldsymbol{\tau} + \sigma_k w] \overset{\mu\to 0}{\longrightarrow} \mathbb{Q}(w) \tag{2.59}$$

for any $w \in \mathbb{R}$. We recall that $\mathbb{E}\boldsymbol{\tau}$ is the mean of $\boldsymbol{w}_k^\star$, and that, in view of Eq. (2.51), the standard deviation σ_k vanishes as $\sqrt{\mu}$. Accordingly, the term $\mathbb{E}\boldsymbol{\tau} + \sigma_k w$ in Eq. (2.59) represents a (μ-dependent) threshold that lies, for small values of μ, in a small neighborhood of the mean. As a result, Eq. (2.59) reveals that the normal approximation is asymptotically exact when: (i) the steady-state variable, $\boldsymbol{w}_k^\star$, deviates "slightly" from its mean; and (ii) the probability converges to some value $\mathbb{Q}(w)$ (i.e., it does not vanish) as $\mu \to 0$. Similar regimes, which arise in various forms in statistics, are usually referred to as *normal deviations* or *small deviations* regimes. In contrast, when one focuses on evaluating the probability that the steady-state variable, $\boldsymbol{w}_k^\star$, deviates from its mean by a constant (rather than vanishing) amount, it is commonplace to talk of a *large deviations* regime.

The large deviations framework is particularly appropriate in the detection framework, for at least two reasons. First, setting a constant threshold γ is appealing because both error probabilities will vanish as the step-size goes to zero; see Eq. (2.56). Second, even allowing a μ-dependent threshold, it is not possible to enforce small deviations around the means under both hypotheses because $\mathbb{E}_0\boldsymbol{\tau} \neq \mathbb{E}_1\boldsymbol{\tau}$ (for example, if the threshold collapses to $\mathbb{E}_0\boldsymbol{\tau}$, we guarantee a small deviation under $\mathcal{H}_0$, but we get a *large* deviation under $\mathcal{H}_1$).

2.5.5 LARGE DEVIATIONS

In order to overcome the limitations of normal approximations, a large deviations analysis has been performed in [59], and the following conclusion has been established:

$$\boxed{\alpha_k \doteq e^{-(1/\mu)\,\Phi_0} = e^{-(1/\mu)\,[\Phi_0 + o(1)]}, \qquad \beta_k \doteq e^{-(1/\mu)\,\Phi_1} = e^{-(1/\mu)\,[\Phi_1 + o(1)]}} \tag{2.60}$$

where $o(1)$ denotes a quantity that vanishes as μ goes to zero, and, accordingly, the notation $\doteq$ means equality to the leading exponential order as μ goes to zero [62]. The quantities Φ_0 and Φ_1, which will be referred to as the *error exponents*, are *independent* of the step-size μ. A preliminary examination of result (2.60) shows several insightful ramifications about the performance of distributed detection over adaptive networks.

Fundamental scaling law. Eq. (2.60) reveals the fundamental scaling law for distributed detection with diffusion adaptation: as μ decreases, the error probabilities go to zero exponentially as functions of $1/\mu$.

Adaptation/learning trade-off. Recalling that smaller values of μ entail less adaptation, we observe from Eq. (2.60) that a better detection quality costs in terms of adaptation speed. This result matches what we have found in the *estimation* setting.

Benefits of cooperation. As done for the estimation setting, two useful comparisons with known detectors can be easily done for the case of *doubly-stochastic* combination policies. For this case, the error exponents admit a special form, namely, $\Phi_0 = N\Omega_0$ and $\Phi_1 = N\Omega_1$, where the quantities Ω_0 and Ω_1 are the error exponents *per single agent*, which are independent of the number of agents and of the network connectivity [59]. They depend solely upon the detection threshold and upon the distribution of the local statistic, $\boldsymbol{\tau}_k(i)$. As a result, we see that the overall error exponents, Φ_0 and Φ_1, increase linearly in the number of agents. This implies that cooperation offers *exponential gains* in terms of detection performance.

Centralized (adaptive) detector. Eq. (2.60) applies to the centralized case as well because the centralized adaptive detector can be regarded as a distributed detector with a *fully connected* network. Therefore, it can be concluded from Eq. (2.60) that in the small-μ regime the distributed diffusion solution exhibits a detection performance governed by the *same error exponents* of the centralized system.

Example 2.9 (Comparison with decaying step-size solutions). It is useful to contrast the results about DAD with those pertaining to distributed detection algorithms with decaying step-size [49–51]. The result in Eq. (2.47) reveals that, under stationary conditions, the agents' detection statistics (i.e., the diffusion outputs $\boldsymbol{w}_k(i)$) converge in distribution to a certain *random* variable. In contrast, in the decaying step-size case, the detection statistic will collapse, as time elapses, into a *deterministic* value (e.g., the Kullback-Leibler divergence). Such convergence to a deterministic value reflects the continuously improving performance as time elapses with decaying step-sizes. In particular, under stationary conditions, the error probabilities for decaying step-size algorithms decay exponentially *as functions of the time index i*; see, e.g. [49–51]. Accordingly, such probabilities might reach "astronomically" small values as time elapses. In our *adaptive* setting (with constant step-size), the situation is different because the error probabilities stabilize as time elapses. The possibility of reducing the error probabilities is now related to decreasing the step-size. In fact, the result in Eq. (2.60) reveals that the error probabilities decay exponentially *as functions of the (inverse of the) step-size* μ, and *not* as functions of the time index *i*. In accordance with the exponential decay w.r.t. $1/\mu$, small variations of the step-size can lead to substantial variations in the error probabilities. ■

We now illustrate how the error exponents can be computed from the statistical properties of the local statistic $\boldsymbol{\tau}_k(i)$. In particular, we shall see that the fundamental statistical descriptor enabling a large-deviations analysis is the logarithmic moment generating function (LMGF) of $\boldsymbol{\tau}_k(i)$, namely,

$$\psi(t) \triangleq \ln \mathbb{E} e^{t\boldsymbol{\tau}_k(i)}, \qquad \text{assuming } \psi(t) < +\infty \ \forall t \in \mathbb{R}. \tag{2.61}$$

As done before, we are not specifying the particular hypothesis that is in force because it suffices to assume that the data come from the same distribution. Then, when we need to distinguish the Type-I and Type-II error probabilities and error exponents, a subscript $h \in \{0, 1\}$ will be appended to denote that the pertinent quantities are computed under the distribution corresponding to the particular hypothesis h.

In [59] it has been proved that computation of the error exponents involves two steps. First, one introduces the following function, which depends only on the local LMGF, $\psi(t)$, and on the *limiting* combination weights, $\{p_\ell\}$:

$$\boxed{\phi(t) = \sum_{\ell=1}^{N} \int_0^{p_\ell t} \frac{\psi(\zeta)}{\zeta} d\zeta.} \tag{2.62}$$

Second, one computes the Fenchel-Legendre transform of $\phi(t)$ [12,13,59,61]:

$$\boxed{\Phi \triangleq \sup_{t\in\mathbb{R}}[\gamma t - \phi(t)] = \gamma\theta - \phi(\theta), \quad \text{where } \theta : \phi'(\theta) = \gamma.} \tag{2.63}$$

Example 2.10 (A Gaussian example). Assume that the local detection statistics, $\boldsymbol{\tau}_k(i)$, are Gaussian random variables, with mean $\mathbb{E}\boldsymbol{\tau}$ and variance σ_τ^2. Because the diffusion output is a linear combination of Gaussian variables, from Eqs. (2.49) and (2.50) we conclude that $\boldsymbol{w}_k^\star$ is Gaussian as well with mean $\mathbb{E}\boldsymbol{\tau}$ and variance σ_k^2. Therefore, analytical expressions for the exact probabilities are available.

Preliminarily, it is convenient to use the following representation for the variance σ_k^2 in Eq. (2.51):

$$\sigma_k^2 = \mu\sigma_{\lim}^2[1 - z_k(\mu)], \quad \text{where } \sigma_{\lim}^2 \triangleq \frac{\sigma_\tau^2}{2}\sum_{\ell=1}^{N} p_\ell^2, \; z_k(\mu) = 1 - \frac{\sigma_k^2}{\mu\sigma_{\lim}^2} \xrightarrow{\mu\to 0} 0. \tag{2.64}$$

Consider now a threshold $\gamma > \mathbb{E}\boldsymbol{\tau}$. We can write:

$$\mathbb{P}[\boldsymbol{w}_k^\star > \gamma] = \mathbb{Q}\left(\frac{\gamma - \mathbb{E}\boldsymbol{\tau}}{\sigma_k}\right) \approx \frac{1}{\sqrt{2\pi \dfrac{(\gamma - \mathbb{E}\boldsymbol{\tau})^2}{\mu\sigma_{\lim}^2[1 - z_k(\mu)]}}} \exp\left\{-\frac{(\gamma - \mathbb{E}\boldsymbol{\tau})^2}{2\mu\sigma_{\lim}^2[1 - z_k(\mu)]}\right\}, \tag{2.65}$$

where we have used the well-known approximation $\mathbb{Q}(x) \approx \frac{1}{\sqrt{2\pi x^2}} e^{-\frac{x^2}{2}}$, which holds for $x > 0$.

Let us now see where a large deviations analysis will lead. Noticing that the LMGF of a Gaussian random variable of mean a and variance b is equal to $e^{at+bt^2/2}$, and using Eqs. (2.62) and (2.63), some straightforward calculations give:

$$\boxed{\Phi = \frac{(\gamma - \mathbb{E}\boldsymbol{\tau})^2}{2\sigma_{\lim}^2} \Leftrightarrow \mathbb{P}[\boldsymbol{w}_k^\star > \gamma] \doteq \exp\left\{-\frac{(\gamma - \mathbb{E}\boldsymbol{\tau})^2}{2\mu\sigma_{\lim}^2}\right\}} \tag{2.66}$$

where, we recall, the symbol $\doteq$ denotes equality to the leading order in the exponent as $\mu \to 0$. Therefore, comparing Eq. (2.66) with Eq. (2.65), we see that the large deviations analysis corresponds to neglecting the square-root factor as well as the corrections $z_k(\mu)$, appearing in Eq. (2.65), which implies the following two conclusions: (i) the large deviations analysis matches the leading exponential order of the true probability in Eq. (2.65); and (ii) the dependencies of the true probability on the agent index, k, are skipped by the large deviations analysis. Are these dependencies important? This question is examined in the next section. ■

2.5.6 REFINED LARGE DEVIATIONS ANALYSIS: EXACT ASYMPTOTICS

A large deviations analysis presents a well-known limitation: it inherently neglects subexponential terms. For example, assume that network agents 1 and 2 exhibit the following (asymptotic) Type-I error probabilities:

$$\alpha_1 = e^{-(1/\mu)\,\Phi_0},$$
$$\alpha_2 = 10\,e^{-(1/\mu)\,\Phi_0} = e^{-(1/\mu)\,\Phi_0 + \ln 10} = e^{-(1/\mu)\,[\Phi_0 + o(1)]}. \tag{2.67}$$

These two probabilities have the same exponent, Φ_0 (they are equal *at the leading order in the exponent*), and, hence, they are considered equivalent from a large deviations perspective. Nonetheless, we shall always have that $\alpha_2 = 10\,\alpha_1$, namely, that agent 2 has a Type-I error probability that is 10 times larger than that of agent 1. This is because the factor 10 acts as a *subexponential* term. On the other hand, there is no doubt that an accurate prediction of the detection performance would require taking into account such subexponential corrections. A refined analysis should therefore look for two functions, $\mathscr{A}_k$ and $\mathscr{B}_k$, which are able to approximate the Type-I and Type-II error probabilities, α_k and β_k, in the following stronger sense:

$$\boxed{\frac{\alpha_k}{\mathscr{A}_k} \xrightarrow{\mu\to 0} 1, \qquad \frac{\beta_k}{\mathscr{B}_k} \xrightarrow{\mu\to 0} 1.} \tag{2.68}$$

Such types of approximations are known in a framework that is commonly referred to as *exact asymptotics* [12,63].

The exact asymptotics for detection over diffusion networks have been obtained in [61], where the following result has been established[4]:

$$\boxed{\begin{aligned} \mathscr{A}_k &= \sqrt{\frac{\mu}{2\pi\,\theta_0^2\,\phi_0''(\theta_0)}}\,\exp\left\{-\frac{1}{\mu}\,[\Phi_0 + c_{k,0}(\mu)]\right\} \\ \mathscr{B}_k &= \sqrt{\frac{\mu}{2\pi\,\theta_1^2\,\phi_1''(\theta_1)}}\,\exp\left\{-\frac{1}{\mu}\,[\Phi_1 + c_{k,1}(\mu)]\right\} \end{aligned}} \tag{2.69}$$

with the functions $c_{k,0}(\mu)$ and $c_{k,1}(\mu)$ converging to zero as $\mu \to 0$, and being in particular of order $O(\mu)$. The detailed procedure to compute such functions is illustrated in Appendix A.1.

Despite an apparent complexity, the formulas in Eq. (2.69) exhibit a remarkable and revealing structure that allows capturing important connections with the physical behavior of the adaptive distributed detectors. Let us elucidate some of these features. We start by the leading order in the exponent. Because $c_{k,0}(\mu)$ and $c_{k,1}(\mu)$ vanish, it is immediately seen from Eq. (2.69) that:

[4] More accurately, the result holds provided that the local statistic, $\tau_k(i)$, is not lattice [64]; see Theorem 3 in [61] for additional details.

$$\mu \ln \mathscr{A}_k \xrightarrow{\mu \to 0} -\Phi_0, \quad \mu \ln \mathscr{B}_k \xrightarrow{\mu \to 0} -\Phi_1, \tag{2.70}$$

which matches perfectly the leading-order behavior prescribed by Eq. (2.60).

More importantly, we see from Eq. (2.69) that the functions $c_{k,0}(\mu)$ and $c_{k,1}(\mu)$, along with the square-root terms, give a shape to the overall *subexponential* corrections, namely, to the $o(1)$ terms appearing in Eq. (2.60). The square-root correction is a typical subexponential refinement arising in the framework of exact asymptotics, and is a consequence of a local CLT (see [63]). Because this correction is related to the network topology only through the Perron eigenvector, it is *independent* of the agent index k, and is therefore applicable to all agents. In contrast, the corrections $c_{k,0}(\mu)$ and $c_{k,1}(\mu)$ do depend on the agent index k and take into account the entire network topology and combination weights.

As a result, we conclude that Eq. (2.69) furnishes a detailed and revealing assessment of the *universal behavior of distributed detection over adaptive networks*: as functions of $1/\mu$, the error (log-) probability curves corresponding to different agents not only stay nearly parallel to each other, but they are also ordered following a criterion dictated by the corrections $c_{k,0}(\mu)$ and $c_{k,1}(\mu)$. As we shall show later (see rightmost panel in Fig. 2.4 further ahead), this criterion reflects the degree of connectivity of each agent. Depending on the combination weights, the more connected an agent is, the lower its error probability will be, and the corrections $c_{k,0}(\mu)$ and $c_{k,1}(\mu)$ are sufficiently rich to capture this behavior.

Example 2.11 (The Gaussian example (revisited)). Let us apply the exact asymptotics framework to Example 2.10. We seek an approximation in the following form:

$$\mathbb{P}[\boldsymbol{w}_k^\star > \gamma] \approx \sqrt{\frac{\mu}{2\pi\,\theta^2\,\phi''(\theta)}}\,\exp\left\{-\frac{1}{\mu}\,[\Phi + c_k(\mu)]\right\}. \tag{2.71}$$

By applying the procedure detailed in Appendix A.1, we end up with the following result:

$$\theta = \frac{(\gamma - \mathbb{E}\boldsymbol{\tau})^2}{\sigma_{\text{lim}}^2}, \quad \phi''(\theta) = \sigma_{\text{lim}}^2, \quad c_k(\mu) = \frac{(\gamma - \mathbb{E}\boldsymbol{\tau})^2}{2\sigma_{\text{lim}}^2}[z_k(\mu) + z_k^2(\mu)], \tag{2.72}$$

where $z_k(\mu)$ has been defined in Eq. (2.64). Using now the error exponent Φ already computed in Example 2.10, and applying Eq. (2.71), the approximation in Eq. (2.66) can be refined to:

$$\boxed{\mathbb{P}[\boldsymbol{w}_k^\star > \gamma] \approx \frac{1}{\sqrt{2\pi\frac{(\gamma - \mathbb{E}\boldsymbol{\tau})^2}{\mu\sigma_{\text{lim}}^2}}}\exp\left\{-\frac{(\gamma - \mathbb{E}\boldsymbol{\tau})^2}{2\mu\sigma_{\text{lim}}^2}[1 + z_k(\mu) + z_k^2(\mu)]\right\}.} \tag{2.73}$$

Let us now compare this formula with the true probability in Eq. (2.65). We see immediately that: (i) the ratio between the square-root factors tends to one; and (ii) the arguments of the exponential functions match in view of the well-known Taylor approximation $(1 - z)^{-1} \approx 1 + z + z^2$ for $z \ll 1$. We observe how the exact asymptotics analysis is able to recover the dependencies on the individual agents (index k) that were ignored by the large deviations analysis. ■

Example 2.11 helps highlight the following important aspect that concerns the normal approximation in Eq. (2.58) and the refined large deviations in Eq. (2.69). In Example 2.11, all the statistical properties of the steady-state diffusion output are determined completely by its first two moments, $\mathbb{E}\boldsymbol{\tau}$ and σ_k^2, because $\boldsymbol{w}_k^\star$ is normally distributed. Then, we have two obvious observations. First, there is no need to refer to a normal "approximation", because the steady-state random variable *is* actually Gaussian, and, hence, Eqs. (2.58) will be exact in this particular case. Second, the exact asymptotics in Eq. (2.73) cannot but depend only on the first two moments of $\boldsymbol{w}_k^\star$ (again, because $\boldsymbol{w}_k^\star$ is Gaussian).

However, the situation changes under *general statistical models*, when the error probabilities will depend on higher moments or on the moment generating function. In these more challenging situations, the normal approximation in Eq. (2.58) will not take into account these higher order dependencies, because normal approximations use only the first two moments. In contrast, the exact asymptotics in Eq. (2.69) will be able to exploit higher order dependencies because they use information contained in the moment generating function.

2.5.7 DAD AT WORK

One illustrative example of detection over adaptive networks is offered in Fig. 2.3, leftmost panel. The network topology is the same adopted for the estimation example, and is displayed in Fig. 2.1. The considered detection problem is a non-Gaussian detection problem, where the (scalar) observations, $\boldsymbol{x}_k(i)$, follow a shifted Laplace distribution with unit-scale parameter, namely:

$$\boxed{\begin{aligned}\mathcal{H}_0 &: \boldsymbol{x}_k(i) \text{ has probability density function } f_0(x) = e^{-\frac{1}{2}|x|}\\ \mathcal{H}_1 &: \boldsymbol{x}_k(i) \text{ has probability density function } f_1(x) = e^{-\frac{1}{2}|x-\rho|}\end{aligned}} \tag{2.74}$$

for some $\rho > 0$. The local statistic adopted by the agents is the log-likelihood ratio of the observations, namely,

$$\boldsymbol{\tau}_k(i) = \ln \frac{f_1(\boldsymbol{x}_k(i))}{f_0(\boldsymbol{x}_k(i))} = |\boldsymbol{x}_k(i)| - |\boldsymbol{x}_k(i) - \rho|. \tag{2.75}$$

The plot shows the time evolution of the detection statistics, $\boldsymbol{w}_k(i)$, for two agents, namely, for $k = 2$ and $k = 4$. We see that the hypothesis initially in force is $\mathcal{H}_0$. Accordingly, after an initial transient, the online detection statistics (oscillating curves) fluctuate around some negative value (which is the divergence $\mathcal{D}_{01}$). Then, at one-half of the observation window, the hypothesis in force becomes $\mathcal{H}_1$, and, after a short transient, the online detection statistics react promptly to such variation and *are able to adapt* their behavior to the new hypothesis, namely, they oscillate around some positive value (which is the divergence $\mathcal{D}_{10}$). The detection threshold is set to $\gamma = 0$. Accordingly, each agent will declare $\mathcal{H}_1$ when such a threshold is exceeded (we see that some wrong decisions are occasionally made during the algorithm evolution).

With reference to the same example, in the rightmost panel of Fig. 2.3 we present a quantitative illustration of the detection performance of the adaptive network, which is obtained by estimating the time-dependent error probability in Eq. (2.43) by the Monte Carlo simulation. Such a "running" error probability is displayed as a function of time, and is expressed in a logarithmic scale for better reading. With respect to the example of the leftmost panel, where the step-size was set to a value $\mu = 0.1$, we add another example, for the case $\mu = 0.05$. After transient phases that correspond either to an initialization phase or to time instants where the true hypothesis drifts, we see that the error probability reaches a steady-state behavior. Specifically, the curves corresponding to agents 2 and 4 become stable. It is interesting to examine the features of the steady-state behavior of the error probability in this detection example, in comparison to what happens for the mean-square-deviation in the estimation

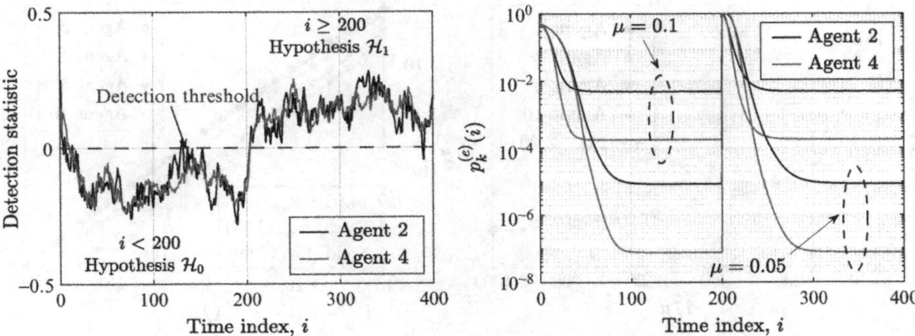

FIG. 2.3

Illustration of the distributed adaptive detection problem. The detection problem in Eq. (2.74), with the observations following a shifted Laplace distribution with unit-scale parameter and with $\rho = 0.6$; see Eq. (2.74). Leftmost panel: One realization of the diffusion algorithm (2.41), namely, of the online detection statistics corresponding to agents 2 and 4. The network topology is shown in Fig. 2.1, and the combination rule is the Metropolis rule in Eq. (2.9). The step-size is set to $\mu = 0.1$. The underlying hypothesis changes when $i = 200$, and the *adaptive* algorithm is able to track such variation promptly. Rightmost panel: Empirical (Monte Carlo) evaluation of the *time-dependent* error probability, $p_k^{(e)}(i)$, appearing in Eq. (2.43), for two values of the step-size.

example; see the rightmost panel of Fig. 2.2. Two important similarities emerge because, even in the detection case, for smaller step-sizes the adaptation becomes slower and the quality of inference increases (i.e., the detection error probability decreases). However, there are also remarkable differences between estimation and detection, as regards the performance of distinct agents. In the estimation case we have observed that the performance levels of different agents tend to equalize as $\mu \to 0$. In contrast, we see from the rightmost panel of Fig. 2.3 that, in the detection case, the curves pertaining to the different agents do not collapse into one and the same curve as μ decreases. In particular, also in the detection case the performance of agent 2 is worse (i.e., higher error probability) than agent 4. However, differently from what happens in the estimation case, the discrepancies do not disappear as μ goes to zero. Such a behavior is perfectly consistent with the theoretical analysis. Indeed, we have already explained that the subexponential correction terms, $c_{k,0}(\mu)$ and $c_{k,1}(\mu)$, appearing in Eq. (2.69), embody the differences existing among distinct agents as regards the impact of their connectivity features on the error probabilities. Moreover, we notice that, as compared with the estimation case (where we reduced the step-size by one order of magnitude, passing from $\mu = 0.1$ to $\mu = 0.01$), a smaller variation of the step-size (reduction by one half, passing from $\mu = 0.1$ to $\mu = 0.05$) implies a substantial variation of the error probability, complying with the *exponential* decay of the latter with the inverse of the step-size, $1/\mu$.

Finally, in the rightmost panel of Fig. 2.4 we check the validity of the two theoretical approximations presented in this chapter, namely, the normal approximation in Eq. (2.58), and the exact asymptotics in Eq. (2.69). We note in passing that, for the chosen distributions, detection statistics, and threshold, the Type-I and Type-II error probabilities can be shown to be equal, namely, we have $\alpha_k = \beta_k$. The error probabilities of four distinct agents ($k = 2, 4, 5, 10$) are displayed as functions of the inverse of the step-size, $1/\mu$. The logarithmic scale adopted on the vertical axis emphasizes the exponential decay

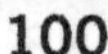

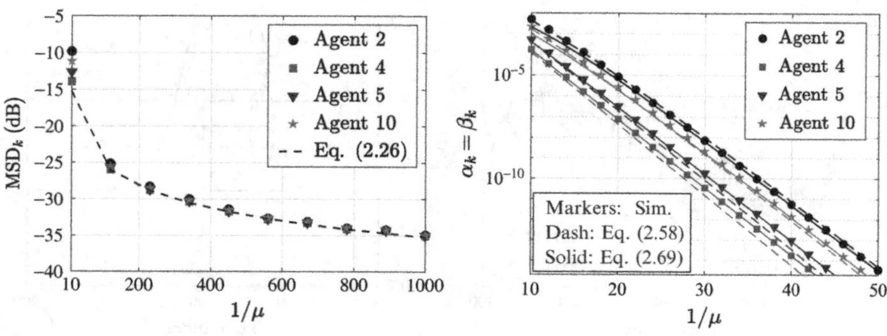

FIG. 2.4

DAE vs. DAD. Comparison between the steady-state performance of DAE (leftmost panel) and DAD (rightmost panel) as a function of the (inverse of the) step-size μ. The leftmost panel shows the steady-state mean-square-deviations, MSD_k, for $k = 2, 4, 5, 10$, along with the theoretical prediction (*dashed curve*) provided by Eq. (2.26). The rightmost panel shows the empirical steady-state error probabilities ($\alpha_k = \beta_k$ in this example), evaluated through Monte Carlo importance sampling, along with: (i) the theoretical normal approximation in Eq. (2.58), *dashed curves*; and (ii) the exact asymptotics in Eq. (2.69), *solid curves*. The setting for the estimation and the detection problem are the same as those considered in Fig. 2.2 and in Fig. 2.3, respectively.

of the error probabilities predicted by the large deviations analysis. First, we observe that the normal approximation turns out to be accurate, especially for relatively small values of $1/\mu$. This implies that the variances in Eq. (2.50) contain useful information about the different detection performance at different agents. However, as indicated by the theory, the predictions obtained with the normal approximation cannot be assumed true as the step-size decreases. On the other hand, the empirical probability converges toward the exact asymptotics (and not toward the normal approximation) as the step-size decreases. The dependency between the network structure and the detection performance at different agents is correctly embodied in the exact asymptotics, as witnessed by the correct ordering of the curves. In contrast to what happens for the normal approximation, the predictions offered by the exact asymptotic for less connected agents seem less accurate in the leftmost part of the plot. In a sense, the two theoretical approximations complement each other.

Finally, the rightmost panel of Fig. 2.4 shows that very low values of the error probabilities are achievable.[5] This observation has some important implications. In Example 2.9 we have remarked that distributed *nonadaptive* implementations such as the running consensus exhibit error probabilities that decay to zero exponentially fast as time elapses. In contrast, the price for adaptation paid by the diffusion algorithm is that the error probabilities tend to some steady-state value as time elapses. But we have just shown that such steady-state error probabilities can reach extremely low values, which means that the *adaptive* algorithm is able to learn well.

[5]The simulation of the system for accurate evaluation of such low probabilities has been carried out using importance sampling techniques, as detailed in [65].

2.6 UNIVERSAL SCALING LAWS: ESTIMATION VERSUS DETECTION

We conclude our analysis by highlighting the universal features of estimation and detection over adaptive networks as well as by illustrating some useful ramifications of the presented theory.

As we have shown in the previous sections, a number of results are available to characterize the estimation performance as well as the detection performance of adaptive networks. Such results are even more revealing if examined in conjunction, because a comparative analysis allows highlighting commonalities as well as distinctive features between estimation and detection. Moreover, we will also compare the results illustrated in the previous sections against known laws for estimation and detection in traditional (centralized or decentralized, nonadaptive) inferential systems.

The (steady-state) inference performance of adaptive networks is summarized in Fig. 2.4. The leftmost panel displays the mean-square-deviation, MSD_k, as a function of the inverse of the step-size, $1/\mu$, and with reference to four agents ($k = 2, 4, 5, 10$). The empirical data points (markers) obtained via Monte Carlo simulation are compared against the theoretical curve given in Eq. (2.26), which is represented with a dashed line. The rightmost panel displays the empirical steady-state error probabilities ($\alpha_k = \beta_k$ in this example), evaluated through Monte Carlo importance sampling, along with: (i) the theoretical normal approximation in Eq. (2.58), dashed curves; and (ii) the exact asymptotics in Eq. (2.69), solid curves. The settings for the estimation and for the detection problem are the same as those considered in Fig. 2.2 and in Fig. 2.3, respectively. A number of important findings are revealed by a detailed analysis of the figure.

Universal scaling laws. In the estimation context, we observe from Eq. (2.26) that the mean-square-deviation scales proportionally to μ (i.e., inversely with $1/\mu$) as μ goes to zero. In a detection context, we observe from Eq. (2.60) that the error probabilities decay exponentially fast as functions of $1/\mu$, as μ goes to zero. These scaling laws are general and represent the *universal* scaling laws governing errors of estimation and detection over adaptive networks.

Adaptation/learning trade-off. Recalling that smaller values of μ mean a lower degree of adaptation, we observe that, in both cases, reaching a better inference quality costs in terms of adaptation speed. This trade-off between tracking speed and steady-state learning performance is well known in the adaptive filtering literature [6].

Equivalence among agents? We see from the leftmost panel in Fig. 2.4 that the different error curves collapse into the limiting theoretical curve in Eq. (2.26), as the step-size goes to zero. Accordingly, we can conclude that all agents reach, for sufficiently small step-sizes, *the same estimation performance*. Let us switch to the analysis of the detection setting, rightmost panel in Fig. 2.4. We see that the error probability curves stay nearly parallel (in the log-scale representation), which confirms that they are equivalent *at the leading order in the exponent*. In this sense we recognize an analogy with the estimation setting, because *for what concerns the exponential rate of decay, the performance is equalized across agents*. On the other hand, a fundamental contrast emerges between estimation and detection: differently from what happens in the former problem, in the latter problem the performance of distinct agents *does not equalize as μ goes to zero*. Actually, the error-probability curves of distinct agents are ordered so as to reflect the network structure, namely, the differences across agents are related to the degree of connectivity of the agents within the network, as can be readily verified by joint inspection of the rightmost panel in Fig. 2.4 and of the network topology shown in Fig. 2.1.

Where these discrepancies come from? Given that the operational conditions (e.g., the network deployment, the diffusion algorithm, and so on) are essentially the same for both the estimation and the detection setting, one might wonder why in the former case the inference performance equalizes across agents while in the latter case it does not. The theoretical machinery developed so far provides a clear explanation of such a phenomenon. Let us compare the performance of agents j and k.

For the estimation setting we use Eq. (2.25). Because here we want to emphasize that the $O(\mu^2)$ corrections appearing in Eq. (2.25) can depend on the particular agent, we denote the (possibly) different corrections pertaining to agents j and k by $\tilde{c}_j(\mu)$ and $\tilde{c}_k(\mu)$, respectively. Thus we get:

$$\frac{\mathrm{MSD}_j}{\mathrm{MSD}_k} = \frac{\mu M \sum_{\ell=1}^{N} p_\ell^2 \sigma_{v,\ell}^2 + \tilde{c}_j(\mu)}{\mu M \sum_{\ell=1}^{N} p_\ell^2 \sigma_{v,\ell}^2 + \tilde{c}_k(\mu)} \approx 1. \tag{2.76}$$

We see from Eq. (2.76) that, in the relative comparison between the performance of the two agents, the impact of the higher order terms progressively disappears as the step-size decreases (because these terms are of order $O(\mu^2)$), which implies the aforementioned performance equalization across agents.

For the detection setting we use Eq. (2.69) to obtain:

$$\frac{\alpha_j}{\alpha_k} \approx \exp\left\{\frac{1}{\mu}\left[c_{k,0}(\mu) - c_{j,0}(\mu)\right]\right\}, \quad \frac{\beta_j}{\beta_k} \approx \exp\left\{\frac{1}{\mu}\left[c_{k,1}(\mu) - c_{j,1}(\mu)\right]\right\}. \tag{2.77}$$

Now, the ratios appearing in Eq. (2.77) do not tend to 1 in general, because we know that the corrections $c_{\ell,h}(\mu)$, for $\ell = 1, 2, \ldots, N$ and $h = 0, 1$, are of order $O(\mu)$. We conclude that, in general, the error probabilities of different agents do *not* equalize as the step-size decreases.

Classic inference scaling laws. We have observed from Eqs. (2.26) and (2.60) that the scaling laws governing errors of estimation and detection over adaptive networks behave very differently. The significance and elegance of this result for adaptive distributed networks lie in revealing an intriguing analogy with other more traditional inferential schemes. As a first example, consider a *classic* (i.e., centralized, nonadaptive) inferential system with N i.i.d. data points. It is known that the optimal estimation error decays as $1/N$ while the error probabilities of the best detector decay exponentially fast to zero with N [1,12]. Another important case is that of distributed (nonadaptive) inference problems with a fusion center, which has been described in Section 2.2.2. The fundamental limits for such a problem have been examined in the context of rate-constrained multiterminal inference, and, more specifically, with reference to the so-called CEO problem; see Example 2.3. In this case, given a total bit-rate R, the squared estimation error vanishes as $1/R$ while the detection performance scales exponentially with R [38,39]. Thus, at an abstract level, reducing the step-size corresponds to increasing the number of independent observations in the first system or increasing the bit-rate in the second system. The above comparisons furnish at least two interesting interpretations for the inverse of the step-size, $1/\mu$. First, $1/\mu$ represents the *cost of information* used by the network for inference purposes, much as the number of data N or the bit-rate R in the considered examples. Indeed, $1/\mu$ measures the cost of adaptation, i.e., the *adaptation time*, because we know that the adaptation time scales roughly as $1/\mu$; see last statement of point (i) in Section 2.4.2. Second, $1/\mu$ measures the number of data samples that, at a certain time i, receive "enough" credit by the diffusion algorithm, and, hence, $1/\mu$ takes on the

Table 2.2 Scaling Laws of Detection and of Estimation for Different Types of Inference Problems

Problem Type	Cost	Estimation	Detection
Centralized and nonadaptive	N (no. of data)	mse $\sim 1/N$	Err. prob. $\sim e^{-N}$
Distributed + FC and nonadaptive	R (bit-rate)	mse $\sim 1/R$	Err. prob. $\sim e^{-R}$
Fully distributed and adaptive	$1/\mu$ (adapt. time)	mse $\sim \mu$	Err. prob. $\sim e^{-1/\mu}$

The symbol $\sim$ means "scales as" whereas the common symbol mse denotes the pertinent mean-square-error estimation performance for all the addressed problems. FC, fusion center.

meaning of an effective *amount of information*; see Eq. (2.53). Such a notion can be useful to compare the adaptive system (which operates on infinitely long sequences of *streaming* data) against the other two systems of the considered examples, which operate on a *fixed* amount of data (N) or bits (R). A summary of the aforementioned analogies and comparisons is provided in Table 2.2.

APPENDIX

A.1 PROCEDURE TO EVALUATE EQ. (2.69)

This appendix collects the steps needed to evaluate Eq. (2.69). In order to simplify the notation, we suppress the subscripts denoting the particular hypothesis in force.

In the forthcoming computation, we shall employ the LMGF of the *steady-state* random variable $\boldsymbol{w}_k^\star$, which is:

$$\phi_k(t;\mu) \triangleq \ln \mathbb{E} e^{t\boldsymbol{w}_k^\star} = \sum_{m=0}^{\infty}\sum_{\ell=1}^{S} \psi\left(\mu\,(1-\mu)^m\, b_{\ell k}(m+1)\; t\right), \tag{A.1}$$

where the latter series representation has been proved in [61]. Moreover, in [61] it is also proved that the derivatives of the LMGF can be computed via term-by-term differentiation. Finally, we remark that, depending on the particular application, the formulas in the forthcoming listing might need to be evaluated numerically, and, for practical purposes, the infinite summations must be obviously truncated.

1. Find the solution θ to the stationary equation appearing in Eq. (2.63):

$$\phi'(\theta) = \frac{1}{\theta}\sum_{\ell=1}^{N} \psi(p_\ell \theta) = \gamma. \tag{A.2}$$

2. Using Eq. (A.1), compute the correction term $c_k(\mu)$ as:

$$c_k(\mu) = [\phi(\theta) - \mu\,\phi_k(\theta/\mu;\mu)] + \frac{[\phi'(\theta) - \phi_k'(\theta/\mu;\mu)]^2}{2\,\phi''(\theta)}. \tag{A.3}$$

REFERENCES

[1] Shao H. Mathematical statistics. New York: Springer-Verlag; 2003.
[2] Lehmann EL, Casella G. Theory of point estimation. New York: Springer; 1998.
[3] Lehmann EL, Romano JP. Testing statistical hypotheses. New York: Springer-Verlag; 2005.
[4] Kay SM. Fundamentals of statistical signal processing: estimation theory. Upper Saddle River, NJ: Prentice-Hall; 1993.
[5] Kailath T, Sayed AH, Hassibi B. Linear estimation. Upper Saddle River, NJ: Prentice-Hall; 2000.
[6] Sayed AH. Adaptive filters. NJ: Wiley; 2008.
[7] Kay SM. Fundamentals of statistical signal processing: detection theory. Upper Saddle River, NJ: Prentice-Hall; 1998.
[8] Poor HV. An introduction to signal detection and estimation. Springer-Verlag; 1988.
[9] Kassam SA. Signal detection in non-Gaussian noise. New York: Springer-Verlag; 1987.
[10] Huber PJ. Robust statistics. NJ: Wiley; 2004.
[11] Györfi L, Kohler M, Krzyżak A, Walk H. A distribution-free theory of nonparametric regression. New York: Springer; 2002.
[12] Dembo A, Zeitouni O. Large deviations techniques and applications. New York: Springer-Verlag; 1998.
[13] den Hollander F. Large deviations. Providence: American Mathematical Society; 2008.
[14] Lam WM, Reibman AR. Design of quantizers for decentralized estimation systems. IEEE Trans Commun 1993;41(11):1602–5.
[15] Gubner JA. Distributed estimation and quantization. IEEE Trans Inf Theory 1993;39(4):1456–9.
[16] Zhang K, Li XR. Optimal sensor data quantization for best linear unbiased estimation fusion. In: 2004 43rd IEEE Conference on Decision and Control (CDC), vol. 3; 2004. p. 2656–61.
[17] Venkitasubramaniam P, Tong L, Swami A. Quantization for maximin ARE in distributed estimation. IEEE Trans Signal Process 2007;55(7):3596–605.
[18] Marano S, Matta V, Willett P. Distributed estimation in large wireless sensor networks via a locally optimum approach. IEEE Trans Signal Process 2008;56(2):748–56.
[19] Luo ZQ. Universal decentralized estimation in a bandwidth constrained sensor network. IEEE Trans Inf Theory 2005;51(6):2210–19.
[20] Xiao JJ, Luo ZQ, Giannakis GB. Performance bounds for the rate-constrained universal decentralized estimators. IEEE Signal Process Lett 2007;14(1):47–50.
[21] Predd JB, Kulkarni SB, Poor HV. Distributed learning in wireless sensor networks. IEEE Signal Process Mag 2006;23(4):56–69.
[22] Marano S, Matta V, Willett P. Nearest-neighbor distributed learning by ordered transmissions. IEEE Trans Signal Process 2013;61(21):5217–30.
[23] Mergen G, Naware V, Tong L. Asymptotic detection performance of type-based multiple access over multiaccess fading channels. IEEE Trans Signal Process 2007;55(3):1081–92.
[24] Marano S, Matta V, Tong L, Willett P. A likelihood-based multiple access for estimation in sensor networks. IEEE Trans Signal Process 2007;55(11):5155–66.
[25] Viswanathan R, Varshney PK. Distributed detection with multiple sensors. I. Fundamentals. Proc IEEE 1997;85(1):54–63.
[26] Blum RS, Kassam SA, Poor HV. Distributed detection with multiple sensors. II. Advanced topics. Proc IEEE 1997;85(1):64–79.
[27] Tsitsiklis JN. Decentralized detection. Adv Stat Signal Process 1993;2:297–344.
[28] Longo M, Lookabaugh TD, Gray RM. Quantization for decentralized hypothesis testing under communication constraints. IEEE Trans Inf Theory 1990;36(2):241–55.
[29] Varshney PK. Distributed detection and data fusion. New York: Springer-Verlag; 1997.

[30] Chamberland JF, Veeravalli VV. Decentralized detection in sensor networks. IEEE Trans Signal Process 2003;51(2):407–16.
[31] Chamberland JF, Veeravalli VV. Wireless sensors in distributed detection applications. IEEE Signal Process Mag 2007;24(3):16–25.
[32] Chen B, Tong L, Varshney PK. Channel-aware distributed detection in wireless sensor networks. IEEE Signal Process Mag 2006;23(4):16–26.
[33] Saligrama V, Alanyali M, Savas O. Distributed detection in sensor networks with packet losses and finite capacity links. IEEE Trans Signal Process 2006;54(11):4118–32.
[34] Rago C, Willett P, Bar-Shalom Y. Censoring sensors: a low-communication-rate scheme for distributed detection. IEEE Trans Aerosp Electron Syst 1996;32(2):554–68.
[35] Marano S, Matta V, Willett PK. Distributed detection with censoring sensors under physical layer secrecy. IEEE Trans Signal Process 2009;57(5):1976–86.
[36] Han TS, Amari S. Statistical inference under multiterminal data compression. IEEE Trans Inf Theory 1998;44(6):2300–24.
[37] Zhang Z, Berger T. Estimation via compressed information. IEEE Trans Inf Theory 1988;34(2):198–211.
[38] Berger T, Zhang Z, Viswanathan H. The CEO problem [multiterminal source coding]. IEEE Trans Inf Theory 1996;42(3):887–902.
[39] Viswanathan H, Berger T. The quadratic Gaussian CEO problem. IEEE Trans Inf Theory 1997;43(5): 1549–59.
[40] Tsitsiklis J, Bertsekas D, Athans M. Distributed asynchronous deterministic and stochastic gradient optimization algorithms. IEEE Trans Autom Control 1986;31(9):803–12.
[41] Xiao L, Boyd S. Fast linear iterations for distributed averaging. Syst Control Lett 2004;53(1):65–78.
[42] Boyd S, Ghosh A, Prabhakar B, Shah D. Randomized gossip algorithms. IEEE Trans Inf Theory 2006;52(6):2508–30.
[43] Nedic A, Ozdaglar A. Distributed subgradient methods for multi-agent optimization. IEEE Trans Autom Control 2009;54(1):48–61.
[44] Dimakis AG, Kar S, Moura JMF, Rabbat MG, Scaglione A. Gossip algorithms for distributed signal processing. Proc IEEE 2010;98(11):1847–64.
[45] Braca P, Marano S, Matta V. Enforcing consensus while monitoring the environment in wireless sensor networks. IEEE Trans Signal Process 2008;56(7):3375–80.
[46] Braca P, Marano S, Matta V, Willett P. Asymptotic optimality of running consensus in testing binary hypotheses. IEEE Trans Signal Process 2010;58(2):814–25.
[47] Braca P, Marano S, Matta V, Willett P. Consensus-based Page's test in sensor networks. Signal Process 2011;91(4):919–30.
[48] Kar S, Moura JMF. Convergence rate analysis of distributed gossip (linear parameter) estimation: fundamental limits and tradeoffs. IEEE J Sel Top Signal Process 2011;5(4):674–90.
[49] Bajovic D, Jakovetic D, Xavier J, Sinopoli B, Moura JMF. Distributed detection via Gaussian running consensus: large deviations asymptotic analysis. IEEE Trans Signal Process 2011;59(9): 4381–96.
[50] Bajovic D, Jakovetic D, Moura JMF, Xavier J, Sinopoli B. Large deviations performance of consensus+ innovations distributed detection with non-Gaussian observations. IEEE Trans Signal Process 2012;60(11): 5987–6002.
[51] Jakovetic D, Moura JMF, Xavier J. Distributed detection over noisy networks: large deviations analysis. IEEE Trans Signal Process 2012;60(8):4306–20.
[52] Sayed AH, Tu SY, Chen J, Zhao X, Towfic ZJ. Diffusion strategies for adaptation and learning over networks. IEEE Signal Process Mag 2013;30(3):155–71.
[53] Sayed AH. Adaptation, learning, and optimization over networks. In: Foundations and trends in machine learning, vol. 7. Boston-Delft: NOW Publishers; 2014. p. 311–801.

[54] Sayed AH. Adaptive networks. Proc IEEE 2014;102(4):460–97.
[55] Lopes CG, Sayed AH. Diffusion least-mean squares over adaptive networks: formulation and performance analysis. IEEE Trans Signal Process 2008;56(7):3122–36.
[56] Chen J, Sayed AH. On the learning behavior of adaptive networks—Part I: transient analysis. IEEE Trans Inf Theory 2015;61(6):3487–517.
[57] Chen J, Sayed AH. On the learning behavior of adaptive networks—Part II: performance analysis. IEEE Trans Inf Theory 2015;61(6):3518–48.
[58] Zhao X, Sayed AH. Performance limits for distributed estimation over LMS adaptive networks. IEEE Trans Signal Process 2012;60(10):5107–24.
[59] Matta V, Braca P, Marano S, Sayed AH. Diffusion-based adaptive distributed detection: steady-state performance in the slow adaptation regime. IEEE Trans Inf Theory 2016;62(8):4710–32.
[60] Roberts SW. Control chart tests based on geometric moving averages. Technometrics 1959;1(3):239–50.
[61] Matta V, Braca P, Marano S, Sayed AH. Distributed detection over adaptive networks: refined asymptotics and the role of connectivity. IEEE Trans Signal Inf Process Netw 2016;2(4):442–60.
[62] Cover T, Thomas J. Elements of information theory. New York: John Wiley & Sons; 1991.
[63] Bahadur RR, Rao RR. On deviations of the sample mean. Ann Math Stat 1960;31(4):1015–27.
[64] Feller W. An introduction to probability and its applications, vol. 2. New York: John Wiley & Sons; 1971.
[65] Matta V, Braca P, Marano S, Sayed AH. Detection over diffusion networks: asymptotic tools for performance prediction and simulation. In: 24th European signal processing conference (EUSIPCO); 2016. p. 1503–7.

CHAPTER

MULTITASK LEARNING OVER ADAPTIVE NETWORKS WITH GROUPING STRATEGIES 3

Jie Chen[a*], **Cédric Richard**[b†], **Shang Kee Ting**[‡], **Ali H. Sayed**[c§]

Center of Intelligent Acoustics and Immersive Communications, School of Marine Science and Technology, Northwestern Polytechnical University, Xi'an, China Laboratoire Lagrange, Université Côte d'Azur, CNRS, OCA, Nice, France† DSO National Laboratories, Singapore, Singapore‡ School of Engineering, École Polytechnique Fédérale de Lausanne (EPFL), Lausanne, Switzerland§*

3.1 INTRODUCTION

A large variety of applications are network-structured and require adaptation to time-varying dynamics. Sensor networks, vehicular networks, communication networks, and power grids are some typical examples.

While centralized strategies can extract information from aggregated data more accurately, they nevertheless become prohibitive in large data scenarios and rely on a risky fusion-based architecture where failure of the single central processor can make this solution unreliable. Distributed strategies are more robust and can be designed to process data in an online streaming fashion, thus avoiding the need to steer large amounts of raw information. Signal processing over networks has provided a powerful and convenient set of tools for such scenarios, allowing for efficient in-network learning and adaptation. Several strategies have been proposed in the literature, including incremental [1–4], consensus [5–7], and diffusion strategies [8–15]. Diffusion strategies are particularly attractive because they are scalable, robust, and enable continuous learning and adaptation in response to data drifts [16–18].

The working hypothesis for these earlier studies is that the nodes cooperate with each other to monitor a single process or to estimate a common parameter vector. We shall refer to problems of this type as single-task problems. Reaching consensus among the agents is critical for successful inference in these problems. Due to the increased heterogeneity in models and data types, there has been growing interest in multitask problems. Over multitask networks, rather than promote consensus among all agents, the agents are allowed to track node-specific interests that happen to share some dependency relation with the interest of other agents. In this way, even though the objectives may be different,

[a]The work of J. Chen was supported in part by NSFC grant 61671382.
[b]The work of C. Richard was supported in part by IDEX UCAJEDI project (ANR-15-IDEX-0001).
[c]The work of A.H. Sayed was supported in part by NSF grant CCF-1524250.

Cooperative and Graph Signal Processing. https://doi.org/10.1016/B978-0-12-813677-5.00003-1

the agents can still benefit from cooperation. In [19,20], the authors describe distributed node-specific estimation algorithms over fully connected networks or tree networks. In [21], the authors formalize the problem of adaptation and learning over multitask networks. They devise a set of distributed online algorithms based on diffusion adaptation strategy. Extensions to asynchronous networks are considered in [22]. In [23], the performance of a single-task diffusion implementation is analyzed when it operates in a multitask environment. An unsupervised clustering strategy that allows each agent to automatically select the neighboring agents with which it can collaborate is also introduced. In this scenario, the only available information is that clusters of nodes with common interests may exist in the network but nodes do not know which other nodes share the same interest. Other useful works have also addressed variations of this scenario in [24–27]. In [28], the authors use multitask diffusion adaptation as described in [21] with a node clustering strategy for studying the relation between the tremor intensity and brain connectivity of Parkinson's patients. In [29], the authors derive a distributed strategy that allows each node in the network to locally adapt the intensity of cooperation with other nodes. The authors in [30] promote cooperation between clusters with ℓ_1-norm coregularizers. The authors in [31,32] examine an alternative way to model relations between tasks by assuming that they all share a common latent feature representation. Variations of this scenario are addressed in [33]. In another scenario, it is assumed that there are parameters of global interest to all nodes in the network, a collection of parameters of common interest within subgroups of nodes, and a set of parameters of local interest at each node [34–36]. In [37,38], the optimum parameter vectors to be estimated by agents are related according to a set of constraints.

An inspection of the literature on diffusion adaptation over networks shows that, in most existing works, single-task and multitask oriented algorithms fuse information from neighboring agents via weighted combinations of estimated parameter vectors. These combinations assign the same scaling weight to all entries in the combined iterates. There are situations, however, where some groups of entries within the iterate vectors should be weighted differently than other groups of entries within the same iterates. Consider an example where the top half of the entries of the parameter vectors to estimate are common across all agents while the bottom half entries are randomly distributed without obvious relationship. Uniformly combining estimates may cause a large estimation error due to the presence of significantly different entries.

Considering groups of variables rather than variables individually can be beneficial for estimation accuracy if structural relationships between variables exist (e.g., spatial, hierarchical, or related to the physics of the problem). Group-sparsity inducing estimators are typical examples that benefit from such prior information. In this chapter, we build on this principle to show how diffusion LMS can be extended to deal with structured criteria involving groups of variables.

This chapter is organized as follows. Section 3.2 presents the network model and provides a brief review of diffusion LMS. The group diffusion LMS algorithm is shown in Section 3.3. Its stochastic behavior is analyzed for known groups of variables and fixed combination coefficients. Section 3.4 introduces unsupervised strategies for grouping the variables and setting the combination coefficients of the group diffusion LMS. In Section 3.5, experiments are conducted to validate the algorithms and theoretical findings. Section 3.6 concludes this chapter.

Notation. The normal font x denotes scalars. Boldface small letters $\boldsymbol{x}$ denote vectors. All vectors are column vectors. Boldface capital letters $\boldsymbol{X}$ denote matrices. The (k, ℓ)th entry of a matrix is denoted by $(\cdot)_{k\ell}$, and the (k, ℓ)th block of a block matrix is denoted by $[\,\cdot\,]_{k\ell}$. The superscript $(\cdot)^\top$ represents the transpose of a matrix or vector. The notation $\|\cdot\|$ denotes the ℓ_2-norm of its matrix or vector argument while $\|\cdot\|_{b,\infty}$ denotes the block maximum norm of its block vector or matrix argument. The spectral radius of a square matrix is denoted by $\rho(\cdot)$. The matrix trace is denoted by trace$(\cdot)$. The operator col$\{\cdot\}$ stacks its vector arguments on top of each other to generate a connected vector. The operator diag$\{\cdot\}$ formulates a (block) diagonal matrix with its arguments. An identity matrix of size $N \times N$ is denoted by $\boldsymbol{I}_N$. The Kronecker product is denoted by $\otimes$, and expectation is denoted by $\mathbb{E}\{\cdot\}$. We denote by $\mathcal{N}_k$ the set of node indices in the neighborhood of node k, including k itself, and $|\mathcal{N}_k|$ its set cardinality.

3.2 NETWORK MODEL AND DIFFUSION LMS

3.2.1 NETWORK MODEL

Let us consider a connected network $G = (\mathcal{V}, \mathcal{E})$ defined by a set $\mathcal{V} = \{1, 2, \ldots, N\}$ of N agents, along with a set $\mathcal{E}$ of edges that are two-element subsets of $\mathcal{V}$. We address the problem of estimating an $L \times 1$ unknown vector at each node k from streaming data collected over the network. At each time instant n, node k has access to time sequences $\{d_k(n), \boldsymbol{x}_{k,n}\}$, where $d_k(n)$ denotes the reference signal and $\boldsymbol{x}_{k,n}$ represents an $L \times 1$ regression vector with covariance matrix $\boldsymbol{R}_{x,k} = \mathbb{E}\{\boldsymbol{x}_{k,n}\boldsymbol{x}_{k,n}^\top\} > 0$. We assume that the data are related via the linear model:

$$d_k(n) = \boldsymbol{w}_k^{\star\top} \boldsymbol{x}_{k,n} + z_k(n) \tag{3.1}$$

for all k, with $\boldsymbol{w}_k^\star$ an unknown parameter vector at node k, and $z_k(n)$ a zero-mean i.i.d. noise of variance $\sigma_{z,k}^2$ that is independent of every other signal. For determining the parameter vectors $\boldsymbol{w}_k^\star$, we consider the mean-square error criterion at each node k defined as:

$$J_k(\boldsymbol{w}_k) = \mathbb{E}\big\{|d_k(n) - \boldsymbol{x}_{k,n}^\top \boldsymbol{w}_k|^2\big\}. \tag{3.2}$$

We shall refer to scenarios where all nodes estimate the same parameter vector, that is, $\boldsymbol{w}_1^\star = \cdots = \boldsymbol{w}_N^\star$, as *single-task* problems. Collaboration among nodes with standard distributed strategies can enhance the estimation performance over the network. On the contrary, we shall refer to cases where nodes may estimate distinct parameter vectors, namely, cases where the $\{\boldsymbol{w}_k^\star\}_{k=1}^N$ are not necessarily the same, as *multitask* problems. Still, we assume that similarities exist in some sense among these parameter vectors. Otherwise the estimation problem would be node-independent and would reduce to the noncooperative setting.

3.2.2 A BRIEF REVIEW OF DIFFUSION LMS

Before introducing the diffusion strategy at the group level, we provide a brief review of standard diffusion LMS derived for single-task scenarios. The goal of this algorithm is to minimize the following global cost function in a distributed manner for an enhanced estimation performance over a noncooperative strategy:

$$J^{\text{glob}}(\boldsymbol{w}) = \sum_{k=1}^{N} J_k(\boldsymbol{w}). \tag{3.3}$$

We denote the minimizer of Eq. (3.3) by $w^\star$. Minimizing Eq. (3.3) over w with J_k defined by the mean-square error (3.2) is equivalent to minimizing the following alternative cost [12,13]:

$$J^{\text{glob}'}(w) = J_k(w) + \sum_{\ell \neq k} \|w - w^\star\|^2_{R_{x,\ell}}. \tag{3.4}$$

To bypass the unknown second-order statistics $R_{x,\ell}$, one can rely on the Rayleigh-Ritz characterization of eigenvalues to approximate the weighted norm in Eq. (3.4) by a scaled unweighted norm [12,13], say as,

$$\|w - w^\star\|^2_{R_{x,\ell}} \approx b_{\ell k} \, \|w - w^\star\|^2 \tag{3.5}$$

for some nonnegative coefficients $b_{\ell k}$. This leads to the following modified cost function at node k:

$$J^{\text{glob}''}(w) = J_k(w) + \sum_{\ell \neq k} b_{\ell k} \, \|w - w^\star\|^2. \tag{3.6}$$

Calculating the gradient vector of Eq. (3.6), restricting communication to immediate neighbors, and using approximation (3.5) along with the arguments from [13], we arrive at the adapt-then-combine (ATC) strategy without raw data exchange [9]:

$$\psi_{k,n} = w_{k,n-1} + \mu \, x_{k,n} \, [d_k(n) - w^\top_{k,n-1} x_{k,n}], \tag{3.7a}$$

$$w_{k,n} = \sum_{\ell \in \mathcal{N}_k} a_{\ell k} \, \psi_{\ell,n}, \tag{3.7b}$$

where μ is a small positive step size. The combine-then-adapt (CTA) form can be derived in a similar way; it is sufficient for our purposes to continue with the ATC form (3.7). The coefficients $\{a_{\ell k}\}$ in the above algorithm are given by:

$$a_{kk} = 1 - \mu \sum_{\ell \in \mathcal{N}_k \backslash \{k\}} b_{\ell k}, \tag{3.8}$$

$$a_{\ell k} = \mu \, b_{\ell k}, \quad \ell \in \mathcal{N}_k \backslash \{k\}, \tag{3.9}$$

$$a_{\ell k} = 0, \quad \ell \notin \mathcal{N}_k. \tag{3.10}$$

In practice, the coefficients $\{a_{\ell k}\}$ are usually treated as free weighting parameters to be chosen by the designer. That is, it is not necessary to worry about selecting the coefficients $\{b_{\ell k}\}$. It is sufficient to select the $\{a_{\ell k}\}$ as nonnegative convex combination coefficients satisfying:

$$a_{\ell k} \geq 0, \quad \sum_{\ell \in \mathcal{N}_k} a_{\ell k} = 1, \quad a_{\ell k} = 0 \text{ if } \ell \notin \mathcal{N}_k. \tag{3.11}$$

The selection of the $\{a_{\ell k}\}$ has a significant impact on the performance of the algorithm for both single and multitask scenarios [13–15,23,25].

3.3 GROUP DIFFUSION LMS

3.3.1 MOTIVATION

It is explained in [13] how Eq. (3.5) leads to the fusion (3.7b) of local estimates in the neighborhood of each node. Note now that all the entries of the intermediate estimate $\boldsymbol{\psi}_{\ell,n}$ are scaled by the same weight $a_{\ell k}$. Fig. 3.1 illustrates one possible limitation of uniform scaling of the entries and why grouping can be useful in some important situations. For example, in the figure, adjacent nodes k and ℓ are estimating parameter vectors $\boldsymbol{w}_k^\star$ and $\boldsymbol{w}_\ell^\star$ whose entries are grouped into three separate sets: both vectors have the same entries in the first group, they significantly differ in the second group due to sensor failure, for instance, and only differ slightly in the third group due to sensor drift. It is not suitable to view this scenario either as a single-task problem or as a multitask problem with a single set of combination weights $a_{\ell k}$. A small combination weight may not be sufficient to promote the closeness of entries in the first and third groups whereas a large combination weight may lead to a large estimation bias caused by the second group.

This example motivates us to introduce a grouping strategy. More generally, let M be a positive integer less than or equal to L, and let $\{\mathcal{G}_m\}_{m=1}^M$ be a partition of the set of indexes $\mathcal{G} = \{1,\ldots,L\}$, namely,

$$\bigcup_{m=1}^{M} \mathcal{G}_m = \mathcal{G}, \quad \mathcal{G}_m \cap \mathcal{G}_{m'} = \varnothing \text{ if } m \neq m'. \tag{3.12}$$

We also let $\boldsymbol{w}_{\mathcal{G}_m}$ or $[\boldsymbol{w}]_{\mathcal{G}_m}$ denote a subvector of $\boldsymbol{w}$ indexed by $\mathcal{G}_m$. In the case of Fig. 3.1B, these are the subvectors that correspond to the groups $\mathcal{G}_1, \mathcal{G}_2$, and $\mathcal{G}_3$. We can then assign larger combination weights

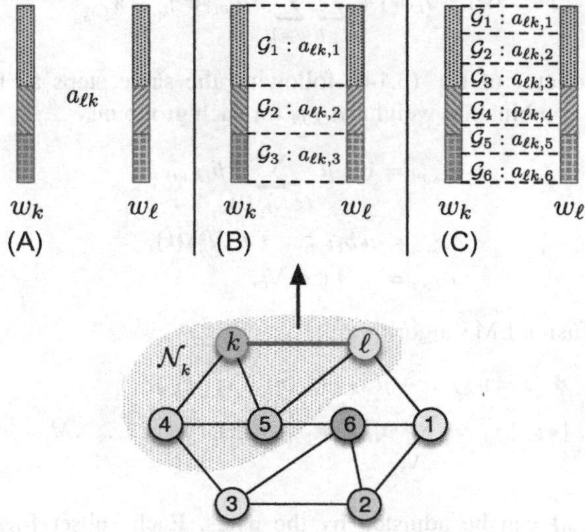

FIG. 3.1

(A) Parameter vector structures for nodes k and ℓ: three sets of entries have different levels of similarity, encoded by grey levels. (B) A scenario with three groups; (C) a second scenario with six groups.

to the first group, smaller or even null-valued weights to the second group, and medium-value weights to the third group. Such a grouping strategy ends up exploiting the structure of the parameter vectors more fully. However, because information on the internal group structures may not be available beforehand, one possible strategy is to split parameter vectors into a number of groups of preset lengths and assign a combination coefficient to each group, as illustrated in Fig. 3.1C. In the sequel, we shall describe an unsupervised adaptive strategy to estimate the parameter vectors in these scenarios in an online manner. Note that the parameter vector entries within each group need not be necessarily contiguous. In the scope of this chapter, we shall only focus on homogeneous groups of entries across the network, namely, we shall assume that the parameter vectors at all nodes possess the same grouping structure across the network. While heterogeneous group models are able to represent more complex application scenarios, this will require further notation and a more complex algorithm development.

3.3.2 GROUP DIFFUSION LMS ALGORITHM

We now motivate the group diffusion LMS from the single-task derivation by approximating the second-order statistics $\boldsymbol{R}_{x,\ell}$ in an alternative manner. Inspecting Eq. (3.5), we now assign a scaling factor to each group of entries instead of using a single factor, i.e., we now use

$$\|\boldsymbol{w} - \boldsymbol{w}^\star\|^2_{\boldsymbol{R}_{x,\ell}} \approx \sum_{m=1}^{M} b_{\ell k,m} \|\boldsymbol{w}_{\mathcal{G}_m} - \boldsymbol{w}^\star_{\mathcal{G}_m}\|^2, \tag{3.13}$$

where $b_{\ell k,m}$ is the nonnegative weight for group m. The global cost (3.6) is then relaxed as follows:

$$J^{\text{glob}''}(\boldsymbol{w}) = J_k(\boldsymbol{w}) + \sum_{\ell \neq k} \sum_{m=1}^{M} b_{\ell k,m} \|\boldsymbol{w}_{\mathcal{G}_m} - \boldsymbol{w}^\star_{\mathcal{G}_m}\|^2. \tag{3.14}$$

Calculating the gradient vector of Eq. (3.14), following the same steps as for diffusion LMS and introducing the following combination weights $a_{\ell k,m}$ for each group m:

$$a_{kk,m} = 1 - \mu \sum_{\ell \in \mathcal{N}_k \backslash \{k\}} b_{\ell k,m}, \tag{3.15}$$

$$a_{\ell k,m} = \mu \, b_{\ell k,m}, \quad \ell \in \mathcal{N}_k \backslash \{k\}, \tag{3.16}$$

$$a_{\ell k,m} = 0, \quad \ell \notin \mathcal{N}_k, \tag{3.17}$$

we arrive at the group diffusion LMS algorithm:

$$\boldsymbol{\psi}_{k,n} = \boldsymbol{w}_{k,n-1} + \mu \boldsymbol{x}_{k,n} (d_k(n) - \boldsymbol{x}_{k,n}^\top \boldsymbol{w}_{k,n-1}), \tag{3.18a}$$

$$[\boldsymbol{w}_{k,n}]_{\mathcal{G}_m} = \sum_{\ell \in \mathcal{N}_k} a_{\ell k,m} [\boldsymbol{\psi}_{\ell,n}]_{\mathcal{G}_m}, \quad \text{for } m = 1, \ldots, M. \tag{3.18b}$$

Parameters $\bigcup_{m=1}^{M} \{a_{\ell k,m}\}$ can be adjusted by the users. Each subset $\{a_{\ell k,m}\}$ now forms a left-stochastic matrix $\boldsymbol{A}_m$, i.e., for $m = 1, \ldots, M$

$$a_{\ell k,m} \geq 0, \quad \sum_{\ell \in \mathcal{N}_k} a_{\ell k,m} = 1, \quad a_{\ell k,m} = 0 \text{ if } \ell \notin \mathcal{N}_k. \tag{3.19}$$

Appropriate selection of these coefficients can enhance the performance of diffusion LMS, especially for scenarios with structural relationships within groups. In Section 3.4.3, we will introduce an unsupervised strategy to adjust these weights in an online manner. One earlier version of the group diffusion strategy (3.18a) and (3.18b) was introduced in [39] and applied there to the problem of A/D converter tuning. In that application, the combination weights $\{a_{\ell k,m}\}$ were selected proportionally to the SNR conditions within relevant frequency bands.

3.3.3 NETWORK BEHAVIOR

We now study the behavior of the group diffusion LMS algorithm (3.18) with constant combination weights $a_{\ell k,m}$ that satisfy conditions (3.19). To proceed, we collect the information from across the network into block vectors and matrices. In particular, we denote by $\boldsymbol{w}_n$ and $\boldsymbol{w}^\star$ the stacked weight estimate vector and the stacked optimum weight vector, respectively:

$$\boldsymbol{w}_n = \mathrm{col}\{\boldsymbol{w}_{1,n}, \ldots, \boldsymbol{w}_{N,n}\}, \tag{3.20}$$

$$\boldsymbol{w}^\star = \mathrm{col}\{\boldsymbol{w}_1^\star, \ldots, \boldsymbol{w}_N^\star\}. \tag{3.21}$$

We consider the case where the $\boldsymbol{w}_k^\star$ are distinct. The weight error vector $\tilde{\boldsymbol{w}}_{k,n}$ for each node k at iteration n is defined by:

$$\tilde{\boldsymbol{w}}_{k,n} = \boldsymbol{w}_{k,n} - \boldsymbol{w}_k^\star. \tag{3.22}$$

These error vectors $\tilde{\boldsymbol{w}}_{k,n}$ are also stacked on top of each other to get the vector:

$$\tilde{\boldsymbol{w}}_n = \mathrm{col}\{\tilde{\boldsymbol{w}}_{1,n}, \ldots, \tilde{\boldsymbol{w}}_{N,n}\}. \tag{3.23}$$

We assume that the regression vectors $\boldsymbol{x}_{k,n}$ arise from a zero-mean random process that is temporally (over n) stationary, white, and independent over space (over k) with $\boldsymbol{R}_{x,k} = \mathbb{E}\{\boldsymbol{x}_k(n)\boldsymbol{x}_k^\top(n)\} > 0$. This independence assumption is widely used in the analysis of adaptive learning systems [40, App. 24.A], [14, Chs. 10-11].

Mean weight behavior analysis

Subtracting optimum vectors $\boldsymbol{w}_k^\star$ from both sides of the adaptation equation (3.18a), and using

$$d_k(n) - \boldsymbol{x}_{k,n}^\top \boldsymbol{w}_{k,n-1} = z_k(n) - \boldsymbol{x}_{k,n}^\top \tilde{\boldsymbol{w}}_{k,n-1} \tag{3.24}$$

gives

$$\boldsymbol{\psi}_{k,n} - \boldsymbol{w}_k^\star = \tilde{\boldsymbol{w}}_{k,n-1} - \mu\, \boldsymbol{x}_{k,n}\boldsymbol{x}_{k,n}^\top \tilde{\boldsymbol{w}}_{k,n-1} + \mu\, \boldsymbol{x}_{k,n} z_k(n). \tag{3.25}$$

Before establishing the relation between the weight error vectors $\tilde{\boldsymbol{w}}_n$ and $\tilde{\boldsymbol{w}}_{n-1}$, it is convenient to introduce the $N \times N$ block matrix

$$\mathcal{A} = \begin{pmatrix} \mathcal{A}_{11} & \cdots & \mathcal{A}_{1N} \\ \vdots & \ddots & \vdots \\ \mathcal{A}_{N1} & \cdots & \mathcal{A}_{NN} \end{pmatrix}. \tag{3.26}$$

Each block $\mathcal{A}_{\ell k}$ is an $L \times L$ diagonal matrix whose ith diagonal entry is $a_{\ell k,m}$, where m refers to the subset of indexes $\mathcal{G}_m$ to which index i belongs. In the single-task case, expression (3.26) reduces to matrix $\mathcal{A} = A \otimes I_N$ considered in [14, Ch. 8] for analyzing the convergence behavior of diffusion LMS, with $(A)_{\ell k} = a_{\ell k}$.

Matrix $\mathcal{A}$ can also be expressed as follows:

$$\mathcal{A} = (A_1 \otimes J_1) + \cdots + (A_M \otimes J_M), \tag{3.27}$$

where J_m is an $L \times L$ diagonal matrix with diagonal entries defined as:

$$(J_m)_{ii} = 1, \quad \text{if } i \in \mathcal{G}_m, \tag{3.28}$$

$$(J_m)_{ii} = 0, \quad \text{otherwise.} \tag{3.29}$$

Because the weights $a_{\ell k,m}$ satisfy condition (3.19), i.e., each matrix A_m is left-stochastic, matrix $\mathcal{A}$ is also left-stochastic.

With the above matrix $\mathcal{A}$, it can be verified that:

$$\tilde{w}_n = \mathcal{A}^\top (\psi_n - w^\star) + (\mathcal{A}^\top - I)\, w^\star, \tag{3.30}$$

where $\psi_n = \text{col}\{\psi_{1,n}, \ldots, \psi_{N,n}\}$. Using Eq. (3.25), we write:

$$\tilde{w}_n = \mathcal{A}^\top (I - \mu\, \mathcal{R}_{x,n})\tilde{w}_{n-1} + \mu\, \mathcal{A}^\top p_{xz,n} + (\mathcal{A}^\top - I) w^\star \tag{3.31}$$

with $\mathcal{R}_{x,n} = \text{diag}\{x_{1,n}x_{1,n}^\top, \ldots, x_{N,n}x_{N,n}^\top\}$ and $p_{xz,n} = \{x_{1,n}z_1(n), \ldots, x_{N,n}z_N(n)\}$. Taking the expectation of both sides of Eq. (3.31) and using the independence assumption, we arrive at the mean behavior equation of the group diffusion LMS algorithm:

$$\mathbb{E}\{\tilde{w}_n\} = \mathcal{A}^\top (I - \mu\, \mathcal{R}_x)\, \mathbb{E}\{\tilde{w}_{n-1}\} + (\mathcal{A}^\top - I)\, w^\star \tag{3.32}$$

with $\mathcal{R}_x = \text{diag}\{R_{x,1}, \ldots, R_{x,N}\}$. We shall now provide a condition on μ to guarantee the stability of Eq. (3.32).

The convergence of Eq. (3.32) is determined by the stability of $\mathcal{A}^\top (I - \mu\, \mathcal{R}_x)$. Algorithm parameters should be chosen to satisfy the mean stability condition:

$$\rho\big(\mathcal{A}^\top (I - \mu\, \mathcal{R}_x)\big) < 1, \tag{3.33}$$

where $\rho(\cdot)$ denotes spectral radius of its matrix argument. Let us first focus on matrix $\mathcal{A}$. Let $x = \text{col}\{x_1, \ldots, x_N\}$ be an arbitrary $L \times 1$ block vector whose individual entries $\{x_k\}$ are vectors of size $L \times 1$ each. Considering Eq. (3.27), using that $\sum_{i=1}^N a_{ji,m} = 1$ with $a_{ji,m} \geq 0$, and Jensen's inequality, we have:

$$\begin{aligned}
\left\| \left(\sum_{m=1}^M A_m^\top \otimes J_m \right) x \right\|^2 &= \sum_{i=1}^N \left\| \sum_{m=1}^M \sum_{j=1}^N a_{ji,m} J_m x_j \right\|^2 \\
&\leq \sum_{i=1}^N \sum_{m=1}^M \sum_{j=1}^N a_{ji,m} \left\| J_m x_j \right\|^2 \\
&= \sum_{m=1}^M \sum_{j=1}^N \left\| J_m x_j \right\|^2.
\end{aligned} \tag{3.34}$$

Matrix $\boldsymbol{J}_m$ is actually an orthogonal projection matrix that sets to 0 in Eq. (3.34) the entries of $\boldsymbol{x}_j$ that are not indexed by $\mathcal{G}_m$. Because $\{\mathcal{G}_m\}_{m=1}^{M}$ is a partition of the set of indexes, we have:

$$\sum_{m=1}^{M}\sum_{j=1}^{N}\|\boldsymbol{J}_m\boldsymbol{x}_j\|^2 = \sum_{j=1}^{N}\|\boldsymbol{x}_j\|^2 = \|\boldsymbol{x}\|^2. \tag{3.35}$$

We conclude that

$$\left\|\mathcal{A}^\top\right\| = \left\|\sum_{m=1}^{M}\boldsymbol{A}_m^\top \otimes \boldsymbol{J}_m\right\|^2 \leq 1. \tag{3.36}$$

We know that the spectral radius of any matrix $\boldsymbol{X}$ satisfies $\rho(\boldsymbol{X}) \leq \|\boldsymbol{X}\|$, for any induced norm. Applying this to $\mathcal{A}^\top(\boldsymbol{I}-\mu\,\mathcal{R}_x)$, we have:

$$\rho(\mathcal{A}^\top(\boldsymbol{I}-\mu\,\mathcal{R}_x)) \leq \left\|\mathcal{A}^\top\right\|\left\|\boldsymbol{I}-\mu\,\mathcal{R}_x\right\| \tag{3.37}$$

$$\leq \left\|\boldsymbol{I}-\mu\,\mathcal{R}_x\right\|. \tag{3.38}$$

It then follows that the group diffusion LMS asymptotically converges in the mean, for any initial condition, if the step size satisfies:

$$0 < \mu < \frac{2}{\max_k \lambda_{\max}(\boldsymbol{R}_{x,k})}. \tag{3.39}$$

Setting $n \longrightarrow \infty$ in Eq. (3.32) leads to the asymptotic mean bias expression:

$$\tilde{\boldsymbol{w}}_\infty = \left[\boldsymbol{I}-\mathcal{A}^\top(\boldsymbol{I}-\mu\,\mathcal{R}_x)\right]^{-1}(\mathcal{A}^\top-\boldsymbol{I})\,\boldsymbol{w}^\star. \tag{3.40}$$

Mean-square error behavior analysis

We shall now perform a mean-square error analysis of the group diffusion LMS. The purpose of this analysis is to evaluate how the variance $\mathbb{E}\{\|\tilde{\boldsymbol{w}}_n\|^2\}$ evolves with time. This analysis is based on the energy conservation framework used in [13,23,32], which starts from the weight-error vector recursion in the compact form:

$$\tilde{\boldsymbol{w}}_n = \boldsymbol{B}_n\,\tilde{\boldsymbol{w}}_{n-1} - \boldsymbol{g}_n - \boldsymbol{r} \tag{3.41}$$

with the transition matrix:

$$\boldsymbol{B}_n = \mathcal{A}^\top(\boldsymbol{I}-\mu\,\mathcal{R}_{x,n}) \tag{3.42}$$

the stochastic driving term:

$$\boldsymbol{g}_n = \mu\mathcal{A}^\top\boldsymbol{p}_{xz,n} \tag{3.43}$$

and the constant driving term:

$$\boldsymbol{r} = (\mathcal{A}^\top - \boldsymbol{I})\,\boldsymbol{w}^\star. \tag{3.44}$$

The expected values of the stochastic quantities (3.42) and (3.43) are given by:

$$\boldsymbol{B} = \mathcal{A}^\top(\boldsymbol{I}-\mu\,\mathcal{R}_x), \tag{3.45}$$

$$\boldsymbol{g} = \boldsymbol{0}_{NL}. \tag{3.46}$$

We define the matrix

$$\boldsymbol{K} = \mathbb{E}\{\boldsymbol{B}_n^\top \otimes \boldsymbol{B}_n^\top\} \tag{3.47}$$

and approximate it by $\boldsymbol{K} \approx \boldsymbol{B}^\top \otimes \boldsymbol{B}^\top$ for sufficiently small step sizes. We also define:

$$\boldsymbol{G} = \mathbb{E}\{\boldsymbol{g}_n \boldsymbol{g}_n^\top\} = \mu^2 \boldsymbol{\mathcal{A}}^\top \operatorname{diag}\{\sigma_{z,1}^2 \boldsymbol{R}_{x,1}, \ldots, \sigma_{z,n}^2 \boldsymbol{R}_{x,N}\} \boldsymbol{\mathcal{A}}. \tag{3.48}$$

We skip the derivations here and refer instead to [24,33]. Following similar arguments, the following statements can be justified.

Theorem 3.1 (Mean-square stability). *Consider the data model (3.1) and assume the independence assumption holds. The group diffusion strategy (3.18) is mean-square stable when the matrix* $\boldsymbol{K}$ *defined by Eq. (3.47), or its approximation, is stable. This condition is satisfied by sufficiently small step sizes.* ■

Theorem 3.2 (Network learning curve). *Consider the same setting of Theorem 3.1 and let* $\zeta_n \triangleq \mathbb{E}\{\|\tilde{\boldsymbol{w}}_n\|^2/N\}$ *denote the average network mean-square deviation (MSD) at time n. Then, the learning curve of the network corresponds to the evolution of* ζ_n *with time and is described by the following recursion over* $n \geq 0$:

$$\begin{aligned}\zeta_{n+1} = \zeta_n + \Big[& \left(vec\{\boldsymbol{G}^\top\}\right)^\top \boldsymbol{K}^n \boldsymbol{\sigma}_I + \|\boldsymbol{r}\|^2_{\boldsymbol{K}^n \boldsymbol{\sigma}_I} - \|\tilde{\boldsymbol{w}}_0\|^2_{(\boldsymbol{I}_{(NL)^2} - \boldsymbol{K})\boldsymbol{K}^n \boldsymbol{\sigma}_I} \\ & -2\left[\boldsymbol{\gamma}_n^\top + (\boldsymbol{B}\,\mathbb{E}\{\tilde{\boldsymbol{w}}_n\})^\top \otimes \boldsymbol{r}^\top\right] \boldsymbol{\sigma}_I \Big],\end{aligned} \tag{3.49}$$

$$\boldsymbol{\gamma}_{n+1} = \boldsymbol{K}^\top \boldsymbol{\gamma}_n + (\boldsymbol{K} - \boldsymbol{I}_{(NL)^2})^\top (\boldsymbol{B}\,\mathbb{E}\{\tilde{\boldsymbol{w}}_n\} \otimes \boldsymbol{r}) \tag{3.50}$$

with $\boldsymbol{\sigma}_I = vec\{\frac{1}{N}\boldsymbol{I}_{NL}\}$, $\zeta_0 = \frac{1}{N}\|\tilde{\boldsymbol{w}}_0\|^2$, $\boldsymbol{\gamma}_0 = \boldsymbol{0}_{(NL)^2 \times 1}$. ■

Theorem 3.3 (Steady-state MSD). *Consider the same setting of Theorem 3.1. The steady-state MSD of the group diffusion strategy (3.18) is given by:*

$$\zeta_\infty = \left(vec\{\boldsymbol{G}^\top\}\right)^\top \boldsymbol{\sigma} + \boldsymbol{r}^\top \boldsymbol{\Sigma}(\boldsymbol{r} - 2\boldsymbol{B}\tilde{\boldsymbol{w}}_\infty) \tag{3.51}$$

with $\tilde{\boldsymbol{w}}_\infty$ *determined by Eq. (3.40), and* $vec\{\boldsymbol{\Sigma}\} = \boldsymbol{\sigma} = \frac{1}{N}(\boldsymbol{I}_{(NL)^2} - \boldsymbol{K})^{-1} vec\{\boldsymbol{I}_{NL}\}$. ■

Although Theorems 3.2 and 3.3 provide closed-form expressions for the network MSD and steady-state MSD, it may not be practical to evaluate Eqs. (3.49)–(3.51) due to the size of the matrices involved, which have dimensions $(NL)^2 \times (NL)^2$. In what follows, we derive equivalent but more compact expressions with matrices of size $NL \times NL$ (see Proof of Corollary 3.1).

Corollary 3.1 (Alternative transient MSD expression). *Consider the same setting of Theorem 3.1. The MSD learning curve of the group diffusion strategy (3.18), provided by Theorem 3.2, can be equivalently expressed as follows:*

$$\begin{aligned}\zeta_{n+1} = \zeta_n + \frac{1}{N} trace \Big(& [\boldsymbol{G} + \boldsymbol{r}\boldsymbol{r}^\top] \boldsymbol{B}^{n\top} \boldsymbol{B}^n \\ & - \tilde{\boldsymbol{w}}_0 \tilde{\boldsymbol{w}}_0^\top \left[\boldsymbol{B}^{n\top}\boldsymbol{B}^n - \boldsymbol{B}^{n+1\top}\boldsymbol{B}^{n+1}\right] - 2\boldsymbol{\Gamma}_n - 2\boldsymbol{B}\,\mathbb{E}\{\tilde{\boldsymbol{w}}_n\}\boldsymbol{r}^\top \Big),\end{aligned} \tag{3.52}$$

$$\boldsymbol{\Gamma}_{n+1} = \boldsymbol{B}\boldsymbol{\Gamma}_n\boldsymbol{B}^\top + \boldsymbol{B}\boldsymbol{r}\,(\boldsymbol{B}^2\,\mathbb{E}\{\tilde{\boldsymbol{w}}_n\})^\top - \boldsymbol{r}(\boldsymbol{B}\,\mathbb{E}\{\tilde{\boldsymbol{w}}_n\})^\top \tag{3.53}$$

with $\zeta_0 = \frac{1}{N}\|\tilde{\boldsymbol{w}}_0\|^2$ *and* $\boldsymbol{\Gamma}_0 = \boldsymbol{0}_{NL}$. ■

Corollary 3.2 (Alternative steady-state MSD expression). *Consider the same setting of Theorem 3.1. The steady-state MSD of the group diffusion strategy (3.18), provided by Theorem 3.3, can be equivalently expressed as follows:*

$$\zeta_\infty = \sum_{n=0}^{\infty} \boldsymbol{B}^n \left(\boldsymbol{G} + (\boldsymbol{r} - 2\boldsymbol{B}\tilde{\boldsymbol{w}}_\infty)\boldsymbol{r}^\top\right)(\boldsymbol{B}^n)^\top. \tag{3.54}$$

■

Expression (3.54) is obtained by performing a series expansion of Eq. (3.51).

3.4 GROUPING STRATEGIES

In many practical cases, information about the group structure is not available beforehand. It is thus necessary to devise grouping strategies to endow agents with the ability to partition the estimated parameter vectors and to associate appropriate combination weights to each group.

3.4.1 FIXED GROUPING STRATEGY

A simple strategy is to split parameter vectors into a number of contiguous groups with preset lengths, possibly equal, and then assign a combination coefficient to each group, as illustrated in Fig. 3.1C. Splitting the parameter vector into subvectors can improve the performance for some applications, especially when there exist correlations among adjacent entries. However, this strategy may fail with particular configurations. For instance, consider the case where only the odd entries of the parameter vectors show some correlation. No matter how the group length is set, the algorithm will not be able to benefit from a uniform grouping strategy except perhaps if the group size is set to one. This motivates us to derive smart adaptive grouping strategies.

3.4.2 ADAPTIVE GROUPING STRATEGY

Adaptive grouping can be viewed as a clustering problem where we need to assign a label to each entry of a parameter vector. Before proceeding with the derivation, it is important to keep in mind that because we are considering algorithms with linear complexity (LMS-type algorithms) within the context of online learning and distributed adaptation, grouping/clustering should neither be performed in a centralized manner nor significantly increase the computational complexity. In other words, deriving a grouping strategy with quadratic complexity would not make much sense in this context. This constraint rules out most clustering algorithms used in machine learning and data analysis, e.g., hierarchical clustering, k-means or spectral clustering. In what follows, we introduce a simple but efficient strategy. As this strategy is time-independent, we shall omit the time index in notation for the sake of simplicity.

We start by introducing the following quantity that characterizes the deviation between the intermediate estimates defined in Eq. (3.18a) at nodes k and ℓ:

$$\delta_{k\ell,i} = \left|(\boldsymbol{\psi}_k)_i - (\boldsymbol{\psi}_\ell)_i\right| \quad \text{for} \quad \ell \in \mathcal{N}_k \tag{3.55}$$

for $i = 1, \ldots, L$. By averaging pairwise quantities $\delta_{k\ell,i}$ within the neighborhood, we then associate with each node k with the following L quantities:

$$\delta_{k,i} = \frac{1}{|\mathcal{N}_k|} \sum_{\ell \in \mathcal{N}_k} \delta_{k\ell,i} \quad \text{for} \quad i = 1, \ldots, L. \tag{3.56}$$

Each $\delta_{k,i}$ shows how the ith entry of $\boldsymbol{\psi}_k$ deviates from those of its neighbors. Entries with similar $\delta_{k,i}$ can be assigned to the same group because they have a similar average contrast level with respect to their neighbors. In this way, groups of entries with small (resp., large) contrast will lead the multitask diffusion algorithm to adopt a consensus (resp., noncooperative) strategy over these entries. Note that each node k can calculate $\delta_{k,i}$ in Eq. (3.56) after collecting the estimates from its neighbors, after the adaptation step (3.18a). We now propose the following steps to generate the groups:

1. Sort $\{\delta_{k,i}\}_{i=1}^{L}$ in ascending order to obtain the ordered sequence $\{\overline{\delta}_{k,i}\}_{i=1}^{L}$;
2. Generate the difference sequence $\Delta_{k,i} = \overline{\delta}_{k,i+1} - \overline{\delta}_{k,i}$, and determine the $K-1$ largest values $\Delta_{k,i}$ to split $\{\overline{\delta}_{k,i}\}_{i=1}^{L}$ into K sections where the largest changes occur;
3. Form a group with the original entries $\delta_{k,i}$ that are in a same section. Repeat this operation with the K sections determined in Step 2.

The computational complexity of this procedure is dominated by the sorting operation in Step 1. A complexity of $\mathcal{O}(L \log L)$ can be achieved with an efficient sorting algorithm. This grouping strategy derives from a vertex clustering algorithm commonly used in the literature [41]. Indeed, consider a fully connected graph with L vertices associated to the L entries of the local estimate $\boldsymbol{\psi}_k$. The edge between vertices (i.e., entries) i and j is assigned a weight equal to $|\delta_{k,i} - \delta_{k,j}|$. The above grouping procedure is then equivalent to generating a minimum spanning tree (MST) with Prim's algorithm on this graph [42], and then grouping the vertices into K clusters by cutting the most significant edges.

Before concluding this section, observe that Step 2 does not take relative differences into consideration for small $\overline{\delta}_{k,i}$. We suggest determining the cutting positions by considering the $K-1$ largest values of the normalized sequence $\xi_{k,i}$, with $\xi_{k,i} = 0$ if $\overline{\delta}_{k,i} < \tau$ and $\xi_{k,i} = (\overline{\delta}_{k,i+1} - \overline{\delta}_{k,i})/\overline{\delta}_{k,i}$ if $\overline{\delta}_{k,i} \geq \tau$, where τ denotes a given threshold.

3.4.3 ADAPTIVE COMBINATION STRATEGY

We now derive an adaptive combination strategy for group diffusion LMS. Motivated by Chen et al. [23] and Zhao and Sayed [43], it consists of adjusting the combination weights $a_{\ell k,m}$ in an online manner via instantaneous MSD minimization. Let us denote by $\tilde{\boldsymbol{w}}_{k,n}$ the weight error vector $\boldsymbol{w}_{k,n} - \boldsymbol{w}_k^{\star}$ after the combination step (3.18b). Considering groups $\mathcal{G}_m$, the instantaneous MSD at each agent k can be expressed as a function of $a_{\ell k,m}$ as follows:

$$\begin{aligned}
\mathbb{E}\{\|\tilde{\boldsymbol{w}}_{k,n}\|^2\} &= \sum_{m=1}^{M} \mathbb{E}\Big\{\big\|[\boldsymbol{w}_k^{\star}]_{\mathcal{G}_m} - \sum_{\ell \in \mathcal{N}_k} a_{\ell k,m}\, [\boldsymbol{\psi}_{\ell,n}]_{\mathcal{G}_m}\big\|^2\Big\} \\
&= \sum_{m=1}^{M} \sum_{\ell \in \mathcal{N}_k} \sum_{p \in \mathcal{N}_k} a_{\ell k,m}\, a_{pk,m} \left(\boldsymbol{\Psi}_{k,n}^{(m)}\right)_{\ell p},
\end{aligned} \tag{3.57}$$

where matrix $\Psi_{k,n}^{(m)}$ is the covariance matrix of the weight error for group m at node k and time instant n, with (ℓ, p)th entry given by:

$$\left(\Psi_{k,n}^{(m)}\right)_{\ell p} = \begin{cases} \mathbb{E}\left\{[\boldsymbol{w}_k^\star - \boldsymbol{\psi}_{\ell,n}]_{\mathcal{G}_m}^\top [\boldsymbol{w}_k^\star - \boldsymbol{\psi}_{p,n}]_{\mathcal{G}_m}\right\}, & \ell, p \in \mathcal{N}_k, \\ 0, & \text{otherwise.} \end{cases} \tag{3.58}$$

To make the problem tractable, we approximate $\Psi_{k,n}^{(m)}$ by an instantaneous value and we drop its off-diagonal entries. In addition, because $\boldsymbol{w}_k^\star$ is unknown, we approximate it by $\hat{\boldsymbol{w}}_k^\star$ as shown in Eq. (3.61). The instantaneous MSD minimization then leads to the optimization problem:

$$\begin{aligned} \min_{\boldsymbol{a}_{k,m}} \quad & \sum_{\ell=1}^{N}\sum_{m=1}^{M} a_{\ell k,m}^2 \left\| [\hat{\boldsymbol{w}}_k^\star - \boldsymbol{\psi}_{\ell,n}]_{\mathcal{G}_m} \right\|^2 \\ \text{subject to} \quad & \mathbf{1}_N^\top \boldsymbol{a}_{k,m} = 1, \quad a_{\ell k,m} \geq 0, \\ & a_{\ell k,m} = 0 \qquad \text{if } \ell \notin \mathcal{N}_k, \end{aligned} \tag{3.59}$$

where $\boldsymbol{a}_{k,m} = [a_{1k,m}, \ldots, a_{Nk,m}]^\top$. The above objective function promotes weak information exchange via small $a_{\ell k,m}$ if the estimate of group $\mathcal{G}_m$ at node ℓ is far from its counterpart at node k. The solution of Eq. (3.59) is given by:

$$a_{\ell k,m} = \frac{\|[\hat{\boldsymbol{w}}_k^\star - \boldsymbol{\psi}_{\ell,n}]_{\mathcal{G}_m}\|^{-2}}{\sum_{j \in \mathcal{N}_k} \|[\hat{\boldsymbol{w}}_k^\star - \boldsymbol{\psi}_{j,n}]_{\mathcal{G}_m}\|^{-2}}, \quad \text{for } \ell \in \mathcal{N}_k. \tag{3.60}$$

We now introduce an instantaneous approximation $\hat{\boldsymbol{w}}_{k,n}^\star$ for $\boldsymbol{w}_k^\star$ at each node k and time instant n. In order to reduce the MSD bias that may result from an inappropriate cooperation between nodes performing distinct estimation tasks, a possible strategy is to use the local one-step ahead approximation:

$$\hat{\boldsymbol{w}}_{k,n}^\star = \boldsymbol{\psi}_{k,n} + \mu_k' \boldsymbol{q}_{k,n}, \tag{3.61}$$

where $\boldsymbol{q}_{k,n} = [d_k(n) - \boldsymbol{x}_{k,n}^\top \boldsymbol{\psi}_{k,n}]\boldsymbol{x}_{k,n}$ is the instantaneous approximation of the negative gradient of $J_k(\boldsymbol{w})$ at $\boldsymbol{\psi}_{k,n}$. Substituting this expression into Eq. (3.60) leads to the combination rule:

$$a_{\ell k,m}(n) = \frac{\|[\boldsymbol{\psi}_{k,n} + \mu_k' \boldsymbol{q}_{k,n} - \boldsymbol{\psi}_{\ell,n}]_{\mathcal{G}_m}\|^{-2}}{\sum_{j \in \mathcal{N}_k} \|[\boldsymbol{\psi}_{k,n} + \mu_k' \boldsymbol{q}_{k,n} - \boldsymbol{\psi}_{j,n}]_{\mathcal{G}_m}\|^{-2}} \tag{3.62}$$

for $\ell \in \mathcal{N}_k$ and $m = 1, \ldots, M$. Furthermore, we observed in our experiments that the normalized gradient $\boldsymbol{q}_{k,n} \leftarrow \boldsymbol{q}_{k,n}/(\|\boldsymbol{q}_{k,n}\| + \epsilon)$ with ϵ a small positive constant can increase the robustness of the resulting strategy.

3.5 SIMULATIONS

In this section, we shall first report simulation results that illustrate the theoretical findings, and then simulate the adaptive grouping and combination weight adjustment algorithms. All agents were initialized with zero parameter vector $\boldsymbol{w}_{k,0} = \mathbf{0}_L$ for all k. Simulated curves were obtained by averaging over 100 Monte-Carlo runs.

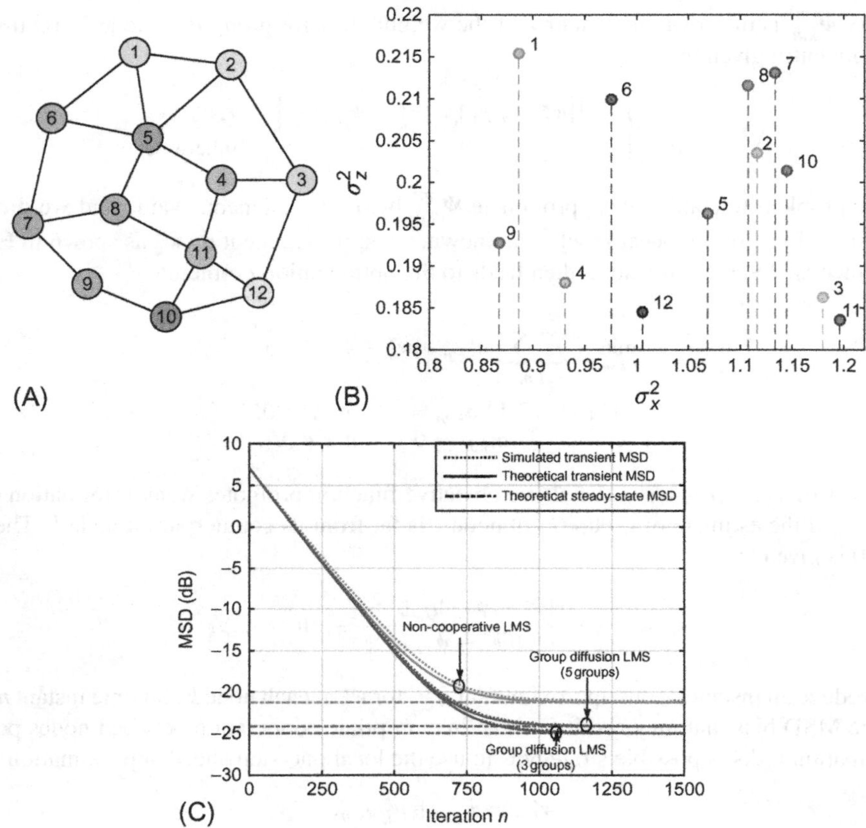

FIG. 3.2

Validation of the mean-square error behavior analysis. (A) Network. (B) Variance of regressors and noise at each node. (C) MSD learning curves.

3.5.1 MODEL VALIDATION

We considered the network with $N = 12$ nodes shown in Fig. 3.2A. The optimum parameter vectors $\{w_k^\star\}_{k=1}^N$ consisted of $L = 15$ entries. The first six entries were common across all nodes, that is,

$$[w_1^\star]_{\mathcal{G}_1} = \cdots = [w_N^\star]_{\mathcal{G}_1} \quad \text{with} \quad \mathcal{G}_1 = \{1, \ldots, 6\}. \tag{3.63}$$

These entries were sampled from a uniform distribution $\mathcal{U}(-1, 1)$. The next four entries were uniformly sampled from $\mathcal{U}(-1, 1)$ for each node, so that there was no relationship between the entries of this group, that is,

$$[w_k^\star]_i = u_{ki} \quad \text{for} \quad i \in \mathcal{G}_2 = \{7, \ldots, 10\} \tag{3.64}$$

with independent u_{ki} sampled from $\mathcal{U}(-1, 1)$. The last five entries were set to:

$$[\mathbf{w}_k^\star]_i = [\mathbf{w}^o]_i + u_{ki} \quad \text{for} \quad i \in \mathcal{G}_3 = \{11, \ldots, 15\} \tag{3.65}$$

with $[\mathbf{w}^o]_i$ uniformly sampled from $\mathcal{U}(-1, 1)$, identical for all nodes, and i.i.d. perturbations u_{ki} sampled from $\mathcal{U}(-0.1, 0.1)$. In the sequel, we shall refer to the generative model (3.63)–(3.65) by $\mathcal{S}_1$ for short. Observe that nodes did not know this setup beforehand. Input vectors $\boldsymbol{x}_n$ were zero-mean $L \times 1$ random vectors governed by a Gaussian distribution with covariance matrix $\boldsymbol{R}_{x,k} = \sigma_{x,k}^2 \boldsymbol{I}_L$. The noises $z_k(n)$ were i.i.d. zero-mean Gaussian random variables, independent of any other signal with variances $\sigma_{z,k}^2$. Variances $\sigma_{x,k}^2$ and $\sigma_{z,k}^2$ used in this experiment were sampled from $\mathcal{U}(0.8, 1.2)$ and $\mathcal{U}(0.18, 0.22)$, respectively. Their values are depicted on the signal-noise variance plot shown in Fig. 3.2B.

First, we illustrate the theoretical model with constant combination coefficients as characterized in Section 3.3.3. The following algorithm settings were considered:

- Noncooperative LMS. This is the limit case where the combination coefficient matrix $\boldsymbol{A}$ is set to $\boldsymbol{I}$.
- Diffusion LMS. This is the limit case where there exists only one group. A uniform combination matrix $\boldsymbol{A}$ with $a_{\ell k} = |\mathcal{N}_k|^{-1}$ was used for this experiment. As for the noncooperative LMS, this setting was used as a baseline.
- Group diffusion LMS with three groups. In this setting, the groups were set according to the generative model. For the first group, $\boldsymbol{A}_1$ was set to a uniform combination matrix with $a_{\ell k,1} = |\mathcal{N}_k|^{-1}$. For the second group, the combination matrix was set to $\boldsymbol{A}_2 = \boldsymbol{I}$. For the third group, the combination matrix was set to an intermediate version $\boldsymbol{A}_3 = 0.5\boldsymbol{I} + 0.5\boldsymbol{A}_1$.
- Group diffusion LMS with five groups. We split the L entries into five groups of $L/5$ consecutive entries. A uniform combination matrix $\boldsymbol{A}_1$ with $a_{\ell k,1} = |\mathcal{N}_k|^{-1}$ was used for the first group. Matrix $\boldsymbol{A}_2$ was generated with the Metropolis rule, that is, $a_{\ell k,2} = \max\{|\mathcal{N}_k|, |\mathcal{N}_\ell|\}^{-1}$ for $k \in \mathcal{N}_k \backslash \{k\}$, $a_{kk,2} = 1 - \sum_{\ell \in \mathcal{N}_k \backslash \{k\}} a_{\ell k,2}$, otherwise $a_{\ell k,2} = 0$, for the second group. The identity matrix was used for the third and fourth groups, namely, $\boldsymbol{A}_3 = \boldsymbol{A}_4 = \boldsymbol{I}$. For the fifth group, the combination matrix was set to $\boldsymbol{A}_5 = 0.5\boldsymbol{I} + 0.5\boldsymbol{A}_1$.

This experimental setup was considered to test the theoretical models rather than reveal the performance gain using a grouping strategy. The step size was set to $\mu = 0.005$. The resulting MSD curves are illustrated in Fig. 3.2C. The theoretical transient and steady-state MSD were evaluated using Eqs. (3.52)–(3.54). The theoretical curves are generally consistent with the Monte Carlo simulated curves. It can be observed that single-task diffusion LMS algorithm had a large MSD due to the bias caused by averaging over the entries of the group $\mathcal{G}_2$. Group diffusion LMS with three groups performed the best, because its three groups correspond to the generative model and we associated reasonable combination coefficients to each group. Group diffusion LMS with five groups performed slightly worse than with the three-group setting because the fourth group overlaps $\mathcal{G}_2$ and $\mathcal{G}_3$ of the generative model. This simulation confirms that a grouping strategy should improve the performance, and also suggests that it should be adaptive.

3.5.2 PERFORMANCE OF THE ADAPTIVE GROUPING STRATEGY

The aim of this section is to compare the static and adaptive grouping strategies along with the adaptive method for setting the combination coefficients. The following algorithms and settings were considered:

- Noncooperative LMS. This noncooperative algorithm was used as a reference for the performance comparison.
- Multitask diffusion LMS in [23]. The number of groups considered with this algorithm is $M = 1$. Nodes are, however, endowed with the adaptive combination function that allows them to adapt the combination weights $a_{\ell k}$ in an online way. This algorithm was also used as a baseline for performance comparison to illustrate the need for an adaptive grouping strategy.
- Group diffusion LMS with preset groups. First, we considered the same groups as the generative model. Next we uniformly split the parameter vectors into M contiguous groups of the same size. With this algorithm, nodes are endowed with the adaptive combination function only.
- Group diffusion LMS with adaptive grouping strategy. With this algorithm, nodes are endowed with the adaptive variables grouping function and the adaptive combination function.

First, we considered the generative model $\mathcal{S}_1$ used for model validation in Section 3.5.1, namely, Eqs. (3.63)–(3.65). The second generative model we considered, denoted by $\mathcal{S}_2$, consisted of a partition into two groups of the parameter vectors entries. The first group involved all the odd entries as follows:

$$[\boldsymbol{w}_1^\star]_{\mathcal{G}_1} = \ldots = [\boldsymbol{w}_N^\star]_{\mathcal{G}_1} \quad \text{with} \quad \mathcal{G}_1 = \{1, 3, \ldots, 15\}, \text{ for } \forall k \tag{3.66}$$

and the second group involved all even entries as follows:

$$[\boldsymbol{w}_k^\star]_i = u_{ki} \quad \text{for } i \in \mathcal{G}_2 = \{2, 4, \ldots, 14\}, \text{ for } \forall k \tag{3.67}$$

with u_{ki} randomly drawn from $\mathcal{U}(-1, 1)$.

Fig. 3.3 illustrates the MSD convergence behavior of the algorithms enumerated above. The noncooperative LMS algorithm can be considered as a baseline for this comparative test because it does not rely on any cooperation. The multitask diffusion LMS considered in [23] reached a slightly larger MSD than the noncooperative LMS. This algorithm is able to adjust the combination weights $a_{\ell k}$ in an adaptive manner, but it cannot take possible group structures into account. It processed the parameter vectors as if they were significantly different and inhibited cooperation between nodes. This result reveals the need for a grouping strategy. By setting the group structure in accordance with the generative model, the group diffusion LMS with preset groups achieved the lowest MSD for both $\mathcal{S}_1$ and $\mathcal{S}_2$.

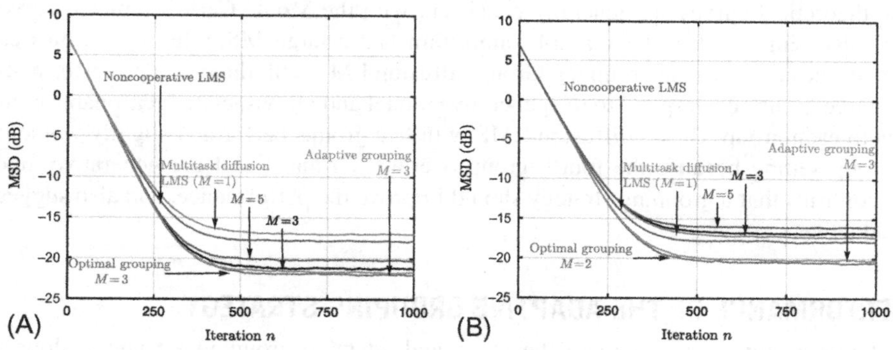

FIG. 3.3

Comparison of MSD learning curves. (A) $\mathcal{S}_1$; (B) $\mathcal{S}_2$.

With $M = 3$ preset uniform groups, the group diffusion LMS also led to a significant performance improvement over the noncooperative LMS for $\mathcal{S}_1$, showing that preset groups can be beneficial. With $M = 5$ groups, this algorithm still outperformed the noncooperative algorithm. A larger MSD than in the case $M = 3$ was, however, observed, which shows that increasing the number of groups may not always be beneficial. It is worth noting that the group diffusion LMS with preset groups of sizes $M = 3$ and $M = 5$ led to unfavorable performance with $\mathcal{S}_2$. The limits of this strategy involving preset uniform groups of entries have already been discussed in Section 3.4.1. Finally, the proposed group diffusion LMS with adaptive grouping and adaptive combination coefficients was run with $M = 3$ groups. Note that M was thus voluntarily overestimated for $\mathcal{S}_2$. For experimental setups $\mathcal{S}_1$ and $\mathcal{S}_2$, it performed almost as well as when using the ground truth groups. Fig. 3.4 and 3.6 show the group structures at

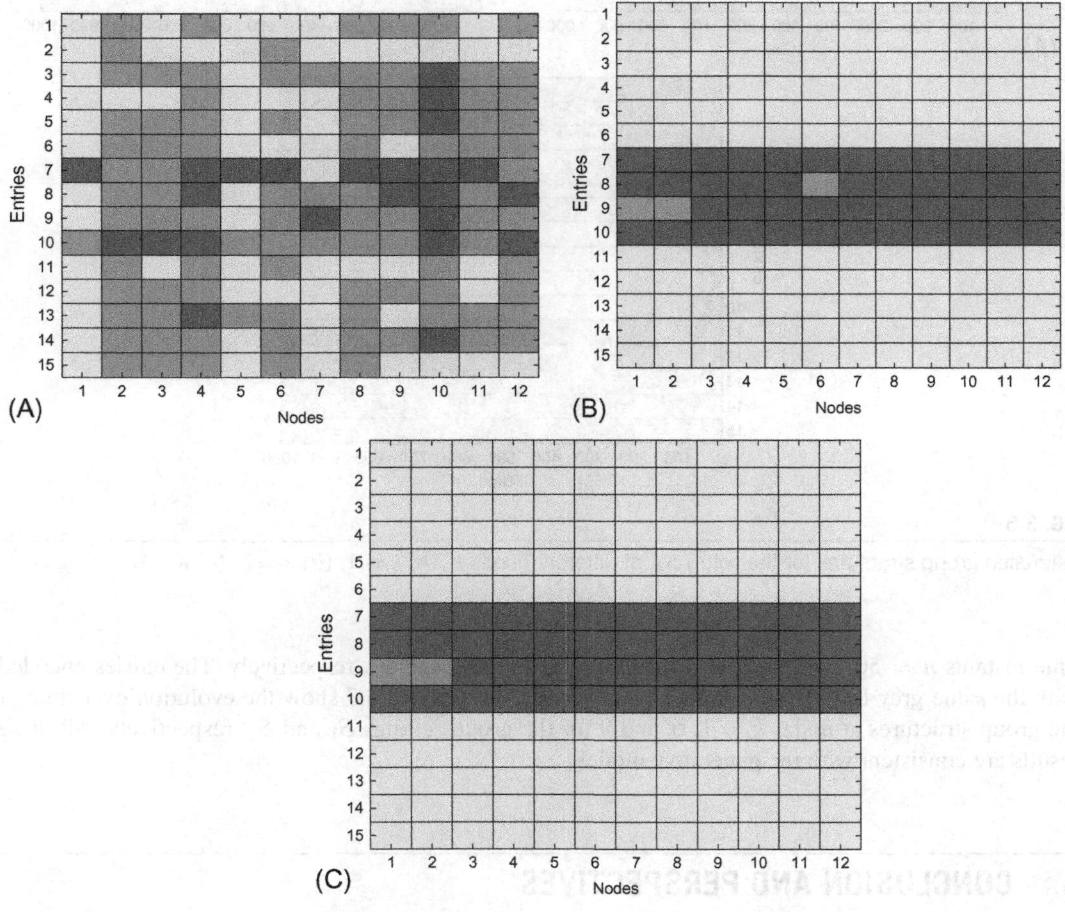

FIG. 3.4

Estimated group structures for the setup $\mathcal{S}_1$, at different time instants n. (A) $n = 50$; (B) $n = 250$; (C) $n = 1000$.

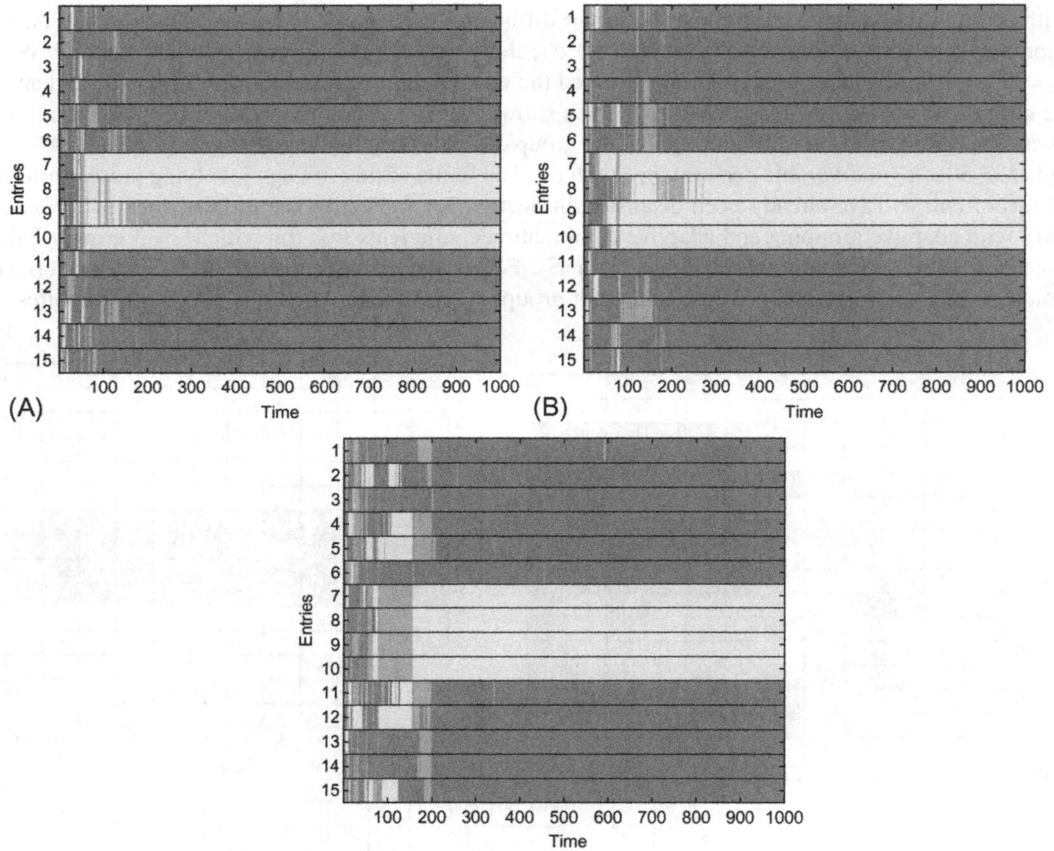

FIG. 3.5

Estimated group structures for the setup $\mathcal{S}_1$, at different nodes k. (A) $k = 1$; (B) $k = 6$; (C) $k = 9$.

time instants $n = 50$, 250, and 1000 for group settings $\mathcal{S}_1$ and $\mathcal{S}_2$, respectively. The entries encoded with the same grey level belong to the same group. Figs. 3.5 and 3.7 show the evolution over time of the group structures at nodes $k = 1$, 6, and 9 for the group settings $\mathcal{S}_1$ and $\mathcal{S}_2$, respectively. All these results are consistent with the generative models.

3.6 CONCLUSION AND PERSPECTIVES

In this paper, we introduced an adaptive grouping procedure into diffusion adaptation to take advantage of structural similarities among parameter vectors to estimate. Simulation results illustrated the effectiveness of the grouping strategy and of the adaptive information fusion rule.

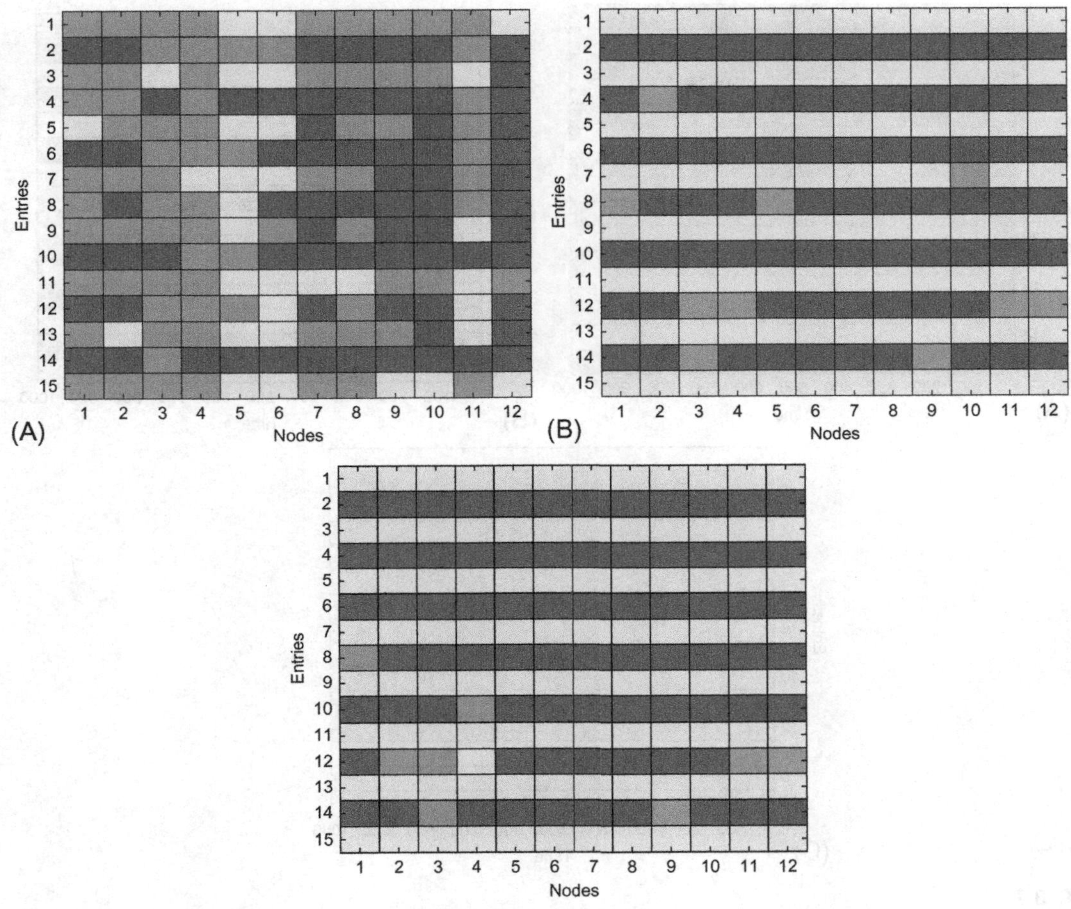

FIG. 3.6

Estimated group structures for the setup $\mathcal{S}_2$, at different time instants n. (A) $n = 50$; (B) $n = 250$; (C) $n = 1000$.

PROOF OF COROLLARY 3.1

The results below are based on the following properties of the Kronecker product:

$$\text{vec}(\boldsymbol{XYZ}) = (\boldsymbol{Z}^\top \otimes \boldsymbol{X})\text{vec}(\boldsymbol{Y}) \quad \text{and} \quad \text{trace}(\boldsymbol{XY}) = \left(\text{vec}(\boldsymbol{Y}^\top)\right)^\top \text{vec}(\boldsymbol{X}).$$

Then, the first term in Eq. (3.49) can be rewritten as follows:

$$\left(\text{vec}\{\boldsymbol{G}^\top\}\right)^\top \boldsymbol{K}^n \boldsymbol{\sigma}_I = \frac{1}{N}\text{trace}\left(\boldsymbol{G}\left[\boldsymbol{B}^{n\top}\boldsymbol{B}^n\right]\right).$$

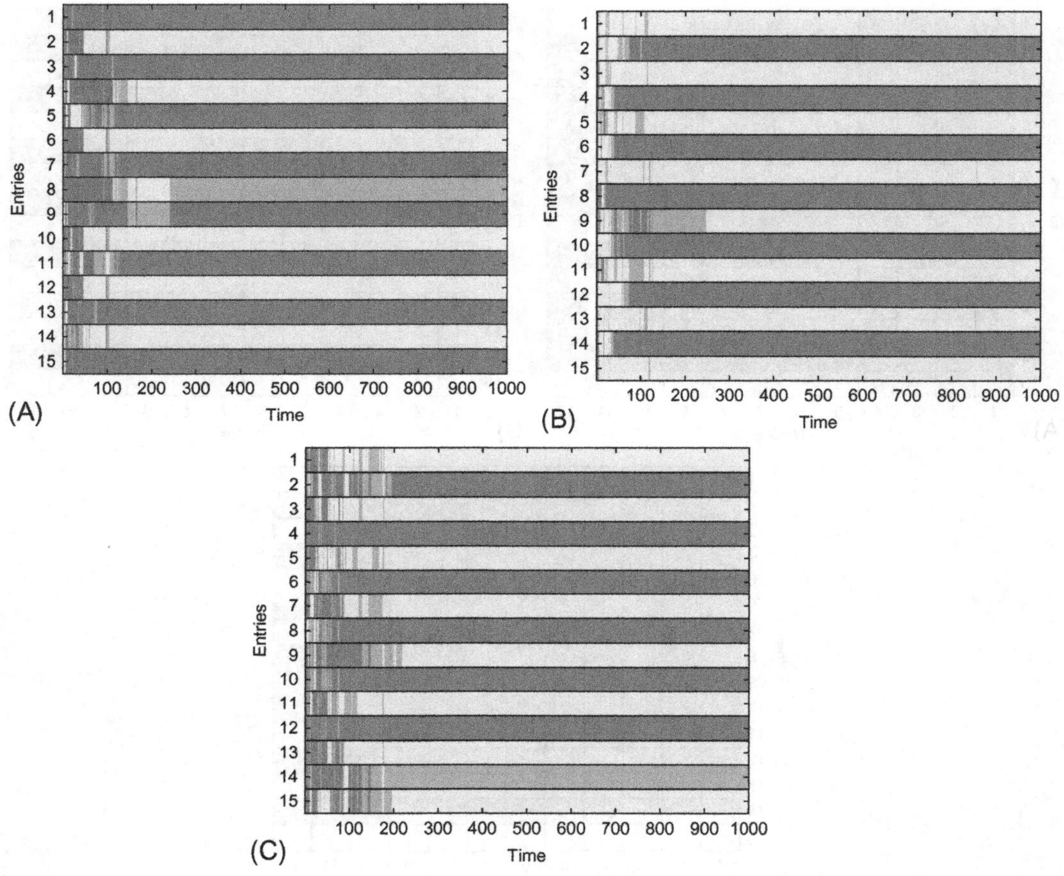

FIG. 3.7

Estimated group structures for the setup $\mathcal{S}_2$, at different nodes k. (A) $k = 1$; (B) $k = 6$; (C) $k = 9$.

The second term is given by:

$$\|\boldsymbol{r}\|^2_{\boldsymbol{K}^n\boldsymbol{\sigma}_I} = \frac{1}{N}\text{trace}\Big(\boldsymbol{r}\boldsymbol{r}^\top\,[\boldsymbol{B}^{n\top}\boldsymbol{B}^n]\Big).$$

The third term can be expressed as follows:

$$\|\tilde{\boldsymbol{w}}_0\|^2_{(\boldsymbol{I}_{(NL)^2}-\boldsymbol{K})\boldsymbol{K}^n\boldsymbol{\sigma}_I} = \frac{1}{N}\text{trace}\Big(\tilde{\boldsymbol{w}}_0\tilde{\boldsymbol{w}}_0^\top\,[\boldsymbol{B}^{n\top}\boldsymbol{B}^n - \boldsymbol{B}^{n+1\top}\boldsymbol{B}^{n+1}]\Big).$$

The last term can be rewritten as:

$$\Big(\boldsymbol{\gamma}_n^\top + [\boldsymbol{B}\,\mathbb{E}\{\tilde{\boldsymbol{w}}_n\}]^\top \otimes \boldsymbol{r}^\top\Big)\boldsymbol{\sigma}_I = \text{trace}(\boldsymbol{\Gamma}_n^\top) + \boldsymbol{r}^\top\boldsymbol{B}\,\mathbb{E}\{\tilde{\boldsymbol{w}}_n\},$$

where $\boldsymbol{\Gamma}_n$ is defined hereafter. With these expressions, we obtain the following update equation that now depends on $\boldsymbol{B}$ rather than $\boldsymbol{K}$:

$$\zeta_{n+1} = \zeta_n + \frac{1}{N}\text{trace}\Big([\boldsymbol{G} + \boldsymbol{r}\boldsymbol{r}^\top]\boldsymbol{B}^{n\top}\boldsymbol{B}^n \\ -\tilde{\boldsymbol{w}}_0\tilde{\boldsymbol{w}}_0^\top\left[\boldsymbol{B}^{n\top}\boldsymbol{B}^n - \boldsymbol{B}^{n+1\top}\boldsymbol{B}^{n+1}\right] - 2\boldsymbol{\Gamma}_n - 2\boldsymbol{B}\,\mathbb{E}\{\tilde{\boldsymbol{w}}_n\}\boldsymbol{r}^\top\Big).$$

The matrix form $\boldsymbol{\Gamma}_{n+1}$ of $\boldsymbol{\gamma}_{n+1}$ is updated as follows:

$$\boldsymbol{\Gamma}_{n+1} = \boldsymbol{B}\boldsymbol{\Gamma}_n\boldsymbol{B}^\top + \boldsymbol{B}\boldsymbol{r}\,(\boldsymbol{B}^2\,\mathbb{E}\{\tilde{\boldsymbol{w}}_n\})^\top - \boldsymbol{r}(\boldsymbol{B}\,\mathbb{E}\{\tilde{\boldsymbol{w}}_n\})^\top.$$

REFERENCES

[1] Bertsekas DP. A new class of incremental gradient methods for least squares problems. SIAM J Optimiz 1997;7(4):913–26.

[2] Rabbat MG, Nowak RD. Quantized incremental algorithms for distributed optimization. IEEE J Sel Topics Areas Commun 2005;23(4):798–808.

[3] Blatt D, Hero AO, Gauchman H. A convergent incremental gradient method with constant step size. SIAM J Optimiz 2007;18(1):29–51.

[4] Lopes CG, Sayed AH. Incremental adaptive strategies over distributed networks. IEEE Trans Signal Process 2007;55(8):4064–77.

[5] Nedic A, Ozdaglar A. Distributed subgradient methods for multi-agent optimization. IEEE Trans Autom Control 2009;54(1):48–61.

[6] Kar S, Moura JMF. Distributed consensus algorithms in sensor networks: link failures and channel noise. IEEE Trans Signal Process 2009;57(1):355–69.

[7] Srivastava K, Nedic A. Distributed asynchronous constrained stochastic optimization. IEEE J Sel Top Signal Process 2011;5(4):772–90.

[8] Lopes CG, Sayed AH. Diffusion least-mean squares over adaptive networks: formulation and performance analysis. IEEE Trans Signal Process 2008;56(7):3122–36.

[9] Cattivelli FS, Sayed AH. Diffusion LMS strategies for distributed estimation. IEEE Trans Signal Process 2010;58(3):1035–48.

[10] Chen J, Sayed AH. Diffusion adaptation strategies for distributed optimization and learning over networks. IEEE Trans Signal Process 2012;60(8):4289–305.

[11] Chen J, Sayed AH. Distributed Pareto optimization via diffusion strategies. IEEE J Sel Top Signal Process 2013;7(2):205–20.

[12] Sayed AH, Tu SY, Chen J, Zhao X, Towfic ZJ. Diffusion strategies for adaptation and learning over networks. IEEE Sig Process Mag 2013;30(3):155–71.

[13] Sayed AH. Diffusion adaptation over networks. In: Chellapa R, Theodoridis S, editors. E-reference signal processing, vol. 3. Elsevier; 2014. p. 323–454.

[14] Sayed AH. Adaptation, learning, and optimization over networks. In: Foundations and trends in machine learning, vol. 7. Boston-Delft: NOW Publishers; 2014. p. 311–801.

[15] Sayed AH. Adaptive networks. Proc IEEE 2014;102(4):460–97.

[16] Chen J, Sayed AH. On the learning behavior of adaptive networks—Part I: Transient analysis. IEEE Trans Inf Theory 2015;61(6):3487–517.

[17] Chen J, Sayed AH. On the learning behavior of adaptive networks—Part II: Performance analysis. IEEE Trans Inf Theory 2015;61(6):3518–48.

[18] Tu SY, Sayed AH. Diffusion strategies outperform consensus strategies for distributed estimation over adaptive networks. IEEE Trans Signal Process 2012;60(12):6217–34.

[19] Bertrand A, Moonen M. Distributed adaptive node-specific signal estimation in fully connected sensor networks—Part I: Sequential node updating. IEEE Trans Signal Process 2010;58(10):5277–91.

[20] Bertrand A, Moonen M. Distributed adaptive estimation of node-specific signals in wireless sensor networks with a tree topology. IEEE Trans Signal Process 2011;59(5):2196–10.

[21] Chen J, Richard C, Sayed AH. Multitask diffusion adaptation over networks. IEEE Trans Signal Process 2014;62(16):4129–44.

[22] Nassif R, Richard C, Ferrari A, Sayed AH. Multitask diffusion adaptation over asynchronous networks. IEEE Trans Signal Process 2016;64(11):2835–50.

[23] Chen J, Richard C, Sayed AH. Diffusion LMS over multitask networks. IEEE Trans Signal Process 2015;63(11):2733–48.

[24] Zhao X, Sayed AH. Distributed clustering and learning over networks. IEEE Trans Signal Process 2015;63(13):3285–300.

[25] Chen J, Richard C, Sayed AH. Adaptive clustering for multitask diffusion networks. In: Proceedings of European signal processing conference (EUSIPCO), Nice, France; 2015. p. 200–4.

[26] Monajemi S, Sanei S, Ong SH, Sayed AH. Adaptive regularized diffusion adaptation over multitask networks. In: Proceedings of IEEE international workshop on machine learning for signal processing (MLSP), Boston, USA; 2015. p. 1–5.

[27] Khawatmi S, Zoubir AM, Sayed AH. Decentralized clustering over adaptive networks. In: Proceedings of European signal processing conference (EUSIPCO), Nice, France; 2015. p. 2696–700.

[28] Monajemi S, Eftaxias K, Sanei S, Ong SH. An informed multitask diffusion adaptation approach to study tremor in Parkinson's disease. IEEE J Sel Top Signal Process 2016;10(7):1306–14.

[29] Wang Y, Tay WP, Hu W. Multitask diffusion LMS with optimized inter-cluster cooperation. In: Proceedings of the IEEE statistical signal processing workshop (SSP), Palma de Mallorca, Spain; 2016. p. 1–5.

[30] Nassif R, Richard C, Ferrari A, Sayed AH. Proximal multitask learning over networks with sparsity-inducing coregularization. IEEE Trans Signal Process 2016;64(23):6329–44.

[31] Chen J, Richard C, Hero AO, Sayed AH. Diffusion LMS for multitask problems with overlapping hypothesis subspaces. In: Proceedings of the IEEE international workshop on machine learning for signal processing (MLSP), Reims, France; 2014. p. 1–6.

[32] Chen J, Richard C, Sayed AH. Multitask diffusion adaptation over networks with common latent representations. IEEE J Sel Top Signal Process 2017;11(3):563–79.

[33] Hua J, Li C, Shen H. Distributed learning of predictive structures from multiple tasks over networks. IEEE Trans Ind Elect 2017;64(5):4246–56.

[34] Bogdanović N, Plata-Chaves J, Berberidis K. Distributed diffusion-based LMS for node-specific parameter estimation over adaptive networks. In: Proceedings of the IEEE international conference on acoustics, speech and signal processing (ICASSP), Florence, Italy; 2014. p. 7223–7.

[35] Plata-Chaves J, Bogdanović N, Berberidis K. Distributed diffusion-based LMS for node-specific adaptive parameter estimation. IEEE Trans Signal Process 2015;63(13):3448–60.

[36] Plata-Chaves J, Bahari HH, Moonen M, Bertrand A. Unsupervised diffusion-based LMS for node-specific parameter estimation over wireless sensor networks. In: Proceedings of the IEEE international conference on acoustics, speech and signal processing (ICASSP), Shanghai, China; 2016. p. 4159–63.

[37] Nassif R, Richard CA, Ferrari AA, Sayed AH. Diffusion LMS for multitask problems with local linear equality constraints. IEEE Trans Signal Process 2017;65(19):4979–93.

[38] Hua F, Nassif R, Richard CA, Wang H. Penalty-based multitask estimation with non-local linear equality constraints. In: Proceedings of the IEEE international workshop on computational advances in multi-sensor adaptive processing (CAMSAP), Curaçao; 2017. p. 433–7.
[39] Ting SK. Adaptive techniques for mitigating circuit imperfections in high performance A/D converters. Ph.D. thesis, Electrical Engineering Department, UCLA; 2014.
[40] Sayed AH. Adaptive filters. John Wiley & Sons; 2008.
[41] Asano T, Bhattacharya B, Keil M, Yao F. Clustering algorithms based on minimum and maximum spanning trees. In: Proceedings of the 4th annual symposium on computational geometry (SCG), Urbana-Champaign, USA; 1988. p. 252–7.
[42] Rosen K. Discrete mathematics and its applications. 7th ed. McGraw-Hill Science; 2011.
[43] Zhao X, Sayed AH. Clustering via diffusion adaptation over networks. In: Proceedings of the international workshop on cognitive information processing (CIP), Baiona, Spain; 2012. p. 1–6.

CHAPTER

BAYESIAN APPROACH TO COLLABORATIVE INFERENCE IN NETWORKS OF AGENTS

4

Kamil Dedecius*, Petar M. Djurić†

Department of Adaptive Systems, Institute of Information Theory and Automation, The Czech Academy of Sciences, Prague, Czech Republic Department of Electrical and Computer Engineering, Stony Brook University, Stony Brook, NY, United States†*

4.1 INTRODUCTION

With the rapid development of devices with high computational performance, the probabilistically consistent and versatile Bayesian methods have become a popular standard in many applications of signal processing [1]. Their main difference in comparison to traditional approaches consists in the representation of the unknown variables of interest. They are described by probability distributions whose location statistics (the mean, mode, or median) express the probable locations of these variables while the dispersion statistics of the distribution (e.g., the variance) quantify the associated uncertainty. An important facet of Bayesian theory is the universality of its methods. According to Bayes' theorem, one updates the initial knowledge of a considered variable, represented by a prior distribution, by following the same generic steps regardless of whether the underlying task is linear or nonlinear regression, filtering of state-space model parameters, or estimation of hierarchical models.

In the last decade, signal processing has faced a host of new challenges related to the fast evolution of spatially distributed systems with components—termed agents, sensors, nodes, or vertices and here referred to as agents—that have relatively high sensing and computational performance, and may communicate with other agents of the network. The applications of these systems range from environment monitoring, disaster relief management, source localization, and precision agriculture to medicine [2–5]. The first algorithms for processing of data acquired by agents were centralized. More specifically, there the agents locally sense the relevant data and send them to a *fusion center*, responsible for evaluation of (nearly) all necessary computations. Subsequently, the results are sent back to the agents, if necessary. In this setting, the fusion center exploits all the data present in the network and thus reaches the best possible estimation performance. The price for this is the high communication and computation demands and the lack of redundancy, making the centralized algorithms prone to failures [6].

Cooperative and Graph Signal Processing. https://doi.org/10.1016/B978-0-12-813677-5.00004-3

Table 4.1 Summary of Notation

Notation	Description
$\mathcal{K} = \{1, \ldots, K\}$	set of agents
K	total number of agents
$k \in \mathcal{K}$	agent index
$\mathfrak{N}_k$	neighborhood of agent k
$\lvert\mathfrak{N}_k\rvert$	cardinality of the set $\mathfrak{N}_k$
$t = 1, 2, \ldots$	discrete time index
$y_{t,k}$	observation of kth agent at time t
$z_{t,k}$	explanatory variable observed by agent k at time t
θ	model parameter
$\widehat{\theta}$	model parameter point estimate
$f(y_{t,k}\lvert z_{t,k}, \theta)$	probability density of observations of agent k
$\pi_k(\theta)$	probability density of θ of agent k
$\eta = \eta(\theta)$	natural parameter
$T(\cdot)$	sufficient statistic
$\xi_{t,k}, \nu_{t,k}$	hyperparameters of agent k at time t
$\zeta_{t,k}$	information of agent k at time t
$a_{kj} \in [0, 1]$	weight assigned by agent k to agent j
$\mathbb{E}[\cdot]$	expectation operator
$\mathcal{D}(\cdot\lvert\lvert\cdot)$	Kullback-Leibler divergence
$\tilde{\pi}_k(\cdot)$	combined posterior density of agent k
$\tilde{\xi}_{t,k}, \tilde{\nu}_{t,k}$	hyperparameters of $\tilde{\pi}_k(\cdot)$
Tr	trace operator
det	determinant
$\mathcal{N}(\mu, \Sigma)$	normal distribution with mean μ and covariance Σ

In order to remove this drawback, fully distributed processing settings have been proposed. First, incremental algorithms have been studied [7–13], where information is passed from agent to agent in a cyclic Hamiltonian path connecting the whole network. Although this removes the need for a fusion center and alleviates the communication and computational burden, the reliability of the system is not improved as each agent and link are single points of failure. A recovery from such a failure by constructing a new path is an NP-hard problem [4]. Then *consensus* [14–20] and *diffusion* strategies [21–26] have been introduced, where the agents share information with their neighbors within a one-hop distance. Both strategies offer significantly more robust solutions. The consensus strategies aim at a general agreement in the value of the estimated variables of interest while the diffusion strategies put emphasis on a local improvement of the estimation quality of each agent. Therefore, while the diffusion algorithms intrinsically exploit a single time scale both for sensing and collaborative data processing, the consensus algorithms usually need multiple iterations between two time instants. This chapter focuses primarily on diffusion strategies.

The existing diffusion algorithms mostly extend their nondistributed counterparts to distributed settings. A majority of them exploit the least squares criterion and its variants, e.g., the least-mean-squares methods [22,27–29], the recursive least-squares methods [21,26], and the Kalman filter [30,31]. There are also algorithms for distributed expectation maximization-based inference of mixture models [32–34] and distributed particle filtering, e.g., in [35,36].

Quite surprisingly, most of the algorithms for distributed inference are *independently* developed from the original nondistributed ones, e.g., the already-mentioned classical least-mean-squares method, recursive least-squares method, and Kalman filter. However, because there is a common underlying principle in Bayesian estimation, a unifying framework has been recently proposed [37]. Within this framework, one can develop methods that can be applied to a wide class of inference tasks with minimal modifications. For instance, the existing recursive least-squares method [21] or the Kalman filter [31] are special cases when particular models and prior distributions are used. In [34], a quasi-Bayesian algorithm for sequential estimation of mixture models was introduced. The components of the models considered there belong to the exponential family of distributions. In [38], a diffusion approximate Bayesian computation method was presented. The method extends the particle filtering principles to cases of unknown or intractable models. The foundations of these approaches are described in the present chapter.

4.2 BAYESIAN INFERENCE OVER NETWORKS

We consider a network to be represented by a connected undirected graph consisting of a set of vertices termed agents. The agents are interconnected by a set of edges, which determine the network topology. The set of agents is denoted by $\mathcal{K} = \{1, 2, \ldots, K\}$, where K is the number of agents in the network. (A summary of notation for this chapter is presented in Table 4.1.) Each agent $k \in \mathcal{K}$ may communicate only with agents in its close neighborhood $\mathfrak{N}_k$, here defined as a set of agents within one-hop distance (note that $k \in \mathfrak{N}_k$). The agents independently observe outcomes $y_{t,k}$ of a common discrete-time stochastic process $\{Y_t; t = 1, 2, \ldots\}$ with an unknown parameter θ, and a known explanatory variable $z_{t,k}$, if it exists. For instance, $z_{t,k}$ may be a regressor.

For the sake of prediction, filtering, and smoothing, the agents employ a probabilistic model in the form of a probability density $f(y_{t,k}|z_{t,k}, \theta)$, or $f(y_{t,k}|\theta)$ if $z_{t,k}$ is not assumed. The value of θ remains unknown, but its reliable estimation is of main interest in the rest of the chapter. The Bayesian approach to estimating θ proceeds by updating the prior distributions of θ at time t, $\pi_k(\theta|y_{0:t-1,k}, z_{0:t-1,k})$, where $y_{0:t-1,k} \equiv \{y_{0,k}, \ldots, y_{t-1,k}\}$ and $z_{0:t-1,k} \equiv \{z_{0,k}, \ldots, z_{t-1,k}\}$, by using the new observation $y_{t,k}$ via Bayes' theorem,

$$\pi_k(\theta|y_{0:t,k}, z_{0:t,k}) \propto f(y_{t,k}|z_{t,k}, \theta)\pi_k(\theta|y_{0:t-1,k}, z_{0:t-1,k}). \tag{4.1}$$

Note that in writing this equation we assume that the observations are independent given the explanatory variables $z_{t,k}$ and the parameter θ. In this chapter, we assume that the reader is familiar with the principles of Bayesian inference, and skip technical details of derivation of posterior distributions and their computations via Monte Carlo methods, variational approaches, and the like. There is a vast literature on this topic, e.g., [39,40]. Next, we describe a prominent case where the posterior distribution is analytically tractable and which will be used in the sequel.

Suppose that the model $f(y_{t,k}|z_{t,k},\theta)$ belongs to the exponential family of distributions, i.e., that it can be written in the form [41]

$$f(y_{t,k}|z_{t,k},\theta) = h(y_{t,k},z_{t,k})g(\theta)\exp\left\{\eta^\top(\theta)\,T(y_{t,k},z_{t,k})\right\}, \tag{4.2}$$

where $h(y_{t,k},z_{t,k})$ is a known function, $g(\theta)$ is a normalizing log-partition function, $\eta(\theta)$ is a natural parameter, and $T(y_{t,k},z_{t,k})$ is a sufficient statistic that completely summarizes all the information about θ contained in $y_{t,k}$ and $z_{t,k}$. Now, we assume that the prior distribution can be written in the *conjugate* form

$$\begin{aligned}\pi_k(\theta|y_{0:t-1,k},z_{0:t-1,k}) &= \pi_k(\theta|\xi_{t-1,k},\nu_{t-1,k})\\ &= q(\xi_{t-1,k},\nu_{t-1,k})g(\theta)^{\nu_{t-1,k}}\exp\left\{\eta^\top(\theta)\,\xi_{t-1,k}\right\},\end{aligned} \tag{4.3}$$

where $\xi_{t-1,k}$ and $\nu_{t-1,k}$ are the prior *hyperparameters*. The former is of the same dimension as $T(y_{t,k},z_{t,k})$ and the latter is a scalar whereas $q(\xi_{t-1,k},\nu_{t-1,k})$ is a known function. Then Bayes' theorem (4.1) updates the prior hyperparameters according to [37,41]

$$\xi_{t,k} = \xi_{t-1,k} + T(y_{t,k},z_{t,k}), \tag{4.4}$$

$$\nu_{t,k} = \nu_{t-1,k} + 1. \tag{4.5}$$

Naturally, it is possible to write

$$\pi_k(\theta|y_{0:t,k},z_{0:t,k}) = \pi_k(\theta|\xi_{t,k},\nu_{t,k}). \tag{4.6}$$

We point out here that in sequential processing at time t, $\pi_k(\theta|y_{0:t,k},z_{0:t,k})$ is the posterior of θ. This distribution is also the prior of θ for the processing that takes place at time instant $t+1$.

4.2.1 STRATEGIES FOR INFERENCE OVER NETWORKS

There are various types of settings of inference over networks. First and foremost, a question of paramount importance is whether the models and their parameters are the same for all the collaborating agents. If the model parameters have a physical interpretation, then their homogeneity is usually guaranteed. However, if $f_k(\theta|\cdot)$ are black box models that may have different structures, problems arise. For the sake of simplicity, we adopt the assumption that the models are the same for all agents, and that they are all interested in the same θ. The Bayesian treatment of inhomogeneous parameters is studied in [42].

The next question is what kind of information may be shared among network agents. If there are (virtually) no limitations in communication resources, the agents may share their observations $y_{t,k}$ and explanatory variables $z_{t,k}$ (the sharing may be in the form of sufficient statistics $T(y_{t,k},z_{t,k})$), and estimates $\hat{\theta}_k$ potentially accompanied by related statistical properties such as covariance matrices. If the Bayesian approach to inference is employed, then the most legitimate way is to share the posterior distributions $\pi_k(\theta|\cdot)$, or their hyperparameters (e.g., $\xi_{t,k}$ and $\nu_{t,k}$) whenever possible.

We discriminate among three possible strategies:

1. Incorporation of neighbors' (and own) measurements. This step is often called *adaptation* (A) in the literature.
2. Incorporation of *estimates* provided by neighbors. This is known as *combination* (C). Unlike in adaptation, a combination criterion is required that ensures the result to be as close to the original estimates as possible.

3. Incorporation of both measurements and estimates of neighbors (i.e., implementation of both adaptation and combination). There are two strategies, and they are known as ATC (adapt-then-combine) or alternatively CTA (combine-then-adapt). It has been proved that ATC outperforms CTA in terms of estimation quality [4,37].

4.2.2 SHARING OF MEASUREMENTS OR STATISTICS

Let us fix an agent $k \in \mathcal{K}$ and assume that at time t it has access to the neighbors' observations $y_{t,j}$ and explanatory variables $z_{t,j}$, where $j \in \mathfrak{N}_k$. Alternatively, these observations may be surrogated by the sufficient statistics $T(y_{t,j}, z_{t,j})$. Then agent k can improve its knowledge about θ by incorporating them in the same way as its own sufficient statistic. If $\pi_k(\theta|\zeta_{t-1,k})$ is the k's prior distribution of θ at time t, where $\zeta_{t-1,k}$ stands for all the information available to agent k by time $t-1$, including any previously shared information, the distributed variant of the Bayes' theorem (4.1) reads

$$\pi_k(\theta|\zeta_{t,k}) \propto \pi_k(\theta|\zeta_{t-1,k}) \prod_{j \in \mathfrak{N}_k} f(y_{t,j}|z_{t,j}, \theta), \tag{4.7}$$

where we assumed that the observations are conditionally independent. If the models $f(y_{t,j}|z_{t,j}, \theta)$ belong to the exponential family of distributions and the prior distribution $\pi_k(\theta|\zeta_{t-1,k})$ is a conjugate distribution, then Bayes' theorem reduces to the update of the k's hyperparameters according to (see Eqs. (4.4) and (4.5)),

$$\xi_{t,k} = \xi_{t-1,k} + \sum_{j \in \mathfrak{N}_k} T(y_{t,j}, z_{t,j}), \tag{4.8}$$

$$\nu_{t,k} = \nu_{t-1,k} + |\mathfrak{N}_k|, \tag{4.9}$$

where $|\mathfrak{N}_k|$ is the cardinality of the set $\mathfrak{N}_k$, i.e., the number of neighbors plus one.

4.2.3 MERGING OF BAYESIAN ESTIMATORS

Now we fix again an agent $k \in \mathcal{K}$ and assume that the network agents updated their prior distributions by either own or shared measurements, and that their posterior distributions $\pi_j(\theta|\zeta_{t,j})$ are shared with the neighbors. That is, agent k has access to the set $\{\pi_j(\theta|\zeta_{t,j}); j \in \mathfrak{N}_k\}$. Each member of this set may be assigned a nonnegative weight $a_{kj} \leq 1$ expressing the (subjective) probability that the related posterior is true at the moment. The weights thus take values from a corresponding probability simplex and sum to unity. For simplicity, we assume that the weights a_{kj} are constant and either uniform, or selected according to a convenient rule, e.g., [5, Chap. 8]. A model-based Bayesian treatment of the weights can be found in [37].

Keeping all the posterior distributions in the set would, however, quickly lead to an explosion of its size. In order to prevent this situation, we aim to combine the individual posterior distributions to a single distribution $\tilde{\pi}_k(\theta|\cdot)$, which best expresses the information in all of them. The Bayesian theory advocates the use of the Kullback-Leibler divergence, $\mathcal{D}(\cdot||\cdot)$, as a proper dissimilarity (loss) measure [41]. A theoretically consistent combination step is equivalent to seeking the minimizer of

$$
\begin{aligned}
\sum_{j \in \mathfrak{N}_k} a_{kj} \mathcal{D}\left(\tilde{\pi}_k(\theta|\cdot) \,\|\, \pi_j(\theta|\cdot)\right) &= \sum_{j \in \mathfrak{N}_k} a_{kj} \mathbb{E}_{\tilde{\pi}_k}\left[\log \frac{\tilde{\pi}_k(\theta|\cdot)}{\pi_j(\theta|\cdot)}\right] \\
&= \mathbb{E}_{\tilde{\pi}_k}\left[\log \frac{\tilde{\pi}_k(\theta|\cdot)}{\frac{\prod_{j\in\mathfrak{N}_k}[\pi_j(\theta|\cdot)]^{a_{kj}}}{\int \prod_{j\in\mathfrak{N}_k}[\pi_j(\theta|\cdot)]^{a_{kj}} d\theta}}\right] - \log \int \prod_{j \in \mathfrak{N}_k} [\pi_j(\theta|\cdot)]^{a_{kj}} \, d\theta \\
&= \mathcal{D}\left(\tilde{\pi}_k(\theta|\cdot) \,\middle\|\, c \prod_{j \in \mathfrak{N}_k} [\pi_j(\theta|\cdot)]^{a_{kj}}\right) + \text{const.},
\end{aligned}
\tag{4.10}
$$

where c is a proportionality constant assuring that the result is a valid density. The first equality in Eq. (4.10) is due to the definition of the Kullback-Leibler divergence, and the second follows from easy algebra. Because the Kullback-Leibler divergence is minimal if the arguments are equal, the minimizing density has the form

$$
\tilde{\pi}_k(\theta|\cdot) \propto \prod_{j \in \mathfrak{N}_k} [\pi_j(\theta|\cdot)]^{a_{kj}}. \tag{4.11}
$$

There is a notable property of this combination rule. If the posterior distributions belong to the exponential family, then Eq. (4.11) provides an analytically tractable way for obtaining the posterior $\tilde{\pi}_k(\theta|\cdot) = \tilde{\pi}_k(\theta|\tilde{\xi}_{t,k}, \tilde{\nu}_{t,k})$. Then its hyperparameters are given by

$$
\tilde{\xi}_{t,k} = \sum_{j \in \mathfrak{N}_k} a_{kj} \xi_{t,k}, \quad \text{and} \quad \tilde{\nu}_{t,k} = \sum_{j \in \mathfrak{N}_k} a_{kj} \nu_{t,k}. \tag{4.12}
$$

The above equations suggest that the hyperparameters of the resulting distribution of the kth agent are obtained by a linear combination of the hyperparameters of the individual distributions, $\xi_{t,j}$ and $\nu_{t,j}$, $j \in \mathfrak{N}_k$. Recall from Eq. (4.5) that $\xi_{t,j}$ and $\nu_{t,j}$ aggregate the agents' observations.

A natural question is whether it is possible to proceed with swapped arguments of the divergence. We proceed by writing,

$$
\begin{aligned}
\sum_{j \in \mathfrak{N}_k} a_{kj} \mathcal{D}\left(\pi_j(\theta|\cdot) \,\|\, \tilde{\pi}_k(\theta|\cdot)\right) &= \sum_{j \in \mathfrak{N}_k} a_{kj} \mathbb{E}_{\pi_j}\left[\log \frac{\pi_j(\theta|\cdot)}{\tilde{\pi}_k(\theta|\cdot)}\right] \\
&= \sum_{j \in \mathfrak{N}_k} a_{kj} \mathbb{E}_{\pi_j}\left[\log \pi_j(\theta|\cdot)\right] \\
&\quad - \mathbb{E}_{\sum_{j\in\mathfrak{N}_k} a_{kj}\pi_j}\left[\log \sum_{j \in \mathfrak{N}_k} a_{kj}\pi_j(\theta|\cdot)\right] \\
&\quad + \mathbb{E}_{\sum_{j\in\mathfrak{N}_k} a_{kj}\pi_j}\left[\log \frac{\sum_{j\in\mathfrak{N}_k} a_{kj}\pi_j(\theta|\cdot)}{\tilde{\pi}_k(\theta|\cdot)}\right]
\end{aligned}
\tag{4.13}
$$

$$
= \mathcal{D}\left(\sum_{j \in \mathfrak{N}_k} a_{kj}\pi_j(\theta|\cdot) \,\middle\|\, \tilde{\pi}_k(\theta|\cdot)\right) + \text{const.} \tag{4.14}
$$

The first two terms in Eq. (4.13) do not depend on $\tilde{\pi}_k(\theta|\cdot)$, and the minimum of the divergence is achieved by minimization of the last term, which leads to

$$\tilde{\pi}_k(\theta|\cdot) = \sum_{j\in\mathfrak{N}_k} a_{kj}\pi_j(\theta|\cdot). \tag{4.15}$$

The result of the minimization is a mixture of the neighbors' posterior distributions. The number of components is equal to $|\mathfrak{N}_k|$, and with time it will explode, unless a suitable component merging/pruning procedure is implemented, e.g., [43].

Example: Covariance intersection

A nice example of the Kullback-Leibler optimal combination (4.11) is the merging of normal densities, which belong to the exponential family of distributions. Assume that the posterior of $\theta \in \mathbb{R}^n$ is $\pi_j(\theta|\cdot) \sim \mathcal{N}(\mu_{t,j}, \Sigma_{t,j})$. If we drop the time indices for notational simplicity and rewrite the density in the form (4.2), we obtain

$$\begin{aligned}\pi_j\left(\theta|\mu_j, \Sigma_j\right) &= (2\pi)^{-\frac{n}{2}}(\det \Sigma_j)^{-\frac{1}{2}} e^{-\frac{1}{2}(\theta-\mu_j)^\mathsf{T}\Sigma_j^{-1}(\theta-\mu_j)} \\ &\propto \exp\left\{\mathrm{Tr}\left(\begin{bmatrix}\mu_j^\mathsf{T}\Sigma_j^{-1} \\ -\frac{1}{2}\Sigma_j^{-1}\end{bmatrix}^\mathsf{T}\begin{bmatrix}\theta^\mathsf{T} \\ \theta\theta^\mathsf{T}\end{bmatrix}\right) - \frac{1}{2}\mu_j^\mathsf{T}\Sigma_j^{-1}\mu_j\right\},\end{aligned} \tag{4.16}$$

where the natural parameter and the sufficient statistic have the form

$$\eta_j = \begin{bmatrix}\mu_j^\mathsf{T}\Sigma_j^{-1} \\ -\frac{1}{2}\Sigma_j^{-1}\end{bmatrix} \quad \text{and} \quad T(\theta) = \begin{bmatrix}\theta^\mathsf{T} \\ \theta\theta^\mathsf{T}\end{bmatrix}.$$

The combination of all the Gaussian distributions will produce another Gaussian. From Eq. (4.11) it follows that

$$\tilde{\eta}_k = \begin{bmatrix}\tilde{\mu}_k^\mathsf{T}\tilde{\Sigma}_k^{-1} \\ -\frac{1}{2}\tilde{\Sigma}_k^{-1}\end{bmatrix} = \sum_{j\in\mathfrak{N}_k} a_{kj}\eta_j = \sum_{j\in\mathfrak{N}_k} a_{kj}\begin{bmatrix}\mu_j^\mathsf{T}\Sigma_j^{-1} \\ -\frac{1}{2}\Sigma_j^{-1}\end{bmatrix}, \tag{4.17}$$

and a little algebra yields the resulting mean vector and covariance matrix of the resulting Gaussian,

$$\tilde{\mu}_{t,k} = \tilde{\Sigma}_{t,k}\left(\sum_{j\in\mathfrak{N}_k} a_{kj}\Sigma_{t,j}^{-1}\mu_{t,j}\right) \quad \text{and} \quad \tilde{\Sigma}_{t,k} = \left[\sum_{j\in\mathfrak{N}_k} a_{kj}\Sigma_{t,j}^{-1}\right]^{-1}. \tag{4.18}$$

This result is known as *covariance intersection*. It is worth noting that the same rule applies to any other distribution from the exponential family.

Which combination algorithm?

Both the presented algorithms that combine the posterior estimates are optimal in the Kullback-Leibler sense, yet they may lead to significantly different results. The obvious question is which algorithm should be used in a given situation. As mentioned above, the algorithm (4.11) should be used in situations where model and parameter homogeneity are assumed. Further, its analytical tractability under conjugate priors is a very attractive feature. On the other hand, the second algorithm (4.15) is better in situations where the agents use different models and/or parameters. Then, the algorithm provides a (mixture) density that better fits the regions where the neighbors' densities are large enough (for more details, see [42]). As pointed out, the algorithm requires an additional procedure for pruning and merging of the components to prevent the mixture from a rapid growth of its number of components. A specific situation arises if (sequential) Monte Carlo methods are used for estimation or filtering, and the posterior distribution is approximated by samples. Sampling from the mixture (4.15) is equivalent to gathering a relevant number of samples from the neighbors. As this may be communication-intensive, the posteriors may be approximated by a normal mixture at each agent as in the Gaussian particle filter [44–46]. The combination rule (4.15) yields again a normal mixture from which sampling is trivial.

Example

Let us briefly investigate the properties of the two combination algorithms on a simple example. Assume that there are two normal posterior distributions available to agent 1,

$$\begin{aligned} \pi_1(\theta|\cdot) &= \mathcal{N}(0,1) \quad \text{with} \quad a_{11} = 0.5, \\ \pi_2(\theta|\cdot) &= \mathcal{N}(1,1) \quad \text{with} \quad a_{12} = 0.5. \end{aligned}$$

It is straightforward to prove that the combination rule (4.11) yields a normal distribution,

$$\tilde{\pi}_1(\theta|\cdot) = \mathcal{N}(0.5, 1),$$

which is a good compromise between the two original distributions. The combination rule (4.15) yields a mixture

$$\tilde{\pi}_1(\theta|\cdot) = 0.5\mathcal{N}(0,1) + 0.5\mathcal{N}(1,1), \tag{4.19}$$

which preserves the available information about the location of the mean values, but at the cost of higher complexity.

Now, assume two normal distributions that differ only in the variance,

$$\begin{aligned} \pi_1(\theta|\cdot) &= \mathcal{N}(0,100) \quad \text{with} \quad a_{11} = 0.5, \\ \pi_2(\theta|\cdot) &= \mathcal{N}(0,1) \quad \text{with} \quad a_{12} = 0.5. \end{aligned}$$

The combination rule (4.11) produces a normal distribution

$$\tilde{\pi}_1(\theta|\cdot) = \mathcal{N}(0, 1.98),$$

and the second combination rule (4.15) leads to a mixture

$$\tilde{\pi}_1(\theta|\cdot) = 0.5\mathcal{N}(0,100) + 0.5\mathcal{N}(0,1). \tag{4.20}$$

If we assume that agent 1 joined the network with an initial prior while the second agent already performed several updates and has a good knowledge of θ, we see that the former combination rule significantly improves the distribution of agent 1. If agent 2 uses $a_{21} = 0.5$ and $a_{22} = 0.5$ (has the same beliefs as agent 1), the combined distribution of this agent would be the same as that of agent 1.

Hence, a combination of a distribution that reflects high-ignorance with one with low ignorance, does not much affect the latter.

4.3 EXAMPLE: DIFFUSION KALMAN FILTER

The first diffusion Kalman filter was proposed in [30]. Later in [31], it was improved with a covariance intersection-based procedure, which was applied in the combination step. Below we derive the diffusion Kalman filter from a Bayesian viewpoint. More on the diffusion Kalman filter but derived from a different perspective can be found in the chapter on Distributed Kalman and Particle Filtering.

To begin, we assume a hidden Markov model of the form[1]

$$x_t|x_{t-1}, z_t \sim \mathcal{N}\left(A_t x_{t-1} + B_t z_t, Q_t\right), \tag{4.21}$$

$$y_t|x_t \sim \mathcal{N}\left(H_t x_t, R_t\right), \tag{4.22}$$

where x_t is an n-dimensional state vector, y_t is an l-dimensional observation vector, z_t is a known input vector of length m, A_t, B_t, and H_t are matrices of compatible dimensions, Q_t and R_t are $n \times n$ and $\ell \times \ell$ state and observation covariance matrices, respectively.

The observation model (4.22) is normal, so we rewrite it by using the exponential family form (4.2),

$$\begin{aligned} f(y_t|x_t) &\propto \exp\left\{-\frac{1}{2}(y_t - H_t x_t)^\intercal R_t^{-1}(y_t - H_t x_t)\right\} \\ &= \exp\left\{\mathrm{Tr}\Bigg(\underbrace{-\frac{1}{2}\begin{bmatrix}-1\\ x_t\end{bmatrix}\begin{bmatrix}-1\\ x_t\end{bmatrix}^\intercal}_{\eta(x_t)}\underbrace{\begin{bmatrix}y_t^\intercal\\ H_t^\intercal\end{bmatrix}R_t^{-1}\begin{bmatrix}y_t^\intercal\\ H_t^\intercal\end{bmatrix}^\intercal}_{T(y_t)}\Bigg)\right\}. \end{aligned} \tag{4.23}$$

Because the model (4.22) is normal, it is advantageous to set the prior distribution to normal too, as it is conjugate to the model and hence the posterior will be analytically tractable,

$$\begin{aligned} \pi(x_t|y_{0:t-1}, z_{0:t-1}) &= \mathcal{N}(x_t^-, P_t^-), \qquad x_t^- \in \mathbb{R}^n, P_t^- \in \mathbb{R}^{n\times n} \\ &\propto \exp\left\{-\frac{1}{2}(x_t - x_t^-)^\intercal (P_t^-)^{-1}(x_t - x_t^-)\right\} \\ &= \exp\left\{\mathrm{Tr}\Bigg(\underbrace{-\frac{1}{2}\begin{bmatrix}-1\\ x_t\end{bmatrix}\begin{bmatrix}-1\\ x_t\end{bmatrix}^\intercal}_{\eta(x_t)}\underbrace{\begin{bmatrix}(x_t^-)^\intercal\\ I\end{bmatrix}(P_t^-)^{-1}\begin{bmatrix}(x_t^-)^\intercal\\ I\end{bmatrix}^\intercal}_{\xi_t}\Bigg)\right\}, \end{aligned} \tag{4.24}$$

where I is an identity matrix of appropriate dimensions. The minus superscripts denote parameters of the Gaussian where the measurements from time t have not been incorporated yet, whereas the plus signs signify that they have been used. Also, the state transition $x_{t-1} \to x_t$ amounts to updating $\mathcal{N}(x_{t-1}^+, P_{t-1}^+) \to \mathcal{N}(x_t^-, P_t^-)$ according to Eq. (4.21).

Next, we return to the implementation of the Kalman filter over a network, where all the agents use the same model. The matrices A_t, B_t, and Q_t are the same for all the agents whereas $H_{t,k}$, and $R_{t,k}$ are

[1] We temporarily drop the agent's indices for simplicity. Also, instead of θ, the unknowns of interest here are the vectors x_t.

distinctive as are the variables $z_{t,k}$. The adaptation step of the diffusion Kalman filter is similar to the ordinary Kalman filter update. First, the state transition given by Eq. (4.21) is performed as usual,

$$\begin{aligned}\pi_k(x_{t,k}|\zeta_{t-1,k}) &= \int \pi(x_{t,k}|x_{t-1}, z_t)\pi_k(x_{t-1}|\zeta_{t-1,k})dx_{t-1} \\ &= \mathcal{N}_k\left(A_t x^+_{t-1,k} + B_t z_{t,k}, A_t P^+_{t-1,k} A_t^\intercal + Q_t\right) \\ &= \mathcal{N}_k(x^-_{t,k}, P^-_{t,k}),\end{aligned} \tag{4.25}$$

and followed by the Bayesian update by the *neighbors'* measurements

$$\begin{aligned}\pi_k(x_t|\zeta_{t,k}) &\propto \pi_k(x_{t,k}|\zeta_{t-1,k}) \prod_{j\in\mathfrak{N}_k} f(y_{t,j}|x_{t,j}) \\ &= \mathcal{N}_k(x^+_{t,k}, P^+_{t,k}).\end{aligned} \tag{4.26}$$

Taking the exponential family form (4.23) and the conjugate form (4.24) into account, we see that

$$\xi_{t,k} = \xi_{t-1,k} + \sum_{j\in\mathfrak{N}_k} \begin{bmatrix} y_{t,j} \\ H_{t,j}\end{bmatrix} R^{-1}_{t,j} \begin{bmatrix} y_{t,j} \\ H_{t,j}\end{bmatrix}^\intercal, \tag{4.27}$$

$$\nu_{t,k} = \nu_{t-1,k} + |\mathfrak{N}_k|. \tag{4.28}$$

Simple algebra reveals the recursions

$$P^+_{t,k} = \left[(P^-_{t,k})^{-1} + \sum_{j\in\mathfrak{N}_k} H^\intercal_{t,j} R^{-1}_{t,j} H_{t,j}\right]^{-1}, \tag{4.29}$$

$$x^+_{t,k} = x^-_{t,k} + P^+_{t,k} \sum_{j\in\mathfrak{N}_k} H^\intercal_{t,j} R^{-1}_{t,j}\left(y_{t,j} - H_{t,j} x^-_{t,k}\right). \tag{4.30}$$

The above two equations describe the adaptation step.

The combination step operates directly with $\xi_{t,j}$ according to Eq. (4.11). Application of Eq. (4.18) from Section 4.2.3 shows that

$$\tilde{\pi}_k(x_{t,k}|\zeta_{t,k}) = \prod_{j\in\mathfrak{N}_k}\left[\mathcal{N}_j(x^+_{t,j}, P^+_{t,j})\right]^{a_{kj}} = \mathcal{N}_k(\tilde{x}^+_{t,k}, \tilde{P}^+_{t,k}) \tag{4.31}$$

with the hyperparameters

$$\tilde{P}^+_{t,k} = \left[\sum_{j\in\mathfrak{N}_k} a_{kj}\left(P^+_{t,j}\right)^{-1}\right]^{-1}, \tag{4.32}$$

$$\tilde{x}^+_{t,k} = \tilde{P}^+_{t,k}\sum_{j\in\mathfrak{N}_k} a_{kj}\left(P^+_{t,j}\right)^{-1} x^+_{t,j}. \tag{4.33}$$

The algorithm is summarized in Algorithm 4.1.

Algorithm 4.1 DIFFUSION KALMAN FILTER WITH ADAPT-THEN-COMBINE (ATC) STRATEGY

Initialize agents $k = 1, 2, \ldots, K$ with the prior densities $\pi_k(\theta|\xi_{k,0})$. Set the weights a_{kj}. For $t = 1, 2, \ldots$ and each agent k do:

Kalman prediction:

- Update the prior densities $\mathcal{N}_k(x^+_{t-1,k}, P^+_{t-1,k}) \rightarrow \mathcal{N}_k(x^-_{t,k}, P^-_{t,k})$, Eq. (4.25).

Kalman update:

1. Acquire observations $y_{t,j}$ of neighbors $j \in \mathfrak{N}_k$.
2. *Adaptation:* Perform Kalman adaptation according to Eq. (4.26) by updating the hyperparameters via Eq. (4.27).
3. Obtain posterior densities $\pi_j(x_{t,j}|\xi_{t,j})$ of neighbors $j \in \mathfrak{N}_k$ by acquiring their respective hyperparameters $\xi_{t,j}, \nu_{t,j}$.
4. *Combination:* Combine posterior densities according by implementing Eqs. (4.32) and (4.33).

4.3.1 SIMULATION EXAMPLE

We illustrate the performance of the diffusion Kalman filter with a two-dimensional tracking problem. The matrices of the model were time-invariant, and they were defined as follows:

$$A = \begin{bmatrix} 1 & 0 & \Delta & 0 \\ 0 & 1 & 0 & \Delta \\ 0 & 0 & 1 & 0 \\ 0 & 0 & 0 & 1 \end{bmatrix}, \qquad Q = q \cdot \begin{bmatrix} \frac{\Delta^3}{3} & 0 & \frac{\Delta^2}{2} & 0 \\ 0 & \frac{\Delta^3}{3} & 0 & \frac{\Delta^2}{2} \\ \frac{\Delta^2}{2} & 0 & \Delta & 0 \\ 0 & \frac{\Delta^2}{2} & 0 & \Delta \end{bmatrix},$$

$$H = \begin{bmatrix} 1 & 0 & 0 & 0 \\ 0 & 1 & 0 & 0 \end{bmatrix}, \qquad R = r^2 \cdot \begin{bmatrix} 1 & 0 \\ 0 & 1 \end{bmatrix},$$

where $\Delta = 0.1$, $q = 5.0$, $r = 0.1k$, and $k = 1, \ldots, 15$ is the agent's number. That is, the agents had different observation noise covariance matrices. The simulation was started from the origin of the coordinate system. The state vector elements represent the position of the target in the plane and its velocity components.

Figs. 4.1 and 4.2 show the topology of the network and the simulated trajectory with noisy observations of agents 1 and 15, respectively. The Kalman filters were initialized with $P^+_{0,k} = 1000 I_{4\times 4}$ and zero vectors $x^+_{0,k}$. The combination weights a_{kj} were uniform and constant. Four strategies were tested: (1) no cooperation (nocoop), (2) adaptation only (A-only), (3) combination only (C-only), and (4) adapt then combine (ATC). Fig. 4.3 depicts the box plots of the mean square errors (MSEs) of the estimation of all four elements of the state vector. We see that the collaboration among agents improved the estimation performance. Further, the performance of the C-only strategy was superior to that of the A-only strategy.

4.4 CONCLUSION

The Bayesian approach to the inference of unknown parameters of probabilistic models has numerous attractive features. One of the most prominent is its wide applicability. Further, regardless of whether

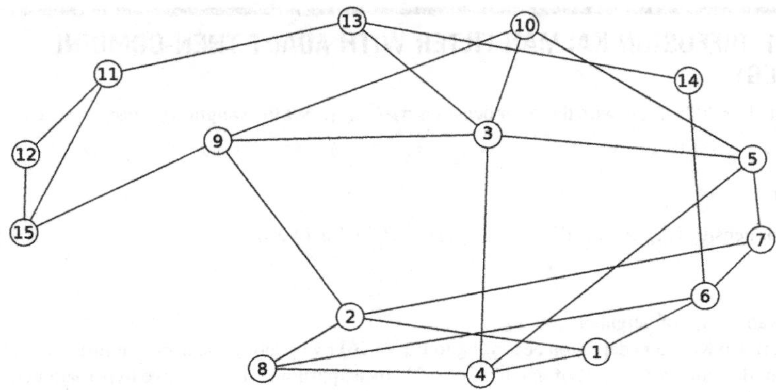

FIG. 4.1

Network layout.

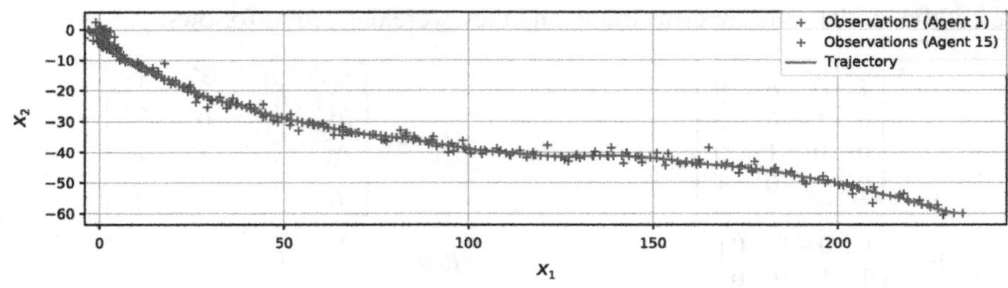

FIG. 4.2

Simulated trajectory and observations of agents 1 and 15.

one deals with linear or nonlinear regression, state-space models, hierarchical models, or any other model type, Bayesian inference relies on the same principles. Unlike in classical (frequentist) statistics, the estimate is represented by a posterior distribution, quantifying not only its expected location but also the uncertainty associated with it. The representation of estimates by posterior distributions remains a cornerstone of the use of Bayesian principles in networked systems, where agents collaborate to improve their own estimates. The main idea is that less-informed agents improve their knowledge while well-informed ones do not reduce it.

There are many open problems that could be investigated. For instance, the determination of combination weights is one, although several methods for choosing them have already been proposed [37,47]. The described combination methods are a small sample from the set of possible approaches, too. For instance, in [48], yet another approach was proposed where the agents fuse received information from neighbors by using mixtures with weights proportional to predictive distributions obtained from the posteriors of the respective agents. Furthermore, the topic of heterogeneous models and/or parameters has attained a huge interest in the last years, but the only Bayesian treatment the

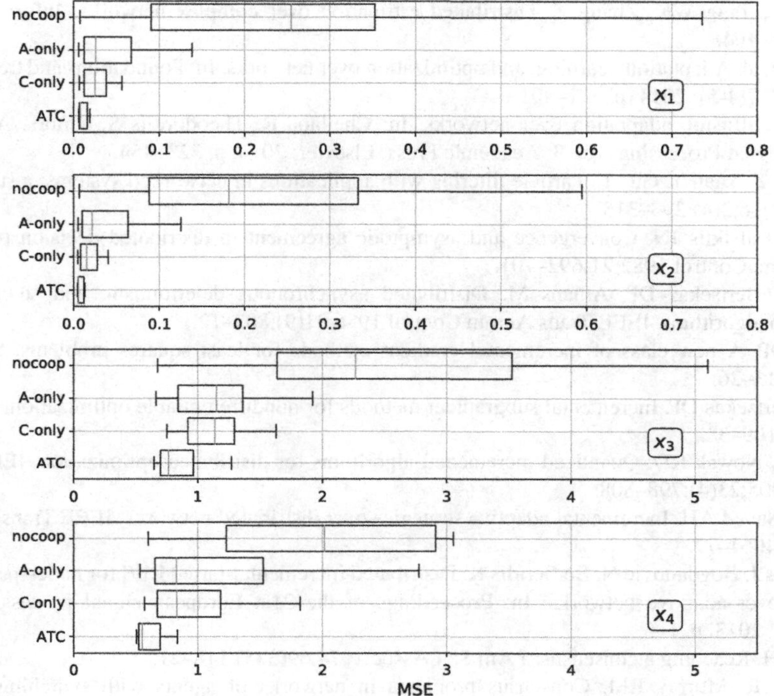

FIG. 4.3

MSEs of various strategies.

authors are aware of was proposed in [42]. Naturally, the robustness to failures and the communication and computational limitations of the agents form other interesting topics for research on collaborative inference in networks of agents.

ACKNOWLEDGMENTS

The work of K. Dedecius was supported by the Czech Science Foundation, project No. 16-09848S. The work of P.M. Djurić was supported by NSF under Award CCF-1618999.

REFERENCES

[1] Candy JV. Bayesian signal processing: classical, modern, and particle filtering methods. John Wiley & Sons; 2016.

[2] Akyildiz IF, Su W, Sankarasubramaniam Y, Cayirci E. A survey on sensor networks. IEEE Commun Mag 2002;40(8):102–14.

[3] Liu Y, Li C, Tang WK, Zhang Z. Distributed estimation over complex networks. Information Sciences 2012;197:91–104.
[4] Sayed AH, et al. Adaptation, learning, and optimization over networks. In: Foundations and trends in machine learning, vol. 7(4-5); 2014. p. 311–801.
[5] Sayed AH. Diffusion adaptation over networks. In: Chellapa R, Theodoridis S, editors. Academic Press Library in Signal Processing, vol. 3. Academic Press, Elsevier; 2014. p. 323–454.
[6] Li W, Wang Z, Yuan Y, Guo L. Particle filtering with applications in networked systems: a survey. Complex Intell Syst 2016;2(4):293–315.
[7] Athans M, Tsitsiklis JN. Convergence and asymptotic agreement in distributed decision problems. IEEE Trans. Autom. Control 1982;21:692–701.
[8] Tsitsiklis J, Bertsekas DP, Athans M. Distributed asynchronous deterministic and stochastic gradient optimization algorithms. IEEE Trans Autom Control 1986;31(9):803–12.
[9] Bertsekas DP. A new class of incremental gradient methods for least squares problems. SIAM J Optim 1997;7(4):913–26.
[10] Nedic A, Bertsekas DP. Incremental subgradient methods for nondifferentiable optimization. SIAM J Optim 2001;12(1):109–38.
[11] Rabbat MG, Nowak RD. Quantized incremental algorithms for distributed optimization. IEEE J Sel Areas Commun 2005;23(4):798–808.
[12] Lopes CG, Sayed AH. Incremental adaptive strategies over distributed networks. IEEE Trans Signal Process 2007;55(8):4064–77.
[13] Plata-Chaves J, Bogdanovic N, Berberidis K. Distributed incremental-based RLS for node-specific parameter estimation over adaptive networks. In: Proceedings of the 21st European signal processing conference (EUSIPCO); 2013. p. 1–5.
[14] DeGroot MH. Reaching a consensus. J Am Stat Assoc 1974;69(345):118–21.
[15] Olfati-Saber R, Murray RM. Consensus problems in networks of agents with switching topology and time-delays. IEEE Trans Autom Control 2004;49(9):1520–33.
[16] Olfati-Saber R, Fax JA, Murray RM. Consensus and cooperation in networked multi-agent systems. Proc IEEE 2007;95(1):215–33.
[17] Schizas ID, Giannakis GB, Luo ZQ. Distributed estimation using reduced-dimensionality sensor observations. IEEE Trans Signal Process 2007;55(8):4284–99.
[18] Schizas ID, Mateos G, Giannakis GB. Distributed LMS for consensus-based in-network adaptive processing. IEEE Trans Signal Process 2009;57(6):2365–82.
[19] Guldogan MB. Consensus Bernoulli filter for distributed detection and tracking using multi-static Doppler shifts. IEEE Signal Process Lett 2014;21(6):672–6.
[20] Hlinka O, Hlawatsch F, Djurić PM. Distributed particle filtering in agent networks: a survey, classification, and comparison. IEEE Signal Process Mag 2013;30(1):61–81.
[21] Cattivelli FS, Lopes CG, Sayed AH. Diffusion recursive least-squares for distributed estimation over adaptive networks. IEEE Trans Signal Process 2008;56(5):1865–77.
[22] Cattivelli FS, Sayed AH. Diffusion LMS strategies for distributed estimation. IEEE Trans Signal Process 2010;58(3):1035–48.
[23] Bertrand A, Moonen M, Sayed AH. Diffusion bias-compensated RLS estimation over adaptive networks. IEEE Trans Signal Process 2011;59(11):5212–24.
[24] Zhao X, Tu SY, Sayed AH. Diffusion adaptation over networks under imperfect information exchange and non-stationary data. IEEE Trans Signal Process 2012;60(7):3460–75.
[25] Dedecius K, Sečkárová V. Dynamic diffusion estimation in exponential family models. IEEE Signal Process Lett 2013;20(11):1114–7.
[26] Arablouei R, Dogancay K, Werner S, Huang YF. Adaptive distributed estimation based on recursive least-squares and partial diffusion. IEEE Trans Signal Process 2014;62(14):3510–22.

[27] Lopes CG, Sayed AH. Diffusion least-mean squares over adaptive networks: formulation and performance analysis. IEEE Trans Signal Process 2008;56(7):3122–36.

[28] Chen J, Richard C, Hero AO, Sayed AH. Diffusion LMS for multitask problems with overlapping hypothesis subspaces. In: Proceedings of the IEEE international workshop on machine learning for signal processing; 2014. p. 1–6.

[29] Plata-Chaves J, Bogdanović N, Berberidis K. Distributed diffusion-based LMS for node-specific adaptive parameter estimation. IEEE Trans Signal Process 2015;63(13):3448–60.

[30] Cattivelli FS, Sayed AH. Diffusion strategies for distributed Kalman filtering and smoothing. IEEE Trans Autom Control 2010;55(9):2069–84.

[31] Hu J, Xie L, Zhang C. Diffusion Kalman filtering based on covariance intersection. IEEE Trans Signal Process 2012;60(2):891–902.

[32] Towfic ZJ, Chen J, Sayed AH. Collaborative learning of mixture models using diffusion adaptation. In: Proceedings of the IEEE international workshop on machine learning for signal processing; 2011. p. 1–6.

[33] Pereira SS, Pages-Zamora A, López-Valcarce R. A diffusion-based distributed EM algorithm for density estimation in wireless sensor networks. In: Proceedings of the IEEE international conference on acoustics, speech and signal processing; 2013. p. 4449–53.

[34] Dedecius K, Reichl J, Djurić PM. Sequential estimation of mixtures in diffusion networks. IEEE Signal Process Lett 2015;22(2):197–201.

[35] Bruno MG, Dias SS. Collaborative emitter tracking using Rao-Blackwellized random exchange diffusion particle filtering. EURASIP J Adv Signal Process 2014;2014(1):19.

[36] Dedecius K. Adaptive approximate filtering of state-space models. In: Proceedings of the European signal processing conference (EUSIPCO); 2015. p. 2236–40.

[37] Dedecius K, Djurić PM. Sequential estimation and diffusion of information over networks: a Bayesian approach with exponential family of distributions. IEEE Trans Signal Process 2017;65(7):1795–809.

[38] Dedecius K, Djurić PM. Diffusion filtration with approximate Bayesian computation. In: Proceedings of the IEEE international conference on acoustics, speech and signal processing; 2015. p. 3207–11.

[39] Gelman A, Carlin JB, Stern HS, Rubin DB. Bayesian data analysis. Chapman & Hall/CRC; 2003.

[40] Robert C. The Bayesian choice: from decision-theoretic foundations to computational implementation. Springer Science & Business Media; 2007.

[41] Bishop CM. Pattern recognition and machine learning. New York, NY: Springer; 2006.

[42] Dedecius K, Sečkárová V. Factorized estimation of partially shared parameters in diffusion networks. IEEE Trans Signal Process 2017;65(19):5153–63.

[43] Frühwirth-Schnatter S. Finite mixture and Markov switching models. Springer Science & Business Media; 2006.

[44] Kotecha JH, Djurić PM. Gaussian particle filtering. IEEE Trans Signal Process 2003;51(10):2592–601.

[45] Kotecha JH, Djurić PM. Gaussian sum particle filtering. IEEE Trans Signal Process 2003;51(10):2602–12.

[46] Hlinka O, Sluciak O, Hlawatsch F, Djurić PM, Rupp M. Likelihood consensus and its application to distributed particle filtering. IEEE Trans Signal Process 2012;60(8):4334–49.

[47] Takahashi N, Yamada I, Sayed AH. Diffusion least-mean squares with adaptive combiners: formulation and performance analysis. IEEE Trans Signal Process 2010;58(9):4795–810.

[48] Djurić PM, Dedecius K. Bayesian estimation of unknown parameters over networks. In: Proceedings of the 24th European signal processing conference (EUSIPCO); 2016. p. 1508–12.

CHAPTER

MULTIAGENT DISTRIBUTED OPTIMIZATION

5

Michael G. Rabbat*, **Alejandro Ribeiro**†

McGill University and Facebook AI Research, Montreal, QC, Canada Department of Electrical and Systems Engineering, University of Pennsylvania, Philadelphia, PA, United States*†

5.1 INTRODUCTION

In multiagent distributed optimization, each agent (or node) has a local objective function, and all agents share the same goal of minimizing the sum (or average) of these local objective functions. To achieve this goal the agents must exchange information over a network that restricts communications to be between adjacent agents.

Problems of this form are motivated by several applications in which data collection is inherently distributed, the most popular of which are decentralized control systems [1–3] and sensor networks [4–6]. In both of these cases, nodes acquire local observations about a phenomenon of interest and they want to integrate that information into a globally optimal control action or a globally optimal estimate. While it is possible for nodes to communicate their observations to a central node or fusion center, this is undesirable because it introduces a single point of failure and a communication bottleneck that can affect the responsiveness of the system. It is therefore more convenient, and often more communication-efficient [7], for nodes to only communicate statistics or some other summary information (e.g., gradients of an objective function based on their local observations) while computing the global estimate or control action in a distributed manner over the network.

Distributed optimization is also motivated in wireless and wired network resource allocation problems [8–10]. In these scenarios, nodes in the network want to define local operating points such as power allocation, band choice, transmit opportunities, and routes. These local decisions must be consistent with the global operation of the network motivating the formulation of optimal aggregate costs that are minimized in a distributed manner. A third motivation for distributed optimization problems arises from large-scale machine learning problems [11–13]. Here, data is not collected over a network but a very large dataset is intentionally split across multiple servers to facilitate computations.

Irrespective of the application, distributed optimization methods decouple the global objective by introducing local variables that are recoupled by consensus constraints (Section 5.2). The consensus constraints explicitly require equality of variables at adjacent nodes, which, by transitivity, implies equality of the variables at all nodes. However, the fact that the constraints are explicit for neighboring nodes only makes it possible to compute descent directions in a distributed manner. This observation is

Cooperative and Graph Signal Processing. https://doi.org/10.1016/B978-0-12-813677-5.00005-5

the basis for all distributed optimization methods, although the way in which the constraints are handled may differ significantly.

The decentralized gradient descent (DGD) method [14–17] incorporates consensus constraints as a penalty function (Section 5.3). The basis for the algorithm is the observation that the gradient of the resulting penalized function can be computed in a distributed manner. Alternatively, distributed dual descent [18,19] and the alternating direction method of multipliers (ADMM) [4,20–24] work with the Lagrangian dual function (Section 5.4). As is the case of DGD, the reason why these algorithms work is that gradients of the dual function can be computed in a distributed manner. A feature common to all these algorithms is that they operate on first-order information only. This has motivated interest in second-order methods, one of which we survey in Section 5.5. Whether operating in the primal or dual domain, distributed optimization methods can be thought of as methods that work iteratively through descent or minimization steps in their local functions followed by information aggregation steps that try to move the local variables toward minimizers of the global objective.

5.2 DISTRIBUTED OPTIMIZATION

We consider a group of agents $i = 1, \ldots, n$ each of which has access to a local convex function $f_i : \mathbb{R}^p \to \mathbb{R}$ defined over a column vector $x \in \mathbb{R}^p$. The agents are interested in finding the minimum of the aggregate function $f : \mathbb{R}^p \to \mathbb{R}$ defined as the one that takes values $f(x) := \sum_{i=1}^{n} f_i(x)$, and they therefore collaborate to solve the optimization problem

$$X^* := \underset{x\in\mathbb{R}^p}{\operatorname{argmin}} f(x) = \underset{x\in\mathbb{R}^p}{\operatorname{argmin}} \sum_{i=1}^{n} f_i(x). \tag{5.1}$$

Here X^* denotes the *set* of global minimizers of f (in general, the minimizer may not be unique), which we assume is a nonempty set. Observe that because the functions f_i are convex, the same holds true for the aggregate function f. We also assume that the set of optimal values in Eq. (5.1) is nonempty so that the optimization problem is well defined. Table 5.1 summarizes the main notation used throughout this chapter.

Table 5.1 Notation Used in This Chapter

Notation	Description
$x_i(k) \in \mathbb{R}^p$	optimization variable at agent i after iteration k
$\mathbf{x}(k) \in \mathbb{R}^{np}$	(stacked) optimization variable for all agents
n	number of agents
f_i	local objective at agent i
v	Lagrange multiplier vector
$\lambda_j(A)$	jth largest eigenvalue of matrix A
A	adjacency matrix
B	incidence matrix
D	degree matrix
L	Laplacian matrix

Methods for solving the convex program in Eq. (5.1) when all the functions f_i are available at a central location are well developed. The assumptions in distributed optimization are that only agent i has access to function f_i and that it is impossible or inadvisable to group all information at a central location. Instead, we assume a pattern of local connectivity described by a graph $\mathcal{G} = (V, E)$ with node set $V = \{1, \ldots, n\}$ and a symmetric edge set E containing $2m$ elements. Edges consist of pairs (i,j) where the presence of a pair in the set E signifies the ability of agent i to communicate with agent j. That the edge set is symmetric means that having $e = (i,j) \in E$ implies $e' = (j,i) \in E$. If we think of a pair of edges $e = (i,j)$ and $e' = (j,i)$ as defining a single connection between agents i and j, it follows that there are m connections in the graph. When there is an edge $e(i,j) \in E$, we say that agents i and j are adjacent, or are neighbors, in the graph and we let $n(i) := \{j : (i,j) \in E\}$ denote the set of all neighbors of agent i.

Given that grouping information is precluded by the problem formulation and that communication can only happen between neighboring agents, our goal is to design an algorithm that operates through iterative message exchanges between i and j and so that agents eventually find a point close to the optimal set X^*. To do so we begin by introducing local variables $x_i \in \mathbb{R}^p$ that we group in the vector $\mathbf{x} = [x_1^\top, \ldots, x_n^\top]^\top \in \mathbb{R}^{np}$ and consider the problem of finding

$$\tilde{\mathbf{x}} \in \operatorname*{argmin}_{\mathbf{x} \in \mathbb{R}^{np}} F(\mathbf{x}) := \sum_{i=1}^{n} f_i(x_i). \tag{5.2}$$

The optimization problem in Eq. (5.2) is clearly not equivalent to Eq. (5.1). In fact, the variables x_i are decoupled in Eq. (5.2) and the solution is equivalent to simply letting the agents operate in isolation to find the locally optimal argument $\tilde{x}_i := \operatorname{argmin}_{x \in \mathbb{R}^p} f_i(x_i)$.

This is easily remedied, at least in theory, by adding constraints of the form $x_i = x_j$ for all pairs of neighboring nodes to formulate the problem of finding

$$\begin{aligned} \mathbf{x}^* \in \operatorname*{argmin}_{\mathbf{x} \in \mathbb{R}^{np}} & \sum_{i=1}^{n} f_i(x_i) \\ \text{s.t.} \quad & x_i = x_j, \quad \text{for all } i = 1, 2, \ldots, n, \text{ and } j \in n(i). \end{aligned} \tag{5.3}$$

If the graph is connected, i.e., if there is a chain of edges connecting any node to any other node, then the problem (5.3) is equivalent to the problem in Eq. (5.1) in the sense that a solution to Eq. (5.3) is of the form $\mathbf{x}^* = [x^{*T}, \ldots, x^{*T}]^T$ where $x^* \in X^*$. This must be the case because any argument that is feasible in Eq. (5.3) must be such that $x_k = x_l$ for any arbitrary pair of nodes k and l, not just neighbors, because the graph is connected.

We can think of problem (5.1) as a *centralized* problem that needs aggregation of information to find the optimal variable x^*. The problem in Eq. (5.2) is simply a collection of *local* problems that can each be solved independently from each other, and different from Eq. (5.1). The problem in Eq. (5.3) is a *distributed collaborative* formulation in which each agent i wants to locally determine a variable x_i that is optimal for the aggregate function f without ever having explicit access to the functions f_k of other agents $k \neq i$. While it may seem that the distinction between Eqs. (5.1) and (5.3) is just a matter of semantics, we will see that this is not the case. The structure of Eq. (5.3) makes it possible to locally compute descent directions of properly relaxed formulations as we explain in Section 5.2.2.

Example 5.1 (Distributed-least squares estimation). To illustrate some of the concepts mentioned above, consider a distributed formulation of the linear least-squares problem. Suppose that the local data at agent i consists of a vector $h_i \in \mathbb{R}^p$ and a scalar $y_i \in \mathbb{R}$, such that

$$y_i = h_i^\top x + w_i,$$

where $x \in \mathbb{R}^p$ is an unobserved parameter vector to be estimated (the same across all agents), and w_i is unobserved white noise. The local objective at agent i is $f_i(x) = (y_i - h_i^\top x)^2$. Thus, the (centralized) optimization problem we would like to solve, analogous to Eq. (5.1), is the familiar least-squares problem,

$$\underset{x \in \mathbb{R}^p}{\text{minimize}} \sum_{i=1}^{n} (y_i - h_i^\top x)^2 = \|y - Hx\|_2^2,$$

where we have stacked the individual values y_i into a vector $y \in \mathbb{R}^n$, and H is a $n \times p$ matrix, the ith row of which is $h_i^\top$.

In the unconstrained distributed reformulation (5.2), each agent locally seeks to find $x_i \in \mathbb{R}^p$ that minimizes $(y_i - h_i^\top x_i)^2$. For $p > 1$, this local problem is underdetermined and there are infinitely many solutions. However, once the variables x_i are coupled through the consensus constraints, as in Eq. (5.3), if $n \geq p$ and the columns of H are linearly independent (so the rank of H is p), then there is a unique consensus solution. ■

5.2.1 MATRIX REPRESENTATIONS OF GRAPHS

For convenience of representation and to understand convergence properties, we will rewrite the constraints $x_i = x_j$ for all i and $j \in n(i)$ that appear in Eq. (5.3) utilizing matrix representations of graphs. The representations we use here are weighted versions of the edge incidence matrix and the Laplacian matrix.

To define these matrices, associate a nonnegative weight w_{ij} with each edge $e = (i,j)$ of the graph. The weights are symmetric and normalized so that they sum up to one,

$$w_{ij} = w_{ji}, \qquad \sum_{j:(i,j)\in E} w_{ij} = 1; \tag{5.4}$$

i.e., the $n \times n$ matrix W with entries $w_{i,j}$ is symmetric and doubly stochastic. Recall that because we are considering symmetric graphs, the presence of the edge $e = (i,j)$ implies the presence of the edge $e' = (j,i)$. Thus, the symmetry condition in Eq. (5.4) simply extends the symmetry of the graph to the edge weights.

Next we will construct the *weighted edge-vertex incidence matrix*, which is useful for expressing the constraint that we wish the solution to be a consensus. Rather than using both edges $e = (i,j)$ and $e' = (j,i)$, which would lead to redundant constraints, let us adopt the convention that we will only use those edges $(i,j) \in E$ for which $i < j$. The weights in Eq. (5.4) are used to construct a weighted edge incidence matrix $B \in \mathbb{R}^{m\times n}$, with entries given by

$$B_{e,v} = B_{(i,j),v} = \begin{cases} \sqrt{w_{ij}} & \text{if } v = i, \\ -\sqrt{w_{ij}} & \text{if } v = j, \\ 0 & \text{otherwise.} \end{cases} \tag{5.5}$$

Rows of the matrix B represent edges of the graph and columns represent nodes. Moreover, the matrix is sparse: each row contains only two (out of n) nonzero entries. The incidence matrix B is used to define the Laplacian as the matrix $L \in \mathbb{R}^{n \times n}$ given by

$$L = B^T B. \tag{5.6}$$

It is easy to verify that the diagonal components of L are $L_{ii} = 1$ and that off-diagonal components are $L_{ij} = -w_{ij}$ for all $e = (i,j) \in E$, and $L_{ij} = 0$ for $(i,j) \notin E$. The symmetry of the graph implies that the Laplacian is a symmetric matrix.

In the distributed optimization methods that we discuss below, the spectral properties of the Laplacian are important. Begin by observing that because the weights sum to one, we must have $L1 = 0$, where 1 denotes the vector of all ones and 0 is a vector of all zeros. This means that 0 is an eigenvalue of L associated with eigenvector 1. In particular, this fact implies that the Laplacian is not full rank. It can be shown that all other eigenvalues are nonzero for connected graphs. We will use λ to denote the smallest nonzero eigenvalue of L and Λ to denote its largest eigenvalue. Observe that $\Lambda \leq 2$ because the weights are normalized to sum to 1. The condition number of the graph is defined as $\rho := \Lambda/\lambda$. It is important to emphasize that the condition number is a property of the graph that is affected by the choice of weights. Proper selection of weights is important to reduce the condition number of the graph.

5.2.2 CONSTRAINT RELAXATIONS

The edge incidence matrix allows concise writing of the constraints in Eq. (5.3). Indeed, let $\mathbf{I} \in \mathbb{R}^{p \times p}$ be the identity matrix and consider a p-dimensional Kronecker product extension of B that we write as $\mathbf{B} = B \otimes \mathbf{I}$. It follows that the constraints in Eq. (5.3) are equivalent to $\mathbf{Bx} = \mathbf{0}$ and that we can rewrite Eq. (5.3) as

$$\mathbf{x}^* := \underset{\mathbf{x} \in \mathbb{R}^{np}}{\operatorname{argmin}}\ F(\mathbf{x}) = \sum_{i=1}^{n} f_i(x_i), \quad \text{s.t. } \mathbf{Bx} = \mathbf{0}. \tag{5.7}$$

For each edge $e = (i,j) \in E$, the constraint $\mathbf{Bx} = \mathbf{0}$ has one row of the form $w_{ij}(x_i - x_j) = 0$ (based on the definition of B). This is, of course, the same as having $x_i = x_j$ except that the weights can be used to (de)emphasize some specific constraints in the algorithm.

The challenge in solving Eq. (5.3) lies in the coupling introduced by the problem constraints. It is then natural that we attempt to overcome this problem by introducing constraint relaxations. The first relaxation that we consider is to replace the constraint $\mathbf{Bx} = \mathbf{0}$ by a quadratic penalty. This results in the optimization problem

$$\mathbf{x}^*_\alpha := \underset{x \in \mathbb{R}^{np}}{\operatorname{argmin}}\ \alpha F(\mathbf{x}) + \frac{1}{2}\|\mathbf{Bx}\|^2, \tag{5.8}$$

where α is a coefficient to control the relative importance of the objective $F(\mathbf{x})$ and the constraint $\mathbf{Bx} = \mathbf{0}$. The problems in Eqs. (5.7) and (5.8) are not equivalent but they are close for large α. It is ready to see that the norm of the difference between $\mathbf{x}_\alpha^*$ and $\mathbf{x}^*$ is of order $\|\mathbf{x}_\alpha^* - \mathbf{x}^*\| = O(1/\alpha)$ [16]. We say that Eq. (5.8) is a primal relaxation of Eq. (5.7) and we will see that it leads to the distributed gradient descent (DGD) method that we cover in Section 5.3.

Alternatively, we can introduce a dual constraint relaxation. To do that, define the nonnegative Lagrange multiplier $\boldsymbol{v}$ associated with the constraint $\mathbf{Bx} = \mathbf{0}$ and define the Lagrangian

$$\mathcal{L}(\mathbf{x}, \boldsymbol{v}) := F(\mathbf{x}) + \boldsymbol{v}^\top \mathbf{Bx}. \tag{5.9}$$

Further define the Lagrangian minimizer variable $\mathbf{x}(\boldsymbol{v}) := \text{argmin}_{x \in \mathbb{R}^{np}} \mathcal{L}(\mathbf{x}, \boldsymbol{v})$ and the dual function as the corresponding Lagrangian minimum

$$g(\boldsymbol{v}) := \min_{x \in \mathbb{R}^{np}} \mathcal{L}(\mathbf{x}, \boldsymbol{v}) = \mathcal{L}(\mathbf{x}(\boldsymbol{v}), \boldsymbol{v}) = F(\mathbf{x}(\boldsymbol{v})) + \boldsymbol{v}^T \mathbf{Bx}(\boldsymbol{v}). \tag{5.10}$$

The dual problem is now defined as the maximization of the dual function and the optimal Lagrange multiplier is the corresponding maximizing argument

$$\boldsymbol{v}^* := \underset{\boldsymbol{v} \in \mathbb{R}^{2m}}{\text{argmax}}\, g(\boldsymbol{v}). \tag{5.11}$$

The importance of the Lagrangian maximizers $\mathbf{x}(\boldsymbol{v})$ is that they can be interpreted as being optimal with respect to a *linear* penalization of the constraint $\mathbf{Bx} = \mathbf{0}$, as opposed to the quadratic penalty as in the case of primal relaxations such as Eq. (5.8). The importance of the optimal dual variable $\boldsymbol{v}^*$ is that when this variable is used as a penalty coefficient, the Lagrangian maximizer $\mathbf{x}(\boldsymbol{v}^*)$ can be used to recover the optimal primal variable $\mathbf{x}^*$ under certain technical conditions [25]. Specifically, for any convex program it is known that $\mathbf{x}^*$ is a subset of $\mathbf{x}(\boldsymbol{v}^*)$,

$$\mathbf{x}^* \subseteq \mathbf{x}(\boldsymbol{v}^*) = \underset{x \in \mathbb{R}^{np}}{\text{argmin}}\, \mathcal{L}(\mathbf{x}, \boldsymbol{v}^*) = \underset{x \in \mathbb{R}^{np}}{\text{argmin}}\, f(\mathbf{x}) + \boldsymbol{v}^{*T} \mathbf{Bx}. \tag{5.12}$$

Thus, if, e.g., the Lagrangian maximizer $x(\boldsymbol{v}^*)$ is unique we know that $\mathbf{x}^*$ is also a singleton that must be equal to $x(\boldsymbol{v}^*)$. In cases when $x(\boldsymbol{v}^*)$ is not unique, it is still possible to recover $\mathbf{x}^*$ from the knowledge of the optimal dual variable $\boldsymbol{v}^*$. In any event, the important point here is that the problem of computing $\mathbf{x}^*$ has been transformed to the problem of determining $\boldsymbol{v}^*$. This latter process can be distributed and forms the basis for the dual methods that we discuss in Section 5.4.

5.3 DISTRIBUTED GRADIENT DESCENT

DGD is an established distributed method to solve Eq. (5.3) that relies on the solution of Eq. (5.8) through gradient descent. To explain this, define $f_\alpha(\mathbf{x}) := \alpha f(\mathbf{x}) + (1/2)\|\mathbf{Bx}\|^2$ and introduce an iteration index k so that $\mathbf{x}_k$ represents the iteration k estimate of the optimal argument $\mathbf{x}_\alpha^*$ of Eq. (5.8). Further let $\mathbf{g}_k := \nabla f_\alpha(\mathbf{x}_k)$ denote the gradient of the objective at iteration k and observe that we can write it as $\mathbf{g}_k = \alpha \nabla f(\mathbf{x}_k) + \mathbf{B}^T\mathbf{B}\mathbf{x}_k$. Further observe that it follows from Eq. (5.6) that the matrix product $\mathbf{B}^T\mathbf{B}$ reduces to $\mathbf{L} = \mathbf{B}^T\mathbf{B}$ where $\mathbf{L} = L \otimes \mathbf{I}$ is a p-dimensional Kronecker product extension of the Laplacian L. We can then write the gradient $\mathbf{g}_k$ as

$$\mathbf{g}_k := \nabla f_\alpha(\mathbf{x}_k) = \alpha \nabla f(\mathbf{x}_k) + \mathbf{L}\mathbf{x}_k. \tag{5.13}$$

We can now introduce a possibly time varying step size ϵ_k and utilize the gradient expression in Eq. (5.13) to define the gradient descent recursion

$$\mathbf{x}_{k+1} = \mathbf{x}_k - \epsilon_k \mathbf{g}_k = \mathbf{x}_k - \epsilon_k\big[\alpha \nabla f(\mathbf{x}_k) + \mathbf{L}\mathbf{x}_k\big]. \tag{5.14}$$

The important observation here is that Eq. (5.14) can be implemented in a distributed manner. Formally, recall that the variable $\mathbf{x}_k = [x_{1,k}; \dots; x_{n,k}] \in \mathbb{R}^{np}$ contains n components $x_{i,k} \in \mathbb{R}^p$ associated with each of the n nodes. We similarly decompose the gradient $\mathbf{g}_k = [g_{1,k}; \dots; g_{n,k}] \in \mathbb{R}^{np}$ in n components $g_{i,k} \in \mathbb{R}^p$ so that $g_{i,k}$ denotes the gradient of $f_\alpha(\mathbf{x})$ with respect to $\mathbf{x}_i$ evaluated at $\mathbf{x}_k$. The following proposition explains that it is possible for node i to compute $g_{i,k}$ while utilizing information that is either available locally or at some neighboring node $j \in n(i)$.

Proposition 5.1. *Recall the definition* $\mathbf{x}_k = [x_{1,k}; \dots; x_{n,k}] \in \mathbb{R}^{np}$ *and write the gradient* $\mathbf{g}_k = \nabla f_\alpha(\mathbf{x}_k)$ *as* $\mathbf{g}_k = [g_{1,k}; \dots; g_{n,k}] \in \mathbb{R}^{np}$ *where* $g_{k,i} = \nabla_{x_i} f_\alpha(\mathbf{x}_k)$ *is the local gradient of* f_α *with respect to the local variable* $x_i \in \mathbb{R}^p$. *The local gradient is explicitly given by*

$$g_{i,k} = \alpha \nabla f_i(x_{i,k}) + x_{i,k} - \sum_{j \in n(i)} w_{ij} x_{j,k}. \tag{5.15}$$

Proof. Using the definition of the function f_α, it follows that the gradient $g_{k,i} = \nabla_{x_i} f_\alpha(\mathbf{x}_k)$ can be written as $g_{k,i} = \alpha \nabla_{x_i} f(\mathbf{x}_k) + \nabla_{x_i} \|\mathbf{B}\mathbf{x}\|^2/2$. Begin by recalling that $f(\mathbf{x}_k) = \sum_{i=1}^n f_i(x_i)$ to conclude that its gradient with respect to x_i is $\nabla_{x_i} f(\mathbf{x}_k) = \nabla_{x_i} f_i(x_{i,k})$ because this is the only term that depends on x_i in the sum. To compute the gradient $\nabla_{x_i} \|\mathbf{B}\mathbf{x}\|^2/2$ observe that the quadratic form $\|\mathbf{B}\mathbf{x}\|^2/2$ can be written as $(1/2)\mathbf{x}^T\mathbf{B}^T\mathbf{B}\mathbf{x} = \mathbf{x}^T\mathbf{L}\mathbf{x}$. Thus, the variable x_i appears only as a square x_i^2 because of the diagonal elements $L_{ii} = 1$ and as a component of the cross products with neighbors $-w_{ij}x_ix_j$ because of the nonzero diagonal elements $L_{ij} = w_{ij}$. We can then conclude that $\nabla_{x_i}(\|\mathbf{B}\mathbf{x}\|^2/2) = x_{i,k} - \sum_{j \in n(i)} w_{ij} x_{j,k}$. Substituting $\nabla_{x_i} f(\mathbf{x}_k) = \nabla_{x_i} f_i(x_i)$ and $\nabla_{x_i}(\|\mathbf{B}\mathbf{x}\|^2/2) = x_{i,k} - \sum_{j \in n(i)} w_{ij} x_{j,k}$ into $g_{k,i} = \alpha \nabla_{x_i} f(\mathbf{x}_k) + \nabla_{x_i} \|\mathbf{B}\mathbf{x}\|^2/2$ the result in Eq. (5.15) follows. ■

That the gradient component $g_{i,k}$ in Eq. (5.15) can be computed locally at node i follows because the variable x_i and the function f_i are available at i whereas the variables $x_{j,k}$ can be obtained from neighbors. Because in gradient descent the component $g_{i,k}$ updates the variable $x_{i,k}$, it follows that Eq. (5.14) is equivalent to the n iterations

$$x_{i,k+1} = x_{i,k} - \epsilon_k g_{i,k} = x_{i,k} - \epsilon_k \left(\alpha \nabla f_i(x_{i,k}) + x_{i,k} - \sum_{j \in n(i)} w_{ij} x_{j,k} \right) \quad \forall i = 1, \dots, n. \tag{5.16}$$

Because the gradient component $g_{i,k}$ can be computed locally, the gradient iteration in Eq. (5.16) can be implemented locally as well. This yields the DGD method that we summarize in Algorithm 5.1. The core of the algorithm is in lines 3 and 4 that represent the implementation of Eqs. (5.15) and (5.16). Implementation of these steps requires exchanging variables with neighboring nodes, which is undertaken in line 2. The algorithm is repeated continuously, as indicated in line 1, or until a convergence criterion is satisfied.

Algorithm 5.1 DISTRIBUTED GRADIENT DESCENT (DGD) METHOD AT NODE i

Require: Initial iterate $x_{i,0}$, weights w_{ij}, penalty coefficient α, step sizes ϵ_k.
for $k = 0, 1, 2, \ldots$ **do**
  Exchange iterates $x_{i,k}$ with neighbors $j \in n(i)$.
  Evaluate gradient $g_{i,k} = \alpha \nabla f_i(x_{i,k}) + x_{i,k} - \sum_{j \in n(i)} w_{ij} x_{j,k}$.
  Update local iterate: $x_{i,k+1} = x_{i,k} - \epsilon_k g_{i,k}$.
end for

Because Eq. (5.16) is equivalent to Eq. (5.14) and the latter is a gradient descent method for f_α it follows that $\mathbf{x}_k$ converges to $\mathbf{x}_\alpha^*$ if the conditions that guarantee convergence of gradient descent methods are satisfied. If the function f_α is strongly convex this can be accomplished by choosing a fixed step size $\epsilon_k = \epsilon$ that is sufficiently small. The convergence rate is linear with a constant that is controlled by the condition number of f_α which, in turn, is a combination of the condition number of f and the condition number of L [16].

Remark 5.1. If we consider constant step sizes $\epsilon_k = \epsilon$ and we define $\tilde{w}_{ii} = (1 - \epsilon)$, $\tilde{w}_{ij} = \epsilon w_{ij}$, and $\tilde{\alpha} = \epsilon\alpha$, it is easy to see that Eq. (5.16) can be alternatively written as

$$x_{i,k+1} = \sum_{j \in n(i) \cup i} \tilde{w}_{ij} x_{j,k} - \tilde{\alpha} \nabla f_i(x_{i,k}), \tag{5.17}$$

where we observe the sum of the weights $\tilde{w}_{ij}$ is $\sum_{j \in n(i) \cup i} \tilde{w}_{ij} = 1$. The expression in Eq. (5.17) is the original formulation of DGD, which has an interesting interpretation. Because the sum of the weights $\tilde{w}_{ij}$ is one, we interpret $\sum_{j \in n(i) \cup i} \tilde{w}_{ij} x_{j,k}$ as an average of local and neighboring variables. We further reinterpret $\tilde{\alpha}$ as a step size and think of DGD as the sequential application of an averaging step followed by a local gradient descent step.

5.4 DUAL METHODS

5.4.1 DUAL ASCENT

We will proceed to develop a distributed algorithm for solving Eq. (5.7) based on the dual ascent method. In the case of symmetric communication graphs $\mathcal{G}$ that we focus on here, there is redundancy in having two separate constraint equations for edges $e = (i,j)$ and $e' = (j,i)$. Instead, we will work with the simplified setting where we only impose one constraint for each edge (e.g., keeping the version where $i < j$). Now the edge incidence matrix $B \in \mathbb{R}^{m \times n}$ can be defined analogously to Eq. (5.5), but with

$$B_{e,i} = B_{(i,j),i} = \sqrt{w_{ij}} \quad \text{and} \quad B_{e,j} = B_{(i,j),j} = -\sqrt{w_{ij}}.$$

Recall that the Lagrangian function for Eq. (5.7) is

$$\mathcal{L}(\boldsymbol{x}, \boldsymbol{v}) = \sum_{i=1}^{n} f_i(x_i) + \boldsymbol{v}^\top \boldsymbol{B}\boldsymbol{x}, \tag{5.18}$$

where $\boldsymbol{v} \in \mathbb{R}^{mp}$ is a vector of Lagrange multipliers and $\boldsymbol{B} = B \otimes I_p$. By Lagrangian duality theory we know that solutions to Eq. (5.7) lie at saddle points of $\mathcal{L}$ (with suitably chosen Lagrange multipliers) [26].

The dual ascent method seeks to find such a saddle point. Let $\boldsymbol{v}^0 \in \mathbb{R}^{mp}$ denote an arbitrary initial Lagrange multiplier vector. Then for $k = 1, 2, \ldots$, the dual ascent method repeats the steps

$$\boldsymbol{x}^k = \underset{\boldsymbol{x} \in \mathbb{R}^{np}}{\operatorname{argmin}}\, \mathcal{L}(\boldsymbol{x}, \boldsymbol{v}^{k-1}), \tag{5.19}$$

$$\boldsymbol{v}^k = \boldsymbol{v}^{k-1} - \eta_k \boldsymbol{B}\boldsymbol{x}^k, \tag{5.20}$$

where $\{\eta_k\}_{k \geq 1}$ is a sequence of positive step sizes.

Let $L_{i,+} = \{e_l = (i_l, j_l) \in E \mid j_l = i\}$ and $L_{i,-} = \{e_l = (i_l, j_l) \in E \mid i_l = i\}$ denote the sets of edges containing i, where i is either greater than or less than the index of its neighbor. Based on the definition of B, the Lagrangian function can be written as

$$\begin{aligned}\mathcal{L}(\boldsymbol{x}, \boldsymbol{v}) &= \sum_{i=1}^{n} f_i(x_i) - \sum_{l=1}^{m} v_l(x_{i_l} - x_{j_l}) \\ &= \sum_{i=1}^{n} \left[f_i(x_i) + x_i \left(\sum_{l \in L_{i,+}} v_l - \sum_{l \in L_{i,-}} v_l \right) \right].\end{aligned}$$

Thus, when the Lagrange multipliers $\boldsymbol{v}^k$ are fixed, the function $\mathcal{L}(\boldsymbol{x}, \boldsymbol{v}^k)$ is separable in the per-node vectors $x_1, \ldots, x_n$, and thus node i can solve for x_i^k in Eq. (5.19) locally via

$$x_i^k = \underset{x \in \mathbb{R}^p}{\operatorname{argmin}}\, f_i(x_i) + x_k \left(\sum_{l \in L_{i,+}} v_l^{k-1} - \sum_{l \in L_{i,-}} v_l^{k-1} \right), \tag{5.21}$$

without coordinating with other nodes. Moreover, node i's subproblem only depends on a subset of the Lagrange multipliers—those associated with edges incident to node i. Similarly, the update of the Lagrange multipliers v_l^k in Eq. (5.20) associated with edge $e_l = (i_l, j_l)$ only depends on the values $x_{i_l}^k$ and $x_{j_l}^k$ at nodes i_l and j_l.

This suggests the distributed algorithm defined in Algorithm 5.2, which we refer to as *distributed dual ascent*, where each node i maintains a copy of its local variable x_i as well as copies of the Lagrange multipliers v_l for those edges incident to i (i.e., for $l \in L_{i,+} \cup L_{i,-}$).

Algorithm 5.2 DISTRIBUTED DUAL ASCENT

Require: Initialize all Lagrange multipliers v_l^0 such that if $e_l = (i_l, j_l)$, then both nodes i_l and j_l know the value of v_l^0.

for $k = 1, 2, \ldots$, at each node i in parallel **do**

  Locally solve for x_i^k as in Eq. (5.21);

  Send the new value x_i^k to all neighbors and receive values x_j^k from each of its neighbors;

  Compute new multipliers v_l^k according to Eq. (5.20).

end for

Distributed dual ascent only involves communication in line 3, and this communication is local because the only nodes that need to maintain a copy of v_l are those at the ends of the edge e_l. Hence, the communication structure of this algorithm respects the underlying communication network topology [27,28].

Moreover, the dual ascent algorithm is guaranteed to converge to the minimizer $\boldsymbol{x}^\star$ of Eq. (5.7) under the assumption that the functions f_i are strictly convex, in addition to some regularity conditions [25]. Unfortunately these conditions are rather strong and one would prefer to have a convergent algorithm in more general settings. This can be accomplished by modifying the Lagrangian function as discussed next.

5.4.2 ALTERNATING DIRECTION METHOD OF MULTIPLIERS

Notice that the linear equality constraints in Eq. (5.7) enter into the Lagrangian function (5.18) as a term that is linear in the variables $\boldsymbol{x}$. To overcome limitations of the dual ascent method, a common approach is to modify the Lagrangian by adding a quadratic term. Specifically, for $\rho > 0$, the *augmented Lagrangian* is defined as

$$\mathcal{L}_\rho(\boldsymbol{x}, \boldsymbol{v}) = \sum_{i=1}^{n} f_i(x_i) + \boldsymbol{v}^\top \boldsymbol{B}\boldsymbol{x} + \frac{\rho}{2} \sum_{(i,j)\in E} \|x_i - x_j\|_2^2 . \tag{5.22}$$

Observe that, for any feasible point $\boldsymbol{x}$, we have $x_i = x_j$ for all i and j, and so the additional penalty terms in the augmented Lagrangian vanish. Nevertheless, the quadratic terms make the augmented Lagrangian strongly convex, even when the objective functions are not necessarily strongly convex, and this greatly aids in developing methods with convergence guarantees for a broader class of objective functions [25,29].

One algorithm based on the augmented Lagrangian is the *method of multipliers*, [29] which, for $k = 1, 2, \ldots$, repeats the steps

$$\boldsymbol{x}^k = \operatorname*{argmin}_{\boldsymbol{x}\in\mathbb{R}^{np}} \mathcal{L}_\rho(\boldsymbol{x}, \boldsymbol{v}^{k-1}), \tag{5.23}$$

$$\boldsymbol{v}^k = \boldsymbol{v}^{k-1} + \rho \boldsymbol{B}\boldsymbol{x}^k. \tag{5.24}$$

However, the quadratic terms also introduce coupling between different variables x_i and x_j, and consequently the augmented Lagrangian is not separable in $x_1, \ldots, x_n$. Thus, the nodes must coordinate in order to perform the $\boldsymbol{x}^k$ update in Eq. (5.23).

A common approach to arrive at a method that does not involve coupling across all nodes is to update the per-node blocks $x_1, \ldots, x_n$ of $\boldsymbol{x}$ one at a time. This leads to the celebrated *alternating direction method of multipliers* (ADMM) [30]. To describe the ADMM more precisely, let us denote by $\mathcal{L}_{i,\rho}(x_i, \boldsymbol{x}_{-i}, \boldsymbol{v})$ the augmented Lagrangian where we separate out the variable x_i at node i from the rest of the network $\boldsymbol{x}_{-i}$; i.e., $\boldsymbol{x}_{-i} = [x_j : j \neq i]$ is the length-$(n-1)$ vector obtained from $\boldsymbol{x}$ by removing the ith entry.

One instance of the ADMM is shown in Algorithm 5.3. Each node i maintains a copy of its local variable, x_i, its multipliers v_l for those edges incident to i as well as copies of the local variables x_j for its neighbors in the network.

Algorithm 5.3 ALTERNATING DIRECTION METHOD OF MULTIPLIERS

Require: Initialize all Lagrange multipliers v_l^0 such that if $e_l = (i_l, j_l)$, then both nodes i_l and j_l know the value of v_l^0. Also, each node i initializes x_i^0 and sends this to each of its neighbors.

```
1: for k = 1, 2, ..., do
2:     for each node i = 1, ..., n do
3:         Update x_i^k = argmin_{x_i ∈ R^p} L_{i,ρ_k}(x_i, x̃_{-i}^{k-1}, v^{k-1}), where x̃_{-i}^{k-1} contains the latest values of nodes other than i;
           specifically, for j < i these will have already been updated to x_j^k, and for j > i they will be equal to x_j^{k-1};
4:         Communicate x_i^k to all neighbors of i;
5:     end for
6:     Compute new multipliers v_l^k according to Eq. (5.24).
7: end for
```

The description of the ADMM given in Algorithm 5.3 has the nodes update sequentially according to their indices, $1, 2, \ldots$, in the inner loop (lines 2–5). Of course, indices have been assigned to the nodes in an arbitrary way at this point, so this order is arbitrary. It is possible to change the order at each outer loop, or more generally to allow the order to be random as long as asymptotically all nodes perform updates equally often.

It is known that the ADMM converges to an optimal solution at a linear rate when the functions f_i are convex [31,32]. Moreover, versions of the ADMM where the node that updates at each step of the inner loop is chosen randomly are also known to converge at the same rate [33–36].

Note that, from the perspective of node i,

$$\mathcal{L}_{i,\rho}(x_i, \tilde{\boldsymbol{x}}_{-i}^{k-1}, \boldsymbol{v}^{k-1}) = f_i(x_i) + x_i\left(\sum_{l \in L_{i,+}} v_l - \sum_{l \in L_{i,-}} v_l\right) + \frac{\rho}{2}\sum_{j \in N_i}\|x_i - x_j\|_2^2 + c_i(\boldsymbol{x}_{-i}, \boldsymbol{v}),$$

where the terms encapsulated into $c_i(\boldsymbol{x}_{-i}, \boldsymbol{v})$ only involve components $x_{i'}$ of $\boldsymbol{x}$ for nodes i' that are not neighbors of i and multipliers v_l associated with edges not incident to node i. Hence, the update at node i in line 3 of Algorithm 5.3 is equivalent to

$$x_i^k = \underset{x_i \in \mathbb{R}^p}{\operatorname{argmin}} f_i(x_i) + x_i\left(\sum_{l \in L_{i,+}} v_l - \sum_{l \in L_{i,-}} v_l\right) + \frac{\rho}{2}\sum_{j \in N_i}\|x_i - x_j\|_2^2,$$

which depends on information that is available at node i after it has communicated with its immediate neighbors. Hence, these updates can be implemented locally without the need to communicate or coordinate across the network.

In practice, the optimization subproblem arising in the update of x_i^k is typically not solved exactly. Instead, an approximate solution may be used [37], or surrogate subproblems may be solved (e.g., using a linear [38] or quadratic [39] approximation of f_i) without sacrificing the convergence properties of the ADMM. Another practical issue that arises when applying the ADMM is to find a suitable choice of the penalty parameter ρ, which may be constant or may vary from iteration to iteration (typically nondecreasing). Theoretical conditions relating the value of ρ to the rate of convergence, and hence providing guidelines for its selection, are available in the case when the functions f_i are quadratic [40] or strongly convex [41].

Finally, we remark that there has recently been interest in exploring convergence properties of the ADMM when the functions f_i are not necessarily convex [42,43]. In such settings, convergence to local solutions may still be guaranteed when the parameter ρ is chosen to be sufficiently large.

5.5 SECOND-ORDER METHODS

Instead of solving Eq. (5.8) with a gradient descent algorithm as in DGD, we can solve Eq. (5.8) using Newton's method. To implement Newton's method we need to compute the Hessian $\mathbf{H}_k := \nabla^2 f_\alpha(\mathbf{x}_k)$ of f_α evaluated at $\mathbf{x}_k$ so as to determine the Newton step $\mathbf{d}_k := -\mathbf{H}_k^{-1}\mathbf{g}_k$. Start by differentiating f_α twice in order to write $\mathbf{H}_k$ as

$$\mathbf{H}_k := \nabla^2 f_\alpha(\mathbf{x}_k) = \alpha \mathbf{G}_k + \mathbf{L}, \tag{5.25}$$

where the matrix $\mathbf{G}_k \in \mathbb{R}^{np \times np}$ is a block diagonal matrix formed by blocks $\mathbf{G}_{ii,k} \in \mathbb{R}^{p \times p}$ containing the Hessian of the ith local function,

$$\mathbf{G}_{ii,k} = \nabla^2 f_i(x_{i,k}). \tag{5.26}$$

It follows from Eqs. (5.25) and (5.26) that the Hessian $\mathbf{H}_k$ is block sparse with blocks $\mathbf{H}_{ij,k} \in \mathbb{R}^{p \times p}$ having the sparsity pattern of $\mathbf{L}$, which is the sparsity pattern of the graph. The diagonal blocks are of the form $\mathbf{H}_{ii,k} = \mathbf{I} + \alpha \nabla^2 f_i(x_{i,k})$ and the off-diagonal blocks are not null only when $j \in n(i)$ in which case $\mathbf{H}_{ij,k} = w_{ij}\mathbf{I}$.

While the Hessian $\mathbf{H}_k$ is sparse, the inverse of $\mathbf{H}_k$ is not, and we need the inverse to compute the Newton direction $\mathbf{d}_k := \mathbf{H}_k^{-1}\mathbf{g}_k$. To overcome this problem we will split $\mathbf{H}_k$ into the sum of two matrices, one that is block diagonal and the other containing the remaining elements of $\mathbf{H}_k$, and then we will rely on a Taylor's expansion of the inverse. This splitting technique is inspired from the Taylor's expansion used in [44,45]. To be precise, write $\mathbf{H}_k = \mathbf{D}_k - \mathbf{J}_k$ where the matrix $\mathbf{D}_k$ is defined as

$$\mathbf{D}_k := \alpha \mathbf{G}_k + \mathbf{I}, \tag{5.27}$$

and $\mathbf{J}_k$ is a matrix we will determine below. The block diagonal matrix $\mathbf{G}_k$ is positive definite because the local functions are assumed to be strongly convex, and thus the matrix $\mathbf{D}_k$ is block diagonal and positive definite. The ith diagonal block $\mathbf{D}_{ii,k} \in \mathbb{R}^p$ of $\mathbf{D}_k$ can be computed and stored by node i as $\mathbf{D}_{ii,k} = \alpha \nabla^2 f_i(x_{i,k}) + \mathbf{I}$. To have $\mathbf{H}_k = \mathbf{D}_k - \mathbf{J}_k$ we must define $\mathbf{J}_k := \mathbf{D}_k - \mathbf{H}_k$. Considering the definitions of $\mathbf{H}_k$ and $\mathbf{D}_k$ in Eqs. (5.25) and (5.27), it follows that

$$\mathbf{J}_k := \mathbf{J} = \mathbf{I} - \mathbf{L}. \tag{5.28}$$

Observe that $\mathbf{J}$ is independent of time and only depends on the Laplacian matrix $\mathbf{L}$. As in the case of the Hessian $\mathbf{H}_k$, the matrix $\mathbf{J}$ is block sparse with blocks $\mathbf{J}_{ij} \in \mathbb{R}^{p \times p}$ having the sparsity pattern of $\mathbf{L}$, which is the sparsity pattern of the graph. Node i can compute the off diagonal blocks $\mathbf{J}_{ij} = w_{ij}\mathbf{I}$ using information about its neighbors' weights. Notice that the diagonal blocks are $\mathbf{J}_{ii} = 0$.

Proceed now to factor $\mathbf{D}_k^{1/2}$ from both sides of the splitting relationship to write $\mathbf{H}_k = \mathbf{D}_k^{1/2}(\mathbf{I} - \mathbf{D}_k^{-1/2}\mathbf{J}\mathbf{D}_k^{-1/2})\mathbf{D}_k^{1/2}$. When we consider the Hessian inverse $\mathbf{H}^{-1}$, we can use the Taylor series $(\mathbf{I} - \mathbf{X})^{-1} = \sum_{j=0}^{\infty} \mathbf{X}^j$ with $\mathbf{X} = \mathbf{D}_k^{-1/2}\mathbf{J}\mathbf{D}_k^{-1/2}$ to write

$$\mathbf{H}_k^{-1} = \mathbf{D}_k^{-1/2} \sum_{l=0}^{\infty} \left(\mathbf{D}_k^{-1/2}\mathbf{J}\mathbf{D}_k^{-1/2}\right)^l \mathbf{D}_k^{-1/2}. \tag{5.29}$$

Observe that the sum in Eq. (5.29) converges if the absolute value of all the eigenvalues of the matrix $\mathbf{D}^{-1/2}\mathbf{B}\mathbf{D}^{-1/2}$ are strictly less than 1. It can be seen that this is true [46]. When the series converge, we can use truncations of this series to define approximations to the Newton step, as we explain in the following section.

Remark 5.2. The Hessian decomposition $\mathbf{H}_k = \mathbf{D}_k - \mathbf{J}$ with the matrices $\mathbf{D}_k$ and $\mathbf{J}$ in Eqs. (5.27) and (5.28), respectively, is not the only valid decomposition that we can use in this approach. Any decomposition of the form $\mathbf{H}_k = \mathbf{D}_k \pm \mathbf{J}_k$ is valid if $\mathbf{D}_k$ is positive definite and the eigenvalues of the matrix $\mathbf{D}_k^{-1/2}\mathbf{J}_k\mathbf{D}_k^{-1/2}$ are in the interval $(-1, 1)$. For example, an alternative decomposition is given by the matrices $\mathbf{D}_k = \alpha\mathbf{G}_k$ and $\mathbf{J} = \mathbf{L}$. This decomposition has the advantage of separating the effects of the function in $\mathbf{D}_k$ and the effects of the network in $\mathbf{J}$. The decomposition in Eqs. (5.27) and (5.28) exhibits a faster convergence of the series in Eq. (5.29) because the matrix $\mathbf{D}_k$ in Eq. (5.27) accumulates more weight in the diagonal than the matrix $\mathbf{D}_k = \alpha\mathbf{G}_k$. The study of alternative decompositions is an interesting area for future work.

5.5.1 DISTRIBUTED APPROXIMATIONS OF THE NEWTON STEP

Network Newton (NN) is defined as a family of algorithms that relies on truncations of the series in Eq. (5.29). The Lth member of this family, NN-L, considers the first $L + 1$ terms of the series to define the approximate Hessian inverse

$$\hat{\mathbf{H}}_k^{(L)^{-1}} := \mathbf{D}_k^{-1/2} \sum_{l=0}^{L} \left(\mathbf{D}_k^{-1/2}\mathbf{J}\mathbf{D}_k^{-1/2}\right)^l \mathbf{D}_k^{-1/2}. \tag{5.30}$$

NN-L uses the approximate Hessian $\hat{\mathbf{H}}_k^{(L)^{-1}}$ as a curvature correction matrix that is used in lieu of the exact Hessian inverse $\mathbf{H}^{-1}$ to estimate the Newton step; i.e., instead of descending along the Newton step $\mathbf{d}_k := -\mathbf{H}_k^{-1}\mathbf{g}_k$ we descend along the NN-L step $\mathbf{d}_k^{(L)} := -\hat{\mathbf{H}}_k^{(L)^{-1}}\mathbf{g}_k$, which we intend as an approximation of $\mathbf{d}_k$. Using the explicit expression for $\hat{\mathbf{H}}_k^{(L)^{-1}}$ in Eq. (5.30) we write the NN-L step as

$$\mathbf{d}_k^{(L)} = -\mathbf{D}_k^{-1/2} \sum_{l=0}^{L} \left(\mathbf{D}_k^{-1/2}\mathbf{J}\mathbf{D}_k^{-1/2}\right)^l \mathbf{D}_k^{-1/2}\mathbf{g}_k, \tag{5.31}$$

where, we recall, the vector $\mathbf{g}_k$ is the gradient of objective function f_α defined. The NN-L update formula can then be written as

$$\mathbf{y}_{k+1} = \mathbf{x}_k + \epsilon\, \mathbf{d}_k^{(L)}, \tag{5.32}$$

where ϵ is a properly selected step size. The algorithm defined by recursive application of Eq. (5.32) can be implemented in a distributed manner because the truncated series in Eq. (5.30) has a local structure controlled by the parameter L. To explain this statement better, define the components $\mathbf{d}_{i,k}^{(L)} \in \mathbb{R}^p$ of the NN-L step $\mathbf{d}_k^{(L)} = [\mathbf{d}_{1,k}^{(L)}; \ldots ; \mathbf{d}_{n,k}^{(L)}]$. A distributed implementation of Eq. (5.32) requires that node i computes $\mathbf{d}_{i,k}^{(L)}$ so as to implement the local descent $\mathbf{x}_{i,k+1} = \mathbf{x}_{i,k} + \epsilon \mathbf{d}_{i,k}^{(L)}$. The key observation here is that the step component $\mathbf{d}_{i,k}^{(L)}$ can indeed be computed through local operations. Specifically, begin by noting that as per the definition of the NN-L descent direction in Eq. (5.31), the sequence of NN descent directions satisfies

$$\mathbf{d}_k^{(l+1)} = \mathbf{D}_k^{-1}\mathbf{J}\mathbf{d}_k^{(l)} - \mathbf{D}_k^{-1}\mathbf{g}_k = \mathbf{D}_k^{-1}\left(\mathbf{J}\mathbf{d}_k^{(l)} - \mathbf{g}_k\right). \tag{5.33}$$

Then observe that because the matrix $\mathbf{J}$ has the sparsity pattern of the graph, this recursion can be decomposed into local components

$$\mathbf{d}_{i,k}^{(l+1)} = \mathbf{D}_{ii,k}^{-1}\left(\sum_{j\in n(i)\cup\{i\}} \mathbf{J}_{ij}\mathbf{d}_{j,k}^{(l)} - \mathbf{g}_{i,k}\right). \tag{5.34}$$

The matrix $\mathbf{D}_{ii,k} = \alpha\nabla^2 f_i(x_{i,k}) + 2(1 - w_{ii})\mathbf{I}$ is stored and computed at node i. The gradient component $\mathbf{g}_{i,k} = (1-w_{ii})\mathbf{x}_{i,k} - \sum_{j\in n(i)} w_{ij}\mathbf{x}_{j,k} + \alpha\nabla f_i(\mathbf{x}_{i,k})$ is also stored and computed at i. Node i can also evaluate the values of the matrix blocks $\mathbf{J}_{ii} = (1 - w_{ii})\mathbf{I}$ and $\mathbf{J}_{ij} = w_{ij}\mathbf{I}$. Thus, if the NN-$k$ step components $\mathbf{d}_{j,k}^{(l)}$ are available at neighboring nodes j, node i can then determine the NN-$(k+1)$ step component $\mathbf{d}_{i,k}^{(l+1)}$ upon being communicated that information.

The expression in Eq. (5.34) represents an iterative computation embedded inside the NN-L recursion in Eq. (5.32). For each time index k, we compute the local component of the NN-0 step $\mathbf{d}_{i,k}^{(0)} = -\mathbf{D}_{ii,k}^{-1}\mathbf{g}_{i,k}$. Upon exchanging this information with neighbors we use Eq. (5.34) to determine the NN-1 step components $\mathbf{d}_{i,k}^{(1)}$. These can be exchanged and plugged in Eq. (5.34) to compute $\mathbf{d}_{i,k}^{(2)}$. Repeating this procedure L times, nodes ends up having determined their NN-L step component $\mathbf{d}_{i,k}^{(L)}$.

The resulting NN-L method is summarized in Algorithm 5.4. The descent iteration in Eq. (5.32) is implemented in line 11. Implementation of this descent requires access to the NN-L descent direction $\mathbf{d}_{i,k}^{(L)}$, which is computed by the loop in lines 6–10. Line 6 initializes the loop by computing the NN-0 step $\mathbf{d}_{i,k}^{(0)} = -\mathbf{D}_{ii,k}^{-1}\mathbf{g}_{i,k}$. The core of the loop is in line 9, which corresponds to the recursion in Eq. (5.34). Line 8 stands for the variable exchange that is necessary to implement line 9. After L iterations through this loop, the NN-L descent direction $\mathbf{d}_{i,k}^{(L)}$ is computed and can be used in line 11. Both lines 6 and 9 require access to the local gradient component $\mathbf{g}_{i,k}$. This is evaluated in line 5 after receiving the prerequisite information from neighbors in line 4. Lines 1 and 3 compute the blocks $\mathbf{J}_{ii,k}$, $\mathbf{J}_{ij,k}$, and $\mathbf{D}_{ii,k}$ that are also necessary in lines 6 and 9.

Algorithm 5.4 NETWORK NEWTON-*L* METHOD AT NODE *i*

Require: Initial iterate $\mathbf{x}_{i,0}$. Weights w_{ij}. Penalty coefficient α.

J matrix blocks: $\mathbf{J}_{ii} = (1 - w_{ii})\mathbf{I}$ and $\mathbf{J}_{ij} = w_{ij}\mathbf{I}$

for $k = 0, 1, 2, \ldots$ **do**

  D matrix block: $\mathbf{D}_{ii,k} = \alpha \nabla^2 f_i(x_{i,k}) + 2(1 - w_{ii})\mathbf{I}$

  Exchange iterates $x_{i,k}$ with neighbors $j \in n(i)$.

  Gradient: $\mathbf{g}_{i,k} = (1 - w_{ii})x_{i,k} - \sum_{j \in n(i)} w_{ij}\mathbf{x}_{j,k} + \alpha \nabla f_i(x_{i,k})$

  Compute NN-0 descent direction $\mathbf{d}_{i,k}^{(0)} = -\mathbf{D}_{ii,k}^{-1}\mathbf{g}_{i,k}$

  for $l = 0, \ldots, L-1$ **do**

    Exchange elements $\mathbf{d}_{i,k}^{(l)}$ of the NN-l step with neighbors

    NN-$(l+1)$ step: $\mathbf{d}_{i,k}^{(l+1)} = \mathbf{D}_{ii,k}^{-1}\left[\sum_{j \in n(i), j=i} \mathbf{J}_{ij}\mathbf{d}_{j,k}^{(l)} - \mathbf{g}_{i,k}\right]$

  end for

  Update local iterate: $\mathbf{x}_{i,k+1} = x_{i,k} + \epsilon\, \mathbf{d}_{i,k}^{(L)}$.

end for

Remark 5.3. By trying to approximate the Newton step, NN-L ends up reducing the number of iterations required for convergence. Furthermore, the larger L is, the closer the NN-L step gets to the Newton step, and the faster NN-L converges. We justify these assertions both analytically and numerically in [46]. It is important to observe, however, that reducing the number of iterations reduces the computational cost but not necessarily the communication cost. In DGD, each node i shares its decision vector $x_{i,k} \in \mathbb{R}^p$ with each of its neighbors $j \in n(i)$. In NN-L, node i exchanges not only the decision vector $x_{i,k} \in \mathbb{R}^p$ with its neighboring nodes, but it also communicates iteratively the local components of the descent directions $\{\mathbf{d}_{i,k}^{(l)}\}_{l=0}^{L-1} \in \mathbb{R}^p$ so as to compute the descent direction $\mathbf{d}_{i,k}^{(L)}$. Therefore, at each iteration, node i sends $|\mathcal{N}_i|$ vectors of size p to the neighboring nodes in DGD while in NN-L it sends $(L+1)|\mathcal{N}_i|$ vectors of the same size. Unless the original problem is well conditioned, NN-L also reduces total communication cost until convergence, even though the cost of each individual iteration is larger [46]. However, the use of large L is unwarranted because the added benefit of better approximating the Newton step does not compensate for the increase in communication cost.

5.6 PRACTICAL CONSIDERATIONS

5.6.1 SYNCHRONOUS VERSUS ASYNCHRONOUS METHODS

All the distributed optimization methods described above are *synchronous*: nodes perform their updates and communicate in a coordinated manner. That is, all nodes perform local computations in parallel, and then they communicate. The iteration counter, k, can be viewed as incrementing in a consistent way across all the nodes. Synchronous methods have many advantages: they are relatively straightforward to describe and implement and they are also easy to debug because their execution is deterministic. However, they also have a drawback: *all nodes progress at the pace of the slowest node* [47]. This can be highly undesirable when one or a few nodes are significantly slower than the others, e.g., if their processor speed is slower, computing the gradient of their local objective function is significantly more complex than the others, or their communication link supports a much lower rate than others. This motivates the use of *asynchronous* methods.

In an asynchronous multiagent method, we allow nodes to proceed with computation and communication at their own pace. The hope is that this will allow the overall system to make progress toward the solution significantly faster than synchronous methods when there is one or more slow node.

The reader should take note that the term "asynchronous" has sometimes been used to mean different things in the distributed optimization literature. For example, in the context of the ADMM, it has been used to mean that the order in which nodes perform their local updates is randomized [33,35]. In the context of gossip and consensus algorithms, it is sometimes used to mean that a single pair of neighboring nodes updates and communicates in one iteration, rather than all pairs of neighboring nodes [48]. When communication takes place over a wireless network, it may also be taken to mean that one node updates and broadcasts information to its neighbors in each iteration [49].

More generally, and most relevant to the current discussion, asynchronous methods typically refer to those in which nodes may perform updates at different rates, messages may be received after some delay, and the communication topology may change over time [50,51]. For this general model, there is still no *de facto* distributed method with strong convergence guarantees available in the literature that addresses all the challenges listed above in the setting considered in this chapter. Recent progress has been made in a variety of directions, including accounting for (individually) heterogeneous update rates [52], delays [53–55], and time-varying networks [56,57]. A promising framework for addressing these challenges is the push-sum approach, which we briefly describe next.

5.6.2 PUSH-SUM METHODS

In the discussion of the distributed gradient method in Section 5.3, the weight matrix **W** played an important role. There, we assumed that the network was symmetric and that **W** was stochastic, which together imply that **W** is doubly stochastic: its row and columns sum to unity. In many practical situations, ensuring that the communication network is symmetric or that the weight matrix **W** is doubly stochastic can be difficult or impossible, especially in asynchronous implementations [47]. At a high level, this is because ensuring that the weights **W** are symmetric and stochastic for each update requires that nodes coordinate with all their neighbors, and a fundamental reason to use asynchronous methods is to reduce the amount of coordination required. In the asynchronous setting there are two alternatives one may consider.

1. Each node may be responsible for assigning weights to the messages it has received when performing an update. This corresponds to giving each node control of its row of the matrix **W**, in which case we arrive at an algorithm where **W** is row-stochastic (but not column-stochastic).
2. Each node may be responsible for assigning weights to the messages it sends. This corresponds to giving each node control of its column of the matrix **W**, and leads to algorithms where **W** is column-stochastic (but not row-stochastic).

The second approach, where **W** is column-stochastic, results in methods associated with the name "push-sum" in the literature, which turn out to have a number of practical advantages. The name push-sum comes from the fact that, when nodes update, they assign weights and then send (i.e., *push*) these messages to their neighbors. The messages wait in a queue at each neighbor until that node is ready to update, and when a node updates, it *sums* all the messages it has received.

Unsurprisingly, column-stochastic matrices are related to Markov chains. In asynchronous push-sum methods, because the node or nodes performing an update vary over time, we have a sequence of time matrices $\mathbf{W}_1, \mathbf{W}_2, \ldots$, one for each update time, where the columns of nonupdating nodes are equal to the corresponding column of the identity matrix (because nothing changes with respect to those nodes' states). Moreover, it is the product of such matrices that appears in the analysis of such methods, after unrolling updates such as Eq. (5.16) over multiple steps.

Products of column-stochastic matrices are also column-stochastic, and the ergodicity theory of Markov chains provides conditions under which products of column-stochastic matrices converge to a limiting column-stochastic matrix where all columns are identical [58]. Let $\mathbf{x}_0$ denote an initial state vector across the network, and let $\mathbf{W}_\infty = \ldots \mathbf{W}_3\mathbf{W}_2\mathbf{W}_1$ denote a product of column-stochastic matrices. When the product converges to an ergodic limit, so $\mathbf{W}_\infty$ is a column-stochastic matrix where all columns are identical, then $\mathbf{W}_\infty = \boldsymbol{\pi}\mathbf{1}^\top$, where $\boldsymbol{\pi}$ is the stationary distribution of the process. Consequently, $\mathbf{y} = \mathbf{W}_\infty \mathbf{x}_0 = (\mathbf{1}^\top \mathbf{x}_0)\boldsymbol{\pi}$ is such that each entry of $\mathbf{y}$ is a scaled version of the sum $\mathbf{1}^\top \mathbf{x}_0$. If we take $\mathbf{x}_0 = \mathbf{1}$, the vector of all ones, then we get $\mathbf{y} = n\boldsymbol{\pi}$. Of course, if we only run a finite number of such iterations then the convergence will be approximate.

Push-sum methods operate by running two copies of this iteration in parallel. One copy may take as input, e.g., the gradient at each node and output $\mathbf{y}_j \approx \pi_j \sum_{i=1}^n \nabla f_i(\mathbf{x}_i^k)$ at node j. The second copy is initialized with the value 1 at every node, and it outputs $z_j \approx n\pi_j$ at node j. Then, by dividing each element of $\mathbf{y}_j$ by z_j, we recover the average gradient. Thus, even though column-stochastic matrices do not necessarily lead to uniform stationary distributions, we can compensate to de bias the algorithm. (Note: If we had instead used row-stochastic matrices, it would not be possible to unmix the contributions of each node, and the resulting algorithm would be equivalent to minimizing $\sum_{i=1}^n \pi_i f_i(x)$, where the contribution to the cost function would depend on the stationary distribution π, and hence on the structure of the matrices $\mathbf{W}_k$.)

Push-sum methods were originally developed to solve the problem of distributed averaging [59]. They were first adapted for distributed optimization in [60], and follow-up work has developed push-sum based first-order methods with extrapolation to converge at faster rates when the functions f_i are strongly convex [61,62], as well as methods that are suited to run over time-varying and directed graphs [56,57]. Developing push-sum distributed optimization methods to handle the full complexity of the asynchronous scenario remains an open and active area of research.

5.7 CONCLUSION

In this chapter we have introduced the problem of consensus-based multiagent convex optimization. We described two different approaches to the design of multiagent algorithms. In the first approach, consensus is achieved by adding a penalty to the separable objective function. The penalty contains one term for each edge in the communication graph, and these terms grow quadratically with the difference between the optimization variables at neighboring agents. The standard DGD algorithm can be viewed as performing gradient descent on this penalized objective, and thus this is one point of departure for developing other decentralized methods.

A second, alternative approach is to explicitly enforce a consensus by introducing constraints to ensure that the optimization variables at neighboring agents are identical. Addressing the constraints via duality leads to approaches such as dual ascent and the ADMM. Examining the Lagrangian function of the constrained problem also reveals that this approach can be seen as imposing a linear penalty.

Multiagent convex optimization has received considerable attention in the research community during the past 15 years, and we now have a solid understanding of theoretical tools and analyses of these methods, including ways to accelerate rates of convergence of first-order methods [63–65] and distributed second-order methods (e.g., as discussed in Section 5.5). We believe the time is ripe for implementations of these methods as well as their use in applications. The discussion of practical issues in Section 5.6, such as dealing with asynchrony and directed communication topologies, addresses some of the issues we have observed to appear in practical implementations, which are not always considered in theoretical studies of multiagent optimization algorithms.

Throughout this chapter we focused on static communication topologies. Another issue that is likely to arise in practical applications (e.g., when agents communicate over a wireless networks) is that the communication topology will vary with time. The issue of time-varying topologies has been studied extensively in the literature, and there are several results guaranteeing asymptotic convergence; see [66] for a recent survey. In general, in the time-varying setting it is not required that the communication topology be connected at every iteration. Rather, convergence to an optimal solution can still be guaranteed even if the communication topology is not connected at any iteration, as long as an aggregate topology (taking the union of edges over some number $b > 0$ of consecutive iterations, so-called *b-connectivity*) is sufficiently well connected. Worst-case rates of convergence in this general setting are typically much slower than in the case of static connected graphs, mainly due to the potentially very slow convergence of averaging in the time-varying setting. However, in the special case where the graph at each iteration can be modeled as being an *independent and identically distributed* (i.i.d.) sample from a family of random graphs, where the expected graph is connected, then convergence rate guarantees (in expectation) that depend on spectral properties of the expected graph can be recovered [67]. An interesting direction for future work could be to consider a regime between these two extremes to better understand what convergence guarantees can be obtained in the setting of time-varying topologies under stronger assumptions than b-connectivity, but without assuming i.i.d. topologies.

Finally, although we have focused on multiagent algorithms for convex optimization in this chapter, there has also been work on algorithms for nonconvex optimization [68–72]. In this case, convergence is only guaranteed to a stationary point.

REFERENCES

[1] Bullo F, Cortés J, Martinez S. Distributed control of robotic networks: a mathematical approach to motion coordination algorithms. Princeton University Press; 2009.

[2] Cao Y, Yu W, Ren W, Chen G. An overview of recent progress in the study of distributed multi-agent coordination. IEEE Trans Ind Inf 2013;9:427–38.

[3] Lopes CG, Sayed AH. Diffusion least-mean squares over adaptive networks: formulation and performance analysis. IEEE Trans Signal Process 2008;56(7):3122–36.

[4] Schizas ID, Ribeiro A, Giannakis GB. Consensus in ad hoc WSNS with noisy links—Part I: Distributed estimation of deterministic signals. IEEE Trans Signal Process 2008;56(1):350–64.

[5] Khan UA, Kar S, Moura JM. Diland: an algorithm for distributed sensor localization with noisy distance measurements. IEEE Trans Signal Process 2010;58(3):1940–7.

[6] Rabbat M, Nowak R. Distributed optimization in sensor networks. In: Proceedings of the 3rd international symposium on information processing in sensor networks. ACM; 2004. p. 20–7.

[7] Rabbat M, Nowak R. Distributed optimization in sensor networks. In: Proceedings of the ACM/IEEE international conference on information processing in sensor networks (IPSN), Berkeley, CA, USA; 2004. p. 20–7.

[8] Ribeiro A. Ergodic stochastic optimization algorithms for wireless communication and networking. IEEE Trans Signal Process 2010;58(12):6369–86.

[9] Ribeiro A. Optimal resource allocation in wireless communication and networking. EURASIP J Wirel Commun Netw 2012;2012(1):1–19.

[10] Rabbat MG, Nowak RD. Decentralized source localization and tracking [wireless sensor networks]. In: IEEE international conference on acoustics, speech, and signal processing, 2004. ICASSP'04, vol. 3. IEEE; 2004. p. iii-921–iii-924.

[11] Bekkerman R, Bilenko M, Langford J. Scaling up machine learning: parallel and distributed approaches. Cambridge University Press; 2011.

[12] Tsianos KI, Lawlor S, Rabbat MG. Consensus-based distributed optimization: practical issues and applications in large-scale machine learning. In: 2012 50th annual Allerton conference on communication, control, and computing (Allerton); 2012. p. 1543–50.

[13] Cevher V, Becker S, Schmidt M. Convex optimization for big data: scalable, randomized, and parallel algorithms for big data analytics. IEEE Signal Process Mag 2014;31(5):32–43.

[14] Nedic A, Ozdaglar A. Distributed subgradient methods for multi-agent optimization. IEEE Trans Autom Control 2009;54(1):48–61.

[15] Jakovetic D, Xavier J, Moura JM. Fast distributed gradient methods. IEEE Trans Autom Control 2014;59(5):1131–46.

[16] Yuan K, Ling Q, Yin W. On the convergence of decentralized gradient descent. arXiv preprint arXiv:13107063; 2013.

[17] Shi W, Ling Q, Wu G, Yin W. Extra: an exact first-order algorithm for decentralized consensus optimization. arXiv preprint arXiv:14046264; 2014.

[18] Duchi JC, Agarwal A, Wainwright MJ. Dual averaging for distributed optimization: convergence analysis and network scaling. IEEE Trans Autom Control 2012;57(3):592–606.

[19] Tsianos KI, Lawlor S, Rabbat MG. Push-sum distributed dual averaging for convex optimization. In: 2012 IEEE 51st annual conference on decision and control (CDC); 2012. p. 5453–8.

[20] Ling Q, Ribeiro A. Decentralized linearized alternating direction method of multipliers. In: 2014 IEEE international conference on acoustics, speech and signal processing (ICASSP); 2014. p. 5447–51.

[21] Boyd S, Parikh N, Chu E, Peleato B, Eckstein J. Distributed optimization and statistical learning via the alternating direction method of multipliers. In: Foundations and trends® in machine learning, vol. 3(1); 2011. p. 1–122.

[22] Shi W, Ling Q, Yuan K, Wu G, Yin W. On the linear convergence of the ADMM in decentralized consensus optimization. IEEE Trans Signal Process 2014;62(7):1750–61.

[23] Mota JF, Xavier JM, Aguiar PM, Puschel M. D-ADMM: a communication-efficient distributed algorithm for separable optimization. IEEE Trans Signal Process 2013;61(10):2718–23.

[24] Chang TH, Hong M, Wang X. Multi-agent distributed optimization via inexact consensus ADMM. IEEE Trans Signal Process 2015;63(2):482–97.

[25] Bertsekas D. Nonlinear programming. 2nd ed. Athena Scientific; 1999.

[26] Boyd S, Vandenberghe L. Convex optimization. Cambridge University Press; 2004.

[27] Rabbat M, Nowak R, Bucklew J. Generalized consensus algorithms in networked systems with erasure links. In: Proceedings of the IEEE workshop on signal processing advances in wireless communications, New York; 2005. p. 1088–92.

[28] Schizas I, Ribeiro A, Giannakis G. Consensus in ad hoc WSNs with noisy links—Part I: Distributed estimation of deterministic signals. IEEE Trans Signal Process 2008;56(1):350–64.

[29] Nocedal J, Wright S. Numerical optimization. 2nd ed. Springer; 2006.

[30] Boyd S, Parikh N, Chu E, Peleato B, Eckstein J. Distributed optimization and statistical learning via the alternating direction method of multipliers. In: Foundations and trends in machine learning, vol. 3(1); 2010. p. 1–122.

[31] Hong M, Luo Z. On the linear convergence of the alternating direction method of multipliers. Math. Program. 2017;162(1–2):165–99.

[32] Shi W, Ling Q, Yuan K, Wu G, Yin W. On the linear convergence of the ADMM in decentralized consensus optimization. IEEE Trans Signal Process 2014;62(7):1750–61.

[33] Wei E, Ozdaglar A. On the $o(1/k)$ convergence of asynchronous distributed alternating direction method of multipliers. In: Proceedings of the IEEE global conference on signal and information processing, Austin, TX, USA; 2013.

[34] Wei E, Ozdaglar A. On the $o(1/k)$ convergence of asynchronous distributed alternating direction method of multipliers; 2013. Preprint, available at http://arxiv.org/abs/1307.8254.

[35] Iutzeler F, Bianchi P, Ciblat P, Hachem W. Asynchronous distributed optimization using a randomized alternating direction method of multipliers. In: Proceedings of the IEEE conference on decision and control, Florence, Italy; 2013.

[36] Iutzeler F, Bianchi P, Ciblat P, Hachem W. Explicit convergence rate of a distributed alternating direction method of multipliers. IEEE Trans Signal Process 2016;61(4):892–904.

[37] Eckstein J, Yao W. Approximate ADMM algorithms derived from Lagrangian splitting. Comput Optim Appl 2017;68(2):363–405.

[38] Ling Q, Shi W, Wu G, Ribeiro A. DLM: decentralized linearized alternating direction method of multipliers. IEEE Trans Signal Process 2014;63(15):4051–64.

[39] Mokhtari A, Shi W, Ling Q, Ribeiro A. DQM: decentralized quadratically approximated alternating direction method of multipliers. IEEE Trans Signal Process 2016;64(19):5158–73.

[40] Teixeira A, Ghadimi E, Shames I, Sandberg H, Johansson M. The ADMM algorithm for distributed quadratic problems: parameter selection and constraint preconditioning. IEEE Trans Signal Process 2016;64(2):290–305.

[41] Nishihara R, Lessard L, Recht B, Packard A, Jordan M. A general analysis of the convergence of ADMM. In: Proceedings of the international conference on machine learning, Lille, France; 2015.

[42] Magnússon S, Weeraddana P, Rabbat M, Fischione C. On the convergence of alternating direction Lagrangian methods for nonconvex structured optimization problems. IEEE Trans Control Netw Syst 2016;3(3):296–309.

[43] Hong M, Luo Z, Razaviyayn M. Convergence analysis of alternating direction method of multipliers for a family of nonconvex problems. SIAM J Optim 2016;26(1):337–64.

[44] Zargham M, Ribeiro A, Ozdaglar A, Jadbabaie A. Accelerated dual descent for network flow optimization. IEEE Trans Autom Control 2014;59(4):905–20.

[45] Zargham M, Ribeiro A, Jadbabaie A. Accelerated dual descent for constrained convex network flow optimization. 2013 IEEE 52nd annual conference on decision and control (CDC); 2013. p. 1037–42.

[46] Mokhtari A, Ling Q, Ribeiro A. Network Newton—Part II: Convergence rate and implementation; 2015.

[47] Tsianos K, Lawlor S, Rabbat M. Consensus-based distributed optimization: practical issues and applications in large-scale machine learning. In: 50th Allerton conference on communication, control, and computing; 2012.

[48] Dimakis A, Kar S, Moura J, Rabbat M, Scaglione A. Gossip algorithms for distributed signal processing. Proc IEEE 2010;98(11):1847–64.

[49] Srivastava K, Nedić A. Distributed asynchronous constrained stochastic optimization. IEEE J Sel Top Signal Process 2011;5(4):772–90.

[50] Tsitsiklis J, Bertsekas D, Athans M. Distributed asynchronous deterministic and stochastic gradient optimization algorithms. IEEE Trans Autom Control 1986;31(9):803–12.

[51] Bertsekas DP, Tsitsiklis JN. Parallel and distributed computation: numerical methods. 1st ed. Upper Saddle River, NJ, USA: Prentice-Hall, Inc.; 1989.

[52] Tsianos K, Rabbat M. Asynchronous decentralized optimization in heterogeneous systems. In: Proceedings of the IEEE conference on decision and control, Los Angeles, CA, USA; 2014.
[53] Tsianos K, Rabbat M. Distributed consensus and optimization under communication delays. In: Proceedings of the Allerton conference on communication, control, and computing, Monticello, IL, USA; 2011.
[54] Tsianos K, Rabbat M. Distributed dual averaging for convex optimization under communication delays. In: Proceedings of the American control conference, Montreal, Canada; 2012.
[55] Wu T, Yuan K, Ling Q, Yin W, Sayed A. Decentralized consensus optimization with asynchrony and delays. In: Proceedings of the Asilomar conference on signals, systems, and computers, Pacific Grove, CA, USA; 2016.
[56] Nedić A, Olshevsky A. Distributed optimization over time-varying directed graphs. IEEE Trans Autom Control 2015;60(3):601–15.
[57] Nedić A, Olshevsky A. Stochastic gradient-push for strongly convex functions on time-varying directed graphs. IEEE Trans Autom Control 2016;61(12):3936–47.
[58] Seneta E. Nonnegative matrices and Markov chains. George Allen & Ullwin Ltd; 1973.
[59] Kempe D, Dobra A, Gehrke J. Gossip-based computation of aggregate information. In: Proceedings of the 44th annual IEEE symposium on foundations of computer science; 2003.
[60] Tsianos K, Rabbat M. Push-sum distributed dual averaging for convex optimization. In: Proceedings of the IEEE conference on decision and control, Maui, HI, USA; 2012.
[61] Zeng J, Yin W. ExtraPush for convex smooth decentralized optimization over directed networks; 2015. Available on arXiv at http://arxiv.org/abs/1511.02942.
[62] Xi C, Khan U. DEXTRA: a fast algorithm for optimization over directed graphs. IEEE Trans Autom Control 2017;62(10):4980–93.
[63] Jakovetic D, Xavier J, Moura J. Fast distributed gradient methods. IEEE Trans Autom Control 2014;59(5):1131–46.
[64] Shi W, Ling Q, Wu G, Yin W. EXTRA: an exact first order algorithm for decentralized consensus optimization. SIAM J Optim 2015;25(2):944–66.
[65] Nedić A, Olshevsky A, Shi W. Achieving geometric convergence for distributed optimization over time-varying graphs; 2016. Accepted at *SIAM J Optim* 2017, available on arXiv at https://arxiv.org/abs/1607.03218.
[66] Nedić A, Olshevsky A, Rabbat MG. Network topology and communication-computation tradeoffs in decentralized optimization; 2017. Preprint, available at http://arxiv.org/abs/1709.08765.
[67] Boyd S, Ghosh A, Prabhakar B, Shah D. Randomized gossip algorithms. IEEE Trans Inf Theory 2006;52(6):2508–30.
[68] Bianchi P, Fort G, Hachem W. Performance of a distributed stochastic approximation algorithm. IEEE Trans Inf Theory 2013;59(11):7405–18.
[69] Zhu M, Martínez S. An approximate dual subgradient algorithm for distributed non-convex constrained optimization. IEEE Trans Autom Control 2013;58(6):1534–9.
[70] Zhang S, Choromanska A, LeCun Y. Deep learning with elastic averaged SGD. In: Proceedings of the advances in neural information processing systems; 2015.
[71] Di Lorenzo P, Scutari G. NEXT: in-network nonconvex optimization. IEEE Trans Signal Inf Process Netw 2016;2(2):120–36.
[72] Tatarenko T, Touri B. Non-convex distributed optimization. IEEE Trans Autom Control 2017;62(8):3744–57.

CHAPTER

DISTRIBUTED KALMAN AND PARTICLE FILTERING

6

Ali H. Sayed*, Petar M. Djurić†, Franz Hlawatsch‡

School of Engineering, École Polytechnique Fédérale de Lausanne (EPFL), Lausanne, Switzerland Department of Electrical and Computer Engineering, Stony Brook University, Stony Brook, NY, United States† Institute of Telecommunications, TU Wien, Vienna, Austria‡*

6.1 DISTRIBUTED SEQUENTIAL STATE ESTIMATION

Distributed strategies can be applied to solving state-space filtering and smoothing problems [1–18]. In these applications, agents interconnected by communication links seek to track the state of some underlying state-space model based on their own measurements and information exchanged with their neighbors.

Many natural and man-made systems can be modeled as dynamic systems described by state-space models [19–21]. The *state* of a dynamic system comprises certain system-related quantities of interest, such as the location and velocity of a moving vehicle or the concentration of a pollutant in a chemical plume. The state is time-varying and unknown. Each agent acquires local measurements and benefits from sharing information with neighboring agents. In this chapter, we consider distributed time-sequential methods for estimating the state from recurring measurements acquired by spatially distributed agents, where each agent is able to communicate only with a limited set of spatially close neighboring agents [1–8,10–18].

The objective of this chapter is to provide an overview of distributed state-space estimation for linear and nonlinear models, with emphasis on filtering and prediction problems while noting that similar techniques can be applied to smoothing problems [12]. In the first part of our treatment, we motivate and examine the diffusion Kalman filter for one-step prediction by following the general approach outlined in [12,15,18]. Similar diffusion strategies for fixed-lag and fixed-point smoothing appear in [12]. We focus on the diffusion Kalman form because it has been shown in [15,18,22,23] that diffusion implementations have superior stability and tracking performance when optimizing aggregate costs in response to streaming data. In the second part of the chapter, we describe distributed particle filtering, where we focus on both consensus-based and diffusion-based approaches. The methods we present can be used in a wide range of applications including surveillance, localization and navigation, environmental and agricultural monitoring, target tracking, exploration, search and rescue, the Internet of Things, and logistics.

Cooperative and Graph Signal Processing. https://doi.org/10.1016/B978-0-12-813677-5.00006-7

6.1.1 THE SETUP

The problem of interest is the estimation of a state $\boldsymbol{x}_n \in \mathbb{R}^N$, where $n \in \{1, 2, \dots\}$ is a discrete time index, from sequences of measurements $\{z_{1,k}, z_{2,k}, \dots, z_{n,k}\}$, with $z_{i,k} \in \mathbb{R}^P$, observed up to the current time n at K different agents $k \in \{1, 2, \dots, K\}$. More specifically, we are interested in estimating $\boldsymbol{x}_n$ in a *time-sequential* and *distributed* manner. Time-sequential processing means that the estimation of $\boldsymbol{x}_n$ at time n recursively reuses relevant results from the estimation of $\boldsymbol{x}_{n-1}$ at time $n-1$ while incorporating the new measurements $z_{n,k}$, $k = 1, 2, \dots, K$. Distributed processing means that the estimation of $\boldsymbol{x}_n$ is done locally at the individual agents, which exchange relevant information in a spatially local manner. Such a distributed (decentralized) mode of operation has important advantages over a centralized mode [16,17]. For example, it does not require a fusion center collecting all the measurements, which would constitute a "single point of failure." It also does not need communication between distant points or the use of complex routing strategies. The distributed solution is robust to network node and link failures and is able to adapt to changing network topologies. Moreover, the computational complexity and communication cost per agent scale well with the network size (number of agents).

The topology of local information exchange is defined by a decentralized network of agents in which each agent k is able to communicate with a certain set of "neighbor" agents $\mathfrak{N}_k \subseteq \{1, 2, \dots, K\}$. This set $\mathfrak{N}_k$ also includes agent k. Usually, the neighbors of an agent are located in its proximity. Practical examples of decentralized agent networks include robotic networks, networks of unmanned terrestrial, aerial, or underwater vehicles, and networks of cameras. We assume that the agent network is connected, i.e., each agent is connected to any other agent via one or multiple "communication hops."

Because the measurements are dispersed among the agents rather than available at a single processing unit, an important aspect of any distributed estimation algorithm is the dissemination of agent-related (and possibly other) information through the network via local communication. Various dissemination schemes are available, such as consensus [24,25], gossip [26,27], and diffusion [12,15,18]. A performance benchmark for a distributed algorithm is the performance that would be achieved by a centralized algorithm that has direct access to all the measurements. The amount of communication that is required for good performance is an important property of a distributed algorithm because communication increases the power consumption and processing delay of the agents.

A summary of the notation used in this chapter is presented in Table 6.1.

6.2 DISTRIBUTED KALMAN FILTERING

In the first part of our treatment, we motivate and examine the diffusion Kalman filter for one-step prediction by following the general approach outlined in [12,15,18]. We start with a brief description of the network topology and an introduction of the data model.

6.2.1 NETWORK TOPOLOGY

We consider a network consisting of K agents connected according to some graph topology. A left-stochastic $K \times K$ matrix $\boldsymbol{A} = (a_{\ell k})$ is associated with the topology, where each entry $a_{\ell k}$ denotes a

Table 6.1 Summary of Notation

Notation	Description
$n \in \{1, 2, \ldots\}$	discrete time index
$\boldsymbol{x}_n \in \mathbb{R}^N$	state vector at time k
$k \in \{1, 2, \ldots, K\}$	agent index
K	number of agents
$\boldsymbol{z}_{n,k} \in \mathbb{R}^p$	measurement vector of agent k at time n
$\boldsymbol{z}_n \triangleq (\boldsymbol{z}_{n,1}^{\mathrm{T}}\ \boldsymbol{z}_{n,2}^{\mathrm{T}} \cdots \boldsymbol{z}_{n,K}^{\mathrm{T}})^{\mathrm{T}}$	vector of all agent measurements at time n
$\boldsymbol{z}_{1:n} \triangleq (\boldsymbol{z}_1^{\mathrm{T}}\ \boldsymbol{z}_2^{\mathrm{T}} \cdots \boldsymbol{z}_n^{\mathrm{T}})^{\mathrm{T}}$	all-agents measurement sequence up to time n
$\mathfrak{N}_k \subseteq \{1, 2, \ldots, K\}$	set of neighboring agents of agent k (including agent k)
$d_k = \lvert\mathfrak{N}_k\rvert$	cardinality of set $\mathfrak{N}_k$
$\boldsymbol{A}$	a left-stochastic matrix
$\boldsymbol{J}$	adjacency matrix
$\mathbb{1}_K$	$K \times 1$ vector with elements equal to 1
δ_{nj}	Kronecker delta function
$\propto$	equal up to a constant factor
$\boldsymbol{F}_n$ and $\boldsymbol{G}_n$	state matrices of the linear state-space model
$\boldsymbol{H}_{n,k}$	local data matrix of the linear state-space model
$\boldsymbol{Q}_n$	covariance matrix of the state noise process
$\boldsymbol{R}_{n,k}$	covariance matrix of the measurement noise process
$\boldsymbol{\Pi}_0$	covariance matrix of the intitial state
$\boldsymbol{I}_N$	$N \times N$ identity matrix
$\mathbf{g}_n(\boldsymbol{x}_{n-1}, \boldsymbol{u}_n)$	state-evolution function at time k
$\boldsymbol{u}_n$	state noise at time n
$f(\boldsymbol{x}_n \mid \boldsymbol{x}_{n-1})$	state-transition probability density function
$\mathbf{h}_{n,k}(\boldsymbol{x}_n, \boldsymbol{v}_{n,k})$	measurement function of agent k at time n
$\boldsymbol{v}_{n,k}$	measurement noise of agent k at time n
$f(\boldsymbol{z}_{n,k} \mid \boldsymbol{x}_n)$	local likelihood function of agent k
$f(\boldsymbol{z}_n \mid \boldsymbol{x}_n)$	global (all-agents) likelihood function
$f(\boldsymbol{x}_n \mid \boldsymbol{z}_{1:n})$	posterior probability density function
$\mathbb{E}[\cdot]$	expectation
$\boldsymbol{x}_n^{(m)}$	mth particle of the state x_n
$w_n^{(m)}$	weight of the particle $x_n^{(m)}$
M	number of particles
$q(\boldsymbol{x}_n \mid \boldsymbol{x}_{n-1}^{(m)})$	proposal probability density function
$\ln(\cdot)$	natural logarithm
$\mathcal{N}(\cdot; \boldsymbol{\mu}, \boldsymbol{\Sigma})$	Gaussian probability density function

nonnegative scaling factor that will be used to scale some of the data transitioning from agent ℓ to agent k (likewise for $a_{k\ell}$). The left-stochasticity of $\boldsymbol{A}$ means that it satisfies

$$\boldsymbol{A}^{\mathrm{T}}\mathbb{1}_K = \mathbb{1}_K, \tag{6.1}$$

so that the entries on each column of $\boldsymbol{A}$ add up to one (the symbol $\mathbb{1}_K$ represents a $K \times 1$ vector of ones). We assume that the graph is strongly connected. The connectedness condition means that, for any two agents ℓ and k, there always exist paths with nonzero scaling weights in $\boldsymbol{A}$ linking ℓ and k in either direction; the path from ℓ to k does not need to agree with the reverse path from k to ℓ. Strong-connectedness additionally means that there exists at least one agent in the graph with $a_{kk} > 0$. That is, at least one agent places some trust or confidence on its local data. These conditions are satisfied by most graphs of interest. It follows from the strong-connectedness condition that the left-stochastic matrix $\boldsymbol{A}$ is primitive [15].

6.2.2 LINEAR STATE-SPACE MODELS

We consider a linear state-space model defined by

$$\boldsymbol{x}_{n+1} = \boldsymbol{F}_n\boldsymbol{x}_n + \boldsymbol{G}_n\boldsymbol{u}_n, \quad n = 0, 1, \ldots, \tag{6.2}$$

$$z_{n,k} = \boldsymbol{H}_{n,k}\boldsymbol{x}_n + \boldsymbol{v}_{n,k}, \quad n = 0, 1, \ldots;\ k = 1, 2, \ldots, K, \tag{6.3}$$

where $\boldsymbol{F}_n$ and $\boldsymbol{G}_n$ are state matrices, and $\boldsymbol{H}_{n,k}$ is a local data matrix. The noise vectors $\boldsymbol{u}_n$ and $\boldsymbol{v}_{n,k}$ are assumed to be realizations of zero-mean white processes that are uncorrelated with covariance matrices denoted by $\boldsymbol{Q}_n$ and $\boldsymbol{R}_{n,k}$, respectively, that is,

$$\mathbb{E}\left[\begin{pmatrix} \boldsymbol{u}_n \\ \boldsymbol{v}_{n,k} \end{pmatrix}\begin{pmatrix} \boldsymbol{u}_j \\ \boldsymbol{v}_{j,k} \end{pmatrix}^{\mathrm{T}}\right] \triangleq \begin{pmatrix} \boldsymbol{Q}_n & \boldsymbol{0} \\ \boldsymbol{0} & \boldsymbol{R}_{n,k} \end{pmatrix}\delta_{nj}, \tag{6.4}$$

where δ_{nj} denotes the Kronecker delta: it is equal to one when $n = j$ and zero otherwise. The measurement noise vectors $\boldsymbol{v}_{n,k}$ are also assumed to be independent of each other across space (i.e., for different k), that is,

$$\mathbb{E}\left[\boldsymbol{v}_{n,k}\boldsymbol{v}_{j,\ell}^{\mathrm{T}}\right] = \boldsymbol{R}_{n,k}\delta_{k\ell}\delta_{nj}. \tag{6.5}$$

The initial state vector, $\boldsymbol{x}_0$, is assumed to have zero mean and covariance matrix

$$\mathbb{E}\left[\boldsymbol{x}_0\boldsymbol{x}_0^{\mathrm{T}}\right] = \boldsymbol{\Pi}_0 > \boldsymbol{0}, \tag{6.6}$$

and it is uncorrelated with $\boldsymbol{u}_n$ and $\boldsymbol{v}_{n,k}$, for all n and k. The notation $\boldsymbol{\Pi}_0 > \boldsymbol{0}$ signifies that $\boldsymbol{\Pi}_0$ is a positive definite matrix. We further assume that $\boldsymbol{R}_{n,k} > \boldsymbol{0}$. The parameter matrices $\{\boldsymbol{F}_n, \boldsymbol{G}_n, \boldsymbol{Q}_n, \boldsymbol{\Pi}_0, \boldsymbol{H}_{n,k}, \boldsymbol{R}_{n,k}\}$ are considered known by each agent k. Note that no Gaussian assumptions are made.

6.2.3 NONCOOPERATIVE FILTERING

Assume initially that each agent k in the network acts individually and uses solely its local data stream $z_{n,k}, n = 1, 2, \ldots$, to track the state vector $\boldsymbol{x}_n$. In this case, agent k can run the well-known Kalman filter in any of its various forms (covariance, measurement and time-update, or information form) to carry out this task [19,28,29]. For instance, let $\hat{\boldsymbol{x}}_{n,k|j}^{\mathrm{ind}}$ denote the linear least-mean-squares error (LLMSE) estimate of x_n by agent k using all the local measurements up to time j, i.e., $\{z_{1,k}, z_{2,k}, \ldots, z_{j,k}\}$. The superscript "ind" is used to emphasize that these estimators are based on individual (noncooperative) behavior. We denote the corresponding estimation error and error-covariance matrix by

$$\tilde{x}^{\text{ind}}_{n,k|j} \triangleq x_n - \hat{x}^{\text{ind}}_{n,k|j}, \tag{6.7}$$

$$P^{\text{ind}}_{n,k|j} \triangleq \mathbb{E}\left[\tilde{x}^{\text{ind}}_{n,k|j}\left(\tilde{x}^{\text{ind}}_{n,k|j}\right)^{\mathrm{T}}\right]. \tag{6.8}$$

Then, it is well known [28,29] that predicted (when $j = n - 1$) and/or filtered (when $j = n$) state estimates can be computed by agent k by one of several forms, listed below.

Covariance form of the noncooperative Kalman filter

start with $\hat{x}^{\text{ind}}_{0,k|-1} = 0, P^{\text{ind}}_{0,k|-1} = \Pi_0$.

repeat at every agent k for $n \geq 0$:

$$\begin{aligned}
e^{\text{ind}}_{n,k} &= z_{n,k} - H_{n,k}\hat{x}^{\text{ind}}_{n,k|n-1}\\
R^{\text{ind}}_{e,n,k} &= R_{n,k} + H_{n,k}P^{\text{ind}}_{n,k|n-1}H^{\mathrm{T}}_{n,k}\\
K^{\text{ind}}_{p,n,k} &= F_n P^{\text{ind}}_{n,k|n-1}H^{\mathrm{T}}_{n,k}\left(R^{\text{ind}}_{e,n,k}\right)^{-1}\\
\hat{x}^{\text{ind}}_{n+1,k|n} &= F_n\hat{x}^{\text{ind}}_{n,k|n-1} + K^{\text{ind}}_{p,n,k}e^{\text{ind}}_{n,k}\\
P^{\text{ind}}_{n+1,k|n} &= F_n P^{\text{ind}}_{n,k|n-1}F^{\mathrm{T}}_n - K^{\text{ind}}_{p,n,k}R^{\text{ind}}_{e,n,k}\left(K^{\text{ind}}_{p,n,k}\right)^{\mathrm{T}} + G_n Q_n G^{\mathrm{T}}_n
\end{aligned} \tag{6.9}$$

end

Measurement- and time-update form of the noncooperative Kalman filter

start with $\hat{x}^{\text{ind}}_{0,k|-1} = 0, P^{\text{ind}}_{0,k|-1} = \Pi_0$.

repeat at every agent k for $n \geq 0$:

(measurement-update)

$$\begin{aligned}
e^{\text{ind}}_{n,k} &= z_{n,k} - H_{n,k}\hat{x}^{\text{ind}}_{n,k|n-1}\\
R^{\text{ind}}_{e,n,k} &= R_{n,k} + H_{n,k}P^{\text{ind}}_{n,k|n-1}H^{\mathrm{T}}_{n,k}\\
\hat{x}^{\text{ind}}_{n,k|n} &= \hat{x}^{\text{ind}}_{n,k|n-1} + P^{\text{ind}}_{n,k|n-1}H^{\mathrm{T}}_{n,k}\left(R^{\text{ind}}_{e,n,k}\right)^{-1}e^{\text{ind}}_{n,k}\\
P^{\text{ind}}_{n,k|n} &= P^{\text{ind}}_{n,k|n-1} - P^{\text{ind}}_{n,k|n-1}H^{\mathrm{T}}_{n,k}\left(R^{\text{ind}}_{e,n,k}\right)^{-1}H_{n,k}P^{\text{ind}}_{n,k|n-1}
\end{aligned} \tag{6.10}$$

(time-update)

$$\begin{aligned}
\hat{x}^{\text{ind}}_{n+1,k|n} &= F_n\hat{x}^{\text{ind}}_{n,k|n}\\
P^{\text{ind}}_{n+1,k|n} &= F_n P^{\text{ind}}_{n,k|n}F^{\mathrm{T}}_n + G_n Q_n G^{\mathrm{T}}_n
\end{aligned}$$

end

Information form of the noncooperative Kalman filter

start with $\hat{\boldsymbol{x}}^{\text{ind}}_{k,0|-1} = \mathbf{0}$, $\left(\boldsymbol{P}^{\text{ind}}_{k,0|-1}\right)^{-1} = \boldsymbol{\Pi}_0^{-1}$.

repeat at every agent k for $n \geq 0$:

$$\begin{aligned}
&\left(\boldsymbol{P}^{\text{ind}}_{n,k|n}\right)^{-1} = \left(\boldsymbol{P}^{\text{ind}}_{n,k|n-1}\right)^{-1} + \boldsymbol{H}^{\mathrm{T}}_{n,k}\boldsymbol{R}^{-1}_{n,k}\boldsymbol{H}_{n,k} \\
&\left(\boldsymbol{P}^{\text{ind}}_{n,k|n}\right)^{-1}\hat{\boldsymbol{x}}^{\text{ind}}_{n,k|n} = \left(\boldsymbol{P}^{\text{ind}}_{n,k|n-1}\right)^{-1}\hat{\boldsymbol{x}}^{\text{ind}}_{n,k|n-1} + \boldsymbol{H}^{\mathrm{T}}_{n,k}\boldsymbol{R}^{-1}_{n,k}\boldsymbol{z}_{n,k} \\
&\hat{\boldsymbol{x}}^{\text{ind}}_{n,k+1|n} = \boldsymbol{F}_n\hat{\boldsymbol{x}}^{\text{ind}}_{n,k|n} \\
&\boldsymbol{P}^{\text{ind}}_{n+1,k|n} = \boldsymbol{F}_n\boldsymbol{P}^{\text{ind}}_{n,k|n}\boldsymbol{F}^{\mathrm{T}}_n + \boldsymbol{G}_n\boldsymbol{Q}_n\boldsymbol{G}^{\mathrm{T}}_n
\end{aligned} \tag{6.11}$$

end

Clearly, these noncooperative forms assume that the agents act individually. However, because all the agents are tracking the same state vector $\boldsymbol{x}_n$, it is expected that the mean-square-error performance can improve through cooperation by having neighboring agents share with each other information about their own estimators. We consider one initial form of cooperation before extending it to arrive at the diffusion Kalman filter later in Section 6.2.7.

6.2.4 INCREMENTAL COOPERATION

Consider an arbitrary agent k and let $\mathfrak{N}_k$ denote its neighborhood, i.e., the collection of all agents ℓ that are connected to agent k by edges and that can share information with it. This set includes also the agent k. The set comprises all agents ℓ for which $a_{\ell k} > 0$; its cardinality is denoted by $d_k = |\mathfrak{N}_k|$.

In addition to its own measurement vector $z_{n,k}$, at each time instant n, agent k has also access to the measurement vectors $\{z_{n,\ell}\}$ from its neighbors, $\ell \in \mathfrak{N}_k \setminus \{k\}$. In this way, at every time instant n, agent k has access to $|\mathfrak{N}_k|$ measurement vectors (its own measurements and the ones collected from its neighbors) and denoted by $\{z_{n,k_1}, z_{n,k_2}, \ldots, z_{n,k_{d_k}}\}$. In this notation, we are referring to the neighbors of agent k (including agent k itself) by the indices $\{k_1, k_2, \ldots, k_{d_k}\}$. Let $\hat{\boldsymbol{x}}^{\text{loc}}_{n,k|j}$ denote the LLMSE estimate of $\boldsymbol{x}_n$ obtained by agent k (including agent k) by using these local (or neighborhood) measurements from time 0 up to time j. The superscript "loc" is used to emphasize that these estimators are based on local neighborhood measurements. We denote the corresponding estimation error and error-covariance matrix by

$$\tilde{\boldsymbol{x}}^{\text{loc}}_{n,k|j} \triangleq \boldsymbol{x}_n - \hat{\boldsymbol{x}}^{\text{loc}}_{n,k|j}, \tag{6.12}$$

$$\boldsymbol{P}^{\text{loc}}_{n,k|j} \triangleq \mathbb{E}\left[\tilde{\boldsymbol{x}}^{\text{loc}}_{n,k|j}(\tilde{\boldsymbol{x}}^{\text{loc}}_{n,k|j})^{\mathrm{T}}\right]. \tag{6.13}$$

Then, as shown in listing (6.14), the predicted (when $j = n-1$) and filtered (when $j = n$) state estimates can be computed by agent k by running multiple measurement updates, one for each neighbor of k [28,29]. In this listing, we are denoting the state estimate at the end of the incremental loop through the neighborhood by $\boldsymbol{\psi}_{n,k}$, which is equal to the filtered estimate $\hat{\boldsymbol{x}}^{\text{loc}}_{n,k|n}$ that results from the d_k measurement updates. This equality is highlighted by the symbol $(\star)$ placed next to one of the equations in listing (6.14). The corresponding error covariance matrix of $\hat{\boldsymbol{x}}^{\text{loc}}_{n,k|n}$ is denoted by $\boldsymbol{P}^{\text{loc}}_{n,k|n}$. Although the variables $\boldsymbol{\psi}_{n,k}$ and $\hat{\boldsymbol{x}}^{\text{loc}}_{n,k|n}$ are identical in the incremental implementation, we keep separate notations for both symbols because they will be distinct and will play different roles in the diffusion form derived further ahead in Section 6.2.7. In particular, the $(\star)$ assignment will be replaced by Eq. (6.48).

Measurement- and time-update incremental form

start with $\hat{\boldsymbol{x}}^{\text{loc}}_{0,k|-1} = \mathbf{0}, \boldsymbol{P}^{\text{loc}}_{0,k|-1} = \boldsymbol{\Pi}_0$.

repeat at every agent k for $n \geq 0$:

(incremental measurement-updates)

set auxiliary variables:

$\boldsymbol{\psi}^{(0)}_{n,k} \leftarrow \hat{\boldsymbol{x}}^{\text{loc}}_{n,k|n-1}$ ($N \times 1$ vector)

$\boldsymbol{P}^{(0)}_{n,k} \leftarrow \boldsymbol{P}^{\text{loc}}_{n,k|n-1}$ ($N \times N$ matrix)

for each neighbor $m = 1, 2, \ldots, d_k$ **do:**

$$\begin{aligned}
\boldsymbol{e}_{n,k_m} &= \boldsymbol{z}_{n,k_m} - \boldsymbol{H}_{n,k_m}\boldsymbol{\psi}^{(m-1)}_{n,k} \\
\boldsymbol{R}_{e,n,k_m} &= \boldsymbol{R}_{n,k_m} + \boldsymbol{H}_{n,k_m}\boldsymbol{P}^{(m-1)}_{n,k}\boldsymbol{H}^{\mathrm{T}}_{n,k_m} \\
\boldsymbol{\psi}^{(m)}_{n,k} &= \boldsymbol{\psi}^{(m-1)}_{n,k} + \boldsymbol{P}^{(m-1)}_{n,k}\boldsymbol{H}^{\mathrm{T}}_{n,k_m}\boldsymbol{R}^{-1}_{e,n,k_m}\boldsymbol{e}_{n,k_m} \\
\boldsymbol{P}^{(m)}_{n,k} &= \boldsymbol{P}^{(m-1)}_{n,k} - \boldsymbol{P}^{(m-1)}_{n,k}\boldsymbol{H}^{\mathrm{T}}_{n,k_m}\boldsymbol{R}^{-1}_{e,n,k_m}\boldsymbol{H}_{n,k_m}\boldsymbol{P}^{(m-1)}_{n,k}
\end{aligned} \tag{6.14}$$

end

$\boldsymbol{P}^{\text{loc}}_{n,k|n} \leftarrow \boldsymbol{P}^{(d_k)}_{n,k}$

$\boldsymbol{\psi}_{n,k} \leftarrow \boldsymbol{\psi}^{(d_k)}_{n,k}$

$\hat{\boldsymbol{x}}^{\text{loc}}_{n,k|n} \leftarrow \boldsymbol{\psi}^{(d_k)}_{n,k} \quad (\star)$

(time-update)

$\hat{\boldsymbol{x}}^{\text{loc}}_{n+1,k|n} = \boldsymbol{F}_n\hat{\boldsymbol{x}}^{\text{loc}}_{n,k|n}$

$\boldsymbol{P}^{\text{loc}}_{n+1,k|n} = \boldsymbol{F}_n\boldsymbol{P}^{\text{loc}}_{n,k|n}\boldsymbol{F}^{\mathrm{T}}_n + \boldsymbol{G}_n\boldsymbol{Q}_n\boldsymbol{G}^{\mathrm{T}}_n$

end

It can be verified that the following relations hold for the variables in the incremental form (6.14); see the proof in [12, Appendix B]:

$$\left(\boldsymbol{P}^{\text{loc}}_{n,k|n}\right)^{-1} = \left(\boldsymbol{P}^{\text{loc}}_{n,k|n-1}\right)^{-1} + \boldsymbol{S}_{n,k} \tag{6.15}$$

with

$$\boldsymbol{S}_{n,k} \triangleq \sum_{\ell \in \mathcal{N}_k} \boldsymbol{H}^{\mathrm{T}}_{n,\ell}\boldsymbol{R}^{-1}_{n,\ell}\boldsymbol{H}_{n,\ell}, \tag{6.16}$$

and

$$\left(\boldsymbol{P}^{\text{loc}}_{n,k|n}\right)^{-1}\hat{\boldsymbol{x}}^{\text{loc}}_{n,k|n} = \left(\boldsymbol{P}^{\text{loc}}_{n,k|n-1}\right)^{-1}\hat{\boldsymbol{x}}^{\text{loc}}_{n,k|n-1} + \boldsymbol{q}_{n,k} \tag{6.17}$$

with

$$q_{n,k} \triangleq \sum_{\ell \in \mathcal{N}_k} H_{n,\ell}^{\mathrm{T}} R_{n,\ell}^{-1} z_{n,\ell}. \tag{6.18}$$

Recursions (6.14) compute the optimal local estimate $\hat{x}_{n,k|n}^{\mathrm{loc}}$ of agent k by incorporating solely the measurements $\{z_{n,\ell}\}$ from the neighborhood $\mathcal{N}_k$. This is a limited form of cooperation because the recursions (6.14) are not exploiting the fact that besides the measurements $\{z_{n,\ell}\}$, the neighbors of agent k also have their own estimates of x_n. These estimates are given by $\hat{x}_{n,\ell|n}^{\mathrm{loc}}$. It is expected that by additionally exploiting the neighborhood estimates, agent k should be able to achieve two objectives: (a) enhance its state estimation accuracy and (b) have its local estimate $\hat{x}_{n,k|n}^{\mathrm{loc}}$ approach the desired global estimate $\hat{x}_{n|n}$ of the state vector x_n (which would be produced by an optimum centralized estimator that has access to all the measurements in the network). We shall use this measurement to motivate and derive the diffusion Kalman filter. First, however, we take a brief digression and review a well-known data fusion construction [28,29], which will be used to form *approximate* local estimates.

6.2.5 DATA FUSION

A problem that is encountered often in applications deals with the need to fuse together data collected from several sources in order to enhance the accuracy of the estimation process. This is similar to the scenario we are facing in the network context, except that the graph topology and the time evolution of the data add a new level of complexity in the form of cross-correlations.

Thus, assume for a moment that we have a collection of K agents that is distributed over some region in space without paying particular attention to whether they are connected by a communication link or not. All agents are interested in estimating the same parameter vector of size $N \times 1$, denoted generically by x, which is assumed to be zero-mean and to have a positive definite covariance matrix $\Pi_x = \mathbb{E}[xx^{\mathrm{T}}]$. For example, the agents could be tracking a moving object with the goal of estimating its speed and direction of motion. Each agent k collects a measurement vector z_k of size $p \times 1$ that is related to the desired x via the linear measurement model (observe that in the discussion in this section we are not dealing with sequential processing of the data and, therefore, the time index is dropped),

$$z_k = H_k x + v_k, \tag{6.19}$$

where H_k is the model matrix that maps x to z_k at agent k and v_k is zero-mean measurement noise with a positive definite covariance matrix $R_k = \mathbb{E}[v_k v_k^{\mathrm{T}}]$.

Assume initially that the measurements from all agents can be collected centrally at some fusion center. For example, each agent k could transmit its measurement vector z_k and its model parameters $\{H_k, R_k\}$ to the fusion center. The data collected at the fusion center thus satisfy the linear model:

$$\underbrace{\begin{pmatrix} z_1 \\ z_2 \\ \vdots \\ z_K \end{pmatrix}}_{\triangleq z} = \underbrace{\begin{pmatrix} H_1 \\ H_2 \\ \vdots \\ H_K \end{pmatrix}}_{\triangleq H} x + \underbrace{\begin{pmatrix} v_1 \\ v_2 \\ \vdots \\ v_K \end{pmatrix}}_{\triangleq v}. \tag{6.20}$$

We assume that the noises $\{v_k\}$ across all agents k are *uncorrelated* with each other so that the covariance matrix of the aggregate noise vector v is

$$R_v = \text{blkdiag}\{R_1, R_2, \ldots, R_K\}. \tag{6.21}$$

Now, it is well known that the LLMSE estimate of x that is based on z is given by [28,29]

$$\hat{x} = \left(\Pi_x^{-1} + H^{\mathrm{T}} R_v^{-1} H\right)^{-1} H^{\mathrm{T}} R_v^{-1} z \tag{6.22}$$

with the resulting minimum mean-square-error (MMSE) matrix equal to

$$P \triangleq \mathbb{E}[\tilde{x}\tilde{x}^{\mathrm{T}}] = \left(\Pi_x^{-1} + H^{\mathrm{T}} R_v^{-1} H\right)^{-1}, \tag{6.23}$$

where $\tilde{x} = x - \hat{x}$ denotes the estimation error. Note that P is a matrix of size $N \times N$.

The solution (6.22) requires all agents to transmit their *raw* data $\{z_k, H_k, R_k\}$ to the fusion center. A more efficient fusion method, one that reduces the communications overhead, can be derived by allowing the agents to perform some local processing and to share the results of this processing step rather than their raw data. Specifically, assume that each agent estimates x using first its own data z_k. We denote the resulting estimator by $\hat{x}_k$ and it is given by

$$\hat{x}_k = \left(\Pi_x^{-1} + H_k^{\mathrm{T}} R_k^{-1} H_k\right)^{-1} H_k^{\mathrm{T}} R_k^{-1} z_k. \tag{6.24}$$

The corresponding MMSE matrix is

$$P_k = \left(\Pi_x^{-1} + H_k^{\mathrm{T}} R_k^{-1} H_k\right)^{-1}. \tag{6.25}$$

Next, assume that the agents share the processed data $\{\hat{x}_k, P_k\}$ with the fusion center instead of the raw data $\{z_k, H_k, R_k\}$. We now show that the desired global quantities $\{\hat{x}, P\}$ can be recovered directly from the locally generated data $\{\hat{x}_k, P_k\}$.

To begin with, observe that we can rework expression (6.23) for the global MMSE matrix as follows:

$$P^{-1} = \Pi_x^{-1} + \sum_{k=1}^{K} H_k^{\mathrm{T}} R_k^{-1} H_k = \sum_{k=1}^{K} P_k^{-1} - (K-1)\Pi_x^{-1}. \tag{6.26}$$

This expression allows the fusion center to determine P^{-1} directly from knowledge of the quantities $\{P_k^{-1}, \Pi_x^{-1}\}$. Note further that the global estimate expression (6.22) can be rewritten as

$$P^{-1}\hat{x} = H^{\mathrm{T}} R_v^{-1} z = \sum_{k=1}^{K} H_k^{\mathrm{T}} R_k^{-1} z_k = \sum_{k=1}^{K} P_k^{-1} \hat{x}_k. \tag{6.27}$$

Therefore, we arrive at the following alternative method to fuse the data from multiple agents:

$$P^{-1} = \sum_{k=1}^{K} P_k^{-1} - (K-1)\Pi_x^{-1}, \tag{6.28}$$

$$P^{-1}\hat{x} = \sum_{k=1}^{K} P_k^{-1} \hat{x}_k. \tag{6.29}$$

Observe that the individual estimates are scaled by the inverses of their MMSE matrices. It is useful to note that the derivation of these expressions requires the noises across the agents to be uncorrelated with each other, so that $\boldsymbol{R}_v$ is block diagonal.

6.2.6 APPROXIMATE FUSION RELATIONS

Let us now return to the network context and use the fusion expressions (6.28) and (6.29) to motivate the diffusion Kalman filter. Let $\hat{\boldsymbol{x}}_{n|n}$ denote the filtered estimate of $\boldsymbol{x}_n$ that is based on the measurements $z_{j,k}$ from across all agents $k = 1, 2, \ldots, K$ at all times j up to time n. Let $\boldsymbol{P}_{n|n}$ denote the corresponding error covariance matrix, i.e.,

$$\boldsymbol{P}_{n|n} \triangleq \mathbb{E}[\tilde{\boldsymbol{x}}_{n|n}\tilde{\boldsymbol{x}}_{n|n}^{\mathrm{T}}], \quad \text{where} \quad \tilde{\boldsymbol{x}}_{n|n} \triangleq \boldsymbol{x}_n - \hat{\boldsymbol{x}}_{n|n}. \tag{6.30}$$

These are the global quantities that we are interested in evaluating in a distributed manner. We can employ the fusion expressions (6.28) and (6.29) to approximate these global variables in terms of the noncooperative variables $\{\hat{\boldsymbol{x}}^{\mathrm{ind}}_{n,k|n}, \boldsymbol{P}^{\mathrm{ind}}_{n,k|n}\}$ as follows:

$$\boldsymbol{P}^{-1}_{n|n} \approx \sum_{k=1}^{K} \left(\boldsymbol{P}^{\mathrm{ind}}_{k,n|n}\right)^{-1} - (K-1)\boldsymbol{\Pi}_n^{-1}, \tag{6.31}$$

$$\boldsymbol{P}^{-1}_{n|n}\hat{\boldsymbol{x}}_{n|n} \approx \sum_{k=1}^{K} \left(\boldsymbol{P}^{\mathrm{ind}}_{n,k|n}\right)^{-1} \hat{\boldsymbol{x}}^{\mathrm{ind}}_{n,k|n}, \tag{6.32}$$

where $\boldsymbol{\Pi}_n \triangleq \mathbb{E}\left[\boldsymbol{x}_n\boldsymbol{x}_n^{\mathrm{T}}\right]$ denotes the covariance matrix of $\boldsymbol{x}_n$, which satisfies the recursion

$$\boldsymbol{\Pi}_{n+1} = \boldsymbol{F}_n\boldsymbol{\Pi}_n\boldsymbol{F}_n^{\mathrm{T}} + \boldsymbol{G}_n\boldsymbol{Q}_n\boldsymbol{G}_n^{\mathrm{T}}. \tag{6.33}$$

Relations (6.31) and (6.32) are approximate expressions while relations (6.28) and (6.29) are exact fusion formulae. To see why, recall that the global estimate $\hat{\boldsymbol{x}}_{n|n}$ is based on the measurements $z_{j,k}$ from time $j = 0$ up to time $j = n$ across all agents k. If we now appeal to the state-space model described by Eqs. (6.2) and (6.3), it is easy to recognize that these measurements cannot generally be related directly to the unknown $\boldsymbol{x}_n$ via a linear model of the form (6.19). Exact fusion formulae in this case are cumbersome and require the use of error cross-covariance matrices (see, e.g., [30–32]). The above approximations, however, are sufficient for our purposes and they are commonly used in the literature on data fusion methods, including the following simpler form:

$$\boldsymbol{P}^{-1}_{n|n} \approx \sum_{k=1}^{K} \left(\boldsymbol{P}^{\mathrm{ind}}_{k,n|n}\right)^{-1}, \tag{6.34}$$

where the term involving $\boldsymbol{\Pi}_n^{-1}$ is ignored. Expression (6.32) amounts to estimating $\hat{\boldsymbol{x}}_{n|n}$ by using a covariance-weighted combination of the local estimates, $\hat{\boldsymbol{x}}^{\mathrm{ind}}_{n,k|n}$ while expression (6.34) estimates the error covariance matrix by pooling together the individual error covariances.

In principle, the same fusion expressions (6.28) and (6.29) could also be used to approximate the same global variables $\{\hat{\boldsymbol{x}}_{n|n}, \boldsymbol{P}_{n|n}\}$ in terms of the incremental (neighborhood) estimates $\{\boldsymbol{\psi}_{n,k}, \boldsymbol{P}^{\mathrm{loc}}_{n,k|n}\}$ defined in Section 6.2.4, say as:

$$P_{n|n}^{-1}\hat{x}_{n|n} \approx \sum_{k=1}^{K}\left(P_{n,k|n}^{\text{loc}}\right)^{-1}\psi_{n,k}, \tag{6.35}$$

$$P_{n|n}^{-1} \approx \sum_{k=1}^{K}\left(P_{n,k|n}^{\text{loc}}\right)^{-1}, \tag{6.36}$$

where we recall that $\psi_{n,k} = \hat{x}_{n,k|n}^{\text{loc}}$. However, observe now that the construction of the estimates $\psi_{n,k}$ across agents relies on shared measurements. For example, the measurements from both agent k and its neighbors influence the value of $\psi_{n,k}$; likewise, some of these same measurements influence the value of other estimates because they belong to the neighborhoods of other agents as well. This data redundancy is not present in the computation of the individual estimates $\hat{x}_{n,k|n}^{\text{ind}}$, where each agent k relies solely on its local sequential data $z_{n,k}$.

The presence of redundant information in the intermediate estimates $\psi_{n,k}$ is due to the graph topology, which defines the neighborhoods over the network. We can exploit this topology to replace the approximations (6.35) and (6.36) by more revealing relations that bring forth the graph structure. To do so, we let J denote the $K \times K$ adjacency matrix of the network graph. This matrix consists of unit and zero entries representing the connectivity of the agents, namely,

$$J_{\ell k} = \begin{cases} 1, & \text{if } a_{\ell k} > 0, \\ 0, & \text{otherwise.} \end{cases} \tag{6.37}$$

Although unnecessary in the final statement of the algorithm, we shall assume for the sake of the derivation that there exists a vector s such that $Js = \mathbb{1}_K$. For example, if J happens to be invertible, then this vector s exists, is unique, and is given by $s = J^{-1}\mathbb{1}_K$. In general, however, it is not always guaranteed that a vector s satisfying $Js = \mathbb{1}_K$ exists. It is nevertheless possible to verify that the matrix J can be made invertible by flipping some of its diagonal entries (from zero to one and from one to zero). By turning a diagonal entry of J from zero to one, we are in effect allowing the agent at that location to rely on its local measurement. On the other hand, by turning a diagonal entry of J from one to zero, we are disabling this feature. This action is tolerable because this measurement is not discarded by the network as it is still used by the neighbors of the agent to update their estimates (as long as at least one entry of the row of J with a diagonal entry equal to zero has a value equal to one). We denote the entries of s by γ_k, $k = 1, 2, \ldots, K$, or $s = \text{col}\{\gamma_k\}$ so that

$$Js = \mathbb{1}_K \iff \sum_{k=1}^{K}\gamma_k J_{\ell k} = 1, \quad \text{for every } \ell = 1, 2, \ldots, K. \tag{6.38}$$

In a manner similar to Eqs. (6.32) and (6.34), we first use the fusion expressions (6.28) and (6.29) to approximate the incremental estimators $\{\psi_{n,k}, P_{n,k|n}^{\text{loc}}\}$ by using the noncooperative variables $\{\hat{x}_{n,k|n}^{\text{ind}}, P_{n,k|n}^{\text{ind}}\}$ as follows:

$$\left(P_{n,k|n}^{\text{loc}}\right)^{-1}\psi_{n,k} \approx \sum_{\ell=1}^{K} J_{\ell k}\left(P_{n,\ell|n}^{\text{ind}}\right)^{-1}\hat{x}_{n,\ell|n}^{\text{ind}}, \tag{6.39}$$

$$\left(P_{n,k|n}^{\text{loc}}\right)^{-1} \approx \sum_{\ell=1}^{K} J_{\ell k}\left(P_{n,\ell|n}^{\text{ind}}\right)^{-1}. \tag{6.40}$$

Recall that the individual estimates $\hat{x}^{\text{ind}}_{n,k|n}$ rely on independent streams of data and that $J_{\ell k} = 1$ if information can flow from agent ℓ to agent k, and $J_{\ell k} = 0$ otherwise.

Using the above relations we can now relate the global variables $\{\hat{x}_{n|n}, P_{n|n}\}$ to the local variables $\{\psi_{n,k}, P^{\text{loc}}_{n,k|n}\}$ and replace Eqs. (6.35) and (6.36). Indeed, using the entries γ_k, we note that

$$\sum_{k=1}^{K} \gamma_k \left(P^{\text{loc}}_{n,k|n}\right)^{-1} \psi_{n,k} \overset{(6.39)}{\approx} \sum_{\ell=1}^{K} \underbrace{\left(\sum_{k=1}^{K} \gamma_k J_{\ell k}\right)}_{=1} \left(P^{\text{ind}}_{n,\ell|n}\right)^{-1} \hat{x}^{\text{ind}}_{n,\ell|n}$$

$$= \sum_{\ell=1}^{K} \left(P^{\text{ind}}_{n,\ell|n}\right)^{-1} \hat{x}^{\text{ind}}_{n,\ell|n} \overset{(6.32)}{\approx} P^{-1}_{n|n} \hat{x}_{n|n} \tag{6.41}$$

and, similarly,

$$\sum_{k=1}^{K} \gamma_k \left(P^{\text{loc}}_{n,k|n}\right)^{-1} \overset{(6.40)}{\approx} \sum_{\ell=1}^{K} \left(P^{\text{ind}}_{n,\ell|n}\right)^{-1} \overset{(6.34)}{\approx} P^{-1}_{n|n}. \tag{6.42}$$

In summary, we arrive at the following approximate relations between the global variables $\{\hat{x}_{n|n}, P_{n|n}\}$ and the neighborhood variables $\{\psi_{n,k}, P^{\text{loc}}_{n,k|n}\}$:

$$P^{-1}_{n|n} \hat{x}_{n|n} \approx \sum_{k=1}^{K} \gamma_k \left(P^{\text{loc}}_{n,k|n}\right)^{-1} \psi_{n,k}, \tag{6.43}$$

$$P^{-1}_{n|n} \approx \sum_{k=1}^{K} \gamma_k \left(P^{\text{loc}}_{n,k|n}\right)^{-1}. \tag{6.44}$$

These relations replace Eqs. (6.35) and (6.36), where the graph structure is represented through the scalars γ_k.

Remark (Covariance-intersection method). Expressions (6.43) and (6.44) have a form similar to another commonly used fusion technique that relies on the use of the covariance-intersection (CI) method [33,34] and which avoids the need for error cross-covariance matrices. In the CI method, the global variables $\{\hat{x}_{n|n}, P_{n|n}\}$ are estimated according to Eqs. (6.43) and (6.44) where the scalars γ_k are instead treated as *design* parameters. Specifically, they are chosen as nonnegative convex combination coefficients satisfying $\sum_{k=1}^{K} \gamma_k = 1$, $\gamma_k \geq 0$, and their values are selected to optimize the resulting $P_{n|n}$, say, by minimizing the trace or determinant of $P_{n|n}$, i.e.,

$$\min_{\{\gamma_k\}} \text{Tr}(P_{n|n}) \quad \text{or} \quad \min_{\{\gamma_k\}} \det(P_{n|n}). \tag{6.45}$$

6.2.7 DIFFUSION COOPERATION

Continuing with Eqs. (6.43) and (6.44), one difficulty with these expressions is that computation of the global estimator $\hat{x}_{n|n}$ requires access to the local estimators $\{\psi_{n,\ell}\}$ from across the entire network.

We can obtain a decentralized solution that relies solely on local interactions as follows. Substituting Eq. (6.44) into Eq. (6.43) gives

$$\hat{\boldsymbol{x}}_{n|n} \approx \left[\sum_{k=1}^{K} \gamma_k \left(\boldsymbol{P}^{\text{loc}}_{n,k|n}\right)^{-1}\right]^{-1} \sum_{k=1}^{K} \gamma_k \left(\boldsymbol{P}^{\text{loc}}_{n,k|n}\right)^{-1} \boldsymbol{\psi}_{n,k}, \tag{6.46}$$

which has the form of a convex weighted average, namely,

$$\hat{\boldsymbol{x}}_{n|n} \approx \sum_{k=1}^{K} \boldsymbol{\Gamma}_k \boldsymbol{\psi}_{n,k} \tag{6.47}$$

with nonnegative-definite coefficient matrices $\boldsymbol{\Gamma}_k$ that add up to the identity matrix, i.e., $\sum_{k=1}^{K} \boldsymbol{\Gamma}_k = \boldsymbol{I}_N$. The result (6.47) suggests one useful approximation by which the local estimators $\{\boldsymbol{\psi}_{n,\ell}\}$ can be fused within the neighborhood of every agent k to obtain a local version for $\hat{\boldsymbol{x}}_{n|n}$, which we shall denote by $\hat{\boldsymbol{x}}_{n,k|n}$ (we are removing the "loc" superscript). This local version is obtained by limiting the convex combination operation (6.47) to the neighborhood of each agent, and by replacing the matrix combination weights $\boldsymbol{\Gamma}_\ell$ by convex combination scalars $a_{\ell k}$. In this way, the computation (6.47) is replaced locally by

$$\hat{\boldsymbol{x}}_{n,k|n} \leftarrow \sum_{\ell \in \mathcal{N}_k} a_{\ell k} \boldsymbol{\psi}_{n,\ell}. \tag{6.48}$$

Comparing with the incremental listing (6.14), we see that the above calculation amounts to replacing the assignment $\hat{\boldsymbol{x}}^{\text{loc}}_{n,k|n} \leftarrow \boldsymbol{\psi}_{n,k}$ (marked by ($\star$) in Eq. 6.14) by a local convex combination step. In view of this substitution, the error covariances of the local estimators $\hat{\boldsymbol{x}}_{n,k|n}$ computed by Eq. (6.48) are not given anymore by the matrices $\boldsymbol{P}_{n,k|n}$ in the listings (6.49) and (6.52) below (where again we removed the superscript "loc"). In summary, we arrive at the listings (6.49) and (6.52) for the diffusion Kalman filter in two equivalent forms: the measurement and time-update form and the information form.

More generally, and in view of Eq. (6.47), a more enhanced fusion of the local estimators $\{\boldsymbol{\psi}_{n,\ell}\}$ is possible by employing convex combination matrices in Eq. (6.48) rather than scalars; for example, these combination matrices could be defined in terms of the inverses $\boldsymbol{P}^{-1}_{n,\ell|n}$ as suggested by Eq. (6.46). This construction, however, would entail added complexity and would require sharing of additional information regarding these inverses. The implementation (6.49) from [12] employs scalar combination coefficients $\{a_{\ell k}\}$ in order to reduce the complexity of the resulting algorithm. Reference [35] studies the alternative fusion of the estimators $\{\boldsymbol{\psi}_{n,\ell}\}$ in the diffusion Kalman filter by exploiting information about the inverses $\boldsymbol{P}^{-1}_{n,\ell|n}$ and using covariance-intersection combinations over neighborhoods in a manner similar to Eq. (6.45).

Diffusion Kalman filter (measurement- and time-update form)

start with $\hat{\boldsymbol{x}}_{0,k|-1} = \boldsymbol{0}, \boldsymbol{P}_{0,k|-1} = \boldsymbol{\Pi}_0$.

repeat at every agent k for $n \geq 0$:

(measurement-updates)

set auxiliary variables:

$$\boldsymbol{\psi}_{n,k}^{(0)} \leftarrow \hat{\boldsymbol{x}}_{n,k|n-1} \quad (N \times 1 \text{ vector})$$

$$\boldsymbol{P}_{n,k}^{(0)} \leftarrow \boldsymbol{P}_{n,k|n-1} \quad (N \times N \text{ matrix})$$

for each neighbor $\ell = 1, 2, \ldots, d_k$ do :

$$\begin{aligned}
\boldsymbol{e}_{n,k_\ell} &= \boldsymbol{z}_{n,k_\ell} - \boldsymbol{H}_{n,k_\ell}\boldsymbol{\psi}_{n,k}^{(\ell-1)} \\
\boldsymbol{R}_{e,n,k_\ell} &= \boldsymbol{R}_{n,k_\ell} + \boldsymbol{H}_{n,k_\ell}\boldsymbol{P}_{n,k}^{(\ell-1)}\boldsymbol{H}_{n,k_\ell}^{\mathrm{T}} \\
\boldsymbol{\psi}_{n,k}^{(\ell)} &= \boldsymbol{\psi}_{n,k}^{(\ell-1)} + \boldsymbol{P}_{n,k}^{(\ell-1)}\boldsymbol{H}_{n,k_\ell}^{\mathrm{T}}\boldsymbol{R}_{e,n,k_\ell}^{-1}\boldsymbol{e}_{n,k_\ell} \\
\boldsymbol{P}_{n,k}^{(\ell)} &= \boldsymbol{P}_{n,k}^{(\ell-1)} - \boldsymbol{P}_{n,k}^{(\ell-1)}\boldsymbol{H}_{n,k_\ell}^{\mathrm{T}}\boldsymbol{R}_{e,n,k_\ell}^{-1}\boldsymbol{H}_{n,k_\ell}\boldsymbol{P}_{n,k}^{(\ell-1)}
\end{aligned} \tag{6.49}$$

end

$$\boldsymbol{P}_{n,k|n} \leftarrow \boldsymbol{P}_{n,k}^{(d_k)}$$

$$\boldsymbol{\psi}_{n,k} \leftarrow \boldsymbol{\psi}_{n,k}^{(d_k)}$$

(combination step)

$$\hat{\boldsymbol{x}}_{n,k|n} = \sum_{\ell \in \mathcal{N}_k} a_{\ell k}\boldsymbol{\psi}_{\ell,n}$$

(time-update)

$$\hat{\boldsymbol{x}}_{n+1,k|n} = \boldsymbol{F}_n\hat{\boldsymbol{x}}_{n,k|n}$$

$$\boldsymbol{P}_{n+1,k|n} = \boldsymbol{F}_n\boldsymbol{P}_{k,n|n}\boldsymbol{F}_n^{\mathrm{T}} + \boldsymbol{G}_n\boldsymbol{Q}_n\boldsymbol{G}_n^{\mathrm{T}}$$

end

We also note that the combination step in Eq. (6.52) is similar to the update used in the consensus Kalman filter derived in [5] with one main difference. The consensus Kalman filter employs a specific combination construction at the local level of the form:

$$\hat{\boldsymbol{x}}_{n,k|n} = (1 + (1 - d_k)\epsilon)\boldsymbol{\psi}_{n,k} + \sum_{\ell \in \mathcal{N}_k \backslash \{k\}} \epsilon\,\boldsymbol{\psi}_{n,\ell}, \tag{6.50}$$

where we recall that $d_k = |\mathcal{N}_k|$, and $\epsilon > 0$ is a small weight that is *equal* for all the neighbors. In comparison, the diffusion implementation (6.52), which can be written as

$$\hat{\boldsymbol{x}}_{n,k|n} = a_{kk}\boldsymbol{\psi}_{n,k} + \sum_{\ell \in \mathcal{N}_k \backslash \{k\}} a_{\ell k}\boldsymbol{\psi}_{n,\ell}, \tag{6.51}$$

allows more freedom in assigning generally different weights $a_{\ell k}$ to different neighbors [10,12].

Diffusion Kalman filter (information form)

start with $\hat{x}_{0,k|-1} = \mathbf{0}, (P_{0,k|-1})^{-1} = \Pi_0^{-1}$.

repeat at every agent k for $n \geq 0$:

$$\begin{aligned}
S_{n,k} &= \sum_{\ell \in \mathcal{N}_k} H_{n,\ell}^{\mathsf{T}} R_{n,\ell}^{-1} H_{n,\ell} \\
q_{n,k} &= \sum_{\ell \in \mathcal{N}_k} H_{n,\ell}^{\mathsf{T}} R_{n,\ell}^{-1} z_{n,\ell} \\
(P_{n,k|n})^{-1} &= (P_{n,k|n-1})^{-1} + S_{n,k} \\
\psi_{n,k} &= \hat{x}_{n,k|n-1} + P_{n,k|n}(q_{n,k} - S_{n,k}\hat{x}_{n,k|n-1}) \\
\hat{x}_{n,k|n} &= \sum_{\ell \in \mathcal{N}_k} a_{\ell k} \psi_{n,\ell} \\
\hat{x}_{n+1,k|n} &= F_n \hat{x}_{n,k|n} \\
P_{n+1,k|n} &= F_n P_{n,k|n} F_n^{\mathsf{T}} + G_n Q_n G_n^{\mathsf{T}}
\end{aligned} \tag{6.52}$$

end

6.2.8 PERFORMANCE ANALYSIS

In view of the approximations (6.44) and (6.48), it is important to examine the mean-square error performance of the resulting diffusion Kalman filter. In this section, we examine how close the local estimates $\{\hat{x}_{n,k|n}\}$ get to the state variable x_n by evaluating the size of the mean-square error in the steady state after the filter has had sufficient time to learn.

We start by collecting all state estimation errors from across the network into an extended $K \times 1$ block vector (whose individual entries are of size $N \times 1$ each):

$$\tilde{\mathcal{X}}_{n|n} \triangleq \begin{pmatrix} \tilde{x}_{1,n|n} \\ \tilde{x}_{n,2|n} \\ \vdots \\ \tilde{x}_{n,K|n} \end{pmatrix}, \quad \text{where } \tilde{x}_{n,k|n} \triangleq x_n - \hat{x}_{n,k|n}, \tag{6.53}$$

and introduce supporting block-diagonal and Kronecker product matrices:

$$\mathcal{U}_{n-1} \triangleq \mathbb{1}_K \otimes u_{n-1}, \tag{6.54}$$

$$\mathcal{V}_n \triangleq \text{blkcol}\{v_{n,1|n}, v_{2,n|n}, \ldots, v_{n,K|n}\}, \tag{6.55}$$

$$\mathcal{P}_{n|n} \triangleq \text{blkdiag}\{P_{n,1|n}, P_{n,2|n}, \ldots, P_{n,K|n}\}, \tag{6.56}$$

$$\mathcal{H}_n \triangleq \text{blkdiag}\{H_{n,1}, H_{n,2}, \ldots, H_{n,K}\}, \tag{6.57}$$

$$\mathcal{S}_n \triangleq \text{blkdiag}\{S_{n,1}, S_{n,2}, \ldots, S_{n,K}\}, \tag{6.58}$$

$$\mathcal{R}_n \triangleq \text{blkdiag}\{R_{n,1}, R_{n,2}, \ldots, R_{n,K}\}, \tag{6.59}$$

$$\mathcal{J} \triangleq J \otimes I_N, \tag{6.60}$$

$$\mathcal{A} \triangleq A \otimes I_N, \tag{6.61}$$

where $\otimes$ denotes the Kronecker product operation. Subtracting $\boldsymbol{x}_n$ from both sides of the expression for $\boldsymbol{\psi}_{n,k}$ in Eq. (6.52), and using $\boldsymbol{z}_{n,\ell} = \boldsymbol{H}_{n,\ell}\boldsymbol{x}_n + \boldsymbol{v}_{n,\ell}$, gives

$$\begin{aligned}\tilde{\boldsymbol{\psi}}_{n,k} &= \tilde{\boldsymbol{x}}_{n,k|n-1} - \boldsymbol{P}_{n,k|n}(\boldsymbol{q}_{n,k} - \boldsymbol{S}_{n,k}\hat{\boldsymbol{x}}_{n,k|n-1}) \\ &= \tilde{\boldsymbol{x}}_{n,k|n-1} - \boldsymbol{P}_{n,k|n}\sum_{\ell\in\mathfrak{N}_k}\boldsymbol{H}_{n,\ell}^{\mathrm{T}}\boldsymbol{R}_{n,\ell}^{-1}(\boldsymbol{z}_{n,\ell} - \boldsymbol{H}_{n,\ell}\hat{\boldsymbol{x}}_{n,\ell|n-1}) \\ &= (\boldsymbol{I}_N - \boldsymbol{P}_{n,k|n}\boldsymbol{S}_{n,k})\tilde{\boldsymbol{x}}_{n,k|n-1} - \boldsymbol{P}_{n,k|n}\sum_{\ell\in\mathfrak{N}_k}\boldsymbol{H}_{n,\ell}^{\mathrm{T}}\boldsymbol{R}_{n,\ell}^{-1}\boldsymbol{v}_{n,\ell},\end{aligned} \tag{6.62}$$

where $\tilde{\boldsymbol{\psi}}_{n,k} \triangleq \boldsymbol{x}_n - \boldsymbol{\psi}_{n,k}$. Using the combination step from Eq. (6.52) gives

$$\tilde{\boldsymbol{x}}_{n,k|n} = \sum_{\ell\in\mathfrak{N}_k} a_{\ell k}\left((\boldsymbol{I}_N - \boldsymbol{P}_{n,\ell|n}\boldsymbol{S}_{n,\ell})\tilde{\boldsymbol{x}}_{n,\ell|n-1} - \boldsymbol{P}_{n,\ell|n}\sum_{i\in\mathfrak{N}_\ell}\boldsymbol{H}_{n,i}^{\mathrm{T}}\boldsymbol{R}_{n,i}^{-1}\boldsymbol{v}_{n,i}\right). \tag{6.63}$$

Using $\tilde{\boldsymbol{x}}_{n,\ell|n-1} = \boldsymbol{F}_{n-1}\tilde{\boldsymbol{x}}_{n-1,\ell|n-1} + \boldsymbol{G}_{n-1}\boldsymbol{u}_{n-1}$, we arrive at the recursion

$$\begin{aligned}\tilde{\boldsymbol{x}}_{n,k|n} = \sum_{\ell\in\mathfrak{N}_k} a_{\ell k}\Bigg(&(\boldsymbol{I}_N - \boldsymbol{P}_{n,\ell|n}\boldsymbol{S}_{n,\ell})\boldsymbol{F}_{n-1}\tilde{\boldsymbol{x}}_{n-1,\ell|n-1} \\ &+(\boldsymbol{I}_N - \boldsymbol{P}_{n,\ell|n}\boldsymbol{S}_{n,\ell})\boldsymbol{G}_{n-1}\boldsymbol{u}_{n-1} - \boldsymbol{P}_{n,\ell|n}\sum_{i\in\mathfrak{N}_\ell}\boldsymbol{H}_{n,i}^{\mathrm{T}}\boldsymbol{R}_{n,i}^{-1}\boldsymbol{v}_{n,i}\Bigg).\end{aligned} \tag{6.64}$$

This relation shows that the network error vector evolves according to the following dynamics:

$$\begin{aligned}\tilde{\boldsymbol{\mathcal{X}}}_{n|n} &= \mathcal{A}^{\mathrm{T}}(\boldsymbol{I}_N - \mathcal{P}_{n|n}\mathcal{S}_n)(\boldsymbol{I}_K \otimes \boldsymbol{F}_{n-1})\tilde{\boldsymbol{\mathcal{X}}}_{n-1|n-1} \\ &\quad + \mathcal{A}^{\mathrm{T}}(\boldsymbol{I}_{KN} - \mathcal{P}_{n|n}\mathcal{S}_n)(\boldsymbol{I}_K \otimes \boldsymbol{G}_{n-1})\boldsymbol{\mathcal{U}}_{n-1} - \mathcal{A}^{\mathrm{T}}\mathcal{P}_{n|n}\mathcal{J}^{\mathrm{T}}\mathcal{H}_n^{\mathrm{T}}\mathcal{R}_n^{-1}\boldsymbol{\mathcal{V}}_n,\end{aligned} \tag{6.65}$$

which we rewrite more compactly as

$$\tilde{\boldsymbol{\mathcal{X}}}_{n|n} = \mathcal{F}_n\tilde{\boldsymbol{\mathcal{X}}}_{n-1|n-1} + \mathcal{G}_n\boldsymbol{\mathcal{U}}_{n-1} - \mathcal{D}_n\boldsymbol{\mathcal{V}}_n, \tag{6.66}$$

where

$$\mathcal{F}_n \triangleq \mathcal{A}^{\mathrm{T}}(\boldsymbol{I}_{KN} - \mathcal{P}_{n|n}\mathcal{S}_n)(\boldsymbol{I}_K \otimes \boldsymbol{F}_{n-1}), \tag{6.67}$$

$$\mathcal{G}_n \triangleq \mathcal{A}^{\mathrm{T}}(\boldsymbol{I}_{KN} - \mathcal{P}_{n|n}\mathcal{S}_n)(\boldsymbol{I}_K \otimes \boldsymbol{G}_{n-1}), \tag{6.68}$$

$$\mathcal{D}_n \triangleq \mathcal{A}^{\mathrm{T}}\mathcal{P}_{n|n}\mathcal{J}^{\mathrm{T}}\mathcal{H}_n^{\mathrm{T}}\mathcal{R}_n^{-1}. \tag{6.69}$$

If we now let $\mathcal{P}_{\tilde{\mathcal{X}},n}$ denote the covariance matrix of the network error vector,

$$\mathcal{P}_{\tilde{\boldsymbol{\mathcal{X}}},n} \triangleq \mathbb{E}\left[\tilde{\boldsymbol{\mathcal{X}}}_{n|n}\tilde{\boldsymbol{\mathcal{X}}}_{n|n}^{\mathrm{T}}\right], \tag{6.70}$$

it then follows from Eq. (6.66) that this matrix satisfies the Lyapunov recursion

$$\mathcal{P}_{\tilde{\boldsymbol{\mathcal{X}}},n} = \mathcal{F}_n\mathcal{P}_{\tilde{\boldsymbol{\mathcal{X}}},n-1}\mathcal{F}_n^{\mathrm{T}} + \mathcal{G}_n(\mathbb{1}_K\mathbb{1}_K^{\mathrm{T}} \otimes \boldsymbol{Q}_{n-1})\mathcal{G}_n^{\mathrm{T}} + \mathcal{D}_n\mathcal{R}_n\mathcal{D}_n^{\mathrm{T}}. \tag{6.71}$$

In order to analyze the stability and performance of the diffusion filter, we shall consider here the following conditions; see [28] for definitions of the notions of stabilizability and detectability.

Assumption. *It is assumed that the model parameters $\{F, G, H_k, Q, R_k\}$ do not depend on the time index, n. We further collect the measurement matrices in the neighborhood of agent k into the block column matrix:*

$$H_k^{\mathrm{loc}} \triangleq \mathrm{blkcol}\{H_{k_1}, H_{k_2}, \ldots, H_{k_{d_k}}\}, \tag{6.72}$$

and assume that the pair $(F, GQ^{1/2})$ is stabilizable and the pair (F, H_k^{loc}) is detectable. ■

Under the stabilizability and detectability conditions, it is known from the convergence properties of the discrete Riccati recursions that each entry $P_{n,k|n}$ of $\mathcal{P}_{n|n}$ converges to the quantity P_k^+ defined by [28]:

$$(P_k^+)^{-1} = (P_k)^{-1} + (H_k^{\mathrm{loc}})^{\mathrm{T}} (R_k^{\mathrm{loc}})^{-1} H_k^{\mathrm{loc}} = (P_k)^{-1} + S_k, \tag{6.73}$$

where

$$R_k^{\mathrm{loc}} \triangleq \mathrm{blkdiag}\{R_{k_1}, R_{k_2}, \ldots, R_{k_{d_k}}\}, \tag{6.74}$$

and P_k is the unique stabilizing solution of the following discrete algebraic Riccati equation (DARE):

$$P_k = F P_k^+ F^{\mathrm{T}} + G Q G^{\mathrm{T}} = F P_k F^{\mathrm{T}} + G Q G^{\mathrm{T}} - K_{p,k} R_{e,k} K_{p,k}^{\mathrm{T}}, \tag{6.75}$$

where

$$K_{p,k} \triangleq F P_k (H_k^{\mathrm{loc}})^{\mathrm{T}} R_{e,k}^{-1}, \quad R_{e,k} \triangleq R_k^{\mathrm{loc}} + H_k^{\mathrm{loc}} P_k (H_k^{\mathrm{loc}})^{\mathrm{T}}. \tag{6.76}$$

Accordingly, the quantities $\{\mathcal{P}_{n|n}, \mathcal{F}_n, \mathcal{G}_n, \mathcal{D}_n\}$ converge to steady-state values denoted by [12]:

$$\mathcal{P}^+ \triangleq \lim_{n\to\infty} \mathcal{P}_{n|n} \text{ (block diagonal)}, \tag{6.77}$$

$$\mathcal{F} \triangleq \lim_{n\to\infty} \mathcal{F}_n = \mathcal{A}^{\mathrm{T}}(I_{KN} - \mathcal{P}^+\mathcal{S})(I_K \otimes F), \tag{6.78}$$

$$\mathcal{G} \triangleq \lim_{n\to\infty} \mathcal{G}_n = \mathcal{A}^{\mathrm{T}}(I_{KN} - \mathcal{P}^+\mathcal{S})(I_K \otimes G), \tag{6.79}$$

$$\mathcal{D} \triangleq \lim_{n\to\infty} \mathcal{D}_n = \mathcal{A}^{\mathrm{T}}\mathcal{P}^+\mathcal{J}^{\mathrm{T}}\mathcal{H}^{\mathrm{T}}\mathcal{R}^{-1}, \tag{6.80}$$

where $\mathcal{S}$, $\mathcal{H}$, and $\mathcal{R}$ are written without the time subscript n because their entries are now time-independent. Furthermore, the stabilizability and detectability conditions ensure that (see Lemma 1 in [12]):

$$\mathcal{X} \triangleq (I_{KN} - \mathcal{P}^+\mathcal{S})(I_K \otimes F) \text{ is block diagonal and stable.} \tag{6.81}$$

Note that the limiting matrix $\mathcal{F}$ in Eq. (6.78) has the form

$$\mathcal{F} = \mathcal{A}^{\mathrm{T}}\mathcal{X}. \tag{6.82}$$

We will verify soon that, under some conditions, $\mathcal{F}$ is also stable for any left-stochastic matrix $\mathcal{A}$ and, in view of Lemma A.1 in Appendix A.1, the error covariance matrix $\mathcal{P}_{\tilde{\mathcal{X}},n}$ in Eq. (6.71) converges to the unique solution of the Lyapunov equation:

$$\mathcal{P}_{\tilde{\mathcal{X}}} = \mathcal{F}\mathcal{P}_{\tilde{\mathcal{X}}}\mathcal{F}^{\mathrm{T}} + \mathcal{G}(\mathbb{1}_K\mathbb{1}_K^{\mathrm{T}} \otimes \mathcal{Q})\mathcal{G}^{\mathrm{T}} + \mathcal{D}\mathcal{R}\mathcal{D}^{\mathrm{T}}. \tag{6.83}$$

Using the vec notation (which stacks the columns of a matrix on top of each other), we can solve for $\mathcal{P}_{\tilde{\mathcal{X}}}$ and write

$$\mathrm{vec}(\mathcal{P}_{\tilde{\mathcal{X}}}) = (\boldsymbol{I}_{(KN)^2} - \mathcal{F}^{\mathrm{T}} \otimes \mathcal{F})^{-1}\mathrm{vec}(\mathcal{G}(\mathbb{1}_K\mathbb{1}_K^{\mathrm{T}} \otimes \mathcal{Q})\mathcal{G}^{\mathrm{T}} + \mathcal{D}\mathcal{R}\mathcal{D}^{\mathrm{T}}). \tag{6.84}$$

In this way, the steady-state mean-square error at any agent k will be given by

$$\lim_{n\to\infty} \mathbb{E}\left[\|\tilde{\boldsymbol{x}}_{n,k|n}\|^2\right] = \mathrm{Tr}(\mathcal{P}_{\tilde{\mathcal{X}}}\mathcal{I}_k), \tag{6.85}$$

where $\mathcal{I}_k$ is a $K \times K$ block diagonal matrix with blocks of size $N \times N$; it contains the identity matrix at block (k,k) and zeros everywhere else. That is, it holds that $\mathcal{I}_k = \boldsymbol{e}_k\boldsymbol{e}_k^{\mathrm{T}} \otimes \boldsymbol{I}_N$, where $\boldsymbol{e}_k$ is the basis column vector of size $K \times 1$ with unit entry at location k and zeros everywhere else. Consequently, the average mean-square error across the network is

$$\mathrm{MSE}_{\mathrm{av}} = \frac{1}{K}\mathrm{Tr}(\mathcal{P}_{\tilde{\mathcal{X}}}). \tag{6.86}$$

We verify the above claims by showing that the matrix $\mathcal{F}$ is stable and that, in view of this fact, the Lyapunov recursion (6.71) converges to the unique solution of the Lyapunov equation (6.83). These facts follow from the results of Lemmas A.1 and A.2.

6.3 DISTRIBUTED PARTICLE FILTERING

In the second part of this chapter, we discuss distributed particle filtering. We will focus on consensus-based methods in Section 6.3.4 and on diffusion-based methods in Section 6.3.5. Before that, we present the underlying state-space model, which generalizes Eqs. (6.2) and (6.3), and we discuss some basics of sequential Bayesian estimation and review the particle filter (PF). We adopt a Bayesian framework for estimation, which means that both the state $\boldsymbol{x}_n$ and the measurements $z_{n,k}$ are modeled as random vectors [36].

6.3.1 STATE-SPACE MODEL

Differently from Section 6.2, the state-space model is now allowed to be nonlinear. The temporal dynamics or evolution of the state $\boldsymbol{x}_n$ is characterized by the state-evolution model (a.k.a. state-transition model or system model)

$$\boldsymbol{x}_{n+1} = \boldsymbol{g}_n(\boldsymbol{x}_n, \boldsymbol{u}_n), \quad n = 0, 1, \ldots, \tag{6.87}$$

which extends Eq. (6.2). In Eq. (6.87), $\mathbf{g}_n(\cdot,\cdot)$ is a known, generally nonlinear function, and $\boldsymbol{u}_n$ is state noise that is statistically independent and identically distributed (i.i.d.) across time n and also statistically independent of the state sequence. The probability density function (pdf) of $\boldsymbol{u}_n$ is assumed known. Eq. (6.87) then determines the *state-transition pdf* $f(\boldsymbol{x}_n|\boldsymbol{x}_{n-1})$ for $n = 1, 2, \ldots$.

Furthermore, the statistical dependence of the measurement $\mathbf{z}_{n,k}$ of agent $k \in \{1, 2, \ldots, K\}$ on the state $\boldsymbol{x}_n$ is characterized by a measurement model

$$\mathbf{z}_{n,k} = \mathbf{h}_{n,k}(\boldsymbol{x}_n, \boldsymbol{v}_{n,k}), \quad n = 1, 2, \ldots; \quad k = 1, 2, \ldots, K, \tag{6.88}$$

which extends Eq. (6.3). In Eq. (6.88), $\mathbf{h}_{n,k}(\cdot,\cdot)$ is a known, generally nonlinear function, and $\boldsymbol{v}_{n,k}$ is measurement noise that is statistically independent across time n and across the agents k, and also statistically independent of the state sequence and of the state noise. The pdf of $\boldsymbol{v}_{n,k}$ is assumed known. Eq. (6.88) then determines the *local likelihood function* of agent k, $f(\mathbf{z}_{n,k}|\boldsymbol{x}_n)$, for $n = 1, 2, \ldots$. Because $\boldsymbol{v}_{n,k}$ is assumed independent of $\boldsymbol{v}_{n,k'}$ for $k' \neq k$, the measurements $\mathbf{z}_{n,k}$ at different agents k are conditionally independent given the state $\boldsymbol{x}_n$. Thus, the *global (all-agents) likelihood function* $f(\mathbf{z}_n|\boldsymbol{x}_n)$, with $\mathbf{z}_n \triangleq (\mathbf{z}_{n,1}^{\mathrm{T}}\, \mathbf{z}_{n,2}^{\mathrm{T}} \cdots \mathbf{z}_{n,K}^{\mathrm{T}})^{\mathrm{T}}$, factorizes into the local likelihood functions, i.e.,

$$f(\mathbf{z}_n|\boldsymbol{x}_n) = \prod_{k=1}^{K} f(\mathbf{z}_{n,k}|\boldsymbol{x}_n). \tag{6.89}$$

The above independence assumptions can be relaxed; for example, the state noise process $\boldsymbol{u}_n$ and the measurement noise processes $\boldsymbol{v}_{n,k}$ may be allowed to be dependent [37]. Further, the conditional independence of the measurements can be removed too [38,39]. Also, the noises in the state and measurement equations do not have to be independent across time [40,41]. However, the independence assumptions stated above are required for the consensus-based distributed PF (DPF) methods presented below. We also note that the case where the state-evolution function $\mathbf{g}_n(\boldsymbol{x}_n, \boldsymbol{u}_n)$ is linear in both $\boldsymbol{x}_n$ and $\boldsymbol{u}_n$ and the measurement function $\mathbf{h}_{n,k}(\boldsymbol{x}_n, \boldsymbol{v}_{n,k})$ is linear in both $\boldsymbol{x}_n$ and $\boldsymbol{v}_{n,k}$ was considered in Section 6.2 (see Eqs. (6.2) and (6.3)). Furthermore, when the state noise $\boldsymbol{u}_n$ in Eq. (6.87) and the measurement noise $\boldsymbol{v}_{n,k}$ in Eq. (6.88) have Gaussian distributions, these distributions are completely characterized by their first- and second-order moments. Finally, we point out that the network may be dynamic in that the set of neighboring nodes may vary with time (i.e., we may have $\mathfrak{N}_{n,k}$ instead of $\mathfrak{N}_k$). However, in the rest of the chapter, we will continue to consider the network to be static.

The above state-evolution and measurement models along with the associated statistical assumptions imply two further conditional independence properties. First, the current state $\boldsymbol{x}_n$ is conditionally independent of all the past measurements, $\mathbf{z}_{1:n-1,k} \triangleq (\mathbf{z}_{1,k}^{\mathrm{T}}\, \mathbf{z}_{2,k}^{\mathrm{T}} \cdots \mathbf{z}_{n-1,k}^{\mathrm{T}})^{\mathrm{T}}$, given the previous state $\boldsymbol{x}_{n-1}$, i.e.,

$$f(\boldsymbol{x}_n|\boldsymbol{x}_{n-1}, \mathbf{z}_{1:n-1,k}) = f(\boldsymbol{x}_n|\boldsymbol{x}_{n-1}), \tag{6.90}$$

for all $k = 1, 2, \ldots, K$. Second, the current measurements $\mathbf{z}_{n,k}$ are conditionally independent of all the past measurements, $\mathbf{z}_{1:n-1,k'}$, given the current state $\boldsymbol{x}_n$, i.e.,

$$f(\mathbf{z}_{n,k}|\boldsymbol{x}_n, \mathbf{z}_{1:n-1,k'}) = f(\mathbf{z}_{n,k}|\boldsymbol{x}_n), \tag{6.91}$$

for any $k' \in \{1, 2, \ldots, K\}$, including in particular $k' = k$.

6.3.2 SEQUENTIAL BAYESIAN ESTIMATION

We address the problem of sequential estimation of the state $\boldsymbol{x}_n$ from the total (all-agents) measurement sequence $\boldsymbol{z}_{1:n} \triangleq (\boldsymbol{z}_1^{\mathrm{T}}\ \boldsymbol{z}_2^{\mathrm{T}} \cdots \boldsymbol{z}_n^{\mathrm{T}})^{\mathrm{T}}$. In the Bayesian context, this essentially amounts to calculating the *posterior pdf* $f(\boldsymbol{x}_n|\boldsymbol{z}_{1:n})$, from which optimal Bayesian estimators and quantities characterizing their performance can be derived. In particular, the MMSE estimator [36] is given by the first moment of $f(\boldsymbol{x}_n|\boldsymbol{z}_{1:n})$, i.e.,

$$\hat{\boldsymbol{x}}_n^{\mathrm{MMSE}} = \mathbb{E}[\boldsymbol{x}_n|\boldsymbol{z}_{1:n}] = \int_{\mathbb{R}^N} \boldsymbol{x}_n f(\boldsymbol{x}_n|\boldsymbol{z}_{1:n})\, \mathrm{d}\boldsymbol{x}_n, \quad n = 1, 2, \ldots. \tag{6.92}$$

In a centralized scenario, based on the conditional independence properties (6.90) and (6.91), the current posterior pdf $f(\boldsymbol{x}_n|\boldsymbol{z}_{1:n})$ can be calculated sequentially (recursively) from the previous posterior pdf $f(\boldsymbol{x}_{n-1}|\boldsymbol{z}_{1:n-1})$ and the current all-agents measurement vector $\boldsymbol{z}_n$. This recursion consists of a prediction step calculating $f(\boldsymbol{x}_n|\boldsymbol{z}_{1:n-1})$ from $f(\boldsymbol{x}_{n-1}|\boldsymbol{z}_{1:n-1})$ and a subsequent update or correction step calculating $f(\boldsymbol{x}_n|\boldsymbol{z}_{1:n})$ from $f(\boldsymbol{x}_n|\boldsymbol{z}_{1:n-1})$. The prediction step involves the state-transition pdf $f(\boldsymbol{x}_n|\boldsymbol{x}_{n-1})$, and the update step involves the global likelihood function $f(\boldsymbol{z}_n|\boldsymbol{x}_n)$ and, thus, the all-agents measurement vector $\boldsymbol{z}_n$. Unfortunately, the prediction and update steps as well as the calculation of the MMSE estimate in Eq. (6.92) are usually computationally infeasible. A prominent exception is the case of the linear-Gaussian state-space model, where optimal sequential Bayesian state estimation reduces to the Kalman filter considered in Section 6.2.3 [19,28,29].

Several computationally feasible approximations to optimal sequential Bayesian state estimation have been proposed in the centralized case; these include the extended Kalman filter [19,36,42], the Gaussian sum filter [43], the unscented (sigma-point) Kalman filter [44,45], the cubature Kalman filter [46], and the PF [20,47–49]. In particular, particle filters are well suited to any nonlinear state-space model and any distributions of the state and measurement noises, and they perform well in situations where Kalman filter-based methods perform poorly. This generally comes at the cost of an increased computational complexity. Below, we will discuss distributed methods for particle filtering in decentralized agent networks. We start with a review of the basic PF.

6.3.3 REVIEW OF THE PARTICLE FILTER

A PF performs an approximation of optimal sequential Bayesian state estimation that is based on Monte Carlo simulation and importance sampling [20,47–49]. In a centralized scenario, the posterior pdf $f(\boldsymbol{x}_n|\boldsymbol{z}_{1:n})$ is represented in an approximative manner by M randomly drawn samples or *particles* $\boldsymbol{x}_n^{(m)}$ and corresponding weights $w_n^{(m)}$, where $m = 1, 2, \ldots, M$. Specifically, at each time step n, propagation of the posterior pdf (i.e., $f(\boldsymbol{x}_{n-1}|\boldsymbol{z}_{1:n-1}) \to f(\boldsymbol{x}_n|\boldsymbol{z}_{1:n})$) is replaced by propagation of the particles and weights (i.e., $\{(\boldsymbol{x}_{n-1}^{(m)}, w_{n-1}^{(m)})\}_{m=1}^M \to \{(\boldsymbol{x}_n^{(m)}, w_n^{(m)})\}_{m=1}^M$). As in the case of optimal sequential Bayesian estimation considered in Section 6.3.2, this propagation consists of a prediction step and an update or correction step.

In the prediction step at time n, for each preceding particle $\boldsymbol{x}_{n-1}^{(m)}$, a new particle $\boldsymbol{x}_n^{(m)}$ is sampled from a suitably chosen proposal pdf (a.k.a. importance pdf) $q(\boldsymbol{x}_n|\boldsymbol{x}_{n-1}^{(m)})$. In the simplest case, $q(\boldsymbol{x}_n|\boldsymbol{x}_{n-1}^{(m)})$ is chosen as $f(\boldsymbol{x}_n|\boldsymbol{x}_{n-1}^{(m)})$, i.e., the state-transition pdf $f(\boldsymbol{x}_n|\boldsymbol{x}_{n-1})$ conditioned on $\boldsymbol{x}_{n-1} = \boldsymbol{x}_{n-1}^{(m)}$; the resulting PF algorithm is known as the sequential importance resampling (SIR) filter. However, more sophisticated "adapted" proposal pdfs that also involve the current measurement z_n can result in improved estimation performance [49,50] (see the section "Distributed proposal adaptation").

In the update step at time n, for each particle $\boldsymbol{x}_n^{(m)}$, a nonnormalized weight is calculated according to

$$\tilde{w}_n^{(m)} = w_{n-1}^{(m)} \frac{f(z_n|\boldsymbol{x}_n^{(m)}) f(\boldsymbol{x}_n^{(m)}|\boldsymbol{x}_{n-1}^{(m)})}{q(\boldsymbol{x}_n^{(m)}|\boldsymbol{x}_{n-1}^{(m)})}, \quad m = 1, 2, \ldots, M. \tag{6.93}$$

For the SIR filter, this simplifies to

$$\tilde{w}_n^{(m)} = w_{n-1}^{(m)} f(z_n|\boldsymbol{x}_n^{(m)}), \quad m = 1, 2, \ldots, M. \tag{6.94}$$

Subsequently, the weights are normalized according to $w_n^{(m)} = \tilde{w}_n^{(m)} / \sum_{m'=1}^{M} \tilde{w}_n^{(m')}$. The set of particles and normalized weights $\{(\boldsymbol{x}_n^{(m)}, w_n^{(m)})\}_{m=1}^{M}$ constitutes a Monte Carlo representation of the posterior pdf $f(\boldsymbol{x}_n|z_{1:n})$. From $\{(\boldsymbol{x}_n^{(m)}, w_n^{(m)})\}_{m=1}^{M}$, a corresponding approximation of the MMSE state estimate $\hat{\boldsymbol{x}}_n^{\text{MMSE}}$ in Eq. (6.92) can be computed as the weighted sample mean, i.e.,

$$\hat{\boldsymbol{x}}_n = \sum_{m=1}^{M} w_n^{(m)} \boldsymbol{x}_n^{(m)}. \tag{6.95}$$

Finally, if a suitable criterion is satisfied (as discussed in [20,51]), the set $\{(\boldsymbol{x}_n^{(m)}, w_n^{(m)})\}_{m=1}^{M}$ is resampled to avoid an effect known as particle degeneracy. In the simplest case [51], the resampled particles are obtained by sampling with replacement from the set $\{(\boldsymbol{x}_n^{(m)}, w_n^{(m)})\}_{m=1}^{M}$, where $\boldsymbol{x}_n^{(m)}$ is sampled with probability $w_n^{(m)}$. This results in M resampled particles $\boldsymbol{x}_n^{(m)}$. The weights are redefined as $w_n^{(m)} = 1/M$.

This recursive algorithm is initialized at time $n = 0$ by M particles $\boldsymbol{x}_0^{(m)}$, $m = 1, 2, \ldots, M$, which are drawn from a suitable prior pdf $f(\boldsymbol{x}_0)$. The initial weights are equal, i.e., $w_0^{(m)} = 1/M$ for all m.

In a distributed implementation based on a decentralized network of agents, each agent runs a local instance of a PF, hereafter briefly referred to as "local PF." The local PF at agent k observes only the local measurements $z_{n,k}$ directly; however, it receives from the neighbor agents indirect information about the measurements of the other agents in the network and also provides to the neighboring agents indirect information about its own measurements. The type of information that is exchanged between neighboring agents depends on the specific method used for distributed particle filtering. A DPF is based on the PF algorithm summarized above. However, it modifies that algorithm to account for the fact that each agent runs its own local PF, and it employs some networkwide distributed scheme (such as consensus, gossip, or diffusion) to disseminate and fuse local information provided by the agents. Several types of DPF methods have been proposed, including [13,17,52–64]. In what follows, we discuss two classes of DPF methods that rely on consensus and diffusion schemes for distributed information dissemination and fusion.

6.3.4 CONSENSUS-BASED METHODS

In a DPF that emulates the performance of the basic PF, the local PF at agent k attempts to track a particle representation $\{(\boldsymbol{x}_{n,k}^{(m)}, w_{n,k}^{(m)})\}_{m=1}^{M}$ of the global posterior pdf $f(\boldsymbol{x}_n|z_{1:n})$. According to Eq. (6.93), this requires the *global* likelihood function (GLF) $f(z_n|\boldsymbol{x}_n)$ (evaluated at the current local particles, i.e., $\boldsymbol{x}_n = \boldsymbol{x}_{n,k}^{(m)}$), rather than merely the *local* likelihood function (LLF) of agent k, $f(z_{n,k}|\boldsymbol{x}_n)$. Because agent k by itself is only able to calculate the LLF, calculation of the GLF requires information provided by the other agents.

DPF based on likelihood consensus

A recently proposed class of DPF algorithms performs a distributed calculation of the GLF using the *likelihood consensus* (LC) scheme [13,56]. To develop that scheme, we first take the logarithm of Eq. (6.89), which yields

$$\ln f(z_n|\boldsymbol{x}_n) = \sum_{k=1}^{K} \ln f(z_{n,k}|\boldsymbol{x}_n), \tag{6.96}$$

where ln denotes the natural logarithm. We can thus write the GLF as

$$f(z_n|\boldsymbol{x}_n) = \exp\big(\ln f(z_n|\boldsymbol{x}_n)\big) = \exp\big(K\lambda(\boldsymbol{x}_n; z_n)\big) \tag{6.97}$$

with

$$\lambda(\boldsymbol{x}_n; z_n) \triangleq \frac{1}{K} \ln f(z_n|\boldsymbol{x}_n) = \frac{1}{K} \sum_{k=1}^{K} \ln f(z_{n,k}|\boldsymbol{x}_n), \tag{6.98}$$

where Eq. (6.96) has been used in the last step. Hence, the GLF $f(z_n|\boldsymbol{x}_n)$ has been rewritten in terms of the *average* of the K log-LLFs $\ln f(z_{n,k}|\boldsymbol{x}_n)$, $k = 1, 2, \ldots, K$. Because the log-LLFs $\ln f(z_{n,k}|\boldsymbol{x}_n)$ are functions of the unknown state $\boldsymbol{x}_n \in \mathbb{R}^N$, rather than simply numbers, we cannot directly use a distributed averaging scheme such as the average consensus algorithm to calculate $\lambda(\boldsymbol{x}_n; z_n)$.

The solution provided by the LC scheme is based on an approximate (finite-order) function expansion of the log-LLFs $\ln f(z_{n,k}|\boldsymbol{x}_n)$. Let $\{\varphi_r(\boldsymbol{x}_n)\}_{r=1}^{R}$ denote a system of R functions that is known to all agents. Possible choices of these functions include monomials [13], orthogonal polynomials [65], Fourier basis functions [56], and spline functions [66]. We consider the approximations of the log-LLFs given by

$$\ln f(z_{n,k}|\boldsymbol{x}_n) \approx \sum_{r=1}^{R} \alpha_{n,k,r}(z_{n,k})\, \varphi_r(\boldsymbol{x}_n), \quad k = 1, 2, \ldots, K \tag{6.99}$$

with suitable expansion coefficients $\alpha_{n,k,r}(z_{n,k})$, $r = 1, 2, \ldots, R$. Note that these expansion coefficients involve the local measurements $z_{n,k}$, and thus they are different at different agents k. Inserting Eq. (6.99) into Eq. (6.98) and changing the order of summation yields

$$\lambda(\boldsymbol{x}_n; z_n) \approx \tilde{\lambda}(\boldsymbol{x}_n; z_n) \triangleq \sum_{r=1}^{R} a_{n,r}(z_n)\, \varphi_r(\boldsymbol{x}_n), \tag{6.100}$$

where

$$a_{n,r}(\mathbf{z}_n) \triangleq \frac{1}{K}\sum_{k=1}^{K} \alpha_{n,k,r}(\mathbf{z}_{n,k}). \tag{6.101}$$

Using in Eq. (6.97) the approximation $\tilde{\lambda}(\boldsymbol{x}_n;\boldsymbol{z}_n)$ instead of $\lambda(\boldsymbol{x}_n;\boldsymbol{z}_n)$, we obtain a corresponding approximation of the GLF, namely,

$$f(\boldsymbol{z}_n|\boldsymbol{x}_n) \approx \tilde{f}(\boldsymbol{z}_n|\boldsymbol{x}_n) \triangleq \exp(K\tilde{\lambda}(\boldsymbol{x}_n;\boldsymbol{z}_n)). \tag{6.102}$$

Distributed calculation of the GLF

Within the approximation given by Eqs. (6.100) and (6.102), the distributed calculation of the GLF $f(\boldsymbol{z}_n|\boldsymbol{x}_n)$ at time n amounts to the distributed calculation of the R numbers $a_{n,r}(\mathbf{z}_n)$, $r = 1, 2, \ldots, R$, each of which is an average of the respective local expansion coefficients $\alpha_{n,k,r}(\mathbf{z}_{n,k})$, $k = 1, 2, \ldots, K$. Note that because the measurements $\mathbf{z}_{n,k}$ have been observed and thus are fixed, we now have to average numbers instead of functions. Thus, the average expansion coefficients $a_{n,r}(\mathbf{z}_n)$, $r = 1, 2, \ldots, R$ can be calculated in a distributed manner via R instances of an average consensus algorithm [24,25]. In addition (see Eq. (6.102)), each agent also needs to know the total number of agents, K. Distributed algorithms for counting the number of agents are available (e.g., [67]).

In iteration i of the rth instance of the average consensus algorithm, which is used to calculate $a_{n,r}(\mathbf{z}_n)$, each agent k updates an "internal state" $\zeta_{k,r}$ according to

$$\zeta_{k,r}^{(i)} = \sum_{k' \in \mathfrak{N}_k} \omega_{k,k'}^{(i)} \zeta_{k',r}^{(i-1)}. \tag{6.103}$$

Here, $\{\omega_{k,k'}^{(i)}\}_{k' \in \mathfrak{N}_k}$ is a set of weights whose choice is discussed in [24,25,68]. Hence, $\zeta_{k,r}^{(i)}$ is a linear combination of the preceding (i.e., at iteration $i-1$) internal state of agent k and the preceding internal states of the neighbor agents k'. The iteration is initialized by choosing the initial internal states as $\zeta_{k,r}^{(0)} = \alpha_{n,k,r}(\mathbf{z}_{n,k})$. Then, for a suitable choice of the weights, and using our assumption that the agent network is connected, it can be shown [25] that the internal state $\zeta_{k,r}^{(i)}$ of each agent k converges to the desired average $a_{n,r}(\mathbf{z}_n)$, i.e.,

$$\lim_{i\to\infty} \zeta_{k,r}^{(i)} = a_{n,r}(\mathbf{z}_n) = \frac{1}{K}\sum_{k'=1}^{K} \alpha_{n,k',r}(\mathbf{z}_{n,k'}). \tag{6.104}$$

For a finite number $i_{\max}$ of iterations, complete convergence cannot be achieved; this means that the internal states $\zeta_{k,r}^{(i_{\max})}$ will be (slightly) different from $a_{n,r}(\mathbf{z}_n)$, and also (slightly) different across the agents k.

Several standard choices of the weights $\omega_{k,k'}^{(i)}$ are available [24,68]. A popular choice is given by the Metropolis weights [24]

$$\omega_{k,k'} = \begin{cases} \dfrac{1}{\max\{|\mathcal{N}_k|, |\mathcal{N}_{k'}|\}}, & k' \neq k, \\ 1 - \sum_{k'' \in \mathcal{N}_k \setminus \{k\}} \omega_{k,k''}, & k' = k. \end{cases} \tag{6.105}$$

Note that these weights do not depend on the iteration index i. Their calculation at agent k requires that agent k knows both $|\mathcal{N}_k|$ and $|\mathcal{N}_{k'}|$ for all $k' \in \mathcal{N}_k$. Certain other choices of the weights require less knowledge [24].

In each iteration of the consensus scheme, agent k has to broadcast its internal states $\zeta_{k,r}^{(i)}$, $r = 1, 2, \ldots, R$ to its neighbors $k' \in \mathcal{N}_k \setminus \{k\}$. Furthermore, agent k receives the internal states $\zeta_{k',r}^{(i)}$, $r = 1, 2, \ldots, R$ from its neighbors $k' \in \mathcal{N}_k \setminus \{k\}$. Thus, during one time step n, each agent k has to broadcast a total of $i_{\max} R$ real numbers to its neighbors.

Calculation of the local expansion coefficients

The local expansion coefficients $\alpha_{n,k,r}(z_{n,k})$, $r = 1, 2, \ldots, R$ arising in the log-LLF approximation (6.99) are calculated *locally* at the respective agent k. This can be done by means of a least-squares (LS) fit between the log-LLF $\ln f(z_{n,k}|\boldsymbol{x}_n)$ and the approximating expansion $\sum_{r=1}^{R} \alpha_{n,k,r}(z_{n,k})\, \varphi_r(\boldsymbol{x}_n)$. In view of the way the local PF operates, the approximation does not need to be good for all possible states $\boldsymbol{x}_n$ but only in those regions of the state-space where the current particles $\boldsymbol{x}_{n,k}^{(m)}$ are located. Therefore, the LS fit at agent k minimizes, with respect to the vector of expansion coefficients $\boldsymbol{\alpha}_{n,k} \triangleq \big(\alpha_{n,k,1}(z_{n,k}) \cdots \alpha_{n,k,R}(z_{n,k})\big)^{\mathrm{T}}$, the LS approximation error of the expansion (6.99) evaluated at the particles $\boldsymbol{x}_{n,k}^{(m)}$, $m = 1, 2, \ldots, M$, i.e.,

$$\sum_{m=1}^{M} \left(\ln f\big(z_{n,k}|\boldsymbol{x}_{n,k}^{(m)}\big) - \sum_{r=1}^{R} \alpha_{n,k,r}(z_{n,k})\, \varphi_r\big(\boldsymbol{x}_{n,k}^{(m)}\big) \right)^2 = \|\boldsymbol{\xi}_{n,k} - \boldsymbol{\Phi}_{n,k}\boldsymbol{\alpha}_{n,k}\|^2 \to \min_{\boldsymbol{\alpha}_{n,k}}. \tag{6.106}$$

Here, $\boldsymbol{\xi}_{n,k} \triangleq \Big(\ln f\big(z_{n,k}|\boldsymbol{x}_{n,k}^{(1)}\big) \cdots \ln f\big(z_{n,k}|\boldsymbol{x}_{n,k}^{(M)}\big)\Big)^{\mathrm{T}}$ and $\boldsymbol{\Phi}_{n,k}$ is an $M \times R$ matrix with columns $\boldsymbol{\phi}_{n,k,r} \triangleq \Big(\varphi_r\big(\boldsymbol{x}_{n,k}^{(1)}\big) \cdots \varphi_r\big(\boldsymbol{x}_{n,k}^{(M)}\big)\Big)^{\mathrm{T}}$. Assuming that $M \geq R$ (i.e., there are at least as many particles as expansion coefficients) and that $\boldsymbol{\Phi}_{n,k}$ has full rank, the solution to the LS problem (6.106) is given by [69]

$$\hat{\boldsymbol{\alpha}}_{n,k} = \boldsymbol{\Phi}_{n,k}^{\#}\, \boldsymbol{\xi}_{n,k}, \quad \text{with } \boldsymbol{\Phi}_{n,k}^{\#} \triangleq \big(\boldsymbol{\Phi}_{n,k}^{\mathrm{T}} \boldsymbol{\Phi}_{n,k}\big)^{-1} \boldsymbol{\Phi}_{n,k}^{\mathrm{T}}. \tag{6.107}$$

Numerical aspects of computing $\hat{\boldsymbol{\alpha}}_{n,k}$ in Eq. (6.107) are discussed in [69,70]. The elements $\hat{\alpha}_{n,k,r}$ of $\hat{\boldsymbol{\alpha}}_{n,k}$ are then used to initialize the LC according to $\zeta_{k,r}^{(0)} = \hat{\alpha}_{n,k,r}$.

A summary of the LC-based DPF algorithm is provided in listing (6.108).

LC-based DPF

---

each agent starts with its own set of particles $\boldsymbol{x}_{0,k}^{(m)} \sim f(\boldsymbol{x}_0)$ and identical weights $w_{0,k}^{(m)} = 1/M$, where $m = 1, 2, \ldots, M$.

repeat at every agent k for $n \geq 1$:

(propose a new set of particles)

$\boldsymbol{x}_{n,k}^{(m)} \sim q(\boldsymbol{x}_n | \boldsymbol{x}_{n-1,k}^{(m)}), \ m = 1, 2, \ldots, M$

(calculate an approximation $\tilde{f}_k(\boldsymbol{z}_n|\boldsymbol{x}_n)$ of the GLF $f(\boldsymbol{z}_n|\boldsymbol{x}_n)$)

- calculate the coefficient vector $\hat{\boldsymbol{\alpha}}_{n,k}$ according to Eq. (6.107)
- for $r = 1, 2, \ldots, R$:

  –initialize the internal LC state: $\zeta_{k,r}^{(0)} = \hat{\alpha}_{n,k,r}$

  –for $i = 1, 2, \ldots, i_{\max}$, update the internal LC state $\zeta_{k,r}^{(i)}$ according to Eq. (6.103)

  –calculate $\tilde{f}_k(\boldsymbol{z}_n|\boldsymbol{x}_n) = \exp\Big(K \sum_{r=1}^{R} \zeta_{k,r}^{(i_{\max})} \varphi_r(\boldsymbol{x}_n)\Big)$ (cf. Eqs. (6.102), (6.100)) (6.108)

(calculate the weights of the proposed particles)

- calculate nonnormalized weights:

$$\tilde{w}_{n,k}^{(m)} = w_{n-1,k}^{(m)} \frac{\tilde{f}_k(\boldsymbol{z}_n|\boldsymbol{x}_{n,k}^{(m)}) f(\boldsymbol{x}_{n,k}^{(m)}|\boldsymbol{x}_{n-1,k}^{(m)})}{q(\boldsymbol{x}_{n,k}^{(m)}|\boldsymbol{x}_{n-1,k}^{(m)})}, \ m = 1, 2, \ldots, M$$

- normalize weights:

$$w_n^{(m)} = \tilde{w}_n^{(m)} / \sum_{m'=1}^{M} \tilde{w}_n^{(m')}$$

(compute the state estimate)

$$\hat{\boldsymbol{x}}_{n,k} = \sum_{m=1}^{M} w_{n,k}^{(m)} \boldsymbol{x}_{n,k}^{(m)}$$

(if necessary, perform resampling [20,51])

end

Exponential family

An alternative LC scheme is possible if the LLFs of all the agents k (viewed as conditional pdfs of $\boldsymbol{z}_{n,k}$) belong to the exponential family of distributions [71], i.e.,

$$f(\boldsymbol{z}_{n,k}|\boldsymbol{x}_n) = c_{n,k}(\boldsymbol{z}_{n,k}) \exp(\boldsymbol{b}_{n,k}^{\mathrm{T}}(\boldsymbol{x}_n) \boldsymbol{d}_{n,k}(\boldsymbol{z}_{n,k}) - g_{n,k}(\boldsymbol{x}_n)), \quad k = 1, 2, \ldots, K, \tag{6.109}$$

with some functions $c_{n,k}(\cdot) \in \mathbb{R}_+$, $\boldsymbol{b}_{n,k}(\cdot) \in \mathbb{R}^q$, $\boldsymbol{d}_{n,k}(\cdot) \in \mathbb{R}^q$, and $g_{n,k}(\cdot) \in \mathbb{R}$, where $q \geq N$. Inserting this expression into Eq. (6.89) yields for the GLF

$$f(\boldsymbol{z}_n|\boldsymbol{x}_n) = C_n(\boldsymbol{z}_n) \exp(K G_n(\boldsymbol{x}_n; \boldsymbol{z}_n)), \tag{6.110}$$

where $C_n(z_n) \triangleq \prod_{k=1}^{K} c_{n,k}(z_{n,k})$ and

$$G_n(\boldsymbol{x}_n; z_n) \triangleq \frac{1}{K} \sum_{k=1}^{K} (\boldsymbol{b}_{n,k}^{\mathrm{T}}(\boldsymbol{x}_n)\boldsymbol{d}_{n,k}(z_{n,k}) - g_{n,k}(\boldsymbol{x}_n)). \tag{6.111}$$

The normalization factor $C_n(z_n)$ does not depend on the state $\boldsymbol{x}_n$ and is therefore irrelevant due to weight normalization (see Section 6.3.3).

As an alternative to expanding the log-LLFs as in Eq. (6.99), we may use separate finite-order function expansions of the state-dependent functions involved in Eq. (6.111) [13], i.e.,

$$\boldsymbol{b}_{n,k}(\boldsymbol{x}_n) \approx \sum_{r=1}^{R_b} \boldsymbol{\beta}_{n,k,r}\, \varphi_r(\boldsymbol{x}_n), \quad g_{n,k}(\boldsymbol{x}_n) \approx \sum_{r=1}^{R_g} \gamma_{n,k,r}\, \psi_r(\boldsymbol{x}_n), \quad k = 1, 2, \ldots, K. \tag{6.112}$$

Here, the coefficients $\boldsymbol{\beta}_{n,k,r}$ are vectors of the same dimension as $\boldsymbol{b}_{n,k}(\boldsymbol{x}_n)$. The coefficients $\boldsymbol{\beta}_{n,k,r}$ and $\gamma_{n,k,r}$ can again be calculated by means of LS fitting, similarly to the section "Calculation of the local expansion coefficients"; note that one separate LS fit is required for each component of $\boldsymbol{\beta}_{n,k,r}$. Inserting Eq. (6.112) into Eq. (6.111) and changing the order of summation yields

$$G_n(\boldsymbol{x}_n; z_n) \approx \tilde{G}_n(\boldsymbol{x}_n; z_n) \triangleq \sum_{r=1}^{R_b} B_{n,r}(z_n)\varphi_r(\boldsymbol{x}_n) - \sum_{r=1}^{R_g} \Gamma_{n,r}\psi_r(\boldsymbol{x}_n) \tag{6.113}$$

with

$$B_{n,r}(z_n) \triangleq \frac{1}{K} \sum_{k=1}^{K} \boldsymbol{\beta}_{n,k,r}^{\mathrm{T}} \boldsymbol{d}_{n,k}(z_{n,k}), \quad \Gamma_{n,r} \triangleq \frac{1}{K} \sum_{k=1}^{K} \gamma_{n,k,r}. \tag{6.114}$$

Using in Eq. (6.110) the approximation $\tilde{G}_n(\boldsymbol{x}_n; z_n)$ instead of $G_n(\boldsymbol{x}_n; z_n)$ yields an approximation of the GLF.

The R_b numbers $B_{n,r}(z_k)$, $r = 1, 2, \ldots, R_b$ and the R_g numbers $\Gamma_{n,r}$, $r = 1, 2, \ldots, R_g$ are seen in Eq. (6.114) to be averages of the local quantities $a_{n,k,r}(z_{n,k}) \triangleq \boldsymbol{\beta}_{n,k,r}^{\mathrm{T}} \boldsymbol{d}_{n,k}(z_{n,k})$ and $\gamma_{n,k,r}$, respectively. Hence, they can be calculated in a distributed manner via $R_b + R_g$ instances of an average consensus algorithm, similarly to the section "Distributed calculation of the GLF." Note that this distributed calculation assumes that each agent k knows its own functions $\boldsymbol{b}_{n,k}(\cdot), \boldsymbol{d}_{n,k}(\cdot)$, and $g_{n,k}(\cdot)$, but not the respective functions of the other agents $k' \neq k$.

This distributed calculation of (an approximation of) the GLF $f(z_n|\boldsymbol{x}_n)$ may be preferable over the general scheme presented in the sections "DPF based on likelihood consensus," "Distributed calculation of the GLF," and "Calculation of the local expansion coefficients" if it is easier to approximate the functions $\boldsymbol{b}_{n,k}(\boldsymbol{x}_n)$ and $g_{n,k}(\boldsymbol{x}_n)$ than the LLFs $\ln f(z_{n,k}|\boldsymbol{x}_n)$. In particular, there is a reduction in the number of real numbers each agent has to broadcast to its neighbors if $R_b + R_g < R$, where R is the order of function expansion used in the general scheme.

Gaussian measurement noise

An important special case of an exponential-family LLF arises with additive Gaussian measurement noise [13]. Here, the measurement model of agent k in Eq. (6.88) is specialized according to

$$z_{n,k} = \boldsymbol{m}_{n,k}(\boldsymbol{x}_n) + \boldsymbol{v}_{n,k}, \quad n = 1,2,\ldots; \quad k = 1,2,\ldots,K, \tag{6.115}$$

where $\boldsymbol{m}_{n,k}(\boldsymbol{x}_n)$ is a generally nonlinear function and the additive measurement noise $\boldsymbol{v}_{n,k}$ satisfies the independence properties formulated in Section 6.3.1 and, in addition, is Gaussian with zero mean and covariance matrix $\boldsymbol{C}_{n,k}$. It follows that the LLF of agent k is given, up to a normalization factor, by

$$f(z_{n,k}|\boldsymbol{x}_n) \propto \exp\Big(-\frac{1}{2}(z_{n,k} - \boldsymbol{m}_{n,k}(\boldsymbol{x}_n))^{\mathrm{T}} \boldsymbol{C}_{n,k}^{-1}(z_{n,k} - \boldsymbol{m}_{n,k}(\boldsymbol{x}_n))\Big). \tag{6.116}$$

This is easily verified to be an exponential-family LLF (6.109) with

$$\boldsymbol{b}_{n,k}(\boldsymbol{x}_n) = \boldsymbol{m}_{n,k}(\boldsymbol{x}_n), \tag{6.117}$$

$$\boldsymbol{d}_{n,k}(z_{n,k}) = \boldsymbol{C}_{n,k}^{-1} z_{n,k}, \tag{6.118}$$

$$g_{n,k}(\boldsymbol{x}_n) = \frac{1}{2}\boldsymbol{m}_{n,k}^{\mathrm{T}}(\boldsymbol{x}_n)\boldsymbol{C}_{n,k}^{-1}\boldsymbol{m}_{n,k}(\boldsymbol{x}_n). \tag{6.119}$$

As in the section "Exponential family," we may use separate function expansion approximations of $\boldsymbol{b}_{n,k}(\boldsymbol{x}_n) = \boldsymbol{m}_{n,k}(\boldsymbol{x}_n)$ and $g_{n,k}(\boldsymbol{x}_n)$. However, an alternative approach is possible because, according to Eq. (6.119), $g_{n,k}(\boldsymbol{x}_n)$ is a function of $\boldsymbol{m}_{n,k}(\boldsymbol{x}_n)$ [13]. Indeed, we may insert a function expansion approximation of $\boldsymbol{b}_{n,k}(\boldsymbol{x}_n) = \boldsymbol{m}_{n,k}(\boldsymbol{x}_n)$, i.e.,

$$\boldsymbol{m}_{n,k}(\boldsymbol{x}_n) \approx \tilde{\boldsymbol{m}}_{n,k}(\boldsymbol{x}_n) \triangleq \sum_{r=1}^{R_b} \boldsymbol{\beta}_{n,k,r}\, \varphi_r(\boldsymbol{x}_n), \tag{6.120}$$

into Eq. (6.119) to obtain an "induced" function expansion approximation of $g_{n,k}(\boldsymbol{x}_n)$,

$$\begin{aligned} g_{n,k}(\boldsymbol{x}_n) \approx \tilde{g}_{n,k}(\boldsymbol{x}_n) &\triangleq \frac{1}{2}\tilde{\boldsymbol{m}}_{n,k}^{\mathrm{T}}(\boldsymbol{x}_n)\boldsymbol{C}_{n,k}^{-1}\tilde{\boldsymbol{m}}_{n,k}(\boldsymbol{x}_n) \\ &= \frac{1}{2}\sum_{r_1=1}^{R_b}\sum_{r_2=1}^{R_b} \boldsymbol{\beta}_{n,k,r_1}^{\mathrm{T}}\, \boldsymbol{C}_{n,k}^{-1}\, \boldsymbol{\beta}_{n,k,r_2}\varphi_{r_1}(\boldsymbol{x}_n)\varphi_{r_2}(\boldsymbol{x}_n). \end{aligned} \tag{6.121}$$

Using any one-to-one mapping of the double index $(r_1, r_2) \in \{1,2,\ldots,R_b\} \times \{1,2,\ldots,R_b\}$ to a single index $r \in \{1,2,\ldots,R_g\}$, with $R_g = R_b^2$, we can rewrite Eq. (6.121) as (cf. Eq. (6.112))

$$\tilde{g}_{n,k}(\boldsymbol{x}_n) = \sum_{r=1}^{R_g} \gamma_{n,k,r}\, \psi_r(\boldsymbol{x}_n), \tag{6.122}$$

where $\gamma_{n,k,r} = \frac{1}{2}\boldsymbol{\beta}_{n,k,r_1}^{\mathrm{T}}\, \boldsymbol{C}_{n,k}^{-1}\, \boldsymbol{\beta}_{n,k,r_2}$ and $\psi_r(\boldsymbol{x}_n) = \varphi_{r_1}(\boldsymbol{x}_n)\varphi_{r_2}(\boldsymbol{x}_n)$. It can be easily verified that the resulting approximate GLF (cf. Eq. (6.89)) can be written as $\tilde{f}(z_k|\boldsymbol{x}_k) \propto \exp(-\frac{1}{2}\tilde{Q}_n(z_n,\boldsymbol{x}_n))$ with $\tilde{Q}_n(z_n,\boldsymbol{x}_n) \triangleq \sum_{k=1}^{K}(z_{n,k} - \tilde{\boldsymbol{m}}_{n,k}(\boldsymbol{x}_n))^{\mathrm{T}}\boldsymbol{C}_{n,k}^{-1}(z_{n,k} - \tilde{\boldsymbol{m}}_{n,k}(\boldsymbol{x}_n))$. This equals the true GLF $f(z_k|\boldsymbol{x}_k)$, except that the means $\boldsymbol{m}_{n,k}(\boldsymbol{x}_n)$ are replaced by their approximations $\tilde{\boldsymbol{m}}_{n,k}(\boldsymbol{x}_n)$.

Distributed Gaussian particle filter

The Gaussian PF (GPF) [72] uses a Gaussian approximation of the posterior pdf $f(\boldsymbol{x}_n|\boldsymbol{z}_{1:n})$. This Gaussian approximation is derived from a weighted particle set, which is calculated in a similar way as in the PF except that no resampling is required. Note that the measurement model is still allowed to be non-Gaussian, as in Eq. (6.88).

In a distributed GPF, the local GPF at agent k propagates a Gaussian approximation $\mathcal{N}(\boldsymbol{x}_n; \boldsymbol{\mu}_{n,k}, \boldsymbol{\Sigma}_{n,k})$ of the global posterior pdf $f(\boldsymbol{x}_n|\boldsymbol{z}_{1:n})$. At time n, particles $\boldsymbol{x}_{n,k}^{(m)}$, $m = 1, 2, \ldots, M$ are drawn from the preceding Gaussian approximation $\mathcal{N}(\boldsymbol{x}_{n-1}; \boldsymbol{\mu}_{n-1,k}, \boldsymbol{\Sigma}_{n-1,k})$. Then, the prediction and update steps are performed as described in Section 6.3.3. From the resulting weighted particles $\{(\boldsymbol{x}_{n,k}^{(m)}, w_{n,k}^{(m)})\}_{m=1}^{M}$, a new Gaussian approximation $\mathcal{N}(\boldsymbol{x}_n; \boldsymbol{\mu}_{n,k}, \boldsymbol{\Sigma}_{n,k})$ is determined according to

$$\boldsymbol{\mu}_{n,k} = \sum_{m=1}^{m} w_{n,k}^{(m)} \boldsymbol{x}_{n,k}^{(m)} \quad \text{and} \quad \boldsymbol{\Sigma}_{n,k} = \sum_{m=1}^{M} w_{n,k}^{(m)} \boldsymbol{x}_{n,k}^{(m)} \boldsymbol{x}_{n,k}^{(m)\mathrm{T}} - \boldsymbol{\mu}_{n,k} \boldsymbol{\mu}_{n,k}^{\mathrm{T}}. \tag{6.123}$$

The weighted sample mean $\boldsymbol{\mu}_{n,k}$ also constitutes the local approximation $\hat{\boldsymbol{x}}_{n,k}$ of the MMSE state estimate $\hat{\boldsymbol{x}}_n^{\mathrm{MMSE}}$ in Eq. (6.92), i.e., $\hat{\boldsymbol{x}}_{n,k} = \boldsymbol{\mu}_{n,k}$.

Just as in the conventional DPF (cf. Eq. (6.93)), the update step of the local GPFs requires the GLF evaluated at the current local particles, i.e., $f(\boldsymbol{z}_n|\boldsymbol{x}_n)$ for $\boldsymbol{x}_n = \boldsymbol{x}_{n,k}^{(m)}$, $m = 1, 2, \ldots, M$. An approximation of $f(\boldsymbol{z}_n|\boldsymbol{x}_{n,k}^{(m)})$ can be calculated in a distributed manner by the LC as explained earlier. However, the following variation yields a significant reduction of computational complexity at the cost of some increase in local communications [13]. Inspired by the parallel GPF implementation proposed in [73], the idea is to "distribute" the M particles over the K agents, such that each local GPF uses a significantly reduced set of particles, and to combine the results of the local GPFs via additional average consensus operations. The local GPF at agent k uses only $M_k < M$ particles, where $\sum_{k=1}^{K} M_k = M$. In the local weight update step at agent k, for each particle $\boldsymbol{x}_{n,k}^{(m)}$ from the reduced local particle set, $m \in \{1, 2, \ldots, M_k\}$, a nonnormalized weight $\tilde{w}_{n,k}^{(m)}$ is calculated according to Eq. (6.93). This requires the GLF evaluated at the M_k particles, i.e., $f(\boldsymbol{z}_n|\boldsymbol{x}_{n,k}^{(m)})$, $m = 1, 2, \ldots, M_k$, which was previously calculated (approximately) by the LC. Furthermore, the sum of the resulting nonnormalized weights is calculated, i.e., $\tilde{W}_{n,k} = \sum_{m=1}^{M_k} \tilde{w}_{n,k}^{(m)}$.

Next, from the weighted particles $\{(\boldsymbol{x}_{n,k}^{(m)}, \tilde{w}_{n,k}^{(m)})\}_{m=1}^{M_k}$, a local nonnormalized mean vector and a local nonnormalized correlation matrix are calculated at each agent k as

$$\boldsymbol{\mu}'_{n,k} = \sum_{m=1}^{M_k} \tilde{w}_{n,k}^{(m)} \boldsymbol{x}_{n,k}^{(m)}, \quad \mathbf{R}'_{n,k} = \sum_{m=1}^{M_k} \tilde{w}_{n,k}^{(m)} \boldsymbol{x}_{n,k}^{(m)} \boldsymbol{x}_{n,k}^{(m)\mathrm{T}}. \tag{6.124}$$

Finally, the local means and correlations from all the agents are combined into a global mean and covariance:

$$\bar{\boldsymbol{\mu}}_{n,k} = \frac{1}{W_{n,k}} \sum_{k=1}^{K} \boldsymbol{\mu}'_{n,k}, \quad \bar{\boldsymbol{\Sigma}}_{n,k} = \frac{1}{W_{n,k}} \sum_{k=1}^{K} \mathbf{R}'_{n,k} - \bar{\boldsymbol{\mu}}_{n,k} \bar{\boldsymbol{\mu}}_{n,k}^{\mathrm{T}}, \tag{6.125}$$

where $W_{n,k} \triangleq \sum_{k=1}^{K} \tilde{W}_{n,k}$. Approximations of the sums $\sum_{k=1}^{K} \boldsymbol{\mu}'_{n,k}$, $\sum_{k=1}^{K} \mathbf{R}'_{n,k}$, and $\sum_{k=1}^{K} \tilde{W}_{n,k}$ can be calculated in a distributed manner by means of an average consensus algorithm (again assuming that the number of agents, K, is known to each agent). Subsequently, the division by $W_{n,k}$ and the subtraction of $\bar{\boldsymbol{\mu}}_{n,k}\bar{\boldsymbol{\mu}}_{n,k}^{\mathrm{T}}$ in Eq. (6.125) are performed locally at each agent. This results in local approximations $\boldsymbol{\mu}_{n,k}$ and $\boldsymbol{\Sigma}_{n,k}$ of, respectively, $\bar{\boldsymbol{\mu}}_{n,k}$ and $\bar{\boldsymbol{\Sigma}}_{n,k}$ at each agent k. The state estimate at agent k, $\hat{\boldsymbol{x}}_{n,k}$, is taken to be $\boldsymbol{\mu}_{n,k}$.

Distributed proposal adaptation

The choice of the proposal pdf $q(\boldsymbol{x}_n; \boldsymbol{x}_{n-1}^{(m)})$ used in the weight update step (6.93) strongly affects the performance of a PF. The proposal pdf should be similar to the posterior pdf $f(\boldsymbol{x}_n|\boldsymbol{z}_{1:n})$ [20]. Using the state-transition pdf $f(\boldsymbol{x}_n|\boldsymbol{x}_{n-1}^{(m)})$ as the proposal pdf, as is done in the SIR filter, does not cater to this goal; in particular, it does not take into account the current measurement $\boldsymbol{z}_n$. The design of a proposal pdf that exploits the measurements is known as *proposal adaptation* [49,50]. In a DPF, the proposal pdfs used by the local PFs should be adapted to the total (all-agents) measurement $\boldsymbol{z}_n = (\boldsymbol{z}_{n,1}^{\mathrm{T}}\, \boldsymbol{z}_{n,2}^{\mathrm{T}} \cdots \boldsymbol{z}_{n,K}^{\mathrm{T}})^{\mathrm{T}}$.

A distributed method for this type of "global" proposal adaptation can again be based on the average consensus principle [56]. In this method, each agent first calculates a "predistorted" local posterior pdf. A Gaussian approximation of the global posterior pdf is then obtained by fusing all the predistorted local posterior pdfs; this approximate global posterior pdf is used by the local PFs as a proposal pdf. The method is inspired by [52] but employs a different predistortion that enables the use of a Gaussian filter for calculating the predistorted local posterior pdfs. To derive the method, we note that the global posterior pdf can be developed as $f(\boldsymbol{x}_n|\boldsymbol{z}_{1:n}) = f(\boldsymbol{x}_n|\boldsymbol{z}_{1:n-1}, \boldsymbol{z}_n) \propto f(\boldsymbol{z}_n|\boldsymbol{x}_n, \boldsymbol{z}_{1:n-1}) f(\boldsymbol{x}_n|\boldsymbol{z}_{1:n-1}) = f(\boldsymbol{z}_n|\boldsymbol{x}_n) f(\boldsymbol{x}_n|\boldsymbol{z}_{1:n-1})$, where Bayes' rule and Eq. (6.91) have been used. Inserting Eq. (6.89), we further obtain

$$f(\boldsymbol{x}_n|\boldsymbol{z}_{1:n}) \propto \left(\prod_{k=1}^{K} f(\boldsymbol{z}_{n,k}|\boldsymbol{x}_n) \right) f(\boldsymbol{x}_n|\boldsymbol{z}_{1:n-1}). \tag{6.126}$$

We now define a predistorted, nonnormalized "local pseudoposterior pdf" at agent k as

$$\tilde{f}(\boldsymbol{x}_n|\boldsymbol{z}_{1:n-1}, \boldsymbol{z}_{n,k}) \triangleq f(\boldsymbol{z}_{n,k}|\boldsymbol{x}_n) \big(f(\boldsymbol{x}_n|\boldsymbol{z}_{1:n-1})\big)^{1/K}. \tag{6.127}$$

Furthermore, we consider the network-wide product of all the local pseudoposterior pdfs,

$$\prod_{k=1}^{K} \tilde{f}(\boldsymbol{x}_n|\boldsymbol{z}_{1:n-1}, \boldsymbol{z}_{n,k}) = \left(\prod_{k=1}^{K} f(\boldsymbol{z}_{n,k}|\boldsymbol{x}_n) \right) f(\boldsymbol{x}_n|\boldsymbol{z}_{1:n-1}) \propto f(\boldsymbol{x}_n|\boldsymbol{z}_{1:n}), \tag{6.128}$$

where Eq. (6.126) has been used in the last step. According to Eq. (6.128), the product of all the local pseudoposterior pdfs equals the global posterior pdf up to a factor. This fact can be used to calculate a Gaussian approximation of the global posterior pdf $f(\boldsymbol{x}_n|\boldsymbol{z}_{1:n})$—note that $f(\boldsymbol{x}_n|\boldsymbol{z}_{1:n})$ involves all the measurements and indeed would be a desirable proposal pdf—by a distributed evaluation of the leftmost side of Eq. (6.128). To make this possible, we use Gaussian approximations of the local pseudoposterior pdfs, i.e.,

$$\tilde{f}(\boldsymbol{x}_n|\boldsymbol{z}_{1:n-1}, \boldsymbol{z}_{n,k}) \approx \mathcal{N}(\boldsymbol{x}_n; \tilde{\boldsymbol{\mu}}_{n,k}, \tilde{\boldsymbol{\Sigma}}_{n,k}), \tag{6.129}$$

and of the global posterior pdf, i.e.,

$$f(\boldsymbol{x}_n|\boldsymbol{z}_{1:n}) \approx q(\boldsymbol{x}_n;\boldsymbol{z}_k) \triangleq \mathcal{N}(\boldsymbol{x}_n;\boldsymbol{\mu}_n,\boldsymbol{\Sigma}_n). \tag{6.130}$$

Note that Eq. (6.130) also defines the global proposal pdf $q(\boldsymbol{x}_n;\boldsymbol{z}_n)$. Inserting the approximations (6.129) and (6.130) into the factorization $f(\boldsymbol{x}_n|\boldsymbol{z}_{1:n}) \propto \prod_{k=1}^{K}\tilde{f}(\boldsymbol{x}_n|\boldsymbol{z}_{1:n-1},\boldsymbol{z}_{n,k})$ (see Eq. (6.128)) yields

$$\mathcal{N}(\boldsymbol{x}_n;\boldsymbol{\mu}_n,\boldsymbol{\Sigma}_n) \propto \prod_{k=1}^{K}\mathcal{N}(\boldsymbol{x}_n;\tilde{\boldsymbol{\mu}}_{n,k},\tilde{\boldsymbol{\Sigma}}_{n,k}). \tag{6.131}$$

Using this relation and the rules for a product of Gaussian densities [52,74], the global mean $\boldsymbol{\mu}_n$ and global covariance $\boldsymbol{\Sigma}_n$—which determine $q(\boldsymbol{x}_n;\boldsymbol{z}_n) = \mathcal{N}(\boldsymbol{x}_n;\boldsymbol{\mu}_n,\boldsymbol{\Sigma}_n)$—can be calculated from the local means $\tilde{\boldsymbol{\mu}}_{n,k}$ and local covariances $\tilde{\boldsymbol{\Sigma}}_{n,k}$ according to

$$\boldsymbol{\mu}_n = \boldsymbol{\Sigma}_n\sum_{k=1}^{K}(\tilde{\boldsymbol{\Sigma}}_{n,k})^{-1}\tilde{\boldsymbol{\mu}}_{n,k}, \quad \boldsymbol{\Sigma}_n = \left(\sum_{k=1}^{K}(\tilde{\boldsymbol{\Sigma}}_{n,k})^{-1}\right)^{-1}. \tag{6.132}$$

Here, the sums $\sum_{k=1}^{K}(\tilde{\boldsymbol{\Sigma}}_{n,k})^{-1}\tilde{\boldsymbol{\mu}}_{n,k}$ and $\sum_{k=1}^{K}(\tilde{\boldsymbol{\Sigma}}_{n,k})^{-1}$ can be computed in a distributed manner by means of an average consensus algorithm.

The calculation of the local mean $\tilde{\boldsymbol{\mu}}_{n,k}$ and local covariance $\tilde{\boldsymbol{\Sigma}}_{n,k}$, as defined in Eq. (6.129), at agent k is based on the observation that Eq. (6.127) can be interpreted as the update step of a Bayesian filter using the predistorted predicted posterior pdf $\big(f(\boldsymbol{x}_n|\boldsymbol{z}_{1:n-1})\big)^{1/K}$ instead of the true predicted posterior pdf $f(\boldsymbol{x}_n|\boldsymbol{z}_{1:n-1})$. Each agent first calculates a Gaussian approximation of the predicted posterior pdf,

$$f(\boldsymbol{x}_n|\boldsymbol{z}_{1:n-1}) \approx \mathcal{N}(\boldsymbol{x}_n;\boldsymbol{\mu}'_{n,k},\boldsymbol{\Sigma}'_{n,k}), \tag{6.133}$$

where $\boldsymbol{\mu}'_{n,k} = \frac{1}{M}\sum_{m=1}^{M}\boldsymbol{x}^{(m)\prime}_{n,k}$ and $\boldsymbol{\Sigma}'_{n,k} = \frac{1}{M}\sum_{m=1}^{M}\boldsymbol{x}^{(m)\prime}_{n,k}\big(\boldsymbol{x}^{(m)\prime}_{n,k}\big)^{\mathrm{T}} - \boldsymbol{\mu}'_{n,k}\big(\boldsymbol{\mu}'_{n,k}\big)^{\mathrm{T}}$. Here, $\{\boldsymbol{x}^{(m)\prime}_{n,k}\}_{m=1}^{M}$ is a set of temporary particles that are randomly drawn from $f(\boldsymbol{x}_n|\boldsymbol{x}^{(m)}_{n-1,k})$, where $\{\boldsymbol{x}^{(m)}_{n-1,k}\}_{m=1}^{M}$ is the set of particles resulting from the preceding filtering step at time $n-1$. The approximation (6.133) implies

$$\big(f(\boldsymbol{x}_n|\boldsymbol{z}_{1:n-1})\big)^{1/K} \approx \mathcal{N}(\boldsymbol{x}_n;\boldsymbol{\mu}'_{n,k},K\boldsymbol{\Sigma}'_{n,k}). \tag{6.134}$$

Using the Gaussian approximations (6.129) and (6.134) in Eq. (6.127) gives

$$\mathcal{N}(\boldsymbol{x}_k;\tilde{\boldsymbol{\mu}}_{n,k},\tilde{\boldsymbol{\Sigma}}_{n,k}) = f(\boldsymbol{z}_{n,k}|\boldsymbol{x}_k)\,\mathcal{N}(\boldsymbol{x}_k;\boldsymbol{\mu}'_{n,k},K\boldsymbol{\Sigma}'_{n,k}). \tag{6.135}$$

This can be calculated by the update step of a Gaussian filter with input mean $\boldsymbol{\mu}'_{n,k}$, input covariance $K\boldsymbol{\Sigma}'_{n,k}$, and measurement $\boldsymbol{z}_{n,k}$. This Gaussian filter update step is performed locally at each agent and produces $\tilde{\boldsymbol{\mu}}_{n,k}$ and $\tilde{\boldsymbol{\Sigma}}_{n,k}$. Examples of a Gaussian filter include the extended Kalman filter [19,36,42], the unscented Kalman filter [44,45], the cubature Kalman filter [46], and the filters described in [75].

6.3.5 DIFFUSION-BASED METHODS

In this subsection, we address diffusion cooperation among agents as in Section 6.2.7 but implemented with particle filtering [60]. However, now the considered state-space models are nonlinear and the pdfs of $\boldsymbol{u}_n$ in Eq. (6.87) and $\boldsymbol{v}_{n,k}$ in Eq. (6.88) are assumed to be known up to proportionality constants. We reiterate that these pdfs can be of any form.

The diffusion-based DPF presented here follows the spirit of cooperation from Section 6.2.7 in the sense that the agents share moments (or parameters of approximating distributions) of the unknown state. We assume that the agents at time $n = 0$ start with the generation of their own particles $\boldsymbol{x}_{0,k}^{(m)}$, $m = 1, 2, \ldots, M$, which are drawn from a prior pdf $f(\boldsymbol{x}_0)$, i.e.,

$$\boldsymbol{x}_{0,k}^{(m)} \sim f(\boldsymbol{x}_0), \quad m = 1, 2, \ldots, M;\ k = 1, 2, \ldots, K. \tag{6.136}$$

Thus, before the tracking of the state starts, each agent has a representation of the prior pdf of $\boldsymbol{x}_0$ given by $\left\{\boldsymbol{x}_{0,k}^{(m)}, w_{0,k}^{(m)} = \frac{1}{M}\right\}_{m=1}^{M}$. Similarly, before processing the measurement vector $\boldsymbol{z}_{n,k}$, the local PF of agent k has a representation of the posterior pdf of $\boldsymbol{x}_{n-1,k}$ given by $\left\{\boldsymbol{x}_{n-1,k}^{(m)}, w_{n-1,k}^{(m)} = \frac{1}{M}\right\}_{m=1}^{M}$, which serves as a prior for the state $\boldsymbol{x}_n$.

Next, we explain how we obtain $\left\{\boldsymbol{x}_{n,k}^{(m)}, w_{n,k}^{(m)} = \frac{1}{M}\right\}_{m=1}^{M}$ from $\left\{\boldsymbol{x}_{n-1,k}^{(m)}, w_{n-1,k}^{(m)} = \frac{1}{M}\right\}_{m=1}^{M}$. The PF of each agent k propagates the particles $\boldsymbol{x}_{n-1,k}^{(m)}$ to particles that represent possible values of the state at time n. The propagation is carried out by drawing samples from the proposal pdf $q(\boldsymbol{x}_n|\boldsymbol{x}_{n-1,k}^{(m)})$, i.e.,

$$\tilde{\boldsymbol{x}}_{n,k}^{(m)} \sim q(\boldsymbol{x}_n|\boldsymbol{x}_{n-1,k}^{(m)}), \quad m = 1, 2, \ldots, M;\ k = 1, 2, \ldots, K. \tag{6.137}$$

As described in Section 6.3.3, the PF subsequently computes the weights of these particles. This is accomplished by

$$\tilde{w}_{n,k}^{(m)} \propto \frac{f(\boldsymbol{z}_{n,k}|\tilde{\boldsymbol{x}}_{n,k}^{(m)}) f(\tilde{\boldsymbol{x}}_{n,k}^{(m)}|\boldsymbol{x}_{n-1,k}^{(m)})}{q(\tilde{\boldsymbol{x}}_{n,k}^{(m)}|\boldsymbol{x}_{n-1,k}^{(m)})}, \tag{6.138}$$

where $\sum_{m=1}^{M} \tilde{w}_{n,k}^{(m)} = 1$. With the particles $\tilde{\boldsymbol{x}}_{n,k}^{(m)}$ and their weights $\tilde{w}_{n,k}^{(m)}$, the agent k has a local approximation of the posterior pdf of $\boldsymbol{x}_n$ given by $\left\{\tilde{\boldsymbol{x}}_{n,k}^{(m)}, \tilde{w}_{n,k}^{(m)}\right\}_{m=1}^{M}$. The cooperation among the agents requires that they exchange these approximating local posterior pdfs and fuse them to form new local posteriors. The best approach in terms of accuracy would be that each agent broadcasts all the particles and weights to its neighbors, but that would be quite costly in terms of communication load because the number of particles and weights is usually large. An alternative is to use the particles and weights to construct a parametric distribution, that is, estimate the parameters of an approximating distribution of the weighted particles. Then, the agents would only exchange the parameters of the approximating distribution.

In the sequel, the approximating distributions are multivariate Gaussians. Thus, the agents proceed by computing the means and covariance matrices of the approximating Gaussians. First, the agents obtain the means by

$$\tilde{\boldsymbol{\mu}}_{n,k} = \sum_{m=1}^{M} \tilde{w}_{n,k}^{(m)} \tilde{\boldsymbol{x}}_{n,k}^{(m)}, \quad k = 1, 2, \ldots, K, \tag{6.139}$$

and then the covariance matrices by

$$\tilde{\boldsymbol{\Sigma}}_{n,k} = \sum_{m=1}^{M} \tilde{w}_{n,k}^{(m)} \left(\tilde{\boldsymbol{x}}_{n,k}^{(m)} - \tilde{\boldsymbol{\mu}}_{n,k}\right) \left(\tilde{\boldsymbol{x}}_{n,k}^{(m)} - \tilde{\boldsymbol{\mu}}_{n,k}\right)^{\mathrm{T}}, \quad k = 1, 2, \ldots, K. \tag{6.140}$$

In the next step, the agents exchange the means and covariance matrices with their neighbors and follow up with merging all the Gaussian posteriors into one Gaussian. To that end, the agents use coefficients $a_{\ell k}$ that quantify how much they trust their neighbors (defined earlier in Section 6.2.1, and here with $a_{\ell k}$ representing how much agent k trusts agent ℓ). We note that $\sum_{\ell \in \mathcal{N}_k} a_{\ell k} = 1$. In the chapter on Bayesian Approach to Inference from this book (see its Subsection 4.2.3), it is explained that the Gaussian after merging has a covariance matrix and mean that can be expressed in terms of the individual covariance matrices and means of the agents by

$$\boldsymbol{\Sigma}_{n,k} = \left(\sum_{\ell \in \mathcal{N}_k} a_{\ell k} \left(\tilde{\boldsymbol{\Sigma}}_{n,k}\right)^{-1} \right)^{-1} \tag{6.141}$$

and

$$\boldsymbol{\mu}_{n,k} = \boldsymbol{\Sigma}_{n,k} \left(\sum_{\ell \in \mathcal{N}_k} a_{\ell k} \left(\tilde{\boldsymbol{\Sigma}}_{n,k}\right)^{-1} \tilde{\boldsymbol{\mu}}_{n,k} \right), \tag{6.142}$$

respectively. More specifically, this choice of $\boldsymbol{\mu}_{n,k}$ and $\boldsymbol{\Sigma}_{n,k}$ minimizes the weighted sum of Kullback-Leibler distances $\sum_{\ell \in \mathcal{N}_k} a_{\ell k} \mathcal{D}(\mathcal{N}_{n,k} \| \tilde{\mathcal{N}}_{n,\ell})$, where $\mathcal{N}_{n,k}$ is the resulting Gaussian after merging, and $\tilde{\mathcal{N}}_{n,\ell}$, $\ell \in \mathcal{N}_k$ are the Gaussians of agent k and its neighbors.

At this point, the agents have more than one way of proceeding. A simple way is that the agents use the newly obtained Gaussian and that each draws a set of particles that represent it, i.e.,

$$\boldsymbol{x}_{n,k}^{(m)} \sim \mathcal{N}(\boldsymbol{x}_n; \boldsymbol{\mu}_{n,k}, \boldsymbol{\Sigma}_{n,k}), \quad m = 1, 2, \ldots, M; \; k = 1, 2, \ldots, K. \tag{6.143}$$

After generation of these particles, the agents have a discrete representation, $\left\{\boldsymbol{x}_{n,k}^{(m)}, w_{n,k}^{(m)} = \frac{1}{M}\right\}_{m=1}^{M}$, of the posterior pdf at time n that serves as a prior pdf of $\boldsymbol{x}_{n+1}$.

Another approach would be to avoid fusion of the local posteriors by Eqs. (6.141) and (6.142) and instead draw particles directly from the mixture of Gaussians with components $\tilde{\mathcal{N}}(\boldsymbol{x}_n; \tilde{\boldsymbol{\mu}}_{n,\ell}, \tilde{\boldsymbol{\Sigma}}_{n,\ell})$ and weights $a_{\ell k}$, i.e.,

$$\boldsymbol{x}_{n,k}^{(m)} \sim \sum_{\ell \in \mathcal{N}_k} a_{\ell k} \tilde{\mathcal{N}}(\boldsymbol{x}_n; \tilde{\boldsymbol{\mu}}_{n,\ell}, \tilde{\boldsymbol{\Sigma}}_{n,\ell}), \quad m = 1, 2, \ldots, M; \quad k = 1, 2, \ldots, K. \tag{6.144}$$

Once done, the posterior pdf is again represented by $\left\{\boldsymbol{x}_{n,k}^{(m)}, w_{n,k}^{(m)} = \frac{1}{M}\right\}_{m=1}^{M}$.

An alternative direction is to first resample the original local particles $\tilde{\boldsymbol{x}}_{n,k}^{(m)}$ according to their weights $\tilde{w}_{n,k}^{(m)}$. Let the resampled particles be denoted by $\bar{\boldsymbol{x}}_{n,k}^{(m)}$. In the next step, these particles are rescaled and shifted so that they correspond to particles of a Gaussian distribution with mean $\boldsymbol{\mu}_{n,k}$ in Eq. (6.142) and covariance $\boldsymbol{\Sigma}_{n,k}$ in Eq. (6.141). It is not difficult to show that this can be achieved by

$$x_{n,k}^{(m)} = Q_{n,k}^{\mathrm{T}} \tilde{Q}_{n,k}^{-T} \left(\bar{x}_{n,k}^{(m)} - \tilde{\mu}_{n,k} \right) + \mu_{n,k}, \tag{6.145}$$

where $Q_{n,k}$ is defined by $\Sigma_{n,k} = Q_{n,k}^{\mathrm{T}} Q_{n,k}$ and $\tilde{Q}_{n,k}$ by $\tilde{\Sigma}_{n,k} = \tilde{Q}_{n,k}^{\mathrm{T}} \tilde{Q}_{n,k}$. The particles obtained by Eq. (6.145) have all equal weights, and thus they represent the prior pdf for the state x_{n+1}. This approach is described in the listing (6.146). Note that the particles generated by Eqs. (6.143) and (6.144) are all different whereas many of the ones obtained by Eq. (6.145) may be replicated.

Diffusion particle filter (adapt and then combine form)

each agent starts with its own set of particles $x_{0,k}^{(m)} \sim f(x_0), m = 1, 2, \ldots, M.$

repeat at every agent k for $n \geq 0$:

(propose a new set of particles)

$\tilde{x}_{n,k}^{(m)} \sim q(x_n | x_{n-1,k}^{(m)}), \quad m = 1, 2, \ldots, M$

(compute the weights of the proposed particles)

$\tilde{w}_{n,k}^{(m)} \propto \frac{f(z_{n,k} | \tilde{x}_{n,k}^{(m)}) f(\tilde{x}_{n,k}^{(m)} | x_{n-1,k}^{(m)})}{q(\tilde{x}_{n,k}^{(m)} | x_{n-1,k}^{(m)})}$

(find the parameters of a Gaussian distribution that approximates the posterior)

$\tilde{\mu}_{n,k} = \sum_{m=1}^{M} \tilde{w}_{n,k}^{(m)} \tilde{x}_{n,k}^{(m)}$

$$\tilde{\Sigma}_{n,k} = \sum_{m=1}^{M} \tilde{w}_{n,k}^{(m)} \left(\tilde{x}_{n,k}^{(m)} - \tilde{\mu}_{n,k} \right) \left(\tilde{x}_{n,k}^{(m)} - \tilde{\mu}_{n,k} \right)^{\mathrm{T}} \tag{6.146}$$

(combine own posterior with the posteriors of the neighbors)

$\Sigma_{n,k} = \left(\sum_{\ell \in \mathcal{N}_k} a_{\ell k} \left(\tilde{\Sigma}_{n,\ell} \right)^{-1} \right)^{-1}$

$\mu_{n,k} = \Sigma_{n,k} \left(\sum_{\ell \in \mathcal{N}_k} a_{\ell k} \left(\tilde{\Sigma}_{n,\ell} \right)^{-1} \tilde{\mu}_{n,\ell} \right)$

(resample the local particles according to their weights)

$\left\{ \left(\tilde{x}_{n,k}^{(m)}, \tilde{w}_{n,k}^{(m)} \right) \right\}_{m=1}^{M} \xrightarrow{\text{resampling}} \left\{ \left(\bar{x}_{n,k}^{(m)}, w_{n,k}^{(m)} = \frac{1}{M} \right) \right\}_{m=1}^{M}$

(rescale and shift the resampled local particles)

$x_{n,k}^{(m)} = Q_{n,k}^{\mathrm{T}} \tilde{Q}_{n,k}^{-\mathrm{T}} \left(\bar{x}_{n,k}^{(m)} - \tilde{\mu}_{n,k} \right) + \mu_{n,k}$, where $Q_{n,k}$ and $\tilde{Q}_{n,k}$ are defined by

$\Sigma_{n,k} = Q_{n,k}^{\mathrm{T}} Q_{n,k}$ and $\tilde{\Sigma}_{n,k} = \tilde{Q}_{n,k}^{\mathrm{T}} \tilde{Q}_{n,k}$, respectively

end

The performance of the diffusion cooperation can be improved if the calculation of the particle weights of each agent involves the measurements of the respective neighbors. This requires that the agents engage in two rounds of communication, one when they exchange their measurements with their neighbors and the other when they exchange the parameters of their posterior pdfs. In this case, the calculation of the weights in Eq. (6.138) is replaced by

$$\tilde{w}_{n,k}^{(m)} \propto \frac{\left(\prod_{\ell \in \mathcal{N}_k} f(z_{n,\ell} | \tilde{x}_{n,k}^{(m)}) \right) f(\tilde{x}_{n,k}^{(m)} | x_{n-1,k}^{(m)})}{q(\tilde{x}_{n,k}^{(m)} | x_{n-1,k}^{(m)})}. \tag{6.147}$$

Everything else in the described process remains the same.

In summary, the diffusion-based particle filtering algorithm described in this subsection relies on approximating the local posterior pdfs by Gaussians whose parameters are then exchanged with the neighbors. Upon receiving the Gaussian parameters from its neighbors, each agent either uses them to create a new Gaussian that serves as the final posterior pdf or simply exploits them directly as in Eq. (6.143) or Eq. (6.144) to generate particles that represent the support of the final posterior pdf and the prior pdf of $\boldsymbol{x}_{n+1}$. However, we reiterate that the underlying approximating distributions of $\left\{\tilde{\boldsymbol{x}}_{n,k}^{(m)}, \tilde{w}_{n,k}^{(m)}\right\}_{m=1}^{M}$ used by the agents do not have to be Gaussians.

6.4 CONCLUSIONS

In this chapter, we discussed the problem of distributed filtering in a network of agents. The filtering amounts to estimating a hidden state process based on measurements that are acquired by the agents and contain information about the state process. The agents cooperate by communicating relevant information with their neighbors. Two main types of filtering methods were addressed. The first is distributed Kalman filtering where the state-space models are linear and the state and measurement noise processes are zero-mean and white vector processes with known covariance matrices. The second type is distributed particle filtering where the models are nonlinear and the pdfs characterizing the state and measurement noise processes are known up to proportionality constants. The distributed Kalman filtering method was based on a diffusion strategy, with a discussion on the relation and differences relative to a consensus-type implementation. The distributed particle filtering methods were based on both consensus and diffusion strategies.

ACKNOWLEDGMENTS

The work of A. H. Sayed was supported by the NSF under grants ECCS-1407712 and CCF-1524250. The author is grateful to the IEEE for allowing reproduction of material from [12] in this book chapter. The work of P. M. Djurić was supported by the NSF under grant CCF-1618999. The work of F. Hawatsch was supported by the Austrian Science Fund (FWF) under grant P27370-N30 and by the Czech Science Foundation (GAČR) under grant 17-19638S.

A.1 APPENDIX

Lemma A.1 (Convergence of Lyapunov recursion). *Consider a general Lyapunov recursion of the form*

$$\boldsymbol{Z}_{n+1} = \boldsymbol{C}_n \boldsymbol{Z}_n \boldsymbol{C}_n^T + \boldsymbol{B}_n, \tag{A.1}$$

where the matrix sequences $\{\boldsymbol{B}_n, \boldsymbol{C}_n\}$ converge uniformly to $\{\boldsymbol{B}, \boldsymbol{C}\}$ as $n \to \infty$ with $\boldsymbol{C}$ being a stable matrix (all its eigenvalues lie strictly inside the unit circle). Then, the sequence $\boldsymbol{Z}_n$ converges to the unique solution $\boldsymbol{Z}$ of the Lyapunov equation

$$Z = CZC^T + B. \tag{A.2}$$

Proof. Using the vec notation we have

$$\text{vec}(Z_{n+1}) = (C_n \otimes C_n)\text{vec}(Z_n) + \text{vec}(B_n) \stackrel{\Delta}{=} T_n\text{vec}(Z_n) + d_n, \tag{A.3}$$

where we introduced the quantities $T_n = (C_n \otimes C_n)$ and $d_n = \text{vec}(B_n)$. We know from the assumptions in the lemma that d_n converges to $d = \text{vec}(B)$ and T_n converges to $T = (C \otimes C)$, which is a stable matrix because C is stable. It follows that $\text{vec}(Z_n)$ converges to $\text{vec}(Z) = (I - C \otimes C)^{-1}d$, which establishes that the limiting Z is the solution to the Lyapunov equation (A.2). We can establish this convergence result more formally as follows by adjusting the proof given for Theorem 1 in App. E of [12]. Let $z_n = \text{vec}(Z_n)$, $z = \text{vec}(Z)$, $T_n - T = \epsilon\Delta_n$, and $d_n - d = \epsilon\delta_n$, for some arbitrary scalar $\epsilon > 0$ and quantities $\{\delta_n, \Delta_n\}$. Using $z = Tz + d$ and $z_{n+1} = T_n z_n + d_n$, we get

$$z_{n+1} - z = T(z_n - z) + \epsilon\Delta_n(z_n - z) + \epsilon\Delta_n z + \epsilon\delta_n. \tag{A.4}$$

Now, because T is a stable matrix, there exists a submultiplicative matrix norm, denoted by $\|\cdot\|_\rho$, such that $\|T\|_\rho = c < 1$ [76]. Using this norm, and the triangle inequality, we have

$$\|z_{n+1} - z\|_\rho \le \|T\|_\rho\|z_n - z\|_\rho + \epsilon\|\Delta_n\|_\rho\|z_n - z\|_\rho + \epsilon\|\Delta_n\|_\rho\|z\|_\rho + \epsilon\|\delta_n\|_\rho. \tag{A.5}$$

In the limit, as $n \to \infty$, we can choose $\|\Delta_n\|_\rho \le 1$ and $\|\delta_n\|_\rho \le 1$, which implies that

$$\|z_{n+1} - z\|_\rho \le (\|T\|_\rho + \epsilon)\|z_n - z\|_\rho + \epsilon(\|z\|_\rho + 1), \text{ as } n \to \infty. \tag{A.6}$$

But because $\|T\|_\rho < 1$, we can select ϵ small enough such that $\|T\|_\rho + \epsilon < 1$ and, hence,

$$\lim_{n\to\infty} \|z_{n+1} - z\|_\rho = \epsilon(\|z\|_\rho + 1)/(1 - \|T\|_\rho - \epsilon). \tag{A.7}$$

Because ϵ can be chosen arbitrarily small, it follows that $\|z_{n+1} - z\| \to 0$ as $n \to \infty$. ■

Lemma A.2 (Stability of $\mathcal{F}$). *For any left-stochastic matrix $\mathcal{A}$ and block diagonal matrix $\mathcal{D}$ with the 2-induced norms of its blocks strictly bounded by one, it holds that the matrix product $\mathcal{B} = \mathcal{A}^T\mathcal{D}$ is stable. Consequently, when $\|(I_N - P_k^+ S_k)F\|_2 < 1$, the matrix $\mathcal{F}$ defined below is stable:*

$$\mathcal{F} \stackrel{\Delta}{=} \mathcal{A}^T\mathcal{X}, \quad \mathcal{X} \stackrel{\Delta}{=} (I_{KN} - \mathcal{P}^+\mathcal{S})(I_K \otimes F). \tag{A.8}$$

Proof. This argument adjusts the proof given for Lemma 2 in [12], where the ρ-norm should be replaced by the block-maximum norm defined as follows [77]. Let $x = \text{col}\{x_1, x_2, \ldots, x_K\}$ denote a $K \times 1$ *block* column vector whose individual entries are themselves vectors of size $N \times 1$ each. The block-maximum norm of x is denoted by $\|x\|_{b,\infty}$ and is defined as $\|x\|_{b,\infty} = \max_{1\le k\le K} \|x_k\|$. This vector norm induces a block-maximum matrix norm. Let $\mathcal{A}$ denote an arbitrary $K \times K$ block matrix with individual block entries of size $N \times N$ each. Then, the block-maximum norm of $\mathcal{A}$ is defined as

$$\|\mathcal{A}\|_{b,\infty} \triangleq \max_{x \neq 0} \frac{\|\mathcal{A}x\|_{b,\infty}}{\|x\|_{b,\infty}}. \tag{A.9}$$

The block-maximum norm has several useful properties; see [77]. In particular, when A is $K \times K$ left-stochastic and $\mathcal{A} = A \otimes I_N$, then it can be verified that $\|\mathcal{A}^T\|_{b,\infty} = 1$. Likewise, when $\mathcal{D}$ is block diagonal with the 2-induced norm of its diagonal blocks bounded by one, it is easy to check that $\|\mathcal{D}\|_{b,\infty} < 1$. Consequently, for $\mathcal{B} = \mathcal{A}^T\mathcal{D}$ we get

$$\rho(\mathcal{B}) \leq \|\mathcal{B}\|_{b,\infty} \leq \|\mathcal{A}^T\|_{b,\infty}\|\mathcal{D}\|_{b,\infty} < 1, \tag{A.10}$$

and we conclude that $\mathcal{B}$ is stable. Alternatively, we note that

$$\|\mathcal{B}^n\|_{b,\infty} \leq \left(\|\mathcal{A}^T\|_{b,\infty}\right)^n \left(\|\mathcal{D}\|_{b,\infty}\right)^n \longrightarrow 0, \text{ as } n \to \infty. \tag{A.11}$$

The matrix $\mathcal{F}$ in Eq. (A.8) has a form that fits into this formulation with $\mathcal{D} = (I_{KN} - \mathcal{P}^+\mathcal{S})(I_K \otimes F) = \mathcal{X}$, which is block diagonal. Moreover, because by assumption $\|(I_{KN} - P_k^+ S_k)F\|_2 < 1$, it holds that $\|\mathcal{D}\|_{b,\infty} = \|\mathcal{X}\|_{b,\infty} < 1$. ■

REFERENCES

[1] Zhu Y, You Z, Zhao J, Zhang K, Li XR. The optimality for the distributed Kalman filtering fusion with feedback. Automatica 2001;37(9):1489–93.

[2] Coates M. Distributed particle filters for sensor networks. In: Proceedings of the 3rd ACM/IEEE international symposium on information processing in sensor networks; 2004. p. 99–107.

[3] Spanos DP, Olfati-Saber R, Murray RM. Approximate distributed Kalman filtering in sensor networks with quantifiable performance. In: Proceedings of the 4th ACM/IEEE international symposium on information processing in sensor networks; 2005. p. 133–9.

[4] Ribeiro A, Giannakis GB, Roumeliotis SI. SOI-KF: distributed Kalman filtering with low-cost communications using the sign of innovations. IEEE Trans Signal Process 2006;54(12):4782–95.

[5] Olfati-Saber R. Distributed Kalman filtering for sensor networks. In: 46th IEEE conference on decision and control; 2007. p. 5492–8.

[6] Khan UA, Moura JM. Distributing the Kalman filter for large-scale systems. IEEE Trans Signal Process 2008;56(10):4919–35.

[7] Cattivelli FS, Lopes CG, Sayed AH. Diffusion strategies for distributed Kalman filtering: formulation and performance analysis. In: Proceedings of the IEEE IAPR workshop on cognitive information processing; 2008. p. 36–41.

[8] Cattivelli FS, Sayed AH. Diffusion mechanisms for fixed-point distributed Kalman smoothing. In: 16th European signal processing conference; 2008. p. 1–5.

[9] Carli R, Chiuso A, Schenato L, Zampieri S. Distributed Kalman filtering based on consensus strategies. IEEE J Sel Areas Commun 2008;26(4):622–33.

[10] Cattivelli F, Sayed AH. Diffusion distributed Kalman filtering with adaptive weights. In: Proceedings of Asilomar conference on signals, systems and computers; 2009. p. 908–12.

[11] Olfati-Saber R. Kalman-consensus filter: optimality, stability, and performance. In: Proceedings of the 48th IEEE conference on decision and control, held jointly with the 28th Chinese control conference; 2009. p. 7036–42.

[12] Cattivelli FS, Sayed AH. Diffusion strategies for distributed Kalman filtering and smoothing. IEEE Trans Autom Control 2010;55(9):2069–84.

[13] Hlinka O, Slučiak O, Hlawatsch F, Djurić PM, Rupp M. Likelihood consensus and its application to distributed particle filtering. IEEE Trans Signal Process 2012;60(8):4334–49.
[14] Djurić PM, Wang Y. Distributed Bayesian learning in multiagent systems: improving our understanding of its capabilities and limitations. IEEE Signal Process Mag 2012;29(2):65–76.
[15] Sayed AH. Adaptation, learning, and optimization over networks. In: Foundations and trends® in machine learning, vol. 7(4–5); 2014. p. 311–801.
[16] Zhao F, Guibas LJ. Wireless sensor networks: an information processing approach. Amsterdam, The Netherlands: Morgan Kaufmann; 2004.
[17] Hlinka O, Hlawatsch F, Djurić PM. Distributed particle filtering in agent networks: a survey, classification, and comparison. IEEE Signal Process Mag 2013;30(1):61–81.
[18] Sayed AH. Adaptive networks. Proc IEEE 2014;102(4):460–97.
[19] Anderson BDO, Moore JB. Optimal filtering. Englewood Cliffs, NJ: Prentice Hall; 1979.
[20] Arulampalam MS, Maskell S, Gordon N, Clapp T. A tutorial on particle filters for online nonlinear/non-Gaussian Bayesian tracking. IEEE Trans Signal Process 2002;50(2):174–88.
[21] Li XR, Jilkov VP. Survey of maneuvering target tracking. Part I. Dynamic models. IEEE Trans Aerosp Electron Syst 2003;39(4):1333–64.
[22] Tu SY, Sayed AH. Diffusion strategies outperform consensus strategies for distributed estimation over adaptive networks. IEEE Trans Signal Process 2012;60(12):6217–34.
[23] Towfic ZJ, Chen J, Sayed AH. Excess-risk of distributed stochastic learners. IEEE Trans Inf Theory 2016;62(10):5753–85.
[24] Xiao L, Boyd S, Lall S. A scheme for robust distributed sensor fusion based on average consensus. In: Proceedings of the 4th ACM/IEEE international symposium on information processing in sensor networks, Los Angeles, CA; 2005. p. 63–70.
[25] Olfati-Saber R, Fax JA, Murray RM. Consensus and cooperation in networked multi-agent systems. Proc IEEE 2007;95(1):215–33.
[26] Aysal TC, Yildiz ME, Sarwate AD, Scaglione A. Broadcast gossip algorithms for consensus. IEEE Trans Signal Process 2009;57(7):2748–61.
[27] Dimakis AG, Kar S, Moura JMF, Rabbat MG, Scaglione A. Gossip algorithms for distributed signal processing. Proc IEEE 2010;98(11):1847–64.
[28] Kailath T, Sayed AH, Hassibi B. Linear estimation. Upper Saddle River, NJ: Prentice Hall; 2000.
[29] Sayed AH. Adaptive filters. Wiley; 2008.
[30] Chang KC, Saha RK, Bar-Shalom Y. On optimal track-to-track fusion. IEEE Trans Aerosp Electron Syst 1997;33(4):1271–6.
[31] Atherton DP, Bather JA, Briggs AJ. Data fusion for several Kalman filters tracking a single target. IEE Proc Radar Sonar Navig 2005;152(5):372–6.
[32] Mitchell HB. Multi-sensor data fusion: an introduction. Springer; 2007.
[33] Julier SJ, Uhlmann JK. A non-divergent estimation algorithm in the presence of unknown correlations. In: Proceedings of the American control conference, vol. 4; 1997. p. 2369–73.
[34] Uhlmann JK. Covariance consistency methods for fault-tolerant distributed data fusion. Inf Fusion 2003;4(3):201–15.
[35] Hu J, Xie L, Zhang C. Diffusion Kalman filtering based on covariance intersection. IEEE Trans Signal Process 2012;60(2):891–902.
[36] Kay SM. Fundamentals of statistical signal processing: estimation theory. Upper Saddle River, NJ: Prentice Hall; 1993.
[37] Djurić PM, Khan M, Johnston DE. Particle filtering of stochastic volatility modeled with leverage. IEEE J Sel Top Signal Process 2012;6(4):327–36.
[38] Hlinka O, Hlawatsch F. Distributed particle filtering in the presence of mutually correlated sensor noises. In: Proceedings of IEEE international conference on acoustics, speech and signal processing; 2013. p. 6269–73.

[39] Moldaschl M, Gansterer WN, Hlinka O, Meyer F, Hlawatsch F. Distributed decorrelation in sensor networks with application to distributed particle filtering. In: Proceedings of IEEE international conference on acoustics, speech and signal processing. 2014. p. 6117–21.
[40] Urteaga I, Djurić PM. Sequential estimation of hidden ARMA processes by particle filtering: Part I. IEEE Trans Signal Process 2017;65(2):482–93.
[41] Urteaga I, Djurić PM. Sequential estimation of hidden ARMA processes by particle filtering: Part II. IEEE Trans Signal Process 2017;65(2):494–504.
[42] Tanizaki H. Nonlinear filters: estimation and applications. Berlin, Germany: Springer; 1996.
[43] Alspach D, Sorenson H. Nonlinear Bayesian estimation using Gaussian sum approximations. IEEE Trans Autom Control 1972;17(4):439–48.
[44] Julier SJ, Uhlmann JK. Unscented filtering and nonlinear estimation. Proc IEEE 2004;92(3):401–22.
[45] van der Merwe R. Sigma-point Kalman filters for probabilistic inference in dynamic state-space models. Ph.D. thesis, OGI School of Science and Engineering, Oregon Health and Science University, Hillsboro, OR; 2004.
[46] Arasaratnam I, Haykin S. Cubature Kalman filters. IEEE Trans Autom Control 2009;54(6):1254–69.
[47] Gordon NJ, Salmond DJ, Smith AFM. Novel approach to nonlinear/non-Gaussian Bayesian state estimation. IEE Proc F (Radar Signal Process) 1993;140(2):107–13.
[48] Djurić PM, Kotecha JH, Zhang J, Huang Y, Ghirmai T, Bugallo MF, Míguez J. Particle filtering. IEEE Signal Process Mag 2003;20(5):19–38.
[49] Cappé O, Godsill SJ, Moulines E. An overview of existing methods and recent advances in sequential Monte Carlo. Proc IEEE 2007;95(5):899–924.
[50] Bugallo MF, Elvira V, Martino L, Luengo D, Míguez J, Djurić PM. Adaptive importance sampling: the past, the present, and the future. IEEE Signal Process Mag 2017;34(4):60–79.
[51] Li T, Bolic M, Djurić PM. Resampling methods for particle filtering: classification, implementation, and strategies. IEEE Signal Process Mag 2015;32(3):70–86.
[52] Oreshkin BN, Coates MJ. Asynchronous distributed particle filter via decentralized evaluation of Gaussian products. In: Proceedings of FUSION; 2010.
[53] Farahmand S, Roumeliotis SI, Giannakis GB. Set-membership constrained particle filter: distributed adaptation for sensor networks. IEEE Trans Signal Process 2011;59(9):4122–38.
[54] Üstebay D, Coates M, Rabbat M. Distributed auxiliary particle filters using selective gossip. In: Proceedings of IEEE international conference on acoustics, speech and signal processing; 2011. p. 3296–9.
[55] Mohammadi A, Asif A. Consensus-based distributed unscented particle filter. In: Proceedings of IEEE statistical signal processing workshop; 2011. p. 237–40.
[56] Hlinka O, Hlawatsch F, Djurić PM. Consensus-based distributed particle filtering with distributed proposal adaptation. IEEE Trans Signal Process 2014;62(12):3029–41.
[57] Bordin CJ, Bruno MGS. Consensus-based distributed particle filtering algorithms for cooperative blind equalization in receiver networks. In: Proceedings of IEEE international conference on acoustics, speech and signal processing; 2011. p. 3968–71.
[58] Bandyopadhyay S, Chung SJ. Distributed estimation using Bayesian consensus filtering. In: Proceedings of American control conference; 2014. p. 634–41.
[59] Mohammadi A, Asif A. Distributed consensus + innovation particle filtering for bearing/range tracking with communication constraints. IEEE Trans Signal Process 2015;63(3):620–35.
[60] Wang H, Djurić PM. Diffusion in networks by cooperative particle filtering. In: Proceedings of the IEEE workshop on computational advances in multi-sensor adaptive processing; 2017.
[61] Yu JY, Coates MJ, Rabbat MG, Blouin S. A distributed particle filter for bearings-only tracking on spherical surfaces. IEEE Signal Process Lett 2016;23(3):326–30.
[62] Li J, Nehorai A. Distributed particle filtering via optimal fusion of Gaussian mixtures. IEEE Trans Signal Inf Process Netw 2018;4(2):280–92.

[63] Vázquez MA, Míguez J. A robust scheme for distributed particle filtering in wireless sensors networks. Signal Process 2017;131:190–201.

[64] Savic V, Wymeersch H, Zazo S. Belief consensus algorithms for fast distributed target tracking in wireless sensor networks. Signal Process 2014;95:149–60.

[65] Gautschi W. Orthogonal polynomials: computation and approximation. Oxford, UK: Oxford University Press; 2004.

[66] Unser M. Splines: a perfect fit for signal and image processing. IEEE Signal Process Mag 1999;16(6):22–38.

[67] Mosk-Aoyama D, Shah D. Fast distributed algorithms for computing separable functions. IEEE Trans Inf Theory 2008;54(7):2997–3007.

[68] Xiao L, Boyd S. Fast linear iterations for distributed averaging. Syst Control Lett 2004;53(1):65–78.

[69] Björck Å. Numerical methods for least squares problems. Philadelphia, PA: SIAM; 1996.

[70] Lawson CL, Hanson RJ. Solving least squares problems. Philadelphia, PA: SIAM; 1995.

[71] Bishop CM. Pattern recognition and machine learning. New York, NY: Springer; 2006.

[72] Kotecha JH, Djurić PM. Gaussian particle filtering. IEEE Trans Signal Process 2003;51(10):2592–601.

[73] Bolić M, Athalye A, Hong S, Djurić PM. Study of algorithmic and architectural characteristics of Gaussian particle filters. J Signal Process Syst 2010;61(2):205–18.

[74] Gales MJF, Airey SS. Product of Gaussians for speech recognition. Comput Speech Lang 2006;20(1):22–40.

[75] Ito K, Xiong K. Gaussian filters for nonlinear filtering problems. IEEE Trans Autom Control 2000;45(5): 910–27.

[76] Horn RA, Johnson CR. Matrix analysis. Cambridge University Press; 1990.

[77] Sayed AH. Diffusion adaptation over networks. In: Chellapa R, Theodoridis S, editors. Academic Press library in signal processing, vol. 3. Academic Press, Elsevier; 2014. p. 323–454.

CHAPTER

GAME THEORETIC LEARNING 7

Ceyhun Eksin*, Brian Swenson†, Soummya Kar†, Alejandro Ribeiro‡
*Department of Industrial and Systems Engineering, Texas A&M University, College Station, TX, United States**
Department of Electrical and Computer Engineering, Carnegie Mellon University, Pittsburgh, PA, United States†
Department of Electrical and Systems Engineering, University of Pennsylvania, Philadelphia, PA, United States‡

7.1 INTRODUCTION

A prevalent idea in game theory is that a Nash equilibrium or rational behavior is a result of dynamic processes of adaptation. As Arrow succinctly puts it: "The attainment of equilibrium process requires a disequilibrium process." This reflects the belief that individuals/players in a game learn to act rationally through repetitive decision-making in the same situation. Based on this idea, there is an active research area known as *learning in games* that focuses on developing individual-based adaptive dynamics. The main objective is to show that bounded-rational decision-making rules can, over time, lead to a Nash equilibrium.

A game consists of a set of players, each with a set of available actions, and a utility function that determines its preferences. At a Nash equilibrium action profile, each player's action is optimal with respect to its utility function, given the actions of the other players in the game. In the setup of the learning in games, players repetitively select actions based on past observations according to a simple decision-making rule. Once players select their actions, the game is played, and the players acquire new information through their observations or communications. In this setup, one can interpret the motivation of reaching rational behavior as the players' aspiration to learn the eventual actions of other players, and play optimally with respect to them.

Traditionally, the learning in games literature has considered perfect global information access, that is, at the end of each play individuals observe actions of other players or their realized payoff, which depends on others' action profiles. A more realistic setting is that the players have partial and noisy access to game parameters and the history of game play. Considering the objective of justifying the study of the Nash equilibrium concept in economics, the development of simple decision-making rules based on partial and noisy information access is of value.

In addition, such partial and noisy information-based individual decision-making rules are potentially applicable to the design of agent behavior in distributed engineering systems. Next, we present several examples of engineering systems where game theory is relevant to motivate game-theoretic learning algorithms robust to partial and noisy information access. For each system, we define the key elements of a game, that is, the players and their action sets and preferences.

Cooperative and Graph Signal Processing. https://doi.org/10.1016/B978-0-12-813677-5.00007-9

Power control in wireless networks [1,2]. Users of a common radio channel are the players. Users compete to maximize their signal-to-interference-noise ratio (SINR) by selecting their power levels.
Cognitive MIMO radio networks [3]. A cognitive radio system aims to increase the usage of frequency resources by having secondary users that are allowed to utilize the channels in the absence of primary users. In this system a game is played among secondary users with the goal to maximize the information rate by selecting an appropriate transmission covariance matrix given power limitations.
Target assignment [4]. The players are a team of mobile robots. Each robot selects a subset of available targets with the goal of maximizing the team's collective damage.
Demand response management in energy systems [5,6]. End users of a smart grid are the players. The smart grid operator employs a pricing strategy that depends on the total consumption of its users. Each user makes hourly consumption decisions to minimize the cost of energy.
HVAC control in smart buildings [7]. A smart-energy building consists of several autonomous units. HVAC units of a building are the players in a game that collectively would like to minimize power consumption while keeping room temperature at desired levels.

In all these examples, the Nash equilibrium (NE) is a model of normative, and sometimes, descriptive individual behavior. Yet, how players reach the Nash equilibrium is of concern. It is impractical to assume perfect global information access where players observe or share the past actions of every other player. Further, there often exist uncertainties in preference-relevant parameters. With these considerations in mind, in this chapter we consider generalizations of the traditional learning in games setup where individuals have local interactions over a network and have noisy information about their preferences.

Specifically, individual preferences represented by utility functions depend on their own action, the actions of other players, and a state of the world unknown to players. The goal is for each player to learn about the strategies of other players and the state of the world by repetitively playing the game and exchanging information with its neighbors in the network (see Fig. 7.1 for an illustration). A player's learning problem would have been a standard inference problem had the other players selected actions from a stationary distribution. However, this is not the case. The fact that all players are trying to infer about each other's future strategy creates a nonstationary process and makes the inference problem atypical. In particular, a player's self-actions affect other players' information, and hence their future actions. In turn, this affects the information that the player receives and its forecasts about other's future

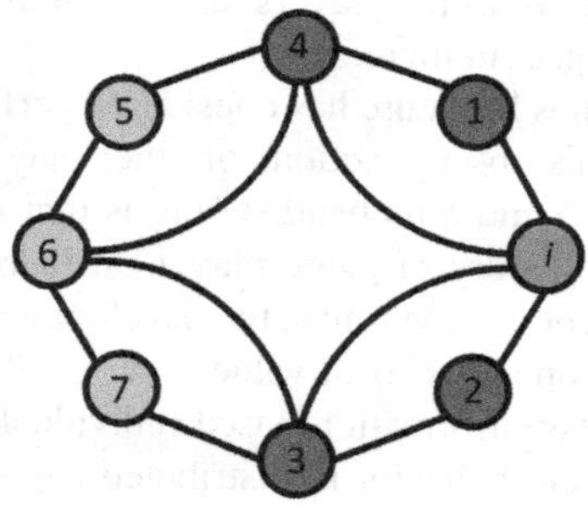

FIG. 7.1

Network learning of player i. Player i exchanges information with its neighbors $\{1, 2, 3, 4\}$ to learn about the future actions of all the players $\{1, 2, \ldots, 7\}$.

play. The networked information access adds delays, further complicating the complex feedback effects between own forecasts and actions and the future behavior of other players.

This chapter is organized as follows: We first present preliminaries of game theory and the setup of learning in games (Section 7.2). In Sections 7.3 and 7.4, we respectively consider networked learning dynamics in complete and incomplete information games. The difference between the complete and incomplete information games is that in the latter, each player has differing information about the state of the world.

7.2 LEARNING IN GAMES PRELIMINARIES

The basic elements of a game include a set of players $\mathcal{N} = \{1, 2, \ldots, N\}$, and for each player $i \in \mathcal{N}$ an action set $\mathcal{A}_i$ and a utility function $u_i : \mathcal{A} \to \mathbb{R}$ where $\mathcal{A} = \prod_{i=1}^{n} \mathcal{A}_i$ is the action profile set (see Table 7.1 for notational conventions used throughout the chapter). Accordingly, we define a complete information game with the tuple $\Gamma := (\mathcal{N}, \mathcal{A}, \{u_i\}_{i\in\mathcal{N}})$. We use $-i$ to denote all players other than player i. We sometimes refer to the action profile $a \in \mathcal{A}$ as $a = (a_i, a_{-i})$ where $a_{-i} \in \mathcal{A}_{-i}$ are the actions of players other than i.

Example 7.1 (Target assignment game). Two robots ($\mathcal{N} = \{1, 2\}$) would like to coordinate on covering the left and right doors of a room ($\mathcal{A}_i = \{L, R\}$ for $i \in \mathcal{N}$). The utility function is given by $u_i(L, L) = u_i(R, R) = 0$ for $i \in \mathcal{N}$, $u_1(R, L) = 2$ and $u_1(L, R) = 2$, and $u_2(R, L) = 1$ and $u_2(L, R) = 4$. The utility function values are given in normal form in Table 7.2, where row and column players are players 1 and 2, respectively. ■

When action profile $a \in \mathcal{A}$ yields a higher utility than the action profile $a' \in \mathcal{A}$, $u_i(a) > u_i(a')$, player i prefers a over a'. A best response action of player i to the actions of other players a_{-i} belongs to the best response set defined as

Table 7.1 Notation

Notation	Description
$\mathbb{R}$:	real numbers
$\mathbf{1}$:	indicator function
t:	continuous time
$E[\cdot]$:	expectation operator
$\mathcal{N}$:	set of players
$\Delta(\cdot)$:	set of probability distributions over a set
y_n:	value of variable y at time n
1_N:	$N \times 1$ ones vector
(x, y):	tuple of x and y
n:	discrete time
$\mathcal{X}$:	a set
i, j:	player indices
y_i:	a variable belonging to i
$\{x, y\}$:	set of x and y

Table 7.2 Normal Form of the Target Assignment Game

		Robot 2	
		Left	Right
Robot 1	Left	(0,0)	(2,4)
	Right	(2,1)	(0,0)

$$BR_i(a_{-i}) := \operatorname*{argmax}_{a_i \in \mathcal{A}_i} u_i(a_i, a_{-i}). \tag{7.1}$$

We define the joint best response function as $BR : \mathcal{A} \to \mathcal{A}$ with its ith element given by BR_i.

An action profile $a^* \in \mathcal{A}$ is a pure Nash equilibrium when no player has a profitable unilateral deviation, that is, for all $i = 1, 2, \ldots, n$

$$u_i(a_i^*, a_{-i}^*) \geq u_i(a_i, a_{-i}^*) \quad \forall\, a_i \in \mathcal{A}_i. \tag{7.2}$$

In the target assignment game, there are two pure Nash equilibria (Left, Right) and (Right, Left).

We can also represent a pure NE action profile $a^* \in \mathcal{A}$ as the fixed point of the best response set, where for every player i

$$a_i^* \in BR_i(a_{-i}^*). \tag{7.3}$$

Players can also randomize among their actions, implementing mixed strategies. A mixed strategy σ_i is a probability distribution over the action space of player i denoted as $\Delta(\mathcal{A}_i)$ where $\sigma_i(a_i)$ is the probability that player i selects action $a_i \in \mathcal{A}_i$. We denote the mixed strategy profile of the players as $\sigma = \{\sigma_1, \ldots, \sigma_n\} \in \Delta(\mathcal{A})$. The expected utility of player i given the strategy profile $\sigma \in \Delta(\mathcal{A})$ is defined as

$$u_i(\sigma) = \sum_{a \in \mathcal{A}} u_i(a)\sigma(a). \tag{7.4}$$

Accordingly, a (mixed) NE is a strategy profile $\sigma \in \Delta(\mathcal{A})$ such that for any player $i = 1, 2, \ldots, N$

$$u_i(\sigma_i, \sigma_{-i}) \geq u_i(\sigma_i', \sigma_{-i}) \quad \forall \sigma_i' \in \Delta(\mathcal{A}_i). \tag{7.5}$$

As in the fixed point definition in Eq. (7.3), we can equivalently represent the above NE definition using the joint best response mapping,

$$\sigma \in BR(\sigma). \tag{7.6}$$

We define the set of NE strategies as $NE := \{\sigma \in \Delta(\mathcal{A}) : \sigma \in BR(\sigma)\}$. This set is nonempty for games with finite action spaces [8].

The complete information game $\Gamma(\mathcal{N}, \mathcal{A}, \{u_i\}_{i \in \mathcal{N}})$ assumes players know their utility functions exactly. However, this is not realistic in various scenarios. For instance, in the target assignment game considered in Example 7.1, the robots might have noisy information about their own utility and/or the utility of the other player. In the following, we introduce incomplete information games to model such scenarios of uncertainty.

The utility function depends on preference relevant states of the world, which we denote as θ belonging to a space Θ. That is, $u_i : \mathcal{A} \times \Theta \to \mathbb{R}$. In addition to the player set, action sets, and utility functions, a characterizing element of the incomplete information game is the information (type) of each player denoted by t_i, which belongs to a type space $\mathcal{T}_i$. Define the set of information of players as $\mathcal{T} := \prod_{i=1}^{n} \mathcal{T}_i$. Assume the game has a common prior $\pi \in \Delta(\Theta \times \mathcal{T})$ over the state and information space. Define the interim belief of player i as $p_i \in \Delta(\Theta \times T)$ as the conditional probability distribution $p_i(\cdot) := \pi(\cdot \mid t_i)$. A game of incomplete information is defined by the tuple $\mathcal{I} := (\mathcal{N}, \mathcal{A}, \Theta, \{u_i\}_{i\in\mathcal{N}}, \pi, \mathcal{T})$.

Example 7.2 (Target assignment game incomplete information). Consider the setup in Example 7.1. Assume there are two rooms $\theta \in \{A, B\}$. Each room has a Left and Right door. When the room is A, players have the same utility function as in Example 7.1. When the room is B, the players' payoffs are flipped. The utility function values are given in normal form in Table 7.3. Assume that robots have a common uniform prior about which room they are in. Further, each robot receives a private signal $x_i = \{A, B\}$ for $i = 1, 2$ that reveals the correct room with some positive probability. The incomplete information game is characterized by the tuple $(\mathcal{N}, \mathcal{A}, \Theta, \{u_i\}_{i\in\mathcal{N}}, \pi, \{x_i\}_{i\in\mathcal{N}})$ where the prior π is the uniform distribution over $\Theta = \{A, B\}$, and the type of the player i is its private signal x_i. ■

The strategy of player i in a game of incomplete information is a mapping from its type to the action space, i.e., $\sigma_i : \mathcal{T}_i \to \mathcal{A}_i$. The expected utility of player i from selecting action $a_i \in \mathcal{A}_i$ against others' strategy profile σ_{-i} is given by

$$E[u_i(a_i, \sigma_{-i}, \theta) \mid t_i] = \sum_{\theta\in\Theta, t_{-i}\in\mathcal{T}_{-i}} u_i(a_i, \sigma_{-i}(t_{-i}), \theta) p_i(\theta, t_{-i}), \tag{7.7}$$

where we make the utility function's dependence on the state of the world θ explicit by writing $u_i(a, \theta) : \mathcal{A} \times \Theta \to \mathbb{R}$. A strategy profile $\sigma^* : \mathcal{T} \to \mathcal{A}$ is a Bayesian-Nash equilibrium (BNE) of $\mathcal{I}$ if and only if for all $i \in \mathcal{N}$, $a_i \in \mathcal{A}_i$, and $t_i \in \mathcal{T}_i$

$$E[u(\sigma_i^*, \sigma_{-i}^*, \theta) \big| t_i] \geq E[u(a_i, \sigma_{-i}^*, \theta) \big| t_i]. \tag{7.8}$$

That is, σ^* is a BNE if and only if each agent plays a best response to other players' strategies given its interim belief at any possible realization of its information.

When players only know the common prior π and nothing else, the game is a complete information game with utility functions defined as expectations of u_i computed with respect to the prior π. That is, define the expected utility of i with respect to prior as $\hat{u}_i(a) := \sum_{\theta\in\Theta} u_i(a, \theta)\pi(\theta)$. Then the complete information game is given by the tuple $\Gamma(\pi) = (\mathcal{N}, \mathcal{A}, \Theta, \{u_i\}_{i\in\mathcal{N}}, \pi) = (\mathcal{N}, \mathcal{A}, \{\hat{u}_i\}_{i\in\mathcal{N}})$.

Table 7.3 Target Assignment Game With Two Rooms

	A				B		
		Robot 2				Robot 2	
		Left	Right			Left	Right
Robot 1	Left	(0,0)	(2,4)	Robot 1	Left	(0,0)	(2,1)
	Right	(2,1)	(0,0)		Right	(2,4)	(0,0)

The incomplete information game $\mathcal{I}$ with a finite set of types $t \in \mathcal{T}$ can be viewed as another complete information game Γ' with each possible information (type) representing a player. That is, define $\mathcal{N}' = \bigcup_{i \in \mathcal{N}} \mathcal{T}_i$ and the action space of player $t_i \in \mathcal{T}_i$ as $\mathcal{A}_{t_i} = \mathcal{A}_i$, with utility function $\hat{u}_{t_i}(\sigma, \theta) = E[u_i(\sigma, \theta)|t_i]$ with the expectation operator defined as in Eq. (7.7). Then in the complete information game $\Gamma' = (\mathcal{N}', \{\mathcal{A}_{t_i}, \hat{u}_{t_i}\}_{t_i \in \mathcal{N}'})$ each type t_i is a different player taking an action a_i from $\mathcal{A}_i$. Hence, the strategy profile of player $i \in \mathcal{N}$, $\sigma_i : \mathcal{T}_i \to \mathcal{A}_i$ in the incomplete information game $\mathcal{I}$ is a list of actions of types in $\mathcal{T}_i$ in the complete information game Γ'.

Learning dynamics

Consider the complete information game with common prior π on the state of the world $\theta \in \Theta$, $\Gamma(\pi) = (\mathcal{N}, \mathcal{A}, \Theta, \{u_i\}_{i \in \mathcal{N}}, \pi)$, which we will refer as the stage game. The game $\Gamma(\pi)$ is repeatedly played over stages $n = 1, 2, \ldots$. At each stage n, there is a set of actions realized that we denote as $a_n = (a_{1,n}, \ldots, a_{N,n})$ where $a_{i,n} \in \mathcal{A}_i$. After each stage, each player acquires new information from the most recent stage game and possibly other exogenous information sources. A learning algorithm in game theory is a collection of individual assessment and behavior rules that respectively determines how players process information and select the next stage action.

Formally, an assessment rule of player i denoted by $\mu_{i,n}$ maps i's information to a probability distribution on the future play of other players and the state of the world $\Delta(\mathcal{A}_{-i} \times \Theta)$. A behavior rule denoted by $\Phi_{i,n}$ maps the player's assessment to an individual action.

Below we provide two simple examples of assessment and behavior rules for the target assignment game of Example 7.1.

Stubborn players. Suppose each robot observes the past action of the other robot. A simple assessment rule is to assume the other robot is going to switch doors in the next stage. According to this assessment rule, robot 1 thinks that if robot 2 selects Left (Right) door at stage n then robot 2 will select Right (Left) door at stage $n + 1$. That is, $\mu_{1,n}(L) = R$ and $\mu_{1,n}(R) = L$. The simplest behavior rule is to do nothing with an assessment and continue to select the same action at each stage. For instance, $\Phi_{i,n} = L$ for all $i = 1, 2$. In this example, the robots are stubborn, and there are no learning dynamics because of their behavior rule.

Cournot best response dynamics. The canonical example of learning dynamics is best response dynamics. In a general N-person game, each player observes the previous action of the other players $a_{-i,n}$. The assessment rule is that each player will repeat its previous action. That is, suppose $a_{-i,n} = a'_{-i} \in \mathcal{A}_{-i}$ then $\mu_{i,n}(a_{-i,n}) = a'_{-i}$. Player i's behavior rule is to take the action that maximizes its utility function, given its assessment of other players' next actions. That is, $\Phi_{i,n}(\mu_{i,n}(a_{-i,n})) = BR_i(\mu_{i,n}(a_{-i,n}))$ where $BR_i(\cdot)$ is as defined in Eq. (7.1).

The eventual action of robots in the target assignment game under the best response dynamics is dependent on the initial actions chosen. Suppose both players select Left door initially, then the best response dynamics lead to a cycle between action profiles where the action profile $a_{i,n} = (L, L)$ is repeatedly followed by $a_{i,n+1} = (R, R)$. However, if robots initially select different actions, i.e., play action profile (L, R) or (R, L), then they continue to select their initial action. This is because both action profiles are equilibria, and at the NE action profile, players best respond to other player's actions as per the definition in Eq. (7.3).

We note that, in both the examples above, the information that players receive at the end of each stage is the other players' actions. In a networked communication setting, this information is not readily available to the players at the end of each stage.

Important classes of games

Restrictions to the structure of the utility functions are common when studying the convergence of the learning dynamics. In this chapter, we will reference the following classes of games: (i) potential, (ii) congestion, (iii) quadratic, (iv) zero-sum, and (v) supermodular games. These classes of games are relevant to applications in distributed engineering systems. In our exposition below, we omit the payoff relevant state of the world θ for notational brevity.

Potential games. Suppose there exists a "potential function" $u : \mathcal{A} \mapsto \mathbb{R}$ such that for all pairs of action profiles $a = (a_i, a_{-i})$ and $a' = (a_i', a_{-i})$, and players i, the *local* payoffs satisfy

$$u_i(a_i, a_{-i}) - u_i(a_i', a_{-i}) = u(a_i, a_{-i}) - u(a_i', a_{-i}). \tag{7.9}$$

The existence of the potential function u implies an "alignment of interests" among players because the joint action that maximizes u is a pure NE action profile of the game defined by the utilities u_i, $i \in \mathcal{N}$ [9].

Any technological system with a global objective can be modeled using a potential game. The target assignment problem [4] and consensus problems are only two examples among many.

Congestion games. Let $R = \{1, \ldots, m\}$ denote a set of resources. For each $i \in \mathcal{N}$, let $\mathcal{A}_i \subseteq 2^R$, where 2^R denotes the power set of R. In particular, an action choice a_i indicates a subset of resources being utilized by player i.

In a congestion game, the cost associated with using a resource is dependent on the total number of players using the same resource. For each $r \in R$, $a \in \mathcal{A}$, let $N_r(a) \in \mathbb{N}$ denote the number of players using resource r under the action profile a. More generally, for a subset of players $\mathcal{K} \subseteq \mathcal{N}$, the number of players in $\mathcal{K}$ utilizing resource r given $\{a_j\}_{j\in\mathcal{K}}$, is given by

$$N_r(\{a_j\}_{j\in\mathcal{K}}) := \sum_{j\in\mathcal{K}} \mathbf{1}(r \in a_j), \tag{7.10}$$

where $\mathbf{1}(r \in a_j) = 1$ if $r \in a_j$ and $\mathbf{1}(r \in a_j) = 0$ otherwise. Given a subset of players $\mathcal{K}$, and a corresponding set of actions $\{a_j\}_{j\in\mathcal{K}}$, we represent the number of players using each resource by $\#N(\{a_j\}_{j\in\mathcal{K}})$, where $\#N : \prod_{j\in\mathcal{K}} \mathcal{A}_j \to \mathbb{N}^m$ is a mapping with the rth entry in $\#N(\{a_j\}_{j\in\mathcal{K}})$ given by $N_r(\{a_j\}_{j\in\mathcal{K}})$.

For $r \in R$ and $k \in \mathbb{N}$, let $c_r(k)$ be the cost associated with using resource r given state θ_r, when there are precisely k players simultaneously using the resource. For $a_i \in \mathcal{A}_i$ and $N_r(a_{-i}) \in \mathbb{N}$, let the utility of player i be given by

$$u_i(a_i, a_{-i}) = -\sum_{r\in a_i} c_r(N_r(a_{-i}) + N_r(a_i)), \tag{7.11}$$

where we have written $N_r(a) = N_r(a_{-i}) + N_r(a_i)$ explicitly to emphasize dependence of the utility on the player i's action a_i and the actions of other players a_{-i}. Note that within the class of congestion games, players do not need to know the full action profile $a = (a_1, \ldots, a_n) \in \mathcal{A}$ precisely to compute their utility. It is sufficient for each player to have knowledge of $\#N(a_{-i}) \in \mathbb{N}^m$ and their own action $a_i \in \mathcal{A}_i$.

Congestion games are a subclass of the potential games. Internet congestion control [10] and transportation network problems can be modeled using congestion games.

Quadratic games. The quadratic utility function is the sum of a quadratic term and a bilinear term,

$$u_i(a_i, a_{-i}) := -\frac{1}{2}a_i^2 + \sum_{j \in \mathcal{N}\setminus\{i\}} \beta_{ij} a_i a_j, \tag{7.12}$$

where $\beta_{ij} \in \mathbb{R}$ for all $i \in \mathcal{N}, j \in -i$ are real valued constants. Notice that because $\partial^2 u_i / \partial a_i^2 = -1 < 0$, the payoff function in Eq. (7.12) is strictly concave with respect to the self action a_i of agent i. Quadratic utility functions are ubiquitous in stochastic optimal control [11,12] and distributed estimation [13,14].

Zero-sum games. There are two players $\mathcal{N} = \{1, 2\}$ where $u_1(a) = -u_2(a)$ for any action profile $a \in \mathcal{A}$.

Supermodular games. A utility function $u_i : \mathbb{R}^N \to \mathbb{R}$ is supermodular if

$$u_i(\min\{x, y\}) + u_i(\max\{x, y\}) \geq u_i(x) + u_i(y) \tag{7.13}$$

for all $x, y \in \mathbb{R}^N$, where $\min(\{x, y\})$ denotes the component-wise minimum and $\max(\{x, y\})$ denotes the component-wise maximum of x and y. The function is strictly supermodular if the inequality is strict for any incomparable pair of vectors x and y. If u_i is twice differentiable, strict supermodularity is equivalent to requiring that $\partial^2 f / \partial x_i \partial x_j > 0$ for all $1 \leq i < j \leq n$. Positive cross-partial derivatives mean that when a player increases its action, it incentivizes other players to increase their actions; see [15] for more on the theory of supermodular games.

Examples of supermodular games include currency attacks [16], power control problems in wireless networks [1], and arms race models [17].

7.3 NETWORK LEARNING ALGORITHMS FOR COMPLETE INFORMATION GAMES

In this section we study network-based learning in games with complete information. Recall that we represent a complete information game using the tuple $\Gamma = (\mathcal{N}, (\mathcal{A}, u_i)_{i \in \mathcal{N}})$. In traditional game-theoretic learning setups, players are assumed to be capable of having instantaneous access to all information required by the learning process. This can be an unrealistic and impractical assumption in many real-world setups. A more realistic assumption is to suppose that players are equipped with an overlaid communication graph through which they may disseminate all information relevant to the learning process. Formally, we say a learning algorithm is *network-based* if the following hold:

Assumption 7.1. *Players are endowed with a preassigned communication graph $G = (\mathcal{N}, E)$, where the vertices $\mathcal{N}$ correspond to the players, and the edge set E consists of communication links between pairs of players that can communicate directly. The graph G is connected. We define the neighbors of player i as $\mathcal{N}_i := \{j \in \mathcal{N} : (i, j) \in E\}$.* ■

Assumption 7.2. *Players directly observe only their own actions.* ■

Assumption 7.3. *A player may exchange information with immediate neighbors $\mathcal{N}_i$, as defined by G, at most once for each iteration or round of the repeated play.* ■

Remark 7.1. For clarity of presentation, in this chapter we focus on network-based learning algorithms with synchronous information dissemination schemes. However, it is straightforward to reformulate the above assumptions to allow for asynchronous information dissemination schemes. For instance, an asynchronous version of the Cournot best response dynamics entails each player randomly waking up to take best-response actions. Random activation of the players guarantees that players update at different times. Considering again the target assignment game example (Example 7.2), in the asynchronous best-response algorithm the robots converge to a Nash equilibrium of the game regardless of the initial action profile, in contrast to the synchronous version, which cycles starting from certain action profiles. Network-based learning in an asynchronous setting has been considered, for example, in [18,19].

As an example of a network-based learning setup, consider the following traffic routing scenario.

Example 7.3. Suppose a group of drivers wishes to navigate a traffic grid. Each driver's vehicle is equipped with an onboard computer and is connected to neighboring vehicles in the group via an ad hoc vehicular network. Each vehicle is equipped with a model of the traffic grid and knows its starting point and destination. Prior to physically engaging in the commute, the drivers wish to compute an equilibrium routing strategy. In particular, it is desired that, by repeatedly exchanging information with neighboring vehicles in the network, the group of vehicles can negotiate on a equilibrium routing strategy that can be recommended to the drivers.

The strategic problem of routing traffic can be modeled as a congestion game by letting each vehicle be represented as a player. The set of actions for each player is the set of route choices available to the vehicle. The utility function of each player is determined by travel time of the vehicle from starting point to destination, as determined by the onboard traffic grid model.[1] An equilibrium routing strategy defines a route choice (possibly probabilistic) for each vehicle that optimizes the utility (travel time) of each player (vehicle). ■

A prototypical network-based learning algorithm proceeds as follows. As in a centralized learning algorithm, each player chooses a sequence of actions $(a_{i,n})_{n\geq 1}$.

Initialize

(i) Each player i chooses an arbitrary initial action, $a_{i,1}$.

Iterate ($n \geq 1$)

(ii) Players engage in one round of information exchange with immediate neighbors.
(iii) Using the information obtained from neighbors, each player i chooses a new action $a_{i,n+1}$.

We say an algorithm converges to NE if the action sequence $(a_{i,n})_{n\geq 1}$ (or some function thereof) converges to the set of NE.

Remark 7.2. We emphasize that in a network-based algorithm, players are not assumed to be capable of viewing their instantaneous payoff information (see Assumption 7.2). Consequently, network-based learning differs significantly from distributed *payoff-based* learning schemes studied in the game-theoretic learning literature, e.g., [20]. In payoff-based learning setups, it is assumed that players physically engage in a game, and at the end of each round of game play, players may observe the payoff received. In this setting, information is implicitly transmitted by physical interaction in the game, and the challenge is to design algorithms that rely solely on the implicitly transmitted information.

[1] More precisely, we let the utility be the negative of the travel time so that players prefer to maximize their utility function.

In the network-based setting, players engage in a form of "virtual" game play in which no payoff measurements are available (e.g., Example 7.3). That is, information cannot be transmitted by the physical interaction in the game—all information dissemination must be carried out directly by the algorithm. The challenge here is to design algorithms that handle information dissemination in an efficient and practical manner.

The fictitious play (FP) algorithm is a canonical game-theoretic learning process [21,22]. In this section we will review methods for implementing the FP algorithm in network-based settings.

Our focus on FP here is motivated by several factors. First, FP is a prototypical game-theoretic learning algorithm—algorithms that rely on an underlying "myopic best response" structure are intimately related with classical FP dynamics, e.g., [23–26]. Such algorithms are used in a variety of settings including cognitive radio [27,28], deep learning [29], learning optimal strategies in large games [29–31], and distributed traffic routing [32,33]. Second, FP has important robustness properties that are crucial for network-based implementation of a learning algorithm [23,34]. These same robustness properties also allow for asynchronous implementation [34,35] as well as low-complexity randomized implementation [23,36]. Finally, we note that FP is accompanied with convergence rate estimates that can be useful when considering practical engineering applications in signal processing and control [37,38].

We note that, in addition to FP, a variety of network-based algorithms for learning in complete information games have been studied in the literature. For example, [39] studies a network-based algorithm for NE seeking in a two-network zero-sum game, [40] studies an algorithm for finding NE in a spatial spectrum access game, and [41] studies a network-based, regret-based reinforcement learning algorithm for tracking the polytope of correlated equilibria in time-varying games. The work [42] presents a method for designing games with a prescribed local dependence. The works [18,19] study gossip-based algorithms for computing NE in a network-based setting in games with continuous-action spaces. In the following section we introduce classical FP and review salient properties of the learning process relevant to engineering and network-based applications.

7.3.1 CLASSICAL FICTITIOUS PLAY

Consider the following dynamical system,

$$\dot{\sigma} \in BR(\sigma) - \sigma, \tag{7.14}$$

where σ is an absolutely continuous mapping from $\mathbb{R}$ to $\Delta(\mathcal{A})$.[2] Note that the rest points of this system coincide with the set of Nash equilibria as defined in Eq. (7.6). The dynamical system (7.14) is known as *continuous-time fictitious play* [37,69][3]. The discrete-time analog of this process is given by the system

---

[2]The existence of such a σ and the general well posedness of the differential inclusion (7.14) is discussed in [43].

[3]In the literature, the nomenclature used to refer to (7.14) is somewhat varied. Instead of continuous-time fictitious play, the system (7.14) is sometimes referred to as "best response dynamics" [53]. We prefer to use the term "continuous-time fictitious play" here to emphasize that the classical fictitious play algorithm is merely an Euler discretization of (7.14). We also note that authors sometimes use the term continuous-time fictitious play to refer to the closely related non-autonomous systems $\sigma(t) \in \frac{1}{t}\left(BR(\sigma(t)) - \sigma(t)\right)$ [37,69]. Since solutions of this later system are identical to solutions of (7.14) after a time change [37], we prefer to simply work directly with the autonomous system (7.14).

$$a_{n+1} \in BR(\sigma_n), \tag{7.15}$$

$$\sigma_n = \frac{1}{n}\sum_{s=1}^{n} \mathbf{1}_{a_s}, \tag{7.16}$$

where $\mathbf{1}_a$ denotes the vertex of $\Delta(\mathcal{A})$ placing weight 1 on the action $a \in \mathcal{A}$. Note that the discrete-time process $(\sigma_n)_{n\geq 1}$ may be expressed recursively as

$$\sigma_{n+1} - \sigma_n \in \frac{1}{n+1}\left(BR(\sigma_n) - \sigma_n\right). \tag{7.17}$$

This may be seen as an Euler discretization of the differential inclusion (7.14) in which the step size is given by the diminishing sequence $(\frac{1}{n+1})_{n\geq 1}$. As $n \to \infty$ the continuous-time interpolation of Eq. (7.17) more closely approximates a solution of Eq. (7.14). (This is made precise in [43].) In light of this relationship, Eq. (7.14) can be thought of as the *mean-field system* associated with Eq. (7.17).

We refer to an absolutely continuous mapping $\sigma : \mathbb{R} \to \Delta(\mathcal{A})$ satisfying Eq. (7.14) for almost every $t \in \mathbb{R}$ as a *continuous-time FP process* and we refer to a sequence $(\sigma_n)_{n\geq 1}$ satisfying Eq. (7.17) as a *discrete-time FP process*, or when there is no risk of confusion, simply as an FP process.

Our primary focus in this chapter will be on discrete-time learning algorithms. However, being a mean-field approximation, continuous-time FP serves an important role, both as a means of gaining intuition about the discrete-time system and as a means of rigorously deriving the asymptotic properties of the discrete-time system [43].

The following theorem states the fundamental Nash convergence result for FP.

Theorem 7.1. *Suppose* Γ *is a potential game, a zero-sum game, or a generic* $2 \times m$ *game. Then continuous-time FP converges to the set of Nash equilibria of* Γ *in the sense that* $\lim_{t\to\infty} d(\sigma(t), NE) = 0$. *Likewise, discrete-time FP converges to the set of NE in the sense that* $\lim_{n\to\infty} d(\sigma_n, NE) = 0$. ■

The class of potential games (see Eq. 7.9) is particularly important in the study of decentralized learning algorithms [44]; such games have an extremely broad range of applications in the field of multiagent systems [32,33,42,45–51]. In many engineering applications involving potential games, it can be preferable to converge to pure equilibria rather than mixed equilibria [4,52]. This issue has been studied in depth in the continuous-time setting. The following theorem demonstrates that while it is possible for (continuous-time) FP to converge to mixed equilibria, this is an exceptional occurrence.

Before stating the theorem, we remark that when we say a property holds for almost every game or almost every initial condition, we mean that the complement of the set where the property holds has Lebesgue measure zero [53,54].

Theorem 7.2. *For almost every potential game, and for almost every initial condition, continuous-time FP converges to a pure-strategy equilibrium.* ■

In order to apply game-theoretic learning algorithms in practical applications, it is important to understand the transient properties of the algorithm such as the rate of convergence. Given a game Γ, we say that a continuous-time FP process $(\sigma(t))_{t\geq 0}$ on Γ with initial condition $\sigma(0) = \sigma_0$ converges at an exponential rate if there exists a constant $c = c(\Gamma, \sigma_0)$ such that

$$d(\sigma(t), NE) \leq ce^{-t}.$$

The following result shows that the rate of convergence of (continuous-time) FP is generally exponential [37,38].

Theorem 7.3. *(i) In every zero-sum game and for every initial condition, the rate of convergence of (continuous-time) FP is exponential. (ii) For almost every potential game and for almost every initial condition, the rate of convergence of (continuous-time) FP is exponential.* ■

We note that while the results in Theorems 7.2 and 7.3 have been shown to hold for continuous-time FP, they have not yet been formally extended to the discrete-time case.[4]

Finally, we remark on an important robustness property of fictitious play that plays a critical role in network-based learning. In order to discuss the robustness property, we must first define the notion of an ϵ-best response. The ϵ-best response map of player i, $BR_\epsilon : \prod_{j\neq i} \Delta(\mathcal{A}_j) \to \Delta(\mathcal{A}_i)$ is given by

$$BR(\sigma_{-i}) := \left\{ \sigma_i \in \Delta(\mathcal{A}_i) : u_i(\sigma_i, \sigma_{-i}) \geq \max_{\sigma_i' \in \Delta(\mathcal{A}_i)} u_i(\sigma_i', \sigma_{-i}) - \epsilon \right\},$$

and the (joint) best response map $BR : \Delta(\mathcal{A}) \to \Delta(\mathcal{A})\Delta(\mathcal{A}_i)$ is given by

$$BR(\sigma) = BR_1(\sigma_{-1}) \times \cdots \times BR_N(\sigma_N).$$

A (discrete-time) process $(\sigma_n)_{n\geq 1}$ is said to be a *weakened* fictitious play process if

$$\sigma_{n+1} - \sigma_n \in \frac{1}{n+1} \left(BR_{\epsilon_n}(\sigma_n) - \sigma_n \right), \tag{7.18}$$

where $(\epsilon_n)_{n\geq 1}$ is a sequence of positive numbers representing "perturbations" to the FP process. Note that classical FP is a special case of Eq. (7.18) in which $\epsilon_n = 0 \ \forall n \geq 1$. The following theorem shows that FP is robust to the perturbation introduced by the sequence $(\epsilon_n)_{n\geq 1}$, so long as ϵ_n decays to zero.

Theorem 7.4. *Γ is a potential game, a zero-sum game, or a generic $2 \times m$ game. Suppose $(\sigma_n)_{n\geq 1}$ is a weakened FP process on Γ satisfying $\lim_{n\to\infty} \epsilon_n = 0$. Then the FP process converges to the set of Nash equilibria in the sense that $\lim_{n\to\infty} d(\sigma_n, NE) = 0$.* ■

This theorem follows, roughly, by studying the asymptotic properties of continuous-time FP and using the continuous-time process as a mean-field approximation of the discrete-time process [23,34].

The above robustness property allows for a breadth of practical applications, including low complexity implementations of FP (e.g., [36] or the actor-critic algorithm in [23]) and asynchronous implementations of FP [34]. Also, as we will see in the next section, it plays a critical role in ensuring convergence in network-based implementations of FP.

7.3.2 NETWORK-BASED FICTITIOUS PLAY

In a network-based setting, players are unable to directly observe the game action history or their own payoff history (see Remark 7.2). Information relevant to the learning process must be gradually disseminated using the overlaid communication graph. As a prototype of a network-based learning algorithm, in this section we consider the network-based implementation of FP.

---

[4] We note that modifications of the discrete-time FP process, such as the use of "inertia" [22,55], can be used to ensure theoretic convergence of discrete-time FP to pure equilibria.

Due to the network-based information dissemination scheme, players may not have perfect knowledge of the "true" empirical distribution[5] σ_n at all times. For each $i, j \in \mathcal{N}$, let $\hat{\sigma}^i_{j,n}$ denote an estimate that player i maintains of $\sigma_{j,n}$. Let $\hat{\sigma}^i_n = (\hat{\sigma}^i_{j,n})_{j \in \mathcal{N}}$ denote player i's estimate of σ_n. Players will form their estimates σ^i_n by exchanging information with neighbors. Let W denote a weight matrix to be used in the information sharing protocol. We denote its element in the ith row and jth column by $W_{i,j}$. We will assume the following.

Assumption 7.4. *The weight matrix W is an $N \times N$ matrix that is doubly stochastic, aperiodic, and irreducible, with sparsity conforming to the communication graph G.* ∎

A prototypical network-based implementation of FP is given below.

Initialize

(i) Each player i chooses an arbitrary initial action $a_i(1)$. The initial empirical distribution for player i is given by $\sigma_{i,1} = \mathbf{1}_{a_{i,1}}$. Each player initializes her local estimate of the empirical distribution as $\hat{\sigma}^i_{i,1} = \mathbf{1}_{a_{i,1}}$ and $\hat{\sigma}^i_{j,1} = 0, j \neq i$.

Iterate ($n \geq 1$)

(ii) Each player i chooses their next-stage action according to the rule

$$a_{i,n+1} \in BR_i(\hat{\sigma}^i_{-i,n}).$$

(iii) Each player i updates their personal empirical distribution as $\sigma_{i,n+1} = \sigma_{i,n} + \frac{1}{n+1}\left(\mathbf{1}_{a_{i,n+1}} - \sigma_{i,n}\right)$.

(iv) Each player i updates their estimate $\hat{\sigma}^i(n+1)$ according to

$$\hat{\sigma}^i_{j,n+1} = \sum_{k \in \mathcal{N}_i} W_{i,k}\left(\hat{\sigma}^k_{j,n} + (\sigma_{j,n+1} - \sigma_{j,n})\chi_{k=j}\right),$$

where $\chi_{k=j} = 1$ if $k = j$ and $\chi_{k=j} = 0$ otherwise.

Step (ii) is the behavioral rule in which agents best-respond given the assessment of others' behavior. Steps (iii) and (iv) constitute the assessment rule of the learning algorithm. In step (iv), player i does a weighted averaging of its neighbors' estimates of the empirical frequency of player j. If player j is a neighbor of player i, then player j appends the effect of its current action $(\sigma_{j,n+1} - \sigma_{j,n})$ to its estimate of self-empirical frequency. The effect of the current action on the empirical estimate is equal to $(\mathbf{1}_{a_{i,n+1}} - \sigma_{i,n})$ weighted by $1/(n+1)$ by step (iii).

The following theorem gives the convergence result for network-based FP [56].

Theorem 7.5. *Let Γ be a potential game. Then the network-based FP algorithm converges to the set of Nash equilibria in the sense that $\lim_{n\to\infty} d(\sigma_n, NE) = 0$. Furthermore, each player achieves asymptotic strategy learning in the sense that $\lim_{n\to\infty} d(\hat{\sigma}^i_n, NE) = 0$, $\forall i \in \mathcal{N}$; that is, players' estimates of σ_n converge to the set of Nash equilibria.* ∎

[5] An element of the sequence $(\sigma_n)_{n\geq 1}$, as defined in Eq. (7.16), is commonly referred to as the empirical distribution of player i, because σ_n tracks the empirical frequency with which player i uses each action.

The convergence of network-based FP follows from the robustness result given in Theorem 7.4. More precisely, the distributed information dissemination protocol used in step (iv) ensures that $\lim_{n\to\infty} \|\hat{\sigma}_n^i - \sigma_n\| = 0$. Using the Lipschitz continuity of u_i, this ensures that the network-based FP process is in fact a weakened FP process, and convergence follows from the robustness result in Theorem 7.4.

Remark 7.3. We note that the distributed information dissemination scheme used in the above network-based FP algorithm (in particular, see step (iv)) is chosen as an example of a viable information dissemination scheme. Using the robustness result in Theorem 7.4, we see that any distributed information dissemination scheme and protocol for approximating σ_n can be used so long as it ensures that $\lim_{n\to\infty} \|\hat{\sigma}_n^i - \sigma(n)\| = 0$. See, e.g., [57–59] for examples of other information dissemination protocols. Furthermore, using asynchronous variants of FP [34,35], one may formulate similar variants of network-based FP relying on asynchronous communication schemes, e.g., [57].

7.4 NETWORK LEARNING ALGORITHMS FOR INCOMPLETE INFORMATION GAMES

We consider learning dynamics for settings where players have differing information about a payoff-relevant state of the world. Players start with an initial common prior estimate about the state of the world, that is, the initial stage game is $\Gamma(\pi) = (\mathcal{N}, \mathcal{A}, \Theta, \{u_i\}_{i\in\mathcal{N}}, \pi)$. They then make noisy private observations about the state of the world. The private signal of player i is denoted by x_i belonging to some space $\mathcal{X}_i$. For instance, the private signal x_i comes from a Gaussian distribution with mean θ and a finite variance. We denote the beginning information of player i by $h_{i,1}$. After receiving the private signal, players have different posterior distributions about the state of the world and the other players' signals. That is, they are in a game of incomplete information defined by the tuple $\Xi := (\mathcal{N}, \mathcal{A}, \Theta, \{u_i\}_{i\in\mathcal{N}}, \pi, \mathcal{X})$ with types of players given by $\{x_i\}_{i\in\mathcal{N}}$.

In a game of incomplete information, the rational behavior is a BNE as per Eq. (7.8). Note that a BNE strategy of a player is a function that maps any possible realization of the signal x_i to an action while an NE strategy corresponds to a distribution over the action space. In contrast to the learning in complete information games, in a game of incomplete information learning a BNE involves each player learning the other players' action at every possible realization of the signal. Given the complete information view of the incomplete information game Ξ where we treat each possible type as a separate player (see the discussion at the end of Section 7.2), the learning algorithms discussed in the previous section can be applied in the following way. Assume x_i belongs to a finite space $\mathcal{X}_i$. For each possible signal profile $\{x_i\}_{i\in\mathcal{N}} \in \mathcal{X}$ run the fictitious play algorithm where the player i learns σ_{x_i}. Then the BNE strategy of player i is given by the $\sigma_i = \{\sigma_{x_i}\}_{x_i\in\mathcal{X}_i}$. In this case, the computational complexity of the learning of a BNE increases with the cardinality of the signal space.

Another impracticality in learning a BNE is that there is a single realization of the private signal in the first stage prior to the learning process. Hence, there is no need to learn the BNE for other possible private signal realizations if the players can learn the NE for the realized signal profile. In the learning process described above, players do not make use of the realization of the signal to infer about the payoff-relevant state of the world. A refined learning goal for the game of incomplete information is for players to aggregate information about the state of the world, i.e., learn the conditional probability distribution $\pi(\theta \mid \{x_i\}_{i\in\mathcal{N}})$, and play the NE of the complete information game with common aggregate distribution $\pi(\theta \mid \{x_i\}_{i\in\mathcal{N}})$ on θ, i.e., $\Gamma(\pi(\theta \mid \{x_i\}_{i\in\mathcal{N}})) = (\mathcal{N}, \mathcal{A}, \Theta, \{u_i\}_{i\in\mathcal{N}}, \pi(\theta \mid \{x_i\}_{i\in\mathcal{N}}))$.

For instance, in Example 2 each private signal carries information about the state of the world. So instead of trying to learn contingent plans for any possible realization of the signal profile, i.e., learn the BNE, robots may try to infer each other's realized signal. Once robots learn each others' signals, they will have a common estimate of the state of the world, and can learn to predict the other player's action exactly. This information aggregation objective is not realizable with an algorithm designed for a complete information game because there is not a mechanism to update estimates of the state of the world.

In the following we present two approaches to information aggregation in incomplete information games. In the first approach, players at each stage observe their neighbors' actions, and play according to the BNE strategy profile of the stage game with types defined as the individuals' current information. We consider the convergence properties of this Bayesian learning approach, and present quadratic games with Gaussian signals as a special case that allows for tractable assessment and behavior. Second, we present a modification to the network-based fictitious play algorithm described in the previous section that allows for players to learn the state of the world and reach an NE.

7.4.1 BAYESIAN LEARNING IN NETWORKS

Players are in a game of incomplete information at each stage. At the beginning of each stage n, players observe the past actions of their neighbors, denoted by $a_{\mathcal{N}_i,n-1}$. At each stage, the information of each player represents its type in the incomplete information game. We assume the state of the world θ and the private signals $\{x_i\}_{i\in\mathcal{N}}$ come from some probability distribution π over the space $\Omega := \Theta \times \mathcal{X}$. We represent the information of player i until time n by $h_{i,n}$. The strategy of player i at stage 1, denoted by $\sigma_{i,1}$, maps any possible information that player i can have to the action space $\mathcal{A}_i$. Together with the probability distribution π and the private signals, the strategy profile of the players at stage 1, σ_1, determines the realized action profile a_1, which then determines the information of players at stage $n = 2$. We denote the strategy profile of players across time by $\sigma = (\sigma_1, \sigma_2, \dots)$ where $\sigma_n = (\sigma_{1,n}, \dots, \sigma_{N,n})$ is the strategy profile at stage n. We can recursively define the partial history of player i generated by the strategy profile σ at stage n denoted by $h^{\sigma}_{i,n}$ as follows. The strategy profile of players at time $n-1$ maps their information to an action profile, that is, $a_{i,n-1} = \sigma_{i,n-1}(h^{\sigma}_{i,n-1})$ for $i \in \mathcal{N}$. Then, given the communication network G, the realized action profile $a_{n-1} \in \mathcal{A}$ determines the information of players in the next stage where player i observes its neighbors' actions $a_{\mathcal{N}_i,n-1}$, that is, $h^{\sigma}_{i,n} = \{h^{\sigma}_{i,n-1}, a_{\mathcal{N}_i,n-1}\}$ and $h_{i,1} = x_i$.

Given the knowledge of the strategy profile σ and prior π, players have a common prior distribution π^{σ} over the space of the state, signals, and the sequence of actions, i.e., over the path of play $\Theta \times \mathcal{X} \times \mathcal{A}^{\mathbb{N}}$. The expectation operator with respect to π^{σ} is denoted by E. Given the notation above, the definition of an equilibrium strategy profile follows.

Definition 7.1. *Players follow a Markov Perfect Bayesian Equilibrium (MPBE) strategy profile σ when for all i, n, and mappings from the path of play to the space of actions that are measurable with respect to player i's information,*[6] *i.e., functions $\sigma'_{i,n} : \Omega \to \mathcal{A}_i$, we have*

$$E[u_i(\sigma_{i,n}, \sigma_{-i,n}, \theta) \mid h^{\sigma}_{i,n}] \geq E[u_i(\sigma'_{i,n}, \sigma_{-i,n}, \theta) \mid h^{\sigma}_{i,n}]. \tag{7.19}$$

■

[6] To be formal, we need to define $\mathcal{B}$ the Borel sigma algebra of the space of exogenous variables Ω. At stage 1, let $\mathcal{H}_{i,1}$ be the smallest sub sigma-algebra of $\mathcal{B}$ that makes x_i measurable. For $n \geq 2$, we can define player i's information $\mathcal{H}^{\sigma}_{i,n}$ and its Markovian strategy $\sigma_{i,n} : \Omega \to \mathcal{A}_i$ that is measurable with respect to $\mathcal{H}^{\sigma}_{i,n}$ (see [60] for a formal definition).

According to this definition, the players follow BNE of the incomplete information game whose information is induced by the actions generated by the BNE equilibrium strategies before them. That is, the incomplete information game at stage t is given by the tuple $\Xi_n := (\mathcal{N}, \mathcal{A}, \Theta, \{u_i\}_{i\in\mathcal{N}}, \pi_\sigma, \{h^\sigma_{i,n}\}_{i\in\mathcal{N}})$. Accordingly, the players are said to be Bayesian in that they form beliefs about the future play of other opponents and the state of the world following Bayes' rule. In addition, each player uses its belief to maximize its current utility, that is, players are myopic. We note that in this model players only observe the actions of their neighbors and do not share their past experiences, signals, or beliefs with other players. Furthermore, players do not observe their realized payoffs.

Remark 7.4. In the previous section, we defined network-based algorithms assuming the existence of a communication graph and once per stage information exchange among the neighbors in the graph. In this model, we assume players observe the actions of their neighbors on a network. Observing actions is a form of information exchange among the actor and the observer, in that, while the actor does not actively send a message, the observer is receiving a message from its neighbor. In the case that players do not have such observational capabilities, this model also allows for players to actively share their actions. In sum, the Bayesian learning algorithm is network-based, as defined in Section 7.3.

Given the MPBE behavior model, the question we would like to answer is whether players learn to coordinate in a supermodular game, and whether they are able to aggregate their information on the state of the world.

We assume the action space $\mathcal{A}_i$ is a compact subset of $\mathbb{R}$ and u_i is continuous in all of its arguments. We restrict attention to symmetric[7] and strictly supermodular games. Our first result is that an MPBE exists for this class of games—see Proposition 1 in [60]. Given the existence of the equilibrium model, we have that players are able to coordinate their actions, and asymptotically reach consensus in strategies and payoffs [60].

Theorem 7.6. *Let σ be a MPBE. For all $i, j \in \mathcal{N}$,*

- *$\sigma_{i,n} \to \sigma_{j,n}$ almost surely as n goes to infinity.*
- *$u_i(\sigma_n, \theta) \to u_j(\sigma_n, \theta)$ almost surely as n goes to infinity.*

■

The second result above (consensus in payoffs) is proven as a corollary to the first result (consensus in actions). The second result means that prior to the start of the game, all players expect to receive similar payoffs eventually regardless of the informativeness of their signals or their position in the network. Even though players agree on their ex ante asymptotic payoffs, they might disagree on their ex post conditional expected payoffs, as exemplified in [60]. The consensus in actions result can be generalized to include endogenous private signals and time-varying directed networks [60].

Next, we consider optimal information aggregation in the following quadratic game, which is an important special case of the model considered above

$$u_i(a, \theta) = -(1 - \lambda)(a_i - \theta)^2 - \lambda(a_i - \bar{a}_{-i})^2, \tag{7.20}$$

where $\lambda \in (0, 1)$ is a constant and $\bar{a}_{-i} = \sum_{j\neq i} a_j/(n-1)$ is the average action of other players. The utility function above is strictly supermodular because $\lambda > 0$. The first term is the estimation loss

[7] For all $i, j \in N$, $\mathcal{A}_i = \mathcal{A}_j$ and $u_i(a, \theta) = u_j(a', \theta)$ if $a_i = a'_j$ and a_{-i} is a permutation of a'_{-j}.

measured by the distance between the realized state and player i's action. The second term represents the player's preference to act in conformity with the rest of the population.

The quadratic utility function above yields a unique MPBE strategy profile σ. The uniqueness follows by showing that the best response mapping is a contraction mapping using the fact that $\lambda \in (0,1)$. This implies uniqueness of BNE in a stage game. Because the MPBE is a sequence of stage game BNE, all stages have unique equilibria. Thus, MPBE must be unique.

Before we state the information aggregation result, we provide an example where information aggregation fails even if players' actions are in consensus.

Example 7.4. Assume the players $i \in \{1,2\}$ have the utility function above. The state of the world $\theta \in \{-1,1\}$ and players have common uniform prior. Each player receives an initial private signal "Heads" (H) or "Tails" (T), that is, $\mathcal{X} = \{H,T\}$. The joint distribution of the private signals is as follows: if $\theta = 1$, players with equal probability observe the signal profile $(x_1,x_2) = (H,H)$ or $(x_1,x_2) = (T,T)$. Otherwise if $\theta = -1$, players with equal probability observe the signal profiles $(x_1,x_2) = (H,T)$ or $(x_1,x_2) = (T,H)$. Given the joint signal profile distribution, both payers receive the signal H or T with equal probability regardless of θ's realization. Thus, the individual private signals are completely uninformative. Hence their expectation of θ is zero. Further, each player knows the other player's expectation of θ is also zero. As a result, the initial equilibrium action of players is zero, that is, $\sigma_{i,1}(x_i) = 0$ for $x_i = \{H,T\}$ and $i = \{1,2\}$. The first stage actions reveal no information to the other players. Therefore, both players continue to take the same action in the following stages.

Next, consider the complete information setting where players observe the complete private signal profile. The private signal profile (x_1,x_2) reveals the realized state θ. Hence, both players select θ as their action for all $n \geq 1$.

Comparing the two scenarios above, we observe that even though in both settings players select the same action for all stages, in the first setting the consensus action is not the action they would choose if they had access to the other player's signal. ■

This example shows that consensus need not imply information aggregation. However, note that the parameters of the joint distribution of the signal profile are fine tuned to achieve this behavior in the first setting. Any infinitesimal change to these parameters breaks the symmetry in the informativeness of the signals. The following theorem shows that this example is not generic [60].

Theorem 7.7. *Denote the space all probability measures over the state and signal profile, i.e., Ω, by $\mathbf{P}$. Let σ^P denote the unique MPBE of the quadratic game given prior $P \in \mathbf{P}$. For a generic $P \in \mathbf{P}$ and all i, $\sigma^P_{i,n} \to E^P[\theta \mid h^\sigma_\infty]$ almost surely.*[8] ■

The theorem states that players eventually select actions as if they have access to each others' private signals for any generic joint prior distribution of the private signals. That is, Example 7.4 belongs to a meager set in the space of probability measures over Ω.

A special case of the utility function above is when $\lambda = 0$. In this case, estimation is the only concern. The equilibrium action is the expectation of the state conditional on the information. This problem is an instance of social learning. Bayesian social learning literature considers the information aggregation when there are no payoff externalities among players, that is, the actions of players do not affect each other's payoffs directly. This theorem extends some of the results in Bayesian social learning in networks [61–63] to the case when payoff externalities exist.

[8] A generic $P \in \mathbf{P}$ belongs to a residual subset of $(\mathbf{P}, d)$ where d denotes the total variation distance between measures.

Asymptotic properties aside, computing a BNE of a stage game is not straightforward, even in quadratic games. In the following, we provide a tractable local filter for players to compute an MPBE strategy when the private signals come from a Gaussian distribution.

Quadratic Gaussian network games

We consider the general form of the quadratic utility function in Eq. (7.12) with an additional bilinear term in the state of the world and self-action,

$$u_i(a_i, a_{-i}) := -\frac{1}{2}a_i^2 + \sum_{j\in -i}\beta_{ij}a_i a_j + \delta_i a_i \theta, \tag{7.21}$$

where $\beta_{ij} \in \mathbb{R}$ for $i \in \mathcal{N}, j \in -i$, and δ_i for $i \in \mathcal{N}$ are real valued constants. We assume the private signal profile $x := (x_1, \dots, x_N)$ comes from a Gaussian distribution with mean $\theta 1_N$ and a diagonal covariance matrix C.

Given its private signal, player i's posterior distribution (belief) on the signal profile x and the state of the world θ is Gaussian. Specifically, the expectation of private signals and the state (x, θ) for $i = 1, \dots, N$ is given by

$$E[(x, \theta) \mid x_i] = 1_{N+1}\mathbf{e}_i^T x, \tag{7.22}$$

where 1_{N+1} is an $(N+1)\times 1$ vector of ones, and $\mathbf{e}_i$ is the canonical vector of size $(N+1)\times 1$ with one in the ith element and zero elsewhere.

In the following we use induction to show that if at time n players' beliefs are Gaussian with expectations that are a linear combination of the private signal profile, then at time n their strategies are linear in their expectation of the signal profile, and their beliefs remain Gaussian at time $n+1$. We start with the following induction step: Suppose that at time n player i has Gaussian beliefs on (x, θ) given its information $h_{i,n}$ with expectations that can be expressed as a linear combination of the signal profile,

$$E[x \mid h_{i,n}] = L_{i,n}x, \qquad E[\theta \mid h_{i,n}] = k_{i,n}^T x, \tag{7.23}$$

where $L_{i,n} \in \mathbb{R}^{N\times N}$ and $k_{i,n} \in \mathbb{R}^{N\times 1}$.

Computing the MPBE strategy at time n is equivalent to computing the BNE of the incomplete information stage game. We can represent the BNE definition in Eq. (7.8) by using the fixed point definition in Eq. (7.3), where the BNE strategy $\sigma_{i,n}$ at stage n satisfies,

$$\sigma_{i,n}(h_{i,n}) = \operatorname*{argmax}_{a_i} E[u_i(a_i, \sigma_{-i,n}(h_{-i,n}), \theta) \mid h_{i,n}] \tag{7.24}$$

for all $h_{i,n}$ and $i \in \mathcal{N}$. The right side of the above equation can be simplified by taking the derivative of the quadratic utility function in Eq. (7.21) with respect to a_i, equating it to zero, and solving for a_i,

$$\sigma_{i,n}(h_{i,n}) = \sum_{j\in -i}\beta_{ij}E[\sigma_{j,n}(h_{j,n}) \mid h_{i,n}] + \delta_i E[\theta \mid h_{i,n}]. \tag{7.25}$$

The second step of the induction is that player i's strategy at time n is a linear combination of its expectation of the private signals, that is,

$$\sigma_{i,n}(h_{i,n}) = v_{i,n}^T E[x \mid h_{i,n}], \tag{7.26}$$

where $v_{i,n} \in \mathbb{R}^{N\times 1}$ is the strategy constant. When we substitute the above equation to Eq. (7.25), we obtain the following set of linear equations,

$$v_{i,n}^T E[x\,|h_{i,n}] = \sum_{j\in -i} \beta_{ij} E[v_{j,n}^T E[x|h_{j,n}]|h_{i,n}] + \delta E[\theta\,|h_{i,n}]. \tag{7.27}$$

Using the linearity of the expectation in Eq. (7.23), we can represent the above equation as

$$v_{i,n}^T L_{i,n} x = \sum_{j\in -i} \beta_{ij} v_{j,n}^T L_{j,n} L_{i,n} x + \delta k_{i,n}^T x. \tag{7.28}$$

Recall that a BNE strategy satisfies the equation in Eq. (7.25) for all realizations of the history. This implies that the above equation needs to hold for any realization of x. This we can ensure by equating the coefficients that multiply each component of x above, yielding the following set of equalities,

$$L_{i,n}^T v_{i,n} = \sum_{j\in -i} \beta_{ij} L_{i,n}^T L_{j,n}^T v_{j,n} + \delta k_{i,n} \tag{7.29}$$

for all $i \in \mathcal{N}$. Solving the set of equalities above for $\{v_{i,n}\}_{i\in\mathcal{N}}$ determines BNE strategies that are linear in the expectation of the private signals as per Eq. (7.26). Together with the linear strategy constants $\{v_{i,n}\}_{i\in\mathcal{N}}$, the realized action at stage n can be expressed as a linear combination of the private signals using the induction step in Eq. (7.23),

$$a_{i,n} = \sigma_{i,n}(h_{i,n}) = v_{i,n}^T L_{i,n} x. \tag{7.30}$$

After selecting its action, player i observes its neighbors' actions $a_{\mathcal{N}_i,n}$, which as per the above equation can be written as a linear combination of the private signals,

$$a_{\mathcal{N}_i,n} = H_{i,n}^T x := \begin{pmatrix} v_{j_{i,1},n}^T L_{j_{i,1},n} \\ \vdots \\ v_{j_{i,d(i)},n}^T L_{j_{i,d(i)},n} \end{pmatrix} x, \tag{7.31}$$

where $\mathcal{N}_i = (j_{i,1}, \ldots, j_{i,d(i)})$. This means that player i's observations are Gaussian distributed. Hence, player i can form its posterior estimate on the private signals and the state using a Kalman filter. In particular, player i propagates its expectation of the private signals as follows,

$$E[x \mid h_{i,n+1}] = E[x \mid h_{i,n}] + K_{i,n}(a_{\mathcal{N}_i,n} - E[a_{\mathcal{N}_i,n} \mid h_{i,n}]), \tag{7.32}$$

where $K_{i,n} \in \mathbb{R}^{N\times|\mathcal{N}_i|}$ is the Kalman filter gain. The computation of the gain matrix at each stage requires keeping track of the variance-covariance matrix $M_{i,n}$ (see [64] for details). Note that because the observations of player i can be expressed as a linear combination the private signals (7.31), its posterior distribution on the private signals remain Gaussian with mean estimates that are also linear combinations of the private signals. This completes the induction argument.

The computation of equilibrium strategies, i.e., forming the set of linear stage BNE equations in Eq. (7.29) and solving for $v_{i,n}$ as well as forming estimates on its observations $E[a_{\mathcal{N}_i,n} \mid h_{i,n}]$, requires each player to keep track of the mean estimate constants of the entire population $\{L_{i,n}, k_{i,n}\}_{i\in\mathcal{N}}$. For this, even though player i does not observe other players' information, it needs to keep track of how they process their information. In a sense, each player simulates the population's progress at each stage. Details of this computation are presented in [64].

In Fig. 7.2, we present an outline of the quadratic network game (QNG) filter that allows player i to compute its MPBE actions. A few remarks are in order. The QNG filter is designed for any quadratic game. That is, the game does not have to be supermodular, implying that we do not have nonnegativity assumption on the cross-product constants β_{ij}. The Gaussian distribution of the private signals together with the existence of linear strategies yield that all players' beliefs on the private signals remain Gaussian at all stages. The Gaussian beliefs separate BNE action computation at each stage to estimation and decision-making, i.e., assessment and behavior, components.

The QNG filter requires a high level of computational prowess from each player. In particular, each player needs to solve the N^2 equalities for N^2 variables in Eq. (7.29). Further, each player needs to keep track of how other players form their beliefs on the private signals to be able to form these sets of equalities. This process is not computationally demanding in comparison to solving N^2 equalities, as it only involves $2N$ addition operations, yet it assumes that players know the network structure.

In the following we consider a numerical implementation of the QNG filter.

Simulation

A network of $N = 50$ autonomous robots wants to move in coordination and at the same time follow a target direction $\theta \in [0°, 10°]$. An action determines the direction of movement on a two-dimensional space. The estimation and coordination goals of the robots can be captured by the beauty contest game with utility functions given in Eq. (7.20). We let $\lambda = 0.5$.

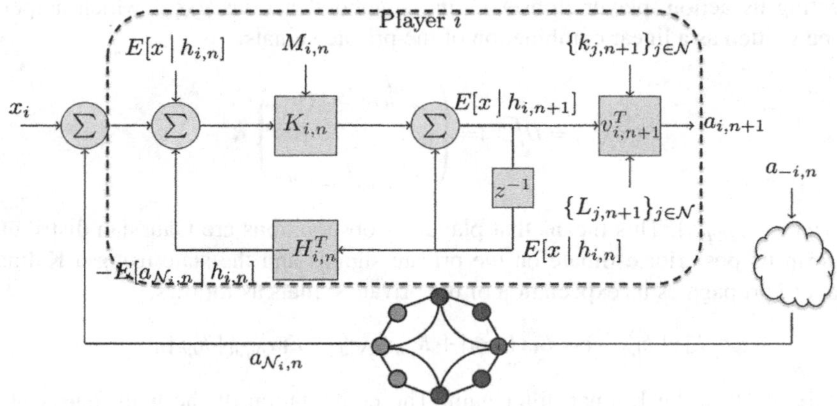

FIG. 7.2

Quadratic network game (QNG) filter. We illustrate the process after actions of stage n are realized. The *dashed box* encloses player i's belief formation and decision-making. After player i observes its neighbor's action $a_{\mathcal{N}_i,n}$, it updates its estimate of the realized signal profile according to Eq. (7.32). In forming estimates of other's actions, player i needs to construct the observation matrix in Eq. (7.31) and the gain matrix $K_{i,n}$. Given the expectation of the private signals at time $n+1$, player i takes its action following the linear strategy in Eq. (7.26). In order to compute the linear strategy constant $v_{i,n+1}$, it constructs and solves the set of linear equations in Eq. (7.29). All players follow the same process simultaneously at each stage.

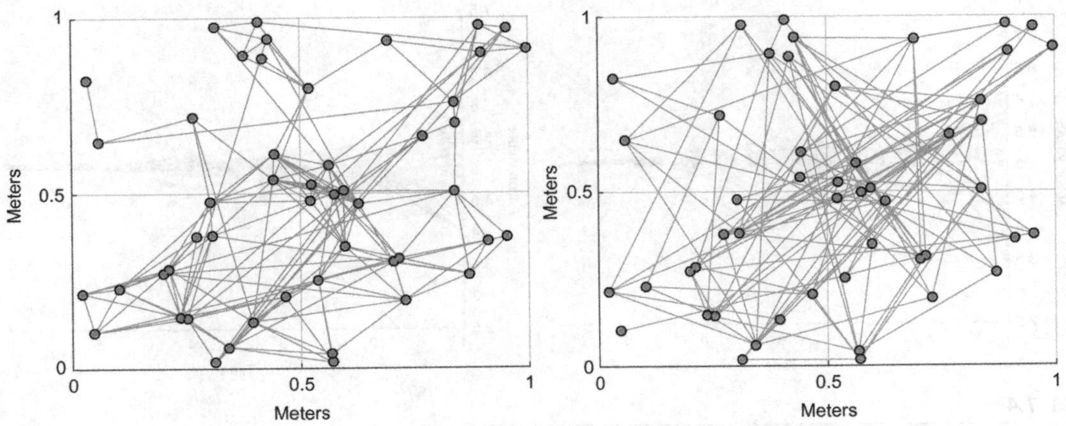

FIG. 7.3

Geometric (left) and random (right) networks. A geometric random network is created by placing the players randomly on a 1 × 1 meter square and connecting pairs with a distance less than 0.25 meter between them. In the random network, there exists a link between any pair of players with probability 0.1.

We evaluate the QNG filter in geometric and random networks (see Fig. 7.3). The geometric network has a diameter[9] of 6 whereas the random network has a diameter of 5. The target direction is chosen to be $\theta = 5°$. Players receive noisy private signals Gaussian with mean θ and standard deviation equal to 1°. The direction of movement of each player over stages is depicted in Fig. 7.4. We observe that players' movement directions converge to the best estimates in the target direction in a finite number of steps. That is, at the end of the convergence time T, we have $E[\theta \mid h_{i,T}] = E[\theta \mid x]$ for all $i \in \mathcal{N}$. Further, convergence time is in the order of the diameter for both the networks. This means that players learn the sufficient statistic to calculate the best estimate in the amount of time it takes for information to propagate on the network.

7.4.2 NETWORK-BASED FICTITIOUS PLAY FOR INCOMPLETE INFORMATION GAMES

Given the computational and memory demands mentioned above, the QNG filter, while being conceptually appealing and having desirable asymptotic properties for supermodular quadratic games, is not practical. In the following, we provide an adaptation of the network-based FP algorithm discussed in Section 7.3 to incomplete information games.

Assume each player i has a local belief $\pi_{i,n}$ that assigns probabilities to different realizations of the state of the world θ. We assume players have a separate state learning process that updates their local belief $\pi_{i,n}$ in addition to the steps of the network-based FP algorithm at stage n. We will be agnostic to the specifics of the state learning process as long as the players' local beliefs converge to a common belief π on the state of the world. In particular, we make the following assumption.

[9]Diameter is the longest shortest path among all pairs of nodes.

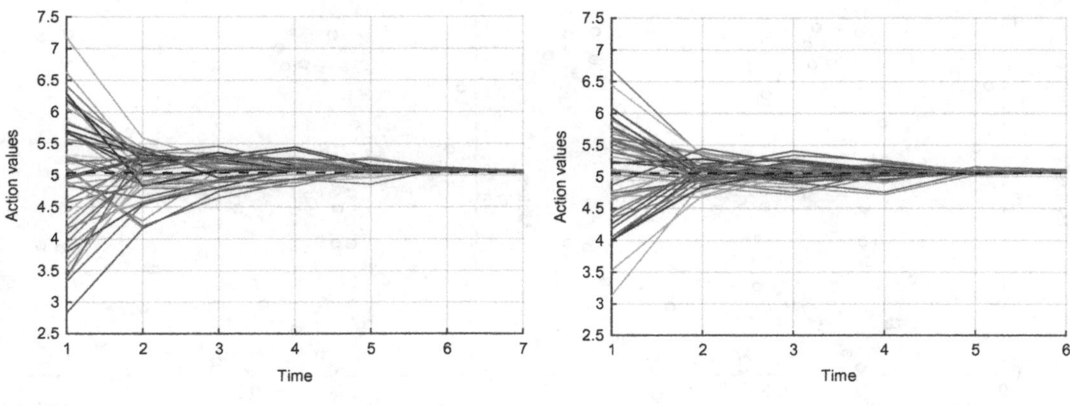

FIG. 7.4

Player actions for geometric (left) and random (right) networks in Fig. 7.3. Action consensus happens in the order of the diameter of the corresponding networks.

Assumption 7.5. *For all players $i \in \mathcal{N}$, the local beliefs converge to a common belief π at a rate faster than* $\log n/n$,

$$TV(\pi_{i,n}, \pi) = O(\log n/n), \tag{7.33}$$

where $TV(\cdot)$ is the total variation distance between the distributions.[10] ■

For example, the state learning process can involve information exchange among players where they exchange their beliefs and implement a consensus-like updating procedure, or it can entail Bayesian updates based on repeated noisy private signal observations about the state of the world.

Given this assumption, we have the following convergence result for the modified network-based FP [65].

Theorem 7.8. *Let $\Gamma(\pi) = (\mathcal{N}, \mathcal{A}, \Theta, \{u_i\}, \pi)$ be a potential game. Assume players follow the network-based fictitious play algorithm with individual state learning processes that satisfy the above assumption. Then, $\lim_{n\to\infty} d(\sigma_n, NE) = 0$.* ■

This theorem can also be seen as another application of the robustness result given in Section 7.3. In particular, the uncertainty on the state and networked interactions perturbs the best response actions of players, resulting in a weakened FP process.

The network-based FP for incomplete information is computationally less demanding than the QNG filter. However, in terms of information exchange, the QNG filter is less demanding. The QNG filter only requires the observation of neighbors' actions while step (iv) of the network-based FP necessitates exchanging entire estimates of other players' empirical distributions with neighbors. In addition, the state learning process might require additional information exchange or observations.

In the following, we present an implementation of the network-based FP for incomplete information games.

[10] TV is the maximum absolute difference between the respective probabilities assigned to elements B of the Borel set $\mathcal{B}(\Theta)$ of the space Θ, i.e., $TV(\pi_{i,n}, \pi) := \sup_{B \in \mathcal{B}(\Theta)} |\pi_{i,n}(B) - \pi(B)|$.

Simulation

We consider the same simulation setup as the numerical example of the QNG filter. We discretize the action space in order to implement the network-based FP, where $\mathcal{A}_i = \{0°, 1°, 2°, \ldots, 10°\}$. In the setup, each robot receives an initial noisy signal related to the target direction $\theta, x_i = \theta + \epsilon_i$ where ϵ_i is drawn from a zero mean normal distribution with standard deviation equal to 1°. Robots learn about the state by averaging their neighbors' estimates with initial estimates equal to the private signal x_i.[11]

We consider two types of networks. The first network is a geometric network generated by randomly placing the robots on a 1 unit × 1 unit square and drawing an edge between pairs that are less than 0.25 units away. The second network is a random network. We observe that the convergence of the algorithm is faster in the random network (46 steps) than the geometric network (113 steps) in Fig. 7.5.

In comparison to the QNG filter, the network-based FP for incomplete information takes longer in terms of the number of stages. As can be expected, the computation time of each stage is much faster for the network-based FP algorithm. In sum, the total simulation time until convergence takes longer for the QNG filter.

7.5 SUMMARY AND DISCUSSION

Game theoretic learning entails a population of players sequentially predicting each others' future actions, selecting individual actions based on their predictions, and making new observations based on the history of play. The goal of each player is to successfully predict others' upcoming actions and play

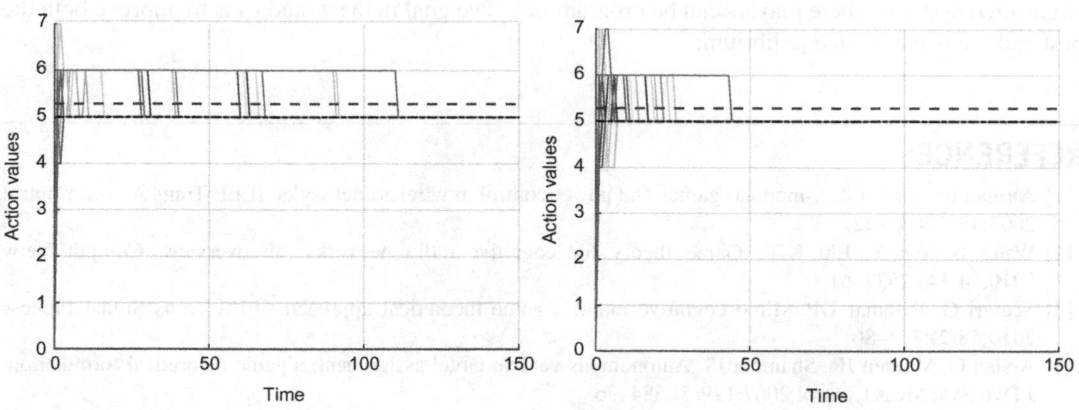

FIG. 7.5

Robots' actions for geometric (left) and small-world (right) networks. *Solid lines* correspond to each robots' actions over time. The *dotted dashed line* is equal to the value of the state of the world $\theta = 5°$ and the *dashed line* is the optimal estimate of the state given all the signals, which is equal to 5.3°. Agents reach consensus in the movement direction 5° faster in the small-world network than the geometric network.

[11] Note that the averaging dynamics in connected networks do not satisfy the assumption of convergence in TV distance in Eq. (7.33). That is, the numerical setup does not necessarily meet the assumption of Theorem 7.8.

optimally with respect to those actions. The discussion in this chapter presumed that players can only exchange messages over a communication/observation network, and cannot measure their individual payoffs. In discussing network-based learning algorithms, we considered two important subcategories of games: complete and incomplete information games. We presented network-based fictitious play algorithms for both categories of games. We also considered Bayesian learning in the case of incomplete information games, and discussed its asymptotic and computational properties.

Many research topics relevant to game theoretic learning have not been explored in this chapter. We discuss two of these active research areas below.

Payoff-based learning algorithms

In these algorithms, players measure their realized utility at each stage; see [22] for a general discussion. Most prominent payoff-based algorithms include log-linear learning that guarantees convergence to a local maximum of the potential function of a potential game [66], and "no-regret" based algorithms that have the desirable property of convergence to broader solution concepts than Nash equilibria, such as correlated equilibria [67,68].

Utility design

In this chapter, we primarily considered convergence to Nash equilibria. When considering social welfare or global objectives in technological settings, convergence to Nash equilibria can be undesired as an NE can be highly inefficient. In response to the inefficiencies of Nash equilibria, besides designing learning algorithms that avoid convergence to inefficient Nash equilibria, another research area focuses on the design of utility functions [42,52,66]. The design of utility functions is particularly relevant to engineered systems where players can be programmed. The goal in these studies is to improve both the best and the worst Nash equilibrium.

REFERENCES

[1] Altman E, Altman Z. S-modular games and power control in wireless networks. IEEE Trans Autom Control 2003;48(5):839–42.

[2] Wang B, Wu Y, Liu KJR. Game theory for cognitive radio networks: an overview. Comput Netw 2010;54(14):2537–61.

[3] Scutari G, Palomar DP. Mimo cognitive radio: a game theoretical approach. IEEE Trans Signal Process 2010;58(2):761–80.

[4] Arslan G, Marden JR, Shamma JS. Autonomous vehicle-target assignment: a game-theoretical formulation. J Dyn Syst Meas Control 2007;129(5):584–96.

[5] Mohsenian-Rad A, Wong VW, Jatskevich J, Schober R, Leon-Garcia A. Autonomous demand-side management based on game-theoretic energy consumption scheduling for the future smart grid. IEEE Trans Smart Grid 2010;1(3):320–31.

[6] Yang P, Tang G, Nehorai A. A game-theoretic approach for optimal time-of-use electricity pricing. IEEE Trans Power Syst 2013;28(2):884–92.

[7] Forouzandehmehr N, Perlaza SM, Han Z, Poor HV. A satisfaction game for heating, ventilation and air conditioning control of smart buildings. In: 2013 IEEE Global communications conference (GLOBECOM). IEEE; 2013. p. 3164–9.

[8] Nash JF, et al. Equilibrium points in n-person games. Proc Natl Acad Sci U S A 1950;36(1):48–9.

[9] Monderer D, Shapley LS. Potential games. Games Econ Behav 1996;14(1):124–43.
[10] Alpcan T, Basar T. A game-theoretic framework for congestion control in general topology networks. In: 2002, Proceedings of the 41st IEEE conference on decision and control, vol. 2. IEEE; 2002. p. 1218–24.
[11] Lamperski A, Doyle JC. On the structure of state-feedback LQG controllers for distributed systems with communication delays. In: Proceedings of the 50th IEEE decision and control and European control conference (CDC-ECC), Orlando, FL, USA; 2011. p. 6901–6.
[12] Nayyar A, Mahajan A, Teneketzis D. Optimal control strategies in delayed sharing information structures. IEEE Trans Autom Control 2011;56(7):1606–20.
[13] Cattivelli F, Sayed A. Analysis of spatial and incremental LMS processing for distributed estimation. IEEE Trans Signal Process 2011;59(4):1465–80.
[14] Xiao J, Ribeiro A, Zhi-Quan L, Giannakis G. Distributed compression-estimation using wireless sensor networks. IEEE Signal Process Mag 2006;23:27–41.
[15] Topkis DM. Supermodularity and complementarity. Princeton University Press; 1998.
[16] Vives X. Complementarities and games: new developments. J Econ Lit 2005:437–79.
[17] Milgrom P, Roberts J. Rationalizability, learning, and equilibrium in games with strategic complementarities. Econometrica 1990;58(6):1255–77.
[18] Salehisadaghiani F, Pavel L. Distributed Nash equilibrium seeking: a gossip-based algorithm. Automatica 2016;72:209–16.
[19] Koshal J, Nedic A, Shanbhag UV. A gossip algorithm for aggregative games on graphs. In: Proceedings of the 51st IEEE conference on decision and control; 2012. p. 4840–5.
[20] Marden JR, Young HP, Arslan G, Shamma JS. Payoff-based dynamics for multi-player weakly acyclic games. SIAM J Control Optim 2009;48(1):373–96.
[21] Fudenberg D, Levine DK. The theory of learning in games, vol. 2. MIT Press; 1998.
[22] Young HP. Strategic learning and its limits, vol. 2002. Oxford University Press; 2004.
[23] Leslie DS, Collins EJ. Generalised weakened fictitious play. Games Econ Behav 2006;56(2):285–98.
[24] Fudenberg D, Levine DK. Consistency and cautious fictitious play. J Econ Dyn Control 1995;19(5):1065–89.
[25] Coucheney P, Gaujal B, Mertikopoulos P. Penalty-regulated dynamics and robust learning procedures in games. Math Oper Res 2014;40(3):611–33.
[26] Wang Y, Saad W, Han Z, Poor HV, Başar T. A game-theoretic approach to energy trading in the smart grid. IEEE Trans Smart Grid 2014;5(3):1439–50.
[27] Dabcevic K, Betancourt A, Marcenaro L, Regazzoni CS. A fictitious play-based game-theoretical approach to alleviating jamming attacks for cognitive radios. In: IEEE ICASSP; 2014. p. 8158–62.
[28] Wang B, Wu Y, Liu KR. Game theory for cognitive radio networks: an overview. Comput Netw 2010;54(14):2537–61.
[29] Heinrich J, Lanctot M, Silver D. Fictitious self-play in extensive-form games. In: Proceedings of the 32nd international conference on machine learning, ICML; 2015. p. 805–13.
[30] Ganzfried S, Sandholm T. Computing an approximate jam/fold equilibrium for 3-player no-limit Texas hold'em tournaments. In: Proceedings of the 7th international joint conference on autonomous agents and multiagent systems, vol. 2. International Foundation for Autonomous Agents and Multiagent Systems; 2008. p. 919–25.
[31] Sandholm T. The state of solving large incomplete-information games, and application to poker. AI Mag 2010;31(4):13–32.
[32] Lambert TJ, Epelman MA, Smith RL. A fictitious play approach to large-scale optimization. Oper Res 2005;53(3):477–89.
[33] Garcia A, Reaume D, Smith RL. Fictitious play for finding system optimal routings in dynamic traffic networks. Transp Res B Methodol 2000;34(2):147–56.
[34] Swenson B, Kar S, Xavier J, Leslie DS. Robustness properties in fictitious-play-type algorithms. SIAM J Control Optim 2017;55(5):3295–318.

[35] Swenson B, Kar S, Xavier J. On asynchronous implementations of fictitious play for distributed learning. In: 2015 49th Asilomar conference on signals, systems and computers. IEEE; 2015. p. 1119–24.
[36] Swenson B, Kar S, Xavier J. Single sample fictitious play. IEEE Trans Autom Control 2017;62(11):6026–31.
[37] Harris C. On the rate of convergence of continuous-time fictitious play. Games Econ Behav 1998;22(2):238–59.
[38] Swenson B, Kar S. On the exponential rate of convergence of fictitious play in potential games; 2017. arXiv preprint arXiv:170708055.
[39] Gharesifard B, Cortes J. Distributed convergence to Nash equilibria in two-network zero-sum games. Automatica 2013;49(6):1683–92.
[40] Chen X, Huang J. Spatial spectrum access game: Nash equilibria and distributed learning. In: Proceedings of the thirteenth ACM international symposium on mobile ad hoc networking and computing. ACM; 2012. p. 205–14.
[41] Gharehshiran ON, Krishnamurthy V, Yin G. Distributed tracking of correlated equilibria in regime switching noncooperative games. IEEE Trans Autom Control 2013;58(10):2435–50.
[42] Li N, Marden JR. Designing games for distributed optimization. IEEE J Sel Top Signal Process 2013;7(2):230–42.
[43] Benaïm M, Hofbauer J, Sorin S. Stochastic approximations and differential inclusions. SIAM J Control Optim 2005;44(1):328–48.
[44] Marden JR, Arslan G, Shamma JS. Cooperative control and potential games. IEEE Trans Syst Man Cybern B Cybern 2009;39(6):1393–407.
[45] Scutari G, Barbarossa S, Palomar DP. Potential games: a framework for vector power control problems with coupled constraints. In: Proceedings of the IEEE international conference on acoustics, speech and signal processing, vol. 4. IEEE; 2006. p. 241–4.
[46] Xu Y, Anpalagan A, Wu Q, Shen L, Gao Z, Wang J. Decision-theoretic distributed channel selection for opportunistic spectrum access: strategies, challenges and solutions. IEEE Commun Surv Tutorials 2013;15(4):1689–713.
[47] Zhu M, Martínez S. Distributed coverage games for energy-aware mobile sensor networks. SIAM J Control Optim 2013;51(1):1–27.
[48] Ding C, Song B, Morye A, Farrell JA, Roy-Chowdhury AK. Collaborative sensing in a distributed PTZ camera network. IEEE Trans Image Process 2012;21(7):3282–95.
[49] Nie N, Comaniciu C. Adaptive channel allocation spectrum etiquette for cognitive radio networks. Mob Netw Appl 2006;11(6):779–97.
[50] Chu X, Sethu H. Cooperative topology control with adaptation for improved lifetime in wireless ad hoc networks. In: Proceedings of IEEE international conference on computer communications. IEEE; 2012. p. 262–70.
[51] Srivastava V, Neel JO, MacKenzie AB, Menon R, DaSilva LA, Hicks JE, et al. Using game theory to analyze wireless ad hoc networks. IEEE Commun Surv Tutorials 2005;7(4):46–56.
[52] Marden JR, Wierman A. Overcoming limitations of game-theoretic distributed control. In: Proceedings of the 48th IEEE conference on decision and control, 2009 held jointly with the 2009 28th Chinese control conference. CDC/CCC 2009. IEEE; 2009. p. 6466–71.
[53] Swenson B, Murray R, Kar S. On best-response dynamics in potential games; submitted for publication. https://arxiv.org/abs/1707.06465.
[54] Swenson B, Murray R, Kar S. Regular potential games; submitted for publication. https://arxiv.org/abs/1707.06466.
[55] Marden JR, Arslan G, Shamma JS. Joint strategy fictitious play with inertia for potential games. IEEE Trans Autom Control 2009;54(2):208–20.
[56] Swenson B, Kar S, Xavier J. Empirical centroid fictitious play: an approach for distributed learning in multi-agent games. IEEE Trans Signal Process 2015;63(15):3888–901.

[57] Dimakis AG, Kar S, Moura JM, Rabbat MG, Scaglione A. Gossip algorithms for distributed signal processing. Proc IEEE 2010;98(11):1847–64.
[58] Kar S, Moura JM. Convergence rate analysis of distributed gossip (linear parameter) estimation: fundamental limits and tradeoffs. IEEE J Sel Top Signal Process 2011;5(4):674–90.
[59] Chen J, Sayed AH. Diffusion adaptation strategies for distributed optimization and learning over networks. IEEE Trans Signal Process 2012;60(8):4289–305.
[60] Molavi P, Eksin C, Ribeiro A, Jadbabaie A. Learning to coordinate in social networks. Oper Res 2015;64(3):605–21.
[61] Djuric PM, Wang Y. Distributed Bayesian learning in multiagent systems. IEEE Signal Process Mag 2012;29:65–76.
[62] Rosenberg D, Solan E, Vieille N. Informational externalities and emergence of consensus. Games Econ Behav 2009;66(2):979–94.
[63] Mueller-Frank M. A general framework for rational learning in social networks. Theor Econ 2013;8:1–40.
[64] Eksin C, Molavi P, Ribeiro A, Jadbabaie A. Bayesian quadratic network game filters. IEEE Trans Signal Process 2014;62(9):2250–64.
[65] Eksin C, Ribeiro A. Distributed fictitious play for multi-agent systems in uncertain environments. IEEE Trans Autom Control 2017;63(4):1177–84.
[66] Marden JR, Shamma JS. Game theory and distributed control. In: Handbook of game theory 2012;4:861–900.
[67] Hart S, Mas-Colell A. A simple adaptive procedure leading to correlated equilibrium. Econometrica 2000;68(5):1127–50.
[68] Hart S, Mas-Colell A. A general class of adaptive strategies. J Econ Theory 2001;98(1):26–54.
[69] Krishna V, Sjöström T. On the convergence of fictitious play. Math Oper Res 1998;23(2):479–511.

PART

SIGNAL PROCESSING ON GRAPHS

2

CHAPTER

GRAPH SIGNAL PROCESSING 8

José M.F. Moura[a]*
*Department of Electrical and Computer Engineering, Carnegie Mellon University, Pittsburgh, PA, United States**

8.1 INTRODUCTION

It is by now a cliché that data is everywhere, and we are and will be inundated by data. According to a 2014 study by IDC [1], in 2020 alone, all digital data created, replicated, and consumed will amount to 44 zetabytes. A zetabyte is 10^{21} bytes, so we will produce yearly data quantified by the number 44 followed by 21 zeros. To have a more physical feeling for this staggering amount of data, we compare it to a coarse estimate of the entire collection of books, reports, written texts, maps, images, audio recordings, and videos in the US Library of Congress (LOC) (charged with maintaining a copy of every book ever printed in the United States). This estimate may vary, but we consider it to be three petabytes [2]. Then, the 44 zetabytes of data produced worldwide in the year 2020 will be roughly equivalent to 15 million LOCs. IDC has recently updated this estimate to 163 zetabytes by 2025 [3,4]. But these data will be very different from the data we are traditionally concerned with in disciplines such as statistics, signal or image processing, computer vision, or machine learning. Beyond traditional time series, speech, audio, radio, radar, biomedical signals, or images and videos, these data arise from the 11 billion internet-of-things (IOT) devices in 2016 that are estimated to triplicate to 30 billion in 2020, from the activity of billions of cell phone users of the many service providers; from public and private urban transportation systems; in health care from the digital records of patients, providers, visits, exams, tests, results, costs, insurance, hospital procedures; from interactions among social network agents, corporate financial data, hyperlinked blogs, or tweeters, metabolic networks, protein interaction networks, just to mention some examples. These *Big Data* are often characterized by three, five, or seven V: *V*ariety, *V*olume, *V*elocity, *V*eracity, *V*ariability, *V*alue, and *V*isualization. In many contexts, these data are produced by scattered sources such as the thousands of webcams monitoring traffic in urban centers, i.e., the data are *distributed*. Besides numerical, the data can be Boolean, ordinal, or categorical. Finally, the data is unlikely to fit neatly in a table; in other words, the data are *un*structured. The goal of this chapter is to introduce data analytic tools to process these data that are much like the classical tools used with time series, images, or videos, but now applied to the variety of *un*structured *Big Data* of today.

[a]This work is partially supported by NSF grant CCF # 1513936.

Cooperative and Graph Signal Processing. https://doi.org/10.1016/B978-0-12-813677-5.00008-0

While this chapter focuses on the analytics of these unstructured data, which we will call *Data Science*, we start by contrasting the goal of the chapter and Data Science with two other close topics, namely, *Network Science* and *Network Processes*.

Network Science. In the last decade and a half, graphs have permeated the study of complex social, biological, and technological systems [5–8] of high dimensional data, or of the dynamics of networked processes such as epidemics, to refer to a few application domains. These studies often lead to first representing the application data by a graph, and then addressing the questions of interest by focusing on and analyzing the structure of the graph. This is the domain of *Network Science* [9] that has received considerable attention [6–8]. For an illustrative Network Science approach to problems, consider the rise and dominance of the Medici elite family in 15th century Florence through marriage, economics, and the patronage of the Medicean networks [10]. Marriage network data in [10] is partially shown [11] by the graph in Fig. 8.1 where nodes are families and edges are marital linkages among families. Reference [10] shows that the "in-betweenness centrality" [12] for the Medici family marriage network is $C_b = 0.362$ while for oligarch families it is $C_b = 0.184$ (see footnote 31 in [10]). Similarly, for economic relations, it is $C_b = 0.429$ for Medici families and $C_b = 0.198$ for other elite families [10]. Further, as noted in footnote 32 in [10], the Medici family is more like a star or spoke network, with sparse ties among families tied by marriage or economic relations to the Medici patriarch (average degree of these neighbors of the Medici being 2). This contrasts with oligarch families where the

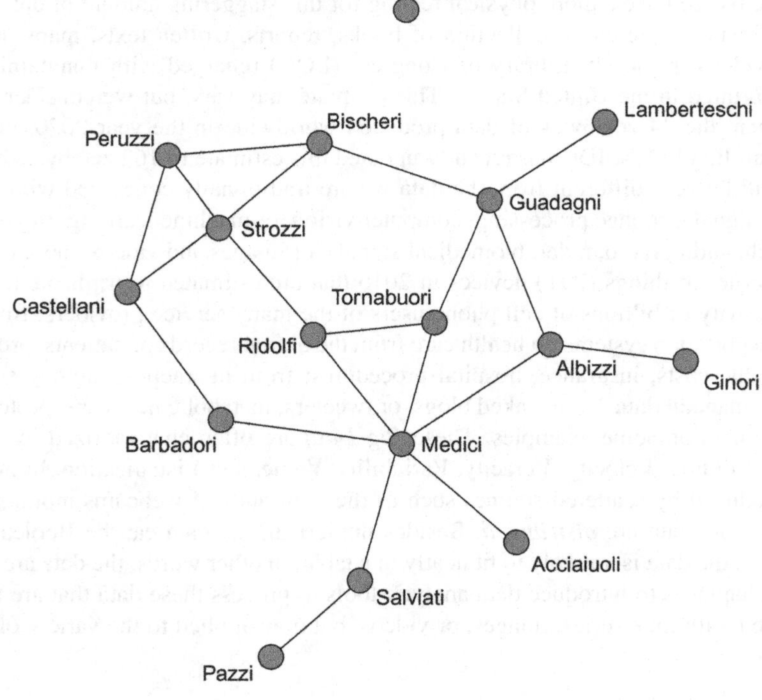

FIG. 8.1

Fifteenth-century Florentine marriages [11].

average degree of their neighbors could be as high as 6.5. This centrality explains the dependence of Medici partners on the Medici that played a significant intermediary role as facilitator of transactions among them. The point here is that, in Network Science, a graph represents the data, and, once the graph is available, the questions of interest are not addressed by analyzing the data, but rather by analyzing the structure of the graph. Network Science focuses on topological and structural properties of the network with quantities of interest such as in- and out-degrees of nodes, degree distribution, number of connected components, size of the giant component, graph diameter, and average shortest path length as well as parameters such as the clustering coefficient that measures network cohesiveness or various other centrality measures. Network models include the Erdös Rényi random graphs, small world graphs, preferential attachment and scale-free random graphs [13], regular graphs, and hub networks [6–8].

Network Processes. The spread of epidemics, fake news, memes, diffusion of opinions, and voting patterns are all examples of *network processes* of great interest in the social and life sciences, for example. These processes collect the state of all the agents (nodes) in the network. Their dynamics as the process evolves in time are governed by local rules of interaction among the agents. For example, in epidemics, a healed node becomes infected by coming into contact with another infected node [14–18]. Studies of these processes consider the emergence of global behaviors from these local interactions. These are difficult questions because the network state space is very large—for an interacting system of N agents and a binary state for each agent, the cardinality of this state space is 2^N. For a small network of $N = 200$, this is $2^{200} = 10^{60}$. This is a staggering large number that makes the analysis of general processes over general networks an almost impossible task. Often, the literature studies these systems for large time, large scale or network size (fluid dynamics) limit, restricting attention to global process parameters such as the fraction of nodes in a given state, say, the fraction of infected nodes, or the fraction of nodes adopting a particular opinion, or fraction of voters voting in the same candidate or party. Further, the literature frequently analyzes these systems under simple conditions that abstract out the network topology—full mixing or complete network where an agent can interact with any other agent. Corresponding limiting models lead to systems of ordinary differential equations governing the dynamics of the global quantities of interest. Our recent work in this area studies the dynamics of network processes for arbitrary network topologies. References [19,20] introduce the scaled network process under which the network process is a reversible Markov process [21], and for which we find its equilibrium probability distribution (time asymptotics) that explicitly exhibits the network topology through its adjacency matrix. From this equilibrium distribution, we can address global questions such as which configuration, i.e., network state, in the long run is more likely to occur, or structural questions such as which parts or substructures of the network are more likely to be infected. These questions depend on both the topology of the network and the local rules of interaction among agents.

In [22,23], we consider the qualitative behavior of these processes in the limit of large networks. Under this limit and appropriate conditions, the dynamics of global statistics, for example the fraction of infected nodes, follow nonlinear ordinary differential equations (NLODE). The qualitative behavior of these equations, requiring establishing the basins of attraction for the limit points of NLODEs, addresses issues such as which opinion or virus may survive when several compete for attention or multiple viruses cohabit the network [22,24].

Data Science. To remove the *un* from *un*structured, we assume that the data is indexed by the N nodes of a graph $G = (V, E)$. The edges in E of G represent dependencies among the data. The nodes in V can be thought of as the sources of data, for example, agents in a social network, individuals in a population, tweeters, or bloggers in networks of hyperlinked blogs [25]. The data at each node of V can itself be a time series, an image, a video, or other features collected or produced by the agent or characterizing the node. For example, we may be interested in a college cohort, say from Carnegie Mellon. We may abstract a network where the nodes are the students in the class and the edges express direct relations among them, for example, their friendships or degree of acquaintance. The data can be their grades, extracurricular activities, time spent studying, the particular dormitory where they live. These data relate to each student, and so the data are indexed by the nodes of the friendship network. We may then pursue the question of predicting the grade point average of current juniors in the fall of their senior year. Coupled to the available data associated with the lifestyle and course grades of each student, the underlying cohort network graph should lead to better predictions than treating the data independently of the social relations among students. We consider a second example. Fig. 8.2 on the left represents a small fragment (20,000 users) of a much larger network of subscribers of a cell phone service provider in a given month, say, month m. Nodes are users and their level of activity (who calls whom) is represented by edges among them. The data in this case indicate simply whether a user maintains service in a given month or drops the service; with reference to the left of Fig. 8.2, this binary valued data is given by the two colors labeling the nodes of the graph—blue users keep their service and red users drop or churn the service. Given the past history of subscribers who have churned in each month, the service provider is interested in predicting which subscribers are at risk of churning, i.e., which blue nodes in month m become red in month $m+1$. Because the churn rate is very small, this detection problem is extremely challenging, like finding a needle in a haystack. A successful solution (high probability of detecting churners while keeping the probability of false alarm very small) requires taking into account the underlying graph structure. In [26], we find a set of network features that are input to a classifier, leading to a probability of churn detection of 71% with a positive false alarm rate of 0.08%; see the right of Fig. 8.2. In other words, if the user base is 100 million and the average monthly churn rate is 1%, of the 1 million expected churners, the detector in [26] correctly finds about 700,000 of these potential churners, missing 300,000 of them while misclassifying only 80,000 as churners of the remaining 99,000,000 subscribers. The analytics of the data indexed by nodes of a graph G is the purview of *Data Science* that we consider in this chapter, in particular, we study graph signal processing (GSP) that extends to graph-based data the methods developed over the last 60 years for time and image signals. We introduce GSP by building it from classical discrete signal processing (DSP). To do this, we will provide in section 8.3 a quick abridged review of DSP concepts. But first we briefly review related literature.

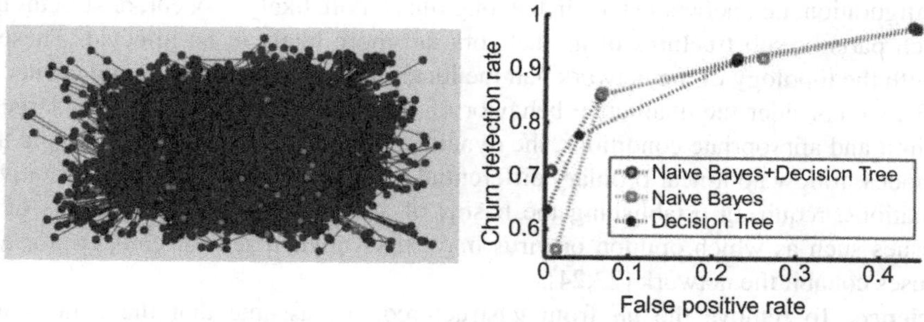

FIG. 8.2

Left: Snippet of a cell phone service provider user network in a given month: *blue dots* (dark gray in print versions) keep the service and *red dots* (light gray in print versions) drop the service in the month of service. Right: Churn detection error performance from [26].

8.2 BRIEF REVIEW OF THE LITERATURE

GSP studies data indexed by graphs. Other approaches that process data indexed by graphs include Markov random fields and, in particular, Gauss-Markov random fields [27–32], graphical models [33–37], data reduction approaches such as [38–41], spectral graph clustering [42–45], or geometric diffusion [46–48], to mention a few. Work on extending wavelets to irregularly spaced data and in sensor networks [49–51] is closer to the perspective of GSP. Reference [52] studies compressed sensing for networked data and [53–56] explore a number of issues such as wavelets on graphs, denoising, and sampling on graphs.

In [57–60], we developed algebraic signal processing for time signals and space signals. The algebraic signal processing theory introduces a signal model as a triplet of an algebra of filters, a signal module, and a linear mapping. We showed that this signal model can be defined under certain conditions by an appropriate definition of the shift filter. The algebraic signal processing leads in a principled way to GSP as we developed in [61–63]. An approach to GSP that uses the graph Laplacian is in [64].

To avoid a too mathematical introduction to GSP, we introduce GSP by illustrating how to cast DSP in the context of GSP. To do this, we start with a brief refresher on DSP in the next section.

8.3 DSP: A QUICK REFRESHER

Discrete signal processing (DSP) [65–69] studies time or image signals.[1] We briefly review time signals and their representations, filtering, delay and impulse response, frequency, spectrum, Fourier, z-transform, and related concepts.

We start with N (complex valued) numbers $\mathcal{S} = \{s_{\alpha_0}, \ldots, s_{\alpha_{N-1}}\}$. To define a time signal, we need an ordering of these numbers so that we know which number precedes and which number succeeds a given number. Let such ordering be given by the N-tuple $s = (s_0, \ldots, s_{N-1})$. Tuples are ordered. We refer to the indexing parameter n as time, and time takes values $n = 0, \ldots, N-1$. Entry s_n of s is the signal sample s_n at time n that precedes the sample s_{n+1} and succeeds sample s_{n-1}. The N-tuple is usually referred to as the (finite) discrete time signal or sequence. Common (abuse of) notation refers to the signal as simply the samples s_n or $s[n]$, where the time interval $n = 0, \ldots, N-1$ is taken for granted.

Remark. To avoid details with the finiteness of time, we assume the signal is extended to the left of $n = 0$ and right of $n = N-1$. Of the many possible extensions, we consider periodic extensions whereby we reduce $s_N = s_0$ and more generally $s_n = s_{n \bmod N}$. ■

The z-transform provides a second useful representation of the signal. The z-transform of the (finite) signal s is then

$$s(z) = s_0 z^0 + s_1 z^{-1} + \cdots + s_{N-1} z^{-(N-1)}.$$

The z-transform is not to be interpreted as a polynomial in the complex variable z. The symbol z^{-1} is a place holder and stands for the delay. The power z^{-n} is the nth delay.

[1] We consider here only linear DSP with a finite number N of samples.

When we develop GSP, we work with vector notation $\mathbf{s} = [s_0 \cdots s_{N-1}]^\top$, so, we recast DSP in vector notation also. The time signal is now interpreted as a vector in $\mathbb{C}^N$. A vector collects *all* signal samples; time is hidden or represented by the index of the coordinates of the vector.

Because $\mathbb{C}^N$ is a vector space, it has a basis with respect to which we can represent any vector as a linear combination of the vectors in the basis. We choose first the standard basis $E = \{e_0, \ldots, e_{N-1}\}$. This is an ordered basis. Using this standard basis in the vector space $\mathbf{C}^N$, the signal $\mathbf{s}$ is written as

$$\mathbf{s} = s_0 \underbrace{\begin{bmatrix} 1 \\ 0 \\ \vdots \\ 0 \end{bmatrix}}_{\mathbf{e}_1} + \cdots + s_{N-1} \underbrace{\begin{bmatrix} 0 \\ \vdots \\ 0 \\ 1 \end{bmatrix}}_{\mathbf{e}_N},$$

where the basis vector e_n is represented in the standard basis by the standard vector $\mathbf{e}_{n+1}$.[2] This equation explicitly indicates that, in the vector representation of the signal, the sampled signal at time $n = 0$ is $s_0\,[1\,0 \cdots 0]^\top$, and more generally at time n it is $s_n \mathbf{e}_n$. The vectors of the standard basis are the (unit) impulses; for example, $\mathbf{e}_n$ is the impulse signal centered at time n, zero everywhere except at time n where it is one. So, the representation above gives the signal s as a linear combination of the unit impulses $\mathbf{e}_n$. This representation in matrix-vector form is

$$\mathbf{s} = \underbrace{[\mathbf{e}_1\,\mathbf{e}_2\,\cdots\,\mathbf{e}_{N-1}]}_{\mathbf{E}=\mathbf{I}_N} \underbrace{\begin{bmatrix} s_0 \\ s_1 \vdots \\ s_{N-1} \end{bmatrix}}_{\mathbf{s}_E},$$

where $\mathbf{E}$ is the matrix of the standard basis vectors that as expected is $\mathbf{I}_N$, the N-dimensional identity matrix. Using the notation in [70], the vector of the representation (called component vector) with respect to the standard basis is $\mathbf{s}_E = \mathbf{s}$.

Because the signal[3] s is an element of the vector space $\mathbf{C}^N$, it can be represented as a linear combination of the vectors of any other ordered basis $B = \{b_0, \ldots, b_{N-1}\}$. These basis' vectors are themselves time signals. To determine $\mathbf{s}_B$ that represents the signal s in the new basis B, let the component vectors representing the vectors of the basis B in the standard basis E be $\mathbf{b}_n$, $n = 0, \ldots, N-1$, and let

$$\mathbf{B} = [\mathbf{b}_0 \cdots \mathbf{b}_{N-1}].$$

The component vector $\mathbf{b}_n$ collects the (ordered) samples of the signal b_n. Then, the signal s is represented in the basis B as

$$\mathbf{s} = \mathbf{B}\,\mathbf{s}_B. \tag{8.1}$$

[2]Representing a basis vector e_n by a vector $\mathbf{e}_{n+1}$ is definitely confusing notation. It will become clear why when we consider, for example, the z-transform of signals.

[3]We emphasize that a signal s is an ordered set of (complex valued) numbers or (complex valued) N-tuple. The vector representation $\mathbf{s}$ of s assumes a reference basis in $\mathbf{C}^N$.

On the left of Eq. (8.1), we have the component vector $\mathbf{s}$ of the signal s with respect to the standard basis in $\mathbf{C}^N$; on the right, we have the component vector $\mathbf{s}_B$ of the same signal s, now with respect to the basis B. Then,

$$\mathbf{s}_B = \mathbf{B}^{-1}\,\mathbf{s}. \tag{8.2}$$

For example, if $B = \left\{\exp\left(j\frac{2\pi}{N}kn\right), k = 0,\ldots,N-1\right\}_{0\leq n\leq N-1}$ is the Fourier basis of N complex exponentials, then $\mathbf{s}_B$ is the vector of Fourier coefficients of the signal s and Eqs. (8.1) and (8.2) are the Fourier synthesis and Fourier analysis expressions of s (inverse Fourier transform and Fourier transform of the signal).

Filters process signals. The simplest filter is the *delay*

$$s_{n-1} = z^{-1} s_n.$$

Consider a signal s_{in} and its delayed version s_{out}. Let their component vectors be $\mathbf{s}_{\text{in}}$ and $\mathbf{s}_{\text{out}}$, respectively. Because delay is a linear operation, we have

$$\mathbf{s}_{\text{out}} = \mathbf{A}\mathbf{s}_{\text{in}}. \tag{8.3}$$

The matrix $\mathbf{A}$ representing the periodic shift is

$$\mathbf{A} = \begin{bmatrix} 0 & 0 & 0 & \cdots & 0 & 1 \\ 1 & 0 & 0 & \cdots & 0 & 0 \\ 0 & 1 & 0 & \cdots & 0 & 0 \\ \vdots & \vdots & \ddots & \ddots & \ddots & 0 \\ 0 & 0 & \cdots & 1 & 0 & 0 \\ 0 & 0 & \cdots & 0 & 1 & 0 \end{bmatrix}. \tag{8.4}$$

This matrix is the cyclic shift. By replacing it in Eq. (8.3),

$$\begin{aligned} \mathbf{s}_{\text{out}} &= \begin{bmatrix} 0 & 0 & 0 & \cdots & 0 & 1 \\ 1 & 0 & 0 & \cdots & 0 & 0 \\ 0 & 1 & 0 & \cdots & 0 & 0 \\ \vdots & \vdots & \ddots & \ddots & \ddots & 0 \\ 0 & 0 & \cdots & 1 & 0 & 0 \\ 0 & 0 & \cdots & 0 & 1 & 0 \end{bmatrix} \mathbf{s}_{\text{in}} \\ &= \begin{bmatrix} 0 & 0 & 0 & \cdots & 0 & 1 \\ 1 & 0 & 0 & \cdots & 0 & 0 \\ 0 & 1 & 0 & \cdots & 0 & 0 \\ \vdots & \vdots & \ddots & \ddots & \ddots & 0 \\ 0 & 0 & \cdots & 1 & 0 & 0 \\ 0 & 0 & \cdots & 0 & 1 & 0 \end{bmatrix} \begin{bmatrix} s_0 \\ s_1 \\ \vdots \\ s_{N-1} \end{bmatrix} \\ &= \begin{bmatrix} s_{N-1} \\ s_0 \\ \vdots \\ s_{N-2} \end{bmatrix}. \end{aligned}$$

A generic filter h is given by its z-transform

$$h(z) = h_0 z^0 + h_1 z^{-1} + \cdots + h_{N-1} z^{-(N-1)}.$$

In vector notation, and with respect to the standard basis E, the filter is represented by the matrix $\mathbf{H}$, a polynomial in the cyclic shift

$$\begin{aligned}\mathbf{H} &= h(\mathbf{A}) \\ &= h_0\mathbf{A}^0 + h_1\mathbf{A}^1 + \cdots + h_{N-1}\mathbf{A}^{(N-1)} \\ &= h_0\mathbf{I} + h_1\mathbf{A} + \cdots + h_{N-1}\mathbf{A}^{(N-1)} \\ &= \begin{bmatrix} h_0 & h_{N-1} & \cdots & h_2 & h_1 \\ h_1 & h_0 & h_{N-1} & & h_2 \\ \vdots & h_1 & h_0 & \ddots & \vdots \\ h_{N-2} & & \ddots & \ddots & h_{N-1} \\ h_{N-1} & h_{N-2} & \cdots & h_1 & h_0 \end{bmatrix}.\end{aligned}$$

This is a circulant matrix. It is of course specified by the filter coefficients that can also be read out from the first row of the filter $\mathbf{H}$.

Filters are *shift invariant*. This can be seen from the z-transform representation

$$z \cdot h(z) = h(z) \cdot z$$

or from the matrix representation

$$\mathbf{A} \cdot h(\mathbf{A}) = h(\mathbf{A}) \cdot \mathbf{A}.$$

Either of these equations states that filtering first by $h(z)$ (or $h(\mathbf{A})$) a signal and then delaying the filtered output leads to the same signal as delaying first the signal by z (or $\mathbf{A}$) and then filtering by $h(z)$ (or $h(\mathbf{A})$) the delayed signal. Finally, we observe that, from the Cayley-Hamilton Theorem [71,72], $\mathbf{A}$ satisfies its characteristic polynomial $\Delta(\mathbf{A})$, where $\Delta(\lambda)$ is the determinant of $\lambda\mathbf{I}-\mathbf{A}$. The characteristic polynomial $\Delta(\mathbf{A})$ has degree N, so, in DSP, as described so far, linear filters are (matrix) polynomials with degree at most $N-1$.

We consider the *Fourier Transform.* Without presenting details, we observe that the shift and filters are circulant matrices. As such, they are diagonalizable and diagonalized by the (inverse) of the discrete Fourier transform matrix (DFT). Because the DFT matrix is unitary and symmetric, its inverse is simply its conjugate DFT*. In other words,

$$\begin{bmatrix} 0 & 0 & 0 & \cdots & \cdots & 1 \\ 1 & 0 & 0 & \cdots & \cdots & 0 \\ 0 & 1 & 0 & \cdots & \cdots & 0 \\ \vdots & \vdots & \ddots & \ddots & \ddots & \vdots \\ 0 & 0 & \cdots & 1 & 0 & 0 \\ 0 & 0 & \cdots & \cdots & 1 & 0 \end{bmatrix} = \mathrm{DFT}^* \cdot \Lambda \cdot \mathrm{DFT} \tag{8.5}$$

$$= \frac{1}{\sqrt{N}} \begin{bmatrix} 1 & 1 & 1 & \cdots & 1 \\ 1 & e^{j\frac{2\pi}{N}} & e^{j\frac{2\pi}{N}2} & \cdots & e^{j\frac{2\pi}{N}(N-1)} \\ 1 & e^{j\frac{2\pi}{N}2} & e^{j\frac{2\pi}{N}4} & \cdots & e^{j\frac{2\pi}{N}2(N-1)} \\ \vdots & & \vdots & & \vdots \\ 1 & e^{j\frac{2\pi}{N}(N-1)} & e^{j\frac{2\pi}{N}(N-1)2} & \cdots & e^{j\frac{2\pi}{N}(N-1)(N-1)} \end{bmatrix} \tag{8.6}$$

$$\cdot \begin{bmatrix} 1 & & & \\ & e^{-j\frac{2\pi}{N}} & & \\ & & \ddots & \\ & & & e^{-j\frac{2\pi}{N}(N-1)} \end{bmatrix} \tag{8.7}$$

$$\cdot \frac{1}{\sqrt{N}} \begin{bmatrix} 1 & 1 & 1 & \cdots & 1 \\ 1 & e^{-j\frac{2\pi}{N}} & e^{-j\frac{2\pi}{N}2} & \cdots & e^{-j\frac{2\pi}{N}(N-1)} \\ 1 & e^{-j\frac{2\pi}{N}2} & e^{-j\frac{2\pi}{N}4} & \cdots & e^{-j\frac{2\pi}{N}2(N-1)} \\ \vdots & & \vdots & & \vdots \\ 1 & e^{-j\frac{2\pi}{N}(N-1)} & e^{-j\frac{2\pi}{N}(N-1)2} & \cdots & e^{-j\frac{2\pi}{N}(N-1)(N-1)} \end{bmatrix}, \tag{8.8}$$

where $(\cdot)^*$ stands for complex conjugate. We remark:

1. **Spectral components**: The columns of the inverse DFT* of the DFT matrix

$$\mathbf{v}_n = \frac{1}{\sqrt{N}} \begin{bmatrix} 1 \\ e^{j\frac{2\pi}{N}n} \\ e^{j\frac{2\pi}{N}2n} \\ \vdots \\ e^{j\frac{2\pi}{N}(N-1)n} \end{bmatrix}, \quad n = 0, \ldots, N-1$$

   are the N eigenvectors of $\mathbf{A}$ and form a complete orthonormal basis for $\mathbf{C}^N$. These vectors are the spectral components.
2. **Frequencies**: The diagonal entries of Λ are the eigenvalues of the time shift $\mathbf{A}$. In Physics and in operator theory, these eigenvalues are the frequencies of the signal. In DSP it is more common to call frequencies

$$\begin{aligned} \Omega_n &= \frac{1}{2\pi j} \ln \lambda_n \\ &= -\frac{1}{2\pi j} \ln e^{-j\frac{2\pi}{N}n} \\ &= \frac{1}{N} n, \quad n = 0, \ldots, N-1. \end{aligned}$$

The N (time) frequencies Ω_n are all distinct, positive, equally spaced, and increasing from 0 to $\frac{N-1}{N}$. The spectral components are the complex exponential sinusoidal functions. For example, corresponding to the zero frequency is the DC spectral component (a vector whose entries are constant and all equal to $\frac{1}{\sqrt{N}}$).

This completes our brief review of DSP. It recasts well known DSP facts in a vector formulation that is appropriate for immediate generalization to GSP, as we show in the next section.

8.4 GRAPH SIGNAL PROCESSING

As we observed in Section 8.3, DSP [65–69] studies time or image signals. GSP[4] [61,62,73] generalizes DSP to graph signals, i.e., signals whose samples are indexed by nodes $v \in V$ of a generic directed, undirected, or mixed graph $G = (V, E)$, rather than necessarily by time instants or pixels of an image as with DSP. We follow the presentation in [61,62]. GSP as introduced in these references and in this chapter recovers DSP when we apply GSP to time signals or images. This is satisfying because when we extend a theory such as DSP to a broader context such as GSP, the new theory should revert to the original theory when applied to the original problem. In other words, GSP as described here if applied to time signals or images should recover the DSP presented in Section 8.3. This is not the case with other versions of GSP.

The main points in Section 8.3 when presenting DSP can be summarized as:

1. **Set of numbers**: Start with a set of N (complex valued) unordered numbers.
2. **Graph signal**: Order the numbers to get a signal (ordered tuple).
3. **Shift**: Define a shift.
4. **Shift invariance**: Assume filters are shift invariant.
5. **Graph spectral analysis**: Diagonalize the shift.

8.4.1 GRAPH SIGNALS

We first discuss graph signals. We start with an unordered set of complex valued numbers, possibly repeated, $\mathcal{S} = \left\{s_{\alpha_0}, \ldots, s_{\alpha_{N-1}}\right\} \subset \mathbb{C}$. We assume that a graph $G = (V, E)$ is given. In G, $V = \{0, \ldots, N-1\}$ is the set of N nodes and E is the set of edges. We further assume that each of the N data in $\mathcal{S}$ is assigned to a node of the graph G. The data in $\mathcal{S}$ is now ordered by the assumed ordering of the nodes of the graph and is given by an N-tuple $s = (s_0, \ldots, s_n, \ldots, s_{N-1})$. We can think of $\mathcal{S}$ as a scrambled version of the samples of the graph signal s. The component s_n of the N-tuple graph signal s, the nth sample of graph signal s, is now indexed by node $n \in V$ of the graph G.

The graph G can be arbitrary, i.e., it may be directed, undirected, or mixed (having both directed and undirected edges). For example, a directed edge from node i to node j captures dependency of the data s_j in node j, or sample s_j, on the data s_i in node i. This dependency will be made clear in Section 8.4.2.

As an example, we consider a time signal and its associated graph. Assume the time signal $s = (s_0, \ldots, s_n, \ldots, s_{N-1})$ where, see Section 8.3, the samples are indexed by time. We construct a graph, see Fig. 8.3, where the N nodes are the time instants 0 through $N-1$ and the label of the nodes are the numerical values of the time samples s_0 through s_{N-1}. Because sample s_{n+1} at time $n+1$ follows the sample s_n at time n, node n is connected to node $n+1$ by a directed edge. Assuming a cyclic or periodic

[4]Like for DSP, we consider here only linear graph signal processing and finite graph signals, i.e., graphs with a finite number N of nodes.

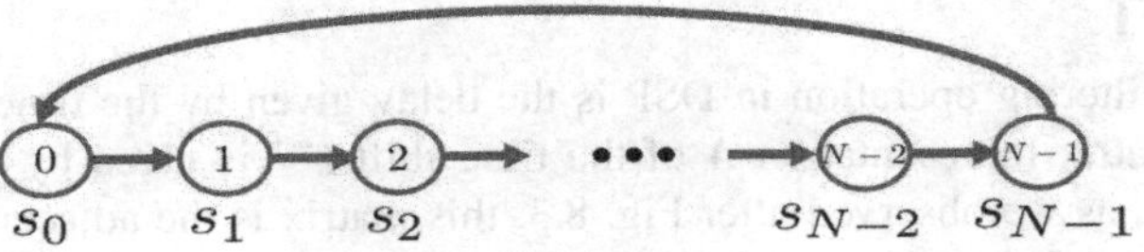

FIG. 8.3

Time cyclic graph.

signal extension $s_n = s_{n \bmod N}$, see remark at the beginning of Section 8.3, node $N-1$ is connected back to node 0. This is a directed cyclic graph and it is associated with time signals.

One way to define a graph is through its adjacency matrix $\mathbf{A}$ [74]. The rows and columns of the adjacency matrix are labeled by its nodes from 0 to $N-1$. Entry $A_{ij} \neq 0$ of $\mathbf{A}$ is a weighted in-edge to node i from node j (or equivalently a weighted out-edge from node j to node i). Node j is an in-neighbor of node i, and node i is an out-neighbor of node j. The set of in-neighbors of i is its in-neighborhood $\mathcal{N}_i$, and similarly for its out-neighborhood. If $A_{ij} = A_{ji} \neq 0$, the edge (i,j) is undirected. If $A_{ij} = 0$, there is no in-edge to node i from node j. The graph may have cycles. Because A_{ij} can be arbitrarily valued, the graph adjacency matrix is a weighted matrix. Returning to time signals and the associated graph in Fig. 8.3, we note that the adjacency matrix of the time cyclic graph is the cyclic shift matrix in Eq. (8.4).

Because the graph signal assigns to each node $n \in V$ one of the complex numbers in $\mathcal{S}$, a graph signal s is a map

$$s: V \to \mathbb{C}.$$

As mentioned in Section 8.3, we work with vectors rather than tuples. With respect to the standard basis in $\mathbb{C}^N$, the graph signal is given by the component vector

$$\mathbf{s} = \begin{bmatrix} s_0 \\ \vdots \\ s_{N-1} \end{bmatrix} \in \mathbb{C}^N.$$

As an example, the impulse graph signal centered at node n is given by

$$\delta_n = \begin{bmatrix} 0 \\ \vdots \\ 1 \\ \vdots \\ 0 \end{bmatrix},$$

which is zero everywhere, except at node n where it is one.

8.4.2 GRAPH SHIFT

The most elementary filtering operation in DSP is the delay given by the time shift z^{-1}. We saw in Section 8.3 that the matrix representation $\mathbf{A}$ of the time shift z^{-1} is given by the cyclic shift matrix given in Eq. (8.4). But, as we observed after Fig. 8.3, this matrix is the adjacency matrix of the time graph illustrated in Fig. 8.3. In other words, the shift matrix $\mathbf{A}$ representing the time shift z^{-1} is the adjacency matrix of the graph associated with the time signal.

We adopt here (in reverse) this insight and propose as graph shift $\mathbf{A}$ for graph signals s indexed by nodes of an arbitrary graph $G = (V, E)$ the adjacency matrix of the graph. In other words, the shifted version of the sample s_n of the graph signal is given by a weighted sum of the samples of the graph signal at the in-neighbors of node n

$$\begin{aligned} s_n &= \sum_{i=0}^{N-1} A_{ni} s_i \\ &= \sum_{i \in \mathcal{N}_n} A_{ni} s_i, \end{aligned}$$

where the weights A_{ni} are the entries of the adjacency matrix $\mathbf{A}$ and $\mathcal{N}_n$ is the in-neighborhood of node n. Collecting all the data in the signal vector $\mathbf{s}$, we get that the output $\mathbf{s}_{\text{out}}$ of the graph shift $\mathbf{A}$ to the input $\mathbf{s}_{\text{in}} = \mathbf{s}$ is given by

$$\mathbf{s}_{\text{out}} = \mathbf{A}\mathbf{s}_{\text{in}}, \tag{8.9}$$

where, we emphasize again, the graph shift is the adjacency $\mathbf{A}$ of the graph.

Remark. The samples of the graph signal are indexed by the nodes of the graph. Relabeling the nodes changes the graph signal component vector and the shift matrix. Let π be a permutation of the nodes of the graph. Then, for example, Eq. (8.9) becomes

$$\pi\, \mathbf{s}_{\text{out}} = \pi\, \mathbf{A}\pi^T \pi \mathbf{s}_{\text{in}},$$

where recall $\pi^{-1} = \pi^T$. Relabeling the nodes, correspondingly permutes the entries of the graph signal and the shift matrix is conjugated by the same permutation π, i.e., the rows of $\mathbf{A}$ are reordered by π and the columns reordered by the inverse of π. ∎

8.4.3 GRAPH FILTERS AND GRAPH CONVOLUTION

A graph filter h is represented by a matrix $\mathbf{H}$, and *graph filtering* or *graph filtering convolution* by h is matrix-vector multiplication

$$\mathbf{s}_{\text{out}} = \mathbf{H} \cdot \mathbf{s}_{\text{in}}.$$

Shift Invariance. In DSP, shift invariant filters play an important role. In this chapter, we restrict attention to shift invariant filters. Shift invariance means that we can commute the operations of shifting and filtering. In other words, the filter is shift invariant if first shifting the input signal s_{in} and then filtering the shifted input by the graph filter h or first filtering s_{in} by h and then shifting the output signal leads to the same graph signal. In matrix-vector product form

$$\underbrace{\mathbf{A}\cdot\underbrace{(\mathbf{H}\cdot\mathbf{s}_{in})}_{\text{filter input}}}_{\text{shift filtered input}} = \underbrace{\mathbf{H}\cdot\underbrace{(\mathbf{A}\cdot\mathbf{s}_{in})}_{\text{shift input}}}_{\text{filter shifted input}}$$

Because this is true for every graph signal, we conclude that for shift invariant filters

$$\mathbf{A}\cdot\mathbf{H} = \mathbf{H}\cdot\mathbf{A}, \tag{8.10}$$

and the graph filter h and the graph shift commute.

As explained in [58,61,62,72], if $\Delta(\mathbf{A}) = m(\mathbf{A})$, where $\Delta(\mathbf{A})$ and $m(\mathbf{A})$ are the characteristic polynomial and the minimal polynomial of the shift $\mathbf{A}$, then the shift invariant filters h are polynomials of the shift. In other words, if Eq. (8.10) holds for graph filter $\mathbf{H}$, then

$$\mathbf{H} = h(\mathbf{A}),$$

where $h(\cdot)$ is, by the Cayley-Hamilton Theorem [75], a polynomial of degree at most $N-1$. Also, if the shift is diagonalizable, shift invariant filters and the shift share the same (complete set of orthonormal) eigenvectors. This means that if $\mathbf{v}_n$ is a spectral component (eigenvector of $\mathbf{A}$) with corresponding eigenvalue λ_n

$$\mathbf{H}\cdot\mathbf{v}_n = h(\lambda_n)\,\mathbf{v}_n,$$

i.e., although graph filtering by $\mathbf{H}$ is in general a matrix-vector multiplication, graph filtering of a spectral component reduces to scalar multiplication by the polynomial filter evaluated at the eigenvalue λ_n. In words, the spectral components are *invariant* to graph filters.

8.4.4 GRAPH FOURIER TRANSFORM AND GRAPH SPECTRAL DECOMPOSITION

We now consider the graph Fourier transform (GFT). For simplicity, we focus on shifts $\mathbf{A}$ with N distinct eigenvalues λ_n, $n = 0, \ldots, N-1$. These shifts $\mathbf{A}$ are diagonalizable and their characteristic and minimal polynomials are equal, $\Delta(\mathbf{A}) = m(\mathbf{A})$. As argued in Section 8.4.3, shift invariant filters are then polynomials of the shift. The general case of $k < N$ distinct eigenvalues and nondiagonalizable shifts is considered in [61–63] and fully treated in [76]; we refer the reader to these references for full details.

In an analogy with classical DSP and Section 8.3, see Eq. (8.5), the inverse of the graph Fourier transform is the matrix of eigenvectors of the shift

$$\begin{aligned}\mathbf{A} &= \underbrace{\left[\mathbf{v}_1\cdots\mathbf{v}_{N-1}\right]}_{\mathbf{GFT}^{-1}}\operatorname{diag}\left[\lambda_1,\ldots,\lambda_{N-1}\right]\underbrace{\left[\mathbf{v}_1\cdots\mathbf{v}_{N-1}\right]^{-1}}_{\mathbf{GFT}} &(8.11)\\ &= \mathbf{GFT}^{-1}\,\Lambda\,\mathbf{GFT} &(8.12)\\ &= \mathbf{GFT}^{H}\,\Lambda\,\mathbf{GFT}, &(8.13)\end{aligned}$$

where $\mathbf{GFT}^{-1}$ is the matrix of eigenvectors of $\mathbf{A}$ and diag is a diagonal matrix. Because $\mathbf{A}$ is diagonalizable, it has a complete eigenbasis. So the graph Fourier transform $\mathbf{GFT}$ is unitary and $\mathbf{GFT}^{-1} = \mathbf{GFT}^H$.

In the sequel, we also use the following representation for the GFT.

$$\begin{aligned}\mathbf{GFT} = \mathbf{W}^H &= \begin{bmatrix} \mathbf{w}_1^H \\ \vdots \\ \mathbf{w}_{N-1}^H \end{bmatrix} \\ &= \begin{bmatrix} \mathbf{v}_1^H \\ \vdots \\ \mathbf{v}_{N-1}^H \end{bmatrix},\end{aligned}$$

because the $\mathbf{GFT}$ is unitary.

The N distinct eigenvalues $\{\lambda_n\}_{0\leq n\leq N-1}$ of $\mathbf{A}$ are the graph frequencies and the eigenvectors $\{\mathbf{v}_n\}_{0\leq n\leq N-1}$ of $\mathbf{A}$ that are the columns of $\mathbf{GFT}^H$ are the graph spectral components.

Given a signal s, with vector component $\mathbf{s}$, its graph Fourier transform is then

$$\begin{aligned}\hat{\mathbf{s}} &= \mathbf{GFT} \cdot \mathbf{s} \\ &= \begin{bmatrix} \mathbf{v}_1^H \cdot \mathbf{s} \\ \vdots \\ \mathbf{v}_{N-1}^H \cdot \mathbf{s} \end{bmatrix}.\end{aligned}$$

This is the analysis formula of the graph Fourier transform. The synthesis of the original signal from its graph Fourier decomposition is

$$\begin{aligned}\mathbf{s} &= \mathbf{GFT}^H \cdot \hat{\mathbf{s}} \\ &= \sum_{n=0}^{N-1} \hat{s}_n \mathbf{v}_n \\ &= \sum_{n=0}^{N-1} \left(\mathbf{v}_n^H \cdot \mathbf{s}\right) \mathbf{v}_n \\ &= \sum_{n=0}^{N-1} \langle \mathbf{v}_n^H, \mathbf{s}\rangle \, \mathbf{v}_n,\end{aligned} \tag{8.14}$$

where $\mathbf{v}_n$ are the spectral components. In Eq. (8.14), $\langle \cdot, \cdot \rangle$ is the dot or scalar product. This is the graph signal equivalent expansion of the time domain Fourier series representation of the signal.

Remark. The vectors $\{\mathbf{w}_n\}_{0\leq n\leq N-1}$ are the left graph eigenvectors of the graph shift $\mathbf{A}$. Because the graph shift $\mathbf{A}$ is assumed to be diagonalizable, the left and right eigenvectors $\mathbf{w}_n$ and $\mathbf{v}_n$ can be chosen to be the same. ∎

Graph Filter Frequency Response. Given a shift invariant filter h, its graph frequency response is

$$\left(h\left(\lambda_0\right),\ldots,h\left(\lambda_{N-1}\right)\right),$$

the polynomial filter h evaluated at the graph frequencies $\lambda_n, n = 0,\ldots,N-1$.

Graph Convolution Theorem. Following [61], see Equation (27) therein, we have the following result for filtering with shift invariant filters

$$\begin{aligned}
\mathbf{s}_{\text{out}} &= \mathbf{H}\cdot\mathbf{s}_{\text{in}} \\
&= h(\mathbf{A})\cdot\mathbf{s}_{\text{in}} \\
&= h\left(\mathbf{GFT}^H\Lambda\,\mathbf{GFT}\right)\cdot\mathbf{s}_{\text{in}} \\
&= \underbrace{\mathbf{GFT}^H\cdot\underbrace{\left[h(\Lambda)\cdot\underbrace{\mathbf{GFT}\cdot\mathbf{s}_{\text{in}}}_{\text{GFT of s}}\right]}_{\text{Filtering in spectral domain}}}_{\text{Inverse GFT}}.
\end{aligned} \tag{8.15}$$

This is the graph signal processing version of the Convolution Theorem of classical DSP. Filtering can be accomplished in the node domain as a matrix-vector multiplication. In alternative, as per Eq. (8.15), (1) we first GFT the input graph signal by $\hat{\mathbf{s}}_{\text{in}} = \mathbf{GFT}\cdot\mathbf{s}_{\text{in}}$, then (2) graph filter in the spectral domain as a pointwise multiplication of $\hat{\mathbf{s}}_{\text{in}}$, the GFT of the graph signal $\mathbf{s}_{\text{in}}$, with the graph frequency response $h(\Lambda)$ of the graph filter h, followed, finally, by (3) the inverse GFT of the graph filtered output in the graph spectral domain as $\mathbf{GFT}^H\cdot\left[h(\Lambda)\cdot\mathbf{GFT}\cdot\mathbf{s}_{\text{in}}\right]$.

Graph Frequency Ordering. To define the concepts of low, band, and high pass signals and filters, we need to order the graph frequencies. Because the underlying graph most likely will contain direct edges, the graph shift $\mathbf{A}$ is in general nonsymmetric and hence the graph frequencies may be complex valued. Because the complex numbers $\mathbb{C}$ is not an ordered field, there remains then the question of how to order the graph frequencies. In [63], we chose to order the graph frequencies through the total variation of the spectral components.

For time signals, the total variation of a signal $\mathbf{s}$ is

$$\text{TV}(\mathbf{s}) = \sum_{n=0}^{N-1}\left|s_n - s_{n-1 \bmod N}\right|.$$

The total variation TV($\mathbf{s}$) of the time signal $\mathbf{s}$ measures how the signal varies over time. In [63], see equation below Fig. 3, we compute the TV for the time spectral components as

$$\text{TV}\left(\mathbf{v}_n\right) = \left|1-\cos\frac{2\pi n}{N}\right| + \left|\sin\frac{2\pi n}{N}\right|.$$

This expression gives that $\text{TV}\left(\mathbf{v}_n\right) = \text{TV}\left(\mathbf{v}_{N-n}\right)$. The total variation for the DC-spectral component of time signals is zero. This agrees with our intuition—because the DC-spectral component of a time signals is a constant signal, its total variation across time is zero. This corresponds to the lowest time frequency $\Omega_0 = 0$. We can also compute from the above expression that the $\text{TV}\left(\mathbf{v}_n\right)$ is strictly increasing from $n = 0$ to $n = \frac{N}{2}$, if N is even, or $\frac{N-1}{2}$ if N is odd. So, with time signals, the frequencies can be ordered by ordering the associated spectral components according to their TV.

We use this method of ordering time frequencies to order the graph frequencies. From [63], the graph total variation of spectral component $\mathbf{v}$ is

$$\mathrm{TV_G}(\mathbf{v}) = \left\| \mathbf{v} - \mathbf{A}^{\mathrm{norm}} \mathbf{v} \right\|_1,$$

where the normalized graph shift is

$$\mathbf{A}^{\mathrm{norm}} = \frac{1}{|\lambda_{\max}|} \mathbf{A},$$

and $\lambda_{\max}$ is the eigenvalue of $\mathbf{A}$ with largest magnitude, i.e., $|\lambda_{\max}| \geq |\lambda_n|, \forall n$. This normalization is used to guarantee that the shifted signal is properly scaled for comparison with the original nonshifted graph signal. It follows that, for spectral component $\mathbf{v}$,

$$\mathrm{TV_G}(\mathbf{v}) = \left| 1 - \frac{\lambda}{|\lambda_{\max}|} \right| \|\mathbf{v}\|_1,$$

where $\| \cdot \|_1$ is the vector ℓ_1 norm. From here, it follows Theorem 2 in [63] that states that, given two distinct complex eigenvalues or graph frequencies λ_m and λ_n of the graph shift $\mathbf{A}$ with eigenvectors or spectral components $\mathbf{v}_m$ and $\mathbf{v}_n$, respectively, the total variation of these spectral components satisfies

$$\mathrm{TV}_G(\mathbf{v}_m) < \mathrm{TV}_G(\mathbf{v}_n)$$

if the graph frequency λ_m is located closer to the value $|\lambda_{\max}|$ on the complex plane than the graph frequency λ_n. Using these results we can order the graph frequencies and from this ordering define low, band, and high pass graph signals and graph filters; see [63] where we illustrate these concepts with the dataset of hyperlinked political blogs [25].

8.4.5 FURTHER TOPICS AND APPLICATIONS

There is significant work that extends GSP to many other areas in signal and data processing. Sampling is considered in, for example [55,77–80], and interpolation in [73]. Uncertainty principles are in [81–83]. Reference [84] considers approximation of signals supported on graphs. References [85,86] extend classical multirate signal processing theory to graphs. In [87–89], the authors consider interpolation of graph signals, spectral estimation of graph processes, and blind identification of graph filters. A recent topic of interest is graph learning from data; see a sample of work related to this topic in [90–94].

There is an increasing set of applications of GSP. In our original work [61,62,73], we illustrated the application of GSP in filter design for denoising, for classification, for lossy compression of graph signals, and for analysis of graph signals, for detection of malfunctioning in sensor networks. Reference [95] applies GSP to image inpainting and other problems of graph signal recovery by variation minimization. In [96], GSP is applied to semisupervised multiresolution classification using adaptive graph filtering with application to indirect bridge structural health monitoring and [97] applies GSP to fast resampling of three-dimensional point clouds via graphs. In [98], multiresolution representations for piecewise-smooth signals on graphs are developed. In recent work [99], we apply GSP to convolutional neural networks (CNN), replacing the image or regular lattice-like convolution step in CNN by graph convolution, leading to noticeable increased performance over other existing approaches.

In this chapter, we studied GSP assuming the graph shift is the adjacency matrix of the graph. There is a healthy literature that develops GSP with other definitions for the graph shift. In [64], GSP is developed adopting the graph Laplacian as shift and several subsequent works have adopted this shift. In [100,101], variations on the adjacency matrix as shift are discussed with different trade-offs with respect to adopting the adjacency matrix as shift as done here and in our original work [61]. We note that, following the algebraic signal processing [57–60,102], there is no "optimal" or "best choice" of shift. Different shifts lead to different signal models, and the question is which of these is better suited and leads to better results in the context of a specific application.

A final note on the spectral decomposition of graph signals. In [76], we show that in many applications the shift has repeated eigenvalues or is not diagonalizable. The graph Fourier transform is then dependent on choices of basis associated with corresponding invariant or spectral subspaces. So, care should be taken to completely specify these choices before claiming results on graph Fourier decompositions. Also, the concepts of frequency and spectral components are then to be handled with care. Further, if the shift is not diagonalizable, we work with the Jordan form of the shift that is numerically sensitive and the spectral components are (oblique) projections on subspaces of dimension larger than one. We refer to [76] for further details.

8.5 CONCLUSION

This chapter considered GSP that extends classical DSP to data that is indexed by the nodes of a graph. We studied linear GSP with an underlying graph with a finite number of nodes. We introduced the main concepts of classic DSP but for graph signals, i.e., data indexed by the nodes of the graph G. The shift plays an important role in GSP. We worked here with the shift that is identified with the adjacency matrix of the graph. Shift invariant filters are polynomials of the shift. We discussed the graph Fourier transform that introduces the graph frequency, and, by a suitable ordering of the graph frequencies, we were able to define low, band, and high pass graph signals and graph filters.

REFERENCES

[1] Turner V. The digital universe of opportunities: rich data and the increasing value of the internet of things. tech. rep.. IDC; 2014. https://www.emc.com/leadership/digital-universe/2014iview/executive-summary.htm.

[2] Johnston L. A "Library of Congress" worth of data: it's all in how you define it; 2012. https://blogs.loc.gov/thesignal/2012/04/a-library-of-congress-worth-of-data-its-all-in-how-you-define-it/ [Recovered 17 December 2017].

[3] Reinsel D, Gantz J, Rydning J. Data age 2025: the evolution of data to life-critical don't focus on big data; focus on the data that's big. tech. rep.. IDC; 2017. An IDC White Paper, Sponsored by Seagate. https://www.seagate.com/files/www-content/our-story/trends/files/Seagate-WP-DataAge2025-March-2017.pdf.

[4] Weissberger A. IDC directions 2016: IOT (internet of things) outlook vs current market assessment. IEEE Communications Society, ComSoc blog; 2016. http://techblog.comsoc.org/2016/03/09/idc-directions-2016-iot-internet-of-things-outlook-vs-current-market-assessment/.

[5] Börner K, Sanyal S, Vespignani A. Network science. Ann Rev Inf Sci Technol 2007;41(1):537–607.

[6] Newman M. Networks: an introduction. Oxford University Press; 2010.

[7] Jackson MO. Social and economic networks. Princeton University Press; 2010.
[8] Easley D, Kleinberg J. Networks, crowds, and markets: reasoning about a highly connected world. Cambridge University Press; 2010.
[9] National Research Council. Network science. Washington, DC: The National Academies Press; 2005. ISBN 978-0-309-10026-7. https://doi.org/10.17226/11516. https://www.nap.edu/catalog/11516/network-science.
[10] Padgett JF, Ansell CK. Robust action and the rise of the Medici, 1400-1434. Am J Sociol 1993;98(6):1259–319.
[11] Bbuuggzz. 15th century Florentine marriages data from Padgett and Ansell.pdf. File licensed under the Creative Commons Attribution-Share Alike 3.0 Unported license; 2013. https://commons.wikimedia.org/wiki/File:15th_Century_Florentine_Marriges_Data_from_Padgett_and_Ansell.pdf [Retrieved 8 December 2017].
[12] Freeman LC. Centrality in social networks conceptual clarification. Soc Networks 1978;1(3):215–39.
[13] Barabási AL. Network science. Cambridge University Press; 2016.
[14] Barrat A, Barthelemy M, Vespignani A. Dynamical processes on complex networks. Cambridge University Press; 2008.
[15] Liggett TM. Interacting particle systems, vol. 276. Springer Science & Business Media; 2012.
[16] Kipnis C, Landim C. Scaling limits of interacting particle systems, vol. 320. Springer Science & Business Media; 2013.
[17] Pastor-Satorras R, Castellano C, Van Mieghem P, Vespignani A. Epidemic processes in complex networks. Rev Mod Phys 2015;87(3):925.
[18] Lanchier N. Interacting particle systems. In: Stochastic modeling. Springer; 2017. p. 235–44.
[19] Zhang J, Moura JMF. Diffusion in social networks as SIS epidemics: beyond full mixing and complete graphs. IEEE J Sel Top Signal Process 2014;8(4):537–51.
[20] Zhang J, Moura JMF. Role of subgraphs in epidemics over finite-size networks under the scaled SIS process. J Complex Networks 2015;3(4):584–605.
[21] Kelly FP. Reversibility and stochastic networks. Cambridge University Press; 2011.
[22] Santos A, Moura JMF, Xavier JM. Bi-virus SIS epidemics over networks: qualitative analysis. IEEE Trans Netw Sci Eng 2015;2(1):17–29.
[23] Santos AA, Kar S, Moura JMF, Xavier J. Thermodynamic limit of interacting particle systems over dynamical networks. In: 2016 50th Asilomar conference on signals, systems and computers. IEEE; 2016. p. 997–1000.
[24] Santos A, Moura JMF, Xavier JM. Sufficient condition for survival of the fittest in a bi-virus epidemics. In: 2015 49th Asilomar conference on signals, systems and computers. IEEE; 2015. p. 1323–7.
[25] Adamic LA, Glance N. The political blogosphere and the 2004 U.S. election: divided they blog. In: Proceedings of the 3rd international workshop on link discovery. ACM; 2005. p. 36–43.
[26] Deri JA, Moura JMF. Churn detection in large user networks. In: 2014 IEEE international conference on acoustics, speech and signal processing (ICASSP). IEEE; 2014. p. 1090–4.
[27] Kindermann R, Snell JL. Markov random fields and their applications, vol. 1. American Mathematical Society; 1980.
[28] Rue H, Held L. Gaussian Markov random fields: theory and applications. CRC Press; 2005.
[29] Besag J. Spatial interaction and the statistical analysis of lattice systems. J R Stat Soc Ser B Methodol 1974:192–236.
[30] Chellappa R, Jain A, editors. Markov random fields. Theory and application. Boston, MA: Academic Press; 1993.
[31] Moura JMF, Balram N. Recursive structure of noncausal Gauss-Markov random fields. IEEE Trans Inf Theory 1992;38(2):334–54.
[32] Balram N, Moura JMF. Noncausal Gauss Markov random fields: parameter structure and estimation. IEEE Trans Inf Theory 1993;39(4):1333–55.

[33] Lauritzen SL. Graphical models, vol. 17. Clarendon Press; 1996.
[34] Wainwright MJ, Jordan MI. Graphical models, exponential families, and variational inference. In: Foundations and trends® in machine learning, vol. 1(1–2); 2008. p. 1–305.
[35] Koller D, Friedman N. Probabilistic graphical models: principles and techniques. MIT Press; 2009.
[36] Jordan M, Sudderth E, Wainwright M, Willsky A. Major advances and emerging developments of graphical models [from the guest editors]. IEEE Signal Process Mag 2010;27(6):17–138.
[37] Sudderth EB, Ihler AT, Isard M, Freeman WT, Willsky AS. Nonparametric belief propagation. Commun ACM 2010;53(10):95–103.
[38] Tenenbaum JF, Silva V, Langford JC. A global geometric framework for nonlinear dimensionality reduction. Science 2000;290:2319–23.
[39] Roweis S, Saul L. Nonlinear dimensionality reduction by locally linear embedding. Science 2000;290:2323–6.
[40] Belkin M, Niyogi P. Laplacian eigenmaps for dimensionality reduction and data representation. Neural Comput 2003;15(6):1373–96.
[41] Donoho DL, Grimes C. Hessian eigenmaps: locally linear embedding techniques for high-dimensional data. Proc Natl Acad Sci U S A 2003;100(10):5591–6.
[42] Belkin M, Niyogi P. Using manifold structure for partially labeled classification. In: Advances in neural information processing systems (NIPS); 2002. p. 953–60.
[43] Belkin M, Niyogi P. Laplacian eigenmaps and spectral techniques for embedding and clustering. In: Advances in neural information processing systems; 2002. p. 585–91.
[44] Ng AY, Jordan MI, Weiss Y. On spectral clustering: analysis and an algorithm. In: Advances in neural information processing systems; 2002. p. 849–56.
[45] Schaeffer SE. Graph clustering. Comput Sci Rev 2007;1(1):27–64.
[46] Coifman RR, Lafon S, Lee A, Maggioni M, Nadler B, Warner FJ, Zucker SW. Geometric diffusions as a tool for harmonic analysis and structure definition of data: diffusion maps. Proc Nat Acad Sci 2005;102(21):7426–31.
[47] Coifman RR, Lafon S, Lee A, Maggioni M, Nadler B, Warner FJ, Zucker SW. Geometric diffusions as a tool for harmonic analysis and structure definition of data: multiscale methods. Proc Natl Acad Sci U S A 2005;102(21):7432–7.
[48] Coifman RR, Maggioni M. Diffusion wavelets. Appl Comput Harmon Anal 2006;21(1):53–94.
[49] Ganesan D, Greenstein B, Estrin D, Heidemann J, Govindan R. Multiresolution storage and search in sensor networks. ACM Trans Storage 2005;1:277–315.
[50] Wagner R, Cohen A, Baraniuk RG, Du S, Johnson DB. An architecture for distributed wavelet analysis and processing in sensor networks. In: Information processing in sensor networks (IPSN); 2006. p. 243–50.
[51] Wagner R, Delouille V, Baraniuk RG. Distributed wavelet de-noising for sensor networks. In: Proc. IEEE conference on decision and control (CDC); 2006. p. 373–9.
[52] Haupt J, Bajwa WU, Rabbat M, Nowak R. Compressed sensing for networked data. IEEE Signal Process Mag 2008;25(2):92–101.
[53] Narang SK, Ortega A. Local two-channel critically sampled filter-banks on graphs. In: IEEE international conference on image processing (ICIP); 2010. p. 333–6.
[54] Hammond DK, Vandergheynst P, Gribonval R. Wavelets on graphs via spectral graph theory. J Appl Comput Harmon Anal 2011;30(2):129–50.
[55] Narang SK, Ortega A. Downsampling graphs using spectral theory. In: IEEE international conference on acoustics, speech and signal processing (ICASSP); 2011. p. 4208–11.
[56] Narang SK, Ortega A. Perfect reconstruction two-channel wavelet filter banks for graph structured data. IEEE Trans Signal Process 2012;60(6):2786–99.
[57] Püschel M, Moura JMF. The algebraic approach to the discrete cosine and sine transforms and their fast algorithms. SIAM J Comp 2003;32(5):1280–316.

[58] Püschel M, Moura JMF. Algebraic signal processing theory; 2006. 67 p. http://arxiv.org/abs/cs.IT/0612077.

[59] Püschel M, Moura JMF. Algebraic signal processing theory: foundation and 1-D time. IEEE Trans Signal Process 2008;56(8):3572–85.

[60] Püschel M, Moura JMF. Algebraic signal processing theory: 1-D space. IEEE Trans Signal Process 2008;56(8):3586–99.

[61] Sandryhaila A, Moura JMF. Discrete signal processing on graphs. IEEE Trans Signal Process 2013;61(7):1644–56.

[62] Sandryhaila A, Moura JMF. Big data analysis with signal processing on graphs: representation and processing of massive data sets with irregular structure. IEEE Signal Process Mag 2014;31(5):80–90.

[63] Sandryhaila A, Moura JMF. Discrete signal processing on graphs: frequency analysis. IEEE Trans Signal Process 2014;62(12):3042–54.

[64] Shuman DI, Narang SK, Frossard P, Ortega A, Vandergheynst P. The emerging field of signal processing on graphs: extending high-dimensional data analysis to networks and other irregular domains. IEEE Signal Process Mag 2013;30:83–98.

[65] Oppenheim AV, Schafer RW. Digital signal processing. Englewood Cliffs, NJ: Prentice-Hall; 1975.

[66] Oppenheim AV, Willsky AS. Signals and systems. Englewood Cliffs, NJ: Prentice-Hall; 1983.

[67] Siebert WM. Circuits, signals, and systems. Cambridge, MA: The MIT Press; 1986.

[68] Oppenheim AV, Schafer RW. Discrete-time signal processing. Englewood Cliffs, NJ: Prentice-Hall; 1989.

[69] Mitra SK. Digital signal processing. A computer-based approach. New York: McGraw Hill; 1998.

[70] Bendersky E. Change of basis in linear algebra (posted on web); 2015. https://eli.thegreenplace.net/2015/change-of-basis-in-linear-algebra/ [Accessed 11 December 2017].

[71] Gantmacher FR. Matrix theory. New York: Chelsea; 1959. p. 21.

[72] Lancaster P, Tismenetsky M. The theory of matrices: with applications. Elsevier; 1985.

[73] Narang SK, Gadde A, Ortega A. Signal processing techniques for interpolation in graph structured data. In: 2013 IEEE international conference on acoustics, speech and signal processing (ICASSP). IEEE; 2013. p. 5445–9.

[74] Chung FRK. Spectral graph theory. American Mathematical Society; 1996.

[75] Gantmacher FR, Brenner JL. Applications of the theory of matrices. Courier Corporation; 2005.

[76] Deri JA, Moura JMF. Spectral projector-based graph Fourier transforms. IEEE J Sel Top Signal Process 2017;11(6):785–95. https://doi.org/10.1109/JSTSP.2017.2731599.

[77] Anis A, Gadde A, Ortega A. Towards a sampling theorem for signals on arbitrary graphs. In: 2014 IEEE international conference on acoustics, speech and signal processing (ICASSP). IEEE; 2014. p. 3864–8.

[78] Chen S, Varma R, Sandryhaila A, Kovačević J. Discrete signal processing on graphs: sampling theory. IEEE Trans Signal Process 2015;63(24):6510–23.

[79] Puy G, Tremblay N, Gribonval R, Vandergheynst P. Random sampling of bandlimited signals on graphs. Appl Comput Harmon Anal 2018;44(2):446–75.

[80] Anis A, Gadde A, Ortega A. Efficient sampling set selection for bandlimited graph signals using graph spectral proxies. IEEE Trans Signal Process 2016;64(14):3775–89.

[81] Agaskar A, Lu YM. Uncertainty principles for signals defined on graphs: bounds and characterizations. In: 2012 IEEE international conference on acoustics, speech and signal processing (ICASSP). IEEE; 2012. p. 3493–6.

[82] Agaskar A, Lu YM. A spectral graph uncertainty principle. IEEE Trans Inf Theory 2013;59(7):4338–56.

[83] Pasdeloup B, Alami R, Gripon V, Rabbat M. Toward an uncertainty principle for weighted graphs. In: 2015 23rd European signal processing conference (EUSIPCO). IEEE; 2015. p. 1496–500.

[84] Zhu X, Rabbat M. Approximating signals supported on graphs. In: IEEE international conference on acoustics, speech and signal processing (ICASSP); 2012. p. 3921–24.

[85] Teke O, Vaidyanathan P. Extending classical multirate signal processing theory to graphs—Part I: Fundamentals. IEEE Trans Signal Process 2017;65(2):409–22.

[86] Teke O, Vaidyanathan P. Extending classical multirate signal processing theory to graphs—Part II: M-channel filter banks. IEEE Trans Signal Process 2017;65(2):423–37.

[87] Segarra S, Marques AG, Leus G, Ribeiro A. Interpolation of graph signals using shift-invariant graph filters. In: 2015 23rd European signal processing conference (EUSIPCO). IEEE; 2015. p. 210–4.

[88] Marques AG, Segarra S, Leus G, Ribeiro A. Stationary graph processes and spectral estimation; 2016. arXiv preprint arXiv:160304667.

[89] Segarra S, Mateos G, Marques AG, Ribeiro A. Blind identification of graph filters. IEEE Trans Signal Process 2017;65(5):1146–59.

[90] Mei J, Moura JMF. Fitting graph models to big data. In: 2015 49th Asilomar conference on signals, systems and computers. IEEE; 2015. p. 387–90.

[91] Mei J, Moura JMF. Signal processing on graphs: estimating the structure of a graph. In: 2015 IEEE international conference on acoustics, speech and signal processing (ICASSP). IEEE; 2015. p. 5495–9.

[92] Mei J, Moura JMF. Signal processing on graphs: performance of graph structure estimation. In: 2016 IEEE international conference on acoustics, speech and signal processing (ICASSP). IEEE; 2016. p. 6165–9.

[93] Mei J, Moura JMF. Signal processing on graphs: causal modeling of unstructured data. IEEE Trans Signal Process 2017;65(8):2077–92.

[94] Egilmez HE, Pavez E, Ortega A. Graph learning from data under Laplacian and structural constraints. IEEE J Sel Top Signal Process 2017;11(6):825–41.

[95] Chen S, Sandryhaila A, Moura JMF, Kovačević J. Signal recovery on graphs: variation minimization. IEEE Trans Signal Process 2015;63(17):4609–24.

[96] Chen S, Cerda F, Rizzo P, Bielak J, Garrett JH, Kovačević J. Semi-supervised multiresolution classification using adaptive graph filtering with application to indirect bridge structural health monitoring. IEEE Trans Signal Process 2014;62(11):2879–93.

[97] Chen S, Tian D, Feng C, Vetro A, Kovačević J. Fast resampling of 3D point clouds via graphs. IEEE Trans Signal Process 2017;66(3):666–81.

[98] Chen S, Singh A, Kovačević J. Multiresolution representations for piecewise-smooth signals on graphs. https://arxiv.org/abs/1803.02944; March 2018.

[99] Du J, Zhang S, Wu G, Moura JMF, Kar S. Topology adaptive graph convolutional networks; 2017. ArXiv preprint arXiv:1710.10370.

[100] Girault B, Gonçalves P, Fleury E. Translation on graphs: an isometric shift operator. IEEE Signal Process Lett 2015;22(12):2416–20. https://doi.org/10.1109/LSP.2015.2488279.

[101] Gavili A, Zhang XP. On the shift operator, graph frequency, and optimal filtering in graph signal processing. IEEE Trans Signal Process 2017;65(23):6303–18. https://doi.org/10.1109/TSP.2017.2752689.

[102] Püschel M, Moura JMF. Algebraic signal processing theory: Cooley-Tukey type algorithms for DCTs and DSTs. IEEE Trans Signal Process 2008;56(4):1502–21.

CHAPTER

SAMPLING AND RECOVERY OF GRAPH SIGNALS

9

Paolo Di Lorenzo*, **Sergio Barbarossa***, **Paolo Banelli**†

*Department of Information Engineering, Electronics, and Telecommunications, Sapienza University of Rome, Rome, Italy** *Department of Engineering, University of Perugia, Perugia, Italy*†

9.1 INTRODUCTION

In a large number of applications involving sensor, transportation, communication, social, or biological networks, the observed data can be modeled as signals defined over graphs, or graph signals for short. As a consequence, over the last few years, there has been a surge of interest in developing novel analysis methods for graph signals, thus leading to the research field known as graph signal processing (GSP); see, e.g., [1,2]. The goal of GSP is to extend classical processing tools to the analysis of signals defined over an irregular discrete domain, represented by a graph, and one interesting aspect is that such methods typically come to depend on the graph topology; see, e.g., [2–6].

A fundamental task in GSP is to infer the values of a graph signal by interpolating the samples collected from a known set of vertices. In the GSP literature, this learning task is known as *interpolation from samples*, and emerges whenever cost constraints limit the number of vertices that we can directly observe. This arises in several applications such as semisupervised learning of categorical data [7], environmental monitoring [8], and missing value prediction as in matrix completion problems [9]. Interpolation methods on graphs rely on the implicit assumption that nodes close to each other have similar values, i.e., the graph encodes similarity among the values observed over the vertices. For instance, in an item-item graph in a recommendation system, a user would rate two similar items with similar ratings [10]. In the same way, predicting the functions of proteins based on a protein network relies on some notion of closeness among the nodes [11]. In other words, the signals of interest must be smooth functions over the graph. In GSP, the smoothness assumption is typically formalized in terms of (approximate) bandlimitedness over a graph Fourier basis, and enables recovery of the signal after sampling over a selected subset of vertices.

A first seminal contribution to sampling theory in GSP is given by [12], where a sufficient condition for unique recovery is stated for a given sampling set; the approach was then extended in [13,14]. Most of the works on graph sampling theory assume that a portion of the graph Fourier basis is explicitly known. For example, the work in [6] provides conditions that guarantee unique reconstruction of signals spanned over a subset of vectors composing the graph Fourier basis, proposing also a greedy method to select the sampling set in order to minimize the effect of sample noise in the worst case. Reference [15] exploited a smart partitioning of the graph in local sets, and proposed iterative methods to reconstruct bandlimited graph signals from sampled data. The work in [16] creates a conceptual link between

Cooperative and Graph Signal Processing. https://doi.org/10.1016/B978-0-12-813677-5.00009-2

the uncertainty principle and the sampling of graph signals, and proposes several optimality criteria (e.g., the mean-square error) to select the sampling set in the presence of noise. Another valid approach is the so-called aggregation sampling [17], which involves successively shifting a signal using the adjacency matrix and aggregating the values at a given node. Greedy sampling strategies with provable performance guarantees were proposed in [18] in a Bayesian reconstruction setting.

If the size of the graph signal is very large as, e.g., in web-scale graphs [19], complexity becomes a crucial issue, such that in many cases we cannot assume to know or efficiently compute the graph Fourier basis. Some works have then proposed sampling methods that do not require such previous knowledge. For instance, the work in [20] proposes efficient methods to select the sampling set based on powers of the variation operator to approximate the bandwidth of the graph signal. There are also alternative approaches that do not consider graph spectral information and rely only on vertex-domain characteristic, e.g., maximum graph cuts [21] and spanning trees [22]. Finally, there exist randomized sampling strategies, e.g., [23–25]. The work in [23] provides an efficient design of sampling probability distribution over the nodes, deriving bounds on the reconstruction error in the presence of noise and/or approximatively bandlimited signals. Reference [24] exploits compressive sampling arguments to derive random sampling strategies with variable density, thus also proposing a fast technique to estimate the optimal sampling distribution accurately. Last, the work in [25] proposes a sampling strategy tailored for large-scale data based on random walks on graphs.

The sampling strategies described so far involve batch methods for sampling and recovery of graph signals. In many applications such as, e.g., transportation networks, brain networks, or communication networks, the observed graph signals are typically time-varying. This requires the development of effective methods capable of learning and tracking dynamic graph signals from a carefully designed, possibly time-varying, sampling set. Some previous works have considered this specific learning task, see, e.g., [26–29]. Specifically, [26] proposed an LMS estimation strategy enabling adaptive learning and tracking from a limited number of smartly sampled observations. The LMS method in [26] was then extended to the distributed setting in [27]. The work in [28] proposed a kernel-based reconstruction framework to accommodate time-evolving signals over possibly time-evolving topologies, leveraging spatiotemporal dynamics of the observed data. Finally, reference [29] proposes a distributed method for tracking bandlimited graph signals, assuming perfect observations and a fixed sampling strategy.

In this chapter, we review some of the recent advances related to sampling and recovery of signals defined over graphs. Due to space limitations, this review will be limited to some specific contributions. The structure of the chapter is explained in the sequel. Section 9.2 defines the adopted notation and recalls some background on GSP. In Section 9.3, we illustrate the conditions for perfect recovery of bandlimited graph signals from samples collected according to design criteria proposed to mitigate the effect of noise or model mismatching. Finally, Section 9.4 illustrates algorithms and optimal sampling strategies for adaptive recovery and tracking of dynamic graph signals, where both sampling set and signal values are allowed to vary with time.

9.2 NOTATION AND BACKGROUND

In this section, we first introduce the notation that we will use throughout the chapter. Then, we briefly recall some basic notions from GSP that will be instrumental for the derivations and arguments of the following sections.

Notation. We indicate scalars by normal letters (e.g., a); vector variables with bold lowercase letters (e.g., $\boldsymbol{a}$) and matrix variables with bold uppercase letters (e.g., $\mathbf{A}$). Scalars a_i and a_{ij} correspond to the ith entry of $\boldsymbol{a}$ and the ijth entry of $\mathbf{A}$, respectively. We indicate by $\|\boldsymbol{a}\|_2$ and $\|\mathbf{A}\|_2$ the ℓ_2 norm and the spectral norm of the vector $\boldsymbol{a}$ and matrix $\mathbf{A}$, respectively. If $\mathbf{A}$ is rectangular, we denote by $\sigma_i(\mathbf{A})$ the ith singular value of $\mathbf{A}$; if $\mathbf{A}$ is square, $\lambda_i(\mathbf{A})$ represents the ith eigenvalue of $\mathbf{A}$. The trace operator of matrix $\mathbf{A}$ is indicated with $\text{Tr}(\mathbf{A})$; $\text{diag}(\boldsymbol{a})$ is a diagonal matrix having $\boldsymbol{a}$ as main diagonal; $\text{rank}(\mathbf{A})$ denotes the rank of matrix $\mathbf{A}$; $\det(\mathbf{A})$ represents the determinant of $\mathbf{A}$, whereas $\text{pdet}(\mathbf{A})$ is the pseudo-determinant of $\mathbf{A}$, i.e., the product of all nonzero eigenvalues of $\mathbf{A}$. The superscript ^{H} denotes the Hermitian operator, i.e., the conjugate transposition of a vector or matrix, whereas $\mathbf{A}^\dagger$ denotes the pseudoinverse of matrix $\mathbf{A}$. $\mathbb{E}\{\cdot\}$ represents the expectation operator. A set of elements is denoted by a calligraphic letter (e.g., $\mathcal{S}$), and $|\mathcal{S}|$ represents the cardinality of set $\mathcal{S}$, i.e., the number of elements of $\mathcal{S}$. The symbols $\cup$, $\cap$, and $\setminus$ denote union, intersection, and difference among sets, respectively. Given a set $\mathcal{S}$, we denote its complement set as $\mathcal{S}^c$, i.e., $\mathcal{V} = \mathcal{S} \cup \mathcal{S}^c$ and $\mathcal{S} \cap \mathcal{S}^c = \emptyset$. $\mathbf{1}$ denotes the vector of all ones whereas $\mathbf{1}_\mathcal{S}$ is the set indicator vector whose ith entry is equal to one, if $i \in \mathcal{S}$, or zero otherwise.

Background. We consider a graph $\mathcal{G} = (\mathcal{V}, \mathcal{E})$ consisting of a set of N nodes $\mathcal{V} = \{1, 2, \ldots, N\}$, along with a set of weighted edges $\mathcal{E} = \{a_{ij}\}_{i,j \in \mathcal{V}}$, such that $a_{ij} > 0$, if there is a link from node j to node i, or $a_{ij} = 0$, otherwise. The adjacency matrix $\mathbf{A}$ of a graph is the collection of all the weights $a_{ij}, i, j = 1, \ldots, N$. The combinatorial Laplacian matrix is defined as $\mathbf{L} = \text{diag}(\mathbf{1}^T\mathbf{A}) - \mathbf{A}$. A signal $\boldsymbol{x}$ over a graph $\mathcal{G}$ is defined as a mapping from the vertex set to the set of complex numbers, i.e., $\boldsymbol{x} : \mathcal{V} \to \mathbb{C}$. The graph $\mathcal{G}$ is endowed with a graph-shift operator $\mathbf{S}$ defined as an $N \times N$ matrix whose entry (i, j), denoted with S_{ij}, can be nonzero only if $i = j$ or the link $(j, i) \in \mathcal{E}$. The sparsity pattern of matrix $\mathbf{S}$ captures the local structure of $\mathcal{G}$; common choices for $\mathbf{S}$ are the adjacency matrix [2], the Laplacian [1], and its generalizations [20]. We assume that $\mathbf{S}$ is diagonalizable, i.e., there exists an $N \times N$ matrix $\mathbf{U} = [\boldsymbol{u}_1, \ldots, \boldsymbol{u}_N]$ and an $N \times N$ diagonal matrix $\mathbf{\Lambda}$ that can be used to decompose $\mathbf{S}$ as $\mathbf{S} = \mathbf{U}\mathbf{\Lambda}\mathbf{U}^{-1}$. When $\mathbf{S}$ is normal, i.e., when $\mathbf{S}\mathbf{S}^H = \mathbf{S}^H\mathbf{S}$, matrix $\mathbf{U}$ is unitary and $\mathbf{U}^{-1} = \mathbf{U}^H$.

Recovery of a signal from its sampled version is possible under the assumption that $\boldsymbol{x}$ admits a *sparse representation*. The basic idea when addressing the problem of sampling graph signals is to suppose that $\mathbf{S}$ plays a key role in explaining the signal of interest. More specifically, we assume that $\boldsymbol{x}$ can be expressed as a linear combination of a subset of the columns of $\mathbf{U}$, i.e.,

$$\boldsymbol{x} = \mathbf{U}\boldsymbol{s}, \tag{9.1}$$

where $\boldsymbol{s} \in \mathbb{C}^N$ is either exactly or approximately sparse. In this context, vectors $\{\boldsymbol{u}_i\}_{i=1}^N$ are interpreted as the graph Fourier basis, and $\{s_i\}_{i=1}^N$ are the corresponding graph signal frequency coefficients, i.e.,

$$\boldsymbol{s} = \mathbf{U}^H\boldsymbol{x} \tag{9.2}$$

takes the role of the Graph Fourier Transform (GFT) of signal $\boldsymbol{x}$. As an example, in many cases the graph signal exhibits clustering features, i.e., it is a smooth function over each cluster (e.g., in semisupervised learning [7]), but it may vary arbitrarily from one cluster to the other. In such a case, if the columns of $\mathbf{U}$ are chosen to represent clusters, the only nonzero (or approximately nonzero) entries of $\boldsymbol{s}$ are the ones associated with the clusters. In the case of *undirected* graphs, $\mathbf{U}$ may be composed from the eigenvectors of the Laplacian, which have well-known clustering properties [30].

The localization properties of graph signals in vertex and frequency domains will play an important role in the ensuing arguments. To introduce such properties, we first define the matrix $\mathbf{U}_\mathcal{F} \in \mathbb{C}^{N \times |\mathcal{F}|}$, which represents the collection of all the columns of $\mathbf{U}$ associated with a subset of frequency indices $\mathcal{F} \subseteq \{1, \ldots, N\}$. Then, we introduce the $N \times N$ band-limiting operator

$$\mathbf{B}_{\mathcal{F}} = \mathbf{U}_{\mathcal{F}} \mathbf{U}_{\mathcal{F}}^{H}. \tag{9.3}$$

The role of $\mathbf{B}_{\mathcal{F}}$ is to project a vector $\boldsymbol{x}$ onto the subspace spanned by the columns of $\mathbf{U}_{\mathcal{F}}$. Thus, we say that a vector $\boldsymbol{x}$ is perfectly localized over the frequency set $\mathcal{F}$ (or $\mathcal{F}$-bandlimited) if

$$\mathbf{B}_{\mathcal{F}} \boldsymbol{x} = \boldsymbol{x} = \mathbf{U}_{\mathcal{F}} \boldsymbol{s}_{\mathcal{F}}, \tag{9.4}$$

where $\mathbf{B}_{\mathcal{F}}$ is given in Eq. (9.3), and the second equality comes from Eq. (9.1) where we have exploited the sparsity of $\boldsymbol{s}$, which is different from zero (and equal to $\boldsymbol{s}_{\mathcal{F}} \in \mathbb{C}^{|\mathcal{F}|}$) only in the frequency support $\mathcal{F}$. Similarly, given a subset of vertices $\mathcal{S} \subseteq \mathcal{V}$, we define the $N \times N$ vertex-limiting operator

$$\mathbf{D}_{\mathcal{S}} = \text{diag}\{\mathbf{1}_{\mathcal{S}}\}. \tag{9.5}$$

Thus, we say that a vector $\boldsymbol{x}$ is perfectly localized over the subset $\mathcal{S} \subseteq \mathcal{V}$ (or $\mathcal{S}$-vertex-limited) if $\mathbf{D}_{\mathcal{S}} \boldsymbol{x} = \boldsymbol{x}$, with $\mathbf{D}_{\mathcal{S}}$ defined as in Eq. (9.5). We also denote by $\mathcal{B}_{\mathcal{F}}$ the set of all $\mathcal{F}$-bandlimited signals, and by $\mathcal{D}_{\mathcal{S}}$ the set of all $\mathcal{S}$-vertex-limited signals. The operators $\mathbf{D}_{\mathcal{S}}$ and $\mathbf{B}_{\mathcal{F}}$ are self-adjoint and idempotent, and represent orthogonal projectors onto the sets $\mathcal{D}_{\mathcal{S}}$ and $\mathcal{B}_{\mathcal{F}}$, respectively. Differently from continuous-time signals, a graph signal can be perfectly localized in *both* vertex and frequency domains. This property is formally stated in the following theorem [16, Th. 2.1].

Theorem 9.1. *There is a graph signal* $\boldsymbol{x}$ *perfectly localized over both vertex set* $\mathcal{S}$ *and frequency set* $\mathcal{F}$ *(i.e.,* $\boldsymbol{x} \in \mathcal{B}_{\mathcal{F}} \cap \mathcal{D}_{\mathcal{S}}$*) if and only if the operator* $\mathbf{B}_{\mathcal{F}} \mathbf{D}_{\mathcal{S}} \mathbf{B}_{\mathcal{F}}$ *has an eigenvalue equal to one; in such a case,* $\boldsymbol{x}$ *is the eigenvector of* $\mathbf{B}_{\mathcal{F}} \mathbf{D}_{\mathcal{S}} \mathbf{B}_{\mathcal{F}}$ *associated with the unit eigenvalue.* ■

Perfect localization onto the sets $\mathcal{S}$ and $\mathcal{F}$ can be equivalently expressed in terms of the operator $\mathbf{D}_{\mathcal{S}} \mathbf{U}_{\mathcal{F}}$ [16], i.e., it holds if and only if

$$\|\mathbf{D}_{\mathcal{S}} \mathbf{U}_{\mathcal{F}}\|_2 = \|\mathbf{B}_{\mathcal{F}} \mathbf{D}_{\mathcal{S}} \mathbf{B}_{\mathcal{F}}\|_2 = 1. \tag{9.6}$$

In the following section, we will illustrate the theory behind sampling and recovery of signals defined over graphs.

9.3 SAMPLING AND RECOVERY

Let us consider the observation of an $\mathcal{F}$-bandlimited graph signal over the sampling set $\mathcal{S}$. The observation model can be cast as:

$$\boldsymbol{y}_{\mathcal{S}} = \mathbf{P}_{\mathcal{S}}^{T} \boldsymbol{x} = \mathbf{P}_{\mathcal{S}}^{T} \mathbf{U}_{\mathcal{F}} \boldsymbol{s}_{\mathcal{F}}, \tag{9.7}$$

where $\boldsymbol{y}_{\mathcal{S}} \in \mathbb{C}^{|\mathcal{S}|}$ is the observation vector over the vertex set $\mathcal{S}$, and $\mathbf{P}_{\mathcal{S}} \in \mathbb{R}^{N \times |\mathcal{S}|}$ is a sampling matrix whose columns are indicator functions for nodes in $\mathcal{S}$, and such that the orthogonal projector over $\mathcal{D}_{\mathcal{S}}$ is given by $\mathbf{D}_{\mathcal{S}} = \mathbf{P}_{\mathcal{S}} \mathbf{P}_{\mathcal{S}}^{T}$ [cf. Eq. (9.5)]. The problem of recovering a bandlimited graph signal from its samples is then equivalent to the problem of properly selecting the sampling set $\mathcal{S}$, and then recovering $\boldsymbol{x}$ from $\boldsymbol{y}_{\mathcal{S}}$ by inverting the system of equations in Eq. (9.7). This approach is known as *selection sampling* and was addressed, for example, in [6,12,13,16].

In the sequel, we will first consider the conditions for perfect recovery of bandlimited graph signals. Then, we will illustrate the effect of noise and model mismatching on the reconstruction performance. Also, because the identification of the sampling set $\mathcal{S}$ plays a key role in the conditions for signal recovery and in the reconstruction performance, we will illustrate optimization strategies to design the sampling set. Finally, we will illustrate results of numerical simulations carried out over synthetic and realistic data.

9.3.1 SAMPLING AND PERFECT RECOVERY OF BANDLIMITED GRAPH SIGNALS

We will now address the fundamental problem of assessing the conditions and the means for perfect recovery of $\boldsymbol{x}$ from $\boldsymbol{y}_{\mathcal{S}}$. To this aim, we introduce the operator $\mathbf{D}_{\mathcal{S}^c} = \mathbf{I} - \mathbf{D}_{\mathcal{S}}$, which projects onto the complement vertex set $\mathcal{S}^c = \mathcal{V} \setminus \mathcal{S}$. Starting from Eq. (9.7), the necessary and sufficient conditions for perfect recovery are stated in the following Theorem [16, Th. 4.1].

Theorem 9.2. *Any $\mathcal{F}$-bandlimited graph signal $\boldsymbol{x}$ can be perfectly recovered from its samples collected over the vertex set $\mathcal{S}$ if and only if*

$$\|\boldsymbol{D}_{\mathcal{S}^c}\boldsymbol{U}_{\mathcal{F}}\|_2 < 1, \tag{9.8}$$

i.e., if there are no $\mathcal{F}$-bandlimited signals that are perfectly localized on $\mathcal{S}^c$.

Proof. From Eq. (9.7), a sufficient condition for signal recovery is the existence of the pseudoinverse matrix $\mathbf{Q} = (\mathbf{P}_{\mathcal{S}}^T\mathbf{U}_{\mathcal{F}})^{\dagger} = (\mathbf{U}_{\mathcal{F}}^H\mathbf{D}_{\mathcal{S}}\mathbf{U}_{\mathcal{F}})^{-1}\mathbf{U}_{\mathcal{F}}^H\mathbf{P}_{\mathcal{S}}$. Because $\mathbf{U}_{\mathcal{F}}^H\mathbf{D}_{\mathcal{S}}\mathbf{U}_{\mathcal{F}} = \mathbf{I} - \mathbf{U}_{\mathcal{F}}^H\mathbf{D}_{\mathcal{S}^c}\mathbf{U}_{\mathcal{F}}$, we obtain that $\mathbf{Q}$ exists if $\|\mathbf{U}_{\mathcal{F}}^H\mathbf{D}_{\mathcal{S}^c}\mathbf{U}_{\mathcal{F}}\|_2 = \|\mathbf{D}_{\mathcal{S}^c}\mathbf{U}_{\mathcal{F}}\|_2 < 1$, i.e., if Eq. (9.8) holds true. Conversely, if $\|\mathbf{D}_{\mathcal{S}^c}\mathbf{U}_{\mathcal{F}}\|_2 = 1$, there exist bandlimited signals that are perfectly localized over $\mathcal{S}^c$ [cf. Eq. (9.6)]. Thus, if we sample one of these signals over $\mathcal{S}$, it would be impossible to recover $\boldsymbol{x}$ from those samples. This proves that condition (9.8) is also necessary. ■

Theorem 9.2 and its proof also suggest the reconstruction formula:

$$\hat{\boldsymbol{x}} = \mathbf{U}_{\mathcal{F}}(\mathbf{P}_{\mathcal{S}}^T\mathbf{U}_{\mathcal{F}})^{\dagger}\boldsymbol{y}_{\mathcal{S}} = \mathbf{U}_{\mathcal{F}}(\mathbf{U}_{\mathcal{F}}^H\mathbf{D}_{\mathcal{S}}\mathbf{U}_{\mathcal{F}})^{-1}\mathbf{U}_{\mathcal{F}}^H\mathbf{P}_{\mathcal{S}}\boldsymbol{y}_{\mathcal{S}}, \tag{9.9}$$

which guarantees reconstruction of the bandlimited graph signal $\boldsymbol{x}$ if condition (9.8) holds true, and has computational complexity equal to $O(N|\mathcal{F}|^2)$. The above reconstruction formula is also known as consistent reconstruction [31] because it keeps the observed samples unchanged.

Let us consider now the implications of condition (9.8) of Theorem 9.2 on the sampling strategy. To fulfill Eq. (9.8), we need to guarantee that there exist no signals that are perfectly localized over the vertex set $\mathcal{S}^c$ and the frequency set $\mathcal{F}$. Because, in general, we have

$$\boldsymbol{y}_{\mathcal{S}} = \mathbf{P}_{\mathcal{S}}^T\mathbf{U}_{\mathcal{F}}\boldsymbol{s}_{\mathcal{F}} + \mathbf{P}_{\mathcal{S}^c}^T\mathbf{U}_{\mathcal{F}}\boldsymbol{s}_{\mathcal{F}}, \tag{9.10}$$

we need to guarantee that $\mathbf{P}_{\mathcal{S}}^T\mathbf{U}_{\mathcal{F}}\boldsymbol{s}_{\mathcal{F}} \neq \mathbf{0}$ for any nontrivial vector $\boldsymbol{s}_{\mathcal{F}}$, which requires $\mathbf{P}_{\mathcal{S}}^T\mathbf{U}_{\mathcal{F}}$ to be full column rank, i.e.,

$$\text{rank}(\mathbf{P}_{\mathcal{S}}^T\mathbf{U}_{\mathcal{F}}) = \text{rank}(\mathbf{U}_{\mathcal{F}}) = |\mathcal{F}|. \tag{9.11}$$

Of course, a necessary condition to satisfy Eq. (9.11) is that

$$|\mathcal{S}| \geq |\mathcal{F}|. \tag{9.12}$$

However, condition (9.12) is not sufficient because $\mathbf{P}_{\mathcal{S}}^T\mathbf{U}_{\mathcal{F}}$ may loose rank, depending on graph topology and samples' location. As a particular case, if the graph is not connected, the vertices can be labeled so that the Laplacian (adjacency) matrix can be written as a block diagonal matrix, with a number of blocks equal to the number of connected components. Correspondingly, each eigenvector of $\mathbf{L}$ can be expressed as a vector having all zero elements, except for the entries corresponding to the connected component. This implies that, if there are no samples over the vertices corresponding to the nonnull entries of the eigenvectors with index included in $\mathcal{F}$, $\mathbf{P}_{\mathcal{S}}^T\mathbf{U}_{\mathcal{F}}$ looses rank. More generally, even if the graph is connected, there may easily occur situations where matrix $\mathbf{P}_{\mathcal{S}}^T\mathbf{U}_{\mathcal{F}}$ is not rank-deficient but it is ill-conditioned, depending on graph topology and samples' location. This case is particularly dangerous when the true signal is only approximately bandlimited (which is the case for most signals in practice) or when the samples are noisy. In such cases, not all sampling sets of given size are equally good, and it becomes fundamental to understand which is the best sampling set that achieves the smallest reconstruction error.

The optimal samples' location depends on the network topology. From the above theory, it turns out that this dependency is strictly related to the structure of the eigenvectors of the Laplacian matrix. If the graph is circulant,[1] the eigenvectors of its Laplacian matrix are the Fourier vectors, so that they have constant modulus. However, as soon as a graph departs from the circularity conditions, the Laplacian eigenvectors tend to be more localized or, equivalently, to have some zero entries. It is not infrequent to have eigenvectors that have only two coefficients different from zero [32]. Indeed, as illustrated in [33], there is a strict relation between symmetries of a graph and sparsity of its Laplacian eigenvectors. This means that as soon as a graph topology departs from a circulant structure, we need to take care of the strategy to be adopted to select the samples' location to guarantee a proper reconstruction of the overall graph signal from a subset of observations. In the sequel, we will first illustrate the effect of noise and model mismatching on graph signal reconstruction, and then we will describe sampling strategies satisfying several optimization criteria.

9.3.2 THE EFFECT OF NOISE AND MODEL MISMATCHING

Let us consider first the reconstruction of bandlimited signals from noisy samples, where the observation model is given by:

$$y_{\mathcal{S}} = \mathbf{P}_{\mathcal{S}}^T(\boldsymbol{x}+\boldsymbol{v}) = \mathbf{P}_{\mathcal{S}}^T\mathbf{U}_{\mathcal{F}}s_{\mathcal{F}} + \mathbf{P}_{\mathcal{S}}^T\boldsymbol{v}, \tag{9.13}$$

where $\boldsymbol{v}$ is a zero-mean noise vector with covariance matrix $\mathbf{R}_v = \mathbb{E}\{\boldsymbol{v}\boldsymbol{v}^H\}$. To design an interpolator in the presence of noise, we consider the best linear unbiased estimator (BLUE), which is given by [34]:

$$\hat{\boldsymbol{x}} = \mathbf{U}_{\mathcal{F}}\left(\mathbf{U}_{\mathcal{F}}^H\mathbf{P}_{\mathcal{S}}\left(\mathbf{P}_{\mathcal{S}}^T\mathbf{R}_v\mathbf{P}_{\mathcal{S}}\right)^{-1}\mathbf{P}_{\mathcal{S}}^T\mathbf{U}_{\mathcal{F}}\right)^{-1}\mathbf{U}_{\mathcal{F}}^H\mathbf{P}_{\mathcal{S}}\left(\mathbf{P}_{\mathcal{S}}^T\mathbf{R}_v\mathbf{P}_{\mathcal{S}}\right)^{-1}y_{\mathcal{S}}. \tag{9.14}$$

The estimator in Eq. (9.14) minimizes the least square error and, if noise is Gaussian in Eq. (9.13), it coincides with the minimum variance unbiased estimator, which attains the Cramér-Rao lower bound. It is immediate to see that Eq. (9.14) is an unbiased estimator, i.e., $\mathbb{E}\{\hat{\boldsymbol{x}}\} = \boldsymbol{x}$. Furthermore, the mean square error (MSE) is given by [34]:

---

[1] A graph is circulant if there exists some ordering of nodes for which the adjacency matrix (or, equivalently, the Laplacian matrix) of the graph is circulant.

$$\text{MSE} = \mathbb{E}\|\hat{x} - x\|^2 = \text{Tr}\left\{\left(\mathbf{U}_{\mathcal{F}}^H \mathbf{P}_{\mathcal{S}} \left(\mathbf{P}_{\mathcal{S}}^T \mathbf{R}_v \mathbf{P}_{\mathcal{S}}\right)^{-1} \mathbf{P}_{\mathcal{S}}^T \mathbf{U}_{\mathcal{F}}\right)^{-1}\right\}. \tag{9.15}$$

As a particular case, if noise is spatially uncorrelated, i.e., $\mathbf{R}_v = \text{diag}\{r_1^2, \ldots, r_N^2\}$, and letting $\boldsymbol{u}_{\mathcal{F},i}^H$ be the ith row of matrix $\mathbf{U}_{\mathcal{F}}$, we obtain:

$$\text{MSE} = \text{Tr}\left\{\left(\mathbf{U}_{\mathcal{F}}^H \mathbf{D}_{\mathcal{S}} \mathbf{R}_v^{-1} \mathbf{U}_{\mathcal{F}}\right)^{-1}\right\} = \text{Tr}\left\{\left(\sum_{i \in \mathcal{S}} \boldsymbol{u}_{\mathcal{F},i} \boldsymbol{u}_{\mathcal{F},i}^H / r_i^2\right)^{-1}\right\}. \tag{9.16}$$

This illustrates how, in the presence of uncorrelated noise, the design of the sampling set should minimize the trace of the inverse of matrix $\mathbf{U}_{\mathcal{F}}^H \mathbf{D}_{\mathcal{S}} \mathbf{R}_v^{-1} \mathbf{U}_{\mathcal{F}}$.

So far we assumed that the true signal $\boldsymbol{x}$ is perfectly bandlimited, i.e., $\boldsymbol{x} \in \mathcal{B}_{\mathcal{F}}$. However, in most applications, the signals are only approximately bandlimited. In such a case, the recovery formula in Eq. (9.9) applied to such signals leads to a reconstruction error, which is analyzed next. In general, an approximately bandlimited graph signal can be expressed as

$$\boldsymbol{x} = \boldsymbol{x}_{\mathcal{F}} + \Delta \boldsymbol{x}, \tag{9.17}$$

where $\boldsymbol{x}_{\mathcal{F}} = \mathbf{B}_{\mathcal{F}} \boldsymbol{x}$ is the bandlimited component, whereas $\Delta \boldsymbol{x} = \mathbf{B}_{\mathcal{F}^c} \boldsymbol{x}$ represents the model mismatch. Sampling the signal over the vertex set $\mathcal{S}$ and using Eq. (9.9) as a recovery formula, then an upper bound (i.e., a worst case) on the reconstruction error is given by [31]:

$$\|\hat{\boldsymbol{x}} - \boldsymbol{x}\| \leq \frac{\|\Delta \boldsymbol{x}\|}{\cos(\theta_{\max})}, \tag{9.18}$$

where $\theta_{\max}$ represents the maximum angle between the subspaces $\mathcal{B}_{\mathcal{F}}$ and $\mathcal{D}_{\mathcal{S}}$, which is defined as:

$$\cos(\theta_{\max}) = \inf_{\|z\|=1} \|\mathbf{D}_{\mathcal{S}} \boldsymbol{z}\|_2 \quad \text{subject to } \mathbf{B}_{\mathcal{F}} \boldsymbol{z} = \boldsymbol{z}. \tag{9.19}$$

In particular, from Eq. (9.19), it is easy to see that $\cos(\theta_{\max}) > 0$ if condition (9.8) holds true. Intuitively, the bound in Eq. (9.18) says that, for the worst-case error to be minimum, the sampling and reconstruction subspaces should be as aligned as possible. Therefore, for approximatively bandlimited signals, an optimal sampling set should be selected in order to maximize the smallest maximum angle between the subspaces $\mathcal{B}_{\mathcal{F}}$ and $\mathcal{D}_{\mathcal{S}}$. Interestingly, from Eqs. (9.19) and (9.3), it appears clear that

$$\cos(\theta_{\max}) = \sigma_{\min}(\mathbf{D}_{\mathcal{S}} \mathbf{U}_{\mathcal{F}}) \tag{9.20}$$

Thus, in the presence of model mismatching, the design of the sampling set should maximize the minimum singular value of matrix $\mathbf{D}_{\mathcal{S}} \mathbf{U}_{\mathcal{F}}$ or, equivalently, the minimum eigenvalue of matrix $\mathbf{U}_{\mathcal{F}}^H \mathbf{D}_{\mathcal{S}} \mathbf{U}_{\mathcal{F}}$.

In the next section, we will illustrate the strategies used to optimize the selection of the sampling set.

9.3.3 SAMPLING STRATEGIES

As previously mentioned, when sampling graph signals, besides choosing the right number of samples, whenever possible it is also fundamental to have a strategy indicating *where* to sample as the samples' location plays a key role in the performance of reconstruction algorithms. In principle, in the ideal case (9.7), any sampling set $\mathcal{S}$ that satisfies condition (9.8) enables unique reconstruction through the interpolation formula in Eq. (9.9). However, in the presence of noise or model mismatching, from Eqs. (9.16) and (9.18)–(9.20), it is clear that the quality of reconstruction is strongly affected by a careful design of the sampling set $\mathcal{S}$. Different costs can then be defined to measure the reconstruction error and are based on optimal design of experiments [35]. For instance, if we seek for the optimal sampling set $\mathcal{S}^{\text{opt}}$ of size M, as the set that minimizes the mean squared error in Eq. (9.16), we have:

$$\mathcal{S}^{\text{A-opt}} = \underset{|\mathcal{S}|=M}{\arg\min}\, \text{Tr}\left\{\left(\mathbf{U}_{\mathcal{F}}^{H}\mathbf{D}_{\mathcal{S}}\mathbf{R}_{v}^{-1}\mathbf{U}_{\mathcal{F}}\right)^{-1}\right\}. \tag{9.21}$$

This is analogous to the so-called *A-optimal design* [35], and is equivalent to the one proposed in [16]. Similarly, if we aim to design the optimal sampling set of size M to minimize the worst-case reconstruction error in the presence of model mismatching [cf. Eqs. (9.18)–(9.20)], we have:

$$\mathcal{S}^{\text{E-opt}} = \underset{|\mathcal{S}|=M}{\arg\max}\; \sigma_{\min}(\mathbf{D}_{\mathcal{S}}\mathbf{U}_{\mathcal{F}}), \tag{9.22}$$

which is equivalent to the so-called *E-optimal design* [35]. The above criterion is equivalent to the one proposed in [6] and, in general, it is useful to find a stable sampling set that satisfies condition (9.8). To select the optimal sampling set, we should solve one of the problems in Eq. (9.21) or Eq. (9.22), which entails the selection of an M-element subset of $\mathcal{V}$ that optimizes the adopted design criterion. This is a finite combinatorial optimization problem (which is known to be NP-hard [36]) whose solution in general requires an exhaustive search over all the possible combinations. Because the number of possible subsets grows factorially as $|\mathcal{V}|$ increases, a brute-force approach quickly becomes infeasible also for graph signals of moderate dimensions. To cope with this issue, in the sequel, we will introduce lower complexity methods based on: i) greedy approaches, and ii) convex relaxations.

Algorithm 9.1 GREEDY SELECTION OF GRAPH SAMPLES

Input Data : $\mathbf{U}_{\mathcal{F}}, M$;
Output Data : $\mathcal{S}$, the sampling set.
Function : initialize $\mathcal{S} \equiv \emptyset$
  while $|\mathcal{S}| < M$
    $s = \arg\max_{j} f(\mathcal{S} \cup \{j\})$;
    $\mathcal{S} \leftarrow \mathcal{S} \cup \{s\}$;
  end

Greedy sampling

In this section, we will consider a numerically efficient albeit suboptimal greedy algorithm to tackle the problem of selecting the sampling set. The greedy approach is described in Algorithm 9.1. The simple idea underlying such a method is to iteratively add to the sampling set those vertices of the graph that lead to the largest increment of an adopted performance metric, i.e., a specific set function $f(\mathcal{S}) : 2^{\mathcal{V}} \to \mathbb{R}$. We will set $f(\mathcal{S}) = -\text{Tr}\left\{(\mathbf{U}_{\mathcal{F}}^{H}\mathbf{D}_{\mathcal{S}}\mathbf{R}_{v}^{-1}\mathbf{U}_{\mathcal{F}})^{-1}\right\}$ if we use an A-optimality design as in Eq. (9.21), or $f(\mathcal{S}) = \sigma_{\min}(\mathbf{D}_{\mathcal{S}}\mathbf{U}_{\mathcal{F}})$ if we consider an E-optimality design as in Eq. (9.22). In fact, because Algorithm 9.1 starts from the empty set, when $|\mathcal{S}| < |\mathcal{F}|$, matrix $\mathbf{U}_{\mathcal{F}}^H \mathbf{D}_{\mathcal{S}} \mathbf{R}_v^{-1} \mathbf{U}_{\mathcal{F}}$ is inevitably rank-deficient, and its inverse does not exist. In this case, considering an A-optimality criterion, we can use $f(\mathcal{S}) = -\text{Tr}\{(\mathbf{U}_{\mathcal{F}}^H \mathbf{D}_{\mathcal{S}} \mathbf{R}_v^{-1} \mathbf{U}_{\mathcal{F}})^\dagger\}$, which becomes equivalent to Eq. (9.21) when condition (9.8) is satisfied.

In general, the performance of the greedy strategy will be suboptimal with respect to an exhaustive search procedure. Nevertheless, if the set function $f(\mathcal{S})$ satisfies some structural properties, the greedy Algorithm 9.1 can be proved to be close to optimality. In particular, submodularity plays a similar role in combinatorial optimization to convexity in continuous optimization and shares other features of concave functions [37].

Definition 9.1. A set function $f : 2^{\mathcal{V}} \to \mathbb{R}$ is submodular if and only if the derived set functions $f_a : 2^{\mathcal{V}\backslash\{a\}} \to \mathbb{R}$

$$f_a(\mathcal{S}) = f(\mathcal{S} \cup \{a\}) - f(\mathcal{S}) \tag{9.23}$$

are monotone decreasing, i.e., if for all subsets $a, \mathcal{A}, \mathcal{B} \subseteq \mathcal{V}$ it holds that

$$\text{if } \mathcal{A} \subseteq \mathcal{B} \;\Rightarrow\; f_a(\mathcal{A}) \geq f_a(\mathcal{B}).$$

■

Intuitively, submodularity is a diminishing returns property where adding an element to a smaller set gives a larger gain than adding one to a larger set. The maximization of monotone increasing submodular functions is still NP-hard, but the greedy heuristic can be used to obtain a solution that is provably close to optimality, with a solution having objective value within $1 - 1/e$ of the optimal combinatorial solution [38].

Unfortunately, both set functions in Eqs. (9.21) and (9.22) are not submodular functions [39].[2] Thus, even if the design criteria in Eqs. (9.21) and (9.22) are useful to minimize the effect of noise [cf. Eq. (9.15)] and model mismatching [cf. Eqs. (9.18)–(9.20)], respectively, we do not have theoretical performance guarantees when applying Algorithm 9.1 to solve such problems. Nevertheless, in the literature of experimental design, a further design criterion is often considered as a surrogate for Eq. (9.21) [or Eq. (9.22)], which writes as:

$$\begin{aligned}
\mathcal{S}^{\text{D-opt}} &= \underset{|\mathcal{S}|=M}{\arg\max} \;\; \log\det\left(\mathbf{U}_{\mathcal{F}}^H \mathbf{D}_{\mathcal{S}} \mathbf{R}_v^{-1} \mathbf{U}_{\mathcal{F}}\right) \\
&= \underset{|\mathcal{S}|=M}{\arg\max} \log\det\left(\sum_{i\in\mathcal{S}} \boldsymbol{u}_{\mathcal{F},i} \boldsymbol{u}_{\mathcal{F},i}^H / r_i^2\right).
\end{aligned} \tag{9.24}$$

This is analogous to the so-called *D-optimal design* [35], and is equivalent to one of the methods proposed in [16] for graph signals sampling. This design strategy aims at maximizing the volume of the parallelepiped built with the selected rows $\{\boldsymbol{u}_{\mathcal{F},i}^H\}_{i\in\mathcal{S}}$ of matrix $\mathbf{U}_{\mathcal{F}}$ (weighted by the inverse of the noise variances $\{r_i^2\}_{i\in\mathcal{S}}$), and the rationale is to design a well-suited basis for the graph signal that we want to estimate. Interestingly, the set function $f(\mathcal{S}) = \log\det(\mathbf{U}_{\mathcal{F}}^H \mathbf{D}_{\mathcal{S}} \mathbf{R}_v^{-1} \mathbf{U}_{\mathcal{F}})$ is a monotone increasing

[2] Interestingly, in a Bayesian recovery setting [18], the negative of the MSE function was proved to be approximately submodular.

submodular function [40]. Thus, in this case, the greedy approach in Algorithm 9.1 can be used to solve Eq. (9.24) with provable performance guarantees. In the implementation of Algorithm 9.1, when $|\mathcal{S}| < |\mathcal{F}|$ and matrix $\mathbf{U}_{\mathcal{F}}^H \mathbf{D}_{\mathcal{S}} \mathbf{R}_v^{-1} \mathbf{U}_{\mathcal{F}}$ is rank-deficient, we can use $f(\mathcal{S}) = \log \text{pdet}(\mathbf{U}_{\mathcal{F}}^H \mathbf{D}_{\mathcal{S}} \mathbf{R}_v^{-1} \mathbf{U}_{\mathcal{F}})$, which is equivalent to Eq. (9.24) when sampling condition (9.8) on perfect recovery is satisfied.

Convex relaxation

Another possible algorithmic solution to problems such as Eqs. (9.21), (9.22), (9.24), is to resort to convex relaxation techniques; see, e.g., [41–43]. To this aim, let us introduce the indicator vector $\boldsymbol{d} = \{d_i\}_{i=1}^N$, such that the ith entry is binary and given by $d_i = 1$ if node i belongs to the sampling set $\mathcal{S}$, and $d_i = 0$ otherwise. Using the indicator vector $\boldsymbol{d}$, we can build a general sampling design problem that can be cast as:

$$\begin{aligned} &\min_{\boldsymbol{d}} \; f(\boldsymbol{d}) \\ &\text{s.t.} \quad \mathbf{1}^T \boldsymbol{d} = M, \\ &\qquad \boldsymbol{d} \in \{0,1\}^N, \end{aligned} \tag{9.25}$$

where $f(\boldsymbol{d}) = \text{Tr}\left\{(\mathbf{U}_{\mathcal{F}}^H \text{diag}(\boldsymbol{d}) \mathbf{R}_v^{-1} \mathbf{U}_{\mathcal{F}})^{-1}\right\}$ for the A-optimal design [cf. Eq. (9.21)], $f(\boldsymbol{d}) = -\sigma_{\min}(\text{diag}(\boldsymbol{d}) \mathbf{U}_{\mathcal{F}})$ for the E-optimal design [cf. Eq. (9.22)], and $f(\boldsymbol{d}) = -\log\det(\mathbf{U}_{\mathcal{F}}^H \text{diag}(\boldsymbol{d}) \mathbf{R}_v^{-1} \mathbf{U}_{\mathcal{F}})$ for the D-optimal design [cf. Eq. (9.24)].

Problem (9.25) has still combinatorial complexity due to the integer nature of the optimization variable $\boldsymbol{d}$. Nevertheless, we can simply relax the indicator variable $\boldsymbol{d}$ to be a real vector belonging to the hypercube $[0, 1]^N$, thus leading to the following formulation:

$$\begin{aligned} &\min_{\boldsymbol{d} \in [0,1]^N} \; f(\boldsymbol{d}) \\ &\text{s.t.} \quad \mathbf{1}^T \boldsymbol{d} = M. \end{aligned} \tag{9.26}$$

It is now easy to check that problem (9.26) is convex for all objective functions $f(\boldsymbol{d})$ defined by the design criteria in Eqs. (9.21), (9.22), (9.24), and its global solution can be found using efficient numerical methods [44]. Of course, because Eq. (9.26) is a relaxed version of Eq. (9.25), its real solution $\boldsymbol{d}^*$ might need a further selection/thresholding step in order to generate a valid integer vector, as required by Eq. (9.25). For instance, a possible solution is to select the M sampling nodes as the ones associated with the M largest entries of $\boldsymbol{d}^*$. Finally, one can also formulate the sampling design problem in the opposite way with respect to Eq. (9.25). In particular, we might be interested in searching for the optimal indicator vector $\boldsymbol{d}$ that minimizes the number of collected samples, i.e., the ℓ_0 norm of the vector $\boldsymbol{d}$, under a performance requirement on the function $f(\boldsymbol{d})$, e.g., the MSE in Eq. (9.16). This category of design problems takes the name of sparse sensing [42,43] and, using similar relaxation arguments as before, such criteria lead to convex optimization problems.

In the next section, we will illustrate some numerical results aimed at assessing the performance of the described sampling and recovery strategies.

9.3.4 NUMERICAL RESULTS

In the sequel, we consider the application of the described sampling and recovery methods to two real graphs: a power network and a road network.

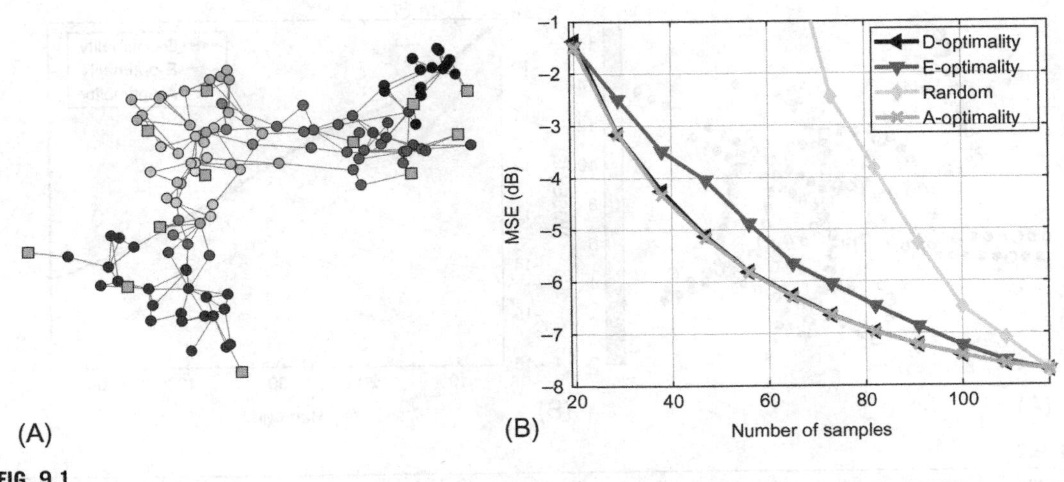

FIG. 9.1

Sampling and recovery over the IEEE 118 Bus graph. (A) Graph topology and sampling set; (B) MSE versus number of samples.

Sampling over power grids. The first example involves the IEEE 118 Bus Test Case, i.e., a portion of the American Electric Power System (in the Midwestern United States) as of December 1962. The graph is composed of 118 nodes (i.e., buses); its topology (i.e., transmission lines connecting buses) is depicted in Fig. 9.1A [45] and the signal at each node encodes the entries of the eigenvector of the Laplacian matrix associated with the second smallest eigenvalue (these entries highlight the presence of three distinct clusters in the network). As illustrated in [46], the dynamics of the power generators give rise to smooth graph signals, so that the bandlimited assumption is justified, although in an approximate sense. In our example, we randomly generate a low pass signal with $|\mathcal{F}| = 12$ and we take a number of samples equal to $|\mathcal{S}| = 12$. The squares in Fig. 9.1A correspond to the samples selected using the greedy Algorithm 9.1 and the A-optimality design in Eq. (9.21). It is interesting to see how the method distributes samples over the clusters and puts the samples, within each cluster, quite far apart from each other. Finally, we compare the reconstruction performance obtained by the considered greedy sampling strategies [cf. Eqs. (9.21), (9.22), and (9.24)] and by random sampling. To this aim we consider graph signal recovery in the presence of an uncorrelated zero mean Gaussian random noise with unit variance, and considering $|\mathcal{F}| = 20$. Thus, Fig. 9.1B reports the MSE in Eq. (9.16) versus the number of samples collected over the graph. As expected, the MSE decreases as the number of samples increases. We can also notice how random sampling performs quite poorly whereas the A-optimal design (and the D-optimal design) outperforms all other strategies.

Traffic flow prediction over road networks. The second example considers sampling of a portion of the road network in the neighborhood of Mazzini Square, which is in the city of Rome, Italy. We have placed landmarks (nodes of the graph) over the streets in a regular fashion, and connected adjacent landmarks on the same lane and at the junctions, thus obtaining the graph topology depicted in Fig. 9.2A. The signal lying on the vertices of the graph represents the flow (number of vehicles per unit of time) of cars passing through the landmark during a period of 30 seconds, and was obtained using

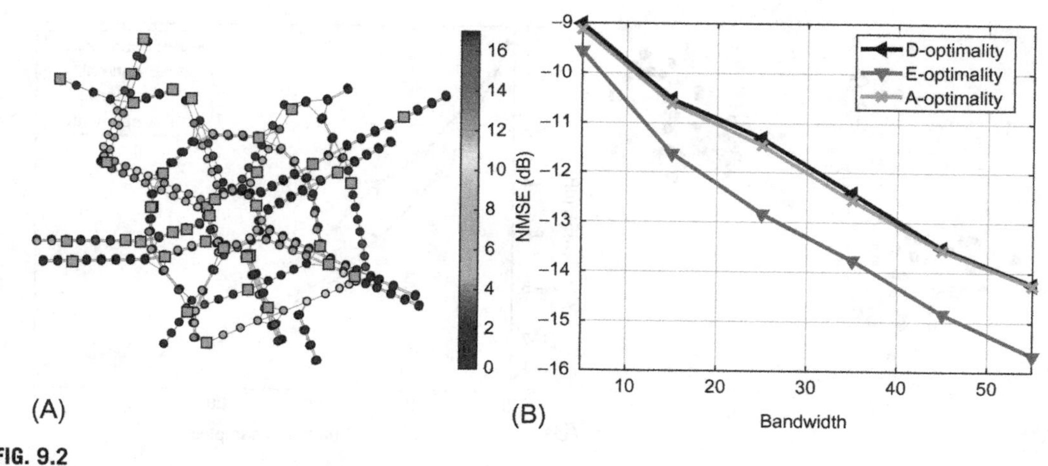

FIG. 9.2

Sampling and recovery of vehicular flows over road networks. (A) Graph topology and sampling set; (B) NMSE versus bandwidth.

a realistic simulator of urban mobility, namely, SUMO [47]. The similarity of values of the signal over adjacent nodes makes the signal smooth, but only approximatively bandlimited. In this sense, there is a mismatching between the observed signal and the bandlimited model used for processing. The goal is to infer the traffic situation over all the road networks from a small number of collected samples. Thus, we consider a bandwidth equal to $|\mathcal{F}| = 30$, and we take a number of samples equal to $|\mathcal{S}| = 40$. The green squares in Fig. 9.2A correspond to the samples selected using the convex relaxation in Eq. (9.26) and the E-optimality design in Eq. (9.22). It is interesting to see how the method distributes almost uniformly the samples over the streets and the junctions. Finally, we compare the reconstruction performance obtained by the sampling strategies based on convex relaxation [cf. Eqs. (9.21), (9.22), and (9.24)] in the presence of model mismatching. To this aim, Fig. 9.2B reports the normalized MSE (NMSE), i.e., NMSE $= \|\hat{\boldsymbol{x}} - \boldsymbol{x}\|^2/\|\boldsymbol{x}\|^2$, versus the graph signal bandwidth, and selecting $|\mathcal{S}| = |\mathcal{F}|$. From Fig. 9.1B, as expected, the NMSE decreases if we use a larger bandwidth. Furthermore, in this case, the E-optimal design outperforms all other strategies, as expected from Eqs. (9.18)–(9.20).

9.3.5 ℓ_1-NORM RECONSTRUCTION OF GRAPH SIGNALS

Let us consider now a different observation model where a bandlimited graph signal $\boldsymbol{x} \in \mathcal{B}_{\mathcal{F}}$ is observed everywhere, but a subset of nodes $\mathcal{S}$ is strongly corrupted by noise, i.e.,

$$\boldsymbol{y} = \boldsymbol{x} + \mathbf{D}_{\mathcal{S}}\boldsymbol{v}, \tag{9.27}$$

where the noise is arbitrary but bounded, i.e., $\|\boldsymbol{v}\|_1 < \infty$. This model is relevant, for example, in sensor networks where a subset of sensors can be damaged or highly interfered. The problem in this case is whether it is possible to recover the graph signal $\boldsymbol{x}$ exactly, i.e., irrespective of noise. Even though this is not a sampling problem, the solution is still related to sampling theory. Clearly, if the signal $\boldsymbol{x}$

is bandlimited and if the indexes of the noisy observations are known, the answer is simple: $\boldsymbol{x} \in \mathcal{B}_{\mathcal{F}}$ can be perfectly recovered from the noisy-free observations, i.e., by completely discarding the noisy observations if the sampling theorem condition (9.8) holds true. But, of course, the challenging situation occurs when the location of the noisy observations is not known. In such a case, we may resort to an ℓ_1-norm minimization, by formulating the problem as follows [16]:

$$\hat{\boldsymbol{x}} = \arg\min_{\boldsymbol{x} \in \mathcal{B}} \|\boldsymbol{y} - \boldsymbol{x}\|_1. \tag{9.28}$$

We provide next some theoretical bounds on the cardinality of $\mathcal{S}$ and $\mathcal{F}$ enabling perfect recovery of the bandlimited graph signal using Eq. (9.28). To this purpose, we recall the following lemma from [16].

Lemma 9.1. *Let us define* $\mu := \max_{j \in \mathcal{F}\ i \in \mathcal{V}} |u_j(i)|$, *where* $u_j(i)$ *is the ith entry of the jth vector of the graph Fourier basis. If for some unknown* $\mathcal{S}$, *we have*

$$|\mathcal{S}| < \frac{1}{2\mu^2\,|\mathcal{F}|}, \tag{9.29}$$

then the ℓ_1*-norm reconstruction method (9.28) recovers* $\boldsymbol{x} \in \mathcal{B}$ *perfectly, i.e.,* $\hat{\boldsymbol{x}} = \boldsymbol{x}$, *for any arbitrary noise* $\boldsymbol{v}$ *present on at most* $|\mathcal{S}|$ *vertices.* ■

An example of ℓ_1 reconstruction based on Eq. (9.28) is useful to grasp some interesting features. We consider the IEEE 118 bus graph in Fig. 9.1A. The signal is assumed to be bandlimited with a spectral content limited to the first $|\mathcal{F}|$ eigenvectors of the Laplacian matrix. In Fig. 9.3, we report the behavior of the NMSE associated with the ℓ_1-norm estimate in Eq. (9.28) versus the number of noisy samples, considering different values of bandwidth $|\mathcal{F}|$. As we can notice from Fig. 9.3, for any value of $|\mathcal{F}|$,

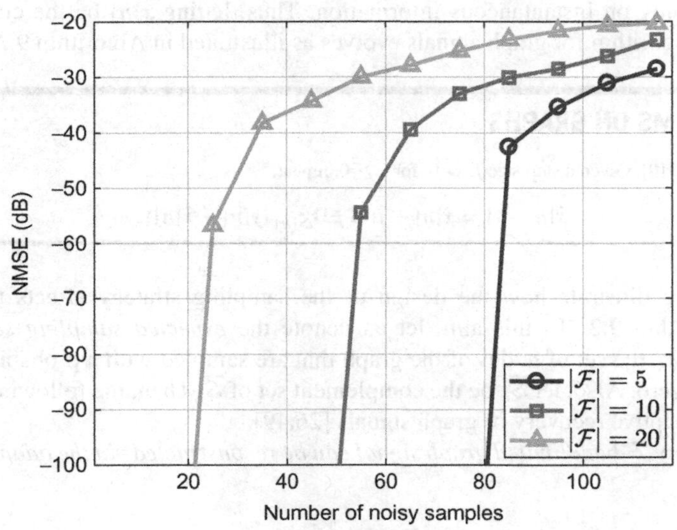

FIG. 9.3

ℓ_1-Norm reconstruction: NMSE versus number of noisy samples.

there exists a threshold value such that, if the number of noisy samples is lower than the threshold, the reconstruction of the signal is error free. As expected, a smaller signal bandwidth allows perfect reconstruction with a larger number of noisy samples.

9.4 ADAPTIVE SAMPLING AND RECOVERY

In this section, we consider processing methods capable of learning and tracking dynamic graph signals from a carefully designed, possibly time-varying, sampling set. To this aim, let us assume that, at each time n, noisy samples of the signal are taken over a (randomly) time-varying subset of vertices, according to the following model:

$$\boldsymbol{y}[n] = \mathbf{D}_{\mathcal{S}[n]}(\boldsymbol{x} + \boldsymbol{v}[n]) = \mathbf{D}_{\mathcal{S}[n]}\mathbf{U}_{\mathcal{F}}\boldsymbol{s}_{\mathcal{F}} + \mathbf{D}_{\mathcal{S}[n]}\boldsymbol{v}[n], \tag{9.30}$$

where $\mathbf{D}_{\mathcal{S}[n]} = \text{diag}\{d_i[n]\}_{i=1}^N \in \mathbb{R}^{N\times N}$ [cf. Eq. (9.5)], with $d_i[n]$ denoting a random sampling binary coefficient, which is equal to 1 if $i \in \mathcal{S}[n]$, and 0 otherwise (i.e., $\mathcal{S}[n]$ represents the *instantaneous*, random sampling set at time n); and $\boldsymbol{v}[n] \in \mathbb{C}^N$ is zero-mean, spatially and temporally independent observation noise, with covariance matrix $\mathbf{R}_v = \text{diag}\{r_1^2, \ldots, r_N^2\}$. The estimation task consists in recovering the vector $\boldsymbol{x}$ (or, equivalently, its GFT $\boldsymbol{s}_{\mathcal{F}}$) from the noisy, streaming, and partial observations $\boldsymbol{y}[n]$ in Eq. (9.30). Following an LMS approach [48], from Eq. (9.30), the optimal estimate for $\boldsymbol{s}_{\mathcal{F}}$ can be found as the vector that solves the following optimization problem:

$$\min_{\boldsymbol{s}} \mathbb{E}\|\mathbf{D}_{\mathcal{S}[n]}(\boldsymbol{y}[n] - \mathbf{U}_{\mathcal{F}}\boldsymbol{s})\|^2, \tag{9.31}$$

where in Eq. (9.31) we have exploited the fact that $\mathbf{D}_{\mathcal{S}[n]}$ is an idempotent matrix for any fixed n [cf. Eq. (9.5)]. An LMS-type solution optimizes Eq. (9.31) by means of a stochastic steepest-descent procedure, relying only on instantaneous information. Thus, letting $\hat{\boldsymbol{x}}[n]$ be the current estimates of vector $\boldsymbol{x}$, the LMS algorithm for graph signals evolves as illustrated in Algorithm 9.2 [26].

Algorithm 9.2 LMS ON GRAPHS

Start with random $\hat{\boldsymbol{x}}[0]$. Given a step size $\mu > 0$, for $n \geq 0$, repeat:

$$\hat{\boldsymbol{x}}[n+1] = \hat{\boldsymbol{x}}[n] + \mu\, \mathbf{B}_{\mathcal{F}}\mathbf{D}_{\mathcal{S}[n]}(\boldsymbol{y}[n] - \hat{\boldsymbol{x}}[n]).$$

In the sequel, we illustrate how the design of the sampling strategy affects the reconstruction capability of Algorithm 9.2. To this aim, let us denote the *expected sampling set* by $\overline{\mathcal{S}} = \{i = 1, \ldots, N \,|\, p_i > 0\}$, i.e., the set of nodes of the graph that are sampled with a probability $p_i = \mathbb{E}\{d_i[n]\}$ strictly greater than zero. Also, let $\overline{\mathcal{S}}_c$ be the complement set of $\overline{\mathcal{S}}$. Then, the following results illustrate the conditions for adaptive recovery of graph signals [26,49].

Theorem 9.3. *Any $\mathcal{F}$-bandlimited graph signal can be reconstructed via the adaptive Algorithm 9.2 if and only if*

$$\left\|\boldsymbol{D}_{\overline{\mathcal{S}}_c}\boldsymbol{U}_{\mathcal{F}}\right\|_2 < 1, \tag{9.32}$$

i.e., if there are no $\mathcal{F}$-bandlimited signals that are perfectly localized on $\overline{\mathcal{S}}_c$. ■

Differently from batch sampling and recovery of graph signals, see, e.g., [6,12,13,16,17], condition (9.32) depends on the *expected* sampling set. In particular, it implies that there are no $\mathcal{F}$-bandlimited signals that are perfectly localized over the set $\overline{\mathcal{S}}_c$. As a consequence, the adaptive Algorithm 9.2 with probabilistic sampling does not need to collect all the data necessary to reconstruct one shot the graph signal at each iteration, but can learn acquiring the needed information over time. The only important thing required by condition (9.32) is that a sufficiently large number of nodes is sampled in *expectation*.

We now illustrate the mean-square performance of Algorithm 9.2. The main results are summarized in Theorem 9.4 [49].

Theorem 9.4. *Assume spatial and temporal independence of the random variables extracted by the sampling process $\{d_i[n]\}_{i,n}$. Then, for any initial condition, Algorithm 9.2 is mean-square error stable if the sampling probability vector $\boldsymbol{p}$ and the step size μ satisfy Eq. (9.32) and*

$$0<\mu<\frac{2\lambda_{\min}\left(\boldsymbol{U}_{\mathcal{F}}^{H}\,diag(\boldsymbol{p})\boldsymbol{U}_{\mathcal{F}}\right)}{\lambda_{\max}^{2}\left(\boldsymbol{U}_{\mathcal{F}}^{H}\,diag(\boldsymbol{p})\boldsymbol{U}_{\mathcal{F}}\right)}.$$

Furthermore, under a small step-size assumption, the MSE writes as:

$$\begin{aligned} MSE &= \lim_{n\to\infty}\ \mathbb{E}\|\hat{\boldsymbol{x}}[n]-\boldsymbol{x}[n]\|^2 \\ &= \frac{\mu}{2}\,Tr\left[\left(\boldsymbol{U}_{\mathcal{F}}^{H}\,diag(\boldsymbol{p})\boldsymbol{U}_{\mathcal{F}}\right)^{-1}\boldsymbol{U}_{\mathcal{F}}^{H}\,diag(\boldsymbol{p})\boldsymbol{R}_{v}\boldsymbol{U}_{\mathcal{F}}\right] \end{aligned} \tag{9.33}$$

and the convergence rate α is well approximated by

$$\alpha = 1-2\mu\lambda_{\min}\left(\boldsymbol{U}_{\mathcal{F}}^{H}\,diag(\boldsymbol{p})\boldsymbol{U}_{\mathcal{F}}\right). \tag{9.34}$$

■

The results of Theorem 9.4 are instrumental to devise optimal probabilistic sampling strategies for Algorithm 9.2, which are described in the sequel.

9.4.1 PROBABILISTIC SAMPLING STRATEGIES

We consider a sampling design that seeks for the probability vector $\boldsymbol{p}$ that minimizes the total sampling rate over the graph, i.e., $\mathbf{1}^T\boldsymbol{p}$, while guaranteeing a target performance in terms of MSE in Eq. (9.33) and of convergence rate in Eq. (9.34) [49]. Then, the optimization problem can be cast as:

$$\begin{aligned} &\min_{\boldsymbol{p}}\quad \mathbf{1}^T\boldsymbol{p} \\ &\text{s.t.}\quad \lambda_{\min}\left(\mathbf{U}_{\mathcal{F}}^{H}\,\text{diag}(\boldsymbol{p})\mathbf{U}_{\mathcal{F}}\right)\geq\frac{1-\bar{\alpha}}{2\mu}, \\ &\qquad \text{Tr}\left[\left(\mathbf{U}_{\mathcal{F}}^{H}\,\text{diag}(\boldsymbol{p})\mathbf{U}_{\mathcal{F}}\right)^{-1}\mathbf{U}_{\mathcal{F}}^{H}\,\text{diag}(\boldsymbol{p})\mathbf{R}_{v}\mathbf{U}_{\mathcal{F}}\right]\leq\frac{2\gamma}{\mu}, \\ &\qquad \mathbf{0}\leq\boldsymbol{p}\leq\boldsymbol{p}^{\max}. \end{aligned} \tag{9.35}$$

The first constraint imposes that the convergence rate of the algorithm is larger than a desired value, i.e., α in Eq. (9.34) is smaller than a target value, say, e.g., $\bar{\alpha}\in(0,1)$. Furthermore, as illustrated in [49], the first constraint on the convergence rate also guarantees adaptive signal reconstruction, i.e., condition

(9.32) holds true. The second constraint guarantees a target mean-square performance, i.e., the MSE in Eq. (9.33) must be less than or equal to a prescribed value, say, e.g., $\gamma > 0$. Finally, the last constraint limits the probability vector to lie in the box $p_i \in [0, p_i^{\max}]$, for all i, with $0 \leq p_i^{\max} \leq 1$ denoting an upper bound on the sampling probability at each node that might depend on external factors such as, e.g., limited energy, processing and/or communication resources, node or communication failures, etc.

Unfortunately, problem (9.35) is nonconvex due to the presence of the nonconvex constraint on the MSE. To handle the nonconvexity of Eq. (9.35), we exploit an upper bound of the MSE function in Eq. (9.33), given by:

$$\text{MSE}(\boldsymbol{p}) \leq \overline{\text{MSE}}(\boldsymbol{p}) \triangleq \frac{\mu}{2} \frac{\text{Tr}\left(\mathbf{U}_{\mathcal{F}}^H \,\text{diag}(\boldsymbol{p}) \mathbf{R}_v \mathbf{U}_{\mathcal{F}}\right)}{\lambda_{\min}\left(\mathbf{U}_{\mathcal{F}}^H \,\text{diag}(\boldsymbol{p}) \mathbf{U}_{\mathcal{F}}\right)}, \quad \text{for all } \boldsymbol{p} \in \mathbb{R}^N. \tag{9.36}$$

Of course, replacing the MSE function with the bound (9.36), the second constraint in Eq. (9.35) is always satisfied. Furthermore, the function in Eq. (9.36) has the nice property to be pseudoconvex because it is the ratio between a convex and concave function, which are both differentiable and positive for all $\boldsymbol{p}$ satisfying the other constraints [50]. Thus, exploiting the upper bound (9.36), we can formulate a surrogate optimization problem for the selection of the probability vector $\boldsymbol{p}$, which can be cast as:

$$\begin{aligned}
&\min_{\boldsymbol{p}} \ \mathbf{1}^T \boldsymbol{p} \\
&\text{subject to} \\
&\qquad \lambda_{\min}\left(\mathbf{U}_{\mathcal{F}}^H \,\text{diag}(\boldsymbol{p}) \mathbf{U}_{\mathcal{F}}\right) \geq \frac{1-\bar{\alpha}}{2\mu}, \\
&\qquad \frac{\text{Tr}\left(\mathbf{U}_{\mathcal{F}}^H \,\text{diag}(\boldsymbol{p}) \mathbf{C}_v \mathbf{U}_{\mathcal{F}}\right)}{\lambda_{\min}\left(\mathbf{U}_{\mathcal{F}}^H \,\text{diag}(\boldsymbol{p}) \mathbf{U}_{\mathcal{F}}\right)} \leq \frac{2\gamma}{\mu}, \\
&\qquad \mathbf{0} \leq \boldsymbol{p} \leq \boldsymbol{p}^{\max}.
\end{aligned} \tag{9.37}$$

Because the sublevel sets of pseudoconvex functions are convex sets [50], it is straightforward to see that the approximated problem (9.37) is convex, and its global solution can be found using efficient numerical tools [44].

9.4.2 DISTRIBUTED ADAPTIVE RECOVERY

The implementation of Algorithm 9.2 would require collecting all the data $\{y_i[n]\}_{i:d_i[n]=1}$, for all n, in a single processing unit that performs the computation. In many practical systems, data are collected in a distributed network, and sharing local information with a central processor might be either unfeasible or not efficient, owing to the large volume of data, time-varying network topology, and/or privacy issues. Motivated by these observations, in this section we extend the LMS strategy in Algorithm 9.2 to a distributed setting where the nodes perform the reconstruction task via online in-network processing, only exchanging data between neighbors defined over a sparse (but connected) communication network, which is described by the graph $\mathcal{G}_c = (\mathcal{V}, \mathcal{E}_c)$. Proceeding as in [27] to derive distributed solution methods for problem (9.31), let us introduce local copies $\{\boldsymbol{s}_i\}_{i=1}^N$ of the global variable $\boldsymbol{s}$, and recast problem (9.31) in the following equivalent form:

$$\min_{\{s_i\}_{i=1}^N} \sum_{i=1}^{N} \mathbb{E}\left|d_i[n]\left(y_i[n] - \boldsymbol{u}_{\mathcal{F},i}^H \boldsymbol{s}_i\right)\right|^2 \qquad (9.38)$$
$$\text{subject to} \quad \boldsymbol{s}_i = \boldsymbol{s}_j \quad \text{for all } i = 1,\ldots,N, j \in \mathcal{N}_i,$$

where $\boldsymbol{u}_{\mathcal{F},i}^H$ is the ith row of matrix $\mathbf{U}_{\mathcal{F}}$ (supposed to be known at node i, or computable in distributed fashion, see, e.g., [51]), and $\mathcal{N}_i = \{j | a_{ij} > 0\}$ is the local neighborhood of node i. To solve problem (9.31), we consider an Adapt-Then-Combine (ATC) diffusion strategy [27], and the resulting algorithm is reported in Algorithm 9.3. The first step in Eq. (9.39) is an adaptation step where the intermediate estimate $\boldsymbol{\psi}_i[n]$ is updated adopting the current observation taken by node i, i.e., $y_i[n]$. The second step is a diffusion step where the intermediate estimates $\boldsymbol{\psi}_j[n]$, from the (extended) spatial neighborhood $\overline{\mathcal{N}}_i = \mathcal{N}_i \bigcup\{i\}$, are combined through the weighting coefficients $\{w_{ij}\}$. Several possible combination rules have been proposed in the literature, such as the Laplacian or the Metropolis-Hastings weights, see, e.g., [52–54]. Finally, given the estimate $\boldsymbol{s}_i[n]$ of the GFT at node i and time n, the last step produces the estimate $x_i[n+1]$ of the graph signal value at node i [cf. Eq. (9.30)]. Here, we assume that graphs $\mathcal{G}$ (i.e., the one used for GSP) and $\mathcal{G}_c$ (i.e., the one describing the communication pattern among nodes) might have in general distinct topologies. We remark that both graphs play an important role in the proposed distributed processing strategy (9.39). First, the processing graph determines the structure of the regression data $\boldsymbol{u}_{\mathcal{F},i}$ used in the adaptation step of Eq. (9.39). In fact, $\{\boldsymbol{u}_{\mathcal{F},i}^H\}_i$ are the rows of the matrix $\mathbf{U}_{\mathcal{F}}$, whose columns are the eigenvectors of the Laplacian matrix associated with the set of support frequencies $\mathcal{F}$. Then, the topology of the communication graph determines how information is spread all over the network through the diffusion step in Eq. (9.39). This illustrates how, when reconstructing graph signals in a distributed manner, we have to take into account both the processing and communication aspects of the problem.

Algorithm 9.3 DIFFUSION LMS ON GRAPHS

Start with random $\boldsymbol{s}_i[0]$, for all $i \in \mathcal{V}$. Given combination weights $\{w_{ij}\}_{i,j}$, step-sizes $\mu_i > 0$, for each time $n \geq 0$ and for each node i, repeat:

$$\begin{aligned}
\boldsymbol{\psi}_i[n] &= \boldsymbol{s}_i[n] + \mu_i d_i[n] \boldsymbol{u}_{\mathcal{F},i}(y_i[n] - \boldsymbol{u}_{\mathcal{F},i}^H \boldsymbol{s}_i[n]) && \text{(adaptation)} \\
\boldsymbol{s}_i[n+1] &= \sum_{j \in \overline{\mathcal{N}}_i} w_{ij} \boldsymbol{\psi}_j[n] && \text{(diffusion)} \\
x_i[n+1] &= \boldsymbol{u}_{\mathcal{F},i}^H \boldsymbol{s}_i[n+1] && \text{(reconstruction)}
\end{aligned} \qquad (9.39)$$

9.4.3 NUMERICAL RESULTS

In this section, we first illustrate the performance of the probabilistic sampling method in Eq. (9.37) over the IEEE 118 bus graph. Then, we consider an application to dynamic inference of brain activity.

Optimal probabilistic sampling. As a first example, let us consider an application to the IEEE 118 Bus Test Case in Fig. 9.1A. The spectral content of the graph signal is assumed to be limited to the

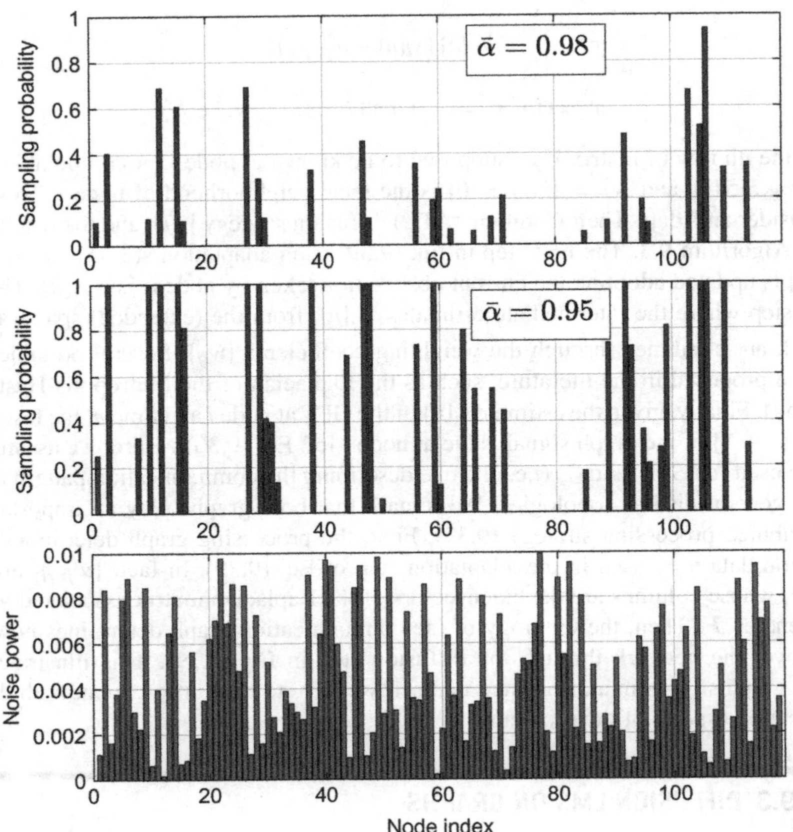

FIG. 9.4

Optimal probabilistic sampling over the IEEE 118 graph.

first ten eigenvectors of the Laplacian matrix of the graph. The observation noise in Eq. (9.30) is zero-mean Gaussian with a diagonal covariance matrix $\mathbf{R}_v$, where each element is illustrated in Fig. 9.4 (bottom). The other parameters are $\mu = 0.1$, and $\gamma = 10^{-3}$. Then, in Fig. 9.4 (top and middle), we plot the optimal probability vector obtained solving (9.37) for two different values of $\bar{\alpha}$. In all cases, the constraints on the MSE and convergence rate are attained strictly. From Fig. 9.4 (top and middle), we notice how the method increases the sampling rate if we require a faster convergence (i.e., a smaller value of $\bar{\alpha}$); it also finds a very sparse probability vector and usually avoids assigning large sampling probabilities to nodes having large noise variances. Interestingly, with the proposed formulation, sparse sampling patterns are obtained thanks to the optimization of the sampling probabilities, which are already real numbers, without resorting to any relaxation of complex integer optimization problems [cf. Eq. (9.25)].

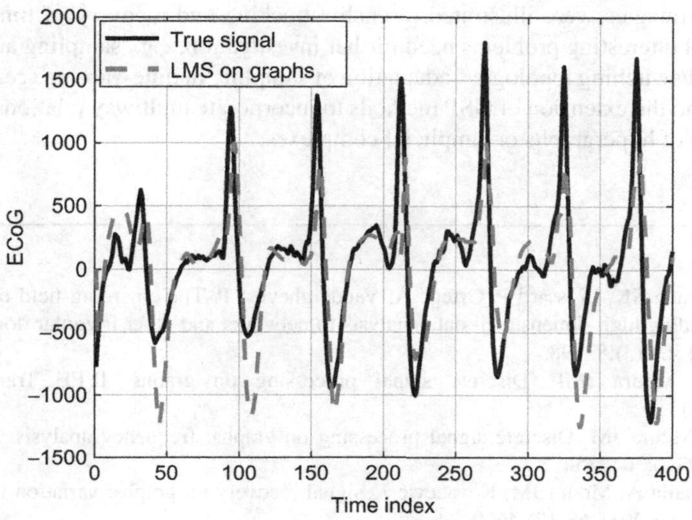

FIG. 9.5

True ECoG and estimate across time.

Inference of brain activity. The last example presents test results on electrocorticography (ECoG) data, captured through experiments conducted in an epilepsy study [55]. Data were collected over a period of five days where the electrodes recorded 76 ECoG time series, consisting of voltage levels measured in different regions of the brain. Two temporal intervals of interest were picked for analysis, namely, the preictal and ictal intervals. In the sequel, we focus on the ictal interval. Further details about data acquisition and preprocessing are provided in [55]. The GFT matrix $\mathbf{U}_{\mathcal{F}}$ is learnt from the first 200 samples of ictal data, using the method proposed in [56], and imposing a bandwidth equal to $|\mathcal{F}| = 30$. In Fig. 9.5, we illustrate the true behavior of the ECoG present at an unobserved electrode chosen at random over the first 400 samples of ictal data, along with an estimate carried out using Algorithm 9.2 (with $\mu = 1.5$). The sampling set is fixed over time (i.e., $p_i = 1$ for all i), and chosen according to the E-optimal design in Eq. (9.24), selecting 32 samples. As we can notice from Fig. 9.5, the method is capable of efficiently inferring and tracking the unknown dynamics of ECoG data at unobserved regions of the brain.

9.5 CONCLUSIONS

In this chapter, we have reviewed some of the methods recently proposed to sample and interpolate signals defined over graphs. First, we have recalled the conditions for perfect recovery under a bandlimited assumption. Second, we have illustrated sampling strategies, based on greedy methods or convex relaxations, aimed at reducing the effect of noise or aliasing on the recovered signal. Then, we considered ℓ_1-norm reconstruction, which allows perfect recovery of graph signals in the presence of a strong impulsive noise over a limited number of nodes. Finally, adaptive methods based on (possibly distributed) LMS strategies were illustrated to enable tracking and recovery of time-varying signals over graphs. Several interesting problems need further investigation, e.g., sampling and recovery in the presence of directed/switching topologies, adaptation of sampling in time-varying scenarios, distributed implementations, and the extension of GSP methods to incorporate multiway relationships among data, e.g., under the form of hypergraphs or simplicial complexes.

REFERENCES

[1] Shuman DI, Narang SK, Frossard P, Ortega A, Vandergheynst P. The emerging field of signal processing on graphs: extending high-dimensional data analysis to networks and other irregular domains. IEEE Signal Process Mag 2013;30(3):83–98.

[2] Sandryhaila A, Moura JMF. Discrete signal processing on graphs. IEEE Trans Signal Process 2013;61:1644–56.

[3] Sandryhaila A, Moura JM. Discrete signal processing on graphs: frequency analysis. IEEE Trans Signal Process 2014;62(12):3042–54.

[4] Chen S, Sandryhaila A, Moura JM, Kovacevic J. Signal recovery on graphs: variation minimization. IEEE Trans Signal Process 2015;63(17):4609–24.

[5] Zhu X, Rabbat M. Approximating signals supported on graphs. In: IEEE international conference on acoustics, speech and signal processing; 2012. p. 3921–4.

[6] Chen S, Varma R, Sandryhaila A, Kovačević J. Discrete signal processing on graphs: sampling theory. IEEE Trans Signal Process 2015;63:6510–23.

[7] Gadde A, Anis A, Ortega A. Active semi-supervised learning using sampling theory for graph signals. In: Proceedings of the 20th ACM SIGKDD international conference on knowledge discovery and data mining. ACM; 2014. p. 492–501.

[8] Janssen S, Dumont G, Fierens F, Mensink C. Spatial interpolation of air pollution measurements using CORINE land cover data. Atmos Environ 2008;42(20):4884–903.

[9] Candès EJ, Recht B. Exact matrix completion via convex optimization. Found Comput Math 2009;9(6):717.

[10] Gomez-Uribe CA, Hunt N. The Netflix recommender system: algorithms, business value, and innovation. ACM Trans Manag Inf Syst 2016;6(4):13.

[11] Yamanishi Y, Vert JP, Kanehisa M. Protein network inference from multiple genomic data: a supervised approach. Bioinformatics 2004;20(suppl_1):i363–70.

[12] Pesenson IZ. Sampling in Paley-Wiener spaces on combinatorial graphs. Trans Am Math Soc 2008;360(10):5603–27.

[13] Narang S, Gadde A, Ortega A. Signal processing techniques for interpolation in graph structured data. In: IEEE international conference on acoustics, speech, and signal processing; 2013. p. 5445–9.

[14] Anis A, Gadde A, Ortega A. Towards a sampling theorem for signals on arbitrary graphs. In: 2014 IEEE international conference on acoustics, speech and signal processing (ICASSP). IEEE; 2014. p. 3864–8.

[15] Wang X, Liu P, Gu Y. Local-set-based graph signal reconstruction. IEEE Trans Signal Process 2015;63(9):2432–44.

[16] Tsitsvero M, Barbarossa S, Di Lorenzo P. Signals on graphs: uncertainty principle and sampling. IEEE Trans Signal Process 2016;64(18):4845–60.

[17] Marques AG, Segarra S, Leus G, Ribeiro A. Sampling of graph signals with successive local aggregations. IEEE Trans Signal Process 2016;64(7):1832–43.

[18] Chamon LF, Ribeiro A. Greedy sampling of graph signals; 2017. arXiv:170401223.

[19] Tremblay N, Puy G, Gribonval R, Vandergheynst P. Compressive spectral clustering. In: International conference on machine learning; 2016. p. 1002–11.

[20] Anis A, Gadde A, Ortega A. Efficient sampling set selection for bandlimited graph signals using graph spectral proxies. IEEE Trans Signal Process 2016;64(14):3775–89.
[21] Narang SK, Ortega A. Local two-channel critically sampled filter-banks on graphs. In: 17th IEEE international conference on image processing. IEEE; 2010. p. 333–6.
[22] Nguyen HQ, Do MN. Downsampling of signals on graphs via maximum spanning trees. IEEE Trans Signal Process 2015;63(1):182–91.
[23] Chen S, Varma R, Singh A, Kovačević J. Signal recovery on graphs: fundamental limits of sampling strategies. IEEE Trans Signal Inf Process Netw 2016;2(4):539–54.
[24] Puy G, Tremblay N, Gribonval R, Vandergheynst P. Random sampling of bandlimited signals on graphs. Appl Comput Harmon Anal 2018;44(2):446–75.
[25] Tremblay N, Amblard P-O, Barthelmé S. Graph sampling with determinantal processes. In: European signal processing conference (Eusipco); 2017.
[26] Di Lorenzo P, Barbarossa S, Banelli P, Sardellitti S. Adaptive least mean squares estimation of graph signals. IEEE Trans Signal Inf Process Netw 2016;2(4):555–68.
[27] Di Lorenzo P, Banelli P, Barbarossa S, Sardellitti S. Distributed adaptive learning of graph signals. IEEE Trans Signal Process 2017;65(16):4193–208.
[28] Romero D, Ioannidis VN, Giannakis GB. Kernel-based reconstruction of space-time functions on dynamic graphs. IEEE J Sel Top Signal Process 2017;11(6):856–69.
[29] Wang X, Wang M, Gu Y. A distributed tracking algorithm for reconstruction of graph signals. IEEE J Sel Top Signal Process 2015;9(4):728–40.
[30] Godsil C, Royle GF. Algebraic graph theory, vol. 207. Springer Science & Business Media; 2013.
[31] Eldar YC. Sampling with arbitrary sampling and reconstruction spaces and oblique dual frame vectors. J Fourier Anal Appl 2003;9(1):77–96.
[32] Teke O, Vaidyanathan P. Uncertainty principles and sparse eigenvectors of graphs. IEEE Trans Signal Process 2017;65(20):5406–20.
[33] Banerjee A, Jost J. On the spectrum of the normalized graph Laplacian. Linear Algebra Appl 2008;428(11-12):3015–22.
[34] Kay SM. Fundamentals of statistical signal processing. Prentice Hall PTR; 1993.
[35] Winer BJ, Brown DR, Michels KM. Statistical principles in experimental design, vol. 2. McGraw-Hill New York; 1971.
[36] Nemhauser GL, Wolsey LA. Integer and combinatorial optimization. In: Interscience series in discrete mathematics and optimization. John Wiley & Sons; 1988.
[37] Lovász L. Submodular functions and convexity. In: Mathematical programming the state of the art. Springer; 1983. p. 235–57.
[38] Nemhauser GL, Wolsey LA, Fisher ML. An analysis of approximations for maximizing submodular set functions-i. Math Program 1978;14(1):265–94.
[39] Summers TH, Cortesi FL, Lygeros J. Correction to “On Submodularity and Controllability in Complex Dynamical Networks”. Available at: http://www.utdallas.edu/~tyler.summers/papers/TCNS_Correction.pdf.
[40] Shamaiah M, Banerjee S, Vikalo H. Greedy sensor selection: leveraging submodularity. In: 2010 49th IEEE conference on decision and control. IEEE; 2010. p. 2572–7.
[41] Joshi S, Boyd S. Sensor selection via convex optimization. IEEE Trans Signal Process 2009;57(2):451–62.
[42] Chepuri SP, Leus G. Sparsity-promoting sensor selection for non-linear measurement models. IEEE Trans Signal Process 2015;63(3):684–98.
[43] Chepuri SP, Leus G. Sparse sensing for distributed detection. IEEE Trans Signal Process 2016;64(6):1446–60.
[44] Boyd S, Vandenberghe L. Convex optimization. Cambridge University Press; 2004.
[45] Sun K, Zheng DZ, Lu Q. A simulation study of OBDD-based proper splitting strategies for power systems under consideration of transient stability. IEEE Trans Power Syst 2005;20(1):389–99.

[46] Pasqualetti F, Zampieri S, Bullo F. Controllability metrics, limitations and algorithms for complex networks. IEEE Trans Control Netw Syst 2014;1(1):40–52.
[47] Behrisch M, Bieker L, Erdmann J, Krajzewicz D. Sumo-simulation of urban mobility: an overview. In: International conference on advances in system simulation. ThinkMind; 2011.
[48] Sayed AH. Adaptive filters. John Wiley & Sons; 2011.
[49] Di Lorenzo P, Banelli P, Isufi E, Barbarossa S, Leus G. Adaptive graph signal processing: algorithms and optimal sampling strategies; to appear in IEEE Transactions on Signal Processing, 2018. arXiv:170903726.
[50] Avriel M, Diewert WE, Schaible S, Zang I. Generalized concavity. SIAM; 2010.
[51] Di Lorenzo P, Barbarossa S. Distributed estimation and control of algebraic connectivity over random graphs. IEEE Trans Signal Process 2014;62(21):5615–28.
[52] Barbarossa S, Sardellitti S, Di Lorenzo P. Distributed detection and estimation in wireless sensor networks, vol. 2. Academic Press Library in Signal Processing; 2014. p. 329–408.
[53] Xiao L, Boyd S, Kim SJ. Distributed average consensus with least-mean-square deviation. J Parallel Distrib Comput 2007;67(1):33–46.
[54] Cattivelli FS, Sayed AH. Diffusion LMS strategies for distributed estimation. IEEE Trans Signal Process 2010;58:1035–48.
[55] Kramer MA, Kolaczyk ED, Kirsch HE. Emergent network topology at seizure onset in humans. Epilepsy Res 2008;79(2):173–86.
[56] Gavish M, Donoho DL. Optimal shrinkage of singular values. IEEE Trans Inf Theory 2017;63(4):2137–52.

CHAPTER

BAYESIAN ACTIVE LEARNING ON GRAPHS 10

Kwang-Sung Jun*, Robert Nowak†

*Wisconsin Institute for Discovery, Madison, WI, United States** *Department of Electrical and Computer Engineering, University of Wisconsin-Madison, Madison, WI, United States†*

10.1 BAYESIAN ACTIVE LEARNING ON GRAPHS

Suppose we are given a graph where each node is associated with a categorical label. An edge between two nodes encodes their similarity. Most node labels are, however, unknown to us. We consider a label prediction task on such a graph where node labels can be queried with some cost.

There are many applications where the data is represented as a graph or constructed to be so. In social networks or citation networks, the graph naturally arises where each friendship or citation relationship is represented as an edge and carries important information between two nodes. For example, in document classification problems, two documents tend to be of the same topic when one cites the other. Even when there is no graph information in the data, a graph can be constructed based on known or computed (dis-)similarities between data points.[1] The advantage of the graph construction is that it captures the manifold structure of the data, which often results in low generalization error. For example, handwritten digits can be recognized efficiently through graph-based learning algorithms [2]. In all these examples, the edge existence or weights often carries how strongly two nodes are related, which can be used to make label predictions on unlabeled nodes. Note that this is precisely an instance of so-called *transduction* (or *semisupervised learning* [1], more broadly) as the test set covariates (graph information in our case) are available during training the model, unlike the standard supervised learning setting.

The goal of this chapter is to perform *active learning*.[2] That is, if we have the freedom to choose which node labels to query, can we judiciously select informative nodes in order to minimize the prediction error on the entire graph while querying as few labels as possible? The active learning problem in the *graph* setting was first considered in [4]. Since then, there has been a flurry of research activities [5–10].

Before defining the problem formally, let us consider a few introductory examples where active learning can be effective. Consider a star graph with n nodes where there exists a center node and the

[1] Popular construction methods include the k-nearest-neighbor graph; see [1, Section 6.2] for details.

[2] We refer to Settles [3] for a complete survey.

Cooperative and Graph Signal Processing. https://doi.org/10.1016/B978-0-12-813677-5.00010-9

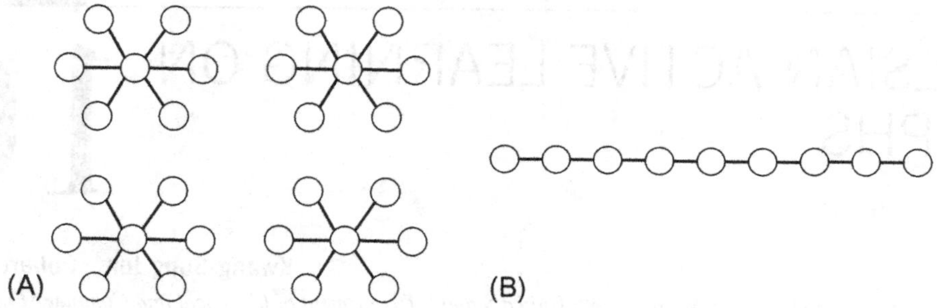

FIG. 10.1
Introductory graph examples: (A) stars; (B) linear chain.

rest of the nodes are all "dongles" that have one and only one edge to the center node. Construct a larger graph with k star graphs (total kn nodes); see Fig. 10.1A. Assume that the labels of the dongle nodes are each $+1$ or -1 at random and the center node's label is decided by the majority label of the dongles; in other words, the center node is "representative" of the rest. In this case, it might be wise to query the labels of the k center nodes first because then afterwards we can predict the labels of each component with error no larger than 0.5. This shows the importance of using the graph structure.

As another example, consider a linear chain graph with n nodes where the labels are ± 1; see Fig. 10.1B. Assume that there exists only one edge in the graph that connects -1 and $+1$ (i.e., cut), and the location of the cut is not known. One can see that performing a binary search takes $O(\log(n))$ time to find the cut after which we incur no prediction error. This is better than randomly choosing which node labels to observe, which takes $O(n)$ time in most cases. This example shows the importance of adaptivity; one can use what labels have been observed to narrow down the cut location faster.

With these introductory examples in mind, we now formally define the active learning problem and present a popular label-generation model that encodes our assumption on the labels.

Notations

Denote by $\mathbf{1}$ and $\mathbf{0}$ the vector of all 1s and all 0s, respectively. Let $\mathbb{1}\{A\}$ be 1 if A is true and 0 otherwise. A boldfaced lowercase and uppercase letter are vectors (e.g., $\mathbf{y}$) and matrices (e.g., $\mathbf{M}$), respectively, where each component is denoted without boldfacing (e.g., y_i and M_{ij}). We use $\mathbf{M}_{\cdot i}$ ($\mathbf{M}_{i\cdot}$) to refer to the ith column (row). One exception is $\mathbf{Y} \in \{\pm 1\}^n$ that denotes the *random* vector of node labels, which must not be interpreted as a matrix; this will be clear from the context as $\mathbf{Y}$ appears in probability statements only. Denote by $\mathbf{I}$ the identity matrix. $\boldsymbol{\ell}$ refers to the set of labeled nodes, and $\mathbf{u}$ is the set of unlabeled nodes. We denote by $\mathbf{Y}_{\boldsymbol{\ell}}$ a subvector of length $|\boldsymbol{\ell}|$ that slices the components of $\mathbf{Y}$ by the indices in $\boldsymbol{\ell}$. Similarly, we denote by $\mathbf{M}_{\boldsymbol{\ell}\mathbf{u}}$ the submatrix sliced from $\mathbf{M}$ by taking rows and columns indicated by $\boldsymbol{\ell}$ and $\mathbf{u}$.

10.1.1 PROBLEM DEFINITION

We are given a weighted undirected graph $G = (N, E)$ with nodes $N = \{1, \ldots, n\}$, edges E, and weights $\mathbf{W} \in \mathbb{R}^{n\times n}$ where $W_{ij} = W_{ji} \geq 0, \forall i \leq j$, and $W_{ij} = 0$ if there is no edge between i and j. We assume

$W_{ii} = 0, \forall i$. The node labels $Y_1, \ldots, Y_n \in \{\pm 1\}$ are jointly drawn from a distribution.[3] These labels are, however, mostly unknown to the algorithm. Let $\ell^{(1)} \subseteq N$ be the set of nodes whose labels are known initially. The algorithm must perform a sequential task specified in the following protocol.

For time step $t = 1, 2, \ldots$

1. PREDICT: Make a label prediction $\hat{Y}_i$ on each unlabeled node $i \notin \ell^{(t)}$. Let $\hat{Y}_i := Y_i, \forall i \in \ell^{(t)}$. An algorithm suffers an error rate $\epsilon_t = \frac{1}{n}\sum_{i=1}^{n} \mathbb{1}\{\hat{Y}_i \neq Y_i\}$, which is *unknown* to the algorithm.
2. QUERY: Select an unlabeled node q and query its label. Receive the label Y_q. Update $\ell^{(t+1)} = \ell^{(t)} \cup \{q\}$.

The goal is to achieve low error rates $\{\epsilon_t\}$ for time steps of interest. Depending on the application, one might aim to have the lowest error at a specific time step t' or low error on average. In general, we would like to enjoy low error rate in the "small sample size" regime as the motivation of active learning is to use a small budget to build a good enough model. Among various approaches for performing PREDICT and QUERY, we focus on those motivated by a specific Bayesian model we describe in the following section.

Note that the protocol above is different from the so-called *batch* active learning where the query budget k is given and an algorithm must choose k nodes and query their labels *altogether* rather than querying one at a time sequentially. Still, a batch active learning algorithm can be used to perform the above *sequential* active learning task.

10.1.2 BINARY MARKOV RANDOM FIELD: A BAYESIAN MODEL OF GRAPH LABELS

Consider the following probabilistic model for the random variable $\mathbf{Y} \in \{\pm 1\}^n$:

$$\mathbb{P}(\mathbf{Y} = \mathbf{y}) = \frac{1}{Z}\exp\left(-\frac{\beta}{2}\sum_{i<j} W_{ij}(y_i - y_j)^2\right), \tag{10.1}$$

where Z is the normalization factor and $\beta > 0$ is a strength parameter. The model above prefers labelings $\mathbf{y} \in \{\pm 1\}^n$ that vary smoothly along the edges; i.e., a larger weight W_{ij} implies a higher likelihood of $y_i = y_j$. We refer to the model above as the binary Markov random field (**BMRF**).

One can rewrite the above in a simpler form. Let $\mathbf{L} := \mathbf{D} - \mathbf{W}$ be the graph Laplacian where $\mathbf{D}$ is a diagonal matrix with $D_{ii} = \sum_{j=1}^{n} W_{ij}$. Then, Eq. (10.1) can be written as

$$\mathbb{P}(\mathbf{Y} = \mathbf{y}) = \frac{1}{Z}\exp\left(-\frac{\beta}{2}\mathbf{y}^\top \mathbf{L}\mathbf{y}\right). \tag{10.2}$$

We fix $\beta = 1$ for the rest of the chapter for simplicity. One can obtain results for $\beta \neq 1$ by replacing $\mathbf{L}$ with $\beta\mathbf{L}$.

Note that BMRF would be equivalent to the Gaussian random field (GRF) if we relax the labels to belong to real values: $\mathbf{Y} \in \mathbb{R}^n$. Indeed, Eq. (10.1) takes the same form as a Gaussian distribution with covariance $\mathbf{L}^{-1}$. For this reason, some researchers refer to BMRF as the "Gaussian Markov random field" or the "binary Gaussian random field."

[3] A multiclass generalization is straightforward via the one-vs-the-rest reduction, which is elaborated in Section 10.3.4.

Note that, while $\mathbf{L}$ is a singular matrix (verify that $\mathbf{L} \cdot \mathbf{1} = \mathbf{0}$), one can simply add $\delta\mathbf{I}$ to $\mathbf{L}$ for some $\delta > 0$, which does not change the original BMRF model after normalization.

10.1.3 PREDICTIONS BY BMRF

Consider performing the task PREDICT with BMRF. Let $\boldsymbol{\ell}$ be the set of labeled nodes and $\mathbf{u} = \{1, \ldots, n\} \backslash \boldsymbol{\ell}$ be the set of unlabeled nodes. Ideally, one should compute the marginal $\mathbb{P}(Y_i = 1 \mid \mathbf{Y}_{\boldsymbol{\ell}} = \mathbf{y}_{\boldsymbol{\ell}})$ for each $i \in \mathbf{u}$ and threshold it at $\frac{1}{2}$, which is called the Bayes decision rule and is optimal when the labels are truly drawn from BMRF. Unfortunately, there is no known polynomial time algorithm for computing the marginal; a naive computation involves summing over exponentially many possible labelings of $\mathbf{Y}_{\mathbf{u}} \in \{\pm 1\}^{|\mathbf{u}|}$. This motivates us to look for approximate computations.

A popular and efficient approximation is label propagation (LP), proposed by Zhu et al. [2]. LP predicts the unknown labels $\mathbf{Y}_{\mathbf{u}}$ as follows:

$$\hat{\mathbf{y}}_{\mathbf{u}} := -(\mathbf{L}_{\mathbf{uu}})^{-1}\mathbf{L}_{\mathbf{u}\boldsymbol{\ell}}\mathbf{y}_{\boldsymbol{\ell}}. \tag{10.3}$$

For brevity, we use the notation $\mathbf{L}_{\mathbf{uu}}^{-1}$ for $(\mathbf{L}_{\mathbf{uu}})^{-1}$ hereafter. The predicted value $\hat{y}_i$ for $i \in \mathbf{u}$, which is a real value in $[-1, 1]$, provides a "soft label" that encodes not only the predicted label but also the strength of the belief. For final predictions, one should take the sign of $\hat{y}_i$.

The LP prediction is motivated by relaxing the binary labels $\mathbf{Y}$ of a BMRF model to continuous. Then, the BMRF model becomes a Gaussian distribution, and the LP prediction is equivalent to the posterior mean of $\mathbf{Y}_{\mathbf{u}}$ conditioned on $\mathbf{Y}_{\boldsymbol{\ell}} = \mathbf{y}_{\boldsymbol{\ell}}$.

10.2 ACTIVE LEARNING IN BMRF: EXPECTED ERROR MINIMIZATION

Active learning asks which node labels one should choose to observe in order to minimize the prediction error rate ϵ_t for time steps of interest. Suppose for now that we make the optimal predictions. That is, given a set of labeled nodes $\boldsymbol{\ell} \subset N$ with labels $\mathbf{y}_{\boldsymbol{\ell}}$, we follow the Bayes decision rule:

$$\hat{Y}_i^*(\mathbf{Y}_{\boldsymbol{\ell}} = \mathbf{y}_{\boldsymbol{\ell}}) := \arg\max_{y \in \{\pm 1\}} \mathbb{P}(Y_i = y \mid \mathbf{Y}_{\boldsymbol{\ell}} = \mathbf{y}_{\boldsymbol{\ell}}), \tag{10.4}$$

where $\mathbf{Y}_{\boldsymbol{\ell}} = \mathbf{y}_{\boldsymbol{\ell}}$ indicates that we condition on the event that the random variable $\mathbf{Y}_{\boldsymbol{\ell}}$ is equal to $\mathbf{y}_{\boldsymbol{\ell}}$. Note $\hat{Y}_i^*(\mathbf{Y}_{\boldsymbol{\ell}} = \mathbf{y}_{\boldsymbol{\ell}}) = Y_i$ for $i \in \boldsymbol{\ell}$ trivially. We hereafter use $\hat{Y}_i^*$ and omit $(\mathbf{Y}_{\boldsymbol{\ell}} = \mathbf{y}_{\boldsymbol{\ell}})$ when it is clear from the context. Because the BMRF fully specifies how the labels are generated, given a set of labeled nodes $\boldsymbol{\ell} \subset N$, the expected error rate of the prediction is a well-defined mathematical quantity. At time t, a natural strategy is to query the node that minimizes the expected error at time $t+1$. Even though this greedy choice does not necessarily allow us to minimize the expected error at any time later than $t+1$, it is still an attractive heuristic that leads to a tractable method. We refer to this query strategy as *expected error minimization* (**EEM**).

Define the unlabeled nodes $\mathbf{u} := N \backslash \boldsymbol{\ell}$. Let us study what the *expected* error rate of the Bayes decision rule looks like after querying $q \in \mathbf{u}$. We define the expected error after knowing the label Y_q as follows, which we call the *lookahead risk* of node q:

$$R_{+q}(\mathbf{Y}_\ell = \mathbf{y}_\ell) := \mathbb{E}_{Y_q}\mathbb{E}_{\mathbf{Y}_{\mathbf{u}\setminus\{q\}}}\left[\frac{1}{n}\sum_{i=1}^{n}\mathbb{1}\{\hat{Y}_i^* \neq Y_i\}\middle| Y_q, \mathbf{Y}_\ell = \mathbf{y}_\ell\right], \tag{10.5}$$

where $\hat{Y}_i^*$ depends on Y_q as well as $\mathbf{Y}_\ell$. Notice that we take the expectation over Y_q as well because we do not know the label of node q yet. We use $R_{+q}(\mathbf{y}_\ell)$ as a shorthand for $R_{+q}(\mathbf{Y}_\ell = \mathbf{y}_\ell)$.

EEM chooses the query that minimizes the lookahead risk:

$$q^* = \arg\min_{q \in N\setminus\ell} R_{+q}(\mathbf{y}_\ell). \tag{10.6}$$

Define $\mathbb{P}_{\mathbf{y}_\ell}(\cdot) := \mathbb{P}(\cdot|\mathbf{Y}_\ell = \mathbf{y}_\ell)$ for brevity and the *zero-one risk*

$$\begin{aligned} R(Y_q = y, \mathbf{y}_\ell) &:= \mathbb{E}_{\mathbf{Y}_{\mathbf{u}\setminus\{q\}}}\left[\sum_{i=1}^{n}\frac{1}{n}\mathbb{1}\{\hat{Y}_i^* \neq Y_i\}\middle| Y_q = y, \mathbf{Y}_\ell = \mathbf{y}_\ell\right] \\ &= \frac{1}{n}\sum_{i=1}^{n}\left(1 - \max_{y' \in \{\pm 1\}} \mathbb{P}_{Y_q = y, \mathbf{y}_\ell}(Y_i = y')\right). \end{aligned} \tag{10.7}$$

Then,

$$R_{+q}(\mathbf{y}_\ell) = \sum_{y \in \{\pm 1\}} R(Y_q = y, \mathbf{y}_\ell)\mathbb{P}_{\mathbf{y}_\ell}(Y_q = y). \tag{10.8}$$

Notice that the key quantity is the posterior marginal distribution $\mathbb{P}_{Y_q = y, \mathbf{y}_\ell}(Y_i = y')$ in Eq. (10.7) and $\mathbb{P}_{\mathbf{y}_\ell}(Y_q = y)$ in Eq. (10.8). An efficient computation of the posterior marginal would lead to an algorithm for PREDICT due to Eq. (10.4) and also to an algorithm for QUERY due to Eq. (10.6). However, there is no known polynomial time algorithm to compute the marginal, as mentioned before. Such a difficulty calls for various approximate methods for EEM.

10.3 ALGORITHMS TO APPROXIMATE EEM

Approximate EEM methods fall into two main categories: (i) relax the binary labels in BMRF to continuous (now a GRF model) in which the lookahead risk has a closed-form solution, or (ii) keep the binary labels but approximate the marginal in Eq. (10.8).

10.3.1 CONTINUOUS RELAXATION

The first category is to relax the binary labels in BMRF to continuous. We now pretend the labels are real-valued, ignoring the fact that we actually observe binary labels only. This relaxation results in a GRF model as discussed before. In GRF, it is standard to make a prediction $\hat{Y}_i$ for node i as $\hat{y}_i$ defined in Eq. (10.3), which is the posterior mean. Now that the zero-one error $\mathbb{1}\{\hat{Y}_i \neq Y_i\}$ does not make sense, considering alternative error notions for continuous labels leads to new algorithms.

V-optimality (VOpt)

V-optimality (**VOpt**) proposed by Ji and Han [6] chooses the squared error $(\hat{Y}_i - Y_i)^2$, which is easy to work with and natural in GRFs. Then, the lookahead risk for the GRF is defined as follows, which has a closed form solution:

$$R^{\mathrm{V}}_{+q} := \mathbb{E}_{Y_q}\mathbb{E}_{\mathbf{Y}_{\mathbf{u}\setminus\{q\}}}\left[\sum_{i=1}^{n}\left(\hat{Y}_i - Y_i\right)^2 \middle| Y_q, \mathbf{Y}_\ell = \mathbf{y}_\ell\right] = \operatorname{tr}\left(\mathbf{L}^{-1}_{(\mathbf{u}\setminus\{q\})\,(\mathbf{u}\setminus\{q\})}\right), \tag{10.9}$$

where $\operatorname{tr}(\cdot)$ is the trace of a matrix. In GRF, $\hat{y}_i$ is the optimal prediction for the squared error, and so the risk is now the sum of the posterior variances, thus the name VOpt where V means variance.

One noticeable feature of VOpt is that the lookahead risk is independent of $\mathbf{y}_\ell$. That is, it depends on "which nodes" are labeled but not "what labels" these nodes possess. Such a method is said to be *nonadaptive*. One can see that this is not ideal to solve the linear chain case described in the introduction. However, nonadaptive methods fit well the batch active learning problem as they do not require knowing the label of the queries.

Σ-optimality (SOpt)

Σ-optimality (**SOpt**) proposed by Ma et al. [8] makes a nonintuitive choice of error called survey error: $(\sum_{i=1}^{n}\hat{Y}_i - \sum_{i=1}^{n}Y_i)^2$. This measures the squared difference of the estimated proportion of the positive labels, which seems to be necessary but not sufficient for achieving a low classification error. The SOpt lookahead risk is defined as follows:

$$R^{\Sigma}_{+q} := \mathbb{E}_{Y_q}\mathbb{E}_{\mathbf{Y}_{\mathbf{u}\setminus\{q\}}}\left[\left(\sum_{i=1}^{n}\hat{Y}_i - \sum_{i=1}^{n}Y_i\right)^2 \middle| Y_q, \mathbf{Y}_\ell = \mathbf{y}_\ell\right] = \mathbf{1}^\top L^{-1}_{(\mathbf{u}\setminus\{q\})\,(\mathbf{u}\setminus\{q\})}\mathbf{1}, \tag{10.10}$$

where $\hat{Y}_i = \hat{y}_i$; see Eq. (10.3). Surprisingly, SOpt turns out to perform well in practice, especially so in earlier rounds of active learning as SOpt prefers querying "influential" nodes (such as hubs) much more than other algorithms do. Specifically, VOpt, as pointed out by Ma et al. [8], tends to query outliers as they have a large variance in GRF while in practice outliers do not convey much information. It is easy to see that SOpt is nonadaptive as well.

Comparison

To see the difference between VOpt and SOpt, define $\mathbf{C} := \mathbf{L}^{-1}_{\mathbf{uu}}$ and $\mathbf{C}' := \mathbf{L}^{-1}_{(\mathbf{u}\setminus\{q\})\,(\mathbf{u}\setminus\{q\})}$. The one-step covariance update rule (or Schur complement) says

$$\begin{pmatrix}\mathbf{C}' & 0\\ 0 & 0\end{pmatrix} = \mathbf{C} - \frac{\mathbf{C}_{\cdot q}\mathbf{C}_{q\cdot}}{C_{qq}}, \tag{10.11}$$

where we assume that the node q corresponds to the last row and column of $\mathbf{C}$ for ease of exposition. This allows us to not only compute the lookahead risk faster when $\mathbf{L}^{-1}_{\mathbf{uu}}$ is available but also understand the behavior of VOpt and SOpt. As $\mathbf{C}$ is the covariance matrix of the Gaussian distribution, we rewrite

$C_{ij} = \rho_{ij}\sigma_i\sigma_j$, where $\sigma_i = \sqrt{C_{ii}}$ and ρ_{ij} is the correlation coefficient between i and j. Then, one can rewrite the query selection q^* made by VOpt and SOpt as follows [8]:

$$\text{VOpt: } q^* = \arg\max_{q\in\mathbf{u}} \frac{\sum_{i\in\mathbf{u}} C_{qi}^2}{C_{qq}} = \arg\max_{q\in\mathbf{u}} \sum_{i\in\mathbf{u}} \rho_{qi}^2\sigma_i^2, \tag{10.12}$$

$$\text{SOpt: } q^* = \arg\max_{q\in\mathbf{u}} \frac{(\sum_{i\in\mathbf{u}} C_{qi})^2}{C_{qq}} = \arg\max_{q\in\mathbf{u}} \left(\sum_{i\in\mathbf{u}} \rho_{qi}\sigma_i\right)^2. \tag{10.13}$$

Both optimalities involve the term $\sum_{i\in\mathbf{u}} \rho_{qi}^2\sigma_i^2$, which means they favor querying nodes that correlate highly with those having large variance. SOpt, however, has extra cross terms: $\sum_{i,j\in\mathbf{u}:i\neq j}(\rho_{qi}\sigma_i)\cdot(\rho_{qj}\sigma_j)$. As pointed out by [8], by the Cauchy-Schwarz inequality, these cross terms are maximized when they are equal. Thus, SOpt additionally favors nodes that have consistent global influence (e.g., hubs).

10.3.2 APPROXIMATION OF THE MARGINAL

Another direction for approximating the lookahead risk is to avoid relaxing BMRF to GRF in the first place. We now stick to the lookahead risk (10.8) that is based on the zero-one error but try to approximate the marginal $\mathbb{P}_{\mathbf{y}_\ell}(Y_q = y)$. We then choose the query minimizing the lookahead risk (10.6) where the marginal is approximated.

Zhu-Lafferty-Ghahramani (ZLG)

Zhu et al. [4] proposed a simple approximation that we call Zhu-Lafferty-Ghahramani (**ZLG**). Note that the LP prediction $\hat{\mathbf{y}}_u$ is always in $[-1, 1]$ due to the property of the harmonic function (see [2] for detail). ZLG simply takes the following shifted and scaled version of $\hat{y}_i$ as the marginal probability of node $i \in \mathbf{u}$:

$$\mathbb{P}_{\mathbf{y}_\ell}(Y_i = 1) \approx \tfrac{1}{2}(\hat{y}_i + 1).$$

Note that $\frac{1}{2}(\hat{y}_i + 1)$ coincides with the probability of a random walk on the graph arriving at a positive node before arriving at a negative node, as pointed out by Zhu et al. [2]. This way, ZLG can be seen as approximating the marginal of node i being positive with "how close i is to positive nodes" measured by the random walk.

Two-step approximation (TSA)

Jun and Nowak [11] proposed a two-step approximation (TSA) to the posterior marginal distribution $\mathbb{P}_{\mathbf{y}_\ell}(Y_k)$. The key lies in the following log probability ratio approximation: $\log \frac{\mathbb{P}(Y_k=1,\mathbf{Y}_\ell=\mathbf{y}_\ell)}{\mathbb{P}(Y_k=-1,\mathbf{Y}_\ell=\mathbf{y}_\ell)} \approx \log \frac{\mu(Y_k=1,\mathbf{Y}_\ell=\mathbf{y}_\ell)}{\mu(Y_k=-1,\mathbf{Y}_\ell=\mathbf{y}_\ell)}$ for some $\mu(\cdot)$. Define the sigmoid function $\sigma(z) := (1+\exp(-z))^{-1}$. Then,

$$\begin{aligned}
\mathbb{P}(Y_k = 1 \mid \mathbf{Y}_\ell = \mathbf{y}_\ell) &= \frac{\mathbb{P}(Y_k = 1, \mathbf{Y}_\ell = \mathbf{y}_\ell)}{\mathbb{P}(Y_k = 1, \mathbf{Y}_\ell = \mathbf{y}_\ell) + \mathbb{P}(Y_k = -1, \mathbf{Y}_\ell = \mathbf{y}_\ell)} \\
&= \sigma(\log \mathbb{P}(Y_k = 1, \mathbf{Y}_\ell = \mathbf{y}_\ell) - \log \mathbb{P}(Y_k = -1, \mathbf{Y}_\ell = \mathbf{y}_\ell)) \\
&\approx \sigma(\log \mu(Y_k = 1, \mathbf{Y}_\ell = \mathbf{y}_\ell) - \log \mu(Y_k = -1, \mathbf{Y}_\ell = \mathbf{y}_\ell)).
\end{aligned} \tag{10.14}$$

We construct $\mu(Y_k = y_k, \mathbf{Y}_\ell = \mathbf{y}_\ell)$ as a two-step upperbound on $\mathbb{P}(Y_k = y_k, \mathbf{Y}_\ell = \mathbf{y}_\ell)$ as follows. We describe the details in Appendix.

Lemma 10.1. *Let $\bar{\mathbf{u}} := \mathbf{u}\backslash\{k\}$ be the set of unlabeled nodes except k. For $y_k \in \{\pm 1\}$,*

$$\begin{aligned}\log \mathbb{P}(Y_k = y_k, \mathbf{Y}_\ell = \mathbf{y}_\ell) &\le -y_k L_{k\ell}\mathbf{y}_\ell + \frac{1}{2}(L_{\bar{u}k}y_k + L_{\bar{u}\ell}\mathbf{y}_\ell)^\top L_{\bar{u}\bar{u}}^{-1}(L_{\bar{u}k}y_k + L_{\bar{u}\ell}\mathbf{y}_\ell) + C' \\ &=: \log \mu(Y_k = y_k, \mathbf{Y}_\ell = \mathbf{y}_\ell)\end{aligned}$$

for some C' that is independent of y_k. ■

Let $f_k := \log \mu(Y_k = 1, \mathbf{Y}_\ell = \mathbf{y}_\ell) - \log \mu(Y_k = -1, \mathbf{Y}_\ell = \mathbf{y}_\ell)$ be the decision value for node k. f_k can be written compactly as

$$f_k = -2\mathbf{L}_{k\ell}\mathbf{y}_\ell + 2\mathbf{L}_{k\bar{\mathbf{u}}}\mathbf{L}_{\bar{\mathbf{u}}\bar{\mathbf{u}}}^{-1}\mathbf{L}_{\bar{\mathbf{u}}\ell}\mathbf{y}_\ell. \tag{10.15}$$

All in all, we can rewrite Eq. (10.14) as

$$\mathbb{P}_{\mathbf{y}_\ell}(Y_k = 1) \approx \sigma(f_k).$$

Note that one can compute the decision value f_k for all $k \in \mathbf{u}$ at once as follows. Denote by $\circ$ the Hadamard product and $[f_k]_{k\in\mathbf{u}}$ a vector whose kth component has value f_k. Using the one-step covariance update rule (10.11), one can show that

$$[f_k]_{k\in\mathbf{u}} = 2 \cdot \left[\frac{1}{(\mathbf{L}_{\mathbf{uu}}^{-1})_{kk}}\right]_k \circ (-\mathbf{L}_{\mathbf{uu}}^{-1}\mathbf{L}_{\mathbf{u}\ell}\mathbf{y}_\ell). \tag{10.16}$$

Comparison

Let $\sigma^{\text{Linear}}(z) := \frac{1}{2}(z+1)$ that is valid for $z \in [-1, 1]$. Then, we can write the approximations of ZLG and TSA as follows:

$$\begin{aligned}\text{ZLG:}\quad & [\mathbb{P}(Y_k = 1 \mid \mathbf{Y}_\ell = \mathbf{y}_\ell)]_k \approx \sigma^{\text{Linear}}(\hat{\mathbf{y}}_\mathbf{u}), \\ \text{TSA:}\quad & [\mathbb{P}(Y_k = 1 \mid \mathbf{Y}_\ell = \mathbf{y}_\ell)]_k \approx \sigma\left(2 \cdot \left[\frac{1}{(\mathbf{L}_{\mathbf{uu}}^{-1})_{kk}}\right]_k \circ \hat{\mathbf{y}}_\mathbf{u}\right),\end{aligned}$$

where we apply σ^{Linear} and σ elementwise. Both methods utilize $\hat{\mathbf{y}}_\mathbf{u}$, the decision value used in LP, and the classification rule based on the TSA marginal happens to be equivalent to LP. Besides using a different sigmoid function, TSA further weights $\hat{\mathbf{y}}_\mathbf{u}$ by $1/(\mathbf{L}_{\mathbf{uu}}^{-1})_{kk}$ where $(\mathbf{L}_{\mathbf{uu}}^{-1})_{kk}$ is always positive. $(\mathbf{L}_{\mathbf{uu}}^{-1})_{kk}$ can be interpreted as the variance of node k in the GRF context. The larger the variance of a node, the closer its decision value is to 0 and the closer the marginal probability is to 1/2. Such variance information is not utilized in ZLG.

One difference is that replacing occurrences of $\mathbf{L}$ with $\beta\mathbf{L}$ in ZLG (and VOpt/SOpt) results in no effect of β whereas in TSA there *exists* an effect of β; the smaller the β, the closer the marginal probabilities is to 1/2. An edge connecting two nodes with different labels is called a *cut*. We have observed that β changes the balance between exploration (searching for new parts of the graph that contain a cut) and exploitation (nailing down on the exact cut). However, parameter tuning in active learning is hard in general and is left as an open problem.

10.3.3 CASE STUDY: LINEAR CHAIN

We compare all the methods presented in this chapter using a simple linear chain example. In Fig. 10.2, we consider a linear chain example where there are 18 nodes with edges between node i and $i+1$ with weight 1 for all $1 \leq i \leq 17$. Labels for nodes 1 and 11 are known to the algorithm a priori. We denote labeled nodes by ✓ where initial labels are in gray, the first two queries are in black, and the last two are with daggers. Symbols +/− indicate the predicted labels by LP after four queries. The method BMRF refers to the EEM strategy based on the exact marginal computations, which we manage to compute thanks to the small graph size.

For the first query, an algorithm sees that there is at least one cut between nodes 1 and 11. ZLG drills into this region and spends its next four queries in nailing down the cut. Consequently, it does not query any node to the right side of node 11 and incurs a large error. In other words, ZLG "exploits" a known cut region but does not "explore" for unknown cut regions. To see why, the posterior marginal $\mathbb{P}(Y_k = -1 \mid \mathbf{Y}_\ell = \mathbf{y}_\ell)$ for $k = 12, \ldots, 18$ under BMRF with exact computation is (0.88, 0.79, 0.72, 0.67, 0.63, 0.60, 0.57) and under TSA is (0.88, 0.73, 0.66, 0.62, 0.60, 0.58, 0.57), which shows a similar trend. When computing the lookahead risk, the exact marginal and TSA both pick node 16. However, under ZLG the marginals are $(1, 1, \ldots, 1)$ for nodes 12–18, which means the expected errors on these nodes are already 0 and thus querying one of them does not help.

In SOpt, the first two queries include exploration query (node 16). Then, the next two queries include node 3, which does not reduce the error rate; node 8 would have a reduced error. SOpt selects queries by which *nodes* have been labeled, ignoring what *labels* they have. In fact, this nonadaptivity is a common characteristic of many graph-based active learning algorithms [5,7–9]. Due to the lack of exploration queries, SOpt is suboptimal in this example. VOpt shares the same issue, so we omit it here. In contrast, the exact computation of EEM (BMRF) and TSA naturally balance between exploration and exploitation.

10.3.4 IMPLEMENTATION

We now describe how to implement active learning methods efficiently. At time step t, VOpt and SOpt have to compute the inverse Laplacian $\mathbf{L}_{\mathbf{uu}}^{-1}$, spending $O(n^2)$ time by applying the one-step covariance update rule (10.11) on the previous inverse Laplacian. Both methods can then find q^* efficiently using Eq. (10.12) or (10.13), which spends $O(n^2)$ time.

Node	1	2	3	4	5	6	7	8	9	10	11	12	13	14	15	16	17	18	Error rate
True label	+	+	+	+	+	+	+	+	+	−	−	−	+	+	+	+	+	+	
ZLG	✓	+	+	+	+	✓	+	✓	✓†	✓†	✓	−	−	−	−	−	−	−	0.33
SOpt	✓	+	✓†	+	+	✓	+	+	−	−	✓	−	✓†	+	+	✓	+	+	0.06
BMRF	✓	+	+	+	+	✓	+	✓†	+	−	✓	−	✓†	+	+	✓	+	+	0.00
TSA	✓	+	+	+	+	✓	+	✓†	+	−	✓	−	✓†	+	+	✓	+	+	0.00

FIG. 10.2

A linear chain example and various QUERY algorithms.

While ZLG and TSA compute the inverse Laplacian $\mathbf{L}_{\mathbf{uu}}^{-1}$ in the same way, computing the lookahead risk needs some work as we also need to compute $\mathbf{L}_{(\mathbf{u}\backslash\{q\})(\mathbf{u}\backslash\{q\})}^{-1}$ for each $q \in \mathbf{u}$. A naive computation of these quantities would require total $O(n^3)$ time. Fortunately, there exists an efficient computation for ZLG and TSA.

Using the following lemma, one can compute the "lookahead marginal" of ZLG without $\mathbf{L}_{(\mathbf{u}\backslash\{q\})(\mathbf{u}\backslash\{q\})}^{-1}$. This leads to computing the lookahead risk (10.8) for ZLG in $O(n^2)$ time.

Lemma 10.2. *Let* $\mathbf{g} := -\mathbf{L}_{\mathbf{uu}}^{-1}\mathbf{L}_{\mathbf{u}\ell}\mathbf{y}_\ell$ *be the LP prediction for the nodes* $\mathbf{u}$*. Let* $\overline{\mathbf{u}} = \mathbf{u}\backslash\{q\}$ *and* $\overline{\ell} = \ell \cup \{q\}$*. Define* $\mathbf{g}^{+q,y}$ *to be the LP predictions for nodes* $\overline{\mathbf{u}}$ *after observing* $Y_q = y$*; i.e.,* $\mathbf{g}^{+q,y} := -\mathbf{L}_{\overline{\mathbf{u}}\overline{\mathbf{u}}}^{-1}\mathbf{L}_{\overline{\mathbf{u}}\overline{\ell}}\mathbf{y}_{\overline{\ell}}$*. Define* $\mathbf{C} := \mathbf{L}_{\mathbf{uu}}^{-1}$*. Then,*

$$\mathbf{g}^{+q,y} = \mathbf{g} + \frac{y - g_q}{C_{qq}}\mathbf{C}_{\cdot q},$$

where the equality ignores the component in the RHS corresponding to the node q. ■

TSA has a similar way to achieve the same time complexity as ZLG.

Lemma 10.3. *Consider the same assumptions as Lemma 10.2. Define* $\mathbf{f} := 2\left[\frac{1}{C_{kk}}\right]_k \circ (-\mathbf{L}_{\mathbf{uu}}^{-1}\mathbf{L}_{\mathbf{u}\ell}\mathbf{y}_\ell) = 2\left[\frac{1}{C_{kk}}\right]_k \circ \mathbf{g}$ *to be the TSA's decision values for nodes* $\mathbf{u}$*. Define* $\mathbf{f}^{+q,y}$ *to be the TSA decision values for nodes* $\overline{\mathbf{u}}$ *after observing* $Y_q = y$*; i.e.,* $\mathbf{f}^{+q,y} := 2\left[\frac{1}{(\mathbf{L}_{\overline{\mathbf{u}}\overline{\mathbf{u}}}^{-1})_{kk}}\right]_k \circ (-\mathbf{L}_{\overline{\mathbf{u}}\overline{\mathbf{u}}}^{-1}\mathbf{L}_{\overline{\mathbf{u}}\overline{\ell}}\mathbf{y}_{\overline{\ell}})$*. Then,* $\forall k \in \mathbf{u}$*,*

$$\mathbf{f}^{+q,y} = 2\left[\left(C_{kk} - \frac{C_{kq}^2}{C_{qq}}\right)^{-1}\right]_k \circ \left(\mathbf{g} + \frac{y - g_q}{C_{qq}}\mathbf{C}_{\cdot q}\right),$$

where the equality ignores the component in the RHS corresponding to the node q. ■

A multiclass extension

Note that there is no need to handle multiclass explicitly in VOpt and SOpt as nonadaptive methods do not depend on the observed labels. Let K be the number of classes and assume $y \in \{1, \ldots, K\}$. For ZLG and TSA, one can handle the multiclass case by instantiating one algorithm for each class in the so-called one-vs-the-rest manner (total K runs). After computing each one-vs-the-rest marginal (binary), we compute the multiclass marginal distribution (now a multinomial) by normalizing the binary marginals. Finally, the multiclass zero-one risk is a trivial extension of Eq. (10.7):

$$R(Y_q = y, \mathbf{y}_\ell) = \frac{1}{n}\sum_{i=1}^{n}\left(1 - \max_{y' \in \{1,\ldots,K\}} \mathbb{P}_{Y_q=y,\mathbf{y}_\ell}(Y_i = y')\right). \tag{10.17}$$

10.4 EXPERIMENTS

We run simulations on both synthetic and real-world datasets to compare various Bayesian active learning methods. Throughout, all methods start from one labeled node that is chosen uniformly at random. We break ties uniformly at random when there exist ties for choosing a query.[4] For TSA, we use $\beta = 1$.

Toy Data. The first toy dataset is a linear chain with 15 nodes where each edge has weight 1. We choose an edge uniformly at random and make it a cut; i.e., assign a positive label on one side and a negative label on the other side. We repeat the experiment 50 times where we assign new labels before each trial. We plot the accuracy versus the number of queries in Fig. 10.3A with the confidence bounds on the accuracies in gray. After 10 queries, we observe a cluster of methods that outperforms the rest. This cluster consists of methods that are equipped with exploitation queries and thus able to nail down the exact cut. The rest are *nonadaptive* methods that are blind to observed labels. This experiment confirms the importance of the exploitation queries.

The second toy dataset is the 10-by-10 grid graph; see Fig. 10.3B. We assign positive labels to a 3-by-3 box at the bottom left and another one at the top right, and negative labels to the rest. Then, for each negative node adjacent to a positive node, we assign a positive label with probability 1/2. This makes the boundary "jittered". We repeat the experiment 50 times where we assign new jittered labels before each trial. We show the result in Fig. 10.3C. Overall, there is no absolute winner. For very

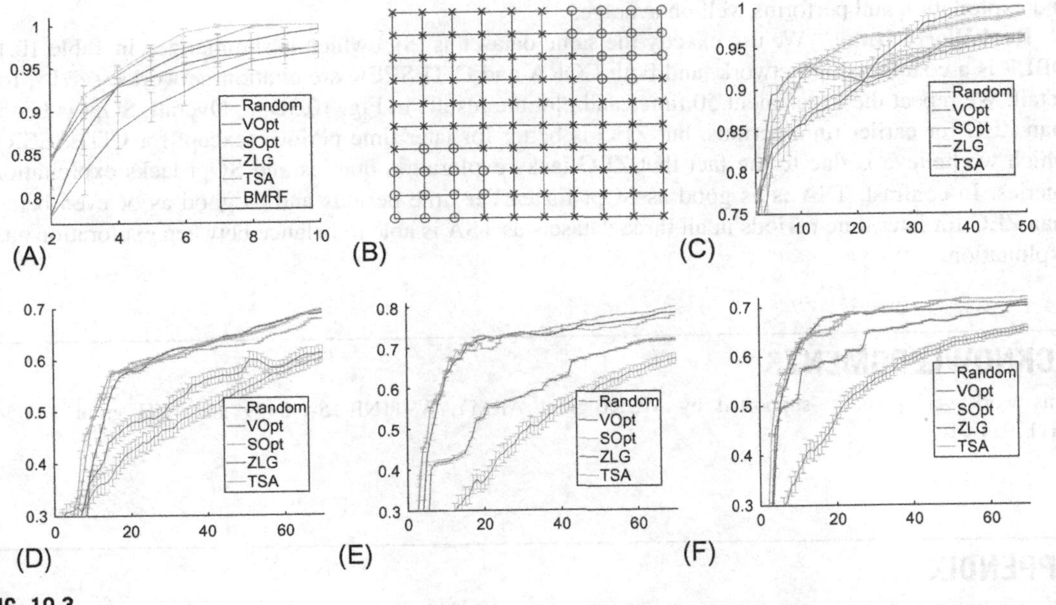

FIG. 10.3

Experiment results. Plots show accuracy versus the number of queries. In *gray* are error bars. (A) Linear chain; (B) Jittered box dataset; (C) Jittered box; (D) DBLP; (E) CORA; (F) CITESEER.

[4]This happens quite often for ZLG in earlier time steps where all the observed node labels are the same.

Table 10.1 Real-World Dataset Summary

Name	$\|N\|$	$\|E\|$	The Number of Classes
DBLP	1711	2898	4
CORA	2485	5069	7
CITESEER	2109	3665	6

early time periods, both VOpt and SOpt perform better than the rest because they perform exploration only—rough locations of the two positive boxes are discovered fast. On the other hand, ZLG incurs very low accuracy in the first half for the following two reasons: (i) before discovering a positive node, every node has the same lookahead risk and ZLG resorts to tie-breaking uniformly at random, and (ii) after discovering the first positive node, ZLG drills down the exact boundary of it while completely not knowing the existence of the other positive box. In the end, however, ZLG becomes the best because it does not waste queries on exploration. TSA, our method, balances between exploration and exploitation and performs well on average.

Real-World Data We use exactly the same dataset as [8],[5] which is summarized in Table 10.1. DBLP is a coauthorship network, and both CORA and CITESEER are citation networks; see [8] for detail. We repeat the experiment 50 times and plot the results in Fig. 10.3D–F. Overall, SOpt is better than ZLG for earlier time periods, but ZLG is better for later time periods (except for CITESEER), which we believe is due to the fact that ZLG lacks exploration queries and SOpt lacks exploitation queries. In contrast, TSA is as good as SOpt for earlier time periods and as good as or even better than ZLG for later time periods in all three datasets as TSA is able to balance between exploration and exploitation.

ACKNOWLEDGMENTS

This work was partially supported by MURI grant ARMY W911NF-15-1-0479 and NIH grant 1 U54 AI117924-01.

APPENDIX

We provide proofs of the lemmas in this appendix.

[5]The dataset can be downloaded from http://www.autonlab.org/autonweb/21763.

A.1 PROOF OF LEMMA 10.1

Let $A := L_{kk} + \mathbf{y}_\ell^\top \mathbf{L}_{\ell\ell}\mathbf{y}_\ell$ and $g(\mathbf{y}_{\bar{u}}) := -\left(\frac{1}{2}\mathbf{y}_{\bar{u}}^\top \mathbf{L}_{\bar{u}\bar{u}}\mathbf{y}_{\bar{u}} + y_k \mathbf{L}_{k\bar{u}}\mathbf{y}_{\bar{u}} + \mathbf{y}_\ell^\top \mathbf{L}_{\ell\bar{u}}\mathbf{y}_{\bar{u}}\right)$. With these definitions, one can show that

$$\log \mathbb{P}(Y_k = y_k, \mathbf{Y}_\ell = \mathbf{y}_\ell) = -\log(Z) - \frac{1}{2}A - y_k \mathbf{L}_{k\ell}\mathbf{y}_\ell + \log\left(\sum_{\mathbf{y}_{\bar{u}}} \exp(g(\mathbf{y}_{\bar{u}}))\right).$$

Note that the last term is the log-sum-exp function that is similar to the max operator. This intuition leads to our *first upper bound*, which simply upper bounds individual terms in the summation by the largest term:

$$\log\left(\sum_{\mathbf{y}_{\bar{u}}} \exp(g(\mathbf{y}_{\bar{u}}))\right) \le \max_{\mathbf{y}_{\bar{u}} \in \{\pm 1\}^{|\bar{u}|}} g(\mathbf{y}_{\bar{u}}) + |\bar{u}| \log 2.$$

We now have an integer optimization problem, which is hard in general. We relax the domain of $\mathbf{y}_{\bar{u}}$ to real, which leads to our *second upper bound*:

$$\max_{\mathbf{y}_{\bar{u}} \in \{\pm 1\}^{|\bar{u}|}} g(\mathbf{y}_{\bar{u}}) \le \max_{\mathbf{y}_{\bar{u}} \in \mathbb{R}^{|\bar{u}|}} g(\mathbf{y}_{\bar{u}}) = \max_{\mathbf{y}_{\bar{u}} \in \mathbb{R}^{|\bar{u}|}} -\left(\frac{1}{2}\mathbf{y}_{\bar{u}}^\top \mathbf{L}_{\bar{u}\bar{u}}\mathbf{y}_{\bar{u}} + (y_k \mathbf{L}_{k\bar{u}} + \mathbf{y}_\ell^\top \mathbf{L}_{\ell\bar{u}})\mathbf{y}_{\bar{u}}\right).$$

We now have a concave quadratic maximization problem. We find the closed form solution as follows. By equating the objective function's derivative to zero, $-\mathbf{L}_{\bar{u}\bar{u}}\mathbf{y}_{\bar{u}} - y_k \mathbf{L}_{k\bar{u}} - \mathbf{y}_\ell^\top \mathbf{L}_{\ell\bar{u}} = 0$. This leads to the solution $\mathbf{y}_{\bar{u}}^* := -\mathbf{L}_{\bar{u}\bar{u}}^{-1}(\mathbf{L}_{\bar{u}k} y_k + \mathbf{L}_{\bar{u}\ell}\mathbf{y}_\ell) = -\mathbf{L}_{\bar{u}\bar{u}}^{-1}\begin{pmatrix}\mathbf{L}_{\bar{u}k} & \mathbf{L}_{\bar{u}\ell}\end{pmatrix}\begin{pmatrix} y_k \\ \mathbf{y}_\ell \end{pmatrix}$. By plugging in $\mathbf{y}_{\bar{u}}^*$ into the objective function, we have

$$\begin{aligned}
&\max_{\mathbf{y}_{\bar{u}} \in \mathbb{R}^{|\bar{u}|}} -\left(\frac{1}{2}\mathbf{y}_{\bar{u}}^\top \mathbf{L}_{\bar{u}\bar{u}}\mathbf{y}_{\bar{u}} + (y_k \mathbf{L}_{k\bar{u}} + \mathbf{y}_\ell^\top \mathbf{L}_{\ell\bar{u}})\mathbf{y}_{\bar{u}}\right) \\
&= \max_{\mathbf{y}_{\bar{u}} \in \mathbb{R}^{|\bar{u}|}} -\left(\frac{1}{2}\mathbf{y}_{\bar{u}}^\top \mathbf{L}_{\bar{u}\bar{u}} + y_k \mathbf{L}_{k\bar{u}} + \mathbf{y}_\ell^\top \mathbf{L}_{\ell\bar{u}}\right)\mathbf{y}_{\bar{u}} \\
&= \left(-\frac{1}{2}(y_k \mathbf{L}_{k\bar{u}} + \mathbf{y}_\ell^\top \mathbf{L}_{\ell\bar{u}}) + y_k \mathbf{L}_{k\bar{u}} + \mathbf{y}_\ell^\top \mathbf{L}_{\ell\bar{u}}\right)\mathbf{L}_{\bar{u}\bar{u}}^{-1}(\mathbf{L}_{\bar{u}k} y_k + \mathbf{L}_{\bar{u}\ell}\mathbf{y}_\ell) \\
&= \frac{1}{2}(y_k \mathbf{L}_{k\bar{u}} + \mathbf{y}_\ell^\top \mathbf{L}_{\ell\bar{u}})\mathbf{L}_{\bar{u}\bar{u}}^{-1}(\mathbf{L}_{\bar{u}k} y_k + \mathbf{L}_{\bar{u}\ell}\mathbf{y}_\ell).
\end{aligned}$$

Altogether,

$$\begin{aligned}
\log \mathbb{P}(Y_k = y_k, \mathbf{Y}_\ell = \mathbf{y}_\ell) \le& -\log(Z) - \frac{1}{2}A - y_k \mathbf{L}_{k\ell}\mathbf{y}_\ell \\
&+ \frac{1}{2}(y_k \mathbf{L}_{k\bar{u}} + \mathbf{y}_\ell^\top \mathbf{L}_{\ell\bar{u}})\mathbf{L}_{\bar{u}\bar{u}}^{-1}(\mathbf{L}_{\bar{u}k} y_k + \mathbf{L}_{\bar{u}\ell}\mathbf{y}_\ell) + |\bar{u}| \log 2.
\end{aligned}$$

A.2 PROOF OF LEMMAS 10.2 AND 10.3

We prove Lemma 10.2 only because the same proof technique can be applied to Lemma 10.3; we refer to [11] for a precise proof of Lemma 10.3.

The original proof can be found in [4]. Let $y_0 \in \{\pm 1\}$. The proof starts by adding a node named 0 with label y_0 to the graph and adding an edge between node 0 and q with weight w_0 while leaving Y_q unobserved. The new node is a "dongle" attached to node q. The intuition is that as w_0 approaches ∞, we effectively assign the label y_0 to Y_q. Let $\mathbf{L}^+ := \mathbf{D}^+ - \mathbf{W}^+$ be the graph Laplacian of this augmented graph. Denote by $\mathbf{e}_q$ the indicator vector that has 1 for dimension corresponding to node q. We define $\mathbf{g}^{+0}$ as the decision vector in the augmented graph as follows (later, we will take w_0 to infinity to get $\mathbf{g}^{+q,y}$):

$$\begin{aligned}\mathbf{g}^{+0} &= (\mathbf{L}^+_{\mathbf{uu}})^{-1}\mathbf{W}^+_{\mathbf{u}(\ell\cup\{0\})}\mathbf{y}_{\ell\cup\{0\}} \\ &= (w_0\mathbf{e}_q\mathbf{e}_q^\top + \mathbf{D}_{\mathbf{uu}} - \mathbf{W}_{\mathbf{uu}})^{-1}\cdot(w_0 y_0\mathbf{e}_q + \mathbf{W}_{\mathbf{u}\ell}\mathbf{y}_\ell) \\ &= (w_0\mathbf{e}_q\mathbf{e}_q^\top + \mathbf{L}_{\mathbf{uu}})^{-1}\cdot(w_0 y_0\mathbf{e}_q + \mathbf{W}_{\mathbf{u}\ell}\mathbf{y}_\ell).\end{aligned}$$

Let $\mathbf{C} := \mathbf{L}_{\mathbf{uu}}^{-1}$. Applying the matrix inversion lemma, $(w_0\mathbf{e}_q\mathbf{e}_q^\top + \mathbf{L}_{\mathbf{uu}})^{-1} = \mathbf{C} - \frac{\mathbf{C}_{\cdot q}\mathbf{C}_{q\cdot}}{w_0^{-1}+C_{qq}}$. Then,

$$\begin{aligned}\mathbf{g}^{+0} &= \left(\mathbf{C} - \frac{\mathbf{C}_{\cdot q}\mathbf{C}_{q\cdot}}{w_0^{-1}+C_{qq}}\right)\cdot(w_0 y_0\mathbf{e}_q + \mathbf{W}_{\mathbf{u}\ell}\mathbf{y}_\ell) \\ &= w_0 y_0\mathbf{C}_{\cdot q} + \mathbf{C}\mathbf{W}_{\mathbf{u}\ell}\mathbf{y}_\ell - \frac{w_0 y_0\mathbf{C}_{\cdot q}C_{qq}}{w_0^{-1}+C_{qq}} - \frac{\mathbf{C}_{\cdot q}\mathbf{C}_{q\cdot}\mathbf{W}_{\mathbf{u}\ell}\mathbf{y}_\ell}{w_0^{-1}+C_{qq}} \\ &\overset{(a)}{=} \frac{y_0\mathbf{C}_{\cdot q} + w_0 y_0\mathbf{C}_{\cdot q}C_{qq}}{w_0^{-1}+C_{qq}} + \mathbf{g} - \frac{w_0 y_0\mathbf{C}_{\cdot q}C_{qq}}{w_0^{-1}+C_{qq}} - \frac{\mathbf{C}_{\cdot q}\mathbf{C}_{q\cdot}\mathbf{W}_{\mathbf{u}\ell}\mathbf{y}_\ell}{w_0^{-1}+C_{qq}} \\ &= \frac{y_0\mathbf{C}_{\cdot q}}{w_0^{-1}+C_{qq}} + \mathbf{g} - \frac{\mathbf{C}_{\cdot q}\mathbf{C}_{q\cdot}\mathbf{W}_{\mathbf{u}\ell}\mathbf{y}_\ell}{w_0^{-1}+C_{qq}} \\ &\overset{w_0\to\infty}{\longrightarrow} \frac{y_0\mathbf{C}_{\cdot q}}{C_{qq}} + \mathbf{g} - \frac{\mathbf{C}_{\cdot q}\mathbf{C}_{q\cdot}\mathbf{W}_{\mathbf{u}\ell}\mathbf{y}_\ell}{C_{qq}} \\ &\overset{(b)}{=} \mathbf{g} + \frac{y_0 - g_q}{C_{qq}}\cdot\mathbf{C}_{\cdot q},\end{aligned}$$

where (a) and (b) is by $\mathbf{g} = \mathbf{C}\mathbf{W}_{\mathbf{u}\ell}\mathbf{y}_\ell$. As a sanity check, one can see that $g_q^{+0} = y_0$. We obtain $\mathbf{g}^{+q,y}$ by simply removing the coordinate in $\mathbf{g}^{+0}$ corresponding to the node q. This concludes the proof.

REFERENCES

[1] Zhu X. Semi-supervised learning literature survey. Tech. Rep. 1530, Computer Sciences, University of Wisconsin-Madison; 2005.

[2] Zhu X, Ghahramani Z, Lafferty J. Semi-supervised learning using Gaussian fields and harmonic functions. In: Proceedings of the international conference on machine learning (ICML); 2003. p. 912–9.

[3] Settles B. Active learning literature survey. Computer Sciences Technical Report 1648, University of Wisconsin-Madison; 2009.

[4] Zhu X, Lafferty J, Ghahramani Z. Combining active learning and semi-supervised learning using Gaussian fields and harmonic functions. In: ICML workshop on the continuum from labeled to unlabeled data in machine learning and data mining; 2003. p. 58–65.
[5] Guillory A, Bilmes JA. Label selection on graphs. In: Advances in neural information processing systems (NIPS); 2009. p. 691–9.
[6] Ji M, Han J. A variance minimization criterion to active learning on graphs. In: Proceedings of the international conference on artificial intelligence and statistics (AISTATS); 2012. p. 556–64.
[7] Gu Q, Han J. Towards active learning on graphs: an error bound minimization approach. In: Proceedings of the IEEE international conference on data mining (ICDM); 2012. p. 882–7.
[8] Ma Y, Garnett R, Schneider J. Sigma-optimality in active learning on Gaussian random fields. In: Advances in neural information processing systems (NIPS); 2013. p. 2751–9.
[9] Gadde A, Anis A, Ortega A. Active semi-supervised learning using sampling theory for graph signals. In: Proceedings of the 20th ACM SIGKDD international conference on knowledge discovery and data mining; 2014. p. 492–501.
[10] Gadde A, Ortega A. A probabilistic interpretation of sampling theory of graph signals. In: Proceedings of the IEEE international conference on acoustics, speech and signal processing, ICASSP; 2015. p. 3257–61.
[11] Jun KS, Nowak R. Graph-based active learning: a new look at expected error minimization. In: IEEE global conference on signal and information processing (GlobalSIP) symposium on non-commutative theory and applications; 2016. p. 1325–9.

CHAPTER

DESIGN OF GRAPH FILTERS AND FILTERBANKS

11

Nicolas Tremblay*, **Paulo Gonçalves**†, **Pierre Borgnat**‡

CNRS, GIPSA-lab, Univ. Grenoble Alpes, Grenoble, France ENS de Lyon, Univ Lyon 1, CNRS, Inria, LIP & IXXI, Univ Lyon, Lyon, France† ENS de Lyon, Univ Lyon 1, CNRS, Laboratoire de Physique & IXXI, Univ Lyon, Lyon, France‡*

11.1 GRAPH FOURIER TRANSFORM AND FREQUENCIES

Basic operations in graph signal processing consist of processing signals indexed on graphs either by filtering them or by changing their domain of representation in order to better extract or analyze the important information they contain. The aim of this chapter is to review general concepts underlying such filters and representations of graph signals. We first recall the different Graph Fourier Transforms that have been developed in the literature, and show how to introduce a notion of frequency analysis for graph signals by looking at their variations. Then, we move to the introduction of graph filters that are defined like the classical equivalent for one-dimensional signals or two-dimensional images, as linear systems that operate on each frequency of a signal. Some examples of filters and their implementations are given. Finally, as alternate representations of graph signals, we focus on multiscale transforms that are defined from filters. Continuous multiscale transforms such as spectral wavelets on graphs are reviewed as well as the versatile approaches of filterbanks on graphs. Several variants of graph filterbanks are discussed for structured as well as arbitrary graphs with a focus on the central point of the choice of the decimation or aggregation operators.

11.1.1 INTRODUCTION

Graph signal processing (GSP) has been introduced recently using at least two complementary formalisms: on one hand, the discrete signal processing on graphs [1] (see also Chapter 8) that emphasizes the adjacency matrix as a shift operator on graphs and develops an equivalent of discrete signal processing (DSP) for signals on graphs; and on the other hand, the approaches rooted in graph spectral analysis, which rely on the spectral properties of a Laplacian matrix on a graph [2–4]. Both approaches yield a harmonic analysis of graph signals via the definition of a Graph Fourier Transform (GFT): an operator projecting signals in the spectral domain of the chosen matrix. While the technical details vary, and some interpretations in the vertex domain may differ, the fundamental objective of both approaches is to decompose a signal onto components of different frequencies and to design filters

Cooperative and Graph Signal Processing. https://doi.org/10.1016/B978-0-12-813677-5.00011-0

that can extract or modify parts of a graph signal according to these frequencies, e.g., providing notions of low pass, band pass, or high pass filters for graph signals. In this section, both approaches (along with other variations) are seen as specific instances of a general guideline for defining GFT and its associated frequency analysis.

Notations

Vectors are written in bold with small letters and matrices in bold and capital letters. Let $\mathcal{G} = (\mathcal{V}, \mathcal{E}, \mathbf{A})$ be a graph with $\mathcal{V}$ the set of N nodes, $\mathcal{E}$ the set of edges, and $\mathbf{A}$ the weighted adjacency matrix in $\mathbb{R}^{N\times N}$. If $\mathbf{A}_{ij} = 0$, there is no connection from node i to node j, otherwise, $\mathbf{A}_{ij}$ is the weight of the edge starting from i and pointing[1] to j. If an undirected edge exists between i and j, then $\mathbf{A}_{ij} = \mathbf{A}_{ji}$. We restrict ourselves to adjacency matrices with positive or null entries: $\mathbf{A}_{ij} \geq 0$. Also, the symbol $\mathbf{I}$ denotes the identity matrix (its dimension should be clear from the context), and $\boldsymbol{\delta}_i$ is a vector whose ith entry is equal to 1 while all other entries are equal to 0. Finally, we will denote by $|\mathcal{E}|$ the number of edges in the graph.

11.1.2 GRAPH FOURIER TRANSFORM

Definition 11.1 (Graph signal). A graph signal is a vector $\mathbf{x} \in \mathbb{R}^N$ whose component x_i is considered to be defined on vertex i. ■

GFT is defined via a choice of reference operator[2] admitting a spectral decomposition. Representing a graph signal in this spectral domain is interpreted as a GFT. We review the standard properties and the various definitions proposed in the literature.

Consider a matrix $\mathbf{R} \in \mathbb{R}^{N\times N}$. To be admissible as a reference operator for the graph, it is often required that for any pair of nodes $i \neq j$, $\mathbf{R}_{ij}$ and $\mathbf{R}_{ji}$ are equal to zero if i and j are not connected, as this will help for efficient implementations of filters (see Section 11.2.4). We assume furthermore that $\mathbf{R}$ is diagonalizable in $\mathbb{C}$. In fact, if $\mathbf{R}$ is not diagonalizable, one needs to consider Jordan's decomposition, which is beyond this chapter's scope. We refer the reader to [1] and Chapter 8 for technical details on how to handle this case. Nevertheless, in practice, we claim that it often suffices to consider only diagonalizable operators because: (i) diagonalizable matrices in $\mathbb{C}$ are dense in the space of matrices; and (ii) graphs under consideration are generally measured (should they model social, sensor, or biological networks) with some inherent noise. Thus, if one ends up unluckily with a nondiagonalizable matrix, a small perturbation within the noise level will make it diagonalizable, provided the graphs have no specific regularities that need to be kept. Still, we may assume with only a small loss of generality that the reference operator $\mathbf{R}$ has a spectral decomposition:

$$\mathbf{R} = \mathbf{U}\boldsymbol{\Lambda}\mathbf{U}^{-1} \tag{11.1}$$

with $\mathbf{U}$ and $\boldsymbol{\Lambda}$ in $\mathbb{C}^{N\times N}$. The columns of $\mathbf{U}$, denoted as $\mathbf{u}_k$, are the right eigenvectors of $\mathbf{R}$ while the rows of $\mathbf{U}^{-1}$, denoted as $\mathbf{v}_k^\top$, are its left eigenvectors. $\boldsymbol{\Lambda}$ is the diagonal matrix of the eigenvalues $\{\lambda_k\}$.

[1] In the literature, the converse convention is sometimes chosen (e.g., in [1]), hence the $\mathbf{A}^\top$ occasionally appearing in this chapter.

[2] In the literature, this reference operator is often noted $\mathbf{S}$ for "shift." Nevertheless, the shift interpretation is essentially valid if one considers $\mathbf{S}$ to be the adjacency matrix (see discussion in Section 11.2.1). In a general setting, we prefer to denote by $\mathbf{R}$ the reference operator.

The GFT is defined as the transformation of a graph signal from the canonical "node" basis to its representation in the eigenvector basis:

Definition 11.2 (Graph Fourier Transform). For a given diagonalizable reference operator $\mathbf{R} = \mathbf{U}\mathbf{\Lambda}\mathbf{U}^{-1}$ associated with a graph $\mathcal{G}$, the GFT of a graph signal $\mathbf{x} \in \mathbb{R}^N$ is:

$$\mathbb{F}_{\mathcal{G}}\mathbf{x} \doteq \hat{\mathbf{x}} \doteq \mathbf{U}^{-1}\mathbf{x}. \tag{11.2}$$

■

The GFT's coefficients are simply the projections on the left eigenvectors of $\mathbf{R}$:

$$\forall k = 1, \dots, N \quad \left(\mathbb{F}_{\mathcal{G}}\,\mathbf{x}\right)_k = \hat{\mathbf{x}}_k = \mathbf{v}_k^T\mathbf{x}. \tag{11.3}$$

Moreover, the GFT is invertible: $\mathbf{U}\,\hat{\mathbf{x}} = \mathbf{U}\mathbf{U}^{-1}\mathbf{x} = \mathbf{x}$. In general, the complex Fourier modes $\mathbf{u}_k$ are not orthogonal to each other. However, when $\mathbf{R}$ is symmetric, the following additional properties hold true.

The special case of symmetric reference operators. If in addition to being real, $\mathbf{R}$ is also symmetric, then $\mathbf{U}$ and $\mathbf{\Lambda}$ are real matrices, and $\mathbf{U}$ may be found orthonormal, that is: $\mathbf{U}^{-1} = \mathbf{U}^\top$. In this case, $\mathbf{v}_k = \mathbf{u}_k$, the GFT of $\mathbf{x}$ is simply $\hat{\mathbf{x}} = \mathbf{U}^\top\mathbf{x}$ with coefficients $\hat{\mathbf{x}}_k = \mathbf{u}_k^\top\mathbf{x}$, and the Parseval relation holds: $||\hat{\mathbf{x}}||_2 = ||\mathbf{x}||_2$. Hereafter, when a symmetric operator $\mathbf{R}$ is encountered, one should have these properties in mind.

Finally, the interpretation of the graph Fourier modes $\mathbf{u}_k$ in terms of oscillations and frequencies will be the scope of Section 11.1.3. In the following, we list possible choices of reference operators, all diagonalizable with different $\mathbf{U}$ and $\mathbf{\Lambda}$, thus all defining different possible GFTs.

GFT for undirected graphs

Undirected graphs are characterized by symmetric adjacency matrices: $\forall(i,j)\ \mathbf{A}_{ij} = \mathbf{A}_{ji}$. This does not necessarily mean that $\mathbf{R}$ has to be chosen symmetric as well, as we will see with the example $\mathbf{R} = \mathbf{L_{rw}}$. The following choices of $\mathbf{R}$ are the most common in the undirected case.

The combinatorial Laplacian [symmetric]. The first choice for $\mathbf{R}$, advocated in [2–4], is to use the graph's combinatorial Laplacian, having properties studied in [5]. It is defined as $\mathbf{L} = \mathbf{D} - \mathbf{A}$ where $\mathbf{D}$ is the diagonal matrix of nodes' strengths, defined as $\mathbf{D}_{ii} = \mathbf{d}_i = \sum_j \mathbf{A}_{ij}$. If the adjacency matrix is binary (i.e., unweighted), this strength reduces to the degree of each node. The advantages of using $\mathbf{L}$ are twofold. (i) It is an intuitive manner to define the GFT: $\mathbf{L}$ being the discretized version of the continuous Laplacian operator which admits the Fourier modes as eigenmodes, it is fair to use by analogy the eigenvectors of $\mathbf{L}$ as graph Fourier modes. Moreover, this choice is associated with a complete theory of vector calculus (e.g., gradients) for graph signals [6] that is useful to solve partial differential equations on graphs. (ii) $\mathbf{L}$ has well-known mathematical properties [5], giving ways to characterize the graph or functions and processes on the graph (see also [7]). Most prominently, it is semidefinite positive (SDP) and its eigenvalues, being all positive or null, will serve in the following to bind eigenvectors with a notion of frequency.

The normalized Laplacian [symmetric]. A second choice for $\mathbf{R}$ is the normalized Laplacian $\mathbf{L_n} = \mathbf{D}^{-\frac{1}{2}}\mathbf{L}\mathbf{D}^{-\frac{1}{2}} = \mathbf{I} - \mathbf{D}^{-\frac{1}{2}}\mathbf{A}\mathbf{D}^{-\frac{1}{2}}$. An interesting property of this choice of Laplacian is that all its eigenvalues lie between 0 and 2 [5].

The adjacency matrix, or deformed Laplacian [symmetric]. Another choice for $\mathbf{R}$ is the adjacency matrix[3] $\mathbf{A}^\top$, as advocated in [1]. One readily sees that the eigenbasis $\mathbf{U}$ of $\mathbf{A}^\top$ and the eigenbasis of the deformed Laplacian $\mathbf{L_d} = \mathbf{I} - \frac{\mathbf{A}^\top}{||\mathbf{A}||_2}$, where $||.||_2$ is the operator 2-norm, are the same. Therefore, the corresponding GFTs are equivalent and, for consistency in the presentation, we will use $\mathbf{R} = \mathbf{L_d}$ here.
The random walk Laplacian [not symmetric]. The random walk Laplacian is yet another Laplacian reading: $\mathbf{L_{rw}} = \mathbf{D}^{-1}\mathbf{L} = \mathbf{I} - \mathbf{D}^{-1}\mathbf{A}$, where $\mathbf{D}^{-1}\mathbf{A}$ serves also to describe a uniform random walk on the graph. Even though $\mathbf{L_{rw}}$ is not symmetric, we know it is diagonalizable in $\mathbb{R}$. In fact, if $\mathbf{u}_k$ is an eigenvector of $\mathbf{L_n}$ with eigenvalue λ_k, then $\mathbf{D}^{-\frac{1}{2}}\mathbf{u}_k$ is an eigenvector of $\mathbf{L_{rw}}$ with the same eigenvalue. Thus, $\mathbf{L_{rw}}$ has the same eigenvalues as $\mathbf{L_n}$, and its Fourier basis $\mathbf{U}$ is real but not orthonormal.
Other possible definitions of the reference operator. For instance, the consensus operator (of the form $\mathbf{I} - \sigma\mathbf{L}$ with some suitable σ) [8], a geometric Laplacian [9], or some other deformed Laplacian one may think of, are valid alternatives.
All these operators imply a different spectral domain (different $\mathbf{U}$ and $\mathbf{\Lambda}$) and, provided one has a nice frequency interpretation (which is the object of Section 11.1.3), they all define possible GFTs. In the graph signal processing literature, the first three operators ($\mathbf{L}$, $\mathbf{L_n}$, and $\mathbf{L_d}$) are the most widely used.

GFT for directed graphs

For directed graphs, the adjacency matrix is no longer symmetric—$\mathbf{A}_{ij}$ is not necessarily equal to $\mathbf{A}_{ji}$—which does not automatically imply that the reference operator $\mathbf{R}$ is not symmetric (e.g., the case $\mathbf{R} = \mathbf{Q}$ below). This case is of great interest in some applications where the graph is naturally directed, such as hyperlink graphs (there is a directed edge between website i and website j if there is a hyperlink in website i directing to j). In directed graphs, the degree of node i is separated in its out-degree, $d_{\text{out}} = \sum_j A_{ij}$, and its in-degree, $d_{\text{in}} = \sum_j A_{ji}$.
Some straightforward approaches [not symmetric]. It is possible to readily transpose the previous notions to the directed case, choosing either $\mathbf{D_{out}}$ or $\mathbf{D_{in}}$ to replace $\mathbf{D}$ in the different formulations: e.g., $\mathbf{L} = \mathbf{D_{in}} - \mathbf{A}^\top$ as in [10], $\mathbf{L_{rw}} = \mathbf{I} - \mathbf{D}_{\mathbf{out}}^{-1}\mathbf{A}$ as in [11], $\mathbf{L_n} = \mathbf{I} - \mathbf{D}_{\mathbf{out}}^{-\frac{1}{2}}\mathbf{A}\mathbf{D}_{\mathbf{out}}^{-\frac{1}{2}}$. A notable choice is to directly use $\mathbf{L_d} = \mathbf{I} - \frac{\mathbf{A}^\top}{||\mathbf{A}||_2}$ as in [1] (we recall that $\mathbf{L_d}$ and $\mathbf{R} = \mathbf{A}^\top$ are equivalent for they share the same eigenvectors and thus, define the same GFT). These matrices are no longer strictly speaking Laplacians as they are no longer SDP, but one may nonetheless consider them as reference operators defining possible GFTs. Note that these definitions entail choosing (rather arbitrarily) either $\mathbf{D_{out}}$ or $\mathbf{D_{in}}$ in their formulations, with the notable exception of $\mathbf{L_d}$. In fact, $\mathbf{L_d}$ naturally generalizes to the directed case and is a classical choice of $\mathbf{R}$ in this context [1].
Chung's directed Laplacian [symmetric]. A less common approach in the graph signal processing community is the one provided by the directed Laplacians introduced by Chung [12]. To define these Laplacians, let us first introduce the random walk transition matrix (or operator) defined as $\mathbf{P} = \mathbf{D}_{\mathbf{out}}^{-1}\mathbf{A}$. It admits a stationary probability[4] $\boldsymbol{\pi} \in \mathbb{R}_+^N$ such that $\boldsymbol{\pi}^\top\mathbf{P} = \boldsymbol{\pi}^\top$. Writing $\mathbf{\Pi} = \text{diag}(\boldsymbol{\pi})$, Chung defines the following two directed Laplacians:

---

[3]We use $\mathbf{A}^\top$ instead of $\mathbf{A}$ for consistency purposes with [1], whose convention for directed edges in the adjacency matrix is converse to ours: what they call $\mathbf{A}$ is what we call $\mathbf{A}^\top$. Without any influence in the undirected case, it has an impact in the directed case.

[4]Assuming the random walk is ergodic, i.e., irreducible and nonperiodic.

$$\mathbf{Q} = \mathbf{\Pi} - \frac{\mathbf{\Pi P} + \mathbf{P}^\top \mathbf{\Pi}}{2}, \tag{11.4}$$

$$\mathbf{Q_n} = \mathbf{\Pi}^{-\frac{1}{2}} \mathbf{Q} \mathbf{\Pi}^{-\frac{1}{2}} = \mathbf{I} - \frac{\mathbf{\Pi}^{\frac{1}{2}} \mathbf{P} \mathbf{\Pi}^{-\frac{1}{2}} + \mathbf{\Pi}^{-\frac{1}{2}} \mathbf{P}^\top \mathbf{\Pi}^{\frac{1}{2}}}{2}. \tag{11.5}$$

Both the combinatorial $\mathbf{Q}$ and the normalized $\mathbf{Q_n}$ directed Laplacians verify the properties of Laplacian matrices: SDP, negative (or null) entries everywhere except on the diagonal and real symmetric. It is easy to see that Eqs. (11.4) and (11.5) generalize the definitions of the undirected case because for an undirected graph, $\mathbf{\Pi} = \mathbf{D}$, $\mathbf{\Pi P} = \mathbf{A}$; hence $\mathbf{Q}$ is the combinatorial Laplacian $\mathbf{L}$, and $\mathbf{Q_n}$ is its normalized version $\mathbf{L_n}$.

Other possible definitions of the reference operator. The previous definitions of $\mathbf{L_{rw}}$ and $\mathbf{L_d}$ for undirected graphs may also be generalized to the directed Laplacian framework to obtain:

$$\mathbf{Q_{rw}} = \mathbf{I} - \frac{\mathbf{P} + \mathbf{\Pi}^{-1} \mathbf{P}^\top \mathbf{\Pi}}{2} \quad \text{and} \quad \mathbf{Q_d} = \mathbf{I} - \frac{\mathbf{\Pi P} + \mathbf{P}^\top \mathbf{\Pi}}{||\mathbf{\Pi P} + \mathbf{P}^\top \mathbf{\Pi}||_2}. \tag{11.6}$$

Additional notes. Other GFTs for directed graphs were proposed via the Hermitian Laplacian as introduced in [13], which generalizes $\mathbf{A}^\top$. A very different approach is to construct a Graph Fourier basis directly from an optimization scheme, requiring some notion of smoothness, or generalization of it; see [14,15]. We will not consider this recent approach here.

11.1.3 FREQUENCIES OF GRAPH SIGNALS

To complement the notion of GFT, one needs to introduce some frequency analysis of the Fourier modes on the graph. The general way of doing so is to compute how fast a mode oscillates on the graph, and the tool of preference is to compute their variations across all edges of the graph. Let us first note the following facts:

- **In the undirected case,** $\mathbf{L}$ is semidefinite positive (SDP). In fact, one may write:

$$V_\mathbf{L}(\mathbf{x}) = \mathbf{x}^\top \mathbf{L} \mathbf{x} = \frac{1}{2} \sum_{(i,j) \in \mathcal{E}} \mathbf{A}_{ij} (x_i - x_j)^2 \geq 0. \tag{11.7}$$

  This function is also called the Dirichlet form. Similarly, $\mathbf{L_n}$ is also SDP:

$$V_{\mathbf{L_n}}(\mathbf{x}) = \mathbf{x}^\top \mathbf{L_n} \mathbf{x} = \frac{1}{2} \sum_{(i,j) \in \mathcal{E}} \mathbf{A}_{ij} \left(\frac{x_i}{\sqrt{d_i}} - \frac{x_j}{\sqrt{d_j}} \right)^2 \geq 0. \tag{11.8}$$

  As far as we know, the Dirichlet forms of $\mathbf{L_{rw}}$ and $\mathbf{L_d}$ do not have such a nice formulation as a sum of local quadratic variations over all edges of the graph. They are nevertheless SDP because: (i) $\mathbf{L_{rw}}$ and $\mathbf{L_n}$ have the same spectrum; (ii) the symmetry of $\mathbf{L_d}$ implies real eigenvalues, and the maximum eigenvalue of $\mathbf{A}/||\mathbf{A}||_2$ being 1 by definition of the norm, the minimum eigenvalue of $\mathbf{L_d}$ is 0.

- **In the directed case**, all directed Laplacians $\mathbf{Q}$, $\mathbf{Q_n}$, $\mathbf{Q_{rw}}$, and $\mathbf{Q_d}$ are SDP due to similar arguments. $\mathbf{Q}$ and $\mathbf{Q_n}$ also have Dirichlet forms in terms of a sum of local quadratic variations, e.g.:

$$V_{\mathbf{Q}}(\mathbf{x}) = \mathbf{x}^{\top}\mathbf{Q}\mathbf{x} = \frac{1}{2}\sum_{(i,j)\in\mathcal{E}} \pi_i \mathbf{P}_{ij}\left(x_i - x_j\right)^2 \geq 0. \tag{11.9}$$

The other reference operators $\mathbf{L} = \mathbf{D_{in}} - \mathbf{A}^{\top}$, $\mathbf{L_{rw}} = \mathbf{I} - \mathbf{D}_{\mathbf{out}}^{-1}\mathbf{A}$, $\mathbf{L_n} = \mathbf{I} - \mathbf{D}_{\mathbf{out}}^{-\frac{1}{2}}\mathbf{A}\mathbf{D}_{\mathbf{out}}^{-\frac{1}{2}}$, $\mathbf{L_d} = \mathbf{I} - \frac{\mathbf{A}^{\top}}{\|\mathbf{A}\|_2}$ are not SDP as their eigenvalues may be complex. Nevertheless, the real part of their eigenvalues is always nonnegative. This is quite clear for $\mathbf{L}_d$. For $\mathbf{L} = \mathbf{D_{in}} - \mathbf{A}^{\top}$, as the sum of row i of $\mathbf{A}^{\top}$ is equal to $\mathbf{d_{in}}(i)$, Gershgorin circle theorem ensures that the real part of all eigenvalues of $\mathbf{L}$ are nonnegative. For $\mathbf{L_{rw}}$: as $\mathbf{P} = \mathbf{D}_{\mathbf{out}}^{-1}\mathbf{A}$ is a stochastic matrix, the Perron-Frobenius theorem ensures that its eigenvalues are in the disk of radius 1 in the complex plane, hence the real part of $\mathbf{L_{rw}}$'s eigenvalues are nonnegative. As $\mathbf{L_{rw}}$ and $\mathbf{L_n}$ have the same set of eigenvalues, it is also true for $\mathbf{L_n}$.

To sum up, all the reference operators considered are either SDP (with real nonnegative eigenvalues) or have eigenvalues whose real component is nonnegative.

Definition 11.3 (Graph frequency). Let $\mathbf{R}$ be a reference operator. If its eigenvalues are real, the generalized graph frequency ν of a graph Fourier mode $\mathbf{u}_k$ is:

$$\nu(\mathbf{u}_k) = \lambda_k \geq 0. \tag{11.10}$$

If its eigenvalues are complex, two different definitions of the generalized graph frequency ν of a graph Fourier mode $\mathbf{u}_k$ exist:

$$\nu(\mathbf{u}_k) = \mathrm{Re}(\lambda_k) \geq 0 \quad \text{or} \quad \nu(\mathbf{u}_k) = |\lambda_k| \geq 0. \tag{11.11}$$

■

Remark. In the case of complex eigenvalues, it is a matter of choice whether we consider the imaginary part of the eigenvalues or not. There is no current consensus on this question. A second remark deals with the case of a multiple eigenvalue λ_k, i.e., if there are several eigenvectors associated to the same λ_k, then, only one frequency $\nu(\lambda_k)$ is defined for the associated eigenspace.

Justification: the link between frequency and variation. Two types of variation measures have been considered in the literature to show the consistency between this definition of graph frequencies and a notion of oscillation over the graph. The first one is based on the quadratic forms of the Laplacian operators. For instance, in the undirected case with the combinatorial Laplacian, Eq. (11.7) applied to any normalized Fourier mode $\mathbf{u}_k$ defined from $\mathbf{L}$ reads:

$$V_{\mathbf{L}}(\mathbf{u}_k) = \mathbf{u}_k^{\top}\mathbf{L}\mathbf{u}_k = \frac{1}{2}\sum_{(i,j)\in\mathcal{E}} \mathbf{A}_{ij}\left(u_k(i) - u_k(j)\right)^2 = \lambda_k \|\mathbf{u}_k\|_2^2 = \lambda_k. \tag{11.12}$$

The larger the local quadratic variations of $\mathbf{u}_k$, the larger its frequency λ_k. Eqs. (11.8) and (11.9) (as well as its counterpart for $\mathbf{Q_n}$) enable making this variation-frequency link for $\mathbf{L_n}$, $\mathbf{Q}$, and $\mathbf{Q_n}$. The second general type of variation that has been defined [16] is the total variation between a signal and its shifted version on the graph (where "shifting" a signal is understood as applying the adjacency matrix to it). For instance, in the case of $\mathbf{L_d}$, the associated variation reads[5]:

$$V_{\mathbf{L_d}}(\mathbf{x}) = \left\| \mathbf{x} - \frac{1}{|\mu_{\max}|}\mathbf{A}^\top \mathbf{x} \right\|_2 = ||\mathbf{L_d x}||_2 , \tag{11.13}$$

where μ_i designate the eigenvalues of $\mathbf{A}$ and $\mu_{\max}$ the one of maximum magnitude. The variation of the graph Fourier mode $\mathbf{u}_k$ from $\mathbf{L_d}$ thus reads:

$$V_{\mathbf{L_d}}(\mathbf{u}_k) = ||\mathbf{L_d u}_k||_2 = |\lambda_k| \, ||\mathbf{u}_k||_2 = |\lambda_k|. \tag{11.14}$$

The larger the total variation of $\mathbf{u}_k$, that is, the further is $\mathbf{u}_k$ from its shifted version along the graph, the larger its frequency.[6] For $\mathbf{L} = \mathbf{D_{in}} - \mathbf{A}^\top$, a similar approach detailed in [10] links the variation of $\mathbf{u}_k$ to its frequency $|\lambda_k|$.

It may also happen that for some operator $\mathbf{R}$, none of these two types of variations (quadratic forms or total variation) show natural. Then, one may use the variation $V_{\mathbf{R'}}$ based on another related operator $\mathbf{R'}$ to define frequencies. For instance, in [11,17], the authors considered the random walk Laplacians $\mathbf{R} = \mathbf{P} = \mathbf{D}_{\mathbf{out}}^{-1}\mathbf{A}$ as the reference operator to define the GFT while the directed combinatorial Laplacian $\mathbf{R'} = \mathbf{Q}$ is used to measure the variations. With these choices, they showed that $V_{\mathbf{Q}}(\mathbf{u}_k)$ is equal to $\mathrm{Re}(\lambda_k)$ up to a normalization constant. Another example of such a case is in [18] where one of the reference operators to build filters is the isometric translation introduced in [19] while the variational operators are built upon the combinatorial Laplacian. Note finally that other notions of variations such as the Hub authority score can be drawn from the literature. We refer the reader to [20] for details on that as well as a complementary discussion on GFTs and their related variations.

Definition 11.3 associates graph frequencies only to graph Fourier modes. For an arbitrary signal, we have the following definition of its frequency analysis:

Definition 11.4 (Frequency analysis). The frequency analysis of any graph-signal $\mathbf{x}$ on $\mathcal{G}$ is given by its components $(\mathbb{F}_{\mathcal{G}}\mathbf{x})_k$ at frequency $\nu(\mathbf{u}_k)$, as given in Definition 11.3. ■

11.1.4 IMPLEMENTATION AND ILLUSTRATION

Implementation. The implementation of the GFT requires diagonalizing $\mathbf{R}$, costing $\mathcal{O}(N^3)$ operations in general and $\mathcal{O}(N^2)$ memory space to store $\mathbf{U}$, and applying $\mathbf{U}$ to a signal $\mathbf{x}$ to obtain $\hat{\mathbf{x}} = \mathbf{U}^{-1}\mathbf{x}$ costs $\mathcal{O}(N^2)$ operations. These costs are prohibitive for large graphs ($N \gtrsim 10^4$ nodes). Recent works investigate how to reduce these costs, by tolerating an approximation on $\hat{\mathbf{x}}$. In the cases where $\mathbf{R}$ is symmetric, the authors in [21,22] suggest approximating $\mathbf{U}$ by a product of $\mathcal{O}(N \log N)$ Givens rotations [23], using a truncated Jacobi algorithm. The resulting approximated fast GFT requires $\mathcal{O}(N^2 \log^2 N)$ operations to compute the Givens rotations, and $\mathcal{O}(N \log N)$ operations to compute the approximate GFT of $\mathbf{x}$. The difficulty in designing fast GFTs boils down to the difficulty of deciphering

[5] In [16], the ℓ_1 norm is used but the ℓ_2 norm can be used equivalently: this is a matter of how one wants to normalize the eigenvectors. In this chapter, we consider the classical Euclidean norm, hence ℓ_2.

[6] There is a direct correspondence between λ_i, the eigenvalues of $\mathbf{L_d}$, and μ_i: $\lambda_i = 1 - \mu_i/|\mu_{\max}|$. We thus recover the results in [16]: the closer is μ_k from $|\mu_{\max}|$ in the complex plane, the smaller the total variation of the associated Fourier mode $\mathbf{u}_k$.

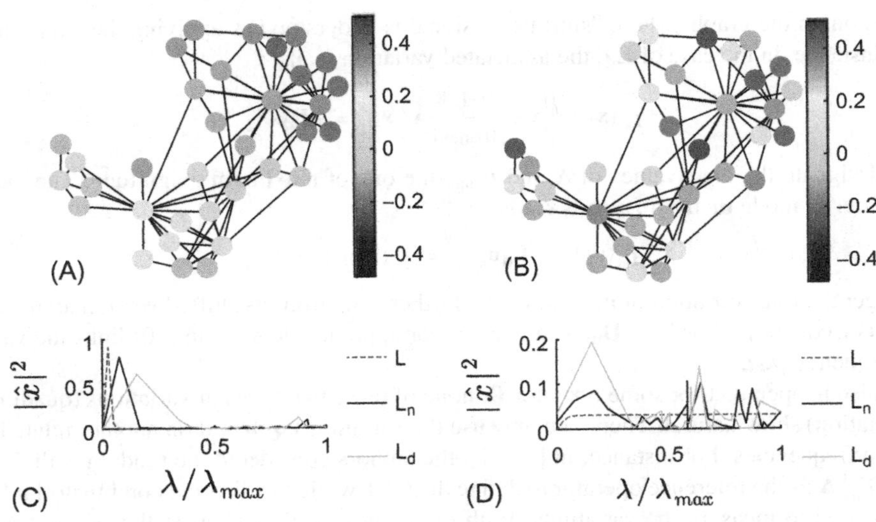

FIG. 11.1

Two graph signals and their GFTs. Plots (A) and (B) represent, respectively, a low-frequency and a high-frequency graph signal on the binary Karate club graph [24]. Plots (C) and (D) are their corresponding GFTs computed for three reference operators: **L**, $\mathbf{L_n}$, and $\mathbf{L_d}$ (equivalent to the GFT defined via the adjacency matrix).

eigenvalues that are very close to one another. This difficulty disappears once we consider smooth filtering operations that are much easier to efficiently approximate, as we will see in the next section.

Illustrations. To illustrate the GFT and the notion of frequency, we show in Fig. 11.1 two graph signals on the Karate club graph [24], corresponding to instances of a low-frequency and a high-frequency signal, respectively. We also show their GFTs, computed for three different choices of $\mathbf{R}$: $\mathbf{L}$, $\mathbf{L_n}$, and $\mathbf{L_d}$. We see that the choice of normalization and the choice of whether to take explicitly the degree matrix into account in the definition of $\mathbf{R}$ has a quantitative impact on the GFTs. Nevertheless, qualitatively, a graph signal that varies slowly (resp. rapidly) along any path of the graph is low frequency (resp. high frequency). In Fig. 11.2, we show how the GFT is not only sensitive to the graph signal but also to the underlying graph structure. In fact, we observe that for a given graph signal, modifications in the graph structure (here adding three links of weight ω) induce modifications in the signal's GFT.

11.2 GRAPH FILTERS

In this section, we assume that the reference operator $\mathbf{R}$ of the graph on which we wish to design filters is diagonalizable in $\mathbb{C}$ as in the previous section. The order of the eigenvalues and eigenvectors is chosen as frequency increasing. That is, given a choice of frequency definition (either $\nu(\lambda_k) = \mathrm{Re}(\lambda_k) \in \mathbb{R}^+$ or $\nu(\lambda_k) = |\lambda_k| \in \mathbb{R}^+$), one has $\nu(\mathbf{u}_1) \leq \nu(\mathbf{u}_2) \leq \cdots \leq \nu(\mathbf{u}_N)$.

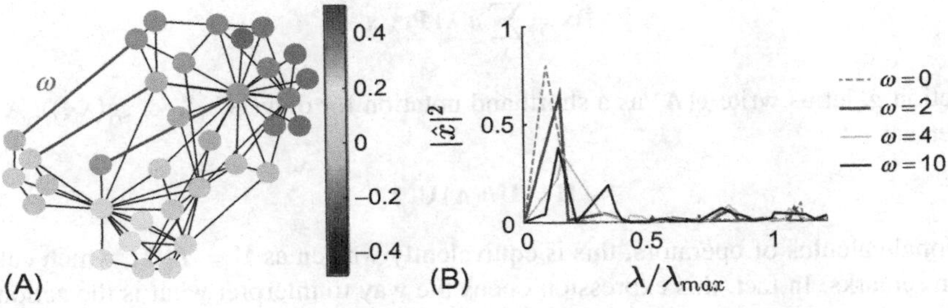

FIG. 11.2

Sensitivity of the GFT to the graph's topology. Plot (A) represents the same low-frequency graph signal as in Fig. 11.1A, but the underlying graph structure is altered by adding three edges with same weight ω (the three new edges are on the left side of the graph). Plot (B) represents the variation of its GFT (here choosing $\mathbf{R} = \mathbf{L}_n$) with respect to the edges' weight ω.

11.2.1 DEFINITION OF GRAPH FILTERS

The reference operator **R** has an eigendecomposition as in Eq. (11.1), and it can also be written as a sum of projectors on all its eigenspaces:

$$\mathbf{R} = \sum_{\lambda} \lambda \, \mathbf{Pr}_{\lambda}, \tag{11.15}$$

where the sum is on all different eigenvalues λ and $\mathbf{Pr}_{\lambda}$ is the projector on the eigenspace associated with eigenvalue λ, i.e.:

$$\mathbf{Pr}_{\lambda} = \sum_{\lambda_k = \lambda} \mathbf{u}_k \mathbf{v}_k^{\top}.$$

Definition 11.5 (Wide-sense definition of a graph filter).
The most general definition of a graph filter is an operator that acts separately on all the eigenspaces of **R**, depending on their eigenvalue λ. Mathematically, any function:

$$h : \mathbb{C} \to \mathbb{R} \tag{11.16}$$

$$\lambda \to h(\lambda) \tag{11.17}$$

defines a graph filter **H** such that:

$$\mathbf{H} = \sum_{\lambda} h(\lambda) \, \mathbf{Pr}_{\lambda} = \sum_{k} h(\lambda_k) \mathbf{u}_k \mathbf{v}_k^{\top}. \tag{11.18}$$

■

To each eigenspace of **R** with eigenvalue λ is associated a filtering weight $h(\lambda)$ that attenuates or increases the importance of this eigenspace in the decomposition of the signal of interest. In fact, one may write the action of **H** on a signal **x** as:

$$\mathbf{H}\mathbf{x} = \sum_{\lambda} h(\lambda)\, \mathbf{Pr}_{\lambda}\, \mathbf{x}. \tag{11.19}$$

For any function g, let us write $g(\mathbf{\Lambda})$ as a shorthand notation for $\mathrm{diag}(g(\lambda_1), \ldots, g(\lambda_N))$. A graph filter can be written as:

$$\mathbf{H} = \mathbf{U}\, h(\mathbf{\Lambda})\, \mathbf{U}^{-1}. \tag{11.20}$$

Using functional calculus of operators, this is equivalently written as $\mathbf{H} = h(\mathbf{R})$, which calls for some interpretation remarks. In fact, this expression opens the way to interpret what is the action of a graph filter in the vertex domain. Firstly, note that applying $\mathbf{R}$ to a graph signal is in fact a *local* computation on the graph: on each node, the resulting transformed signal is a weighted sum of the values of the original signal on its (direct) neighbors. Therefore, in the node space, a filter $\mathbf{H} = h(\mathbf{R})$ can be interpreted as an operator that weights the information on the signal transmitted through edges of the graph, the same way classical filters are built on the basic operation of time-shift. This fundamental analogy shift/reference operator was first made in [1] and explains why in the GSP literature, reference operators are often called "shift operators".

Now, to illustrate the notion of graph filtering in the spectral domain, consider the graph signal $\mathbf{x} = \sum_k \alpha_k \mathbf{u}_k$ and $\mathbf{y} = \mathbf{H}\mathbf{x}$. The kth Fourier component of $\mathbf{x}$ being, by construction, $\hat{\mathbf{x}}_k = \mathbf{v}_k^\top \mathbf{x} = \alpha_k$, its filtered version reads:

$$\mathbf{y} = \mathbf{H}\mathbf{x} = \sum_k h(\lambda_k)\alpha_k \mathbf{u}_k. \tag{11.21}$$

Hence, the kth Fourier component of the filtered signal is $\hat{\mathbf{y}}_k = h(\lambda_k)\,\alpha_k$, recalling the classical interpretation of filtering as a multiplication in the Fourier domain. We call $h(\lambda)$ the frequency response of the filter.

In many cases, it is more convenient and natural to restrict ourselves to functions h that associate the same real number to all values of $\lambda \in \{\lambda \text{ s.t. } \nu(\lambda) = \nu \in \mathbb{R}^+\}$. That is, considering any two eigenvalues λ_1 and λ_2 associated with the same frequency $\nu(\lambda_1) = \nu(\lambda_2)$, we restrict ourselves to functions h such that $h(\lambda_1) = h(\lambda_2)$. This entails the following narrowed-sense definition of a graph filter.

Definition 11.6 (Narrowed-sense definition of a graph filter). Any function:

$$h : \mathbb{R}^+ \to \mathbb{R} \tag{11.22}$$

$$\nu \to h(\nu) \tag{11.23}$$

defines a narrowed-sense graph filter $\mathbf{H}$ such that:

$$\mathbf{H} = \sum_{\lambda} h(\nu(\lambda))\, \mathbf{Pr}_{\lambda} = \mathbf{U}h(\nu(\mathbf{\Lambda}))\mathbf{U}^{-1}. \tag{11.24}$$

■

Remark. If the eigenvalues are real, both Definitions 11.5 and 11.6 are equivalent because $\nu(\lambda) = \lambda \in \mathbb{R}^+$.

Examples of narrowed-sense filters.

- The constant filter equal to c: $h(\nu) = c$. In this case, $h(\nu(\mathbf{\Lambda})) = c\mathbf{I}$ and $\mathbf{H} = c\mathbf{I}$: all frequencies are allowed to pass, and no component is filtered out.

- The Kronecker delta in ν^*: $h(\nu) = \delta_{\nu,\nu^*}$. If there exists one (or several) eigenvalues λ of $\mathbf{R}$ such that $\nu(\lambda) = \nu^*$, then $\mathbf{H} = \sum_{\lambda \text{ s.t. } \nu(\lambda)=\nu^*} \mathbf{Pr}_\lambda$. If not, then $\mathbf{H} = \mathbf{0}$. For this filter, only the frequency ν^* is allowed to pass.
- The ideal low-pass with cut-off frequency ν_c: $h(\nu) = 1$ if $\nu \leq \nu_c$, and 0 otherwise. In this case: $\mathbf{H} = \sum_{\lambda,\text{s.t.}\nu(\lambda)\leq\nu_c} \mathbf{Pr}_\lambda$, *i.e.*, only frequencies up to ν_c are allowed to pass.
- The heat kernel $h(\nu) = \exp^{-\nu/\nu_0}$: the weight associated with ν is exponentially decreasing with the frequency ν. Actually $\mathbf{y} = \mathbf{H}\,\mathbf{x}_0$ is the solution of the graph diffusion (or heat) equation (see [2]) at time $t = 1/\nu_0$ with initial condition $\mathbf{x}_0$.

11.2.2 PROPERTIES OF GRAPH FILTERS

From now on, in order to simplify notations and concepts in this introductory chapter on graph filtering, we will restrict ourselves to symmetric reference operators, such as $\mathbf{L}$, $\mathbf{L_n}$, or $\mathbf{L_d}$ in the undirected case, or the directed Laplacians $\mathbf{Q}$ or $\mathbf{Q_n}$ in the directed case. In this case, the eigenvalues are real such that the frequency definition is straightforward $\nu(\lambda) = \lambda$, both filter definitions are equivalent such that the frequency response reads:

$$h : \mathbb{R}^+ \to \mathbb{R} \tag{11.25}$$

$$\lambda \to h(\lambda), \tag{11.26}$$

and one may find a real orthonormal graph Fourier basis $\mathbf{U}$ such that $\mathbf{U}^{-1} = \mathbf{U}^\top$. A filtering operator associated with h thereby reads:

$$\mathbf{H} = \mathbf{U}h(\mathbf{\Lambda})\mathbf{U}^\top \in \mathbb{R}^{N\times N}. \tag{11.27}$$

All results presented in the following may be (carefully) generalized to the unsymmetric case.

Definition 11.7. We write $\mathcal{C}_p$ the set of finite-order polynomials in $\mathbf{R}$:

$$\mathcal{C}_p = \left\{ \mathbf{H} \text{ s.t. } \mathbf{H} = \sum_{i=0}^{n} a_i \mathbf{R}^i,\ \{a_i\}_{i=0,\dots,n} \in \mathbb{R}^{n+1}, n \in \mathbb{N}\backslash\{+\infty\} \right\}. \tag{11.28}$$

■

Proposition 11.1. *$\mathcal{C}_p$ is equal to the set of graph filters.*

Proof. Consider $\mathbf{H} \in \mathcal{C}_p$. Then, defining $h(\lambda) = \sum_{i=0}^{n} a_i\lambda^i$, one has: $\mathbf{H} = \sum_{i=0}^{n} a_i\mathbf{R}^i = \mathbf{U}\sum_{i=0}^{n} a_i\mathbf{\Lambda}^i\mathbf{U}^\top = \mathbf{U}h(\mathbf{\Lambda})\mathbf{U}^\top$, *i.e.*, $\mathbf{H}$ is a filter. Now, consider $\mathbf{H}$ a filter, *i.e.*, there exists h such that $\mathbf{H} = \mathbf{U}h(\mathbf{\Lambda})\mathbf{U}^\top$. Consider the polynomial $\sum_{i=0}^{N-1} a_i\lambda^i$ that interpolates through all pairs $(\lambda_i, h(\lambda_i))$. The maximum degree of such a polynomial is $N-1$ as there are maximum N points to interpolate, and may be smaller if eigenvalues have multiplicity larger than one. Thereby, one may write: $\mathbf{H} = \mathbf{U}h(\mathbf{\Lambda})\mathbf{U}^\top = \mathbf{U}\sum_{i=0}^{N-1} a_i\mathbf{\Lambda}^i\mathbf{U}^\top = \sum_{i=0}^{N-1} a_i\mathbf{R}^i$. Writing $n = N-1$, this means that $\mathbf{H} \in \mathcal{C}_p$. ■

Consequence. An equivalent definition of a graph filter is a polynomial in $\mathbf{R}$.

Definition 11.8. We write $\mathcal{C}_d$ the set of all diagonal operators in the graph Fourier space:

$$\mathcal{C}_d = \{\mathbf{H}, \text{ s.t. } \mathbf{U}^\top\mathbf{H}\mathbf{U} \text{ is diagonal}\}. \tag{11.29}$$

■

Proposition 11.2. *The set of graph filters is included in $\mathcal{C}_d$. Both sets are equal if all eigenspaces of* **R** *are simple (i.e., all eigenvalues are of multiplicity one).* ■

Proof. By definition of graph filters, they are included in $\mathcal{C}_d$. Now, in general, an element of $\mathcal{C}_d$ is not necessarily a graph filter. In fact, given $\mathbf{H} \in \mathcal{C}_d$, all diagonal entries of $\mathbf{U}^\top \mathbf{H} \mathbf{U}$ may be chosen independently, which is not the case for the diagonal entries of $h(\mathbf{\Lambda})$ corresponding to the same eigenspace. Thus, both sets are equal if all eigenspaces are of dimension one. ■

Consequence. In the case where all eigenvalues are simple, an equivalent definition of a graph filter is a diagonal matrix in the graph Fourier basis.

Definition 11.9. We write $\mathcal{C}_c$ the set of matrices that commute with **R**:

$$\mathcal{C}_c = \{\mathbf{H} \text{ s.t. } \mathbf{RH} = \mathbf{HR}\}. \tag{11.30}$$

■

Proposition 11.3. *$\mathcal{C}_p \subseteq \mathcal{C}_c$. The equality holds if all eigenvalues of* **R** *are simple.* ■

Proof. A polynomial in **R** necessarily commutes with **R**, thus $\mathcal{C}_p \subseteq \mathcal{C}_c$. Now, in general, an element in $\mathcal{C}_c$ is not necessarily in $\mathcal{C}_p$. In fact, if $\mathbf{H} \in \mathcal{C}_c$, then $\mathbf{\Lambda Y} = \mathbf{Y \Lambda}$ with $\mathbf{Y} = \mathbf{U}^\top \mathbf{HU}$. Suppose $\mathbf{\Lambda} = \lambda \mathbf{I}$. In this case, commutativity does not constrain **Y** at all, thereby **H** is not necessarily in $\mathcal{C}_p$. Now, if we suppose that all eigenvalues have multiplicity one, *i.e.*, all diagonal entries of $\mathbf{\Lambda}$ are different, the only solution for **Y** is to be a diagonal matrix, *i.e.*, $\mathcal{C}_c = \mathcal{C}_d$. We showed in Proposition 11.2 that the set of graph filters is equal to $\mathcal{C}_d$ if all eigenvalues have multiplicity one, therefore: $\mathcal{C}_c = \mathcal{C}_p$ if all eigenvalues have multiplicity one. ■

Consequence. In the case where all eigenvalues are simple, an equivalent definition of a graph filter is a linear operator that commutes with **R**.

11.2.3 SOME DESIGNS OF GRAPH FILTERS

From the previous sections, it should now be clear that the frequency response of a filter **H** only alters the frequencies $\nu(\lambda)$ corresponding to the discrete set of eigenvalues of **R** (see Eq. 11.18). More generally though, the frequency response of a graph filter can be defined over a continuous range of λ's, leading to the notion of *universal filter* design, i.e., a filter whose frequency response $h(\lambda)$ is designed for all λ's and not only adapted to the specific eigenvalues of **R**. On the contrary, a *graph-dependent filter* design depends specifically on these eigenvalues.

FIR filters. From the results of Section 11.2.2, a natural class of graph filters is given in the form of finite impulse response (FIR) filters, as by Eq. (11.28) with a polynomial of finite order, which realizes a weighted moving average (MA) filtering of a signal. Also, one can design any universal filter by fitting the desired response $h(\lambda)$ with a polynomial $\sum_{i=0}^{n} a_i \lambda^i$. The larger n is, the closer the filter can approximate the desired shape. If the approximation is done only using the $h(\lambda_k)$, the design is graph-dependent, else it is universal if fitting some function $h(\lambda)$.

Let us then go back to the interpretation of **R** as a graph shift operator in the node space, (as studied or recalled for instance in [1,20,25,26]), and see how it operates for FIR filters. Applied to a graph signal, the terms $\mathbf{R}^i$ in Eq. (11.28), act as an i-hop local computation on the graph: on each node, the resulting filtered signal is a weighted sum of the values of the original signal lying in its ith neighborhood, that is, nodes attainable with a path of length i along the graph. Then, as for classical signals, FIR filters only imply a finite neighborhood of each node, and this will translate in Section 11.2.4 into distributed, fast implementations of these filters. Still, FIR filters are usually poor at approximating filters with sharp changes of desired frequency response, as illustrated, for instance, in Fig. 1 of [25].

ARMA filters. A more versatile approximation of $h(\lambda)$ can be obtained with a rational design [25,27,28]:

$$h(\lambda) = \frac{\sum_{i=0}^{q} b_i \lambda^i}{1 + \sum_{i=1}^{p} a_i \lambda^i} = \frac{p_q(\lambda)}{p_p(\lambda)}. \tag{11.31}$$

Such a rational filter is called an auto-regressive moving average filter of order (p, q) and is commonly noted ARMA(p, q). Again, it is known from classical DSP that an ARMA design, being a IIR (infinite impulse response) filter, is more adaptable at approximating various shapes of filters, especially with sharp changes in the frequency response. The filtering relation $\mathbf{y} = \mathbf{H}\mathbf{x}$ for ARMA filters can be written in the node domain as: $(1 + \sum_{i=1}^{p} a_i \mathbf{R}^i)\mathbf{y} = (\sum_{i=0}^{q} b_i \mathbf{R}^i)\mathbf{x}$. This ARMA filter expression will lead to the distributed implementation, discussed later in Section 11.2.4. For instance, for an ARMA(1,0) (i.e., an AR(1)) one will have to use: $\mathbf{y} = -a_1 \mathbf{R}\mathbf{y} + b_0 \mathbf{x}$.

Example. A first design of low pass graph filtering is given by the simplest least-square denoising problem, where the Dirichlet form $\mathbf{x}^T \mathbf{R}\mathbf{x}$ is used as a Tikhonov regularization promoting smoothness on the graph. Using the (undirected) Laplacian and assuming one observes $\mathbf{y}$, the filter is given by:

$$\mathbf{x}_* = \arg\min_{\mathbf{x}} \frac{1}{2} \|\mathbf{x} - \mathbf{y}\|_2^2 + \gamma \mathbf{x}^\top \mathbf{L}\mathbf{x} \tag{11.32}$$

The solution is then given in the spectral domain (for $\mathbf{L}$) by $(\mathbb{F}\mathbf{x}_*)_k = h_{AR(1)}(\lambda_k)(\mathbb{F}\mathbf{y})_k$ with $h_{AR(1)}(\lambda) = 1/(1 + \gamma\lambda)$. It turns out to be a (universal) AR(1) filter. ■

Design of coefficients. To design the coefficients of ARMA filters, the classical approach is to find the set of coefficients a_i and b_i to approximate the desired $h(\lambda)$ as a rational function. However, as recalled in [25,27], the usual designs in DSP are not easily transposed to the GSP framework because the frequency response is given in terms of the λ's, and not in terms of $j\omega$ or $e^{j\omega}$.

Henceforth, it has been studied in [25,27] how to approximate the filter coefficients in a universal manner (i.e., with no specific reference to the graph spectrum) using a Shank's method: 1) Determine the a_i by first finding a polynomial approximation $P_h(\lambda)$ of $h(\lambda)$, and solve the system of equations $p_p(\lambda)P_h(\lambda) = p_q(\lambda)$ to identify the a_i's. Then 2) solve the least-square problem to minimize $\int_\lambda |p_q(\lambda)/p_p(\lambda) - h(\lambda)|^2 d\lambda$ w.r.t. λ to find the b_i's.

A second method is to approximate the filter response in a graph-dependent design, on the specific frequencies λ_k only. To do so, the method in [29], instead of fitting the polynomial ratio, solves the following optimization problem:

$$\min_{\mathbf{a},\mathbf{b}} \sum_{k=0}^{N-1} \left| h(\lambda_k) \left(1 + \sum_{i=1}^{p} a_i \lambda_k^i \right) - \sum_{i=0}^{q} b_i \lambda_k^i \right|^2. \tag{11.33}$$

Using again a polynomial approximation $P_h(\lambda)$, the solution derives from the least-square solution (see details in [29]).

AR filters to model random processes. We do not discuss much in this chapter about random processes on graphs; for that, see the framework to study stationary random processes on graphs in [26,30], and Chapter 12. Still, a remark can be made for the parametric modeling of random processes. As introduced in [1], one can model a process on a graph as the output of a graph filter, generally taken as an ARMA filter. Here, we discuss the case of AR filters. The linear prediction is written as:

$$\tilde{\mathbf{x}} = \sum_{i=1}^{p} a_i \mathbf{R}^i \mathbf{x}. \tag{11.34}$$

The coefficients of this filter can directly be obtained by numerical inversion of $\mathbf{x} = \mathbf{Ba}$ where $\mathbf{B} = (\mathbf{Rx}, \mathbf{R}^2\mathbf{x}, \ldots, \mathbf{R}^p\mathbf{x})$. The solution is given by the pseudoinverse: $\mathbf{a}^\# = (\mathbf{B}^\top\mathbf{B})^{-1}\mathbf{B}^\top\mathbf{x}$, which minimizes the squared error $||\mathbf{x} - \mathbf{Ba}||_2^2$, w.r.t to $\mathbf{a}$. Another possibility would be to estimate the coefficients of the AR model by means of the orthogonality principle leading to the Yule-Walker (YW) equations, as follows:

$$\mathbb{E}\left\{(\mathbf{R}^k\mathbf{x})^\top\mathbf{Rx})\right\} = \sum_{i=1}^{p} a_i \mathbb{E}\left\{(\mathbf{R}^i\mathbf{x})^\top\mathbf{R}^k\mathbf{x}\right\} = 0. \tag{11.35}$$

Depending on the structure of the reference operator $\mathbf{R}$, we have:

1. If $\mathbf{R}$ is symmetric ($\mathbf{R}^\top = \mathbf{R}$, as often for undirected graph), the autocorrelation function involved in this YW system is: $\rho_\mathbf{x}(m,n) = \mathbb{E}\left\{(\mathbf{R}^m\mathbf{x})^\top\mathbf{R}^n\mathbf{x}\right\} = \gamma_\mathbf{x}(m+n)$;
2. When $\mathbf{R}$ is unitary (for instance with $\mathbf{R}$ as the isometric operator from [19]), the corresponding autocorrelation will be $\rho_\mathbf{x}(m,n) = \mathbb{E}\left\{(\mathbf{R}^m\mathbf{x})^\top\mathbf{R}^n\mathbf{x}\right\} = \gamma_\mathbf{x}(n-m)$ and the usual techniques to solve the YW system can be used.

Experimentally, it was found that the isometric operator of [19] or a consensus operator is the one that offers a more exact and stable modeling [18].

11.2.4 IMPLEMENTATIONS OF GRAPH FILTERS

Given a frequency response h, how to efficiently filter a graph signal $\mathbf{x}$?

The direct approach. It consists in first diagonalizing $\mathbf{R}$ to obtain $\mathbf{U}$ and $\mathbf{\Lambda}$, then computing the filter matrix $\mathbf{H} = \mathbf{U}h(\mathbf{\Lambda})\mathbf{U}^\top$, and finally left-multiplying the graph signal $\mathbf{x}$ by $\mathbf{H}$ to obtain its filtered version. The overall computational cost of this procedure is $\mathcal{O}(N^3)$ arithmetic operations due to the diagonalization and $\mathcal{O}(N^2)$ memory space as the GFT $\mathbf{U}$ is dense in general.

The polynomial approximate filtering approach. More efficiently though, we can first quickly estimate $\lambda_{\min}$ and $\lambda_{\max}$ (for instance via the power method) and second, look for a polynomial that best approximates $h(\lambda)$ on the whole interval $[\lambda_{\min}, \lambda_{\max}]$. Let us call $\tilde{h}_i$ the coefficients of this approximate polynomial. We have:

$$\mathbf{Hx} = \mathbf{U}h(\mathbf{\Lambda})\mathbf{U}^\top\mathbf{x} \simeq \mathbf{U}\sum_{i=0}^{p} \tilde{h}_i \mathbf{\Lambda}^i\mathbf{U}^\top\mathbf{x} = \sum_{i=0}^{p} \tilde{h}_i \mathbf{R}^i\mathbf{x}. \tag{11.36}$$

The number of required arithmetic operations is $\mathcal{O}(p|\mathcal{E}|)$, where p is a trade-off between precision and computational cost. Also, the easier is h approachable by a low-order polynomial, the better the approximation. At the same time, the larger p is, the more accurate the approximation is. The authors of [31] recommend Chebyshev polynomials as an approximation basis as they are known to be optimal in the ∞-norm sense. In some circumstances, other choices can be preferred. For instance, when approximating the ideal low pass, Chebyshev polynomials yield Gibbs oscillations around the cut-off frequency that turn penalizing for smooth filters. In that case, other choices are possible [32], such as the Jackson-Chebyshev polynomials that attenuate such unwanted oscillations.

The Lanczos approximate filtering approach. In [33], and based on works by Gallopoulos and Saad [34], the authors propose an approximate filtering approach based on Lanczos iterations. Given a signal $\mathbf{x}$, the Lanczos algorithm computes an orthonormal basis $\mathbf{V}_p \in \mathbb{R}^{N \times p}$ of the Krylov subspace associated with $\mathbf{x}$: $K_p(\mathbf{L}, \mathbf{x}) = \text{span}(\mathbf{x}, \mathbf{L}\mathbf{x}, \ldots, \mathbf{L}^{p-1}\mathbf{x})$, as well as a small tridiagonal matrix $\mathbf{H}_p \in \mathbb{R}^{p \times p}$ such that: $\mathbf{V}_p^* \mathbf{L} \mathbf{V}_p = \mathbf{H}_p$. The approximate filtering then reads:

$$\mathbf{H}\mathbf{x} \simeq ||\mathbf{x}||_2 \mathbf{V}_p h(\mathbf{H}_p)\delta_1. \tag{11.37}$$

At fixed p, this approach has a typical complexity in $\mathcal{O}(p|\mathcal{E}|)$, possibly raised to $\mathcal{O}(p|\mathcal{E}| + Np^2)$ if a reorthonomalization is needed to stabilize the algorithm; that is a cost that is comparable to that of the polynomial approximation approach. Theoretically, the quality of approximation is similar (see [33] for details). In practice, however, it has been observed that if the spectrum is regularly spaced, polynomial approximations should be preferred while the Lanczos method has an edge over others in the case of irregularly spaced spectra. This is understandable as Krylov subspaces are also used for diagonalization purposes (see for instance, chapter 6 of [35]) and thus naturally adapt to the underlying spectrum.

Distributed implementation of ARMA filters. The ARMA filters being defined through a rational fraction are IIR filters. Henceforth, the polynomials approach to distribute and fasten the computation are not the most efficient ones. The methods developed in [27,28] yield a distributed implementation of ARMA filters. The first point is to remember that the rational filter of Eq. (11.31) can be implemented from its partial fraction decomposition as a sum of polynomial fractions of order 1 only. Then, the distributed implementation of filtering $\mathbf{x}$ can be done by studying the first-order recursion [27]:

$$\mathbf{y}(t+1) = c\mathbf{M}\mathbf{y}(t) + d\mathbf{x}, \tag{11.38}$$

where $\mathbf{y}(t+1)$ is the filter output at iteration t. The operator $\mathbf{M}$ is chosen equal to $(\lambda_{\max} - \lambda_{\min})\mathbf{I} - \mathbf{R}$, with the same eigenvectors as $\mathbf{R}$, and with a minimal spectral radius that ensures good convergence properties for the recursion. The coefficients c and d are chosen in $\mathbb{C}$ so that, with $r = -d/c$ and $\rho = 1/c$, the proposed recursion reproduces the effect of the following ARMA(1,0) filter: $h(\lambda) = r/(\lambda - \rho)$. The coefficients r and ρ are the residue and the pole of the rational function, respectively.

Because $\mathbf{M}$ is local in the graph, the recursive application of Eq. (11.38) is local and the algorithm is then naturally distributed on the graph, with a memory and operation at each recursion in $\mathcal{O}(K|\mathcal{E}|)$ for K filters in parallel to compute the output of an ARMA(K, K) filter. This approach is shown in [27,28] to converge efficiently. Moreover, in [25], the behavior of this design is also studied in time-varying settings, when the graph and signal are possibly time varying; it is then shown that the recursion can remain stable and usable as a distributed implementation of IIR filters.

Illustration. We show in Fig. 11.3 an example of a filtering operation on a graph signal. We consider here the case of a Tikhonov denoising (see Eq. 11.32), i.e., with a frequency response equal to $h(\lambda) = 1/(1 + \gamma\lambda)$.

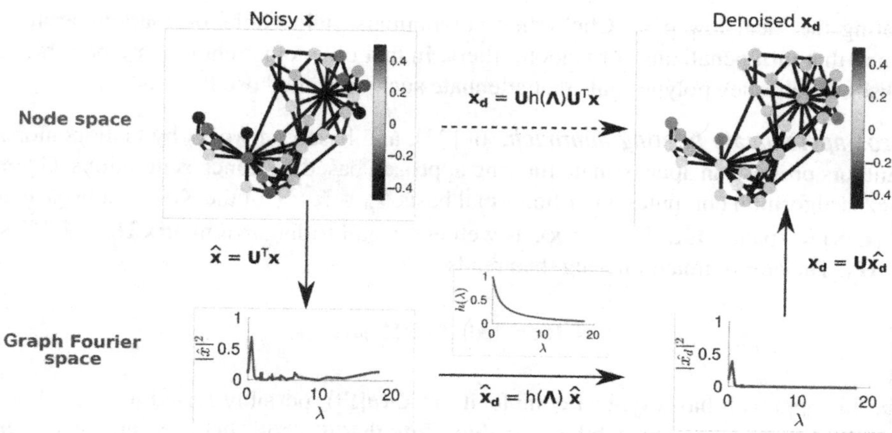

FIG. 11.3

Illustration of graph filters: a denoising toy experiment. The input signal **x** is a noisy version (additive Gaussian noise) of the low-frequency graph signal displayed in Fig. 11.1. We show here the filtering operation in the graph Fourier domain associated with $\mathbf{R} = \mathbf{L_n}$.

11.3 FILTERBANKS AND MULTISCALE TRANSFORMS ON GRAPHS

To process and filter signals on graphs, it is attractive to develop the equivalent of multiscale transforms such as wavelets, i.e., ways to decompose a graph signal on components at different scales, or frequency ranges. The road to multiscale transforms on graphs has originally been tackled in the vertex domain [36] to analyze data on networks. Thereafter, a general design of multiscale transforms was based on the diffusion of signals on the graph structure, leading to the powerful framework of diffusion wavelets [37,38]. These latter works were already based on diffusion operators, usually a Laplacian or a random walk operator, whose powers are decomposed in order to obtain a multiscale orthonormal basis. The objective was to build a kind of equivalent of discrete wavelet transforms for graph signals.

In this chapter, we focus on two other constructions of multiscale transforms on graphs that are more related to the GFT. The frequency analysis and filters described here are: (1) the method of [3], which develops an analog of continuous wavelet transform on graphs, (2) approaches that combine filters on graphs with graph decompositions through decimation (pioneered in [4] with decimation of bipartite graphs) or aggregation of nodes; in a nutshell, these methods are very close to the filter banks implementation of discrete wavelets [39].

11.3.1 CONTINUOUS MULTISCALE TRANSFORMS

In the following, we work with undirected graphs and $\mathbf{R} = \mathbf{L_n} = \mathbf{I} - \mathbf{D}^{-\frac{1}{2}}\mathbf{A}\mathbf{D}^{-\frac{1}{2}}$, whose eigenvalues are contained in the interval [0, 2]. The generalization to other operators can be done using the guidelines of previous sections.

The first continuous multiscale transform based on the GFT was introduced via the spectral graph wavelet transform [3] (using similar concepts as in diffusion polynomial frames [40]). These wavelets were defined by analogy to the classical wavelets in the following sense. Classically, a wavelet family $\{\psi_{s,\tau}(t)\}$ centered around time τ and at scale s is the translated and dilated version of a mother wavelet $\psi(t)$, generally defined as a zero-mean, square integrable function. Mathematically it is expressed for all $s \in \mathbb{R}^+$ and $\tau \in \mathbb{R}$ as:

$$\psi_{s,\tau}(t) = \frac{1}{s}\psi\left(\frac{t-\tau}{s}\right) \tag{11.39}$$

or equivalently, in the frequency domain, with $\mathbb{F}$ the continuous Fourier transform:

$$\mathbb{F}[\psi_{s,\tau}](\omega) = \mathbb{F}[\psi](s\omega)\mathbb{F}[\delta_\tau](\omega) = \hat{\psi}(s\omega)e^{-i\omega\tau}. \tag{11.40}$$

Then, for a signal x, the wavelet coefficient at scale s and instant τ is given by the inner product $W_{s,\tau}x = \langle x, \psi_{s,\tau}\rangle$.

By analogy, transposing Eq. (11.40) with GFT, a spectral graph wavelet $\psi_{s,a}$ at scale s and node a reads[7]:

$$\psi_{s,a} = \mathbf{U}h(s\mathbf{\Lambda})\mathbf{U}^\top\delta_a, \tag{11.41}$$

where the graph filter $h(\lambda)$ plays the role of the wavelet band pass filter $\hat{\psi}(\omega)$. The shifted scaled wavelet identifies to the impulse response of $h(s\lambda)$ to a Dirac localized on node a. In particular, the shape of the filter originally proposed in [3] was:

$$h^{\mathrm{SGW}}(\lambda) = \begin{cases} \lambda_1^{-\alpha}\lambda^{\alpha} \text{ for } \lambda < \lambda_1 \\ q(\lambda) \text{ for } \lambda_1 \leq \lambda \leq \lambda_2 \\ \lambda_2^{\beta}\lambda^{-\beta} \text{ for } \lambda > \lambda_2 \end{cases}$$

with α, β, λ_1, and λ_2 four parameters, and $q(\lambda)$ the unique cubic polynomial interpolation that preserves continuity and the derivative's continuity. Several properties on the obtained wavelets may be theoretically derived, for instance the notion of locality (the fact that a wavelets' energy is mostly contained around the node on which it is centered). Given a selection of scales $\mathcal{S} = (s_1, \ldots, s_m)$ and a graph signal $\mathbf{x}$, the signal wavelet coefficient associated with the node a and the scale $s \in \mathcal{S}$ reads: $W_{s,a}\mathbf{x} = \psi_{s,a}^\top\mathbf{x}$. Then, a question that naturally arises is that of invertibility of the wavelet transform: can one recover any signal $\mathbf{x}$ from its wavelet coefficients? As defined here, the wavelet transform is not invertible as it does not take into account—due to the zero-mean constraint of the wavelets—the signal's information associated with the null frequency, i.e., associated with the first eigenvector $\mathbf{u}_1$. To enable invertibility, one may simply add any low pass filter $h_0(\lambda)$ to the set of filters $\{h^{\mathrm{SGW}}(s\lambda)\}_{s\in\mathcal{S}}$ of the wavelet transform. We write ϕ_a their associated atoms:

$$\phi_a = \mathbf{U}h_0(\mathbf{\Lambda})\mathbf{U}^\top\delta_a. \tag{11.42}$$

[7]Note that the scale parameter stays continuous but the localization parameter is discretized to the set of nodes a of the graph.

The following theorem derives:

Theorem 11.1 (Theorem 5.6 in [3]). *Given a set of scales $\mathcal{S}$, the set of atoms $\{\{\psi_{s,a}\}_{s\in\mathcal{S}} \cup \{\phi_a\}\}_{a\in\mathcal{V}}$ forms a frame with bounds A, B given by:*

$$A = \min_{\lambda\in[0,\lambda_{max}]} G(\lambda), \tag{11.43}$$

$$B = \max_{\lambda\in[0,\lambda_{max}]} G(\lambda), \tag{11.44}$$

where $G(\lambda) = (h_0(\lambda))^2 + \sum_{s\in\mathcal{S}} (h^{\text{SGW}}(s\lambda))^2$. ■

In theory, invertibility is guaranteed provided that A is different from 0. Nevertheless, in practice, one should strive to design filter shapes (wavelet and low pass filters) and to choose a set of scales such that A is as close as possible to B in order to deal with well-conditioned inverses. Doing so, we obtain the so-called tight (or snug) frames, i.e., frames such that $A = B$ (or $A \approx B$). An approach is to use classical dyadic decompositions using band limited filters such as in Table 1 of [41]. Another desirable property of such frames is their discriminatory power: the ability to discern different signals only by considering their wavelet coefficients. For a filterbank to be discriminative, each filter element needs to take into account information from a similar number of eigenvalues of the Laplacian. The eigenvalues of an arbitrary graph being unevenly spaced on $[0, \lambda_{\text{max}}]$, one needs to compute or estimate the exact density of the spectrum of the graph under consideration [42].

11.3.2 DISCRETE MULTISCALE TRANSFORMS

A second general way to generate multiscale transforms is via a succession of filtering and decimation operations, as in Fig. 11.4. This scheme is usually cascaded as in Fig. 11.5, and each level of the cascade represents a scale of description of the input signal. Thereby, as soon as decimation enters into the process, we talk about "discrete" multiscale transforms, as the scale parameter can no longer be continuously varied. For details on this particular approach to multiscale transforms in classical signal processing, we refer, e.g., to the book by Strang and Nguyen [39]. In the following, we directly consider the graph-based context. Let us first settle notations:

- The decimation operator may be generally defined by partitioning the set of nodes $\mathcal{V}$ into two sets $\mathcal{V}_0$ and $\mathcal{V}_1$. As this subdivision is a partition, we have $\mathcal{V}_0 \cup \mathcal{V}_1 = \mathcal{V}$ and $\mathcal{V}_0 \cap \mathcal{V}_1 = \emptyset$. Moreover, let us define $\downarrow_{\mathcal{V}_i}$ the downsampling operator associated with $\mathcal{V}_i$: given any graph signal $\mathbf{x}$, $\mathbf{y}_i = \downarrow_{\mathcal{V}_i} \mathbf{x}$ is the reduction of $\mathbf{x}$ to $\mathcal{V}_i$. We also define the upsampling operator $\uparrow_{\mathcal{V}_i} = \downarrow_{\mathcal{V}_i}^{\top}$. Given $\mathbf{y}_i$ a signal defined on $\mathcal{V}_i$, $\uparrow_{\mathcal{V}_i} \mathbf{y}_i$ is the zero-padded version of $\mathbf{y}_i$ on the whole graph. The combination of both operators reads: $\uparrow_{\mathcal{V}_i}\downarrow_{\mathcal{V}_i} = \text{diag}(\mathcal{I}_{\mathcal{V}_i})$, where $\mathcal{I}_{\mathcal{V}_i}$ is the indicator function of $\mathcal{V}_i$. Moreover, we define:

$$\mathbf{J} = 2 \uparrow_{\mathcal{V}_0}\downarrow_{\mathcal{V}_0} - \mathbf{I} = \text{diag}(\mathcal{I}_{\mathcal{V}_0}) - \text{diag}(\mathcal{I}_{\mathcal{V}_1}). \tag{11.45}$$

- We define two analysis filters: a low pass graph filter $\mathbf{H}_0$ and a high pass graph filter $\mathbf{H}_1$, as well as two synthesis graph filters $\mathbf{G}_0$ and $\mathbf{G}_1$. All filters are associated with their frequency responses $h_0(\lambda)$, $h_1(\lambda)$, $g_0(\lambda)$ and $g_1(\lambda)$.

The signal $\mathbf{y}_0 = \downarrow_{\mathcal{V}_0} \mathbf{H}_0\mathbf{x}$ is called the approximation of $\mathbf{x}$, whereas $\mathbf{y}_1 = \downarrow_{\mathcal{V}_1} \mathbf{H}_1\mathbf{x}$ is generally understood as the necessary details to recover $\mathbf{x}$ from its approximation.

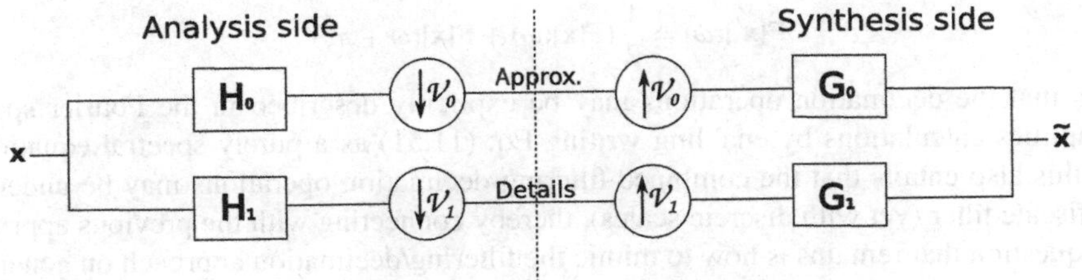

FIG. 11.4

A filterbank seen as a succession of filtering and decimating operators.

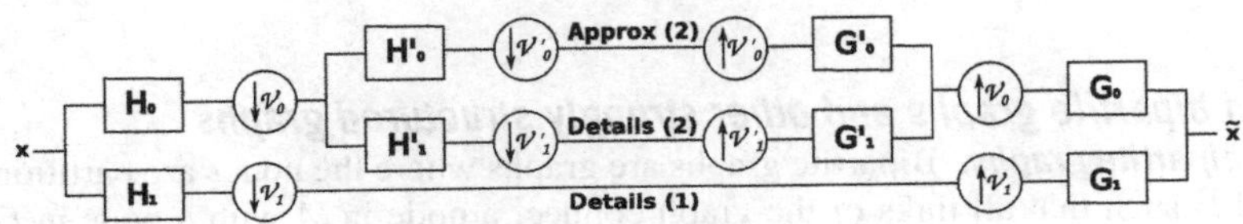

FIG. 11.5

A cascaded filterbank (here two levels).

Given the scheme of Fig. 11.4, one writes the processed signal $\tilde{\mathbf{x}}$ as:

$$\tilde{\mathbf{x}} = (\mathbf{G}_0 \uparrow_{\mathcal{V}_0} \downarrow_{\mathcal{V}_0} \mathbf{H}_0 + \mathbf{G}_1 \uparrow_{\mathcal{V}_1} \downarrow_{\mathcal{V}_1} \mathbf{H}_1)\,\mathbf{x} \tag{11.46}$$

$$= \frac{1}{2}(\mathbf{G}_0\mathbf{H}_0 + \mathbf{G}_1\mathbf{H}_1)\,\mathbf{x} + \frac{1}{2}(\mathbf{G}_0\mathbf{J}\mathbf{H}_0 - \mathbf{G}_1\mathbf{J}\mathbf{H}_1)\,\mathbf{x}. \tag{11.47}$$

When designing such discrete filterbanks, and in order to enable perfect reconstruction ($\forall \mathbf{x} \in \mathbb{R}^N$ $\tilde{\mathbf{x}} = \mathbf{x}$), one deals with two main equations linking all four filters and the matrix $\mathbf{J}$:

$$\mathbf{G}_0\mathbf{H}_0 + \mathbf{G}_1\mathbf{H}_1 = 2\mathbf{I}, \tag{11.48}$$

$$\mathbf{G}_0\mathbf{J}\mathbf{H}_0 - \mathbf{G}_1\mathbf{J}\mathbf{H}_1 = \mathbf{0}. \tag{11.49}$$

Left-multiplying by $\mathbf{U}^\top$ and right-multiplying by $\mathbf{U}$, one obtains equivalently:

$$g_0(\mathbf{\Lambda})h_0(\mathbf{\Lambda}) + g_1(\mathbf{\Lambda})h_1(\mathbf{\Lambda}) = 2\mathbf{I}, \tag{11.50}$$

$$g_0(\mathbf{\Lambda})\mathbf{U}^\top\mathbf{J}\mathbf{U}h_0(\mathbf{\Lambda}) - g_1(\mathbf{\Lambda})\mathbf{U}^\top\mathbf{J}\mathbf{U}h_1(\mathbf{\Lambda}) = \mathbf{0}. \tag{11.51}$$

Eq. (11.50) is purely spectral, and may be seen as a set of N equations:

$$\forall \lambda_i \quad g_0(\lambda_i)h_0(\lambda_i) + g_1(\lambda_i)h_1(\lambda_i) = 2. \tag{11.52}$$

On the other hand, Eq. (11.51) is not so simple due to the decimation operation and needs to be investigated in detail.

In 1D classical signal processing (equivalent to the undirected circle graph), the decimation operator samples one every two nodes. Moreover, given $\mathbf{x}' = \uparrow_2\downarrow_2 \mathbf{x}$, one classically has the following aliasing phenomenon (Theorem 3.3 of [39]):

$$\mathbb{F}[\mathbf{x}'](\omega) = \frac{1}{2}\left(\mathbb{F}[\mathbf{x}](\omega) + \mathbb{F}[\mathbf{x}](\omega + \pi)\right). \tag{11.53}$$

This means that the decimation operations may be explicitly described in the Fourier space, which greatly simplifies calculations by enabling writing Eq. (11.51) as a purely spectral equation as well. Moreover, this also entails that the combined filtering-decimation operations may be understood as a global multiscale filter (yet with discrete scales), thereby connecting with the previous approach. Now, the central question that remains is how to mimic the filtering/decimation approach on general graphs?

In Section 11.3.2, it will be shown that for very well-structured graphs such as bipartite graphs or m-cyclic graphs, generalizations of the decimation operator may be defined and their effect explicitly described as graph filters. Then the case of the arbitrary graph is studied in Section 11.3.2 where the "one every two nodes" paradigm is not transposable directly; several approaches to do that will then be reviewed.

Filterbanks on bipartite graphs and other strongly structured graphs

Filterbanks on bipartite graphs. Bipartite graphs are graphs where the nodes are partitioned in two sets of nodes $\mathcal{A}$ and $\mathcal{B}$ such that all links of the graph connect a node in $\mathcal{A}$ with a node in $\mathcal{B}$. On bipartite graphs, the "one-every-two-node" paradigm has a natural extension: decimation ensembles are set to $\mathcal{V}_0 = \mathcal{A}$ and $\mathcal{V}_1 = \mathcal{B}$. Leveraging the fact that bipartite graphs' spectra are symmetrical[8] around the value 1, Narang and Ortega [4] show the bipartite graph spectral folding phenomenon:

$$\forall\lambda \quad \mathbf{Pr}_\lambda \mathbf{J} = \mathbf{J}\mathbf{Pr}_{2-\lambda} \tag{11.54}$$

(with $\mathbf{Pr}_\lambda$ as in Eq. 11.15). This means that for any filter $\mathbf{H} = \mathbf{U}h(\mathbf{\Lambda})\mathbf{U}^\top$ one has:

$$\mathbf{G}\mathbf{J} = \mathbf{U}g(\mathbf{\Lambda})\mathbf{U}^\top\mathbf{J} = \mathbf{J}\mathbf{U}g(2\mathbf{I} - \mathbf{\Lambda})\mathbf{U}^\top. \tag{11.55}$$

Eq. (11.51) therefore boils down to a second set of N purely spectral equations:

$$\forall\lambda_i \quad g_0(2-\lambda_i)h_0(\lambda_i) - g_1(2-\lambda_i)h_1(\lambda_i) = 0. \tag{11.56}$$

Eqs. (11.52) and (11.56) give us $2N$ equations linking the $4N$ parameters of the four filters to ensure perfect reconstruction. The other $2N$ degrees of liberty are free to be used to design filterbanks with other desirable properties of filterbanks, such as (bi-)orthogonality, compact-supportness of the atoms, and of course also to adapt to the specific application for which these filters are designed.

Filterbanks on other regular structures. Extending these ideas, several authors have proposed similar approaches to define filterbanks on other regular structures such as M-block cyclic graphs [43] or circulant graphs [44,45]. In any case, writing decimation operations exactly as graph filters requires regular structures on graphs inducing at least some regularity in the spectrum one may take advantage of. All these approaches lead to exact reconstruction procedures. However, arbitrary graphs do not have such regularities, and other approaches are required.

Filterbanks on arbitrary graphs

In order to extend the filterbanks approach to arbitrary graphs, one needs to either generalize the decimation operator or to bypass decimation via aggregation operators. We discuss in this last part some solutions that were proposed in the literature. A complementary and thoughtful discussion on

[8] That is, if λ is an eigenvalue of $\mathcal{L}$, then so is $2 - \lambda$.

graph decimation, graph aggregation, and graph reconstruction can be found in sections III and IV of [46]. For that, the two key questions are:

- How to generalize the decimation operator on arbitrary graphs? We will see that generalized decimation operators either try to mimic the classical decimation and attempt to sample "one every two nodes", or aggregate nodes to form supernodes according in general to some graph-cut objective function.
- How to build the new coarser-scale graph from the decimated nodes (or aggregated supernodes)? In fact, after each decimation, if one wants to cascade the filterbank, a new coarse-grain graph has to be built in order to define the next level's graph filters. The nodes (resp. supernodes) are set thanks to decimation (resp. aggregation), but how do we link them together?

Graph decimation. The first work to generalize filterbanks on arbitrary graphs is due to Narang and Ortega [4] and consists in decomposing the graph into an edge-disjoint collection of bipartite subgraphs, and then to apply the scheme presented in Section 11.3.2 on each of the subgraphs. In this collection, each subgraph has the same node set, and the union of all subgraphs sums to the original graph. To perform this decomposition (which is not unique), the same authors propose a coloring-based algorithm, called Harary's decomposition. Sakiyama and Tanaka [47] also used this decomposition as one of their design's cornerstones. Unsatisfied by the NP-completeness of the coloring problem (even though heuristics exist), Nguyen and Do [48] propose another decomposition method based on maximum spanning trees.

The bipartite paradigm's main advantage comes from the fact that decimation has an explicit formulation in the graph's Fourier space, thereby enabling exact filter designs depending on the given task. In our opinion, when applied to arbitrary graphs, its main drawback comes from the nonunicity of the bipartite subgraphs decomposition as well as the seeming arbitrariness of such a decomposition: from a graph signal point-of-view, what is the meaning of a bipartite decomposition? Letting go of this paradigm and slightly changing the general filterbank design presented in Section 11.3.2, other generalized graph decimations have been proposed. For instance, in [49], the authors propose separating the graph in two sets $\mathcal{V}_0, \mathcal{V}_1$ according to its max cut, i.e., maximizing $\sum_{i\in\mathcal{V}_0}\sum_{j\in\mathcal{V}_1}\mathbf{W}_{ij}$. In [46], the authors suggest similarly partitioning the graph into two sets according to the polarity of the last eigenvector (i.e., the eigenvector associated with the highest frequency). In [50], the authors use an original approach based on random forests to sample nodes, where they have a probabilistic version of "equally spaced" nodes on the graph.

Graph aggregation. Another paradigm in graph reduction is graph aggregation, where, instead of *selecting* nodes as in decimation, one *aggregates* entire regions of the graph in "supernodes." In general, these methods are based on first clustering the nodes in a partition $\mathcal{P} = \{\mathcal{V}_1, \mathcal{V}_2, \ldots, \mathcal{V}_J\}$. Each of these subsets will define a supernode of the coarse graph; this reduced graph thus contains J supernodes. Once a rule is chosen to connect these supernodes together (the object of the next paragraph), the coarse graph is fully defined, and the method may be iterated to obtain a multiresolution of the initial graph's structure. All these methods differ mainly on the choice of the algorithm or the objective function to find this partition. For instance, one may find methods based on random walks [51], on short time diffusion distances [52], on the algebraic distance [53], etc. Other multiresolution approaches may also be found in [54–56]. One may also find many approaches from the network science community in the field of community detection [57,58], and in particular multiscale community detection [59–61]. All these methods are concerned with providing a multiresolution description of the graph structure, but do not consider any graph signal. Recently, GSP filterbanks have been proposed to define a multiscale representation of graph signals based on these approaches. In [62], we proposed such an approach where we defined a generalized Haar filterbank: instead of averaging and differentiating over pairs of nodes as in the classical Haar filterbank, we average and differentiate over the subsets $\mathcal{V}_j$ of a partition in subgraphs. In [63], the authors proposed a similar approach and defined other types of filterbanks such as the hierarchical graph Laplacian eigentransform. Another similar Haar filterbank may be found in [38]. All these methods are independent of exactly which aggregation algorithm one chooses to find the partition. Let us also cite methods that provide multiresolution approaches without necessarily defining low pass and high pass filters explictly in the graph Fourier domain [38,64,65]. Finally, let us point out that these methods may be extended to graph partitions $\mathcal{P}$ containing overlaps, as in [66].

Coarse graph reconstruction. Once one decided how to decimate nodes, or how to partition them in supernodes, how should one connect these nodes together in order to form a consistent reduced graph? In order to satisfy constraints such as interlacement (coarsely speaking, that the spectrum of the reduced graph is representative of the spectrum of the initial graph) and sparsity, Shuman et al. [46] propose a Kron reduction followed by a sparsification step. The Laplacian of the reduced graph is thus defined as the Schur complement of the initial graph's Laplacian relative to the unsampled nodes. The sparsification step is performed via a sparsifier based on effective resistances by Spielman and Srivastava [67] that approximately preserve the spectrum. In [50], the authors propose another approach to the intuitive idea that the initial and coarse graph should have similar spectral properties: they looked for the coarse Laplacian matrix that satisfies an intertwining relation. In [48], the authors connect nodes according to the set of nested bipartite graphs obtained by their maximum spanning tree algorithm. In aggregation methods [62,63], there is an inherent natural way of connecting supernodes: the weight of the link between supernodes i and j is equal to the sum of the weights of the links connecting nodes in $\mathcal{V}_i$ to nodes in $\mathcal{V}_j$.

Illustrations. We show in Fig. 11.6 an example of multiresolution analysis of a given graph signal on the Minnesota traffic graph, using a method of successive graph aggregation to compute the details and approximations at different scales. The specific method for this illustration is the one detailed in [62].

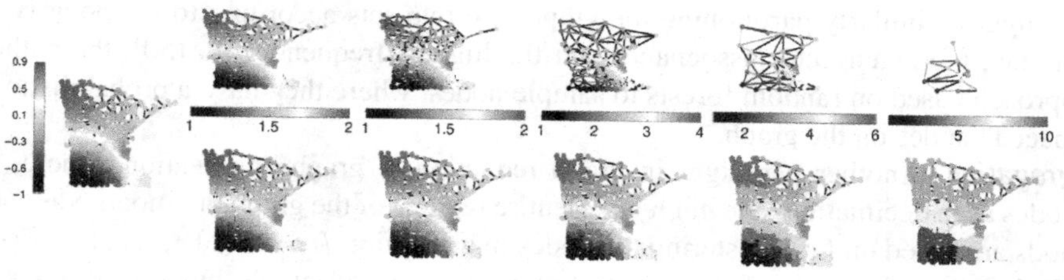

FIG. 11.6

An example of multiresolution analysis of a graph signal (from [62]). Left: original smooth graph signal (sum of the five lowest Fourier modes normalized by its maximum absolute value) defined on the Minnesota traffic graph. The *vertical scale bar* of this figure is valid for all graph signals represented on this figure. Top row: successive approximations of the graph signal. The *horizontal scale bar* on the bottom of each figure corresponds to the weights of the links of the corresponding coarsened graph. Lower row: for each of the successive approximations, we represent the upsampled reconstructed graph signal obtained from the corresponding approximations.

11.4 CONCLUSION

The purpose of this chapter was to introduce the reader to a basic understanding of a GFT. We stressed how it can be generally introduced for undirected or directed graphs by choosing a reference operator whose spectral domain will define the frequency domain for graph signals. Then, we led the reader to the more elaborated designs of graph filters and multiscale transforms on graphs. The first section is voluntarily introductory and almost self-contained. Indeed, we endeavored to delineate a general guideline that shows, in an original manner, that there is no major discrepancy between choosing a Laplacian, an Adjacency, or a random walk operator as long as one chooses accordingly the appropriate notion of frequency to analyze graph signals. Then, after a proper definition of graph filters, our objective was to review the literature and to propose guidelines and pointers to the relevant results on graph filters and related multiscale transforms. This last part being written as a review, we beg for the reader's indulgence as many details are skipped and some works are only reported here in a sketchy manner.

ACKNOWLEDGMENTS

Work supported by the ANR grant Graphsip ANR-14-CE27-0001-02 and ANR-14-CE27-0001-03, the LabEx PERSYVAL-Lab (ANR-11-LABX-0025-01) funded by the French program Investissement d'avenir, and the LabEx MILyon (ANR-10-LABX-0070).

REFERENCES

[1] Sandryhaila A, Moura JMF. Discrete signal processing on graphs. IEEE Trans Signal Process 2013;61(7): 1644–56.

[2] Shuman DI, Narang SK, Frossard P, Ortega A, Vandergheynst P. The emerging field of signal processing on graphs: extending high-dimensional data analysis to networks and other irregular domains. IEEE Signal Process Mag 2013;30(3):83–98. https://doi.org/10.1109/MSP.2012.2235192.

[3] Hammond DK, Vandergheynst P, Gribonval R. Wavelets on graphs via spectral graph theory. Appl Comput Harmon Anal 2011;30(2):129–50.

[4] Narang SK, Ortega A. Perfect reconstruction two-channel wavelet filter banks for graph structured data. IEEE Trans Signal Process 2012;60(6):2786–99. https://doi.org/10.1109/TSP.2012.2188718.

[5] Chung FRK. Spectral graph theory, vol. 92. American Mathematical Society; 1997.

[6] Grady L, Polimeni JR. Discrete calculus: applied analysis on graphs for computational science. Springer; 2010.

[7] Barrat A, Barthlemy M, Vespignani A. Dynamical processes on complex networks. Cambridge University Press; 2008.

[8] Kar S, Moura JMF. Consensus + innovations distributed inference over networks: cooperation and sensing in networked systems. IEEE Signal Process Mag 2013;30(3):99–109. https://doi.org/10.1109/MSP.2012.2235193.

[9] Rabiei H, Richard F, Coulon O, Lefevre J. Local spectral analysis of the cerebral cortex: new gyrification indices. IEEE Trans Med Imaging 2017;36(3):838–48. https://doi.org/10.1109/TMI.2016.2633393.

[10] Singh R, Chakraborty A, Manoj BS. Graph Fourier transform based on directed Laplacian. In: 2016 international conference on signal processing and communications (SPCOM); 2016. p. 1–5. https://doi.org/10.1109/SPCOM.2016.7746675.

[11] Sevi H, Rilling G, Borgnat P. Multiresolution analysis of functions on directed networks. In: Wavelets and sparsity XVII; 2017.

[12] Chung F. Laplacians and the Cheeger inequality for directed graphs. Ann Comb 2005;9(1):1–19. http://link.springer.com/article/10.1007/s00026-005-0237-z.

[13] Yu G, Qu H. Hermitian Laplacian matrix and positive of mixed graphs. Appl Math Comput 2015;269(Supplement C):70–6. https://doi.org/10.1016/j.amc.2015.07.045.

[14] Sardellitti S, Barbarossa S, Lorenzo PD. On the graph Fourier transform for directed graphs. IEEE J Sel Top Signal Process 2017;11(6):796–811. https://doi.org/10.1109/JSTSP.2017.2726979.

[15] Shafipour R, Khodabakhsh A, Mateos G, bibinfoauthorNikolova E. A digraph Fourier transform with spread frequency components. In: Proceedings of IEEE global conference on signal and information processing; 2017.

[16] Sandryhaila A, Moura JMF. Discrete signal processing on graphs: frequency analysis. IEEE Trans Signal Process 2014;62(12):3042–54. https://doi.org/10.1109/TSP.2014.2321121.

[17] Sevi H, Rilling G, Borgnat P. Analyse fréquentielle et filtrage sur graphes dirigés. In: 26e Colloque sur le Traitement du Signal et des Images. GRETSI-2017; 2017. p. id. 220.

[18] Ben Alaya S, Gonçalves P, Borgnat P. Linear prediction on graphs based on autoregressive models. In: Graph signal processing workshop; 2016.

[19] Girault B, Goncalves P, Fleury E. Translation on graphs: an isometric shift operator. IEEE Signal Process Lett 2015;22(12).

[20] Anis A, Gadde A, Ortega A. Efficient sampling set selection for bandlimited graph signals using graph spectral proxies. IEEE Trans Signal Process 2016;64(14):3775–89. https://doi.org/10.1109/TSP.2016.2546233.

[21] LeMagoarou L, Gribonval R, Tremblay N. Approximate fast graph Fourier transforms via multi-layer sparse approximations. IEEE Trans Signal Inf Process Netw 2017. https://doi.org/10.1109/TSIPN.2017.2710619.

[22] LeMagoarou L, Tremblay N, Gribonval R. Analyzing the approximation error of the fast graph Fourier transform. In: Proceedings of the Asilomar conference on signals, systems, and computers; 2017.

[23] Givens W. Computation of plane unitary rotations transforming a general matrix to triangular form. J Soc Ind Appl Math 1958;6(1):26–50. http://www.jstor.org/stable/2098861.

[24] Zachary W. An information flow model for conflict and fission in small groups. J Anthropol Res 1977;33(4):452–73.

[25] Isufi E, Loukas A, Simonetto A, Leus G. Autoregressive moving average graph filtering. IEEE Trans Signal Process 2017;65(2):274–88. https://doi.org/10.1109/TSP.2016.2614793.

[26] Marques AG, Segarra S, Leus G, Ribeiro A. Stationary graph processes and spectral estimation. IEEE Trans Signal Process 2017;65(22):5911–26. https://doi.org/10.1109/TSP.2017.2739099.

[27] Loukas A, Simonetto A, Leus G. Distributed autoregressive moving average graph filters. IEEE Signal Process Lett 2015;22(11):1931–5. https://doi.org/10.1109/LSP.2015.2448655.

[28] Shi X, Feng H, Zhai M, Yang T, Hu B. Infinite impulse response graph filters in wireless sensor networks. IEEE Signal Process Lett 2015;22(8):1113–7.

[29] Liu J, Isufi E, Leus G. Autoregressive moving average graph filter design. In: 6th joint WIC/IEEE symposium on information theory and signal processing in the Benelux; 2016.

[30] Perraudin N, Vandergheynst P. Stationary signal processing on graphs. IEEE Trans Signal Process 2017;65(13):3462–77. https://doi.org/10.1109/TSP.2017.2690388.

[31] Shuman D, Vandergheynst P, Frossard P. Chebyshev polynomial approximation for distributed signal processing. In: 2011 International conference on distributed computing in sensor systems and workshops (DCOSS); 2011. p. 1–8. https://doi.org/10.1109/DCOSS.2011.5982158.

[32] Tremblay N, Puy G, Gribonval R, Vandergheynst P. Compressive spectral clustering. In: Proceedings of the 33rd international conference on machine learning (ICML), vol. 48. JMLR: W&CP; 2016. p. 1002–11. http://jmlr.csail.mit.edu/proceedings/papers/v48/tremblay16.pdf.

[33] Susnjara A, Perraudin N, Kressner D, Vandergheynst P. Accelerated filtering on graphs using Lanczos method; 2015. arXiv preprint arXiv:150904537. https://arxiv.org/abs/1509.04537.

[34] Gallopoulos E, Saad Y. Efficient solution of parabolic equations by Krylov approximation methods. SIAM J Sci Stat Comput 1992;13(5):1236–64. https://doi.org/10.1137/0913071.

[35] Saad Y. Numerical methods for large eigenvalue problems. 2nd ed. Classics in Applied Mathematics 66. SIAM; 2011. ISBN 1-61197-072-5 978-1-61197-072-2.

[36] Crovella M, Kolaczyk E. Graph wavelets for spatial traffic analysis. In: INFOCOM 2003. Twenty-second annual joint conference of the IEEE computer and communications. IEEE societies, vol. 3; 2003. p. 1848–57.

[37] Coifman RR, Maggioni M. Diffusion wavelets. Appl Comput Harmon Anal 2006;21(1):53–94.

[38] Gavish M, Nadler B, Coifman RR. Multiscale wavelets on trees, graphs and high dimensional data: theory and applications to semi supervised learning. In: Proceedings of the 27th international conference on machine learning (ICML-10); 2010. p. 367–74.

[39] Strang G, Nguyen T. Wavelets and filter banks. SIAM; 1996.

[40] Maggioni M, Mhaskar HN. Diffusion polynomial frames on metric measure spaces. Appl Comput Harmon Anal 2008;24(3):329–53. https://doi.org/10.1016/j.acha.2007.07.001.

[41] Leonardi N, Van De Ville D. Tight wavelet frames on multislice graphs. IEEE Trans Signal Process 2013;61(13):3357–67. https://doi.org/10.1109/TSP.2013.2259825.

[42] Shuman DI, Wiesmeyr C, Holighaus N, Vandergheynst P. Spectrum-adapted tight graph wavelet and vertex-frequency frames. IEEE Trans Signal Process 2015;63(16):4223–35. https://doi.org/10.1109/TSP.2015.2424203.

[43] Teke O, Vaidyanathan PP. Extending classical multirate signal processing theory to graphs; Part I: Fundamentals. IEEE Trans Signal Process 2017;65(2):409–22. https://doi.org/10.1109/TSP.2016.2617833.

[44] Ekambaram VN, Fanti GC, Ayazifar B, Ramchandran K. Circulant structures and graph signal processing. In: 2013 IEEE international conference on image processing; 2013. p. 834–8. https://doi.org/10.1109/ICIP.2013.6738172.

[45] Kotzagiannidis MS, Dragotti PL. Splines and wavelets on circulant graphs. CoRR 2016;abs/1603.04917. http://arxiv.org/abs/1603.04917.

[46] Shuman DI, Faraji MJ, Vandergheynst P. A multiscale pyramid transform for graph signals. IEEE Trans Signal Process 2016;64(8):2119–34. https://doi.org/10.1109/TSP.2015.2512529.

[47] Sakiyama A, Tanaka Y. Oversampled graph Laplacian matrix for graph filter banks. IEEE Trans Signal Process 2014;62(24):6425–37. https://doi.org/10.1109/TSP.2014.2365761.

[48] Nguyen HQ, Do MN. Downsampling of signals on graphs via maximum spanning trees. IEEE Trans Signal Process 2015;63(1):182–91. https://doi.org/10.1109/TSP.2014.2369013.

[49] Narang SK, Ortega A. Local two-channel critically sampled filter-banks on graphs. In: 2010 17th IEEE international conference on image processing (ICIP); 2010. p. 333–6.

[50] Avena L, Castell F, Gaudillière A, Mélot C. Intertwining wavelets or multiresolution analysis on graphs through random forests; 2017.

[51] Lafon S, Lee AB. Diffusion maps and coarse-graining: a unified framework for dimensionality reduction, graph partitioning, and data set parameterization. IEEE Trans Pattern Anal Mach Intell 2006;28(9):1393–403. https://doi.org/10.1109/TPAMI.2006.184.

[52] Livne OE, Brandt A. Lean algebraic multigrid (LAMG): fast graph Laplacian linear solver. SIAM J Sci Comput 2012;34(4):B499–522. https://doi.org/10.1137/110843563.

[53] Ron D, Safro I, Brandt A. Relaxation-based coarsening and multiscale graph organization. Multiscale Model Simul 2011;9(1):407–23. https://doi.org/10.1137/100791142.

[54] Dhillon I, Guan Y, Kulis B. Weighted graph cuts without eigenvectors a multilevel approach. IEEE Trans Pattern Anal Mach Intell 2007;29(11):1944–57. https://doi.org/10.1109/TPAMI.2007.1115.

[55] Karypis G, Kumar V. A fast and high quality multilevel scheme for partitioning irregular graphs. SIAM J Sci Comput 1998;20(1):359–92. https://doi.org/10.1137/S1064827595287997.

[56] Murtagh F, Contreras P. Algorithms for hierarchical clustering: an overview. Data Min Knowl Disc 2012;2(1):86–97. https://doi.org/10.1002/widm.53.
[57] Newman ME. Modularity and community structure in networks. Proc Natl Acad Sci U S A 2006;103(23): 8577–82.
[58] Fortunato S. Community detection in graphs. Phys Rep 2010;486(3–5):75–174.
[59] Reichardt JA, Bornholdt S. Statistical mechanics of community detection. Phys Rev E 2006;74(1):016110. https://doi.org/10.1103/PhysRevE.74.016110.
[60] Schaub MT, Delvenne JC, Yaliraki SN, Barahona M. Markov dynamics as a zooming lens for multiscale community detection: non clique-like communities and the field-of-view limit. PLOS One 2012;7(2):e32210.
[61] Tremblay N, Borgnat P. Graph wavelets for multiscale community mining. IEEE Trans Signal Process 2014;62(20):5227–39. https://doi.org/10.1109/TSP.2014.2345355.
[62] Tremblay N, Borgnat P. Subgraph-based filterbanks for graph signals. IEEE Trans Signal Process 2016;64(15):3827–40. https://doi.org/10.1109/TSP.2016.2544747.
[63] Irion J, Saito N. Applied and computational harmonic analysis on graphs and networks. In: Wavelets and sparsity XVI, vol. 9597; 2015. p. 95971F. https://doi.org/10.1117/12.2186921.
[64] Lee AB, Nadler B, Wasserman L. Treelets: an adaptive multi-scale basis for sparse unordered data. Ann Appl Stat 2008;2(2):435–71. http://www.jstor.org/stable/30244207.
[65] Mishne G, Talmon R, Cohen I, Coifman RR, Kluger Y. Data-driven tree transforms and metrics. IEEE Trans Signal Inf Process Netw 2017. https://doi.org/10.1109/TSIPN.2017.2743561.
[66] Szlam AD, Maggioni M, Coifman RR, Bremer Jr JC. Diffusion-driven multiscale analysis on manifolds and graphs: top-down and bottom-up constructions. In: Proc SPIE, vol. 5914; 2005. p. 59141D.
[67] Spielman DA, Srivastava N. Graph sparsification by effective resistances. SIAM J Comput 2011;40(6): 1913–26. https://doi.org/10.1137/080734029.

CHAPTER

STATISTICAL GRAPH SIGNAL PROCESSING: STATIONARITY AND SPECTRAL ESTIMATION

12

Santiago Segarra*, Sundeep Prabhakar Chepuri†, Antonio G. Marques‡, Geert Leus†
Massachusetts Institute of Technology, Cambridge, MA, United States Delft University of Technology, Delft, The Netherlands† Department of Signal Theory and Communications, King Juan Carlos University, Madrid, Spain‡*

12.1 RANDOM GRAPH PROCESSES

12.1.1 INTRODUCTION

Most of the tools in graph signal processing are deterministic in nature, e.g., graph signal denoising using diffusion [1], sampling and reconstruction of graph signals [2–7], graph filter design [8–11], and so on. Only recently, statistical signal processing methods tailored to graph signals have been introduced. As we know from classical signal processing focusing on spatiotemporal signals, statistical methods allow one to take statistical information into account when designing optimal sampling and reconstruction schemes, e.g., Wiener filtering for denoising, interpolation, prediction, and so on [12]. This generally leads to a better average performance compared to deterministic methods. Key to the majority of statistical methods is the concept of weak stationarity, which means that the first- and second-order statistics of the random process do not change over space and/or time. The extension of this concept to graph signals is not trivial due to the fact that these signals have an irregular structure, which is generally characterized by a so-called graph shift (a generalization of the shift in time and/or space). This is what will be discussed in the current chapter.

The first works discussing stationary graph processes observe that in contrast to a shift in time and/or space, a graph shift is not energy preserving [13,14]. Hence, these papers base their definition of a weakly stationary graph process on a new isometric graph shift. However, this new shift cannot be carried out by means of local operations and hence the connection between stationarity and locality is lost. Therefore, in this chapter, we present definitions based on the original graph shift, allowing for stationarity tests and estimation schemes based on local information. Stationary graph processes are also characterized by a power spectral density (PSD) and this chapter provides a rigorous treatment of various PSD estimators, including nonparametric and parametric methods. Our treatment of stationary graph processes is based on the comprehensive study presented in [15]. Graph stationarity was also studied in [16], where the analysis is carried out using the Laplacian matrix as the graph shift operator. In this chapter, the proposed framework is also extended to random processes that are jointly stationary

Cooperative and Graph Signal Processing. https://doi.org/10.1016/B978-0-12-813677-5.00012-2

in the time and vertex domain [17]. This paves the way for statistical tools for random processes over two domains: the regular time domain and the irregular graph domain.

The field of compressive sensing has recently been extended to compressive covariance sensing [18], which is based on the idea that the covariance matrix or PSD of a spatiotemporal process can be estimated from compressed measurements without any prior assumptions on sparsity or smoothness. A special case of compressive covariance sensing occurs when the compression is realized by subsampling (below the Nyquist rate), also known as sparse covariance sensing. This allows one to design statistical signal processing tools from only a subset of measurements. The last part of the chapter explains how these ideas can be extended to random graph processes, where the covariance matrix does not have any apparent structure, as for spatiotemporal processes [19]. We demonstrate how the covariance matrix—and thus the PSD—of a graph process can be estimated from a subset of the nodes without the use of priors. Again, nonparametric as well as parametric methods are considered and we additionally show how to select the nodes in a greedy fashion.

12.1.2 CHAPTER ORGANIZATION

The definition of weakly stationary graph processes is presented in Section 12.2 along with discussions about the relation with the classical definition in time. Section 12.2.1 introduces the notion of power spectral density (PSD) followed by a recollection of relevant examples and useful properties. The characterization of stationarity for graph processes that also vary over time is presented in Section 12.2.2. Because stationary processes are easier to understand in the frequency domain, Section 12.3 is devoted to the study of different methods for *spectral estimation*, which can also be used to improve the estimate of the covariance matrix itself. These include both nonparametric and parametric approaches. Finally, in Section 12.4 we discuss methods to estimate the PSD and the covariance of random graph processes using only observations from a subset of nodes. We also develop a low-complexity and near-optimal method to select the nodes in a greedy manner.

12.1.3 NOTATION

Let $\mathcal{G} = (\mathcal{N}, \mathcal{E})$ be a directed graph or network with a set of N nodes $\mathcal{N}$ and directed edges $\mathcal{E}$ such that $(i,j) \in \mathcal{E}$ if there exists an edge from node i to node j. We associate with $\mathcal{G}$ the graph shift operator (GSO) $\mathbf{S}$, defined as an $N \times N$ matrix whose entry $S_{j,i} \neq 0$ only if $i = j$ or if $(i,j) \in \mathcal{E}$ [9,20]. The sparsity pattern of $\mathbf{S}$ captures the local structure of $\mathcal{G}$, but we make no specific assumptions on the values of the nonzero entries of $\mathbf{S}$; hence the GSO can represent the adjacency matrix, the Laplacian, or other graph-related matrices. In this chapter we assume that $\mathbf{S}$ is *normal* to guarantee the existence of a unitary matrix $\mathbf{V} = [\mathbf{v}_1, \mathbf{v}_2, \ldots, \mathbf{v}_N]$ and a diagonal matrix $\mathbf{\Lambda}$ such that $\mathbf{S} = \mathbf{V}\mathbf{\Lambda}\mathbf{V}^H$. We use $\mathbf{x} = [x_1, \ldots, x_N]^T \in \mathbb{R}^N$ to denote a generic graph signal and $\tilde{\mathbf{x}} := \mathbf{V}^H\mathbf{x}$ to denote its frequency representation, with $\mathbf{V}^H$ being the graph Fourier transform (GFT) [9]. Finally, we use $\mathbf{H} : \mathbb{R}^N \to \mathbb{R}^N$ to denote a linear shift-invariant graph filter of the form

$$\mathbf{H} = \sum_{l=0}^{L-1} h_l \mathbf{S}^l = \mathbf{V}\text{diag}(\tilde{\mathbf{h}})\mathbf{V}^H = \mathbf{V}\text{diag}\big(\mathbf{\Psi}_L \mathbf{h}\big)\mathbf{V}^H, \tag{12.1}$$

where $\tilde{\mathbf{h}}$ denotes the frequency response of the filter $\mathbf{H}$, $\boldsymbol{\Psi}_L$ is an $N \times L$ Vandermonde matrix with entries $\Psi_{k,l} = \lambda_k^{l-1}$, and $\mathbf{h}$ is a vector collecting the polynomial coefficients. The notation $\circ$, $\otimes$, and $\odot$ denote the elementwise, Kronecker, and Khatri-Rao matrix products, respectively. The notation $\oplus$ stands for the Kronecker sum.

12.2 WEAKLY STATIONARY GRAPH PROCESSES

We extend three equivalent definitions of weak stationarity in time to the graph domain, the most common being the invariance of the first and second moments to time shifts. We will see that under certain conditions those definitions can be rendered equivalent for the graph domain as well. Intuitively, stating that a *graph* process is stationary is an inherently incomplete assertion because we need to declare which graph we are referring to. Hence, the proposed definitions depend on the GSO $\mathbf{S}$, so that a process $\mathbf{x}$ can be stationary in $\mathbf{S}$ but not in $\mathbf{S}' \neq \mathbf{S}$.

Defining a standard zero-mean white random process $\mathbf{n}$ as one with mean $\mathbb{E}[\mathbf{n}] = \mathbf{0}$ and covariance $\mathbb{E}\left[\mathbf{n}\mathbf{n}^H\right] = \mathbf{I}$, we state our first definition of graph stationarity.

Definition 12.1. Given a normal shift operator $\mathbf{S}$, a zero-mean random process $\mathbf{x}$ is weakly stationary with respect to $\mathbf{S}$ if it can be written as the response of a linear shift-invariant graph filter $\mathbf{H} = \sum_{l=0}^{N-1} h_l \mathbf{S}^l$ to a zero-mean white input $\mathbf{n}$. ■

The definition states that stationary graph processes can be written as the output of graph filters when excited with a white input. This generalizes the well-known fact that stationary processes in time can be expressed as the output of linear time-invariant systems with white noise as input. If we write $\mathbf{x} = \mathbf{H}\mathbf{n}$, the covariance matrix $\mathbf{C}_\mathbf{x} := \mathbb{E}\left[\mathbf{x}\mathbf{x}^H\right]$ of the process $\mathbf{x}$ is given by

$$\mathbf{C}_\mathbf{x} = \mathbb{E}\left[(\mathbf{H}\mathbf{n})(\mathbf{H}\mathbf{n})^H\right] = \mathbf{H}\mathbb{E}\left[\mathbf{n}\mathbf{n}^H\right]\mathbf{H} = \mathbf{H}\mathbf{H}^H, \tag{12.2}$$

which shows that the color of $\mathbf{x}$ is determined by the filter $\mathbf{H}$. We can think of Definition 12.1 as a constructive definition of stationarity because it describes how a stationary process can be generated. Alternatively, one can define stationarity from a descriptive perspective by imposing requirements on the moments of the random graph process in either the vertex or the frequency domain.

Definition 12.2. Given a normal shift operator $\mathbf{S}$, a zero-mean random process $\mathbf{x}$ is weakly stationary with respect to $\mathbf{S}$ if the following two equivalent properties hold

(a) For any set of nonnegative integers a, b, and $c \leq b$ it holds that

$$\mathbb{E}\left[(\mathbf{S}^a\mathbf{x})\left(\left(\mathbf{S}^H\right)^b\mathbf{x}\right)^H\right] = \mathbb{E}\left[(\mathbf{S}^{a+c}\mathbf{x})\left(\left(\mathbf{S}^H\right)^{b-c}\mathbf{x}\right)^H\right]. \tag{12.3}$$

(b) Matrices $\mathbf{C}_\mathbf{x}$ and $\mathbf{S}$ are simultaneously diagonalizable. ■

The statements in Definition 12.2a and b can indeed be shown to be equivalent [15]. These two statements generalize known definitions of stationarity in time. Definition 12.2a generalizes the requirement that the second moment of a stationary process must be invariant to time shifts whereas Definition 12.2b extends the requisite that the covariance of time stationary processes must be circulant.

In Definition 12.2a we require the correlation to be invariant to *how* we shift our signal—namely, forward $\mathbf{Sx}$ or backward $\mathbf{S}^H\mathbf{x}$—as long as the total number of shifts remains constant. Indeed, in both the left and right hand sides of Eq. (12.3) the signal is shifted a total of $a+b$ times. This generalizes what happens to stationary signals in time, where correlation depends on the total number of shifts but not on the particular time instants. More specifically, when $\mathbf{S}$ is a directed cycle we have that $\mathbf{S}^H = \mathbf{S}^{-1}$. Also, notice that for the directed cycle $\mathbf{S}^N = \mathbf{I}$. Then, if we set $a = 0$, $b = N$ and $c = l$, Eq. (12.3) boils down to $\mathbb{E}\left[\mathbf{xx}^H\right] = \mathbb{E}\left[\mathbf{S}^l\mathbf{x}(\mathbf{S}^l\mathbf{x})^H\right]$, which is the definition of a stationary signal in time. Intuitively, accumulating the same number of shifts in both sides of Eq. (12.3) is necessary because the operator $\mathbf{S}$, in general, does not preserve the energy. Thus, requiring $\mathbb{E}\left[\mathbf{xx}^H\right] = \mathbb{E}\left[\mathbf{S}^l\mathbf{x}(\mathbf{S}^l\mathbf{x})^H\right]$ for stationarity with respect to a general GSO would be infeasible. Definition 12.2a strikes the right balance of being valid for general normal GSOs while particularizing to the accepted classical definition when $\mathbf{S}$ represents the domain of time signals.

Definition 12.2b characterizes stationarity from a graph frequency perspective by requiring the covariance $\mathbf{C_x}$ to be diagonalized by the GFT matrix $\mathbf{V}$. When particularized to time, Definition 12.2b requires $\mathbf{C_x}$ to be diagonalized by the Fourier matrix and, therefore, must be circulant. This fact is exploited in classical signal processing to define the PSD of a stationary process as the eigenvalues of the circulant covariance matrix, motivating the PSD definition in Section 12.2.1.

Thus far, we have presented three extensions of the concept of stationarity into the realm of graph processes, two of which are equivalent and, hence, grouped in Definition 12.2. At this point, the attentive reader might have a natural inquiry. Are Definitions 12.1 and 12.2 equivalent for general graphs, as they are for stationarity in time? In fact, it can be shown that Definitions 12.1 and 12.2 are equivalent for any graph $\mathbf{S}$ that is normal and whose eigenvalues are all distinct [15]. Fig. 12.1 presents a concise summary of the definitions discussed in this section.

Coexisting approaches. Stationary graph processes were first defined and analyzed in [13]. The fundamental problem identified there is that GSOs do not preserve energy in general and therefore cannot be isometric [21]. This problem is addressed in [14] with the definition of an isometric graph shift that preserves the eigenvector space of the Laplacian GSO but modifies its eigenvalues.

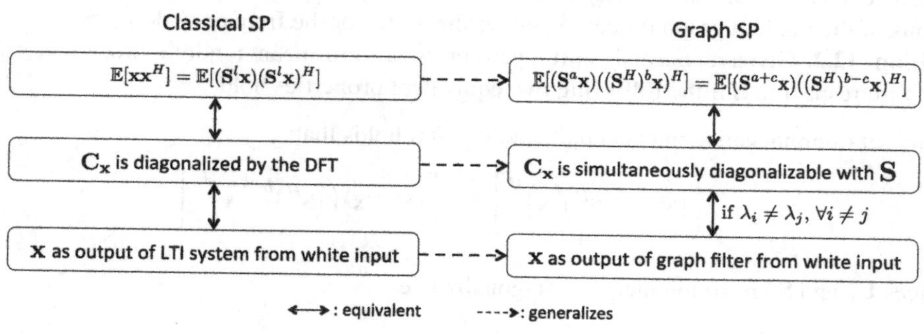

FIG. 12.1

Equivalent definitions of a weakly stationary graph process. Three equivalent definitions for weak stationarity in time and their corresponding extensions to the graph domain. In graphs, two of the definitions are always equivalent and the third one is equivalent for shifts with distinct eigenvalues.

A stationary graph process is then defined as one whose probability distributions are invariant with respect to multiplications with the isometric shift. One drawback of this approach is that the isometric shift is a complex-valued operator and has a sparsity structure (if any) different from **S**. By contrast, the vertex-based definition in Eq. (12.3) is based on the original GSO **S**, which is local and real-valued. As a result, Eq. (12.3) provides intuition on the relations between stationarity and locality, which can be leveraged to develop stationarity tests or estimation schemes that work with local information. Graph stationarity was also studied in [16] where the requirement of having a covariance matrix diagonalizable by the eigenvectors of the Laplacian GSO is adopted as a definition. This condition is shown to be equivalent to statistical invariance with respect to the translation operator introduced in [22]. When the shift **S** coincides with the Laplacian of the graph and the eigenvalues of **S** are all distinct, Definitions 12.1 and 12.2 are equivalent to those in [13,16]. Hence, the definitions presented here differ from [16] in that we consider general normal shifts instead of Laplacians and that we see Definition 12.1 as a definition, not a property. These are mathematically minor differences that are important in practice though; see [15,23] for more details.

12.2.1 POWER SPECTRAL DENSITY

Stationarity reduces the degrees of freedom of a random graph process, thus facilitating its description and understanding. It follows from Definition 12.2b that one can express the remaining degrees of freedom in the frequency domain via the notion of PSD, as defined next.

Definition 12.3. The PSD of a random process $\mathbf{x}$ that is stationary with respect to $\mathbf{S} = \mathbf{V}\mathbf{\Lambda}\mathbf{V}^H$ is the nonnegative $N \times 1$ vector $\mathbf{p}$

$$\mathbf{p} := \text{diag}\left(\mathbf{V}^H\mathbf{C_x}\mathbf{V}\right). \tag{12.4}$$

■

Observe that because $\mathbf{C_x}$ is diagonalized by $\mathbf{V}$ (see Definition 12.2b) the matrix $\mathbf{V}^H\mathbf{C_x}\mathbf{V}$ is diagonal and it follows that the PSD in Eq. (12.4) corresponds to the eigenvalues of the positive semidefinite covariance matrix $\mathbf{C_x}$. Thus, Eq. (12.4) is equivalent to

$$\mathbf{C_x} = \mathbf{V}\text{diag}(\mathbf{p})\mathbf{V}^H. \tag{12.5}$$

Zero-mean white noise is an example of a random process that is stationary with respect to *any* graph shift $\mathbf{S}$. The PSD of white noise with covariance $\mathbb{E}[\mathbf{n}\mathbf{n}^H] = \sigma^2\mathbf{I}$ is $\mathbf{p} = \sigma^2\mathbf{1}$. Also notice that, by definition, any random process $\mathbf{x}$ is stationary with respect to the shift $\mathbf{S} = \mathbf{C_x}$ defined by its covariance matrix, with corresponding PSD $\mathbf{p} = \text{diag}(\mathbf{\Lambda})$. This can be exploited in the context of network topology inference. Given a set of graph signals $\{\mathbf{x}_r\}_{r=1}^R$ it is common to infer the underlying topology by building a graph $\mathcal{G}_{corr}$ whose edge weights correspond to cross-correlations among the entries of the signals. In that case, the process generating those signals is stationary in the shift given by the adjacency of $\mathcal{G}_{corr}$; see [23] for details. A random process $\mathbf{x}$ is also stationary with respect to the shift given by its precision matrix, which is defined as the (pseudo-)inverse $\mathbf{\Theta} = \mathbf{C_x}^\dagger$. The PSD, in this case, is $\mathbf{p} = \text{diag}(\mathbf{\Lambda})^\dagger$. This is particularly important when $\mathbf{x}$ is a Gaussian Markov Random Field (GMRF) whose Markovian dependence is captured by the unweighted graph $\mathcal{G}_{MF}$. It is well known [24, Ch. 19] that in these cases $\Theta_{i,j}$ can be nonzero only if (i,j) is either a link of $\mathcal{G}_{MF}$, or an element in the diagonal. Thus, *any* GMRF is stationary with respect to the *sparse* shift $\mathbf{S} = \boldsymbol{\Theta}$, which captures the conditional dependence between the elements of $\mathbf{x}$.

Two important properties that hold for random processes in time can be shown to be true as well for the PSD of graph processes.

Property 12.1. Let $\mathbf{x}$ be stationary in $\mathbf{S}$ with covariance $\mathbf{C_x}$ and PSD $\mathbf{p_x}$. Consider a filter $\mathbf{H}$ with frequency response $\tilde{\mathbf{h}}$ and define $\mathbf{y} := \mathbf{Hx}$. Then, the process $\mathbf{y}$:

(a) Is stationary in $\mathbf{S}$ with covariance $\mathbf{C_y} = \mathbf{HC_xH}^H$.
(b) Has a PSD given by $\mathbf{p_y} = |\tilde{\mathbf{h}}|^2 \circ \mathbf{p_x}$, where $|\cdot|^2$ is applied elementwise. ■

Property 12.2. Given a process $\mathbf{x}$ stationary in $\mathbf{S} = \mathbf{V\Lambda V}^H$ with PSD $\mathbf{p}$, define the GFT process as $\tilde{\mathbf{x}} = \mathbf{V}^H\mathbf{x}$. Then, it holds that $\tilde{\mathbf{x}}$ is uncorrelated and its covariance matrix is

$$\mathbf{C}_{\tilde{\mathbf{x}}} := \mathbb{E}\left[\tilde{\mathbf{x}}\tilde{\mathbf{x}}^H\right] = \mathbb{E}\left[(\mathbf{V}^H\mathbf{x})(\mathbf{V}^H\mathbf{x})^H\right] = \text{diag}(\mathbf{p}). \tag{12.6}$$

■

Property 12.1 is a statement of the spectral convolution theorem for graph signals. Property 12.2 is fundamental to motivate the analysis and modeling of stationary graph processes in the frequency domain, which we undertake in the remainder of this chapter. It also shows that if a process $\mathbf{x}$ is stationary in the shift $\mathbf{S} = \mathbf{V\Lambda V}^H$, then the GFT $\mathbf{V}^H$ provides the Karhunen-Loève expansion of the process.

The concept of stationarity and, consequently, that of PSD can be extended to processes defined jointly in a graph and over time. Before we review this extension in the ensuing section, we discuss requirements on the first moment of stationary graph processes.

The mean of stationary graph processes. While Definitions 12.1 and 12.2 assume that the random process $\mathbf{x}$ has mean $\bar{\mathbf{x}} := \mathbb{E}[\mathbf{x}] = \mathbf{0}$, traditional stationary time processes are allowed to have a (nonzero) constant mean $\bar{\mathbf{x}} = \alpha\mathbf{1}$, with α being an arbitrary scalar. Stationary graph processes, by contrast, are required to have a first-order moment of the form $\bar{\mathbf{x}} = \alpha\mathbf{v}_k$, i.e., a scaled version of an eigenvector of $\mathbf{S}$. This choice: (i) takes into account the structure of the underlying graph; (ii) maintains the validity of Property 12.1; and (iii) encompasses the case $\mathbf{v}_k = \mathbf{1}$ when $\mathbf{S}$ is either the adjacency matrix of a directed cycle or the Laplacian of any graph, recovering the classical first-order requirement for weak stationarity.

12.2.2 JOINT TIME AND GRAPH STATIONARITY

In many real-world network applications, observations are taken periodically, giving rise to a *sequence* $\mathbf{X} = [\mathbf{x}_1, \mathbf{x}_2, \ldots, \mathbf{x}_T] \in \mathbb{R}^{N\times T}$ of graph signals. Each signal has size N the number of nodes in the network and there are T of those signals. Up to this point, we have been focusing on the statistical variation across the vertices of the network. That is, we took one particular column of $\mathbf{X}$ and analyzed the statistical relations between the signal values at different vertices. The purpose of this section is to carry out this analysis jointly across rows and columns of $\mathbf{X}$. The ultimate goal is to present the conditions under which a random process is considered to be *jointly* stationary in both the vertex and the time domain [17].

The first step to analyze the statistical properties of the vertex-time process $\mathbf{X}$, which whenever convenient will be represented as $\mathbf{x} = \text{vec}(\mathbf{X})$, is to identify its graph support. As shown in Fig. 12.2,

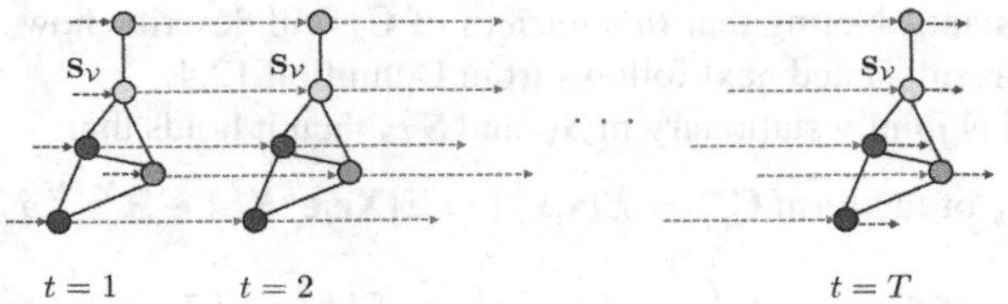

FIG. 12.2

Support of a vertex-time process. The shift $\mathbf{S}_{\mathcal{V}}$ captures the dependence across the nodes of the underlying network. *Solid lines* represent the edges in $\mathbf{S}_{\mathcal{V}}$. *Dashed lines* represent connections between the same node at two consecutive time instants.

for every time instant one can plot a graph that accounts for the graph support of the corresponding column of $\mathbf{X}$. With this representation, a horizontal path in the picture represents a particular node at different time instants. To account for the time variation, node n at time t is the origin of a link toward its successor (node n at time $t+1$) as well as the destination of a link from its predecessor (node n at time $t-1$). Suppose that the spatial graph $\mathbf{S}_{\mathcal{V}} = \mathbf{V}_{\mathcal{V}}\mathbf{\Lambda}_{\mathcal{V}}\mathbf{V}_{\mathcal{V}}^H$ is the same for all columns, as is the case in Fig. 12.2, and let us use $\mathbf{S}_{\mathcal{T}} = \mathbf{V}_{\mathcal{T}}\mathbf{\Lambda}_{\mathcal{T}}\mathbf{V}_{\mathcal{T}}^H$, the adjacency of the directed cycle, to denote the support of the time domain. Then, it holds that the graph support of $\mathbf{X}$, which will be denoted as $\mathbf{S}_{\mathcal{J}}$, is given by the Cartesian product [25] of $\mathbf{S}_{\mathcal{V}}$ and $\mathbf{S}_{\mathcal{T}}$. Mathematically, this implies that the joint shift $\mathbf{S}_{\mathcal{J}} \in \mathbb{R}^{NT\times NT}$ can be written as

$$\mathbf{S}_{\mathcal{J}} = \mathbf{S}_{\mathcal{T}} \oplus \mathbf{S}_{\mathcal{V}} = \mathbf{I}_{\mathcal{T}} \otimes \mathbf{S}_{\mathcal{V}} + \mathbf{S}_{\mathcal{T}} \otimes \mathbf{I}_{\mathcal{V}}, \tag{12.7}$$

where $\mathbf{I}_{\mathcal{T}}$ and $\mathbf{I}_{\mathcal{V}}$ are identity matrices of appropriate size. Using basic properties of the Kronecker product, it follows from Eq. (12.7) that the eigendecomposition of the joint shift is given by $\mathbf{S}_{\mathcal{J}} = (\mathbf{V}_{\mathcal{T}} \otimes \mathbf{V}_{\mathcal{V}})(\mathbf{\Lambda}_{\mathcal{T}} \oplus \mathbf{\Lambda}_{\mathcal{V}})(\mathbf{V}_{\mathcal{T}} \otimes \mathbf{V}_{\mathcal{V}})^H$, revealing that the GFT associated with $\mathbf{S}_{\mathcal{J}}$ is $\mathbf{V}_{\mathcal{T}} \otimes \mathbf{V}_{\mathcal{V}}$, the Kronecker product of the GFTs associated[1] with $\mathbf{S}_{\mathcal{T}}$ and $\mathbf{S}_{\mathcal{V}}$ [26].

Once the graph support of the joint process and its corresponding GFT have been identified, for $\mathbf{X}$ to be jointly stationary in $\mathbf{S}_{\mathcal{V}}$ and $\mathbf{S}_{\mathcal{T}}$ it suffices to particularize the definitions presented in the previous section for the shift $\mathbf{S}_{\mathcal{J}}$, giving rise to the following result.

Definition 12.4. A process $\mathbf{X}$ is *jointly stationary* in $\mathbf{S}_{\mathcal{V}}$ and $\mathbf{S}_{\mathcal{T}}$ if the covariance matrix $\mathbf{C}_{\mathbf{x}} = \mathbb{E}[\text{vec}(\mathbf{X})\text{vec}(\mathbf{X})^T]$ can be written as $\mathbf{C}_{\mathbf{x}} = (\mathbf{V}_{\mathcal{T}} \otimes \mathbf{V}_{\mathcal{V}})\text{diag}(\mathbf{p}_{\mathbf{x}})(\mathbf{V}_{\mathcal{T}} \otimes \mathbf{V}_{\mathcal{V}})^H$. ∎

Clearly, the nonnegative vector $\mathbf{p}_{\mathbf{x}}$ of length NT stands for the PSD of $\mathbf{X}$. If the eigenvalues of $\mathbf{S}_{\mathcal{J}}$ are nonrepeated, the definition is equivalent to requiring $\mathbf{C}_{\mathbf{x}}$ to be written as a (positive semidefinite) graph filter on the shift operator $\mathbf{S}_{\mathcal{J}}$.

While Definition 12.4 describes the *spectral* properties of the covariance of a jointly stationary process, it is also of interest to understand its implications in the vertex and time domains. To that end, recall that $\mathbf{e}_i$ represents the ith canonical vector, the signal $\mathbf{x}_t = \mathbf{X}\mathbf{e}_t \in \mathbb{R}^N$ collects the values of the process at time instant t, and the signal $\chi_n = \mathbf{X}^T\mathbf{e}_n \in \mathbb{R}^T$ collects the values of the process at node n

[1] Recall that the fact of $\mathbf{S}_{\mathcal{T}}$ being the directed cycle implies that the GFT $\mathbf{V}_{\mathcal{T}}^H$ is the $T \times T$ DFT matrix so that $[\mathbf{V}_{\mathcal{T}}]_{k,k'} = \frac{1}{\sqrt{T}}\exp(j\frac{2\pi}{T}kk')$.

for the different time instants. Noting that *submatrices* of $\mathbf{C_x}$ will describe how *subsets* of the elements of $\mathbf{X}$ are correlated, the result stated next follows from Definition 12.4.

Property 12.3. If $\mathbf{X}$ is jointly stationary in $\mathbf{S}_\mathcal{V}$ and $\mathbf{S}_\mathcal{T}$, then it holds that:

1. Any submatrix of $\mathbf{C_x}$ of the form $\mathbf{C}^{\mathcal{V}}_{t,t'} = \mathbb{E}[\mathbf{x}_t\mathbf{x}_{t'}^T] = \mathbb{E}[\mathbf{X}\mathbf{e}_t\mathbf{e}_{t'}^T\mathbf{X}^T] \in \mathbb{R}^{N\times N}$ is jointly diagonalizable with $\mathbf{S}_\mathcal{V}$.
2. Any submatrix of $\mathbf{C_x}$ of the form $\mathbf{C}^{\mathcal{T}}_{n,n'} = \mathbb{E}[\boldsymbol{\chi}_n\boldsymbol{\chi}_{n'}^T] = \mathbb{E}[\mathbf{X}^T\mathbf{e}_n\mathbf{e}_{n'}^T\mathbf{X}] \in \mathbb{R}^{T\times T}$ is jointly diagonalizable with $\mathbf{S}_\mathcal{T}$ and, hence, it is circulant. ■

The statement in Property 12.3.2 is equivalent to saying that $\mathbb{E}[X_{n,t}X_{n',t'}] = \mathbb{E}[X_{n,t+a}X_{n',t'+a}]$, which is the classical requirement for a *multivariate* time series to be considered stationary [27, Sec. 2.1.3]. Particularizing the results in Property 12.3 for $t = t'$ and $n = n'$ yields the subsequent property.

Property 12.4. If $\mathbf{X}$ is jointly stationary in $\mathbf{S}_\mathcal{V}$ and $\mathbf{S}_\mathcal{T}$, then it holds that:

1. All the graph signals $\mathbf{x}_t = \mathbf{X}\mathbf{e}_t$ are stationary in $\mathbf{S}_\mathcal{V}$
2. All the time-varying signals $\boldsymbol{\chi}_n = \mathbf{X}^T\mathbf{e}_n$ are stationary in $\mathbf{S}_\mathcal{T}$. ■

The result above is not an equivalence. That is, there may be processes that satisfy the two conditions stated in Property 12.4 but do not possess the structure in Definition 12.4. Even if those processes cannot be considered jointly stationary, they are likely to arise in practice, so that the design of signal processing schemes that leverage their structure is of interest.

Remark. Definition 12.4 is also valid if the joint shift $\mathbf{S}_\mathcal{J}$ is defined as either the Kronecker product or the strong product [25] between graphs $\mathbf{S}_\mathcal{V}$ and $\mathbf{S}_\mathcal{T}$. The reason is that for any of these three graph products, the eigenbasis of the joint shift is $\mathbf{V}_\mathcal{T} \otimes \mathbf{V}_\mathcal{V}$ [25,26].

Jointly stationary and separable processes

We close this section by elaborating on a *subclass* of jointly stationary processes of particular relevance. To that end, let matrix $\mathbf{H}_\mathcal{V}$ be a generic graph filter in the shift $\mathbf{S}_\mathcal{V}$ and, similarly, $\mathbf{H}_\mathcal{T}$ a generic linear time-invariant filter. Those filters are used in the following definition.

Definition 12.5. Let $\mathbf{X}$ be a process jointly stationary in $\mathbf{S}_\mathcal{V}$ and $\mathbf{S}_\mathcal{T}$. Then, the process $\mathbf{X}$ is called *separable* if it can be written as $\mathbf{X} = \mathbf{H}_\mathcal{V}\mathbf{W}\mathbf{H}_\mathcal{T}^T$, where $\mathbf{W} \in \mathbb{R}^{N\times T}$ is a zero-mean white process with $\mathbb{E}[W_{i,j}W_{i,j}] = 1$ and $\mathbb{E}[W_{i,j}W_{i',j'}] = 0$ for all $(i,j) \neq (i',j')$. ■

From the previous definition one can view the jointly stationary and separable process $\mathbf{X}$ as one generated by processing each of the columns of $\mathbf{W}$ with the same graph filter and, then, each of the resultant rows with the same linear time-invariant filter. Note that one can also apply first the time-invariant filter $\mathbf{H}_\mathcal{T}$ and then the graph filter $\mathbf{H}_\mathcal{V}$. Upon defining $\mathbf{C}_{\mathbf{x},\mathcal{V}} = \mathbf{H}_\mathcal{V}\mathbf{H}_\mathcal{V}^T$, $\mathbf{C}_{\mathbf{x},\mathcal{T}} = \mathbf{H}_\mathcal{T}\mathbf{H}_\mathcal{T}^T$, $\mathbf{p}_{\mathbf{x},\mathcal{V}} = \mathrm{diag}(\mathbf{V}_\mathcal{V}^H\mathbf{C}_{\mathbf{x},\mathcal{V}}\mathbf{V}_\mathcal{V})$ and $\mathbf{p}_{\mathbf{x},\mathcal{V}} = \mathrm{diag}(\mathbf{V}_\mathcal{V}^H\mathbf{C}_{\mathbf{x},\mathcal{T}}\mathbf{V}_\mathcal{T})$, it is easy to show that the following properties hold.

Property 12.5. Let $\mathbf{X} = \mathbf{H}_\mathcal{V}\mathbf{W}\mathbf{H}_\mathcal{T}^T$ be a jointly stationary and separable process in $\mathbf{S}_\mathcal{V}$ and $\mathbf{S}_\mathcal{T}$. Then, it holds that:

1. The correlation of $\mathbf{X}$ can be factorized as $\mathbf{C_x} = \mathbf{C}_{\mathbf{x},\mathcal{T}} \otimes \mathbf{C}_{\mathbf{x},\mathcal{V}}$.
2. The PSD of $\mathbf{x}$ can be written as $\mathbf{p_x} = \mathbf{p}_{\mathbf{x},\mathcal{T}} \otimes \mathbf{p}_{\mathbf{x},\mathcal{V}}$. ■

The factorable structure of the correlation implies that, for any given (t,t'), the covariance $\mathbf{C}^{\mathcal{V}}_{t,t'} = \mathbb{E}[\mathbf{x}_t\mathbf{x}_{t'}^T]$ is a *scaled* version of $\mathbf{C}_{\mathbf{x},\mathcal{V}}$. In other words, after a trivial scaling, the covariance of any of the columns of the separable process $\mathbf{X}$ is the same. Similarly, it holds that $\mathbf{C}^{\mathcal{V}}_{n,n'} = \mathbb{E}[\chi_n \chi^T_{n'}]$ is a scaled version of $\mathbf{C}_{\mathbf{x},\mathcal{T}}$ for all (n, n'). The fact of the PSD being factorable reveals that the number of degrees of freedom of the PSD of a jointly stationary and separable process is $N + T$, which contrasts with the NT degrees of freedom of a generic jointly stationary process. This more parsimonious description of the PSD vector—equivalently, of the covariance matrix—can be exploited when designing spectral (covariance) estimation schemes for processes obeying Definition 12.5.

12.3 POWER SPECTRAL DENSITY ESTIMATORS

We can exploit the fact that $\mathbf{x}$ is a stationary graph process in $\mathbf{S} = \mathbf{V}\text{diag}(\mathbf{\Lambda})\mathbf{V}^H$ to design efficient estimators of the covariance $\mathbf{C}_{\mathbf{x}}$. In particular, instead of estimating $\mathbf{C}_{\mathbf{x}}$ directly, which has $N(N+1)/2$ degrees of freedom, one can estimate $\mathbf{p}$ first, which only has N degrees of freedom, and then leverage that $\mathbf{C}_{\mathbf{x}} = \mathbf{V}\text{diag}(\mathbf{p})\mathbf{V}^H$.

Motivated by this, the focus of this section is on estimating $\mathbf{p}$, the PSD of a stationary random graph process $\mathbf{x}$, using as input either *one* or a few realizations $\{\mathbf{x}_r\}_{r=1}^R$ of $\mathbf{x}$. To illustrate the developments in Sections 12.3 and 12.4, we will use as a running example a random process defined on the well-known Zachary's Karate club network [28] (Figs. 12.3 and 12.4). As shown in Fig. 12.4, this graph consists of 34 nodes or members of the club and 78 undirected edges symbolizing friendships among members.[2]

12.3.1 NONPARAMETRIC PSD ESTIMATORS

Nonparametric estimators—as opposed to their parametric counterparts—do not assume any specific generating model on the process $\mathbf{x}$. This more agnostic view of $\mathbf{x}$ comes with the price of needing, in general, to observe more graph signals to achieve satisfactory performance. In this section, we extend to the graph setting the periodogram, the correlogram, and the least-squares (LS) estimator, which are classical unbiased nonparametric estimators. Moreover, for the special case where the observations are Gaussian, we derive the Cramér-Rao lower bound. We also discuss the windowed average periodogram, which attains a better performance when a few observations are available by introducing bias in a controlled manner while drastically reducing the variance.

Periodogram, correlogram, and LS estimator

From Eq. (12.6) it follows that one may express the PSD as $\mathbf{p} = \mathbb{E}\left[|\mathbf{V}^H\mathbf{x}|^2\right]$. That is, the PSD is given by the expected value of the squared frequency components of the random process. This leads to a natural approach for the estimation of $\mathbf{p}$ from a finite set of R realizations of the process $\mathbf{x}$. Indeed, we compute the GFT $\tilde{\mathbf{x}}_r = \mathbf{V}^H\mathbf{x}_r$ of each observed signal $\mathbf{x}_r$ and estimate $\mathbf{p}$ as

$$\hat{\mathbf{p}}_{\text{pg}} := \frac{1}{R}\sum_{r=1}^{R}|\tilde{\mathbf{x}}_r|^2 = \frac{1}{R}\sum_{r=1}^{R}\left|\mathbf{V}^H\mathbf{x}_r\right|^2. \tag{12.8}$$

[2]The process to assess the performance of the different PSD estimators was created using the generating filter $\mathbf{H} = \sum_{l=0}^{3} h_l\mathbf{S}^l$ where $\mathbf{S}$ was set as the Laplacian matrix and the filter coefficients as $\mathbf{h} = [1, -0.15, 0.075, -10^{-4}]^T$ (cf. Definition 12.1). The coefficients were chosen for the filter to be of low order and to have a low pass behavior, as can be appreciated from the "*True PSD*" curves in Fig. 12.3, where most of the energy is concentrated in the low frequencies.

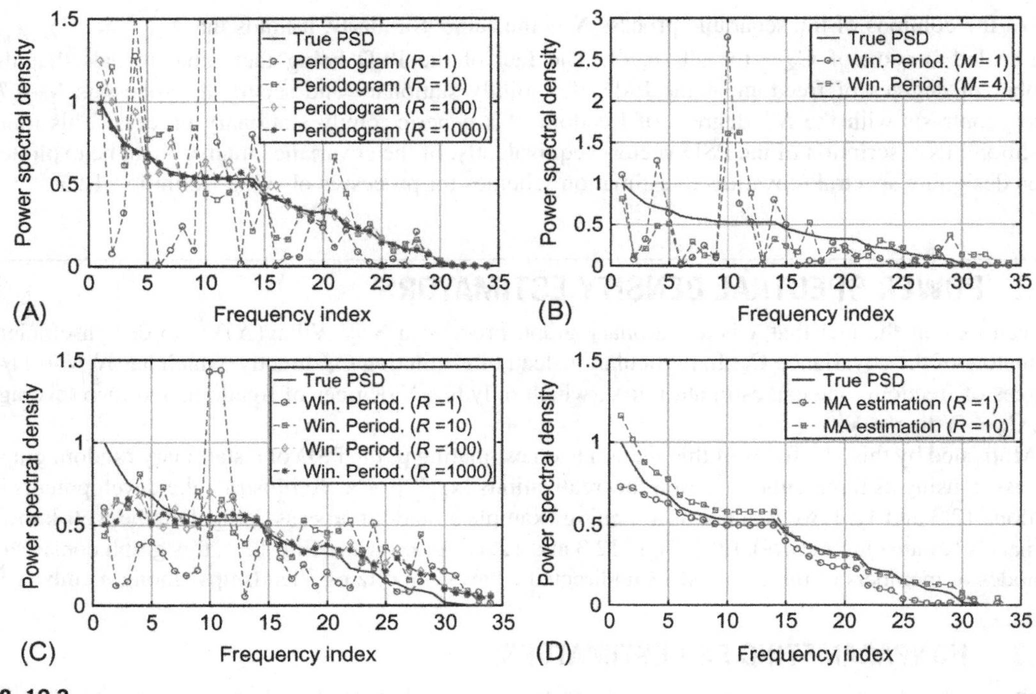

FIG. 12.3

Power spectral density estimation. All estimators are based on the same random process defined on the Karate club network [28]. (A) Periodogram estimation with different numbers of observations. (B) Windowed average periodogram from a single realization and a different number of windows. (C) Windowed average periodogram for four windows and a varying number of realizations. (D) Parametric MA estimation for 1 and 10 realizations.

The estimator in Eq. (12.8) is termed *periodogram* due to its evident similarity with its homonym in classical estimation. It is simple to show that $\hat{\mathbf{p}}_{\text{pg}}$ is an unbiased estimator, that is, $\mathbb{E}\left[\hat{\mathbf{p}}_{\text{pg}}\right] = \mathbf{p}$. A more detailed analysis of the performance of $\hat{\mathbf{p}}_{\text{pg}}$, for the case where the observations are Gaussian, is given in Proposition 12.1.

An alternative nonparametric estimation scheme, denominated *correlogram*, can be devised by starting from the definition of $\mathbf{p}$ in Eq. (12.4). Namely, one may substitute $\mathbf{C}_\mathbf{x}$ in Eq. (12.4) by the *sample covariance* $\hat{\mathbf{C}}_\mathbf{x} = (1/R)\sum_{r=1}^{R} \mathbf{x}_r\mathbf{x}_r^H$ computed based on the available observations to obtain

$$\hat{\mathbf{p}}_{\text{cg}} := \text{diag}\left(\mathbf{V}^H\hat{\mathbf{C}}_\mathbf{x}\mathbf{V}\right) := \text{diag}\left[\mathbf{V}^H\left[\frac{1}{R}\sum_{r=1}^{R}\mathbf{x}_r\mathbf{x}_r^H\right]\mathbf{V}\right]. \tag{12.9}$$

Notice that the matrix $\mathbf{V}^H\hat{\mathbf{C}}_\mathbf{x}\mathbf{V}$ is in general, not diagonal because the eigenbasis of $\hat{\mathbf{C}}_\mathbf{x}$ differs from $\mathbf{V}$, the eigenbasis of $\mathbf{C}_\mathbf{x}$. Nonetheless, we keep only the diagonal elements $\mathbf{v}_i^H\hat{\mathbf{C}}_\mathbf{x}\mathbf{v}_i$ for $i = 1, \ldots, N$ as our PSD estimator. It can be shown that the correlogram $\hat{\mathbf{p}}_{\text{cg}}$ in Eq. (12.9) and the periodogram $\hat{\mathbf{p}}_{\text{pg}}$ in Eq. (12.8) lead to identical estimators, as is the case in classical signal processing.

The correlogram can also be interpreted as an LS estimator. The decomposition in Eq. (12.5) allows a linear parameterization of the covariance matrix $\mathbf{C}_{\mathbf{x}}$ as

$$\mathbf{C}_{\mathbf{x}}(\mathbf{p}) = \sum_{i=1}^{N} p_i \mathbf{v}_i \mathbf{v}_i^H \tag{12.10}$$

This linear parameterization will also be useful for the sampling schemes developed in Section 12.4. Vectorizing $\mathbf{C}_{\mathbf{x}}$ in Eq. (12.10) results in a set of N^2 equations in $\mathbf{p}$

$$\mathbf{c}_{\mathbf{x}} = \text{vec}(\mathbf{C}_{\mathbf{x}}) = \sum_{i=1}^{N} p_i \text{vec}(\mathbf{v}_i \mathbf{v}_i^H) = \mathbf{G}_{\text{np}}\mathbf{p}, \tag{12.11}$$

where $\text{vec}(\mathbf{v}_i \mathbf{v}_i^H) = \mathbf{v}_i^* \otimes \mathbf{v}_i$. Relying on the Khatri-Rao product, we then form the $N^2 \times N$ matrix $\mathbf{G}_{\text{np}}$ as

$$\mathbf{G}_{\text{np}} := [\mathbf{v}_1^* \otimes \mathbf{v}_1, \ldots, \mathbf{v}_N^* \otimes \mathbf{v}_N] = \mathbf{V}^* \odot \mathbf{V}.$$

Using the sample covariance matrix $\hat{\mathbf{C}}_{\mathbf{x}}$ as an estimate of $\mathbf{C}_{\mathbf{x}}$, we can *match* the estimated covariance vector $\hat{\mathbf{c}}_{\mathbf{x}} = \text{vec}(\hat{\mathbf{C}}_{\mathbf{x}})$ to the true covariance vector $\mathbf{c}_{\mathbf{x}}$ in the LS sense as

$$\hat{\mathbf{p}}_{\text{ls}} = \underset{\mathbf{p}}{\text{argmin}}\, \|\hat{\mathbf{c}}_{\mathbf{x}} - \mathbf{G}_{\text{np}}\mathbf{p}\|_2^2 = (\mathbf{G}_{\text{np}}^H \mathbf{G}_{\text{np}})^{-1} \mathbf{G}_{\text{np}}^H \hat{\mathbf{c}}_{\mathbf{x}}. \tag{12.12}$$

In other words, the LS estimator minimizes the squared error $\text{tr}[(\hat{\mathbf{C}}_{\mathbf{x}} - \mathbf{C}_{\mathbf{x}}(\mathbf{p}))^T(\hat{\mathbf{C}}_{\mathbf{x}} - \mathbf{C}_{\mathbf{x}}(\mathbf{p}))]$. From expression (12.12) it can be shown that the ith element of $\hat{\mathbf{p}}_{\text{ls}}$ is $\mathbf{v}_i^H \hat{\mathbf{C}}_{\mathbf{x}} \mathbf{v}_i$. Combining this with Eq. (12.9), we get that the LS estimator $\hat{\mathbf{p}}_{\text{ls}}$ and the correlogram $\hat{\mathbf{p}}_{\text{cg}}$—and hence the periodogram as well—are all identical estimators.

The estimators derived in this subsection do not assume any data distribution and are well suited for cases where the data probability density function is not available. In what follows, we provide performance bounds for these estimators under the condition that the observed signals are Gaussian.

Mean squared error and the Cramér-Rao bound

Suppose that the data consists of realizations from a sequence of independent and identically distributed (i.i.d.) Gaussian random vectors $\{\mathbf{x}_r\}_{r=1}^R$, where for each r, the vector $\mathbf{x}_r \sim \mathcal{N}(\mathbf{0}, \mathbf{C}_{\mathbf{x}}(\mathbf{p}))$. Under this setting, we can characterize the variance, hence the mean squared error (MSE), of the periodogram estimator (as well as the equivalent correlogram and LS estimators). In the following proposition, we present expressions for its bias and variance.

Proposition 12.1. Let $\{\mathbf{x}_r\}_{r=1}^R$ be independent samples of the process $\mathbf{x}$ stationary in $\mathbf{S}$ with PSD $\mathbf{p}$. Then, the bias $\mathbf{b}_{\text{pg}}$ of the periodogram estimator in Eq. (12.8) is zero,

$$\mathbf{b}_{\text{pg}} := \mathbb{E}\left[\hat{\mathbf{p}}_{\text{pg}}\right] - \mathbf{p} = \mathbf{0}. \tag{12.13}$$

Further define the covariance of the periodogram as $\boldsymbol{\Sigma}_{\text{pg}} := \mathbb{E}\left[(\hat{\mathbf{p}}_{\text{pg}} - \mathbf{p})(\hat{\mathbf{p}}_{\text{pg}} - \mathbf{p})^H\right]$. If the process $\mathbf{x}$ is Gaussian and $\mathbf{S}$ is symmetric, then $\boldsymbol{\Sigma}_{\text{pg}}$ can be written as

$$\boldsymbol{\Sigma}_{\text{pg}} := \mathbb{E}\left[(\hat{\mathbf{p}}_{\text{pg}} - \mathbf{p})(\hat{\mathbf{p}}_{\text{pg}} - \mathbf{p})^H\right] = (2/R)\text{diag}^2(\mathbf{p}). \tag{12.14}$$

■

As was mentioned before, Proposition 12.1 states that the periodogram is an unbiased estimator, i.e., $\mathbb{E}\left[\hat{\mathbf{p}}_{\text{pg}}\right] = \mathbf{p}$, as expected given its classical counterpart. While Eq. (12.13) is valid for any distribution, observe that the covariance expression in Eq. (12.14) requires the process $\mathbf{x}$ to be Gaussian. This requirement stems from the fact that the derivation of $\boldsymbol{\Sigma}_{\text{pg}}$ involves fourth-order moments of $\mathbf{x}$. This is natural because an analogous limitation arises for time signals [29, Sec. 8.2]. Notice also that the PSD estimates of different frequencies are uncorrelated, because Eq. (12.14) reveals that $\boldsymbol{\Sigma}_{\text{pg}}$ is a diagonal matrix. A proof of the above result along with generalizations for the cases in which $\mathbf{S}$ is not necessarily symmetric (but normal) can be found in [15].

The MSE of the periodogram, defined as $\text{MSE}(\hat{\mathbf{p}}_{\text{pg}}) := \mathbb{E}\left[\|(\hat{\mathbf{p}}_{\text{pg}} - \mathbf{p})\|_2^2\right]$, can be readily computed using the result in Proposition 12.1

$$\text{MSE}(\hat{\mathbf{p}}_{\text{pg}}) = \|\mathbf{b}_{\text{pg}}\|_2^2 + \text{tr}[\boldsymbol{\Sigma}_{\text{pg}}] = (2/R)\|\mathbf{p}\|_2^2. \tag{12.15}$$

As becomes apparent from Eq. (12.15), the periodogram is expected to yield large relative errors when only a few observations R are available. In Fig. 12.3A we show the periodogram estimation for different numbers of observations R. Notice that, indeed, when $R = 1$ the estimation is very poor. Nonetheless, when increasing R the estimation tends to the true PSD. A method that can achieve better performance for lower values of R—windowed average periodogram—will be introduced after showing that the periodogram is an efficient estimator.

The Cramér-Rao bound provides a lower bound on the covariance of unbiased estimators when the available data records are finite. The Cramér-Rao bound matrix is equal to the inverse of the Fisher information matrix, $\mathbf{F}$, and it is given by $\mathbf{F} = (R/2)\,\text{diag}^{-2}(\mathbf{p})$; see, e.g., [30, Ch. 6.13]. The efficiency of the periodogram follows readily by comparing $\mathbf{F}^{-1}$ with Eq. (12.14).

Windowed average periodogram

When only one or just a few observations of the process $\mathbf{x}$ are available, the periodogram and correlogram yield large errors [cf. Eq. (12.15)]. A way to overcome this roadblock is to artificially generate multiple signals from the few available ones. Bartlett and Welch methods are classical examples of this procedure because they utilize windows to generate multiple samples of the process, even if only a single realization is given [31, Sec. 2.7]. Intuitively, a long signal is partitioned into pieces where each piece can be considered as a different signal. This operation introduces bias in the estimator but reduces variance to the point that the overall MSE can be improved. The frequency counterparts of such classical methods are filter banks, where the signal is partitioned in the Fourier domain. Both the windowed average periodogram—including Bartlett and Welch methods—and the filter banks can be extended for the estimation of graph processes. In this section, we only focus on the former, but extensions of this analysis as well as a full derivation of filter-bank estimators can be found in [15].

The application of a window $\mathbf{w}$ to a signal[3] $\mathbf{x}$ entails a component-wise multiplication to produce the signal $\mathbf{x_w} = \text{diag}(\mathbf{w})\mathbf{x}$, where we assume that windows are normalized to have energy $\|\mathbf{w}\|_2^2 = N$. We may leverage the definition of the GFT to write

$$\tilde{\mathbf{x}}_\mathbf{w} = \mathbf{V}^H\mathbf{x_w} = \mathbf{V}^H\text{diag}(\mathbf{w})\mathbf{x} = \mathbf{V}^H\text{diag}(\mathbf{w})\mathbf{V}\tilde{\mathbf{x}} =: \tilde{\mathbf{W}}\tilde{\mathbf{x}}, \tag{12.16}$$

where we implicitly defined $\tilde{\mathbf{W}} := \mathbf{V}^H\text{diag}(\mathbf{w})\mathbf{V}$ as the dual of the windowing operator in the frequency domain. For time signals the frequency representation of a window is its Fourier transform and the dual operator of windowing is the convolution between the spectra of the window and the signal. This parallelism is lost for graph signals. Nonetheless, Eq. (12.16) can be used to design windows with small spectral distortion, i.e., windows for which $\tilde{\mathbf{W}} \approx \mathbf{I}$. Recall that our objective is to generate multiple signals from only one, thus instead of focusing on a single window we consider a bank of M windows $\mathcal{W} = \{\mathbf{w}_m\}_{m=1}^M$ and use it to construct the windowed signals $\mathbf{x}_m := \text{diag}(\mathbf{w}_m)\mathbf{x}$. Based on these windowed signals, we build the *windowed average periodogram* as

$$\hat{\mathbf{p}}_\mathcal{W} := \frac{1}{M}\sum_{m=1}^{M}\left|\mathbf{V}^H\mathbf{x}_m\right|^2 = \frac{1}{M}\sum_{m=1}^{M}\left|\mathbf{V}^H\text{diag}(\mathbf{w}_m)\mathbf{x}\right|^2. \tag{12.17}$$

The name given to $\hat{\mathbf{p}}_\mathcal{W}$ becomes apparent when comparing Eq. (12.17) with Eq. (12.8). Indeed, the former is almost equivalent to the latter with the caveat that the M signals considered in Eq. (12.17) are not independent. As a consequence, the variance decreases slower than $1/M$ with the number of windows, this being the rate found in Proposition 12.1 for the averaging of R independent signals. Moreover, the dependence between the different $\mathbf{x}_m$ introduces a distortion (bias) in the estimator. To state these effects more formally, we construct the dual operators associated with each window $\tilde{\mathbf{W}}_m := \mathbf{V}^H\text{diag}(\mathbf{w}_m)\mathbf{V}$ [cf. Eq. (12.16)], and use them to define the power *spectrum mixing* matrix of windows m and m' as the componentwise product $\tilde{\mathbf{W}}_{m,m'} := \tilde{\mathbf{W}}_m \circ \tilde{\mathbf{W}}_{m'}^*$. Based on the spectrum mixing matrices, the following proposition presents the bias and covariance of $\hat{\mathbf{p}}_\mathcal{W}$.

Proposition 12.2. Let $\hat{\mathbf{p}}_\mathcal{W}$ be the windowed average periodogram computed based on a window bank $\mathcal{W} = \{\mathbf{w}_m\}_{m=1}^M$ and single observation $\mathbf{x}$ of a stationary process in $\mathbf{S}$. Then, the bias of $\hat{\mathbf{p}}_\mathcal{W}$ is given by

$$\mathbf{b}_\mathcal{W} := \mathbb{E}\left[\hat{\mathbf{p}}_\mathcal{W}\right] - \mathbf{p} = \left(\frac{1}{M}\sum_{m=1}^{M}\tilde{\mathbf{W}}_{m,m} - \mathbf{I}\right)\mathbf{p}. \tag{12.18}$$

Furthermore, if $\mathbf{x}$ is Gaussian and $\mathbf{S}$ is symmetric, the trace of the covariance $\mathbf{\Sigma}_\mathcal{W} := \mathbb{E}\left[(\hat{\mathbf{p}}_\mathcal{W} - \mathbb{E}\left[\hat{\mathbf{p}}_\mathcal{W}\right])(\hat{\mathbf{p}}_\mathcal{W} - \mathbb{E}\left[\hat{\mathbf{p}}_\mathcal{W}\right])^H\right]$ is given by

$$\text{tr}[\mathbf{\Sigma}_\mathcal{W}] = \frac{2}{M^2}\sum_{m=1,\, m'=1}^{M}\text{tr}\left[(\tilde{\mathbf{W}}_{m,m'}\mathbf{p})(\tilde{\mathbf{W}}_{m,m'}\mathbf{p})^H\right]. \tag{12.19}$$

■

Expression (12.18) reveals that the bias of $\hat{\mathbf{p}}_\mathcal{W}$ is given by the discrepancy between the average spectrum mixing of the windows—depending on both the window silhouette $\mathbf{w}_m$ and the underlying

[3] To keep notation simple, in this subsection we use $\mathbf{x}$ to denote a realization of process $\mathbf{x}$.

graph through $\mathbf{V}$—and the identity matrix. Notice that even if the individual spectrum mixing matrices $\tilde{\mathbf{W}}_{m,m}$ are far from the identity, a small bias can still be achieved by controlling their average. The covariance expression in Eq. (12.19) can be further decomposed into a term akin to Eq. (12.14) plus another one that quantifies the added effect of dependency between the windowed signals; see [15] for more details. Furthermore, as done in Eq. (12.15), we can use Proposition 12.2 to obtain a closed-form expression for the MSE of $\hat{\mathbf{p}}_{\mathcal{W}}$ that can then guide the design criteria for optimal window banks. However, the associated optimization problems are nonconvex. Although some basic developments in this area are presented in [15], the efficient design of optimal windows is still an open problem.

In Fig. 12.3B we illustrate the windowed average periodogram estimation for $M = 4$ random windows for a single observation of the random process on the Karate club network. Notice that the estimation is better than the one obtained by the (regular) periodogram, i.e., $M = 1$. In Fig. 12.3C we present the windowed average periodogram estimation for $M = 4$ but with an increasing number of observations. Notice that for a low number of observations ($R = 1$ and $R = 10$) the estimation improves that of the periodogram [cf. Fig. 12.3A]. Nonetheless, it can be seen that this estimator is biased because there is still a residual error, even for large values of R.

12.3.2 PARAMETRIC PSD ESTIMATORS

A stationary graph process $\mathbf{x}$ can always be represented as the response of a graph filter $\mathbf{H}$ when applied to a white input [cf. Definition 12.1]. The cases where $\mathbf{H}$ depends on just a few parameters—much less than N—ultimately result in a further reduction of the degrees of freedom of the process $\mathbf{x}$. In particular, we may obtain a parametric description of the PSD of $\mathbf{x}$ as a function of the few coefficients of $\mathbf{H}$. In this section, we leverage this reduction in degrees of freedom to design PSD estimators. We discuss in detail the case where $\mathbf{H}$ corresponds to a moving average (MA) model, and then briefly review the constructions for an autoregressive (AR) model. For the combined ARMA model, the developments for the MA and AR processes can be mimicked; see [15] for more details on the ARMA model.

Moving average graph processes

Consider a vector of coefficients $\boldsymbol{\beta} = [\beta_0, \ldots, \beta_{L-1}]^T$, for $L \ll N$, and assume that the stationary process $\mathbf{x}$ is generated as $\mathbf{x} = \mathbf{H}(\boldsymbol{\beta})\mathbf{n}$ where $\mathbf{n}$ is white and $\mathbf{H}(\boldsymbol{\beta}) = \sum_{l=0}^{L-1} \beta_l \mathbf{S}^l$. From this generative model, it immediately follows that the covariance of $\mathbf{x}$ can be written as a function of $\boldsymbol{\beta}$, i.e., $\mathbf{C}_\mathbf{x}(\boldsymbol{\beta}) = \mathbf{H}(\boldsymbol{\beta})\mathbf{H}^H(\boldsymbol{\beta})$. Regarding the PSD of $\mathbf{x}$, from the definition in Eq. (12.4) we have that $\mathbf{p}(\boldsymbol{\beta}) = \operatorname{diag}\left(\mathbf{V}^H\mathbf{C}_\mathbf{x}(\boldsymbol{\beta})\mathbf{V}\right)$, from where it follows that the PSD of $\mathbf{x}$ is equal to the squared magnitude of the frequency representation of the filter. The dependence of $\mathbf{C}_\mathbf{x}$ and $\mathbf{p}$ on $\boldsymbol{\beta}$ are explicitly stated below

$$\mathbf{C}_\mathbf{x}(\boldsymbol{\beta}) = \sum_{l=0,l'=0}^{L-1} \beta_l \mathbf{S}^l \beta_{l'} (\mathbf{S}^H)^{l'}, \quad \mathbf{p}(\boldsymbol{\beta}) = |\tilde{\mathbf{h}}(\boldsymbol{\beta})|^2 = |\boldsymbol{\Psi}_L \boldsymbol{\beta}|^2. \tag{12.20}$$

The covariance and PSD expressions in Eq. (12.20) correspond to the natural graph counterparts of MA time processes generated by FIR filters; see [15] for discussions on the relevance of these processes.

The estimation of $\boldsymbol{\beta}$ can now be pursued in either the graph or frequency domain through covariance or PSD fitting, respectively. More specifically, in the graph domain, we compute the sample covariance $\hat{\mathbf{C}}_\mathbf{x}$ and use a matrix distortion function $D_\mathrm{C}(\hat{\mathbf{C}}_\mathbf{x}, \mathbf{C}_\mathbf{x}(\boldsymbol{\beta}))$ to measure the dissimilarity between $\hat{\mathbf{C}}_\mathbf{x}$ and

$\mathbf{C_x}(\boldsymbol{\beta})$. Alternatively, in the frequency domain, we compute the periodogram $\hat{\mathbf{p}}_{\text{pg}}$ as in Eq. (12.8) and use a vector distortion function $D_{\mathbf{p}}(\hat{\mathbf{p}}_{\text{pg}}, |\boldsymbol{\Psi}_L\boldsymbol{\beta}|^2)$ to compare the periodogram $\hat{\mathbf{p}}_{\text{pg}}$ with the PSD $|\boldsymbol{\Psi}_L\boldsymbol{\beta}|^2$. We then select the coefficients $\boldsymbol{\beta}$ that lead to the minimal distortion, as specified below, for either the graph or the frequency domain

$$\hat{\boldsymbol{\beta}} := \underset{\boldsymbol{\beta}}{\text{argmin}}\ D_{\mathbf{C}}(\hat{\mathbf{C}}_{\mathbf{x}}, \mathbf{C_x}(\boldsymbol{\beta})), \quad \hat{\boldsymbol{\beta}} := \underset{\boldsymbol{\beta}}{\text{argmin}}\ D_{\mathbf{p}}(\hat{\mathbf{p}}_{\text{pg}}, |\boldsymbol{\Psi}_L\boldsymbol{\beta}|^2). \tag{12.21}$$

Notice that both the functional forms of $\mathbf{C_x}(\boldsymbol{\beta})$ and $\mathbf{p}(\boldsymbol{\beta})$ in Eq. (12.20) are indefinite quadratics in $\boldsymbol{\beta}$. Hence, the optimization problems in Eq. (12.21) will not be convex in general. In the particular case where the distortion $D_{\mathbf{p}}$ is given by the squared ℓ_2 norm of the difference, i.e., $D_{\mathbf{p}}(\hat{\mathbf{p}}_{\text{pg}}, |\boldsymbol{\Psi}_L\boldsymbol{\beta}|^2) = \|\hat{\mathbf{p}}_{\text{pg}} - |\boldsymbol{\Psi}_L\boldsymbol{\beta}|^2\|_2^2$, efficient (phase-retrieval) solvers with probabilistic guarantees are available [32,33]. Alternative tractable formulations of Eq. (12.21) are discussed in [15], one of which is described next.

When $\mathbf{S}$ is symmetric, the expression (12.20) reduces to

$$\mathbf{C_x} = \sum_{k=0}^{Q-1} b_k\mathbf{S}^k, \quad p_n = \sum_{k=0}^{Q-1} b_k\lambda_n^k. \tag{12.22}$$

Here, $Q := \min\{2L-1, N\}$ unknown expansion coefficients $\{b_k\}_{k=0}^{Q-1}$ are collected in the vector $\mathbf{b} = [b_0, b_1, \ldots, b_{Q-1}]^T \in \mathbb{R}^Q$. By ignoring the structure in $\mathbf{b}$, i.e., the relation between $\mathbf{b}$ and $\boldsymbol{\beta}$, we arrive at a linear parameterization of $\mathbf{C_x}$ using the set of Q symmetric matrices $\{\mathbf{S}^0, \mathbf{S}, \ldots, \mathbf{S}^{Q-1}\}$ as a basis. Vectorizing $\mathbf{C_x}$ in Eq. (12.22), we obtain

$$\mathbf{c_x} = \text{vec}(\mathbf{C_x}) = \sum_{k=0}^{Q-1} b_k \text{vec}(\mathbf{S}^q) = \mathbf{G}_{\text{ma}}\mathbf{b}, \tag{12.23}$$

where we implicitly defined the matrix $\mathbf{G}_{\text{ma}} := \left[\text{vec}(\mathbf{S}^0), \ldots, \text{vec}(\mathbf{S}^{Q-1})\right]$. Because $\mathbf{C_x}$ depends linearly on $\mathbf{b}$—as opposed to quadratically on $\boldsymbol{\beta}$—we may efficiently solve Eq. (12.21) for some choices of $D_{\mathbf{C}}$. For example, the LS estimate of $\mathbf{b}$ is given by $\hat{\mathbf{b}} = (\mathbf{G}_{\text{ma}}^H\mathbf{G}_{\text{ma}})^{-1}\mathbf{G}_{\text{ma}}^H\hat{\mathbf{c}}_{\mathbf{x}}$. We illustrate the implementation of this relaxation in Fig. 12.3D. Notice that the PSD estimation is quite faithful even for $R = 1$, and it slightly improves for $R = 10$.

Autoregressive graph processes

A stationary process can be better and better approximated as an MA process by increasing the order of the associated FIR filter. However, the merits of the parametric estimators depend on having a small number of parameters describing the generating process. For some stationary processes, an AR model using an infinite impulse response filter leads to a more parsimonious description. For example, consider the diffusion process driven by the graph filter $\mathbf{H} = \sum_{l=0}^{\infty}\alpha^l\mathbf{S}^l$, where α represents the diffusion rate. For small enough α, the filter can be rewritten as $\mathbf{H} = (\mathbf{I} - \alpha\mathbf{S})^{-1}$, with frequency response $\tilde{\mathbf{h}} = \text{diag}(\mathbf{I} - \alpha\boldsymbol{\Lambda}^{-1})$. Thus, $\mathbf{H}$ can be viewed as a single-pole AR filter, leading to a more meager description. More generally, an AR filter of order M can be described as $\mathbf{H} = \alpha_0\prod_{m=1}^{M}(\mathbf{I} - \alpha_m\mathbf{S})^{-1}$ for some vector of parameters $\boldsymbol{\alpha} = [\alpha_0, \ldots, \alpha_M]^T$. Correspondingly, the frequency response of this filter is given by $\tilde{\mathbf{h}} = \alpha_0\,\text{diag}\big(\prod_{m=1}^{M}(\mathbf{I} - \alpha_m\boldsymbol{\Lambda})^{-1}\big)$. If we define the graph process $\mathbf{x} = \mathbf{Hn}$ with $\mathbf{n}$ white, we may leverage the previous expressions to obtain explicit formulas for the covariance and PSD of $\mathbf{x}$ as a function of the parameters $\boldsymbol{\alpha}$,

$$\mathbf{C_x}(\boldsymbol{\alpha}) = \alpha_0^2 \prod_{m=1}^{M} (\mathbf{I} - \alpha_m \mathbf{S})^{-1} (\mathbf{I} - \alpha_m \mathbf{S})^{-H}, \quad \mathbf{p}(\boldsymbol{\alpha}) = \alpha_0^2 \operatorname{diag}\left(\prod_{m=1}^{M} |\mathbf{I} - \alpha_m \boldsymbol{\Lambda}|^{-2} \right). \tag{12.24}$$

The mechanism to obtain the corresponding parametric PSD estimator is equivalent to the one explained for MA processes, where $\mathbf{C_x}(\boldsymbol{\beta})$ and $\mathbf{p}(\boldsymbol{\beta})$ in Eq. (12.20) are replaced by $\mathbf{C_x}(\boldsymbol{\alpha})$ and $\mathbf{p}(\boldsymbol{\alpha})$ in Eq. (12.24). The associated optimization problems [cf. Eq. (12.21)] will be nonconvex in general and become intractable for large orders M.

Yule-Walker schemes [31, Sec. 3.4] tailored to graph signals may be of help, as discussed next. The all-pole filter $\mathbf{H}^{-1}(\boldsymbol{\alpha}) = \prod_{k=1}^{M} (\mathbf{I} - \alpha_k \mathbf{S})$ can be alternatively expressed as $\mathbf{H}^{-1}(\mathbf{a}) = \mathbf{I} - \sum_{k=1}^{M} a_k \mathbf{S}^k$, where $\mathbf{a} = [a_1, a_2, \ldots, a_M]^T$. Thus, the AR signal satisfies the equations

$$\mathbf{x} = \sum_{k=1}^{M} a_k \mathbf{S}^k \mathbf{x} + \mathbf{n}. \tag{12.25}$$

In other words, the graph signal $\mathbf{x}$ depends *linearly* on the M shifted graph signals $\{\mathbf{S}^k \mathbf{x}\}_{k=1}^{M}$ according to the above AR model. As a result, the covariance matrix of $\mathbf{x}$ and its vectorized form can be expressed as

$$\mathbf{C_x} = \sum_{k=1}^{M} a_k \mathbf{S}^k \mathbf{C_x} + \mathbf{C_{nx}}, \quad \mathbf{c_x} = \operatorname{vec}(\mathbf{C_x}) = \sum_{k=1}^{M} a_k \operatorname{vec}(\mathbf{S}^k \mathbf{C_x}) + \operatorname{vec}(\mathbf{C_{nx}}) \approx \mathbf{G}_{\text{ar}} \mathbf{a}, \tag{12.26}$$

where we have defined $\mathbf{G}_{\text{ar}} := [\operatorname{vec}(\mathbf{SC_x}), \ldots, \operatorname{vec}(\mathbf{S}^M \mathbf{C_x})]$ and where we have assumed that $\mathbf{C_{nx}} = \mathbb{E}[\mathbf{nx}^H]$ is a small error term. Note that in contrast to the previous linear equations for the nonparametric (12.11) and MA (12.23) models, the system matrix $\mathbf{G}_{\text{ar}}$ now explicitly depends on the unknown covariance $\mathbf{C_x}$. Still, when the sample covariance matrix $\hat{\mathbf{C}}_\mathbf{x}$ is available, we can solve Eq. (12.26) through LS as $\hat{\mathbf{a}} = (\hat{\mathbf{G}}_{\text{ar}}^H \hat{\mathbf{G}}_{\text{ar}})^{-1} \hat{\mathbf{G}}_{\text{ar}}^H \hat{\mathbf{c}}_\mathbf{x}$, where $\hat{\mathbf{G}}_{\text{ar}}$ is defined as $\mathbf{G}_{\text{ar}}$ replacing $\mathbf{C_x}$ by $\hat{\mathbf{C}}_\mathbf{x}$.

12.4 NODE SUBSAMPLING FOR PSD ESTIMATION

Compression or data reduction is preferred for large-scale graph processes as the size of the datasets often inhibits a direct computation of the second-order statistics. In this section, we focus on recovering the second-order statistics of stationary graph processes from subsampled graph signals. We refer to this problem as *graph covariance sampling* [19].

The fact that we reconstruct the power spectrum instead of the graph signal itself enables us to sparsely sample the nodes, even in the absence of any spectral priors such as smoothness, sparsity, or band-limitedness with known support. The proposed concept basically generalizes the field of compressive covariance sensing [18] to the graph setting, which is not trivial. This is because, for weakly stationary signals with a regular support or signals supported on a circulant graph, the covariance matrix has a clear structure (e.g., Toeplitz, circulant) that enables an elegant subsampler design. However, for second-order stationary graph signals residing on arbitrary graphs, the covariance matrix in general does not admit *any* clear structure that can be easily exploited.

12.4.1 THE SAMPLING PROBLEM

Consider the problem of estimating the graph power spectrum of the weakly stationary graph signal $\mathbf{x} \in \mathbb{R}^N$ from a set of $K \ll N$ linear observations stacked in $\mathbf{y} \in \mathbb{R}^K$, given by

$$\mathbf{y} = \mathbf{\Phi x}, \tag{12.27}$$

where $\mathbf{\Phi}$ is a known $K \times N$ selection matrix with Boolean entries, i.e., $\mathbf{\Phi} \in \{0, 1\}^{K \times N}$ and where several realizations of $\mathbf{y}$ may be available. The matrix $\mathbf{\Phi}$ is referred to as the *subsampling* or *sparse sampling* matrix. Such a sparse sampling scheme generally results in a reduction in the storage and processing costs. Moreover, for applications where nodes correspond to sensing devices—such as weather stations in climatology and electroencephalography probes in brain networks—it also leads to smaller hardware and communications costs.

The covariance matrices $\mathbf{C_x} = \mathbb{E}[\mathbf{xx}^H] \in \mathbb{R}^{N \times N}$ and $\mathbf{C_y} = \mathbb{E}[\mathbf{yy}^H] \in \mathbb{R}^{K \times K}$ contain the second-order statistics of $\mathbf{x}$ and $\mathbf{y}$, respectively. In practice, a sample covariance matrix is computed based on R signal observations. More precisely, suppose that R observations of the uncompressed and compressed graph signals are available, denoted by the vectors $\{\mathbf{x}_r\}_{r=1}^R$ and $\{\mathbf{y}_r\}_{r=1}^R$, respectively. Then forming the sample covariance matrix, $\hat{\mathbf{C}}_\mathbf{x} = (1/R)\sum_{r=1}^R \mathbf{x}_r\mathbf{x}_r^H$, from R snapshots of $\mathbf{x}$ costs $\mathcal{O}(N^2R)$ while forming the sample covariance matrix, $\hat{\mathbf{C}}_\mathbf{y} = (1/R)\sum_{r=1}^R \mathbf{y}_r\mathbf{y}_r^H$, from R snapshots of $\mathbf{y}$ *only* costs $\mathcal{O}(K^2R)$. Therefore, when $K \ll N$, there will clearly be a significant reduction in the storage and processing costs due to compression.

12.4.2 COMPRESSED LS ESTIMATOR

In this section, we will extend the previously derived LS estimators (for nonparametric as well as parametric PSD estimation) to the case where only compressed graph signals are available. The reason we only focus on those estimators is not because they lead to the best performance, but because they can be used to design the best subset of nodes to sample.

Let us condense the linearly structured covariance matrix $\mathbf{C_x}$ for the nonparametric case (see Eq. 12.10), the parametric MA case with symmetric shifts (see Eq. 12.22), and the parametric AR case (see Eq. 12.26), in a single expression as

$$\mathbf{C_x}(\boldsymbol{\theta}) = \sum_{i=1}^{L} \theta_i \mathbf{Q}_i; \quad \boldsymbol{\theta} = [\theta_1, \ldots, \theta_L]^T, \tag{12.28}$$

where for the nonparametric case, we have $L = N$, $\boldsymbol{\theta} := \mathbf{p}$, and $\mathbf{Q}_i := \mathbf{v}_i\mathbf{v}_i^H$, for the MA case with symmetric shifts, we have $L = Q$, $\boldsymbol{\theta} := \mathbf{b}$, and $\mathbf{Q}_i := \mathbf{S}^{i-1}$, and for the AR case, we have $L = M$, $\boldsymbol{\theta} := \mathbf{a}$, and $\mathbf{Q}_i := \mathbf{S}^{i-1}\mathbf{C_x}$.

Using the compression scheme described in Eq. (12.27), the covariance matrix $\mathbf{C_y}$ of the subsampled graph signal $\mathbf{y}$ can be related to $\mathbf{C_x}$ as

$$\mathbf{C_y}(\boldsymbol{\theta}) = \mathbf{\Phi C_x \Phi}^T = \sum_{i=1}^{L} \theta_i \mathbf{\Phi Q}_i \mathbf{\Phi}^T. \tag{12.29}$$

This means that the expansion coefficients of $\mathbf{C_y}$ with respect to the set $\{\boldsymbol{\Phi}\mathbf{Q}_1\boldsymbol{\Phi}^T, \ldots, \boldsymbol{\Phi}\mathbf{Q}_L\boldsymbol{\Phi}^T\}$ are the *same* as those of $\mathbf{C_x}$ with respect to the set $\{\mathbf{Q}_1, \ldots, \mathbf{Q}_L\}$, and they are preserved under linear compression. It is not clear at this point whether these expansion coefficients, which basically characterize the graph power spectrum, can be uniquely recovered from $\mathbf{C_y}(\boldsymbol{\theta})$.

Vectorizing $\mathbf{C_y}$ as $\mathbf{c_y} = \text{vec}(\mathbf{C_y}) = (\boldsymbol{\Phi} \otimes \boldsymbol{\Phi})\text{vec}(\mathbf{C_x}) \in \mathbb{R}^{K^2}$ we obtain

$$\mathbf{c_y} = (\boldsymbol{\Phi} \otimes \boldsymbol{\Phi})\mathbf{G}\boldsymbol{\theta}, \tag{12.30}$$

where $\mathbf{G} = [\text{vec}(\mathbf{Q}_1), \ldots, \text{vec}(\mathbf{Q}_L)]$. When only a finite number of observations are available, we use the compressed sample data covariance matrix $\hat{\mathbf{C}}_\mathbf{y}$ instead of $\mathbf{C_y}$, leading to the approximation $\hat{\mathbf{c}}_\mathbf{y} \approx (\boldsymbol{\Phi} \otimes \boldsymbol{\Phi})\mathbf{G}\boldsymbol{\theta}$.

The parameter $\boldsymbol{\theta}$ is identifiable from this system of equations if $(\boldsymbol{\Phi} \otimes \boldsymbol{\Phi})\mathbf{G}$ has full column rank, which requires $K^2 \geq L$. Assuming that this is the case, the graph power spectrum (thus the second-order statistics of $\mathbf{x}$) can be estimated in closed form via LS as

$$\hat{\boldsymbol{\theta}} = [(\boldsymbol{\Phi} \otimes \boldsymbol{\Phi})\mathbf{G}]^\dagger \hat{\mathbf{c}}_\mathbf{y}. \tag{12.31}$$

It can be shown that a full row rank (wide) matrix $\boldsymbol{\Phi} \in \mathbb{R}^{K \times N}$ yields a full column rank matrix $(\boldsymbol{\Phi} \otimes \boldsymbol{\Phi})\mathbf{G}$ if and only if the matrix $(\boldsymbol{\Phi} \otimes \boldsymbol{\Phi})\mathbf{G}$ is tall, i.e., $K^2 \geq L$, and $\text{null}(\boldsymbol{\Phi} \otimes \boldsymbol{\Phi}) \cap \text{range}(\mathbf{G}) = \{\mathbf{0}\}$. When this is the case, we can recover the graph power spectrum by observing *only* $\mathcal{O}(\sqrt{L})$ nodes.

An important remark is required at this point with respect to the parametric AR model. Note from Eq. (12.26) that in this case the matrix $\mathbf{G}$ depends itself on the *uncompressed* covariance matrix $\mathbf{C_x}$, which is unknown. Hence, Eq. (12.31) cannot be directly applied. One option is to simply assume we roughly know it and although this is not going to lead to a good estimate, it might be good enough for designing a suboptimal sampling scheme (see Section 12.4.3). Another option is to restrict ourselves to particular subsampling schemes that preserve the linear structure in Eq. (12.26) but for the compressed data instead of the uncompressed data; see [19] for more details.

12.4.3 SPARSE SAMPLER DESIGN

We have seen so far that the design of the subsampling matrix $\boldsymbol{\Phi}$ is crucial for the reconstruction of the covariance of the random graph process. In this subsection, we design a sparse subsampling matrix $\boldsymbol{\Phi}$ to ensure that the observation matrix $(\boldsymbol{\Phi} \otimes \boldsymbol{\Phi})\mathbf{G}$ has full column rank and the solution for $\boldsymbol{\theta}$ has a small error.

Algorithm 12.1 GREEDY ALGORITHM

1. **Require** $\mathcal{X} = \emptyset, K$.
2. **for** $k = 1$ to K
3. $\quad s^* = \underset{s \notin \mathcal{X}}{\text{argmax}}\, f(\mathcal{X} \cup \{s\})$
4. $\quad \mathcal{X} \leftarrow \mathcal{X} \cup \{s^*\}$
5. **end**
6. **Return** $\mathcal{X}$

Consider a structured sparse sampling matrix $\boldsymbol{\Phi}(\mathbf{z}) \in \{0, 1\}^{K \times N}$, such that the entries of this matrix are determined by a binary *component selection* vector $\mathbf{z} = [z_1, \ldots, z_N]^T \in \{0, 1\}^N$, where $z_i = 1$ indicates that the ith node is selected by $\boldsymbol{\Phi}(\mathbf{z})$.

Uniqueness and sensitivity of the LS solution developed in the previous subsection depends on the spectrum (i.e., the set of eigenvalues) of the matrix

$$\mathbf{T}(\mathbf{z}) = [(\boldsymbol{\Phi}(\mathbf{z}) \otimes \boldsymbol{\Phi}(\mathbf{z}))\mathbf{G}]^T [(\boldsymbol{\Phi}(\mathbf{z}) \otimes \boldsymbol{\Phi}(\mathbf{z}))\mathbf{G}] = \mathbf{G}^T (\mathrm{diag}(\mathbf{z}) \otimes \mathrm{diag}(\mathbf{z}))\,\mathbf{G}.$$

More specifically, the performance of LS is better if the spectrum of the matrix $(\boldsymbol{\Phi} \otimes \boldsymbol{\Phi})\mathbf{G}$ is more uniform, i.e., its condition number is close to unity [34]. Thus, a sparse sampler $\mathbf{z}$ can be obtained by solving

$$\underset{\mathbf{z}\in\{0,1\}^N}{\mathrm{argmax}} \quad f(\mathbf{z}) \quad \text{s.t.} \quad \|\mathbf{z}\|_0 = K, \tag{12.32}$$

with either $f(\mathbf{z}) = -\mathrm{tr}[\mathbf{T}^{-1}(\mathbf{z})]$, $f(\mathbf{z}) = \lambda_{\min}(\mathbf{T}(\mathbf{z}))$, or $f(\mathbf{z}) = \log\det[\mathbf{T}(\mathbf{z})]$. These functions balance the spectrum of $\mathbf{T}(\mathbf{z})$.

Although the above problem can be solved using standard convex relaxation techniques [35], due to the involved complexity of solving the relaxed convex problem and keeping in mind the large-scale problems that might arise in the graph setting, we will now focus on the optimization problem (12.32) with $f(\mathbf{z}) = \log\det[\mathbf{T}(\mathbf{z})]$ as it can be solved near optimally using a low-complexity greedy algorithm. To do so, we introduce the concept of submodularity, a notion based on the property of diminishing returns. This is useful for solving discrete combinatorial optimization problems of the form (12.32) (see e.g., [36]). Submodularity can be formally defined as follows.

Definition 12.6. Given two sets $\mathcal{X}$ and $\mathcal{Y}$ such that for every $\mathcal{X} \subseteq \mathcal{Y} \subseteq \mathcal{N}$ and $s \in \mathcal{N}\backslash\mathcal{Y}$, the set function $f : 2^N \to \mathbb{R}$ defined on the subsets of $\mathcal{N}$ is said to be submodular, if it satisfies $f(\mathcal{X} \cup \{s\}) - f(\mathcal{X}) \geq f(\mathcal{Y} \cup \{s\}) - f(\mathcal{Y})$. ■

Suppose the submodular function is monotone nondecreasing, i.e., $f(\mathcal{X}) \leq f(\mathcal{Y})$ for all $\mathcal{X} \subseteq \mathcal{Y} \subseteq \mathcal{N}$ and normalized, i.e., $f(\emptyset) = 0$, then a greedy maximization of such a function as summarized in Algorithm 12.1 is *near optimal* with an approximation factor of $(1 - 1/e)$; see [37].

To use this framework, we have to rewrite $f(\mathbf{z}) = \log\det[\mathbf{T}(\mathbf{z})]$ as a set function

$$f(\mathcal{X}) = \log\det\left[\sum_{(i,j)\in\mathcal{X}\times\mathcal{X}} \mathbf{g}_{i,j}\mathbf{g}_{i,j}^T\right], \tag{12.33}$$

where the index set $\mathcal{X}$ is related to the component selection vector $\mathbf{z}$ as $\mathcal{X} = \{m \mid z_m = 1, m = 1, \ldots, N\}$ and the column vectors $\mathbf{g}_{i,j}$ correspond to the rows of $\mathbf{G}$ as $\mathbf{G} = [\mathbf{g}_{1,1}, \mathbf{g}_{1,2}, \ldots, \mathbf{g}_{N,N}]^T$. We use such an indexing because the sampling matrix $\boldsymbol{\Phi} \otimes \boldsymbol{\Phi}$ results in a Kronecker structured (row) subset selection.

Modifying this set function slightly to

$$f(\mathcal{X}) = \log\det\left[\sum_{(i,j)\in\mathcal{X}\times\mathcal{X}} \mathbf{g}_{i,j}\mathbf{g}_{i,j}^T + \epsilon\mathbf{I}\right] - N\log\epsilon, \tag{12.34}$$

we obtain a normalized, nondecreasing, submodular function on the set $\mathcal{X} \subset \mathcal{N}$. Here, $\epsilon > 0$ is a small constant. In Eq. (12.34), $\epsilon\mathbf{I}$ is needed to carry out the first few iterations of Algorithm 12.1 and $-N\log\epsilon$ ensures that $f(\emptyset)$ is zero. It is worth mentioning that the greedy algorithm is linear in K while computing Eq. (12.34) dominates the computational complexity. Finally, random subsampling (i.e., $\mathbf{z}$ having random 0 or 1 entries) is not suitable as it might not always result in a full-column rank model matrix.

In Fig. 12.4, we illustrate the PSD estimation based on the observations from a subset of nodes for 100 realizations of the random process on the Karate club network. For the nonparametric model, the selected graph nodes obtained from Algorithm 12.1 are indicated with a black circle in Fig. 12.4A. Based on the observations from these 20 selected graph nodes, the PSD estimate obtained using LS is shown in Fig. 12.4B. It can be seen that the PSD estimate based on the observations from a subset of nodes fits reasonably well to the true PSD.

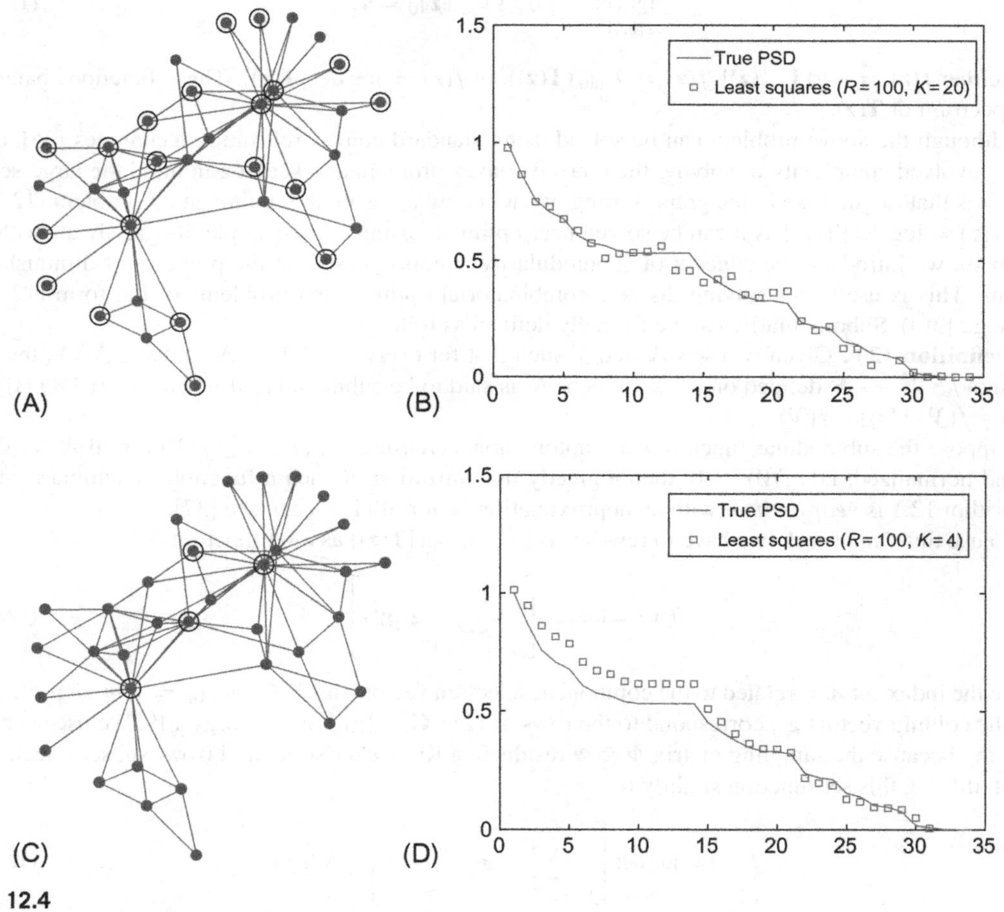

FIG. 12.4

PSD estimation from a subset of nodes. Estimators are based on a random process defined on the Karate club network [28]. (A) Graph sampling for nonparametric PSD estimation. Here, 20 out of 34 nodes are observed. The sampled nodes are highlighted by the *circles* around the nodes. (B) Nonparametric PSD estimation based on observations from 20 nodes and 100 data snapshots. (C) Graph sampling for parametric MA PSD estimation. Here, 4 out of 34 nodes are observed. (D) Parametric MA PSD estimation based on observations from 4 nodes and 100 data snapshots.

For the parametric MA model, wherein the PSD is parametrized with $Q = 7$ MA parameters, the selected graph nodes obtained from Algorithm 12.1 are shown in Fig. 12.4C and the reconstructed PSD using LS is shown in Fig. 12.4D. In this case, we sample only 4 out of 34 graph nodes and yet obtain a PSD estimate that fits very well to the true PSD.

12.5 DISCUSSION AND THE ROAD AHEAD

In this chapter, we have introduced the concept of weakly stationary graph processes and their related power spectral density. We discussed the links between the different definitions as well as the relations with classical signal processing. Furthermore, we extended this idea to processes that are jointly stationary in the vertex and time domain, where the subclass of separable processes is of particular importance due to their more parsimonious description. The chapter has also focused on estimating the PSD and the covariance using nonparametric as well as parametric methods. Equivalences and differences with classical PSD techniques for spatiotemporal signals have been established. Finally, we presented different techniques to estimate the PSD and the covariance from only a subset of the nodes without any loss of identifiability. This can be viewed as a particular instance of sparse covariance sampling. In this context, we also proposed a greedy method to select the best nodes to sample in order to guarantee a satisfying estimation of the PSD and the covariance.

While this chapter only covers weakly stationary graph processes, a definition of strict stationarity is still open. One option could be to define a strictly stationary graph process as the output of filtering i.i.d. noise. Ergodicity is also a concept that we did not discuss in this chapter. Ergodicity in a graph signal processing context would mean that the statistics of the graph process could be derived from successive graph shifts of a single realization (observed at one or multiple nodes) [38]. Due to the finite length of graph signals, this will entail certain problems and exact estimates of the statistics (even asymptotically) will rarely be possible. How to model nonstationary graph processes in an intuitively pleasing way is another unexplored area. A way forward in this direction could be the introduction of so-called node-varying graph filters [39], where the variation of the filter taps can be expanded in a particular basis. Filtering white or i.i.d. noise using such filters leads to a nonstationary graph process that is parametrized by a limited number of coefficients. Yet, other parametrized graph filter structures could be employed as a model for nonstationary graph processes, e.g., edge-variant graph filters [40] or median graph filters [41,42]. Finally, in this chapter, we limited ourselves to normal graph shift operators that are endowed with a unitary matrix of eigenvectors. Stationarity for nonnormal graph shifts (whether diagonalizable or not) is a topic for future research. Some of the concepts discussed in this chapter can be easily extended to nonorthonormal and/or generalized eigenvectors but others require more research.

REFERENCES

[1] Zhang F, Hancock ER. Graph spectral image smoothing using the heat kernel. Pattern Recogn 2008;41(11):3328–42.

[2] Pesenson I. Sampling in Paley-Wiener spaces on combinatorial graphs. Trans Am Math Soc 2008;360(10):5603–27.

[3] Chen S, Sandryhaila A, Moura J, Kovačević J. Signal recovery on graphs: variation minimization. IEEE Trans Signal Process 2015;63(17):4609–24.
[4] Chen S, Varma R, Sandryhaila A, Kovačević J. Discrete signal processing on graphs: sampling theory. IEEE Trans Signal Process 2015;63(24):6510–23.
[5] Anis A, Gadde A, Ortega A. Towards a sampling theorem for signals on arbitrary graphs. In: IEEE international conference on acoustics, speech and signal processing (ICASSP); 2014. p. 3864–8.
[6] Marques AG, Segarra S, Leus G, Ribeiro A. Sampling of graph signals with successive local aggregations. IEEE Trans Signal Process 2016;64(7):1832–43.
[7] Segarra S, Marques AG, Leus G, Ribeiro A. Reconstruction of graph signals through percolation from seeding nodes. IEEE Trans Signal Process 2016;64(16):4363–78.
[8] Shuman DI, Vandergheynst P, Frossard P. Distributed signal processing via Chebyshev polynomial approximation. CoRR 2011;abs/1111.5239.
[9] Sandryhaila A, Moura JMF. Discrete signal processing on graphs: frequency analysis. IEEE Trans Signal Process 2014;62(12):3042–54.
[10] Shi X, Feng H, Zhai M, Yang T, Hu B. Infinite impulse response graph filters in wireless sensor networks. IEEE Signal Process Lett 2015;22(8):1113–7.
[11] Isufi E, Loukas A, Simonetto A, Leus G. Autoregressive moving average graph filtering. IEEE Trans Signal Process 2017;65(2):274–88.
[12] Hayes MH. Statistical digital signal processing and modeling. John Wiley and Sons; 2009.
[13] Girault B. Stationary graph signals using an isometric graph translation. In: European signal processing conference (EUSIPCO); 2015. p. 1516–20.
[14] Girault B, Gonçalves P, Fleury E. Translation on graphs: an isometric shift operator. IEEE Signal Process Lett 2015;22(12):2416–20.
[15] Marques AG, Segarra S, Leus G, Ribeiro A. Stationary graph processes and spectral estimation. IEEE Trans Signal Process 2017;65(22):5911–26.
[16] Perraudin N, Vandergheynst P. Stationary signal processing on graphs. IEEE Trans Signal Process 2017;65(13):3462–77.
[17] Perraudin N, Loukas A, Grassi F, Vandergheynst P. Towards stationary time-vertex signal processing. In: IEEE international conference on acoustics, speech and signal processing (ICASSP); 2017. p. 3914–8.
[18] Romero D, Ariananda DD, Tian Z, Leus G. Compressive covariance sensing: structure-based compressive sensing beyond sparsity. IEEE Signal Process Mag 2016;33(1):78–93.
[19] Chepuri SP, Leus G. Graph sampling for covariance estimation. IEEE Trans Signal Inf Process Netw 2017;3(3):451–66.
[20] Sandryhaila A, Moura JMF. Discrete signal processing on graphs. IEEE Trans Signal Process 2013;61(7):1644–56.
[21] Gavili A, Zhang XP. On the shift operator, graph frequency and optimal filtering in graph signal processing. IEEE Trans Signal Process 2017;65(23):6303–18.
[22] Shuman DI, Ricaud B, Vandergheynst P. Vertex-frequency analysis on graphs. Appl Comput Harmon Anal 2016;40(2):260–91.
[23] Segarra S, Marques AG, Mateos G, Ribeiro A. Network topology inference from spectral templates. IEEE Trans Signal Inf Process Netw 2017;3(3):467–83.
[24] Murphy KP. Machine learning: a probabilistic perspective. MIT Press; 2012.
[25] Imrich W, Klavzar S. Product graphs: structure and recognition. Wiley; 2000.
[26] Sandryhaila A, Moura J. Big data analysis with signal processing on graphs: representation and processing of massive data sets with irregular structure. IEEE Signal Process Mag 2014;31(5):80–90.
[27] Lütkepohl H. New introduction to multiple time series analysis. Springer; 2007.
[28] Zachary WW. An information flow model for conflict and fission in small groups. J Anthropol Res 1977;33(4):452–73.

[29] Hayes MH. Statistical digital signal processing and modeling. John Wiley & Sons; 2009.
[30] Scharf LL. Statistical signal processing. Reading, MA: Addison-Wesley; 1991.
[31] Stoica P, Moses RL. Spectral analysis of signals. Upper Saddle River, NJ: Pearson/Prentice Hall; 2005.
[32] Fienup JR. Phase retrieval algorithms: a comparison. Appl Opt 1982;21(15):2758–69.
[33] Candes EJ, Li X, Soltanolkotabi M. Phase retrieval via Wirtinger flow: theory and algorithms. IEEE Trans Inf Theory 2015;61(4):1985–2007.
[34] Golub GH, Van Loan CF. Matrix computations. Johns Hopkins Studies in the Mathematical Sciences. Baltimore, MD: Johns Hopkins University Press; 1996.
[35] Chepuri SP, Leus G. Sparse sensing for statistical inference. Found Trends Signal Process 2016;9(3–4):233–368.
[36] Krause A. Optimizing sensing: theory and applications. Ph.D. dissertation, School of Computer Science. Pittsburgh, PA: Carnegie Mellon University; 2008.
[37] Nemhauser GL, Wolsey LA, Fisher ML. An analysis of approximations for maximizing submodular set functions—I. Math Program 1978;14(1):265–94.
[38] Gama F, Ribeiro A. Weak law of large numbers for stationary graph processes. In: IEEE international conference on acoustics, speech and signal processing (ICASSP); 2017. p. 4124–8.
[39] Segarra S, Marques AG, Ribeiro A. Optimal graph-filter design and applications to distributed linear network operators. IEEE Trans Signal Process 2017;65(15):4117–31.
[40] Coutino M, Isufi E, Leus G. Distributed edge-variant graph filters. In: IEEE international workshop on computational advances in multi-sensor adaptive processing (CAMSAP); 2017.
[41] Segarra S, Marques AG, Arce GR, Ribeiro A. Center-weighted median graph filters. In: Global conference on signal and information processing (GlobalSIP); 2016. p. 336–40.
[42] Segarra S, Marques AG, Arce G, Ribeiro A. Design of weighted median graph filters. In: IEEE international workshop on computational advances in multi-sensor adaptive processing (CAMSAP); 2017.

CHAPTER

INFERENCE OF GRAPH TOPOLOGY

13

Gonzalo Mateos*, **Santiago Segarra**†, **Antonio G. Marques**‡

Department of Electrical and Computer Engineering, University of Rochester, Rochester, NY, United States Institute for Data, Systems, and Society, Massachusetts Institute of Technology, Cambridge, MA, United States*† *Department of Signal Theory and Communications, King Juan Carlos University, Madrid, Spain*‡

13.1 INTRODUCTION

Coping with the challenges found at the intersection of Network Science and Big Data necessitates fundamental breakthroughs in modeling, identification, and controllability of distributed network processes often conceptualized as signals defined on graphs [1–3]. For instance, graph-supported signals can model vehicle trajectories over road networks [4]; economic activity observed over a network of production flows between industrial sectors [5,6]; infectious states of individuals susceptible to an epidemic disease spreading on a social network [7]; gene expression levels defined on top of gene regulatory networks [8–10]; brain activity signals supported on brain connectivity networks [11–13]; and media cascades that diffuse on online social networks [14,15], to name a few. There is an evident mismatch between our scientific understanding of signals defined over regular domains (time or space) and graph-valued signals. Knowledge about time series was developed over the course of decades and boosted by real needs in areas such as communications, speech, or control. On the contrary, the prevalence of network-related signal processing problems and the access to quality network data are recent events [1].

Under the assumption that the signals are related to the topology of the graph where they are supported, the goal of graph signal processing (GSP) is to develop algorithms that fruitfully leverage this relational structure, and can make inferences about these relationships when they are only partially observed [5,10,16]. Most GSP efforts to date assume that the underlying network is known, and then analyze how the graph's algebraic and spectral characteristics impact the properties of the graph signals of interest. However, such an assumption is often untenable in practice and arguably most graph construction schemes are largely informal, distinctly lacking an element of validation. In studies of, e.g., functional brain connectivity or regulation among genes, inference of nontrivial pairwise interactions between signal elements (i.e., blood oxygen-level dependent time series per voxel or gene expression levels, respectively) is often the goal per se.

Cooperative and Graph Signal Processing. https://doi.org/10.1016/B978-0-12-813677-5.00013-4

In this chapter we present a framework to leverage information available from graph signals to learn the underlying graph topology. The unknown graph represents direct relationships between signal elements, which one aims to recover from observable indirect relationships generated by a diffusion process on the graph. The fresh look advocated here leverages concepts from convex optimization and stationarity of graph signals in order to identify the graph-shift operator (a matrix representation of the graph) given only its eigenvectors. These *spectral templates* can be obtained, e.g., from the sample covariance of independent graph signals diffused on the sought network. The novel idea is to find a graph-shift operator that, while being consistent with the provided spectral information, endows the network with certain desired properties such as sparsity or minimum-energy edge weights. The focus of the chapter is on inferring *undirected* graphs. The recovery of directed graphs from diffused observations is briefly discussed in the last section of the chapter.

13.2 GRAPH INFERENCE: A HISTORICAL OVERVIEW

Network topology inference is a prominent problem in Network Science [10,17]. Because networks typically encode similarities between nodes, several topology inference approaches construct graphs whose edge weights correspond to nontrivial correlations or coherence measures between signal profiles at incident nodes. In this vein, informal (but widely popular) scoring methods rely on ad hoc thresholding of user-defined score functions such a Pearson product-moment correlation, Spearman rank correlation, or mutual information. Formal hypothesis-testing methods to assess nontrivial correlations have been proposed as well [10, Ch. 7.3.1]. Because in graph inference settings one performs a hypothesis test per node pair, the problem of multiple testing is prevalent and often addressed via, e.g., false discovery rate (FDR) procedures.

Acknowledging that the observed correlations can be due to latent network effects, alternative statistical methods rely on inference of full partial correlations to eliminate potential confounding variables [10, Ch. 7.3.2]. Under Gaussianity assumptions, this line of work has well-documented connections with covariance selection [18] and sparse precision matrix estimation [19–23] as well as high-dimensional sparse linear regression [24]. Extensions to directed graphs include structural equation models (SEMs) [14,25,26], Granger causality [17,27], or their nonlinear (e.g., kernelized) variants [28,29].

Recent GSP-based network inference frameworks postulate instead that the network exists as a latent underlying structure, and that observations are generated as a result of a network process defined in such a graph [30–35]. For instance, network structure is estimated in [33] to unveil unknown relations among nodal time series adhering to an autoregressive model involving graph filter dynamics. Different from [32–35] that operate on the graph domain, the goal here is to identify graphs that endow the given observations with desired spectral (frequency-domain) characteristics. Two works have recently explored this approach by identifying a matrix representation of the network given its eigenvectors [30, 31], and rely on observations of stationary graph signals [36–38]. Different from [34,35,39,40] that infer structure from signals assumed to be smooth over the sought graph, here the measurements are assumed related to the graph via filtering (e.g., modeling the diffusion of an idea or the spreading of a disease). Smoothness models can be subsumed as special cases encountered when diffusion filters have a low-pass frequency response.

13.3 GRAPH INFERENCE FROM DIFFUSED SIGNALS

A weighted and undirected graph $\mathcal{G}$ consists of a node set $\mathcal{N}$ of cardinality N, an edge set $\mathcal{E}$ of unordered pairs of elements in $\mathcal{N}$, and edge weights $A_{ij} \in \mathbb{R}$ such that $A_{ij} = A_{ji} \neq 0$ for all $(i,j) \in \mathcal{E}$. The edge weights A_{ij} are collected as entries of the symmetric adjacency matrix $\mathbf{A}$ and the node degrees in the diagonal matrix $\mathbf{D} := \text{diag}(\mathbf{A1})$. These are used to form the combinatorial Laplacian matrix $\mathbf{L}_c := \mathbf{D}-\mathbf{A}$ and the normalized Laplacian $\mathbf{L} := \mathbf{I} - \mathbf{D}^{-1/2}\mathbf{A}\mathbf{D}^{-1/2}$. More broadly, one can define a generic graph-shift operator (GSO) $\mathbf{S} \in \mathbb{R}^{N\times N}$ as any matrix whose off-diagonal sparsity pattern is equal to that of the adjacency matrix of $\mathcal{G}$ [2]. Although the choice of $\mathbf{S}$ can be adapted to the problem at hand, most existing works set it to either $\mathbf{A}$, $\mathbf{L}_c$, or $\mathbf{L}$. See Table 13.1 for a summary of relevant notation.

13.3.1 STRUCTURE OF A NETWORK DIFFUSION PROCESS

The main focus of this chapter is on identifying graphs that explain the structure of a random signal. Formally, let $\mathbf{x} = [x_1, \ldots, x_N]^T \in \mathbb{R}^N$ be a graph signal in which the ith element x_i denotes the signal value at node i of an unknown graph $\mathcal{G}$ with symmetric shift operator $\mathbf{S}$. Further suppose that we are given a zero-mean white signal $\mathbf{w}$ with covariance matrix $\mathbb{E}\left[\mathbf{w}\mathbf{w}^T\right] = \mathbf{I}$. We say that $\mathbf{S}$ *represents the structure of* the signal $\mathbf{x}$ *if* there exists a diffusion process in the GSO $\mathbf{S}$ that produces the signal $\mathbf{x}$ from the white signal $\mathbf{w}$, that is,

$$\mathbf{x} = \alpha_0 \prod_{l=1}^{\infty}(\mathbf{I} - \alpha_l \mathbf{S})\mathbf{w} = \sum_{l=0}^{\infty} \beta_l \mathbf{S}^l \mathbf{w}. \tag{13.1}$$

Table 13.1 Notation

Notation	Description
$\mathbf{x}$	vector with ith entry x_i
$\mathbf{X}$	matrix with (i,j)th entry X_{ij}
$\mathcal{I}$	set
$\mathbf{X}_{\mathcal{I}}$	submatrix of $\mathbf{X}$ formed by the rows indexed by $\mathcal{I}$
$(\cdot)^T$	matrix transpose
$(\cdot)^\dagger$	matrix pseudoinverse
$\text{vec}(\cdot)$	matrix vectorization operator
$\sigma_{\min}(\cdot)$	minimum singular value of argument matrix
$\otimes$	Kronecker product
$\odot$	Khatri-Rao (columnwise Kronecker) product
$\text{tr}\{\cdot\}$	matrix trace
$\Vert\mathbf{x}\Vert_p$	vector ℓ_p-norm
$\Vert\mathbf{X}\Vert_p$	vector ℓ_p-norm of the vectorized form of $\mathbf{X}$
$\Vert\mathbf{X}\Vert_{M(p)}$	induced matrix ℓ_p-norm
$\Vert\mathbf{X}\Vert_F := \sqrt{\text{tr}\{\mathbf{X}^T\mathbf{X}\}}$	matrix Frobenius norm
$\text{diag}(\mathbf{x})$	diagonal matrix with (i,i)th entry x_i
$\mathbf{I}$	identity matrix
$\mathbf{0}$	all-zero vector
$\mathbf{1}$	all-one vector

While $\mathbf{S}$ encodes only one-hop interactions, each successive application of the shift percolates (correlates) the original information across an iteratively increasing neighborhood; see e.g. [41]. The product and sum representations in Eq. (13.1) are common and equivalent models for the generation of random signals. Indeed, any process that can be understood as the linear propagation of a white input through a static, undirected graph can be written in the form in Eq. (13.1). These include, for example, processes generated by the so-called *diffusion* Laplacian *kernels* [42].

The justification to say that $\mathbf{S}$ is the structure of $\mathbf{x}$ is that we can think of the edges of $\mathbf{S}$ as direct (one-hop) relationships between the elements of the signal. The diffusion described by Eq. (13.1) generates indirect relationships. In this context, the network topology inference problem is to recover the fundamental relationships described by $\mathbf{S}$ from a set $\mathcal{X} := \{\mathbf{x}_r\}_{r=1}^R$ of R independent snapshots of the random signal $\mathbf{x}$. We show next that this is an underdetermined problem.

Because we focus on the inference of undirected graphs, the shift operator $\mathbf{S}$ is symmetric and diagonalizable. Hence, upon defining the orthogonal eigenvector matrix $\mathbf{V} := [\mathbf{v}_1, \ldots, \mathbf{v}_N]$ and the eigenvalue matrix $\mathbf{\Lambda} := \text{diag}(\boldsymbol{\lambda})$ with $\boldsymbol{\lambda} := [\lambda_1, \ldots, \lambda_N]^T$, it holds that

$$\mathbf{S} = \mathbf{V}\mathbf{\Lambda}\mathbf{V}^T = \mathbf{V}\text{diag}(\boldsymbol{\lambda})\mathbf{V}^T. \tag{13.2}$$

Further observe that while the diffusion expressions in Eq. (13.1) are polynomials on the GSO of possibly infinite degree, the Cayley-Hamilton theorem implies that they are equivalent to polynomials of a degree smaller than N. Upon defining the vector of coefficients $\mathbf{h} := [h_0, \ldots, h_{L-1}]^T$ and the graph filter $\mathbf{H} \in \mathbb{R}^{N \times N}$ as $\mathbf{H} := \sum_{l=0}^{L-1} h_l \mathbf{S}^l$, the generative model in Eq. (13.1) can be rewritten as

$$\mathbf{x} = \left(\sum_{l=0}^{L-1} h_l \mathbf{S}^l \right) \mathbf{w} = \mathbf{H}\mathbf{w} \tag{13.3}$$

for some particular $\mathbf{h}$ and L. Because a graph filter $\mathbf{H}$ is a polynomial on $\mathbf{S}$ [2], graph filters are linear graph-signal operators that have the *same eigenvectors* as the shift (i.e., the operators $\mathbf{H}$ and $\mathbf{S}$ commute).

More important for the arguments in Sections 13.3 and 13.4, the filter representation in Eq. (13.3) can be used to show that *the eigenvectors of* $\mathbf{S}$ *are also eigenvectors of the covariance matrix* $\mathbf{C}_x := \mathbb{E}\left[\mathbf{x}\mathbf{x}^T\right]$. To that end, notice that because $\mathbf{w}$ is white the said covariance is given by

$$\mathbf{C}_x = \mathbb{E}\left[\mathbf{H}\mathbf{w}(\mathbf{H}\mathbf{w})^T\right] = \mathbf{H}\mathbb{E}\left[\mathbf{w}\mathbf{w}^T\right]\mathbf{H}^T = \mathbf{H}\mathbf{H}^T. \tag{13.4}$$

If we further use the spectral decomposition of the shift in Eq. (13.2) to express the filter as $\mathbf{H} = \sum_{l=0}^{L-1} h_l (\mathbf{V}\mathbf{\Lambda}\mathbf{V}^T)^l = \mathbf{V}(\sum_{l=0}^{L-1} h_l \mathbf{\Lambda}^l)\mathbf{V}^T$, we can write the covariance matrix as

$$\mathbf{C}_x = \mathbf{V} \left(\sum_{l=0}^{L-1} h_l \mathbf{\Lambda}^l \right)^2 \mathbf{V}^T. \tag{13.5}$$

A consequence of Eq. (13.5) is that the *eigenvectors* of the shift $\mathbf{S}$ and the covariance $\mathbf{C}_x$ are the same. Alternatively, one can say that the difference between $\mathbf{C}_x$ in Eq. (13.5), which includes indirect relationships between components, and $\mathbf{S}$ in Eq. (13.2), which includes exclusively direct relationships, is only on their *eigenvalues*. While the diffusion in Eq. (13.1) obscures the eigenvalues of $\mathbf{S}$, the eigenvectors $\mathbf{V}$ remain present in $\mathbf{C}_x$ as templates of the original spectrum. Additional motivation for the model in (13.5) is provided in Remark 13.1.

Identity (13.5) also shows that the problem of finding a GSO that generates $\mathbf{x}$ from a white input $\mathbf{w}$ with unknown coefficients [cf. Eq. (13.1)] is *underdetermined*. As long as the matrices $\mathbf{S}$ and $\mathbf{C}_x$ have the same eigenvectors, filter coefficients that generate $\mathbf{x}$ through a diffusion process on $\mathbf{S}$ exist. In fact, the covariance matrix $\mathbf{C}_x$ itself is a GSO that can generate $\mathbf{x}$ through a diffusion process and so is the precision matrix $\mathbf{C}_x^{-1}$. To sort out this ambiguity, which amounts to selecting the eigenvalues of the shift, we assume that the GSO of interest is optimal in some sense. This pursuit is the subject of the next section, where we formally state the graph inference problem.

Remark 13.1 (Graph stationarity meets topology inference). Recently, a group of works has generalized the definition of stationarity to graph processes [36–38]. In a nutshell, a graph signal is stationary in a particular GSO $\mathbf{S}$ if either the signal can be expressed as the output of a graph filter with white inputs [36, Def. 2], or if its covariance matrix is simultaneously diagonalizable with $\mathbf{S}$ [36, Def. 3]. These are precisely the conditions in Eqs. (13.3) and (13.5), respectively. Hence, our problem of identifying a GSO that explains the fundamental structure of $\mathbf{x}$ is equivalent to the problem of identifying a shift $\mathbf{S}$ in which the signal $\mathbf{x}$ is stationary. Notice that this is equivalent to saying that the mapping between the sought shift $\mathbf{S}$ and the covariance matrix $\mathbf{C}_x$ is given by a (matrix) polynomial, including $\mathbf{C}_x = \mathbf{S}$ and $\mathbf{C}_x = \mathbf{S}^{-1}$ as particular cases. Advances dealing with identification of undirected networks from diffused nonstationary graph signals are outlined in Section 13.5.

13.3.2 OPTIMAL GRAPH SHIFT OPERATOR

Given estimates $\hat{\mathbf{V}}$ of the filter eigenvectors (e.g., obtained from observations $\mathcal{X} := \{\mathbf{x}_r\}_{r=1}^R$ via the eigenvectors of the *sample covariance* $\hat{\mathbf{C}}_x = \frac{1}{R}\sum_{r=1}^R \mathbf{x}_r\mathbf{x}_r^T$), recovery of $\mathbf{S}$ amounts to selecting its eigenvalues $\mathbf{\Lambda}$ and to that end we assume that the shift of interest is optimal in some sense. At the same time, we should account for the discrepancies between $\hat{\mathbf{V}}$ and the actual eigenvectors of $\mathbf{S}$, due to finite sample size constraints and noise corrupting the observations in $\mathcal{X}$. Accordingly, we seek a shift operator $\mathbf{S}$ that: (a) is optimal with respect to (often convex) criteria $f(\mathbf{S})$; (b) belongs to a convex set $\mathcal{S}$ that specifies the desired type of shift operator (e.g., the adjacency $\mathbf{A}$ or Laplacian $\mathbf{L}$); and (c) is close to $\hat{\mathbf{V}}\mathbf{\Lambda}\hat{\mathbf{V}}^T$ as measured by a convex matrix distance $d(\cdot,\cdot)$. Formally, one can solve

$$\mathbf{S}^* := \underset{\mathbf{\Lambda},\mathbf{S}\in\mathcal{S}}{\operatorname{argmin}}\ f(\mathbf{S}), \quad \text{s.t. } d(\mathbf{S}, \hat{\mathbf{V}}\mathbf{\Lambda}\hat{\mathbf{V}}^T) \leq \epsilon, \tag{13.6}$$

which is a convex optimization problem provided $f(\mathbf{S})$ is convex, and ϵ is a tuning parameter chosen based on a priori information on the imperfections. Within the scope of the signal model (13.1), the formulation (13.6) entails a general class of network topology inference problems parametrized by the choices in (a)–(c) above. Following a formal statement of the problem, we briefly outline the spectrum of alternatives for points (a)–(c), while concrete choices are made for the analysis and numerical tests in Sections 13.4 and 13.5.

Problem statement. Given a set $\mathcal{X} := \{\mathbf{x}_r\}_{r=1}^R$ of R independent samples of the random signal $\mathbf{x}$ adhering to Eq. (13.1), estimate the optimal description of the structure of $\mathbf{x}$ in the form of the graph-shift operator $\mathbf{S}^* \in \mathcal{S}$ defined in Eq. (13.6).

Criteria. The selection of $f(\mathbf{S})$ allows incorporating physical characteristics of the desired graph into the formulation while being consistent with the spectral templates $\hat{\mathbf{V}}$. For instance, the matrix pseudonorm $f(\mathbf{S}) = \|\mathbf{S}\|_0$, which counts the number of nonzero entries in $\mathbf{S}$, can be used to minimize the number of edges toward identifying sparse graphs (e.g., of direct relations among signal elements); $f(\mathbf{S}) = \|\mathbf{S}\|_1$ is a convex proxy for the aforementioned edge cardinality function. Alternatively, the Frobenius norm $f(\mathbf{S}) = \|\mathbf{S}\|_F$ can be adopted to minimize the energy of the edges in the graph, or $f(\mathbf{S}) = \|\mathbf{S}\|_\infty$ can be selected to obtain shifts $\mathbf{S}$ associated with graphs of uniformly low edge weights. This can be meaningful when identifying graphs subject to capacity constraints.

Constraints. The constraint $\mathbf{S} \in \mathcal{S}$ in Eq. (13.6) incorporates a priori knowledge about $\mathbf{S}$. If we let $\mathbf{S} = \mathbf{A}$ represent the adjacency matrix of an undirected graph with nonnegative weights and no self-loops, we can explicitly write $\mathcal{S}$ as follows

$$\mathcal{S}_{\mathrm{A}} := \left\{\mathbf{S} \,|\, S_{ij} \geq 0, \quad \mathbf{S} \in \mathcal{M}_N, \quad S_{ii} = 0, \quad \sum_j S_{j1} = 1\right\}. \tag{13.7}$$

The first condition in $\mathcal{S}_{\mathrm{A}}$ encodes the nonnegativity of the weights whereas the second condition incorporates that $\mathcal{G}$ is undirected, hence, $\mathbf{S}$ must belong to the set $\mathcal{M}_N$ of real and symmetric $N \times N$ matrices. The third condition encodes the absence of self-loops, thus, each diagonal entry of $\mathbf{S}$ must be null. Finally, the last condition fixes the scale of the admissible graphs by setting the weighted degree of the first node to 1, and rules out the trivial solution $\mathbf{S} = \mathbf{0}$. Other GSOs such as the normalized Laplacian $\mathbf{L}$ can be accommodated in this framework via minor adaptations to $\mathcal{S}$; see [30].

The form of the convex matrix distance $d(\cdot,\cdot)$ depends on the particular application. For instance, if $\|\mathbf{S} - \hat{\mathbf{V}}\mathbf{\Lambda}\hat{\mathbf{V}}^T\|_F$ is chosen, the focus is more on the similarities across the entries of the shifts while $\|\mathbf{S} - \hat{\mathbf{V}}\mathbf{\Lambda}\hat{\mathbf{V}}^T\|_{\mathcal{M}(2)}$ focuses on their spectrum.

13.3.3 JOINT INFERENCE OF MULTIPLE GRAPHS

So far we have dealt with the identification of a *single* graph from signal observations under the assumption that the available observations are stationary on the sought shift. However, many contemporary setups involve *multiple* related networks, each with a subset of available observations. This is the case in multihop communication networks in dynamic environments where links are created or destroyed as nodes change their position; in neuroscience where observations for different patients are available and the objective is to estimate their brain functional networks, and in gene-to-gene networks where the goal is to identify pairwise interactions between genes when measurements for different tissues of the same patient are available. Looking at the joint identification of multiple shifts can be useful even if the interest is only in one of the networks because joint formulations exploit additional sources of information and, hence, are likely to give rise to better solutions. Although noticeably less than its single-network counterpart, joint inference of multiple networks has attracted attention, especially for the case of (Gaussian) Markov random fields, which give rise to sparse precision matrices [43–47], and in the context of dynamic (time-varying) graphs [48–50]. All the aforementioned works consider that the multiple graphs share a common node set while being allowed to have different edge sets, a structure often referred to as a multilayer graph [51]. Given the previous motivation, in this section we extend the problem formulation in Eq. (13.6) to encompass the case of joint network inference.

To formally state our joint network topology inference problem, consider a scenario with K different graphs $\{\mathcal{G}^{(k)}\}_{k=1}^K$ defined over the same set $\mathcal{N}$ of nodes, but with possibly different sets of edges and weights. This implies the existence of K different GSOs $\{\mathbf{S}^{(k)}\}_{k=1}^K$, all represented by $N \times N$ matrices, whose sparsity pattern and nonzero values may be different across k. Suppose that, associated with each of the graphs, we have access to a set of graph signals collecting information attached to the nodes. Formally, we use $\mathbf{X}^{(k)} := [\mathbf{x}_1^{(k)}, \ldots, \mathbf{x}_{R_k}^{(k)}] \in \mathbb{R}^{N \times R_k}$ to denote the matrix containing the R_k graph signals associated with graph $\mathcal{G}^{(k)}$. As done for the case of a single graph, we assume that $\{\mathbf{x}_i^{(k)}\}_{i=1}^{R_k}$ are independent realizations of a random process $\mathbf{x}^{(k)}$ whose structure is represented by $\mathbf{S}^{(k)}$ [cf. Eq. (13.1)]. In order to justify the concept of joint inference, we further assume that graphs k and k' are *similar*, which we encode as some matrix distance $d(\mathbf{S}^{(k)}, \mathbf{S}^{(k')})$ being small. Hence, the problem of joint network inference amounts to finding $\{\mathbf{S}^{(k)}\}_{k=1}^{K}$ when given the signals $\{\mathbf{X}^{(k)}\}_{k=1}^{K}$ generated as in Eq. (13.1) and under the assumption of similarity between different graphs.

Mimicking the development in Section 13.3.2, given estimates $\{\hat{\mathbf{V}}^{(k)}\}_{k=1}^{K}$ of the eigenvectors of each of the sought GSOs (e.g., obtained from the sample covariances of the different sets of observations $\{\mathbf{X}^{(k)}\}_{k=1}^{K}$), recovery of the GSOs boils down to selecting the optimal eigenvalues $\{\mathbf{\Lambda}^{(k)}\}_{k=1}^{K}$. In this respect, problem (13.6) can be extended to obtain

$$\{\mathbf{S}^{*(k)}\}_{k=1}^{K} := \underset{\{\mathbf{\Lambda}^{(k)}, \mathbf{S}^{(k)} \in \mathcal{S}\}_{k=1}^{K}}{\operatorname{argmin}} \sum_{k=1}^{K} \gamma_k f(\mathbf{S}^{(k)}) + \sum_{k<k'} \nu_{k,k'} d_1(\mathbf{S}^{(k)}, \mathbf{S}^{(k')}) \quad \text{s.t } d_2(\mathbf{S}^{(k)}, \hat{\mathbf{V}}^{(k)} \mathbf{\Lambda}^{(k)} \hat{\mathbf{V}}^{(k)T}) \leq \epsilon, \quad \text{for all } k, \tag{13.8}$$

where the weights γ_k and $\nu_{k,k'}$ respectively encode the relative importance of the optimality criterion on each GSO and the level of similarity between pairs of GSOs. As discussed in Section 13.3.2, the optimal criterion f and the constraint set $\mathcal{S}$ can be selected to promote or enforce specific properties in the sought graphs. Specific to formulation (13.8), we can also choose the distance function d_1 to promote different modes of similarity between GSOs. For example, choosing as distance d_1 the norms $\|\mathbf{S}^{(k)} - \mathbf{S}^{(k')}\|_0$ or $\|\mathbf{S}^{(k)} - \mathbf{S}^{(k')}\|_1$ would promote the pair of shifts to have the same sparsity pattern and weights whereas selecting d_1 as $\|\mathbf{S}^{(k)} - \mathbf{S}^{(k')}\|_{\text{F}}$ would promote similar weights regardless of the sparsity pattern.

For more details and results about the problem of jointly inferring *multiple* graph shift operators, including provable guarantees of the associated algorithms, we refer the reader to [52]. For concreteness, the remainder of this chapter focuses on the problem of inferring a *single* graph.

13.4 ROBUST NETWORK TOPOLOGY INFERENCE

This section deals with robust network inference problems from imperfect (noisy or incomplete) spectral templates.

13.4.1 NOISY SPECTRAL TEMPLATES

We first address the case where knowledge of an approximate version of the spectral templates $\hat{\mathbf{V}} = [\hat{\mathbf{v}}_1, \ldots, \hat{\mathbf{v}}_N]$ is available, e.g., from the eigenvectors of a *sample* covariance matrix $\hat{\mathbf{C}}_x$. For the particular case of sparse shifts, adopting $f(\mathbf{S}) = \|\mathbf{S}\|_1$ as the criterion in Eq. (13.6) and $d(\mathbf{S}, \hat{\mathbf{V}}\mathbf{\Lambda}\hat{\mathbf{V}}^T) = \|\mathbf{S} - \hat{\mathbf{V}}\mathbf{\Lambda}\hat{\mathbf{V}}^T\|_F$ yields

$$\mathbf{S}_1^* := \underset{\mathbf{\Lambda}, \mathbf{S} \in \mathcal{S}}{\operatorname{argmin}} \|\mathbf{S}\|_1, \quad \text{s.t. } \|\mathbf{S} - \hat{\mathbf{V}}\mathbf{\Lambda}\hat{\mathbf{V}}^T\|_F \leq \epsilon. \tag{13.9}$$

Note that $\|\mathbf{S}\|_1$ in Eq. (13.9) refers to the ℓ_1 norm of the vectorized version of $\mathbf{S}$. Moreover, further uncertainties can be introduced in the definition of the feasible set $\mathcal{S}$, e.g., in the scale of the admissible graphs for the case of $\mathcal{S} = \mathcal{S}_A$ (cf. Proposition 13.1 and [30] for additional details).

To assess the effect of the noise in recovering the sparsest $\mathbf{S}$ (henceforth denoted as $\mathbf{S}_0^*$, the solution of Eq. (13.9) when $f(\mathbf{S}) = \|\mathbf{S}\|_0$), some additional notation must be introduced. Define $\hat{\mathbf{W}} := \hat{\mathbf{V}} \odot \hat{\mathbf{V}} \in \mathbb{R}^{N^2 \times N}$, where $\odot$ denotes the Khatri-Rao product. Let $\mathbf{s}_0^* := \text{vec}(\mathbf{S}_0^*)$, denote by $\mathcal{D}$ the diagonal indices such that $(\mathbf{s}_0^*)_{\mathcal{D}} = \text{diag}(\mathbf{S}_0^*)$ and partition its complement $\mathcal{D}^c$ into $\mathcal{K}$ and $\mathcal{K}^c$, with the former indicating the positions of the nonzero entries of $\mathbf{s}_{0\mathcal{D}^c}^* := (\mathbf{s}_0^*)_{\mathcal{D}^c}$, where matrix *calligraphic subscripts* select rows. Denoting by † the matrix pseudoinverse, we define $\hat{\mathbf{M}} := (\mathbf{I} - \hat{\mathbf{W}}\hat{\mathbf{W}}^\dagger)_{\mathcal{D}^c} \in \mathbb{R}^{N^2 - N \times N^2}$, i.e., the orthogonal projector onto the kernel of $\hat{\mathbf{W}}^T$ constrained to the off-diagonal elements in $\mathcal{D}^c$. With $\mathbf{e}_1$ denoting the first canonical basis vector, we construct

$$\hat{\mathbf{R}} := [\hat{\mathbf{M}}, \mathbf{e}_1 \otimes \mathbf{1}_{N-1}] \in \mathbb{R}^{N^2 - N \times N^2 + 1} \tag{13.10}$$

by horizontally concatenating $\hat{\mathbf{M}}$ and a column vector of size $|\mathcal{D}^c|$ with ones in the first $N-1$ positions and zeros elsewhere. Further, we drop the nonnegativity constraint in $\mathcal{S}_A$—to obtain $\tilde{\mathcal{S}}_A$—and incorporate the scale ambiguity by augmenting $d(\mathbf{S}, \mathbf{S}')$ as $\tilde{d}(\mathbf{S}, \mathbf{S}') = (d(\mathbf{S}, \mathbf{S}')^2 + (\sum_j S_{j1} - 1)^2)^{1/2}$. With this notation, the following result on robust recovery of network topologies holds (see [30] for a proof).

Proposition 13.1. *Assuming that there exists at least one $\mathbf{S}'$ such that $\tilde{d}(\mathbf{S}_0^*, \mathbf{S}') \leq \epsilon$, the solution $\hat{\mathbf{s}}_1^* := \text{vec}(\hat{\mathbf{S}}_1^*)$ to Eq. (13.9) for $\mathcal{S} = \tilde{\mathcal{S}}_A$ with scale ambiguity satisfies*

$$\|\hat{\mathbf{s}}_1^* - \mathbf{s}_0^*\|_1 \leq C\epsilon, \quad \textit{with } C = 2C_1 + 2C_2C_3 \tag{13.11}$$

if the two following conditions are satisfied:

(1) $\text{rank}(\hat{\mathbf{R}}_{\mathcal{K}}) = |\mathcal{K}|$; *and*
(2) *There exists a constant $\delta > 0$ such that*

$$\psi := \|\mathbf{I}_{\mathcal{K}^c}(\delta^{-2}\hat{\mathbf{R}}\hat{\mathbf{R}}^T + \mathbf{I}_{\mathcal{K}^c}^T\mathbf{I}_{\mathcal{K}^c})^{-1}\mathbf{I}_{\mathcal{K}}^T\|_{M(\infty)} < 1. \tag{13.12}$$

Constants C_1, C_2, and C_3 are given by

$$C_1 = \frac{\sqrt{|\mathcal{K}|}}{\sigma_{\min}(\hat{\mathbf{R}}_{\mathcal{K}}^T)}, \quad C_2 = \frac{1 + \|\hat{\mathbf{R}}^T\|_2 C_1}{1 - \psi}, \quad C_3 = \|\hat{\mathbf{R}}^\dagger\|_2 N, \tag{13.13}$$

where $\sigma_{\min}(\cdot)$ denotes the minimum singular value of the argument matrix. ■

When given noisy versions $\hat{\mathbf{V}}$ of the spectral templates of our target GSO, Proposition 13.1 quantifies the effect that the noise has on the recovery. More precisely, the recovered shift is guaranteed to be at a maximum distance from the desired shift bounded by the tolerance ϵ times a constant, which depends on $\hat{\mathbf{R}}$ and the support $\mathcal{K}$. This also implies that as the number of observed signals increases we recover the true GSO. In particular, as the number of observed signals increases, the sample covariance

$\hat{\mathbf{C}}_x$ tends to the covariance $\mathbf{C}_x$ and, for the cases where the latter has no repeated eigenvalues, the noisy eigenvectors $\hat{\mathbf{V}}$ tend to the eigenvectors $\mathbf{V}$ of the desired shift; see, e.g., [53, Th. 3.3.7]. In particular, with better estimates $\hat{\mathbf{V}}$ the tolerance ϵ in Eq. (13.9) needed to guarantee feasibility can be made smaller, entailing a smaller discrepancy between the recovered $\mathbf{S}_1^*$ and the sparsest shift $\mathbf{S}_0^*$. In the limit when $\hat{\mathbf{V}} = \mathbf{V}$ and under no additional uncertainties, the tolerance ϵ can be made zero and Eq. (13.11) guarantees perfect recovery under conditions (1) and (2).

13.4.2 INCOMPLETE SPECTRAL TEMPLATES

Thus far we have assumed that an estimate of the *entire set* of eigenvectors $\mathbf{V} = [\mathbf{v}_1, \ldots, \mathbf{v}_N]$ is known. However, there are scenarios where only some of the eigenvectors (say K out of N) are available. This would be the case when, e.g., the given signal ensemble is bandlimited and $\mathbf{V}$ is found as the eigenbasis of the low-rank $\mathbf{C}_x$. More generally, if $\mathbf{C}_x$ contains repeated eigenvalues there is a rotation ambiguity in the definition of the associated eigenvectors. Hence, in this case, we keep the eigenvectors that can be unambiguously characterized and, for the remaining ones, we include the rotation ambiguity as an additional constraint in our optimization problem.

Formally assume that the K first eigenvectors $\mathbf{V}_K = [\mathbf{v}_1, \ldots, \mathbf{v}_K]$ are those that are known. For simplicity of exposition, suppose as well that $\mathbf{V}_K$ is estimated to be error free. Then, the network topology inference problem with incomplete spectral templates can be formulated as

$$\bar{\mathbf{S}}_1^* := \underset{\mathbf{S} \in \mathcal{S}, \mathbf{S}_{\bar{K}}, \boldsymbol{\lambda}}{\operatorname{argmin}} \ \|\mathbf{S}\|_1, \quad \text{s.t. } \mathbf{S} = \mathbf{S}_{\bar{K}} + \textstyle\sum_{k=1}^{K} \lambda_k \mathbf{v}_k \mathbf{v}_k^T, \quad \mathbf{S}_{\bar{K}} \mathbf{V}_K = \mathbf{0}, \tag{13.14}$$

where we already particularized the objective to the ℓ_1-norm convex relaxation. The formulation in Eq. (13.14) enforces $\mathbf{S}$ to be partially diagonalized by the known spectral templates, $\mathbf{V}_K$, while its remaining component $\mathbf{S}_{\bar{K}}$ is forced to belong to the orthogonal complement of range($\mathbf{V}_K$). Notice that, as a consequence, the rank of $\mathbf{S}_{\bar{K}}$ is at most $N - K$. An advantage of using only partial information of the eigenbasis as opposed to the whole $\mathbf{V}$ is that the set of feasible solutions in Eq. (13.14) is larger than that in Eq. (13.6). This is particularly important when the templates do not come from a prescribed GSO but, rather, one has the freedom to choose $\mathbf{S}$ provided it satisfies certain spectral properties (see [41] for examples in the context of distributed estimation).

GSO recovery guarantees can be derived for problem (13.14) [30]. To formally state these, define $\mathbf{W}_K := \mathbf{V}_K \odot \mathbf{V}_K$ and $\boldsymbol{\Upsilon} := [\mathbf{I}_{N^2}, \mathbf{0}_{N^2 \times N^2}]$. Also, define matrices $\mathbf{B}^{(i,j)} \in \mathbb{R}^{N \times N}$ for $i < j$ such that $B_{ij}^{(i,j)} = 1$, $B_{ji}^{(i,j)} = -1$, and all other entries are zero. Based on this, we denote by $\mathbf{B} \in \mathbb{R}^{\binom{N}{2} \times N^2}$ a matrix whose rows are the vectorized forms of $\mathbf{B}^{(i,j)}$ for all $i, j \in \{1, 2, \ldots, N\}$ where $i < j$. In this way, $\mathbf{Bs} = \mathbf{0}$ when $\mathbf{s}$ is the vectorized form of a symmetric matrix. Further, we define the following matrices

$$\mathbf{P}_1 := \begin{bmatrix} \mathbf{I} - \mathbf{W}_K \mathbf{W}_K^\dagger \\ \mathbf{I}_{\mathcal{D}} \\ \mathbf{B} \\ \mathbf{0}_{NK \times N^2} \\ (\mathbf{e}_1 \otimes \mathbf{1}_N)^T \end{bmatrix}^T, \quad \mathbf{P}_2 := \begin{bmatrix} \mathbf{W}_K \mathbf{W}_K^\dagger - \mathbf{I} \\ \mathbf{0}_{N \times N^2} \\ \mathbf{0}_{\binom{N}{2} \times N^2} \\ \mathbf{I} \otimes V_K^T \\ \mathbf{0}_{1 \times N^2} \end{bmatrix}^T, \tag{13.15}$$

and $\mathbf{P} := [\mathbf{P}_1^T, \mathbf{P}_2^T]^T$. With this notation and denoting by $\mathcal{J}$ the indices of the support of $\mathbf{s}_0^* = \text{vec}(\mathbf{S}_0^*)$, the following result is proved in [30].

Proposition 13.2. *Whenever $\mathcal{S} = \mathcal{S}_{\text{A}}$ and assuming problem (13.14) is feasible, $\bar{\mathbf{S}}_1^* = \mathbf{S}_0^*$ if the two following conditions are satisfied:*

(1) $\text{rank}([\mathbf{P}_{1\mathcal{J}}^T, \mathbf{P}_2^T]) = |\mathcal{J}| + N^2$; *and*
(2) *There exists a constant $\delta > 0$ such that*

$$\eta := \|\boldsymbol{\Upsilon}_{\mathcal{J}^c}(\delta^{-2}\mathbf{P}\mathbf{P}^T + \boldsymbol{\Upsilon}_{\mathcal{J}^c}^T\boldsymbol{\Upsilon}_{\mathcal{J}^c})^{-1}\boldsymbol{\Upsilon}_{\mathcal{J}}^T\|_{M(\infty)} < 1. \tag{13.16}$$

■

The proposition provides sufficient conditions for the relaxed problem in Eq. (13.14) to recover the sparsest graph, even when not all the eigenvectors are known. In practice it is observed that for a smaller number K of known spectral templates, the value of η in Eq. (13.16) tends to be larger, indicating a less favorable setting for recovery.

Notice that scenarios that combine the settings in Sections 13.4.1 and 13.4.2, i.e., where the knowledge of the K templates is imperfect, can be handled by combining the formulations in problems (13.9) and (13.14). This can be achieved upon implementing the following modifications to problem (13.14): considering the shift $\mathbf{S}'$ as a new optimization variable, replacing the first constraint in problem (13.14) with $\mathbf{S}' = \mathbf{S}_{\bar{K}} + \sum_{k=1}^{K} \lambda_k \mathbf{v}_k \mathbf{v}_k^T$, and adding $d(\mathbf{S}, \mathbf{S}') \leq \epsilon$ as a new constraint [cf. (13.9)].

Laplacian graph shift operators. Counterparts to the optimizations in problems (13.9) and (13.14) as well as for the recovery guarantees in Propositions 13.1 and 13.2 can be derived for the case of (normalized) Laplacian operators. This requires changing the definition of $\mathcal{S}$ and accounting for the fact that the Laplacian has a zero eigenvalue; see the numerical tests in the following section and [30] for further details.

13.4.3 NUMERICAL TESTS

Here we test the proposed topology inference methods on different synthetic and real-world graphs. A comprehensive performance evaluation is carried out through comparisons with state-of-the-art methods and a test case that illustrates how our framework can promote sparsity on a given network.

Comparison with baseline statistical methods. First we analyze the performance of the topology inference algorithm (13.9) (henceforth referred to as SpecTemp) in comparison with two workhorse statistical methods, namely, (thresholded) correlation networks [10, Ch. 7.3.1] and graphical lasso [19]. The goal is to recover the adjacency matrix of an undirected and unweighted graph with no self-loops from the observation of filtered graph signals $\mathcal{X} := \{\mathbf{x}_r\}_{r=1}^R$. For the implementation of SpecTemp, we use the eigendecomposition of the sample covariance $\hat{\mathbf{C}}_x$ in order to extract noisy spectral templates $\hat{\mathbf{V}}$. We then solve problem (13.9) for $\mathcal{S} = \mathcal{S}_{\text{A}}$, where ϵ is selected as the smallest value that admits a feasible solution. For the correlation-based method, we keep the absolute value of the sample correlation of the observed signals, force zeros on the diagonal, and set all values below a certain threshold to zero. This threshold is determined during a training phase; see [30] for additional details. Lastly, for graphical lasso we follow the implementation in [19] that finds $\mathbf{S}$ by solving

$$\max_{\mathbf{S} \succeq 0} \quad \log\det\mathbf{S} - \text{tr}\left\{\hat{\mathbf{C}}_x\mathbf{S}\right\} - \rho\|\mathbf{S}\|_1, \tag{13.17}$$

where the tuning parameter ρ is selected during the training phase. We then force zeros on the diagonal and keep the absolute values of each entry.

We test the recovery of adjacency matrices $\mathbf{S} = \mathbf{A}$ of Erdős-Rényi (ER) random graphs with $N = 20$ nodes and edge probability $p = 0.2$. We vary the number of observed signals from 10^1 to 10^6 in powers of 10. Each signal is generated by passing white Gaussian noise through a graph filter $\mathbf{H}$. Two different types of filters are considered. As a first type we consider a *general* filter $\mathbf{H}_1 = \mathbf{V}\text{diag}(\hat{\mathbf{h}}_1)\mathbf{V}^T$, where the entries of $\hat{\mathbf{h}}_1$ are independent and chosen randomly between 0.5 and 1.5. The second type is a *specific* filter of the form $\mathbf{H}_2 = (\delta_{\mathcal{H}}\mathbf{I} + \mathbf{S})^{-1/2}$, where the constant $\delta_{\mathcal{H}}$ is chosen so that $\delta_{\mathcal{H}}\mathbf{I} + \mathbf{S}$ is positive definite to ensure that $\mathbf{H}_2$ is real and well defined. According to Eq. (13.4), this implies that the precision matrix of the filtered signals is given by $\mathbf{C}_x^{-1} = \mathbf{H}_2^{-2} = \delta_{\mathcal{H}}\mathbf{I} + \mathbf{S}$, which coincides with $\mathbf{S}$ in the off-diagonal elements. For each combination of filter type and number of observed signals, we generate 10 ER graphs that are used for training and 20 ER graphs that are used for testing. Based on the 10 training graphs, the optimal threshold for the correlation method and parameter ρ for graphical lasso [cf. Eq. (13.17)] are determined and then used for the recovery of the 20 testing graphs. Given that for SpecTemp we are fixing ϵ beforehand, no training is required.

As a figure of merit we use the F-measure, i.e., the harmonic mean of edge precision and edge recall, that solely takes into account the support of the recovered graph while ignoring the weights. In Fig. 13.1 we plot the performance of the three methods as a function of the number of filtered graph signals observed for filters $\mathbf{H}_1$ and $\mathbf{H}_2$, where each point is the mean F-measure over the 20 testing graphs. When considering a general graph filter $\mathbf{H}_1$, SpecTemp clearly outperforms the other two. For instance, when 10^5 signals are observed, our average F-measure is 0.81 while the measures for correlation and graphical lasso are 0.29 and 0.25, respectively. Moreover, of the three methods, the proposed approach in Eq. (13.9) is the only consistent one, i.e., achieving perfect recovery with

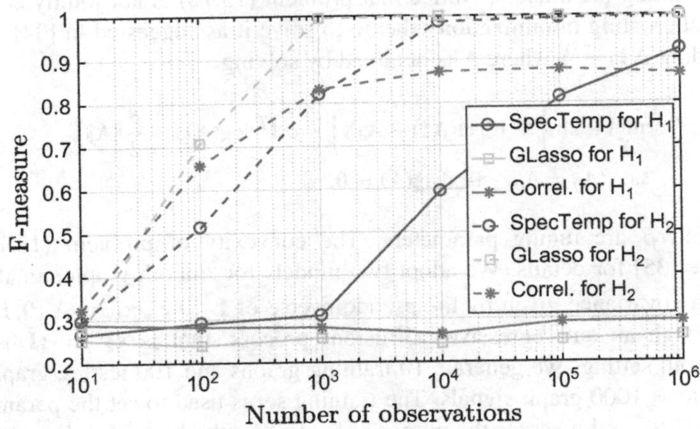

FIG. 13.1

Comparison with baseline statistical methods. Performance comparison between the proposed SpecTemp approach (13.9), graphical lasso [19], and correlation-based recovery. For general filters, SpecTemp outperforms the competing alternatives.

increasing number of observed signals. Although striking at first glance, the deficient performance of the baseline statistical methods was expected. For general filters $\mathbf{H}_1$, neither the correlation nor the precision matrices are sparse or share the support of the GSO to be recovered $\mathbf{S}$. When analyzing the specific case of graph filters $\mathbf{H}_2$, where the precision matrix exactly coincides with the desired graph-shift operator, graphical lasso outperforms both SpecTemp and the correlation-based method. This is not surprising because graphical lasso was designed for the recovery of sparse precision matrices and is optimal (in the maximum-likelihood sense) for Gaussian signals. Notice, however, that for a large number of observations, SpecTemp, without assuming any specific filter model, also achieves perfect recovery and yields an F-measure equal to 1. Consequently, if a practitioner knows a priori that the sought graph is (close to) the precision matrix and Gaussian signal assumptions are tenable, then graphical lasso will be the preferred method. However, for the general case in which this information is unavailable, SpecTemp is a more prudent alternative.

Comparison with GSP methods. Here we compare the network topology inference approach (13.6) with the state-of-the art algorithms in [34,35], which are designed to identify the Laplacian of a graph from observations of smooth graph signals. We select $f(\mathbf{S}) = \|\mathbf{S}\|_1$ and $d(\mathbf{S}_1, \mathbf{S}_2) = \|\mathbf{S}_1 - \mathbf{S}_2\|_F^2$ in Eq. (13.6), resulting in the SpecTemp formulation in Eq. (13.9). We study the recovery of the combinatorial Laplacian $\mathbf{S} = \mathbf{L}_c$ of the Barabási-Albert preferential attachment graphs [54], with $N = 20$ generated from $m_0 = 4$ initially placed nodes, where each new node is connected to $m = 3$ existing ones. Collecting the observations $\{\mathbf{x}_r\}_{r=1}^R$ as the columns of matrix $\mathbf{X}$, the approach in [34] seeks to recover $\mathbf{S} = \mathbf{L}_c$ by solving

$$\begin{aligned} \min_{\mathbf{S},\mathbf{Y}} \quad & \|\mathbf{X} - \mathbf{Y}\|_F^2 + \alpha \operatorname{tr}\left\{\mathbf{Y}^T \mathbf{S} \mathbf{Y}\right\} + \beta \|\mathbf{S}\|_F^2 \\ \text{s.t.} \quad & \operatorname{tr}\{\mathbf{S}\} = N, \quad S_{ij} = S_{ji} \leq 0, \; i \neq j, \quad \mathbf{S}\mathbf{1} = \mathbf{0}, \end{aligned} \tag{13.18}$$

where α and β are tuning parameters. Notice that problem (13.18) is not jointly convex in $\mathbf{S}$ and $\mathbf{Y}$, thus, we adopt an alternating minimization scheme to solve it as suggested in [34]. By contrast, [35] recovers $\mathbf{S}$ as $\mathbf{S} = \operatorname{diag}(\mathbf{A1}) - \mathbf{A}$ where $\mathbf{A}$ is obtained by solving

$$\begin{aligned} \min_{\mathbf{A}} \quad & 2\operatorname{tr}\left\{\mathbf{X}^T [\operatorname{diag}(\mathbf{A1}) - \mathbf{A}]\mathbf{X}\right\} - \alpha\, \mathbf{1}^T \log(\mathbf{A1}) + \frac{\beta}{2}\|\mathbf{A}\|_F^2 \\ \text{s.t.} \quad & A_{ij} = A_{ji} \geq 0, \; \operatorname{diag}(\mathbf{A}) = \mathbf{0}, \end{aligned} \tag{13.19}$$

where, again, α and β are tuning parameters. The convexity of problem (13.19) facilitates its implementation; see [35] for details. We adopt two models for smooth graph signals: i) multivariate normal signals with covariance given by the pseudoinverse of $\mathbf{L}_c$, i.e., $\mathbf{x}_1 \sim \mathcal{N}(\mathbf{0}, \mathbf{L}_c^\dagger)$; and ii) white signals filtered through an autoregressive (diffusion) process, that is $\mathbf{x}_2 = (\mathbf{I} + \mathbf{L}_c)^{-1}\mathbf{w}$, where $\mathbf{w} \sim \mathcal{N}(\mathbf{0}, \mathbf{I})$. For both settings we generate 10 training graphs and 100 testing graphs, and for every graph we generate $R = 1000$ graph signals. The training set is used to set the parameters α and β in Eqs. (13.18) and (13.19), and it serves the purpose of selecting the best ϵ [cf. Eq. (13.9)]. To increase the difficulty of the recovery task, every signal $\mathbf{x}$ is perturbed as $\hat{\mathbf{x}} = \mathbf{x} + \sigma\, \mathbf{x} \circ \mathbf{z}$, for $\sigma = 0.1$ and $\mathbf{z} \sim \mathcal{N}(\mathbf{0}, \mathbf{I})$, where $\circ$ denotes the elementwise product. We assess the performance via the F-measure, the ℓ_2 relative error of recovery of the edges, and the ℓ_2 relative error of recovery of the degrees. The performance achieved by each method in the testing sets is summarized in Table 13.2. In all but one

Table 13.2 Comparison With GSP Methods

	Inverse Laplacian			Diffusion		
Barabási-Albert	**Proposed**	**Kalofolias**	**Dong et al.**	**Proposed**	**Kalofolias**	**Dong et al.**
F-measure	**0.926**	0.855	0.873	**0.945**	0.845	0.894
Edge error	**0.143**	0.173	0.209	**0.135**	0.154	0.235
Degree error	**0.108**	0.124	0.169	0.109	**0.092**	0.188

Performance comparison between the proposed SpecTemp approach (13.9), Kalofolias (13.19), *and Dong et al.* (13.18). *Bold values indicate the best performance according to the different metrics.*

case, SpecTemp attains the highest F-measures and the lowest errors for both signal models. Similar results were found for ER graphs; see [30].

Network deconvolution. The network deconvolution problem is the identification of an adjacency matrix $\mathbf{S} = \mathbf{A}$ that encodes direct dependencies when given an adjacency $\mathbf{T}$ that includes indirect relationships. The problem is a generalization of channel deconvolution and can be solved by making $\mathbf{T} = \mathbf{S}(\mathbf{I} - \mathbf{S})^{-1}$ [55]. This solution assumes a diffusion as in Eq. (13.1) but for the particular case of a single-pole-single-zero graph filter. A more general approach is to assume that $\mathbf{T}$ can be written as a polynomial of $\mathbf{S}$ but be agnostic to the form of the filter. This leads to problem formulation (13.9) with $\mathbf{V}$ given by the eigenvectors of $\mathbf{T}$. Note that here matrix $\mathbf{T}$ is not necessarily an empirical covariance matrix.

In this context, our goal is to identify the structural properties of proteins from a mutual information graph of the co-variation between the constitutional amino-acids [56]; see [55] for details. For example, for a particular protein, we want to recover the structural graph in Fig. 13.2A (left) when given the graph of mutual information in Fig. 13.2A (right). Notice that the structural contacts along the first four subdiagonals of the graphs were intentionally removed to assess the capability of the methods in detecting the contacts between distant amino acids. The graph recovered by network deconvolution [55] is illustrated in Fig. 13.2B (left), whereas the one recovered using SpecTemp is depicted in Fig. 13.2B (right). Comparing both recovered graphs, SpecTemp leads to a sparser graph that follows more closely the desired structure to be recovered. To quantify this latter assertion, in Fig. 13.2C we plot the fraction of the real contact edges recovered for each method as a function of the number of edges considered, as done in [55]. For example, if for a given method the 100 edges with the largest weight in the recovered graph contain 40% of the edges in the ground truth graph, we say that the 100 top edge predictions achieve a fraction of recovered edges equal to 0.4. As claimed in [55], network deconvolution improves the estimation when compared to raw mutual information data. Nevertheless, from Fig. 13.2C it follows that SpecTemp outperforms network deconvolution. Notice that when $\epsilon = 0$ [cf. Eq. (13.9)] we are forcing the eigenvectors of $\mathbf{S}$ to coincide exactly with those of the matrix of mutual information $\mathbf{S}'$. However, because $\mathbf{S}'$ is already a valid adjacency matrix, we end up recovering $\mathbf{S} = \mathbf{S}'$. By contrast, for larger values of ϵ, the additional flexibility in the choice of eigenvectors allows us to recover shifts $\mathbf{S}$ that more closely resemble the ground truth. For example, when considering the top 200 edges, the mutual information and network deconvolution methods recover 36% and 43% of the desired edges, respectively, while our method for $\epsilon = 1$ achieves a recovery of 53%. In Fig. 13.2D we present this same analysis for a different protein and similar results can be appreciated.

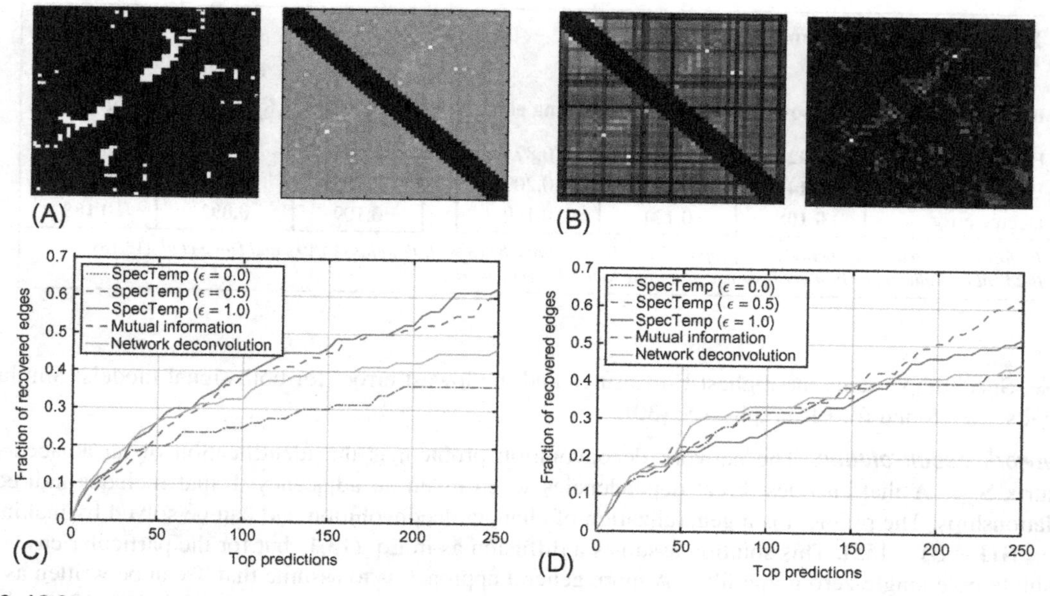

FIG. 13.2

Identifying the structural properties of proteins. (A) Real and (B) inferred contact networks between amino-acid residues for protein BPT1 BOVIN. Ground truth contact network (A, left), mutual information of the covariation of amino-acid residues (A, right), contact network inferred by network deconvolution (B, left), contact network inferred the proposed SpecTemp approach (13.9) (B, right). (C) Fraction of the real contact edges between amino acids recovered for each method as a function of the number of edges considered. (D) Counterpart of (C) for protein YES HUMAN.

13.5 NONSTATIONARY DIFFUSION PROCESSES

We now deal with more general *nonstationary* signals $\mathbf{x}$ that adhere to linear diffusion dynamics (13.1) in $\mathcal{G}$, but where the input covariance $\mathbf{C}_w = \mathbb{E}\left[\mathbf{w}\mathbf{w}^T\right]$ can be arbitrary. In other words, we relax the assumption of $\mathbf{w}$ being white, which led to the stationary signal model dealt with so far (cf. Remark 13.1). Such a broader model is, for instance, relevant to (geographically) correlated sensor network data, or to models of opinion dynamics where (even before engaging in discussion) the network agents can be partitioned into communities according to their standing on the subject matter.

For generic (nonidentity) input covariance matrix $\mathbf{C}_w$, we face the challenge that the signal covariance [cf. Eq. (13.4)]

$$\mathbf{C}_x = \mathbf{H}\mathbf{C}_w\mathbf{H}^T \tag{13.20}$$

is no longer simultaneously diagonalizable with $\mathbf{S}$. This rules out using the eigenvectors of the sample covariance $\hat{\mathbf{C}}_x$ as spectral templates of $\mathbf{S}$. Still, as argued following Eq. (13.3) the eigenvectors of the GSO coincide with those of the graph filter $\mathbf{H}$ that govern the underlying diffusion dynamics. This motivates using snapshot observations $\mathcal{X} := \{\mathbf{x}_r\}_{r=1}^R$ together with additional information on the excitation input $\mathbf{w}$ (either realizations of the graph signal, sparsity assumptions, or its covariance matrix $\mathbf{C}_w$ [26]) to *identify the filter* $\mathbf{H}$, with the ultimate goal of estimating its eigenvectors $\mathbf{V}$ [57]. These spectral templates are then used as inputs to the GSO identification problem (13.6), exactly as in the stationary setting of Sections 13.3 and 13.4. Accordingly, focus is henceforth placed on the graph filter (i.e., system) identification task.

13.5.1 LINEAR GRAPH FILTER IDENTIFICATION

Consider $m = 1, \ldots, M$ diffusion processes on $\mathcal{G}$, and assume that the observed nonstationary signal $\mathbf{x}_m$ corresponds to an input $\mathbf{w}_m$ diffused by an unknown graph filter $\mathbf{H} = \sum_{l=0}^{L-1} h_l \mathbf{S}^l$, which encodes the structure of the network via $\mathbf{S}$. In this section, we show how additional knowledge about *realizations* of the input signals $\mathbf{w}_m$ can be used to identify $\mathbf{H}$ and, as a byproduct, its eigenvectors $\mathbf{V}$.

Input-output signal realization pairs. Suppose first that realizations of M input-output pairs $\{\mathbf{w}_m, \mathbf{x}_m\}_{m=1}^M$ are available, which can be arranged in the data matrices $\mathbf{W} = [\mathbf{w}_1, \ldots, \mathbf{w}_M]$ and $\mathbf{X} = [\mathbf{x}_1, \ldots, \mathbf{x}_M]$. The goal is to identify a symmetric filter $\mathbf{H} \in \mathcal{M}_N$ such that the observed signals $\mathbf{x}_m$ and the predicted ones $\mathbf{H}\mathbf{w}_m$ are close in some sense. In the absence of measurement noise this simply amounts to solving a system of M linear matrix equations

$$\mathbf{x}_m = \mathbf{H}\mathbf{w}_m, \quad m = 1, \ldots, M. \tag{13.21}$$

When noise is present, using the workhorse least-squares (LS) criterion the filter can be estimated as

$$\hat{\mathbf{H}} = \underset{\mathbf{H} \in \mathcal{M}_N}{\operatorname{argmin}} \sum_{m=1}^{M} \|\mathbf{x}_m - \mathbf{H}\mathbf{w}_m\|_2^2. \tag{13.22}$$

Because $\mathbf{H}$ is symmetric, the free optimization variables in Eq. (13.22) correspond to, say, the lower triangular part of $\mathbf{H}$, meaning the entries on and below the main diagonal. These $N_H := N(N+1)/2$ nonredundant entries can be conveniently arranged in the so-termed half-vectorization of $\mathbf{H}$, i.e., a vector $\text{vech}(\mathbf{H}) \in \mathbb{R}^{N_H}$ from which one can recover $\text{vec}(\mathbf{H}) \in \mathbb{R}^{N^2}$ via duplication. Indeed, there exists a unique duplication matrix $\mathbf{D}_N \in \{0, 1\}^{N^2 \times N_H}$ such that one can write $\mathbf{D}_N \text{vech}(\mathbf{H}) = \text{vec}(\mathbf{H})$. The Moore-Penrose pseudoinverse of $\mathbf{D}_N$, denoted as $\mathbf{D}_N^\dagger$, possesses the property $\text{vech}(\mathbf{H}) = \mathbf{D}_N^\dagger \text{vec}(\mathbf{H})$. With this notation in place, several properties of the solution $\hat{\mathbf{H}}$ of Eq. (13.22) are stated next.

Proposition 13.3. *Regarding the graph filter problem (13.22), it holds that:*

(a) *The entries of the symmetric solution $\hat{\mathbf{H}}$ are given by*

$$\text{vech}(\hat{\mathbf{H}}) = \left[(\mathbf{W}^T \otimes \mathbf{I}_N)\mathbf{D}_N\right]^\dagger \text{vec}(\mathbf{X}). \tag{13.23}$$

(b) $\text{rank}\left((\mathbf{W}^T \otimes \mathbf{I}_N)\mathbf{D}_N\right) \leq N_H - (N - \text{rank}(\mathbf{W}) + 1)(N - \text{rank}(\mathbf{W}))/2.$

(c) *The minimizer of Eq. (13.22) is unique if and only if* $\text{rank}(\mathbf{W}) = N$.

■

Proposition 13.3 asserts that if the excitation input set $\{\mathbf{w}_m\}_{m=1}^M$ is sufficiently rich, i.e., if $M \geq N$ and the excitation signals are linearly independent, the entries of the diffusion filter $\mathbf{H}$ can be found as the solution of an LS problem. Interestingly, the fact of $\mathbf{H}$ having only $N_H = N(N+1)/2$ different entries cannot be exploited to reduce the number of input signals required to identify $\mathbf{H}$. The reason being that the matrix $(\mathbf{W}^T \otimes \mathbf{I}_N)\mathbf{D}_N$ is rank deficient if $\mathbf{W}^T$ has a nontrivial nullspace. Symmetry, however, can be exploited to enhance the estimation performance in overdetermined scenarios with noisy observations.

Once $\hat{\mathbf{H}}$ is estimated using Eq. (13.23), the next step is to decompose the filter as $\hat{\mathbf{H}} = \hat{\mathbf{V}}\hat{\mathbf{\Lambda}}\hat{\mathbf{V}}^T$ and use $\hat{\mathbf{V}}$ as input for the GSO identification problem (13.6). Note that obtaining such an eigendecomposition is always possible because filter estimates $\hat{\mathbf{H}} \in \mathcal{M}_N$ are constrained to be symmetric.

13.5.2 QUADRATIC GRAPH FILTER IDENTIFICATION

In a number of applications, realizations of the excitation input $\mathbf{w}_m$ may be challenging to acquire, but information about the *statistical* description of $\mathbf{w}_m$ could still be available. To be specific, assume that the excitation inputs are zero mean and their covariance $\mathbf{C}_{w,m} = \mathbb{E}\left[\mathbf{w}_m\mathbf{w}_m^T\right]$ is known. Further suppose that for each input $\mathbf{w}_m$, we have access to a set of observations $\{\mathbf{x}_m^{(r)}\}_{r=1}^{R_m}$, which are then used to *estimate the output covariance* as $\hat{\mathbf{C}}_{x,m} = \frac{1}{R_m}\sum_{r=1}^{R_m}\mathbf{x}_m^{(r)}(\mathbf{x}_m^{(r)})^T$. Because under Eq. (13.3) the true *covariance* is $\mathbf{C}_{x,m} = \mathbb{E}\left[\mathbf{x}_m\mathbf{x}_m^T\right] = \mathbf{H}\mathbf{C}_{w,m}\mathbf{H}^T$ [cf. Eq. (13.20)], the aim is to identify a filter $\mathbf{H}$ such that matrices $\hat{\mathbf{C}}_{x,m}$ and $\mathbf{H}\mathbf{C}_{w,m}\mathbf{H}^T$ are close.

Assuming for now perfect knowledge of the signal covariances, the above rationale suggests studying the solutions of the following system of matrix quadratic equations

$$\mathbf{C}_{x,m} = \mathbf{H}\mathbf{C}_{w,m}\mathbf{H}^T, \quad m = 1,\ldots,M. \tag{13.24}$$

To gain some initial insights, let us focus first on one of the matrix equations in Eq. (13.24) (or alternatively, suppose that $M = 1$). Given the eigendecomposition of the symmetric and positive semidefinite (PSD) covariance matrix $\mathbf{C}_{x,m} = \mathbf{V}_{x,m}\mathbf{\Lambda}_{x,m}\mathbf{V}_{x,m}^T$, the *principal square root* of $\mathbf{C}_{x,m}$ is the unique symmetric and PSD matrix $\mathbf{C}_{x,m}^{1/2}$, which satisfies $\mathbf{C}_{x,m} = \mathbf{C}_{x,m}^{1/2}\mathbf{C}_{x,m}^{1/2}$. It is given by $\mathbf{C}_{x,m}^{1/2} = \mathbf{V}_{x,m}\mathbf{\Lambda}_{x,m}^{1/2}\mathbf{V}_{x,m}^T$, where $\mathbf{\Lambda}_{x,m}^{1/2}$ stands for a diagonal matrix with the nonnegative square roots of the eigenvalues of $\mathbf{C}_{x,m}$.

With this notation in place, let us introduce the matrices $\mathbf{C}_{wxw,m} := \mathbf{C}_{w,m}^{1/2}\mathbf{C}_{x,m}\mathbf{C}_{w,m}^{1/2}$ and $\mathbf{H}_{ww,m} := \mathbf{C}_{w,m}^{1/2}\mathbf{H}\mathbf{C}_{w,m}^{1/2}$. Clearly, $\mathbf{C}_{wxw,m}$ is both symmetric and PSD. Regarding the transformed filter $\mathbf{H}_{ww,m}$, note that by construction we have that $\mathbf{H}_{ww,m}$ is symmetric. Moreover, if $\mathbf{H}$ is assumed to be PSD, then so will be $\mathbf{H}_{ww,m}$. These properties will be instrumental toward characterizing the solutions of the matrix quadratic equation $\mathbf{C}_{x,m} = \mathbf{H}\mathbf{C}_{w,m}\mathbf{H}^T$ in Eq. (13.24), which can be recovered from the solutions $\mathbf{H}_{ww,m}$ of

$$\mathbf{C}_{wxw,m} = \mathbf{C}_{w,m}^{1/2}\mathbf{C}_{x,m}\mathbf{C}_{w,m}^{1/2} = \mathbf{C}_{w,m}^{1/2}\mathbf{H}\mathbf{C}_{w,m}\mathbf{H}\mathbf{C}_{w,m}^{1/2} = \mathbf{H}_{ww,m}\mathbf{H}_{ww,m}. \tag{13.25}$$

Positive semidefinite graph filters. Suppose that $\mathbf{H}$ is PSD (henceforth denoted by $\mathbf{H} \in \mathcal{M}_N^{++}$), so that $\mathbf{H}_{ww,m}$ in Eq. (13.25) is PSD as well. Such filters arise, for example, in heat diffusion processes of the form $\mathbf{x} = (\sum_{l=0}^{\infty}\beta^l\mathbf{L}_c^l)\mathbf{w}$ with $\beta > 0$, where the Laplacian GSO $\mathbf{L}_c$ is PSD and the filter coefficients $h_l = \beta^l$ are all positive. In this setting, the solution of Eq. (13.25) is unique and given by the principal square root

$$\mathbf{H}_{ww,m} = \mathbf{C}_{wxw,m}^{1/2}. \tag{13.26}$$

Consequently, if $\mathbf{C}_{w,m}$ is nonsingular then the definition of $\mathbf{H}_{ww,m}$ can be used to recover $\mathbf{H}$ via

$$\mathbf{H} = \mathbf{C}_{w,m}^{-1/2}\mathbf{C}_{wxw,m}^{1/2}\mathbf{C}_{w,m}^{-1/2}. \tag{13.27}$$

The previous arguments demonstrate that the assumption $\mathbf{H} \in \mathcal{M}_N^{++}$ gives rise to a strong identifiability result. Indeed, if $\{\mathbf{C}_{x,m}\}_{m=1}^{M}$ are known perfectly, the graph filter is identifiable even for $M = 1$.

However, in pragmatic settings where only empirical covariances are available, the observation of multiple ($M > 1$) diffusion processes can improve the performance of the system identification task. Given empirical covariances $\{\hat{\mathbf{C}}_{x,m}\}_{m=1}^{M}$ respectively estimated with enough samples R_m to ensure that they are full rank, we define $\hat{\mathbf{C}}_{wxw,m} := \mathbf{C}_{w,m}^{1/2}\hat{\mathbf{C}}_{x,m}\mathbf{C}_{w,m}^{1/2}$ for each m. The quadratic equation (13.27) motivates solving the LS problem

$$\hat{\mathbf{H}} = \operatorname*{argmin}_{\mathbf{H}\in\mathcal{M}_N^{++}} \sum_{m=1}^{M} \|\hat{\mathbf{C}}_{wxw,m}^{1/2} - \mathbf{C}_{w,m}^{1/2}\mathbf{H}\mathbf{C}_{w,m}^{1/2}\|_F^2. \tag{13.28}$$

Whenever the number of samples R_m—and accordingly the accuracy of the empirical covariances $\hat{\mathbf{C}}_{x,m}$—differs significantly across diffusion processes $m = 1,\ldots,M$, it may be prudent to introduce nonuniform coefficients to downweigh those residuals in Eq. (13.28) with inaccurate covariance estimates.

Symmetric graph filters. Consider now a more general setting whereby $\mathbf{H}$ is only assumed to be symmetric, and once more let us start by considering only one of the M equations in Eq. (13.24) to gain insights. With the unitary matrix $\mathbf{V}_{wxw,m}$ denoting the eigenvectors of $\mathbf{C}_{wxw,m} = \mathbf{C}_{w,m}^{1/2}\mathbf{C}_{x,m}\mathbf{C}_{w,m}^{1/2}$ and with $\mathbf{b}_m \in \{-1,1\}^N$ being a binary (signed) vector, one can conclude that the set of solutions to $\mathbf{C}_{wxw,m} = \mathbf{H}_{ww,m}\mathbf{H}_{ww,m}$ [cf. Eq. (13.25)] is given by

$$\mathbf{H}_{ww,m} = \mathbf{C}_{wxw,m}^{1/2}\mathbf{V}_{wxw,m}\text{diag}(\mathbf{b}_m)\mathbf{V}_{wxw,m}^T. \tag{13.29}$$

That is, while for the case where $\mathbf{H}$ was PSD we had that the solution was unique and given by $\mathbf{C}_{wxw,m}^{1/2}$, when $\mathbf{H}$ is symmetric, any matrix obtained by changing the sign of one (or more) of the eigenvalues of $\mathbf{C}_{wxw,m}^{1/2}$ is also a feasible solution. Leveraging this and provided that the input covariance matrix $\mathbf{C}_{w,m}$ is nonsingular, it follows that all symmetric solutions to $\mathbf{C}_{x,m} = \mathbf{H}\mathbf{C}_{w,m}\mathbf{H}^T$ are described by the set

$$\mathcal{H}_m^{\text{sym}} = \left\{\mathbf{H} \in \mathcal{M}_N | \mathbf{H} = \mathbf{C}_{w,m}^{-1/2}\mathbf{C}_{wxw,m}^{1/2}\mathbf{V}_{wxw,m}\text{diag}(\mathbf{b}_m)\mathbf{V}_{wxw,m}^T\mathbf{C}_{w,m}^{-1/2} \text{ and } \mathbf{b}_m \in \{-1,1\}^N\right\}. \tag{13.30}$$

Which, as pointed out earlier, confirms that in the absence of the PSD assumption, the problem for $M = 1$ is nonidentifiable. Inspection of $\mathcal{H}_m^{\text{sym}}$ shows that there are 2^N possible solutions to the quadratic equation (13.20), which are parameterized by the binary vector $\mathbf{b}_m$. For the PSD setting the solution is unique and corresponds to $\mathbf{b}_m = \mathbf{1}$.

For $M > 1$, the set of feasible solutions to the system of Eq. (13.24) is naturally given by

$$\mathcal{H}_{1:M}^{\text{sym}} = \bigcap_{m=1}^{M} \mathcal{H}_m^{\text{sym}}.$$

Upon defining the tall matrices $\mathbf{A}_m := (\mathbf{C}_{x,m}^{-1/2}\mathbf{V}_{wxw,m}) \odot (\mathbf{C}_{x,m}^{-1/2}\mathbf{C}_{wxw,m}^{1/2}\mathbf{V}_{wxw,m}) \in \mathbb{R}^{N^2 \times N}$ and using those to rewrite $\mathcal{H}_m^{\text{sym}}$ in Eq. (13.30), the feasible set $\mathcal{H}_{1:M}^{\text{sym}}$ can be compactly written as

$$\mathcal{H}_{1:M}^{\text{sym}} = \bigcap_{m=1}^{M} \left\{\mathbf{H} \in \mathcal{M}_N | \text{vec}(\mathbf{H}) = \mathbf{A}_m \mathbf{b}_m \text{ and } \mathbf{b}_m \in \{-1, 1\}^N\right\}. \tag{13.31}$$

Because $\mathcal{H}_{1:M}^{\text{sym}}$ is the intersection of M finite size sets, it is conceivable that for larger values of M, the cardinality of $\mathcal{H}_{1:M}^{\text{sym}}$ will be reduced to render the problem identifiable (up to an unavoidable sign ambiguity because if $\mathbf{H} \in \mathcal{M}_N$ is a solution of Eq. (13.24), so is $-\mathbf{H}$).

Imperfect observations: In pragmatic settings where only empirical covariances $\{\hat{\mathbf{C}}_{x,m}\}_{m=1}^{M}$ are available, $\{\mathbf{A}_m\}_{m=1}^{M}$ cannot be perfectly estimated. Hence, the equalities $\text{vec}(\mathbf{H}) = \mathbf{A}_m \mathbf{b}_m$ must be relaxed and $\hat{\mathbf{A}}_m := (\mathbf{C}_{x,m}^{-1/2}\hat{\mathbf{V}}_{wxw,m}) \odot (\mathbf{C}_{x,m}^{-1/2}\hat{\mathbf{C}}_{wxw,m}^{1/2}\hat{\mathbf{V}}_{wxw,m})$ must be used in lieu of $\mathbf{A}_m$. Provided that covariances are estimated with enough samples to ensure full rankness, our approach is to solve the LS problem

$$\min_{\{\mathbf{b}_m\}_{m=1}^{M}} \sum_{m,m'} \|\hat{\mathbf{A}}_m \mathbf{b}_m - \hat{\mathbf{A}}_{m'} \mathbf{b}_{m'}\|_2^2 \quad \text{s.t. } \mathbf{b}_m \in \{-1, 1\}^N \text{ for all } m. \tag{13.32}$$

Note that both terms within the ℓ_2 norm in Eq. (13.32) should equal $\text{vec}(\mathbf{H})$ in a noiseless setting. Thus, we are minimizing this discrepancy across the M processes considered. If the accuracy of the empirical covariances $\hat{\mathbf{C}}_{x,m}$ differs significantly across diffusion processes $m = 1, \ldots, M$, it may be prudent to introduce nonuniform coefficients to weigh the residuals, taking into account those accuracies. While the objective in Eq. (13.32) is convex in the unknowns $\{\mathbf{b}_m\}_{m=1}^{M}$, the binary constraints render the optimization nonconvex. An efficient algorithm to find a solution to Eq. (13.32) with theoretical guarantees is discussed next.

Algorithmic approach: Here we explain how the graph filter identification task can be tackled using a semidefinite relaxation (SDR) [58]. This convexification technique has been successfully applied to a wide variety of nonconvex quadratically constrained quadratic programs (QCQP) in applications such as MIMO detection [59] and transmit beamforming [60]. To that end, we first cast the filter identification (perfect-observations based) problem as a Boolean quadratic program (BQP).

Proposition 13.4. *Consider the unknown binary vectors $\mathbf{b}_m \in \{-1, 1\}^N$ and the known observation matrices $\mathbf{A}_m \in \mathbb{R}^{N^2 \times N}$, and use those to define*

$$\mathbf{b} := [\mathbf{b}_1^T, \ldots, \mathbf{b}_M^T]^T \in \{-1, 1\}^{NM}, \tag{13.33}$$

$$\mathbf{\Psi} := \begin{bmatrix} \mathbf{A}_1 & -\mathbf{A}_2 & \mathbf{0} & \cdots & \mathbf{0} & \mathbf{0} \\ \mathbf{0} & \mathbf{A}_2 & -\mathbf{A}_3 & \cdots & \mathbf{0} & \mathbf{0} \\ \vdots & \vdots & \vdots & \ddots & \vdots & \vdots \\ \mathbf{0} & \mathbf{0} & \mathbf{0} & \cdots & \mathbf{A}_{M-1} & -\mathbf{A}_M \end{bmatrix} \in \mathbb{R}^{N^2(M-1) \times NM}. \tag{13.34}$$

If $\text{rank}(\mathbf{\Psi}) = NM - 1$ *then the symmetric diffusion filter* $\mathbf{H}^*$ *can be exactly recovered (up to a sign) as* $\text{vec}(\mathbf{H}^*) = \mathbf{A}_1 \mathbf{b}_1^*$, *where* $\mathbf{b}_1^*$ *is the first* $N \times 1$ *subvector of the solution to the following BQP problem*

$$\mathbf{b}^* = \underset{\mathbf{b}\in\{-1,1\}^{NM}}{\operatorname{argmin}} \ \mathbf{b}^T\mathbf{\Psi}^T\mathbf{\Psi}\mathbf{b}. \tag{13.35}$$

■

Problem (13.35) offers a natural formulation for the pragmatic setting whereby $\{\mathbf{C}_{x,m}\}_{m=1}^M$ are replaced by sample estimates $\{\hat{\mathbf{C}}_{x,m}\}_{m=1}^M$, and one would aim at minimizing Eq. (13.32). Given a solution of Eq. (13.35) with $\hat{\mathbf{\Psi}}$ replacing $\mathbf{\Psi}$, the entries of $\mathbf{H} \in \mathcal{H}_N$ can be estimated as

$$\operatorname{vec}(\hat{\mathbf{H}}) = \frac{1}{M}\sum_{m=1}^{M} \hat{\mathbf{A}}_m \mathbf{b}_m^*. \tag{13.36}$$

Even though the BQP is a classical NP-hard combinatorial optimization problem [58], via SDR one can obtain near-optimal solutions with provable approximation guarantees. To derive the SDR of Eq. (13.35), first introduce the $NM \times NM$ symmetric PSD matrices $\mathbf{W} := \mathbf{\Psi}^T\mathbf{\Psi}$ and $\mathbf{B} := \mathbf{b}\mathbf{b}^T$. By construction, the binary matrix $\mathbf{B}$ has rank one and its diagonal entries are $B_{ii} = b_i^2 = 1$. Conversely, any matrix $\mathbf{B} \in \mathbb{R}^{NM\times NM}$ that satisfies $\mathbf{B} \succeq 0$, $B_{ii} = 1$, and $\operatorname{rank}(\mathbf{B}) = 1$ necessarily has the form $\mathbf{B} = \mathbf{b}\mathbf{b}^T$, for some $\mathbf{b} \in \{-1,1\}^{NM}$. Using these definitions, one can write $\mathbf{b}^T\mathbf{W}\mathbf{b} = \operatorname{trace}(\mathbf{b}^T\mathbf{W}\mathbf{b}) = \operatorname{trace}(\mathbf{W}\mathbf{b}\mathbf{b}^T) = \operatorname{trace}(\mathbf{W}\mathbf{B})$ and accordingly problem (13.35) is equivalent to

$$\begin{aligned} &\min_{\mathbf{B}} \operatorname{trace}(\mathbf{W}\mathbf{B}) \\ &\text{s.t. } \mathbf{B} \succeq \mathbf{0}, \quad \operatorname{rank}(\mathbf{B}) = 1, \quad B_{ii} = 1, \quad i = 1,\ldots,NM. \end{aligned} \tag{13.37}$$

The only source of nonconvexity in problem (13.37) is the rank constraint, and dropping it yields the convex SDR. The resultant problem is a semidefinite program (SDP), which can be solved using an off-the-shelf interior-point method [61]. It is immediate that if the solution of the relaxed problem (denoted as $\mathbf{B}^*$) is a rank-one matrix, then $\mathbf{B}^* = \mathbf{b}^*(\mathbf{b}^*)^T$ solves the original BQP as well. However, in general it holds that $\operatorname{rank}(\mathbf{B}^*) \neq 1$ and an algorithm to generate a feasible solution of problem (13.35) from $\mathbf{B}^*$ must be put forth. An effective scheme that comes with theoretical guarantees is to adopt the so-termed Gaussian randomization procedure, which is the one that achieved the best performance in [62]. For general details on this method we refer the readers to [58,63].

Nonsymmetric graph filters. We finally describe the more challenging case of inferring directed graphs (see [67] for further details). As in the previous sections, we start by assuming that perfect knowledge of the signal covariances is available and focus on one of the M matrix equations in (13.24). After doing this, the interest is on characterizing the set of real and possibly asymmetric $\mathbf{H}$ solving $\mathbf{C}_{x,m} = \mathbf{H}\mathbf{C}_{w,m}\mathbf{H}^T$. This problem is strongly related to that of finding the square roots of $\mathbf{C}_{x,m}$. Indeed, with $\mathbf{C}_{x,m}^{1/2}$ denoting the principal square root of $\mathbf{C}_{x,m}$ and with $\mathbf{U}$ denoting an $N \times N$ *orthogonal* matrix (such that $\mathbf{U}_m\mathbf{U}_m^T = \mathbf{I}$), it is not difficult to show that any square matrix $\mathbf{H}_{w,m}$ such that $\mathbf{H}_{w,m}\mathbf{H}_{w,m}^T = \mathbf{C}_{x,m}$ is of the form $\mathbf{H}_{w,m} = \mathbf{C}_{x,m}^{1/2}\mathbf{U}_m$. This can be leveraged to show the following result.

Lemma 13.1. *Let $\mathbf{U}_m$ be an orthogonal matrix. Then, if $\mathbf{C}_{w,m}$ and $\mathbf{C}_{x,m}$ are full rank, the set $\mathcal{H}_m^{nsym}$ containing all the (possibly asymmetric) matrices $\mathbf{H}$ that solve Eq. (13.24) for a particular m is given by*

$$\mathcal{H}_m^{nsym} = \left\{\mathbf{H} | \mathbf{H} = \mathbf{C}_{x,m}^{1/2}\mathbf{U}_m\mathbf{C}_{w,m}^{-1/2} \text{ and } \mathbf{U}_m\mathbf{U}_m^T = \mathbf{I}\right\}. \tag{13.38}$$

A simple substitution suffices to show that every $\mathbf{H}$ of the form in Eq. (13.38) solves Eq. (13.24). Conversely, given an $\mathbf{H}$ that solves Eq. (13.24), form the matrix $\mathbf{U}_m = \mathbf{C}_{x,m}^{-1/2}\mathbf{H}\mathbf{C}_{w,m}^{1/2}$. Observe that if $\mathbf{U}_m$ is orthogonal then $\mathbf{H} \in \mathcal{H}_m^{nsym}$. Orthogonality of $\mathbf{U}_m$ follows because $\mathbf{U}_m\mathbf{U}_m^T = \mathbf{C}_{x,m}^{-1/2}\mathbf{H}\mathbf{C}_{w,m}\mathbf{H}^T\mathbf{C}_{x,m}^{-1/2} = \mathbf{I}$, where the last equality comes from the fact that $\mathbf{H}$ solves Eq. (13.24).

It is instructive to compare $\mathcal{H}_m^{nsym}$ with its counterpart for the symmetric case $\mathcal{H}_m^{sym}$. Note that for the symmetric case, the solution set was spanned by unitary matrices of the form diag($\mathbf{b}_m$) with $\mathbf{b}_m \in \{-1, 1\}^N$. Because there are only 2^N of such matrices, the cardinality of $\mathcal{H}_m^{sym}$ was at most 2^N. By contrast, the solution set (13.38) is spanned by matrices of the form $\mathbf{U}_m \in \mathcal{U}_N$, where $\mathcal{U}_N$ denotes the Stiefel manifold containing all $N \times N$ real orthogonal matrices. Because $\mathcal{U}_N$ contains an infinite number of elements, so does $\mathcal{H}_m^{nsym}$. In fact, it is known that $\mathcal{U}_N$ has dimension $N(N-1)/2$ [64]. Hence, if the matrices $\mathbf{C}_{x,m}$ and $\mathbf{C}_{w,m}$ are both nonsingular, it follows that $\mathcal{H}_m^{nsym}$ in Eq. (13.38) is also a manifold with dimension $N(N-1)/2$, the same than that of $\mathcal{U}_N$.

For the general case of $M > 1$, the set of feasible solutions is just given by the intersection of Eq. (13.38) for all diffusing processes, i.e.,

$$\mathcal{H}_{1:M}^{nsym} = \bigcap_{m=1}^{M} \mathcal{H}_m^{nsym} = \bigcap_{m=1}^{M} \left\{\mathbf{H} | \mathbf{H} = \mathbf{C}_{x,m}^{1/2}\mathbf{U}_m\mathbf{C}_{w,m}^{-1/2} \text{ and } \mathbf{U}_m\mathbf{U}_m^T = \mathbf{I}\right\}.$$

To gain insights on the size of $\mathcal{H}_{1:M}^{nsym}$ note that for each additional m one has N^2 new constraints (corresponding to the N^2 observations provided by $\mathbf{C}_{x,m}$ and $\mathbf{C}_{w,m}$) and only $N(N-1)/2$ new degrees of freedom (those corresponding to $\mathbf{U}_m$). Hence, it is conceivable that as M increases, the problem becomes identifiable (up to the unresolvable sign ambiguity).

Imperfect observations: As before, when only empirical covariances are available, the equalities in Eq. (13.38) must be relaxed and $\{\mathbf{C}_{x,m}\}_{m=1}^M$ must be replaced with $\{\hat{\mathbf{C}}_{x,m}\}_{m=1}^M$. Provided that the latter can be estimated with enough samples to ensure full rankness, our approach is to solve the following least-squares (LS) problem with manifold constraints

$$\min_{\{\mathbf{U}_m\}_{m=1}^M} \sum_{m,m'} \|\hat{\mathbf{C}}_{x,m}^{1/2}\mathbf{U}_m\mathbf{C}_{w,m}^{-1/2} - \hat{\mathbf{C}}_{x,m'}^{1/2}\mathbf{U}_{m'}\mathbf{C}_{w,m'}^{-1/2}\|_F^2 \quad \text{s.t. } \mathbf{U}_m \in \mathcal{U}_N \text{ for all } m. \tag{13.39}$$

As pointed out after presenting problem (13.32), the counterpart of the above optimization for the symmetric case, we note that: (i) in a noiseless setting both terms within the Frobenius norm in Eq. (13.39) should equal $\mathbf{H}$, hence, the goal is to minimize the discrepancy across the M observed processes; and (ii) weighted versions of the objective are pertinent if either the accuracy of the empirical covariances $\hat{\mathbf{C}}_{x,m}$ or the conditioning number of the input covariances $\mathbf{C}_{w,m}$ vary significantly across $m = 1, \ldots, M$. Regarding the algorithms to solve problem (13.39), even though the objective function is convex in $\{\mathbf{U}_m\}_{m=1}^M$, the set $\mathcal{U}_N$ is not. As a result, optimization methods tailored to the Stiefel manifold must be used [65,66], which is the subject of the next section.

Before doing so, we note that, as discussed after the definition of $\mathcal{H}_{1:M}^{nsym}$, it is expected that as the number M of observed processes increases, recoverability of problem (13.39) improves, provided that the excitation input matrices $\mathbf{C}_{w,m}^{1/2}$ bear sufficiently new information. By increasing M in one unit, we are introducing a new matrix variable $\mathbf{U}_m$ that lives in a lower-dimensional manifold of size $N(N-1)/2$. But, on the other hand, we are incorporating M new matrix cost terms, each of them with N^2 scalar terms. Hence, the overall effect should boost recovery, as confirmed in [67].

Algorithmic approach: While a number of gradient methods are available for manifold constrained optimization [65,66,68], these can only guarantee convergence to a stationary point, which depends on the particular initialization. To partially address this issue, we propose a two-step algorithm whereby: (i) we first find a judicious filter initialization $\mathbf{H}^{(0)}$ by solving a biconvex recovery problem; and (ii) we refine such an initialization by solving a modified version of problem (13.39) using a gradient method projected onto the Stiefel manifold $\mathcal{U}_N$.

To be more specific, let us describe first how the initialization in step (i) is obtained. To that end, we introduce the auxiliary variables $\mathbf{H}_L$ and $\mathbf{H}_R$ and reformulate the recovery filter problem as a biconvex optimization with linear constraints $\mathbf{H}_L = \mathbf{H}_R = \mathbf{H}$, namely

$$\{\mathbf{H}_L^{(0)}, \mathbf{H}_R^{(0)}\} = \operatorname*{arg\,min}_{\mathbf{H}_L,\mathbf{H}_R} \sum_{m=1}^{M} \|\hat{\mathbf{C}}_{x,m} - \mathbf{H}_L \mathbf{C}_{w,m} \mathbf{H}_R\|_F^2 \qquad \text{s.t. } \mathbf{H}_L = \mathbf{H}_R. \tag{13.40}$$

Problem (13.40) can be tackled using an alternating LS scheme, or a more sophisticated scheme based on the Alternating Direction Method of Multipliers (ADMM) [69], which has been applied to a wide variety of linearly constrained convex and nonconvex problems.

Once the solution to this first problem (denoted as $\mathbf{H}^{(0)}$) has been obtained, in the second step we consider the following slightly modified version of Eq. (13.39)

$$\begin{aligned} \min_{\mathbf{H},\{\mathbf{U}_m\}_{m=1}^M} \; & \sum_m \|\mathbf{H} - \hat{\mathbf{C}}_{x,m}^{1/2} \mathbf{U}_m \mathbf{C}_{w,m}^{-1/2}\|_F^2 \\ \text{s.t. } & \mathbf{U}_m \in \mathcal{U}_N \text{ for all } m, \end{aligned} \tag{13.41}$$

which is again a quadratic convex objective with nonconvex manifold constraints. Problem (13.41) is then solved using a gradient method that, at each iteration, projects the iterates of each $\mathbf{U}_m$ onto the Stiefel manifold $\mathcal{U}_N$. For the first gradient iteration, the filter $\mathbf{H}$ is initialized as $\mathbf{H} = \mathbf{H}^{(0)}$, the output of the first step. Because multiple local optima exist, we run the first step with I_1 random initializations and, for each of those, we solve problem (13.41). We then collect all of the obtained outputs and select as the final solution the one leading to the sparsest shift. We refer to [67] for additional details.

13.5.3 NUMERICAL TESTS

Here we illustrate the recovery of two real-world graphs in order to assess the performance of some of the proposed network topology inference algorithms from nonstationary diffusion processes.

Brain graph. Consider a brain graph $\mathcal{G}$ with $N = 66$ nodes or neural regions and edge weights given by the density of anatomical connections between regions [70]. Denoting by $\mathbf{S} = \mathbf{A}$ the weighted adjacency of the brain graph, we consider two types of filters $\mathbf{H}_1 = \sum_{l=0}^{2} h_l \mathbf{A}^l$ and $\mathbf{H}_2 = (\mathbf{I} + \alpha\mathbf{A})^{-1}$, where the coefficients h_l and α are drawn uniformly on [0, 1]. We then generate M random input-output pairs $\{\mathbf{x}_m, \mathbf{w}_m\}_{m=1}^M$ (cf. Section 13.5.1), where signals are filtered by either $\mathbf{H}_1$ or $\mathbf{H}_2$, and estimate the filter using Eq. (13.23). Problem (13.9) with $\hat{\mathbf{V}}$ given by the eigenvectors of the estimated filter is then solved in order to infer the brain graph. In Fig. 13.3A (top) we plot the recovery error $\|\mathbf{S}_1^* - \mathbf{S}\|_F / \|\mathbf{S}\|_F$ as a function of M for both types of filters. First, notice that the performance is roughly independent of the filter type. More importantly, for $M \geq N$, the optimal filter estimation is unique (cf. Proposition 13.3) and leads to perfect recovery. We also consider the case where the observation of the output signals $\mathbf{x}_m$ is

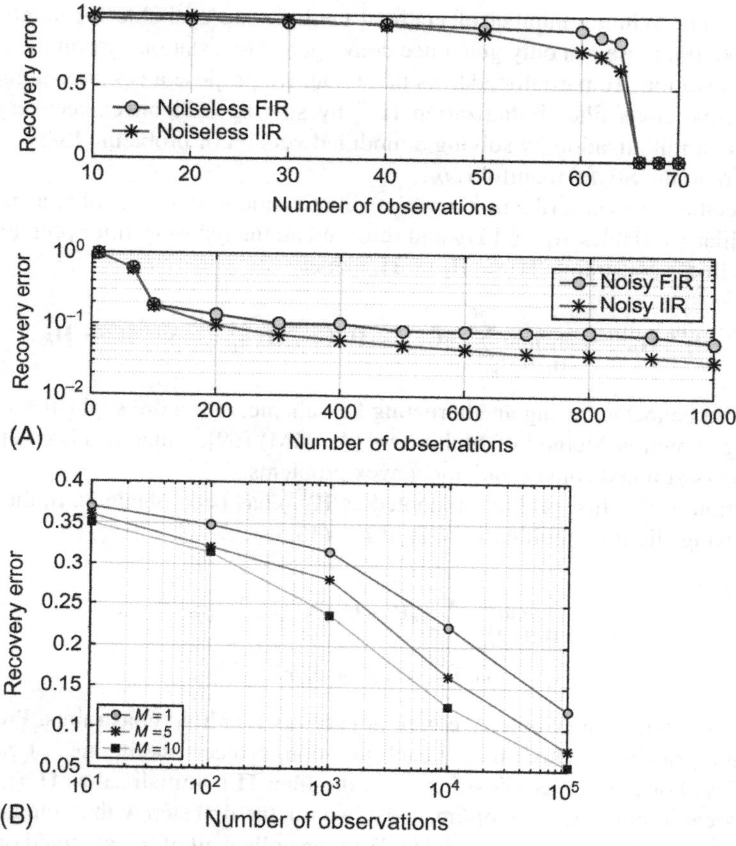

FIG. 13.3

Inference of graph topology from nonstationary signals. (A) Brain network recovery error for FIR and IIR filters versus number of observed signals in noiseless (*top*) and noisy (*bottom*) settings. (B) Error in recovering a social network as a function of the number of opinion profiles observed and parametrized by the number of topics M.

noisy; see Fig. 13.3A (bottom). For this latter case, even though the estimation improves with increasing M, a larger number of observations is needed to guarantee successful recovery of the brain graph.

Social network. We consider the social network of a karate club studied by Zachary [71], represented by a graph $\mathcal{G}$ consisting of $N = 34$ nodes or members of the club and undirected edges symbolizing friendships among them. Denoting by $\mathbf{L}$ the normalized Laplacian of $\mathcal{G}$, we define the GSO $\mathbf{S} = \mathbf{I} - \alpha\mathbf{L}$ with $\alpha = 1/\lambda_{\max}(\mathbf{L})$, modeling the diffusion of opinions between the members of the club. A signal $\mathbf{x}$ can be regarded as a unidimensional opinion of each club member regarding a specific topic, and each application of $\mathbf{S}$ can be seen as an opinion update. Our goal is to recover $\mathbf{L}$—hence, the social structure of the Karate club–from the observations of opinion profiles. We consider M different processes in the graph corresponding, e.g., to opinions on M different topics and assume that an opinion profile $\mathbf{x}_m$ is generated by the diffusion through the network of an initial signal $\mathbf{w}_m$. More precisely, for each topic

$m = 1, \ldots, M$, we model $\mathbf{w}_m$ as a zero-mean process with known covariance $\mathbf{C}_{w,m}$. We are then given a set $\{\mathbf{x}_m^{(r)}\}_{r=1}^R$ of opinion profiles generated from different sources $\{\mathbf{w}_m^{(r)}\}_{r=1}^R$ diffused through a filter of unknown nonnegative coefficients $\boldsymbol{\beta}$. From these R opinion profiles we build an estimate $\hat{\mathbf{C}}_{x,m}$ of the output covariance and, leveraging the fact that $\mathbf{S}$ is PSD and $\boldsymbol{\beta} \geq 0$ (cf. Section 13.5.2), we estimate the unknown filter $\hat{\mathbf{H}}$ by solving Eq. (13.28). Lastly, we use the eigenvectors $\hat{\mathbf{V}}$ of $\hat{\mathbf{H}}$ to solve Eq. (13.9), where $\mathcal{S}$ is modified accordingly for the recovery of a normalized Laplacian; see [30]. In Fig. 13.3B we plot the shift recovery error as a function of the number of observations R and for three different values of M. As R increases, the estimate $\hat{\mathbf{C}}_{x,m}$ becomes more reliable, entailing a better estimation of the underlying filter and, ultimately, leading to more accurate eigenvectors $\hat{\mathbf{V}}$. Hence, we observe a decreasing error with increasing R. Moreover, for a fixed R, the error in the estimation of $\hat{\mathbf{C}}_{x,m}$ can be partially overcome by observing multiple processes, thus, larger values of M lead to smaller errors.

13.6 DISCUSSION

With $\mathbf{S} = \mathbf{V}\boldsymbol{\Lambda}\mathbf{V}^T$ being the eigendecompositon of the shift operator associated with an undirected graph $\mathcal{G}$, we studied the problem of identifying $\mathbf{S}$ (hence the topology of $\mathcal{G}$) using a two-step approach where: (i) we first estimate the eigenvectors $\mathbf{V}$; and (ii) we then use $\mathbf{V}$ as input to find the eigenvalues $\boldsymbol{\Lambda}$ robustly via a convex optimization problem. Under the assumption that observed signals $\mathcal{X} = \{\mathbf{x}_r\}_{r=1}^R$ resulted from diffusion dynamics on the graph or, equivalently, that they were (graph) stationary in $\mathbf{S}$, it was shown that $\mathbf{V}$ could be estimated from the eigenvectors of the sample covariance of $\mathcal{X}$. As a consequence, several well-established methods for topology identification based on Pearson and partial correlations can be viewed as particular instances of the approach here presented. Contrasting with the stationary setting where $\mathbf{S}$ and the covariance matrix of the observed signals are simultaneously diagonalizable, for general (nonstationary) diffusion processes they are not. There is a workaround that entails estimating the unknown diffusion (graph) filter, a polynomial in the shift operator that preserves the sought eigenbasis $\mathbf{V}$. To carry out this initial system identification step, extra information is required on the input signal driving the diffusion process on the graph. Numerical tests showcase the effectiveness of the developed topology inference framework in recovering synthetic and real-world graphs.

ACKNOWLEDGMENTS

The authors want to thank A. Ribeiro, R. Shafipour, Y. Wang, and C. Uhler for their collaboration in the papers that serve as foundation of this chapter, as well as the financial support of the Spanish MINECO grants OMICRON (TEC2013-41604-R) and KLINILYCS (TEC2016-75361-R).

REFERENCES

[1] Shuman DI, Narang SK, Frossard P, Ortega A, Vandergheynst P. The emerging field of signal processing on graphs: extending high-dimensional data analysis to networks and other irregular domains. IEEE Signal Process Mag 2013;30(3):83–98.

[2] Sandryhaila A, Moura JMF. Discrete signal processing on graphs. IEEE Trans Signal Process 2013;61(7):1644–56.

[3] Sandryhaila A, Moura JMF. Big Data analysis with signal processing on graphs. IEEE Signal Process Mag 2014;31(5):80–90.
[4] Deri JA, Moura JMF. New York city taxi analysis with graph signal processing. In: Proceedings of the IEEE global conference on signal and information processing; 2016. p. 1275–9.
[5] Marques AG, Segarra S, Leus G, Ribeiro A. Sampling of graph signals with successive local aggregations. IEEE Trans Signal Process 2016;64(7):1832–43.
[6] Romero D, Ma M, Giannakis GB. Kernel-based reconstruction of graph signals. IEEE Trans Signal Process 2017;65(3):764–78.
[7] Barrat A, Barthelemy M, Vespignani A. Dynamical processes on complex networks. New York, NY: Cambridge University Press; 2008.
[8] Mittler R, Vanderauwera S, Gollery M, Breusegem FV. Reactive oxygen gene network of plants. Trends Plant Sci 2004;9(10):490–8.
[9] Segarra S, Huang W, Ribeiro A. Diffusion and superposition distances for signals supported on networks. IEEE Trans Signal Inf Process Netw 2015;1(1):20–32.
[10] Kolaczyk ED. Statistical analysis of network data: methods and models. New York, NY: Springer; 2009.
[11] Sporns O. Networks of the brain. MIT Press; 2011.
[12] Huang W, Goldsberry L, Wymbs NF, Grafton ST, Bassett DS, Ribeiro A. Graph frequency analysis of brain signals. IEEE J Sel Topics Signal Process 2016;10(7):1189–203.
[13] Ménoret M, Farrugia N, Pasdeloup B, Gripon V. Evaluating graph signal processing for neuroimaging through classification and dimensionality reduction. In: Proceedings of the IEEE global conference on signal and information processing; 2017.
[14] Baingana B, Mateos G, Giannakis GB. Proximal-gradient algorithms for tracking cascades over social networks. IEEE J Sel Topics Signal Process 2014;8:563–75.
[15] Gomez-Rodriguez M, Song L. Diffusion in social and information networks: research problems, probabilistic models and machine learning methods. In: Proceedings of the 21th ACM SIGKDD international conference on knowledge discovery and data mining, Sydney, NSW, Australia, August 10–13; 2015. p. 2315–6.
[16] Chen S, Varma R, Sandryhaila A, Kovačević J. Discrete signal processing on graphs: sampling theory. IEEE Trans Signal Process 2015;63(24):6510–23.
[17] Sporns O. Discovering the human connectome. Boston, MA: MIT Press; 2012.
[18] Dempster AP. Covariance selection. Biometrics 1972;28(1):157–75.
[19] Friedman J, Hastie T, Tibshirani R. Sparse inverse covariance estimation with the graphical lasso. Biostatistics 2008;9(3):432–41.
[20] Banerjee O, Ghaoui LE, d'Aspremont A. Model selection through sparse maximum likelihood estimation for multivariate gaussian or binary data. J Mach Learn Res 2008;9:485–516.
[21] Lake BM, Tenenbaum JB. Discovering structure by learning sparse graph. In: Annual cognitive science conference; 2010. p. 778–83.
[22] Slawski M, Hein M. Estimation of positive definite M-matrices and structure learning for attractive gaussian markov random fields. Linear Algebra Appl 2015;473:145–79.
[23] Egilmez HE, Pavez E, Ortega A. Graph learning from data under Laplacian and structural constraints. IEEE J Sel Topics Signal Process 2017;11(6):825–41.
[24] Meinshausen N, Buhlmann P. High-dimensional graphs and variable selection with the lasso. Ann Stat 2006;34:1436–62.
[25] Cai X, Bazerque JA, Giannakis GB. Inference of gene regulatory networks with sparse structural equation models exploiting genetic perturbations. PLoS Comput Biol 2013;9(5):e1003068. https://doi.org/10.1371/journal.pcbi.1003068.
[26] Shen Y, Baingana B, Giannakis GB. Tensor decompositions for identifying directed graph topologies and tracking dynamic networks. IEEE Trans Signal Process 2017;65(14):3675–87.

[27] Brovelli A, Ding M, Ledberg A, Chen Y, Nakamura R, Bressler SL. Beta oscillations in a large-scale sensorimotor cortical network: directional influences revealed by Granger causality. PNAS 2004;101: 9849–54.

[28] Karanikolas GV, Giannakis GB, Slavakis K, Leahy RM. Multi-kernel based nonlinear models for connectivity identification of brain networks. In: Proceedings of the IEEE international conference on acoustics, speech, signal process, Shanghai, China, March 20–25; 2016.

[29] Shen Y, Baingana B, Giannakis GB. Kernel-based structural equation models for topology identification of directed networks. IEEE Trans Signal Process 2017;65(10):2503–16.

[30] Segarra S, Marques AG, Mateos G, Ribeiro A. Network topology inference from spectral templates. IEEE Trans Signal Inf Process Netw 2017;3(3):467–83.

[31] Pasdeloup B, Gripon V, Mercier G, Pastor D, Rabbat MG. Characterization and inference of graph diffusion processes from observations of stationary signals. IEEE Trans Signal Inf Process Netw 2017. https://doi.org/10.1109/TSIPN.2017.2742940.

[32] Thanou D, Dong X, Kressner D, Frossard P. Learning heat diffusion graphs. IEEE Trans Signal Inf Process Netw 2017;3(3):484–99.

[33] Mei J, Moura JMF. Signal processing on graphs: causal modeling of unstructured data. IEEE Trans Signal Process 2017;65(8):2077–92.

[34] Dong X, Thanou D, Frossard P, Vandergheynst P. Learning Laplacian matrix in smooth graph signal representations. IEEE Trans Signal Process 2016;64(23):6160–73.

[35] Kalofolias V. How to learn a graph from smooth signals. In: International conference on artificial intelligence and statistics (AISTATS). J Mach Learn Res; 2016. p. 920–9.

[36] Marques AG, Segarra S, Leus G, Ribeiro A. Stationary graph processes and spectral estimation. IEEE Trans Signal Process 2017;65(22):5911–26.

[37] Perraudin N, Vandergheynst P. Stationary signal processing on graphs. IEEE Trans Signal Process 2017;65(13):3462–77.

[38] Girault B. Stationary graph signals using an isometric graph translation. In: Proceedings of European signal processing conference; 2015. p. 1516–20.

[39] Chepuri SP, Liu S, Leus G, Hero AO. Learning sparse graphs under smoothness prior. In: Proceedings of the IEEE international conference on acoustics, speech, signal process, New Orleans, LA, March 5–9; 2017. p. 6508–12.

[40] Rabbat MG. Inferring sparse graphs from smooth signals with theoretical guarantees. In: Proceedings of the IEEE international conference on acoustics, speech, signal process, New Orleans, LA, March 5–9; 2017. p. 6533–7.

[41] Segarra S, Marques AG, Ribeiro A. Optimal graph-filter design and applications to distributed linear network operators. IEEE Trans Signal Process 2017;65(15):4117–31.

[42] Smola AJ, Kondor R. Kernels and regularization on graphs. In: Learning theory and kernel machines. Springer; 2003. p. 144–58.

[43] Guo J, Levina E, Michailidis G, Zhu J, et al. Joint estimation of multiple graphical models. Biometrika 2011;98(1):1–15.

[44] Danaher P, Wang P, Witten DM. The joint graphical lasso for inverse covariance estimation across multiple classes. J R Stat Soc Ser B Stat Methodol 2014;76(2):373–97.

[45] Ryali S, Chen T, Supekar K, Menon V. Estimation of functional connectivity in fMRI data using stability selection-based sparse partial correlation with elastic net penalty. NeuroImage 2012;59(4):3852–61.

[46] Honorio J, Samaras D. Multi-task learning of gaussian graphical models. In: Proceedings of the 27th international conference on machine learning (ICML-10); 2010. p. 447–54.

[47] Varoquaux G, Gramfort A, Poline JB, Thirion B. Brain covariance selection: better individual functional connectivity models using population prior. In: Advances in neural information processing systems; 2010. p. 2334–42.
[48] Zhou S, Lafferty J, Wasserman L. Time varying undirected graphs. Mach Learn 2010;80(2–3):295–319.
[49] Kalofolias V, Loukas A, Thanou D, Frossard P. Learning time varying graphs. In: 2017 IEEE international conference on acoustics, speech and signal processing (ICASSP); 2017. p. 2826–30.
[50] Baingana B, Giannakis GB. Tracking switched dynamic network topologies from information cascades. IEEE Trans Signal Process 2017;65(4):985–97. https://doi.org/10.1109/TSP.2016.2628354.
[51] Oselio B, Kulesza A, Hero AO. Multi-layer graph analysis for dynamic social networks. IEEE J Sel Top Signal Process 2014;8(4):514–23. https://doi.org/10.1109/JSTSP.2014.2328312.
[52] Segarra S, Wang Y, Uhler C, Marques AG. Joint inference of networks from stationary graph signals. In: Proceedings of Asilomar conference on signals, systems, computers, Pacific Grove, CA; 2017.
[53] Ortega JM. Numerical analysis: a second course. Classics in applied mathematics. Society for Industrial and Applied Mathematics; 1990.
[54] Bollobás B. Random graphs. Cambridge University Press; 2001.
[55] Feizi S, Marbach D, Medard M, Kellis M. Network deconvolution as a general method to distinguish direct dependencies in networks. Nat Biotech 2013;31(8):726–33.
[56] Marks DS, Colwell LJ, Sheridan R, Hopf TA, Pagnani A, Zecchina R, et al. Protein 3d structure computed from evolutionary sequence variation. PLoS ONE 2011;6(12):e28766.
[57] Shafipour R, Segarra S, Marques AG, Mateos G. Network topology inference from non-stationary graph signals. In: Proceedings of the IEEE international conference on acoustics, speech, signal process; 2017. p. 5870–4.
[58] Luo ZQ, Ma WK, So AMC, Ye Y, Zhang S. Semidefinite relaxation of quadratic optimization problems. IEEE Signal Process Mag 2010;27(3):20–34.
[59] Tan PH, Rasmussen LK. The application of semidefinite programming for detection in CDMA. IEEE J Sel Areas Commun 2001;19(8):1442–9.
[60] Gershman AB, Sidiropoulos ND, Shahbazpanahi S, Bengtsson M, Ottersten B. Convex optimization-based beamforming. IEEE Signal Process Mag 2010;27(3):62–75.
[61] Ye Y. Interior point algorithms-theory and analysis; 1998.
[62] Shafipour R, Segarra S, Marques AG, Mateos G. Identifying the topology of undirected networks from diffused non-stationary graph signals. IEEE Trans Signal Process 2018. https://arxiv.org/abs/1801.03862.
[63] Nesterov Y. Semidefinite relaxation and nonconvex quadratic optimization. Optim Methods Softw 1998;9(1–3):141–60.
[64] Boothby WM. An introduction to differentiable manifolds and Riemannian geometry. Academic Press; 2002.
[65] Absil PA, Mahony R, Sepulchre R. Optimization algorithms on matrix manifolds. Princeton University Press; 2009.
[66] Boumal N, Mishra B, Absil PA, Sepulchre R, et al. Manopt, a matlab toolbox for optimization on manifolds. J Mach Learn Res 2014;15(1):1455–9.
[67] Shafipour R, Segarra S, Marques AG, Mateos G. Topology inference of directed networks from diffused graph signals. In: IEEE Data Science Workshop; 2018.
[68] Manton JH. Optimization algorithms exploiting unitary constraints. IEEE Trans Signal Process 2002;50(3):635–50.
[69] Boyd S, Parikh N, Chu E, Peleato B, Eckstein J. Distributed optimization and statistical learning via the alternating direction method of multipliers. Found Trends Mach Learn 2011;3(1):1–122.
[70] Hagmann P, Cammoun L, Gigandet X, Meuli R, Honey CJ, Wedeen VJ, et al. Mapping the structural core of human cerebral cortex. PLoS Biol 2008;6(7):e159.
[71] Zachary WW. An information flow model for conflict and fission in small groups. J Anthropol Res 1977;33(4):452–73.

CHAPTER

PARTIALLY ABSORBING RANDOM WALKS: A UNIFIED FRAMEWORK FOR LEARNING ON GRAPHS

14

Xiao-Ming Wu*, Zhenguo Li†, Shih-Fu Chang‡

Department of Computing, The Hong Kong Polytechnic University, Kowloon, Hong Kong Special Administrative Region Huawei Noah's Ark Lab, Shatin, Hong Kong Special Administrative Region† Department of Electrical Engineering and Department of Computer Science, Columbia University, New York, NY, United States‡*

14.1 INTRODUCTION

As large amounts of graph-structured data become available in various domains such as social networks, the Internet, and knowledge bases, the importance of learning on graphs is growing. Analyzing graph structures and estimating relations between vertices benefit numerous applications, including web search [1], recommender systems [2,3], question answering systems [4], community detection [5], text analysis [6,7], spam filtering [8,9], image segmentation [10], and biological data mining [11].

In the past two decades, many popular tools have been developed for solving learning tasks on graphs, including ranking [12], semisupervised classification [13,14], and local clustering [15,16]. These mainly include models based on random walks such as personalized PageRank for web-page ranking [1] and hitting and commute times [17,18] for proximity analysis; models based on Laplacian regularization such as the regularized Laplacian kernel matrices for collaborative filtering [3] and the Gaussian harmonic function method for semisupervised learning [13]; kernel learning for pairwise constraint propagation [19,20]; and many other graph-theoretic proximity measures [21,22].

Despite the widespread popularity of some models that are used in various applications, recent studies have shown of their limitations in modeling the graph topology, including the commute and hitting times [2,18], the personalized PageRank [23], and the Gaussian harmonic function method [24]. To prevent the improper use of these models in practice and to reach their full potential, a unified framework is needed for understanding their behavior and connections, and further facilitating model selection and design.

In this chapter, we present a unified framework called partially absorbing random walks (ParWalks) [23]. A ParWalk is a second-order Markov chain that allows partial absorption at each state. We show that by setting the absorbing capacity of each state properly, various popular models, including personalized PageRank, hitting times, the pseudoinverse of the graph Laplacian matrix, and the

Cooperative and Graph Signal Processing. https://doi.org/10.1016/B978-0-12-813677-5.00014-6

Gaussian harmonic function method, could be reproduced by ParWalks. The unified framework enables model comparison and selection and opens the door for model design. In particular, we show that if the absorbing capacity of each state is small, the probabilities of ParWalks absorbed at a query vertex can well encode graph cluster structures [23,25]. This result not only justifies existing models such as hitting times to a fixed vertex, the normalized personalized PageRank, and the pseudoinverse of the graph Laplacian matrix, but also allows us to design new models. Our analysis and models are supported by extensive experiments on both synthetic and real data.

14.2 PARTIALLY ABSORBING RANDOM WALKS

14.2.1 THE PARWALK MODEL

Let us consider a simple diffusion process illustrated in Fig. 14.1A. At the beginning, a unit flow (shaded) is injected at a vertex on the graph. At the first step, some fraction of the flow (right-shaded) is "absorbed" at the vertex while the rest (left-shaded) propagates to its neighbors. Whenever the flow passes a vertex, some fraction of it will get absorbed at that vertex. As this process continues, the amount of flow absorbed will accumulate and there will be less and less flow left running on the graph. After a certain number of steps, the unit flow will mostly be absorbed. Note that if the fraction of flow absorbed at each vertex is the same, this diffusion process is essentially the same as the bookmark-coloring algorithm in [26], though both of them were proposed independently.

We can describe the above diffusion process in terms of random walks. Let us define a discrete-time stochastic process $X = \{X_t : t \geq 0\}$ on the state set $\{1, 2, \ldots, n\}$. Note that for random walks on a graph, the state set is the set of vertices of the graph. The initial state X_0 is given, say $X_0 = i$, the next state

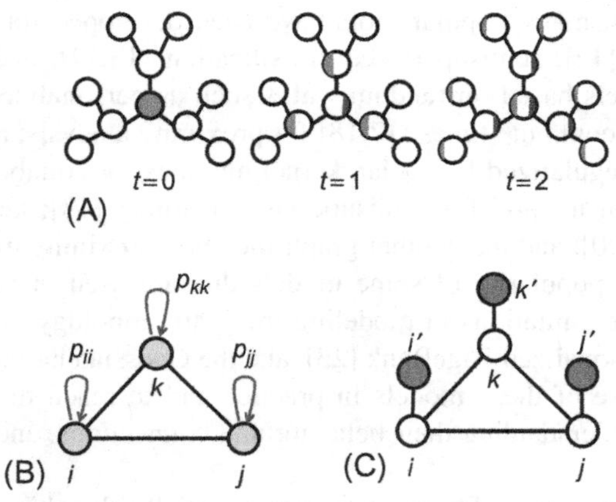

FIG. 14.1

Partially absorbing random walks. (A) A flow diffusion perspective (see text). (B) A second-order Markov chain. (C) An equivalent standard Markov chain with additional sinks.

X_1 is determined by the transition probability $\mathbb{P}(X_1 = j|X_0 = i) = p_{ij}$, and the subsequent states are determined by the following transition probabilities

$$\mathbb{P}(X_{t+2} = j|X_{t+1} = i, X_t = k) = \begin{cases} 1, & j = i, i = k, \\ 0, & j \neq i, i = k, \\ \mathbb{P}(X_{t+2} = j|X_{t+1} = i) = p_{ij}, & i \neq k, \end{cases} \tag{14.1}$$

where $t \geq 0$. Note that the process X is time homogeneous, i.e., the transition probabilities in Eq. (14.1) are independent of t. By the definition, if the previous and current states are the same, the process will remain in the current state forever. Otherwise the next state is conditionally independent of the previous state given the current state, i.e., the process behaves like an ordinary random walk.

As illustrated in Fig. 14.1B, the stochastic process X at each state i will be stuck at i with some probability p_{ii}. Hence, we shall call X the partially absorbing random walk, and p_{ii} the *absorbing capacity* of state i. If $0 < p_{ii} < 1$, then we say that i is a partially absorbing state. If $p_{ii} = 1$, then we say that i is a fully absorbing state. We will refer to both as absorbing states without being confusing. Finally, if $p_{ii} = 0$, then we say that i is a nonabsorbing state. Note that if $p_{ii} \in \{0, 1\}$ for every state $i \in N$, the above process will reduce to a standard Markov chain [27].

We can see that a ParWalk is a second-order Markov chain completely specified by the first-order transition probabilities $\{p_{ij}\}$. It is also not difficult to observe that any ParWalk can be realized as a standard Markov chain by adding a sink vertex (a fully absorbing state) to each vertex (Fig. 14.1C), and with the transition probability from state i to sink i' exactly as p_{ii}.

14.2.2 THE ABSORPTION PROBABILITIES OF PARWALKS

In the following, we will show that the probability of a ParWalk starting from any state i to be absorbed at any state j (within a finite number of steps) can be derived in a closed form.

Before we proceed further, let us first define some notations used in this chapter (see Table 14.1 for a summary of notations). We consider undirected graphs that are connected, weighted, and without self-loops. Denote by $\mathcal{G} = (\mathcal{V}, W)$ a graph with a set $\mathcal{V}$ of n vertices, where $W = [w_{ij}] \in \mathbb{R}^{n\times n}$ ($w_{ii} = 0$) is a symmetric nonnegative affinity matrix that specifies the edge weight. Denote by $D = \mathrm{diag}(d_1, d_2, \ldots, d_n)$ the degree matrix, where $d_i = \sum_j w_{ij}$ is the degree of vertex i. The graph Laplacian [28] is defined as $L := D - W$, and the symmetric normalized graph Laplacian is defined as $L_{\mathrm{sym}} = D^{-\frac{1}{2}} L D^{-\frac{1}{2}}$. Let $\lambda_1, \lambda_2, \ldots, \lambda_n \geq 0$ be arbitrary, and set $\Lambda = \mathrm{diag}(\lambda_1, \lambda_2, \ldots, \lambda_n)$, which we will refer to as a regularizer. Denote by α a positive scalar.

The first order transition probabilities of a ParWalk on the graph $\mathcal{G}$ can be defined as follows:

$$p_{ij} = \begin{cases} \frac{\alpha\lambda_i}{\alpha\lambda_i + d_i}, & i = j, \\ \frac{w_{ij}}{\alpha\lambda_i + d_i}, & i \neq j. \end{cases} \tag{14.2}$$

Clearly, the absorbing capacity p_{ii} of state i is determined by α and λ_i. The larger $\alpha\lambda_i$ is, the larger the absorbing capacity is. Let $A = [a_{ij}] \in \mathbb{R}^{n\times n}$ be the absorption probability matrix, where a_{ij} is the probability of the ParWalk from i to be absorbed at j. We can derive A as follows.

Theorem 14.1. *Suppose $\lambda_i > 0$ for some i. Then $A = (L + \alpha\Lambda)^{-1}\alpha\Lambda$.* ■

Proof. Because $\alpha > 0$, and $\lambda_i > 0$ for some i, the matrix $L + \alpha\Lambda$ is positive definite and hence nonsingular. Moreover, the matrix $D + \alpha\Lambda$ is nonsingular because D is nonsingular. Thus, the matrix

Table 14.1 Notations

Notation	Description
$\mathcal{V}$	set of all vertices in a graph
W	affinity weight matrix
w_{ij}	edge weight between vertices i and j
D	degree matrix
d_i	degree of vertex i
L	graph Laplacian matrix
L_{sym}	normalized graph Laplacian matrix
$L^{\dagger}$	pseudoinverse of the graph Laplacian matrix
Λ	regularizer matrix
λ_i	scalar regularizer of vertex i
I	identity matrix
H	hybrid regularizer matrix combining D and I
A	absorption probability matrix
a_{ij}	probability of a ParWalk starting from state i absorbed at state j
M	normalized absorption probability matrix
p_{ij}	transition probability of a ParWalk from state i to state j
p_{ii}	absorbing capacity of state i
α	balancing scalar

$I-(D+\alpha\Lambda)^{-1}W=(D+\alpha\Lambda)^{-1}(L+\alpha\Lambda)$ is also nonsingular. We can observe that the absorbing probabilities $\{a_{ij}\}$ satisfy the following equations:

$$a_{ii}=\frac{\alpha\lambda_i}{\alpha\lambda_i+d_i}\times 1+\sum_{j\neq i}\frac{w_{ij}}{\alpha\lambda_i+d_i}a_{ji}, \tag{14.3}$$

$$a_{ij}=\sum_{k\neq i}\frac{w_{ik}}{\alpha\lambda_i+d_i}a_{kj},\ i\neq j. \tag{14.4}$$

Upon writing Eqs. (14.3), (14.4) in matrix form, we have

$$(I-(D+\alpha\Lambda)^{-1}W)A=(D+\alpha\Lambda)^{-1}\alpha\Lambda, \tag{14.5}$$

whence $A=(I-(D+\alpha\Lambda)^{-1}W)^{-1}(D+\alpha\Lambda)^{-1}\alpha\Lambda=(L+\alpha\Lambda)^{-1}\alpha\Lambda$. ∎

It is easy to see that as long as there is at least one absorbing state in the graph, a ParWalk starting from any vertex will eventually be absorbed. The following corollary confirms that A is indeed a probability matrix.

Corollary 14.1. *Suppose $\lambda_i>0$ for some i. Then A is a nonnegative matrix with each row summing up to 1.*

Proof. See Appendix. ∎

14.2.3 HIGHER-ORDER PARWALKS

We can naturally generalize the construction of a ParWalk to mth order ($m > 2$), i.e., the process is absorbed at a state only after staying at it for m-consecutive steps. Mathematically, it is defined by the following transition probabilities:

$$\mathbb{P}(X_{t+m+1} = j | X_{t+m} = i_m, \ldots, X_t = i_1) = \begin{cases} 1, & j = i_m = \cdots = i_1, \\ 0, & j \neq i_m, i_m = \cdots = i_1, \\ \mathbb{P}(X_{t+m+1} = j | X_{t+m} = i_m) = p_{i_m j}, & \text{else}. \end{cases} \tag{14.6}$$

Similar to second-order ParWalks, an mth order ParWalk is also completely specified by the first-order transition probabilities $\{p_{i_m j}\}$. The absorption probabilities of an mth order ParWalk, with its first-order transition probabilities defined as in Eq. (14.2), can be derived similarly as a second-order ParWalk, though in a much more complicated form. However, it turns out that an mth order ParWalk does not have additional modeling power than a second-order ParWalk, as explained below.

Suppose that $\tilde{A}$ is the absorption probability matrix of an mth order ParWalk with its first-order transition probabilities $\{p_{ij}\}_{i,j}$ defined as in Eq. (14.2). We can construct a second-order ParWalk X such that its absorption probability matrix A is exactly the same as $\tilde{A}$. The transition probabilities of X can be defined as:

$$\mathbb{P}(X_{t+2} = j | X_{t+1} = i, X_t = k) = \begin{cases} 1, & j = i, i = k, \\ 0, & j \neq i, i = k, \\ p_{ii}^m, & j = i, i \neq k, \\ \frac{p_{ij}}{1-p_{ii}}(1 - p_{ii}^m), & j \neq i, i \neq k. \end{cases} \tag{14.7}$$

To see the rationale behind this definition, note that for $j \neq i$,

$$\mathbb{P}(X_{t+2} = j | X_{t+1} = i) = p_{ij}\frac{1 - p_{ii}^m}{1 - p_{ii}} = p_{ij} + p_{ii}p_{ij} + p_{ii}^2 p_{ij} + \cdots + p_{ii}^{m-1} p_{ij}.$$

By the definition, it is not difficult to see that $A = \tilde{A}$. Also, the second-order ParWalk X can be realized by choosing λ_i' such that $\frac{\alpha\lambda_i'}{\alpha\lambda_i'+d_i} = p_{ii}^m = \left(\frac{\alpha\lambda_i}{\alpha\lambda_i+d_i}\right)^m$.

14.3 A UNIFIED VIEW FOR LEARNING ON GRAPHS

The absorption probabilities of a ParWalk can be used to measure the proximity between any two vertices on a graph, which can be utilized for learning tasks including ranking, semisupervised classification, and clustering. In this section, we show that ParWalks emcompass various popular models for learning on graphs. We first consider the case of ParWalks starting from a fixed vertex, and then consider the case of ParWalks absorbed at a fixed vertex (Fig. 14.2).

14.3.1 PARWALKS STARTING FROM A FIXED STATE

In the following, we will show that the personalized PageRank algorithm [1], the kernel matrix $(L + \alpha I)^{-1}$ used in collaborative filtering [3], and the Gaussian harmonic function method for label propagation [13] and its variants are all special cases of ParWalks starting from a fixed state.

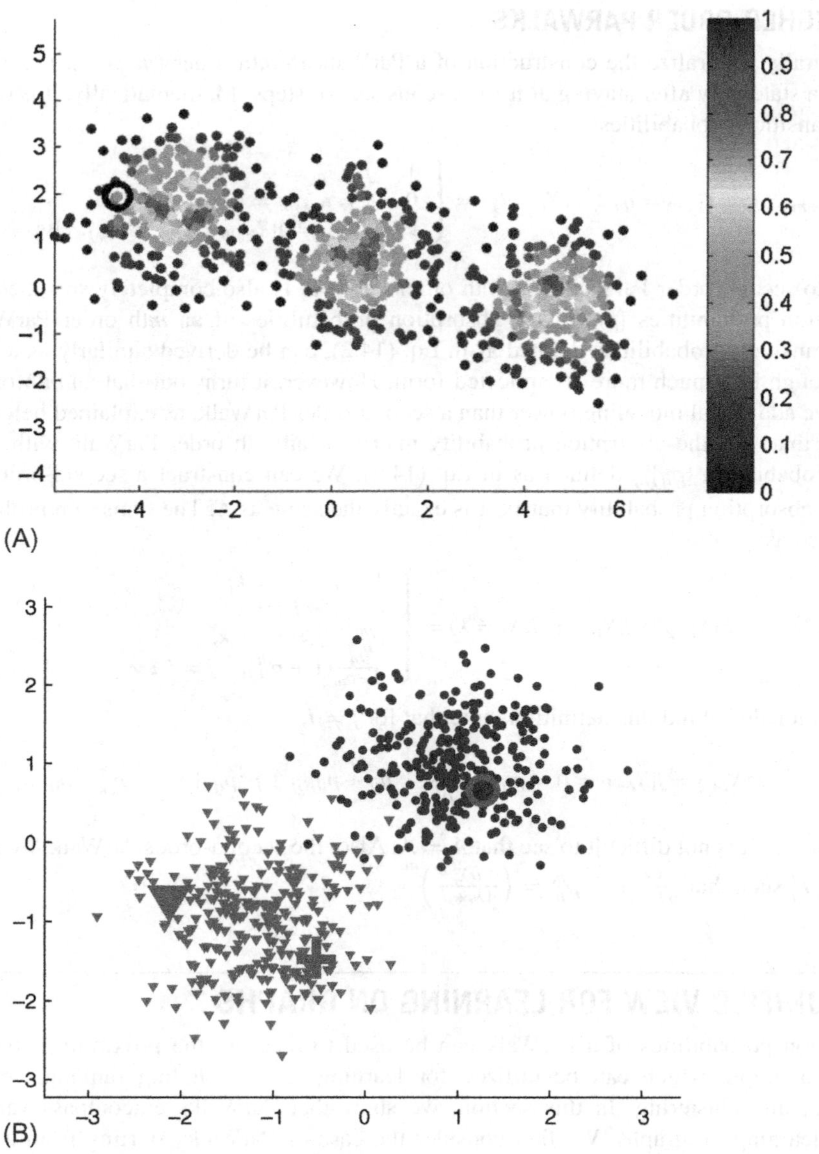

FIG. 14.2

Two synthetic datasets. (A) Three two-dimensional Gaussians of 900 points. The *circle* indicates the query point. The graph construction is described in Section 14.4. The degree of a vertex is indicated by the colormap. Best viewed in color. (B) Two 20-dimensional Gaussians of 600 points with the first two dimensions plotted. The *cross* denotes a query point. The *triangle* and the *circle* denote the labeled point for each Gaussian respectively. For illustration purpose, points within each Gaussian are arranged to appear consecutively in both datasets.

Clearly, the probabilities of a ParWalk starting from a fixed vertex i and absorbed at each vertex on the graph correspond to the ith row of the absorption probability matrix A. The following theorem shows that when the absorbing capacity of each vertex is sufficiently small, the rows of A will converge to a distribution proportional to the regularizer $(\lambda_1, \lambda_2, \ldots, \lambda_n)$, regardless of the graph structure.

Theorem 14.2. *Suppose $\lambda_i > 0$ for all i. Then*

$$\lim_{\alpha \to 0+} (L + \alpha\Lambda)^{-1}\alpha\Lambda = \mathbf{1}\bar{\boldsymbol{\lambda}}^\top, \tag{14.8}$$

where $(\bar{\boldsymbol{\lambda}})_i = \lambda_i / (\sum_{j=1}^{n} \lambda_j)$. In particular, $\lim_{\alpha \to 0+}(L + \alpha I)^{-1}\alpha I = \frac{1}{n}\mathbf{1}\mathbf{1}^\top$.

Proof. See Appendix. ■

Relation with personalized PageRank

In the personalized PageRank [1] algorithm, a random walk at each step returns to some vertex i with probability $0 < \beta < 1$, where β is often referred to as the "teleportation" probability. Denote by $\mathbf{a}$ the stationary distribution of the random walk. Denote by $\mathbf{e}$ the indicator vector of i, i.e., $(\mathbf{e})_i = 1$ and $(\mathbf{e})_j = 0$ for $j \neq i$. Then the equilibrium equation for the random walk can be written as:

$$\mathbf{a}^\top = \beta\mathbf{e}^\top + (1-\beta)\mathbf{a}^\top D^{-1}W \tag{14.9}$$

$$\Longleftrightarrow \mathbf{a}^\top D^{-1}(1-\beta)\left(D - W + \frac{\beta}{1-\beta}D\right) = \beta\mathbf{e}^\top$$

$$\Longleftrightarrow \mathbf{a}^\top = \mathbf{e}^\top\left(L + \frac{\beta}{1-\beta}D\right)^{-1}\frac{\beta}{1-\beta}D. \tag{14.10}$$

By Eq. (14.10), it can be seen that the personalized PageRank vector of i is exactly the ith row of the absorption probability matrix A with $\alpha = \frac{\beta}{1-\beta}$ and $\Lambda = D$. It is also not difficult to check that the absorbing capacity of each state is constant and equal to β. This indicates that a ParWalk will tend to be absorbed at vertices with larger degrees due to their denser connection with the graph. Indeed, by Theorem 14.2, the absorption probabilities will be dominated by the degree of vertices as $\beta \to 0$ $(\alpha \to 0)$, which can be seen from Fig. 14.3A–F. We can also see that even for such a simple dataset, for a wide range of α, there are no clear gaps between clusters in the absorption probabilities.

Relation with the kernel matrix $(L + \alpha I)^{-1}$

The Laplacian regularized kernel matrix $(L + \alpha I)^{-1}$ has been found empirically successful as a proximity measure in ranking and collaborative recommendation [3]. Obviously, it is equal to the absorption probability matrix A with $\Lambda = I$, up to a constant factor $1/\alpha$. By Eq. (14.2), we can see that when α is small, the absorbing capacity $p_{ii} = \frac{\alpha}{\alpha + d_i}$ of vertex i is approximately inversely proportional to its degree d_i.

In contrast to personalized PageRank, setting partial absorption in this way cancels out the bias in d_i and makes absorption probabilities evenly distributed within each cluster, as shown in Fig. 14.3G–L. We can also see that the gaps between clusters become clearer when α becomes smaller. Even when $\alpha \to 0$ and the absorption probabilities converge to a constant (see Theorem 14.2), the gaps still exist, demonstrating that they are desirable for representing the cluster structure.

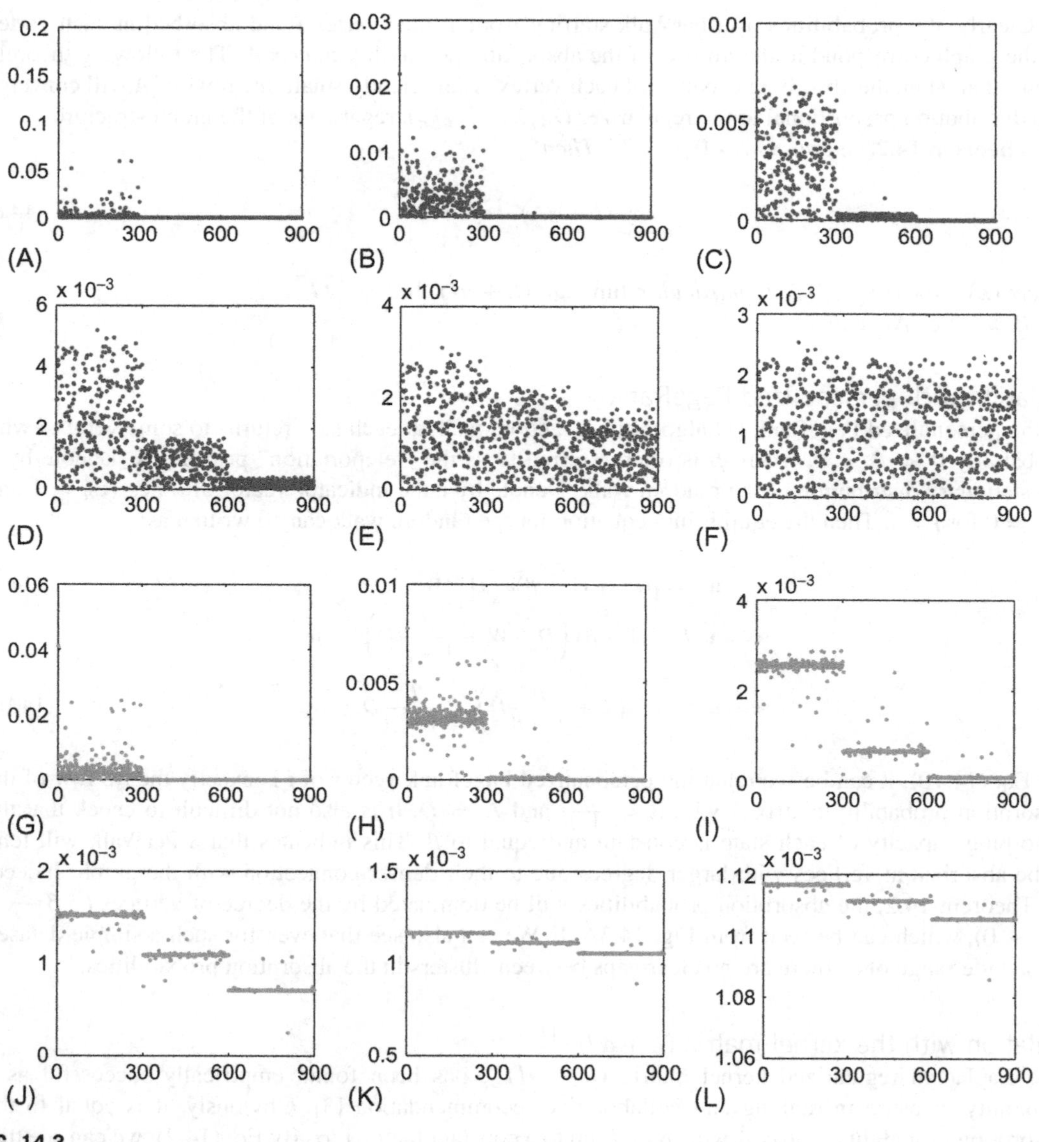

FIG. 14.3

Absorption probabilities of ParWalks starting from a fixed vertex indicated by the black circle in Fig. 14.2A. The horizontal axis is the index of a point, and the vertical axis is the probability of a ParWalk absorbed at that point. (A)–(F) Absorption probabilities of ParWalks with $\Lambda = \alpha D$, $\alpha = 10^{-k}$, $k = 1, 2, \ldots, 6$; (G)–(L) Absorption probabilities of ParWalks with $\Lambda = \alpha I$, $\alpha = 10^{-k}$, $k = 1, 2, \ldots, 6$.

Relations with the harmonic function method and its variants

The seminal Gaussian harmonic function (GHF) method for label propagation [13] can be interpreted in absorbing random walks, thus is naturally a special case of ParWalks. In the GHF method, the labeled vertices are fully absorbing states, and the unlabeled vertices are nonabsorbing states. To classify an unlabeled instance i, one needs to compute and compare the probabilities of a random walk starting from i and absorbed at the labeled vertices of different classes. The regularized GHF method [29] is also a ParWalk with $\lambda_i = 1$ at the labeled vertices while $\lambda_i = 0$ at the unlabeled vertices. The local and global consistency method [14], with the unnormalized Laplacian instead of the normalized Laplacian, is a ParWalk with $\Lambda = I$. A variant of this method is a ParWalk with $\Lambda = D$, which is the same as personalized PageRank. If we add an additional sink to the graph, a variant of the GHF method [30] and a variant of the regularized GHF method [31] can all be included as instances of ParWalks.

From Fig. 14.3, we can see that for ParWalks starting from a fixed state, the choice of Λ is crucial. It seems that setting $\Lambda = I$ is the only way to avoid the bias introduced by Λ. However, it is entirely different for ParWalks ending at a fixed state. We will see in the following that in that case even a random Λ can work very well.

14.3.2 PARWALKS ABSORBED AT A FIXED STATE

The probability vector of ParWalks absorbed at a fixed vertex j is exactly the jth column of the absorption probability matrix A. For simplicity, it is equivalent to consider the symmetric matrix $M = [m_{ij}] \in \mathbb{R}^{n \times n} = (L + \alpha\Lambda)^{-1}$, whose jth column is equal to the jth column of A up to a constant factor $1/\alpha\lambda_j$. In the following, we will show that M can be decomposed into a constant matrix plus a proximity matrix, where the latter converges to a meaningful limit when $\alpha \to 0$.

Denote by $\bar{L} = \Lambda^{-\frac{1}{2}} L \Lambda^{-\frac{1}{2}}$. It is easy to see that $\bar{L}$ is symmetric and positive semidefinite. Note that $\bar{L}$ has the same rank $n-1$ as L (because the graph is connected), and has eigenvalue 0 of multiplicity 1. Let $\bar{L} = U\Gamma U^\top$ be the eigen-decomposition of $\bar{L}$ with eigenvalues $0 = \gamma_1 < \gamma_2 \le \cdots \le \gamma_n$, and orthonormal eigenvector matrix $U = (\mathbf{u}_1, \ldots, \mathbf{u}_n)$, with $\mathbf{u}_1 = (\frac{\sqrt{\lambda_1}}{\sqrt{\sum_i \lambda_i}}, \ldots, \frac{\sqrt{\lambda_n}}{\sqrt{\sum_i \lambda_i}})^\top$. Denote by $\bar{L}^\dagger$ the pseudoinverse of $\bar{L}$, and denote by $\mathbf{1}$ the all-one vector. The following theorem decomposes M.

Theorem 14.3. *$M = C + E$, where $C = \frac{1}{\alpha \sum_i \lambda_i} \mathbf{1}\mathbf{1}^\top$, and $E = \Lambda^{-\frac{1}{2}} \left(\sum_{i=2}^{n} \frac{1}{\gamma_i + \alpha} \mathbf{u}_i \mathbf{u}_i^\top \right) \Lambda^{-\frac{1}{2}}$.*

Proof. By definition,

$$M = (L + \alpha\Lambda)^{-1} = \Lambda^{-\frac{1}{2}} (\Lambda^{-\frac{1}{2}} L \Lambda^{-\frac{1}{2}} + \alpha I)^{-1} \Lambda^{-\frac{1}{2}} = \Lambda^{-\frac{1}{2}} \left(\sum_{i=1}^{n} (\gamma_i + \alpha) \mathbf{u}_i \mathbf{u}_i^\top \right)^{-1} \Lambda^{-\frac{1}{2}}$$

$$= \Lambda^{-\frac{1}{2}} \left(\sum_{i=1}^{n} \frac{1}{\gamma_i + \alpha} \mathbf{u}_i \mathbf{u}_i^\top \right) \Lambda^{-\frac{1}{2}} = \frac{1}{\alpha \sum_i \lambda_i} \mathbf{1}\mathbf{1}^\top + \Lambda^{-\frac{1}{2}} \left(\sum_{i=2}^{n} \frac{1}{\gamma_i + \alpha} \mathbf{u}_i \mathbf{u}_i^\top \right) \Lambda^{-\frac{1}{2}}.$$

■

Corollary 14.2. $\lim_{\alpha \to 0} E = \Lambda^{-\frac{1}{2}} \bar{L}^\dagger \Lambda^{-\frac{1}{2}}$.

Proof. It follows from $\bar{L}^\dagger = \sum_{i=2}^{n} \frac{1}{\gamma_i} \mathbf{u}_i \mathbf{u}_i^\top$. ■

By Theorem 14.3, M is completely determined by E because C is a constant matrix. By Corollary 14.2, when $\alpha \to 0$, E converges to the pseudoinverse of $\bar{L}$ doubly normalized by $\Lambda^{-\frac{1}{2}}$, which encodes the graph structure. In the following, we will show that M (or the column vectors of A) encompasses and relates various well-known proximity measures. To demonstrate their ability in capturing global graph structures, we compare them on a more challenging synthetic dataset two 20-dimensional Gaussians, as shown in Fig. 14.2B. The graph construction and parameter settings are described in Section 14.4.

Relation with the normalized personalized PageRank

The personalized PageRank vector normalized by vertex degree was shown to be highly effective in retrieving local clusters [15]. Suppose that $\mathbf{a}$ is the personalized PageRank vector of some vertex i. We have shown before that $\mathbf{a}$ is equal to the ith row of A with $\Lambda = D$. It is easy to see that the normalized vector, $\mathbf{a}./\mathbf{d}$, is equal to the ith column of A, up to a constant factor $1/\lambda_i$. The comparisons of personalized PageRank and the normalized one are shown in Fig. 14.4A and B. We can see that while the former completely fails in representing the cluster structure, the latter works amazingly well. It is surprising that a simple normalization would make such a difference.

Relation with the hitting times

The hitting time h_{ij} is defined as the average number of steps for a random walk from vertex i to hit vertex j for the first time. Denote by $\mathbf{e}_i$ the indicator vector of i, i.e., $(\mathbf{e}_i)_i = 1$ and $(\mathbf{e}_i)_j = 0$ for $j \neq i$. h_{ij} can be computed as [18]:

$$h_{ij} = d(\mathcal{V})\left\langle \frac{1}{\sqrt{d_j}}\mathbf{e}_j, L_{\text{sym}}^{\dagger}\left(\frac{1}{\sqrt{d_j}}\mathbf{e}_j - \frac{1}{\sqrt{d_i}}\mathbf{e}_i\right)\right\rangle \tag{14.11}$$

$$= d(\mathcal{V})\left(\frac{1}{d_j}\mathbf{e}_j{}^{\top} L_{\text{sym}}^{\dagger}\mathbf{e}_j - \frac{1}{\sqrt{d_i d_j}}\mathbf{e}_i{}^{\top} L_{\text{sym}}^{\dagger}\mathbf{e}_j\right), \tag{14.12}$$

where $d(\mathcal{V})$ is the sum of the degrees of all vertices in the graph. For a fixed vertex j, consider the hitting times of random walks from any vertex i to j, $(h_{ij})_{i=1,\ldots,n}$. By Eq. (14.12), $(h_{ij})_{i=1,\ldots,n}$ is determined by the second term $\frac{1}{\sqrt{d_i d_j}}\mathbf{e}_i{}^{\top} L_{\text{sym}}^{\dagger}\mathbf{e}_j$ (as the first term is constant), which is the (i,j) entry of $D^{-\frac{1}{2}} L_{\text{sym}}^{\dagger} D^{-\frac{1}{2}}$. Note that for $\Lambda = D$, we have $\bar{L} = L_{\text{sym}}$, and by Corollary 14.2, $\lim_{\alpha \to 0} E = D^{-\frac{1}{2}} L_{\text{sym}}^{\dagger} D^{-\frac{1}{2}}$. Also note that smaller h_{ij} indicates that vertices i and j are more similar. Hence, in measuring the proximities between j and other vertices, the hitting times $(h_{ij})_{i=1,\ldots,n}$ are essentially the same as the jth column of M with $\Lambda = D$ when α is sufficiently small.

There is a sharp contrast between the hitting times from a fixed vertex i and the hitting times to a fixed vertex i, as shown in Fig. 14.4C and D. We can see that the former totally fails to capture the cluster structure of high-dimensional data, which verifies the analysis in [18]. In contrast, the latter does very well.

Relation with the pseudoinverse of the graph Laplacian and the commute times

For $\Lambda = I$, we have $\bar{L} = L$, and by Corollary 14.2, $\lim_{\alpha \to 0} E = L^{\dagger}$, where $L^{\dagger}$ is the pseudoinverse of the graph Laplacian matrix. This establishes the equivalence between $L^{\dagger}$ and $M = (L + \alpha I)^{-1}$ (when

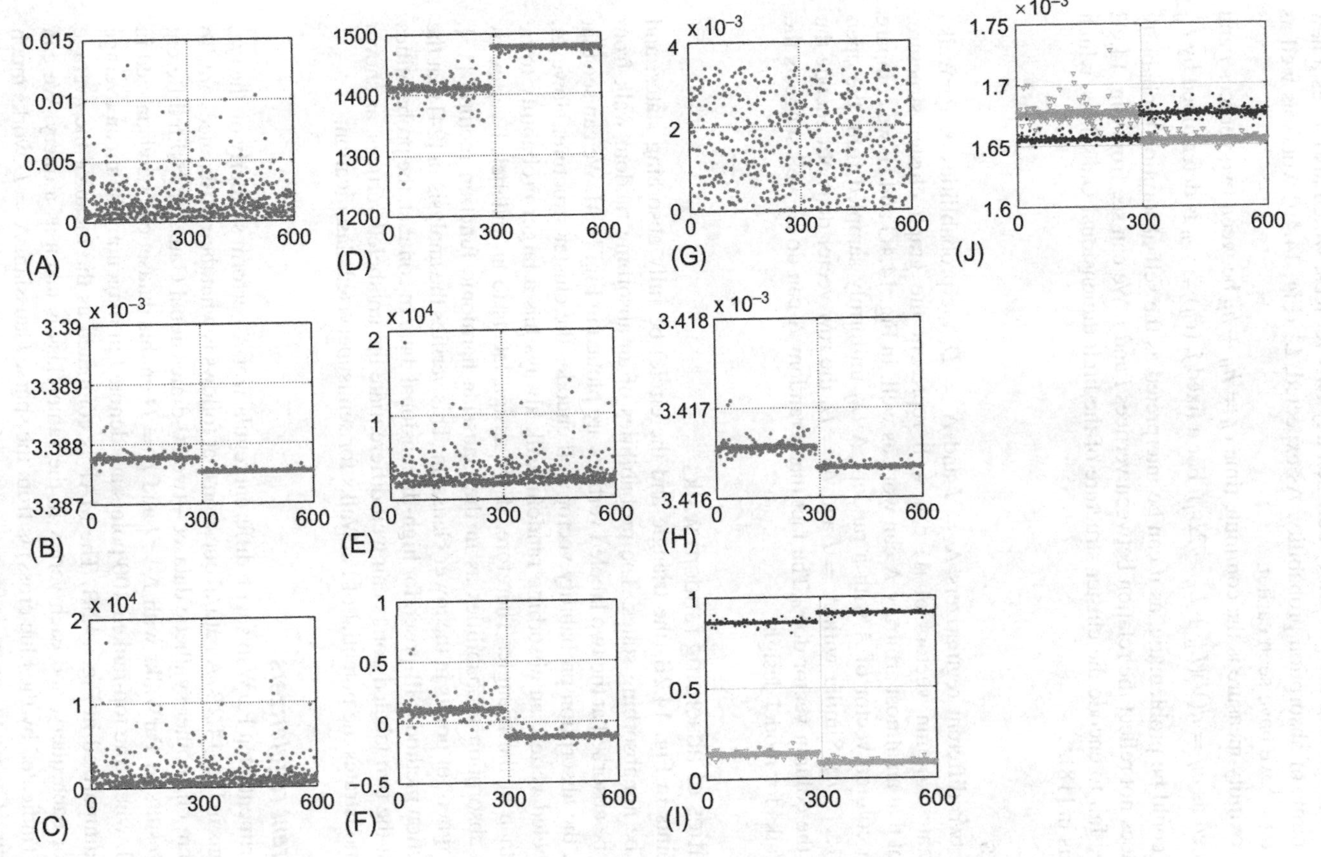

FIG. 14.4

Proximities between other points and the query point i in the two Gaussians in Fig. 14.2B. The query point i is denoted as a blue cross in Fig. 14.2. (A) The personalized PageRank vector of i. (B) The normalized personalized PageRank vector of i. (C) Hitting times of a random walk from i to hit other points. (D) Hitting times of random walks from other points to hit i. (E) Commute times between i and other points. (F) The ith row of $L^{\dagger}$. (G) The probabilities of a ParWalk with a random Λ starting from i to be absorbed at other points. (H) The probabilities of ParWalks with a random Λ starting from other points to be absorbed at i. (I) The probabilities of absorbing random walks starting from other points to be absorbed at the two labeled points in Fig. 14.2B, denoted by dots and triangles, respectively. (J) The probabilities of ParWalks starting from other points to be absorbed at the two labeled points, denoted by dots and *triangles*, respectively.

α is sufficiently small) in measuring the proximity between any two vertices, which also gives a new interpretation of $L^\dagger$ in terms of absorption probability. As expected, $L^\dagger$ (Fig. 14.4F) works as well as the kernel matrix $(L+\alpha I)^{-1}$ we have seen earlier.

A closely related proximity measure is the commute time $c_{ij} = h_{ij} + h_{ji}$ between two vertices i and j, which can be computed as $c_{ij} = d(\mathcal{V})(L^\dagger_{ii} + L^\dagger_{jj} - 2L^\dagger_{ij})$. For a fixed j, $(c_{ij})_{j=1,\dots,n}$ is determined by $L^\dagger_{ii}$ and $L^\dagger_{ij}$. The first term could be problematic, as it can be interpreted as the self-absorption probability a_{ii} of vertex i, which does not reflect the relation between vertices i and j. We can see from Fig. 14.4E that the commute times fail to encode the cluster structure of the high-dimensional Gaussians, which also verifies the analysis in [18].

Random regularizers

We have seen that for two different regularizers $\Lambda = I$ and $\Lambda = D$, the probabilities of ParWalks absorbed at a fixed state (column vectors of A) can well represent the graph cluster structure. Surprisingly, it turns out that an almost arbitrary Λ can work as well. In Fig. 14.4G and H, we compare the ith row and the ith column vector of A with a random Λ (λ_j uniformly sampled from the open interval (0,1), for $j = 1,\dots,n$). Similar with $\Lambda = I$ and $\Lambda = D$, the row vector does not encode the cluster structure while the column vector does. The fact that a random Λ can do so well suggests the great potential of ParWalks for model design.

Absorption probabilities of absorbing random walks

Let the two labeled points in Fig. 14.2B (the triangle and the circle) be fully absorbing states, and other unlabeled points be nonabsorbing states. The probabilities of an absorbing random walk from any unlabeled vertex to be absorbed at the two labeled vertices are plotted in Fig. 14.4I. We can see that for each labeled vertex, the absorption probability vector well encodes the cluster structure. However, starting from any unlabeled vertex, an absorbing random walk always has a larger probability to be absorbed at the circle than at the triangle. Therefore, if we assign labels to an unlabeled vertex by directly comparing the absorption probabilities as in the Gaussian harmonic function method [13], wrong labels will be assigned for points in the lower Gaussian. This verifies the analysis in [24] that the Gaussian harmonic function method is ill-posed for high-dimensional data. In contrast, the probabilities of ParWalks absorbed at the two labeled points can well differentiate the unlabeled vertices, as shown in Fig. 14.4I, which demonstrates the potential of ParWalks for semisupervised classification.

Comparisons of different regularizers

We have shown by experiments that ParWalks with different regularizer Λ perform similarly on the two 20-dimensional Gaussians in Fig. 14.2B. A natural question that follows is what their differences are. To see that, we compare them on another synthetic dataset—two 20-dimensional Gaussians with different variances. The ranking results by ParWalks with $\Lambda = I$ and $\Lambda = D$ on this dataset can be visualized in Fig. 14.5C, D, G, and H, where the top-ranked 400 points are denoted in magenta. The mean average precisions (MAP) are summarized in Fig. 14.5B. The MAP is computed as the mean of the average precision (AP) scores for each query, and each vertex in the same class is taken as a query. We can see that when the query (indicated by the blue cross) is from the sparse Gaussian, $\Lambda = D$ works much better than $\Lambda = I$. In contrast, when the query is from the dense Gaussian, $\Lambda = I$ is better than $\Lambda = D$.

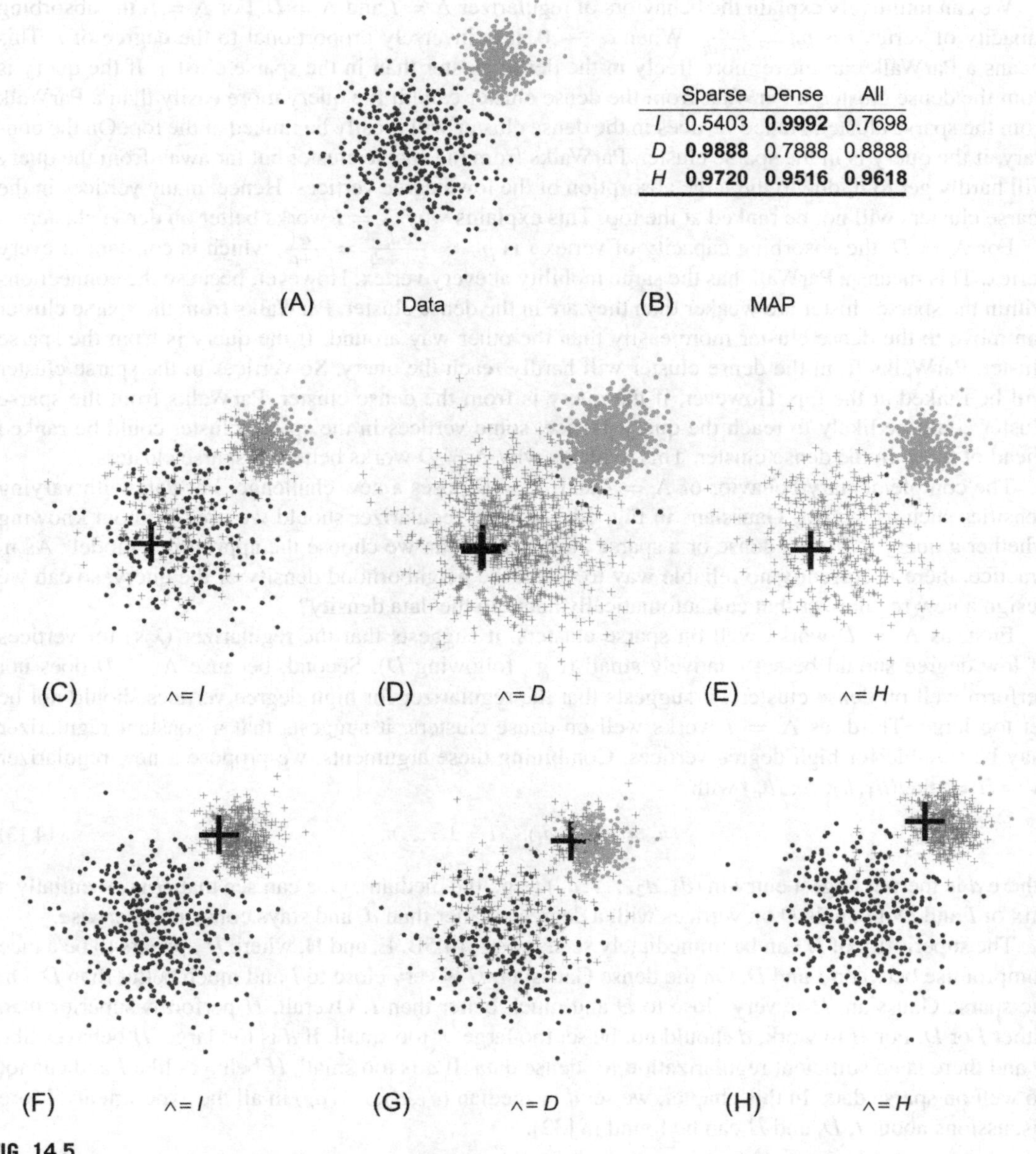

FIG. 14.5

Ranking results by ParWalks with different regularizers. (A) Two 20-dimensional Gaussians with variances 1 and 0.16, respectively, and 400 points in each Gaussian. (B) Mean average precision (MAP). (C)–(E) Ranking w.r.t. the query (denoted by the big *cross*) in the sparse Gaussian. The top 400 ranked points are denoted by small crosses. (F)–(H) Ranking w.r.t. the query (denoted by the big *cross*) in the dense Gaussian. The top 400 ranked points are denoted by small crosses. (A) Data; (B) MAP; (C) $\Lambda = I$; (D) $\Lambda = D$; (E) $\Lambda = H$; (F) $\Lambda = I$; (G) $\Lambda = D$; (H) $\Lambda = H$.

We can intuitively explain the behaviors of regularizer $\Lambda = I$ and $\Lambda = D$. For $\Lambda = I$, the absorbing capacity of vertex i is $p_{ii} = \frac{\alpha}{\alpha+d_i}$. When $\alpha \to 0$, it is inversely proportional to the degree of i. This means a ParWalk can move more freely in the dense cluster than in the sparse cluster. If the query is from the dense cluster, a ParWalk from the dense cluster can hit the query more easily than a ParWalk from the sparse cluster. Hence vertices in the dense cluster will mostly be ranked at the top. On the contrary, if the query is in the sparse cluster, ParWalks from the sparse cluster but far away from the query will hardly get to it, due to the large absorption of the low degree vertices. Hence, many vertices in the sparse clusters will not be ranked at the top. This explains why $\Lambda = I$ works better on dense clusters.

For $\Lambda = D$, the absorbing capacity of vertex i is $p_{ii} = \frac{\alpha d_i}{\alpha d_i + d_i} = \frac{\alpha}{1+\alpha}$, which is constant at every vertex. This means a ParWalk has the same mobility at every vertex. However, because the connections within the sparse cluster are weaker than they are in the dense cluster, ParWalks from the sparse cluster can move to the dense cluster more easily than the other way around. If the query is from the sparse cluster, ParWalks from the dense cluster will hardly reach the query. So vertices in the sparse cluster will be ranked at the top. However, if the query is from the dense cluster, ParWalks from the sparse cluster could be likely to reach the query. Hence, some vertices in the sparse cluster could be ranked ahead of those in the dense cluster. This explains why $\Lambda = D$ works better on sparse clusters.

The complementary behavior of $\Lambda = I$ and $\Lambda = D$ poses a new challenge. For data with varying densities such as the two Gaussians in Fig. 14.5A, what regularizer should we use? Without knowing whether a query is from a dense or a sparse cluster, how can we choose the appropriate model? As in practice, there is currently no reliable way to detect the neighborhood density of the query, so can we design a new regularizer that can automatically adapt to the data density?

First, as $\Lambda = D$ works well on sparse clusters, it suggests that the regularizer (λ_is) for vertices of low degree should be set relatively small (e.g., following D). Second, because $\Lambda = D$ does not perform well on dense clusters, it suggests that the regularizer for high-degree vertices should not be set too large. Third, as $\Lambda = I$ works well on dense clusters, it suggests that a constant regularizer may be suitable for high degree vertices. Combining these arguments, we propose a new regularizer $\Lambda = H = \text{diag}(h_1, h_2, \ldots, h_n)$ with

$$h_i = \min(\hat{d}, d_i), \quad i = 1, \ldots, n, \tag{14.13}$$

where $\hat{d}$ is the τth largest entry in $(d_1, d_2, \ldots, d_n)$ (e.g., the median). One can see that H is essentially a mix of I and D it equals D on vertices with a degree smaller than $\hat{d}$, and stays constant otherwise.

The superiority of H can be immediately seen in Fig. 14.5B, E, and H, where H appears to be a nice compromise between I and D. On the dense Gaussian, H is very close to I and much better than D. On the sparse Gaussian, H is very close to D and much better than I. Overall, H performs superior than either I or D. For H to work, $\hat{d}$ should not be set too large or too small. If $\hat{d}$ is too large, H behaves like D and there is no sufficient regularization for dense data. If $\hat{d}$ is too small, H behaves like I and cannot do well on sparse data. In this chapter, we set $\hat{d} = \text{median}\ (d_1, d_2, \ldots, d_n)$ in all the experiments. More discussions about I, D, and H can be found in [32].

14.4 EXPERIMENTS

In this section, we apply ParWalks for image retrieval and semisupervised classification on real benchmark datasets, and compare their performance with several state-of-the-art models.

14.4.1 IMAGE RETRIEVAL

For image retrieval, we first construct a data graph with each image as a vertex. Then for each query image, we rank the images with respect to the query and compute the average precision (AP). For each dataset, we compute the mean average precision (MAP) on each class and the entire dataset (the average of the MAP on all classes). We compare the performance of models including ParWalks with regularizers $\Lambda = I$, $\Lambda = D$, and $\Lambda = H$, personalized PageRank (PR), and manifold ranking (MR) [12].

Parameter setup

We construct a weighted 20-NN graph for each dataset, including the synthetic data in Figs. 14.2 and 14.5A. The edge weight between vertices i and j is set as $w_{ij} = \exp(-d_{ij}^2/\sigma)$ if i is within j's 20 nearest neighbors or vice versa, otherwise $w_{ij} = 0$. d_{ij} is the Euclidean distance between vertices i and j. We set $\sigma = 0.2 \times s$ with s being the average square distance of each vertex to its 20th nearest neighbor. For ParWalks, we use the same $\alpha = 1e-6$ for different regularizers. For personalized PageRank, we set the teleportation parameter $\beta = 0.15$ as suggested in [1]. For manifold ranking, we set the regularization parameter to 0.99 as suggested in [12].

Datasets

We conduct image retrieval on three real datasets: USPS, MNIST, and CIFAR-10. The USPS[1] dataset contains 9298 images of handwritten digits from 0 to 9 of size 16 × 16, with 1553, 1269, 929, 824, 852, 716, 834, 792, 708, and 821 in each class. We use each instance as a query for image retrieval on all the 9298 images. Some sample images are displayed in Fig. 14.7. Another popular handwritten digit dataset is the MNIST[2] [33] dataset, which contains 70,000 images of handwritten digits from 0 to 9 of size 28 × 28, with 6903, 7877, 6990, 7141, 6824, 6313, 6876, 7293, 6825, and 6958 in each class. It consists of a training set of 60,000 examples and a test set of 10,000 examples. We use each instance in the test set as a query for image retrieval on all the 70,000 images. The CIFAR-10[3] dataset consists of 60,000 tiny color images of size 32x32 in 10 mutually exclusive classes, with 6000 in each class. There are 50,000 training images and 10,000 test images. Each image is represented by a 512-dimensional GIST feature vector [34]. We use each test image as a query for image retrieval on all the 60,000 images. Some sample images are displayed in Fig. 14.6.

Experimental results

The results are displayed in Table 14.2, where $\hat{d}$ denotes the median degree of vertices in each class and the entire graph. From the values of $\hat{d}$, we can see that the data graphs constructed are of varying densities. Some image clusters can be highly dense because images of that class have more similar features, e.g., digit "1" and "9" in USPS (Fig. 14.7A and D) and MNIST, "plane" (Fig. 14.6A) and "ship" in CIFAR. But some image clusters can be rather sparse due to fewer similar features, e.g., digit "2" and "4" in USPS (Fig. 14.7B and C) and MNIST, "dog" (Fig. 14.6B) and "cat" in CIFAR.

[1] https://web.stanford.edu/~hastie/ElemStatLearn/datasets/zip.info.txt.
[2] http://yann.lecun.com/exdb/mnist/.
[3] http://www.cs.toronto.edu/~kriz/cifar.html.

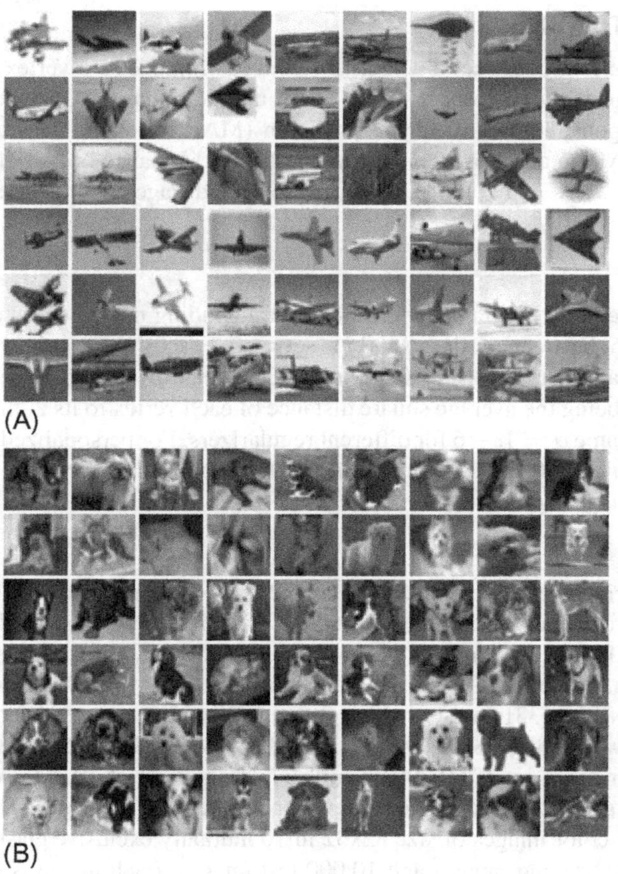

FIG. 14.6

Sample images from the CIFAR-10 dataset. (A) Plane. (B) Dog.

We can draw several observations from the results. First, we can see that ParWalks with regularizers I, D, and H all outperform personalized PageRank and manifold ranking by a large margin on each dataset. Second, regularizers I and D show distinctive yet complementary behaviors. For example, I is much better than D on dense classes, e.g., "plane" and "ship" in CIFAR, digit "1" in MNIST, and digit "9" in USPS. In contrast, D performs much better than I on sparse classes, e.g., digits "2", "4", and "5" in USPS, and "auto", "cat", "dog", "horse", and "truck" in CIFAR. Third, the proposed regularizer H adapts to the data density and combines the strengths of I and D (as highlighted), thus achieving the best overall performance (last column in Table 14.2) on all three datasets. The results demonstrate the superiority of ParWalks and verify our analysis.

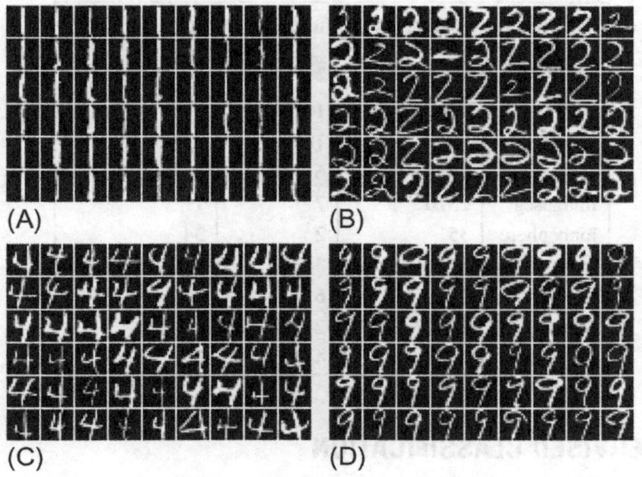

FIG. 14.7

Sample images from the USPS dataset. (A) Digit 1. (B) Digit 2. (C) Digit 4. (D) Digit 9.

Table 14.2 Mean Average Precision on the USPS, MNIST, and CIFAR Datasets

	0	1	2	3	4	5	6	7	8	9	All
USPS											
$\hat{d}$	0.76	18.66	0.04	0.13	0.27	0.05	0.83	1.68	0.17	1.70	0.47
I	0.9805	**0.9882**	0.8760	0.8926	0.6462	0.7781	**0.9401**	0.9194	0.7460	**0.7296**	0.8497
D	0.9819	0.9751	**0.9057**	0.8926	**0.6816**	**0.7972**	0.9231	0.9153	0.7450	0.6959	0.8514
H	0.9797	**0.9871**	**0.9101**	0.8961	**0.6819**	**0.7971**	**0.9408**	0.9167	0.7679	**0.7231**	**0.8601**
PR	0.8860	0.9720	0.6080	0.7639	0.4879	0.5684	0.8374	0.8253	0.6255	0.7022	0.7277
MR	0.9570	0.9871	0.8272	0.8273	0.4671	0.6303	0.9167	0.8225	0.6750	0.7191	0.7829
MNIST											
$\hat{d}$	0.30	11.18	0.07	0.15	0.36	0.15	0.49	1.06	0.11	0.79	0.32
I	0.9877	**0.9759**	0.9269	0.8867	0.7916	0.8004	0.9745	**0.8848**	0.8118	0.6602	0.8700
D	0.9881	0.9249	**0.9324**	0.8744	0.8102	0.8097	0.9706	0.8502	0.8161	0.6573	0.8634
H	0.9868	**0.9746**	**0.9397**	0.8831	0.8002	0.8070	0.9742	**0.8832**	0.8341	0.6613	**0.8744**
PR	0.8867	0.7444	0.6574	0.7006	0.5941	0.5750	0.8303	0.6916	0.5874	0.5916	0.6859
MR	0.9803	0.9436	0.8897	0.8166	0.6355	0.7152	0.9546	0.7883	0.7140	0.6463	0.8084
	plane	**auto**	**bird**	**cat**	**deer**	**dog**	**frog**	**horse**	**ship**	**truck**	**All**
CIFAR											
$\hat{d}$	0.65	0.15	0.33	0.13	0.36	0.15	0.27	0.16	0.51	0.16	0.23
I	**0.2999**	0.2760	**0.1570**	0.1320	0.1703	0.1848	0.2949	0.2243	**0.3195**	0.2493	0.2308
D	0.2387	**0.3049**	0.1454	**0.1562**	0.1581	**0.2141**	0.2901	**0.2488**	0.2835	**0.2741**	0.2314
H	**0.2917**	**0.2945**	**0.1552**	**0.1496**	0.1621	**0.2054**	0.2891	**0.2342**	**0.3128**	**0.2609**	**0.2356**
PR	0.2335	0.2050	0.1418	0.1007	0.2136	0.1403	0.2612	0.1571	0.2655	0.1701	0.1889
MR	0.2296	0.1513	0.1286	0.0821	0.1715	0.1022	0.1924	0.1201	0.2321	0.1124	0.1522

Table 14.3 Statistics of the Nine Datasets

	# Examples	# Classes	# Dimensions
USPS	9298	10	256
YaleB	5760	10	1200
satimage	6435	6	36
imageseg	2310	7	19
ionosphere	351	2	34
iris	150	3	4
protein	116	6	20
spiral	100	2	3
soybean	47	4	35

14.4.2 SEMISUPERVISED CLASSIFICATION

For semisupervised classification, we compare ParWalks with a random Λ with several popular semisupervised classification methods, including the Gaussian harmonic function (GHF) method coupled with and without class mass normalization (CMN) [13], and the local and global consistency (LGC) method [14].

Semisupervised classification with ParWalks

Denote the indices of labeled vertices as $i_1, \ldots, i_m$. To classify an unlabeled vertex k, we assign k to the class of the labeled vertex $\hat{j}$, where $\hat{j} = \arg\max_{j\in\{i_1,\ldots,i_m\}} a_{jk}$. Simply put, we find the labeled vertex from which a ParWalk has the largest probability to be absorbed at vertex k. Our method is based on the analysis in Section 14.3.2 that the column vectors of A encode the cluster structure.

Parameter setup

As in Section 14.4.1, we set $\alpha = 1e - 6$ for ParWalks. We set the regularization parameter in the LGC method as 0.9.[4] We use nine real datasets for this experiment, including USPS, YaleB,[5] and seven popular UCI datasets,[6] as summarized in Table 14.3.

We construct a weighted 20-NN graph for each dataset, except for YaleB, imageseg, and iris, where we build 50-NN, 50-NN, and 25-NN graphs, respectively, to ensure that the graphs are connected. As in Section 14.4.1, the edge weight between vertices i and j is set as $w_{ij} = \exp(-d_{ij}^2/\sigma)$ if i is within j's k nearest neighbors or vice versa, and $w_{ij} = 0$ otherwise. And $\sigma = 0.2 \times r$, where r is the average square distance between each point to its 20th nearest neighbor. For USPS and YaleB, we randomly sample 20 instances as labeled data. For other datasets, we randomly sample 10 instances as labeled data. We make sure that at least one label is sampled for each class. The classification accuracy is averaged over 100 trials.

[4] We also tested other regularization parameters, but found LGC performed best with 0.9.
[5] https://computervisiononline.com/dataset/1105138686.
[6] https://archive.ics.uci.edu/ml/datasets.html.

Table 14.4 Classification Accuracy on the Nine Datasets

	GHF	GHF+CMN	LGC	ParWalks
USPS	0.445	0.775	0.821	**0.880**
YaleB	0.733	0.847	0.884	**0.906**
satimage	0.650	0.741	0.725	**0.781**
imageseg	0.595	0.624	0.638	**0.665**
ionosphere	0.699	0.724	0.731	**0.752**
iris	0.902	0.894	0.903	**0.928**
protein	0.440	0.511	0.477	**0.572**
spiral	0.754	0.726	0.729	**0.835**
soybean	0.889	0.856	0.816	**0.905**

Experimental results

The results are summarized in Table 14.4. We can see clearly that ParWalks with a random Λ consistently outperforms other methods (as highlighted in bold), which justifies our method and demonstrates its superiority. The harmonic function method without the class mass normalization performs poorly on high-dimensional data, e.g., USPS, YaleB, satimage, and ionosphere, which confirms the analysis in [24]. However, as shown in Section 14.3.2 and Fig. 14.5I, the absorption probability vectors actually encode the cluster structure. Hence, a post processing such as the class mass normalization could help improve the performance significantly, as shown in Table 14.4 (GHF+CMN). The results of LGC are better than GHF and GHF+CMN, but not comparable to ParWalks. We refer readers to [35] for more comparisons and discussions.

14.5 CONCLUSIONS

In this chapter, we introduce a stochastic process called partially absorbing random walks (ParWalks). We provide a unified view of a variety of popular models for learning on graphs under this framework. By comparing them, we identify conditions under which ParWalks can reliably capture graph structures, which opens the door for model selection and model design. We also conduct extensive experiments to verify our arguments and proposals. We refer interested readers to [35] for more theoretical analysis and insights. It was recently shown in [36] that ParWalks can be easily scaled up on top of vertex-centric graph engines such as VENUS [37] or PowerGraph [38]. Also, our model was successfully applied to the Huawei App Store for large-scale app push recommendation, and significantly outperformed other state-of-the-art methods [36]. We expect our model and analysis to benefit many more applications in practice.

ACKNOWLEDGMENT

This research received support from the grant 1-ZVJJ funded by the Hong Kong Polytechnic University.

APPENDIX

A.1 PROOFS

Lemma A.1. *Suppose $\lambda_i > 0$ for some i, then any eigenvalue of $(\Lambda + D)^{-1}W$ is of magnitude less than 1.*

Proof. Let $C = (\Lambda + D)^{-1}W$. We can see that all eigenvalues of C are real, because C is similar to the real symmetric matrix $(\Lambda + D)^{-1/2}W(\Lambda + D)^{-1/2}$. Let v be any eigenvalue of C. We claim that $|v| < 1$. Let $\mathbf{u} = (u_1, u_2, \ldots, u_n)^\top$ be an eigenvector associated with v, where $u_i \in \mathbb{R}$ for $i = 1, 2, \ldots, n$. Observe that C is nonnegative and the sum of each row of C is less than or equal to 1. Since $\lambda_i > 0$ for some i, the sum of the ith row of C is less than 1. If $|u_i| = \max_j\{|u_j|\}$, then $|vu_i| = |C(i, :)\mathbf{u}| < |u_i|$, yielding $|v| < 1$. Otherwise, there must be a $k \neq i$, $|u_k| = \max_j\{|u_j|\}$ and $|u_k| > |u_i|$. It is easy to see that $C(k, :)\mathbf{u} \neq vu_k$ if $|v| \geq 1$, which contradicts the assumption that $C\mathbf{u} = v\mathbf{u}$. Hence, we conclude that $|v| < 1$. ■

Lemma A.2. *Suppose $\lambda_i > 0$ for some i. Then,*

$$(\Lambda + L)^{-1}\Lambda = \sum_{t \geq 0}((\Lambda + D)^{-1}W)^t(\Lambda + D)^{-1}\Lambda. \tag{A.1}$$

Proof. $(\Lambda+L)^{-1} = (\Lambda+D-W)^{-1} = (I-(\Lambda+D)^{-1}W)^{-1}(\Lambda+D)^{-1} = \sum_{t\geq 0}((\Lambda+D)^{-1}W)^t(\Lambda+D)^{-1}$, where the last equation follows from Lemma A.1. ■

Proof of Corollary 14.1. *The nonnegativity of A follows directly from Lemma A.2 because each matrix in the summation is nonnegative. Denote by $\mathbf{1}$ the all-one vector and $\mathbf{0}$ the zero vector. Because $L + \alpha\Lambda$ is nonsingular, it suffices to show that $(L + \alpha\Lambda)(A\mathbf{1} - \mathbf{1}) = \mathbf{0}$, which follows by plugging in $A = (L + \alpha\Lambda)^{-1}\alpha\Lambda$ and using the fact that $L\mathbf{1} = \mathbf{0}$.*

Proof of Theorem 14.2. *Note that because $\Lambda^{-1}L$ is similar to the symmetric and positive semidefinite matrix $\bar{L} = \Lambda^{-1/2}L\Lambda^{-1/2}$, they have same real eigenvalues. Let $\bar{L} = UEU^\top$ be the eigen-decomposition of $\bar{L}$ with eigenvalues $0 = \gamma_1 < \gamma_2 \leq \cdots \leq \gamma_n$ ($\gamma_2 > 0$ due to the connectivity of the graph). Then the eigen-decomposition of $\Lambda^{-1}L$ can be written as*

$$\Lambda^{-1}L = VEV^{-1}, \text{ with } V = \Lambda^{-1/2}U \ (V^{-1} = U^\top\Lambda^{1/2}). \tag{A.2}$$

By Eq. (A.2), we have

$$(L + \alpha\Lambda)^{-1}\alpha\Lambda = \left(\frac{1}{\alpha}\Lambda^{-1}L + I\right)^{-1} = VE_\alpha V^{-1}, \tag{A.3}$$

where

$$E_\alpha = diag\left(1, \frac{\alpha}{\gamma_2 + \alpha}, \ldots, \frac{\alpha}{\gamma_n + \alpha}\right). \tag{A.4}$$

Hence,

$$\lim_{\alpha \to 0+}(L + \alpha\Lambda)^{-1}\alpha\Lambda = \Lambda^{-1/2}U(:, 1)U(:, 1)^\top\Lambda^{1/2} = \mathbf{1}\bar{\lambda}^\top, \tag{A.5}$$

where in the last equation we have used the fact that $U(:, 1) = \left(\frac{\sqrt{\lambda_1}}{\sqrt{\sum_i \lambda_i}}, \ldots, \frac{\sqrt{\lambda_n}}{\sqrt{\sum_i \lambda_i}}\right)^\top$.

REFERENCES

[1] Page L, Brin S, Motwani R, Winograd T. The PageRank citation ranking: bringing order to the web. tech. rep.. Stanford InfoLab; 1999.

[2] Brand M. A random walks perspective on maximizing satisfaction and profit. In: Proceedings of the 2005 SIAM international conference on data mining. SIAM; 2005. p. 12–9.

[3] Fouss F, Pirotte A, Renders J, Saerens M. Random-walk computation of similarities between nodes of a graph with application to collaborative recommendation. IEEE Trans Knowl Data Eng 2007;19(3):355–69.

[4] Ferrucci D, Brown E, Chu-Carroll J, Fan J, Gondek D, Kalyanpur AA, et al. Building Watson: an overview of the DeepQA project. AI Mag 2010;31(3):59–79.

[5] Fortunato S. Community detection in graphs. Phys Rep 2010;486(3):75–174.

[6] Sen P, Namata G, Bilgic M, Getoor L, Galligher B, Eliassi-Rad T. Collective classification in network data. AI Mag 2008;29(3):93.

[7] Balmin A, Hristidis V, Papakonstantinou Y. ObjectRank: authority-based keyword search in databases. In: Proceedings of the thirtieth international conference on very large data bases, vol. 30. VLDB Endowment; 2004. p. 564–75.

[8] Gyöngyi Z, Garcia-Molina H, Pedersen J. Combating web spam with trustrank. In: Proceedings of the thirtieth international conference on Very large data bases, vol. 30. VLDB Endowment; 2004. p. 576–87.

[9] Wu B, Chellapilla K. Extracting link spam using biased random walks from spam seed sets. In: Proceedings of the 3rd international workshop on adversarial information retrieval on the web. ACM; 2007. p. 37–44.

[10] Grady L. Random walks for image segmentation. IEEE Trans Pattern Anal Mach Intell 2006;28(11):1768–83.

[11] Parthasarathy S, Tatikonda S, Ucar D. A survey of graph mining techniques for biological datasets. In: Managing and mining graph data. Springer; 2010. p. 547–80.

[12] Zhou D, Weston J, Gretton A, Bousquet O, Schölkopf B. Ranking on data manifolds. In: Advances in neural information processing systems; 2004. p. 169–76.

[13] Zhu X, Ghahramani Z, Lafferty JD. Semi-supervised learning using Gaussian fields and harmonic functions. In: Proceedings of the 20th international conference on machine learning. ACM; 2003. p. 912–9.

[14] Zhou D, Bousquet O, Lal TN, Weston J, Schölkopf B. Learning with local and global consistency. In: Advances in neural information processing systems 16. MIT Press; 2004. p. 321–8.

[15] Andersen R, Chung F, Lang K. Local graph partitioning using PageRank vectors. In: 47th Annual IEEE symposium on foundations of computer science, 2006. FOCS'06. IEEE; 2006. p. 475–86.

[16] Andersen R, Chung F. Detecting sharp drops in PageRank and a simplified local partitioning algorithm. Theory Appl Models Comput 2007:1–12.

[17] Lovász L. Random walks on graphs: a survey. In: Combinatorics, Paul Erdos is Eighty, vol. 2(1); 1993. p. 1–46.

[18] Von Luxburg U, Radl A, Hein M. Hitting and commute times in large random neighborhood graphs. J Mach Learn Res 2014;15:1751–98.

[19] Li Z, Liu J, Tang X. Pairwise constraint propagation by semidefinite programming for semi-supervised classification. In: Proceedings of the 25th international conference on machine learning; 2008. p. 576–83.

[20] Wu XM, So MC, Li Z, Li SY. Fast graph Laplacian regularized kernel learning via semidefinite-quadratic–linear programming. In: Advances in neural information processing systems; 2009. p. 1964–72.

[21] Sarkar P, Moore AW. Random walks in social networks and their applications: a survey. In: Social network data analytics. Springer; 2011. p. 43–77.

[22] Li Z, Fang Y, Liu Q, Cheng J, Cheng R, Lui J. Walking in the cloud: parallel SimRank at scale. Proc VLDB Endowment 2015;9(1):24–35.

[23] Wu XM, Li Z, So MC, Wright J, Chang SF. Learning with partially absorbing random walks. In: Advances in neural information processing systems 25. Curran Associates, Inc.; 2012. p. 3077–85.

[24] Nadler B, Srebro N, Zhou X. Statistical analysis of semi-supervised learning: the limit of infinite unlabelled data. In: Advances in neural information processing systems; 2009. p. 1330–8.
[25] Wu XM, Li Z, Chang SF. Analyzing the harmonic structure in graph-based learning. In: Advances in neural information processing systems; 2013. p. 3129–37.
[26] Berkhin P. Bookmark-coloring algorithm for personalized PageRank computing. Internet Math 2006;3(1):41–62.
[27] Kemeny J, Snell J. Finite markov chains. Springer; 1976.
[28] Chung FR. Spectral graph theory. American Mathematical Society; 1997.
[29] Chapelle O, Zien A. Semi-supervised classification by low density separation. In: Proceedings of the tenth international workshop on artificial intelligence and statistics. Max-Planck-Gesellschaft; 2005. p. 57–64.
[30] Kveton B, Valko M, Rahimi A, Huang L. Semi-supervised learning with max-margin graph cuts. In: International conference on artificial intelligence and statistics; 2010. p. 421–8.
[31] Bengio Y, Delalleau O, Le Roux N. Label propagation and quadratic criterion. In: Semi-supervised learning; 2006. p. 193–216.
[32] Wu XM, Li Z, Chang SF. New insights into Laplacian similarity search. In: Proceedings of the IEEE conference on computer vision and pattern recognition; 2015. p. 1949–57.
[33] LeCun Y, Bottou L, Bengio Y, Haffner P. Gradient-based learning applied to document recognition. Proc IEEE 1998;86(11):2278–324.
[34] Oliva A, Torralba A. Modeling the shape of the scene: a holistic representation of the spatial envelope. Int J Comput Vis 2001;42(3):145–75.
[35] Wu XM. Learning on graphs with partially absorbing random walks: theory and practice. Ph.D. thesis, Columbia University; 2016.
[36] Guo H, Tang R, Ye Y, Li Z, He X. A graph-based push service platform. In: International conference on database systems for advanced applications. Springer; 2017. p. 636–48.
[37] Cheng J, Liu Q, Li Z, Fan W, Lui JC, He C. VENUS: vertex-centric streamlined graph computation on a single PC. In: 2015 IEEE 31st international conference on data engineering (ICDE). IEEE; 2015. p. 1131–42.
[38] Gonzalez JE, Low Y, Gu H, Bickson D, Guestrin C. Powergraph: distributed graph-parallel computation on natural graphs. In: OSDI, vol. 12; 2012. p. 2.

PART 3

DISTRIBUTED COMMUNICATIONS, NETWORKING, AND SENSING

CHAPTER

METHODS FOR DECENTRALIZED SIGNAL PROCESSING WITH BIG DATA

15

Hoi-To Wai*, Anna Scaglione*, Eric Moulines†

*School of Electrical, Computer and Energy Engineering, Arizona State University, Tempe, AZ, United States**

CMAP, Ecole Polytechnique, Palaiseau, France†

15.1 INTRODUCTION

The growing demand for data analytics and the end of Moore's scaling law are fueling the trend toward virtualization of distributed resources. The advances made in decentralized signal processing algorithms are part of this trend as they enable networked machines to solve large-scale inference problems through the implementation of relatively simple local data processing and communication policies [1].

In a decentralized setting, the data are distributed over a network of N agents, modeled as an undirected graph $G = (V, E)$, where $V = \{1, \ldots, N\}$ is the set of agents and the set of edges $E \subseteq V \times V$ describes the network connectivity. In general, each node will only be able to reach directly a set of neighbors, which is a relatively small fraction of the nodes in V. It is thus important to apply decentralized optimization algorithms that rely on *near-neighbor* information exchanges (c.f. Section 15.1.1). Fortunately, a substantial class of machine learning algorithms is amenable to be solved in a decentralized fashion because their formulation can be cast as a constrained optimization problem in the following form:

$$\min_{\boldsymbol{\theta} \in \mathbb{R}^d} F(\boldsymbol{\theta}) \text{ s.t. } \boldsymbol{\theta} \in \mathcal{C}, \text{ where } F(\boldsymbol{\theta}) := \frac{1}{N} \sum_{i=1}^{N} f_i(\boldsymbol{\theta}), \tag{15.1}$$

using N connected agents. Here we consider a smooth optimization setting where $f_i(\boldsymbol{\theta})$ is a continuously differentiable (possibly nonconvex) function held by the ith agent and $\mathcal{C}$ is a closed and bounded convex set in $\mathbb{R}^d$.

A common instance of Eq. (15.1) is the empirical risk minimization (ERM) problem (see Fig. 15.1), where the private risk function $f_i(\boldsymbol{\theta})$ models the loss of $\boldsymbol{\theta}$ incurred over the private data held by agent i, e.g.,

$$f_i(\boldsymbol{\theta}) = \frac{1}{|\Omega_i|} \sum_{k \in \Omega_i} \ell_i(\boldsymbol{\theta}, y_{i,k}), \tag{15.2}$$

Cooperative and Graph Signal Processing. https://doi.org/10.1016/B978-0-12-813677-5.00015-8

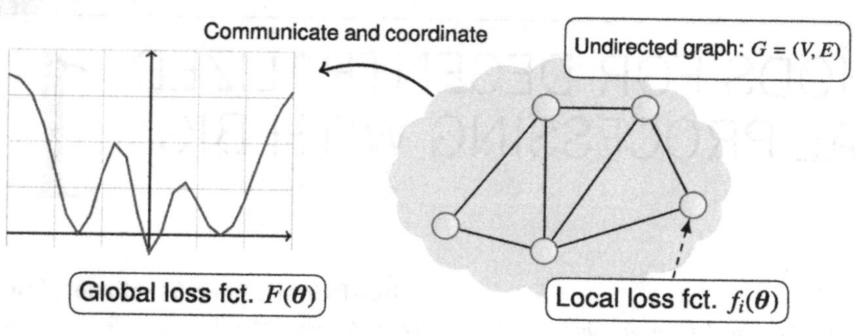

FIG. 15.1

Decentralized signal processing for big data. Illustrating the scenario considered in this chapter. Our focus is to tackle a (possibly nonconvex) optimization problem (15.1) in a decentralized setting, where the private data are held by machines/agents connected on a network G.

where $\ell_i(\boldsymbol{\theta}, \boldsymbol{y}_{i,k})$ quantifies the mismatch between a statistical model parameterized by $\boldsymbol{\theta}$ and the kth data entry, $\boldsymbol{y}_{i,k}$, held by agent i. In this instance, $\mathcal{C}$ corresponds to a regularization constraint imposed on $\boldsymbol{\theta}$ that promotes desirable properties such as sparsity or a low rank, which capture prior knowledge about the solutions that can help overcome the curse of dimensionality in searching for a solution in $\mathbb{R}^d$. Problem (15.1) also covers a number of applications in control theory and signal processing, including system identification [2], matrix completion [3], and sparse learning [4,5]. While in most instances the inclusion of the high-dimensional constraint is a fundamental ingredient to attain good estimation performance, the curse of dimensionality returns to haunt us due to the significant computational cost added when enforcing such constraints through a *projection* step, aimed at ensuring feasible iterates. We refer to this class of algorithms as decentralized projection-based algorithms and review them in Section 15.1.1.

Unlike the projection-based algorithms, *projection-free* optimization methods have been recently studied to address the curse of dimensionality caused by these high-dimensional constraints from the computational standpoint. This chapter aims at giving a friendly and up-to-date introduction to both projection-based algorithms and projection-free decentralized optimization algorithms, discussing in depth their implementation issues and convergence properties. The projection-free decentralized optimization algorithms introduced and studied in this chapter are rooted in the classical centralized Frank-Wolfe (FW) algorithm [6] and are, therefore, referred to as the decentralized FW (DeFW). From the algorithmic perspective, the FW algorithm replaces the costly projection step in projection-based algorithms with a constrained linear optimization, which often admits a computationally efficient solution.

The rest of the chapter is organized as follows. In Section 15.2, we discuss both the classical decentralized projected gradient descent algorithm and the newly proposed DeFW algorithm and study their convergence *rates* for both convex and nonconvex optimization. In Section 15.3 we showcase a concrete machine learning application of the algorithms surveyed, solving a low-rank matrix completion problem on distributed data. We conclude in Section 15.4 by presenting numerical results on both synthetic data and real data corroborating our theoretical claims.

15.1.1 BACKGROUND

There is a vast literature on decentralized optimization algorithms to tackle Eq. (15.1). The first class of them is characterized by a common basic primitive for distributed computation called *average consensus protocol*, with the earliest references dating back to the 1980s [7] (see, e.g., [8] for a review). This basic protocol is combined more or less directly with ideas that stem from the vast literature on iterative solutions of centralized optimization problems. In particular, the rich theory on first-order optimization methods, such as gradient descent, has been a prolific source of inspiration. For example, Nedich and Ozdaglar [9] and Ram et al. [10] proposed one of the first examples of a decentralized projected subgradient descent method; Chen [11] characterized the convergence rate of the method in [10] and a few other proximal gradient-like methods; and Jakovetic et al. [12] studied a decentralized version of the Nesterov-type accelerated gradient descent method. Later Shi et al. [13], Nedić et al. [14], and Qu and Li [15] proposed a decentralized gradient descent method with a linear convergence rate that is extremely efficient. Related works are also [16,17] and the references therein. Along the same line are the Newton-type methods considered in, e.g., [18,19], which take inspiration from the vast literature on second-order methods.

Another class of popular decentralized algorithms is developed from the primal-dual optimization framework. Here, the most popular example is the alternating direction of multiplier method (ADMM) [20]. Interestingly, the decentralized implementations often follow by reformulating the optimization problem into problems with a consensus constraint, after which the standard primal dual optimization procedure can be applied. Compared to the methods developed from first/second-order optimization, these algorithms offer greater flexibility in handling complex constraints. Recent advances in this direction include Duchi et al. [21] and Wei and Ozdaglar [22], which analyzed the convergence rate of the general algorithm; Chang et al. [23] and Simonetto and Jamali-Rad [24] which analyzed the decentralized algorithm for problems with a complex constraint structure; and Hong [25], which analyzed the convergence of primal-dual algorithms for nonconvex optimization. Other decentralized algorithms have also been developed, e.g., [26,27], based on fundamentally different ideas from the ones cited above. The interested readers are referred to Chapter 5 for a comprehensive treatment of the aforementioned decentralized optimization methods.

It is worth mentioning that most of the analyses of decentralized algorithms assume convexity, while nonconvex decentralized optimization has received far less attention (exceptions are, e.g., [16,25–29]). As nonconvex problems are often encountered in control theory, signal processing, and machine learning, it is important to fill this gap.

As mentioned in the introduction, big data analytics often map onto the solution of high-dimensional constrained problems. The constraints are essential to solving the problem and finding the *needle in the haystack*; unfortunately, using a *projection* operator (or more generally the proximal operator) in handling these large problems becomes computationally prohibitive. The goal of this chapter is to introduce and analyze the DeFW algorithm as an attractive alternative for handling such problem. Note that the centralized FW algorithm has already found many applications in optimal control [30], matrix completion [31], image and video colocation [32], electric vehicle charging [33], and traffic assignment [34]; see the overview article [35]. Additionally, the convergence of the centralized FW algorithm has been studied for convex problems [6,35] and a few results shed some light on its convergence properties for nonconvex problems [36–38].

15.1.2 NOTATIONS AND MATHEMATICAL PRELIMINARIES

To prepare the reader for what follows, this section introduces the basic notation as well as some assumptions and definitions used throughout this chapter.

For any $d \in \mathbb{N}$, we use the notation $[d]$ to refer to the set $\{1, \ldots, d\}$. We use boldfaced lower-case letters to denote vectors and boldfaced upper-case letters to denote matrices. For a vector $\boldsymbol{x}$ (or a matrix $\boldsymbol{X}$), the notation $[\boldsymbol{x}]_i$ (or $[\boldsymbol{X}]_{i,j}$) denotes its ith element (or (i,j)th element). The vectorization of a matrix $\boldsymbol{X} \in \mathbb{R}^{m_1 \times m_2}$ is denoted by $\text{vec}(\boldsymbol{X}) = [\boldsymbol{x}_1; \boldsymbol{x}_2; \ldots; \boldsymbol{x}_{m_2}] \in \mathbb{R}^{m_1 m_2}$ such that $\boldsymbol{x}_i$ is the ith column of $\boldsymbol{X}$. The vector $\mathbf{e}_i \in \mathbb{R}^d$ is the ith unit vector such that $[\mathbf{e}_i]_j = 0$ for all $j \neq i$ and $[\mathbf{e}_i]_i = 1$. For some positive finite constants C_1 and nonnegative functions $f(t), g(t)$, the notations $f(t) = \mathcal{O}(g(t))$ indicate $f(t) \leq C_1 g(t)$.

We denote by $\mathbf{E}$ the Euclidean space embedded in $\mathbb{R}^d$ and the Euclidean norm is denoted by $\|\cdot\|_2$. The binary operator $\langle \cdot, \cdot \rangle$ denotes the inner product on $\mathbf{E}$. In addition, $\mathbf{E}$ is equipped with a norm $\|\cdot\|$. Let G, L, μ be nonnegative constants. Consider a function $f: \mathbb{R}^d \to \mathbb{R}$, the function f is G-Lipschitz if for all $\boldsymbol{\theta}, \boldsymbol{\theta}' \in \mathbf{E}$,

$$|f(\boldsymbol{\theta}) - f(\boldsymbol{\theta}')| \leq G \|\boldsymbol{\theta} - \boldsymbol{\theta}'\|. \tag{15.3}$$

Notice that the norm on the right side does not need to be the standard Euclidean norm, e.g., it can be the trace-norm of matrices. The function f is L-smooth if for all $\boldsymbol{\theta}, \boldsymbol{\theta}' \in \mathbf{E}$, $\|\nabla f(\boldsymbol{\theta}') - \nabla f(\boldsymbol{\theta})\|_2 \leq L \|\boldsymbol{\theta}' - \boldsymbol{\theta}\|_2$. The equation above implies that for all $\boldsymbol{\theta}, \boldsymbol{\theta}' \in \mathbf{E}$,

$$f(\boldsymbol{\theta}) - f(\boldsymbol{\theta}') \leq \langle \nabla f(\boldsymbol{\theta}'), \boldsymbol{\theta} - \boldsymbol{\theta}' \rangle + \frac{L}{2} \|\boldsymbol{\theta} - \boldsymbol{\theta}'\|_2^2. \tag{15.4}$$

The function f is μ-strongly convex if for all $\boldsymbol{\theta}, \boldsymbol{\theta}' \in \mathbf{E}$,

$$f(\boldsymbol{\theta}) - f(\boldsymbol{\theta}') \leq \langle \nabla f(\boldsymbol{\theta}), \boldsymbol{\theta} - \boldsymbol{\theta}' \rangle - \frac{\mu}{2} \|\boldsymbol{\theta} - \boldsymbol{\theta}'\|_2^2, \tag{15.5}$$

moreover, f is convex if the equation above is satisfied with $\mu = 0$. Consider Problem (15.1), its constraint set $\mathcal{C} \subseteq \mathbf{E}$ is convex and bounded with the diameter defined as:

$$\rho := \max_{\boldsymbol{\theta}, \boldsymbol{\theta}' \in \mathcal{C}} \|\boldsymbol{\theta} - \boldsymbol{\theta}'\|, \quad \bar{\rho} := \max_{\boldsymbol{\theta}, \boldsymbol{\theta}' \in \mathcal{C}} \|\boldsymbol{\theta} - \boldsymbol{\theta}'\|_2, \tag{15.6}$$

note that ρ is defined with respect to (w.r.t.) the norm $\|\cdot\|$ while $\bar{\rho}$ is defined w.r.t. the Euclidean norm. When the objective function F is μ-strongly convex with $\mu > 0$, the optimal solution to Eq. (15.1) is unique and denoted by $\boldsymbol{\theta}^\star$, we also define

$$\delta := \min_{\boldsymbol{s} \in \partial \mathcal{C}} \|\boldsymbol{s} - \boldsymbol{\theta}^\star\|_2, \tag{15.7}$$

where $\partial \mathcal{C}$ is the boundary set of $\mathcal{C}$. If $\delta > 0$, the solution $\boldsymbol{\theta}^\star$ is in the interior of $\mathcal{C}$.

In the following developments, the effects of the network interactions are captured mathematically by a doubly stochastic matrix, $\boldsymbol{A} \in \mathbb{R}_+^{N \times N}$, that can be interpreted as a weighted adjacency matrix associated with the network graph $G = (V, E)$; that is, $A_{ij} \geq \eta$ for some $\eta > 0$ if $(i,j) \in E$. Because the matrix is doubly stochastic $\boldsymbol{A}^\top \mathbf{1} = \boldsymbol{A}\mathbf{1} = \mathbf{1}$ and it is assumed that the second largest singular value, $\sigma_2(\boldsymbol{A})$, is strictly less than one, which implies that the graph is a single component, this matrix is the *mixing matrix* required by the average consensus protocol [7]. A *communication round* corresponds to the network nodes sharing a message through the network edges once. To keep our discussion simple, we shall focus on the *static* network setting where $\boldsymbol{A}$ is time invariant.

15.2 DECENTRALIZED OPTIMIZATION ALGORITHMS

Next, two possible decentralized methods are described that are appropriate to solve problem instances in the same form as Eq. (15.1). The first algorithm is the classical decentralized projected gradient (DPG) algorithm in [10] and its variants. The second is the decentralized Frank-Wolfe (DeFW) algorithm introduced and studied in [39]. For both of them, the convergence rates are derived as well.

15.2.1 DECENTRALIZED PROJECTED GRADIENT (DPG)

The DPG algorithm emulates the centralized projected gradient descent (PG) [40]. In particular, let $t \in \mathbb{N}$ be the iteration number, the projection can be described as:

$$\boldsymbol{\theta}_{t+1} = \mathcal{P}_{\mathcal{C}}(\boldsymbol{\theta}_t - \gamma_t \nabla F(\boldsymbol{\theta}_t)), \tag{15.8}$$

where $\gamma_t \in (0, 1]$ is a step size and $\mathcal{P}_{\mathcal{C}}(\cdot)$ is the projection operator onto $\mathcal{C}$:

$$\mathcal{P}_{\mathcal{C}}(\boldsymbol{x}) := \arg\min_{\boldsymbol{\theta} \in \mathcal{C}} \|\boldsymbol{\theta} - \boldsymbol{x}\|_2^2. \tag{15.9}$$

To mimic the centralized PG algorithm in the decentralized setting, the agents need to retrieve information about the global gradient $\nabla F(\boldsymbol{\theta}_t)$. The DPG algorithm achieves this using the following recursions—let $\boldsymbol{\theta}_t^i$ be the local iterate held by agent i at iteration t,

$$\text{(Consensus step)} \quad \bar{\boldsymbol{\theta}}_t^i = \sum_{j=1}^{N} A_{ij} \boldsymbol{\theta}_t^j, \tag{15.10a}$$

$$\text{(PG step)} \quad \boldsymbol{\theta}_{t+1}^i = \mathcal{P}_{\mathcal{C}}\left(\bar{\boldsymbol{\theta}}_t^i - \gamma_t \nabla f_i(\bar{\boldsymbol{\theta}}_t^i)\right), \tag{15.10b}$$

where $\bar{\boldsymbol{\theta}}_t^i$ is an auxiliary variable that holds a local approximate of the global average parameter $(1/N)\sum_{j=1}^{N} \boldsymbol{\theta}_t^j$. The consensus step (15.10a) is similar to the average consensus protocol in [7] while the PG step (15.10b) is analogous to the centralized PG algorithm (15.8), with the exception that the global gradient $\nabla F(\boldsymbol{\theta}_t)$ is replaced by the local gradient function $\nabla f_i(\bar{\boldsymbol{\theta}}_t^i)$, evaluated at the approximate global iterate. Despite using the local gradient vector in lieu of the global one, the DPG algorithm achieves convergence because the evaluation of $\nabla f_i(\bar{\boldsymbol{\theta}}_t^i)$ incorporates information about the local functions held by the other agents that propagates through the mixing step. More specifically, the algorithm exhibits sublinear convergence for convex problems with a diminishing step size γ_t.

As seen, the iteration steps of the DPG algorithm are conceptually simple to implement. However, for high-dimensional problems, the projection operation (15.9) can be computationally prohibitive, even when a closed form solution is available for its update. For example, when $\mathcal{C}$ is a trace norm ball for matrices of dimension $m_1 \times m_2$, i.e.,

$$\mathcal{C} = \{\boldsymbol{\theta} \in \mathbb{R}^{m_1 \times m_2} \;:\; \|\boldsymbol{\theta}\|_{\sigma,1} \leq R\}, \tag{15.11}$$

the projection operator admits a closed form solution as:

$$\mathcal{P}_{\mathcal{C}}(\boldsymbol{X}) = \boldsymbol{U} \max\{\boldsymbol{0}, \boldsymbol{\Sigma} - \boldsymbol{\Lambda}^\star\} \boldsymbol{V}^\top, \text{ where } \boldsymbol{X} = \boldsymbol{U}\boldsymbol{\Sigma}\boldsymbol{V}^\top, \tag{15.12}$$

for some diagonal $\boldsymbol{\Lambda}^\star$ such that $\|\mathcal{P}_{\mathcal{C}}(\boldsymbol{X})\|_{\sigma,1} \leq R$. Clearly, the projection step amounts to computing a full singular value decomposition (SVD) of the operand. The associated complexity of such a step grows as $\mathcal{O}(\max\{m_1 m_2^2, m_2 m_1^2\})$ is a cost endured by all the N agents in all iterations. This is highly undesirable for big data applications where $m_1, m_2 \gg 0$. The DPG has served as a prototype algorithm for a number of sophisticated decentralized optimization algorithms, e.g., [12–15]. These algorithms require fewer iterations to convergence, but are equally burdened by the high complexity of the projection step.

Convergence analysis. The convergence properties of the DPG algorithm are known for a general setting with a time-varying mixing matrix (i.e., the matrix $\boldsymbol{A}[t]$ may change at every iteration) [10]. Specifically, it has been established in [10] that the algorithm converges almost surely when the step size is chosen such that $\sum_{t=1}^{\infty} \gamma_t = \infty$ and $\sum_{t=1}^{\infty} \gamma_t^2 < \infty$ (and the time-varying network is connected in an ergodic sense). However, the convergence rate of the DPG algorithm has not been studied in [10]. Here, we describe the convergence rate analysis conducted in [11] for convex problems, whose result can be summarized as follows:

CONVERGENCE OF DPG (CONVEX)

Theorem 15.1 ([11]). *Consider Problem (15.1) and suppose that each of f_i is convex and L-smooth. If we apply the DPG algorithm to solve Eq. (15.10) and choose the step size as $\gamma_t \leq 1/L$, then it holds for all $T \geq 1$ that:*

$$\min_{1 \leq t \leq T} F(\boldsymbol{\theta}_t^i) - \big(\min_{\boldsymbol{\theta} \in \mathcal{C}} F(\boldsymbol{\theta})\big) \leq \frac{D_1 + D_2 \sum_{t=1}^{T} \gamma_t^2}{\sum_{t=1}^{T} \gamma_t}, \tag{15.13}$$

where D_1, D_2 are some finite constants that depend on $\sigma_2(\boldsymbol{A})$. If we set $\gamma_t = C/\sqrt{t}$ for some $C < \infty$, then $\min_{1 \leq t \leq T} F(\boldsymbol{\theta}_t^i) - (\min_{\boldsymbol{\theta} \in \mathcal{C}} F(\boldsymbol{\theta})) = \mathcal{O}(\log T/\sqrt{T})$. Moreover, the algorithm attains consensus, that is $\lim_{t \to \infty} \|\boldsymbol{\theta}_t^i - (1/N) \sum_{j=1}^{N} \boldsymbol{\theta}_t^j\| = 0 \ \forall \ i$.

Theorem 15.1 proves that, in terms of the difference between objective values at iteration T and at an optimal solution (a.k.a. the primal optimality gap), the DPG algorithm converges sublinearly at a rate of $\mathcal{O}(\log T/\sqrt{T})$. The proof of Theorem 15.1 proceeds first in showing that the algorithm attains consensus asymptotically, and then in bounding the optimality gap by the descent lemma in [40] (as f_i is assumed to be a smooth function).

If we relax the assumption that the objective function of Problem (15.1) is convex, little is known about the convergence (rate) of the DPG algorithm. Recent work [41] has shown that a decentralized gradient descent method applied to the unconstrained version of Eq. (15.1) converges at a sublinear rate for nonconvex problems, i.e., $\|\boldsymbol{\theta}_t^i - \bar{\boldsymbol{\theta}}\| = \mathcal{O}(1/\sqrt{t})$, where $\bar{\boldsymbol{\theta}}$ is a stationary point to Eq. (15.1), yet the algorithm considered therein is different from the DPG algorithm in Eq. (15.10). The state of the art in understanding the convergence of the DPG applied to a nonconvex problem can be found in [16] and is quoted below:

CONVERGENCE OF DPG (NONCONVEX)

Theorem 15.2 ([16]). *Consider Problem (15.1) and the DPG algorithm (15.10). Suppose that each of f_i is L-smooth. If we choose the step size such that $\sum_{t=1}^{\infty} \gamma_t = \infty$ and $\sum_{t=1}^{\infty} \gamma_t^2 < \infty$, then the sequence $\{\theta_t^i\}_{t\geq 1}$ satisfies:*

1. *(Consensus)* $\lim_{t\to\infty} \|\theta_t^i - (1/N)\sum_{j=1}^{N} \theta_t^j\| = 0$ *for all* $i \in [N]$.
2. *(Stationary point)* $\lim_{t\to\infty} \|\theta_t^i - \bar{\theta}\| = 0$*, where* $\bar{\theta}$ *is a stationary point of Eq. (15.1).*

Finally, we remark that in a centralized setting, the PG algorithm is known to converge at a linear rate for strongly convex objective functions. Such convergence rate is not observed for the DPG algorithm because the latter requires a diminishing step size to guarantee convergence. An active research area is to develop DPG-like algorithms that achieve linear convergence using a constant step size, e.g., [13–15].

15.2.2 DECENTRALIZED FRANK-WOLFE (DEFW)

The decentralized Frank-Wolfe (DeFW) algorithm [39] is born out of the classical centralized FW algorithm [6], which is briefly reviewed next. Denote by $t \in \mathbb{N}$ the iteration number. Assume that the initial point $\theta_1 \in \mathcal{C}$ is feasible, which may be found using prior knowledge on the constraint set, e.g., for the trace-norm ball constraint in Eq. (15.11), an obvious solution is to pick the zero matrix $\mathbf{0}$ as θ^1. To solve problem (15.1) the *centralized* FW algorithm performs the following iteration:

$$a_t \in \arg\min_{a\in\mathcal{C}} \langle \nabla F(\theta_t), a\rangle, \quad (15.14a)$$

$$\theta_{t+1} = \theta_t + \gamma_t(a_t - \theta_t), \quad (15.14b)$$

where $\gamma_t \in (0, 1]$ is a step size to be determined. Observe that θ_{t+1} is a convex combination of θ_t and a_t, which are both feasible, therefore $\theta_t \in \mathcal{C}$ as $\mathcal{C}$ is convex.

Algorithm 15.1 DECENTRALIZED FRANK-WOLFE (DEFW)

1: **Input**: Initial point $\theta_1^i \in \mathcal{C}$ for $i = 1, \ldots, N$, doubly stochastic weighted adjacency matrix of $G, A \in \mathbb{R}^{n\times n}$.
2: Initialize the slack variables with $\overline{\nabla_0^i F} \leftarrow \mathbf{0}$ and $\nabla f_i(\bar{\theta}_0^i) \leftarrow \mathbf{0}$ for all $i \in [N]$.
3: **for** $t = 1, 2, \ldots$ **do**
4: *Consensus*: approximate the average iterate:

$$\bar{\theta}_t^i \leftarrow \sum_{j=1}^{N} A_{ij}\theta_t^j, \quad \forall\, i \in [N]. \quad (15.15)$$

5: *Aggregation*: approximate the average gradient:

$$\overline{\nabla_t^i F} \leftarrow \sum_{j=1}^{N} A_{ij}(\overline{\nabla_{t-1}^j F} + \nabla f_j(\bar{\theta}_t^j) - \nabla f_j(\bar{\theta}_{t-1}^j)), \ \forall\, i \in [N]. \quad (15.16)$$

6: *Frank-Wolfe Step*: update

$$\boldsymbol{\theta}_{t+1}^{i} \leftarrow (1-\gamma_t)\bar{\boldsymbol{\theta}}_t^{i} + \gamma_t \boldsymbol{a}_t^{i} \text{ where } \boldsymbol{a}_t^{i} \in \arg\min_{\boldsymbol{a}\in\mathcal{C}} \langle \overline{\nabla_t^{i} F}, \boldsymbol{a}\rangle,\ \forall\, i \in [N], \tag{15.17}$$

where $\gamma_t \in (0,1]$ is a diminishing step size.

7: **end for**

8: **Return**: $\bar{\boldsymbol{\theta}}_{t+1}^{i}, \forall\, i \in [N]$.

To this end, one would be tempted to develop a DeFW algorithm in a similar fashion as the DPG algorithm, i.e., simply modifying the centralized FW algorithm by replacing Eq. (15.14b) with an average consensus update while using the local gradient $\nabla f_i(\cdot)$ for the update direction (15.14a). However, as we shall explain later, this procedure may not converge to a meaningful solution of the problem (15.1). As a remedy, we approximate two pieces of the global information about the problem (15.1) that appears in the FW updates (15.14)—the parameter $\boldsymbol{\theta}_t$ and the global gradient $\nabla F(\boldsymbol{\theta}_t) = (1/N)\sum_{i=1}^{N} \nabla f_i(\boldsymbol{\theta}_t)$, neither of which can be obtained locally from a single agent. In a similar spirit to the development of the DPG algorithm, the DeFW algorithm, summarized in Algorithm 15.1, combines the classical average consensus (AC) approach [7] with the FW algorithm iterations. In the pseudocode, the local parameter kept by agent i at iteration t is denoted by $\boldsymbol{\theta}_t^{i}$ and the global average parameter is denoted by

$$\bar{\boldsymbol{\theta}}_t := \frac{1}{N}\sum_{i=1}^{N} \boldsymbol{\theta}_t^{i}. \tag{15.18}$$

The DeFW algorithm produces two auxiliary sequences $\{\bar{\boldsymbol{\theta}}_t^{i}\}_{t\geq 1}$ and $\{\overline{\nabla_t^{i} F}\}_{t\geq 1}$ that correspond to the local approximates of $\bar{\boldsymbol{\theta}}_t$ and $\nabla F(\bar{\boldsymbol{\theta}}_t)$, respectively.

As seen, the DeFW algorithm uses two rounds of near-neighbor communication steps (Steps 4 and 5) to mimic the centralized FW algorithm. In particular, in the *consensus* step, each agent shares its local variable $\boldsymbol{\theta}_t^{i}$ with its neighbor to approximate $\bar{\boldsymbol{\theta}}_t$, while in the *aggregation* step, agent j computes from $\bar{\boldsymbol{\theta}}_t^{j}$ the message:

$$\overline{\nabla_{t-1}^{j} F} + \nabla f_j(\bar{\boldsymbol{\theta}}_t^{j}) - \nabla f_j(\bar{\boldsymbol{\theta}}_{t-1}^{j}) \tag{15.19}$$

and shares it with the neighbors, thereby computing an approximation of $\nabla F(\bar{\boldsymbol{\theta}}_t)$. Finally, in the *FW* step, each agent computes the FW update in a similar fashion as the centralized FW algorithm based on $\bar{\boldsymbol{\theta}}_t^{i}$ and $\overline{\nabla_t^{i} F}$.

Note that during the *aggregation* step, the agents do not transmit plainly the current local gradient $\nabla f_j(\bar{\boldsymbol{\theta}}_t^{j})$. Instead, they exchange a carefully crafted message that is a linear combination of the previous estimate $\overline{\nabla_{t-1}^{j} F}$ and the difference in the successive gradients $\nabla f_j(\bar{\boldsymbol{\theta}}_t^{j}) - \nabla f_j(\bar{\boldsymbol{\theta}}_{t-1}^{j})$. This step is the key ingredient that allows the DeFW algorithm to achieve convergence while requiring a constant number of communication rounds per iteration. In fact, this step is partly inspired by the SAGA method [42] and has also been employed in a few recent works on decentralized optimization [14,15,26].

Compared to the DPG method (15.10), the DeFW algorithm requires an additional *aggregation* step to compute the approximate global gradient while the global gradient is not required in the DPG method. The primary reason for this is the fact that the FW step computation (15.14a) is not *smooth* in general with respect to the gradient $\nabla F(\bar{\boldsymbol{\theta}}_t)$. Concretely, consider $\mathcal{C} = \{\boldsymbol{\theta} \in \mathbb{R}^2 : \|\boldsymbol{\theta}\|_1 \leq 1\}$ and let $\nabla F(\boldsymbol{\theta}) = (1, 1-\epsilon)$ and $\nabla F(\boldsymbol{\theta}') = (1, 1+\epsilon)$ be two gradient vectors for any $\epsilon > 0$, we observe that

$$(-1,0) = \arg\min_{\boldsymbol{a}\in\mathcal{C}} \langle \nabla F(\boldsymbol{\theta}), \boldsymbol{a}\rangle, \quad (0,-1) = \arg\min_{\boldsymbol{a}\in\mathcal{C}} \langle \nabla F(\boldsymbol{\theta}'), \boldsymbol{a}\rangle. \tag{15.20}$$

Therefore, a small perturbation to the gradient direction may lead to a huge difference in the FW direction $\boldsymbol{a}_t$ found. On the other hand, the projection operator in the DPG method is *nonexpansive* such that it tolerates small changes in the gradient direction and retains the information in the gradient after the projection. Now, if the DeFW algorithm proceeds by taking $\boldsymbol{a}_t^i = \arg\min_{\boldsymbol{a}\in\mathcal{C}} \langle \nabla f_i(\bar{\boldsymbol{\theta}}_t^i), \boldsymbol{a}\rangle$ in a similar fashion as in the DPG method, the computed direction $\boldsymbol{a}_t^i$ can be greatly different from that of taking it with respect to the global gradient $\nabla F(\bar{\boldsymbol{\theta}}_t)$. Intuitively, this would prevent convergence to a stationary point of Eq. (15.1) because the computed directions are likely to be completely unrelated to the global gradient that the algorithm is supposed to follow. It is, therefore, necessary to adopt a two-step average consensus procedure to implement the DeFW algorithm.

A main feature of the DeFW algorithm (as well as its centralized counterpart) is that the linear optimization[1] (LO) step (15.14a) can often be solved more efficiently than computing the projection, which justifies the epithet of *projection-free* algorithm. Taking the trace-norm ball example with $\mathcal{C}$ defined in Eq. (15.11) again, the corresponding LO amounts to the following computation:

$$\boldsymbol{a}_t = \arg\min_{\boldsymbol{a}\in\mathcal{C}} \langle \nabla F(\boldsymbol{\theta}_t), \boldsymbol{a}\rangle = -R\boldsymbol{u}_1\boldsymbol{v}_1^\top, \tag{15.21}$$

where $\boldsymbol{u}_1, \boldsymbol{v}_1$ are the top left/right singular vector of the matrix $\nabla F(\boldsymbol{\theta}_t)$. The complexity of such computation is $\mathcal{O}(\max\{m_1, m_2\})$, which is considerably less than the projection step discussed above. As reviewed by Jaggi [35], a number of other constraint types also admit an efficient LO computation when compared to their projection counterpart.

Convergence analysis. The DeFW algorithm can be seen as an inexact/perturbed version of the FW algorithm operating on the global average parameter $\bar{\boldsymbol{\theta}}_t$. The following lemmas, proven in [39], establish this. For some $\alpha \in (0, 1]$, define $t_0(\alpha)$ as the smallest integer such that

$$\sigma_2(\boldsymbol{A}) \leq \left(\frac{t_0(\alpha)}{t_0(\alpha)+1}\right)^{\alpha} \cdot \frac{1}{1+(t_0(\alpha))^{-\alpha}}. \tag{15.22}$$

Notice that $t_0(\alpha)$ is upper bounded by:

$$t_0(\alpha) \leq \left\lceil \frac{1}{\sigma_2(\boldsymbol{A})^{-1/(1+\alpha)} - 1} \right\rceil, \tag{15.23}$$

which is finite as long as $\sigma_2(\boldsymbol{A}) < 1$. Now, the lemmas follows:

Lemma 15.1. *Set the step size $\gamma_t = 1/t^{\alpha}$ in the DeFW algorithm for some $\alpha \in (0, 1]$, then $\bar{\boldsymbol{\theta}}_t^i$ in Eq. (15.15) satisfies:*

$$\max_{i\in[N]} \|\bar{\boldsymbol{\theta}}_t^i - \bar{\boldsymbol{\theta}}_t\|_2 \leq C_p/t^{\alpha}, \ \forall t \geq 1, \quad C_p := (t_0(\alpha))^{\alpha}\sqrt{N}\bar{\rho}. \tag{15.24}$$

■

[1] Notice that Eq. (15.14a) is a convex optimization problem with a linear objective.

Lemma 15.2. *Set the step size $\gamma_t = 1/t^\alpha$ in the DeFW algorithm for some $\alpha \in (0,1]$. Suppose that each of f_i is L-smooth, $\bar{\theta}_t^i$ is updated according to Eq. (15.15), then $\overline{\nabla_t^i F}$ in Eq. (15.16) satisfies*

$$N^{-1}\sum_{i=1}^N \overline{\nabla_t^i F} = N^{-1}\sum_{i=1}^N \nabla f_i(\bar{\theta}_t^i),\ \forall t \geq 1, \tag{15.25}$$

and:

$$\max_{i\in[N]} \|\overline{\nabla_t^i F} - \overline{\nabla_t F}\|_2 \leq C_g/t^\alpha,\ \forall t \geq 1, \tag{15.26}$$

$$C_g := (t_0(\alpha))^\alpha \cdot 2\sqrt{N}(2C_p + \bar{\rho})L. \tag{15.27}$$

■

Notice that the conditions (15.24) and (15.26) show that both the local approximations $\bar{\theta}_t^i$ and $\overline{\nabla_t^i F}$ approach the global averages asymptotically at a rate $\mathcal{O}(1/t^\alpha)$, as controlled by a step-size parameter. Moreover, Eq. (15.24) implies that the iterates produced by DeFW attain consensus asymptotically. Therefore, for each agent, each iteration of the DeFW algorithm updating $\bar{\theta}_t$ is equivalent to running an inexact version of the corresponding update in the FW algorithm. Using the observations above, the following result holds [39]:

CONVERGENCE OF DEFW (CONVEX)

Theorem 15.3 ([39]). *Consider Problem (15.1) and the DeFW algorithm. Set the step size as $\gamma_t = 2/(t+1)$. Suppose that each of f_i is convex and L-smooth, then:*

$$F(\bar{\theta}_t) - F(\theta^\star) \leq \frac{8\bar{\rho}(C_g + LC_p) + 2L\bar{\rho}^2}{t+1}, \tag{15.28}$$

for all $t \geq 1$, where $\theta^\star$ is an optimal solution to Eq. (15.1). Furthermore, if F is μ-strongly convex and the optimal solution $\theta^\star$ lies in the interior of C, i.e., $\delta > 0$ (cf. Eq. 15.7), we have

$$F(\bar{\theta}_t) - F(\theta^\star) \leq \frac{(4\bar{\rho}(C_g + LC_p) + L\bar{\rho}^2)^2}{2\delta^2\mu} \cdot \frac{9}{(t+1)^2}, \tag{15.29}$$

for all $t \geq 1$. In the above, the two constants C_p and C_g are defined in Eqs. (15.24) and (15.27), respectively.

As the consensus error, $\max_{i\in[N]} \|\bar{\theta}_t^i - \bar{\theta}_t\|_2$, decays to zero (cf. Eq. 15.24), the local iterates $\{\bar{\theta}_t^i\}_{t\geq 1}$ share similar convergence guarantees as their centralized counterpart $\{\bar{\theta}_t\}_{t\geq 1}$. We note that the first $\mathcal{O}(1/t)$ convergence rate is analogous to the well-known result for the centralized FW algorithm (see e.g., [35]). The second $\mathcal{O}(1/t^2)$ convergence rate, achievable under the additional assumption of strong convexity, can be proven thanks to a recent observation made in [38]. It is important to note that the upper bound in Eq. (15.28) is smaller than the upper bound in Eq. (15.29) when t is small.

Now, suppose one can relax the convexity assumption on f_i. It is of interest to study the convergence of the FW gap:

$$g_t := \max_{\theta \in C} \langle \nabla F(\bar{\theta}_t), \bar{\theta}_t - \theta \rangle, \tag{15.30}$$

as a measure of the stationarity of the iterate $\bar{\theta}_t$. This is motivated by the observation that when $g_t = 0$, then the iterate $\bar{\theta}_t$ is a stationary point to Eq. (15.1), because $\langle \nabla F(\bar{\theta}_t), \bar{\theta}_t - \theta \rangle \leq 0$ for all $\theta \in C$. Also, define the set of stationary point to Eq. (15.1) as:

$$C^\star = \left\{ \underline{\theta} \in C : \max_{\theta \in C} \langle \nabla F(\underline{\theta}), \underline{\theta} - \theta \rangle = 0 \right\}. \tag{15.31}$$

To proceed, we use the following technical assumption

Assumption 15.1. *The set $C^\star$ is nonempty. Moreover, the function $F(\theta)$ takes a finite number of values over $C^\star$, i.e., the set $F(C^\star) = \{F(\theta) : \theta \in C^\star\}$ is finite.* ■

While verifying that Assumption 15.1 is valid for a given problem is not straightforward, a number of studies (e.g., [36,41]) rely on this technical assumption to shed light on the performance of iterative algorithms solving nonconvex problems. Moreover, it is not unreasonable to assume that Eq. (15.1) has a finite number of stationary points because the set C is bounded; note that Assumption 15.1 is satisfied for this case. Under these assumptions the following theorem holds:

CONVERGENCE OF DEFW (NONCONVEX)

Theorem 15.4 ([39]). *Consider Problem (15.1) and the DeFW algorithm. Set the step size as $\gamma_t = 1/t^\alpha$ for some $\alpha \in (0, 1]$. Suppose each of f_i is L-smooth and G-Lipschitz (possibly nonconvex). It holds that:*

1. *for all $T \geq 6$ that are even, if $\alpha \in [0.5, 1)$,*

$$\min_{t \in [T/2+1, T]} g_t \leq \frac{1}{T^{1-\alpha}} \cdot \frac{1-\alpha}{(1-(2/3)^{1-\alpha})} \cdot \left(G\rho + (L\bar{\rho}^2/2 + 2\bar{\rho}(C_g + LC_p)) \log 2 \right); \tag{15.32}$$

if $\alpha \in (0, 0.5)$,

$$\min_{t \in [T/2+1, T]} g_t \leq \frac{1}{T^\alpha} \cdot \frac{1-\alpha}{(1-(2/3)^{1-\alpha})} \cdot \left(G\rho + \frac{(L\bar{\rho}^2/2 + 2\bar{\rho}(C_g + LC_p))(1-(1/2)^{1-2\alpha})}{1-2\alpha} \right). \tag{15.33}$$

2. *under Assumption 15.1 and $\alpha \in (0.5, 1]$, the sequence of objective values $\{F(\bar{\theta}_t)\}_{t \geq 1}$ converges, $\{\bar{\theta}_t\}_{t \geq 1}$ has limit points, and each limit point is in $C^\star$.*

In the above, the two constants C_p and C_g are defined in Eqs. (15.24) and (15.27), respectively.

Note that setting $\alpha = 0.5$ gives the quickest convergence rate of $\mathcal{O}(1/\sqrt{T})$. This rate is comparable to the best-known rate for centralized projected gradient descent [43]. It is worth mentioning that Theorem 15.4 is even stronger than previously established theorems regarding the convergence of the centralized FW applied to nonconvex problems. For instance, Ref. [36] requires that the local minimizer is unique; Ref. [37] gives the same convergence rate, but uses an adaptive step size.

Lastly, let us comment on the choice of step size for practical convergence. We remark that the results in Theorem 15.4 are given in terms of the *average parameter* $\bar{\boldsymbol{\theta}}_t$, which is the quickest when $\alpha = 0.5$. However, from Lemma 15.1, we notice that the local approximation error for consensus has its slowest decay when $\alpha = 0.5$. Because achieving consensus is also one of our main objectives, this presents a potential tradeoff in the choice of α. Extensive numerical experiments support the rule of thumb of setting $\alpha = 0.75$, which generally yields a good compromise in the performance.

15.2.3 COMMENTS ON DPG AND DEFW ALGORITHMS

From the discussion above it is readily apparent that the DeFW algorithm brings about two significant benefits. It enjoys a state of the art convergence rate for convex problems ($\mathcal{O}(\log(t)/\sqrt{t})$ for DPG versus $\mathcal{O}(1/t)$ for DeFW) as well as a significant computational cost reduction in its implementation when dealing with high-dimensional constrained problems, compared to the DPG. Furthermore, for nonconvex problems, the DeFW algorithm convergence rate is comparable to that of a centralized projected gradient method.

However, the DeFW algorithm entails a stronger requirement on the in-network communication protocol relative to the DPG. In particular, each iteration of the DeFW algorithm requires two communication rounds between the agents while the DPG algorithm only requires one. This is not a matter only of accounting for the number of iterations fairly. When extending the DeFW algorithm to work on time-varying networks (i.e., to work asynchronously), requiring that two consecutive communication rounds are performed successfully adds some coordination complexity. By requiring only one data exchange, the DPG algorithm is naturally extended to work with time varying networks and each iteration can be performed asynchronously. In fact, the convergence results pertaining to the DPG discussed before apply directly in the case of an asynchronous implementation.

15.3 APPLICATION: MATRIX COMPLETION

The goal of this section is to showcase practical applications of the DeFW algorithm to the scalable implementation of big data analytics by applying the DeFW to the matrix completion problem.

The setting consists of a network of agents endowed with incomplete observations of a matrix $\boldsymbol{\theta}_{\text{true}}$ of dimension $m_1 \times m_2$ with $m_1, m_2 \gg 0$. The ith agent has corrupted observations from the *training* set $\Omega_i \subset [m_1] \times [m_2]$ that are expressed as:

$$Y_{k,l} = [\boldsymbol{\theta}_{\text{true}}]_{k,l} + Z_{k,l}, \ \ \forall\, (k,l) \in \Omega_i. \tag{15.34}$$

To recover a low-rank $\boldsymbol{\theta}_{\text{true}}$ (which is generally a high-dimensional search) one can cast the regression problem into the following trace-norm constrained matrix completion (MC) problem:

$$\min_{\theta \in \mathbb{R}^{m_1 \times m_2}} \sum_{i=1}^{N} \sum_{(k,l)\in\Omega_i} \tilde{f}_i([\theta]_{k,l}, Y_{k,l}) \text{ s.t. } \|\theta\|_{\sigma,1} \le R, \tag{15.35}$$

where $\tilde{f}_i : \mathbb{R}^2 \to \mathbb{R}$ is a loss function picked by agent i according to the observations he/she received. Notice that Eq. (15.35) is also related to the low-rank subspace system identification problem described in [2], where $\boldsymbol{Y}$ with $[\boldsymbol{Y}]_{k,l} = Y_{k,l}$, $\boldsymbol{\theta}_{\text{true}}$ are modeled as the measured system response and the ground-truth, low-rank response (see also [44] and references therein).

Similar MC problems were considered in [45–48], where Ling et al. [45] studied a consensus-based optimization method similar to the one described in this chapter while Mackey et al. [46], Yu et al. [47], and Recht and Ré [48] studied the parallel computation setting where the agents are working synchronously in a fully connected network. These works assume that the rank of $\boldsymbol{\theta}_{\text{true}}$ is known in advance and solve the MC problem via matrix factorization. In addition, the algorithms in [45, 46] require that each local observation set Ω_i only have entries taken from a disjoint subset of the columns/rows only. The approach discussed next does not have any of the aforementioned restrictions.

We consider two different statistical models for $Z_{k,l}$ and propose the corresponding optimization problem formulation. When $Z_{k,l}$ is the i.i.d. Gaussian noise of variance σ_i^2, we choose $\tilde{f}_i(\cdot,\cdot)$ to be the square loss function, i.e.,

$$\tilde{f}_i([\theta]_{k,l}, Y_{k,l}) := (1/\sigma_i^2) \cdot (Y_{k,l} - [\theta]_{k,l})^2. \tag{15.36}$$

This yields the classical MC problem in [3]. Even though the square loss function is strongly convex when applied element-wise, the resulting problem (15.35) is not strongly convex as we only have partial observations on the matrix. The next model considers the sparse+low rank matrix completion problem posed in [49], where the observations are contaminated with sparse noise. More specifically, the noise term $Z_{k,l}$ has only a few entries in Ω_i where $Z_{k,l}$ is nonzero. We choose $\tilde{f}_i(\cdot,\cdot)$ to be the negated Gaussian loss, i.e.,

$$\tilde{f}_i([\theta]_{k,l}, Y_{k,l}) := \left(1 - \exp\left(-\frac{([\theta]_{k,l} - Y_{k,l})^2}{\sigma_i}\right)\right), \tag{15.37}$$

where $\sigma_i > 0$ controls the robustness to outliers for the data obtained at the ith agent. Here, $\tilde{f}_i(\cdot,\cdot)$ is a *smoothed* ℓ_0 loss [50] with enhanced robustness to outliers in the data. The resulting MC problem (15.35) is nonconvex.

Note that Eq. (15.35) is a special case of problem (15.1) with $\mathcal{C}$ being the trace-norm ball. Both the DPG and DeFW algorithms introduced in the last section can be applied on Eq. (15.35) directly. In particular, thanks to its projection-free nature, the DeFW algorithm leads to a low complexity implementation of Eq. (15.35). To this end, we also observe a few properties on the communication and storage cost of the DeFW algorithm:

- The gradient surrogate $\nabla_t^i F$ (15.16) is supported only on $\cup_{i=1}^N \Omega_i$. In fact, for all $i \in [N]$, the local gradient

$$\nabla f_i(\bar{\theta}_t^i) = \sum_{(k,l)\in\Omega_i} \tilde{f}_i'([\bar{\theta}_t^i]_{k,l}, Y_{k,l}) \cdot \mathbf{e}_k(\mathbf{e}_l')^\top \tag{15.38}$$

is supported on Ω_i, where $\bar{\theta}_t^i$ is defined in Eq. (15.15). In the above, $\mathbf{e}_k$ ($\mathbf{e}'_l$) is the kth (lth) canonical basis vector for $\mathbb{R}^{m_1}$ ($\mathbb{R}^{m_2}$) and $\tilde{f}'_i(\theta, y)$ is the derivative of $\tilde{f}_i(\theta, y)$ taken with respect to θ. Consequently, the average $\overline{\nabla_t^i F}$ is supported only on $\cup_{i=1}^N \Omega_i$. As $| \cup_{i=1}^N \Omega_i| \ll m_1 m_2$, the size of the message exchanged during the *aggregating* step (Line 5 in DeFW) is relatively small.

- The update direction $\boldsymbol{a}_t^i$ is a rank-one matrix composed of the top singular vectors of $\overline{\nabla_t^i F}$ (cf. Eq. 15.21). Because every iteration in DeFW adds at most N distinct pair of singular vectors to $\bar{\theta}_t$, the rank of $\bar{\theta}_t^i$ is upper bounded by tN if we initialize by $\bar{\theta}_0^i = \mathbf{0}$. We can reduce the communication cost in Line 4 in DeFW by exchanging these singular vectors. Note that $(tN) \cdot (m_1 + m_2)$ entries are stored/exchanged instead of $m_1 \cdot m_2$.
- When the agents are *only* concerned with predicting the entries of θ_{true} in the subset $\Xi \subset [m_1] \times [m_2]$, instead of propagating the singular vectors as described above, the *consensus* step can be carried out by exchanging only the entries of θ_{t+1}^i in $\Xi \cup \left(\cup_{i=1}^N \Omega_i\right)$ without affecting the operations of the DeFW algorithm. In this case, the storage/communication cost is $|\Xi \cup \left(\cup_{i=1}^N \Omega_i\right)|$.

Faster matrix completion. In the recent work [51], the authors have proposed an extension to the DeFW algorithm, tailoring it to the MC problem. The goal is to further speed up the algorithm by relying on even simpler operations that are possible in this specific case. Recall that the DeFW algorithm for the matrix completion problem (15.35) seeks to compute the principal components of $\nabla F(\theta_t)$ at each iteration. Algebraically, such a task can be completed using the simple power method. Leveraging on this observation, the "Fast DeFW" algorithm combines the *consensus* and *aggregation* steps in the DeFW algorithm with a *decentralized power method*. In particular, the computation required at the agents is now reduced to a series of matrix-vector multiplications. Interestingly, the algorithm also hides the private information about the local gradients because what is communicated is the product of a random matrix-vector multiplication.

Choosing the parameters of the decentralized power method appropriately, the Fast DeFW achieves similar convergence rate as the DeFW algorithm while reducing significantly the size of the messages that need to be exchanged. However, a drawback of Fast DeFW is that it does entail a higher number of communication rounds per iteration, demanding tighter agent coordination compared to the DeFW.

15.4 NUMERICAL EXPERIMENTS

Having described the nuts and bolts of the algorithms and how they are expected to perform, this section is dedicated to exposing their performance merits through numerical experiments, all focused on the matrix completion application described in Section 15.3. To simulate the decentralized optimization setting, we artificially construct a network of $N = 50$ agents, where the underlying communication network G is an Erdos-Renyi graph with connectivity of 0.1. The doubly stochastic matrix $\boldsymbol{A}$ used in the average consensus protocol is calculated according to the Metropolis-Hastings rule in [52].

As stated before, the MC goal is to estimate missing entries of an unknown matrix through corrupted partial measurements. We use two datasets—the first dataset is synthetically generated and the unknown matrix θ_{true} is rank-K and has dimensions of $m_1 \times m_2$, where m_1, m_2, K are varied in different experiments; the matrix is generated as $\theta_{\text{true}} = \sum_{i=1}^K \boldsymbol{y}_i \boldsymbol{x}_i^\top / K$ where $\boldsymbol{y}_i, \boldsymbol{x}_i$ have i.i.d. $\mathcal{N}(0, 1)$. The

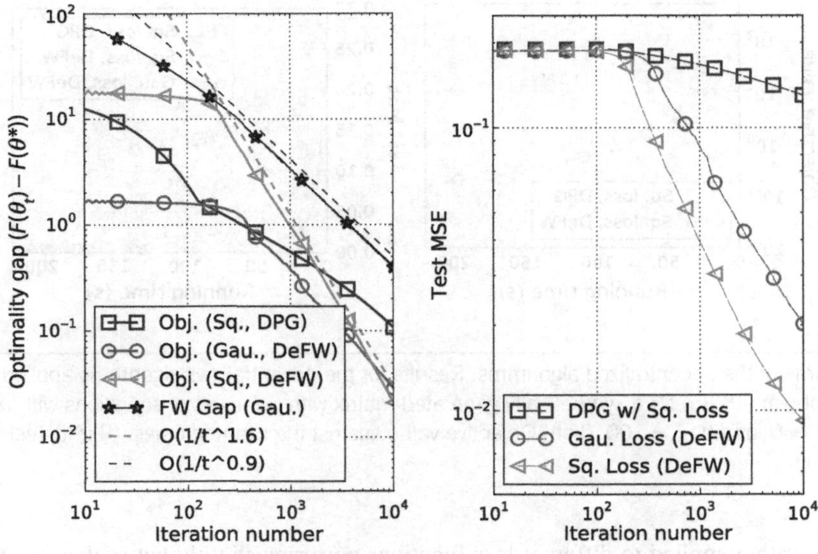

FIG. 15.2

Convergence rates of the decentralized algorithms. Applying decentralized algorithms on the matrix completion problem (15.35) for a synthetically generated matrix with noiseless observations with $m_1 = 100$, $m_2 = 250$, and $K = 5$. (Left) Optimality gap against the iteration number. (Right) Test MSE against the iteration number.

second dataset is the `movielens100k` dataset [53]. The unknown matrix $\boldsymbol{\theta}_{\text{true}}$ consists of the movie ratings of $m_1 = 943$ users on $m_2 = 1682$ movies, and a total of 10^5 entries in $\boldsymbol{\theta}_{\text{true}}$ are available as integers ranging from 1 to 5. The datasets are divided into training and testing sets and the mean square error (MSE) on the *testing set* is evaluated as:

$$\text{Test MSE} = \frac{1}{|\Omega_{\text{test}}|} \sum_{(k,l)\in\Omega_{\text{test}}} |[\boldsymbol{\theta}_{\text{true}}]_{k,l} - [\hat{\boldsymbol{\theta}}]_{k,l}|^2, \tag{15.39}$$

where $\hat{\boldsymbol{\theta}}$ denotes the estimated $\boldsymbol{\theta}$ produced by the algorithm and Ω_{test} contains the set of data points that is missing. The decentralized algorithms are all implemented in MATLAB and tested on an Intel Xeon server. Furthermore, for the DPG algorithm, we set $\gamma_t = 1/\sqrt{t}$ as suggested by the convergence analysis; for the DeFW algorithm with the square loss objective defined in Eq. (15.36), $\gamma_t = 2/(t+1)$; for the DeFW with the Gaussian loss objective (15.37), $\gamma_t = 1/t^{0.75}$. The MATLAB program for the simulations is available at the online repository https://github.com/hoitowai/defw. The experimental results are as follows.

Convergence rates. Our first example considers the synthetic data case with $m_1 = 100$, $m_2 = 250$, and $K = 5$. The number of observations made is $|\Omega_i| = 100$. The observations are noiseless, thus the optimal objective value for Eq. (15.35) is zero. The numerical results are presented in Fig. 15.2, where we plot the optimality gap and test MSE against the iteration number. We observe that the theoretically predicted convergence rates of the decentralized algorithms match the numerical trends. Moreover,

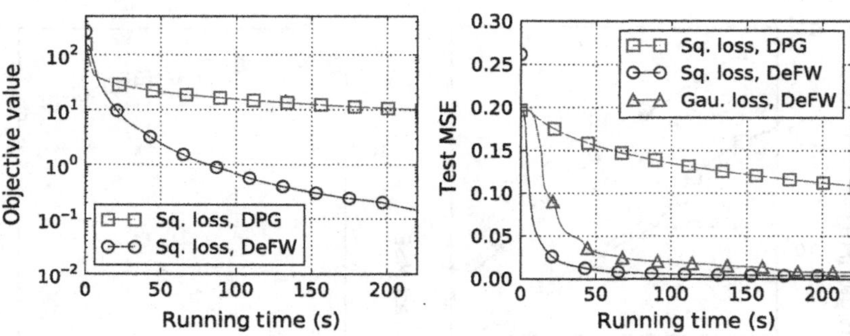

FIG. 15.3

Computation time of the decentralized algorithms. Results for the decentralized algorithms applied to the matrix completion problem (15.35) for a synthetically generated matrix with noiseless observations with $m_1 = 200$, $m_2 = 1000$, $K = 5$, and $|\Omega_i| = 100$. (Left) Objective value against the running times. (Right) Test MSE against the running time.

the DeFW algorithm applied to different loss functions performs slightly better than the theoretically predicted rate bounds. In the squared loss case, a possible conjecture explaining this phenomenon is a hidden strong convexity property such that the optimal solution of Eq. (15.35) is unique when the matrix $\boldsymbol{\theta}_{\text{true}}$ is sufficiently low rank.

Running time. The next example focuses on a high-dimensional synthetic data case with $m_1 = 200$, $m_2 = 1000$, $K = 5$, and $|\Omega_i| = 100$. The observations are also noiseless, leading to $F(\boldsymbol{\theta}^\star) = 0$. Our focus here is on the running time performance of the DPG and DeFW algorithms. As seen in Fig. 15.3, the DeFW algorithm runs with a much smaller computation time to achieve the same level of optimality compared to the DPG algorithm. This is consistent with our claims and is the result of the fact that the DPG algorithm requires a costly projection step per iteration.

Application on real data. Our last example (Fig. 15.4) focuses on the real data case of the `movielens100k` dataset. Here, we consider the case with *noisy* observation such that the added noise is *sparse*, i.e., some of the observed entries can be viewed as outliers. We observe desired convergence for the DeFW algorithm applied to the squared loss and the negated Gaussian loss functions. In particular, the negated Gaussian loss case achieves better test MSE performance due to its robustness to outlier noise. It is also observed that the iterates attain consensus as the iteration number grows.

15.5 CONCLUSIONS AND OTHER EXTENSIONS

This chapter described the classical decentralized projected gradient descent algorithm and the so-called decentralized Frank-Wolf algorithm for projection-free decentralized optimization. We also summarized the known results regarding their convergence rates when solving both convex and nonconvex problems. The numerical analysis of the decentralized matrix completion problem exposes their performance and validates the various claims reported in the chapter. Some open questions lie ahead, particularly the study of the asynchronous version of the DeFW algorithm that is still missing.

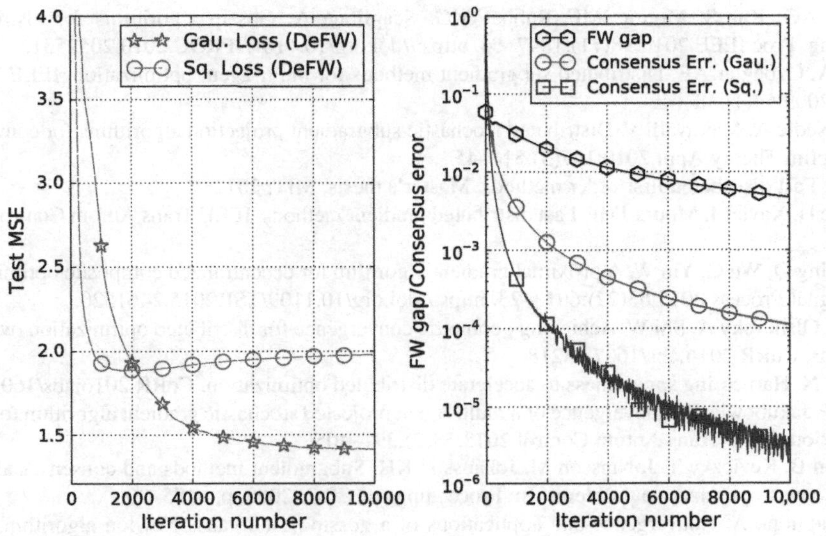

FIG. 15.4

Real data experiment with movielens100k. Decentralized matrix completion (15.35) using the DeFW algorithm. Notice that the DPG algorithm is not considered in the example due to its excessive run-time. (Left) Test MSE against iteration number. (Right) FW gap and consensus error against iteration number.

ACKNOWLEDGMENTS

The authors would like to thank Dr. Jean Lafond for contributing to the development of the DeFW algorithm and the support from NSF under the award number CCF-1553746 and CCF-1531050.

REFERENCES

[1] Sayed AH, Tu SY, Chen J, Zhao X, Towfic ZJ. Diffusion strategies for adaptation and learning over networks: an examination of distributed strategies and network behavior. IEEE Signal Process Mag 2013;30(3):155–71. https://doi.org/10.1109/MSP.2012.2231991.

[2] Liu Z, Vandenberghe L. Interior-point method for nuclear norm approximation with application to system identification. SIAM J Matrix Anal Appl 2010;31(3):1235–56.

[3] Candès EJ, Recht B. Exact matrix completion via convex optimization. Found Comput Math 2009;9(6): 717–72.

[4] Ravazzi C, Fosson SM, Magli E. Randomized algorithms for distributed nonlinear optimization under sparsity constraints. IEEE Trans Signal Process 2016;64(6):1420–34.

[5] Patterson S, Eldar YC, Keidar I. Distributed compressed sensing for static and time-varying networks. IEEE Trans Signal Process 2014;62(19):4931–46.

[6] Frank M, Wolfe P. An algorithm for quadratic programming. Naval Res Logis Quart 1956;3(1–2):95–110.

[7] Tsitsiklis J. Problems in decentralized decision making and computation. Ph.D. thesis, Department of Electrical Engineering and Computer Science, MIT, Boston, MA; 1984.

[8] Dimakis AG, Kar S, Moura JMF, Rabbat MG, Scaglione A. Gossip algorithms for distributed signal processing. Proc IEEE 2010;98(11):1847–64. https://doi.org/10.1109/JPROC.2010.2052531.
[9] Nedich A, Ozdaglar AE. Distributed subgradient methods for multi-agent optimization. IEEE Trans Autom Control 2009;54(1):48–61.
[10] Ram S, Nedić A, Veeravalli V. Distributed stochastic subgradient projection algorithms for convex optimization. J Optim Theory Appl 2010;147(3):516–45.
[11] Chen IA. Fast distributed first-order methods. Master's thesis, MIT; 2012.
[12] Jakovetic D, Xavier J, Moura JMF. Fast distributed gradient methods. IEEE Trans Autom Control 2014;59(5):1131–46.
[13] Shi W, Ling Q, Wu G, Yin W. A proximal gradient algorithm for decentralized composite optimization. IEEE Trans Signal Process 2015;63(22):6013–23. https://doi.org/10.1109/TSP.2015.2461520.
[14] Nedić A, Olshevsky A, Shi W. Achieving geometric convergence for distributed optimization over time-varying graphs. CoRR 2016;abs/1607.03218.
[15] Qu G, Li N. Harnessing smoothness to accelerate distributed optimization. CoRR 2016;abs/1605.07112.
[16] Bianchi P, Jakubowicz J. Convergence of a multi-agent projected stochastic gradient algorithm for non-convex optimization. IEEE Trans Autom Control 2013;58(2):391–405.
[17] Johansson B, Keviczky T, Johansson M, Johansson KH. Subgradient methods and consensus algorithms for solving convex optimization problems. In: Proceedings of CDC; 2008. p. 4185–90.
[18] Li X, Scaglione A. Convergence and applications of a gossip based gauss Newton algorithm. IEEE Trans Signal Process 2013;61(21):5231–46.
[19] Varagnolo D, Zanella F, Cenedese A, Pillonetto G, Schenato L. Newton-Raphson consensus for distributed convex optimization. IEEE Trans Autom Control 2016;61(4):994–1009. https://doi.org/10.1109/TAC.2015.2449811.
[20] Boyd S, Parikh N, Chu E, Peleato B, Eckstein J. Distributed optimization and statistical learning via the alternating direction method of multipliers. Found Trends Mach Learn 2011;3(1):1–122.
[21] Duchi J, Agarwal A, Wainwright MJ. Dual averaging for distributed optimization: convergence analysis and network scaling. IEEE Trans Autom Control 2012;57(3):592–606.
[22] Wei E, Ozdaglar A. On the o(1/k) convergence of asynchronous distributed alternating direction method of multipliers. CoRR 2013;abs/1307.8254.
[23] Chang TH, Nedic A, Scaglione A. Distributed constrained optimization by consensus-based primal-dual perturbation method. IEEE Trans Autom Control 2014;59(6):1524–38.
[24] Simonetto A, Jamali-Rad H. Primal recovery from consensus-based dual decomposition for distributed convex optimization. J Optim Theory Appl 2016;168(1):172–97. https://doi.org/10.1007/s10957-015-0758-0.
[25] Hong M. Decomposing linearly constrained nonconvex problems by a proximal primal dual approach: algorithms, convergence, and applications. CoRR 2016;abs/1604.00543.
[26] Lorenzo PD, Scutari G. NEXT: in-network nonconvex optimization. IEEE Trans Signal Inf Process Netw 2016;2(2):120–136.
[27] Wai HT, Chang TH, Scaglione A. A consensus-based decentralized algorithm for non-convex optimization with application to dictionary learning. In: Proc ICASSP; 2015. p. 3546–550.
[28] Yang Y, Scutari G, Palomar DP, Pesavento M. A parallel stochastic approximation method for nonconvex multi-agent optimization problems. CoRR 2014;abs/1410.5076.
[29] Wai HT, Scaglione A. Consensus on state and time: decentralized regression with asynchronous sampling. IEEE Trans Signal Process 2015;63(11):2972–85.
[30] Wu Z, Teo K. A conditional gradient method for an optimal control problem involving a class of nonlinear second-order hyperbolic partial differential equations. J Math Anal Appl 1983;91(2):376–93.
[31] Jaggi M, Sulovsky M. A simple algorithm for nuclear norm regularized problems. In: ICML; 2010.
[32] Joulin A, Tang K, Fei-Fei L. Efficient image and video co-localization with Frank-Wolfe algorithm. In: ECCV; 2014.

[33] Zhang L, Kekatos V, Giannakis GB. Scalable electric vehicle charging protocols. CoRR 2016;abs/1510.00403v2.
[34] Fukushima M. A modified Frank-Wolfe algorithm for solving the traffic assignment problem. Transp Res B Methodol 1984;18(2):169–77.
[35] Jaggi M. Revisiting Frank-Wolfe: projection-free sparse convex optimization. In: ICML; 2013.
[36] Ghosh S, Lam H. Computing worst-case input models in stochastic simulation. CoRR 2015;abs/1507.05609.
[37] Lacoste-Julien S. Convergence rate of Frank-Wolfe for non-convex objectives. CoRR 2016;abs/1607.00345.
[38] Lafond J, Wai HT, Moulines E. On the online Frank-Wolfe algorithms for convex and non-convex optimizations. CoRR 2016;abs/1510.01171v2.
[39] Wai HT, Lafond J, Scaglione A, Moulines E. Decentralized Frank-Wolfe algorithm for convex and non-convex problems. IEEE Trans Autom Control 2017;62(11):5522–5537.
[40] Bertsekas DP. Nonlinear programming. Athena Scientific; 1999.
[41] Tatarenko T, Touri B. Non-convex distributed optimization. IEEE Trans Autom Control 2017;62(8):3744–57.
[42] Defazio A, Bach F, Lacoste-Julien S. SAGA: a fast incremental gradient method with support for non-strongly convex composite objectives. In: NIPS; 2014.
[43] Ghadimi S, Lan G. Accelerated gradient methods for nonconvex nonlinear and stochastic programming. Math Program 2015;156(1):59–99.
[44] Scaglione A, Pagliari R, Krim H. The decentralized estimation of the sample covariance. In: Proc Asilomar; 2008. p. 1722–6.
[45] Ling Q, Xu Y, Yin W, Wen Z. Decentralized low-rank matrix completion. In: Proc ICASSP; 2012.
[46] Mackey L, Talwalkar A, Jordan MI. Distributed matrix completion and robust factorization. J Mach Learn Res 2015;16:913–60.
[47] Yu HF, Hsieh CJ, Si S, Dhillon I. Scalable coordinate descent approaches to parallel matrix factorization for recommender systems. In: ICDM. IEEE; 2012. p. 765–74.
[48] Recht B, Ré C. Parallel stochastic gradient algorithms for large-scale matrix completion. Math Program Comput 2013;5(2):201–26.
[49] Chandrasekaran V, Sanghavi S, Parrilo PA, Willsky AS. Rank-sparsity incoherence for matrix decomposition. SIAM J Optim 2011;21(2):572–96. https://doi.org/10.1137/090761793.
[50] Mohimani GH, Babaie-Zadeh M, Jutten C. Fast sparse representation based on smoothed L0 norm. In: ICA. Lecture notes in computer science. Springer; 2007. p. 389–96.
[51] Wai HT, Scaglione A, Lafond J, Moulines E. Fast and privacy preserving distributed low-rank regression. In: ICASSP; 2017.
[52] Xiao L, Boyd S. Fast linear iterations for distributed averaging. Syst Control Lett 2004;53(1):65–78. https://doi.org/10.1016/j.sysconle.2004.02.022.
[53] Harper FM, Konstan JA. The MovieLens datasets: history and context. ACM Trans Interact Intell Syst 2016;5(4). https://doi.org/10.1145/2827872.

CHAPTER

THE EDGE CLOUD: A HOLISTIC VIEW OF COMMUNICATION, COMPUTATION, AND CACHING 16

Sergio Barbarossa*, Stefania Sardellitti*, Elena Ceci*, Mattia Merluzzi*
*Department of Information Engineering, Electronics, and Telecommunications, Sapienza University of Rome, Rome, Italy**

16.1 INTRODUCTION

The major goal of next generation (5G) communication networks is to build a communication infrastructure that will enable new business opportunities in diverse sectors, or *verticals*, such as automated driving, e-health, virtual/augmented reality, Internet of Things (IoT), smart grids, and so on [1,2]. These services have very different specifications and requirements in terms of latency, reliability, data rate, number of connected devices, and so on. Thinking of enabling such diverse services using a common communication platform might seem like a crazy idea. But, in reality, if the system is properly designed, reusing a common infrastructure for different purposes might induce a significant economic advantage. The key idea for making this possible is to use *virtualization* [3] and implement *network slicing* [4]. Through virtualization, many network functionalities are implemented in software through virtual machines that can be instantiated and moved upon request [5]. Building on virtualization, network slicing partitions a *physical* network into multiple *virtual* networks, each matched to its specific requirements and constraints, thus enabling operators to provide networks on an as-a-service basis while meeting a wide range of use cases in parallel.

This new reality, sometimes called the fourth Industrial Revolution, can be realized by a new architecture able to meet advanced requirements, especially in terms of latency (below 5 ms), reliability (around 0.99999), coverage (up to 100 devices/m^2), and data rate (more than 10 Gbps). At the physical layer, 5G builds on a significant increase of system capacity by incorporating massive MIMO techniques, dense deployment of radio access points, and wider bandwidth. All these strategies are facilitated by the introduction of millimeter wave (mmWave) communications [6–8]: mmWaves make possible the reduction of the antenna size, thus enabling the use of an array with many elements, as required in massive MIMO. Dense deployment is also facilitated because mmWaves give rise to a stronger intercell attenuation. Finally, increasing the carrier frequency facilitates the usage of wider bandwidths. However, the significant improvement achievable at the physical layer could still be insufficient to meet the challenging and diverse requirements of very low latency and ultrareliability. A further improvement comes from a paradigm shift that puts applications at the center of the system

Cooperative and Graph Signal Processing. https://doi.org/10.1016/B978-0-12-813677-5.00016-X

design. Network Function Virtualization (NFV) and Multiaccess Edge Computing (MEC) [9] are the key tools of this application-centric networking. In particular, MEC plays the key role of bringing cloud-computing resources at the edge of the network, within the Radio Access Network (RAN), in proximity to mobile subscribers [9,10]. MEC is particularly effective to deliver context-aware services or to enable computation offloading from resource-poor mobile devices to fixed servers or to perform intelligent cache prefetching, based on local learning of the most popular contents across space and time.

Given this perspective, the goal of this chapter is to show that graph-based methods can play a significant role in optimizing resource allocation or deriving new learning mechanisms. The organization of this chapter is the following. In Section 16.2 we present the edge cloud architecture and we motivate the holistic approach that looks at $3C$ resources as a common pool of resources to be handled jointly with the goal of achieving, on the user side, a satisfactory quality of experience and, on the network side, a balanced and efficient use of resources. Then, in Section 16.3, we will focus on the joint optimization of computation and communication resources, with specific attention to computation offloading in the edge-cloud. In Section 16.4, we will concentrate on the joint optimization of caching and communication. Differently from storage, which is fundamentally *static*, caching is inherently *dynamic* so that cache memories are prefetched when and where needed, and then released. In both cases of joint optimization, the goal is to bring resources, either computation (virtual machines) or cache, as close as possible to the end user to enable truly low latency and low energy consumption services. After presenting this holistic view, we will present in Section 16.5 some learning mechanisms based on graph signal processing. In particular, we show how to reconstruct the radio environment map (REM), which enables cognitive usage of the radio resources. Then, building again on graph representations, in Section 16.6 we show how to achieve an optimal resource allocation across a network while being robust to link failures. The proposed approach is based on a small perturbation analysis of network topologies affected by sporadic edge failures. Finally, in Section 16.7 we draw some conclusions and suggest some possible further developments.

16.2 HOLISTIC VIEW OF COMMUNICATION, COMPUTATION, AND CACHING

The new infrastructure provided by the next communication networks can be seen as a truly distributed and pervasive computer that provides very different services to mobile users with sufficiently good quality of experience. The physical resources composing this pervasive computer are cache memories, computing machines, and communication channels. The system should serve the end user, either a mobile subscriber or a car or the component of a production process with, ideally, zero latency. This means an end-to-end latency smaller than the user perception capability or than the maximum value ensuring proper control, such as breaking time in automated driving. To enable this vision, at the physical layer, the network will support a much higher system (or area) capacity (bits/sec/km^2). In 5G systems, a 1000-fold increase of system capacity is planned, exploiting mmWave communications, massive MIMO, and dense deployment of access points. However, in spite of this enormous improvement in system capacity, the zero-latency ideal could still be far from obtainment because it is very complicated, if not impossible, to control latency over a wide area network. For this reason, the next step is to bring computation and cache resources as close as possible to the end user, where proximity is actually measured in terms of service time. This creates a new ecosystem, called

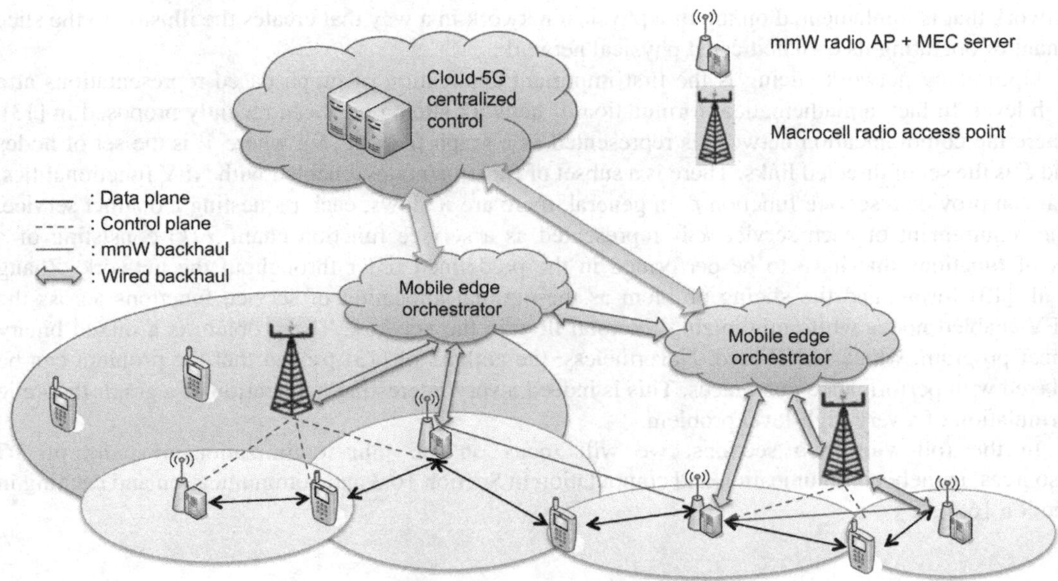

FIG. 16.1

Edge cloud architecture.

edge cloud, whose architecture is sketched in Fig. 16.1. In this system, within a macrocell served by one base station, we have multiple millimeter-wave access points (AP) covering much smaller areas. Each AP is endowed with computation and caching capabilities to enable mobile users to get proximity access to cloud functionalities. This makes it possible to provide cloud services with very low latency and a high data rate while at the same time keeping data traffic and computation as local as possible. Of course, the computing and caching capabilities of local MEC servers are significantly lower than a typical cloud, but they also serve a limited number of requests and, whenever their resources are insufficient, they may interact with nearby MEC servers, under the supervision of a MEC orchestrator. In this system, mobile applications are handled by *virtual machines* (or containers) instantiated at the edge of the network, close to the end user. The edge is either the ensemble of network access points, as in Multiaccess Edge Computing (MEC) [11], or it might even include the mobile terminals as well, as in *fog computing* [12].

Similarly, content moves dynamically when and where it is more convenient to have them. Caching can in fact be seen as a *noncausal* communication, where content moves before it is actually requested to minimize the downloading time. In this framework, it makes sense to allocate 3C resources *jointly*, with the objective of guaranteeing some ultimate user quality of experience.

Assuming such a holistic perspective, the first important question is why use a *common* platform, call it 5G or the generations to follow, to accommodate services having such different requirements, such as IoT, virtual reality, or automated driving. This is indeed one of the main challenges faced by 5G systems. The approach proposed in the 5G roadmap is *network slicing* [5]. A network slice is a virtual network that is implemented on top of a physical network in a way that creates the illusion to the slice tenant of operating its own dedicated physical network.

Optimizing network slicing is the first important application of graph-based representations at a high level. In fact, a mathematical formulation of network slicing has been recently proposed in [13], where the communication network is represented as a graph $\mathcal{G} = (\mathcal{V}, \mathcal{E})$, where $\mathcal{V}$ is the set of nodes and $\mathcal{E}$ is the set of directed links. There is a subset of function nodes, enabled with NFV functionalities, that can provide a service function f. In general, there are K flows, each requesting a distinct service. The requirement of each service k is represented as a service function chain $\mathcal{F}(k)$ consisting of a set of functions that have to be performed in the predefined order throughout the network. Zhang et al. [13] formulated the slicing problem as the optimal allocation of service functions across the NFV-enabled nodes while minimizing the total flow in the network. The problem is a mixed binary linear program, which is NP-hard. Nevertheless, the authors of [13] proved that the problem can be relaxed with performance guarantees. This is indeed a very interesting application of a graph-theoretic formulation of a very high-level problem.

In the following two sections, we will focus on the joint optimization of pairs of 3C resources, namely communication and computation in Section 16.3 and communication and caching in Section 16.4.

16.3 JOINT OPTIMIZATION OF COMMUNICATION AND COMPUTATION

Smartphones have really exploded in their usage and capabilities, placing significant demand upon battery usage. Unfortunately, advancements in battery technology have not kept pace with the demands of users and their smartphones. One approach to overcome the battery energy limitations is to offload computations from mobile devices to fixed devices. Computation offloading may be convenient for the following reasons [14,15]: (i) to save energy and then prolong the battery lifetime of hand-held devices; (ii) to enable simple devices, such as inexpensive sensors, to run sophisticated applications; and (iii) to reduce latency. From a user perspective, one of the parameters most affecting the quality of experience is the end-to-end (E2E) latency, i.e., the time necessary to get the results from running an application. In the case of offloading, this latency includes: (i) the time to send bits from the mobile device to the fixed server to enable the program; (ii) the time to run the application remotely; and (iii) the time to get the result back. It is precisely this E2E latency that couples communication and computation resources and then motivates the joint allocation of these resources. We recall now the approach proposed in [16] and later expanded in [14,17].

We first consider the case where multiple users are served by a single AP/MEC pair. Then, we will move to the more challenging case where multiple users are served by multiple APs and MEC servers. In the first case, the assignment of each UE to a pair of AP and MEC is supposed to be given; in the second case, the assignment is part of the optimization problem. In both cases, for economical reasons associated with promoting their capillary deployment, the computational capabilities of MEC servers are enormously smaller than a typical cloud. This implies that the number of cores per server is very limited or, in other words, that the available cores in an MEC server must operate in a multitasking mode to accommodate the requests of multiple users. This means that a server running K applications for as many mobile users will allocate a certain percentage β_k of its CPU time to the users that are being served concurrently. If F_S denotes the number of CPU cycles/sec that the server can run, the percentage of CPU cycles/sec assigned to the kth user is then $f_k = \beta_k F_S$.

Multiple users served by a single AP/MEC pair. We start by considering K user equipments (UE) assigned to a single AP and a single MEC. The decision to offload a computation from the mobile device to the MEC server depends on the characteristics of the application to be offloaded. Not all applications are equally amenable to offloading. The decision should take into account all sources of energy consumption in a smartphone, such as display, network, CPU, GPS, camera, and so on. Profiling energy consumption of applications running on smartphones, rather than on a general purpose computer, is not an easy task because of *asynchronous power behavior*, where the effect on a component's power state due to a program entity lasts beyond the end of that program entity [18]. The signal processing community could provide a significant contribution to this research field by optimizing app developments taking into account the associated energy profiling for a class of smartphone operating systems, e.g., iOS, Android, and so on, and a class of applications. In this chapter, we do not dig into these aspects. We rather concentrate on the joint optimization of radio and computational resources associated with computation offloading, in a multiuser context. From this point of view, we simplify the classification of applications by identifying a few most significant parameters as relevant for computation offloading. For each user k, we consider: (i) the number b_k of bits to be transmitted from the mobile user to the server to transfer the program execution; and (ii) the number of CPU cycles w_k necessary to run the application to be offloaded. We denote by L_k the E2E latency requested from UE k. The overall latency T_k experienced by the kth UE for offloading an application is the sum of three terms: (i) the time $T_k^{\mathtt{tx}}$ necessary to transmit all bits to the server to enable the transfer of program execution; (ii) the time $T_k^{\mathtt{exe}}$ for the server to run the application; and (iii) the time $T_k^{\mathtt{rx}}$ to get the result back to the UE. In formulas,

$$T_k = T_k^{\mathtt{tx}} + T_k^{\mathtt{exe}} + T_k^{\mathtt{rx}}. \tag{16.1}$$

This equation, in its simplicity, shows that enforcing an E2E latency constraint induces a coupling between communication and computation resources.

From a user-centric perspective, the goal might either be to minimize the E2E latency under a maximum transmit power constraint or, by duality, to minimize the transmit power necessary to guarantee a desired latency. We follow this latter approach, but clearly the two strategies can be interchanged. Let us now express the single contributions in Eq. (16.1) in terms of the parameters to be optimized.

The first contribution is the time $T_k^{\mathtt{tx}}$ to transmit b_k bits from the UE to the AP:

$$T_k^{\mathtt{tx}}(p_k) = \frac{c_k}{r_k(p_k)}, \tag{16.2}$$

where $c_k = b_k/B$, B is the bandwidth and $r_k(p_k)$ is the spectral efficiency over the channel between UE and AP, which is equal to

$$r_k(p_k) = \log_2\,(1 + \alpha_k p_k), \tag{16.3}$$

where p_k is the transmit power of UE k; $\alpha_k = |h_k|^2/(d_k^{\gamma}\sigma_n^2)$ is a an equivalent channel coefficient that incorporates the channel coefficient h_k, the noise variance σ_n^2, the distance d_k between UE and AP, and the channel exponent factor γ. The second contribution in Eq. (16.1) is the execution time at the server, which is equal to $T_k^{\text{exe}} = w_k/f_k$. From the user perspective, the third term in Eq. (16.1) does not imply a transmit power, but only the energy to process the received data. This term is typically much smaller than the first term and in the following derivations we will assume it to be a fixed term incorporated in the overall latency.

We are now ready to formulate the computation offloading optimization problem in terms of the transmit powers p_k and the CPU percentages f_k, $k = 1, \ldots, K$:

$$\begin{aligned}
&\min_{p, f_s} \quad \sum_{k=1}^{K} p_k, \qquad \textbf{[P.1]} \\
&\text{s.t.} \quad \frac{c_k}{\log_2(1 + p_k \alpha_k)} + \frac{w_k}{f_k} \le L_k, \quad k = 1, \ldots, K \\
&\qquad 0 < p_k \le P_T, \quad f_k > 0, \quad k = 1, \ldots, K \\
&\qquad \sum_{k=1}^{K} f_k \le F_S,
\end{aligned} \tag{16.4}$$

where $\mathbf{p} = (p_1, \ldots, p_K)$ and $\mathbf{f}_s = (f_1, \ldots, f_K)$.

This is a convex problem that can be easily solved. In particular, the optimal computational rates can be expressed in closed form as [19]:

$$f_k = \frac{\sqrt{w_k \eta_k}}{\sum_{k=1}^{K} \sqrt{w_k \eta_k}} F_S, \tag{16.5}$$

where η_k are coefficients that depend on the channel coefficients. This simple formula shows how the allocation of computational resources depends not only on computational aspects, but also on the channel state. Note also that the above formula contrasts with the proportional allocation of computational rates that would have been performed in a conventional system, i.e.,

$$f_k = \frac{w_k}{\sum_{k=1}^{K} w_k} F_S. \tag{16.6}$$

A further substantial improvement to computation offloading comes from the introduction of mmWave links. Merging MEC with an underlying mmWave physical layer creates a unique opportunity to bring IT services to the mobile user with very low latency and a very high data rate. This merge is indeed one of the main objectives of the joint Europe/Japan H2020 Project called 5G-MiEdge (Millimeter-wave Edge Cloud as an Enabler for 5G Ecosystem) [20]. The challenge coming from the use of mmWave links is that they are more prone to blocking events [21], which may jeopardize the benefits of computation offloading. A possible way to counteract blocking events in an MEC system using mmWave links was proposed in [19,22].

Multiple users served by multiple APs and multiple MEC servers. Let us consider now a more complex scenario where multiple users may get radio access through multiple APs and multiple MECs. Besides resource allocation, our goal now is to find also the optimal association between UEs, APs, and MEC servers. We consider a system composed of N_b small cell access points, N_c MEC servers, and K mobile UE's. Within the edge-cloud scenario depicted in Fig. 16.1, the association of a mobile user to an access point does not necessarily follow the same principles of current systems, where a mobile user gets access to the base station with the largest signal-to-noise ratio. In the edge-cloud scenario depicted in Fig. 16.1, the association of a UE to a pair of AP and MEC servers depends not only on radio channel parameters, but also on the availability of computational resources at the MEC server. Furthermore, a UE can get radio access from a certain AP, but its application can run elsewhere, not necessarily on the nearest MEC, depending on the availability of computational resources. Actually, because the applications run as virtual machines (VM), we can think of migrating these VMs in order to follow the user. The orchestration of MEC servers in order to provide seamless service continuity to mobile users is an item that has been recently included in the standardization activities of ETSI, within the MEC study group [23]. Migrating VMs is not an easy task because the instantiation of a VM requires times that are too large with respect to some of the latency requirements foreseen in 5G. This has motivated significant research efforts in investigating light forms of virtual machines, named *containers*, that do not need the instantiation of the whole operating system, but only of a restricted kernel [24].

Here, we do not consider the migration of VMs, but we do consider the possibility of letting a UE get access under one AP while having its application run in an MEC located elsewhere. In this case, we need to incorporate in the E2E latency the delay along the backhaul link connecting the AP and MEC. In particular, we denote by T_{Bnm} the latency between access point n and MEC server m.

Following an approach similar to what we proposed in [25], we generalize now the resource allocation problem by incorporating binary variables $a_{knm} \in \{0, 1\}$ that assume a value $a_{knm} = 1$ if user k gets radio access through AP n to have its application running on MEC server m, and $a_{knm} = 0$ otherwise. For the sake of simplicity, we assume that each user is served by a single base station and a single cloud. Our goal now is to find the optimal assignment rule, together with the optimal transmit powers p_k and the computational rates f_{mk} assigned by MEC server m to UE k. As in the previous section, our goal is to minimize the overall UE power consumption, under a latency constraint.

The resulting optimization problem is:

$$
\begin{aligned}
\min_{\mathbf{p},\mathbf{f},\mathbf{a}} \quad & f(\mathbf{p},\mathbf{a}) \triangleq \sum_{k=1}^{K}\sum_{n=1}^{N_b}\sum_{m=1}^{N_c} p_k a_{knm} \qquad (\mathcal{P}) \\
\text{s.t.} \quad & \text{(i) } g_{knm}(p_k, f_{mk}, a_{knm}) \leq L_k, \ \forall\, k, n, m, \\
& \text{(ii) } p_k \leq P_k, \quad p_k \geq 0, \ \forall\, k, \\
& \text{(iii) } h_m(\mathbf{f},\mathbf{a}) \triangleq \sum_{k=1}^{K}\sum_{n=1}^{N_b} a_{knm} f_{mk} \leq F_m, \ \forall\, m, \ \mathbf{f} \geq \mathbf{0}, \\
& \text{(iv) } \sum_{n=1}^{N_b}\sum_{m=1}^{N_c} a_{knm} = 1, \ a_{knm} \in \{0,1\}, \ \forall\, k, n, m,
\end{aligned}
\tag{16.7}
$$

where $\mathbf{f} := (f_{mk})_{\forall m,k}$, $\mathbf{a} := (a_{knm})_{\forall k,n,m}$, and

$$
g_{knm}(p_k, f_{mk}, a_{knm}) \triangleq a_{knm}\left(\frac{c_k}{r_{kn}(p_k)} + \frac{w_k}{f_{mk}} + T_{Bnm}\right)
$$

with $r_{kn}(p_k) = \log_2(1 + \alpha_{kn} p_k)$ denoting the spectral efficiency of UE k accessing AP n and α_{kn} the equivalent channel coefficient between UE k and AP n.

The objective function is the total transmit power consumption from the mobile users. The constraints have the following meaning: (i) the overall latency for each user k must be less than the maximum value L_k; (ii) the total power spent by each user must be lower than a fixed total power budget P_k; (iii) the sum of the computational rates f_{mk} assigned by each server cannot exceed the server computational capability F_m; and (iv) each mobile user should be served by one AP/MEC pair; this is enforced by imposing $\sum_{n=1}^{N_b}\sum_{m=1}^{N_c} a_{knm} = 1$, for each k, together with $a_{knm} \in \{0, 1\}$.

Unfortunately, problem $\mathcal{P}$ is a mixed-binary problem and is, in general, NP-hard. To overcome this difficulty, as we suggested in [26,27], we relax the binary variables a_{knm} to be real variables in the interval $[0, 1]$ and adopt a suboptimal successive convex approximation strategy [25,28], able to converge to local optimal solutions. Additionally, to drive the assignment variables a_{knm} to contain only one value equal to one and all others to zero, for each k, we incorporate a further constraint recently suggested in [13]. The penalty method in [13] is based on the fact that the following problem

$$\begin{aligned} \min_{\mathbf{a}_k} \quad & \| \mathbf{a}_k + \epsilon \mathbf{1} \|_p^p \triangleq \sum_{n=1}^{N_b}\sum_{m=1}^{N_c} (a_{knm} + \epsilon)^p \\ \text{s.t.} \quad & \| \mathbf{a}_k \|_1 = 1, \\ & a_{knm} \in [0, 1], \quad \forall n, m \end{aligned} \tag{16.8}$$

with $\mathbf{a}_k = (a_{knm})_{\forall n,m}$ and $p \in (0, 1)$, $\epsilon > 0$, admits an optimal solution that is binary, i.e., only one element is one and all the others are zero. The optimal solution is $c_{\epsilon,k} = (1+\epsilon)^p + (N_b N_c - 1)\epsilon^p$. Therefore, by relaxing the binary variables a_{knm} so that they belong to the following convex set

$$\mathcal{A} = \left\{ (\mathbf{a}_k)_{k\in\mathcal{I}} \; : \; a_{knm} \in [0, 1], \sum_{n=1}^{N_b}\sum_{m=1}^{N_c} a_{knm} = 1, \; \forall k, n, m \right\},$$

where $\mathcal{I}$ denotes the set of K users, we formulate the following relaxed optimization problem [26]:

$$\begin{aligned} \min_{\mathbf{p},\mathbf{f},\mathbf{a}} \quad & f_{\mathcal{P}_\sigma}(\mathbf{p}, \mathbf{a}) \triangleq f(\mathbf{p}, \mathbf{a}) + \sigma P_\epsilon(\mathbf{a}) \qquad (\mathcal{P}_\sigma) \\ \text{s.t.} \quad & \text{(i) } g_{knm}(p_k, f_{mk}, a_{knm}) \le L_k, \; \forall k, n, m, \\ & \text{(ii) } h_m(\mathbf{f}, \mathbf{a}) \triangleq \sum_{k=1}^{K}\sum_{n=1}^{N_b} a_{knm} f_{mk} \le F_m, \; \forall\, m, \; \mathbf{f} \ge \mathbf{0}, \\ & \text{(iii) } p_k \le P_k, \quad p_k \ge 0, \; \forall k \in \mathcal{I}, \; \mathbf{a} \in \mathcal{A}, \end{aligned} \tag{16.9}$$

where $\sigma > 0$ is the penalty parameter, and

$$P_\epsilon(\mathbf{a}) \triangleq \sum_{k=1}^{K} \| \mathbf{a}_k + \epsilon \mathbf{1} \|_p^p - c_{\epsilon,k}. \tag{16.10}$$

It is important to emphasize that this penalty is differentiable with respect to the unknown variables. Even by relaxing the binary variables $\mathbf{a}$, the problem in Eq. (16.9) is still nonconvex because the objective function and the constraints (i), (ii) are nonconvex. In [26], we proposed a successive convex approximation (SCA) technique, inspired by Scutari et al. [28], to devise an efficient iterative penalty SCA approximation algorithm (PSCA) converging to a local optimal solution of Eq. (16.9). We omit the details here, but we report some numerical results.

To test the effectiveness of the proposed offloading strategy, in Fig. 16.2 we report the optimal total transmit power consumption versus the maximum latency L_k. We consider a network composed of $K = 4$ users, a number of base stations equal to the number of clouds, i.e., $N_b = N_c = 2$. The other parameters are set as follows: $F_1 = 2.7 \cdot 10^9$, $F_2 = 6 \cdot 10^8$, $P_k = 2 \cdot 10^{-1}$, $p = 0.025$. From Fig. 16.2, we may observe that the PSCA algorithm provides results very close to the exhaustive

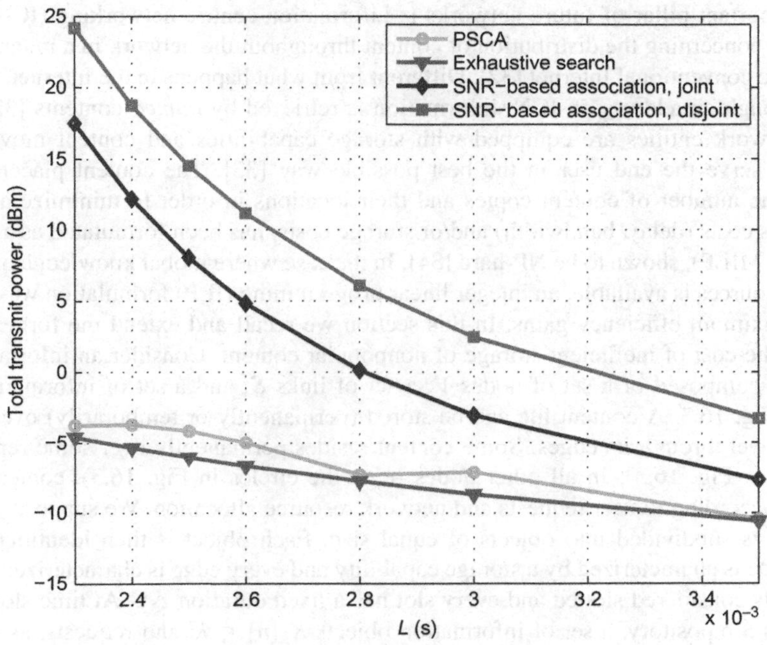

FIG. 16.2

Overall UE transmit power consumption versus L.

search algorithm whose complexity is exponential. Additionally, we consider as a comparison term the SNR-based association method, in both cases where the radio and computational resources are optimized jointly or disjointly. It can be noted that the PSCA algorithm yields considerable power savings compared to methods based on SNR only because it takes advantage of the optimal assignment of each user to a cloud through the most convenient base station.

16.4 JOINT OPTIMIZATION OF CACHING AND COMMUNICATION

Caching popular content in storage disks distributed across the network yields significant advantages in terms of reduction of downloading times and limitation of data traffic. Caching can be seen as a *noncausal* communication where popular content moves throughout the network in the off-peak hours to anticipate the users' requests. Clearly an effective caching strategy builds significantly on the ability to learn and predict users' behaviors. This capability lies at the foundation of *proactive caching* [29] and it motivates the need to merge future networks with big data analytics [30]. An alternative approach to proactive caching based on reinforcement learning to learn file popularity across time and space was recently proposed in [31].

Another important pillar of future networks is information-centric networking (ICN), a relatively novel paradigm concerning the distribution of content throughout the network in a manner much more efficient than the conventional Internet [32]. Different from what happens in the Internet, where content is retrieved through an address, in ICN, information is retrieved by *named* contents [32]. In the ICN framework, network entities are equipped with storage capabilities and content moves throughout the network to serve the end user in the best possible way [33]. The content placement problem, incorporating the number of content copies and their locations in order to minimize a cost function capturing access costs (delay, bandwidth) and/or storage costs, has been formulated as a mixed integer linear program (MILP), shown to be NP-hard [34]. In the case where global knowledge of user requests and network resources is available, an integer linear programming (ILP) formulation was given in [33], yielding the maximum efficiency gains. In this section we recall and extend the formulation of [33] to incorporate the cost of inefficient storage of nonpopular content. Consider an information network $\mathcal{G} = (\mathcal{V}, \mathcal{E}, \mathcal{K})$, composed of a set of nodes $\mathcal{V}$, a set of links $\mathcal{E}$, and a set of information objects $\mathcal{K}$, as depicted in Fig. 16.3. A content file can be stored (permanently or temporarily) over the nodes of this graph or travel through its edges. Some content resides permanently over some repository nodes (e.g., the disks in Fig. 16.3). In all other nodes (e.g., the circles in Fig. 16.3), content may appear and disappear, according to user requests and network resource allocation. We suppose, for simplicity, that all content is subdivided into objects of equal size. Each object is then identified by an index $k \in \mathcal{K}$. Each node is characterized by a storage capability and every edge is characterized by a transport capacity. Time is considered slotted and every slot has a fixed duration $\Delta\tau$. At time slot n, each node $u \in \mathcal{V}$ hosts, as a repository, a set of information objects $K_u[n] \in \mathcal{K}$ and requests, as a consumer, a set of information objects $Q_u[n] \in \mathcal{K}$. Let $\mathbf{q}[n] \in \{0, 1\}^{|\mathcal{V}||\mathcal{K}|}$ be the request arrival process such that $q_u[k, n] = 1$ if node u requests object k at time n, and $q_u[k, n] = 0$ otherwise.

Given this graph, we define a vertex signal over its nodes and an edge signal over its edges. The vertex signal $s_u[k, n]$ is a binary signal defined as:

$$s_u[k, n] = \begin{cases} 1, & \text{if content } k, \text{ at time } n, \text{ is stored on node } u \\ 0, & \text{otherwise} \end{cases}, u \in \mathcal{V}.$$

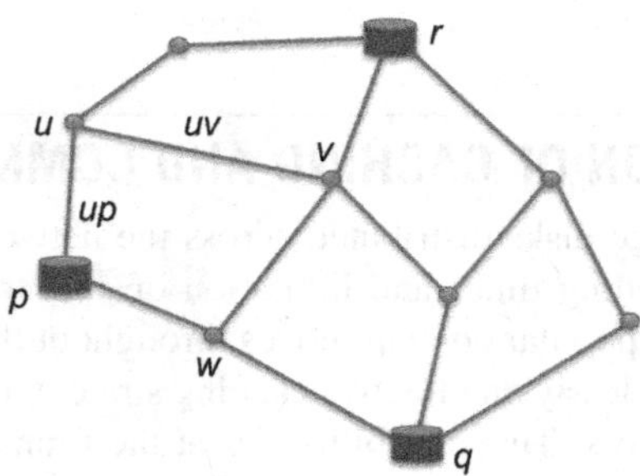

FIG. 16.3

Information network.

The amount of content stored on node u, at time n, is then $S_u[n] := \sum_k s_u[k,n]$. Similarly, we can define an edge signal as a binary signal, defined on each edge, as

$$t_{uv}[k,n] = \begin{cases} 1, & \text{if content } k, \text{ at time } n, \text{ is transported over link } uv \\ 0, & \text{otherwise} \end{cases}, uv \in \mathcal{E}.$$

The amount of content transported over link uv at time n, is then $T_{uv}[n] := \sum_k t_{uv}[k,n]$. Typically, each content may be hosted on every node and moved whenever useful. The storage and capacity constraints limit the variability of both $S_u[n]$ and $T_{uv}[n]$ as

$$0 \le S_u[n] \le S_u, \quad 0 \le T_{uv}[n] \le T_{uv}, \tag{16.11}$$

where S_u is the storage capability of node u, whereas T_{uv} is the transport capacity of link uv. The state of the network, at time slot n, is represented by the vector $\mathbf{x}[n] := [\mathbf{s}[n]; \mathbf{t}[n]]$, with $\mathbf{s}[n] := (s_u[k,n])_{\forall u,k}$ and $\mathbf{t}[n] := (t_{uv}[k,n])_{\forall k, uv \in \mathcal{E}}$.

In principle, a content $k \in \mathcal{K}$ may be cached, at any time slot n, in more then one location. However, there is a cost in keeping content in one place, if it is not utilized. The goal of *dynamic caching* is to find the state vector $\mathbf{x}[n]$ that minimizes an overall cost function that includes the cost for caching and the cost for transportation, under constraints dictated by the storage capability, the transport capacity, and the users' requirements in terms of latency to get access to their desired content.

The fundamental difference between *caching* and *storage* is that storage is intrinsically *static* whereas caching is fundamentally *dynamic*. This means that cached content moves throughout the network, appears in some nodes, and disappears from others. There are only some *repository* nodes (e.g., nodes p, q, and r in Fig. 16.3) that keep a permanent record or have fast access to a content delivery network. The assumption is that each content is hosted in at least one repository node.

The basic question about caching is then to decide, dynamically, depending on the users' requests, when and where to place all content, how to move it, and when to drop content to save memory. The decision for caching an object k at node u at time slot n must result from a trade-off between the cost for storing for a certain amount of time and the cost for transporting the content from its current location to the network access point nearest to the user who requested it.

The cost associated with storing content k on node u during T consecutive time slots in the time window $[n' - T + 1, n']$, is

$$E_{st} = \sum_{n=n'-T+1}^{n'} \sum_{k \in \mathcal{K}} \sum_{u \in \mathcal{V}} s_u[k,n] c_u[k], \tag{16.12}$$

where $c_u[k]$ is the energy cost for keeping content k on node u per unit of time. This unit time cost depends on the popularity of content k in a neighborhood of node u. For instance, we can set

$$c_u[k] = \frac{c_0}{1 + P_u[k]/P_0}, \tag{16.13}$$

where $P_u[k]$ is the popularity of content k at node u and c_0 is the (energy) cost for keeping a content object with zero popularity and P_0 is the popularity level that justifies halving the cost for caching per unit of time, with respect to zero-popularity content. The introduction of the cost coefficients $c_u[k]$ is what makes the formulation *context-aware*. In fact, the popularity $P_u[k]$ may vary across the network.

A fundamental issue in proactive caching is to select the repository nodes where to store the content objects as a function of their popularity. To make this choice *proactive* we associate a probabilistic measure of centrality to each node u, defined as

$$w_u[k] = \frac{\sum_{uv \in \mathcal{E}} B_{uv} P_v[k]}{\sum_{v \in \mathcal{V}} P_v[k]}$$

where B_{uv} is the length of the shortest path between nodes u and v and $P_v[k]$ is the probability that object k is requested by node v. We can store each content object k in the node u where $w_u[k]$ takes its minimum value, i.e. in the node having the average minimum number of hops from the nodes requiring content k. This selection defines the set of information objects $\mathcal{K}_u$ stored at each repository node u and it remains unaltered as long as the content popularity remains approximately constant.

The cost associated with content transportation is

$$E_{tr} = \sum_{n=n'-T+1}^{n'} \sum_{k \in \mathcal{K}} \sum_{uv \in \mathcal{E}} t_{uv}[k,n] c_{uv}[k], \tag{16.14}$$

where $c_{uv}[k]$ is the energy cost for transporting object k over link uv. In general, when user u makes a request of content k, we may associate to that request a maximum delivery time, which we call $D_u[k]$. We also denote by $\mathcal{N}_u$ the neighborhood of node u, i.e., the set of nodes that are one hop away from node u, and by $\mathbf{x}_T := [\mathbf{x}[n'-T+1]; \ldots; \mathbf{x}[n']]$ the state vector during T consecutive time slots.

The dynamic caching optimization problem can then be formulated as

$$\hat{\mathbf{x}}_T = \arg\min_{\mathbf{x}_T} (E_{st}(\mathbf{x}_T) + E_{tr}(\mathbf{x}_T)) \tag{16.15}$$

subject to the following constraints

$$\begin{aligned}
&(a) \quad q_u[k,n] \le s_u[k,n] + \sum_{v \in \mathcal{N}_u} \sum_{j=0}^{D_u[k]} t_{vu}[k, n+j], \\
&(b) \quad s_u[k,n] \le s_u[k,n-1] + \sum_{v \in \mathcal{N}_u} t_{vu}[k, n-1], \\
&(c) \quad t_{vu}[k,n] \le s_v[k,n-1] + \sum_{w \in \mathcal{N}_v} t_{wv}[k, n-1], \\
&(d) \quad s_u[k,n] = 1, \ \forall k \in K_u[n], \ s_u[k,0] = 0, k \notin K_u[n], \\
&(e) \quad S_u[n] \le S_u, \\
&(f) \quad T_{uv}[n] \le T_{uv}, \\
&(g) \quad s_u[k,n] \in \{0,1\}, \ t_{uv}[k,n] \in \{0,1\},
\end{aligned} \tag{16.16}$$

$\forall u \in \mathcal{V}, vu \in \mathcal{E}, k \in \mathcal{K}, n \in [n'-T+1, n']$.

The above constraints reflect the storage and flow constraints [33]:

(a) ensures that if object k is requested by node u at time slot n, then k either is in the cache of node u at time n or needs to be received by node u from a neighbor node $v \in \mathcal{N}_u$ within $D_u[k]$ time slots;

(b) assures that if k is being cached at node u at time n, then k either was in the cache of u at time $n-1$ or was received by node u from a neighbor node $v \in \mathcal{N}_u$ at time $n-1$;

(c) assures that if object k is received by node u from a neighbor node $v \in \mathcal{N}_u$ at time n, then k either was in the cache of v at time $n-1$ or was received by node v from a neighbor node $w \in \mathcal{N}_v$ at time $n-1$;
(d) describes the initial condition constraints that assure that each node u always stores the objects that it hosts as a repository, $\mathcal{K}_u[n]$, and at $n = 0$ nothing else;
(e) and (f) define the storage and transport capacity constraints;
(g) states the binary nature of the network configuration (storage and transport) variables.

To simplify the solution of the above problem, we let the entries of vector $\mathbf{x}_T$ be real variables in [0, 1]. A numerical example resulting from our relaxed formulation is shown in Fig. 16.4 where we illustrate the optimal transport energy versus the arrival request rate. We consider a network composed of $|\mathcal{V}| = 10$ nodes and $|\mathcal{K}| = 4$ information objects to be transported, by setting $T = 25$, $\Delta\tau = 1$s, $T_{uv} = 2$Mb, and $S_u = 4$. We considered, for simplicity, no knowledge of popularity and the same transportation costs over all links. To better evaluate the effect of the transport energy, we neglected the storage energy E_{st} term in the integer linear program (ILP) (16.15), by assuming that only three repository nodes store the information objects for all time. As a benchmark method, we consider the shortest path algorithm, which at each request forwards the desired content along the shortest path. It can be noted that the relaxed ILP method yields a considerable performance gain with respect to the shortest path algorithm; moreover, the improvement grows as the maximum delivery time $D_u[k]$ (set equal for each k) increases, due to the greater degrees of freedom of the algorithm.

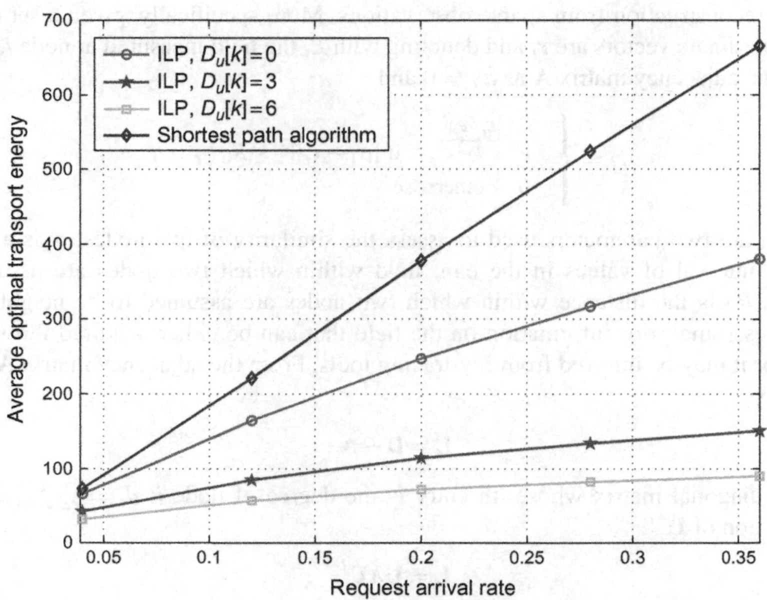

FIG. 16.4

Average optimal transport energy versus the request arrival rate.

16.5 GRAPH-BASED RESOURCE ALLOCATION

Enabling proactive resource allocation strategies is a key feature of 5G networks. Proactivity is rooted in the ability to predict users' behavior. Proactive caching is one example where the prediction is based on learning the popularity matrix. But of course caching is not the only network aspect that can benefit from learning. Radio coverage is one more case where learning maps of the radio environment may be useful to ensure seamless connectivity to moving users, possibly keeping the smallest number of access points active to save energy. This requires prediction of users' mobility and the capability to build radio environment maps (REM) [35]. Building a REM is also a key step to enable cognitive radio [35–37]. Balancing data traffic across the network is another problem that could take advantage of the capability to predict data flows exploiting spatiotemporal correlation (low-rank) [38,39].

16.5.1 RADIO ENVIRONMENT MAP

In this section, we show how graph-based representations can be useful to build a REM from sporadic measurements. Graph-based representations play a key role in many machine learning techniques as a way to formally take into account all similarities among the entities of an interconnected system. In the signal processing community, there is a growing interest in methods for processing signals defined over a graph, or graph signal processing (GSP), for short [40]. We show now an application of GSP to recovering the REM in an urban environment from sporadic measurements collected by mobile devices. The goal is to reconstruct the field over an ideal grid, built according to the city map, starting from observations taken over a subset of nodes. We use a graph-based approach to identify patterns useful for the ensuing reconstruction from sparse observations. More specifically, given a set of N points in space, whose coordinate vectors are $\mathbf{r}_i$ and denoting with E_i the field measured at node i, we define the coefficients of the adjacency matrix $\mathbf{A}$ as $a_{ii} = 0$ and

$$a_{i,j} = \begin{cases} e^{-\frac{|E_i - E_j|^2}{2\sigma^2}}, & \text{if } ||\mathbf{r}_i - \mathbf{r}_j||^2 \leq R_0 \ , i \neq j, \\ 0, & \text{otherwise} \end{cases}$$

where σ and R_0 are two parameters used to assess the similarity of two nodes: σ is a variable used to establish the interval of values in the e.m. field within which two nodes are assumed to sense a *similar* value; R_0 is the distance within which two nodes are assumed to be neighbors. Building matrix $\mathbf{A}$ requires some prior information on the field that can be either acquired through time from measurements or it may be inferred from ray-tracing tools. From the adjacency matrix $\mathbf{A}$, we build the Laplacian matrix

$$\mathbf{L} = \mathbf{D} - \mathbf{A}, \tag{16.17}$$

where $\mathbf{D}$ is the diagonal matrix whose ith entry is the degree of node i: $d_i = \sum_{j=1}^{N} a_{ij}$. Taking the eigendecomposition of $\mathbf{L}$

$$\mathbf{L} = \mathbf{U}\mathbf{\Lambda}\mathbf{U}^T \tag{16.18}$$

we have a way to identify the principal components of the field. It is well known from spectral graph theory [41], in fact, that the eigenvectors associated with the smallest eigenvalues of $\mathbf{L}$ identify clusters, i.e., well-connected components. Hence, the eigenvectors associated with the smallest eigenvalues of the Laplacian matrix built according to the above method are useful to identify patterns in the e.m. field. Denoting with $\mathbf{u}_k$ the eigenvector associated with the kth eigenvalue, the useful signal $\mathbf{x}$ can then be modeled as the superposition of the K principal eigenvectors:

$$\mathbf{x} = \sum_{k=1}^{K} \mathbf{u}_k \, s_k := \mathbf{U}_K \, \mathbf{s} \tag{16.19}$$

with $K < N$ to be determined from measurements and $\mathbf{U}_K := [\mathbf{u}_1, \ldots, \mathbf{u}_K]$.

In the GSP literature, a signal as in Eq. (16.19), with $K < N$, is called a *band-limited* signal over the graph. In general, a real signal is never perfectly band-limited, but it can be approximately band-limited. Having a band-limited model is instrumental to establish the condition for the recovery of the entire signal from a subset of samples [42].

In a real situation, it is typical to have several access points whose radio coverage areas overlap. For each access point, we can build a dictionary using the method described above, using for the e.m. field a ray-tracing algorithm. We denote by $\mathbf{U}_K^{(m)}$ the dictionary built when only AP m is active. At any given time frame, only a few APs are active. Therefore, the overall map can be written as

$$\mathbf{x} = \sum_{m=1}^{M} \sum_{k=1}^{K} \mathbf{u}_k s_k^{(m)} := \sum_{m=1}^{M} \mathbf{U}_K^{(m)} \mathbf{s}^{(m)} := \mathbf{U}\mathbf{s}, \tag{16.20}$$

where M is the number of APs covering the area of interest (not all of them necessarily active at the same time), $\mathbf{U} := (\mathbf{U}_K^{(1)}, \ldots, \mathbf{U}_K^{(M)})$ and $\mathbf{s} := (\mathbf{s}^{(1)}; \ldots; \mathbf{s}^{(M)})$ is sparse. The observed signal typically consists in a limited number of measurements collected along the grid. We may write the observed signal as:

$$\mathbf{y} = \boldsymbol{\Sigma} \sum_{m=1}^{M} \mathbf{U}_K^{(m)} \mathbf{s}^{(m)} = \boldsymbol{\Sigma}\mathbf{U}\mathbf{s}, \tag{16.21}$$

where $\boldsymbol{\Sigma}$ is a diagonal selection matrix whose ith entry is one if node i is observed and zero otherwise. The recovery of the overall radio coverage map can then be formulated as a sparse recovery problem. We used Basis Pursuit (BP), which implies solving the following convex problem:

$$\begin{aligned} \hat{\mathbf{s}} &= \arg\min_{\mathbf{s}} \|\mathbf{s}\|_1 \\ \text{s.t.} \quad & \mathbf{y} = \boldsymbol{\Sigma}\mathbf{U}\mathbf{s} \end{aligned} \tag{16.22}$$

and then we used $\hat{\mathbf{x}} = \mathbf{U}\,\hat{\mathbf{s}}$.

An example of reconstruction using BP is shown in Fig. 16.5. The grid is composed of $N = 547$ nodes and the number M of AP's covering the city area illustrated in the figure is four. The APs are located in the southeast, northeast, northwest, and southwest side of the examined area. The number of measurements is 115. Measurement noise is considered negligible. We assumed a bandwidth $K = 40$, equal for all APs. The background (continuous) color is the map ground-truth, obtained using the ray-tracing tool Remcom Wireless InSite 2.6.3 [43]. The colors on each vertex of the grid represent the reconstructed value. Comparing each node color with the background, we can testify to the goodness of the method to reconstruct the overall map. The normalized mean square error (NMSE), measured as the square norm of the error normalized by the square norm of the true signal, in this example is

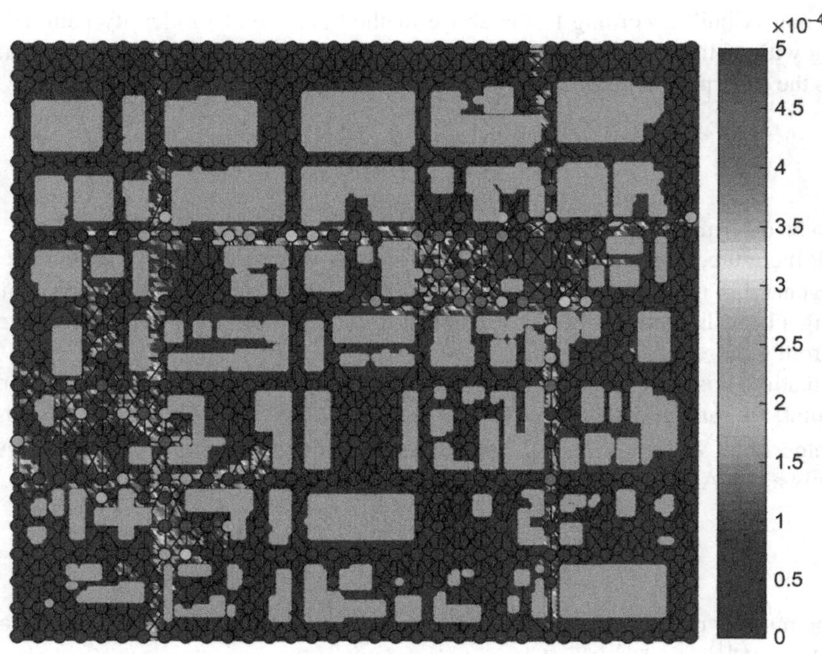

FIG. 16.5

Example of reconstructed e.m. field.

NMSE = 0.018. The quality of the reconstruction depends on the number of measurements and on the assumption on the bandwidth. Clearly, the larger the bandwidth, the better the reconstruction, but the larger is also the number of measurements to be taken to enable the reconstruction. This suggests that the choice of the bandwidth must come from a trade-off between accuracy and complexity.

16.5.2 MATCHING USERS TO 3*C* RESOURCES

In Section 16.3 we motivated the use of a joint allocation of computation and communication resources in computation offloading. We also incorporated the assignment rule between UE, AP, and MEC within the overall optimization problem. The resulting formulation yields better performance than a disjoint formulation; however, it is also computationally demanding because it involves the solution of a mixed-integer programming problem.

A possible way to overcome this difficulty is to simplify the rule for associating UEs to AP and MEC. One possibility is to resort to *matching theory*, a low-complexity tool used to solve the combinatorial problem of matching players from different sets, based on their preferences. Matching theory can be seen as the problem of finding a bipartite graph connecting two sets, depending on the preference lists. Matching theory has already been proposed in [44] for resource allocation in multitiered wireless heterogeneous architectures, with applications to cognitive radio networks, heterogeneous small-cell-based networks, and device-to-device communications (D2D). In [45], a multistage matching game is used in the C-RAN context to assign radio remote heads (RRH), base band units (BBU), and computing resources for computation offloading, aimed at minimizing the refusal ratio, i.e., the proportion of offloading tasks that is not able to meet the deadlines. A well-known matching problem is the college admission game presented in [46], where a deferred-acceptance (DA) algorithm is proved to converge to a stable matching with extremely low complexity. The key initial step of matching theory is to establish a preference rule. For instance, in [47] the users' preferences are defined as the R-factor, which captures both Packet Success Rate (PSR) and wireless delay. However, as pointed out in [47], the complexity of this algorithm increases considerably when dealing with interdependent preferences, i.e., when the preference of a user is affected by the acceptance of the others. This is indeed the case of user association in wireless networks because, continuing in the example defined above, the R factor of a user changes as other users get accepted by the same AP. To overcome this problem, the authors of [47] divide the game into two interdependent subgames:

1. An admission matching game with R-factor guarantees, depending on the maximum delay experienced at each access point;
2. A coalitional game to transfer users, where a coalition is the set of users associated to a certain AP.

In particular, a user assigned to a certain AP a through the first subgame could prefer to be matched to another AP b because the utility functions change as users get admitted. Then, a user k requests to be transferred from a to b if it improves its R-factor. The transfer is accepted if and only if:

1. The access point b does not exceed its quota (maximum number of admitted users);
2. The social welfare (sum of the R-factors of the two coalitions) is increased.

Starting from an initial partition (sets of coalitions) obtained with the deferred acceptance algorithm, the algorithm in [47] converges to a final partition that is also Nash-stable. In the holistic view of 3C resources, other utility functions can be used to take into account all three aspects of 3C: communication, computation, and caching. For instance, additional parameters to be taken into account are the computational load on MEC servers in case of computation offloading and the amount of storage for caching.

One more example where graph theory can be used is load balancing. In fact, especially in view of the dense deployment of access points, there is a high probability that the load, either data rate, computational load, or storage, can be highly unbalanced throughout the network [48]. One possibility to balance the situation is to split the networks in many nonoverlapping clusters. A cluster head is then elected in each cluster and it enforces a balance within the cluster. Then, balancing across clusters is achieved by repeated clustering and balancing steps. A possible way to do clustering is to use spectral clustering, which starts from the creation of a similarity (adjacency) matrix. In this case, as suggested in [49], it could be useful to include in the construction of the adjacency matrix a *dissimilarity* measure that assesses how much two nodes are unbalanced. In this way, the ensuing clustering tends to put together nodes that are close but unbalanced so that the resulting in-cluster balancing will be more effective.

16.6 NETWORK RELIABILITY

The edge-cloud architecture described in Section 16.2 clearly builds on the reliability of the network connectivity. However, in practice, the presence of a link between a pair of nodes is subject to random changes. In a wireless communication system, for instance, it is typical to have random link failures due to fading. With mmWave communications, link failures are typically even more pronounced because of blocking due to obstacles between transmitting and receiving devices. The goal of this section is to build on graph-based representations to assess the effect of random failure on a limited number of edge on macroscopic network parameters, such as, for example, connectivity. We build our study on a small perturbation analysis of the eigendecomposition of the Laplacian matrix describing the graph, as suggested in [50]. An outcome of our analysis is the identification of the most critical links, i.e., those links whose failure has a major effect on some network macroscopic features such as connectivity.

A small perturbation analysis of the eigendecomposition of a matrix is a classical problem that has been studied for a long time, see, e.g., [51,52]. In this section we focus on the small perturbation analysis of the eigendecomposition of a perturbed Laplacian $\boldsymbol{L}+\delta\boldsymbol{L}$, incorporating an original graph Laplacian $\boldsymbol{L}$ plus the addition or deletion of a small percentage of edges. We consider a graph composed of N vertices, so that the dimension of $\boldsymbol{L}$ is $N\times N$. We denote by $\tilde{\lambda}_i=\lambda_i+\Delta\lambda_i$ the perturbed ith eigenvalue and by $\tilde{\boldsymbol{u}}_i=\boldsymbol{u}_i+\Delta\boldsymbol{u}_i$ the associated perturbed eigenvector. If only one link fails, let us say link m, the perturbation matrix can be written as $\delta\boldsymbol{L}(m)=-\boldsymbol{a}_m\boldsymbol{a}_m^T$, where $\boldsymbol{a}_m=[a_{m_1}\cdots a_{m_n}]^T$ is a column vector of size N that has all entries equal to zero, except the two elements $a_m(i_m)=1$ and $a_m(f_m)=-1$, where i_m and f_m are the initial and final vertices of the failing edge m. In case of the addition of a new edge, the perturbation matrix is simply the opposite of the previous expression, i.e., $\delta\boldsymbol{L}(m)=\boldsymbol{a}_m\boldsymbol{a}_m^T$. It is straightforward to see that the perturbation of the Laplacian matrix due to the simultaneous deletion of a small set of edges is simply $\delta\boldsymbol{L}=-\sum_{m\in\mathcal{E}_p}\boldsymbol{a}_m\boldsymbol{a}_m^T$ where $\mathcal{E}_p$ denotes the set of perturbed edges. The perturbed eigenvalues and eigenvectors $\tilde{\lambda}_i$ and $\tilde{\boldsymbol{u}}_i$, in the case where all eigenvalues are distinct and the perturbation affects a few percentage of links, are related to the unperturbed values λ_i and $\boldsymbol{u}_i$ by the following formulas [51]:

$$\tilde{\lambda}_i\simeq\lambda_i+\boldsymbol{u}_i^T\delta\boldsymbol{L}\,\boldsymbol{u}_i, \tag{16.23}$$

$$\tilde{\boldsymbol{u}}_i\simeq\boldsymbol{u}_i+\sum_{j\neq i}\frac{\boldsymbol{u}_j^T\delta\boldsymbol{L}\,\boldsymbol{u}_i}{\lambda_i-\lambda_j}\boldsymbol{u}_j. \tag{16.24}$$

In particular, the perturbations due to the failure of a generic link m on the ith eigenvalue and associated eigenvector are:

$$\begin{aligned}\Delta\lambda_i(m)&=\boldsymbol{u}_i^T\delta\boldsymbol{L}(m)\boldsymbol{u}_i=-\boldsymbol{u}_i^T\boldsymbol{a}_m\boldsymbol{a}_m^T\boldsymbol{u}_i\\&=-||\boldsymbol{a}_m^T\boldsymbol{u}_i||^2=-[u_i(f_m)-u_i(i_m)]^2\end{aligned} \tag{16.25}$$

and

$$\begin{aligned}\Delta\boldsymbol{u}_i(m)&=\sum_{j\neq i}\frac{\boldsymbol{u}_j^T\delta\boldsymbol{L}(m)\boldsymbol{u}_i}{\lambda_i-\lambda_j}\boldsymbol{u}_j=-\sum_{j\neq i}\frac{\boldsymbol{u}_j^T\boldsymbol{a}_m\boldsymbol{a}_m^T\boldsymbol{u}_i}{\lambda_i-\lambda_j}\boldsymbol{u}_j\\&=\sum_{j\neq i}\frac{[u_j(i_m)-u_j(f_m)][u_i(f_m)-u_i(i_m)]}{\lambda_i-\lambda_j}\boldsymbol{u}_j.\end{aligned} \tag{16.26}$$

Within the limits of validity of first order perturbation analysis, the overall perturbation resulting from the deletion of multiple edges is the sum of all the perturbations occurring on single edges:

$$\Delta\lambda_i = \sum_{m \in \mathcal{E}_p} \Delta\lambda_i(m), \tag{16.27}$$

where $\mathcal{E}_p$ denotes the set of perturbed edges. In their simplicity, the above formulas capture some of the most relevant aspects of perturbation and their relation to graph topology. In fact, it is known from spectral graph theory, see e.g., [41], that the entries of the Laplacian eigenvectors associated with the smallest eigenvalues tend to be smooth and assume the same sign over vertices within a cluster while they can vary arbitrarily across different clusters. Taking into account these properties, the above perturbation formulas (16.23)–(16.26) give rise to the following interpretations:

1. the edges whose deletion causes the largest perturbation are intercluster edges;
2. given a connected graph, the eigenvector associated with the null eigenvalue does not induce any perturbation on any other eigenvalue/eigenvector because it is constant;
3. the eigenvector perturbation is larger for quantities associated with eigenvalues very similar to each other (recall that formulas (16.23) and (16.24) hold true only for distinct eigenvalues).

16.6.1 A NEW MEASURE OF EDGE CENTRALITY

Based on the above derivations, we propose a new measure of edge centrality, which we call *perturbation centrality*. We assume a connected undirected graph. If we denote by K the number of clusters in the graph and by $\Delta\lambda_i(m)$ the perturbation of the ith eigenvalue due to the deletion of edge m, we define the topology perturbation centrality of edge m as follows [50]:

$$\mathrm{p}_K(m) := \sum_{i=2}^{K} |\Delta\lambda_i(m)|. \tag{16.28}$$

The summation starts from $i = 2$ simply because, from Eq. (16.23), the perturbation induced by the deletion of any edge on the smallest eigenvalue is null. The above parameter $\mathrm{p}_K(m)$ assigns to each edge the perturbation that its deletion causes to the overall network connectivity, measured as the sum of the K smallest eigenvalues of the Laplacian matrix [41]. This parameter is particularly relevant in case of modular graphs, i.e., graphs evidencing the presence of clusters. In such a case, it is well known from spectral clustering theory [41] that the smallest eigenvalues of the Laplacian carry information about the number of clusters in a graph.

In Fig. 16.6 we report an example of a modular graph, obtained by connecting two clusters through a few edges. The *perturbation centrality* is encoded in the color intensity of each edge. It is interesting to see that the edges with the darkest color are, as expected, the ones connecting the two clusters.

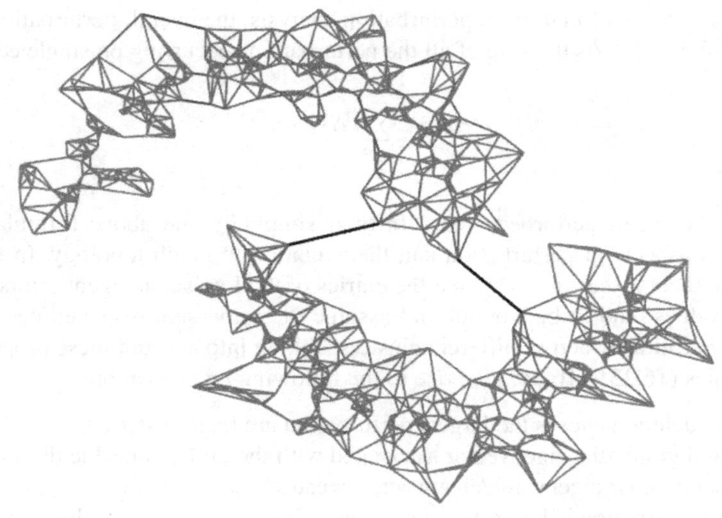

FIG. 16.6

Example of perturbation centrality measure.

16.6.2 APPLICATION: ROBUST INFORMATION TRANSMISSION OVER WIRELESS NETWORKS

Now we apply our statistical analysis to optimize the resource (power) allocation over a wireless network in order to make the network robust against random link failures. We consider a wireless communication network with M links, where each link is subject to a random failure because of fading or blocking. Every edge is characterized by an outage probability $P_{out}(m), m = 1, \ldots, M$. We suppose the failure events over different links to be independent of each other. We consider first a single-input-single-output (SISO) Rayleigh flat fading channel for each link. In such a case, the channel coefficient h is a complex Gaussian random variable (r.v.) with zero mean and circularly symmetric. Hence, the r.v. $\alpha = |h|^2$ has an exponential distribution. Denoting with $F_n(x;\lambda)$ the cumulative distribution function (CDF) of a gamma random variable x of order n, with parameter λ, the CDF of α can then be written as $F_1(\alpha;\lambda)$. We also denote with $C = \log_2(1 + |h|^2\rho)$ the link capacity (in bits/sec/Hz), where $\rho = \frac{P_T(m)}{\sigma_n^2 r_m^2}$ is the signal-to-noise ratio (SNR), $P_T(m)$ is the transmitted power over the mth link, σ_n^2 is the noise variance, and r_m the distance covered by link m. Denoting by R the data rate, the outage probability $P_{out}(m)$ is defined as:

$$\begin{aligned} P_{out}(m) &= Pr\{C < R\} = Pr\{\log_2(1 + |h|^2\rho) < R\} \qquad (16.29)\\ &= Pr\left\{|h|^2 < \frac{2^R - 1}{\rho}\right\} \\ &= \int_0^{\frac{2^R-1}{\rho}} \lambda e^{-\lambda\alpha} d\alpha = F_1\left(\frac{2^R-1}{\rho};\lambda\right) = 1 - e^{-\frac{\lambda}{\rho}(2^R-1)}. \end{aligned}$$

Because the CDF of α is invertible, it is useful to introduce its inverse. In particular, if $y = F_n(x;\lambda)$, we denote its inverse as $x = F_n^{-1}(y;\lambda)$. Expression (16.29) can then be inverted to derive the transmit power $P_T(m)$ as a function of the outage probability:

$$P_T(m) = \frac{\lambda\sigma_n^2 r_m^2(2^R - 1)}{\log(1 - P_{out}(m))} = \frac{\sigma_n^2 r_m^2(2^R - 1)}{F_1^{-1}(P_{out}(m);\lambda)}. \tag{16.30}$$

The small perturbation statistical analysis derived above can be used to formulate a robust network optimization problem. We assess the network robustness, in terms of connectivity, as the ability of the network to give rise to small changes of connectivity as a consequence of a small number of edge failures. The network connectivity is measured by the second smallest eigenvalue of the Laplacian, also known as the graph *algebraic connectivity*. This parameter is known to provide a bound for the graph conductance [53]. Our goal now is to evaluate the transmit powers $P_T(m)$, or equivalently, through Eq. (16.30), the outage probabilities, that minimize the average perturbation of the algebraic connectivity, subject to a cost function on the total transmit power $P_{T_{\max}}$ of the overall network. In formulas, we wish to solve the following optimization problem:

$$\begin{aligned} \min_{\mathbf{P}_{out}} \quad & \sum_{m\in\mathcal{E}} \mathbb{E}\{|\Delta\lambda_2(m)|\} \\ \text{s.t.} \quad & \sum_{m\in\mathcal{E}} P_T(m) \le P_{T_{\max}} \\ & P_{out}(m) \in [0,1],\ \forall m \in \mathcal{E}. \end{aligned}$$

Using Eqs. (16.25) and (16.30), we can rewrite the optimization problem explicitly in terms of the outage probabilities $P_{out}(m)$ as:

$$\begin{aligned} \min_{\mathbf{P}_{out}} \quad & \sum_{m\in\mathcal{E}} P_{out}(m)[u_2(i_m) - u_2(f_m)]^2 \\ \text{s.t.} \quad & \sum_{m\in\mathcal{E}} \frac{r_m^2}{F_1^{-1}(P_{out}(m);\lambda)} \le C_{\max} \qquad (\mathcal{Q}) \\ & P_{out}(m) \in [0,1],\ \forall m \in \mathcal{E}, \end{aligned}$$

where $C_{\max} := \frac{P_{T_{\max}}}{\sigma_n^2(2^R-1)}$.

Problem $(\mathcal{Q})$ is nonconvex because the constraint set is not convex. However, if we perform the change of variable $t_m := 1/F_1^{-1}(P_{out}(m);\lambda) = -\lambda/\log(1 - P_{out}(m)), m = 1,\ldots,M$, the first constraint becomes linear. The objective function becomes nonconvex. However, if we limit the variability of the unknown variables to the set $t_m \ge \lambda/2$, $\forall m$, the objective function becomes convex so that the original problem converts into the following convex problem:

$$\begin{aligned} \min_{\mathbf{t}} \quad & \sum_{m\in\mathcal{E}} F_1\left(\tfrac{1}{t_m};\lambda\right)|\Delta\lambda_2(m)| = \sum_{m\in\mathcal{E}} \left(1 - e^{-\frac{\lambda}{t_m}}\right)|\Delta\lambda_2(m)| \\ \text{s.t.} \quad & \sum_{m\in\mathcal{E}} r_m^2 t_m \le C_{\max} \qquad (\mathcal{Q}_1) \\ & t_m \ge \tfrac{\lambda}{2},\quad \forall m \in \mathcal{E}. \end{aligned}$$

We can now generalize the previous formulation to the multi-input multi-output (MIMO) case, assuming multiple independent Rayleigh fading channels. One fundamental property of MIMO systems is the diversity gain, which makes them more robust against fading with respect to SISO systems [54]. In fact, different performance, can be obtained depending on the number of antennas on the transmitting sides n_T and receiving sides n_R exploiting the diversity gain. In a MIMO system with $n = n_T \times n_R$ statistically independent channels, denoting by h_{ij} the coefficient between the ith transmit and the jth receive antenna, the pdf of the random variable $\alpha := \sum_{i=1}^{n_T} \sum_{j=1}^{n_R} |h_{ij}|^2$ is the Gamma distribution:

$$P_A(\alpha) = \frac{\lambda^n}{(n-1)!} \alpha^{n-1} e^{-\lambda\alpha} \tag{16.31}$$

and we denote by $F_n(\alpha; \lambda)$ its cumulative distribution function (CDF), with parameters n and λ. Proceeding similarly to the SISO case, the optimization problem can be formulated as

$$\begin{aligned} \min_{\mathbf{t}} \quad & \sum_{m \in \mathcal{E}} F_n(\tfrac{1}{t_m}; \lambda) |\Delta\lambda_2(m)| \\ \text{s.t.} \quad & \sum_{m \in \mathcal{E}} r_m^2 t_m \leq C_{\max} \qquad\qquad (\mathcal{Q}_2) \\ & t_m \geq \lambda/(n+1), \quad \forall m \in \mathcal{E}, \end{aligned}$$

where the constraint on the variables t_m has been introduced to make the problem convex. Indeed, problem $\mathcal{Q}_1$ is a special case of problem $\mathcal{Q}_2$, when $n = 1$. An interesting result about the convexity of problem $\mathcal{Q}_2$ is that the bounding region increases with the number of independent channels.

As a numerical example, we considered a connected network composed by two clusters, with a total of $|\mathcal{E}| = 1612$ edges and two bridge edges between the two clusters. For the sake of simplicity, we assumed the same distances r_m over all links. In Fig. 16.7, we compare the expected perturbations of the algebraic connectivity, normalized to the nominal value λ_2, obtained using our optimization

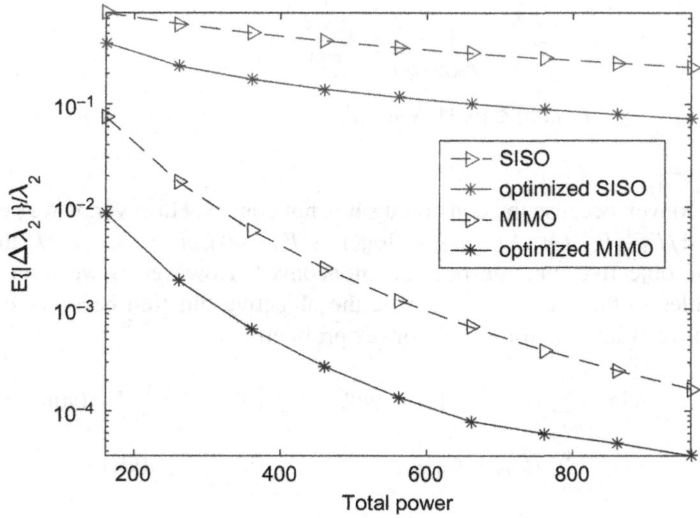

FIG. 16.7

Expected perturbation of algebraic connectivity versus total power.

procedure or using the same power over all links, assuming the same overall power consumption. We report the result for both SISO and MIMO cases. From Fig. 16.7, we can observe a significant gain in terms of the total power necessary to achieve the same expected perturbation of the network algebraic connectivity. We can also see the advantage of using MIMO communications, at least in the case of statistically independent links.

16.7 CONCLUSIONS

In this chapter we have described some of the aspects of the edge-cloud architecture, a framework proposed to bring cloud and communication resources as close as possible to mobile users to reduce latency and achieve a more efficient usage of the available energy. From the edge-cloud perspective, we have motivated a holistic view that aims at optimizing the allocation of communication, computation, and caching resources jointly. Within this framework, graph-based representations play a key role. In this chapter, we considered just a few cases where these representations can provide a valid and innovative tool for an efficient deployment of the edge-cloud system. As happens in most engineering problems, big potentials come with big challenges. One of these is complexity. To take full advantage of graph representations, there is the need for devising efficient distributed computational tools to analyze graph-based signals. Furthermore, we believe that graph representations are only the beginning of the story, as they are built incorporating only pairwise relations. More sophisticated tools may be envisaged by enlarging the horizon to include multiway relations, using, for example, simplicial complexes or hypergraphs, as suggested in [55], or multilayer network representations [56,57]. Furthermore, in this work, we have basically restricted our attention to time-invariant graph representations and to linear models. Clearly, a significant improvement can be expected by enlarging the view to time-varying graphs and nonlinear models [58,59].

ACKNOWLEDGMENTS

The research leading to these results has been jointly funded by the European Commission (EC) H2020 and the Ministry of Internal affairs and Communications (MIC) in Japan under grant agreements Nr. 723171 5G MiEdge in EC and 0159-0149, 0150, 0151 in MIC.

REFERENCES

[1] 5G empowering vertical industries. 5G PPP White Paper; 2016.
[2] Andrews JG, Buzzi S, Choi W, Hanly SV, Lozano A, Soong AC, et al. What will 5G be? IEEE J Sel Areas Commun 2014;32(6):1065–82.
[3] Mijumbi R, Serrat J, Gorricho JL, Bouten N, De Turck F, Boutaba R. Network function virtualization: state-of-the-art and research challenges. IEEE Commun Surv Tutor 2016;18(1):236–62.
[4] Rost P, Mannweiler C, Michalopoulos DS, Sartori C, Sciancalepore V, Sastry N, et al. Network slicing to enable scalability and flexibility in 5G mobile networks. IEEE Commun Mag 2017;55(5):72–9.
[5] Vassilaras S, Gkatzikis L, Liakopoulos N, Stiakogiannakis IN, Qi M, Shi L, et al. The algorithmic aspects of network slicing. IEEE Commun Mag 2017;55(8):112–9.

[6] Heath RW, Gonzalez-Prelcic N, Rangan S, Roh W, Sayeed AM. An overview of signal processing techniques for millimeter wave MIMO systems. IEEE J Sel Topics Signal Process 2016;10(3):436–53.

[7] Xiao M, Mumtaz S, Huang Y, Dai L, Li Y, Matthaiou M, et al. Millimeter wave communications for future mobile networks. IEEE J Sel Areas Commun 2017;35(9):1909–35.

[8] Sakaguchi K, Haustein T, Barbarossa S, Calvanese-Strinati E, Clemente A, Destino G, et al. Where, when, and how mmWave is used in 5G and beyond. IEICE Trans Electron 2017;E100-C(10):790–808.

[9] Taleb T, Samdanis K, Mada B, Flinck H, Dutta S, Sabella D. On multi-access edge computing: a survey of the emerging 5G network edge architecture & orchestration. IEEE Commun Surv Tutor 2017;19(3):1657–81.

[10] Hu YC, Patel M, Sabella D, Sprecher N, Young V. Mobile edge computing: a key technology towards 5G. ETSI White Paper 11; 2015.

[11] Wang S, Zhang X, Zhang Y, Wang L, Yang J, Wang W. A survey on mobile edge networks: convergence of computing, caching and communications. IEEE Access 2017;5:6757–79.

[12] Bonomi F, Milito R, Natarajan P, Zhu J. Fog computing: a platform for Internet of things and analytics. In: Big data and internet of things: a roadmap for smart environments. Springer; 2014. p. 169–86.

[13] Zhang N, Liu YF, Farmanbar H, Chang TH, Hong M, Luo ZQ. Network slicing for service-oriented networks under resource constraints. IEEE J Sel Areas Commun 2017;35(11):2512–21.

[14] Barbarossa S, Sardellitti S, Di Lorenzo P. Communicating while computing: distributed mobile cloud computing over 5G heterogeneous networks. IEEE Signal Process Mag 2014;31(6):45–55.

[15] Wang C, Liang C, Yu FR, Chen Q, Tang L. Computation offloading and resource allocation in wireless cellular networks with mobile edge computing. IEEE Trans Wireless Commun 2017;16(8):4924–38.

[16] Barbarossa S, Sardellitti S, Di Lorenzo P. Joint allocation of computation and communication resources in multiuser mobile cloud computing. In: IEEE workshop SPAWC; 2013. p. 26–30.

[17] Sardellitti S, Scutari G, Barbarossa S. Joint optimization of radio and computational resources for multicell mobile-edge computing. IEEE Trans Signal Inf Process Netw 2015;1(2):89–103.

[18] Pathak A, Hu YC, Zhang M. Where is the energy spent inside my app?: fine grained energy accounting on smartphones with Eprof. In: Proceedings of the 7th ACM European conference on computer systems. ACM; 2012. p. 29–42.

[19] Barbarossa S, Ceci E, Merluzzi M. Overbooking radio and computation resources in mmW-mobile edge computing to reduce vulnerability to channel intermittency. In: 2017 European conference on networks and communications (EuCNC); 2017. p. 1–5.

[20] 5G-MiEdge. Millimeter-wave edge cloud as an enabler for 5G ecosystem. Europe/Japan project co-funded by the European Commission's Horizon 2020 and Japanese Ministry of Internal Affairs and Communications. http://5g-miedge.eu.

[21] Andrews JG, Bai T, Kulkarni MN, Alkhateeb A, Gupta AK, Heath RW. Modeling and analyzing millimeter wave cellular systems. IEEE Trans Commun 2017;65(1):403–30.

[22] Barbarossa S, Ceci E, Merluzzi M, Calvanese-Strinati E. Enabling effective mobile edge computing using millimeterwave links. In: 2017 IEEE international conference on communications workshops (ICC workshops); 2017. p. 367–72.

[23] Mobile edge computing (MEC); end to end mobility aspects. ETSI GR MEC 018 V1.1.1; 2017.

[24] Li W, Kanso A. Comparing containers versus virtual machines for achieving high availability. In: 2015 IEEE international conference on cloud engineering (IC2E); 2015. p. 353–8.

[25] Sardellitti S, Barbarossa S, Scutari G. Distributed mobile cloud computing: joint optimization of radio and computational resources. In: 2014 IEEE Globecom workshops (GC Wkshps); 2014. p. 1505–10.

[26] Sardellitti S, Barbarossa S, Merluzzi M. Optimal association of mobile users to multi-access edge computing resources. IEEE Trans Signal Inf Process Netw; 2017 (submitted).

[27] Sardellitti S, Merluzzi M, Barbarossa S. Optimal association of mobile users to multi-access edge computing resources. In: Proceedings of IEEE international conference on communications (ICC), May 2018, Kansas City, USA.

[28] Scutari G, Facchinei F, Lampariello L. Parallel and distributed methods for constrained nonconvex optimization—Part I: Theory. IEEE Trans Signal Process 2017;65(8):1929–44.
[29] Baştuğ E, Bennis M, Zeydan E, Kader MA, Karatepe IA, Er AS, et al. Big data meets telcos: a proactive caching perspective. J Commun Netw 2015;17(6):549–57.
[30] Zeydan E, Bastug E, Bennis M, Kader MA, Karatepe IA, Er AS, et al. Big data caching for networking: moving from cloud to edge. IEEE Commun Mag 2016;54(9):36–42.
[31] Sadeghi A, Sheikholeslami F, Giannakis GB. Optimal and scalable caching for 5G using reinforcement learning of space-time popularities. Preprint arXiv:1708.06698; 2017.
[32] Jacobson V, Smetters DK, Thornton JD, Plass MF, Briggs NH, Braynard RL. Networking named content. In: Proceedings of the 5th international conference on emerging networking experiments and technologies. ACM; 2009. p. 1–12.
[33] Llorca J, Tulino AM, Guan K, Esteban J, Varvello M, Choi N, et al. Dynamic in-network caching for energy efficient content delivery. In: 2013 Proceedings of IEEE INFOCOM; 2013. p. 245–9.
[34] Krishnan P, Raz D, Shavitt Y. The cache location problem. IEEE/ACM Trans Netw 2000;8(5):568–82.
[35] Bazerque JA, Mateos G, Giannakis GB. Group-Lasso on splines for spectrum cartography. IEEE Trans Signal Process 2011;59(10):4648–63.
[36] Yilmaz HB, Tugcu T, Alagoz F, Bayhan S. Radio environment map as enabler for practical cognitive radio networks. IEEE Commun Mag 2013;51(12):162–9.
[37] Romero D, Kim SJ, Giannakis GB, López-Valcarce R. Learning power spectrum maps from quantized power measurements. IEEE Trans Signal Process 2017;65(10):2547–60.
[38] Mardani M, Giannakis GB. Robust network traffic estimation via sparsity and low rank. In: 2013 IEEE international conference on acoustics, speech, signal process (ICASSP); 2013. p. 4529–33.
[39] Xu J, Deng D, Demiryurek U, Shahabi C, van der Schaar M. Mining the situation: spatiotemporal traffic prediction with big data. IEEE J Sel Top Signal Process 2015;9(4):702–15.
[40] Shuman DI, Narang SK, Frossard P, Ortega A, Vandergheynst P. The emerging field of signal processing on graphs: extending high-dimensional data analysis to networks and other irregular domains. IEEE Signal Proc Mag 2013;30(3):83–98.
[41] Von Luxburg U. A tutorial on spectral clustering. Stat Comput 2007;17(4):395–416.
[42] Tsitsvero M, Barbarossa S, Di Lorenzo P. Signals on graphs: uncertainty principle and sampling. IEEE Trans Signal Process 2016;64(18):4845–60.
[43] https://www.remcom.com/.
[44] Gu Y, Saad W, Bennis M, Debbah M, Han Z. Matching theory for future wireless networks: fundamentals and applications. IEEE Commun Mag 2015;53(5):52–9.
[45] Li T, Magurawalage CS, Wang K, Xu K, Yang K, Wang H. On efficient offloading control in cloud radio access network with mobile edge computing. In: 2017 IEEE 37th international conference on distributed computing systems (ICDCS); 2017. p. 2258–63.
[46] Gale D, Shapley LS. College admissions and the stability of marriage. Am Math Mon 1962;69(1):9–15.
[47] Saad W, Han Z, Zheng R, Debbah M, Poor HV. A college admissions game for uplink user association in wireless small cell networks. In: IEEE INFOCOM 2014—IEEE international conference on computer communications; 2014. p. 1096–104.
[48] Vu TK, Bennis M, Samarakoon S, Debbah M, Latva-aho M. Joint load balancing and interference mitigation in 5G heterogeneous networks. IEEE Trans Wirel Commun 2017;16(9):6032–46.
[49] Samarakoon S, Bennis M, Saad W, Latva-Aho M. Dynamic clustering and sleep mode strategies for small cell networks. In: 2014 11th international symposium on wireless communication systems (ISWCS); 2014. p. 934–8.
[50] Ceci E, Barbarossa S. Small Perturbation Analysis of Network Topologies. In: 2018 IEEE International conference on acoustics, speech, signal process (ICASSP); 2018. p. 4194–98.
[51] Wilkinson JH. The algebraic eigenvalue problem. New York, NY: Oxford University Press, Inc.; 1988.

[52] Stewart G. Introduction to matrix computations. Computer science and applied mathematics. Academic Press; 1973.
[53] Newman M. Networks: an introduction. New York, NY: Oxford University Press, Inc.; 2010.
[54] Barbarossa S. Multiantenna wireless communication systems. Mobile communications series. Artech House; 2003.
[55] Barbarossa S, Tsitsvero M. An introduction to hypergraph signal processing. In: 2016 IEEE international conference on acoustics, speech, signal process (ICASSP); 2016. p. 6425–9.
[56] Kivelä M, Arenas A, Barthelemy M, Gleeson JP, Moreno Y, Porter MA. Multilayer networks. J Complex Netw 2014;2(3):203–71.
[57] Boccaletti S, Bianconi G, Criado R, Del Genio CI, Gómez-Gardenes J, Romance M, et al. The structure and dynamics of multilayer networks. Phys Rep 2014;544(1):1–122.
[58] Shen Y, Baingana B, Giannakis GB. Kernel-based structural equation models for topology identification of directed networks. IEEE Trans Signal Process 2017;65(10):2503–16.
[59] Romero D, Ioannidis VN, Giannakis GB. Kernel-based reconstruction of space-time functions on dynamic graphs. IEEE J Sel Top Signal Process 2017;11(6):856–69.

CHAPTER

APPLICATIONS OF GRAPH CONNECTIVITY TO NETWORK SECURITY

17

Ying Liu*, Wade Trappe*, Andrey Garnaev*
*Wireless Information Network Laboratory (WINLAB), Rutgers, The State University of New Jersey, North Brunswick, NJ, United States**

17.1 INTRODUCTION

Our communication infrastructure is a complex ecosystem of separate yet interconnected systems. It consists of a variety of networks, including the broader Internet, cellular networks, optical backhaul networks, and local area networks. It provides service to almost all aspects of our daily life and is likely to become even more essential to us as we move toward a future involving the Internet of Things (IoT) and vehicular systems. Unfortunately, as these networks become intermingled, ensuring their reliability in the presence of network threats will become increasingly more challenging. While cybersecurity technologies, i.e., technologies designed to prevent the proliferation of malicious code or to prevent unwanted processes from accessing certain parts of a computer system or network, are essential to protecting our use of the communication infrastructure, it is prudent to ensure that our networks are deployed in configurations that allow them to have resilience in spite of cyberattacks being present.

Network resilience is the ability of the network to withstand harm or to return to an acceptable operational condition after it has been harmed by external perturbations. From a network perspective, adversarial resilience happens through two basic mechanisms: properly setting up the connections between entities in a network to withstand threats and allow network protocols to have an avenue for redirecting communications when connections are broken; and through feedback mechanisms that call forth additional redundancy in the network's design to repurpose components to meet new challenges.

Communication networks are built with protocols that allow them to withstand a certain amount of network faults: links can fail due to natural reasons, and network routing protocols are designed to rediscover routes. Meanwhile, transport protocols, such as TCP and store-and-forward protocols from disruption tolerant networking, are designed with buffering and redelivery mechanisms that allow networks to ride out periods of disconnection. Although network protocols go a long way in allowing a network to withstand failures, they generally do not involve changing the network itself, and can fail in their purpose when there are no paths connecting a communication source and destination. Consequently, network management and security functions need to have the ability to

Cooperative and Graph Signal Processing. https://doi.org/10.1016/B978-0-12-813677-5.00017-1

summon additional resources in order to provide enough topological connections to allow network protocols to restore functionality. In general, a network with better connectivity has a stronger ability to recover from attacks.

This chapter will explore the notion of network resilience with a focus on how knowledge regarding a network's topological structure can be leveraged by adversaries to more effectively attack networks and, conversely, how a network's topological structure can be adapted to provide enhanced resilience to attacks against a network. We will begin the chapter with a brief review of graph theory, including the concepts of graph connectivity that will be the basis for most of the attack and defense mechanisms explored in this chapter. We will also explore attacks that are directed at puncturing a network's connectivity, and the impact that such an attack can have. Then, we will explore how one can improve the network connectivity and thereby provide the means for a network to withstand attacks.

17.2 ALGEBRAIC GRAPH THEORY OVERVIEW

Communication networks can be examined as a graph that evolves during the operation of the network, and thus graph theory is an ideal mathematical tool for examining communication networks. Graph theory deals with the structures, properties, and mathematical representation of a graph, and while the theory is built with just two simple components (i.e., edges and vertices), it can describe numerous real-world phenomena. In communication networks, nodes or vertices correspond to communicating entities, such as network routers, and the edges correspond to the existence of a communication link between these nodes.

Studying the relationships in a graph is important for revealing the network's properties, and the various interdependencies in the network determine the graph's local and global connectivity properties: the operation of nodes in one part of a network can depend on the function of nodes in other parts of a network. Due to the interdependence, if a network attack happens in one part of a network, the effects can propagate to other parts of a network. This could also disrupt a wider area than the locality near the attack and may ultimately cause the abrupt collapse of a whole communication network. In the remainder of the chapter, we will explore the mathematical notions associated with a network's resilience. The notation we use is provided in Table 17.1.

A starting point for studying the connectivity of the graph associated with a communication network is the branch of graph theory known as flow theory. Graph theory provides algorithms that can find an available path between a communication source and destination pair, such as Dijkstra, Bellman-Ford, Depth First Search (DFS), Minimum Spanning Tree, and Max Flow Min Cut. While these network algorithms are the graph theoretic basis for many network routing protocols, they do not examine the actual topology of the network and can fail in their purpose if the network's topology does not have sufficient connectivity to support a path between a source and destination. Instead, in order to declare that a network is robust to attacks or faults, one would like to answer the questions: (1) do all pairs of sources and destinations have connections? (2) how many links are needed for each source destination? and (3) which nodes are connected most and likely to be used for forming a path between an arbitrary source and destination? The answers to these questions can be quite complex: for example, it takes $O(VE + V^2 lgV)$ time to find the All-Pairs Shortest Paths (APSP) using Dijkstra's algorithm with Fibonacci heaps, where V and E are corresponding number of nodes and edges in a graph. This value becomes prohibitively large for a high-density network whose $E \approx \Theta(V^2)$, which under this condition becomes $O(V^4)$.

Table 17.1 Mathematical Notation Used in This Chapter

Notation	Description
G	a graph or network topology
V	vertex set
E	edge set
L	Laplacian matrix
λ^i	the ith eigenvalue of Laplacian matrix, L
λ^1	algebraic connectivity or Fiedler value
λ^1_j	node Fiedler value associated with removing node j
$G\backslash i$	the remaining graph of removing node i from a graph G
n	total number of nodes in a network
$\boldsymbol{P}$	a vector of transmission power, P_i, at node i
ϵ	receiving threshold
T_{ij}	throughput on a link from node i to receiver j
d	distance between two nodes
σ^2	natural noise
h	fading channel gain
w	weight on a link in graph, G
$\overline{P}$	total network power
Π	an allocation of $\overline{P}$
$\boldsymbol{I}$	an identity matrix
Γ_γ	stochastic game
γ	discounted factor to replay a game, Γ
$\bar{\lambda_i}$	instantaneous payoff of node Fiedler value
C_h	a jammer's hiding cost
α	probability of being detected in the hiding mode
δ	probability of continuing an attack
$\{\cdot\}^T$	transpose of a vector, $\{\cdot\}$
$\boldsymbol{p}$	a probability vector employed by a scanner, $\boldsymbol{p}^T = (p_1, p_2, \ldots, p_n)$
$\boldsymbol{q}$	a probability vector employed by a jammer, $\boldsymbol{q}^T = (q_1, q_2, \ldots, q_n)$
e	a vertical vector of all ones
$\mathbf{A}$	game matrix
V_γ	game value of Γ_γ
$\mathbb{E}(\cdot)$	expectation of $(\cdot)$
τ	jamming time (slots) spent before a jammer launches a successful attack
t	an attack aims at damaging throughput
c	an attack aims at damaging connectivity
$\mathbf{D}$	game matrix to compare jamming time

While traditional graph theory is quite useful for studying smaller networks, it is often desirable to study networks and their characteristics using specialized branches of graph theory that allow one to reduce the complexity of the problems being investigated. Algebraic graph theory is a branch of graph theory involving the analysis of the spectrum distribution of the Laplacian matrix by eigenvalue decomposition. Because a graph's Laplacian matrix is built from the graph's adjacency matrix, many topological properties of the network can be explored using the eigenvalues of the Laplacian matrix. This is advantageous because it takes $O(V^3)$ complexity to calculate the full eigenvalue decomposition of the Laplacian (or less if one is interested in a restricted set of eigenvalues).

Algebraic graph theory is useful in understanding the resilience of a network as the second smallest eigenvalue of the network's Laplacian matrix, which is known as the Fiedler value or algebraic connectivity, provides a measure of the network's connectivity. The larger the Fiedler value, the more connected the network, and thus the more opportunities for a network to bounce back or redirect its communications when facing a network attack. The Fiedler value and general analysis of a network's spectrum distribution can yield insight into the network's structure, and thereby serve as the basis for introducing additional resilience in a network by judiciously introducing extra edges (or, conversely, give an adversary a tool to identify weak points of a network where an attack can have maximal impact).

17.2.1 FIEDLER VALUE AND GRAPH CONNECTIVITY

One characterization of network connectivity is the network's Fiedler value, which is the second smallest eigenvalue of the Laplacian matrix, $L(V,E)$, of a network's topological graph, $G(V,E)$ where V is the graph vertex set and E is the edge set for V. For the discussion, because algebraic connectivity does not consider time we will ignore the notion of time, t, in the graph representation here, but later our exploration will involve the notion of time and network characteristics evolving with time. The Fiedler value is always nonnegative, and its value is zero if and only if the graph is disconnected, in which case the number of zero eigenvalues of L equals the number of connected components in a graph (e.g., a disconnected network consisting of two subnetworks that are themselves topologically connected would yield two eigenvalues equal to zero). Referring to [1], the Fiedler value, represented by λ^1, of a graph, G can be obtained by the following eigenvalue optimization problem.

$$\begin{aligned} \lambda^1 &= \min y^T L(V,E) y \\ \text{s.t.} y^T y &= 1 \text{ and } y^T \mathbf{1} = 0, \end{aligned} \tag{17.1}$$

where y is a vector that does not equal to $\mathbf{1}$.

The Laplacian matrix of a given graph is defined as follows: Given a graph $G(V,E)$ without self cycles and multiple links between two nodes, the Laplacian matrix L is calculated by

$$L(V,E) = D(V,E) - A(V,E), \tag{17.2}$$

where $D(V,E)$ is a diagonal matrix whose diagonal entry contains the degrees for each node. $A(V,E)$ is the adjacency matrix with each entry being a value of zero or one when nodes are connected to each other. In addition, its diagonal is zero because, for a communication network we assume that $G(V,E)$ has no self cycles. According to Eq. (17.2), the Laplacian matrix has the following properties [2,3]:

- *Observation 1*: L is a symmetric matrix. Its (i,j)th and (j,i)th entries are equal and its diagonal entries contain each node's total degree.
- *Observation 2*: All of its eigenvalues are real because L is symmetric.

- *Observation 3*: L is a positive semidefinite matrix. Thus, it has no negative eigenvalues. Its first smallest eigenvalue is always 0 because the sum of each row or column is zeros. By sorting the eigenvalues, we obtain: $\lambda^0 = 0 \leq \lambda^1 \leq \lambda^2 \leq \cdots \leq \lambda^{n-1}$
- *Observation 4*: The number of zero eigenvalues indicates the number of disconnected components in the graph. If the graph is strongly connected, then the second smallest eigenvalue, λ^1, which is also the Fiedler value, is always larger than zero.
- *Observation 5*: Now consider an attacker aiming to disrupt the functionality of the network. If an attacker kills the links in between nodes (or when a network's links are broken because of natural distances) and thereby produces a new edge set E_1, then the new Fiedler value satisfies $\lambda^1(V, E_1) \leq \lambda^1(V, E)$ where $E_1 \subseteq E$
- *Observation 6*: The Fiedler value's upper bound is limited by the minimum degree of nodes and the total number of nodes that exists in the network. The upper bound approaches the minimum degree value of all the nodes as the size of the network becomes large. The exact relationship is given by [2]

$$\lambda^1(V, E) \leq \frac{|V|}{|V| - 1} \min_v d_v. \tag{17.3}$$

We note that one may examine *Observation 5* a different way. In particular, it implies that the Fiedler value can become larger when adding edges to a graph, and thus a network can become more robust by increasing the Fiedler value.

Because we are interested in network vulnerability when a node is removed from a graph, we now introduce a modified notion of the Fiedler value, which corresponds to the role of a particular node in the network's connectivity as measured by the impact associated with removing all of that node's links (i.e., connections to other nodes in the network). Specifically, we propose a new measure of connectivity, which we term a node's Fiedler value:

Definition. For a graph $G(V, E)$, the *node Fiedler value* associated with node j corresponds to the Fiedler value $\lambda_j^1(V, E_j)$, where E_j corresponds to a revised set of edges for G where all edges containing node j have been removed from E. ■

Nodal Fiedler values have properties that λ_i^1, $i = 1, 2, \ldots, n$, on the graphs that remain after removing each node from the same original graph, G, and are comparable in terms of graph connectivity. Because all the remaining graphs have the same number of nodes, the only differences among them are the number of edges. Thus, according to *Observation 5*, comparison can be performed by adding and removing edges when the number of nodes is equal. This observation provides the foundation for comparing the connectivity impact that an adversary could have on a network by removing a node from the network, such as results from a jamming attack or a sophisticated attack in which an adversary broadcasts instructions to remove a node from the network routing tables. Additionally, this notion will allow us to select the most connectivity-influential node in a network. For example, for a five-node network, the Fiedler value indicates that the star topology is more connected than a line topology with $\lambda^1 = 1$ compared to $\lambda^1 = 0.3820$. For a star topology, removing the central node that connects to all other nodes has more influence on connectivity than removing a leaf node because its Fiedler value becomes zero (in fact, we have a totally disconnected network). To not risk confusion, we will not write superscript 1 in the following text.

17.2.2 MOST CONNECTIVITY-INFLUENTIAL NODE

Our objective is to identify the weakest point in the network in the sense that removal of this node would weaken the network's connectivity the most. To do this, we examine the connectivity of the topologies that remain on a case-by-case basis after removing each node, and discover the nodes in the network whose deletion would have the most harmful impact on the network's algebraic connectivity.

Algorithm 17.1 SELECT A NODE THAT HAS THE MOST HARMFUL IMPACT ON CONNECTIVITY

procedure SELECTNODE(G)
 Step 1: Remove each node, i, in a graph G
 Step 2: Calculate the Fiedler value λ_i, $i = 1, \ldots, n$ for the remaining graph $G \backslash i$
 Step 3: The node with most influence on the connectivity is
 $C = \arg_i \max \lambda_i$
 Step 4: K nodes with most influence on connectivity correspond to the first k largest λ_i
end procedure

17.3 USING CONNECTIVITY TO DIRECT AN INTERFERENCE ATTACK

Each edge in a network graph describes a relationship between entities, such as a communication link in a wireless network, that may be quantified by throughput or in a binary manner. We now consider an application of the Fiedler value for directing an attack against a wireless network. To describe the physical layer connectivity, we will use a Shannon-style formulation to capture the throughput of a physical layer link. Consider a wireless network consisting of n nodes in radio range of each other. We denote a node by $v_i = (x_{1i}, x_{2i})$, $i \in [1, n]$, where (x_{1i}, x_{2i}) is the 2-d coordinate for node v_i. Let $V = \{v_i, i \in [1, n]\}$ be the set of all nodes. We assume that, when each node communicates, it emits the same power in all directions. Due to fading gains, path loss, and mutual interference of the signals, not every signal can reach each receiver. Let $\boldsymbol{P} = (P_1, \ldots, P_n)$ be the transmission power allocation, where P_i is the signal power used by node i. Interference between signals could take place, and its effect depends on the distance between the receiver and the sender. Using a Shannon formulation, the throughput of the received signal by node j is

$$T_{ij,\epsilon}(\boldsymbol{P}) = \begin{cases} 0, & \mathrm{SINR}_{ij}(\boldsymbol{P}) < \epsilon, \\ \ln(1 + \mathrm{SINR}_{ij}(\boldsymbol{P})), & \mathrm{SINR}_{ij}(\boldsymbol{P}) \geq \epsilon, \end{cases}$$

where $\epsilon \geq 0$ is a threshold value for SINR (below this threshold, the link is unsustainable and hence no throughput), and $\mathrm{SINR}_{ij}(\boldsymbol{P}) = (h_i P_i / d_{ij}^2)/(\sigma^2 + \sum_{k \neq i, k \neq j} h_k P_k / d_{kj}^2)$ with σ^2 is the background noise, h_i is the fading channel gain, and

$$d_{ij} = \sqrt{(x_{1i} - x_{1j})^2 + (x_{2i} - x_{2j})^2}$$

is the distance between node v_i and node v_j. (Note, for simplicity of notation and derivations, we are using ln instead of $\log_2$ in the capacity formula.)

To define a communication network's topology, *links* (edges) between nodes have to be established. In this example, we consider *symmetric* communication, i.e., two nodes (say, node i and node j) are considered to be linked if and only if $T_{ij,\epsilon}(\boldsymbol{P}) > 0$ and $T_{ji,\epsilon}(\boldsymbol{P}) > 0$, and because communication is symmetric, the link is considered an undirected edge. Denote the link between node v_i and node v_j by e_{ij}. Let $E(\boldsymbol{P})$ be the set of all links. The graph $G(\boldsymbol{P}) = (V, E(\boldsymbol{P}))$ is simple, i.e., there are no self loops, and there are no multiple links connecting two nodes.

The graph $G(\boldsymbol{P})$, associated with a network, can be represented by the Laplacian matrix as

$$L_{ij}(G(\boldsymbol{P})) = \begin{cases} -1, & i \neq j,\ v_i \text{ and } v_j \text{ are linked}, \\ 0, & i \neq j,\ v_i \text{ and } v_j \text{ are not linked}, \\ -\sum\limits_{k=1,k\neq i}^{n} L_{ik}, & i = j, \end{cases}$$

where $L_{ii}(G(\boldsymbol{P}))$ equals the number of nodes connected with node v_i. We note that it is possible to consider a weighted network by assigning throughput as the weight for each link, in which case the weighted network can be represented by a weighted Laplacian matrix as

$$L_{ij}(G(\boldsymbol{P})) = \begin{cases} -w_{ij}, & i \neq j,\ v_i \text{ and } v_j \text{ are linked}, \\ 0, & ij,\ v_i \text{ and } v_j \text{ are not linked}, \\ -\sum\limits_{k=1,k\neq i}^{n} L_{ik}, & i = j, \end{cases}$$

where $w_{ij} = T_{ij,\epsilon}(\boldsymbol{P}) + T_{ji,\epsilon}(\boldsymbol{P})$ is the total throughput of the symmetric communication between node v_i and node v_j, and $L_{ii}(G(\boldsymbol{P}))$ is the total throughput of the symmetric communication between node v_i and others nodes.

Because $L(G(\boldsymbol{P}))$ is positive semidefinite and symmetric, its eigenvalues are all nonnegative. By ordering the eigenvalues in an increasing way, we have: $0 = \lambda^0(G(\boldsymbol{P})) \leq \lambda^1(G(\boldsymbol{P})) \leq \cdots \leq \lambda^{n-1}(G(\boldsymbol{P}))$. The eigenvector corresponding to the first eigenvalue is always $\boldsymbol{e}^T = (1, \ldots, 1)$. The second eigenvalue $\lambda^1(G(\boldsymbol{P}))$ is the Fiedler value. To emphasize that we consider connectivity based on the fact that there is bidirectional throughput (above a threshold value) for a link, we will use the term *throughput connectivity* and *throughput Fiedler value*. For a fixed power assignment $\boldsymbol{P}$, the throughput Fiedler value can be found as the solution of the following optimization problem

$$\lambda^1(G(\boldsymbol{P})) = \min_{\boldsymbol{y}^T\boldsymbol{y}=1, \boldsymbol{e}^T\boldsymbol{y}=0} \boldsymbol{y}^T L(G(\boldsymbol{P}))\boldsymbol{y}.$$

Let us illustrate the behavior of throughput connectivity by the following example. Let the network consist of five nodes $(0,0)$, $(1,0)$, $(0,1)$, $(1,1)$, and $(2,0.5)$ (Fig. 17.1A), and $h = 1$, $\sigma^2 = 2$, $\epsilon = 0.1, 0.25$ and $\boldsymbol{P} = (10, 20, 15, 25, P_5)$ where P_5 varies from 0.2 to 40. Of course, increasing ϵ yields a decrease in the total throughput (Fig. 17.1B). Throughput connectivity is piece-wise constant versus power (in our case, P_5, see, Fig. 17.1C), while weighted throughput connectivity is piece-wise continuous on P_5 (Fig. 17.1D). Thus, weighted throughput connectivity is more sensitive than throughput connectivity to a variation of the power. In this example, we observe that there is a continuum where throughput connectivity obtains its maximum, and the value of this maximum is not too sensitive to the threshold ϵ (in the considered example they coincide for $\epsilon = 0.1$ and $\epsilon = 0.25$, and are equal to 3). Also, we can observe that there is a reduction of the set where the

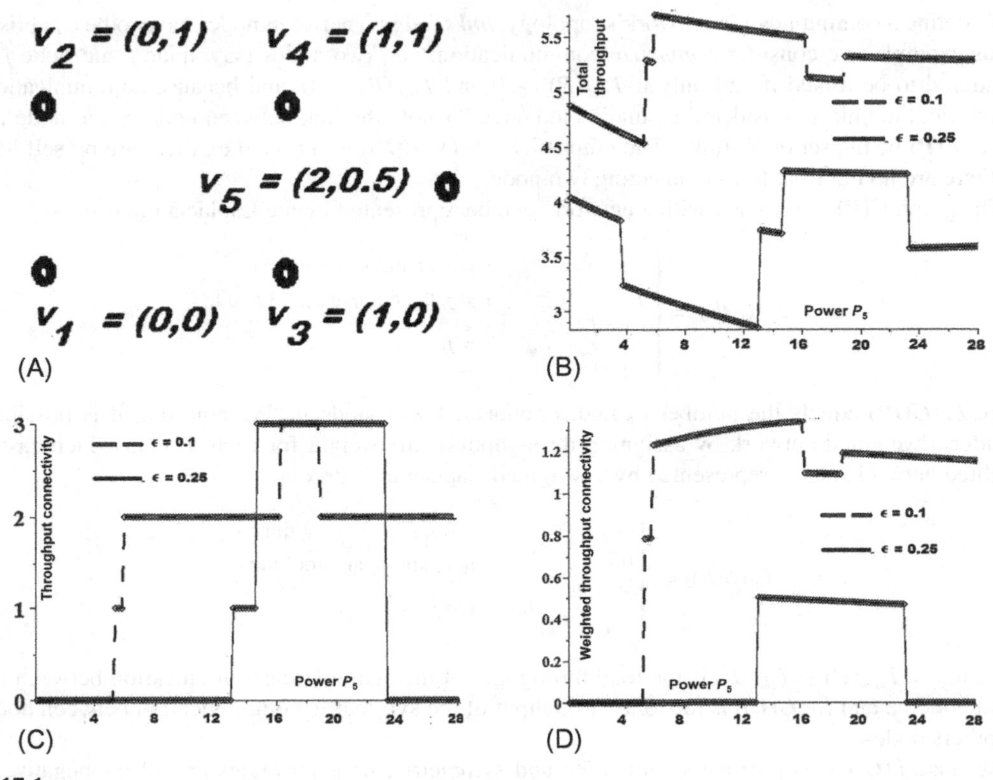

FIG. 17.1

(A) Nodes of the network, (B) Total throughput, (C) Throughput connectivity, and (D) Weighted throughput connectivity as functions on P_5.

throughput connectivity obtains its maximum when reducing the threshold ϵ, but there is no simple monotonic dependence between throughput connectivity and the threshold ϵ. For weighted throughput connectivity, such dependence could be observed as well as the fact that it obtains its maximum for a unique P_5.

17.3.1 IMPROVING CONNECTIVITY AGAINST SELF-INTERFERENCE

The Fiedler value has been used to optimize a network's design, and we now survey a few such works. A greedy heuristic algorithm was presented in [4], which adds edges (from a set of candidate edges) to a graph to maximize its algebraic connectivity. A distributed algorithm for the estimation and control of the connectivity of ad hoc networks for random topologies was suggested in [5], while a steepest-descent algorithm was proposed for control of the algebraic connectivity in [6]. The problem of improving network connectivity by adding a set of relays to increase the number of links between network nodes was considered in [7]. In [8] a genetic algorithm and swarm algorithm were applied for finding the best positions of adding nodes to a network to meet trade off between deployment cost and network connectivity. A decentralized algorithm to increase the connectivity of a multiagent system was suggested in [9]. In [10], a problem of finding the best vertex positioning to maximize the Fiedler value of a weighted graph was studied. Finally other measures of connectivity (such as global message connectivity, worst-case connectivity, network bisection connectivity, and k-connectivity, have been used in network design [11]).

In all these papers, the possibility of establishing a new communication link in a network did not depend on signal interference. Interference, however, has a significant impact because signals sent to establish new communication links also serve as noise for all other links, thereby reducing the network's capacity for maintaining existing communication links. To deal with this problem, we introduced two notions of connectivity: throughput connectivity, which reflects the possibility of establishing communication between nodes for a given power level; and weighted throughput connectivity, which associates with each link a weight corresponding to that link's throughput.

We now use these notions in an adaptive transmission protocol that reallocates transmission power between nodes to alleviate interference. Let Π be the set of feasible transmission power vectors. For example, this could be $\Pi(\overline{P}) = \{\boldsymbol{P} \geq 0 : \sum_{j=1}^{n} P_i = \overline{P}\}$, where $\overline{P}$ is the total power allowed by the network's administrator. Then, the problem of optimal transmission power assignment is given as the following max-min problem:

$$\lambda^1(G(\boldsymbol{P})) = \max_{\boldsymbol{P}\in\Pi(\overline{P})} \min_{\boldsymbol{y}^T\boldsymbol{y}=1, \boldsymbol{e}^T\boldsymbol{y}=0} \boldsymbol{y}^T L(G(\boldsymbol{P}))\boldsymbol{y}. \tag{17.4}$$

This maximization problem of the second smallest eigenvalue of the Laplacian matrix based on its inner parameters is equivalent to the following optimization problem (see [12]):

$$\begin{aligned} &\max_{\boldsymbol{P},z} z, \\ &\text{subject to} \\ &L(G(\boldsymbol{P})) - z\boldsymbol{I} \succ \boldsymbol{0}, \quad \boldsymbol{P} \in \Pi(\overline{P}) \text{ and } z > 0, \end{aligned} \tag{17.5}$$

where $\boldsymbol{I}$ is the $n \times n$ identity matrix, and "$\succ$" represents positive definiteness. Because $L(G(\boldsymbol{P}))$ is symmetric, $L(G(\boldsymbol{P})) - z\boldsymbol{I}$ is also symmetric. Therefore, Eq. (17.5) belongs to the class of semi-definite programming (SDP) problems [13]. It can be solved by SDP optimization tools, such as SDPT3 [14,15], SDPA-M [7,16], and CSDP [17].

Fig. 17.2A illustrates the dependence of throughput connectivity and weighted throughput connectivity versus total power $\overline{P}$ with $\epsilon = 0.1$. It is interesting that these two forms of connectivity are nondecreasing due to the cooperative reallocation of transmission power between the nodes. Meanwhile, as was shown in Fig. 17.1, selfishly increasing the transmission power of just one node can lead to decreasing the network's connectivity. The cooperative throughput case is larger than the case for selfish throughput. This illustrates that throughput connectivity reflects the possibility of establishing communication between nodes for a signal power level while the weighted throughput connectivity associates the throughput as a weight in the associated network graph. Throughput connectivity is less sensitive to a network's parameters than one based on weighted throughput connectivity, which makes the power allocation protocol for maintaining throughput connectivity potentially advantageous because one would not have to exchange channel state information as frequently.

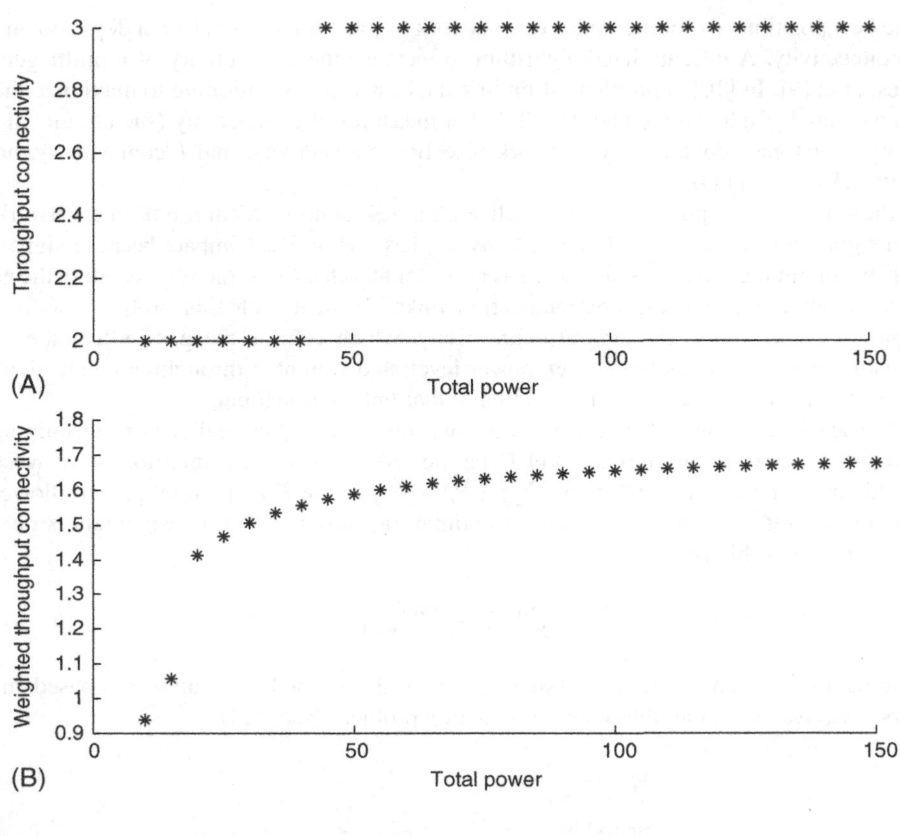

FIG. 17.2

(A) Throughput connectivity and (B) weighted throughput connectivity as functions of $\overline{P}$.

17.4 DYNAMIC GAME FOR IMPROVING CONNECTIVITY AGAINST JAMMING

We now move to applying the Fiedler value to model network interference attacks. A malicious jammer can purposely break links and separate a node from the rest of a network, which can interrupt the normal operation of the network, especially if the node is the hub of several routes. Therefore, investigating the impact of the removal of critical nodes and analyzing the jammer's strategy in choosing a node for an attack is essential to maintain network connectivity in adversarial settings.

Most research about network connectivity traditionally focuses on designing secure routing protocols by which packets can route around holes in networks [18–20]. Those routing protocols usually aim to find the most efficient and free path in a topological graph after an attack happens. The impact of a broken link or removal of a node in a path, and the resulting diffusion of attack damage across the broader network context, has been studied much less, particularly when the connectivity issues appear at the physical layer. Ensuring the robustness of the physical layer typically involves examining links in isolation (e.g., robust error coding), and notably separate from the broader network context.

The robustness of *networks* at the physical layer should examine the network's performance after one or more nodes or links are degraded or removed at the physical layer. For example, an attacker can strategically delete nodes according to his purposes though targeted interference, aimed at greedily removing nodes with a higher degree first or deleting nodes in high-density areas in order to exacerbate the damage.

Game theory is a natural tool for rationalizing a jammer's behavior. Game theory investigates the interactions between players to arrive at equilibrium strategies for both sides [21]. In [22], a survey of works that applied game theory to deal with network security is given. Game theory papers at the physical layer often model the rational behaviors of a jammer or an eavesdropper or cooperative behavior between them to solve the problem of allocating transmission power or increasing the transmission rate. Typically, the utility function being employed is Shannon capacity, SINR (signal to interference and noise ratio), information entropy, or bit error. There is a limited set of works dealing with maintaining the connectivity of the network topology. In [23], the problem of minimizing the probability that the spanning tree is disrupted by an adversary attack was studied. In [24], to identify key players engaged in attacking a network, the Shapley value was applied. In [25], a problem with two types (good and bad) of users was studied by a repeated game, where good users were willing to trade energy for connectivity depending on neighbors' behaviors, while bad users try to destroy connectivity and lure the good users to waste energy.

Here we consider a game where users' throughput and network connectivity are combined in a unified framework. We examine the problem of interference attacks that are intended to harm connectivity and throughput. We arrive at antijamming strategies aimed at coping with interference attacks through a unified stochastic game. In such a framework, an entity trying to protect a network faces a dilemma: (1) the underlying motivations for the adversary can be quite varied, which depends largely on the network's characteristics such as power and distance; (2) the metrics for such an attack can be incomparable (e.g., network connectivity and total throughput). To deal with the problem of such incomparable metrics, we use the attack's expected duration as a unifying metric to compare distinct attack metrics because a longer duration of and unsuccessful attack assumes a higher cost. Based on this common metric, a mechanism of max-min selection for an attack prevention strategy is outlined.

17.4.1 A STOCHASTIC GAME: CONNECTIVITY INTRUSION PREVENTION

As a motivating application, we consider the problem of mitigating a connectivity attack directed against a wireless ad hoc network. We propose strategies to prevent such attacks. We will apply methods of stochastic games in combination with network connectivity, and refer the reader to surveys for applying stochastic games to network security [26–28].

Attack model

Our adversarial model involves a jammer aiming to hurt the network by choosing a node to direct interference against while the network itself aims to reduce the harm this attack has on the network by scanning to detect the interference. The category of the jammer's attack is fixed throughout the entire intrusion, which might either be a jamming attack to minimize throughput or an attack aimed at disrupting connectivity. The jammer uses its knowledge of the network to find a node that could harm the network the most when its communication is blocked. If the victim node determined by a jammer is also simultaneously being scanned by the scanner, the jammer will switch to a silent hiding mode (incurring a cost/penalty), and if he is not caught, he can continue his attack. However, if the node he chooses is not scanned, the jammer performs a jamming attack. We assume a jammer can observe the presence of the scanning authority through some form of detection technique (e.g., side channel information) and thus the jammer will not perform a jamming attack when being scanned because there is a fear of being caught.

Game formulation

Let C_h be the hiding mode cost for the jammer corresponding to its type of attack. Let α be the probability to be detected in the hiding mode, and $1-\alpha$ be the probability not to be detected. Thus, the instantaneous cost to the jammer combines the expected hiding cost and the cost of network penetration in the future if the jammer is not caught. Naturally we assume the instantaneous payoff for the scanning authority equals the instantaneous cost for the jammer. This recursively played zero-sum game Γ_γ can be considered as a single-state stochastic game [29], which can be solved by stationary strategies, as follows:

$$\Gamma_\gamma = \begin{array}{c} \\ 1 \\ 2 \\ \dots \\ n \end{array} \begin{array}{c} \begin{array}{cccc} 1 & 2 & \dots & n \end{array} \\ \begin{bmatrix} C_h+\gamma\Gamma_\gamma & \bar{\lambda}_2 & \dots & \bar{\lambda}_n \\ \bar{\lambda}_1 & C_h+\gamma\Gamma_\gamma & \dots & \bar{\lambda}_n \\ \dots & \dots & \dots & \dots \\ \bar{\lambda}_1 & \bar{\lambda}_2 & \dots & C_h+\gamma\Gamma_\gamma \end{bmatrix} \end{array}, \tag{17.6}$$

where the rows correspond to the scanner's strategies, i.e., chosen nodes to scan, and columns correspond to the jammer's strategies, i.e., chosen nodes to attack.

Let us describe in detail the components of this matrix. Assume the authority has chosen strategy i and the jammer has chosen strategy j. If $i \neq j$, node j is jammed successfully, the jammer suffers the instantaneous cost $\bar{\lambda}_j$, and the game is over. If $i = j$ then the jammer switches to the hiding mode paying the instantaneous cost C_h. With probability α, the jammer will be detected, and the game is over. However, if the jammer is not detected, with probability $1-\delta$, he stops the attempts and exits the game. The game is over. Whereas, with probability δ the jammer keeps playing the game recursively. Therefore, the instantaneous reward for the scanner is $\alpha C_h + (1-\alpha)\left[C_h + \delta \cdot \text{val}(\Gamma_\gamma)\right]$. Then, the conditional probability to continue the jamming attack is $\gamma = (1-\alpha)\delta$, and with this probability the game Γ is played recursively with the expected instantaneous jammer costs accumulated as $C_h + \gamma\Gamma_\gamma$. Because $\gamma < 1$, it can be considered as a discount factor and guarantees the convergence of the solution. Here, employing stochastic games is quite natural because the authority and the jammer have opposing objectives, and it is uncertain how persistent the jammer can be in its malicious attack before it is detected. The routine for the jamming attack is presented in Fig. 17.3.

Cost of connectivity attack

The game (17.6) can be used to model different types of attacks by appropriately assigning parameters. The variable, $\bar{\lambda}_i$, can correspond to either the network's connectivity in a connectivity disruption attack or the network's throughput in a throughput disruption attack.

The Nodal Fielder value, defined in Section 17.2.1, can be used as the cost of a connectivity attack because the Fiedler value is related to the connectivity of a network. The smaller the Fiedler value, the larger the negative impact. These Fiedler values, $\{\lambda_i(G\backslash i)|i = 1, 2, \dots, n\}$, on the remaining graphs

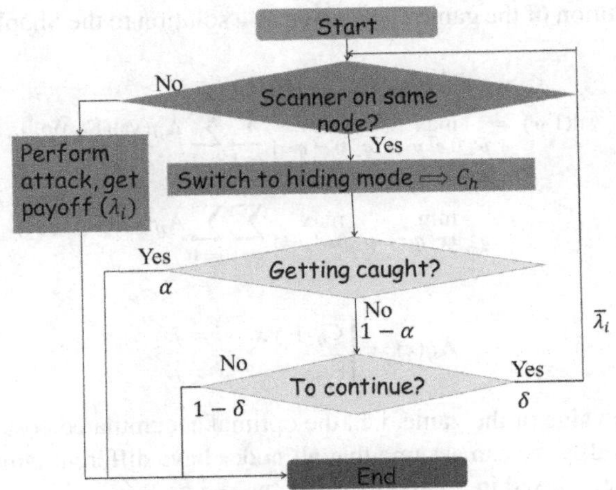

FIG. 17.3

The routine for a connectivity jamming attack in an ad hoc network, C_h is cost of hiding, $\bar{\lambda}_i$ is connectivity damage for removing a node, which is the cost for launching an attack, α is the probability of being caught, and δ is the probability of continuing an attack.

obtained by removing a different node from the same graph, Γ, are comparable. Thus if the jammer aims to reduce network connectivity, the Fiedler value $\{\lambda_i\}$ can be considered as the cost of such an adversary's attack. Namely, the jammer's cost of attack to disrupt connectivity is

$$\bar{\lambda}_i = \lambda_i(G \backslash i). \tag{17.7}$$

If the adversary targets the network's throughput, then the total throughput for the unaffected network can be considered as a cost of such attack. If the adversary has a selected node i for a jamming attack, the total throughput for the rest of the network, or the cost of the throughput jamming attack, is given as:

$$\bar{\lambda}_i = \sum_{l,j=1, j\neq i, l\neq i}^{n} \ln\left(1 + \frac{h_{l,j}P_{l,j}}{\sigma^2 + \sum_{k=1,k\neq j}^{n} h_{k,j}P_{k,j}}\right). \tag{17.8}$$

Defensive strategy

Game Γ_γ has a solution in (mixed) stationary strategies, i.e., strategies that are independent of history and the current time slot. A (mixed) stationary strategy for the scanner is a probability vector $\boldsymbol{p}^T = (p_1, p_2, \ldots, p_n)$, where p_i is the probability to scan node i and $\boldsymbol{e}^T\boldsymbol{p} = 1$. A (mixed) stationary strategy for the jammer is a probability vector $\boldsymbol{q}^T = (q_1, q_2, \ldots, q_n)$, where q_i is the probability to jam node i, and $\boldsymbol{e}^T \boldsymbol{q} = 1$. Solution of the game Γ_γ is given as a solution to the Shapley (-Bellmann) equation game [29]:

$$\begin{aligned} \text{val}(\Gamma_\gamma) &= \max_{p \geq 0, e^T p = 1} \min_{q \geq 0, e^T q = 1} \sum_{i=1}^{n} \sum_{j=1}^{n} \mathbf{A}_{ij}(\text{val}(\Gamma_\gamma)) p_i q_j, \\ &= \min_{q \geq 0, e^T q = 1} \max_{p \geq 0, e^T p = 1} \sum_{i=1}^{n} \sum_{j=1}^{n} \mathbf{A}_{ij}(\text{val}(\Gamma_\gamma)) p_i q_j, \end{aligned} \tag{17.9}$$

$$\mathbf{A}_{ij}(x) = \begin{cases} C_h + \gamma x, & i = j, \\ \bar{\lambda}_j, & i \neq j, \end{cases} \tag{17.10}$$

and $V_\gamma := \text{val}(\Gamma_\gamma)$ is the value of the game, i.e., the optimal accumulated cost to the jammer.

Without loss of generality we can assume that all nodes have different jamming costs, i.e., $\bar{\lambda}_i \neq \bar{\lambda}_j$ for $i \neq j$ and all nodes are indexed in ascending order by $\bar{\lambda}_i$, i.e.,

$$\bar{\lambda}_1 < \bar{\lambda}_2 < \cdots < \bar{\lambda}_n. \tag{17.11}$$

Despite the fact that the stochastic game considered has an $n \times n$ instantaneous payoff matrix, we can obtain the solution explicitly from the following theorem:

Theorem 17.1. *The stochastic game Γ_γ has an equilibrium in stationary strategies $(\boldsymbol{p}, \boldsymbol{q})$ and the value V_γ given as follows:*

(a) *Let*

$$C_h/(1-\gamma) < \bar{\lambda}_1. \tag{17.12}$$

Then $V_\gamma = \bar{\lambda}_1$ and

$$\begin{aligned} p_i &\begin{cases} = 0, & i = 1, \\ \geq \dfrac{\bar{\lambda}_i - \bar{\lambda}_1}{\bar{\lambda}_i - C_h - \gamma \bar{\lambda}_1}, & i \geq 2, \end{cases} \\ q_i &= \begin{cases} 1, & i = 1, \\ 0, & i \geq 2. \end{cases} \end{aligned} \tag{17.13}$$

(b) *Let*

$$\bar{\lambda}_1 \leq C_h/(1-\gamma) < \bar{\lambda}_2. \tag{17.14}$$

Then $V_\gamma = C_h/(1-\gamma)$

$$\begin{aligned} p_i(x) &= \begin{cases} 1, & i = 1, \\ 0, & i \geq 2, \end{cases} \\ q_i(x) &= \begin{cases} 1, & i = 1, \\ 0, & i \geq 2. \end{cases} \end{aligned} \tag{17.15}$$

(c) *Let*

$$\bar{\lambda}_k < C_h/(1-\gamma) \le \bar{\lambda}_{k+1} \tag{17.16}$$

with $\lambda_{n+1} = \infty$, *and* $m \in [1, k]$ *be such that*

$$\varphi_m^{k+1} \le 1 < \varphi_{m+1}^{k+1}, \tag{17.17}$$

with

$$\varphi_s^{k+1} = \sum_{i=1}^{s} \frac{\bar{\lambda}_s - \bar{\lambda}_i}{C_h + \gamma\bar{\lambda}_s - \bar{\lambda}_i} \text{ for } s \le k \tag{17.18}$$

and $\varphi_{k+1}^{k+1} = \infty$. *Note that, by Eq. (17.16),* φ_s^{k+1} *is increasing from zero for* $s = 1$ *to infinity for* $s = k+1$. *Thus, m is uniquely defined by Eq. (17.17).*
Then,

$$p_i = \begin{cases} \dfrac{V_\gamma - \bar{\lambda}_i}{C_h + \gamma V_\gamma - \bar{\lambda}_i}, & i \le m, \\ 0, & i > m, \end{cases}$$
$$q_i = \begin{cases} \dfrac{1/(C_h + \gamma V_\gamma - \bar{\lambda}_i)}{\sum_{j=1}^{m} 1/(C_h + \gamma V_\gamma - \bar{\lambda}_j)}, & i \le m, \\ 0, & i > m, \end{cases} \tag{17.19}$$

and V_γ *is a unique root of the equation*

$$F_m(V_\gamma) := \sum_{i=1}^{m} \frac{V_\gamma - \bar{\lambda}_i}{C_h + \gamma V_\gamma - \bar{\lambda}_i} = 1. \tag{17.20}$$

■

Proof. First note that V_γ, $\boldsymbol{p}$, and $\boldsymbol{q}$ is a solution of the Shapley equation (17.9) if and only if $V_\gamma = v$ and

$$\begin{aligned} &\max v, \\ &\sum_{i=1}^{n} \mathbf{A}_{ij}(V_\gamma) p_i \ge v, \quad i \in \{1, \ldots, n\}, \end{aligned} \tag{17.21}$$

$$\begin{aligned} &\min v, \\ &\sum_{j=1}^{n} \mathbf{A}_{ij}(V_\gamma) q_i \le v, \quad j \in \{1, \ldots, n\}. \end{aligned} \tag{17.22}$$

Taking into account Eq. (17.10) and that $\boldsymbol{p}$ and $\boldsymbol{q}$ are probability vectors, the LP problems (17.21) and (17.22) are equivalent to

$$\max\ \nu,$$
$$(C_h+\gamma V_\gamma-\bar{\lambda}_i)p_i+\bar{\lambda}_i\ge \nu,\quad i\in\{1,\dots,n\}, \tag{17.23}$$

$$\min\ \nu,$$
$$(C_h+\gamma V_\gamma-\bar{\lambda}_i)q_i+\sum_{j=1}^{n}\bar{\lambda}_j q_j\le \nu,\quad j\in\{1,\dots,n\}. \tag{17.24}$$

Then, Eqs. (17.23) and (17.24) imply that $V_\gamma, \boldsymbol{p}$, and $\boldsymbol{q}$ is a solution of the Shapley equation (17.9) if and only if:

$$(C+\gamma V_\gamma-\bar{\lambda}_i)q_i+\sum_{j=1}^{n}\bar{\lambda}_j q_j\begin{cases}=V_\gamma, & p_i>0,\\ \le V_\gamma, & p_i=0,\end{cases} \tag{17.25}$$

$$(C+\gamma V_\gamma-\bar{\lambda}_i)p_i+\bar{\lambda}_i\begin{cases}=V_\gamma, & q_i>0,\\ \ge V_\gamma, & q_i=0.\end{cases} \tag{17.26}$$

Let Eq. (17.12) hold. Then, by Eqs. (17.11), (17.25), and (17.26), there is no i such that $p_i>0$ and $q_i>0$. Also, $q_1=1$ and $p_1=0$. Substituting them into Eqs. (17.25) and (17.26) implies (a). Suppose Eq. (17.12) does not hold. Then, by Eqs. (17.11), (17.25), and (17.26), there is an m such that $p_i>0$ and $q_i>0$.

$$p_i\begin{cases}>0, & i\le m,\\ =0, & i>m\end{cases}\quad\text{and } q_i\begin{cases}>0, & i\le m,\\ =0, & i>m.\end{cases} \tag{17.27}$$

Let $m=1$. Then, by Eq. (17.11), (17.25) and (17.26), the condition (17.14) has to hold, and (b) follows. Now assume Eq. (17.16) holds, then note that

$$\max\left\{\frac{\bar{\lambda}_i-C_h}{\gamma},\bar{\lambda}_i\right\}=\begin{cases}\dfrac{\bar{\lambda}_i-C_h}{\gamma}, & \bar{\lambda}_i\ge\dfrac{C_h}{1-\gamma},\\ \bar{\lambda}_i, & \bar{\lambda}_i\le\dfrac{C_h}{1-\gamma}.\end{cases} \tag{17.28}$$

Because $m>1$, by Eqs. (17.11), (17.25), (17.26), and (17.27) $\boldsymbol{p}$ and $\boldsymbol{q}$ have the form given by Eq. (17.19). Then, V has to be given as a root of Eq. (17.20). It is only left to show that this equation has a unique root. Let $m<k$. Then, by Eqs. (17.11) and (17.28), F_m is increasing in $[\bar{\lambda}_m,\bar{\lambda}_{m+1}]$ such that, by Eq. (17.17), $F_m(\bar{\lambda}_m)=\varphi_m^{k+1}\le 1<\varphi_{m+1}^{k+1}=F_m(\bar{\lambda}_{m+1})$. Thus, V is uniquely defined. Let $m=k$. Then, by Eqs. (17.11) and (17.28), F_k is increasing in $[\bar{\lambda}_k,(\bar{\lambda}_k-C)/\gamma]$, and $F_k((\bar{\lambda}_k-C)/\gamma)>1$, and (c) follows.

From this theorem, if the hiding cost C_h is too large, i.e., $C_h\ge\bar{\lambda}_n$, then all the nodes will be under attack and thus must be scanned, i.e., $p_i>0$ and $q_i>0$ for any i, and the value of the game is the unique root of the equation $F_n(V_\gamma)=1$. Also, the value V_γ of the game is increasing with respect to C_h and γ, including $\gamma=1$. Because the game Γ_0 is a single time-slot game, it is just a matrix game whose solution is given in the following theorem.

Theorem 17.2. *The one time slot matrix game* Γ_0*, which is the limit of the stochastic game* Γ_γ *as* γ *tends to zero, has value* $V_0 = V(C_h)$ *and equilibrium strategies* $\mathbf{p}$ *and* $\mathbf{q}$ *given as:*

(a) *Let*

$$C_h < \bar{\lambda}_1. \tag{17.29}$$

Then $V(C_h) = \bar{\lambda}_1$ *and*

$$p_i \begin{cases} = 0, & i = 1, \\ \geq \dfrac{\bar{\lambda}_i - \bar{\lambda}_1}{\bar{\lambda}_i - C_h}, & i \geq 2, \end{cases}$$
$$q_i(x) = \begin{cases} 1, & i = 1, \\ 0, & i \geq 2. \end{cases} \tag{17.30}$$

(b) *Let*

$$\bar{\lambda}_1 \leq C_h < \bar{\lambda}_2. \tag{17.31}$$

Then $V(C_h) = C_h$ *and*

$$p_i = \begin{cases} 1, & i = 1, \\ 0, & i \geq 2, \end{cases}$$
$$q_i = \begin{cases} 1, & i = 1, \\ 0, & i \geq 2. \end{cases} \tag{17.32}$$

(c) *Let*

$$\bar{\lambda}_k < C_h \leq \bar{\lambda}_{k+1} \tag{17.33}$$

and m be given by Eq. (17.17). Then,

$$V(C_h) = \frac{1 + \sum_{j=1}^{m} \bar{\lambda}_j/(C_h - \bar{\lambda}_j)}{\sum_{j=1}^{m} 1/(C_h - \bar{\lambda}_j)},$$
$$p_i = \begin{cases} \dfrac{V(C_h) - \bar{\lambda}_i}{C_h - \bar{\lambda}_i}, & i \leq m, \\ 0, & i > m, \end{cases}$$
$$q_i = \begin{cases} \dfrac{1/(C_h - \bar{\lambda}_i)}{\sum_{j=1}^{m} 1/(C_h - \bar{\lambda}_j)}, & i \leq m, \\ 0, & i > m. \end{cases} \tag{17.34}$$

■

Theorem 17.2 suggests two procedures to find the value of the stochastic game.

Theorem 17.3. *The value of the stochastic game Γ_γ is given as follows $V_\gamma = \frac{x-C_h}{\gamma}$, where $x \geq C_h$ is a unique root of the equation $\frac{x-C_h}{\gamma} = V(x)$. The unique root of $V(x)$ can be found by*

(a) *an iterative procedure $x_0 = C_h$, $x_{i+1} = \gamma V(x_i) + C_h$, $i = 0, 1, \ldots$ until $|x_{i+1} - x_i| \leq \epsilon$ with ϵ is tolerance.*

(b) *the bisection method because $(x - C_h)/\gamma - V(x) = -C_h/\gamma - v(0) < 0$ for $x = 0$ and $(x - C_h)/\gamma - v(x) > 0$ for enough large x.*

■

17.4.2 COMPARISON OF DIFFERENT ATTACKS

The jammer can deteriorate network performance by reducing either its connectivity or throughput. Motivated by these different categories of malicious activity, the jammer could vary the corresponding optimal strategies. However, the scanner might have no knowledge of the jammer's motivation for an attack, and thus about the strategy employed. The scanner might only know the set of all possible motivations and the corresponding optimal strategies used by the jammer.

Under this situation, the need for comparing these strategies arises because they have different metrics. However, for the game considered, in spite of the difference between attack objectives, the ultimate goal of a jammer is to speed up the process of completing the attack because a longer time commitment involves more cost. Thus, the expected time for a successful attack can be considered as a common metric for all the categories of malicious activity, where the scanner wants to maximize this metric while the jammer aims to minimize it. If we assume the rival chooses a specific category and he continues that type of attack until completing the attack, then the expected jamming time, T, before a successful attack appears can be represented as the following (τ is jamming time slots spent before the jammer launches a successful attack):

$$\mathbb{E}(\tau, \boldsymbol{p}, \boldsymbol{q}) = \sum_{\tau=1}^{\infty} \tau \left[\left(\sum_{i=1}^{n} \gamma p_i q_i \right)^{\tau-1} \left(1 - \sum_{i=1}^{n} \gamma p_i q_i \right) \right] = \frac{1}{1 - \gamma \boldsymbol{p}^T \boldsymbol{q}}, \tag{17.35}$$

where $\boldsymbol{q}$ is a probability vector that represents a category of strategies employed by the jammer. $\boldsymbol{p}$ is a probability vector that represents a category of strategy applied by the scanner to scan the attack. Thus, these strategies depend on the category of malicious activity chosen by the jammer.

Although our approach might be applied to any category of an attack, we focus only on attacks against a network's connectivity and a network's throughput. We denote these metrics (connectivity, throughput) by the symbols "c" and "t". The optimal strategy pair, $(\boldsymbol{p}_c, \boldsymbol{q}_c)$, for dealing with an attack against connectivity was found in the previous section. The optimal strategies $(\boldsymbol{p}_t, \boldsymbol{q}_t)$ for dealing with an attack aiming to harm throughput can be found by substituting connectivity cost λ_i with the remaining throughput expressed in Eq. (17.8) into matrix (17.6).

The scanner wants to maximize the jammer's attacking time in order to force the jammer to make his attack more expensive whereas the jammer wants to minimize it. The scanner does not know what type of attack the jammer intends to use. The jammer does not know what type of attack the scanner intends to focus his defense against. Thus, the rivals face a dilemma of choosing the proper strategies, which can be described by the following zero-sum 2×2 matrix game

$$\mathbf{D} = \begin{matrix} & c & t \\ c & \mathbb{E}(\boldsymbol{p}_c, \boldsymbol{q}_c) & \mathbb{E}(\boldsymbol{p}_c, \boldsymbol{q}_t) \\ t & \mathbb{E}(\boldsymbol{p}_t, \boldsymbol{q}_c) & \mathbb{E}(\boldsymbol{p}_t, \boldsymbol{q}_t) \end{matrix},$$

where the rows correspond to the scanner's strategies, i.e., choosing the attack's category to response, and columns correspond to the jammer's strategies, i.e., choosing the attack's category.

This matrix game has an equilibrium (see [30]) either in pure strategies or in mixed strategies when the rival randomizes his selection. Because the game is zero-sum, then the scanner's equilibrium strategy is also his max-min strategy, i.e., it is the best response strategy for the most dangerous adversary's attack. This result is given in the following two propositions.

Proposition 17.1. *The game has an equilibrium in (pure) strategies if and only if the following conditions hold:*

1. *If $\boldsymbol{p}_t^T \boldsymbol{q}_c \le \boldsymbol{p}_c^T \boldsymbol{q}_c \le \boldsymbol{p}_c^T \boldsymbol{q}_t$ then (c, c) is an equilibrium,*
2. *If $\boldsymbol{p}_c^T \boldsymbol{q}_c \le \boldsymbol{p}_t^T \boldsymbol{q}_c \le \boldsymbol{p}_t^T \boldsymbol{q}_t$ then (t, c) is an equilibrium,*
3. *If $\boldsymbol{p}_t^T \boldsymbol{q}_t \le \boldsymbol{p}_c^T \boldsymbol{q}_t \le \boldsymbol{p}_c^T \boldsymbol{q}_c$ then (c, t) is an equilibrium,*
4. *If $\boldsymbol{p}_c^T \boldsymbol{q}_t \le \boldsymbol{p}_t^T \boldsymbol{q}_t \le \boldsymbol{p}_t^T \boldsymbol{q}_c$ then (t, t) is an equilibrium.*

■

Proposition 17.2. *If there is no equilibrium in pure strategies, the rivals apply randomized strategies. Namely, with probability, x_c (x_t), the scanner should defend against "c" ("t") attack's category, and with probability, y_c (y_t), the jammer applies strategy corresponding to "c" ("t") attack's category, where*

$$\begin{aligned} x_c &= \frac{\mathbb{E}(\boldsymbol{p}_t, \boldsymbol{q}_t) - \mathbb{E}(\boldsymbol{p}_t, \boldsymbol{q}_c)}{\mathbb{E}(\boldsymbol{p}_c, \boldsymbol{q}_c) + \mathbb{E}(\boldsymbol{p}_t, \boldsymbol{q}_t) - \mathbb{E}(\boldsymbol{p}_c, \boldsymbol{q}_t) - \mathbb{E}(\boldsymbol{p}_t, \boldsymbol{q}_c)}, \\ x_t &= 1 - x_c, \\ y_c &= \frac{\mathbb{E}(\boldsymbol{p}_t, \boldsymbol{q}_t) - \mathbb{E}(\boldsymbol{p}_c, \boldsymbol{q}_t)}{\mathbb{E}(\boldsymbol{p}_c, \boldsymbol{q}_c) + \mathbb{E}(\boldsymbol{p}_t, \boldsymbol{q}_t) - \mathbb{E}(\boldsymbol{p}_c, \boldsymbol{q}_t) - \mathbb{E}(\boldsymbol{p}_t, \boldsymbol{q}_c)}, \\ y_t &= 1 - y_c. \end{aligned} \tag{17.36}$$

■

17.4.3 NUMERICAL RESULTS

In this section, numerical results are given to illustrate the impact of network parameters, such as transmission power and SINR's threshold, on maintaining the communication links. In the simulation setting, the network consists of six nodes, i.e., $n = 6$ with background noise $\sigma^2 = 1$. The scanner scans the network to prevent malicious activity. The channel gain matrix, $\boldsymbol{h}$, was randomly generated and given as follows:

$$\boldsymbol{h} = \begin{bmatrix} 0 & 0.3128 & 1.1790 & 1.6488 & 1.6335 & 0.8458 \\ 0.3128 & 0 & 0.4524 & 1.9653 & 0.5215 & 0.1885 \\ 1.1790 & 0.4524 & 0 & 1.4605 & 1.1887 & 1.1970 \\ 1.6488 & 1.9653 & 1.4605 & 0 & 0.0450 & 0.9418 \\ 1.6355 & 0.5215 & 1.1887 & 0.0450 & 0 & 1.3919 \\ 0.8458 & 0.1885 & 1.1970 & 0.9418 & 1.3919 & 0 \end{bmatrix}.$$

For each node, as a power transmission protocol, we consider uniform power allocation, which involves a node allocating the same power to all its neighbors, though each node can have different total power levels. Here we consider that transmission powers (P_1 and P_2) of nodes 1 and 2 vary from 0 to 20, $P_3 = 11$, $P_4 = 10$, $P_5 = 9$, and $P_6 = 8$. The protocol of uniformly allocating transmission power is known to be optimal for independent and identically distributed Gaussian channels [31,32]. Given $\boldsymbol{h}$, $\boldsymbol{P}$, σ^2, and receiving threshold, they imply a network's topology.

Fig. 17.4 illustrates the authority's and the jammer's strategies for a connectivity jamming game as a function of probability, γ, and that the game is continuous. It shows how the jammer tries to avoid being detected by the authority in order to perform a successful attack. Fig. 17.4A indicates the jammer

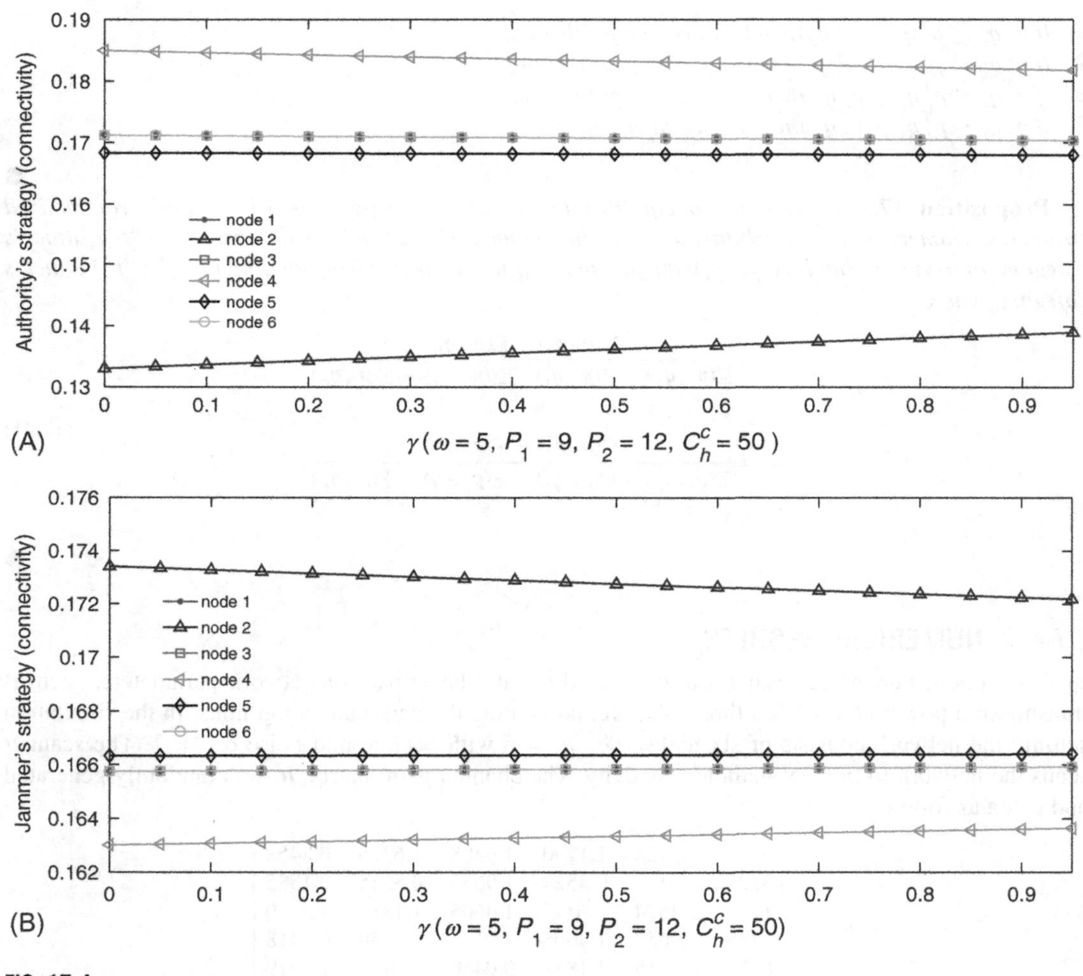

FIG. 17.4

(A) Authority's strategy for connectivity attack and (B) Jammer's strategy for connectivity attack.

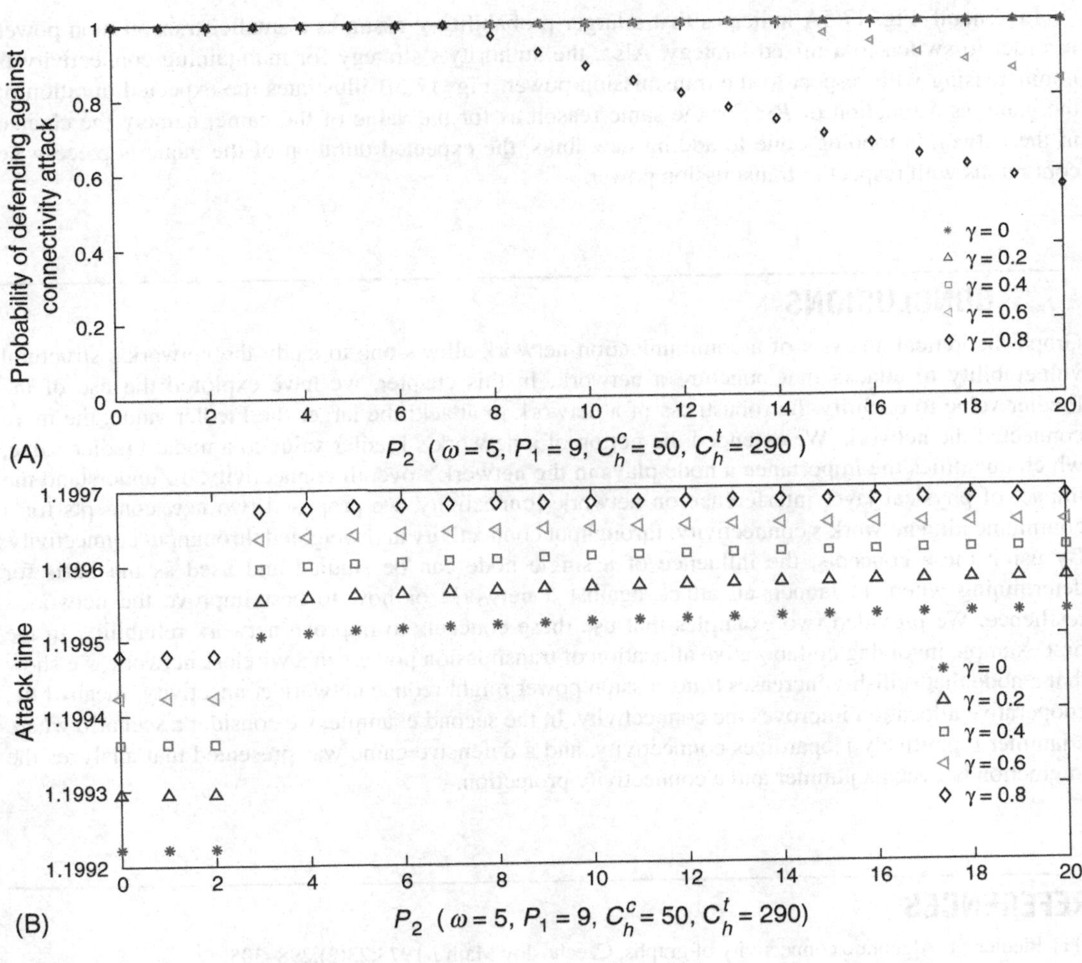

FIG. 17.5

Max-min selection of scanning strategy: (A) the probability that the authority intends to deal with an attack aiming to disrupt connectivity and (B) the expected durations of the game as a function on the transmission power of node 2.

tends to reallocate resources for a new attack as γ becomes larger while Fig. 17.4B illustrates that the best strategy for a jammer is in a reverse order to that of the scanner.

Fig. 17.5A illustrates the probability that the authority intends to deal with an attack aimed at disrupting connectivity as a function of P_2. Under the simulation setting, the figure shows: (1) a scanner should defend against a connectivity attack if an attacker has low probability, γ, to continue its attack; (2) As the transmission power of a legitimate node increases, the scanner switches to defending against a throughput attack.

In general, Fig. 17.5A indicates that a larger probability γ assumes a smaller transmission power in order to switch to a mixed strategy. Also, the authority's strategy for maintaining connectivity is nonincreasing with respect to the transmission power. Fig. 17.5B illustrates the expected duration of the game as a function of P_2. For the same reason as for the value of the game, namely the change in the network's topology due to adding new links, the expected duration of the game is piece-wise continuous with respect to transmission power.

17.5 CONCLUSIONS

Graph theoretical analysis of a communication network allows one to study the network's structural vulnerability to attacks that puncture a network. In this chapter, we have explored the use of the Fiedler value to quantify the robustness of a network to attack: the larger the Fiedler value, the more connected the network. We extended the notion of a network's Fiedler value to a nodal Fiedler value, which quantifies the importance a node plays in the network's overall connectivity. To understand the impact of physical layer interference on network connectivity, we proposed two new concepts for a communication network's connectivity: throughput connectivity and weighted throughput connectivity. By using these concepts, the influence of a single node can be studied and used as the basis for determining where to launch an attack against a network, or how to best improve the network's resilience. We provided two examples that use these concepts to improve network reliability. In the first example involving collaborative allocation of transmission powers in a wireless network, we show that a node that selfishly increases transmission power might reduce network connectivity; meanwhile, cooperative allocation improves the connectivity. In the second example, we consider a scenario where a jammer repetitively jeopardizes connectivity, and a defensive game was presented that analyzes the interaction between a jammer and a connectivity protection.

REFERENCES

[1] Fiedler M. Algebraic connectivity of graphs. Czechoslov Math J 1973;23(2):298–305.

[2] Pandana C, Liu KJR. Robust connectivity-aware energy-efficient routing for wireless sensor networks. IEEE Trans Wirel Commun 2008;7(10):3904–16.

[3] Mohar B, Alavi Y. The Laplacian spectrum of graphs. In: Graph theory, combinatorics, and applications, vol. 2; 1991. p. 871–98.

[4] Ghosh A, Boyd S. Growing well-connected graphs. In: Proceedings of the 45th IEEE conference on decision and control (CDC); 2006. p. 6605–11.

[5] Di Lorenzo P, Barbarossa S. Distributed estimation and control of algebraic connectivity over random graphs. IEEE Trans Signal Process 2014;62(21):5615–28.

[6] Morbidi F. On the control of the algebraic connectivity and clustering of a mobile robotic network. In: Proceedings of the European control conference (ECC); 2013. p. 2801–6.

[7] Ibrahim AS, Seddik KG, Liu KJR. Improving connectivity via relays deployment in wireless sensor networks. In: Proceedings of the IEEE global telecommunications conference (GLOBECOM); 2007. p. 1159–63.

[8] Romoozi M, Babaei H. Improvement of connectivity in mobile ad-hoc networks by adding static nodes based on a realistic mobility model. Int J Comput Sci Issues 2011;8(4):76–83.

[9] De Gennaro M, Jadbabaie A. Decentralized control of connectivity for multi-agent systems. In: Proceedings of the IEEE conference on decision and control (CDC); 2009. p. 3628–33.

[10] Kim Y, Mesbahi M. On maximizing the second smallest eigenvalue of a state-dependent graph Laplacian. In: Proceedings of the American control conference (ACC); 2005. p. 99–103.

[11] Han Z, Swindlehurst AL, Liu KJR. Optimization of MANET connectivity via smart deployment/movement of unmanned air vehicles. IEEE Trans Veh Technol 2009;58(7):3533–46.

[12] Blanco AM, Bandoni JA. Eigenvalue and singular value optimization. In: Mecanica computational; 2003.

[13] Vandenberghe L, Boyd S. Semidefinite programming. SIAM Rev 1996;38(1):49–95.

[14] Tutuncu RH, Toh KC, Todd MJ. Solving semidefinite-quadratic-linear programs using SDPT3. Math Program 2003;95(2):189–217.

[15] Todd MJ, Toh KC, Tutuncu RH. A MATLAB software package for semidefinite programming. Technical report, School of OR and IE, Cornell University Ithacat NY; 1996.

[16] Fujisawa K, Futakata Y, Kojima M, Matsuyama S, Nakamura S, Nakata K, et al. SDPA-M (semidefinite programming algorithm in MATLAB) user's manual-version 6.2, vol. 10. Research Reports on Mathematical and Computing Sciences, Series B: Operation Res., Dep. Math. and Computing Sci., Tokyo Institute of Technol., Japan; 2000.

[17] Borchers B. CSDP, AC library for semidefinite programming. Optim Methods Softw 1999;11(1–4):613–23.

[18] Walters JP, Liang Z, Shi W, Chaudhary V. Wireless sensor network security: a survey. In: Security in distributed, grid, mobile, and pervasive computing, vol. 1; 2007. p. 367.

[19] Karlof C, Wagner D. Secure routing in wireless sensor networks: attacks and countermeasures. Ad Hoc Netw 2003;1(2):293–315.

[20] Yang H, Luo H, Ye F, Lu S, Zhang L. Security in mobile ad hoc networks: challenges and solutions. IEEE Wirel Commun 2004;11(1):38–47.

[21] Roy S, Ellis C, Shiva S, Dasgupta D, Shandilya V, Wu Q. A survey of game theory as applied to network security. In: Proceedings of the IEEE the 43rd Hawaii international conference on system sciences (HICSS); 2010. p. 1–10.

[22] Manshaei MH, Zhu Q, Alpcan T, Bacşar T, Hubaux JP. Game theory meets network security and privacy. ACM Comput Surv (CSUR) 2013;45(3):25.

[23] Gueye A, Walrand J, Anantharam V. Design of network topology in an adversarial environment. In: Decision and game theory for security. Springer; 2010. p. 1–20.

[24] Lindelauf R, Blankers I. Key player identification: a note on weighted connectivity games and the Shapley value. In: Proceedings of the IEEE international conference on advances in social networks analysis and mining (ASONAM); 2010. p. 356–9.

[25] Theodorakopoulos G, Baras JS. A game for ad hoc network connectivity in the presence of malicious users. In: Proceedings of the IEEE global telecommunications conference (GLOBECOM '06); 2006. p. 1–5.

[26] Alpcan KCNT, Başar T. Stochastic games for security in networks with interdependent nodes. In: Proceedings of the IEEE international conference on game theory for networks (GameNets' 09); 2009. p. 697–703.

[27] Wang B, Wu Y, Liu KJR, Clancy TC. An anti-jamming stochastic game for cognitive radio networks. IEEE J Sel Areas Commun 2011;29:877–89.

[28] Garnaev A, Trappe W. Anti-jamming strategies: a stochastic game approach. In: Aguero R, et al, editors. Mobile networks and management. LNICST, vol. 141. Springer; 2015. p. 230–43.

[29] Neyman A, Sorin S. Stochastic games and applications, vol. 570. Springer Science & Business Media; 2003.

[30] Nisan N, Roughgarden T, Tardos E, Vazirani VV. Algorithmic game theory, vol. 1. Cambridge: Cambridge University Press; 2007.

[31] Telatar IE. Capacity of multi-antenna Gaussian channels. European Trans Telecommun 1999;10(6):585–95.

[32] Rhee W, Cioffi JM. Ergodic capacity of multi-antenna Gaussian multiple-access channels. In: Proceedings of the IEEE conference record of the thirty-fifth Asilomar conference on signals, systems and computers, vol. 1; 2001. p. 507–12.

CHAPTER

TEAM METHODS FOR DEVICE COOPERATION IN WIRELESS NETWORKS

18

David Gesbert*, Paul de Kerret*

*Communication Systems Department, EURECOM, Biot, France**

18.1 INTRODUCTION

18.1.1 DEVICE CENTRIC NETWORK OPTIMIZATION

Tens of billions of machines (sensors, robots, computers, tablets, cars, etc.) are expected to be connected to the wireless Internet within the next five to ten years. In the face of such unprecedented demand, future mobile networks must deliver on a large number of criteria such as improved spectral efficiencies, reduced latencies, and better and more consistent throughput experience in the cell as well as extended battery life. From a networking point of view, operators will require highly flexible backhaul architectures that can adapt to large fluctuations in the traffic patterns while maintaining OPEX (including energy) costs low (Table 18.1).

Infrastructure-centric designs have been—and to a large extent still are—the prevailing paradigm in wireless cellular systems such as 4G and 5G. Under this framework, network control and resource optimization tasks are deferred to the infrastructure or *cloud*. One should note the easier path to global network management that stems from such a centralized nature of computation. Nevertheless, pure network-centric designs relying on optical-supported mobile clouds currently envisioned for 5G are powerful yet expensive solutions that come with their own technical and security limitations. Finally, due to cost concerns and the possible lack of efficient preexisting infrastructures, such designs are difficult to implement precisely in those developing markets where universal broadband access could make the biggest difference. In such cases, the quicker, cheaper installation of heterogeneous wireless networks with less stringent requirements on backhaul communications is appealing. In developed user markets and elsewhere, the use of flying radio access networks with base stations carried by autonomous drones [1,2] can provide for an ultraflexible deployment of network coverage where and when it is needed the most (hotspots, concerts, sport events) or also help first responders with connectivity needs in disaster recovery scenarios. In all these examples, there is interest in designing a network of devices that can mutually cooperate or self-organize without the help of a centralized architecture and backhaul. Instead, devices should leverage local computing, communication, and memory capabilities to interact directly so as to help provide the best service possible. Such a device-centric paradigm

Cooperative and Graph Signal Processing. https://doi.org/10.1016/B978-0-12-813677-5.00018-3

Table 18.1 Summary of Notation

Notation	Description
TD	team decision
DM	decision maker
CSI	channel state information
IS	information structure
DHIS	deterministic hierarchical information structure
SHIS	stochastic hierarchical information structure
TX	transmitter
RX	receiver
n	number of decision makers
$U(\bullet)$	joint utility function
$\boldsymbol{h}$	channel state
$\hat{\boldsymbol{h}}^{(j)}$	estimate of the channel state at node j
$w^{(j)}(\bullet)$	decision function at node j
K	number of users
$R(\bullet)$	sum rate
$\mathcal{Q}(\bullet)$	quantizer
$\boldsymbol{\Sigma}_j$	CSI noise covariance matrix at TX j
$\rho^{(j,j')}$	correlation factor between the CSI noise at TX j and TX j'
$\mathcal{N}_{\mathbb{C}}(0,1)$	standard Gaussian distribution with zero mean and unit variance

renders necessary a system protocol architecture where direct communication between devices (D2D) is made possible.

The notion of cooperation has been heavily studied in the context of wireless networks as a tool to extend coverage, improve spectral efficiency and battery autonomy, or manage the interference that stems from frequency reuse [3]. As an example, so-called coordinated multi-point transmission methods have been proposed for inclusion in the 3GPP standards that feature cooperation algorithms between neighboring base stations based on combinations of multiuser MIMO, power control, and advanced resource allocation methods. Such methods are typically studied under a centralized framework enabled by the so-called Cloud RANs where wireless devices at the edge (terminals, base stations) push their observed data into an optical backhaul-supported cloud where servers run optimization algorithms before optimum decisions are sent back to edge devices for application. Interestingly, the application of such cooperation concepts in a device-centric setup has so far been mostly open, due to the challenge posed by the lack of reliable centralized channel state information in such settings.

18.1.2 COOPERATION WITH DECENTRALIZED INFORMATION

In device-centric architectures, wireless devices located at the edge of the network are recast as autonomous agents. These agents run decentralized algorithms that are designed to maximize a global network performance metric, e.g., the average sum throughput or the total user capacity under outage constraints, or minimizing latency toward accessing data content, to name a few examples. Decentralized decision algorithms are needed to guide the devices in their choice of transmission parameters such as power levels, beam design, time frequency resource utilization, routing path, etc. In principle, the coordinated decisions across neighboring devices help overall system performance. A salient feature of device-centric coordination, however, is the lack of reliable observed data (channel measurements, signal-to-noise ratios, etc.) at each decision-making device and the need to build some robustness with respect to this imperfect knowledge. In particular, an agent must make a transmission parameter decision on the basis of mostly local information, which often takes the form of a noisy and partial estimate of the global system state. Furthermore, devices have limited capability to communicate to each other. This prevents the full sharing (centralization) of system state estimates between the agents. Inevitably, a loss is to be expected in any decentralized setting when compared to the solution that would be obtained in a fully centralized setting with ideal backhaul links. The purpose and challenge behind robust device-centric coordination is exactly to minimize this loss.

Here, the device communication and decision-making capability is geared at enabling a *collective network-friendly intelligence*. As such, these smart devices differ profoundly from previously studied problems in cooperative wireless networks, such as those related to frequency agile cognitive radios or (ad hoc) user mobile relaying. The emphasis on the *network utility* and the taking into account of finite *rate and latency constraints* for interdevice communications also differs sharply from classical device cooperation studies, using, e.g., iterative game theoretic approaches [4–6], although useful connections can be made. More precisely, in our setting, the decision makers (DMs) are not conflicting with each other as in a conventional game theoretic sense. In fact it is the decentralized (and noisy) nature of the observed data, based upon which the decisions are made, that hampers the full coordination as opposed to the egoistic nature of the device itself. The theoretical roots behind device-centric coordination are found in the field of Bayesian Game with incomplete information [7] as well as the so-called *team decision theory* [8]. We should, however, raise to the reader's attention the fact that most of the line of work dealing with the use of game theoretic approaches in decentralized wireless resource allocation problems is related to trying to converge to a game equilibrium via an iterative algorithm. Each such iteration entails new observations of some utility or price, allowing the players to ultimately converge toward a coordinated decision state. In contrast, this work focuses on latency-constrained applications that require robust *single-shot* (Bayesian) decision algorithms.

18.1.3 CHAPTER ORGANIZATION AND OBJECTIVES

This chapter is meant as a brief overview of the challenges and promises related to device-centric coordination with application to future wireless networks, especially such networks that will feature one or more decentralized components, i.e., not fully relying on the Cloud-RAN implementation. We first formulate a large class of optimization problems, denoted as "Team Decision (TD) problems," that are well adapted to the context of device-centric coordination. Such problems are hard to crack in their widest generality, as can be inferred from the classical literature on decentralized control [8]. Nevertheless, we point out how the solution to a decentralized coordination problem (and its complexity) critically depends on the associated *information structure (IS)*. The latter describes in quantifiable terms the nature and quality of the observations made locally at each DM and how such local information relates to the true global system state (correlation or noise level). Wireless networks have the advantage that their design is under human control, hence the IS can be shaped in one of many possible ways, for instance by tuning quantization parameters and feedback rates. Key examples of IS designs are highlighted with their advantages toward the construction of coordination algorithms. In the second part of the chapter, we turn to the application of robust TD methods to the problem of decentralized MIMO precoding in wireless networks. Through the prism of this example, we show various strategies for deriving robust algorithms, several of which are rooted in the principle of exploiting approximation models and/or discretization of the observation and decision spaces. Numerical results highlight the benefits of robust coordination over naive coordination or lack of coordination.

18.2 TEAM DECISIONS FRAMEWORK

18.2.1 GENERAL FORMULATION OF TEAM DECISION

We give here a general formulation for a TD problem for application in a large class of device-centric wireless coordination scenarios. The decentralized network of devices as defined as follows. A network of n DMs is considered. In some examples of interest here, the DMs are wireless TXs that seek to optimize one (or possibly several) transmission parameters. We assume the n decisions couple into a resulting network performance index, which is defined below. The decision space at the kth DM is d_k dimensional and can cover a variety of domains such as the selection of a power level, a beam in a continuous or discrete grid of beams, usage of a time-frequency resource unit, a message destination (for point to multipoint networks), and many more. The general TD problem can be formulated as follows

$$(w_1^\star, \ldots, w_n^\star) = \underset{w_1,\ldots,w_n}{\operatorname{argmax}} \mathbb{E}_{h,\hat{h}^{(1)},\ldots,\hat{h}^{(n)}} \left[U\left(h, w_1(\hat{h}^{(1)}), \ldots, w_n(\hat{h}^{(n)})\right) \right], \tag{18.1}$$

where

- $h \in \mathbb{C}^m$ is the state of the system. For instance for a wireless network with n single antenna TXs, K single antenna receivers (RXs) in a flat-fading propagation scenario, the *instantaneous* system (channel) state is characterized by a random channel matrix of size $K \times n$, or equivalently a vector of $m = K \cdot n$ coefficients.
- $\hat{h}^{(j)} \in \mathbb{C}^m$ is the *local* estimate of the system state h, which is available at the jth DM.
- $w_j : \mathbb{C}^m \to \mathcal{A}_j \subset \mathbb{C}^{d_j}$ is the strategy (or policy) adopted by the jth DM. Note that the decision is made to be purely a function of what is locally observed by the jth DM. Hence for an instantaneous observation $\hat{h}^{(j)}$, the decision is $w_j(\hat{h}^{(j)})$.
- $U : \mathbb{C}^m \times \Pi_{j=1}^n \mathbb{C}^{d_j} \to \mathbb{R}$ is the global network *utility* (e.g., throughput) resulting from the policy adopted by the devices.
- $p_{h,\hat{h}^{(1)},\ldots,\hat{h}^{(n)}}$ is the *joint probability distribution* of the true system state and all local estimates. Hence $\mathbb{E}_{h,\hat{h}^{(1)},\ldots,\hat{h}^{(n)}}$ refers to the expectation operator under the joint probability rule $p_{h,\hat{h}^{(1)},\ldots,\hat{h}^{(n)}}$.

Note that while Eq. (18.1) describes a decentralized policy search, the centralized design case is simply a particular case where $\hat{h}^{(j)} = \hat{h}^{(1)}, \forall j = 2, \ldots, n$.

There are several reasons that intuitively explain the decentralized and noisy nature of state information that underpins (18.1). First, devices typically have limited sensing and feedback capabilities.

They can also be mobile with individual velocities, which tends to add varying levels of outdating to the collected information. Finally, direct exchange of channel state information between devices does not come free, or if it does the latency related to exchange may induce further outdating to the channel state information (CSI), making the CSI degradation fundamentally *device-dependent*.

18.2.2 STATIC VERSUS SEQUENTIAL POLICY DESIGN

The TD formulation (18.1) refers to a *static* setting where each of the n DMs designs a policy in order to optimally coordinate with other DMs in the Bayesian sense on the basis of a unique noisy observation of the system state. As predicted by coordination theoretic analysis [9], coordination performance is ultimately limited by the mutual correlation between observations $\hat{\boldsymbol{h}}^{(1)}, \ldots, \hat{\boldsymbol{h}}^{(n)}$ and the correlation between these estimates and the true state $\boldsymbol{h}$. The coordination setup in Eq. (18.1) precludes explicit interaction between devices, i.e., no further exchange of information (local estimates or intermediate decisions) is allowed between the devices, a hypothesis that is consistent with low latency application scenarios. In some cases, the low latency condition can be relaxed and multiple rounds of information exchanges are assumed between DMs. This opens the door to family of so-called *sequential* decision algorithms whereby a device can optimize its policy as a function of messages received from other DMs in the previous round. Eventually and under mild conditions, the algorithm will converge toward a solution near to that obtained in the centralized case and rates of convergence can be analyzed. The rest of this chapter is focused on static (single-shot) decision-making but the reader is referred to [10,11] for an overview of distributed optimization problems in signal processing and communication and to [4–6] for game theoretic approaches.

18.2.3 BEST RESPONSE FORMULATION

The above optimization is formulated in a *Bayesian* manner as a joint policy design problem. Note that by virtue of decentralization, no physical entity in the network has access to the full set of instantaneous informations $\hat{\boldsymbol{h}}^{(1)}, \hat{\boldsymbol{h}}^{(2)}, \ldots, \hat{\boldsymbol{h}}^{(n)}$. However the full knowledge of underlying joint distributions is assumed, so that is possible to compute (and maximize) the network utility in an *expected* sense.

Finding the n policies simultaneously is a daunting task and the complexity of problem (18.1) can be relaxed by adopting the classical game-theoretic *Best Response* optimization approach [12]. The Best Response optimal policy is denoted by $\boldsymbol{w}_j^{\mathrm{BR}}$ and is obtained by iteratively solving:

$$\boldsymbol{w}_j^{\mathrm{BR}} = \operatorname*{argmax}_{\boldsymbol{w}_j} \mathbb{E}_{\boldsymbol{h},\hat{\boldsymbol{h}}^{(1)},\ldots,\hat{\boldsymbol{h}}^{(n)}} \left[U\left(\boldsymbol{h}, \boldsymbol{w}_1^{\mathrm{BR}}, \ldots, \boldsymbol{w}_{j-1}^{\mathrm{BR}}, \boldsymbol{w}_j, \boldsymbol{w}_{j+1}^{\mathrm{BR}}, \ldots, \boldsymbol{w}_n^{\mathrm{BR}}\right)\right], \quad \forall j = 1, \ldots, n, \tag{18.2}$$

where for clarity we have omitted explicitly writing the dependency of the functions. This will be done recurrently in the rest of the chapter but it should always be kept in mind that $\boldsymbol{w}_j$ is only a function of $\hat{\boldsymbol{h}}^{(j)}$ and stands for $\boldsymbol{w}_j(\hat{\boldsymbol{h}}^{(j)})$.

Note, however, that in both the cases of Eqs. (18.2) and (18.1), the formulation calls for an optimization within the space of n functions $\boldsymbol{w}_j(\bullet), j = 1, \ldots, n$. In fact, just like the original formulation in Eq. (18.1), the problem in Eq. (18.2) is to be solved in a central computing location on the basis of probability density information $p_{h,\hat{h}^{(1)},\ldots,\hat{h}^{(n)}}$ alone. Yet the *application* of the policies $w_j^{\mathrm{BR}}(\hat{h}^{(j)}), j = 1, \ldots, n$, is carried out at each DM and remains fundamentally decentralized.

Although simpler than Eq. (18.1), the problem in Eq. (18.2) is generally quite difficult to solve from an algorithm design and complexity point of view. Furthermore, the coordination performance (i.e., the network utility) that can be attained under a decentralized information setting is bound to be less than what can be achieved under a centralized information scenario. The loss of performance due to imperfect sharing of the noisy CSI among the DMs is referred to as the *price of distributedness*. In practice this loss depends on the quality of the channel state estimates made available to the devices. How information about channel states is allocated among the DMs is captured by the notion of *information structure* (IS), which is covered in more detail in Section 18.2.5.

18.2.4 NAIVE AND LOCALLY ROBUST COORDINATION

The original TD problem in Eq. (18.1) seeks robustness with respect to uncertainties along two ways. First, DM j needs to be robust with respect to uncertainties related to its own local information $\hat{h}^{(j)}$. Secondly, as a multiagent problem, this device ought to take into account uncertainties at other DMs with which it seeks to coordinate. Ignoring both local and global uncertainties leads to the following *naive* policy denoted by w_j^{nv} and obtained from:

$$\left(v_1, \ldots, v_{j-1}, w_j^{\mathrm{nv}}(\hat{h}^{(j)}), v_{j+1}, \ldots, v_n\right) = \underset{w_1, \ldots, w_n}{\operatorname{argmax}}\, U\left(\hat{h}^{(j)}, w_1(\hat{h}^{(j)}), \ldots, w_n(\hat{h}^{(j)})\right),$$

where decisions $\left(v_1, \ldots, v_{j-1}, v_{j+1}, \ldots, v_n\right)$ are only auxiliary variables and will not be used in the actual transmission. In the above optimization, DM j optimistically assumes that (i) his local information $\hat{h}^{(j)}$ is perfect (equal to h) and (ii) that *all other DMs* have identical information. Interestingly, it is possible to relax the robustness with respect to the distributed nature of information while retaining robustness with respect to *local* uncertainties. Doing so, the following *locally robust* (LR) policy w_j^{LR} is obtained at DM j:

$$\left(v'_1, \ldots, v'_{j-1}, w_j^{\mathrm{LR}}, v'_{j+1}, \ldots, v'_n\right) = \underset{w_1, \ldots, w_n}{\operatorname{argmax}}\, \mathbb{E}_{h,\hat{h}^{(j)}}\left[U\left(h, w_1(\hat{h}^{(j)}), \ldots, w_n(\hat{h}^{(j)})\right)\right],$$

where this time $\mathbb{E}_{h,\hat{h}^{(j)}}$ accounts for noise in the local information at the jth DM. Here, the DM accounts for local estimation noise, yet *erroneously* assumes that the noise is the same everywhere else. This approach corresponds, in fact, to a conventional robust design in a centralized setting. The performances of naive and LR strategies vary strongly depending on the scenarios. Yet, they are often building blocks of more advanced schemes, as will be seen later on.

18.2.5 INFORMATION STRUCTURES

The IS underpinning the TD problem in Eqs. (18.1) and (18.2) describes how the local information $\hat{h}^{(j)}$ available at the jth DM relates to local estimates at other DMs $\hat{h}^{(j')}, j' \neq j$ as well as to the true global state information vector h. Ultimately the IS is characterized by the joint distribution $p_{h,\hat{h}^{(1)},\ldots,\hat{h}^{(n)}}$, which in turns governs the price of distributedness.

Additive white Gaussian noise model

An intuitive and mathematically tractable model for the decentralized information structures consists in considering that the nodes receive global information that is corrupted by an arbitrarily shaped, device-dependent Gaussian noise. In this case, the estimate at the jth DM is modeled as:

$$\hat{\boldsymbol{h}}^{(j)} \triangleq \sqrt{\mathbf{I} - (\boldsymbol{\Sigma}^{(j)})}\boldsymbol{h} + \sqrt{\boldsymbol{\Sigma}^{(j)}}\boldsymbol{\delta}^{(j)}, \tag{18.3}$$

where $\boldsymbol{\Sigma}^{(j)} \in \mathbb{R}^{m\times m}$ is a diagonal matrix taking its values in [0,1] and whose kth element represents the CSIT quality of the kth channel element at TX j. Furthermore, the CSIT noise error terms $\boldsymbol{\delta}^{(j)} \in \mathbb{C}^m$ have their elements i.i.d. $\mathcal{N}_{\mathbb{C}}(0, 1)$, are independent of the true channel, and are jointly distributed such that

$$\mathbb{E}\left[\boldsymbol{\delta}^{(j)}(\boldsymbol{\delta}^{(j')})^{\mathrm{H}}\right] = (\rho^{(j,j')})^2\mathbf{I}_m \tag{18.4}$$

with the parameters $\rho^{(j,j')} \in [0, 1]$ being the *CSI noise correlation factor.*

The main interest of this model is that it allows modeling *partially centralized CSIT,* thus bridging the gap between fully distributed configuration with *independent* CSIT errors and centralized CSIT. Indeed, the CSIT configuration where

$$\boldsymbol{\Sigma}^{(j)} = \boldsymbol{\Sigma}^{(j')},\ \rho^{(j,j')} = 1, \quad \forall j, j' = 1, \ldots, n \tag{18.5}$$

corresponds to the conventional centralized CSIT configuration [13,14] while taking

$$\rho^{(j,j')} = 0, \quad \forall j, j' = 1, \ldots, n, \quad j' \neq j \tag{18.6}$$

corresponds to the distributed CSIT configuration with independent CSIT noise [15].

Deterministic hierarchical information structure

In some network setups, some wireless nodes may be endowed with greater information gathering capabilities (e.g., high-end devices) either due to practical connectivity constraint (e.g., better connectivity to some devices) or due to a protocol design aiming at minimizing backhaul load by sharing the information only to some devices.

A so-called *Deterministically Hierarchical* Information Structure (DHIS) is obtained when the DMs can be ordered by increasing quality of CSI with DM j having access to the information at DM $j-1$ in addition to some local information. This implies that DM 1 is the least informed one while DM n is the most informed one and knows the information at all preceding DMs. Mathematically, it means that there exists some functions $f_{j,j'}$ such that

$$\hat{\boldsymbol{h}}^{(j')} = f_{j,j'}(\hat{\boldsymbol{h}}^{(j)}), \quad \forall j' < j. \tag{18.7}$$

The advantage of the DHIS is that akin to the information chain in Eq. (18.7), DMs can follow a chain of policies where a better informed DM j can *adapt* its own policies by relying on its knowledge of the decision at the lesser informed DM j' for '$j' < j$. This allows increased coordination between the DMs and simplifies strongly the optimization problem. A remaining difficulty resides in the fact the DM j cannot safely predict the behavior of better-informed devices j' with $j' > j$. Suboptimal solutions exist, however, where for instance DM j may conservatively assume that better-informed ones only have access to the same information $\hat{\boldsymbol{h}}^{(j)}$ that it has itself. This method is discussed in a practical case in Section 18.3.4.

The two nodes case

An interesting subcase of the hierarchical structure above is the two DMs scenario where the first DM has zero prior information (other than the common statistical knowledge). This case is referred to as *Master Slave* information structure. Here, the first DM is the slave: Being deprived of any real time information, its strategy consists of taking a fixed decision that maximized the network utility in an *average sense*. In this setting, we will apply the previous heuristic, which consists in letting DM 1 solve the optimization by assuming that DM 2 has received the same channel information, i.e., no information. DM 1 then solves:

$$(\boldsymbol{w}_1^{\text{stat}}, \boldsymbol{v}_2) = \underset{\boldsymbol{w}_1,\boldsymbol{w}_2}{\operatorname{argmax}}\, \mathbb{E}_{\boldsymbol{h}}\left[U(\boldsymbol{h}, \boldsymbol{w}_1, \boldsymbol{w}_2)\right],$$

where $\boldsymbol{w}_1^{\text{stat}}$ is no longer a policy but a fixed deterministic (yet statistically optimal, hence the subscript "stat") decision. Note that $\boldsymbol{v}_2$ is an auxiliary variable and will not be used in practice: it corresponds to the erroneous estimation at DM 1 of the policy used at DM 2.

Turning to the second DM, his best option is to *adapt itself* to the decision made by the first DM. As such the second DM is a *master* as it attempts to control the situation. The policy is adapted at the second DM as follows:

$$\boldsymbol{w}_2^{\star} = \underset{\boldsymbol{w}_2}{\operatorname{argmax}}\, \mathbb{E}_{\boldsymbol{h},\hat{\boldsymbol{h}}^{(2)}}\left[U\left(\boldsymbol{h}, \boldsymbol{w}_1^{stat}, \boldsymbol{w}_2(\hat{\boldsymbol{h}}^{(2)})\right)\right].$$

Note that the above optimization is meaningful because the second DM has access to the same underlying statistical information as the first DM such that it can also compute $\boldsymbol{w}_1^{\text{stat}}$ before solving for its own transmission strategy w_2. Hence, the master-slave information structure allows nicely decoupling the multiagent coordination problem into a sequence of separated single-agent problems.

Stochastically hierarchical information structure

The deterministic notion of hierarchy above imposes strong constraints on feedback (or information exchange) mechanisms between DMs, which not all practical network scenarios will be compatible with. Interestingly, the restrictive inclusion relation shown in Eq. (18.7) may be relaxed by adopting a stochastic notion of hierarchy. Referring back to the Gaussian information model shown in Eq. (18.3), a *Stochastically Hierarchical* Information Structure (SHIS) is one whereby the following relation holds:

$$\boldsymbol{\Sigma}^{(1)} \geq \boldsymbol{\Sigma}^{(2)} \geq \cdots \geq \boldsymbol{\Sigma}^{(n)}.$$

In other words, there exists a ranking between DMs in terms of the *quality* with which they observe the channel state $\boldsymbol{h}$. The SHIS model is also called *physically degraded* configuration in the information theory community [16]. Because the stochastic hierarchy does not remove the fundamental uncertainties related to local observations at the DM, this information structure does not directly lead to a strong simplification of the optimization problem (18.1). Nevertheless, if exploited properly, it can lead to an improved coordination between DMs. In [17,18], considering decentralized network MIMO precoding with SHIS, a transmission scheme is developed to exploit the stochastic hierarchical structure so as to improve the coordination between the TXs, and hence the performance.

18.3 TEAM DECISION METHODS FOR DECENTRALIZED MIMO PRECODING

In this section, we show how the TD formulation (18.1) unfolds in a particular practical scenario. In this chapter, we illustrate these methods through the prism of the example of *decentralized MIMO precoding*. We first define formally the setting considered and show how it fits in the TD framework introduced earlier. We then present three different methods to tackle the TD problem formulated.

18.3.1 SYSTEM SETTING

We study a so-called network MIMO transmission from n TXs to K RXs where the jth TX is equipped with M_j antennas while the ith RX is equipped with N_i antennas. The ith RX is sent d_i streams *jointly* from all the TXs. The total number of RX antennas, the total number of TX antennas, and the total number of streams are respectively given by

$$N_{\text{tot}} \triangleq \sum_{i=1}^{K} N_i, \quad M_{\text{tot}} \triangleq \sum_{i=1}^{K} M_i, \quad d_{\text{tot}} \triangleq \sum_{i=1}^{K} d_i. \tag{18.8}$$

We always consider that $M_{\text{tot}} \geq K$ such that in a perfect coordination setting (i.e., with ideal CSI) a precoding solution exists that allows for all users to be served at the same time, e.g., via zero-forcing precoding [19,20]. We further assume that the RXs have perfect CSI and that linear filtering is used on both the TX and the RX side, and that the RXs treat interference as noise. The channel from the n TXs to the K RXs is represented by the multiuser channel matrix $\mathbf{H} \in \mathbb{C}^{N_{\text{tot}} \times M_{\text{tot}}}$ where $\mathbf{H}_{i,j} \in \mathbb{C}^{N_i \times M_j}$ denotes the channel matrix from TX j to RX i. For the sake of exposition, we consider in the numerical evaluations that the channel elements are distributed following a standard Rayleigh fading with unit variance.

The transmission is then described as

$$\begin{bmatrix} \mathbf{y}_1 \\ \vdots \\ \mathbf{y}_K \end{bmatrix} = \mathbf{H}\mathbf{x} + \boldsymbol{\eta} = \begin{bmatrix} \mathbf{H}_1\mathbf{x} \\ \vdots \\ \mathbf{H}_K\mathbf{x} \end{bmatrix} + \begin{bmatrix} \eta_1 \\ \vdots \\ \eta_K \end{bmatrix}, \tag{18.9}$$

where $\mathbf{y}_i \in \mathbb{C}^{N_i}$ is the signal received at the ith RX, $\mathbf{H}_i \in \mathbb{C}^{N_i \times M_{\text{tot}}}$ the channel from all TXs to the ith RX, and $\boldsymbol{\eta} \triangleq [\eta_1, \ldots, \eta_n]^{\text{T}} \in \mathbb{C}^{N_{\text{tot}}}$ the normalized Gaussian noise with its elements i.i.d. as $\mathcal{CN}(0,1)$.

INFORMATION STRUCTURE

TX j receives the channel estimate $\hat{\mathbf{H}}^{(j)} \in \mathbb{C}^{N_{\text{tot}} \times M_{\text{tot}}}$ and designs its transmit coefficient $\mathbf{x}_j \in \mathbb{C}^{M_j}$ as a function of $\hat{\mathbf{H}}^{(j)}$, *without any form of information exchange with the other TXs*. To keep the notations consistent with Section 18.2, we use the vectorized version

$$\hat{\boldsymbol{h}}^{(j)} \triangleq \text{vect}\left(\hat{\mathbf{H}}^{(j)}\right) \tag{18.10}$$

and accordingly $\boldsymbol{h} = \text{vect}(\mathbf{H})$ where vect(•) denotes the vectorization operation. For convenience, we will use both $\hat{\boldsymbol{h}}^{(j)}$ and $\hat{\mathbf{H}}^{(j)}$ with the implicit reference to Eq. (18.10).

We will consider in the following the noisy Gaussian CSI model introduced in Section 18.2.5. The estimate at TX j is hence given by

$$\hat{\boldsymbol{h}}^{(j)} \triangleq \sqrt{1-\boldsymbol{\Sigma}^{(j)}}\boldsymbol{h} + \sqrt{\boldsymbol{\Sigma}^{(j)}}\boldsymbol{\delta}^{(j)}, \tag{18.11}$$

where $\boldsymbol{\Sigma}^{(j)} \in \mathbb{R}^{N_{\text{tot}}M_{\text{tot}} \times N_{\text{tot}}M_{\text{tot}}}$ is the covariance matrix of the CSIT noise at TX j, and is assumed to be diagonal in this chapter.

DECENTRALIZED PRECODING

In this distributed CSIT setting, the DM is the TX and the precoding function of TX j is denoted by

$$w_j : \mathbb{C}^{N_{\text{tot}}M_{\text{tot}}} \rightarrow \mathbb{C}^{M_j \times d_{\text{tot}}} \tag{18.12}$$

such that the transmit signal $\mathbf{x}_j$ at TX j, for a given received estimate $\hat{\boldsymbol{h}}^{(j)}$, is equal to

$$\mathbf{x}_j = w_j(\hat{\boldsymbol{h}}^{(j)})s \tag{18.13}$$

with $s \triangleq [s_1^{\text{T}}, \ldots, s_K^{\text{T}}]^{\text{T}} \in \mathbb{C}^{d_{\text{tot}}}$ containing the d_{tot} data symbols to be transmitted to the K users and distributed as i.i.d. $\mathcal{N}_{\mathbb{C}}(0, 1)$. Upon concatenation of all TX's precoding decisions, the multiuser joint precoder $\mathbf{T} \in \mathbb{C}^{M_{\text{tot}} \times d_{\text{tot}}}$ used for the transmission for a given channel realization is equal to

$$\mathbf{T} \triangleq \begin{bmatrix} w_1(\hat{\boldsymbol{h}}^{(1)}) \\ w_2(\hat{\boldsymbol{h}}^{(2)}) \\ \vdots \\ w_n(\hat{\boldsymbol{h}}^{(n)}) \end{bmatrix}. \tag{18.14}$$

We consider a per-TX power constraint such that $\|w_j(\hat{\boldsymbol{h}}^{(j)})\|^2 \leq P_j, \forall j$, with P_j being the power constraint at TX j. It is also useful to introduce the *precoder to user k*, denoted by $\mathbf{T}_k \in \mathbb{C}^{M_{\text{tot}} \times d_k}$, such that

$$\mathbf{x} = \sum_{k=1}^{K} \mathbf{T}_k s_k. \tag{18.15}$$

The decentralized joint MIMO precoding with distributed CSIT setting is illustrated in Fig. 18.1.

NETWORK UTILITY

We are interested in the particular example where the network utility of Eq. (18.1) represents the sum of all users' rates.

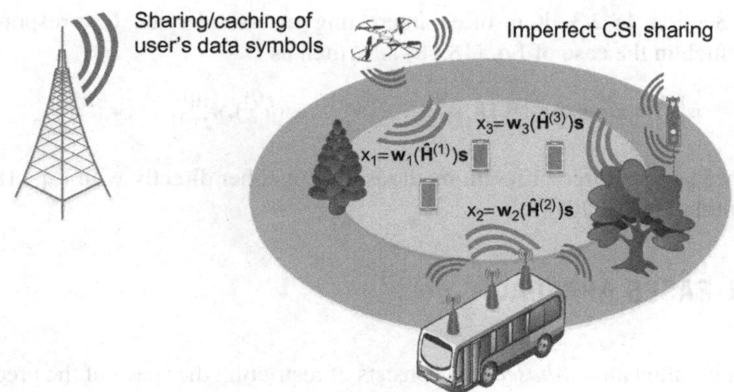

FIG. 18.1

Decentralized MIMO precoding with distributed CSIT. Due to imperfect and heterogeneous backhaul, the transmitting devices receive imperfect *and unequal* channel. Each device computes then its transmit parameters (e.g., beamforming) using *only* this locally available information.

As stated earlier, the received signal at RX k is assumed to be linearly filtered by $\mathbf{G}_k^{\mathrm{H}} \in \mathbb{C}^{d_k \times N_k}$. Due to the assumption of Gaussian signaling, the rate of user k can be written as

$$R_k \triangleq \log_2 \left| \mathbf{I}_{d_k} + \mathbf{T}_k^{\mathrm{H}} \mathbf{H}_k^{\mathrm{H}} \left(\mathbf{I}_{N_k} + \sum_{\ell=1,\ell\neq k}^{K} \mathbf{H}_k \mathbf{T}_\ell \mathbf{T}_\ell^{\mathrm{H}} \mathbf{H}_k^{\mathrm{H}} \right)^{-1} \mathbf{H}_k \mathbf{T}_k \right|. \tag{18.16}$$

Finally, we introduce the average sum rate $\mathbb{E}[R]$ as

$$\mathbb{E}[R] \triangleq \sum_{k=1}^{K} \mathbb{E}[R_k]. \tag{18.17}$$

TEAM DECISION FORMULATION

With distributed CSIT, the TD problem of Eq. (18.1) applied to the case of rate maximizing decentralized precoding can be written as:

$$(\boldsymbol{w}_1^\star, \ldots, \boldsymbol{w}_n^\star) = \underset{(\boldsymbol{w}_1,\ldots,\boldsymbol{w}_n)\in\mathcal{W}}{\operatorname{argmax}} \ \mathbb{E}[\mathrm{R}(\boldsymbol{w}_1(\hat{\boldsymbol{h}}^{(1)}), \ldots, \boldsymbol{w}_n(\hat{\boldsymbol{h}}^{(n)}))], \tag{18.18}$$

where $\mathcal{W}$ is defined as

$$\mathcal{W} \triangleq \left\{(\boldsymbol{w}_1, \ldots, \boldsymbol{w}_n) \mid \boldsymbol{w}_j : \mathbb{C}^{N_{\mathrm{tot}}M_{\mathrm{tot}}} \to \mathbb{C}^{M_j \times d_{\mathrm{tot}}}, \forall \mathbf{x} \in \mathbb{C}^{N_{\mathrm{tot}}M_{\mathrm{tot}}}, \|\boldsymbol{w}_j(\mathbf{x})\|^2 \leq P_j, \forall j\right\}. \tag{18.19}$$

As discussed in Section 18.2.3, it is often interesting to consider the best-response optimization problem (18.2), which in the case of Eq. (18.18) is written as

$$w_j^{\mathrm{BR}} = \operatorname*{argmax}_{w_j} \mathbb{E}\left[\mathrm{R}\left(h, w_1^{\mathrm{BR}}, \ldots, w_{j-1}^{\mathrm{BR}}, w_j(\hat{h}^{(j)}), w_{j+1}^{\mathrm{BR}}, \ldots, w_n^{\mathrm{BR}}\right)\right]. \tag{18.20}$$

In the following, we present three different methods to deal either directly with Eq. (18.18), or with its best-response formulation (18.20).

18.3.2 MODEL-BASED APPROACH

PRINCIPLE

Our first approach is called *model-based* and consists of restricting the space of the precoding functions by introducing a model using some parameters $\boldsymbol{\theta} \in \mathbb{C}^p$ that should typically be optimized in order to maximize the value of the joint utility achieved. How one reduces the infinite dimensional functional space to a finite parametrized space is naturally crucial. Often, the performance of this approach heavily depends on the existence of a good model that governs the devices' optimal decision. Good heuristics can hence emerge from the analysis of the problem in some limiting regimes (e.g., high/low SNR, large antenna settings).

We consider here the model of *regularized Zero-Forcing (ZF)*, which has been shown to be an efficient and robust scheme in the centralized CSIT configuration. In this model, the precoding function at TX j takes the form [19,20]:

$$w_j^{\mathrm{rZF}}(\hat{h}^{(j)}) \triangleq \mathbf{E}_j^{\mathrm{H}} \left((\hat{\mathbf{H}}^{(j)})^{\mathrm{H}}\hat{\mathbf{H}}^{(j)} + \theta_j \mathbf{I}_{M_{\mathrm{tot}}}\right)^{-1} (\hat{\mathbf{H}}^{(j)})^{\mathrm{H}} \frac{\sqrt{P_j}}{\sqrt{\Psi^{(j)}}} \tag{18.21}$$

with parameter $\theta_j > 0$ and where $\mathbf{E}_j^{\mathrm{H}} \in \mathbb{C}^{M_j \times M_{\mathrm{tot}}}$ allows selecting the precoding coefficients effectively used to transmit at TX j and is defined as

$$\mathbf{E}_j^{\mathrm{H}} \triangleq \left[\mathbf{0}_{M_j \times \sum_{j'=1}^{j-1} M_j} \quad \mathbf{I}_{M_j} \quad \mathbf{0}_{M_j \times \sum_{j'=j+1}^{n} M_j}\right]. \tag{18.22}$$

The scalar $\Psi^{(j)}$ corresponds to the power normalization at TX j. Hence, it holds that

$$\Psi^{(j)} \triangleq \|\mathbf{E}_j^{\mathrm{H}} \left((\hat{\mathbf{H}}^{(j)})^{\mathrm{H}}\hat{\mathbf{H}}^{(j)} + \theta_j \mathbf{I}_{M_{\mathrm{tot}}}\right)^{-1} (\hat{\mathbf{H}}^{(j)})^{\mathrm{H}}\|_{\mathrm{F}}^2. \tag{18.23}$$

With this parametrization, the TD optimization problem (18.18) simplifies to

$$(\theta_1^\star, \ldots, \theta_n^\star) = \operatorname*{argmax}_{(\theta_1, \ldots, \theta_n)} \mathbb{E}[\mathrm{R}(w_1^{\mathrm{rZF}}(\hat{h}^{(1)}), \ldots, w_n^{\mathrm{rZF}}(\hat{h}^{(n)}))]. \tag{18.24}$$

Through this model, the TD optimization reduces to the optimization with respect to a vector of deterministic scalars $\theta_1 \ldots \theta_n$. The model can be further simplified by parameterizing using a single common parameter θ. This forces all TXs to use the same regularization coefficient. The simplified optimization then reads as

$$\theta^\star = \operatorname*{argmax}_{\theta} \mathbb{E}[\mathrm{R}(w_1^{\mathrm{rZF}}(\hat{h}^{(1)}), \ldots, w_n^{\mathrm{rZF}}(\hat{h}^{(n)}))], \qquad \text{subject to } \theta_j = \theta, \forall j. \tag{18.25}$$

With single-antenna users and in the regime of a large number of antennas where the number of antennas at each TX grows at the same rate as the number of users, *and when the TXs use the precoding model* (18.21), it is possible to accurately approximate the expectation inside Eq. (18.25) by a deterministic equivalent R_0 [21]. This deterministic equivalent depends only on the statistical information and is obtained from a fixed-point equation [20]. The optimal parameters $\theta_j^\star$ can then be obtained using any nonconvex optimizer. In particular, in the simplified problem with a single parameter (18.25), $\theta^\star$ can be obtained via a simple linear search. We omit the deterministic equivalent expressions that require heavy notations and we refer to [21] for more details.

Note that using the deterministic equivalent is a method to transform the stochastic optimization problem (18.24) into a deterministic one. Yet it would also have been possible to apply any standard method of stochastic optimization to tackle directly Eq. (18.24) (see [22] for an overview of stochastic optimization).

PERFORMANCE EVALUATION AND SIMULATIONS

In Fig. 18.2, we show the performance obtained in a setting with $n = 2$ TXs having each $M_1 = M_2 = 15$ antennas and $K = 30$ single antenna RXs with $\rho^{(1,2)} = \rho^{(2,1)}$ uniformly distributed between [0, 1]. At TX 1, $\mathbf{\Sigma}^{(1)} = 0\mathbf{I}_{N_{\text{tot}}M_{\text{tot}}}$, which indicates that the CSI is perfect at TX 1. At TX 2, $\mathbf{\Sigma}^{(2)} = \sigma^2\mathbf{I}_{N_{\text{tot}}M_{\text{tot}}}$ with σ varyings from 0 to 1, meaning that the CSI at TX 2 varies from perfect to fully inaccurate.

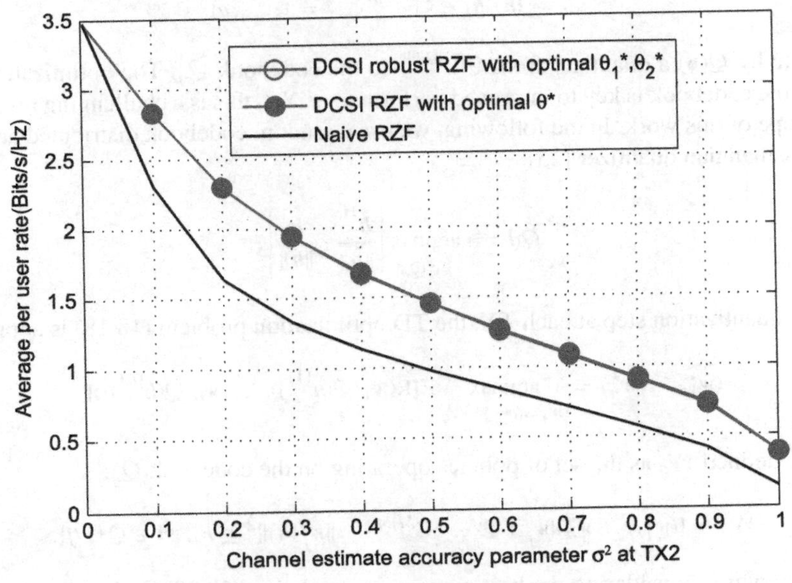

FIG. 18.2

Average rate per user as a function of the CSIT accuracy σ. As the CSIT quality degrades at the second TX, the consistency between the estimates at the TXs degrades and it becomes more important to use an adapted robust precoding scheme.

When TX 2 has quasiperfect CSIT, the optimization of the regularization coefficient does not significantly enhance the system performance when compared to the naive choice of the parameters based on the locally available CSIT. In contrast, as the CSIT configuration becomes more asymmetric, the gap between the proposed TD robust parameter choice and the locally robust parameter choice becomes more important.

18.3.3 DISCRETIZATION-BASED APPROACH

PRINCIPLE

The *discretization-based* approach consists in quantizing the estimate (input) space, thus reducing the dimension of the decision space from an infinite dimensional space to a finite dimensional one. Clearly, good performance can only be obtained with sufficient quantization points, i.e., if the dimension of the approximating space is large. Following the well-known *curse of dimensionality*, the number of quantization points should then grow exponentially with the dimension of the estimate space, such that this approach requires a lot of computing power and an efficient implementation if the dimension of the estimate is large. Yet it has the advantage of being a generic method, independent of any heuristic and adapted to any distribution of the channel and the CSIT noise.

Specifically, let us denote the codebook used at each TX by $\mathcal{Q}^q$, and assume that it contains q instances of the multiuser channel state $\boldsymbol{h}$, i.e.,

$$\mathcal{Q}_q \triangleq \{\boldsymbol{h}_\ell | \boldsymbol{h}_\ell \in \mathbb{C}^{N^{\text{tot}}M^{\text{tot}}}, \ell = 1, \ldots, q\}. \tag{18.26}$$

We then denote by $\mathcal{Q}(\bullet)$ a quantizer from $\mathbb{C}^{N^{\text{tot}}M^{\text{tot}}}$ to the codebook $\mathcal{Q}_q$. The optimization of both the quantizer and the codebook is key to improved performance. Yet, this is a challenging research problem outside the scope of this work. In the following, we use a random codebook distributed according to $p_{\boldsymbol{h}}$ and use a Grassmannian quantizer [23]:

$$\mathcal{Q}(\boldsymbol{h}) \triangleq \operatorname*{argmax}_{\hat{\boldsymbol{h}} \in \mathcal{Q}_q} \left| \frac{\hat{\boldsymbol{h}}^{\text{H}}}{\|\hat{\boldsymbol{h}}\|} \cdot \frac{\boldsymbol{h}}{\|\boldsymbol{h}\|} \right|. \tag{18.27}$$

Following this quantization step at each TX, the TD optimization problem (18.18) is approximated as

$$(\boldsymbol{w}_1^\star, \ldots, \boldsymbol{w}_n^\star) = \operatorname*{argmax}_{(\boldsymbol{w}_1, \ldots, \boldsymbol{w}_n) \in \mathcal{W}_q} \mathbb{E}[\mathrm{R}(\boldsymbol{w}_1(\mathcal{Q}(\hat{\boldsymbol{h}}^{(1)})), \ldots, \boldsymbol{w}_n(\mathcal{Q}(\hat{\boldsymbol{h}}^{(n)})))], \tag{18.28}$$

where we have defined $\mathcal{W}_q$ as the set of policies operating on the codebook $\mathcal{Q}_q$:

$$\mathcal{W}^q \triangleq \{(\boldsymbol{w}_1^q, \ldots, \boldsymbol{w}_n^q) | \boldsymbol{w}_j^q : \mathcal{Q}^q \to \mathbb{C}^{M_j \times d^{\text{tot}}}, \|\boldsymbol{w}_j^q(\hat{\boldsymbol{h}})\|^2 \leq P_j, \forall \hat{\boldsymbol{h}} \in \mathcal{Q}^q, \forall j\}. \tag{18.29}$$

This approach requires considering the best-response formulation (18.20) as the optimization remains otherwise intractable. For each codebook element $\boldsymbol{h}_\ell \in \mathcal{Q}^q$ and each TX j, we then solve

$$\boldsymbol{w}_j^{\text{BR}}(\boldsymbol{h}_\ell) = \operatorname*{argmax}_{\boldsymbol{w}_j} \mathbb{E}[\mathrm{R}\left(\boldsymbol{h}, \boldsymbol{w}_1^{\text{BR}} \circ \mathcal{Q}, \ldots, \boldsymbol{w}_{j-1}^{\text{BR}} \circ \mathcal{Q}, \boldsymbol{w}_j, \boldsymbol{w}_{j+1}^{\text{BR}} \circ \mathcal{Q}, \ldots, \boldsymbol{w}_n^{\text{BR}} \circ \mathcal{Q}\right) | \hat{\boldsymbol{h}}^{(j)} = \boldsymbol{h}_\ell]. \tag{18.30}$$

Optimization (18.30) is a conventional stochastic optimization problem for which many efficient methods can be used. In what follows, sample average approximation (SAA) using Monte Carlo runs is used [22]. The details of the algorithms are skipped and can be found in [24].

PERFORMANCE EVALUATION

In these simulations, we choose $n = 2$ TXs and $K = 2$ RXs with all the nodes having a single antenna. We also choose

$$\mathbf{\Sigma}_1 = 0.5\mathbf{I}_{N_{\text{tot}}M_{\text{tot}}}, \quad \mathbf{\Sigma}_2 = 0.1\mathbf{I}_{N_{\text{tot}}M_{\text{tot}}}, \quad \rho_{1,2} = 0. \tag{18.31}$$

To evaluate the efficiency of the proposed precoding scheme, we compare its performance with the upper bound obtained in the case where both TXs have access to the perfect instantaneous CSI and use the sum-rate maximization algorithm from [25]. We also compare the robust precoding scheme to the conventional decentralized precoding approach where each TX designs its precoder using the robust sum-rate maximization algorithm from [26], which is hence the *locally robust (LR) precoding scheme*. The quantization codebook is designed with $q = 10{,}000$ elements.

In Fig. 18.3, the average sum rate is plotted as a function of the SNR. It can be seen that the discretization approach outperforms the locally robust precoding at any SNR value. The robust precoding performs well at low to medium SNR and, in contrast to the LR precoding, is able to achieve a positive slope by serving only one user at high SNR. The proposed precoding suffers at high SNR from a degradation of the performance due to the quantization noise. This loss is expected to be reduced with more computational power and the optimization of the codebooks and quantizer.

18.3.4 HIERARCHICAL APPROACH

PRINCIPLES

We now consider the DHIS described in Section 18.2.5. Consequently, the TXs can be ordered such that TX j also has access to the CSIT at TX j' for $j' < j$. In this case, the best-response optimization problem (18.20) for a given channel realization $\hat{\boldsymbol{h}}^{(j)}$ simplifies to

$$\boldsymbol{w}_j^{\text{BR}}(\hat{\boldsymbol{h}}^{(j)}) = \operatorname*{argmax}_{\boldsymbol{w}_j} \mathbb{E}_{\boldsymbol{h},\hat{\boldsymbol{h}}^{(j+1)},\ldots,\hat{\boldsymbol{h}}^{(n)} | \hat{\boldsymbol{h}}^{(1)},\ldots,\hat{\boldsymbol{h}}^{(j)}} \left[\mathrm{R}\left(\boldsymbol{w}_1^{\text{BR}}, \ldots, \boldsymbol{w}_{j-1}^{\text{BR}}, \boldsymbol{w}_j, \boldsymbol{w}_{j+1}^{\text{BR}}, \ldots, \boldsymbol{w}_n^{\text{BR}} \right) \right]. \tag{18.32}$$

The key element in Eq. (18.32) is the conditioning on $\hat{\boldsymbol{h}}^{(1)}, \ldots, \hat{\boldsymbol{h}}^{(j)}$, which implies that the uncertainty concerns only the estimates at the TXs having a more accurate estimate, i.e., TX j' with $j' > j$. This deterministic hierarchical assumption strongly simplifies the problem as it allows starting from the most informed TX that knows all the estimates before turning to the decision at the less informed TXs. Yet the remaining difficulty resides in the fact that for $j < n$, TX j must still cope with its lack of knowledge associated with the better informed devices. Fortunately this problem can be circumvented by resorting to a simple heuristic strategy consisting of considering that TX j—when computing its precoding coefficients—assumes that TX j' for $j' > j$, has also received the same channel estimate $\hat{\boldsymbol{h}}^{(j)}$. Following this approximation, the policy $\boldsymbol{w}_j^{\text{HC}}$ at TX j is obtained from

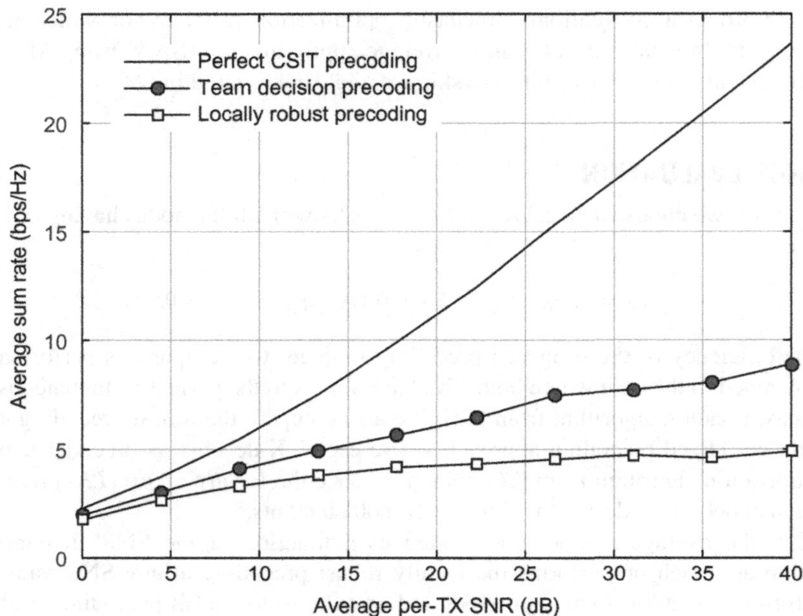

FIG. 18.3

Average sum rate as a function of the per-TX power constraint. The robust precoding solution significantly improve, the sum rate, in particular at high SNR. Higher gains are expected to be possible through the optimization of the codebook at each TX.

$$(\boldsymbol{w}_j^{\mathrm{HC}}, \boldsymbol{v}_{j+1}, \ldots, \boldsymbol{v}_n) = \underset{\boldsymbol{w}_j,\ldots,\boldsymbol{w}_n}{\operatorname{argmax}}\, \mathbb{E}[\mathrm{R}(\boldsymbol{w}_1^{\mathrm{HC}}, \ldots, \boldsymbol{w}_{j-1}^{\mathrm{HC}}, \boldsymbol{w}_j(\hat{\boldsymbol{h}}^{(j)}), \ldots, \boldsymbol{w}_n(\hat{\boldsymbol{h}}^{(j)}))]. \tag{18.33}$$

The auxiliary variables $\boldsymbol{v}_{j+1}, \ldots, \boldsymbol{v}_n$ are not used for the actual transmission due to that fact that TX j' with $j' > j$ will use whatever more accurate information is available locally to improve the precoding decision, i.e., it will solve Eq. (18.33) with its own local CSIT $\hat{\boldsymbol{h}}^{(j')}$.

The optimization problem now reduces to a conventional robust precoder optimization. Indeed, the expression in Eq. (18.33) depends only on the channel estimate $\hat{\boldsymbol{h}}^{(j)}$ such that it is no longer necessary for TX j to *estimate* the information available at the other TXs. Hence, it is possible to adapt the locally robust precoding scheme from the literature to that setting using standard linear algebra (see [27,28] for more details on the computation of the precoder at each TX).

PERFORMANCE EVALUATION

To evaluate the performance of the proposed hierarchical precoding algorithm, the performance is averaged over 1000 channel realizations via Monte Carlo simulations. We consider a simple configuration with $n = 4$ single-antenna TXs and $K = 4$ single-antenna RXs. We furthermore assume that each TX has the same power constraint P. The hierarchical precoding algorithm is compared with the maximum sum rate algorithm from [25] using perfect CSIT at every TX and with a *locally robust* algorithm from the literature [26,29], which is hence applied in a distributed manner at each TX using the CSI locally available. We show in Fig. 18.4, the average sum rate as a function of the per-TX power constraint in the following simple CSI configuration

$$\begin{aligned}\mathbf{\Sigma}^{(1)} &= \mathbf{\Sigma}^{(2)} = 0.25\mathbf{I}_{N_{\text{tot}}M_{\text{tot}}},\\ \mathbf{\Sigma}^{(3)} &= \mathbf{\Sigma}^{(4)} = 0\mathbf{I}_{N_{\text{tot}}M_{\text{tot}}}.\end{aligned} \tag{18.34}$$

It can be seen that the TD robust scheme significantly outperforms the locally robust scheme. In particular, a positive slope is achieved. This follows from the DHIS that allows the TXs having perfect CSIT to adapt to the transmit coefficients of the TXs having less accurate CSIT, thus effectively reducing interference. This simulation confirms hence the intuition that hierarchical CSIT can be beneficial to enforce consistency and allows reaching good performance even when some TXS have very inaccurate CSIT.

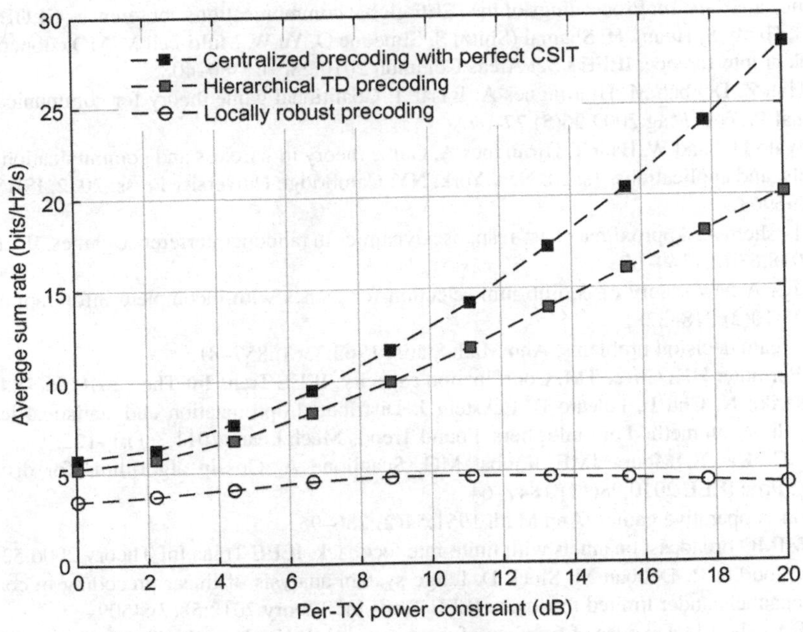

FIG. 18.4

Average sum rate as a function of the available transmit power P. The hierarchical precoding algorithm performs very well as the hierarchical structure allows the TXs with an accurate CSIT to reduce the interference generated by the other TXs, i.e., to compensate for their precoding decisions.

18.4 CONCLUSION

This chapter introduces the challenges related to device-centric coordination where devices only have their own local and noisy versions of the channel state information. We present a few avenues for further research and some initial results for solving the decentralized policy design arising from device-centric coordination. An illustration of the benefits of robust coordination design is given for the example of decentralized MIMO precoding in wireless networks.

ACKNOWLEDGMENT

The authors gracefully acknowledge the support of the European H2020 ERC project *PERFUME* for this work.

REFERENCES

[1] Chen J, Gesbert D. Optimal positioning of flying relays for wireless networks: a LOS map approach. In: Proceedings of the IEEE international conference on communications (ICC); 2017.
[2] Mozaffari M, Saad W, Bennis M, Debbah M. Drone small cells in the clouds: design, deployment and performance analysis. In: Proceedings of the IEEE global communications conference (GLOBECOM); 2015.
[3] Gesbert D, Hanly S, Huang H, Shamai (Shitz) S, Simeone O, Yu W. Multi-cell MIMO cooperative networks: a new look at interference. IEEE J Sel Areas Commun 2010;28(9):1380–408.
[4] Saad W, Han Z, Debbah M, Hjorungnes A, Basar T. Coalitional game theory for communication networks. IEEE Signal Process Mag 2009;26(5):77–97.
[5] Han Z, Niyato D, Saad W, Baar T, Hjrungnes A. Game theory in wireless and communication networks: theory, models, and applications. 1st ed. New York, NY: Cambridge University Press; 2012. ISBN 0521196965, 9780521196963.
[6] Bistritz I, Leshem A. Approximate best-response dynamics in random interference games. IEEE Trans Autom Control 2018;63(6):1549–62.
[7] Harsanyi JC. A new theory of equilibrium selection for games with incomplete information. Games Econ Behav 1995;10(2):318–32.
[8] Radner R. Team decision problems. Ann Math Statist 1962;33(3):857–81.
[9] Cuff PW, Permuter HH, Cover TM. Coordination capacity. IEEE Trans Inf Theo 2010;56(9):4181–206.
[10] Boyd S, Parikh N, Chu E, Peleato B, Eckstein J. Distributed optimization and statistical learning via the alternating direction method of multipliers. Found Trends Mach Learn 2011;3(1):1–122.
[11] Dimakis AG, Kar S, Moura JMF, Rabbat MG, Scaglione A. Gossip algorithms for distributed signal processing. Proc IEEE 2010;98(11):1847–64.
[12] Nash J. Non-cooperative games. Ann Math 1951;54(2):286–95.
[13] Jindal N. MIMO broadcast channels with finite-rate feedback. IEEE Trans Inf Theory 2006;52(11):5045–60.
[14] Wagner S, Couillet R, Debbah M, Slock D. Large system analysis of linear precoding in correlated MISO broadcast channels under limited feedback. IEEE Trans Inf Theory 2012;58(7):4509–37.
[15] de Kerret P, Gesbert D. Degrees of freedom of the network MIMO channel with distributed CSI. IEEE Trans Inf Theory 2012;58(11):6806–24.
[16] Cover T, Thomas A. Elements of information theory. Wiley-Interscience; 2006.
[17] de Kerret P, Gesbert D. Network MIMO: transmitters with no CSI can still be very useful. In: Proceedings of the IEEE international symposium on information theory (ISIT); 2016.

[18] Bazco A, de Kerret P, Gesbert D, Gresset N. Generalized degrees-of-freedom of the 2-user case MISO broadcast channel with distributed CSIT. In: Proceedings of the IEEE international symposium on information theory (ISIT); 2017.

[19] Spencer QH, Swindlehurst AL, Haardt M. Zero-forcing methods for downlink spatial multiplexing in multiuser MIMO Channels. IEEE Trans Signal Process 2004;52(2):461–71.

[20] Couillet R, Debbah M. Random matrix methods for wireless communications. Cambridge University Press; 2011.

[21] Li Q, de Kerret P, Gesbert D, Gresset N. Robust regularized ZF in cooperative broadcast channel under distributed CSIT. IEEE Trans Inf Theory; submitted November 2016 (in review).

[22] Shapiro A, Dentcheva D, Ruszczynski A. Lectures on stochastic programming: modeling and theory. Philadelphia, PA, USA: Society for Industrial and Applied Mathematics; 2014. ISBN 1611973422, 9781611973426.

[23] Dai W, Liu Y, Rider B. Quantization bounds on Grassmann manifolds and applications to MIMO communications. IEEE Trans Inf Theory 2008;54(3):1108–23.

[24] de Kerret P, Gesbert D. Quantized team precoding: a robust approach for network MIMO under general CSI uncertainties. In: Proceedings of the IEEE international workshop on signal processing advances in wireless communications (SPAWC); 2016.

[25] Christensen SS, Agarwal R, Carvalho E, Cioffi JM. Weighted sum-rate maximization using weighted MMSE for MIMO-BC beamforming design. IEEE Trans Wirel Commun 2008;7(12):4792–9.

[26] Fritzsche R, Fettweis GP. Robust sum rate maximization in the multi-cell MU-MIMO downlink. In: Proceedings of the IEEE wireless communications and networking conference (WCNC); 2013.

[27] Fritzsche R, Fettweis G. Distributed robust sum rate maximization in cooperative cellular networks. In: Proceedings of the IEEE workshop on cooperative and cognitive mobile networks (CoCoNet); 2013.

[28] de Kerret P, Fritzsche R, Gesbert D, Salim U. Robust precoding for network MIMO with hierarchical CSIT. In: Proceedings of the IEEE international symposium on wireless communication systems (ISWCS); 2014.

[29] Negro F, Ghauri I, Slock DTM. Sum rate maximization in the noisy MIMO interfering broadcast channel with partial CSIT via the expected weighted MSE. In: Proceedings of the IEEE international symposium on wireless communication systems (ISWCS); 2012.

CHAPTER

COOPERATIVE DATA EXCHANGE IN BROADCAST NETWORKS

19

Alex Sprintson*

*Department of Electrical and Computer Engineering, Texas A&M University, College Station, TX, United States**

19.1 INTRODUCTION

In the last decade, there has been a significant interest in *network coding*, a novel technique to improve the reliability, robustness, and security of communication networks and distributed storage systems. The network coding technique was originally proposed as a way to improve the throughput of data transmission over a multicast network by allowing intermediate network nodes to combine packets from different flows [1–3]. Over the last two decades, many novel network coding techniques have been proposed for wireless, wireline, and distributed storage settings. It was shown that network coding can benefit a large number of applications, including wireless networks, multimedia streaming, disruption tolerant networks, distributed storage, and network security.

Wireless network coding employs innovative coding techniques that leverage the broadcast properties of the wireless medium to improve network performance. The chapter focuses on the *cooperative data exchange (CDE)* problem, which is one of the central problems in this area [4–6]. An instance of Problem CDE includes a set of messages $X = \{x_1, \ldots, x_n\}$ and a set of wireless clients $C = \{c_1, \ldots, c_k\}$, such that each client $c_i \in C$ initially holds a subset X_i of the messages in the set X. The clients share a lossless broadcast channel to transmit a linear combination of the messages. The goal is to enable all clients to decode missing messages while minimizing the total number of transmissions (Table 19.1).

Fig. 19.1A presents an instance of Problem CDE. In this instance, there are three wireless clients $\{c_1, c_2, c_3\}$ that need to obtain five messages $\{x_1, \ldots, x_5\}$. Initially, the clients c_1, c_2, and c_3 hold messages $\{x_1, x_3, x_5\}$, $\{x_2, x_3, x_4\}$ and $\{x_1, x_2, x_4\}$, respectively. Note that each client is missing two messages. The solution without network coding requires transmission of all five messages. Fig. 19.1B shows a coding solution that includes only three transmissions, with every client making exactly one transmission.

This problem arises naturally in many practical settings, including synchronization of the local databases and file systems, gathering data from correlated sources in sensor networks, and distributed data storage. In particular, the use of coding techniques is instrumental for improving the efficiency of wireless device-to-device (D2D) communication [7,8]. With D2D communication, mobile users in a

Cooperative and Graph Signal Processing. https://doi.org/10.1016/B978-0-12-813677-5.00019-5

Table 19.1 Summary of Notation

Notation	Description
$X = \{x_1, \ldots, x_n\}$	set of messages
n	total number of messages
$C = \{c_1, \ldots, c_k\}$	set of clients
k	the number of clients
$\mathbb{F}_q$	the underlining Galois field of size q
X_i	side information set of client c_i
$\overline{X}_i$	set of required messages of client c_i
p_i	packet transmitted at round i
c_{t_i}	the client transmitting at round i
$[\gamma_i^1, \gamma_i^2, \ldots, \gamma_i^n]$	encoding vector of packet p_i
Γ_i	encoding matrix for packets $p_1, \ldots, p_i$
OPT	optimal number of transmissions
$u_i = [u_i^1, u_i^2, \ldots, u_i^n]$	the ith *unit* encoding vector
U_i	a matrix whose rows include unit encoding vectors that correspond to messages in X_i
r_i	number of transmissions made by client c_i
OPT_i	the minimum number of transmissions required when each client has access to messages in X_i and all packets in $\{p_1, \ldots, p_i\}$
$\hat{p}_1, \ldots, \hat{p}_{OPT_i}$	OPT_i transmission packets that are part of the optimal solution
Ω_{i-1}	matrix whose rows correspond to coding coefficients of packets $\hat{p}_1, \ldots, \hat{p}_{OPT_{i-1}}$
γ_i	a random linear combination of rows of U_i
g-robust schedule	a schedule that can tolerate failure of g clients
η	maximum number of transmission made by any client $c_i \in C$
c_{i^*}	the client that makes the largest number of transmissions
Θ_1, Θ_2, and Θ_3	encoding matrices computed in different phases of the robust CDE algorithm
μ	number of packets accessible to a wiretapper
E	set of packets observed by a wiretapper
g-weakly secure schedule	a schedule that can protect any subset of g packets from the wiretapper

physical vicinity of each other can form small ad hoc groups to exchange data, synchronize databases, exchange secret keys, and collaboratively download large files.

This chapter focuses on the design of efficient data exchange algorithms for wireless environments. We consider the setting with perfect channels and reliable clients as well as settings with malicious and faulty clients. In addition, we address the design and analysis of light-weight, information-theoretical secure schemes for protecting user information from eavesdroppers.

Handling faulty and unreliable clients. Wireless nodes are inherently unreliable. In particular, some of the clients may stop transmission at any time due to loss of coverage or limited power supply. Accordingly, we discuss coding schemes that can tolerate a limited number of client failures. Resilience to failures is achieved by adding redundant transmissions that ensure that all valid clients are able to receive all messages in any failure scenario.

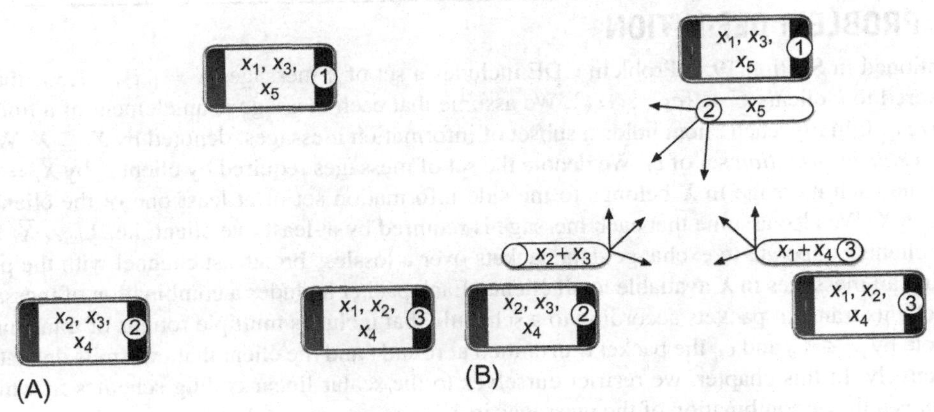

FIG. 19.1

An instance of Problem CDE with three clients.

Information exchange in the presence of wiretappers. Due to their reliance on the wireless spectrum, wireless networks are vulnerable to attack and misuse. In particular, a wiretapper located in the vicinity of the nodes that exchange information might be able to intercept some of the transmitted packets and obtain information about the information messages. We discuss a lightweight, information-theoretic mechanism that prevents the eavesdropper from being able to obtain any meaningful information about the transmitted messages.

RELATED WORK

Cooperative communication at the physical, network, and application layers has been the subject of extensive research in the past few years. Physical layer user cooperation in the form of signal relaying has been shown to result in higher data rates, extended coverage, and robustness to link outages [9–11].

The CDE problem was first formulated in [12] for one-hop broadcast networks. Lower and upper bounds on the minimum required number of transmissions were established in [13]. Courtade and Wesel [5] have considered a setting of the problem where information messages follow a random distribution. The problem has been generalized for arbitrary networks in [5,14–17]. Several works have presented randomized and deterministic solutions for the problem [4,6,18]. References [19–23] studied variations and extensions of Problem CDE. In particular, references [24,25] addressed scenarios with various transmission costs. Coding schemes that provide secrecy and weak security in the presence of an eavesdropper have been considered in [19,20,22,26], respectively.

In information theory, the problem is closely related to the *secure key agreement (SKA)* problem [27] and to the concept of a multivariate mutual information (MMI) measure [28].

A closely related coding problem with side information is the *index coding* problem [29–33]. This problem was originally motivated by satellite broadcast applications with caching clients. However, the index coding setup is centralized and noncooperative with a single transmitter server holding all information messages and passive clients having different demands. A related problem of set reconciliation between two or more similar sets was studied in [34].

19.2 PROBLEM DEFINITION

As mentioned in Section 19.1, Problem CDE includes a set of n messages $X = \{x_1, \dots, x_n\}$ that need be delivered to k clients $C = \{c_1, \dots, c_k\}$. We assume that each message is an element of a finite field $\mathbb{F}_q$ of size q. Initially, each client holds a subset of information messages, denoted by $X_i \subseteq X$. We refer to X_i as a *side information* set of c_i. We denote the set of messages required by client c_i by $\overline{X}_i = X \setminus X_i$. We assume each message in X belongs to the side information set of at least one of the clients, i.e., $\cup_{c_i \in C} X_i = X$. We also assume that each message is required by at least one client, i.e., $\cup_{c_i \in C} \overline{X}_i = X$.

The clients cooperate to exchange data packets over a lossless broadcast channel with the purpose of making all messages in X available to all clients. Each packet includes a combination of messages in X. The clients transmit packets according to a schedule that includes multiple rounds of transmissions. We denote by $p_i \in \mathbb{F}_q$ and c_{t_i} the packet transmitted at round i and the client that transmits data at round i, respectively. In this chapter, we restrict ourselves to the scalar linear coding schemes in which the packet p_i is a linear combination of the messages in X_{t_i}, i.e.,

$$p_i = \sum_{x_j \in X_{t_i}} \gamma_i^j x_j, \tag{19.1}$$

where $\gamma_i^j \in \mathbb{F}_q$ are the *encoding coefficients* of p_i. We refer to $\gamma_i = [\gamma_i^1, \gamma_i^2, \dots, \gamma_i^n]$ as the *encoding vector* of p_i, where $\gamma_i^j = 0$ for $x_j \notin X_{t_i}$. We denote by Γ_i the $i \times n$ matrix whose rows are composed of vectors $\gamma_1, \dots, \gamma_i$. We refer to Γ_i as the *encoding matrix* for packets $p_1, \dots, p_i$.

Note that, in general, p_i can be a combination of messages in X_{t_i} and the packets $\{p_1, \dots, p_{i-1}\}$ previously transmitted over the channel. However, this will not provide any benefit in our setting.

Our goal is to devise a scheme that enables each client $c_i \in C$ to obtain all messages in $\overline{X}_i$ while minimizing the total number of transmissions. We refer to the minimum number of transmissions required to satisfy all clients as OPT.

To formally define our algorithm, we need a few more notations. We denote by $u_i = [u_i^1, u_i^2, \dots, u_i^n]$ the ith *unit* encoding vector that corresponds to the original packet x_i, where $u_i^i = 1$ and $u_i^j = 0$ for $i \neq j$. We also denote by U_i an $|X_i| \times n$ matrix whose rows include vectors u_i that correspond to messages in X_i.

LINEAR FORMULATION AND RELATED PROBLEMS

Courtade et al. [15] have observed that the CDE problem can be formulated as the following integer linear program (ILP) that determines the number of transmissions r_i that need to be made by any client $c_i \in C$.

$$\begin{aligned}
\text{min.} \ & \sum_{c_i \in C} r_i \\
\text{s.t.} \ & \sum_{c_i \in C \setminus \hat{C}} r_i \geq \left| \bigcap_{c_i \in \hat{C}} \overline{X}_i \right|, \quad \forall \hat{C} \subseteq C. \\
& r_i \in \{0, 1\}, \ \forall c_i \in C.
\end{aligned}$$

Note that the objective of the ILP is to minimize the *sum-rate*, i.e., the total number of transmissions made by all clients. For each subset $\hat{C}$ of C we require that at least $|\bigcap_{c_i \in \hat{C}} \overline{X}_i|$ transmissions are made by the clients outside of $\hat{C}$. Note that this is a necessary condition because all clients in $\hat{C}$ are missing all the messages in $|\bigcap_{c_i \in \hat{C}} \overline{X}_i|$. It turns out that this condition is sufficient, i.e., for any feasible solution $\{r_i \,| c_i \in C\}$, of the ILP there exists a feasible network coding scheme that meets the demands of every client. Such a scheme can be obtained through random selection of the network coding coefficients. It is possible to show the correctness of this approach using the general principles of network coding. The interested reader is referred to [6,15] for more details.

Courtade and Wesel [5] have used this formulation to show that if the messages are randomly distributed then the minimum number of transmissions is equal to

$$\left\lceil \frac{1}{k-1} \sum_{i=1}^{k} |\overline{X}_i| \right\rceil \tag{19.2}$$

with probability approaching 1, as the number of messages approaches infinity.

Note that without the integrality constraints, the optimal solution might have fractional values. In general, the cost of a fractional solution can be lower than the cost of an integral one. In practice, the fractional solution can be implemented by dividing messages and packets into small parts. This technique is referred to as *subpacketization* in the literature. In this chapter, we focus on the scalar linear solutions in which packets and messages cannot be split.

The fractional version of the CDE problem is closely related to the *global omniscience* problem in information theory [27,35]. In information-theoretic formulations, the messages are assumed to be generated by discrete memoryless sources, and the side information is obtained by users privately observing a component of the source. The users communicate over the broadcast channel with the goal of all users being able to obtain the entire source, i.e., to become *omniscient*.

19.3 RANDOMIZED ALGORITHM FOR PROBLEM CDE

19.3.1 ALGORITHM DESCRIPTION

This section presents a randomized algorithm for Problem CDE. The algorithm requires that the size of the underlying finite field is at least $k \cdot n$, i.e., $q \geq k \cdot n$. The algorithm identifies an optimal solution with the probability of at least $1 - \frac{n \cdot k}{q}$. We can amplify the probability by repeating the algorithm multiple times.

The algorithm operates in rounds. At the first round, we choose the client with the maximum side information set to make a transmission, i.e.,

$$c_{t_1} = \arg\max_{c_i \in C} |X_i|.$$

At each of the following rounds, $i = 2, 3, \ldots$, we choose the client c_{t_i} that has the maximum number of degrees of freedom, i.e.,

$$c_{t_i} = \arg\max_{c_j \in C} \left\{ \text{rank} \begin{bmatrix} U_j \\ \Gamma_{i-1} \end{bmatrix} \right\},$$

where $\begin{bmatrix} U_j \\ \Gamma_{i-1} \end{bmatrix}$ refers to the matrix obtained by combining rows of U_j and Γ_{i-1}.

The client c_{t_i} broadcasts packet p_i, which is a random combination of messages in its side information set. That is, the packet p_i is formed according to Eq. (19.1), where $\gamma_i^j = 0$ for $x_j \in \bar{X}_{t_i}$; other elements of γ_i are random elements of the field $\mathbb{F}_q$.

19.3.2 ALGORITHM ANALYSIS

We proceed to show that the algorithm finds an optimal solution, i.e., the solution that satisfies the demands of all clients with the minimum number of transmissions. The key idea of the proof is to show that every step of the algorithm does not result in any loss of optimality. For each round of the algorithm, let OPT_i be the minimum number of transmissions required to satisfy all clients given that each client $c_i \in C$ has access to packets already transmitted over the channel, $p_1, \ldots, p_i$, in addition to the packets in X_i already available to it. That is, OPT_i is the minimum number of transmissions required when each client has access to messages in X_i and all packets in $\{p_1, \ldots, p_i\}$.

In Theorem 19.1 we show that with high probability, the value of OPT_i decreases by one during each iteration of the algorithm. If the desired event occurs at each iteration (i.e., the value of OPT_i decreases by one), then after OPT iterations, the algorithm will be able to identify an optimal solution to the problem (also with high probability).

Theorem 19.1. *With probability at least* $1 - \frac{k}{q}$, *it holds that* $OPT_i = OPT_{i-1} - 1$, *provided that the size of the underlying field* $q \geq k$. ■

We proceed to present the proof of Theorem 19.1. Let Γ_{i-1} be the encoding matrix for packets transmitted in rounds $1, 2, \ldots, i-1$. Because all clients can be satisfied with OPT_{i-1} additional transmissions, there exist OPT_{i-1} vectors $\hat{p}_1, \ldots, \hat{p}_{OPT_{i-1}}$ that together with packets $p_1, \ldots, p_{i-1}$ constitute a solution to Problem CDE. We denote by Ω_{i-1} the matrix whose rows correspond to coding coefficients of packets $\hat{p}_1, \ldots, \hat{p}_{OPT_{i-1}}$.

Note that matrix Ω_{i-1} satisfies the following conditions:

Condition I: Each row of Ω_{i-1} is a linear combination of rows in U_j, for some client $c_j \in C$.

Condition II: For each client $c_j \in C$, it holds that the matrix formed by combining rows of Ω_{i-1}, Γ_{i-1}, and U_j is of rank n.

Let c_{t_i} be the client that was chosen to transmit at round i and let γ_i be a random linear combination of rows of U_i.

Lemma 19.1. $OPT_{i-1} > n - rank \begin{bmatrix} U_{t_i} \\ \Gamma_{i-1} \end{bmatrix}$.

Proof. Because c_{t_i} is the client with the maximum degrees of freedom, it holds that

$$\text{rank} \begin{bmatrix} U_{t_i} \\ \Gamma_{i-1} \end{bmatrix} \geq \text{rank} \begin{bmatrix} U_j \\ \Gamma_{i-1} \end{bmatrix} \tag{19.3}$$

for any other client $c_j \in C$.

We consider two cases. In the first case, there exists a client c_j for which Eq. (19.3) holds with strict inequality. In this case, it holds that

$$OPT_{i-1} \geq n - \text{rank}\begin{bmatrix} U_j \\ \Gamma_{i-1} \end{bmatrix} > n - \text{rank}\begin{bmatrix} U_{t_i} \\ \Gamma_{i-1} \end{bmatrix}.$$

In the second case, Eq. (19.3) holds with equality for all clients in C. We note that the optimal solution contains a client c_j that makes at least one transmission. For this client, it holds that

$$OPT_{i-1} > n - \text{rank}\begin{bmatrix} U_j \\ \Gamma_{i-1} \end{bmatrix} = n - \text{rank}\begin{bmatrix} U_{t_i} \\ \Gamma_{i-1} \end{bmatrix}.$$

■

The key observation of the proof is captured by the following lemma:

Lemma 19.2. *With probability at least $1 - \frac{k}{q}$, there exists a row in Ω_{i-1} that can be replaced by γ_i such that both Conditions I and II are still satisfied.*

Proof. By Claim 19.1, $OPT_{i-1} > n - \text{rank}\begin{bmatrix} U_{t_i} \\ \Gamma_{i-1} \end{bmatrix}$. This implies that there exists at least one row $\hat{\omega}$ in Ω_{i-1} that is not required by c_{t_i} to satisfy its requirements. Let Ω_i be the matrix formed from Ω_{i-1} by removing row $\hat{\omega}$. Note that the matrix formed by combining rows of Ω_i, Γ_{i-1}, and U_{t_i} is of rank n.

Now, let $\bar{\Omega}_{i-1}$ be a matrix formed from Ω_{i-1} by replacing row $\hat{\omega}$ by γ_i. Because γ_i is a random linear combination of rows of U_{t_i}, Condition I is satisfied. We note that the matrix formed by combining rows of $\bar{\Omega}_{i-1}$, Γ_{i-1}, and U_{t_i} has rank n.

We now prove that for any client $c_j \in C\setminus\{c_{t_i}\}$, the matrix formed by combining rows of $\bar{\Omega}_{i-1}$, Γ_{i-1}, and U_j is also of rank n with high probability.

Let c_j be an arbitrary client, such that $j \neq t_i$. First, note that the matrix formed by combining rows of Ω_i, Γ_{i-1}, and U_j is of rank at least $n-1$. Note that we only need to consider the case in which its rank is $n-1$.

Let ζ_j be a normal vector to the combined row space of matrices Ω_i, Γ_{i-1}, and U_j. We note that there exists a row vector $u_{\hat{l}}$ in U_{t_i} such that the inner product $\langle u_{\hat{l}}, \zeta_j\rangle \neq 0$. Indeed, if this is not the case, then all vectors of U_{t_i} would belong to the row space of the matrix formed by combining rows of Ω_i, Γ_{i-1}, and U_j, which will result in a contradiction. In particular, this will contradict the fact that the rank of the matrix formed by combining rows of Ω_i, Γ_{i-1} and U_{t_i} is n while the rank of the matrix formed by combining rows of Ω_i, Γ_{i-1}, and U_j is $n-1$.

Next, we use the fact that $\langle u_{\hat{l}}, \zeta_j\rangle \neq 0$ to prove that $\langle \gamma_i, \zeta_j\rangle \neq 0$ with probability at least $1 - \frac{1}{q}$. Because γ_i is a linear combination of vectors in U_{t_i}, we can write it as

$$\gamma_i = \sum_{u_l \in U_{t_i}} \gamma_i^l \cdot u_l.$$

Thus,

$$\langle \gamma_i, \zeta_j\rangle = \sum_{u_l \in U_{t_i}} \gamma_i^l \cdot \langle u_l, \zeta_j\rangle = \gamma_i^{\hat{l}} \cdot \langle u_{\hat{l}}, \zeta_j\rangle + c,$$

where c is an element in $\mathbb{F}_q$. Because $\langle u_{\hat{l}}, \zeta_j\rangle \neq 0$ and because $\gamma_i^{\hat{l}}$ is selected at random from $\mathbb{F}_q$, the probability that $\langle \gamma_i, \zeta_j\rangle = 0$ is bounded by $1/q$.

We proved that the probability that Condition II is satisfied for an arbitrary client is at least $1 - \frac{1}{q}$. Then, by the union bound, the probability that Condition II is satisfied for all clients is at least $1 - \frac{k}{q}$. ■

Lemma 19.2 shows that with probability at least $1 - \frac{k}{q}$ there exists a row $\hat{\omega}$ in Ω_{i-1} that can be replaced with γ_i such that the resulting matrix still represents a broadcast scheme that allows all clients to decode the messages they need with OPT_{i-1} transmissions. This implies, in turn, that matrix Ω_i formed from Ω_{i-1} by removing row $\hat{\omega}$ represents $OPT - 1$ encoding vectors required to complete the information exchange after iteration i. This completes the proof of Theorem 19.1.

Theorem 19.2 implies that the algorithm identifies an optimal solution to Problem CDE with high probability.

Theorem 19.2. *The algorithm computes an optimal solution to Problem CDE with probability at least* $1 - \frac{k \cdot n}{q}$*, provided that the size of the underlying field* $q \geq k \cdot n$*.*

Proof. Theorem 19.1 implies that each iteration of the algorithm reduces the value of OPT_i by one with probability $1 - \frac{k}{q}$. We can use the union bound argument to show that the probability that one of the rounds fails to reduce the value of OPT_i is bounded by

$$\frac{OPT \cdot k}{q} \leq \frac{n \cdot k}{q}.$$

We conclude that the probability of success is at least $1 - \frac{k \cdot n}{q}$. ■

The probability of success can be amplified by selecting a sufficiently large q. For example, for $q \geq 4kn$ the probability of success is at least 3/4. An additional way to increase the probability of success is to execute the algorithm multiple times. This observation is captured in the following corollary.

Corollary 19.1. *For any* $\varepsilon > 0$ *there exists an algorithm that finds an optimal solution to Problem CDE with probability at least* $1 - \varepsilon$ *in time polynomial in the size of the input and* $\log(\varepsilon)$.

19.4 DATA EXCHANGE WITH FAULTY CLIENTS

In the previous section we assumed that all clients are reliable and are able to complete the transmission as specified by the schedule. However, in many practical settings, the clients might leave the system at any time, e.g., due to poor channel conditions, a mobility event, loss of power supply, or an instruction from the user. In this section, we address the problem of constructing a *robust schedule*, i.e., the schedule that enables all remaining clients to decode the packets they need even if a limited number of clients fail or leave the system.

The key idea of our approach is to add a sufficient number of redundant transmissions so no additional transmissions are necessary in the event of a failure. We show that while the original problem (in which all clients are reliable) can be solved in polynomial time, designing an efficient robust schedule is an NP-hard problem even for a single failure scenario [36]. Accordingly, we present an approximation algorithm with the required number of transmissions within a constant factor of the optimum.

Fig. 19.2 presents an example of a reliable schedule. In this example, there are four clients that initially have a subset of messages in the set $\{x_1, x_2, x_3, x_4\}$. A robust schedule includes four transmissions, one from each client. The solution is robust to failure of any two clients. That is, even if one of the clients fails, the remaining clients are still able to decode the messages they need from any two remaining transmissions.

The problem of the design of an efficient robust schedule has a setting similar to that described in Section 19.2, with the only difference being that up to g clients can fail. When a client fails, some of its transmissions might not reach other clients. We are focusing on the worst-case scenario, i.e., any g out of k clients can fail, and all faulty clients do not make any transmissions. We say that a coding

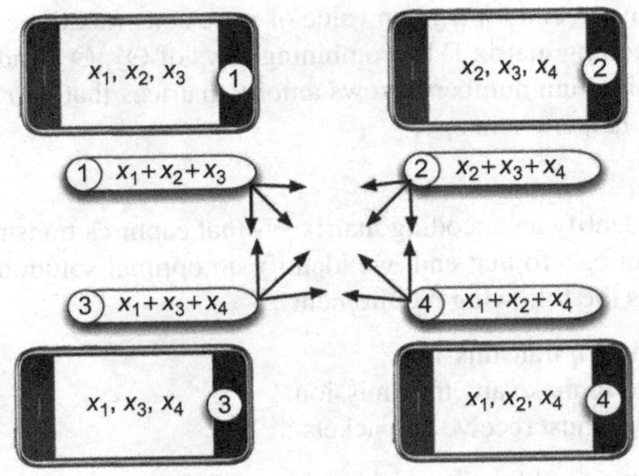

FIG. 19.2

An example of a robust schedule.

scheme is *g-robust* if for any subset $C' \subset C$ of cardinality g, all clients in $C \setminus C'$ can decode all messages in X even if all clients in C' fail to transmit packets required by the coding scheme. Our goal is to find, for the given instance of Problem CDE, a g-robust schedule that minimizes the number of transmissions.

We observe that the necessary and sufficient condition for the existence of a g-robust schedule is that any message $x_i \in X$ is initially held by at least $g + 1$ clients. Indeed, if a message is held by less than g clients, it would not be possible to recover it upon the failure of these clients. Also, a simple suboptimal achievability scheme would include transmitting each message $g + 1$ times.

19.4.1 APPROXIMATION ALGORITHM

Because it is intractable to construct an optimal robust schedule, we present a polynomial-time approximation scheme that computes a coding scheme that is robust to any single failure. The number of transmissions required by the coding scheme is at most two times of the number of transmissions made by the optimal solution.

The key idea of the approximation scheme is to mimic the optimal solution. While we do not know the structure of the optimal solution, we can identify two of its key parameters:

(i) maximum number of transmission η made by any client $c_i \in C$;
(ii) the identity of the client c_{i^*} that makes the largest number of transmissions.

We can identify η and c_{i^*} by performing an exhaustive search among all possible values of $\eta \in \{1, \ldots, n\}$ and among all k clients in C. For each combination of η and c_{i^*} we perform the three-phase procedure (described below) and select the values that yield the minimum possible number of transmissions. This will result in at most $k \cdot n$ invocations of the procedure. We refer to c_{i^*} as a *helper* client.

For a given helper client c_{i^*} and for a given value of η we construct three matrices Θ_1, Θ_2, and Θ_3. Then, we construct an encoding matrix Γ by combining rows of Θ_1, Θ_2, and Θ_3. Finally, we select an encoding matrix with a minimum number of rows among matrices that correspond to different values of η and the selection of a helper client c_{i^*}.

Phase I

The goal of Phase I is to identify an encoding matrix Θ_1 that captures transmissions that are sufficient to handle a failure of client c_{i^*}. To that end, we identify an optimal solution to a modified version of Problem CDE that satisfies the following requirements:

1. Each client makes at most η transmissions;
2. Client c_{i^*} is not allowed to make any transmissions;
3. All clients, including c_{i^*}, must receive all packets in X.

It can be shown that by a modification of Algorithm CDE, it is possible to identify a solution for this variation of the problem with high probability [37].

For a client $c_i \in C$, let Θ_1^i be the encoding matrix that includes coding vectors of packets transmitted by c_i in Phase 1. We also denote by λ_i the number of transmissions made by client c_i in Phase 1 of the algorithm. Note that λ_i is equal to the number of rows of Θ_1^i.

Because the coding vectors represented by Θ_1 are not sufficient for handling failures of any client other than c_{i^*}, we add additional transmissions in Phases II and III of the algorithm.

Phase II

In Phase II of the algorithm, we form encoding matrix Θ_2. Transmissions represented by Θ_2 (in combination with transmissions represented by Θ_1) ensure that the helper node c_{i^*} is able to obtain all packets in the event of a failure of any client $c_i \in C \setminus \{c_{i^*}\}$.

Matrix Θ_2 is constructed as follows. For each submatrix Θ_1^i of Θ_1, we add a corresponding submatrix Θ_2^i of Θ_2, whose purpose is to compensate for the degrees of freedom lost by the helper in the event of failure of client c_i.

In particular, for each $c_i \in C \setminus \{c_{i^*}\}$, we construct matrix Θ_2^i as follows. Let Ψ_i be the vertical concatenation of matrices U_{i^*} and Θ_1, excluding the rows that belong to Θ_1^i. Note the rank of Ψ_i is at least $n - \lambda_i$, because the vertical concatenation of U_{i^*} and Θ_1 is of rank n. As we show below, it is possible to add λ_i or less unit vectors to Ψ_i to make it full rank. These vectors will correspond to uncoded messages transmitted by clients in $C \setminus \{c_i, c_{i^*}\}$.

Indeed, let U be a $n \times n$ diagonal matrix that corresponds to the unit encoding vectors of all messages in X. Then, it is possible to identify $n - \lambda_i$ rows in U that can be added to Ψ_i to complete it to full rank. Each of these rows corresponds to a message in X that is not held by the helper node c_{i^*}. Because each message is held by at least two clients, it is possible to find a client $c_j \in C \setminus \{c_i, c_{i^*}\}$ that can transmit this message.

Note that the total number of rows added to Θ_2 in Phase II of the algorithm is bounded by the number of rows in Θ_1.

Phase III

In Phase III of the algorithm, we construct matrix Θ_3 whose rows correspond to η linear combinations of the messages in the helper's side information set U_{i^*}. Transmissions represented by Θ_3 (in combination with transmissions represented by Θ_1 and Θ_2) ensure that any client $c_i \in C \setminus \{c_{i^*}\}$ can obtain all packets in the case of a failure of any other client $c_j \in C \setminus \{c_{i^*}\}$.

19.4.2 ALGORITHM ANALYSIS

First, we show that with high probability the algorithm yields a solution robust to a failure of one client.

Lemma 19.3. *The coding scheme represented by the concatenation of matrices Θ_1, Θ_2, and Θ_3 is a feasible 1-robust solution to Problem CDE with high probability.*

Proof. As discussed above, the transmissions represented by Θ_1 are sufficient for the scenario with no failure or the failure of the helper c_{i^*}. From the construction of the encoding matrix Θ_2, the transmissions added in Phase II of the algorithm enable the helper to decode all messages in X even if one of the clients fails. It is left to show that transmissions made in Phases I–III are sufficient to satisfy the demands of any client in $c_i \in C \setminus \{c_{i^*}\}$, in the event of a failure of a client $c_j \in C \setminus \{c_i, c_{i^*}\}$.

Let Γ be a concatenation of matrices Θ_1, Θ_2, excluding the rows that correspond to Θ_1^j. Note that Γ captures all packets received by all nodes, including c_i and the helper c_{i^*}, in the event of failure of node c_j. This implies that the concatenation of Γ and U_{i^*} has rank n. On the other hand, the concatenation of Γ and U_i has rank at least $n - \eta_j$. It is easy to verify that the concatenation of Γ and U_i can be completed to the full rank by adding at most $n - \eta_j$ rows from U_{i^*}. This implies that if we add $n - \eta_j$ linear combinations of rows in U_{i^*} to the concatenation of Γ and U_i, then the resulting matrix has rank n with high probability. ■

We proceed to analyze the approximation ratio of the algorithm, i.e., the ratio between the number of transmissions required by the algorithm and the number of transmissions required to identify the optimal 1-robust solution.

Lemma 19.4. *Let OPT' be the minimum number of transmissions required by the optimal algorithm for the 1-robust solution. Then, the minimum number of transmissions required by the algorithm presented in Section 19.4.1 is at most $2 \cdot OPT'$.*

Proof. We bound the number of transmissions required by all phases of the algorithm. We can assume that the value η and the identity of the client c_{i^*} are correct because the algorithm performs an exhaustive search among these two parameters.

In Phase I, the algorithm identifies the minimum number of transmissions required to complete the information exchange if client c_{i^*} cannot transmit but must receive all packets, and each client cannot transmit more than η packets. We note that because the optimal solution only requires $OPT' - \eta$ transmissions to solve this problem, the number of transmissions needed in Phase I is at most $OPT' - \eta$.

Because the number of transmissions in Phase II is less or equal to the number of transmissions made in Phase I, and the number of transmissions made in Phase III is bounded by η, we conclude that the total number of transmissions is at most

$$2(OPT' - \eta) + \eta \leq 2 \cdot OPT'.$$

■

19.5 DATA EXCHANGE IN THE PRESENCE OF AN EAVESDROPPER

In this section, we focus on the setting in which the information transfer is performed in the presence of an eavesdropper that can obtain all packets transmitted over the channel. Specifically, suppose that a CDE algorithm performs μ transmissions represented by the encoding matrix Γ_μ, and the eavesdropper has access to μ packets $E = \{p_1, \ldots, p_\mu\}$.

The goal of the eavesdropper is to obtain information about the messages in X or sparse linear combinations of up to g messages of X, where g is the security parameter of our scheme. We use the concept of *weak information-theoretic security* in which all messages in X are assumed to be uniform independent random variables over $\mathbb{F}_q$. The *g-weak security* requires that

$$I(X_g; E) = 0 \tag{19.4}$$

for any subset X_g of X of size g. That is, the adversary will not be able to extract any information about packets in X_g by observing the set of packets transmitted over the channel.

The goal a g-week secure coding scheme is to construct an encoding matrix Γ that satisfies the following conditions:

1. Each client $c_i \in C$ is able to obtain all messages in $\overline{X}_i$;
2. The scheme has the minimum possible number of transmissions;
3. $(X_g; E) = 0$ for any subset X_g of X of size g.

Fig. 19.3 presents an example of an instance of the problem. In this example the clients c_1, c_2, and c_3 are exchanging messages in the set $X = \{x_1, \ldots, x_5\}$. In the beginning of data exchange process, clients c_1, c_2, and c_3 have subsets of messages $\{x_1, x_2, x_3\}$, $\{x_2, x_3, x_4\}$, and $\{x_3, x_4, x_5\}$, respectively. A possible solution to this problem is for clients c_1, c_2, and c_3 to broadcast linear combinations $x_1 + 2x_2 + x_3$, $x_2 + 2x_3 + x_4$, and $x_3 + 2x_4 + x_5$, respectively (all operations are over $GF(5)$). Note that by observing the transmitted packets, the eavesdropper will not be able to decode any of the original messages. Note also that the eavesdropper will not be able to obtain any linear combination of two messages, so even if it has any single message as a prior side information, it will not be able to decode any additional message in X.

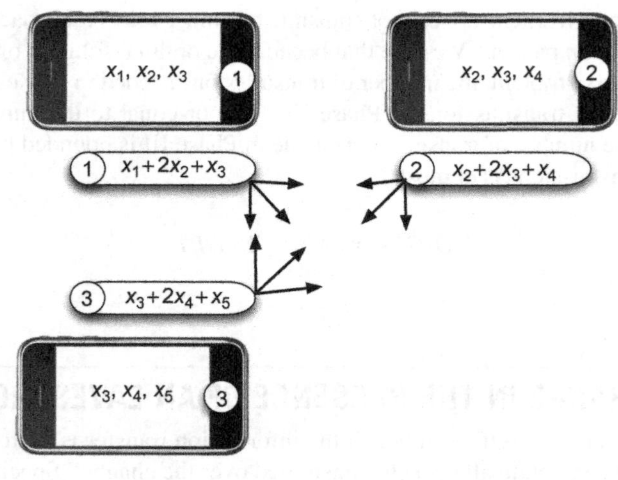

FIG. 19.3

Example of a weakly secure solution against an eavesdropper.

For linear coding schemes, the condition of Eq. (19.4) is equivalent to the following algebraic condition: the Hamming weight of any nonzero linear combination of rows of Γ_μ is at least $g+1$. Clearly, if an adversary can obtain a linear combination of g or less messages in X, this will result in a violation of Eq. (19.4). Conversely, if any linear combination contains at least $g+1$ messages, then Eq. (19.4) will be satisfied because every set of g messages will be protected by a message that does not belong to this set and because this message is a random element of the field $\mathbb{F}_q$.

Our goal is to design a solution to the CDE problem represented by encoding matrix Γ_μ that maximizes the minimum weight of a nonzero vector in the row span of Γ_μ. While the weakly secure scheme does not explicitly require reducing the number of transmissions, solutions with fewer transmissions expose less information to the adversary. To construct a g-weekly secure solution with the maximum possible value of g for a given instance of the problem, the number of transmissions should be identical to that of the optimal solution of Problem CDE.

More specifically, the Singelton bound implies that the maximum weight of a nonzero vector in the row span of a full-rank $\mu \times n$ matrix is at most $n - \mu + 1$. This implies, in turn, that the maximally achievable value of g is $n - OPT$, where OPT is the minimum number of transmissions required to solve the CDE problem. In this section, we show that this bound is tight by proving that the CDE algorithm presented in Section 19.3.1 computes a g-weakly secure scheme for $g = n - OPT$ with high probability, provided that the size of the underlying field $\mathbb{F}_q$ is sufficiently large.

Let Γ_μ be the $\mu \times n$ encoding matrix computed by the CDE algorithm presented in Section 19.3.1. We need to show that with high probability, this is a generator matrix of an MDS code, i.e., any $\mu \times \mu$ submatrix $\hat{\Gamma}$ of Γ_μ is of full rank. Let $\hat{\Gamma}$ be such an arbitrary submatrix of Γ_μ. If there exists a client $c_i \in C$ that initially has messages that correspond to the columns in Γ_μ that do not belong to $\hat{\Gamma}$, then matrix $\hat{\Gamma}$ must be full rank, because c_i needs to decode messages that correspond to the columns of $\hat{\Gamma}$.

If there does not exist such a client, we can add to C an imaginary client $\hat{c}$ whose side information set corresponds to the columns that do not belong to $\hat{\Gamma}$. We denote by $\hat{U}$ the matrix that includes unit encoding vectors that correspond to the side information set of $\hat{c}$. Our goal is to show that the same solution Γ_μ (computed for the original set C of clients) is able to satisfy this client with high probability.

It is sufficient to show that at each iteration i, client $\hat{c}$ gains an additional degree of freedom, i.e.,

$$\operatorname{rank}\begin{pmatrix}\Gamma_{i+1}\\ \hat{U}\end{pmatrix} = \operatorname{rank}\begin{pmatrix}\Gamma_i\\ \hat{U}\end{pmatrix} + 1.$$

Let $c_{t_{i+1}}$ be the client that transmits at round $i+1$. First, we note that the number of degrees of freedom that t_{i+1} has is more than that of $\hat{c}$ in the beginning of the cycle, i.e.,

$$\operatorname{rank}\begin{pmatrix}\Gamma_i\\ U_{t_{i+1}}\end{pmatrix} > \operatorname{rank}\begin{pmatrix}\Gamma_i\\ \hat{U}\end{pmatrix}.$$

This is because clients $c_{t_{i+1}}$ and $\hat{c}$ are missing at most $OPT - i - 1$ and $OPT - i$ degrees of freedom, respectively. Then, there exists at least one row in $U_{t_{i+1}}$ which is not in the row span of the concatenation of Γ_i and $\hat{U}$. Hence, when $c_{t_{i+1}}$ transmits a random linear combination of messages in $U_{t_{i+1}}$, client $\hat{c}$ will gain a degree of freedom with probability at least $1 - 1/q$ (this can be shown by an argument similar to that used in the proof of Lemma 19.2).

We summarize this result by the following theorem:

Theorem 19.3. *The CDE algorithm presented in Section 19.3.1 computes a g-weakly secure coding scheme* Γ_μ *with probability at least* $1 - \binom{n}{OPT}\frac{OPT}{q}$, *for* $g = n - OPT$ *and for a sufficiently large q.*

Proof. As discussed above, for each set of $n - \mu$ columns of Γ_μ, we associate an additional client that has the messages corresponding to these columns (provided that such client does not already exist in C). The number of such clients is bounded by $\binom{n}{OPT}$. The probability that $\hat{c}$ gains a degree of freedom at each iteration is at least $1 - 1/q$. By union bound, the probability that such a client can obtain all messages is at least $1 - OPT/q$. Again, by the union bound, the probability that all additional clients can decode all messages is at least $1 - \binom{n}{OPT}\frac{OPT}{q}$. ■

19.6 CONCLUSION

Cooperative data exchange is a fascinating problem that has interesting connections to many other areas of wireless communication and coding theory. Over the last five years, the problem has received significant attention from the research community. Many interesting extensions and reformulations of the problem have been proposed.

The problem of securing the data against eavesdroppers is closely related to the problem of constructing minimum-distance separable (MDS) codes with generation matrix constraints [38]. This problem gives rise to an intriguing conjecture that states that it is possible to linearly transform a Vandermonde matrix to obtain the constrained generator matrix with high probability. This conjecture, in turn, is closely related to interesting conjectures in algebraic geometry, abstract algebra, and number theory.

REFERENCES

[1] Ahlswede R, Cai N, Li SY, Yeung RW. Network information flow. IEEE Trans Inf Theory 2000;46(4):1204–16.

[2] Li SY, Yeung RW, Cai N. Linear network coding. IEEE Trans Inf Theory 2003;49(2):371–81.

[3] Koetter R, Médard M. An algebraic approach to network coding. IEEE Trans Netw 2003;11(5):782–95.

[4] Sprintson A, Sadeghi P, Booker G, El Rouayheb S. A randomized algorithm and performance bounds for coded cooperative data exchange. In: Proceedings of IEEE ISIT; 2010. p. 1888–92.

[5] Courtade TA, Wesel RD. Coded cooperative data exchange in multihop networks. IEEE Trans Inf Theory 2014;60(2):1136–58.

[6] Milosavljevic N, Pawar S, El Rouayheb S, Gastpar M, Ramhandran K. Deterministic algorithm for the cooperative data exchange problem. In: Proceedings of IEEE ISIT; 2011. p. 410–4.

[7] Neely MJ. Wireless peer-to-peer scheduling in mobile networks. In: Proceedings of CISS; 2012.

[8] Keller L, Le A, Cici B, Seferoglu H, Fragouli C, Markopoulou A. Microcast: cooperative video streaming on smartphones. In: Proceedings of the 10th international conference on mobile systems, applications, and services. MobiSys '12. New York, NY: ACM; 2012. p. 57–70. ISBN 978-1-4503-1301-8. https://doi.org/10.1145/2307636.2307643.

[9] Sendonaris A, Erkip E, Aazhang B. User cooperation diversity—Part I: System description. IEEE Trans Commun 2003;51(11):1927–38.

[10] Laneman JN, Tse DNC, Wornell GW. Cooperative diversity in wireless networks: efficient protocols and outage behavior. IEEE Trans Inf Theory 2004;50(12):3062–80.
[11] Hong YW, Huang WJ, Chiu FH, Kuo CCJ. Cooperative communications in resource-constrained wireless networks. IEEE Signal Process Mag 2007;24(3):47–57.
[12] Rouayheb SE, Chaudhry MAR, Sprintson A. On the minimum number of transmissions in single-hop wireless coding networks. In: Proceedings of information theory workshop (ITW '07), Tahoe City, California, USA; 2007.
[13] El Rouayheb S, Sprintson A, Sadeghi P. On coding for cooperative data exchange. In: Proceedings of IEEE ITW; 2010.
[14] Gonen M, Langberg M. Coded cooperative data exchange problem for general topologies. In: Proceedings of IEEE ISIT; 2012. p. 2606–10.
[15] Courtade TA, Xie B, Wesel RD. Optimal exchange of packets for universal recovery in broadcast networks. In: Proceedings of military communications conference; 2010. p. 2250–5.
[16] Courtade TA, Wesel RD. Efficient universal recovery in broadcast networks. In: Proceedings of 48th annual Allerton conference on communication, control, and computing; 2010. p. 1542–9.
[17] Courtade TA, Wesel RD. Coded cooperative data exchange in multihop networks. IEEE Trans Inf Theory 2014;60(2):1136–58.
[18] Sprintson A, Sadeghi P, Booker G, El Rouayheb S. Deterministic algorithm for coded cooperative data exchange. In: ICST QShine; 2010.
[19] Courtade TA, Wesel RD. Weighted universal recovery, practical secrecy, and an efficient algorithm for solving both. In: Proceedings of 49th annual Allerton conference on communication, control, and computing; 2011. p. 1349–57.
[20] Yan M, Sprintson A, Zelenko I. Weakly secure data exchange with generalized Reed-Solomon codes. In: Proceedings of IEEE ISIT; 2014. p. 1366–70.
[21] Yan M, Sprintson A. On error correcting algorithms for the cooperative data exchange problem. In: Proceedings of 2014 IEEE international symposium on network coding (NetCod 2014), Aalborg, Denmark; 2014.
[22] Courtade TA, Halford TR. Coded cooperative data exchange for a secret key. In: Proceedings of IEEE ISIT; 2014. p. 776–80.
[23] Heidarzadeh A, Sprintson A. Cooperative data exchange with unreliable clients. In: 2015 53rd annual Allerton conference on communication, control, and computing (Allerton); 2015. p. 496–503.
[24] Ozgul D, Sprintson A. An algorithm for cooperative data exchange with cost criterion. In: Proceedings of ITA workshop; 2011.
[25] Tajbakhsh SE, Sadeghi P, Shams R. A generalized model for cost and fairness analysis in coded cooperative data exchange. In: Proceedings of NetCod; 2011.
[26] Yan M, Sprintson A. Algorithms for weakly secure data exchange. In: Proceedings of NetCod; 2013.
[27] Csiszar I, Narayan P. Secrecy capacities for multiple terminals. IEEE Trans Inf Theory 2004;50(12):3047–61. https://doi.org/10.1109/TIT.2004.838380.
[28] Chan C, Al-Bashabsheh A, Ebrahimi JB, Kaced T, Liu T. Multivariate mutual information inspired by secret-key agreement. Proc IEEE 2015;103(10):1883–913. https://doi.org/10.1109/JPROC.2015.2458316.
[29] Birk Y, Kol T. Coding-on-demand by an informed source (ISCOD) for efficient broadcast of different supplemental data to caching clients. IEEE Trans Inf Theory 2006;52(6):2825–30.
[30] Bar-Yossef Z, Birk Y, Jayram TS, Kol T. Index coding with side information. In: Proceedings of the 47th annual IEEE symposium on foundations of computer science (FOCS); 2006. p. 197–206.
[31] El Rouayheb S, Sprintson A, Georghiades C. On the index coding problem and its relation to network coding and matroid theory. IEEE Trans Inf Theory 2010;56(7):3187–95.
[32] Alon N, Hassidim A, Lubetzky E, Stav U, Weinstein A. Broadcasting with side information. In: Proceedings of the 49th annual IEEE symposium on foundations of computer science (FOCS); 2008. p. 823–32.

[33] Blasiak A, Kleinberg R, Lubetzky E. Index coding via linear programming; April 2010. Available at http://arxiv.org/abs/1004.1379.
[34] Minsky Y, Trachtenberg A, Zippel R. Set reconciliation with nearly optimal communication complexity. IEEE Trans Inf Theory 2003;49(9):2213–8.
[35] Chan C, Al-Bashabsheh A, Zhou Q, Ding N, Liu T, Sprintson A. Successive omniscience. IEEE Trans Inf Theory 2016;62(6):3270–89. https://doi.org/10.1109/TIT.2016.2555923.
[36] Yan M, Sprintson A. Approximation algorithms for erasure correcting data exchange. In: 2015 IEEE information theory workshop (ITW), Jerusalem, Israel; 2015.
[37] Ozgul D, Sprintson A. An algorithms for cooperative data exchange with cost criterion. In: Proceedings of 2011 information theory and application workshop (ITA); 2011.
[38] Yan M, Sprintson A, Zelenko I. Weakly secure data exchange with generalized Reed-Solomon codes. In: 2014 IEEE international symposium on information theory; 2014. p. 1366–70.

CHAPTER

COLLABORATIVE SPECTRUM SENSING IN THE PRESENCE OF BYZANTINE ATTACKS

20

Bhavya Kailkhura*, Aditya Vempaty†, Pramod K. Varshney‡

*Center for Applied Scientific Computing, Lawrence Livermore National Laboratory, Livermore, CA, United States**

IBM Thomas J. Watson Research Center, Yorktown Heights, NY, United States†

Department of EECS, Syracuse University, Syracuse, NY, United States‡

20.1 INTRODUCTION

20.1.1 COGNITIVE RADIO NETWORKS

Increasing demand for wireless communication in various areas of human life has brought about an exponential increase in the number of wireless services. This exponential increase has resulted in spectrum scarcity as the electromagnetic spectrum has become too crowded to accommodate the growing number of wireless services. Recently, this problem has attracted a lot of attention in the research community, and several potentially viable solutions have been proposed to mitigate the problem of spectrum scarcity. A survey conducted by the FCC has shown the presence of highly underutilized licensed spectrum bands [1]. These observations have motivated the research community to review the conventional spectrum allocation policies and explore new spectrum allocation policies to alleviate the issue of underutilized spectrum. Dynamic spectrum access (DSA) is an extremely promising idea in the area of nonconventional spectrum allocation. In DSA, the same spectrum band is accessed by several users as opposed to the conventional spectrum access where only the licensed user is permitted to transmit in its spectrum band. The unlicensed (secondary) users access the frequency band in such a way that the interference caused by them stays within the allowed limits. In CRNs [2], the secondary users (CRs) sense the presence of a primary user without interfering with the primary user's communication, and only use the spectrum if the primary user is inferred to be absent. This process is called spectrum sensing, which allows a secondary user to operate when the primary is not present. Clearly, the efficacy of the scheme depends on the accuracy of this inference performance. To ensure high accuracy, collaborative spectrum sensing has been proposed where multiple CRs collaborate with each other by sharing their inferences to make a global inference. This has been shown to have various advantages in terms of spectrum utilization and robustness.

Cooperative and Graph Signal Processing. https://doi.org/10.1016/B978-0-12-813677-5.00020-1

20.1.2 BYZANTINE ATTACKS

As is evident from the process described above, the data fusion scheme is a key component of collaborative spectrum sensing. An adversary interested in using this spectrum would be motivated to disrupt this inference process. While the presence of multiple CRs makes it difficult for a single adversary, these schemes are still quite vulnerable to adversarial attacks. One typical attack on distributed inference is the Byzantine attack, motivated by the *Byzantine generals problem.*

In 1982, Lamport et al. presented the so-called Byzantine generals problem [3] as follows: "a group of generals of the Byzantine army camped with their troops around an enemy city. Communicating only by messengers, the generals must agree upon a common battle plan. However, one or more of them may be traitors who will try to mislead the others. The problem is to find an algorithm to ensure that the loyal generals will reach agreement." This problem is similar in principle to the problem considered in this chapter. It was shown that if the fraction of Byzantine generals is less than $1/3$, there is a way for the loyal generals to reach a consensus agreement, regardless of what the Byzantine generals do. If the fraction is above $1/3$, consensus can no longer be guaranteed. There are many diverse behaviors that a Byzantine entity may engage in with the intent to disrupt network inference, such as a node may lie about connectivity, flood a network with false traffic, attempt to subjugate control information, falsely describe opinions of another node (e.g., peer-to-peer), or capture a strategic subset of devices and collude.

The Byzantine attack on distributed inference problems such as collaborative spectrum sensing are similar in spirit to the Byzantine generals problem described above. While Byzantine attacks may, in general, refer to many types of malicious behavior, our focus in this chapter is on data corruption attacks. In this type of attack, an attacker, after sensing the spectrum, reports falsified (erroneous) data to the network to degrade inference performance. In this chapter, we refer to such a data corruption attacker as a Byzantine and the data thus generated is referred to as Byzantine data. In the distributed spectrum-sensing context, Byzantine CRs can affect global decisions regarding the presence or absence of the primary user (PU) by reporting false sensing data. This might result in a collision of secondary users (SUs) with the PU (if a busy PU is wrongly detected as idle) or in spectrum wastage (if an idle PU is detected as busy). Thus, robust collaborative spectrum sensing in the presence of spectrum sensing data falsification (SSDF) attacks is of utmost importance.

The central goal of this chapter is to discuss the effect of corrupted (or falsified) data on the spectrum-sensing performance of CRNs and robust strategies to ensure reliable overall performance for several practical network architectures. Notations used in this chapter are provided in Table 20.1.

20.2 COLLABORATIVE SPECTRUM SENSING

Next, we give a brief introduction to spectrum sensing in cognitive radio networks (CRNs).

The collaborative spectrum-sensing (CSS) model comprises a group of N CRs that acquire observations regarding a PU, then collaborate with other CRs by sharing their local observations (raw or processed) via their network topology to decide either on hypothesis H_1 (primary user present) or H_0 (primary user absent). These CRs can be honest or malicious.

Typically each CR (secondary user) uses an energy detection scheme for sensing the PU. When the PU is absent (Hypothesis H_0), we assume $P_r = W$ where W is Gaussian distributed as $\mathcal{N}(\mu_0, \sigma_0^2)$.

Table 20.1 Table of Notations

Notation	Description
P_t	PU transmission power
P_r	CR receive power
d	distance between PU and CR
d_0	close in-reference distance
n	path loss exponent
N	number of CRs
μ	mean values of distributions
σ	standard deviation of distributions
i	CR index
α	fraction of Byzantines in CRN
v_i	local decision made by *i*th CR
u_i	decision sent by *i*th CR to FC
P_d	local probability of detection
P_f	local probability of false alarm
$P_d^{H(B)}$	local probability of detection of Honest (Byzantine) CR
$P_f^{H(B)}$	local probability of false alarm of Honest (Byzantine) CR
$P_{a,b}^{H(B)}$	probability that an Honest (Byzantine) CR sends *a* to FC when deciding *b*
α_{blind}	blinding fraction of Byzantines in CRN
κ_i	reputation metric for *i*th CR
T	time window of reputation calculation
$u_0[t]$	global decision at time instant *t*
Λ_i^T	test statistic between expected and observed behavior of *i*th CR
$A = \{a_{ij}\}$	adjacency matrix of CRN graph
$\mathcal{N}_i$	neighborhood of *i*th CR
$D = diag(d_1, \ldots, d_N)$	degree matrix with degrees d_i
z_i^t	*i*th CR's sensed signal at time *t* in peer-to-peer CRNs
ζ_i	deterministic gain of *i*th sensing channel
s^t	deterministic signal at time *t*
Y_i	summary statistic of *i*th CR
M	number of sensing samples
η_i	local SNR at *i*th CR
$x_i(k)$	information shared by *i*th CR with neighbors at consensus iteration *k*
w_i	weight given to *i*th CR's information during consensus
W	Perron matrix
Δ_i	Byzantine attack strength for consensus disruption
P_i	Byzantine attack probability for consensus disruption
$\tilde{w}_i$	tampered weight set by Byzantines
$w_i^{H(B)}$	optimal consensus weights for honest (Byzantine) CRs

When the PU is present (Hypothesis H_1), observations are naturally a function of the distance between the PU and the CR, which affects the power received at the CR. Given P_t, the transmission power of the PU, the power received at a CR located at a distance d from the PU, $P_r(d)$ under hypothesis H_1 (primary present) can be represented as

$$P_r\,[\mathrm{dB}] = P_t\,[\mathrm{dB}] - PL(d)\,[\mathrm{dB}]. \tag{20.1}$$

The links between the PU and CRs are subject to independent and identically distributed (i.i.d.) log-normal shadowing path loss, thus

$$PL(d)\,[\mathrm{dB}] = \widehat{PL}(d) + X_\sigma = \widehat{PL}(d_0) + 10n\log\frac{d}{d_0} + X_\sigma, \tag{20.2}$$

where $PL(d)$ is the path loss as a function of d, the distance between primary and secondary users; X_σ represents a zero-mean Gaussian distributed random variable with standard deviation σ; d_0 denotes the close in-reference distance; and n is the path loss exponent that is equal to the rate at which the path loss increases with distance between primary and secondary users. $\widehat{PL}(d)$ is the mean of $PL(d)$ and can be found using the HATA model [4], which has been suggested by the IEEE 802.22 working group as the path-loss model for a typical CRN environment. Here, it is worth mentioning that IEEE 802.22 is the first standard developed for CRNs coexisting in the TV channels' bands. Assuming a rural environment,

$$\begin{aligned}\widehat{PL}(d) = 27.77 + 46.05\log(f_c) - 4.78(\log(f_c))^2 - 13.82\log(h_t)\\ - (1.1\log(f_c) - 0.7)h_r + (44.9 - 6.55\log(h_t))\log(d),\end{aligned}$$

where f_c is the carrier frequency, h_t and h_r are the effective transmitter and receiver antenna heights (in meters), respectively. The unit of d is kilometers (km).

20.3 SPECTRUM SENSING IN A PARALLEL CRN

20.3.1 SYSTEM MODEL

Network model

Consider a parallel network comprised of a central entity (known as the Fusion Center (FC)) and a set of N CRs that faces the task of determining whether the PU is present. The ith CR makes an observation and compresses it using some local processing. A Byzantine attack on such a system compromises some of the CRs that may then intentionally send falsified local spectrum-sensing decisions to the FC to make the final decision incorrect. We assume that a fraction α of the N CRs that observe the PU have been compromised by an attacker. Let the local decisions made by CRs be $v_i \in \{0,1\}$, and $u_i \in \{0,1\}$, $i = 1,\ldots,N$, denote the decision sent by the CR to the FC. Here $u_i = v_i$, if i is an uncompromised (honest) CR, but for a compromised (Byzantine) CR i, u_i need not be equal to v_i. The FC yields a final decision after processing the local decisions. Observations at the CRs are assumed to be conditionally independent and identically distributed given the hypothesis. Denote the probabilities of detection and false alarm as $P_d^H = P(u_i = 1|H_1)$ and $P_f^H = P(u_i = 1|H_0)$ for honest CRs and $P_d^B = P(u_i = 1|H_1)$ and $P_f^B = P(u_i = 1|H_0)$ for Byzantines.

Byzantine attack model

Assuming that the sensing and decision strategies are the same among all honest CRs and among all Byzantine CRs, let us define the following strategies $P^H_{j,1}$, $P^H_{j,0}$ and $P^B_{j,1}$, $P^B_{j,0}$ ($j \in \{0, 1\}$) for the honest and Byzantine CRs, respectively:

Honest CRs:

$$P^H_{1,1} = 1 - P^H_{0,1} = P^H(x = 1|y = 1) = 1, \tag{20.3}$$

$$P^H_{1,0} = 1 - P^H_{0,0} = P^H(x = 1|y = 0) = 0. \tag{20.4}$$

Byzantine CRs:

$$P^B_{1,1} = 1 - P^B_{0,1} = P^B(x = 1|y = 1), \tag{20.5}$$

$$P^B_{1,0} = 1 - P^B_{0,0} = P^B(x = 1|y = 0). \tag{20.6}$$

$P^H(x = a|y = b)$ ($P^B(x = a|y = b)$) is the probability that an honest (Byzantine) CR sends a to the FC when its actual local decision is b. From now onward, we will refer to Byzantine flipping probabilities simply by ($P_{1,0} := P^B_{1,0}, P_{0,1} := P^B_{0,1}$). We also assume that the FC is not aware of the exact set of Byzantine CRs and considers each CR i to be Byzantine with a certain probability α.

20.3.2 FUNDAMENTAL LIMIT

The objective of Byzantines is to degrade the detection performance of the network by choosing $(P_{1,0}, P_{0,1})$ intelligently. Due to the difficulty in characterizing the exact performance of a general fusion scheme, Kullback-Leibler divergence (KLD) is used to characterize the performance of the detection scheme due to its strong relationship with the global detection performance. The KLD between the distributions $P(u_i|H_1)$ and $P(u_i|H_0)$ is expressed as

$$D(P(u_i|H_1)||P(u_i|H_0)) = \sum_{j\in\{0,1\}} P(u_i|H_1) \log \frac{P(u_i|H_1)}{P(u_i|H_0)}. \tag{20.7}$$

The critical power of the distributed detection network is the minimum fraction of Byzantine CRs needed to make the data from CRs uninformative to the CR, i.e., make KLD at the FC equal to zero. This would result in the FC making a decision based only on the prior information as the network is *blind* to the data from local CRs. This critical power for a Byzantine is characterized as follows.

BOX 20.1 Critical Power

Lemma 20.1 ([5]). *The minimum fraction of Byzantines needed to blind the FC is*

$$\alpha_{blind} = \frac{P^H_d - P^H_f}{(P^B_d - P^B_f) + (P^H_d - P^H_f)}.$$

The optimal attack strategy for the Byzantines is given by the following.

BOX 20.2 Optimal Attacking Strategies

Lemma 20.2 ([5]). *Let us consider $P_d^B = P_d^H$ and $P_f^B = P_f^H$. Optimal attacking strategies, $(P_{1,0}^*, P_{0,1}^*)$, which minimize the KLD are*

$$(P_{1,0}^*, P_{0,1}^*) = \begin{cases} (p_{1,0}, p_{0,1}) & \text{if } \alpha \geq 0.5, \\ (1,1) & \text{if } \alpha < 0.5, \end{cases}$$

where $(p_{1,0}, p_{0,1})$ satisfy $\alpha(p_{1,0} + p_{0,1}) = 1$.

Consider the case where the PU is transmitting at the UHF frequency of 617 MHz with effective transmitter antenna height $h_t = 100$ m and the effective isotropic radiated power (EIRP) is assumed to be 35 dBm. All CRs are assumed to be equipped with a simple energy detector and effective receiver antenna height $h_t = 1$ m. The minimum power for a signal to be detected is assumed to be -94 dBm. Let the noise power equal to -106 dBm and in the log-normal shadowing path-loss model as well as let noise standard deviation $\sigma = \sigma_0 = 11.6$. Assume a large rural city environment where the distance between the PU and SUs is assumed to be equal to 13 km. Also, set the threshold at the local CRs to achieve a maximum probability of misdetection equal to 0.25.

Fig. 20.1 shows the detection performance in terms of the minimum KLD[1] as a function of the fraction of Byzantines for different attack strategies. There are two types of Byzantine attacks: independent malicious Byzantine attacks (IMBA) and cooperative malicious Byzantine attacks (CMBA). In an independent attack, each Byzantine CR attacks the spectrum sensing system independently, relying on its own observation. Because the Byzantine CRs do not know the identity of the other Byzantines in the network, $P_d^B = P_d^H$ and $P_f^B = P_f^H$, which gives $\alpha_{blind} = 0.5$ from Lemma 20.1. This implies that the number of Byzantines needs to be at least 50% to blind the FC when the Byzantines attack the network independently. In a CMBA, Byzantine CRs collaborate to make a decision regarding the true hypothesis and use this information to attack the network. Using collaboration, the Byzantines can reduce the minimum critical power α_{blind} by increasing $(P_d^B - P_f^B)$. Consider Byzantines are colluding using the "L out of M" rule to make their decision, i.e., if more than L out of the M Byzantines make a decision, of say "1", then all the collaborating Byzantines in the network send a "0". The value of L is taken to be $M/2$, which corresponds to the majority rule. As Fig. 20.1 shows, α_{blind} decreases with the collaboration of the Byzantines.

20.3.3 MITIGATION TECHNIQUES

Section 20.3.2 discusses the issue of spectrum sensing from the attacker's perspective. Now, we discuss some counter measures used in practice to protect the network from these Byzantines. Byzantines can be treated as outliers and, therefore, one may use signal processing techniques to mitigate their effects.

A simple and intuitive method to mitigate the effect of Byzantines is to identify them [5]. For identification purposes, one needs to observe the behavior of CRs' sequentially over time. We first discuss some of the schemes proposed in the literature that treat the FC as a watchdog to mitigate the effect of Byzantines.

[1]Minimization is performed over attack strategies $p_1^B := 1 - P_{0,1}$ and $q_1^B := P_{1,0}$.

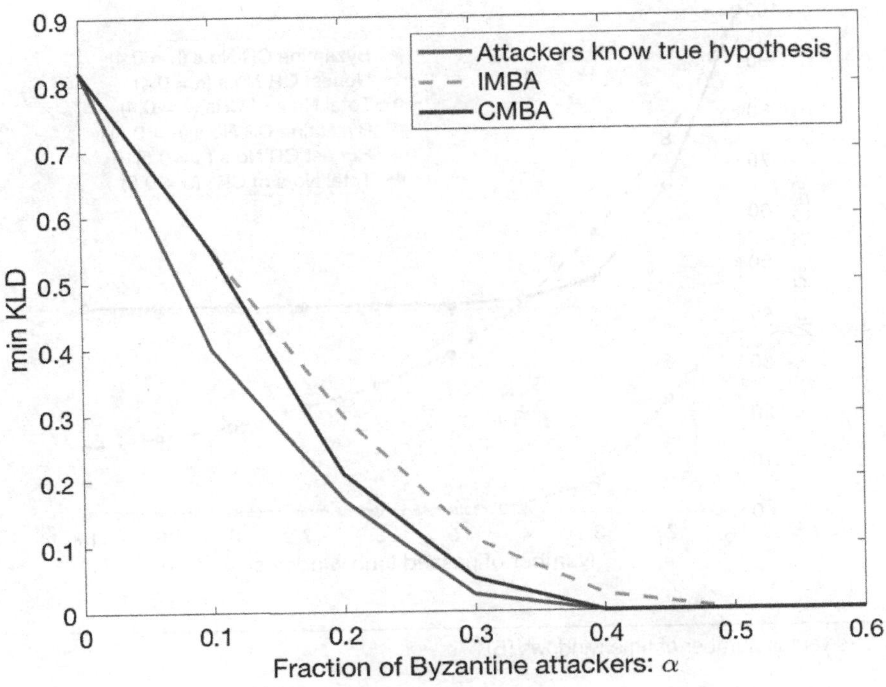

FIG. 20.1

The KLD as a function of the fraction of Byzantines [5].

Reputation-based scheme

A simple and effective scheme to identify the Byzantines is by assigning a reputation to each CR based on the quality of the data they provide [5]. Let us divide the CSS process into time windows consisting of T sensing periods. Next, define a reputation metric κ_i for each CR as the number of mismatches in a time interval T between ith CR's local decision and the global decision made at the FC using the majority rule. The reputation metric is given by

$$\kappa_i = \sum_{t=1}^{T} \mathcal{I}_{(u_i[t] \neq u_0[t])}, \tag{20.8}$$

where $u_i[t]$ is the ith CR's local decision at time instant t, $u_0[t]$ is the global decision made at the FC at time instant t, and $\mathcal{I}_{(S)}$ is the indicator function over the set S. The CRs for which this reputation metric κ_i is greater than a predetermined threshold κ are tagged as Byzantines and removed from the fusion process.

Fig. 20.2 plots the isolation of CRs from information fusion at the FC as a function of the number of time windows when $N = 100, (P_d, P_f) = (0.8, 0.2)$. Each time window consists of $T = 4$ sensing periods. At $\alpha = 0.4$, in a span of only four time windows, the proposed scheme isolates all the

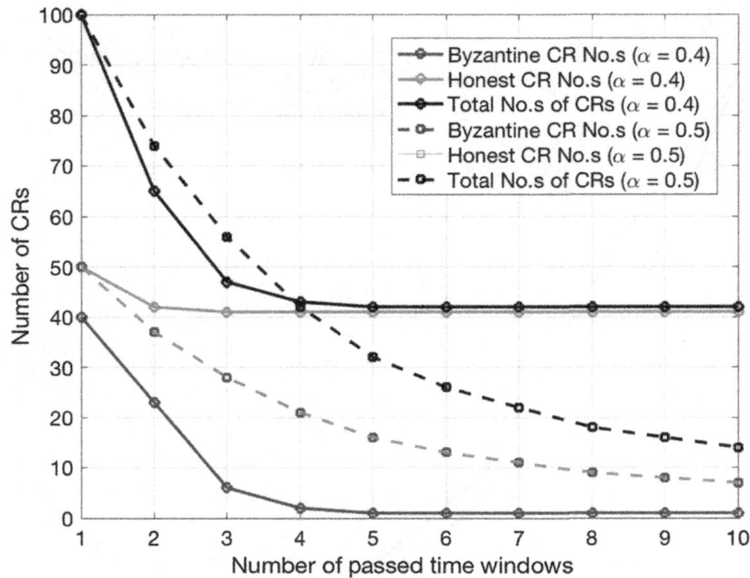

FIG. 20.2

Number of CRs versus number of time windows [5].

Byzantine attackers without cutting off a significant number of honest CRs. However when, $\alpha = 0.5$, honest and Byzantine CRs are eliminated with equal probability. Therefore, this scheme works only when the number of Byzantines in the network is less than 50% of the total number of CRs because the FC uses the majority rule for fusion. If the Byzantines have a majority in number, the above reputation-based scheme identifies the honest CRs as outliers and removes them from the network and thereby worsens the inference performance of the network.

Adaptive learning

An interesting method to improve the performance of the network is to use the information of the identified Byzantines to the network's benefit [6]. More specifically, a three-tier adaptive learning scheme can be used that learns the parameters of the identified Byzantines and uses these learnt parameters in the Chair-Varshney rule [7] to make the final decision. The three-tier scheme can be described as follows: (i) identification of Byzantines in the network, (ii) estimation of parameters of the identified Byzantines at the FC, and (iii) adaptive fusion rule.

The basic idea of the learning-based identification scheme is to compare every CR's observed behavior over time with the expected behavior of an honest CR. The CRs whose observed behavior is too far from the expected behavior are tagged as Byzantines. This scheme works even when the Byzantines are in the majority (>50%) because it does not use the global decision for identification purposes. The behavior of every CR is characterized by the probability of sending a "1" to the FC.

This value is a function of the operating point of the CR (P_f, P_d) and the prior probabilities of the hypotheses, which are assumed to be known at the FC for honest CRs.

At the FC, the expected behavior is estimated for every CR over time by averaging the number of times a particular decision (0 or 1) is made over a time interval of T instants. These probabilities can be updated after every time instant. The test statistic Λ_i^T for the ith CR after time T is the deviation between the expected and observed behavior for every CR. The FC declares a CR as a Byzantine if Λ_i^T is greater than a particular threshold λ. This threshold λ is determined as the minimum value when the Byzantine's operating point is in the region below the $P_d = P_f$ line on the receiver operating characteristics (ROC).

After identifying the Byzantines, their parameters can be estimated by assuming that all the Byzantines have the same operating point. This assumption is typically made in the literature because it is assumed that a single adversary has attacked some of the CRs in the network and reprogrammed them to behave as Byzantines. Therefore, it can be assumed that all these malicious CRs have the same operating point on the ROC. These estimated parameters are used in the Chair-Varshney optimal fusion rule [7] in an adaptive manner to find the global decision. It is important to note that this scheme works for any fraction of Byzantines in the network but assumes the knowledge of honest CR behavior and the primary user statistics.

As can be seen from Fig. 20.3, the Byzantines can be exactly detected without any mismatches when T is between 100 and 150 for both the cases when $\alpha = 0.3$ and $\alpha = 0.7$.

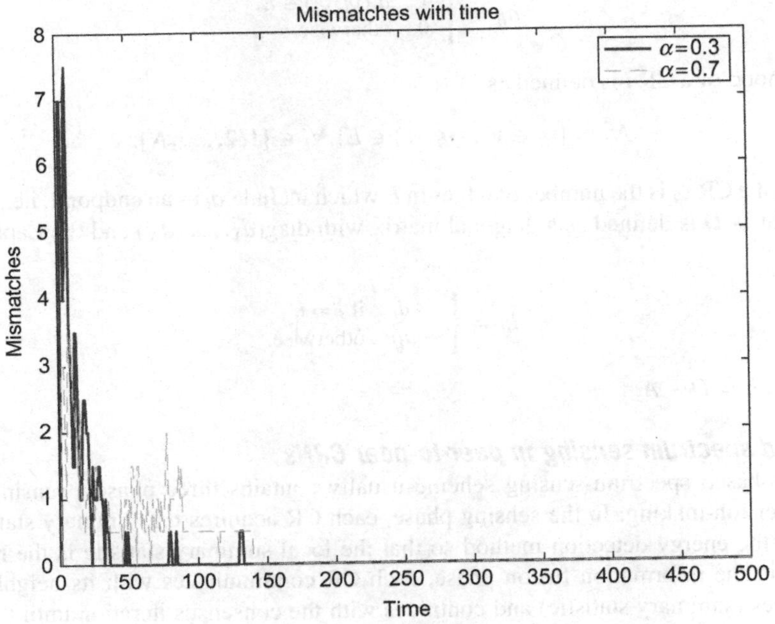

FIG. 20.3

Mismatches versus time when $N = 20, (P_d^H, P_f^H) = (0.9, 0.1)$ [6].

20.4 SPECTRUM SENSING IN A PEER-TO-PEER CRN

In the previous sections, the problem of spectrum sensing with corrupted data in a parallel network with a centralized FC was studied. However, in many scenarios, a centralized FC may not be available or the FC may become an information bottleneck, potentially leading to CRN failure. In such scenarios, it may be desirable to employ alternate peer-to-peer local information exchange in order to reach a global decision. One such decentralized approach for peer-to-peer local information exchange and inference is the use of a consensus algorithm. This section considers the problem of spectrum sensing in peer-to-peer CRNs in the presence of data falsification (Byzantine) attacks. Spectrum-sensing approaches considered in this section are based on fully distributed consensus algorithms where all the CRs exchange information only with their neighbors in the absence of a FC.

20.4.1 SYSTEM MODEL

Network model

Consider a CRN topology modeled as an undirected graph $G = (V, E)$, where $V = \{\upsilon_1, \dots, \upsilon_N\}$ represents the set of CRs in the network with $|V| = N$. The set of communication links in the network correspond to the set of edges E, where $(\upsilon_i, \upsilon_j) \in E$, if and only if there is a communication link between υ_i and υ_j so that, υ_i and υ_j can directly communicate with each other. The adjacency matrix A of the graph is defined as

$$a_{ij} = \begin{cases} 1 & \text{if } (\upsilon_i, \upsilon_j) \in E, \\ 0 & \text{otherwise.} \end{cases}$$

The neighborhood of a CR i is defined as

$$\mathcal{N}_i = \{\upsilon_j \in V : (\upsilon_i, \upsilon_j) \in E\}, \forall i \in \{1, 2, \dots, N\}.$$

The degree d_i of a CR υ_i is the number of edges in E which include υ_i as an endpoint, i.e., $d_i = \sum_{j=1}^{N} a_{ij}$. The degree matrix D is defined as a diagonal matrix with $\text{diag}(d_1, \dots, d_N)$ and the Laplacian matrix L is defined as

$$l_{ij} = \begin{cases} d_i & \text{if } j = i, \\ -a_{ij} & \text{otherwise.} \end{cases}$$

In other words, $L = D - A$.

Decentralized spectrum sensing in peer-to-peer CRNs

The consensus-based spectrum-sensing scheme usually contains three phases: sensing, information fusion, and decision-making. In the sensing phase, each CR acquires the summary statistic about the PU. We adopt the energy detection method so that the local summary statistic is the received signal energy. Next, in the information fusion phase, each CR communicates with its neighbors to update their state values (summary statistic) and continues with the consensus iteration until the whole CRN converges to a steady state, which is the global test statistic. Finally, in the decision-making phase, CRs make their own decisions about the presence of the PU using this global test statistic. In the following, each of these phases is described in more detail.

20.4.2 SENSING PHASE

Consider a CRN with N CRs using the energy detection scheme [8]. For the ith CR, the sensed signal z_i^t at time instant t is given by

$$z_i^t = \begin{cases} n_i^t, & \text{under } H_0, \\ \zeta_i s^t + n_i^t, & \text{under } H_1, \end{cases}$$

where ζ_i is the deterministic gain corresponding to the sensing channel, s^t is the deterministic signal at time instant t, n_i^t is AWGN, i.e., $n_i^t \sim \mathcal{N}(0, \sigma_i^2)$ (where $\mathcal{N}$ denotes the normal distribution) and independent across time. Each CR i calculates a summary statistic Y_i over a spectrum sensing interval of M samples, as

$$Y_i = \sum_{t=1}^{M} |z_i^t|^2,$$

where M is determined by the time-bandwidth product [8]. Because Y_i is the sum of the squares of M i.i.d. Gaussian random variables, it can be shown that $\frac{Y_i}{\sigma_i^2}$ follows a central chi-square distribution with M degrees of freedom (χ_M^2) under H_0, and, a noncentral chi-square distribution with M degrees of freedom and parameter η_i under H_1, i.e.,

$$\frac{Y_i}{\sigma_i^2} \sim \begin{cases} \chi_M^2, & \text{under } H_0, \\ \chi_M^2(\eta_i), & \text{under } H_1, \end{cases}$$

where $\eta_i = E_s|\zeta_i|^2/\sigma_i^2$ is the local SNR at the ith CR and $E_s = \sum_{t=1}^{M} |s^t|^2$ represents the sensed signal energy over M spectrum sensing instants. Note that the local SNR is M times the average SNR at the output of the energy detector, which is $\frac{E_s|\zeta_i|^2}{M\sigma_i^2}$.

20.4.3 INFORMATION FUSION PHASE

In this section, we give a brief introduction to conventional consensus algorithms [9] and explain how consensus is reached using the following two steps.

Step 1: All CRs establish communication links with their neighbors and broadcast their information state regarding the PU, $x_i(0) = Y_i$.

Step 2: Each CR updates its local state information by a local fusion rule (weighted combination of its own value and those received from its neighbors) [9]. Let us denote ith CR's updated information at iteration k by $x_i(k)$. This CR continues to broadcast information $x_i(k)$ and update its local information state until consensus is reached. This process of updating information state can be written in a compact form as

$$x_i(k+1) = x_i(k) + \frac{\epsilon}{w_i} \sum_{j \in \mathcal{N}_i} (x_j(k) - x_i(k)), \tag{20.9}$$

where ϵ is the time step and w_i is the weight given to CR i's information. Using the notation $x(k) = [x_1(k), \ldots, x_N(k)]^T$, network dynamics can be represented in the matrix form as,

$$x(k+1) = Wx(k),$$

where $W = I - \epsilon \operatorname{diag}(1/w_1, \ldots, 1/w_N)L$ is referred to as a Perron matrix. The consensus algorithm is nothing but a local fusion or update rule that fuses the CRs' local information state with information coming from neighbor CRs, and it is well known that every CR asymptotically reaches the same information state for arbitrary initial values [9].

20.4.4 DECISION MAKING PHASE

The final information state x^* after reaching consensus for the above consensus algorithm will be the weighted average of the initial states of all the CRs [9] or $x^* = \sum_{i=1}^N w_i Y_i / \sum_{i=1}^N w_i, \forall i$. Average consensus can be seen as a special case of weighted average consensus with $w_i = w$, $\forall i$. After the whole network reaches a consensus, each CR makes its own decision about the presence (or absence) of the PU using a predefined threshold λ,

$$\text{Decision} = \begin{cases} H_1 & \text{if } x^* > \lambda, \\ H_0 & \text{otherwise,} \end{cases}$$

where weights are given by [10]

$$w_i = \frac{\eta_i/\sigma_i^2}{\sum_{i=1}^N \eta_i/\sigma_i^2}. \tag{20.10}$$

Let us refer to $\Lambda = \sum_{i=1}^N w_i Y_i / \sum_{i=1}^N w_i$ as the final test statistic.

Next, we look into Byzantine attacks on consensus-based spectrum-sensing schemes and discuss the performance degradation of the weighted average consensus-based spectrum-sensing algorithms due to these attacks.

Byzantine attack model

In data falsification attacks, attackers try to manipulate the final test statistic (i.e., $\Lambda = \sum_{i=1}^N w_i Y_i / \sum_{i=1}^N w_i$) in a manner so as to degrade the spectrum-sensing performance. By falsifying initial values Y_i or weights w_i, the attackers can manipulate the final test statistic. Spectrum-sensing performance will be degraded because Byzantine CRs can always set a higher weight to their manipulated information. Thus, the final statistic's value across the whole network will be dominated by the Byzantine CRs' local statistic that will lead to degraded spectrum-sensing performance.

Byzantine CRs tamper with their initial values Y_i and send $\tilde{Y}_i$ such that the spectrum-sensing performance is degraded.

Under H_0:

$$\tilde{Y}_i = \begin{cases} Y_i + \Delta_i & \text{with probability } P_i \\ Y_i & \text{with probability } (1 - P_i) \end{cases}$$

Under H_1:

$$\tilde{Y}_i = \begin{cases} Y_i - \Delta_i & \text{with probability } P_i \\ Y_i & \text{with probability } (1 - P_i) \end{cases}$$

where P_i is the attack probability and $\Delta_i \geq 0$ is a constant value that represents the attack strength, which is zero for honest CRs. As we show later, Byzantine CRs will use a large value of Δ_i so that the final statistic's value is dominated by the Byzantine CR's local statistic, leading to a degraded spectrum-sensing performance.

20.4.5 FUNDAMENTAL LIMIT

The objective of Byzantine CRs is to degrade the spectrum-sensing performance of the network by falsifying their data (Y_i, w_i). To analyze the worst case spectrum-sensing performance, let us assume that Byzantines have an advantage and know the true hypothesis. Consider the case when weights of the Byzantines have been tampered by setting their value at $\tilde{w}_i$ and look at the effect of falsifying the initial values Y_i. One can use the deflection coefficient [11] to characterize the security performance of the detection scheme due to its simplicity and its strong relationship with the global detection performance. Deflection coefficient of the global test statistic is defined as: $\mathcal{D}(\Lambda) = \frac{(\mu_1 - \mu_0)^2}{\sigma_{(0)}^2}$, where $\mu_k = \mathbb{E}[\Lambda | H_k]$, $k = 0, 1$, is the conditional mean and $\sigma_{(k)}^2 = \mathbb{E}[(\Lambda - \mu_k)^2 | H_k], k = 0, 1$, is the conditional variance. Let us define the critical point of the distributed detection network as the minimum fraction of Byzantine CRs needed to make the deflection coefficient of the global test statistic equal to zero. In this case, we say that the network becomes *blind* and denote it by α_{blind}. We assume that the communication between CRs is error-free and our network topology is fixed during the whole consensus process and, therefore, consensus can be reached without disruption [9].

Without loss of generality, assume that the CRs corresponding to the first N_1 indices $i = 1, \ldots, N_1$ are Byzantines and the remaining CRs corresponding to indices $i = N_1 + 1, \ldots, N$ are honest CRs. Let us define $w = [\tilde{w}_1, \ldots, \tilde{w}_{N_1}, w_{N_1+1}, \ldots, w_N]^T$ and $\sum w = \sum_{i=1}^{N_1} \tilde{w}_i + \sum_{i=N_1+1}^{N} w_i$.

BOX 20.3 Critical Power

Lemma 20.3 ([12]). *For data fusion schemes in this consensus-based system, the condition to blind the CRN or equivalently to make the deflection coefficient zero is given by*

$$\sum_{i=1}^{N_1} \tilde{w}_i (2P_i \Delta_i - \eta_i \sigma_i^2) = \sum_{i=N_1+1}^{N} w_i \eta_i \sigma_i^2.$$

Note that, when $w_i = \tilde{w}_i = z, \eta_i = \eta, \sigma_i = \sigma, P_i = P, \Delta_i = \Delta, \forall i$, the blinding condition simplifies to $\frac{N_1}{N} = \frac{1}{2} \frac{\eta \sigma^2}{P \Delta}$ (Fig. 20.4).

Consider a network with four honest CRs and two Byzantine CRs. Sensing channel gains of the CRs are assumed to be $h = [0.8, 0.7, 0.72, 0.61, 0.69, 0.9]$ and weights are given by Eq. (20.10). We

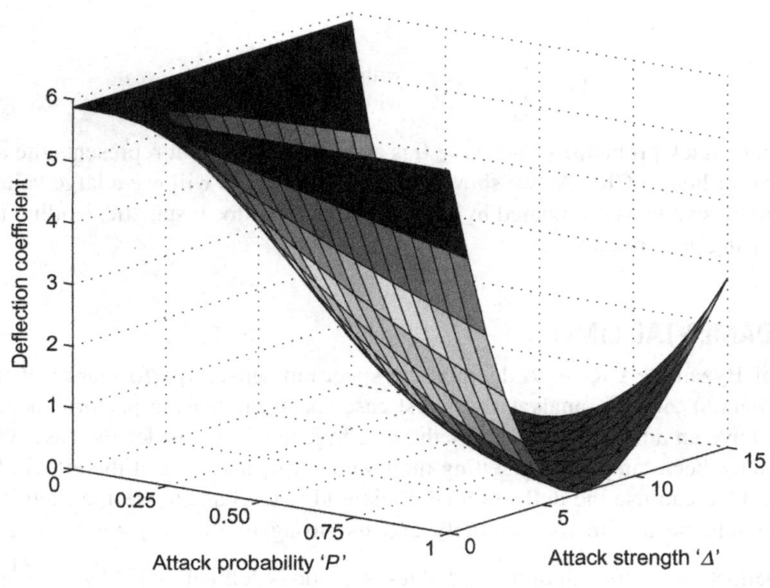

FIG. 20.4

Deflection coefficient as a function of attack parameters [12].

also assume that $M = 12$, $E_s = 5$, and $\sigma_i^2 = 1, \forall i$. Notice that the deflection coefficient is zero when the condition in Lemma 20.3 is satisfied. Another observation to make is that the deflection coefficient can be made zero even when only two out of six CRs are Byzantines. Thus, by appropriately choosing attack parameters (P, Δ), less than 50% of data-falsifying Byzantines can blind the network.

One approach to mitigate the effect of sensing data falsification is to assign weights based on the quality of the data. In other words, a lower weight can be given to the data of the CR identified as a Byzantine. However, to implement this approach in peer-to-peer networks, one has to address the following two issues.

First, in the conventional weighted average consensus algorithm, weight w_i given to CR i's data is controlled or updated by the CR itself. Thus, a Byzantine CR can always set a higher weight to its manipulated information and the final statistics will be dominated by the Byzantine CRs' local statistic that will lead to degraded detection performance. Therefore, conventional consensus algorithms cannot be used in the presence of an attacker.

Second, as will be seen later, the optimal weights given to CRs' sensing data depend on the following unknown parameters: identity of the CRs, which indicates whether the CR is honest or Byzantine, and the underlying statistical distribution of the CRs' data.

In the next section, we discuss a learning-based robust weighted average consensus algorithm that addresses these concerns [12].

20.4.6 MITIGATION TECHNIQUE

To address the first issue discussed in the previous section, which is the optimal weight design, one can employ a consensus algorithm in which the weight for the ith CR's information is controlled (or updated) by the neighbors of the ith CR rather than by the ith CR itself. Note that networks deploying such an algorithm are more robust to weight manipulation because if a Byzantine CR j wants to assign an incorrect weight to the data of its neighbor i in the global test statistic, it has to ensure that all the neighbors of CR i put the same incorrect weight as CR j.

BOX 20.4 Robust Consensus Algorithm

Consider a modified Perron matrix $\hat{W} = I - \epsilon(T \otimes L)$ where L is the original graph Laplacian, $\otimes$ is the element-wise matrix multiplication operator, and T is a transformation given by [12]

$$[T]_{ij} = \begin{cases} \dfrac{\sum_{j \in \mathcal{N}_i} w_j}{l_{ii}} & \text{if } i = j, \\ w_j & \text{otherwise.} \end{cases}$$

Observe that the above transformation T satisfies the condition that weights are controlled (or updated) by neighbors $\mathcal{N}_i$ of CR i rather than by CR i itself.

Based on the above transformation T, the robust distributed consensus algorithm can be written as the following:

$$x_i(k+1) = x_i(k) + \epsilon \sum_{j \in \mathcal{N}_i} w_j(x_j(k) - x_i(k)). \tag{20.11}$$

Consider a network with six CRs where the CRs employ the robust algorithm as given in Eq. (20.11) (with $\epsilon = 0.3$) to reach a consensus. Fig. 20.5 shows the updated state values at each CR as a function of consensus iterations. Assume that the initial data vector is $x(0) = [5, 2, 7, 9, 8, 1]^T$ and the weight vector is $w = [0.65, 0.55, 0.48, 0.95, 0.93, 0.90]^T$. Fig. 20.5 shows the convergence of the proposed algorithm iterations. It is observed that within 20 iterations, consensus has been reached on the global decision statistics, the weighted average of the initial values (states).

Next, to address the second issue discussed in the previous section, one can exploit the statistical distribution of the sensing data and devise techniques to mitigate the influence of Byzantines on the spectrum-sensing system. A three-tier mitigation scheme can be devised where the following three steps are performed at each CR: (1) identification of Byzantine neighbors, (2) estimation of parameters of identified Byzantine neighbors, and (3) adaptation of consensus algorithm (or update weights) using estimated parameters.

The optimal weights for the honest/Byzantine CRs, assuming that the identities of the CRs are known, are given in Lemma 20.4.

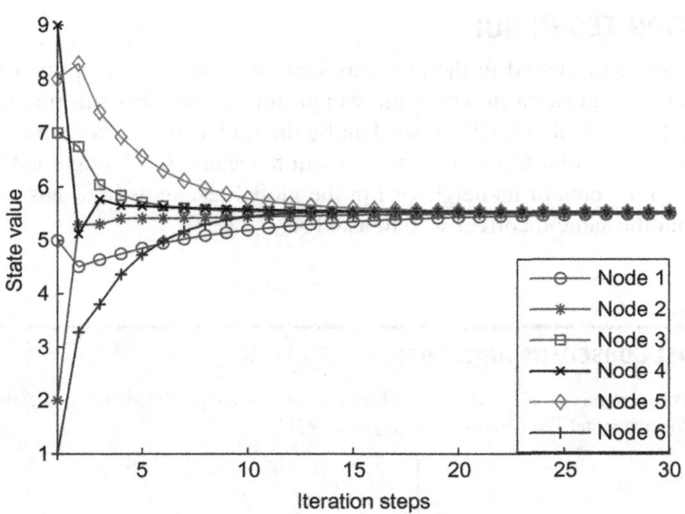

FIG. 20.5

Convergence of the network with six CRs ($\epsilon = 0.3$) [12].

BOX 20.5 Optimal Weights

Lemma 20.4 ([12]). *Optimal weights which maximize the deflection coefficient of the global test statistic,* $\Lambda = \frac{\sum_{i=1}^{N_1} w_i^B \tilde{Y}_i + \sum_{i=N_1+1}^{N} w_i^H Y_i}{\sum w}$ *where* $\sum w = \sum_{i=1}^{N_1} w_i^B + \sum_{i=N_1+1}^{N} w_i^H$, *are given by*

$$w_i^B = \frac{(\eta_i \sigma_i^2 - 2P_i\Delta_i)}{\Delta_i^2 P_i(1-P_i) + 2M\sigma_i^4}, \quad \text{and} \quad w_i^H = \frac{\eta_i}{2M\sigma_i^2}.$$

Assume that $M = 12$, $\eta_i = 3$, $\sigma_i^2 = 0.5$ and the attack parameters are $(P_i, \Delta_i) = (0.5, 9)$. Fig. 20.6 compares the weighted average consensus-based spectrum-sensing scheme with the equal gain combining scheme[2] and the scheme where Byzantines are excluded from the fusion process. It can be clearly seen from the figure that our weighted average consensus scheme performs better than the rest of the schemes.

Notice that the optimal weights for the Byzantines are functions of the attack parameters (P_i, Δ_i), which may not be known to the neighboring CRs in practice. In addition, the parameters of the honest CRs might also not be known. In such cases, techniques based on the expectation maximization (EM) algorithm and maximum likelihood (ML) estimation can be employed to learn the operating parameters (or weights) of the CRs in the network to enable an adaptive design of the local fusion or update rules that are updated after each learning iteration [12].

[2]In equal gain combining scheme, all the CRs (including Byzantines) are given the same weight.

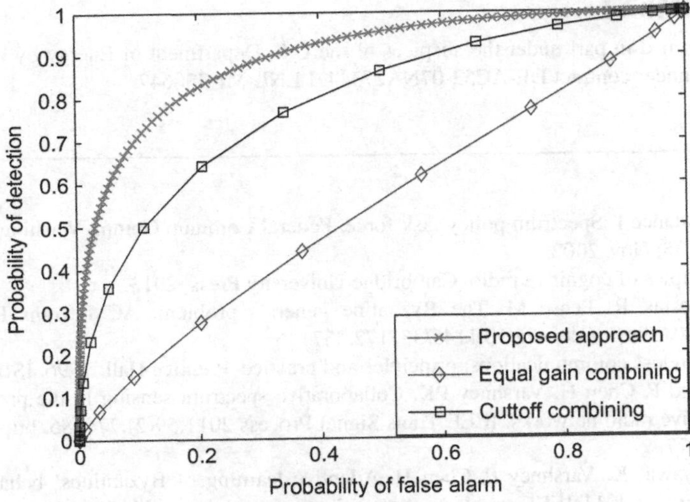

FIG. 20.6

ROC for different protection approaches [12].

20.5 CONCLUSION AND OPEN ISSUES

This chapter discussed the problem of Byzantine attack and defense for CSS in CRNs. To begin, we discussed the vulnerability of CSS to Byzantine attacks in a parallel topology, characterized the critical power analytically, and obtained closed-form expression for an optimal attack. We next discussed two mitigation schemes, (1) reputation-based mitigation, and (2) adaptive learning-based mitigation of Byzantines. It was found that adaptive learning-based mitigation techniques are more robust to SSDF attacks. Next, we considered a scenario where a centralized FC was not available and the CRN was implemented in a peer-to-peer fashion. In such scenarios, when the Byzantine attackers in the network are above a certain fraction, existing consensus-based spectrum-sensing algorithms are ineffective. We discussed a robust distributed weighted average consensus algorithm and a learning technique to estimate the operating parameters (or weights) of the CRs. This enables an adaptive design of the local fusion or update rules to mitigate the effect of data falsification attacks. In a nutshell, the Byzantine attack in CSS is one adversarial action that disrupts the proper functioning of CRNs. More efforts are needed to tackle the unsolved research challenges on Byzantine attacks and defense in CRNs. There are many interesting questions that remain to be explored in the future work such as more sophisticated design across multiple layers of the networking protocol stack, advanced distributed inference at the physical layer, sophisticated network coding schemes for large networks, a variety of cryptographic techniques for different applications, and development of complex Byzantine misbehavior models and methods to detect and mitigate such Byzantines. We envision that the topic of Byzantine attack and defense will remain a fruitful research area in the coming years.

ACKNOWLEDGMENT

This work was performed in part under the auspices of the U.S. Department of Energy by Lawrence Livermore National Laboratory under contract DE-AC52-07NA27344. LLNL-MI-738647.

REFERENCES

[1] Kolodzy P, Avoidance I. Spectrum policy task force. Federal Commun Comm, Washington, DC, Report ET Docket No. 02,135; Nov. 2002.
[2] Biglieri E. Principles of cognitive radio. Cambridge University Press; 2013.
[3] Lamport L, Shostak R, Pease M. The Byzantine generals problem. ACM Trans Program Lang Syst 1982;4(3):382–401. https://doi.org/10.1145/357172.357176.
[4] Rappaport T. Wireless communications: principles and practice. Prentice Hall; 1996. ISBN 0133755363.
[5] Rawat AS, Anand P, Chen H, Varshney PK. Collaborative spectrum sensing in the presence of Byzantine attacks in cognitive radio networks. IEEE Trans Signal Process 2011;59(2):774–86. https://doi.org/10.1109/TSP.2010.2091277.
[6] Vempaty A, Agrawal K, Varshney P, Chen H. Adaptive learning of Byzantines' behavior in cooperative spectrum sensing. In: 2011 IEEE wireless communications and networking conference; 2011. p. 1310–5. https://doi.org/10.1109/WCNC.2011.5779320.
[7] Chair Z, Varshney P. Optimal data fusion in multiple sensor detection systems. IEEE Trans Aerosp Electron Syst 1986(1):98–101.
[8] Digham FF, Alouini MS, Simon MK. On the energy detection of unknown signals over fading channels. IEEE Trans Commun 2007;55(1):21–4.
[9] Olfati-Saber R, Fax JA, Murray RM. Consensus and cooperation in networked multi-agent systems. Proc IEEE 2007;95(1):215–33.
[10] Zhang W, Wang Z, Guo Y, Liu H, Chen Y, Mitola III J. Distributed cooperative spectrum sensing based on weighted average consensus. In: 2011 IEEE global telecommunications conference (GLOBECOM 2011). IEEE; 2011. p. 1–6.
[11] Kay SM. Fundamentals of statistical signal processing. Prentice Hall PTR; 1993.
[12] Kailkhura B, Brahma S, Varshney PK. Data falsification attacks on consensus-based detection systems. IEEE Trans Signal Inf Process Netw 2017;3(1):145–58.

PART

SOCIAL NETWORKS

4

CHAPTER

DYNAMICS OF INFORMATION DIFFUSION AND SOCIAL SENSING

21

Vikram Krishnamurthy*, William Hoiles†

*School of Electrical and Computer Engineering, Cornell Tech, Cornell University, New York, NY, United States**
Biosymetrics, New York, NY, United States†

21.1 INTRODUCTION AND MOTIVATION

Humans can be viewed as social sensors that interact over a social network to provide information about their environment. Examples of information produced by such social sensors include Twitter posts, Facebook status updates, and ratings on online reputation systems such as Yelp and Tripadvisor. Social sensors go beyond physical sensors, for example, user opinions/ratings (such as the quality of a restaurant) are available on Tripadvisor but are difficult to measure via physical sensors. Similarly, future situations revealed by the Facebook status of a user are impossible to predict using physical sensors [1].

Statistical inference using social sensors is an area that has witnessed remarkable progress in the last decade. It is relevant in a variety of applications, including localizing special events for targeted advertising [2,3], marketing [4,5], localization of natural disasters [6], and predicting the sentiment of investors in financial markets [7,8]. For example, Asur and Huberman [9] report that models built from the rate of tweets for particular products can outperform market-based predictors.

21.1.1 CONTEXT: WHY SOCIAL SENSORS?

Social sensors present unique challenges from a statistical estimation point of view. First, social sensors interact with and influence other social sensors. For example, ratings posted on online reputation systems strongly influence the behavior of individuals.[1] Such interactive sensing can result in nonstandard information patterns due to correlations introduced by the structure of the underlying social network. Thus certain events "go viral" [5,12]. Second, due to privacy concerns and time constraints, social sensors typically do not reveal raw observations of the underlying state of nature. Instead, they reveal their decisions (ratings, recommendations, votes), which can be viewed as a low-resolution

[1]It is reported in [10] that 81% of hotel managers regularly check Tripadvisor reviews. Luca [11] reports that a one-star increase in the Yelp rating maps to a 5%–9% revenue increase.

Cooperative and Graph Signal Processing. https://doi.org/10.1016/B978-0-12-813677-5.00021-3

(quantized) function of their raw measurements and interactions with other social sensors. This can result in misinformation propagation, herding, and information cascades. Third, the response of a social sensor may not be consistent with that of a utility maximizer; social sensors are typically risk-averse.

Social sensors are enabled by technological networks. Indeed, social media sites that support interpersonal communication and collaboration using Internet-based social network platforms are growing rapidly. McKinsey estimates that the economic impact of social media on business is potentially greater than $1 trillion because social media facilitates efficient communication and collaboration within and across organizations.

21.1.2 MAIN RESULTS AND ORGANIZATION

There is strong motivation to construct models that facilitate understanding the dynamics of information flow in social networks. This chapter presents a tutorial description of four important aspects of sensing-based information diffusion in social networks from a signal processing perspective.

Information diffusion in large-scale social networks

The first topic considered in this chapter (Section 21.2) is diffusion of information in social networks comprised of a population of interacting social sensors. The states of sensors evolve over time as a probabilistic function of the states of their neighbors and an underlying target process. Several recent papers investigate such information diffusion in real-world social networks. Motivated by marketing applications, Sun et al. [13] studies diffusion (contagion) behavior in Facebook. Using data from 260,000 Facebook *pages* (which advertise products, services, and celebrities), Sun et al. [13] analyzes information diffusion. In [14], the spread of *hashtags* on Twitter is studied. There is a wide range of social phenomena such as diffusion of technological innovations, sentiment, cultural fads, and economic conventions [15,16] where individual decisions are influenced by the decisions of others.

We consider the so called susceptible-infected-susceptible (SIS) model [17] for information diffusion in a social network. It is shown for social networks comprised of a large number of agents how the dynamics of degree distribution can be approximated by the mean field dynamics. Mean field dynamics have been studied in [18] and applied to social networks in [16] and lead to a tractable model for the dynamic social sensors.

We demonstrate using influenza datasets from the U.S. Centers for Disease Control and Prevention (CDC) how Twitter can be used as a real-time social sensor for tracking the spread of influenza. That is, a health network (namely, the Influenza-like Illness Surveillance Network (ILInet)) is sensed by a real-time microblogging social media network (namely, Twitter).

We also review two recent methods for sampling social networks, namely, social sampling and respondent-driven sampling. Respondent-driven sampling is now used by the U.S. Centers for Disease Control and Prevention (CDC) as part of the National HIV Behavioral Surveillance System in health networks.

Bayesian social learning in online reputation systems

The second topic of this chapter (Section 21.3) considers online reputation systems where individuals make recommendations based on their private observations and recommendations of friends. Such interaction of individuals and their social influence is modeled as Bayesian social learning [15,19,20] on a directed acyclic graph. We consider two important classes of such problems; risk-averse social learning in financial systems and data incest in reputation systems. The risk-averse social learning and associated quickest change detection is important in detecting market shocks in high-frequency trading. Data incest (misinformation propagation) arises as a result of correlations in recommendations due to the intersection of multiple paths in the information exchange graph. Necessary and sufficient conditions are given on the structure of information exchange graphs to mitigate data incest. Experimental results on human subjects are presented to illustrate the effect of social influence and data incest on decision making.

The setup differs from classical signal processing where sensors use noisy observations to compute estimates—in social learning, agents use noisy observations together with decisions made by previous agents to estimate the underlying state of nature. Social learning has been used widely in economics, marketing, political science, and sociology to model the behavior of financial markets, crowds, social groups, and social networks; see [15,19–23] and numerous references therein. Related models have been studied in the context of sequential decision-making in information theory [24,25] and statistical signal processing [26,27] in the electrical engineering literature. Social learning can result in unusual behavior such as herding [20], where agents eventually choose the same action irrespective of their private observations. As a result, the actions contain no information about the private observations and so the Bayesian estimate of the underlying random variable freezes. Such behavior can be undesirable, particularly if individuals herd and make incorrect decisions.

Revealed preferences and detection of utility maximizers

The third topic considered in this chapter (Section 21.4) is the principle of revealed preferences arising in microeconomics. It is used as a constructive test to determine: Are social sensors utility optimizers in their response to external influence? The key question considered is as follows: Given a time-series of data $\mathcal{D} = \{(p_t, x_t), t \in \{1, 2, \ldots, T\}\}$ where $p_t \in \mathbb{R}^m$ denotes the external influence and x_t denotes the response of an agent, is it possible to detect whether the agent is a *utility maximizer*?

These issues are fundamentally different to the *model-centric* theme used in the signal processing literature, where one postulates an objective function (typically convex) and then proposes optimization algorithms. In contrast the revealed preference approach is *data centric*—given a dataset, we wish to determine if is consistent with utility maximization.

We present a remarkable result called Afriat's theorem [28,29], which provides a necessary and sufficient condition for a finite dataset $\mathcal{D}$ to have originated from a utility maximizer. Also a multiagent version of Afriat's theorem is presented to determine if the dataset generated by multiple agents is consistent with playing from the equilibrium of a potential game.

Unlike model-centric applications of game theory in signal processing, the revealed preferences approach is data-centric: (1) Given a time series dataset of probe and response signals, how can one detect whether the response signals are consistent with a Nash equilibrium generated by players in a concave potential game? (2) If consistent with a concave potential game, how can the utility function of the players be estimated?

We present three datasets involving social sensors to illustrate Afriat's theorem of revealed preferences. These datasets are: (i) an auction conducted by undergraduate students at Princeton University, (ii) aggregate power consumption in the electricity market of Ontario, and (iii) a Twitter dataset for specific hashtags.

Varian has written several influential papers on Afriat's theorem in the economics literature. These include measuring the welfare effect of price discrimination [30], analyzing the relationship between prices of broadband Internet access and time-of-use service [31], and ad auctions for advertisement position placement on page search results from Google [31,32]. Despite widespread use in economics, surprisingly, revealed preference theory is relatively unknown in the electrical engineering literature.

Social interaction of YouTube consumers

The fourth topic considered in this chapter (Section 21.5) is the engagement dynamics of social sensors to online video content. Specifically, we consider how users interact with video content created on the YouTube social network. YouTube is the largest user-driven video content provider in the world and has become a major platform for disseminating multimedia information. YouTube contains more than 1 billion users who collectively watch millions of hours of YouTube videos and generate billions of views every day (e.g., 150 years of video are watched every day). Additionally, users upload more than 300 hours of video content every minute. YouTube generates billions in revenue through advertising and also shares the revenue with the popular users that upload videos through the Partner program. YouTube is clearly a social media site; however, is YouTube also a social networking site? In classical online social networks the interaction is directly between users—that is, user-user interactions. However, YouTube is unique in that the interaction between users includes video content—that is, the interaction follows users-content-users. In fact the interaction between users is incentivized using the posted videos. In this way it is not merely the interest preferences between users that promote user-user interaction, but also the content of the videos that governs the social interactions between users.

Using real-world data consisting of more than 6 million videos spread over 25 thousand channels, we empirically examine the sensitivity of YouTube metalevel features on the engagement dynamics of users in YouTube. Insight into the dynamics of social sensors in YouTube can be used to predict how users will interact with posted video content. These results are important for designing methods for optimizing user engagement and for improving the efficiency of content distribution networks [33–35]. Estimating the popularity of YouTube videos based on metalevel features is a challenging problem given the diversity of users and content providers. Time-series methods for modeling the YouTube engagement dynamics of videos over time include ARMA time series models [36], multivariate linear regression models [37], and Gompertz models [38,39]. These methods do not utilize any of the metalevel features of the YouTube video to estimate the user engagement dynamics. In [40] a bag-of-words Bernoulli Naive Bayes classifier is applied to perform a binary classification (popular/unpopular) of YouTube videos based on the title alone. The classifier was able to achieve a classification accuracy of 66%. In [33] visual perception and extreme learning machines were applied to the metalevel features of videos and found to be able to accurately estimate the (popular/unpopular) videos with an accuracy of 80%. It was determined that the main metalevel features that impact video engagement include first day view count, number of subscribers, and contrast of the video thumbnail.

The above methods focus on how to estimate user engagement to specific videos; however, they do not consider the social learning dynamics that are present between users and channel owners. The key topics focused on in Section 21.5 are: (i) how user engagement is affected by changes in the metalevel (title, thumbnail, tag) features of the videos, (ii) the causal relationship between channel subscribers and user engagement, (iii) the engagement dynamics of videos over time with exogenous social media events, and (iv) the engagement of users to videos in a channel's video playlist. The insight provided can be used by channel owners to design policies for maximizing user engagement by adjusting video metalevel features, promoting on external social media venues, and periodically adjusting the uploading schedule of videos.

21.1.3 PERSPECTIVE

The unifying theme that underpins the four topics in this chapter stems from statistical signal processing and controlled sensing. These are used to predict global behavior given local behavior: individual social sensors interact with other sensors and we are interested in understanding the behavior of the entire network. Information diffusion, social learning, and revealed preferences are important issues for social sensors. We treat these issues in a highly stylized manner so as to provide easy accessibility to a signal processing audience. The underlying tools used in this chapter are widely used in signal processing, economics, and network science.

Let us briefly discuss how the four themes of this chapter interact; these four themes are depicted in Fig. 21.1.

The network diffusion models are non-Bayesian and describe the behavior of large numbers of social sensors. The mean field dynamics model for the diffusion of information has the form of an averaged stochastic approximation algorithm (which is widely used in adaptive filtering). Note, however, that the stochastic approximation-type equation is a generative model, and not an algorithm.

The Bayesian social learning models in contrast describe highly stylized individual behavior of social sensors. At this level, it is important to model risk-averse human decision-making and the Bayesian social learning model serves as a useful generative model.

Underpinning both the network diffusion and Bayesian social learning model, are utility (cost) functions that the social sensors optimize in order to make decisions. The natural question is: Given real world data, is the behavior of agents consistent with optimizing a utility function? If yes, can the utility function be estimated? Revealed preferences yield a useful set of algorithms that can answer both these questions. More generally, it can be used to detect play from the Nash equilibrium of a potential game. Put simply, revealed preferences provide the data-driven justification for the utility function models.

Finally, detailed analysis of the YouTube data provides for an interesting real-world study of how social sensors interact. It is important to note that while YouTube is clearly a social media site, it is also a social networking site. Classical online social networks (OSNs) are dominated by user-user

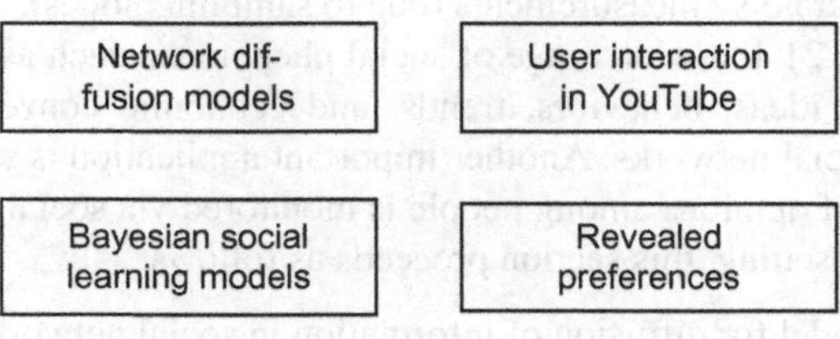

FIG. 21.1

Four themes considered in this chapter.

interactions. However YouTube is unique in that the interaction between users includes video content, that is, the interaction follows users-content-users. The interaction between users in the YouTube social network is incentivized using the posted videos. In addition to the social incentives, YouTube also gives monetary incentives to promote users increasing their popularity. As more users view and interact with a user's video or channel, YouTube will pay the user proportional to the advertisement exposure on the users channel. Therefore, users not only maximize exposure to increase their social popularity, but also for monetary gain, which introduces unique dynamics in the formation of edges in the YouTube social network.

Books and tutorials

The literature in social learning, information diffusion, and revealed preferences is extensive. In each of the following sections, we provide a brief review of relevant works. Seminal books on social networks, social learning, and network science include [15,41–43]. There is a growing literature dealing with the interplay of technological and social networks [44]. Social networks overlaid on technological networks account for a significant fraction of Internet use. As discussed in [44], three key aspects that cut across social and technological networks are the emergence of global coordination through local actions, resource sharing models, and the wisdom of crowds. These themes are addressed in the current chapter in the context of social learning, diffusion, and revealed preferences. Other tutorials include [45,46].

21.2 INFORMATION DIFFUSION IN LARGE SCALE SOCIAL NETWORKS

This section addresses the first topic of the chapter, namely, information diffusion models and their mean field dynamics in social networks. The setting is as follows: The states of individual nodes in the social network evolve over time as a probabilistic function of the states of their neighbors *and* an underlying target process (state of nature). The underlying target process can be viewed as the market conditions or competing technologies that evolve with time and affect the information diffusion. The nodes in the social network are sampled randomly to determine their state. As the adoption of the new technology diffuses through the network, its effect is observed via sentiments (such as tweets) of these selected members of the population. These selected nodes act as social sensors. In signal processing terms, the underlying target process can be viewed as a signal, and the social network can be viewed as a sensor. The key difference compared to classical sensing is that the sensor now is a social network with diffusion dynamics and noisy measurements (due to sampling nodes).

As described in Section 21.1, a wide range of social phenomena such as diffusion of technological innovations, cultural fads, ideas, behaviors, trends, and economic conventions [15,47–49] can be modeled by diffusion in social networks. Another important application is sentiment analysis (opinion mining) where the spread of opinions among people is monitored via social media.

Motivated by the above setting, this section proceeds as follows:

1. We describe the SIS model for diffusion of information in social networks, which has been extensively studied in [16,17,41,42,50].
2. Next, it is shown how the dynamics of the infected degree distribution of the social network can be approximated by the mean field dynamics. The mean field dynamics state that as the number of

agents in the social network goes to infinity, the dynamics of the infected degree distribution converge to that of an ordinary differential (or difference) equation. Such averaging theory results are widely used to analyze adaptive filters. For social networks, they yield a useful tractable model for the diffusion dynamics.

3. We illustrate the diffusion model by using data sets from three related social networks to track the spread of influenza during the period September 1 to December 31, 2009. The friendship network of 744 undergraduate students at Harvard is used together with the outpatient Influenza-like Illness Surveillance Network (ILInet) to monitor the spread of influenza. Then it is shown that Twitter posts related to influenza during this period are correlated with the spread of influenza. Thus in this example, influenza diffuses in a human health network (Harvard friendship network at a local level and ILInet at a global level) and Twitter is used as a social sensor to monitor the spread of influenza.
4. Finally, this section also describes how social networks can be sampled. We review two recent methods for sampling social networks, namely, social sampling and respondent-driven sampling, the latter being used in health networks.

The aim is to estimate the underlying target state that is being sensed by the social network and also the state probabilities of the nodes by sampling measurements at nodes in the social network. In a Bayesian estimation context, this is equivalent to a filtering problem involving estimation of the state of a prohibitively large-scale Markov chain in noise. The mean field dynamics yield a tractable approximation with provable bounds for the information diffusion. Such mean field dynamics have been studied in [18] and applied to social networks in [16,41,50]. For an excellent recent exposition of interacting particle systems comprising agents, each with a finite state space, see [51], where the more apt term "Finite Markov Information Exchange (FMIE) process" is used (Fig. 21.2).

Regarding real datasets, in addition to the case study presented below, for other examples of diffusion datasets and their analysis see [13,14]. A repository of social network datasets can be obtained at [52].

21.2.1 SOCIAL NETWORK MODEL

A social network is modeled as a graph with N vertices:

$$G = (V, E), \quad \text{where } V = \{1, 2, \ldots, N\}, \text{ and } E \subseteq V \times V. \tag{21.1}$$

Here, V denotes the finite set of vertices, and E denotes the set of edges. In social networks, it is customary to use the terminology *network*, *nodes*, and *links* for *graph*, *vertices*, and *edges*, respectively.

We use the notation (m, n) to refer to a link between node m and n. The network may be undirected in which case $(m, n) \in E$ implies $(n, m) \in E$. In undirected graphs, to simplify notation, we use the notation m, n to denote the undirected link between node n and m. If the graph is directed, then $(m, n) \in E$ does not imply that $(n, m) \in E$. We will assume that self loops (reflexive links) of the form i, i are excluded from E.

An important parameter of a social network $G = (V, E)$ is the connectivity of its nodes. Let $\mathcal{N}^{(m)}$ and $D^{(m)}$ denote the neighborhood set and degree (or connectivity) of a node $m \in V$, respectively. That is, with $|\cdot|$ denoting cardinality,

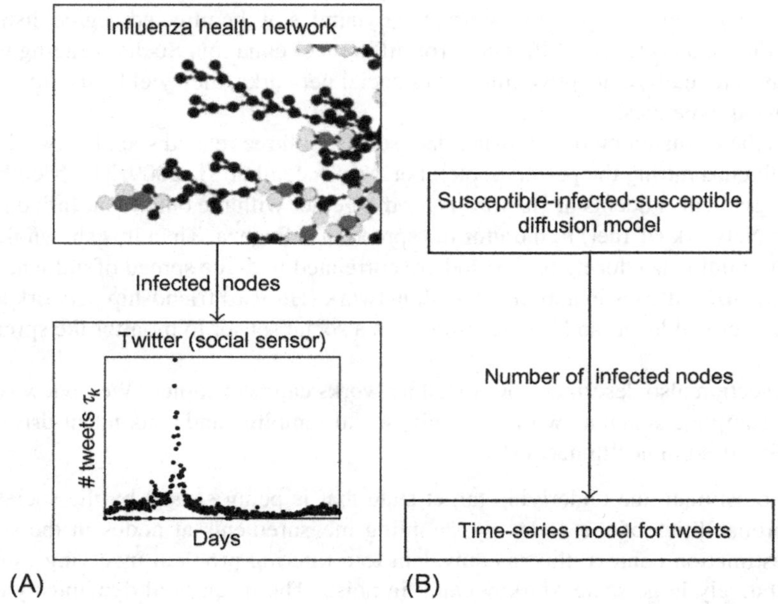

FIG. 21.2

The dynamics of a health network are modeled using the SIS model and linear and nonlinear autoregressive with exogenous input time series models, refer to Section 21.2 for details. (A) Dynamics of health network impact the # of tweets. (B) Model for the number of tweets resulting from the influenza health network.

$$\mathcal{N}^{(m)} = \{n \in V : m, n \in E\}, \quad D^{(m)} = \left|\mathcal{N}^{(m)}\right|. \tag{21.2}$$

For convenience, we assume that the maximum degree of the network is uniformly bounded by some fixed integer $\bar{D}$.

Let $N(d)$ denote the number of nodes with degree d, and let the degree distribution $P(d)$ specify the fraction of nodes with degree d. That is, for $d = 0, 1, \ldots, \bar{D}$,

$$N(d) = \sum_{m \in V} I\left\{D^{(m)} = d\right\}, \quad P(d) = \frac{N(d)}{N}.$$

Here, $I\{\cdot\}$ denotes the indicator function. Note that $\sum_d P(d) = 1$. The degree distribution can be viewed as the probability that a node selected randomly with uniform distribution on V has a connectivity d.

Random graphs generated to have a degree distribution P that is Poisson were formulated by Erdös and Rényi [53]. Several recent works show that large-scale social networks are characterized by connectivity distributions that are different to Poisson distributions. For example, the world wide web has a power law connectivity distribution $P(d) \propto d^{-\gamma}$, where γ ranges between 2 and 3. Such scale-free networks are studied in [54]. In the rest of this chapter, we assume that the degree distribution of the social network is arbitrary but known—allowing arbitrary degree distribution facilities modeling complex networks.

Let $k = 0, 1, \ldots$ denote discrete time. Assume the target process s is a finite state Markov chain with transition probability

$$A_{ss'} = \mathbb{P}\left(s_{k+1} = s' | s_k = s\right). \tag{21.3}$$

In the example of technology diffusion, the target process can denote the availability of competition or market forces that determine whether a node adopts the technology. In the model below, the target state will affect the probability that an agent adopts the new technology.

21.2.2 SIS DIFFUSION MODEL FOR INFORMATION IN SOCIAL NETWORKS

The model we present below for the diffusion of information in the social network is called the SIS model [17,41]. The diffusion of information is modeled by the time evolution of the state of individual nodes in the network. Let $x_k^{(m)} \in \{0, 1\}$ denote the state at time k of each node m in the social network. Here, $x_k^{(m)} = 0$ if the agent at time k is susceptible and $x_k^{(m)} = 1$ if the agent is infected. At time k, the state vector of the N nodes is

$$x_k = \left[x_k^{(1)}, \ldots, x_k^{(N)}\right]' \in \{0, 1\}^N. \tag{21.4}$$

Assume that the process x evolves as a discrete time Markov process with transition law depending on the target state s. If node m has degree $D^{(m)} = d$, then the probability of node m switching from state i to j is

$$\mathbb{P}\left(x_{k+1}^{(m)} = j | x_k^{(m)} = i, x_k^{(i-)}, s_k = s\right) = p_{ij}(d, A_k^{(m)}, s),\ i, j \in \{0, 1\}. \tag{21.5}$$

Here, $A_k^{(m)}$ denotes the number of infected neighbors of node m at time k. That is,

$$A_k^{(m)} = \sum_{n \in N^{(m)}} I\left\{n : x_k^{(m)} = 1\right\}. \tag{21.6}$$

In other words, the transition probability of an agent depends on its degree distribution and the number of active neighbors.

With the above probabilistic model, we are interested in modeling the evolution of infected agents over time. Let $\rho_k(d)$ denote the fraction of infected nodes at each time k with degree d. We call ρ the *infected node distribution*. So with $d = 0, 1, \ldots, \bar{D}$,

$$\rho_k(d) = \frac{1}{N(d)} \sum_{m \in V} I\left\{D^{(m)} = d, x_k^{(m)} = 1\right\}. \tag{21.7}$$

The SIS model assumes that the infection spreads according to the following dynamics:

1. At each time instant k, a single agent, denoted by m, among the N agents is chosen uniformly. Therefore, the probability that the chosen agent m is infected and of degree d is $\rho_k(d)\, P(d)$. The probability that the chosen agent m is susceptible and of degree d is $(1 - \rho_k(d))\, P(d)$.
2. Depending on whether its state $x_k^{(m)}$ is infected or susceptible, the state of agent m evolves according to the transition probabilities specified in Eq. (21.5).

With the Markov chain transition dynamics of individual agents specified above, it is clear that the infected distribution $\rho_k = (\rho_k(1), \ldots, \rho_k(\bar{D}))$ is an $\prod_{d=1}^{\bar{D}} N(d)$ state Markov chain. Indeed, given $\rho_k(d)$, due to the infection dynamics specified above

$$\rho_{k+1}(d) \in \left\{\rho_k(d) - \frac{1}{N(d)}, \ \rho_k(d) + \frac{1}{N(d)}\right\}. \tag{21.8}$$

Our aim below is to specify the transition probabilities of the Markov chain ρ. Let us start with the following statistic that forms a convenient parametrization of the transition probabilities. Given the infected node distribution ρ_k at time k, define $\theta(\rho_k)$ as the probability that at time k a uniformly sampled link in the network points to an infected node. We call $\theta(\rho_k)$ as the *infected link probability*. Clearly

$$\begin{aligned}\theta(\rho_k) &= \frac{\sum_{d=1}^{\bar{D}}(\text{\# of links from infected node of degree } d)}{\sum_{d=1}^{\bar{D}}(\text{\# of links of degree } d)} \\ &= \frac{\sum_{d=1}^{\bar{D}} d\,P(d)\,\rho_k(d)}{\sum_d^{\bar{D}} d\,P(d)}.\end{aligned} \tag{21.9}$$

In terms of the infected link probabilities, the scaled transition probabilities[2] of the process ρ are:

$$\begin{aligned}\bar{p}_{01}(d,\theta_k,s) &\overset{\text{defn}}{=} \frac{1}{P(d)}\,\mathbb{P}\left(\rho_{k+1}(d) = \rho_k(d) + \frac{1}{N(d)}\middle| s_k = s\right) \\ &= (1-\rho_k(d))\sum_{a=0}^{d} p_{01}(d,a,s)\,\mathbb{P}(a \text{ out of l neighbors infected}) \\ &= (1-\rho_k(d))\sum_{a=0}^{d} p_{01}(d,a,s)\binom{d}{a}\theta_k^a(1-\theta_k)^{d-a}, \\ \bar{p}_{10}(d,\theta_k,s) &\overset{\text{defn}}{=} \frac{1}{P(d)}\,\mathbb{P}\left(\rho_{k+1}(d) = \rho_k(d) - \frac{1}{N(d)}\middle| s_k = s\right) \\ &= \rho_k(d)\sum_{a=0}^{d} p_{10}(d,a,s)\binom{d}{a}\theta_k^a(1-\theta_k)^{d-a}.\end{aligned} \tag{21.10}$$

In the above, the notation θ_k is the short form for $\theta(\rho_k)$. The transition probabilities $\bar{p}_{01}$ and $\bar{p}_{10}$ defined above model the diffusion of information about the target state s over the social network. We have the following martingale representation theorem for the evolution of Markov process ρ.

Let $\mathcal{F}_k$ denote the sigma algebra generated by $\{\rho_0, \ldots, \rho_k, s_0, \ldots s_k\}$.

Theorem 21.1. *For $d = 1, 2, \ldots, \bar{D}$, the infected distributions evolve as*

$$\rho_{k+1}(d) = \rho_k(d) + \frac{1}{N}\left[\bar{p}_{01}(d,\theta(\rho_k),s_k) - \bar{p}_{10}(d,\theta(\rho_k),s_k) + w_{k+1}\right], \tag{21.11}$$

---

[2]The transition probabilities are scaled by the degree distribution $P(d)$ for notational convenience. Indeed, because $N(d) = NP(d)$, by using these scaled probabilities we can express the dynamics of the process ρ in terms of the same step size $1/N$ as described in Theorem 21.1. Throughout this chapter, we assume that the degree distribution $P(d)$, $d \in \{1, 2, \ldots, \bar{D}\}$, is uniformly bounded away from zero. That is, $\min_d P(d) > \epsilon$ for some positive constant ϵ.

where w is a martingale increment process, that is $\mathbf{E}\{w_{k+1}|\mathcal{F}_k\} = 0$. Recall s is the finite state Markov chain that models the target process. ■

The above theorem is a well-known martingale representation of a Markov chain [55]—it says that a discrete time Markov process can be obtained by discrete time filtering of a martingale increment process. The theorem implies that the infected distribution dynamics resemble what is commonly called a stochastic approximation (adaptive filtering) algorithm in statistical signal processing: the new estimate is the old estimate plus a noisy update (the "noise" being a martingale increment) that is weighed by a small step size $1/N$ when N is large. Subsequently, we will exploit the structure in Theorem 21.1 to devise a mean field dynamics model that has a state of dimension $\bar{D}$. This is to be compared with the intractable state dimension $\prod_{d=1}^{\bar{D}} N(d)$ of the Markov chain ρ.

21.2.3 MEAN FIELD DYNAMICS OF INFORMATION DIFFUSION

The mean field dynamics state that as the number of agents N grows to infinity, the dynamics of the infected distribution ρ, described by Eq. (21.11), in the social network evolve according to the following deterministic difference equation that is modulated by a Markov chain that depends on the target state evolution s:

For $d = 1, 2, \dots, \bar{D}$,

$$\begin{aligned}
\bar{\rho}_{k+1}(d) &= \bar{\rho}_k(d) + \frac{1}{N}\left[\bar{p}_{01}(d, \theta(\bar{\rho}_k), s_k) - \bar{p}_{10}(d, \theta(\bar{\rho}_k), s_k)\right],\\
\bar{p}_{01}(d, \theta, s_k) &= (1 - \bar{\rho}_k(d)) \sum_{a=0}^{d} p_{01}(d, a, s_k) \binom{d}{a} \theta^a (1-\theta)^{d-a},\\
\bar{p}_{10}(d, \theta, s_k) &= \bar{\rho}_k(d) \sum_{a=0}^{d} p_{10}(d, a, s_k) \binom{d}{a} \theta^a (1-\theta)^{d-a},\\
\theta(\bar{\rho}_k) &= \frac{\sum_{d=1}^{\bar{D}} d\, P(d)\, \bar{\rho}_k(d)}{\sum_{d}^{\bar{D}} d\, P(d)}.
\end{aligned} \tag{21.12}$$

That the above mean field dynamics follow from Eq. (21.11) is intuitive. Such averaging results are well known in the adaptive filtering community where they are deployed to analyze the convergence of adaptive filters. The difference here is that the limit mean field dynamics are not deterministic but Markov modulated. Moreover, the mean field dynamics here constitute a model for information diffusion rather than the asymptotic behavior of an adaptive filtering algorithm. As mentioned earlier, from an engineering point of view, the mean field dynamics yield a tractable model for estimation.

We then have the following exponential bound result for the error of the mean field dynamics approximation.

Theorem 21.2. *For a discrete time horizon of T points, the deviation between the mean field dynamics $\bar{\rho}_k$ in Eq. (21.12) and actual infected distribution in ρ_k Eq. (21.11) satisfies*

$$\mathbb{P}\left\{\max_{0\le k\le T} \|\rho_k - \bar{\rho}_k\|_\infty \ge \epsilon\right\} \le C_1 \exp(-C_2 \epsilon^2 N), \tag{21.13}$$

where C_1 and C_2 are positive constants and $T = O(N)$. ■

The proof of the above theorem follows from [18, Lemma 1] and is presented in [56]. Actually in [18] the mean field dynamics are presented in continuous time as a system of ordinary differential equations. The exponential bound follows from an application of the Azuma-Hoeffding inequality. The above theorem provides an exponential bound (in terms of the number of agents N) for the probability of deviation of the sample path of the infected distribution from the mean field dynamics for any finite time interval T.

The stochastic approximation and adaptive filtering literature [57,58] have several averaging analysis methods for recursions of the form (21.11). The well-studied mean square error analysis [57,58] computes bounds on $\mathbf{E}\|\bar{\rho}_k - \rho_k\|^2$ instead of the maximum deviation in Theorem 21.2. A mean square error analysis of estimating a Markov modulated empirical distribution is given in [59]. Such mean square analysis assumes a finite but small step size $1/N$ in Eq. (21.11).

Related literature

Given the above SIS model, it is appropriate to pause briefly and review related literature. There are several other models for studying the spread of infection and technology in complex networks including susceptible-alert-infected-susceptible (SAIS), and susceptible-exposed-infected-vigilant (SEIV); see [43,60]. SIS models have been extensively studied in [16,17,41,42,46] to model information/infection diffusion, for example, the adoption of a new technology in a consumer market.

Degree-based mean field dynamics approximations for SIS models have been derived in [16,61]. Pair approximations (PA) and approximate master equations (AME) yield more general models for the complex dynamics of large-scale networks [61]. However, the resulting differential/difference equations that characterize the dynamics in PA and AME are no longer polynomial functions of the state. In this more general case, however, a suboptimal filter such as a particle filter can be used to track the infection diffusion.

It is also important to note that the right side of the mean field difference equation (21.12) is a polynomial function of the infected degree distribution $\bar{\rho}$. As a result, when the graph is sampled, resulting in noisy observations of $\bar{\rho}$, one can construct an exact finite dimensional Bayesian filter for the conditional mean estimate of $\bar{\rho}$ at each time k using the filtering algorithms in [62]. We refer the reader to [63] for details and also posterior Cramer-Rao lower bounds for estimating the infected degree distribution in the case of Erdos-Rényi and also power law (scale-free) networks such as Twitter. In comparison, Hamdi et al. [64] provides a stochastic approximation algorithm and analysis on a Hilbert space for tracking the degree distribution of evolving random networks with a duplication-deletion model.

On networks having fixed degree distribution, López-Pintado [16] identified conditions under which a network is susceptible to an epidemic using a mean-field approach and provided a closed form solution for the infection diffusion threshold. The diffusion properties of networks were investigated using stochastic dominance of their underlying degree distributions, as in [65]. We generalize these stochastic dominance results for evolving networks by considering a simple preferential attachment model as this can generate a scale-free network [66].

Finally, Pastor-Satorras and Vespignani [17] study the link between the power law exponent and the diffusion threshold. For the preferential attachment model, Ghoshal et al. [66] study the connection between the parameters that dictates the evolution (node and edge addition probability) and the degree distribution. Krishnamurthy et al. [63] has similar results using stochastic dominance, but the key emphasis is on providing a structured way to study such ordinal sensitivity relationships in large networks.

Numerical example

We simulate the diffusion of information through a network comprising $N = 100$ nodes (with maximum degree $\bar{D} = 17$). It is assumed that at time $k = 0$, 5% of nodes are infected. The mean field dynamics model is investigated in terms of the infected link probability (21.9). The infected link probability $\theta(\rho_k)$ is computed using Eq. (21.12).

Assume each agent is a myopic optimizer and, hence, chooses to adopt the technology only if $c^{(m)} \leq A_k^{(m)}$; $r = 1$. At time k, the costs $c^{(m)}$, $m = 1, 2, \ldots, 100$, are i.i.d. random variables simulated from uniform distribution $U[0, C(s_k)]$. Therefore, the transition probabilities in Eq. (21.5) are

$$p_{01}(d, A_k^{(m)}, s_k) = \mathbb{P}\left(c^{(m)} \leq A_k^{(m)}\right) = \begin{cases} \frac{A_k^{(m)}}{C(s_k)}, & A_k^{(m)} \leq C(s_k), \\ 1, & A_k^{(m)} > C(s_k). \end{cases} \tag{21.14}$$

The probability that a product fails is $p_F = 0.3$, i.e.,

$$p_{10}(d, A_k^{(m)}, s_k) = 0.3.$$

The infected link probabilities obtained from network simulation (21.9) and from the discrete-time mean field dynamics model (21.12) are illustrated in Fig. 21.3. To illustrate that the infected link probability computed from Eq. (21.12) follows the true one (obtained by network simulation), we assume that the value of C jumps from 1 to 10 at time $k = 200$, and from 10 to 1 at time $k = 500$. As can be seen in Fig. 21.3, the mean field dynamics provide an excellent approximation to the true infected distribution.

21.2.4 EXAMPLE: SOCIAL SENSING OF INFLUENZA USING TWITTER

In this section, we utilize datasets from three different social networks (namely, (i) Harvard social network, (ii) influenza datasets from the U.S. Centers for Disease Control and Prevention (CDC), and (iii) Twitter, to show how Twitter can be used as a real time social sensor for detecting outbreaks of influenza.

Twitter as a social sensor

A key advantage of using social media for rapid sensing of disease outbreaks in health networks is that it is low cost and provides rapid results compared with traditional techniques. For example, CDC must contact thousands of hospitals to query the data, which causes a reporting lag of approximately one to two weeks [67]. Using real-time microblogging platforms such as Twitter for disease detection has several advantages: the tweets are publicly available, high tweet posting frequency users often provide metadata (i.e., city, gender, age), and Twitter contains a diverse set of users [67].

Several papers have considered using Twitter data for estimating influenza infection rates. In [68,69] support vector regression supervised learning algorithms is used to relate the volume of Twitter posts that contain specific words (i.e., *flu, swine, influenza*) to the number of confirmed influenza cases in the

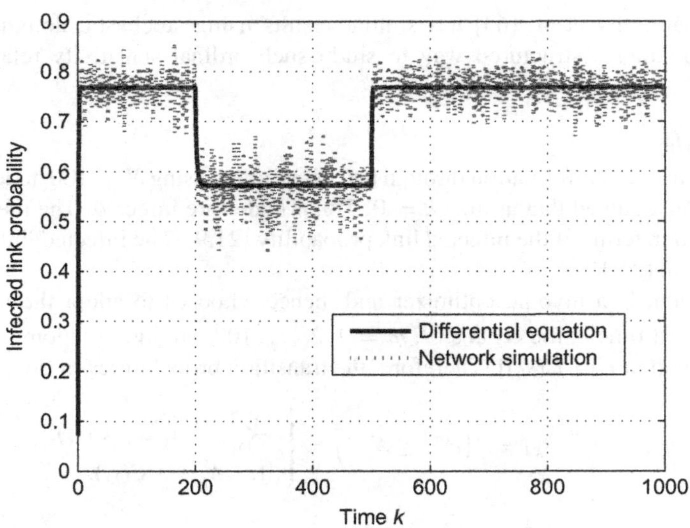

FIG. 21.3

The infected link probability obtained from network simulation compared to the one obtained from the mean field dynamics model (21.12). The transition probabilities in Eq. (21.14) depend only on the number of infected neighbors $A_k^{(m)}$ (the parameters are defined in Scenario 1).

United States as reported by the CDC. Multiple linear regression [70,71] and unsupervised Bayesian algorithms [72] have been used to relate the number of tweets of specific words to the influenza rate reported by the CDC. The detection algorithms [68,70,72] do not consider the dynamics of the disease propagation and the dynamics of information diffusion in the Twitter network. To reduce the effect of information diffusion in the network, Broniatowski et al. [73] proposes a support vector machine (SVM) classifier to detect: (a) if the tweet indicates the user's awareness of influenza or indicates the user is infected, and (b) if the influenza reference is in reference to another person. The classified tweets are then used to train a multiple linear regression model. To account for the diffusion dynamics of Twitter, Achrekar et al. [74,75] utilize an Autoregressive with Exogenous input (ARX) model. The exogenous input is the number of unique Twitter users with influenza-related tweets, and the output is the number of infected users as reported by the CDC.

If the social network is known, then the influenza spread can be formulated in terms of the diffusion model (21.11). Given the population of several hundred million, it is reasonable to adopt the mean field dynamics (21.12). With the influenza infection rate modeled using Eq. (21.12), the results can be used as an exogenous input to an ARX or nonlinear ARX (NARX) models to predict the volume of Twitter messages related to influenza, as illustrated in Fig. 21.2. In this framework, the Twitter messages are used to validate the underlying propagation model of influenza of use for predicting the infection rate and outbreak detection.

Social network influenza dataset

We consider the dataset [76] obtained from a social network of 744 undergraduate students from Harvard. The health of the 744 students was monitored from September 1, 2009 to December 31,

2009 and was reported by the university health services. To construct the social network, students were presented with a background questionnaire. In the questionnaire, students are asked, "Please provide the contact information for two to three Harvard students who you know and who you think would like to participate in this study," and "...provide us with the names and contact information of two to three of your friends...." This information was used to construct the degree distribution and links of the social network. A movie containing the spread of the influenza in the 744 college students over the 122-day sampling period can be viewed as the Youtube video titled, "Social Network Sensors for Early Detection of Contagious Outbreaks" at http://www.youtube.com/watch?v=0TD06g2m8qM. Fig. 21.4A–C displays three illustrative snapshots from this video; in the online version of this chapter, red nodes denote infected students while yellow nodes depict their neighbors in the social network.

Models for influenza diffusion

From the data in the YouTube video for the Harvard students, we observed the following regarding the transition probabilities $p_{ij}(d, A, s)$ defined in Eq. (21.5). As expected, students with a larger number of infected neighbors A contract influenza sooner. The data shows that the transition probabilities were approximately independent of the degree of the node d. Because the data provided was during an actual influenza outbreak, we set the target state of the network (i.e., s) constant. Therefore the transition probabilities depend only on the number of infected neighbors and were estimated as

$$p_{01}(a = 1) = 0.02, \quad p_{01}(a = 2) = 0.15.$$

That is, the dataset reveals that the probability of getting infected given $a = 2$ infected neighbors is substantially higher than with $a = 1$ infected neighbor, as expected. The estimated infected link probability θ_k in Eq. (21.9) versus time (days) k is displayed in Fig. 21.4D. Recall from Section 21.2.3 that the infected link probability θ_k is related to the mean field dynamics equation (21.12). This allows the transition probabilities and θ_k to be used to predict the infection rate dynamics.

Other graph-theoretic measures also play a role in the analysis of the diffusion. Students with high k-coreness[3] are expected to contract influenza earlier. Additionally, students that have high betweenness centrality (i.e., number of shortest paths from all students to all others that pass through that student) contract influenza earlier then students with low betweenness centrality. These observations show that the diffusion of influenza in the network depends strongly on the underlying health network structure. The dynamic model (21.7) accounts for the effects of the degree of nodes; however to account for the effects from betweenness centrality and k-coreness would require a more sophisticated formulation then that presented in Section 21.2.1.

Time series model for influenza tweets

In Section 21.2.4 we illustrated how the mean field dynamics model (21.12) can be used to estimate the influenza infection rate with the model parameters estimated from a sampled set of the entire population. To validate the estimated parameters for the entire network requires that the infection rate be related to an observable response, in this case the number of Twitter mentions of a specific keyword. Two time-series models are considered for relating the infection rate to the number of Twitter mentions. The models are validated using two real-world datasets of Twitter mentions and number of influenza cases in the United States.

[3] k-coreness is the largest subnetwork comprising nodes of degree at least k.

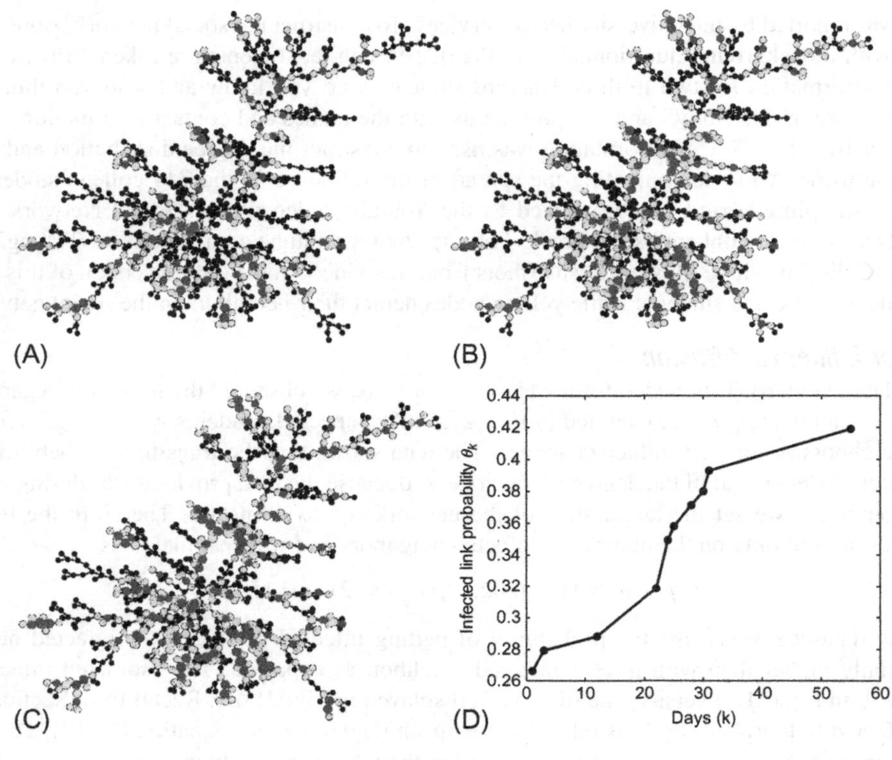

FIG. 21.4

Snapshots from YouTube video of Harvard undergraduate social network propagation of influenza and the estimated infected link probability θ_k (21.9) for October 10 to December 23, 2009. (A) October 10, 2009; (B) November 11, 2009; (C) December 23, 2009; (D) October 10 to December 23, 2009.

The number of influenza cases in the United States is obtained from the CDC,[4] which publishes weekly reports from the ILInet. The data reported by the CDC is comprised of reports from over 3000 health providers nationwide and was obtained for the dates between September 1, 2012 to October 1, 2013. The associated Twitter data for the 122-day period was obtained using the software PeopleBrowsr.[5] The prespecified Twitter search terms used were: *flu, swine, and influenza*. Because our focus is on monitoring influenza dynamics in the United States, we excluded all tweets tagged as originating from outside the United States. The total number of mentions of a specific keyword on each is obtained using PeopleBrowsr.

[4]http://gis.cdc.gov/grasp/fluview/fluportaldashboard.html.
[5]http://gr.peoplebrowsr.com/.

We used two time-series models for the volume of tweets and compared their performance. The first time series model considered is the ARX model defined by:

$$\tau_k = \sum_{i=1}^{n_a} a_i \tau_{k-i} + \sum_{i=0}^{n_b-1} b_i \rho_{k-\Delta-i} + d + v_k. \tag{21.15}$$

In Eq. (21.15), τ_k is the number of influenza-related tweets at k, ρ_k is the exogenous input of the infected influenza patients, $n_a, n_b, a_i, b_i, \Delta$ and d are model parameters with v_k an i.i.d. noise process. Δ models the delay between patient contraction and the respective individual tweeting their symptoms. d models the mean number of tweets related to influenza that are not related to an actual infection.

The second time series model we used is the nonlinear autoregressive exogenous (NARX) model given by:

$$\tau_k = F(\tau_{k-1}, \ldots, \tau_{k-n_a}, \rho_{k-\Delta}, \ldots, \rho_{k-\Delta-n_b}) + v_k. \tag{21.16}$$

In Eq. (21.16) F denotes a nonlinear function that relates the exogenous input and previous tweets to the current number of tweets. Here we consider F as a support vector machine that can be trained using historical data. Note that if F was independent of previous tweets, previous exogenous inputs, and no delay (i.e., $n_b = 0$ and $\Delta = 0$), then Eq. (21.16) would be identical to the SVM classifier used in [68,69] to relate the number of tweets to the number of infected agents.

The number of reported influenza cases, associated Twitter data, and results of the model training and prediction are displayed in Fig. 21.5 for the ARX (21.15) and NARX (21.16) models. As seen from Fig. 21.5A, the dominant word for indicating a possible influenza outbreak is *flu* as compared with *swine* and *influenza*. Notice that there is a lag between the maximum confirmed influenza cases and the # of tweets; however, there is an increase in the number of tweets prior to the peak of infected patients. These dynamics are a result of a combination of infection propagation dynamics and the diffusion of information on Twitter. To account for these dynamics the ARX and NARX models presented in Section 21.2.4 are utilized. The training and prediction accuracy of these models for $n_a = 0, n_b = 2$ (i.e., model input parameters $\rho_{k-\Delta}$ and $\rho_{k-\Delta-1}$) are displayed in Fig. 21.5B. As seen, the NARX (21.16) model provides a superior estimate as compared with the ARX model (21.15). Interestingly there is a $\Delta = 18$ day delay between the maximum number of infected patients and the maximum number of Twitter mentions containing the word *flu*. This is in contrast to the dynamics observed for the 2009 [68] and 2010–2011 [75] influenza outbreaks, which show that the increase in Twitter mentions occurs earlier or at the same time as the number of infected patients increases. This also emphasizes the importance of using the mean field dynamics model for influenza propagation as compared with only using Twitter data for predicting the influenza infection rate. Here we have used the CDC data to estimate the number of infected agents; however, the mean field dynamics model (21.12) could be used to estimate the dynamics of disease propagation and relate this to the observable number of tweets in real time.

To summarize, the above datasets illustrate how Twitter can be used as a sensor for monitoring the spread of influenza in a heath network. The propagation of influenza was modeled according to the SIS model and the dynamics of tweets according to an autoregressive model.

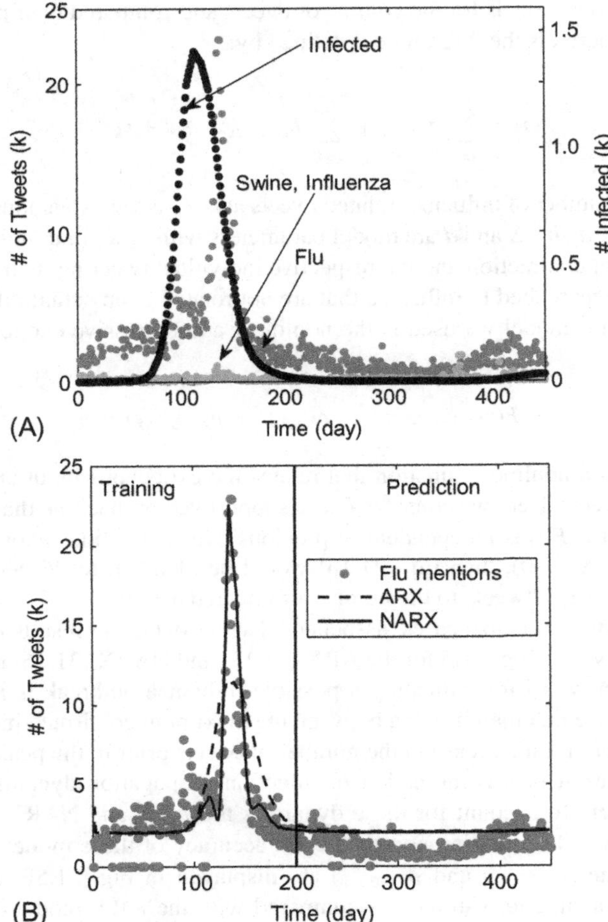

FIG. 21.5

The experimental data is obtained for September 2012 to October 2013 as described in Section 21.2.4. The ARX (21.15) and NARX (21.16) models are utilized to estimate the number of tweets with *flu* given the number of infected influenza patients. (A) Experimental data of # of tweets and # of infected with influenza. (B) ARX model (21.15) and NARX model (21.16) for the # of tweets given the # of infected influenza patients. The ARX and NARX models are trained using the initial 200 days of data. The predictive accuracy of the models is illustrated for the remaining 230 days.

21.2.5 SENTIMENT-BASED SENSING MECHANISM

In the above dataset, samples of influenza-affected individuals were obtained from a Harvard social network. More generally, it is often necessary to sample individuals in a social network to estimate an underlying state of nature such as the sentiment. An important question regarding sensing in a social network is: How can one construct a small but representative sample of a social network with a large number of nodes? In [77] several scale-down and back-in-time sampling procedures are studied. Below we review three sampling schemes. The simplest possible sampling scheme is uniform sampling. We also briefly describe *social sampling* and *respondent-driven sampling*, which are recent methods that have become increasingly popular.

Uniform sampling

Consider the following sampling-based measurement strategy. At each period k, $\alpha(d)$ individuals are sampled[6] independently and uniformly from the population $N(d)$ comprising agents with connectivity degree d. That is, a uniform distributed i.i.d. sequence of nodes, denoted by $\{m_l, l = 1 : \alpha(d)\}$, is generated from the population $N(d)$. The messages $y_k^{(m_l)}$ of these $\alpha(d)$ individuals are recorded. From these independent samples, the empirical sentiment distribution $z_k(d)$ of degree d nodes at each time k is obtained as

$$z_k(d, y) = \frac{1}{\alpha(d)} \sum_{l=1}^{\alpha(d)} I\left\{y_k^{(m_l)} = y\right\}, \quad y = 1, \ldots, Y. \tag{21.17}$$

At each time k, the empirical sentiment distribution z_k can be viewed as noisy observations of the infected distribution ρ_k and target state process s_k.

Social sampling

Social sampling is an extensive area of research; see [78] for recent results. In social sampling, participants in a poll respond with a summary of their friend's responses. This leads to a reduction in the number of samples required. If the average degree of nodes in the network is d, then the savings in the number of samples is by a factor of d because a randomly chosen node summarizes the results from d of its friends. However, the variance and bias of the estimate depend strongly on the social network structure.[7] In [78], a social sampling method is introduced and analyzed where nodes of degree d are sampled with probability proportional to $1/d$. This is intuitive because weighing neighbors' values by the reciprocal of the degree undoes the bias introduced by large degree nodes. It then illustrates this social sampling method and variants on the LiveJournal network (livejournal.com) comprising more than 5 million nodes and 160 million directed edges.

MCMC based respondent-driven sampling (RDS)

Respondent-driven sampling (RDS) was introduced by Heckathorn [79,80] and Lee [81] as an approach for sampling from hidden populations in social networks and has gained enormous popularity in recent years. There are more than 120 RDS studies worldwide involving sex workers and injection drug users [82]. As mentioned in [83], the U.S. Centers for Disease Control and Prevention (CDC) recently selected RDS for a 25-city study of injection drug users that is part of the National HIV Behavioral Surveillance System [84].

RDS is a variant of the well-known method of snowball sampling where current sample members recruit future sample members. The RDS procedure is as follows: A small number of people in the

---

[6]For large population sizes N, sampling with and without replacement are equivalent.

[7]In [78], nice intuition is provided in terms of intent polling and expectation polling. In intent polling, individuals are sampled and asked who they intend to vote for. In expectation polling, individuals are sampled and asked who they think would win the election. For a given sample size, one would believe that expectation polling is more accurate than intent polling because in expectation polling, an individual would typically consider its own intent together with the intents of its friends.

target population serve as seeds. After participating in the study, the seeds recruit other people they know through the social network in the target population. The sampling continues according to this procedure with current sample members recruiting the next wave of sample members until the desired sampling size is reached. Typically, monetary compensation is provided for participating in the data collection and recruitment.

RDS can be viewed as a form of Markov Chain Monte Carlo (MCMC) sampling (see [83] for an excellent exposition). Let $\{m_l, l = 1 : \alpha(d)\}$ be the realization of an aperiodic irreducible Markov chain with state space $N(d)$ comprising nodes of degree d. This Markov chain models the individuals of degree d that are snowball sampled, namely, the first individual m_1 is sampled and then recruits the second individual m_2 to be sampled, who then recruits m_3 and so on. Instead of the independent sample estimator (21.17), an asymptotically unbiased MCMC estimate is then generated as

$$\frac{\sum_{l=1}^{\alpha(d)} \frac{I(y_k^{(m_l)}=y)}{\pi(m_l)}}{\sum_{l=1}^{\alpha(d)} \frac{1}{\pi(m_l)}}, \tag{21.18}$$

where $\pi(m)$, $m \in N(d)$, denotes the stationary distribution of the Markov chain. For example, a reversible Markov chain with prescribed stationary distribution is straightforwardly generated by the Metropolis Hastings algorithm.

In RDS, the transition matrix and, hence, the stationary distribution π in the estimator (21.18) is specified as follows: Assume that edges between any two nodes m and n have symmetric weights $W(m,n)$ (i.e., $W(m,n) = W(n,m)$, equivalently, the network is undirected). In RDS, node m recruits node n with transition probability $W(m,n)/\sum_n W(m,n)$. Then, it can be easily seen that the stationary distribution is $\pi(m) = \sum_{n\in V} W(m,n)/\sum_{m\in V,n\in V} W(m,n)$. Using this stationary distribution, along with the above transition probabilities for sampling agents in Eq. (21.18), yields the RDS algorithm.

It is well known that a Markov chain over a nonbipartite connected undirected network G is aperiodic. Then, the initial seed for the RDS algorithm can be picked arbitrarily, and the above estimator is an asymptotically unbiased estimator.

Note the difference between RDS and social sampling: RDS uses the network to recruit the next respondent whereas social sampling seeks to reduce the number of samples by using people's knowledge of their friends' (neighbors') opinions.

Finally, the reader may be familiar with the DARPA network challenge in 2009 where the locations of 10 red balloons in the continental United States were to be determined using social networking. In this case, the winning MIT Red Balloon Challenge Team used a recruitment-based sampling method. The strategy can also be viewed as a variant of the query incentive network model of [85].

21.2.6 SUMMARY AND EXTENSIONS

This section has discussed the diffusion of information in social networks. Mean field dynamics were used to approximate the asymptotic infected degree distribution. An illustrative example of the spread of influenza was provided. Finally, methods for sampling the population in social networks were reviewed. Below we discuss some related concepts and extensions.

Bayesian filtering problem

Given the sentiment observations described above, how can the infected degree distribution ρ_k and target state s_k be estimated at each time instant? The partially observed state space model with dynamics (21.12) and discrete time observations from sampling the network can be used to obtain Bayesian filtering estimates of the underlying state of nature. Computing the conditional mean estimate s_k, ρ_k given the sentiment observation sequence is a Bayesian filtering problem. In fact, filtering of such jump Markov linear systems have been studied extensively in the signal processing literature [86,87] and can be solved via the use of sequential Markov chain Monte Carlo methods. For example, Sakaki et al. [6] report on how a particle filter is used to localize earthquake events using Twitter as a social sensor.

Reactive information diffusion

A key difference between social sensors and conventional sensors in statistical signal processing is that social sensors are reactive: A social sensor uses additional information gained to modify its behavior. Consider the case where the sentiment-based observation process is made available in a public blog. Then, these observations will affect the transition dynamics of the agents and, therefore, the mean field dynamics.

How does connectivity affect mean field equilibrium?

The papers [16,50] examine the structure of fixed points of the mean field differential equation (21.12) when the underlying target process s is not present (equivalently, s is a one-state process). They consider the case where the agent transition probabilities are parametrized by $p_{01}(d,a) = \mu F(d,a)$ and $p_{10} = p_F$. Then, defining $\lambda = \mu / p_F$, they study how the following two thresholds behave with the degree distribution and diffusion mechanism:

1. *Critical threshold* λ_c: This is defined as the minimum value of λ for which there exists a fixed point of Eq. (21.12) with positive fraction of infected agents, i.e., $\rho_\infty(d) > 0$ for some d and, for $\lambda \leq \lambda_c$, such a fixed point does not exist.
2. *Diffusion threshold* λ_d: Suppose the initial condition ρ_0 for the infected distribution is infinitesimally small. Then, λ_d is the minimum value of λ for which $\rho_\infty(d) > 0$ for some d, and such that, for $\lambda \leq \lambda_d$, $\rho_\infty(d) = 0$ for all d.

Determining how these thresholds vary with degree distribution and diffusion mechanisms is very useful for understanding the long-term behavior of agents in the social network.

21.3 BAYESIAN SOCIAL LEARNING MODELS FOR ONLINE REPUTATION SYSTEMS

In this section we address the second topic of the chapter, namely, Bayesian social learning among social sensors. The motivation can be understood in terms of the following social sensing example. Consider the following interactions in a multiagent social network where agents seek to estimate an underlying state of nature. Each agent visits a restaurant based on reviews on an online reputation website. The agent then obtains a private measurement of the state (e.g., the quality of food in a restaurant) in noise. After that, he reviews the restaurant on the same online reputation website. The

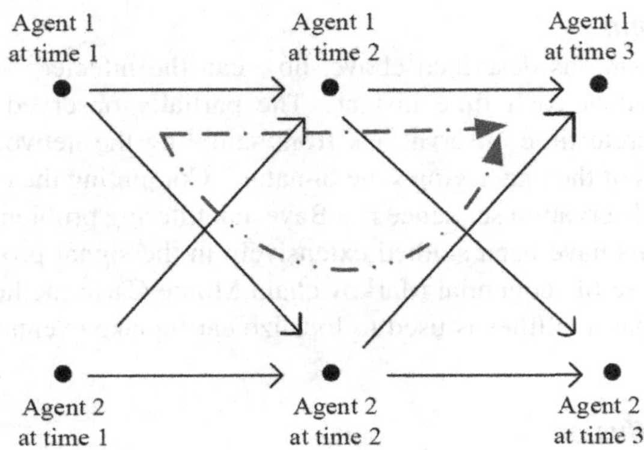

FIG. 21.6

Example of the information flow (communication graph) in a social network with two agents and over three event epochs. The *arrows* represent exchange of information.

information exchange in the social network is modeled by a directed graph. Data incest [88] arises due to loops in the information exchange graph. This is illustrated in the graph of Fig. 21.6. Agents 1 and 2 exchange beliefs (or actions) as depicted in Fig. 21.6. The fact that there are two distinct paths between Agent 1 at time 1 and Agent 1 at time 3 (these paths are denoted by dashed lines) implies that the information of Agent 1 at time 1 is double counted, leading to a data incest event.

How can data incest be removed so that agents obtain a fair (unbiased) estimate of the underlying state? The methodology of this section can be interpreted in terms of the recent *Time* article [89] that provides interesting rules for online reputation systems. These include: (i) review the reviewers, and (ii) censor fake (malicious) reviewers. The data incest removal algorithm proposed in this chapter can be viewed as "reviewing the reviews" of other agents to see if they are associated with data incest or not.

The rest of this section is organized as follows:

1. Section 21.3.1 describes the social learning model that is used to mimic the behavior of agents in online reputation systems. Section 21.3.2 describes how risk averse social learning models apply to detecting market shocks in high frequency financial systems.
2. Sections 21.3.3–21.3.5 deal with modeling data incest and incest removal algorithms for online reputation systems. The information exchange between agents in the social network is formulated on a family of time-dependent directed acyclic graphs. A necessary and sufficient condition is given on the graph structure of information exchange between agents so that a fair rating is achievable.
3. Section 21.3.6 discusses conditions under which treating individual social sensors as Bayesian optimizers is a useful idealization of their behavior. In particular, it is shown that the ordinal behavior of humans can be mimicked by Bayesian optimizers under reasonable conditions.
4. Section 21.3.7 presents a dataset obtained from a psychology experiment to illustrate social learning and data incest patterns.

Related work

Collaborative recommendation systems are reviewed and studied in [90,91]. The books [15,43] study information cascades in social learning. In [92], a model of Bayesian social learning is considered in which agents receive private information about the state of nature and observe actions of their neighbors in a tree-based network. Another type of misinformation caused by influential agents (agents who heavily affect actions of other agents in social networks) is investigated in [21]. Misinformation in the context of this chapter is motivated by sensor networks where the term "data incest" is used [88]. Data incest also arises in belief propagation (BP) algorithms [93,94], which are used in computer vision and error-correcting coding theory. BP algorithms require passing local messages over the graph (Bayesian network) at each iteration. For graphical models with loops, BP algorithms are only approximate due to the overcounting of local messages [95], which is similar to data incest in social learning. With the algorithms presented in this section, data incest can be mitigated from Bayesian social learning over nontree graphs that satisfy a topological constraint. The closest work to the current chapter is [88]. However, in [88], data incest is considered in a network where agents exchange their private belief states—that is, no social learning is considered. Simpler versions of this information exchange process and estimation were investigated in [96–98]. We also refer the reader to [44] for a discussion of recommender systems.

21.3.1 CLASSICAL SOCIAL LEARNING

We briefly review the classical social learning model for the interaction of individuals. Subsequently, we will deal with more general models over a social network.

Consider a multiagent system that aims to estimate the state of an underlying finite state random variable $x \in \mathbb{X} = \{1, 2, \ldots, X\}$ with known prior distribution π_0. Each agent acts once in a predetermined sequential order indexed by $k = 1, 2, \ldots$ Assume at the beginning of iteration k, all agents have access to the public belief π_{k-1} defined in Step (iv) below. The social learning protocol proceeds as follows [15,20]:

(i) *Private Observation*: At time k, agent k records a private observation $y_k \in \mathbb{Y}$ from the observation distribution $B_{iy} = P(y|x = i), i \in \mathbb{X}$. Throughout this section we assume that $\mathbb{Y} = \{1, 2, \ldots, Y\}$ is finite.

(ii) *Private Belief*: Using the public belief π_{k-1} available at time $k - 1$ (Step (iv) below), agent k updates its private posterior belief $\eta_k(i) = P(x_k = i|a_1, \ldots, a_{k-1}, y_k)$ using Bayes formula:

$$\eta_k = \frac{B_{y_k}\pi}{\mathbf{1}_X' B_y \pi}, B_{y_k} = \text{diag}(P(y_k|x = i), i \in \mathbb{X}). \tag{21.19}$$

Here $\mathbf{1}_X$ denotes the X-dimensional vector of ones, η_k is an X-dimensional probability mass function (pmf).

(iii) *Myopic Action*: Agent k takes action $a_k \in \mathcal{A} = \{1, 2, \ldots, A\}$ to minimize its expected cost

$$a_k = \arg\min_{a \in \mathcal{A}} \mathbf{E}\{c(x, a)|a_1, \ldots, a_{k-1}, y_k\} = \arg\min_{a \in \mathcal{A}}\{c_a' \eta_k\}. \tag{21.20}$$

Here $c_a = (c(i, a), i \in \mathbb{X})$ denotes an X dimensional cost vector, and $c(i, a)$ denotes the cost incurred when the underlying state is i and the agent chooses action a. Agent k then broadcasts its action a_k.

(iv) *Social Learning Filter*: Given the action a_k of agent k, and the public belief π_{k-1}, each subsequent agent $k' > k$ performs social learning to update the public belief π_k according to the "social learning filter":

$$\pi_k = T(\pi_{k-1}, a_k), \text{ where } T(\pi, a) = \frac{R_a^\pi \pi}{\sigma(\pi, a)}, \tag{21.21}$$

where $\sigma(\pi, a) = \mathbf{1}'_X R_a^\pi P' \pi$ is the normalization factor of the Bayesian update. In Eq. (21.21), the public belief $\pi_k(i) = P(x_k = i|a_1, \dots a_k)$ and $R_a^\pi = \text{diag}(P(a|x = i, \pi), i \in \mathbb{X})$ has elements

$$P(a_k = a|x_k = i, \pi_{k-1} = \pi) = \sum_{y \in \mathbb{Y}} P(a|y, \pi) P(y|x_k = i),$$

$$P(a_k = a|y, \pi) = \begin{cases} 1 \text{ if } c_a' B_y P' \pi \le c_{\tilde{a}}' B_y P' \pi, \ \tilde{a} \in \mathcal{A}, \\ 0 \text{ otherwise.} \end{cases}$$

The following result, which is well known in the economics literature [15,20], states that if agents follow the above social learning protocol, then after some finite time $\bar{k}$, an *information cascade* occurs.[8]

Theorem 21.3 ([20]). *The social learning protocol leads to an* information cascade *in finite time with probability 1. That is, after some finite time $\bar{k}$ social learning ceases and the public belief $\pi_{k+1} = \pi_k$, $k \ge \bar{k}$, and all agents choose the same action $a_{k+1} = a_k$, $k \ge \bar{k}$.* ■

Instead of reproducing the proof, let us give some insight as to why Theorem 21.3 holds. It can be shown using martingale methods that at some finite time $k = k^*$, the agent's probability $P(a_k|y_k, \pi_{k-1})$ becomes independent of the private observation y_k. Then clearly, $P(a_k = a|x_k = i, \pi_{k-1}) = P(a_k = a|\pi)$. Substituting this into the social learning filter (21.21), we see that $\pi_k = \pi_{k-1}$. Thus after some finite time k^*, the social learning filter hits a fixed point and social learning stops. As a result, all subsequent agents $k > k^*$ completely disregard their private observations and take the same action a_{k^*}, thereby forming an information cascade (and therefore a herd).

21.3.2 RISK-AVERSE SOCIAL LEARNING AND DETECTING MARKET SHOCKS

Here we consider the statistical signal processing problem involving agent-based models of financial markets, which, at a microlevel, are driven by socially aware and risk-averse trading agents. These agents trade (buy or sell) stocks at each trading instant by using the decisions of all previous agents (social learning) in addition to a private (noisy) signal they receive on the value of the stock. We are interested in the following: (1) Modeling the dynamics of these risk-averse agents, (2) Sequential detection of a market shock based on the behavior of these agents. Structural results that characterize social learning under a risk measure, CVaR (Conditional Value-at-risk), are presented and formulation of the Bayesian change point detection problem is provided. The structural results exhibit two

[8] A *herd of agents* takes place at time $\bar{k}$, if the actions of all agents after time $\bar{k}$ are identical, i.e., $a_k = a_{\bar{k}}$ for all time $k > \bar{k}$. An information cascade implies that a herd of agents occurs. Trusov et al. [4] quotes the following anecdote of user influence and herding in a social network: "...when a popular blogger left his blogging site for a two-week vacation, the site's visitor tally fell, and content produced by three invited substitute bloggers could not stem the decline."

interesting properties: (i) Risk-averse agents herd more often than risk-neutral agents. (ii) The stopping set in the sequential detection problem is nonconvex.

It is well documented in behavioral economics [99] and psychology [100] that people prefer a certain but possibly less desirable outcome over an uncertain but potentially larger outcome. To model this risk averse behavior, commonly used risk measures[9] are Value-at-Risk (VaR), Conditional Value-at-Risk (CVaR), Entropic risk measure, and Tail value at risk; see [101]. We consider social learning under CVaR risk measure. CVaR [102] is an extension of VaR that gives the total loss given a loss event and is a coherent risk measure [103]. Below, we choose the CVaR risk measure as it exhibits the following properties [102,103]: (i) It associates higher risk with higher cost; (ii) It ensures that risk is not a function of the quantity purchased but arises from the stock; and (iii) It is convex. CVaR as a risk measure has been used in solving portfolio optimization problems [104,105] and order execution. For an overview of risk measures and their application in finance, see [101].

CVaR social learning model

The market microstructure is modeled as a discrete time dealer market motivated by algorithmic and high-frequency tick-by-tick trading [106]. There is a single traded stock or asset, a market observer, and a countable number of trading agents. The asset has an initial true underlying value $x_0 \in \mathcal{X} = \{1, 2, \dots, X\}$. The market observer does not receive direct information about $x \in \mathcal{X}$ but only observes the public buy/sell actions of agents, $a_k \in \mathcal{A} = \{1(\text{buy}), 2(\text{sell})\}$. The agents themselves receive noisy private observations of the underlying value x and consider this in addition to the trading decisions of the other agents visible in the order book. At a random time, τ^0 determined by the transition matrix P, the asset experiences a jump change in its value to a new value. The aim of the market observer is to detect the change time (global decision) with minimal cost, having access to only the actions of these socially aware agents. Let $y_k \in \mathcal{Y} = \{1, 2, \dots, Y\}$ denote agent k's private observation. The initial distribution is $\pi_0 = (\pi_0(i), i \in \mathcal{X})$ where $\pi_0(i) = \mathbb{P}(x_0 = i)$.

The agent-based model has the following dynamics:

1. *Shock in the asset value*: At time $\tau^0 > 0$, the asset experiences a jump change (shock) in its value due to exogenous factors. The change point τ^0 is modeled by a *phase type (PH) distribution*. The family of all PH distributions forms a dense subset for the set of all distributions [107], i.e., for any given distribution function F such that $F(0) = 0$, one can find a sequence of PH distributions $\{F_n, n \geq 1\}$ to approximate F uniformly over $[0, \infty)$. The PH-distributed time τ^0 can be constructed via a multistate Markov chain x_k with state space $\mathcal{X} = \{1, \dots, X\}$ as follows: Assume state "1" is an absorbing state and denotes the state after the jump change. The states $2, \dots, X$ (corresponding to beliefs $e_2, \dots, e_X$) can be viewed as a single composite state that x resides in before the jump. So $\tau^0 = \inf\{k : x_k = 1\}$ and the transition probability matrix P is of the form

$$P = \begin{bmatrix} 1 & 0 \\ \underline{P}_{(X-1)\times 1} & \bar{P}_{(X-1)\times(X-1)} \end{bmatrix}. \tag{21.22}$$

[9]A risk measure $\varrho : \mathcal{L} \to \mathbb{R}$ is a mapping from the space of measurable functions to the real line, which satisfies the following properties: (i) $\varrho(0) = 0$. (ii) If $S_1, S_2 \in \mathcal{L}$ and $S_1 \leq S_2$ a.s then $\varrho(S_1) \leq \varrho(S_2)$. (iii) if $a \in \mathbb{R}$ and $S \in \mathcal{L}$, then $\varrho(S+a) = \varrho(S)+a$. The risk measure is coherent if in addition ϱ satisfies: (iv) If $S_1, S_2 \in \mathcal{L}$, then $\varrho(S_1+S_2) \leq \varrho(S_1)+\varrho(S_1)$. (v) If $a \geq 0$ and $S \in \mathcal{L}$, then $\varrho(aS) = a\varrho(S)$. The expectation operator is a special case where subadditivity is replaced by additivity.

The distribution of the absorption time to state 1 is

$$\nu_0 = \pi_0(1), \quad \nu_k = \bar{\pi}_0' \bar{P}^{k-1} \underline{P}, \quad k \geq 1, \tag{21.23}$$

where $\bar{\pi}_0 = [\pi_0(2), \dots, \pi_0(X)]'$. The key idea is that by appropriately choosing the pair (π_0, P) and the associated state space dimension X, one can approximate any given discrete distribution on $[0, \infty)$ by the distribution $\{\nu_k, k \geq 0\}$; see [107, pp. 240–243]. The event $\{x_k = 1\}$ means the change point has occurred before time k according to PH distribution (21.23). In the special case when x is a 2-state Markov chain, the change time τ^0 is geometrically distributed.

2. *Agent's Private Observation*: Agent k's private (local) observation denoted by y_k is a noisy measurement of the true value of the asset. It is obtained from the observation likelihood distribution as,

$$B_{xy} = \mathbb{P}(y_k = y | x_k = x). \tag{21.24}$$

3. *Private Belief update*: Agent k updates its private belief using the observation y_k and the prior public belief $\pi_{k-1}(i) = \mathbb{P}(X = i | a_1, \dots, a_{k-1})$ as the following Hidden Markov Model update

$$\eta_k = \frac{B_{y_k} P' \pi_{k-1}}{\mathbf{1}' B_{y_k} P' \pi_{k-1}}, \tag{21.25}$$

where $\mathbf{1}$ denotes the X-dimensional vector of ones.

4. *Agent's trading decision*: Agent k executes an action $a_k \in \mathcal{A} = \{1(\text{buy}), 2(\text{sell})\}$ to myopically minimize its cost. Let $c(i, a)$ denote the cost incurred if the agent takes action a when the underlying state is i. Let the local cost vector be

$$c_a = [c(1, a)\ c(2, a) \dots\ c(X, a)]. \tag{21.26}$$

The costs for different actions are taken as

$$c(i, j) = p_j - \beta_{ij} \quad \text{for } i \in \mathcal{X}, j \in \mathcal{A}, \tag{21.27}$$

where β_{ij} corresponds to the agent's demand. Here demand is the agent's desire and willingness to trade at a price p_j for the stock. Here p_1 is the quoted price for purchase and p_2 is the price demanded in exchange for the stock. We assume that the price is the same during the period in which the value changes. As a result, the willingness of each agent only depends on the degree of uncertainty on the value of the stock.

Remark 21.1. The analysis provided in this paper straightforwardly extends to the case when different agents are facing different prices, as in an order book. For notational simplicity we assume the cost are time invariant.

The agent considers measures of risk in the presence of uncertainty in order to overcome the losses incurred in trading. To illustrate this, let $c(x, a)$ denote the loss incurred with action a while at unknown and random state $x \in \mathcal{X}$. When an agent solves an optimization problem involving $c(x, a)$ for selecting the best trading decision, it will take into account not just the expected loss, but also the "riskiness" associated with the trading decision a. The agent therefore chooses an action a_k to minimize the CVaR measure[10] of trading as

$$
\begin{aligned}
a_k &= \underset{a\in\mathcal{A}}{\operatorname{argmin}}\{\mathrm{CVaR}_\alpha(c(x_k,a))\} \qquad (21.28)\\
&= \underset{a\in\mathcal{A}}{\operatorname{argmin}}\left\{\min_{z\in\mathbb{R}}\left\{z+\frac{1}{\alpha}\mathbb{E}_{y_k}\left[\max\left\{(c(x_k,a)-z),0\right\}\right]\right\}\right\}.
\end{aligned}
$$

Here $\alpha\in(0,1]$ reflects the degree of risk aversion for the agent (the smaller α is, the more risk-averse the agent is). Define

$$\mathcal{H}_k := \sigma\text{-algebra generated by } (a_1,a_2,\ldots,a_{k-1},y_k) \qquad (21.29)$$

$\mathbb{E}_{y_k}$ denotes the expectation with respect to private belief, i.e., $\mathbb{E}_{y_k}=\mathbb{E}[.|\mathcal{H}_k]$ when the private belief is updated after observation y_k.

5. *Social Learning and Public belief update*: Agent k's action is recorded in the order book and hence broadcast publicly. Subsequent agents and the market observer update the public belief on the value of the stock according to the social learning Bayesian filter as follows

$$\pi_k = T^{\pi_{k-1}}(\pi_{k-1},a_k) = \frac{R^{\pi_{k-1}}_{a_k}P'\pi_{k-1}}{\mathbf{1}'R^{\pi_{k-1}}_{a_k}P'\pi_{k-1}}. \qquad (21.30)$$

Here, $R^{\pi_{k-1}}_{a_k}=\operatorname{diag}(\mathbb{P}(a_k|x=i,\pi_{k-1}),i\in\mathcal{X})$, where $\mathbb{P}(a_k|x=i,\pi_{k-1})=\sum_{y\in\mathcal{Y}}\mathbb{P}(a_k|y,\pi_{k-1})\mathbb{P}(y|x_k=i)$ and

$$\mathbb{P}(a_k|y,\pi_{k-1}) = \begin{cases} 1 & \text{if } a_k=\underset{a\in\mathcal{A}}{\operatorname{argmin}}\,\mathrm{CVaR}_\zeta(c(x_k,a));\\ 0 & \text{otherwise.}\end{cases}$$

Note that π_k belongs to the unit simplex $\Pi(X)\stackrel{\Delta}{=}\{\pi\in\mathbb{R}^X:\mathbf{1}'_X\pi=1, 0\le\pi\le1 \text{ for all } i\in\mathcal{X}\}$.

6. *Market Observer's Action*: The market observer (securities dealer) seeks to achieve quickest detection by balancing delay with false alarm. At each time k, the market observer chooses action[11] u_k as

$$u_k\in\mathcal{U}=\{1(\text{stop}),2(\text{continue})\}. \qquad (21.31)$$

Here "Stop" indicates that the value has changed and the dealer incorporates this information before selling new issues to investors. The formulation presented considers a general parametrization of the costs associated with detection delay and false alarm costs. Define

$$\mathcal{G}_k := \sigma\text{-algebra generated by } (a_1,a_2,\ldots,a_{k-1},a_k). \qquad (21.32)$$

[10] For the reader unfamiliar with risk measures, it should be noted that CVaR is one of the "big" developments in risk modeling in finance in the last 15 years. In comparison, the value at risk (VaR) is the percentile loss namely, $\mathrm{VaR}_\alpha(x)=\min\{z:F_x(z)\ge\alpha\}$ for cdf F_x. While CVaR is a coherent risk measure, VaR is not convex and so not coherent. CVaR has other remarkable properties [102]: it is continuous in α and jointly convex in (x,α). For continuous cdf F_x, $\mathrm{CVaR}_\alpha(x)=\mathbf{E}\{X|X>\mathrm{VaR}_\alpha(x)\}$. Note that the variance is not a coherent risk measure.

[11] It is important to distinguish between the "local" decisions a_k of the agents and "global" decisions u_k of the market observer. Clearly the decisions a_k affect the choice of u_k as will be made precise below.

(i) *Cost of Stopping*: The asset experiences a jump change(shock) in its value at time τ^0. If the action $u_k = 1$ is chosen before the change point, a false alarm penalty is incurred. This corresponds to the event $\bigcup_{i\geq 2}\{x_k = i\} \cap \{u_k = 1\}$. Let $\mathcal{I}$ denote the indicator function. The cost of false alarm in state $i, i \in \mathcal{X}$ with $f_i \geq 0$ is thus given by $f_i\mathcal{I}(x_k = i, u_k = 1)$. The expected false alarm penalty is

$$\begin{aligned} C(\pi_k, u_k = 1) &= \sum_{i\in\mathcal{X}} f_i \mathbb{E}\{\mathcal{I}(x_k = i, u_k = 1)|\mathcal{G}_k\} \\ &= \mathbf{f}'\pi_k, \end{aligned} \tag{21.33}$$

where $\mathbf{f} = (f_1, \ldots, f_X)$ and it is chosen with increasing elements, so that states further from "1" incur higher false alarm penalties. Clearly, $f_1 = 0$.

(ii) *Cost of delay*: A delay cost is incurred when the event $\{x_k = 1, u_k = 2\}$ occurs, i.e., even though the state changed at k, the market observer fails to identify the change. The expected delay cost is

$$\begin{aligned} C(\pi_k, u_k = 2) &= d\,\mathbb{E}\{\mathcal{I}(x_k = i, u_k = 1)|\mathcal{G}_k\} \\ &= de_1'\pi_k, \end{aligned} \tag{21.34}$$

where $d > 0$ is the delay cost and e_1 denotes the unit vector with 1 in the first position.

Fig. 21.7 illustrates the above social learning model in which the information exchange between the risk-averse social sensors is sequential.

Market observer's quickest detection objective

The market observer chooses its action at each time k as

$$u_k = \mu(\pi_k) \in \{1(\text{stop}), 2(\text{continue})\}, \tag{21.35}$$

where μ denotes a stationary policy. For each initial distribution $\pi_0 \in \Pi(X)$ and policy μ, the following cost is associated

$$J_\mu(\pi_0) = \mathbb{E}^{\mu}_{\pi_0}\left\{\sum_{k=1}^{\tau-1}\rho^{k-1}C(\pi_k, u_k = 2) + \rho^{\tau-1}C(\pi_k, u_k = 1)\right\}. \tag{21.36}$$

Here $\rho \in [0, 1]$ is the discount factor, which is a measure of the degree of impatience of the market observer. (As long as $\mathbf{f}$ is nonzero, stopping is guaranteed in finite time and so $\rho = 1$ is allowed.)

Given the cost, the market observer's objective is to determine τ^0 with minimum cost by computing an optimal policy μ^* such that

$$J_{\mu^*}(\pi_0) = \inf_{\mu\in\boldsymbol{\mu}} J_\mu(\pi_0). \tag{21.37}$$

The sequential detection problem (21.37) can be viewed as a partially observed Markov decision process (POMDP) where the belief update is given by the social learning filter.

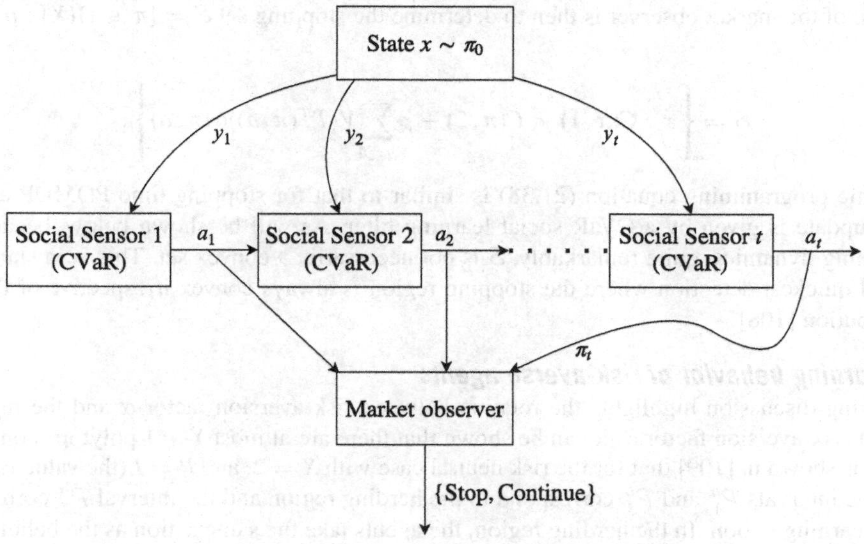

FIG. 21.7

Sequential detection with risk-averse social sensors. Each social sensor receives a noisy observation on the state and chooses an action to minimize its CVaR measure of trading. The social sensors communicate their actions to subsequent sensors. The market observer seeks to determine if there is a change in the value of the underlying asset from the actions of the sensors.

Stochastic dynamic programming formulation

The optimal policy of the market observer $\mu^* : \Pi(X) \to \{1, 2\}$ is the solution of Eq. (21.36) and is given by Bellman's dynamic programming equation as follows:

$$V(\pi) = \min \left\{ C(\pi, 1), C(\pi, 2) + \rho \sum_{a \in \mathcal{A}} V(T^{\pi}(\pi, a))\sigma(\pi, a) \right\}, \tag{21.38}$$

$$\mu^*(\pi) = \operatorname{argmin} \left\{ C(\pi, 1), C(\pi, 2) + \rho \sum_{a \in \mathcal{A}} V(T^{\pi}(\pi, a))\sigma(\pi, a) \right\},$$

where $T^{\pi}(\pi, a) = \frac{R^{\pi}_a P' \pi}{\mathbf{1}' R^{\pi}_a P' \pi}$ is the CVaR-social learning filter and $\sigma(\pi, a) = \mathbf{1}' R^{\pi}_a P' \pi$ is the normalization factor of the Bayesian update. $C(\pi, 1)$ and $C(\pi, 2)$ from Eqs. (21.33) and (21.34) are the market observer's costs. As $C(\pi, 1)$ and $C(\pi, 2)$ are nonnegative and bounded for $\pi \in \Pi(X)$, the stopping time τ is finite for all $\rho \in [0, 1]$.

The aim of the market observer is then to determine the stopping set $\mathcal{S} = \{\pi \in \Pi(X) : \mu^*(\pi) = 1\}$ given by:

$$\mathcal{S} = \left\{ \pi : C(\pi, 1) < C(\pi, 2) + \rho \sum_{a \in \mathcal{A}} V(T^{\pi}(\pi, a))\sigma(\pi, a) \right\}.$$

The dynamic programming equation (21.38) is similar to that for stopping time POMDP except that the belief update is given by a CVaR social learning filter. As will be shown below, because of the social learning dynamics, quite remarkably, $\mathcal{S}$ is not necessarily a convex set. This is in stark contrast to classical quickest detection where the stopping region is always convex irrespective of the change time distribution [108].

Social learning behavior of risk-averse agents

The following discussion highlights the relation between risk-aversion factor α and the regions $\mathcal{P}_l^{\alpha}$. For a given risk-aversion factor α, it can be shown that there are at most $Y + 1$ polytopes on the belief space. It was shown in [109] that for the risk neutral case with $X = 2$, and $P = I$ (the value is a random variable) the intervals $\mathcal{P}_1^{\alpha}$ and $\mathcal{P}_3^{\alpha}$ correspond to the herding region and the interval $\mathcal{P}_2^{\alpha}$ corresponds to the social learning region. In the herding region, the agents take the same action as the belief is frozen. In the social learning region there is observational learning. However, when the agents are optimizing a more general risk measure (CVaR), the social learning region is different for different risk-aversion factors. The social learning region for the CVaR risk measure is shown in Fig. 21.8. It can be observed

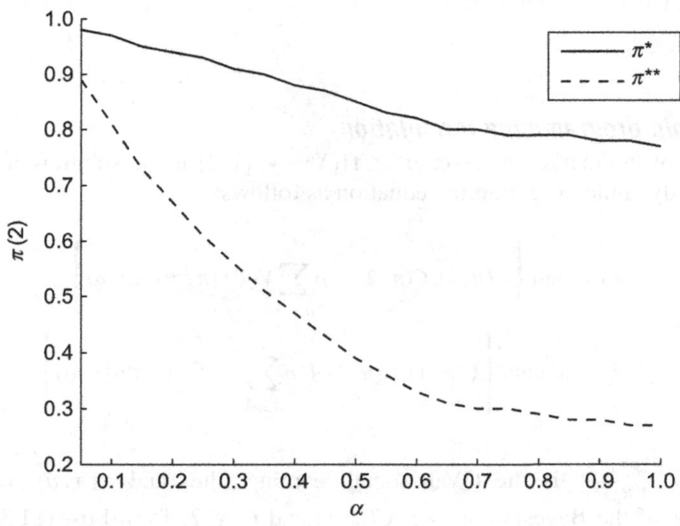

FIG. 21.8

The social learning region for the risk-aversion parameter $\alpha \in (0, 1]$. It can be seen that the curves corresponding to π^{**} and π^* do not intersect and their separation (social learning region) varies with α. Here $P = I$, i.e., the value is a random variable.

from Fig. 21.8 that $\mathcal{P}_1^\alpha$ becomes smaller, $\mathcal{P}_2^\alpha$ becomes smaller, and $\mathcal{P}_3^\alpha$ becomes larger as α decreases. The following parameters were chosen:

$$B = \begin{bmatrix} 0.8 & 0.2 \\ 0.3 & 0.7 \end{bmatrix}, \quad P = \begin{bmatrix} 1 & 0 \\ 0 & 1 \end{bmatrix}, \quad c = \begin{bmatrix} 1 & 2 \\ 3 & 0.5 \end{bmatrix}.$$

This can be interpreted as risk-averse agents showing a larger tendency to go with the crowd rather than "risk" choosing the other action. With the same B and c parameters, but with transition matrix

$$P = \begin{bmatrix} 1 & 0 \\ 0.1 & 0.9 \end{bmatrix}$$

the social learning region is shown in Fig. 21.9. From Fig. 21.9, it is observed that when the state is evolving and when the agents are sufficiently risk averse, the social learning region is very small. It can be interpreted as: agents having a strong risk-averse attitude do not prefer to "learn" from the crowd, but rather face the same consequences, when $P \neq I$.

Nonconvex stopping set for market shock detection

We now illustrate the solution to the Bellman's stochastic dynamic programming equation (21.38), which determines the optimal policy for quickest market shock detection by considering an agent-based model with two states. Clearly the agents (local decision makers) and market observer interact

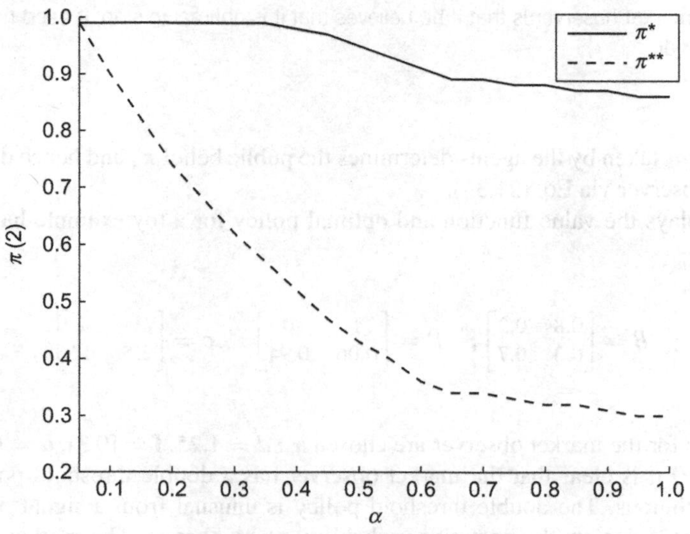

FIG. 21.9

The social learning region for the risk-aversion parameter $\alpha \in (0,1]$. It can be seen that the social learning region is absent when agents are sufficiently risk averse and is larger when the stock value is known to change, i.e., $P \neq I$.

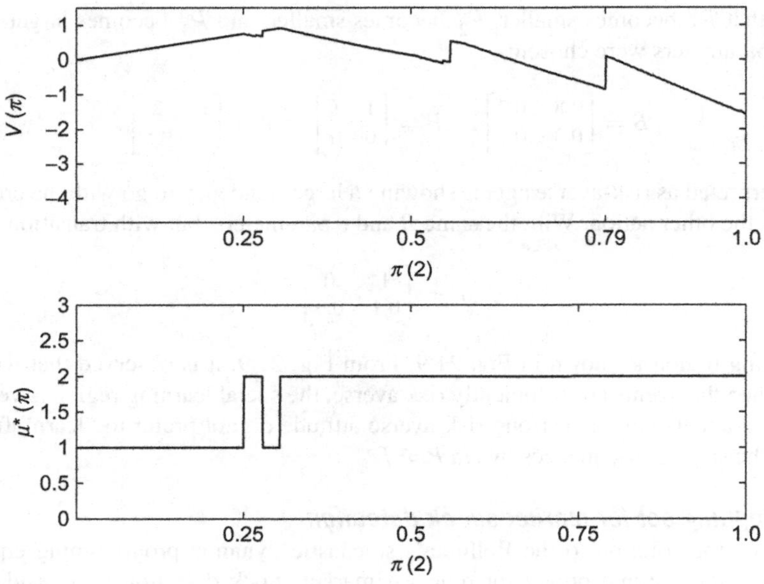

FIG. 21.10

The value function $V(\pi)$ and the double threshold optimal policy $\mu^*(\pi)$ are plotted over $\pi(2)$. The significance of the double threshold policy is that the stopping regions are nonconvex. The implication of a nonconvex stopping set for the market observer is that if he believes that it is optimal to stop, it need not be optimal to stop when his belief is larger.

the local decisions a_k taken by the agents determines the public belief π_k and hence determines decision u_k of the market observer via Eq. (21.35).

Fig. 21.10 displays the value function and optimal policy for a toy example having the following parameters:

$$B = \begin{bmatrix} 0.8 & 0.2 \\ 0.3 & 0.7 \end{bmatrix}, \quad P = \begin{bmatrix} 1 & 0 \\ 0.06 & 0.94 \end{bmatrix}, \quad c = \begin{bmatrix} 1 & 2 \\ 2.5 & 0.5 \end{bmatrix}.$$

The parameters for the market observer are chosen as: $d = 1.25$, $\mathbf{f} = [0\ 3]$, $\alpha = 0.8$, and $\rho = 0.9$.

From Fig. 21.10 it is clear that the market observer has a double threshold policy and the value function is discontinuous. The double threshold policy is unusual from a signal processing point of view. Recall that $\pi(2)$ depicts the posterior probability of no change. The market observer "changes its mind" it switches from no change to change as the posterior probability of change decreases! Thus the global decision (stop or continue) is a nonmonotone function of the posterior, probability obtained from local decisions in the agent-based model. The example illustrates the unusual behavior of the social learning filter.

Summary

In this subsection we provided a Bayesian formulation of the problem of quickest detection of change in the value of a stock using the decisions of socially aware risk averse agents. The quickest detection problem was shown to be nontrivial—the stopping region is in general non-convex when the agents' risk attitude was accounted for by considering a coherent risk measure, CVaR. Results that characterize the structural properties of social learning under the CVaR risk measure were provided and the importance of these results in understanding the global behavior was discussed. It was observed that the behavior of these risk-averse agents is, as expected, different from risk-neutral agents. Risk-averse agents herd sooner and do not prefer to "learn" from the crowd, i.e., the social learning region is smaller the more risk averse the agents. For further structural results on the risk-averse social learning filter, please see [110].

21.3.3 DATA INCEST IN ONLINE REPUTATION SYSTEMS

In comparison to the previous subsections, where the social learning model was formulated on a line graph, we now consider social learning on a family of time-dependent directed acyclic graphs; in such cases, apart from herding, the phenomenon of data incest arises.

Consider an online reputation system comprised of agents $\{1,2,\ldots,S\}$ that aim to estimate an underlying state of nature (a random variable). Let $x \in \mathbb{X} = \{1,2,\ldots,X\}$ represent the state of nature (such as the quality of a restaurant/hotel) with known prior distribution π_0. Let $k = 1,2,3,\ldots$ depict epochs at which events occur. These events involve taking observations, evaluating beliefs, and choosing actions as described below. The index k marks the historical order of events. For simplicity, we refer to k as "time."

It is convenient also to reduce the coordinates of time k and agent s to a single integer index n:

$$n \triangleq s + S(k-1), \quad s \in \{1,\ldots,S\},\ k = 1,2,3,\ldots \tag{21.39}$$

We refer to n as a "node" of a time-dependent information flow graph G_n which we now define. Let

$$G_n = (V_n, E_n), \quad n = 1,2,\ldots \tag{21.40}$$

denote a sequence of time-dependent *directed acyclic graphs (DAGs)*[12] of information flow in the social network until and including time k where $n = s + S(k-1)$. Each vertex in V_n represents an agent s' in the social network at time k' and each edge (n', n'') in $E_n \subseteq V_n \times V_n$ shows that the information (action) of node n' (agent s' at time k') reaches node n'' (agent s'' at time k''). It is clear that G_n is a subgraph of G_{n+1}.

The adjacency matrix A_n of G_n is an $n \times n$ matrix with elements $A_n(i,j)$ given by

$$A_n(i,j) = \begin{cases} 1 & \text{if } (v_j, v_i) \in E, \\ 0 & \text{otherwise} \end{cases}, \quad A_n(i,i) = 0. \tag{21.41}$$

The transitive closure matrix T_n is the $n \times n$ matrix

$$T_n = \operatorname{sgn}((\mathbf{I}_n - A_n)^{-1}), \tag{21.42}$$

[12] A DAG is a directed graph with no directed cycles.

where for matrix M, the matrix sgn(M) has elements

$$\operatorname{sgn}(M)(i,j) = \begin{cases} 0 & \text{if } M(i,j) = 0, \\ 1 & \text{if } M(i,j) \neq 0. \end{cases}$$

Note that $A_n(i,j) = 1$ if there is a single hop path between nodes i and j, In comparison, $T_n(i,j) = 1$ if there exists a path (possible multihop) between i and j.

The information reaching node n depends on the information flow graph G_n. The following two sets will be used to specify the incest removal algorithms below:

$$\mathcal{H}_n = \{m : A_n(m,n) = 1\}, \tag{21.43}$$

$$\mathcal{F}_n = \{m : T_n(m,n) = 1\}. \tag{21.44}$$

Thus $\mathcal{H}_n$ denotes the set of previous nodes m that communicate with node n in a single-hop. In comparison, $\mathcal{F}_n$ denotes the set of previous nodes m whose information eventually arrives at node n. Thus $\mathcal{F}_n$ contains all possible multihop connections by which information from a node m eventually reaches node n.

Example

Consider $S = 2$ two agents with information flow graph for three time points $k = 1, 2, 3$ depicted in Fig. 21.11 characterized by the family of DAGs $\{G_1, \ldots, G_7\}$. The adjacency matrices $A_1, \ldots, A_7$ are constructed as follows: A_n is the upper left $n \times n$ submatrix of A_{n+1} and

$$A_7 = \begin{bmatrix} 0 & 0 & 1 & 1 & 0 & 0 & 1 \\ 0 & 0 & 0 & 1 & 0 & 0 & 0 \\ 0 & 0 & 0 & 0 & 1 & 0 & 0 \\ 0 & 0 & 0 & 0 & 0 & 1 & 0 \\ 0 & 0 & 0 & 0 & 0 & 0 & 1 \\ 0 & 0 & 0 & 0 & 0 & 0 & 1 \\ 0 & 0 & 0 & 0 & 0 & 0 & 0 \end{bmatrix}.$$

Let us explain these matrices. Because nodes 1 and 2 do not communicate, clearly A_1 and A_2 are zero matrices. Nodes 1 and 3 communicate, hence A_3 has a single one, etc. Note that if nodes 1, 3, 4, and 7 are assumed to be the same individual, then at node 7, the individual remembers what happened

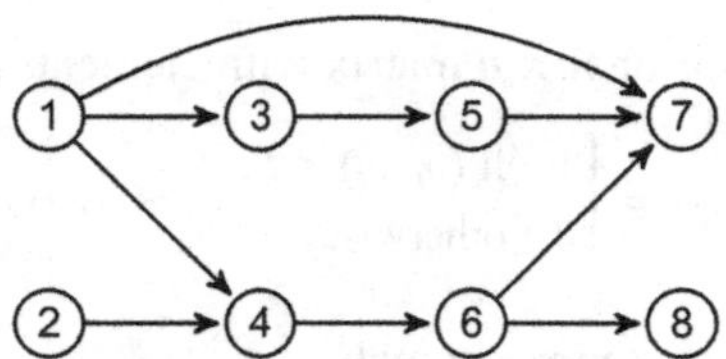

FIG. 21.11

Example of information flow network with $S = 2$ two agents, namely $s \in \{1, 2\}$ and time points $k = 1, 2, 3, 4$. *Circles* represent the nodes indexed by $n = s + S(k - 1)$ in the social network and each edge depicts a communication link between two nodes.

at node 5 and node 1, but not node 3. This models the case where the individual has selective memory and remembers certain highlights. From Eqs. (21.43) and (21.44),

$$\mathcal{H}_7 = \{1, 5, 6\}, \quad \mathcal{F}_7 = \{1, 2, 3, 4, 5, 6\},$$

where $\mathcal{H}_7$ denotes all one-hop links to node 7 while $\mathcal{F}_7$ denotes all multihop links to node 7.

21.3.4 DATA INCEST MODEL AND SOCIAL INFLUENCE CONSTRAINT

Each node n receives recommendations from its immediate friends (one hop neighbors) according to the information flow graph defined above. That is, it receives actions $\{a_m, m \in \mathcal{H}_n\}$ from nodes $m \in \mathcal{H}_n$ and then seeks to compute the associated public beliefs $\pi_m, m \in \mathcal{H}_n$. If node n naively (incorrectly) assumes that the public beliefs $\pi_m, m \in \mathcal{H}_n$ are independent, then it would fuse these as

$$\pi_{n-} = \frac{\prod_{m\in\mathcal{H}_n} \pi_m}{1'_X \prod_{m\in\mathcal{H}_n} \pi_m}. \text{ (WRONG!)} \tag{21.45}$$

This naive data fusion would result in data incest.

Aim

The aim is to provide each node n the true posterior distribution

$$\pi^0_{n-}(i) = P(x = i | \{a_m, m \in \mathcal{F}_n\}) \tag{21.46}$$

subject to the following *social influence constraint*: There exists a fusion algorithm $\mathcal{A}$ such that

$$\pi^0_{n-} = \mathcal{A}(\pi_m, m \in \mathcal{H}_n). \tag{21.47}$$

Discussion. Fair rating and social influence

We briefly pause to discuss Eqs. (21.46) and (21.47). (i) We call π^0_{n-} in Eq. (21.46) the *true* or *fair online rating* available to node n because $\mathcal{F}_n$ defined in Eq. (21.44) denotes all information (multihop links) available to node n. By definition π^0_{n-} is incest free and is the desired conditional probability that agent n needs.

Indeed, if node n combines π^0_{n-} together with its own private observation via social learning, then clearly

$$\eta_n(i) = P(x = i | \{a_m, m \in \mathcal{F}_n\}, y_n), \quad i \in \mathbb{X},$$
$$\pi_n(i) = P(x = i | \{a_m, m \in \mathcal{F}_n\}, a_n), \quad i \in \mathbb{X},$$

are, respectively, the correct (incest free) private belief for node n and the correct after-action public belief. If agent n does not use π^0_{n-}, then incest can propagate; for example, if agent n naively uses Eq. (21.45).

Why should an individual n agree to use π^0_{n-} to combine with its private message? It is here that the social influence constraint (21.47) is important. $\mathcal{H}_n$ can be viewed as the "*social message*," i.e., personal friends of node n because they directly communicate to node n while the associated beliefs can be viewed as the "*informational message*." As described in the remarkable recent paper [111], the social message from personal friends exerts a large social influence[13]—it provides significant incentive (peer pressure) for individual n to comply with the protocol of combining its estimate with π^0_{n-} and thereby prevent incest. Ref. [111] shows that receiving messages from known friends has significantly more influence on an individual than the information in the messages. This study includes a comparison of information messages and social messages on Facebook and their direct effect on voting behavior. To quote [111], "The effect of social transmission on real-world voting was greater than the direct effect of the messages themselves..." In Section 21.3.7, we provide results of an experiment on human subjects that also illustrates social influence in social learning. Ref. [112] is an influential paper in the area of social influence.

21.3.5 INCEST REMOVAL IN ONLINE REPUTATION SYSTEMS

It is convenient to work with the logarithm of the unnormalized belief[14]; accordingly define

$$l_n(i) \propto \log \pi_n(i), \quad l_{n-}(i) \propto \log \pi_{n-}(i), \quad i \in \mathbb{X}.$$

The following theorem shows that the logarithm of the fair rating π^0_{n-} defined in Eq. (21.46) can be obtained as a weighted linear combination of the logarithms of previous public beliefs.

Theorem 21.4 (Fair Rating Algorithm). *Suppose the network administrator runs the following algorithm for an online reputation system:*

$$\begin{aligned} l_{n-}(i) &= w_n' \, l_{1:n-1}(i) \\ \text{where } w_n &= T_{n-1}^{-1} t_n. \end{aligned} \tag{21.48}$$

Then $l_{n-}(i) \propto \log \pi^0_{n-}(i)$. That is, the fair rating $\log \pi^0_{n-}(i)$ defined in Eq. (21.46) is obtained. In Eq. (21.48), w_n is an $n-1$ dimensional weight vector. Recall that t_n denotes the first $n-1$ elements of the nth column of the transitive closure matrix T_n. ■

Theorem 21.4 says that the fair rating π^0_{n-} can be expressed as a linear function of the action log-likelihoods in terms of the transitive closure matrix T_n of graph G_n. This is intuitive because π^0_{n-} can be viewed as the sum of information collected by the nodes such that there are paths between all these nodes and n.

Theorem 21.5 (Achievability of Fair Rating). *Consider the fair rating algorithm specified by Eq. (21.48). With available information ($\pi_m, m \in \mathcal{H}_n$) to achieve the estimates l_{n-} of algorithm* (21.48), *a necessary and sufficient condition on the information flow graph G_n is*

$$A_n(j, n) = 0 \implies w_n(j) = 0. \tag{21.49}$$

(Recall w_n is specified in Eq. (21.48).) ■

---

[13]In a study conducted by social networking site *myYearbook*, 81% of respondents said they had received advice from friends and followers relating to a product purchase through a social site; 74 percent of those who received such advice found it to be influential in their decision. (*Click Z*, January 2010).

[14]The unnormalized belief proportional to $\pi_n(i)$ is the numerator of the social learning filter (21.21).The corresponding unnormalized fair rating corresponding to $\pi^0_{n-}(i)$ is the joint distribution $P(x = i, \{a_m, m \in \mathcal{F}_n\})$. By taking the logarithm of the unnormalized belief, Bayes formula merely becomes the sum of the log likelihood and log prior. This allows us to devise a data incest removal algorithm based on linear combinations of the log beliefs.

Note that the constraint (21.49) is purely in terms of the adjacency matrix A_n, because the transitive closure matrix (21.42) is a function of the adjacency matrix.

21.3.6 ORDINAL DECISIONS AND BAYESIAN SOCIAL SENSORS

The social learning protocol assumes that each agent is a Bayesian utility optimizer. The following discussion puts together ideas from the economics literature to show that under reasonable conditions, such a Bayesian model is a useful idealization of agents' behaviors. This means that the Bayesian social learning follows simple intuitive rules and is, therefore, a useful idealization. (In Section 21.4, we discuss the theory of revealed preferences that yields a nonparametric test on data to determine if an agent is a utility maximizer.)

Humans typically make *monotone* decisions—the more favorable the private observation, the higher the recommendation. Humans make *ordinal* decisions[15] because humans tend to think in symbolic ordinal terms. Under what conditions is the recommendation a_n made by node *n monotone increasing* in its observation y_n and *ordinal*? Recall from the social learning protocol (21.20) that the actions of agents are

$$a_k = \arg\min_{a \in \mathcal{A}} \{c_a' B_{y_k} \pi_k\}.$$

So an equivalent question is: Under what conditions is the argmin increasing in observation y_n? Note that an increasing argmin is an *ordinal* property—that is, $\text{argmin}_a\, c_a' B_{y_n} \pi_{n-}^0$ increasing in y implies $\text{argmin}_a\, \phi(c_a' B_{y_n} \pi_{n-}^0)$ is also increasing in y for any monotone function $\phi(\cdot)$.

The following result gives sufficient conditions for each agent to give a recommendation that is monotone and ordinal in its private observation:

Theorem 21.6. *Suppose the observation probabilities and costs satisfy the following conditions:*

(A1) *B_{iy} are TP2 (totally positive of order 2); that is, $B_{i+1,y} B_{i,y+1} \le B_{i,y} B_{i+1,y+1}$.*
(A2) *$c(x,a)$ is submodular. That is, $c(x,a+1) - c(x,a) \le c(x+1,a+1) - c(x+1,a)$.*

Then

1. *Under (A1) and (A2), the recommendation $a_n(\pi_{n-}^0, y_n)$ made by agent n is increasing and hence ordinal in observation y_n, for any π_{n-}^0.*
2. *Under (A2), $a_n(\pi_{n-}^0, y_n)$ is increasing in belief π_{n-}^0 with respect to the monotone likelihood ratio (MLR) stochastic order*[16] *for any observation y_n.* ■

The proof is in [109]. We can interpret the above theorem as follows. If agents makes recommendations that are monotone and ordinal in the observations and monotone in the prior, then they mimic the Bayesian social learning model. Even if the agent does not exactly follow a Bayesian social learning model, its monotone ordinal behavior implies that such a Bayesian model is a useful idealization.

---

[15] Humans typically convert numerical attributes to ordinal scales before making decisions. For example, it does not matter if the cost of a meal at a restaurant is \$200 or \$205; an individual would classify this cost as "high." Also credit rating agencies use ordinal symbols such as AAA, AA, A.

[16] Given probability mass functions $\{p_i\}$ and $\{q_i\}$, $i = 1, \ldots, X$ then p MLR dominates q if $\log p_i - \log p_{i+1} \le \log q_i - \log q_{i+1}$.

Condition (A1) is widely studied in monotone decision-making; see the classical book by Karlin [113] and Karlin and Rinott [114]; numerous examples of noise distributions are TP2. Indeed in the highly cited paper [115] in the economics literature, observation $y+1$ is said to be more "favorable news" than observation y if Condition (A1) holds.

Condition (A2) is the well known submodularity condition [116–118]. (A2) makes sense in a reputation system for the costs to be well posed. Suppose the recommendations in action set $\mathcal{A}$ are arranged in increasing order and also the states in $\mathbb{X}$ for the underlying state are arranged in ascending order. Then (A2) says: if recommendation $a+1$ is more accurate than recommendation a for state x; then recommendation $a+1$ is also more accurate than recommendation a for state $x+1$ (which is a higher quality state than x).

In the experiment results reported in Section 21.3.7, we found that (A1) and (A2) of Theorem 21.6 are justified.

21.3.7 PSYCHOLOGY EXPERIMENT DATASET

To illustrate social learning, data incest, and social influence, this section presents an actual psychology experiment that was conducted by our colleagues at the department of psychology of the University of British Columbia in September and October, 2013. A total of 36 undergraduate students participated in the experiment for course credit.

Experiment setup

The experimental study involved 1658 individual trials. Each trial comprised two participants who were asked to perform a perceptual task interactively. The perceptual task was as follows: Two arrays of circles denoting left and right were given to each pair of participants. Each participant was asked to judge which array (left or right) had the larger average diameter. The participants answer (left of right) constituted their action. So the action space is $\mathcal{A} = \{0 \text{ (left)}, 1 \text{ (right)}\}$.

The circles were prepared for each trial as follows: two 4×4 grids of circles were generated by uniformly sampling from the radii: $\{20, 24, 29, 35, 42\}$ (in pixels). The average diameter of each grid was computed, and if the means differed by more than 8% or less than 4%, new grids were made. Thus in each trial, the left array and right array of circles differed in the average diameter by 4%–8% (Fig. 21.12).

For each trial, one of the two participants was chosen randomly to start the experiment by choosing an action according to his/her observation. Thereafter, each participant was given access to their partner's previous response (action) and the participant's own previous action prior to making his/her judgment. This mimics the social learning protocol. The participants continued choosing actions according to this procedure until the experiment terminated. The trial terminated when the response of each of the two participants did not change for three successive iterations (the two participants did not necessarily have to agree for the trial to terminate).

In each trial, the actions of participants were recorded along with the time interval taken to choose their action. As an example, Fig. 21.13 illustrates the sample path of decisions made by the two participants in one of the 1658 trials. In this specific trial, the average diameter of the left array of circles was 32.1875 and the right array was 30.5625 (in pixels); so the ground truth was 0 (left).

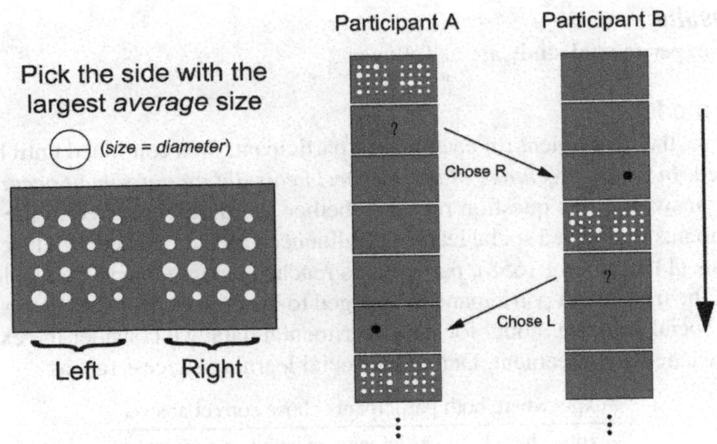

FIG. 21.12

Two arrays of circles were given to each pair of participants on a screen. Their task is to interactively determine which side (left or right) had the larger average diameter. The partner's previous decision was displayed on screen prior to the stimulus.

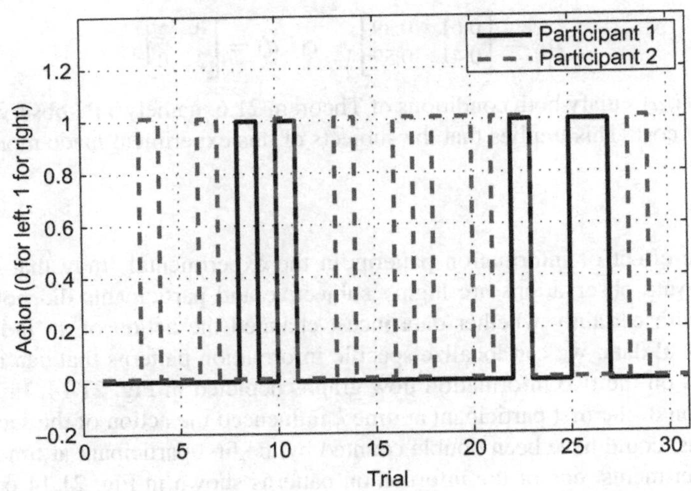

FIG. 21.13

Example of sample path of actions chosen by two participants in a single trial of the experiment. In this trial, both participants eventually chose the correct answer 0 (left).

Experimental results

The results of our experimental study are as follows:

Social learning model

As mentioned above, the experiment for each pair of participants was continued until both participants' responses stabilized. *In what percentage of these experiments did an agreement occur between the two participants?* The answer to this question reveals whether "herding" occurred in the experiments and whether the participants performed social learning (influenced by their partners). The experiments show that in 66% of trials (1102 among 1658), participants reached an agreement; that is, herding occurred. Further, in 32% of the trials, both participants converged to the correct decision after a few interactions.

To construct a social learning model for the experimental data, we consider the experiments where both participants reached an agreement. Define the social learning success rate as

$$\frac{\text{\#expts where both participants chose correct answer}}{\text{\#expts where both participants reached an agreement}}.$$

In the experimental study, the state space is $\mathbb{X} = \{0, 1\}$ where $x = 0$, when the left array of circles has the larger diameter and $x = 1$, when the right array has the larger diameter. The initial belief for both participants is considered to be $\pi_0 = [0.5, 0.5]$. The observation space is assumed to be $\mathbb{Y} = \{0, 1\}$.

To estimate the social learning model parameters (observation probabilities B_{iy} and costs $c(i, a)$), we determined the parameters that best fit the learning success rate of the experimental data. The best fit parameters obtained were[17]

$$B_{iy} = \begin{bmatrix} 0.61 & 0.39 \\ 0.41 & 0.59 \end{bmatrix}, \quad c(i, a) = \begin{bmatrix} 0 & 2 \\ 2 & 0 \end{bmatrix}.$$

Note that B_{iy} and $c(i, a)$ satisfy both conditions of Theorem 21.6, namely TP2 observation probabilities and single-crossing cost. This implies that the subjects of this experiment made monotone and ordinal decisions.

Data incest

Here, we study the effect of information patterns in the experimental study that can result in data incest. Because private observations are highly subjective and participants did not document these, we cannot claim with certainty whether data incest changed the action of an individual. However, from the experimental data, we can localize specific information patterns that can result in incest. In particular, we focus on the two information flow graphs depicted in Fig. 21.14. In the two graphs of Fig. 21.14, the action of the first participant at time k influenced the action of the second participant at time $k + 1$, and thus, could have been double counted by the first participant at time $k + 2$. We found that in 79% of experiments, one of the information patterns shown in Fig. 21.14 occurred (1303 out of 1658 experiments). Further, in 21% of experiments, the information patterns shown in Fig. 21.14 occurred and at least one participant changed his/her decision, i.e., the judgment of participant at

[17] Parameter estimation in social learning is a challenging problem not addressed in this chapter. Due to the formation of cascades in finite time, construction of an asymptotically consistent estimator is impossible because actions after the formation of a cascade contain no information.

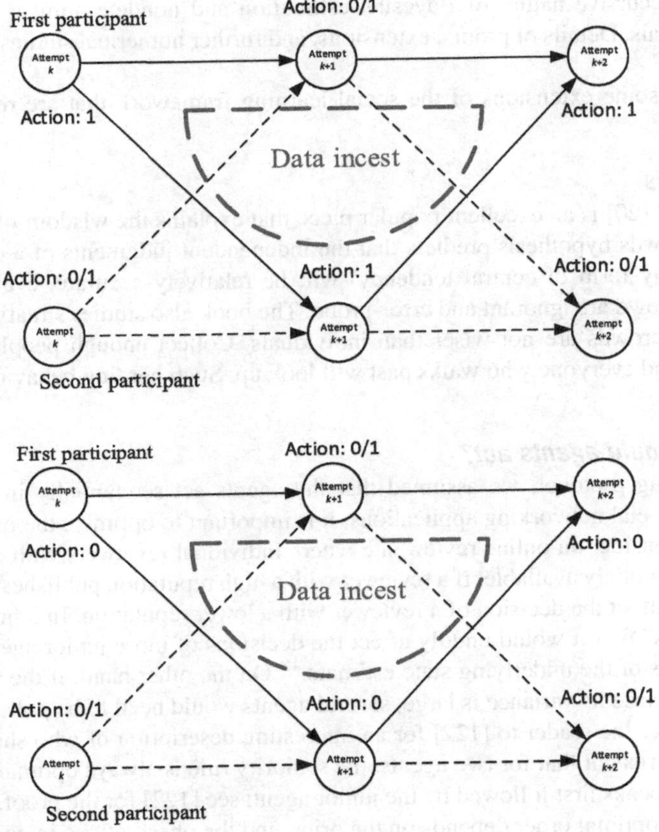

FIG. 21.14

Two information patterns from our experimental studies that can result in data incest.

time $k + 1$ differed from his/her judgments at time $k + 2$ and k. These results show that even for experiments involving two participants, data incest information patterns occur frequently (79%) and cause individuals to modify their actions (21%). It shows that social learning protocols require careful design to handle and mitigate data incest.

21.3.8 SUMMARY AND EXTENSIONS

In this section, we have outlined a controlled sensing problem over a social network in which the administrator controls (removes) data incest and thereby maintains an unbiased (fair) online reputation system. The state of nature could be geographical coordinates of an event (in a target localization problem) or quality of a social unit (in an online reputation system). As discussed above, data incest arises due to the recursive nature of Bayesian estimation and nondeterminism in the timing of the sensing by individuals. Details of proofs, extensions, and further numerical studies are presented in [88, 119].

We summarize some extensions of the social learning framework that are relevant to interactive sensing.

Wisdom of crowds

Surowiecki's book [120] is an excellent popular piece that explains the wisdom-of-crowds hypothesis. The wisdom-of-crowds hypothesis predicts that the independent judgments of a crowd of individuals (as measured by any form of central tendency) will be relatively accurate, even when most of the individuals in the crowd are ignorant and error-prone. The book also studies situations (such as rational bubbles) in which crowds are not wiser than individuals. Collect enough people on a street corner staring at the sky, and everyone who walks past will look up. Such herding behavior is typical in social learning.

In which order should agents act?

In the social learning protocol, we assumed that the agents act sequentially in a predefined order. However, in many social networking applications, it is important to optimize the order in which agents act. For example, consider an online review site where individual reviewers with different reputations make their reviews publicly available. If a reviewer with a high reputation publishes her review first, this review will unduly affect the decision of a reviewer with a lower reputation. In other words, if the most senior agent "speaks" first it would unduly affect the decisions of more junior agents. This could lead to an increase in bias of the underlying state estimate.[18] On the other hand, if the most junior agent is polled first, then because its variance is large, several agents would need to be polled in order to reduce the variance. We refer the reader to [122] for an interesting description of who should speak first in a public debate.[19] It turns out that for two agents, the seniority rule is always optimal for any prior—that is, the senior agent speaks first followed by the junior agent; see [122] for the proof. However, for more than two agents, the optimal order depends on the prior, and the observations in general.

21.4 REVEALED PREFERENCES: ARE SOCIAL SENSORS UTILITY MAXIMIZERS?

We now move on to the third main topic of the chapter, namely, the principle of revealed preferences. The main question addressed is: Given a dataset of decisions made by a social sensor, is it possible to determine if the social sensor is a utility maximizer? More generally, is a dataset from a multiagent

[18] To quote [121]: "In 94% of cases, groups (of people) used the first answer provided as their final answer... Groups tended to commit to the first answer provided by any group member." People with dominant personalities tend to speak first and most forcefully "even when they actually lack competence."

[19] As described in [122], seniority is considered in the rules of debate and voting in the Supreme Court. "In the past, a vote was taken after the newest justice to the Court spoke, with the justices voting in order of ascending seniority largely, it was said, to avoid the pressure from long-term members of the Court on their junior colleagues."

system consistent with play from a Nash equilibrium? If yes, can the behavior of the social sensors be learned using data from the social network?

These questions are fundamentally different to the model-based theme that is widely used in the signal processing literature in which an objective function (typically convex) is proposed and then algorithms are constructed to compute the minimum. In contrast, the revealed preference approach is data-centric—we wish to determine whether the dataset is obtained from a utility maximizer. In simple terms, revealed preference theory seeks to determine if an agent is a utility maximizer subject to budget constraints based on observing its choices over time. In signal processing terminology, such problems can be viewed as set-valued system identification of an argmax nonlinear system. The principle of revealed preferences is widely studied in the microeconomics literature. As mentioned in Section 21.1, Varian has written several influential papers in this area. In this section we will use the principle of revealed preferences on datasets to determine how social sensors behave as a function of external influence. The setup is depicted in the schematic diagram Fig. 21.15.

21.4.1 AFRIAT'S THEOREM FOR A SINGLE AGENT

The theory of revealed preferences was pioneered by Samuelson [123]. Afriat published a highly influential paper [28] on revealed preferences (see also [124]). Given a time-series of data $\mathcal{D} = \{(p_t, x_t), t \in \{1, 2, \ldots, T\}\}$ where $p_t \in \mathbb{R}^m$ denotes the external influence, x_t denotes the response of agent, and t denotes the time index, is it possible to detect if the agent is a *utility maximizer*? An agent is a *utility maximizer* if for every external influence p_t, the chosen response x_t satisfies

$$x_t(p_t) \in \underset{\{p_t'x \le I_t\}}{\arg\max}\, u(x) \tag{21.50}$$

with $u(x)$ a nonsatiated utility function. Nonsatiated means that an increase in any element of response x results in the utility function increasing.[20] As shown by Diewert [125], without local nonsatiation the maximization problem (21.50) may have no solution.

In Eq. (21.50) the budget constraint $p_t'x \le I_t$ denotes the total amount of resources available to the social sensor for selecting the response x to the external influence p_t. For example, if p_t is the electricity price and x_t the associated electricity consumption, then the budget of the social sensor is the available monetary funds for purchasing electricity. In the real-world social sensor datasets provided in this chapter, further insights are provided for the budget constraint.

The celebrated "Afriat's theorem" provides a necessary and sufficient condition for a finite dataset $\mathcal{D}$ to have originated from a utility maximizer. Afriat's theorem has subsequently been expanded and refined, most notably by Diewert [125], Varian [29], and Blundell [126].

Theorem 21.7 (Afriat's Theorem). *Given a dataset $\mathcal{D} = \{(p_t, x_t) : t \in \{1, 2, \ldots, T\}\}$, the following statements are equivalent:*

1. *The agent is a utility maximizer and there exists a nonsatiated and concave utility function that satisfies Eq. (21.50).*

---

[20]The nonsatiated assumption rules out trivial cases such as a constant utility function that can be optimized by any response.

2. *For scalars* u_t *and* $\lambda_t > 0$ *the following set of inequalities has a feasible solution:*

$$u_\tau - u_t - \lambda_t p_t'(x_\tau - x_t) \leq 0 \text{ for } t, \tau \in \{1, 2, \ldots, T\}. \quad (21.51)$$

3. *A nonsatiated and concave utility function that satisfies Eq. (21.50) is given by:*

$$u(x) = \min_{t \in T}\{u_t + \lambda_t p_t'(x - x_t)\}. \quad (21.52)$$

4. *The dataset* $\mathcal{D}$ *satisfies the Generalized Axiom of Revealed Preference (GARP), namely for any* $k \leq T$, $p_t' x_t \geq p_t' x_{t+1} \quad \forall t \leq k - 1 \implies p_k' x_k \leq p_k' x_1$. ■

As pointed out in [29], a remarkable feature of Afriat's theorem is that if the dataset can be rationalized by a nontrivial utility function, then it can be rationalized by a continuous, concave, monotonic utility function. "Put another way, violations of continuity, concavity, or monotonicity cannot be detected with only a finite number of demand observations."

Verifying GARP (statement 4 of Theorem 21.7) on a dataset $\mathcal{D}$ comprising T points can be done using Warshall's algorithm with $O(T^3)$ [29,127] computations. Alternatively, determining if Afriat's inequalities (21.51) are feasible can be done via an LP feasibility test (using, for example, interior point methods [128]). Note that the utility (21.52) is not unique and is ordinal by construction. Ordinal means that any monotone increasing transformation of the utility function will also satisfy Afriat's theorem. Therefore the utility mimics the ordinal behavior of humans, see also Section 21.3.6. Geometrically the estimated utility (21.52) is the lower envelop of a finite number of hyperplanes that is consistent with the dataset $\mathcal{D}$.

Note that GARP is equivalent to the notion of "cyclical consistency" [129]—they state that the responses are consistent with utility maximization if no negative cycles are present. As an example, consider a dataset $\mathcal{D}$ with $T = 2$ observations resulting from a utility maximization agent. Then GARP states that $p_1' x_1 \geq p_1' x_2 \implies p_2' x_2 \leq p_2' x_1$. From Eq. (21.50), the underlying utility function must satisfy $u(x_1) \geq u(x_2) \implies u(x_2) \leq u(x_1)$ where the equality results if $x_1 = x_2$.

Another remarkable feature of Afriat's Theorem is that no parametric assumptions of the utility function of the agent are necessary. To gain insight into the construction of the inequalities (21.51), let us assume the utility function $u(x)$ in Eq. (21.50) is increasing for positive x, concave, and differentiable. If x_t solves the maximization problem (21.50), then from the Karush-Kuhn-Tucker (KKT) conditions there must exist Lagrange multipliers λ_t such that

$$\nabla u(x_t) = \lambda_t \nabla(p_t' x_t - I_t) = \lambda_t p_t$$

is satisfied for all $t \in \{1, 2, \ldots, T\}$. Note that because $u(x)$ is increasing, $\nabla u(x_t) = \lambda_t p_t > 0$, and because p_t is strictly positive, $\lambda_t > 0$. Given that $u(x)$ is a concave differentiable function, it follows that

$$u(x) \leq u(x_t) + \nabla u(x_t)'(x - x_t) \quad \forall t \in \{1, 2, \ldots, T\}.$$

Denoting $u_t = u(x_t)$ and $u_\tau = u(x_\tau)$, and using the KKT conditions and concave differential property, the inequalities (21.51) result. To prove that if the solution of Eq. (21.51) is feasible then GARP is satisfied can be performed using the duality theorem of linear programming as illustrated in [127].

21.4.2 REVEALED PREFERENCES FOR MULTIAGENT SOCIAL SENSORS

We now consider a multiagent version of Afriat's theorem for deciding if a dataset is generated by playing from the equilibrium of a potential game.[21] An example is the control of power consumption in the electrical grid. Consider a corporate network of financial management operators that selects the electricity prices in a set of zones in the power grid. By selecting the prices of electricity the operators are expected to be able to control the power consumption in each zone. The operators wish to supply their consumers with sufficient power; however, given the finite amount of resources the operators in the corporate network must interact. This behavior can be modeled as a game. Recent analysis of energy use scheduling and demand-side management schemes in the energy market have been performed using potential games [131–133]. Another example of potential games are *congestion games* [134–137] in which the utility of each player depends on the amount of resource it and other players use.

Consider the social network presented in Fig. 21.15, given a time series of data from N agents $\mathcal{D} = \{(p_t, x_t^1, \ldots, x_t^n) : t \in \{1, 2, \ldots, T\}\}$ with $p_t \in \mathbb{R}^m$ the external influence, x_t^i the response of agent i, and t the time index, is it possible to detect if the dataset originated from agents that play a potential game?

The characterization of how agents behave as a function of external influence, for example, the price of using a resource, and the responses of other agents in a social network, is key for analysis. Consider the social network illustrated in Fig. 21.15. There are a total of n interacting agents in the network and each can produce a response x_t^i in response to the other agents and an external influence p_t. Without any a priori assumptions about the agents, how can the behavior of the agents in the social network be learned? In the engineering literature the behavior of agents is typically defined a priori using a *utility function* however our focus here is on learning the behavior of agents. The *utility function* captures

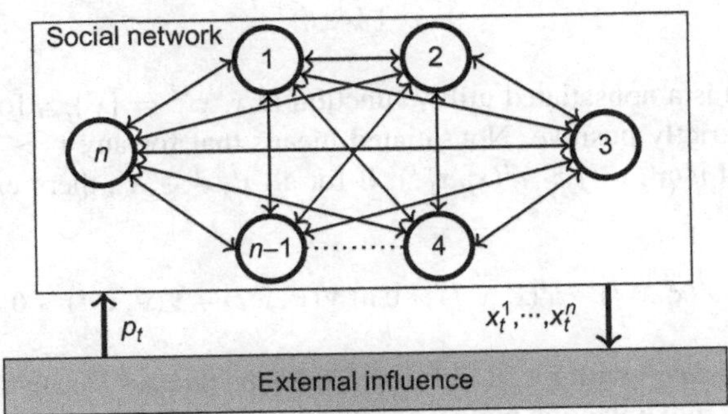

FIG. 21.15

Schematic of a social network containing n agents where $p_t \in \mathbb{R}^m$ denotes the external influence, and $x_t^i \in \mathbb{R}^m$ the response of agent i in response to the external influence and other agents at time t. Note that *dotted line* denotes consumers $4, \ldots, n-1$. The aim is to determine if the dataset $\mathcal{D}$ defined in Eq. (21.53), is consistent with play from a Nash equilibrium.

[21] As in [130], we consider potential games because they are sufficiently specialized so that there exist datasets that fail Afriat's test.

the satisfaction or payoff an agent receives from a set of possible responses, denoted by X. Formally, a utility function $u : X \to \mathbb{R}$ represents a preference relation between responses x_1 and x_2 if and only if for every $x_1, x_2 \in X$, $u(x_1) \leq u(x_2)$ implies x_2 is preferred to x_1. Given a time series of data $\mathcal{D} = \{(p_t, x_t^1, \dots, x_t^n) : t \in \{1, 2, \dots, T\}\}$ with $p_t \in \mathbb{R}^m$ the external influence, x_t^i the response of agent i, and t the time index, is it possible to detect if the series originated from an agent that is a *utility maximizer*?

In a network of social sensors (Fig. 21.15), the responses of agents may be dependent on both the external influence p_t and the responses of the other agents in the network, denoted by x_t^{-i}. The utility function of the agent must now include the responses of other agents—formally if there are n agents, each has a utility function $u^i(x^i, x_t^{-i})$ with x^i denoting the response of agent i, x_t^{-i} the responses of the other $n-1$ agents, and $u^i(\cdot)$ the utility of agent i. Given a dataset $\mathcal{D}$, is it possible to detect if the data is consistent with agents playing a game and maximizing their individual utilities? Deb, following Varian's and Afriat's work, shows that refutable restrictions exist for the dataset $\mathcal{D}$, given by Eq. (21.53), to satisfy Nash equilibrium (21.54) [28,130,138]. These refutable restrictions are, however, satisfied by most $\mathcal{D}$ [130]. The detection of agents engaged in a concave potential game, and generating responses that satisfy Nash equilibrium, provide stronger restrictions on the dataset $\mathcal{D}$ [130,139]. We denote this behavior as *Nash rationality*, defined as follows:

Definition 21.1 ([139–141]). Given a dataset

$$\mathcal{D} = \{(p_t, x_t^1, x_t^2, \dots, x_t^n) : t \in \{1, 2, \dots, T\}\}, \tag{21.53}$$

$\mathcal{D}$ is consistent with *Nash equilibrium* play if there exist utility functions $u^i(x^i, x^{-i})$ such that

$$x_t^i = x_t^{i*}(p_t) \in \underset{\{p_t' x^i \leq I_t^i\}}{\arg\max}\, u^i(x^i, x^{-i}). \tag{21.54}$$

In Eq. (21.54), $u^i(x, x^{-i})$ is a nonsatiated utility function in x, $x^{-i} = \{x^j\}_{j \neq i}$ for $i, j \in \{1, 2, \dots, n\}$, and the elements of p_t are strictly positive. Nonsatiated means that for any $\epsilon > 0$, there exists a x^i with $\|x^i - x_t^i\|_2 < \epsilon$ such that $u^i(x^i, x^{-i}) > u^i(x_t^i, x_t^{-i})$. If for all $x^i, x^j \in X^i$, there exists a concave potential function V that satisfies

$$u^i(x^i, x^{-i}) - u^i(x^j, x^{-i}) > 0 \text{ iff } V(x^i, x^{-i}) - V(x^j, x^{-i}) > 0 \tag{21.55}$$

for all the utility functions $u^i(\cdot)$ with $i \in \{1, 2, \dots, n\}$, then the dataset $\mathcal{D}$ satisfies *Nash rationality*. ■

Just as with the utility maximization budget constraint in Eq. (21.50), the budget constraint $p_t' x^i \leq I_t^i$ in Eq. (21.54) models the total amount of resources available to the social sensor for selecting the response x_t^i to the external influence p_t.

The detection test for Nash rationality (Definition 21.1) has been used in [142] to detect if oil-producing countries are collusive, and in [130] for the analysis of household consumption behavior.

In the following sections, decision tests for utility maximization and a nonparametric learning algorithm for predicting agent responses are presented. Three real-world datasets are analyzed using the nonparametric decision tests and learning algorithm. The datasets are comprised of bidders' auctioning behavior, electrical consumption in the power grid, and on the tweeting dynamics of agents in the social network Twitter, illustrated in Fig. 21.15.

21.4.3 DECISION TEST FOR NASH RATIONALITY

This section presents a nonparametric test for Nash rationality given the dataset $\mathcal{D}$ defined in Eq. (21.53). If the dataset $\mathcal{D}$ passes the test, then it is consistent with play according to a Nash equilibrium of a concave potential game. In Section 21.4.4, a learning algorithm is provided that can be used to predict the response of agents in the social network provided in Fig. 21.15.

The following theorem provides necessary and sufficient conditions for a dataset $\mathcal{D}$ to be consistent with Nash rationality (Definition 21.1). The proof is analogous to Afriat's Theorem when the concave potential function of the game is differentiable [130,139,143].

Theorem 21.8 (Multiagent Afriat's Theorem). *Given a dataset $\mathcal{D}$ (21.53), the following statements are equivalent:*

1. *$\mathcal{D}$ is consistent with Nash rationality (Definition 21.1) for an n-player concave potential game.*
2. *Given scalars v_t and $\lambda_t^i > 0$ the following set of inequalities have a feasible solution for $t, \tau \in \{1, \ldots, T\}$,*

$$v_\tau - v_t - \sum_{i=1}^{n} \lambda_t^i p_t'(x_\tau^i - x_t^i) \leq 0. \tag{21.56}$$

3. *A concave potential function that satisfies Eq. (21.54) is given by:*

$$\hat{V}(x^1, x^2, \ldots, x^n) = \min_{t \in T} \left\{ v_t + \sum_{i=1}^{n} \lambda_t^i p_t'(x^i - x_t^i) \right\}. \tag{21.57}$$

4. *The dataset $\mathcal{D}$ satisfies the Potential Generalized Axiom of Revealed Preference (PGARP) if the following two conditions are satisfied.*
   a. *For every dataset $\mathcal{D}_\tau^i = \{(p_t, x_t^i) : t \in \{1, 2, \ldots, \tau\}\}$ for all $i \in \{1, \ldots, n\}$ and all $\tau \in \{1, \ldots, T\}$, $\mathcal{D}_\tau^i$ satisfies GARP.*
   b. *The responses x_t^i originated from players in a concave potential game.* ■

Note that if only a single agent (i.e., $n = 1$) is considered, then Theorem 21.8 is identical to Afriat's Theorem. Similar to Afriat's Theorem, the constructed concave potential function (21.57) is ordinal—that is, unique up to positive monotone transformations. Therefore several possible options for $\hat{V}(\cdot)$ exist that would produce identical preference relations to the actual potential function $V(\cdot)$. In (4) of Theorem 21.8, the first condition only provides necessary and sufficient conditions for the dataset $\mathcal{D}$ to be consistent with a Nash equilibrium of a game, therefore the second condition is required to ensure consistency with the other statements in the Multiagent Afriat's Theorem. The intuition that connects statements 1 and 3 in Theorem 21.8 is provided by the following result from [141]; for any smooth potential game that admits a concave potential function V, a sequence of responses $\{x^i\}_{i \in \{1,2,\ldots,n\}}$ is generated by a pure-strategy Nash equilibrium if and only if it is a maximizer of the potential function,

$$\begin{aligned} x_t = \{x_t^1, x_t^2, \ldots, x_t^n\} &\in \arg\max V(\{x^i\}_{i \in \{1,2,\ldots,n\}}) \\ \text{s.t.} \quad p_t' x^i &\leq I_t^i \quad \forall i \in \{1, 2, \ldots, n\} \end{aligned} \tag{21.58}$$

for each probe vector $p_t \in \mathbb{R}_+^m$.

The nonparametric test for Nash rationality involves determining if Eq. (21.56) has a feasible solution. Computing parameters v_t and $\lambda_t^i > 0$ in Eq. (21.56) involves solving a linear program with T^2 linear constraints in $(n+1)T$ variables, which has polynomial time complexity [128]. In the special case of one agent, the constraint set in Eq. (21.56) is the dual of the *shortest path problem* in network flows. Using the graph theoretic algorithm presented in [144], the solution of the parameters u_t and λ_t in Eq. (21.51) can be computed with time complexity $O(T^3)$.

21.4.4 LEARNING ALGORITHM FOR RESPONSE PREDICTION

In the previous section a nonparametric test to detect if a dataset $\mathcal{D}$ is consistent with Nash rationality was provided. If the $\mathcal{D}$ satisfies Nash rationality, then the Multiagent Afriat's Theorem can be used to construct the concave potential function of the game for agents in the social network illustrated in Fig. 21.15. In this section we provide a nonparametric learning algorithm that can be used to predict the responses of agents using the constructed concave potential function (21.57).

To predict the response of agent i, denoted by $\hat{x}_\tau^i$, for probe p_τ and budget I_τ^i, the optimization problem (21.58) is solved using the estimated potential function $\hat{V}$ (21.57), p_τ, and I_τ^i. Computing $\hat{x}_\tau^i$ requires solving an optimization problem with linear constraints and concave piece-wise linear objective. This can be solved using the interior point algorithm [128]. The algorithm used to predict the response $\hat{x}_\tau = (\hat{x}_\tau^1, \hat{x}_\tau^2, \ldots, \hat{x}_\tau^n)$ is given below:

Step 1: Select a probe vector $p_\tau \in \mathbb{R}_+^m$, and response budget I_τ^i for the estimation of optimal response $\hat{x}_\tau \in \mathbb{R}_+^{m\times n}$.

Step 2: For dataset $\mathcal{D}$, compute the parameters v_t and λ_t^i using Eq. (21.56).

Step 3: The response $\hat{x}_\tau$ is computed by solving the following linear program given $\{\mathcal{D}, p_\tau, I_\tau^i\}$, and $\{v_t, \lambda_t^i\}$ from Step 2:

$$\begin{aligned} \max \quad & z \\ \text{s.t.} \quad & z \le v_t + \sum_{i=1}^{n} \lambda_t^i p_t'(\hat{x}_\tau^i - x_t^i) \text{ for } t = 1, \ldots, T \\ & p_\tau' \hat{x}_\tau^i \le I_\tau^i \quad \forall i \in \{1, 2, \ldots, n\}. \end{aligned} \tag{21.59}$$

21.4.5 DATASET 1: ONLINE MULTIWINNER AUCTION

This section illustrates how Afriat's Theorem from Section 21.4.1 can be used to determine if bidders in an online multiwinner auction are utility optimizers. Online auctions are rapidly gaining popularity because bidders do not have to gather at the same geographical location. Several researchers have focused on the timing of bids and multiple bidding behavior in Amazon and eBay auctions [145–148]. The analysis of the bidding behavior can be exploited by auctioneers to target suitable bidders and thereby increase profits.

The multiwinner auction dataset was obtained from an experimental study conducted among undergraduate students in electrical engineering at Princeton University on March 25, 2011.[22] The

[22]The experimental data is provided by Leberknight et al. [149].

multiwinner auction consists of bidders competing for questions that will aid them for an upcoming midterm exam. The social network is composed of $n = 12$ bidders where the bidders do not interact with other bidders, they only interact with the external influence, refer to Fig. 21.15. Each bidder is endowed with 500 tokens prior to starting the multiwinner auction. The number of questions being auctioned is not known to the bidders; this prevents the bidders from immediately submitting their entire budget in the final auction. Each auction consists of auctioning a single question at an initial price of 10 tokens and has a duration of 30 min. At the beginning of each auction the bidders are provided with the number of winners, denoted k, that auction will have, and the budget of each bidder. The bids are private information with each bidder only informed when their bid has been outbid. The bidders do not communicate with each other during the auction. At the end of each auction, the first k highest bidders are selected, denoted by $\zeta_1, \zeta_2, \ldots, \zeta_k$. The bidders $\zeta_1, \zeta_2, \ldots, \zeta_k$ are awarded with the question, and pay the second largest bid amount (i.e., bidder ζ_k pays ζ_{k-1}'s bid amount). In the multiwinner auction it is in the self interest of bidders to force other bidders to spend too much, eliminating them from competing in successive auctions.

If the bidding behavior of agents satisfies Afriat's test (21.51) for utility maximization, the next goal is to classify the behavior of bidders into two categories: strategic and frantic. If a bidder fails Afriat's test then they are classified as irrational. Strategic bidders will typically submit a large number of bids and a smaller bid amount when compared to frantic bidders. With this bidding behavior, strategic bidders force the other bidders to spend too much, eliminating them from competing in future auctions. Frantic bidders are, however, only interested in winning the current auction.

To apply Afriat's test (21.51), the external influence p_t, and bidder responses x_t must be defined. The external influence for each bidder is defined by $p_t^i = [p_t^i(1), p_t^i(2)]$ with $p_t^i(1) = \textit{initial bid amount}$ representing the bidders interest level for winning, and $p_t^i(2) = \textit{\# of winners}$ representing the perception of winning where i is the bidder and t the auction. Two datasets are considered for analysis denoted by $\mathcal{D}_1$ and $\mathcal{D}_2$. An identical external influence is used to construct both $\mathcal{D}_1$ and $\mathcal{D}_2$. The responses in $\mathcal{D}_1$ are given by $x_t^i = [x_t^i(1), x_t^i(2)]$ where $x_t^i(1) = \textit{\# of bids}$ and $x_t^i(2) = \textit{mean bid amount}$; and for $\mathcal{D}_2$ the inputs of x_t^i are given by $x_t^i(1) = \textit{\# of bids}$ and $x_t^i(2) = \textit{mean change in bid amount}$. The response $x_t^i(2)$ in $\mathcal{D}_1$ provides the expected bid amount, and $x_t^i(2)$ in $\mathcal{D}_2$ a measure of the statistical dispersion of the bids. The budget I_t^i of each bidder has units of *tokens* multiplied by *# of bids*, and is constrained as the number of tokens and auction duration are finite. The datasets $\mathcal{D}_1$ and $\mathcal{D}_2$ are constructed from $T = 6$ auctions. The nonparametric test (21.51) is applied to each dataset $\mathcal{D}_1$ and $\mathcal{D}_2$ to detect irrational bidders. For dataset $\mathcal{D}_1$ bidder 4 is irrational, and for $\mathcal{D}_2$ bidder 11 is irrational. Note that the classification of irrational behavior is dependent on the choice of response signals used for analysis by the experimentalist.

For utility maximization bidders, an estimate of the utility function of each bidder is required to classify them as strategic or frantic. To estimate the utility function of the bidders, a subset of data from $\mathcal{D}_1$, denoted as $\bar{\mathcal{D}}_1$, is selected such that the preferences of all agents i in $\bar{\mathcal{D}}_1$ are identical. It was determined that $\bar{\mathcal{D}}_1 = \{(p_t^i, x_t^i) : i \in \{1, 3, 5, 7, 12\}\}$. Because the preferences of these bidders are identical, we can consider all the data in $\bar{\mathcal{D}}_1$ as originating from a single representative bidder. This allows an improved estimate of the utility function of these bidders as compared to learning the utility function for each bidder separately. An analogous explanation is used for the construction of $\bar{\mathcal{D}}_2 = \{(p_t^i, x_t^i) : i \in \{1, 2, 3, 4, 7, 8, 9\}\}$ from the dataset $\mathcal{D}_2$. The estimated utility function for $\bar{\mathcal{D}}_1$ is given in Fig. 21.16A and for $\bar{\mathcal{D}}_2$ in Fig. 21.16B. As seen from Fig. 21.16A and B, bidders have a preference to increase the number of bids compared with increasing the mean bid amount or the difference in mean

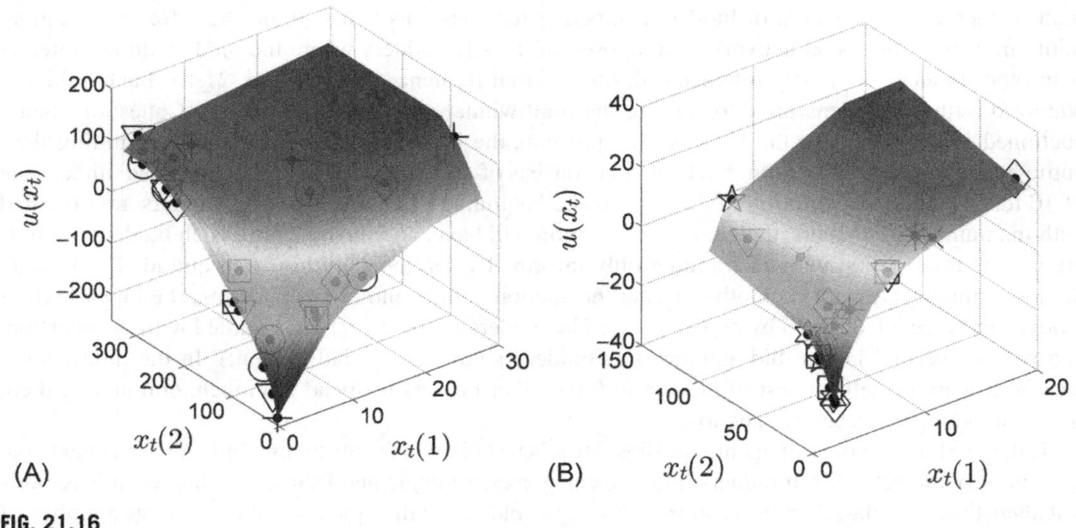

FIG. 21.16

Estimated utility function of bidders is constructed using Eq. (21.52) using the datasets $\bar{\mathcal{D}}_1$ and $\bar{\mathcal{D}}_2$ defined in Section 21.4.5. The *black dots* represent the utility associated with experimentally measured responses, and the *color* in the online version of the chapter (*blue to red*) indicates the utility level. The *black dots* indicate the observed demands, and the shape (i.e., *circle*, *diamond*, etc.) denotes the respective bidder. (A) Estimated utility function $u(x_t)$ using dataset $\bar{\mathcal{D}}_1$ defined in Section 21.4.5. (B) Estimated utility function $u(x_t)$ using dataset $\bar{\mathcal{D}}_2$ defined in Section 21.4.5.

bid amount. This follows logically as $x_t^i(2)$ increases, the bidder will have to pay more tokens to win the question, limiting their ability to bid in future auctions. Interestingly, the bidders show strategic and frantic behavior in both datasets $\bar{\mathcal{D}}_1$ and $\bar{\mathcal{D}}_2$, as seen in Fig. 21.16A and B. This is consistent with the results in [147] that show that bidders change their bidding behavior between successive auctions.

The analysis shows that auctioneers should target bidders that show frantic bidding behavior as they are likely to overspend on items, increasing the revenue of the auctioneer. Such behavior can be detected using the utility maximization test and constructed utility function from Afriat's Theorem.

21.4.6 DATASET 2: ONTARIO ELECTRICAL ENERGY MARKET DATASET

In this section we consider the aggregate power consumption of different zones in the Ontario power grid. A sampling period of $T = 79$ days starting from January 2013 is used to generate the dataset $\mathcal{D}$ for the analysis. All price and power consumption data are available from the *Independent Electricity System Operator*[23] (IESO) website. Each zone is considered as an agent in the corporate network illustrated in Fig. 21.15. The study of corporate social networks was pioneered by Granovetter [150, 151], which shows that the social structure of the network can have important economic outcomes.

[23]http://ieso-public.sharepoint.com/.

Examples include agent's choice of alliance partners, assumption of rational behavior, self-interest behavior, and the learning of other agent's behavior. Here we test for rational behavior (i.e., utility maximization and Nash rationality), and if true then learn the associated behavior of the zones. This analysis provides useful information for constructing demand-side management (DSM) strategies for controlling power consumption in the electricity market. For example, if a utility function exists it can be used in the DSM strategy presented in [152,153].

The zone's power consumption is regulated by the associated price of electricity set by the senior management officer in each respective zone. Because there is a finite amount of power in the grid, each officer must communicate with other officers in the network to set the price of electricity. Here we utilize the aggregate power consumption from each of the $n = 10$ zones in the Ontario power grid and apply the nonparametric tests for utility maximization (21.51) and Nash rationality (21.56) to detect if the zones are demand responsive. If the utility maximization or Nash rationality tests are satisfied, then the power consumption behavior is modeled by constructing the associated utility function (21.52) or concave potential function of the game (21.57).

To perform the analysis, the external influence p_t and response of agents x_t must be defined. In the Ontario power grid the wholesale price of electricity is dependent on several factors such as consumer behavior, weather, and economic conditions. Therefore the external influence is defined as $p_t = [p_t(1), p_t(2)]$ with $p_t(1)$ the average electricity price between midnight and noon, and $p_t(2)$ as the average between noon and midnight with t denoting day. The response of each zone corresponds to the total aggregate power consumption in each respective tie associated with $p_t(1)$ and $p_t(2)$ and is given by $x_t^i = [x_t^i(1), x_t^i(2)]$ with $i \in \{1, 2, \ldots, n\}$. The budget I_t^i of each zone has units of dollars as p_t has units of \$/kWh and x_t^i units of kWh.

We found that the aggregate consumption data of each zone does not satisfy Afriat's utility maximization test (21.51). This points to the possibility that the zones are engaged in a concave potential game—this would not be a surprising result as network congestion games have been shown to reduce peak power demand in distributed demand management schemes [132]. To test if the dataset $\mathcal{D}$ is consistent with Nash rationality, the detection test (21.56) is applied. The dataset for the power consumption in the Ontario power gird is consistent with Nash rationality. Using Eq. (21.59), a concave potential function for the game is constructed. Using the constructed potential function, when do agents prefer to consume power? The *marginal rate of substitution*[24] (MRS) can be used to determine the preferred time for power usage. Formally, the MRS of $x^i(1)$ for $x^i(2)$ is given by

$$\text{MRS}_{12} = \frac{\partial \hat{V} / \partial x^i(1)}{\partial \hat{V} / \partial x^i(2)}.$$

From the constructed potential function we find that $\text{MRS}_{12} > 1$, suggesting that the agents prefer to use power in the time period associated with $x_t(1)$—that is, the agents are willing to give up MRS_{12} kWh of power in the time period associated with $x^i(2)$ for 1 additional kWh of power in the time period associated with $x^i(1)$.

The analysis in this section suggests that the power consumption behavior of agents is consistent with players engaged in a concave potential game. Using the Multiagent Afriat's Theorem, the agents

[24]The amount of one good that an agent is willing to give up in exchange for another good while maintaining the same level of utility.

preference for using power was estimated. This information can be used to improve the DSM strategies presented in [152,153] to control power consumption in the electricity market.

21.4.7 DATASET 3: TWITTER DATA

Does the tweeting behavior of Twitter agents satisfy a utility maximization process? The goal is to investigate how tweets and trend indices[25] impact the tweets of agents in the Twitter social network. The information provided by this analysis can be used in social media marketing strategies to improve a brand and for brand awareness. As discussed in [154], Twitter may rely on a huge amount of agent-generated data, which can be analyzed to provide novel personal advertising to agents.

To apply Afriat's utility maximization test (21.51), we choose the external influence and response as follows. External influence $p_t = [\#Sony, 1/\#Playstation]$ for each day t. The associated response taken by the agents in the network is given by $x_t = [\#Microsoft, \#Xbox]$. Notice that the probe $p_t(2)$ can be interpreted as the frequency of tweets with the word *Playstation* (i.e., the trending index). The dataset $\mathcal{D}$ of external influence and responses is constructed from $T = 80$ days of Twitter data starting from January 1, 2013. The dataset $\mathcal{D}$ satisfies the utility maximization test (21.51). This establishes that utility function exists for agents that is dependent on the number of tweets containing the words *Microsoft* and *Xbox*. The data shows that tweets containing the word *Microsoft* and *Xbox* are dependent on the number of tweets containing *Sony* and trending index of *Playstation*. This dependency is expected as Microsoft produces the game console Xbox, and Sony produces the game console Playstation, both of which have a large number of brand followers (e.g., Xbox has more than 3 million, and Playstation more than 4 million). To gain further insight into the behavior of the agents, Eq. (21.52) from Afriat's Theorem is used to construct a utility function for the agents. Fig. 21.17 shows the constructed utility function of

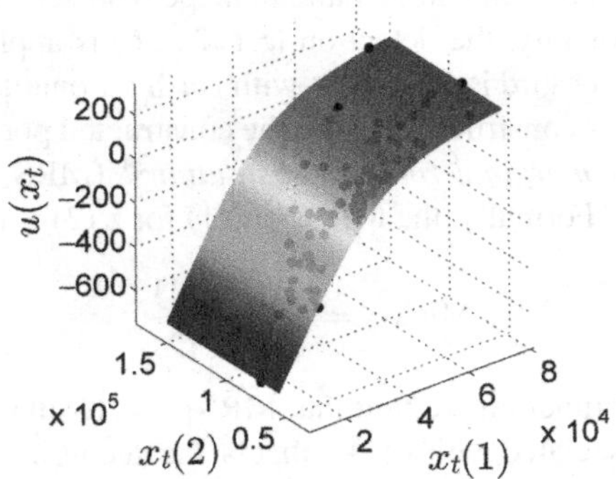

FIG. 21.17

Estimated utility function $u(x_t)$ using dataset $\mathcal{D}$ defined in Section 21.4.7 and constructed using the nonparametric learning algorithm (21.52) from Afriat's theorem.

[25] Here we define the *trend index* as the frequency of tweets containing a particular word [154].

the agents. As seen, agents have a higher utility for using the word *Microsoft* as compared to *Xbox*—that is, agents prefer to use the word *Microsoft* to that of *Xbox*. Interestingly, if we define the response to be $x_t = [\#Microsoft, 1/\#Xbox]$, then the dataset satisfies utility maximization. From the constructed utility function, not shown, the agents prefer to increase the tweets containing the word *Microsoft* compared to increasing the trend index of *Xbox*. If instead $x_t = [1/\#Microsoft, 1/\#Xbox]$, then the dataset satisfies utility maximization and agents prefer to increase the trend index of *Microsoft* compared to that of *Xbox*.

To summarize, the above analysis suggests the following interesting fact: *Xbox* has a lower utility than *Microsoft* in terms of Twitter sentiment. Therefore, online marketing strategies should target the brandname *Microsoft* instead of *Xbox*.

21.4.8 SUMMARY AND EXTENSIONS

The principle of revealed preferences is an active research area with numerous recent papers. We have already mentioned the papers [29,125,126]. Below we summarize some related literature that extends the basic framework of Afriat's theorem.

Afriat's theorem holds for finite datasets and gives an explicit construction of a class of concave utility functions that rationalizes the dataset. Mas-Colell [155] has given sufficient conditions under which, as the data set size T grows to infinity, the underlying utility function of the consumer can be fully identified.

Though the classical Afriat's theorem holds for linear budget constraints $p_t'x \leq I_t$ in Eq. (21.50), an identical formulation holds for certain nonlinear budget constraints as illustrated in [156]. The budget constraints considered in [156] are of the form $\{x \in \mathbb{R}_+^m | g(x) \leq 0\}$ where $g : \mathbb{R}_+^m \to \mathbb{R}$ is an increasing continuous function and $\mathbb{R}_+^m$ denotes the positive orthant. Also Ref. [156] shows how the results in [155] on recoverability of the utility function can be extended to such nonlinear budget constraints. However, learning the utility function from a finite dataset in the case of a nonlinear budget constraint requires sophisticated machine learning algorithms [157]. The machine learning algorithms can only guarantee that the estimated utility function is approximately consistent with the dataset $\mathcal{D}$—that is, the estimated utility is not guaranteed to contain all the preference relations consistent with the dataset $\mathcal{D}$.

In [144], results in statistical learning theory are applied to the principle of revealed preferences to address the question: Is the class of demand functions derived from monotone concave utilities efficiently probably approximately correct (PAC) learnable? It is shown that Lipschitz utility functions are efficiently PAC learnable. In [143], the authors extend the results of [144] and show that for agents engaged in a concave potential game that satisfy Nash rationality, if the underlying potential function satisfies the Lipschitz condition then the potential function of the game is PAC learnable.

In many cases, the responses of agents are observed in noise. Then determining if an agent is a utility maximizer (or a multiagent system's response is consistent with play from a Nash) becomes a statistical decision test. In [158] it is shown how stochastic optimization algorithms can be devised to optimize the probe signals to minimize the type II errors of the decision test subject to a fixed type I error.

Change point detection in utility functions

Ref. [159] extends the classical revealed preference framework to agents with a "dynamic utility function." The utility function jump changes at an unknown time instant by a linear perturbation. Given the dataset of probe and responses of an agent, the objective in [159] is to develop a nonparametric test to detect the change point and utility functions before and after the change, which is henceforth referred to as the change point detection problem.

Such change point detection problems arise in online searches in social media. Online search is currently the most popular method for information retrieval [160] and can be viewed as an agent maximizing the information utility, i.e., the amount of information consumed by an online agent given the limited resource on time and attention. There has been a gamut of research that links Internet search behavior to ground truths such as symptoms of illness, political elections, or major sporting events [161]. Detection of utility change in online searches, therefore, is helpful to identify changes in ground truth. Also, the intrinsic nature of the online search utility function motivates such a study under a revealed preference setting.

The problem of detecting a sudden linear perturbation change in the utility function is motivated by several reasons. First, the linear perturbation assumption provides sufficient selectivity such that the nonparametric test is not trivially satisfied by all datasets but still provides enough degrees of freedom. Second, the linear perturbation can be interpreted as the change in the marginal rate of utility relative to a "base" utility function. In online social media, the linear perturbation coefficients measure the impact of marketing or the measure of severity of the change in ground truth on the utility of the agent. This is similar to the linear perturbation models used to model taste changes [162–164] in microeconomics. Finally, in social networks, linear change in the utility is often used to model the change in utility of an agent based on the interaction with the agent's neighbors [165]. Compared to the taste change model, our model is unique in that we allow the linear perturbation to be introduced at an unknown time.

A related important practical issue that we also consider in this paper is the application of revealed preference framework to high dimensional data ("big-data"). As an example of high dimensional data arising in online social media, we investigate the detection of the utility maximization process inherent in video sharing via YouTube. Detecting utility maximization behavior with such high-dimensional data is computationally demanding. Ref. [159] uses dimensionality reduction to overcome the computational cost associated with high-dimensional data. The high-dimensional data is projected into a lower-dimensional subspace using the Johnson-Lindenstrauss (JL) transform.

21.5 SOCIAL INTERACTION OF CHANNEL OWNERS AND YOUTUBE CONSUMERS

In this section, time-series analysis methods are applied to real-world YouTube data to determine how social sensors interact with YouTube channel owners. Several key results are presented that elucidate the dynamics of social sensors in the YouTube social network. This section contains five main results.

1. Section 21.5.2 illustrates the sensitivity of social sensor engagement to changes in metalevel (title, thumbnail, tags) features of YouTube videos. It is found that metalevel feature optimization causes an increase in user engagement for approximately 50% of videos. Optimization of the title of the video causes a significant improvement of users finding the video from YouTube search results.

Additionally, optimization of the thumbnail causes an increase in users accessing the video from the related video list.[26]

2. In Section 21.5.3 Granger causality is used to show that a causal relationship exists between a channel's subscriber count and the social sensor engagement of videos on the channel. However, this causal relationship is dependent on the category. For example, 80% of the "entertainment" channels satisfy the Granger causality test while only 40% of the "food" channels satisfy the test.
3. In Section 21.5.4 it is determined that for popular gaming YouTube channels with a dominant (constant) upload schedule, deviating from the schedule increases the views and the comment counts of the channel (e.g., increases user engagement). Specifically, when the channel goes off schedule the channel gains views 97% of the time and the channel gains comments 68% of the time.
4. In Section 21.5.5 we illustrate that the social sensor engagement dynamics with YouTube videos can be modeled using a generalized Gompertz model. The generalized Gompertz model accounts for the initial viral increase in views from subscribers, the subsequent linear growth that results from nonsubscribers, and views from exogenous events such as promotion on other popular social media platforms. It is important to account for exogenous events when estimating the efficiency of metalevel optimization procedures.
5. In Section 21.5.6 the generalized Gombertz model is used to study the dynamics of social sensors to video playthroughs (sequence of videos on the same topic). It is illustrated that the early view count dynamics are highly correlated with the view count dynamics of future videos. Both the short-term view count and long-term migration of users to future videos in the playthrough decrease after the initial video in the playthrough is posted. This results even when the channel's subscriber count increases. A possible reason for this decrease is that subsequent videos in the playthrough become repetitive and hence decrease user engagement.

The results in this section are based on the extensive BroadBandTV[27] (BBTV) dataset. Extrapolating these results to other YouTube datasets is an important problem worth addressing by the reader. For example, an extension of this work could involve studying the effect of video characteristics on different traffic sources, for example, the effect of tweets or posts of videos on Twitter or Facebook.

21.5.1 YOUTUBE DATASET

All the results in this section are constructed using the extensive YouTube dataset provided by BBTV.[28] The dataset contains daily samples of metalevel features of YouTube videos and channels on the BBTV platform from April 2007 to May 2015, and has a size of several terabytes. The metalevel features include views, comments, likes, dislikes, shares, and subscribers, which are recorded each day because

[26]In YouTube, the suggested videos refers to the overall list to the right of the video player on the watch page, which is populated with suggestions for what to watch next. A subset of these suggested videos, known as related videos, can also be displayed at the end of a YouTube video.

[27]BroadBandTV is one of the largest YouTube video partners in the world. http://bbtv.com/press/broadbandtv-now-the-largest-multi-platform-network-worldwide.

[28]http://bbtv.com/.

Table 21.1 Dataset Summary

Videos	6 million
Channels	26,000
Average number of videos (per channel)	250
Average age of videos	275 days
Average number of views (per video)	10,000

Table 21.2 YouTube Dataset Categories (Out of 6 Million Videos)

Category	Fraction
Gaming	0.69
Entertainment	0.07
Food	0.07
Music	0.035
Sports	0.017

the video was published. The dataset contains information for more than 6 million videos spread over 25 thousand channels. Table 21.1 shows the statistics summary of the videos present in the dataset.

Table 21.2 shows the summary of the various categories of the videos present in the dataset. The dataset contains a large percentage of gaming videos. Fig. 21.18 shows the fraction of videos as a function of the age of the videos. There is a large fraction of videos uploaded within a year. Also, the dataset captures the exponential growth in the number of videos uploaded to YouTube. Similar to [38], we define three categories of videos based on their popularity: highly popular, popular, and unpopular. Table 21.3 gives a summary of the fraction of videos in the dataset belonging to each category. As can be seen from Table 21.3, the majority of the videos in the dataset belong to the popular category.

A unique feature of the dataset is that it contains information about the "metalevel optimization" for videos. The metalevel optimization is a change in the title, tags, or thumbnail of an existing video in order to increase the popularity. BBTV markets a product that intelligently automates the metalevel optimization. Table 21.4 gives a summary of the statistics of the various metalevel optimization present in the dataset.

21.5.2 SOCIAL SENSOR ENGAGEMENT SENSITIVITY TO METALEVEL OPTIMIZATION

Here we analyze how changing metalevel features after a video is posted impacts the user engagement of the video. Metalevel data plays a significant role in the discovery of content through YouTube searches and in video recommendations through the YouTube related videos.[29] Hence, optimizing the metalevel data to enhance the discoverability and user engagement of videos is of significant

[29]Related and suggested videos appear surrounding the current video being viewed by the user.

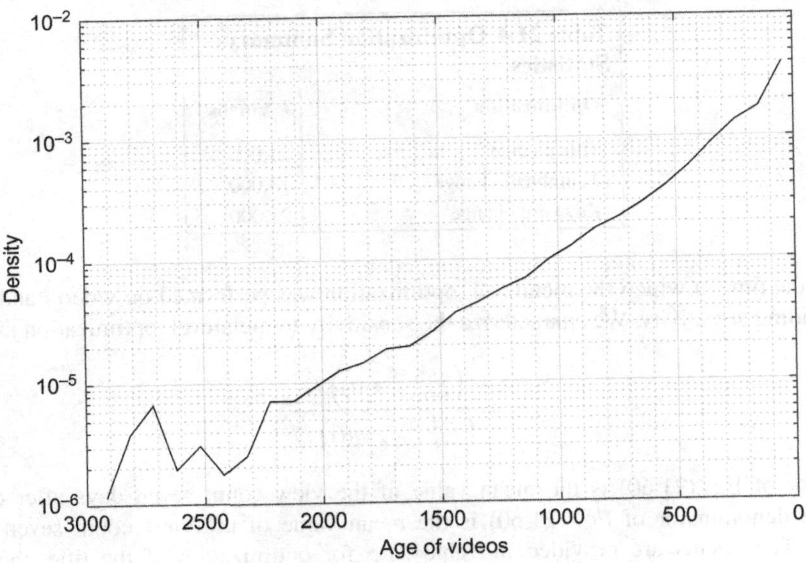

FIG. 21.18

The fraction of videos in the dataset as a function of the age of the videos. There is a significant percentage of newer videos (videos with less age) compared to older videos. Hence, the dataset capture the exponential growth of the number of videos uploaded to YouTube.

Table 21.3 Popularity Distribution of Videos in the Dataset

Criteria	Fraction
Highly popular (total views > 10^4)	0.12
Popular (150 < total views < 10^4)	0.67
Unpopular (total views < 150)	0.21

importance to content providers. Therefore, in this section, we study how optimizing the title, thumbnail, or keywords affects the view count of YouTube videos.

To perform the sensitivity analysis, we utilize the dataset presented in Section 21.5.1, and remove any time-sensitive videos. Time-sensitive videos are those videos that are relevant for a short period of time and the popularity of such videos cannot be improved by optimization. We removed the following two time-sensitive categories of videos: "politics" and "movies and trailers." In addition, we removed videos (from other categories) that contained the following keywords in their video metadata: "holiday," "movie," or "trailers." For example, holiday videos are not watched frequently during off-holiday times.

Table 21.4 Optimization Summary Statistics

Optimization	# Videos
Title change	21,000
Thumbnail change	13,000
Keyword change	21,000

Let $\hat{\tau}_i$ be the time at which the metalevel optimization was performed on video i and let s_i denote the corresponding sensitivity. We characterize the sensitivity to metalevel optimization as follows:

$$s_i = \frac{\left(\sum_{t=\hat{\tau}_i}^{\hat{\tau}_i+6} v_i(t)\right)/7}{\left(\sum_{t=\hat{\tau}_i-6}^{\hat{\tau}_i} v_i(t)\right)/7}. \tag{21.60}$$

The numerator of Eq. (21.60) is the mean value of the view count seven days after optimization. Similarly, the denominator of Eq. (21.60) is the mean value of the view count seven days before optimization. The results are provided in Table 21.5 for optimization of the title, thumbnail, and keywords.

As shown in Table 21.5, at least half the optimizations resulted in an increase in the popularity of the video. In addition, compared to videos with no optimization, the metalevel optimization improves the probability of increased popularity by 45%. This is consistent with YouTube and BBTV recommendations to optimize metalevel features to increase user engagement. However, some classes of videos benefit from optimizing metadata much more than others. The effect may be due to small user channels, which have a limited number of videos and subscribers, gaining by optimizing the metalevel

Table 21.5 Sensitivity to Metalevel Optimization

Optimization	Fraction of Videos With Increased Popularity
Title change	0.52
Thumbnail change	0.533
Keyword change	0.50
No change[a]	0.35

The table shows that in more than 50% the videos, metalevel optimization resulted in an increase in the popularity of the video.

[a] *"No change" was obtained by randomly selecting 10^4 videos that performed no optimization and evaluating s_i three months from the date of posting the video.*

Table 21.6 Sensitivity of Various Traffic Sources to Metalevel Optimization for Videos With Increased Popularity

Optimization	Related	Promoted	Search
Title change	1.13	NA[a]	1.24
Thumbnail change	1.20	NA[a]	1.125
Keyword change	1.10	1.16	1

The title optimization resulted in significant improvement (approximately 25%) from the YouTube search. Similarly, thumbnail optimization improved traffic from the related videos and keyword optimization resulted in increased traffic from related and promoted videos.

[a]*Not enough data available: A binomial test to check for the true hypothesis with 95% confidence interval requires that the sample size, n, should be at least* $\left(\frac{1.96}{0.04}\right)^2 p(1-p)$. *With* $p = 0.5$, $n > 600$.

data of the video compared to hugely popular channels such as Sony or CNN. The highly popular channels (e.g., Sony or CNN) upload videos frequently (even multiple times daily), so video content becomes irrelevant quickly. The question of which class of users gains by optimizing the meta level features of the video is part of our ongoing research.

Table 21.6 summarizes the impact of various metalevel changes on the three major sources of YouTube traffic, i.e., YouTube search,[30] YouTube promoted,[31] and traffic from related videos.[32] For those videos where metalevel optimization increased the popularity (the ratio of the mean value of the views after and before optimization is higher than one), we computed the sensitivity for various traffic sources as in Eq. (21.60). Table 21.6 summarizes the median statistics of the ratio of the traffic sources before and after optimization. The title optimization resulted in significant improvement (approximately 25%) from the YouTube search. Similarly, thumbnail optimization improved traffic from the related videos and keyword optimization resulted in increased traffic from related and promoted videos.

Summary: This section studied the sensitivity of view count with respect to metalevel optimization. The main finding is that metalevel optimization increased the popularity of video in the majority of cases. In addition, we found that optimizing the title improved traffic from the YouTube search. Similarly, thumbnail optimization improved traffic from the related videos and keyword optimization resulted in increased traffic from related and promoted videos.

[30]Video views that resulted users selecting the video from the YouTube search results.

[31]Video views that result from channels paying YouTube to increase their probability of being included at the top of search result lists.

[32]Video views that resulted from users clicking on a thumbnail that was listed on the page of another video they were viewing.

21.5.3 CAUSAL RELATIONSHIP BETWEEN CHANNEL SUBSCRIBERS AND SOCIAL SENSOR ENGAGEMENT

In this section the goal is to detect whether a causal relationship exists between subscriber and viewer counts and how it can be used to estimate the next day subscriber count of a channel. The results are of interest for measuring the popularity of a YouTube channel. Fig. 21.19 displays the subscriber and view count dynamics of a popular movie trailer channel on YouTube. It is clear from Fig. 21.19 that the subscribers spike with a corresponding spike in the view count. In this section we model this causal relationship of the subscribers and view count using the Granger causality test from the econometric literature [166].

The main idea of Granger causality is that if the value(s) of a lagged time-series can be used to predict another time series, then the lagged time-series is said to "Granger cause" the predicted time series. To formalize the Granger causality model, let $s^j(t)$ denote the number of subscribers to a channel j on day t, and $v_i^j(t)$ the corresponding view count for a video i on channel j on day t. The total number of videos in a channel on day t is denoted by $\mathcal{I}(t)$. Define,

$$\hat{v}^j(t) = \sum_{i=1}^{\mathcal{I}(t)} v_i^j(t), \tag{21.61}$$

as the total view count of channel j at time t. The Granger causality test involves testing if the coefficients b_i are nonzero in the following equation, which models the relationship between subscribers and view counts:

$$s^j(t) = \sum_{k=1}^{n_s} a_k^j s^j(t-k) + \sum_{i=k}^{n_v} b_k^j \hat{v}^j(t-k) + \varepsilon^j(t), \tag{21.62}$$

where $\varepsilon^j(t)$ represents normal white noise for channel j at time t. The parameters $\{a_i^j\}_{\{i=1,\ldots,n_s\}}$ and $\{b_i^j\}_{\{i=1,\ldots,n_v\}}$ are the coefficients of the AR model in Eq. (21.62) for channel j, with n_s and n_v denoting the lags for the subscriber and view counts time series respectively. If the time-series $\mathcal{D}^j = \{s^j(t), \hat{v}^j(t)\}_{t\in\{1,\ldots,T\}}$ of a channel j fits the model (21.62), then we can test for a causal relationship between subscribers and view count. In Eq. (21.62), it is assumed that $|a_i| < 1$, $|b_i| < 1$ for stationarity. The causal relationship can be formulated as a hypothesis testing problem as follows:

$$H_0 : b_1 = \cdots = b_{n_v} = 0 \text{ vs. } H_1 : \text{At least one } b_i \neq 0. \tag{21.63}$$

The rejection of the null hypothesis, H_0, implies that there is a causal relationship between subscriber and view counts.

First, we use the Box-Ljung test [167] is to evaluate the quality of the model (21.62) for the given dataset $\mathcal{D}^j$. If satisfied, then the Granger causality hypothesis (21.63) is evaluated using the Wald test [168]. If both hypothesis tests pass then we can conclude that the time series $\mathcal{D}^j$ satisfies Granger causality—that is, the previous day subscriber and view count have a causal relationship with the current subscriber count.

A key question prior to performing the Granger causality test is what percentage of videos in the YouTube dataset (Appendix) satisfy the AR model in Eq. (21.62). To perform this analysis we apply the Box-Ljung test with a confidence of 0.95 (P-value$=0.05$). First, we need to select n_s and n_v, the number of lags for the subscribers and view count time series. For $n_s = n_v = 1$, we found that only 20% of the channels satisfy the model (21.62). When n_s and n_v are increased to 2, the number of channels satisfying the model increases to 63%. For $n_s = n_v = 3$, we found that 91% of the channels satisfy the model (21.62), with a confidence of 0.95 (P-value = 0.05). Hence, in the below analysis we select $n_s = n_v = 3$. It is interesting to note that the mean value of coefficients b_i decrease as i increases, indicating that older view counts have less influence on the subscriber count. Similar results also hold for the coefficients a_i. Hence, as expected, the previous day subscriber count and the previous day view count most influence the current subscriber count.

The next key question is does a causal relationship exist between the subscriber dynamics and the view count dynamics. This is modeled using the hypothesis in Eq. (21.63). To test Eq. (21.63) we use the Wald test with a confidence of 0.95 (P-value = 0.05) and found that approximately 55% of the channels satisfy the hypothesis. For approximately 55% of the channels that satisfy the AR model (21.62), the view count Granger causes the current subscriber count. Interestingly, if different channel categories are accounted for, then the percentage of channels that satisfy Granger causality vary widely as illustrated in Table 21.7. For example, 80% of the entertainment channels satisfy Granger causality while only 40% of the food channels satisfy Granger causality. These results illustrate the importance of channel owners to not only maximize their subscriber count, but to also upload new videos or increase the views of old videos to increase their channel's popularity (i.e., via increasing their subscriber count). Additionally, increasing the number of subscribers will also increase the view count of videos that are uploaded by the channel owners.

21.5.4 VIDEO UPLOAD SCHEDULING AND SOCIAL SENSOR ENGAGEMENT

Here we investigate how the video upload scheduling dynamics of YouTube channels impact social sensor engagement. We find the interesting property that for popular gaming YouTube channels with a dominant (constant) upload schedule, deviating from the schedule increases the views and the comment counts of the channel (e.g., increases user engagement).

Table 21.7 Fraction of Channels Satisfying the Hypothesis: View Count "Granger Causes" Subscriber Count, Split According to Category

Category[a]	Fraction
Gaming	0.60
Entertainment	0.80
Food	0.40
Sports	0.67

[a] *YouTube assigns a category to videos, rather than channels. The category of the channel was obtained as the majority of the category of all the videos uploaded by the channel.*

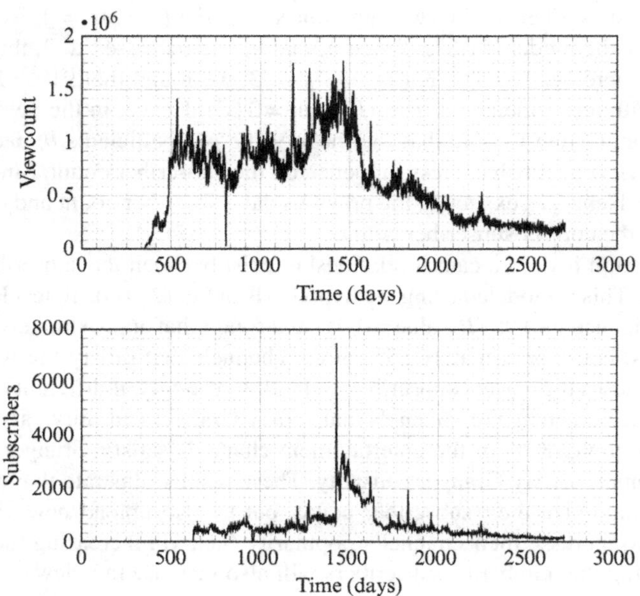

FIG. 21.19

Viewcount and subscribers for the popular movie trailer channel VISOTrailers. The Granger causality test for view counts "Granger causes" subscriber count is true with a *P*-value of 5×10^{-8}.

Creator Academy,[33] in their best practice section, recommends uploading videos on a regular schedule to get repeat views. The reason for a regular upload schedule is to increase user engagement and to rank higher in the YouTube recommendation list. However, we show in this section that going "off the schedule" can be beneficial for a gaming YouTube channel with a regular upload schedule, in terms of the number of views and the number of comments.

From the dataset, we "filtered out" video channels with a *dominant* upload schedules, as follows: The dominant upload schedule was identified by taking the periodogram of the upload times of the channel and then comparing the highest value to the next highest value. If the ratio defined above is greater than two, we say that the channel has a dominant upload schedule. From the dataset containing 25,000 channels, only 6500 channels contain a dominant upload schedule. Some channels, particularly those that contain high amounts of copied videos such as trailers and movie/TV snippets, upload videos on a daily basis. These have been removed from the above analysis. The expectation is that by doing so we concentrate on those channels that contain only user-generated content.

We found that channels with gaming content account for 75% of the 6500 channels with a dominant upload schedule[34] and the main tags associated with the videos were: "game," "gameplay," and

[33]YouTube website for helping with channels.

[34]This could also be due to the fact that gaming videos account for 70% of the videos in the dataset.

"videogame."[35] We computed the average views when the channel goes off the schedule and found that on an average when the channel goes off schedule the channel gains views 97% of the time and the channel gains comments 68% of the time. This suggests that channels with "gameplay" content have a periodic upload schedule and benefit from going off the schedule.

21.5.5 SOCIAL SENSOR ENGAGEMENT DYNAMICS WITH YOUTUBE VIDEOS

Several time-series analysis methods have been employed in the literature to model the view count dynamics of YouTube videos. These include ARMA time series models [36], multivariate linear regression models [37], hidden Markov models [169], normal distribution fitting [170], and parametric model fitting [38,39]. Though all these models provide an estimate of the view count dynamics of videos, we are interested in segmenting view count dynamics of a video resulting from subscribers, nonsubscribers, and exogenous events. Exogenous events are due to video promotion on other social networking platforms such as Facebook or the video being referenced by a popular news organization or celebrity on Twitter. Detecting and accounting for exogenous events are motivated by the need for extracting accurate view counts resulting from exogenous events that provide an estimate of the efficiency of video promotion methods and metalevel feature optimizations.

The view count dynamics of popular videos in YouTube typically show an initial viral behavior due to subscribers watching the content, and then a linear growth resulting from nonsubscribers. The linear growth is due to new users migrating from other channels or to interested users discovering the content either through search or recommendations (we call this phenomenon *migration* similar to [38]). Hence, without exogenous events, the view count dynamics of a video due to subscribers and nonsubscribers can be estimated using piece-wise linear and nonlinear segments. In [38], it is shown that a Gompertz time series model can be modeled on the view count dynamics from subscribers and nonsubscribers, if no exogenous events are present. In this chapter, we generalize the model in [38] to account for views from exogenous events. It should be noted that classical change-point detection methods [171] cannot be used here as the underlying distribution generating the view count is unknown.

To account for the view count dynamics introduced from exogenous events we use the generalized Gompertz model given by:

$$
\begin{aligned}
\bar{v}_i(t) &= \sum_{k=0}^{K_{\max}} w_i^k(t)u(t-t_k), \\
w_i^k(t) &= M_k\left(1-e^{-\eta_k\left(e^{b_k(t-t_k)}-1\right)}\right)+c_k(t-t_k),
\end{aligned}
\tag{21.64}
$$

where $\bar{v}_i(t)$ is the total view count for video i at time t, $u(\cdot)$ is the unit step function, t_0 is the time the video was uploaded, t_k with $k \in \{1,\ldots,K_{\max}\}$ are the times associated with the $K_{\max}$ exogenous events, and $w_i^k(t)$ are Gompertz models that account for the view count dynamics from uploading the video and from the exogenous events. In total there are $K_{\max}+1$ Gompertz models with each having parameters t_k, M_k, η_k, b_k. M_k is the maximum number of requests not including migration for an exogenous event at t_k, η_k and b_k model the initial growth dynamics from event t_k, and c_k accounts for the migration of

[35] We used a topic model to obtain the main tags.

other users to the video. In Eq. (21.64) the parameters $\{M_k, \eta_k, b_k\}_{k=0}$ are associated with the subscriber views when the video is initially posted, the parameters $\{t_k, M_k, \eta_k, b_k\}_{k=1}^{K_{\max}}$ are associated with views introduced from exogenous events, and the views introduced from migration are given by $\{c_k\}_{k=0}^{K_{\max}}$. Each Gompertz model (21.64) captures the initial viral growth when the video is initially available to users, followed by a linearly increasing growth resulting from user migration to the video.

The parameters $\theta_i = \{a_k, t_k, M_k, \eta_k, b_k, c_k\}_{k=0}^{K_{\max}}$ in Eq. (21.64) can be estimated by solving the following mixed-integer nonlinear program:

$$\theta_i \in \arg\min\left\{\sum_{t=0}^{T_i}\left(\bar{v}_i(t) - v_i(t)\right)^2 + \lambda K\right\}$$

$$K = \sum_{k=0}^{K_{\max}} a_k, \quad a_k \in \{0, 1\} \quad k \in \{0, \ldots, K_{\max}\}, \tag{21.65}$$

with T_i the time index of the last recorded views of video v_i, and a_k a binary variable equal to 1 if an exogenous event is present at t_k. Note that Eq. (21.65) is a difficult optimization problem as the objective is nonconvex as a result of the binary variables a_k [172]. In the YouTube social network when an exogenous event occurs this causes a large and sudden increase in the number of views. However, as seen in Fig. 21.20, a few days after the exogenous event occurs, the views only result from migration (i.e., linear increase in total views). Assuming that each exogenous event is followed by a linear increase in views, we can estimate the total number of exogenous events $K_{\max}$ present in a given time series by first using a segmented linear regression method, and then counting the number of segments of connected linear segments with a slope less then $c_{\max}$. The parameter $c_{\max}$ is the maximum slope for the views to be considered to result from viewer migration. Plugging $K_{\max}$ into Eq. (21.65) results in the optimization of a nonlinear program for the unknowns $\{t_k, M_k, \eta_k, b_k, c_k\}_{k=0}^{K_{\max}}$. This optimization problem can be solved using sequential quadratic programming techniques [173].

To illustrate how the Gompertz model (21.64) can be used to detect exogenous events, we apply Eq. (21.64) to the view count dynamics of a video that only contains a single exogenous event. Fig. 21.20 displays the total view count of a video where an exogenous event occurs at time $t = 41$ (i.e., $t_1 = 41$ in Eq. 21.64) days after the video is posted.[36] The initial increase in views for the video for $t \leq 7$ days results from the 2910 subscribers of the channel viewing the video. For $7 \leq t \leq 41$, other users that are not subscribed to the channel migrate to view the video at an approximately constant rate of 13 views/day. At $t = 41$, an exogenous event occurs causing an increase in the views per day. The difference in viewers resulting from the exogenous event is 7174. For $t \geq 43$, the views result primarily from the migration of users to approximately two views/day. Hence, using the generalized Gompertz model (21.64) we can differentiate between subscriber views, views caused by exogenous events, and views caused by migration.

[36]Due to privacy reasons, we cannot detail the specific event. Some of the reasons for the sudden increase in the popularity of the video include: Another user on YouTube mentioning the video, which will encourage viewers from that channel to view the video, resulting in a sudden increase in the number of views. Another possibility is that the channel owner or a YouTube partner such as BBTV did significant promotional initiatives on other social media sites such as Twitter, Facebook, etc., to promote the channel or video.

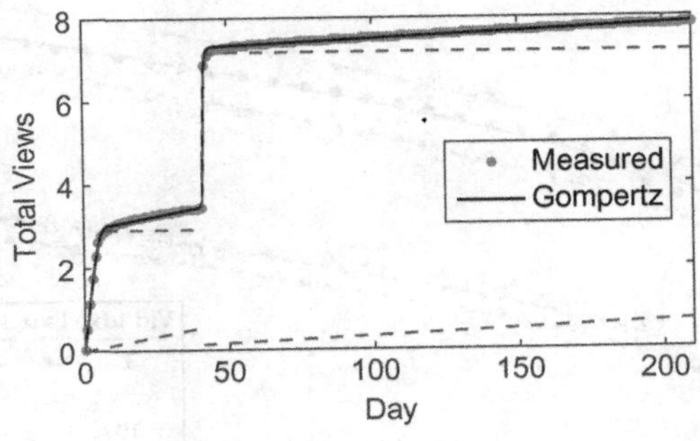

FIG. 21.20

Due to an exogenous event on day 41, there is a sudden increase in the number of views. The total view count fitted by the Gompertz model $\bar{v}_i(t)$ in Eq. (21.64) is shown in black (solid line) with the virality (exponential) and migration (linear) illustrated by the *dashed line.*

21.5.6 SOCIAL SENSOR ENGAGEMENT FOR CHANNEL PLAYTHROUGHS

One of the most popular sequences of YouTube videos is the video game "playthrough." A video game playthrough is a set of videos for which each video has a relaxed and casual focus on the game that is being played and typically contains commentary from the user presenting the playthrough. Unlike YouTube channels such as CNN, BBC, and CBC in which each new video can be considered independent from the others, in a video playthrough the future view count of videos is influenced by the previously posted videos in the playthrough. To illustrate this effect we consider a video playthrough for the game "BioShock Infinite"—a popular video game released in 2013. The channel, popular for hosting such video playthroughs, contains close to 4500 videos and 180 video playthroughs. The channel is highly popular and has garnered a combined view count close to 100 million views with 150 thousand subscribers over a period of three years. Fig. 21.21 illustrates that the early view count dynamics are highly correlated with the view count dynamics of future videos. Both the short-term view count and long-term migration of future videos in the playthrough decrease after the initial video in the playthrough is posted. This results for two reasons: either the viewers purchase the game or the viewers leave as the subsequent playthroughs become repetitive as a result of game quality or video commentary quality. A unique effect with video playthroughs is that although the number of subscribers to the channel hosting the videos in Fig. 21.21 increases over the 600-day period, the linear migration is still maintained after the initial 50 days after the playthrough is published. Additionally, the slope of the migration is related to the early total view count as illustrated in Fig. 21.21B.

21.5.7 SUMMARY AND EXTENSIONS

The application of time-series analysis and machine learning methods to gain insight into the social sensor dynamics on YouTube is an active area of research with several promising outcomes. First, they can be used to reduce the operating cost of content distribution networks. In [35] a two time-scale game-theoretic learning algorithm is constructed to optimally cache videos in the future 5G mobile network based on the dynamics of the social sensors. In [34] the optimal caching decision

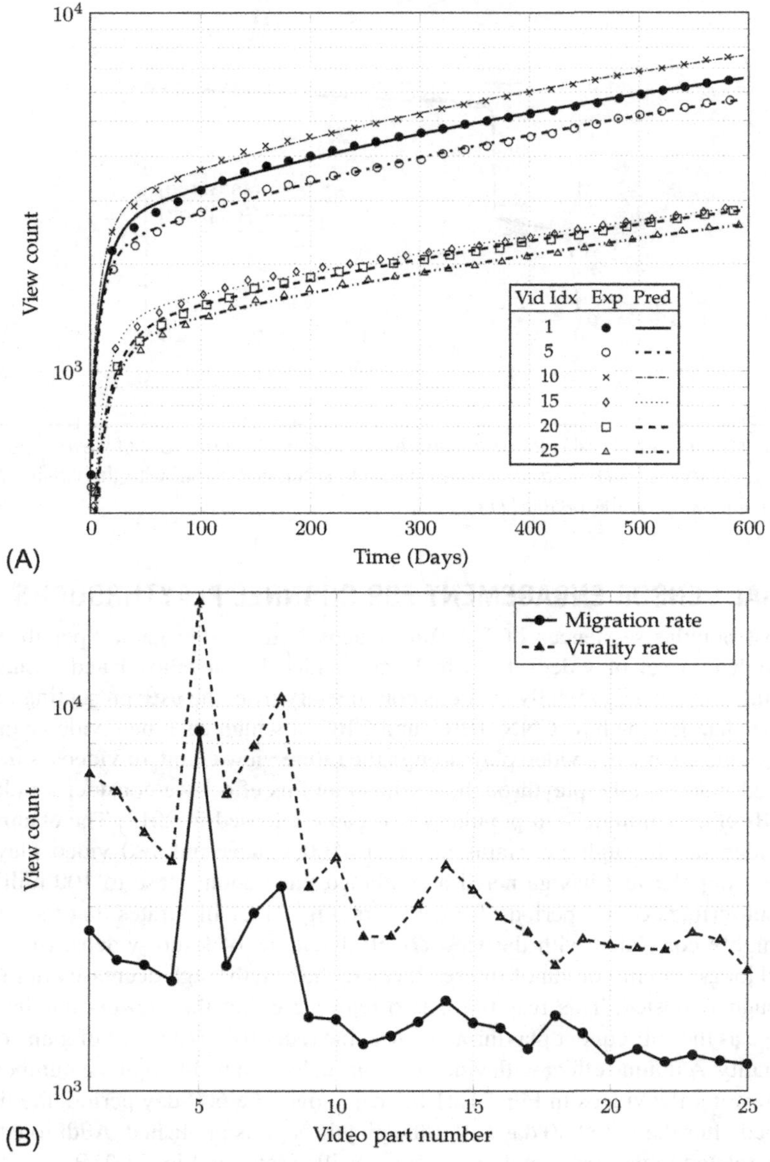

FIG. 21.21

Actual and predicted view count of a playthrough containing 25 YouTube videos for the game "BioShock Infinite." The predictions are computed by fitting a modified Gompertz model (21.64) to the measured view count for each video in the playthrough. (A) Actual and predicted view count of playthrough. We plot the 1st, 5th, 10th, 15th, 20th, and 25th video from the playlist containing 25 videos. In the legend, *Exp* and *Pred* correspond to the actual and predicted value using Eq. (21.64), respectively. Figure shows that the view counts decrease for subsequent videos in the playlist. (B) The virality rate specifies the early views due to subscribers, and the migration rate (in units of views/1000 days) specifies the subsequent linear growth due to nonsubscribers.

is formulated as a mixed-integer linear program that accounts for the dynamics of the social sensors. Second, knowledge of the user dynamics can be used to optimize the metalevel features of videos to maximize user engagement as illustrated in this section.

Significant work remains on the analysis of user dynamics in the YouTube social network. Recall that in the YouTube social network interaction between users and channel owners includes:

1. Commenting on users videos. Commenting is YouTube's version of engagement, and it has some of the most involved, engaged, and dedicated users. Additionally, users can comment on other users' comments, which is very similar to user interaction on blog posting sites except related to the uploaded videos.
2. Subscribing to YouTube channels provides a method of forming relationships between users.
3. Users can directly comment on a YouTube channel without the need to only interact when a video is posted.
4. Users can also interact by embedding videos from another user's channel directly into their own channel to promote exposure or form communities of users.

In addition to the social incentives, YouTube provides monetary incentives to promote users increasing their popularity and engagement. As more users view and interact with a user's video or channel, YouTube will pay the user proportional to the advertisement exposure on the user's channel. Therefore, users not only maximize exposure to increase their social popularity, but also for monetary gain, which introduces unique dynamics in the formation of edges in the YouTube social network. Using the dataset discussed in Section 21.5.1, Fig. 21.22 plots the communication network where an edge indicates comments and responses between users that have interacted at least 1000 times. From Fig. 21.22A, initially there appear to be two clusters of users that have strong interactions, indicating that user preferences play a significant role in forming the edges in these clusters. After a period of three months, Fig. 21.22B illustrates that more users have entered the network; however, there still appear to be two primary clusters of interacting users. At six months, Fig. 21.22C shows a dense interaction between several users in the social network. The dynamics of these interaction links are governed by both the user preferences and the video content that is uploaded by the users. Prior to edge formation, these clustered communities can be detected by applying the homophilic community detection tests introduced in [174]. These tests are designed to cluster users based on their content preferences.

The dynamics of edge formation/destruction and user popularity in the social network (illustrated in Fig. 21.22) are governed by the user-user interaction and the user-content-user interaction. Two key questions to address in the YouTube social network are: How do the social dynamics (subscribing, commenting, video content quality, video category, etc.) impact the popularity of videos and the dynamics of the communication network between users? Answers to this question provide valuable insight into the evolving dynamics of the social network illustrated in Fig. 21.22.

21.6 CLOSING REMARKS

This chapter has discussed four important and interrelated themes regarding the dynamics of social sensors, namely, diffusion models for information in social networks, Bayesian social learning, revealed preferences, and how social sensors interact over YouTube channels. In each case, examples involving real datasets were given to illustrate the various concepts. The unifying theme behind these three topics

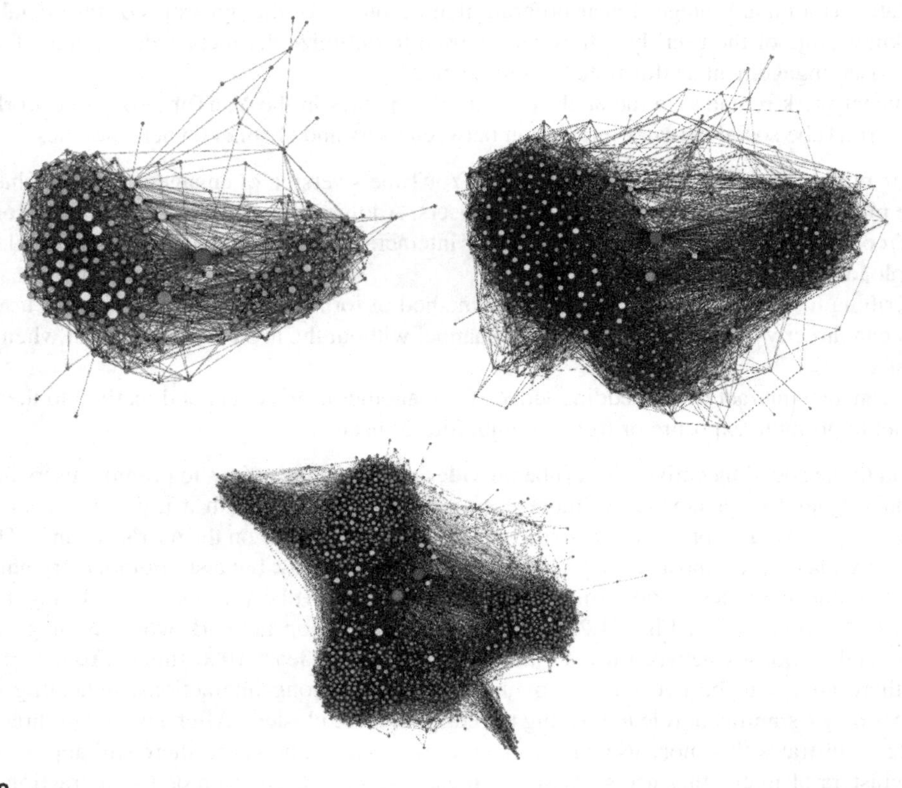

FIG. 21.22

Snapshots of the YouTube social network at 0 months, 3 months, and 6 months constructed using the dataset discussed in Section 21.5.1. Node sizes and color indicate their degree (in online version of chapter, *red* is higher, *blue* is lower). The complete social network is composed of more than 1.13 million users, and contains more than 2.98 million edges. Only users with at least 1000 edges are displayed. (A) Initial: 239 users, 3112 edges. (B) After three months: 503 users, 8558 edges. (C) After six months: 1391 users, 31,701 edges.

stems from predicting global behavior given local behavior: individual social sensors make decisions and learn from other social sensors and we are interested in understanding the behavior of the entire network. In Section 21.2 we showed that the global degree of infected nodes can be determined by mean field dynamics. In Section 21.3, it was shown that despite the apparent simplicity in information flows between social sensors, the global system can exhibit unusual behavior such as herding and data incest. Finally, in Section 21.4 a nonparametric method was used to determine the utility functions of a multiagent system—this can be used to predict the response of the system.

This chapter has dealt with social sensing issues of relevance to a signal processing audience. There are several topics of relevance to social sensors that are omitted due to space constraints, including:

- Coordination of decisions via game-theoretic learning [175–177] or Bayesian game models such as global games [178].
- Consensus formation over social networks and cooperative models of network formation [179,180].
- Small world models [181,182].
- Peer to peer media sharing [183,184].
- Privacy and security modeling [185,186].

REFERENCES

[1] Rosi A, Mamei M, Zambonelli F, Dobson S, Stevenson G, Ye J. Social sensors and pervasive services: approaches and perspectives. In: Proceedings of the 2011 IEEE international conference on pervasive computing and communications workshops (PERCOM workshops). IEEE; 2011. p. 525–30.

[2] Lee R, Sumiya K. Measuring geographical regularities of crowd behaviors for Twitter-based geo-social event detection. In: Proceedings of the 2nd ACM SIGSPATIAL international workshop on location based social networks. ACM; 2010. p. 1–10.

[3] Cheng Z, Caverlee J, Lee K. You are where you tweet: a content-based approach to geo-locating Twitter users. In: Proceedings of the 19th ACM international conference on information and knowledge management. ACM; 2010. p. 759–68.

[4] Trusov M, Bodapati AV, Bucklin RE. Determining influential users in internet social networks. J Market Res 2010;XLVII:643–58.

[5] Leskovec J, Adamic LA, Huberman BA. The dynamics of viral marketing. ACM Trans Web (TWEB) 2007;1(1):5.

[6] Sakaki T, Okazaki M, Matsuo Y. Earthquake shakes Twitter users: real-time event detection by social sensors. In: Proceedings of the 19th international conference on world wide web. New York, NY: ACM; 2010. p. 851–60.

[7] Bollen J, Mao H, Zeng X. Twitter mood predicts the stock market. J Comput Sci 2011;2(1):1–8.

[8] Pang B, Lee L. Opinion mining and sentiment analysis. Found Trends Inf Retr 2008;2(1–2):1–135.

[9] Asur S, Huberman BA. Predicting the future with social media. In: 2010 IEEE/WIC/ACM international conference on web intelligence and intelligent agent technology (WI-IAT), vol. 1. IEEE; 2010. p. 492–9.

[10] Ifrach B, Maglaras C, Scarsini M. Monopoly pricing in the presence of social learning. NET Institute Working Paper No 12-01; 2011.

[11] Luca M. Reviews, reputation, and revenue: the case of Yelp.com. Technical Report 12-016, Harvard Business School; 2011.

[12] Goel S, Watts D, Goldstein D. The structure of online diffusion networks. In: Proceedings of the 13th ACM conference on electronic commerce. ACM; 2012. p. 623–38.

[13] Sun E, Rosenn I, Marlow C, Lento TM. Gesundheit! modeling contagion through Facebook news feed. In: Proceedings of the 3rd international AAAI conference on weblogs and social media, San Jose, CA; 2009. p. 146–53.

[14] Romero DM, Meeder B, Kleinberg J. Differences in the mechanics of information diffusion across topics: idioms, political hashtags, and complex contagion on Twitter. In: Proceedings of the 20th international conference on world wide web, Hyderabad, India; 2011. p. 695–704.

[15] Chamley C. Rational herds: economic models of social learning. Cambridge University Press; 2004.

[16] López-Pintado D. Diffusion in complex social networks. Games Econ Behav 2008;62(2):573–90.

[17] Pastor-Satorras R, Vespignani A. Epidemic spreading in scale-free networks. Phys Rev Lett 2001;86(14):3200.
[18] Benaïm M, Weibull J. Deterministic approximation of stochastic evolution in games. Econometrica 2003;71(3):873–903.
[19] Banerjee A. A simple model of herd behavior. Q J Econ 1992;107(3):797–817.
[20] Bikchandani S, Hirshleifer D, Welch I. A theory of fads, fashion, custom, and cultural change as information cascades. J Polit Econ 1992;100(5):992–1026.
[21] Acemoglu D, Ozdaglar A. Opinion dynamics and learning in social networks. Dyn Games Appl 2011;1(1):3–49.
[22] Lobel I, Acemoglu D, Dahleh M, Ozdaglar A. Preliminary results on social learning with partial observations. In: Proceedings of the 2nd international conference on performance evaluation methodologies and tools. Nantes, France: ACM; 2007. p. 69.
[23] Acemoglu D, Dahleh M, Lobel I, Ozdaglar A. Bayesian learning in social networks. Working Paper 14040, National Bureau of Economic Research; 2008.
[24] Cover T, Hellman M. The two-armed-bandit problem with time-invariant finite memory. IEEE Trans Inf Theory 1970;16(2):185–95.
[25] Hellman ME, Cover TM. Learning with finite memory. Ann Math Stat 1970;41(3):765–82.
[26] Chamley C, Scaglione A, Li L. Models for the diffusion of beliefs in social networks: an overview. IEEE Signal Process Mag 2013;30(3):16–29.
[27] Krishnamurthy V, Poor HV. Social learning and Bayesian games in multiagent signal processing: how do local and global decision makers interact? IEEE Signal Process Mag 2013;30(3):43–57.
[28] Afriat S. The construction of utility functions from expenditure data. Int Econ Rev 1967;8(1):67–77.
[29] Varian H. The nonparametric approach to demand analysis. Econometrica 1982;50(1):945–73.
[30] Varian H. Price discrimination and social welfare. Am Econ Rev 1985:870–5.
[31] Varian H. Revealed preference and its applications. Econ J 2012;122(560):332–8.
[32] Varian H. Online ad auctions. Am Econ Rev 2009:430–4.
[33] Hoiles W, Aprem A, Krishnamurthy V. Engagement and popularity dynamics of YouTube videos and sensitivity to meta-data. IEEE Trans Knowl Data Eng 2017;7:1426–37.
[34] Shahrear T, Hoiles W, Krishnamurthy V. Adaptive scheme for caching YouTube content in a cellular network: machine learning approach. IEEE Access 2017;5:5870–81.
[35] Hoiles W, Gharehshiran ON, Krishnamurthy V, Đào ND, Zhang H. Adaptive caching in the YouTube content distribution network: a revealed preference game-theoretic learning approach. IEEE Trans Cogn Commun Netw 2015;1(1):71–85.
[36] Gürsun G, Crovella M, Matta I. Describing and forecasting video access patterns. In: 2011 Proceedings of INFOCOM. IEEE; 2011. p. 16–20.
[37] Pinto H, Almeida J, Gonçalves M. Using early view patterns to predict the popularity of YouTube videos. In: Proceedings of the sixth ACM international conference on web search and data mining. ACM; 2013. p. 365–74.
[38] Richier C, Altman E, Elazouzi R, Jimenez T, Linares G, Portilla Y. Bio-inspired models for characterizing YouTube viewcout. In: 2014 IEEE/ACM international conference on advances in social networks analysis and mining. IEEE; 2014. p. 297–305.
[39] Richier C, Elazouzi R, Jimenez T, Altman E, Linares G. Forecasting online contents' popularity; 2015. arXiv preprint arXiv:150600178.
[40] Zhang A. Judging YouTube by its Covers. Tech. rep.. Department of Computer Science and Engineering, University of California, San Diego; 2015. http://cseweb.ucsd.edu/jmcauley/cse255/reports/wi15/AngelZhang.pdf.
[41] Vega-Redondo F. Complex social networks, vol. 44. Cambridge University Press; 2007.
[42] Jackson MO. Social and economic networks. Princeton University Press; 2010.

[43] Easley D, Kleinberg J. Networks, crowds, and markets: reasoning about a highly connected world. Cambridge University Press; 2010.

[44] Chen KC, Chuang M, Poor HV. From technological networks to social networks. IEEE J Sel Areas Commun 2013;31(9):548–72.

[45] Krishnamurthy V, Poor HV. A tutorial on interactive sensing in social networks. IEEE Trans Comput Soc Syst 2014;1(1):3–21.

[46] Krishnamurthy V, Gharehshiran ON, Hamdi M. Interactive sensing and decision making in social networks. Found Trends Signal Process 2014;7(1–2):1–196.

[47] Granovetter M. Threshold models of collective behavior. Am J Sociol 1978;83(6):1420–43.

[48] Mossel E, Roch S. On the submodularity of influence in social networks. In: Proceedings of 39th annual ACM symposium on theory of computing. ACM; 2007. p. 128–34.

[49] Chen N. On the approximability of influence in social networks. SIAM J Discret Math 2009;23(3):1400–15.

[50] López-Pintado D. Contagion and coordination in random networks. Int J Game Theory 2006;34(3):371–81.

[51] Aldous D. Interacting particle systems as stochastic social dynamics. Bernoulli 2013;19(4):1122–49.

[52] Leskovec J. SNAP library. http://snap.stanford.edu/data/index.html.

[53] Erdös P, Rényi A. On random graphs, I. Publ Math Debr 1959;6:290–7.

[54] Barabasi A, Reka A. Emergence of scaling in random networks. Science 1999;286(5439):509.

[55] Elliott RJ, Aggoun L, Moore JB. Hidden Markov models—estimation and control. New York: Springer-Verlag; 1995.

[56] Krishnamurthy V. Partially observed Markov decision processes. Cambridge University Press; 2016.

[57] Benveniste A, Metivier M, Priouret P. Adaptive algorithms and stochastic approximations; Applications of mathematics, vol. 22. Springer-Verlag; 1990.

[58] Kushner HJ, Yin G. Stochastic approximation algorithms and recursive algorithms and applications. 2nd ed. Springer-Verlag; 2003.

[59] Yin G, Krishnamurthy V, Ion C. Regime switching stochastic approximation algorithms with application to adaptive discrete stochastic optimization. SIAM J Optim 2004;14(4):1187–215.

[60] Hethcote HW. The mathematics of infectious diseases. SIAM Rev 2000;42(4):599–653.

[61] Porter MA, Gleeson JP. Dynamical systems on networks: a tutorial, vol. 4. Springer; 2016.

[62] Hernández-González M, Basin MV. Discrete-time filtering for nonlinear polynomial systems over linear observations. Int J Syst Sci 2014;45(7):1461–72.

[63] Krishnamurthy V, Bhatt S, Pedersen T. Tracking infection diffusion in social networks: filtering algorithms and threshold bounds. IEEE Trans Signal Inf Process Netw 2017;3(2):298–315.

[64] Hamdi M, Krishnamurthy V, Yin G. Tracking a Markov-modulated stationary degree distribution of a dynamic random graph. IEEE Trans Inf Theory 2014;60(10):6609–25.

[65] Jackson MO, Rogers BW. Relating network structure to diffusion properties through stochastic dominance. BE J Theor Econ 2007;7(1).

[66] Ghoshal G, Chi L, Barabási AL. Uncovering the role of elementary processes in network evolution. Sci Rep 2013;3:2920.

[67] Culotta A. Lightweight methods to estimate influenza rates and alcohol sales volume from twitter messages. Lang Resour Eval 2013;47(1):217–38.

[68] Signorini A, Segre A, Polgreen P. The use of twitter to track levels of disease activity and public concern in the us during the influenza a H1N1 pandemic. PLOS One 2011;6(5):e19467.

[69] Zhang F, Luo J, Li C, Wang X, Zhao Z. Detecting and analyzing influenza epidemics with social media in China. In: Advances in knowledge discovery and data mining. Springer; 2014. p. 90–101.

[70] Culotta A. Towards detecting influenza epidemics by analyzing twitter messages. In: Proceedings of the first workshop on social media analytics. ACM; 2010. p. 115–22.

[71] Culotta A. Lightweight methods to estimate influenza rates and alcohol sales volume from twitter messages. Lang Res Eval 2013;47(1):217–38.

[72] Li J, Cardie C. Early stage influenza detection from twitter; 2013. arXiv preprint arXiv:13097340.
[73] Broniatowski D, Paul M, Dredze M. National and local influenza surveillance through twitter: an analysis of the 2012–2013 influenza epidemic. PLOS One 2013;8(12):e83672.
[74] Achrekar H, Gandhe A, Lazarus R, Yu S, Liu B. Twitter improves seasonal influenza prediction. In: HEALTHINF 2012 international conference on health informatics; 2012. p. 61–70.
[75] Achrekar H, Gandhe A, Lazarus R, Yu S, Liu B. Predicting flu trends using twitter data. In: 2011 IEEE conference on computer communications workshops. IEEE; 2011. p. 702–7.
[76] Christakis N, Fowler J. Social network sensors for early detection of contagious outbreaks. PLOS One 2010;5(9):e12948.
[77] Leskovec J, Faloutsos C. Sampling from large graphs. In: Proceedings of the 12th ACM SIGKDD international conference on knowledge discovery and data mining. Philadelphia: ACM press; 2006. p. 631–6.
[78] Dasgupta A, Kumar R, Sivakumar D. Social sampling. In: Proceedings of the 18th ACM SIGKDD international conference on knowledge discovery and data mining. Beijing: ACM; 2012. p. 235–43.
[79] Heckathorn DD. Respondent-driven sampling: a new approach to the study of hidden populations. Soc Probl 1997;44:174–99.
[80] Heckathorn DD. Respondent-driven sampling II: deriving valid population estimates from chain-referral samples of hidden populations. Soc Probl 2002;49:11–34.
[81] Lee S. Understanding respondent driven sampling from a total survey error perspective. Surv Pract 2009;2(6).
[82] Malekinejad M, Johnston L, Kendall C, Kerr L, Rifkin M, Rutherford G. Using respondent-driven sampling methodology for HIV biological and behavioral surveillance in international settings: a systematic review. AIDS Behav 2008;12(S1):105–30.
[83] Goel S, Salganik MJ. Respondent-driven sampling as Markov chain Monte Carlo. Stat Med 2009;28:2209–29.
[84] Lansky A, Abdul-Quader A, Cribbin M, Hall T, Finlayson T, Garffin R, et al. Developing an HIV behavioral surveillance system for injecting drug users: the National HIV Behavioral Surveillance System. Public Health Rep 2007;122(S1):48–55.
[85] Kleinberg J, Raghavan P. Query incentive networks. In: 46th annual IEEE symposium on foundations of computer science, 2005. FOCS 2005. IEEE; 2005. p. 132–41.
[86] Doucet A, Gordon N, Krishnamurthy V. Particle filters for state estimation of jump Markov linear systems. IEEE Trans Signal Process 2001;49:613–24.
[87] Logothetis A, Krishnamurthy V. Expectation maximization algorithms for MAP estimation of jump Markov linear systems. IEEE Trans Signal Process 1999;47(8):2139–56.
[88] Krishnamurthy V, Hamdi M. Mis-information removal in social networks: dynamic constrained estimation on directed acyclic graphs. IEEE J Sel Top Signal Process 2013;7(2):333–46.
[89] Tuttle B. Fact-checking the crowds: how to get the most out of hotel-review sites. Time Mag July 29, 2013.
[90] Adomavicius G, Tuzhilin A. Toward the next generation of recommender systems: a survey of the state-of-the-art and possible extensions. IEEE Trans Knowl Data Eng 2005;17(6):734–49.
[91] Konstas I, Stathopoulos V, Jose JM. On social networks and collaborative recommendation. In: Proceedings of the 32nd international ACM SIGIR conference on research and development in information retrieval. ACM; 2009. p. 195–202.
[92] Kanoria Y, Tamuz O. Tractable Bayesian social learning on trees. In: Proceedings of the IEEE international symposium on information theory (ISIT); 2012. p. 2721–5.
[93] Pearl J. Fusion, propagation, and structuring in belief networks. Artif Intell 1986;29(3):241–88.
[94] Murphy K, Weiss Y, Jordan M. Loopy belief propagation for approximate inference: an empirical study. In: Proceedings of the fifteenth conference uncertainty in artificial intelligence; 1999. p. 467–75.
[95] Yedidia J, Freeman W, Weiss Y. Constructing free-energy approximations and generalized belief propagation algorithms. IEEE Trans Inf Theory 2005;51(7):2282–312.

[96] Aumann RJ. Agreeing to disagree. Ann Stat 1976;4(6):1236–9.
[97] Geanakoplos J, Polemarchakis H. We can't disagree forever. J Econ Theory 1982;28(1):192–200.
[98] Borkar V, Varaiya P. Asymptotic agreement in distributed estimation. IEEE Trans Autom Control 1982;27(3):650–5.
[99] Cohn RA, Lewellen WG, Lease RC, Schlarbaum GG. Individual investor risk aversion and investment portfolio composition. J Financ 1975;30(2):605–20.
[100] Donkers B, Soest AV. Subjective measures of household preferences and financial decisions. J Econ Psychol 1999;20(6):613–42.
[101] Mitra S, Ji T. Risk measures in quantitative finance. Int J Bus Contin Risk Manag 2010;1(2):125–35.
[102] Rockafellar RT, Uryasev S. Optimization of conditional value-at-risk. J Risk 2000;2:21–41.
[103] Artzner P, Delbaen F, Eber JM, Heath D. Coherent measures of risk. In: Risk management: value at risk and beyond; 2002. p. 145.
[104] Palmquist J, Uryasev S, Krokhmal P. Portfolio optimization with conditional value-at-risk objective and constraints. Department of Industrial & Systems Engineering, University of Florida; 1999.
[105] Lim C, Sherali HD, Uryasev S. Portfolio optimization by minimizing conditional value-at-risk via nondifferentiable optimization. Comput Optim Appl 2010;46(3):391–415.
[106] Cartea A, Jaimungal S. Modelling asset prices for algorithmic and high-frequency trading. Appl Math Finance 2013;20(6):512–47.
[107] Neuts MF. Structured stochastic matrices of MG-1 type and their applications. Dekker; 1989.
[108] Krishnamurthy V. Bayesian sequential detection with phase-distributed change time and nonlinear penalty—a POMDP lattice programming approach. IEEE Trans Inf Theory 2011;57(10):7096–124.
[109] Krishnamurthy V. Quickest detection POMDPs with social learning: interaction of local and global decision makers. IEEE Trans Inf Theory 2012;58(8):5563–87.
[110] Krishnamurthy V, Bhatt S. Sequential detection of market shocks with risk-averse CVAR social sensors. IEEE J Sel Top Signal Process 2016;10(6):1061–72.
[111] Bond R, Fariss C, Jones J, Kramer A, Marlow C, Settle J, et al. A 61-million-person experiment in social influence and political mobilization. Nature 2012;489:295–8.
[112] Kempe D, Kleinberg J, Tardos E. Maximizing the spread of influence through a social network. In: Proceedings of the 9th ACM SIGKDD international conference on knowledge discovery and data mining, Washington, DC; 2003. p. 137–46.
[113] Karlin S. Total positivity, vol. 1. Stanford Univ.; 1968.
[114] Karlin S, Rinott Y. Classes of orderings of measures and related correlation inequalities. I. Multivariate totally positive distributions. J Multivar Anal 1980;10(4):467–98.
[115] Milgrom P. Good news and bad news: representation theorems and applications. Bell J Econ 1981;12(2):380–91.
[116] Topkis DM. Supermodularity and complementarity. Princeton University Press; 1998.
[117] Milgrom P, Shannon C. Monotone comparative statistics. Econometrica 1992;62(1):157–80.
[118] Amir R. Supermodularity and complementarity in economics: an elementary survey. South Econ J 2005;71(3):636–60.
[119] Hamdi M, Krishnamurthy V. Removal of data incest in multi-agent social learning in social networks; 2013. ArXiv e-prints .
[120] Surowiecki J. The wisdom of crowds. New York: Anchor; 2005.
[121] Anderson C, Kilduff GJ. Why do dominant personalities attain influence in face-to-face groups? The competence-signaling effects of trait dominance. J Pers Soc Psychol 2009;96(2):491–503.
[122] Ottaviani M, Sørensen P. Information aggregation in debate: who should speak first? J Public Econ 2001;81(3):393–421.
[123] Samuelson P. A note on the pure theory of consumer's behaviour. Economica 1938:61–71.
[124] Afriat S. Logic of choice and economic theory. Clarendon Press Oxford; 1987.

[125] Diewert W. Afriat and revealed preference theory. Rev Econ Stud 1973:419–25.
[126] Blundell R. How revealing is revealed preference? J Eur Econ Assoc 2005;3(2–3):211–35.
[127] Fostel A, Scarf H, Todd M. Two new proofs of Afriat's theorem. Econ Theory 2004;24(1):211–9.
[128] Boyd S, Vandenberghe L. Convex optimization. Cambridge University Press; 2004.
[129] Varian H. Non-parametric tests of consumer behaviour. Rev Econ Stud 1983;50(1):99–110.
[130] Deb R. Interdependent preferences, potential games and household consumption. MPRA Paper 6818, University Library of Munich, Germany; 2008.
[131] Chapman A, Verbic G, Hill D. A healthy dose of reality for game-theoretic approaches to residential demand response. In: 2013 IREP symposium bulk power system dynamics and control-IX optimization, security and control of the emerging power grid (IREP). IEEE; 2013. p. 1–13.
[132] Ibars C, Navarro M, Giupponi L. Distributed demand management in smart grid with a congestion game. In: 2010 first IEEE international conference on smart grid communications. IEEE; 2010. p. 495–500.
[133] Wu C, Mohsenian-Rad H, Huang J, Wang A. Demand side management for wind power integration in microgrid using dynamic potential game theory. In: 2011 IEEE GLOBECOM workshops; 2011. p. 1199–204.
[134] Rosenthal R. A class of games possessing pure-strategy Nash equilibria. Int J Game Theory 1973;2(1):65–7.
[135] Hayrapetyan A, Tardos E, Wexler T. The effect of collusion in congestion games. In: Proceedings of the thirty-eighth annual ACM symposium on theory of computing. STOC'06. ACM; 2006. p. 89–98. ISBN 1-59593-134-1.
[136] Babaioff M, Kleinberg R, Papadimitriou C. Congestion games with malicious players. In: Proceedings of the 8th ACM conference on electronic commerce. EC'07. New York, NY: ACM; 2007. p. 103–12. ISBN 978-1-59593-653-0.
[137] McGill P, Schumaker MF. Boundary conditions for single-ion diffusion. Biophys J 1996;71:1723–42.
[138] Varian H. Revealed preference. In: Samuelsonian economics and the twenty-first century; 2006. p. 99–115.
[139] Deb R. A testable model of consumption with externalities. J Econ Theory 2009;144(4):1804–16.
[140] Ui T. Correlated equilibrium and concave games. Int J Game Theory 2008;37:1–13.
[141] Neyman A. Correlated equilibrium and potential games. Int J Game Theory 1997;26(2):223–7.
[142] Carvajal A, Deb R, Fenske J, Quah J. Revealed preference tests of the Cournot model. Econometrica 2013;81(6):2351–79.
[143] Hoiles W, Krishnamurthy V. Nonparametric demand forecasting and detection of demand-responsive consumers. IEEE Trans Smart Grid 2015;6(2):695–704.
[144] Beigman E, Vohra R. Learning from revealed preference. In: Proceedings of the 7th ACM conference on electronic commerce. EC'06. New York, NY, USA: ACM; 2006. p. 36–42.
[145] Borle S, Boatwright P, Kadane J. The timing of bid placement and extent of multiple bidding: an empirical investigation using eBay online auctions. Stat Sci 2006:194–205.
[146] Ockenfels A, Roth A. The timing of bids in internet auctions: market design, bidder behavior, and artificial agents. AI Mag 2002;23(3):79.
[147] Pownall R, Wolk L. Bidding behavior and experience in internet auctions. Eur Econ Rev 2013;61:14–27.
[148] DellaVigna S. Psychology and economics: evidence from the field. J Econ Lit 2009;47(2):315–72.
[149] Leberknight C, Inaltekin H, Chiang M, Poor H. The evolution of online social networks: a tutorial survey. IEEE Signal Process Mag 2012;29(2):41–52.
[150] Granovetter M. Economic action and social structure: the problem of embeddedness. Am J Sociol 1985:481–510.
[151] Granovetter M. The impact of social structure on economic outcomes. J Econ Perspect 2005:33–50.
[152] Dave S, Sooriyabandara M, Zhang L. Application of a game-theoretic energy management algorithm in a hybrid predictive-adaptive scenario. In: 2011 2nd IEEE PES international conference and exhibition on innovative smart grid technologies (ISGT Europe); 2011. p. 1–6.

[153] Nguyen H, Song J, Han Z. Demand side management to reduce peak-to-average ratio using game theory in smart grid. In: 2012 IEEE conference on computer communications workshops (INFOCOM WKSHPS); 2012. p. 91–6.
[154] Giummolè F, Orlando S, Tolomei G. A study on microblog and search engine user behaviors: how Twitter trending topics help predict Google hot queries. HUMAN 2013;2(3):195.
[155] Mas-Colell A. On revealed preference analysis. Rev Econ Stud 1978:121–31.
[156] Forges F, Minelli E. Afriat's theorem for general budget sets. J Econ Theory 2009;144(1):135–45.
[157] Lahaie S. Kernel methods for revealed preference analysis. In: ECAI 2010: 19th European conference on artificial intelligence; 2010. p. 439–44.
[158] Krishnamurthy V, Hoiles W. Afriat's test for detecting malicious agents. IEEE Signal Process Lett 2012;19(12):801–4.
[159] Aprem A, Krishnamurthy V. Utility change point detection in online social media: a revealed preference framework. IEEE Trans Signal Process 2017;65(7):1869–80.
[160] Strebel J, Erdem T, Swait J. Consumer search in high technology markets: exploring the use of traditional information channels. J Consum Psychol 2004;14(1-2):96–104.
[161] Zhou X, Ye J, Feng Y. Tuberculosis surveillance by analyzing Google trends. IEEE Trans Biomed Eng 2011;58(8):2247–54.
[162] Adams A, Blundell R, Browning M, Crawford I. Prices versus preferences: taste change and revealed preference. Tech. rep.. IFS Working Papers; 2015.
[163] McFadden DL, Fosgerau M. A theory of the perturbed consumer with general budgets. Tech. rep.. National Bureau of Economic Research; 2012.
[164] Fudenberg D, Iijima R, Strzalecki T. Stochastic choice and revealed perturbed utility. Econometrica 2015;83(6):2371–409.
[165] Chasparis GC, Shamma JS. Control of preferences in social networks. In: 2010 49th IEEE conference on decision and control (CDC). IEEE; 2010. p. 6651–6.
[166] Granger CW. Investigating causal relations by econometric models and cross-spectral methods. Econometrica: J Econ Soc 1969:424–38.
[167] G M Ljung GEPB. On a measure of lack of fit in time series models. Biometrika 1978;65(2):297–303.
[168] Wald A. Sequential analysis. Dover; 1973.
[169] Jiang L, Miao Y, Yang Y, Lan Z, Hauptmann A. Viral video style: a closer look at viral videos on YouTube. In: Proceedings of international conference on multimedia retrieval. ACM; 2014. p. 193.
[170] Figueiredo F, Benevenuto F, Almeida J. The tube over time: characterizing popularity growth of YouTube videos. In: Proceedings of the fourth ACM international conference on web search and data mining. ACM; 2011. p. 745–54.
[171] Tartakovsky A, Nikiforov I, Basseville M. Sequential analysis: hypothesis testing and changepoint detection. CRC Press; 2014.
[172] Burer S, Letchford A. Non-convex mixed-integer nonlinear programming: a survey. Surv Oper Res Manag Sci 2012;17(2):97–106.
[173] Bertsekas DP. Nonlinear programming. Athena Scientific; 1999.
[174] Gharehshiran O, Hoiles W, Krishnamurthy V. Detection of homophilic communities and coordination of interacting meta-agents: a game-theoretic viewpoint. IEEE Trans Signal Inf Process Netw 2016;2(1):84–101.
[175] Hart S, Mas-Colell A. A simple adaptive procedure leading to correlated equilibrium. Econometrica 2000;68(5):1127–50.
[176] Hart S, Mas-Colell A, Babichenko Y. Simple adaptive strategies: from regret-matching to uncoupled dynamics; World Scientific Series in Economic Theory, vol. 4. World Scientific Publishing; 2013.
[177] Namvar O, Krishnamurthy V, Yin G. Distributed tracking of correlated equilibria in regime switching noncooperative games. IEEE Trans Autom Control 2013;58(10):2435–50.

[178] Angeletos G, Hellwig C, Pavan A. Dynamic global games of regime change: learning, multiplicity, and the timing of attacks. Econometrica 2007;75(3):711–56.
[179] Kozma B, Barrat A. Consensus formation on adaptive networks. Phys Rev E 2008;77(1):016102.
[180] Tahbaz-Salehi A, Jadbabaie A. Consensus over ergodic stationary graph processes. IEEE Trans Autom Control 2010;55(1):225–30.
[181] Watts DJ. Networks, dynamics, and the small-world phenomenon. Am J Sociol 1999;105(2):493–527.
[182] Kleinberg J. Navigation in a small world. Nature 2000;406(6798):845.
[183] Halvey MJ, Keane MT. Exploring social dynamics in online media sharing. In: Proceedings of 16th international conference on world wide web, Banff, AB, Canada; 2007. p. 1273–4.
[184] Zhao HV, Lin WS, Liu KJR. Behavior dynamics in media-sharing social networks. Cambridge University Press; 2011.
[185] Lange PG. Publicly private and privately public: social networking on YouTube. J Comput-Mediat Commun 2007;13(1):361–80.
[186] Liu L, Yu E, Mylopoulos J. Security and privacy requirements analysis within a social setting. In: Proceedings of 11th IEEE international requirements engineering conference, Monterey Bay, CA; 2003. p. 151–61.

CHAPTER

ACTIVE SENSING OF SOCIAL NETWORKS: NETWORK IDENTIFICATION FROM LOW-RANK DATA

22

Hoi-To Wai*, Anna Scaglione*, Amir Leshem†

School of Electrical, Computer and Energy Engineering, Arizona State University, Tempe, AZ, United States Faculty of Engineering, Bar-Ilan University, Ramat Gan, Israel†*

22.1 INTRODUCTION

Suppose you are trying to win an argument with a group of friends, advertising a product to a group of individuals, or lobbying with a number of decision-makers. As demonstrated by a number of works [1,2], the "optimal" strategy to maximize the chance of success depends on the *social network structure* and the parameters of the *underlying opinion dynamics*. The desire to fully exploit this network information in decision-making has motivated many studies on *modeling and learning* the social network, as we review in Section 22.1.1.

With the advent of online social networks such as Facebook, Twitter, and LinkedIn, the opportunities for human beings to share information are unprecedented, and the social network of interest now consists of a large number of potential agents. To our rescue, a huge amount of social interaction data has become available simultaneously, enabling researchers to apply sophisticated computation tools to explore social network data. A common approach in this field is first developing a simple statistical model that embeds the latent network structure, then applying statistical inference techniques to learn the social network.

In this chapter we analyze the social interaction data given as *steady-state opinions* of different agents on different topics. For example, the voting records in the US Senate can be seen as a manifestation of steady-state opinions because they represent the results of several prior interactions among the members of the senate and other stakeholders. In this realm, a simple model fails to capture many salient features exhibited by the real data. A prominent feature missing in previous work is that the steady-state opinions have naturally a low-rank structure. The voting records in the senate are a particularly acute example of this phenomenon because senators often choose to side with their own party. Note that learning the social network from low-rank data is difficult, as the social network contains a large number of agents and is a high-dimensional object. Having low-rank data prevents one from learning the social network effectively if one does not choose the appropriate framework.

Cooperative and Graph Signal Processing. https://doi.org/10.1016/B978-0-12-813677-5.00022-5

We address this challenge in this chapter by putting to use a suite of tailored system identification tools and system modeling. In particular, the first part of this chapter is concerned with the modeling of social networks, as described in Section 22.2. Our approach captures the low-rank feature of opinion data by modeling them as the result of the so-called *DeGroot* opinion dynamics model with *stubborn agents*. As suggested by their designation, the stubborn agents are zealots who do not change their opinions even when they enter into contact with others. This modified opinion dynamics model naturally gives rise to the low-rank structure with polarization phenomena described above.

Leveraging these model steady-state equations, the second part of this chapter addresses the the social network *identification* problem by describing the computational techniques and then provides theoretical identifiability guarantees. We call *network identification* the problem of learning *both* the topology and the relative strength of interactions between agents. In Section 22.3, we compare two network identification techniques. One is the popular graphical LASSO method, which exploits a model that ties the latent network to the opinion data correlation. The second one is the model-based learning method that performs a regression on the data using the DeGroot model with stubborn agents. The latter is cast into a *blind* compressive sensing problem and this allows Characterizing rigorously network recoverability guarantees. The identifiability condition is shown to depend implicitly on parameters of the network structure one wishes to recover and on the number of zealots that *excite* the social system. We show that if the number of stubborn agents in the network is proportional to the in-degree of the network, then it is possible to identify the network perfectly. Furthermore, we observe an interesting interplay between the network structure and its identifiability, in which the network can be identified most easily when the network topology is close to a regular graph. Finally, we remark that the model-based learning method can be seen as a general *network RADAR* methodology that is applicable to inferring more general data structures shaped by network interactions, such as gene regulatory networks.

22.1.1 MODELS FOR THE ANALYSIS OF OPINIONS

While it is possible to use nonparametric machine learning approaches to make an inference about group behavior on certain issues and decisions, there are two key reasons to prefer parametric approaches: (1) the parameter estimates provide insights that can be used to interpret the results; (2) the specific use of opinion dynamics models explains a recurrent feature found in real data: certain types of actions or ratings from a possibly vast group of agents tend to exhibit a low rank. This indicates that their decision-making process can be traced back to a belief system that is similar across different groups of agents.

The *social system identification* problem we pose in this chapter draws inspiration from opinion dynamics models that have been developed to interpret the trends observed in society for many years. These models are of interest in the emerging field of graph signal processing, as it pertains to solving social network inference problems. This section reviews briefly the key ones, with emphasis on those that are most relevant to our quest. Of course, unlike classical system identification problems, where laws of physics support the mathematical framework, all these models are not the ground truth. However, the working hypothesis is that whatever can mirror the behavior observed and can be used to make predictions or provide an approximation of the relationships observed, can be viewed as the mathematical law that generates the data structure, just as many other models in statistics that are used to fit empirically the data without a formal derivation that the phenomenon producing the data obeys those laws.

Social network modeling

The interest in modeling mathematically how individuals choose their actions under pressure from society is centuries old. The choice of a social agent is clearly not only a function of what the agent observes (i.e., its private information), but also the result of how the individual internalizes what other peers choose. The research published on social network modeling dates back to the beginning of the 20th century with the focus on explaining the phenomenon of *crowd wisdom* [3]. The *opinions* in the demonstration Galton made in [3] were the estimates of the weight of an ox that Galton averaged over several guesses reaching a remarkably close answer to the true weight. This phenomenon can be mathematically justified if one considers the estimate x_i as a noisy observation $x_i = x^* + w_i$ of the actual weight of the ox x^*, with unbiased error, and applies the law of large numbers to justify the convergence of the average to the true weight. This mathematical finding supported the idea that *voting* is a good mechanism to make decisions because it reduces the *noise* embedded in individual assessments.

The first opinion dynamics model that includes the social network structure explicitly was developed by DeGroot in the early 1970s [4]. The DeGroot model relies on the intuition that agents are influenced by their immediate neighbors and therefore change their opinions by taking a a convex combination of the neighbors' opinions. The DeGroot model considers continuous valued opinions and its analysis has a strong similarity to that of Markov Chains, particularly when the opinions $0 \leq x_i \leq 1$ represent the probability of a binary action, i.e., $x_i = P(a_i = 1)$, such as voting for a bill, or choosing a restaurant. Denoting by $\boldsymbol{x}(t)$ the vector of the agents opinions after the tth interaction, and by $\boldsymbol{A}$ the weighted adjacency matrix, containing in each row the coefficients of the aforementioned convex combination. From a graph signal processing perspective, the matrix $\boldsymbol{A}$ can be interpreted as a graph shift operator (GSO) and the DeGroot model is essentially a first order autoregressive graph filter:

$$\boldsymbol{x}(t) = \boldsymbol{A}\boldsymbol{x}(t-1). \tag{22.1}$$

If the network is connected, the DeGroot model leads to agreement in the opinion. Of course, social learning is affected often by pathological behavior. In some cases the agents may agree on an action that is completely wrong. This is the so-called phenomenon of *group think*, which was well documented by the cognitive studies of Asch [5]. To capture the inability of a group to exercise critical thinking and resist social pressure, in spite of the evidence for a preferable contrary option, Bikhchandani et al. [6] explained the phenomenon by showing how *rational herds* can form, even if the agents are optimally integrating information as Bayesian agents. Other related models include the voter model [7] and the SIS/SIR model [8] pertaining to discrete opinion dynamics.

There are two prevalent ways to model the pathological divisions that are observed in the beliefs or actions of social agents. One is by considering in the models the presence of zealots or stubborn agents, i.e., sociopaths who only trust their own opinion [9–11]. The second approach is embodied by the so-called models of *bounded confidence* that capture the fundamental lack of trust for those who are too different [12]. In particular, Acemoglu et al. [11,13], Yildiz and Scaglione [14] analyzed the behavior of the DeGroot model under the influences of the stubborn agents; Mobilia [9], Mobilia et al. [10], Yildiz et al. [15,16], and Waagen et al. [17] considered the voter model (or discrete opinion dynamics in general) in the presence of the stubborn agents. It is also worthwhile to mention the empirical studies in [18,19] that provide analysis on actual data to verify the opinion dynamics models.

We wish to add that herding phenomena and agreement or disagreement in beliefs are two steady-state distinct regimes. Discrete actions opinion models lead to herding, which means that at the steady state all agents subject to the dynamics end up taking an action with probability one, which, depending on the interaction model, can be exactly the same action or they can split in groups that take the same action. Disagreement in beliefs means, instead, that the agents have different probabilities of taking an action; this is what we are considering in this chapter as the basis for our network identification theory. This is not to say that discrete interaction models are not of interest in general, only that they go beyond the scope of this chapter.

Social network identification

The social network identification problem is a relatively new topic. The research to date can roughly be categorized into statistical inference and dynamics-based learning, as we shall survey below.

Statistical inference methods are widely adopted due to their simplicity and the rich theoretical guarantees. Their working hypothesis is that two agents are not connected directly in a social network if their opinions are statistically independent, when conditioned on the opinions of some other agent. An important consequence of this model is that the *inverse covariance matrix* has the same support as the adjacency matrix of the social network. This forms the basis for the popular graphical LASSO method [20]. Recently, methods have been proposed to analyzing *graph signals* in a similar vein. For example, Segarra et al. [21] has proposed to collect the spectral templates of the network topology from the output of a graph filter. In the class of graphical model methods, Tang et al. [22] proposed a transfer factor graph model to infer links on heterogenous social networks, Tang et al. [23] proposed a trusts evolution model for product review data, Bresler [24] considered the Ising's model and proposed a greedy algorithm that is guaranteed to find the n-nodes graph with $\Omega(\log n)$ samples, Etesami et al. [25] and He et al. [26] studied the Hawkes model and modeled the actions of social agents as an arrival process, and Pouget-Abadie and Horel [27] considered the network inference problem from observing a cascading process. Most of the methods above demand *big* and *high-rank* data, e.g., the data are observations of a random process excited by full-rank, independent random sources.

Model-based (or dynamics-based) learning interprets the interaction data as states of a dynamical system, where the latter is specified by a set of difference/differential equations. This fits into the hypothesis that the states of the agents/nodes in the networked system evolve with a fixed and known rule [28] and allows us to assign physical meaning to the network inference results. Like the statistical learning method described above, the general approach is to apply a linear regression to find the best fit network to the dynamical system. In the case of a sparse network, a sparsity enhancing regularizer, i.e., the popular ℓ_1 norm, may be adopted. These methods have been particularly successful in identifying network structures, e.g., oscillator networks [29], epidemic networks [30], social networks [31,32], and other networks [33,34]. Most of the above works focus on the network identification problem using *transient* data, with the exception of [29,33,35]. In most cases, a rigorous mathematical analysis on the performance of the proposed methods is lacking.

22.1.2 NOTATION

For any natural number $n \in \mathbb{N}$, we denote $[n]$ as the set $\{1, 2, \ldots, n\}$. Vectors (resp. matrices) are denoted by boldfaced letters (resp. capital letters). We denote x_i as the ith element of the vector $\boldsymbol{x}$, $[\boldsymbol{E}]_{\mathcal{S},:}$ (resp. $[\boldsymbol{E}]_{:,\mathcal{S}}$) denotes the submatrix of $\boldsymbol{E} \in \mathbb{R}^{m\times n}$ with only the rows (resp. columns) selected from $\mathcal{S} \subseteq [m]$ (resp. $\mathcal{S} \subseteq [n]$). Vector $\mathbf{e}_k \in \mathbb{R}^n$ is a unit vector with zeros everywhere except for the kth coordinate. The superscript $(\cdot)^\top$ denotes matrix/vector transpose. $\|\cdot\|_2$ denotes the Euclidean norm and $\|\cdot\|_1$ is the

ℓ_1-norm. The vectorization of a matrix $\boldsymbol{X} \in \mathbb{R}^{m_1 \times m_2}$ is denoted by $\text{vec}(\boldsymbol{X}) = [\boldsymbol{x}_1; \boldsymbol{x}_2; \ldots; \boldsymbol{x}_{m_2}] \in \mathbb{R}^{m_1 m_2}$ such that $\boldsymbol{x}_i$ is the ith column of $\boldsymbol{X}$.

22.2 DEGROOT OPINION DYNAMICS

This section reviews the classical DeGroot opinion dynamics that are central to our modeling of the opinion data. To establish our model, let us begin by considering a social network as being represented by a simple directed graph $G = (V, E)$, where $V = [n]$ is the set of agents and $E \subseteq V \times V$ is the edge set. We denote $(i, j) \in E$ if there exists an edge from agent i to j. The graph is also associated with a *trust* matrix between agents, denoted as $\boldsymbol{A} \in \mathbb{R}^{n \times n}$ such that $A_{ij} \geq \eta$ if and only if $(j, i) \in E$, for some $\eta > 0$. The matrix is normalized such that it is stochastic with row sums of ones, i.e., $\boldsymbol{A}\boldsymbol{1} = \boldsymbol{1}$.

There are K different topics discussed by the agents. Each discussion is indexed by $k \in [K]$. The opinion of the ith agent at discrete time t is denoted by a scalar $x_i(t; k)$[1] at time $t \in \mathbb{N}$ during the kth discussion. As the individuals' opinions are constantly influenced by the opinions of others, the DeGroot model postulates that the agents' opinions are updated according to the random process:

$$x_i(t; k) = \sum_{j \in \mathcal{N}_i} A_{ij} x_j(t-1; k). \tag{22.2}$$

In matrix form, the above can be written as

$$\boldsymbol{x}(t; k) = \boldsymbol{A}\boldsymbol{x}(t-1; k), \tag{22.3}$$

where we stack the vectors as $\boldsymbol{x}(t; k) = (x_1(t; k), \ldots, x_n(t; k))^\top$, $[\boldsymbol{A}]_{ij} = A_{ij}$ and $\boldsymbol{A}$ is nonnegative and stochastic, i.e., $\boldsymbol{A}\boldsymbol{1} = \boldsymbol{1}$ for all t, s. See [4] and [36, Chapter 1] for a detailed description.

We remark that it is possible to consider a setting with a time-varying trust matrix of random connectivity. In fact, our analysis can be extended as one replaces $\boldsymbol{A}$ in the previous equation with a random matrix $\boldsymbol{A}(t; k)$, which satisfies the following:

Assumption 22.1. *The matrix $\boldsymbol{A}(t; k)$ is an independently and identically distributed (i.i.d.) random matrix drawn from a distribution satisfying $\mathbb{E}[\boldsymbol{A}(t; k)] = \boldsymbol{A}$ for all $t \in \mathbb{N}, k \in [K]$, where the expectation is taken w.r.t. the realization of $\boldsymbol{A}(t; k)$.*

All the results that follow can be generalized to the stochastic interaction model just described but, for simplicity, we focus on the static case from now on. The random interaction model will be revisited later.

A well-known fact in the distributed control literature [37] is that the agents' opinions in Eq. (22.2) attain consensus as $t \to \infty$, i.e.,

$$\lim_{t \to \infty} \boldsymbol{x}(t; k) = \boldsymbol{1}\boldsymbol{c}^\top \boldsymbol{x}(0; k) \tag{22.4}$$

[1] For example, the ith agent's opinion $x_i(t; k)$ parameterizes a probability mass function of his/her stance on the kth discussion. While this chapter focuses on the case when $x_i(t; k)$ is a scalar, it should be noted that the techniques developed can be easily extended to the vector case.

for some $\boldsymbol{c} \in \mathbb{R}^n$. We have $x_i(\infty; k) = x_j(\infty; k)$ for all i, j for the steady state opinions. However, attaining consensus is clearly not the case in the real-world data, as the agents usually disagree with the others and the opinions exhibit clustering behavior.

22.2.1 EFFECTS OF STUBBORN AGENTS

We consider extending the social network G by appending S *stubborn agents* into the social network. Formally, stubborn agents (a.k.a. zealots) are members of a social network whose opinions cannot be swayed by others. If agent i is stubborn, then $x_i(t; k) = x_i(0; k)$ for all t. Adapting to the DeGroot opinion dynamics, these agents can be characterized by the structure of their respective rows in the trust matrix:

Definition 22.1. *An agent i is stubborn if and only if its corresponding row in the trust matrix* $\boldsymbol{A}$ *is the canonical basis vector; i.e., for all j,*

$$A_{ij} = \begin{cases} 1, & \text{if } j = i, \\ 0, & \text{otherwise.} \end{cases} \tag{22.5}$$

The extended social network G' consists of $n + S$ agents, denoted by $V' = [n + S]$ and the edge set is denoted by $E' \subseteq V' \times V'$. Without loss of generality, we let $V_s := [S]$ be the set of stubborn agents and $V_r := V \setminus V_s = \{1+S, \ldots, n+S\}$ be the set of regular agents. The trust matrix $\boldsymbol{A}$ can thus be partitioned as follows:

$$\boldsymbol{A} = \begin{pmatrix} \boldsymbol{I} & \boldsymbol{0} \\ \boldsymbol{B} & \boldsymbol{D} \end{pmatrix}, \tag{22.6}$$

where $\boldsymbol{B}$ and $\boldsymbol{D}$ are the submatrices of $\boldsymbol{A}$. The matrix $\boldsymbol{B}$ is the network between stubborn and regular agents, and $\boldsymbol{D}$ is the network among the regular agents themselves. See Fig. 22.1 for an illustration of the notations involved. We further impose the following assumptions:

Assumption 22.2. *Each agent in* V_r *has nonzero trust in at least one agent in* V_s.

Assumption 22.3. *The induced subgraph* $G'[V_r] = (V_r, E'(V_r))$ *is strongly connected.*

It can be shown that the two assumptions above imply that the principal submatrix $\boldsymbol{D}$ satisfies $\|\boldsymbol{D}\|_2 < 1$. We are interested in the steady state opinion resulting from Eq. (22.2) at $t \to \infty$, which can be characterized using the observation below.

Observation 22.1 ([14,38]). *Under Assumptions 22.2 and 22.3. Consider Eq. (22.2) and setting* $t \to \infty$, *we have:*

$$\lim_{t\to\infty} \boldsymbol{x}(t; k) = \boldsymbol{A}^{\infty}\boldsymbol{x}(0; k) \quad \text{where } \boldsymbol{A}^{\infty} := \begin{pmatrix} \boldsymbol{I} & \boldsymbol{0} \\ (\boldsymbol{I} - \boldsymbol{D})^{-1}\boldsymbol{B} & \boldsymbol{0} \end{pmatrix}. \tag{22.7}$$

We have the following observations from Eq. (22.7):

- The steady-state opinions depend solely on the stubborn agents and the structure of the network. Unlike the case *without* stubborn agents (cf. Eq. 22.4), information about the network structure $\boldsymbol{D}, \boldsymbol{B}$ is retained in Eq. (22.8).
- The range space of $\boldsymbol{A}^{\infty}$ has a dimension of at most S only. Because the number of stubborn agents S is usually much less than the number of regular agents, this implies that the steady-state opinion

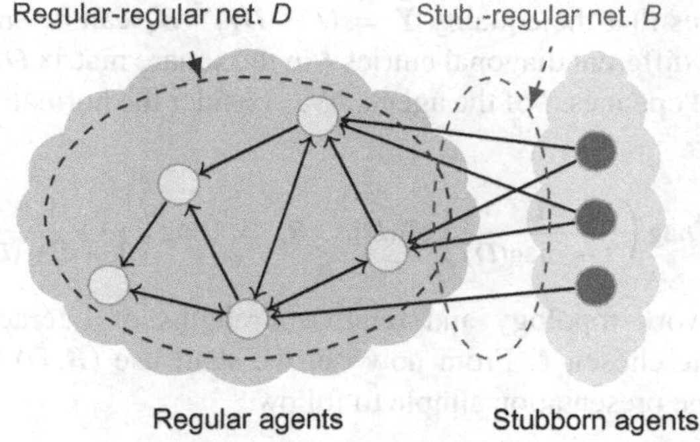

FIG. 22.1

Illustration of the social network structure. The extended social network with *stubborn agents* and the corresponding notations regarding the weighted adjacency matrix.

also lies in a low dimensional space. Importantly, we see that in the presence of zealots the DeGroot opinion dynamics naturally give rise to a *low-rank* structure in the steady-state opinion data.

We now describe an input-output relationship in the steady-state opinions generated from the DeGroot model with stubborn agents. Denote $\boldsymbol{x}^k := \lim_{t\to\infty} \boldsymbol{x}(t;k)$ as the steady-state opinions at the kth discussion and define the partition $\boldsymbol{x}^k := (\boldsymbol{z}^k, \boldsymbol{y}^k)^\top \in \mathbb{R}^{n+S}$ such that $\boldsymbol{z}^k \in \mathbb{R}^S$ and $\boldsymbol{y}^k \in \mathbb{R}^n$ are are the opinions of stubborn agents and regular agents, respectively. The input-output relationship holds:

$$\boldsymbol{y}^k = (\boldsymbol{I} - \boldsymbol{D})^{-1}\boldsymbol{B}\boldsymbol{z}^k, \tag{22.8}$$

where we can take $\boldsymbol{z}^k$ as an "input" to a linear system, $\boldsymbol{y}^k$ is the corresponding "output" and the matrix $(\boldsymbol{I} - \boldsymbol{D})^{-1}\boldsymbol{B}$ is the linear system.

Before concluding the section, we discuss a subtle yet relevant property of the steady-state output model (22.8).

Ambiguity. The input-output pair relationship (22.8) depends on the matrix product $(\boldsymbol{I}-\boldsymbol{D})^{-1}\boldsymbol{B}$, which naturally admits a scaling ambiguity. Precisely, we observe:

Observation 22.2. *Let $\boldsymbol{\ell} \in \mathbb{R}^n$ be bounded such that $\boldsymbol{0} \leq \boldsymbol{\ell} < \boldsymbol{1}$ and define the pair of matrix $(\boldsymbol{B}_{\boldsymbol{\ell}}, \boldsymbol{D}_{\boldsymbol{\ell}})$ such that:*

$$\boldsymbol{D}_{\boldsymbol{\ell}} = \mathrm{Diag}(\boldsymbol{\ell}) + \mathrm{Diag}\left(\frac{\boldsymbol{1}-\boldsymbol{\ell}}{\boldsymbol{1}-\mathrm{diag}(\boldsymbol{D})}\right)\mathrm{off}(\boldsymbol{D}), \quad \boldsymbol{B}_{\boldsymbol{\ell}} = \mathrm{Diag}\left(\frac{\boldsymbol{1}-\boldsymbol{\ell}}{\boldsymbol{1}-\mathrm{diag}(\boldsymbol{D})}\right)\boldsymbol{B}, \tag{22.9}$$

where $\mathrm{off}(\boldsymbol{D}) = \boldsymbol{D} - \mathrm{Diag}(\boldsymbol{D})$ and the fraction inside the bracket is an element-wise division. We have $(\boldsymbol{I}-\boldsymbol{D})^{-1}\boldsymbol{B} = (\boldsymbol{I}-\boldsymbol{D}_{\boldsymbol{\ell}})^{-1}\boldsymbol{B}_{\boldsymbol{\ell}}$.

The observation relies on simple linear algebra and can be found in [39]. In fact, the form of model ambiguity stated above can be understood as losing the information on the *rate of convergence* in the opinion dynamics, which is natural as the data $\boldsymbol{y}^k, \boldsymbol{z}^k$ recorded are the steady-state opinions.

Observation 22.2 implies that the equality $\tilde{Y} = (\boldsymbol{I} - \boldsymbol{D}_{\boldsymbol{\ell}})^{-1}\boldsymbol{B}_{\boldsymbol{\ell}}$ can be satisfied by infinitely many pairs of $(\boldsymbol{B}_{\boldsymbol{\ell}}, \boldsymbol{D}_{\boldsymbol{\ell}})$, each with different diagonal entries $\boldsymbol{\ell}$ in the square matrix $\boldsymbol{D}_{\boldsymbol{\ell}}$. As a remedy, we choose to emphasize the degree of "openness" of the agents. We consider the normalized network $(\boldsymbol{B}_{\boldsymbol{\ell}}, \boldsymbol{D}_{\boldsymbol{\ell}})$ with $\boldsymbol{\ell} = \boldsymbol{0}$, written as:

$$\boldsymbol{D}_0 := \text{Diag}\left(\frac{1}{1 - \text{diag}(\boldsymbol{D})}\right) \text{off}(\boldsymbol{D}), \quad \boldsymbol{B}_0 := \text{Diag}\left(\frac{1}{1 - \text{diag}(\boldsymbol{D})}\right) \boldsymbol{B}. \tag{22.10}$$

Notice that both the network topology and relative strengths of interaction between agents are preserved, regardless of the chosen $\boldsymbol{\ell}$. From now on, we shall use $(\boldsymbol{B}, \boldsymbol{D})$ to denote the normalized network $(\boldsymbol{B}_0, \boldsymbol{D}_0)$ to keep the presentation simple to follow.

22.3 NETWORK IDENTIFICATION

The last section described an opinion dynamics model that gives rise to a low rank data structure for the collection of its steady state opinions.

Our next endeavor is to introduce network identification methods to recover the structure and parameters in the social network. Specifically, given the DeGroot model, we focus on estimating $\boldsymbol{D}, \boldsymbol{B}$ from the observed opinions. We describe a state-of-the-art approach and a model-based learning approach that was introduced in [39], which was designed specifically for the opinion data originated from the DeGroot model. For the latter method, we demonstrate a sufficient condition such that the social network can be perfectly identified, even when the observed data has a significantly lower rank than the number of agents.

22.3.1 GRAPHICAL LASSO

The graphical LASSO (gLASSO) is a popular method introduced in [20] for inferring the latent structure of the random variables (r.v.s) generated from a Gaussian Markov random field (a.k.a. undirected graphical model) [40]. The method relies on the following observation—consider a random vector $\boldsymbol{X} \in \mathbb{R}^n$ generated from a graphical model with $G = (V, E)$ as the underlying graph and X_i is an r.v. associated with node $i \in V$. Assume that the covariance matrix $\boldsymbol{C}_X$ of $\boldsymbol{X}$ is full rank. If $(i, j) \notin E$ and $S \subseteq V$ is a graph cut between them, then X_i, X_j are independent when conditioned on $(X_s)_{s \in S}$. Furthermore, the conditional independence property can be captured by the support of the inverse of the covariance matrix such that $[\boldsymbol{C}_X^{-1}]_{ij} = 0$. In light of this, Friedman et al. [20] proposed the following gLASSO optimization:

$$\min_{\boldsymbol{A} \in \mathbb{R}^{n \times n}} \; -\log\det \boldsymbol{A} + \text{Tr}(\hat{\boldsymbol{C}}_X \boldsymbol{A}) + \rho \|\text{vec}(\boldsymbol{A})\|_1 \;\text{ s.t. }\; \boldsymbol{A} = \boldsymbol{A}^\top, \tag{22.11}$$

where $\rho > 0$ is a regularization parameter, $\hat{\boldsymbol{C}}_X$ is the empirical covariance matrix of $\boldsymbol{X}$, to approximate the inverse of $\boldsymbol{C}_X$ and therefore the connectivity of the graph G. The gLASSO problem (22.11) is essentially a penalized maximum likelihood method for the graphical model. In particular, if we set the regularization parameter as $\rho = \Theta(1/\sqrt{k})$ where k is the number of samples obtained for estimating $\boldsymbol{C}_X$, then the latent graph structure can be recovered in high probability [41].

We note that the covariance matrix computed from steady-state opinions of the DeGroot model does not fit naturally into the graphical model above. In particular, as discussed before, the obtained steady state opinions span only an S-dimensional space and the corresponding covariance matrix has rank at most S. Nevertheless, the gLASSO can still be applied as a heuristic to estimate the latent network structure of the social network. To obtain insights on its performance, let us assume that the initial opinions of the stubborn agents are white, i.e., $\mathbb{E}[z^k(z^k)^\top] = \boldsymbol{I}$. The covariance matrix of the regular agents' opinions can be expressed as:

$$\boldsymbol{C}_y = \mathbb{E}[\boldsymbol{y}^k(\boldsymbol{y}^k)^\top] = (\boldsymbol{I} - \boldsymbol{D})^{-1}\boldsymbol{B}\boldsymbol{B}^\top(\boldsymbol{I} - \boldsymbol{D})^{-\top}. \tag{22.12}$$

Note that $\boldsymbol{C}_y$ is rank-deficient, and its pseudoinverse (denoted by $(\cdot)^\dagger$) is given as:

$$\boldsymbol{C}_y^\dagger = (\boldsymbol{I} - \boldsymbol{D})(\boldsymbol{B}\boldsymbol{B}^\top)^\dagger(\boldsymbol{I} - \boldsymbol{D})^\top. \tag{22.13}$$

Effectively, solving the gLASSO problem (22.11) by setting $\hat{\boldsymbol{C}}_x = (1/K)\sum_{k=1}^K \boldsymbol{y}^k(\boldsymbol{y}^k)^\top$ as the empirical covariance matrix of the regular agents' opinions finds the sparsest positive semidefinite matrix that approximates Eq. (22.13). As $\boldsymbol{I} - \boldsymbol{D}$ is sparse, it is anticipated that the gLASSO method may be able to recover the support of $\boldsymbol{D}$ partially. However, there is no theoretical guarantee to its identifiability condition, even when the covariance $\boldsymbol{C}_y$ is estimated perfectly.

22.3.2 MODEL-BASED LEARNING

Unlike the gLASSO method, the authors considered a model-based learning method in [39] that exploits the structure in the observed opinions directly. In particular, we find that such an approach achieves perfect identification of the social network given a small number of stubborn agents, i.e., when the observed opinion data is low rank.

The method is motivated by the popular technique of sparse recovery using ℓ_1 minimization. Let $(\hat{\boldsymbol{B}}, \hat{\boldsymbol{D}})$ be an estimate of the tuple $(\boldsymbol{B}, \boldsymbol{D})$ and $\hat{\boldsymbol{b}}_i, \hat{\boldsymbol{d}}_i$ be the respective ith row vector. We consider the following ℓ_2 loss function [cf. Eq. 22.8]:

$$J_i(\hat{\boldsymbol{b}}_i, \hat{\boldsymbol{d}}_i) := \sum_{k=1}^K |\hat{\boldsymbol{b}}_i^\top \boldsymbol{z}^k + \hat{\boldsymbol{d}}_i^\top \boldsymbol{y}^k - y_i^k|^2. \tag{22.14}$$

We propose to identify the social network via solving the following problem:

NETWORK IDENTIFICATION PROBLEM

For all $i \in [n]$, we solve

$$\begin{aligned} \min_{\hat{\boldsymbol{b}}_i, \hat{\boldsymbol{d}}_i} \quad & J_i(\hat{\boldsymbol{b}}_i, \hat{\boldsymbol{d}}_i) + \rho \cdot g(\hat{\boldsymbol{b}}_i, \hat{\boldsymbol{d}}_i) \\ \text{s.t.} \quad & \hat{\boldsymbol{b}}_i \geq \boldsymbol{0},\ \hat{\boldsymbol{d}}_i \geq \boldsymbol{0},\ \hat{\boldsymbol{b}}_i^\top \boldsymbol{1} + \hat{\boldsymbol{d}}_i^\top \boldsymbol{1} = 1,\ [\hat{\boldsymbol{d}}_i]_i = 0, \end{aligned} \tag{P1}$$

where $\rho > 0$ is a regularization parameter, the last equality constraint forces the optimization to find the normalized network, and the regularizer $g(\hat{\boldsymbol{b}}_i, \hat{\boldsymbol{d}}_i)$ is chosen based on prior knowledge of the network.

We consider the following choices of the regularization functions $g(\hat{b}_i, \hat{d}_i)$:

$$g^1(\hat{b}_i, \hat{d}_i) := \|\hat{d}_i\|_1 + \mathcal{I}_{\Omega_{b_i}}(\hat{b}_i), \quad g^b(\hat{b}_i, \hat{d}_i) := \|\hat{b}_i\|_1, \tag{22.15}$$

where $\Omega_{b_i} = \{j \in [S] : [\boldsymbol{b}_i]_j = 0\}$ is the index set such that $\boldsymbol{b}_i$ has its zeros on and $\mathcal{I}_{\Omega_{b_i}}(\hat{b}_i)$ is an indicator function such that

$$\mathcal{I}_{\Omega_{b_i}}(\hat{b}_i) = \begin{cases} 0, & \text{if } [\hat{b}_i]_j = 0, \ \forall j \in \Omega_{b_i}, \\ \infty, & \text{otherwise.} \end{cases} \tag{22.16}$$

We refer to the setting when solving Eq. (P1) with $g^1(\cdot)$ as the "*Active Sensing*" case because the latter requires knowledge of the support set Ω_{b_i}. Such knowledge may be gained when the social network is actively probed, i.e., the stubborn agents are introduced intentionally in a social experiment setting. Nevertheless, even when such knowledge is unavailable, one can still infer the network by solving Eq. (P1) with $g^b(\cdot)$ as the regularizer, whose efficacy is confirmed in Section 22.4. Lastly, the network identification problem can be solved efficiently when done on a parallel computer, as the subproblems (P1) are decoupled from each other.

We remark that problem (P1), which is a natural formulation in light of Eq. (22.8), is similar in spirit to the reconstruction formulations considered in [29–34]. In particular, Bazerque et al. [33] studied a similar model to Eq. (22.8) and provided an identifiability condition in terms of the Kruskal rank, which can be difficult to verify. The following discussion establishes a sufficient condition for network identifiability in terms of the number of stubborn agents, S, present.

Network Identifiability Condition. Our next task is to study the network identifiability conditions. Specifically, we wish to derive the smallest possible number of stubborn agents and the corresponding configuration that guarantees perfect identifiability, which, in turn, represents also the lowest possible opinion data rank required. We assume:

Assumption 22.4. *The matrix $\boldsymbol{D} \in \mathbb{R}^{n \times n}$ is sparse and each row of it, $\boldsymbol{d}_i$, satisfies $\|\boldsymbol{d}_i\|_0 \leq d_{\max}$ for all $i \in [n]$.*

Assumption 22.5. *The observation model (22.8) is exact such that opinions are observed without noise and we observe opinions from $K \geq S$ topics.*

Assumption 22.6. *The support of the matrix $\boldsymbol{B}$, $\Omega_B := \{(i,j) : [\boldsymbol{B}]_{ij} = 0\}$, is known.*

In light of Assumptions 22.5 and 22.6, we shall study the the following network identification problem: for all $i \in [n]$,

$$\begin{aligned} \min_{\hat{b}_i, \hat{d}_i} \ \|\hat{d}_i\|_0 \ \text{s.t.} \ & \hat{b}_i \geq \boldsymbol{0}, \ \hat{d}_i \geq \boldsymbol{0}, \ \hat{b}_i^\top \boldsymbol{1} + \hat{d}_i^\top \boldsymbol{1} = 1, \ [\hat{d}_i]_i = 0, \\ & \hat{b}_i^\top \boldsymbol{z}^k + \hat{d}_i^\top \boldsymbol{y}^k = y_i^k, \ \forall k, \ [\hat{b}_i]_j = 0, \ \forall j \in \Omega_{b_i}, \end{aligned} \tag{22.17}$$

the above problem is similar to problem (P1) with the regularizer $g^1(\cdot)$. Analyzing the set of feasible solutions to Eq. (22.17) leads us to study the linear system:

$$\hat{b}_i^\top \boldsymbol{z}^k + \hat{d}_i^\top \boldsymbol{y}^k = y_i^k, \ \forall k \implies \boldsymbol{Z}^\top \hat{b}_i + \boldsymbol{Y}^\top \hat{d}_i = \boldsymbol{y}_i, \tag{22.18}$$

where $Z := (z^1, \ldots, z^K) \in \mathbb{R}^{S\times K}$, $Y := (y^1, \ldots, y^K) \in \mathbb{R}^{n\times K}$ and $y_i = (y_i^1, \ldots, y_i^K)$ is the ith row of Y. From Eq. (22.8), it holds that

$$YZ^{\dagger} = (I - D)^{-1}B, \tag{22.19}$$

where $Z^{\dagger}$ denotes the pseudoinverse of Z. Traditionally, analyzing the identifiability of Eq. (22.18) requires characterizing the *spark* of the resulting "sensing matrix" (see [42]). However, determining the spark of a matrix is nontrivial.

In fact, the system (22.18) is closely related to the problem of *compressed sensing*, as we consider the following alternative representation:

$$\begin{aligned} Z^{\top}\hat{b}_i + Y^{\top}\hat{d}_i = y_i &\Longleftrightarrow Z^{\top}\hat{b}_i + Y^{\top}(\hat{d}_i - e_i) = 0 \\ &\Longleftrightarrow \hat{b}_i + (YZ^{\dagger})^{\top}(\hat{d}_i - e_i) = 0 \\ &\Longleftrightarrow \hat{b}_i + B^{\top}(I - D)^{-\top}(\hat{d}_i - e_i) = 0 \\ &\Longleftrightarrow B^{\top}((I - D)^{-\top}(\hat{d}_i - e_i) + e_i) = b_i - \hat{b}_i \\ &\Longleftrightarrow B^{\top}(I - D)^{-\top}(\hat{d}_i - d_i) = b_i - \hat{b}_i. \end{aligned} \tag{22.20}$$

On the left side of Eq. (22.20), we note that due to the self-trust constraint $[\hat{d}_i]_i = 0$, the number of unknowns in d_i is $n-1$. On the right side of Eq. (22.20), the difference $b_i - \hat{b}_i$ is zero on indices j whenever $[b_i]_j = 0$, as the support of b_i is known to Eq. (22.17); otherwise, the terms on the right side are, in general, unknown.

In light of the above, a sufficient condition for network identification can be obtained by ignoring the rows in the linear system whenever $[b_i]_j \neq 0$. In particular, we require that the matrix obtained by deleting such rows from $B^{\top}(I - D)^{-\top} \in \mathbb{R}^{S\times n}$ to have a null space that consists only of *dense* vector. It follows that one could study the so-called restricted isometry property of such matrix. Before giving our identifiability condition, we provide two further remarks:

- Observe that $B^{\top}(I - D)^{-\top} = B^{\top}(I + D + D^2 + \ldots)^{\top}$; i.e., the sensing matrix (before row deletion) is a perturbed version of $B^{\top}$. When the perturbation induced by D is small, we could study B alone as the sensing matrix.
- There exists a trade-off between $|\Omega_{b_i}|$ and the identifiability performance. Notice that a sensing matrix's performance (e.g., as measured by the so-called restricted isometry property constant) is typically better if the matrix is dense. However, as indicated in Eq. (22.20), we need to ensure that there is a sufficient number of known observations (or zeros) in the right hand side of the underdetermined system (22.20), which is determined by $|\Omega_{b_i}|$.

The second remark prompts us to consider an optimized placement of stubborn agents when the matrix $B^{\top}$ is required to be sparse while maintaining a good sensing performance. As suggested in [43], a good choice is to construct B such that each row in B has a constant number d of nonzero elements. Our construction is stated in the following assumption:

Assumption 22.7. *The support of $B \in \mathbb{R}^{n\times S}$, i.e., Ω_B, is constructed such that each row of it has exactly ℓ nonzero elements, selected randomly and independently.*

The theorem below provides the condition on $d_{\max}$ and S such that the social network can be identified through Eq. (22.17):

A SUFFICIENT CONDITION FOR NETWORK IDENTIFICATION

Define $H(x)$ as the binary entropy function. We have:

Theorem 22.1. *Define* $\alpha := 2d_{\max}/n, b_{min} := \min_{ij\in\Omega_B} B_{ij}, b_{max} := \max_{ij\in\Omega_B} B_{ij}, \beta := S/n$ *and* $\beta' := \beta - \ell/n$. *Under Assumptions 22.4 and 22.7 and the conditions*

$$\ell > \max\left\{4, 1 + \frac{H(\alpha) + \beta' H(\alpha/\beta')}{\alpha \log(\beta'/\alpha)}\right\}, \; b_{min}(2d-3) - 1 - 2b_{max} > 0. \tag{22.21}$$

Then, as $n \to \infty$*, there is a unique optimal solution to Eq. (22.17) that* $(\hat{\boldsymbol{b}}_i, \hat{\boldsymbol{d}}_i) = (\boldsymbol{b}_i, \boldsymbol{d}_i)$ *with probability one. Moreover, the failure probability is bounded as:*

$$Pr\Big((\hat{\boldsymbol{b}}_i, \hat{\boldsymbol{d}}_i) \neq (\boldsymbol{b}_i, \boldsymbol{d}_i), \; \forall\, i \in [n]\Big) \leq \left(\frac{\ell}{\beta}\right)^4 \frac{\ell - 1}{n^2} + \mathcal{O}(n^{2-(\ell-1)(\ell-3)}). \tag{22.22}$$

The proof of Theorem 22.1 can be found in [39]. The claim is proven by treating the unknown entries of $\boldsymbol{B}^\top$ as *erasure bits*, and showing that the sensing matrix with erasure corresponds to a high-quality expander graph.

The first condition in Eq. (22.21) provides a guideline for determining the number of stubborn agents S needed and the role played by the sparsity parameter ℓ for $\boldsymbol{B}$. To gain some intuition, consider the situation where $n \to \infty$, $\beta', \alpha \to 0$ while the ratio β'/α is constant; then, the second condition in Eq. (22.21) can be approximated by

$$\ell > \max\left\{4, 1 + \frac{\beta'}{\alpha}\frac{H(\alpha/\beta')}{\log(\beta'/\alpha)}\right\}, \tag{22.23}$$

where the right hand side is minimized by $\beta'/\alpha \approx 1.27$ and requiring $\ell > 4.362$. Hence, setting $\ell = 5$ so this condition holds, the stubborn agents needed are:

$$S \geq \beta n = 5 + \beta' n \geq 5 + 1.27\alpha n \geq 5 + 2.54 d_{\max} = \Omega(d_{\max}). \tag{22.24}$$

On the other hand, the second condition in Eq. (22.21) indicates that the amount of *relative trust* on the stubborn agents in the paper cannot be too small. This is reasonable because the network identification performance should depend on the degree of influence of the stubborn agents relative to everyone else. Table 22.1 gives a list of the values required for β' and subsequently the required number of stubborn agents can be derived. Note that the number of stubborn agents required is still large. However, as this number only corresponds to a sufficient condition for perfect network identification, in practice the model-based method also provides good performance when this condition is significantly relaxed.

We remark that the probability bound in Eq. (22.22) is associated with the random construction of Ω_B in Assumption 22.7. In particular, when n is finite, this failure probability grows with the size of $\Omega_{\boldsymbol{B}}$, i.e., $\mathcal{O}(\ell^5)$. This indicates a possible tradeoff between the size of $\Omega_{\boldsymbol{B}}$ and the identification accuracy. We conclude this section by showing how the learning method we described can be adapted to establish the identifiability of random graph models and how to deal with randomized interactions.

Random Graph Models. In the previous section we provided a sufficient condition for identifiability based on specific parameters of a given network. Next, we show that for Erdos-Renyi (ER) random

Table 22.1 Evaluating the Minimum β' Required by Eq. (22.21) for the Sufficient Condition of Perfect Network Identification With Different Combinations of ℓ, α

	$\alpha = 0.08$	$\alpha = 0.16$	$\alpha = 0.24$	$\alpha = 0.32$	$\alpha = 0.40$
$\ell = 5$	0.3420	0.5280	0.6730	0.7940	0.8950
$\ell = 6$	0.2340	0.3850	0.5100	0.6190	0.7160
$\ell = 7$	0.1870	0.3190	0.4330	0.5360	0.6290

Note that $\beta' > \alpha$ and the number of stubborn agents required can be evaluated as $S \approx \beta' n + \ell$ and the maximum in-degree required is $\alpha n/2$.

networks, given a certain edge probability, any sufficiently large set of stubborn agents will suffice to identify the network with high probability. Furthermore, we show that the fraction of the number of stubborn agents vanishes compared to the network size, as the network size increases.

For notational simplicity, we shall set $m = n + S$ in the following. Because social networks are inherently directed, we consider the directed ER random graph $D_{m,p}$ where each of the $m(m-1)$ pairs has probability p of being a directed edge. In addition, our interest is in weighted graphs where the weight described the level of influence: therefore, we assume that A_{ij} is a random variable that is 0 if $(i,j) \notin D_{m,p}$ and otherwise has an arbitrary marginal distribution, compatible with the fact that their row sums lie on the simplex. Assume that we have a weighted directional network $G = (V_m, E_m, A)$, where the probability of a link is $p(m)$ depends on m. We would like to randomly select a subset $V_s \subset V_m$ of size s. V_s will be the set of stubborn agents used to identify the network actively. To ensure that the network is connected with high probability, it suffices that [44]

$$p(m) \geq (1+\epsilon)\frac{\log m}{m} \tag{22.25}$$

for any $\epsilon > 0$. Note that $p(m) \to 0$ as $m \to 0$. We would like to ensure that the identifiability conditions in Theorem 22.1 also hold with high probability. In particular, let us fix $\ell = 5$ such that each nonstubborn node is influenced by exactly $\ell = 5$ stubborn nodes. The ER model is consistent with having the stubborn nodes choose independently and at random the nodes they influence, but we need to make sure that there are at least $\ell = 5$ connection in the random ER graph between each stubborn node and randomly selected neighbors that are nonstubborn that can be activated. Hence, in addition to Eq. (22.25), an ER graph meets the assumptions in Theorem 22.1 if, with high probability:

1. The number of stubborn nodes satisfies Eq. (22.24), i.e., it is at least greater than a constant factor multiplied to the maximal in-degree of a nonstubborn node with $s \geq 5 + 2.54 d_{\max}$.
2. In the bipartite graph $G_B = (V_s, V_n, E \cap (V_s \times V_n))$ the in-degree of each nonstubborn node is at least 5.

Because the in-degree of a node is a binomial random variable, $d_i \sim Bin(m, p(m))$, we know that it will concentrate around $m \cdot p(m)$. The expected number of links between each nonstubborn node and the stubborn nodes is $S \cdot p(m)$, where $S = |V_s|$. The maximal in-degree of a nonstubborn node is distributed like the maximum of n independent random variables each having expected value of $m \cdot p(m)$. To bound the tail of $d_{\max}$ and of the probability that a nonstubborn node has less than $\ell = 5$ stubborn nodes in the group, we use the union bound and the following theorem [45]:

Theorem 22.2. *Let Y be Bin(n, p) and let $\alpha > 0$. Then:*

$$\begin{aligned} P(Y > (1+\alpha)pn) &< e^{(\alpha-1)pn}(\alpha)^{-\alpha pn}, \\ P(Y < (1-\alpha)pn) &< e^{-\frac{\alpha^2 pn}{2}}. \end{aligned} \tag{22.26}$$

To facilitate our analysis, we assume that the number of stubborn agents satisfies:

$$s(m) \geq 5 + 2.54 \cdot (1 + \xi(m)) \cdot p(m) \cdot n(m) \tag{22.27}$$

for some $\xi(m) > 0$ that will be determined later. Note that $m = n(m) + s(m)$ and our choice for the number of stubborn nodes is dependent on m. Moreover, we shall set $\epsilon = 1$ in Eq. (22.25).

Our first task is to upper bound the error probability that $s(m) < 5 + 2.54 d_{\max}$. To this end, applying the union bound to all nonstubborn nodes we get:

$$\begin{aligned} P(s(m) < 5 + 2.54 d_{\max}) &\leq n \cdot P\left(\frac{1}{2.54}(s(m) - 5) < d_1\right) \\ &\leq n(m) \cdot P((1 + \xi(m)) \cdot p(m) \cdot n(m) < d_1) \\ &\leq n(m) \cdot e^{(\xi(m)-1)p(m)n(m) - \log(\xi(m))\xi(m)p(m)n(m)}, \end{aligned} \tag{22.28}$$

where we have applied the first inequality in Theorem 22.2 and the fact that d_1 is a binary random variable with $Bin(n(m), p(m))$. Setting $\xi(m) > e$ gives:

$$P(s(m) < 5 + 2.54 d_{\max}) \leq n(m) \cdot e^{-p(m)n(m)} \leq \frac{n(m)}{m^{(n(m)/m)(1+\epsilon)}}. \tag{22.29}$$

To verify the last identifiability condition we will need a larger number of stubborn agents. Notice that the we require each nonstubborn node in the bipartite graph G_B to have at least $\ell = 5$ stubborn neighbors. To this end, we need to upper bound the probability that a nonstubborn node does not have $\ell = 5$ stubborn neighbors. In fact, the expected number of stubborn neighbors of any nonstubborn node in G_B is $s(m) \cdot p(m)$. Now, let $\mathcal{N}_{B,i}$ be the neighborhood set of the nonstubborn node i in G_B, applying Theorem 22.2 and setting $\alpha = (1 - 5/(s(m) \cdot p(m)))$ gives:

$$P\left(|\mathcal{N}_{B,i}| < 5\right) = P\left(|\mathcal{N}_{B,i}| < (1-\alpha)p(m)s(m)\right) \leq e^{-(\alpha)^2 p(m)s(m)/2}, \tag{22.30}$$

where we have used the fact that $|\mathcal{N}_{B,i}|$ is a random variable with $Bin(s, p(m))$.

We want to select s such that the right side of Eqs. (22.29) and (22.30) decays to zero as $m \to \infty$ while satisfying Eq. (22.27). We observe that a possible choice for $s(m)$ is given by

$$s(m) = \frac{m \log\log m}{2\log m}. \tag{22.31}$$

To verify the above choice of $s(m)$, first it can be checked that as $m \to \infty$, the above choice of s will always satisfy Eq. (22.27). Second, as $s/m \to 0$, we have $n/m \to 1$ and thus the right side of Eq. (22.29) decays to zero. Lastly, it is easy to check that

$$\lim_{m\to\infty} s(m) \cdot p(m) = \infty, \tag{22.32}$$

and thus $\alpha = (1 - 5/(s(m) \cdot p(m))) \to 1$. As a result, the right side of Eq. (22.30) also decays to zero. We summarize the above discussion with the following theorem:

Theorem 22.3. *Assume that $G = (V_m, E_m, \boldsymbol{A})$ is a directed weighted ER graph where $p(m) > 2\frac{\log m}{m}$, and we randomly pick a set of $s = \frac{m \log\log m}{2\log m}$ agents as stubborn agents. Then with probability approaching one, the network is connected and there is a set of stubborn agents that can identify the network.*

The theorem is actually pessimistic regarding the number of stubborn agents required. If, for example, the sparsity is fixed and the number of neighbors is a fraction of the network size, then we can use a logarithmic number of stubborn agents. Another important observation is that when we use stubborn agents we make an effort to cause them to influence a large number of nonstubborn agents. In this case they will not follow the general population Erdos-Renyi statistics and our assumptions regarding the size of the set of stubborn agents can be significantly relaxed.

Finally, we emphasize that at the current point the theorem does not prove that with high probability *any* set of stubborn agents of the given size will be able to identify the network. This will require significant strengthening of the current arguments, because the number of subsets of a given size is quite large. An interesting problem will be to identify such a set of stubborn agents for a given network.

Random Opinion Dynamics. So far, our network identification method requires collecting the steady-state opinions $(\boldsymbol{z}^k, \boldsymbol{y}^k)$ resulting from a static opinion dynamics model. A more realistic setting is to consider a randomized opinion dynamics. Importantly, we recall Assumption 22.1 and the following randomized dynamics:

$$\boldsymbol{x}(t+1;k) = \boldsymbol{A}(t;k)\boldsymbol{x}(t;k), \quad \text{where } \boldsymbol{A}(t;k) = \begin{pmatrix} \boldsymbol{I} & \boldsymbol{0} \\ \boldsymbol{B}(t;k) & \boldsymbol{D}(t;k) \end{pmatrix}. \tag{22.33}$$

In the same spirit, we also define $\mathbb{E}[\boldsymbol{B}(t;k)] = \boldsymbol{B}$ and $\mathbb{E}[\boldsymbol{D}(t;k)] = \boldsymbol{D}$. Now, let us reexamine the requirements on the opinion data. From Eq. (22.8), as one wishes that the collected data $(\boldsymbol{y}^k, \boldsymbol{z}^k)$ from the kth discussion to satisfy $\boldsymbol{y}^k = (\boldsymbol{I} - \boldsymbol{D})^{-1}\boldsymbol{B}\boldsymbol{z}^k$. Naturally, this can be obtained by taking the following expectations:

$$\boldsymbol{y}^k = \mathbb{E}[\boldsymbol{y}(\infty;k)|\boldsymbol{z}^k] \quad \text{and} \quad \boldsymbol{z}^k = \mathbb{E}[\boldsymbol{z}(t;k)]. \tag{22.34}$$

However, in practice, this may be difficult to realize as computing the expectation requires taking an average over the ensemble of the sample paths of $\{\boldsymbol{A}(t;k)\}_{\forall t,\forall k}$.

Instead of proceeding with Eq. (22.34), we prove that the randomized opinion dynamics are an ergodic process and replace Eq. (22.34) with a *time average*. To fix the idea, let us consider a noisy observation model on the opinions:

$$\hat{\boldsymbol{x}}(t;k) = \boldsymbol{x}(t;k) + \boldsymbol{n}(t;k). \tag{22.35}$$

Now, suppose that we accrue a time series of the opinions $\{\hat{\boldsymbol{x}}(t;k)\}_{t\in\mathcal{T}_k}$, where $\mathcal{T}_k \subseteq \mathbb{N}$ is an arbitrary sampling set. We define:

$$\hat{\boldsymbol{x}}(\mathcal{T}_k;k) \triangleq \frac{1}{|\mathcal{T}_k|}\sum_{t\in\mathcal{T}_k} \hat{\boldsymbol{x}}(t;k) \approx \mathbb{E}[\boldsymbol{x}(\infty;k)|\boldsymbol{x}(0;k)]. \tag{22.36}$$

Specifically, the temporal opinions samples are collected through random (and possibly noisy) sampling at time instances on the opinions. The following theorem characterizes the performance of Eq. (22.36):

Proposition 22.1. *Consider the estimator in Eq. (22.36) and we denote* $\bar{x}(\infty; k) \triangleq \lim_{t\to\infty} \mathbb{E}\{x(t; k) | x(0; k)\} = A^{\infty} x(0; k)$. *Assume that* $\mathbb{E}\{\|D(t; k)\|_2\} < 1$. *If* $T_o \to \infty$,

1. *then the estimator (22.36) is unbiased:*

$$\mathbb{E}[\hat{x}(\mathcal{T}_k; k) | z(0; k)] = \bar{x}(\infty; k). \tag{22.37}$$

2. *then the estimator (22.36) is asymptotically consistent:*

$$\lim_{|\mathcal{T}_k|\to\infty} \mathbb{E}[\|\hat{x}(\mathcal{T}_k; k) - \bar{x}(\infty; k)\|_2^2 | x(0; k)] = 0. \tag{22.38}$$

For the latter case, we have

$$\mathbb{E}\Big[\|\hat{x}(\mathcal{T}_k; k) - \bar{x}(\infty; k)\|_2^2 \mid x(0; k)\Big] \le \frac{C'}{|\mathcal{T}_k|}\left(\sum_{i=0}^{|\mathcal{T}_k|-1} \lambda^{\min_\ell |t_{\ell+i} - t_\ell|}\right), \tag{22.39}$$

where $C' < \infty$ *is a constant and* $\lambda = \lambda_{max}(D) < 1$, *i.e., the latter term is a geometric series with bounded sum.*

22.4 NUMERICAL EXPERIMENTS

This section includes numerical experiments of the network identification methods we discussed in detail, corroborating the theoretical claims made previously. To set up the experiments, we assume that the DeGroot model is exact and generate the synthetic opinions according to Eq. (22.8). We assume that the identity of the stubborn agents is known a priori and their initial opinions are generated as uniform random variables in [0, 1], except for one of the weights, which is set such that they add up to one. For the network identification problem (P1), we set $\rho = 10^{-1}$ and examine with different choices of regularization $g(\cdot)$.

Throughout this section, we consider the following metrics for performance comparison. First, to compare the network identification performance we evaluate the normalized mean squared error (NMSE), defined as:

$$\text{NMSE} := \|D - \hat{D}\|_F^2 / \|D\|_F^2. \tag{22.40}$$

For the network topology recovery performance, we evaluate the area under ROC (AUROC) and the area under precision-recall curve (AUPR). Notice that if AUROC and AUPR approach 1, then the topology recovery performance is perfect.

Model-based learning versus gLASSO. The first numerical example compares the model-based learning approach to the conventional gLASSO method for learning a sparse graphical model, in terms of the network topology recovery performance. We perform 100 Monte Carlo trials to evaluate the average network topology recovery performance. For each of these trials, the regular agents' network has $n = 100$ agents and is generated as an ER graph with connectivity $p = 0.08$; for the stubborn-regular network (corresponding to the matrix $B \in \mathbb{R}^{n\times S}$), for each regular agent, we randomly select

$\ell = 5$ stubborn agents to connect the regular agent to; note that this construction corresponds to that in Assumption 22.7. Finally, uniformly distributed edge weights are assigned to the edges and the resultant weighted adjacency matrix $\boldsymbol{A}$ is normalized to have its rows sum to one.

The result, presented in Fig. 22.2, compares the average network topology recovery performances against the number of stubborn agents S. We observe that the topology recovery performance improves with the number of stubborn agents, i.e., when the rank of the observed data improves. In particular, a good performance (AUPR/AUROC ≈ 0.99) is observed with $S \approx 40$ using the model-based learning approach (cf. Eq. P1). We also observe that the gLASSO method yields a worse performance than the model-based learning method as the former does not account for the low rankness in the observed data.

Identifiability condition. Our second numerical example examines the impact of the identifiability condition given in Theorem 22.1. We consider a similar setting as in the previous example. In addition, we also compare a random ER graph-like construction for the stubborn-normal agent network, as detailed in the figure's caption.

Fig. 22.3 shows the performance comparison, where we present the NMSE of $\boldsymbol{D}$ and AUROC against the number of stubborn agents S. As the regular-regular network is an ER graph with connectivity $p = 0.08$ and $n = 100$, this setting can be approximated by the case with $\ell = 5, d_{\max} = 8$ in Table 22.1. Theorem 22.1 predicts that with $S \approx 57$ stubborn agents and if Assumption 22.7 holds, then the social network can be perfectly identified by solving Eq. (22.17), where the latter is approximated by Eq. (P1) with $g^1(\cdot)$. The NMSE comparison corroborates with the theorem. In particular, with $S \approx 53$ stubborn agents, the averaged NMSE approaches 10^{-15}, i.e., the machine

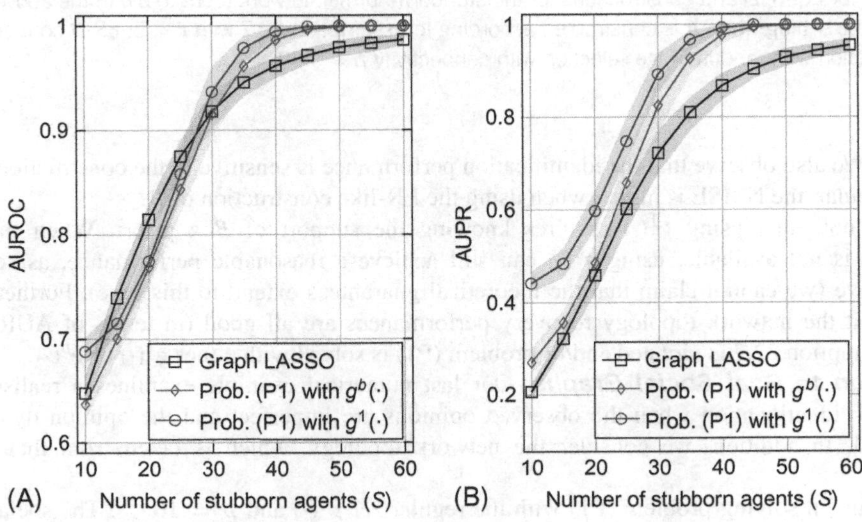

FIG. 22.2

Topology recovery performance. Comparing the topology recovery performance against the number of stubborn agents S. The regular-regular network is a 100-nodes ER graph with connectivity $p = 0.08$. The *shaded area* shows the 5%/95% percentile interval for the AUROC/AUPR performances. The model-based learning (P1) is benchmarked with two choices of regularizers—$g^1(\cdot)$ and $g^b(\cdot)$ from Eq. (22.15).

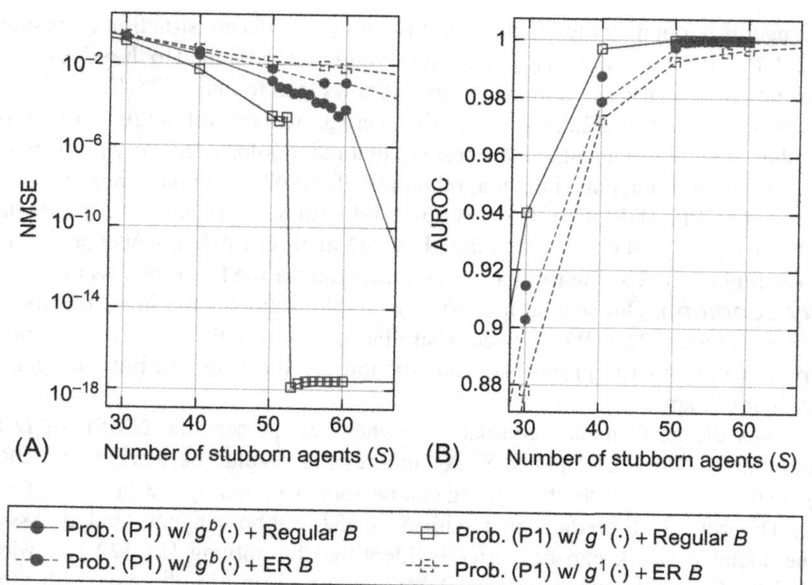

FIG. 22.3

Verifying the identifiability condition. Comparing the network identification performance using different regularizers for Eq. (P1) and constructions for the stubborn-normal network (corr. to $\boldsymbol{B}$). In the above, Regular $\boldsymbol{B}$ refers to the setting when $\boldsymbol{B}$ is constructed according to Assumption 22.7 with $\ell = 5$; ER $\boldsymbol{B}$ corresponds to the construction with random edge selection with connectivity $p = 0.08$.

precision. We also observe that the identification performance is sensitive to the construction model of $\boldsymbol{B}$. In particular, the NMSE is higher when using the ER-like construction of $\boldsymbol{B}$.

Lastly, note that using $g^1(\cdot)$ requires knowing the support of $\boldsymbol{B}$ a priori. When the support knowledge is not available, using $g^b(\cdot)$ can still achieve a reasonable performance, as seen in the figures above (we cannot claim that the theoretical guarantees extend to this case). Furthermore, we observe that the network topology recovery performances are all good (in terms of AUROC) even when Assumption 22.7 is violated and/or problem (P1) is solved with either $g^1(\cdot)$ or $g^b(\cdot)$.

Application to Real Social Graphs. Our last numerical example examines a realistic setting for network identification when the observed opinions are imperfect and the opinion dynamics are randomized. In addition, we consider the network topology, which is taken from that of a real network.

We focus on solving problem (P1) with the regularizer $g^1(\cdot)$ and $\rho = 10^{-10}$. The social network considered is the `ReedCollege` network taken from the `Facebook100` dataset [46], which consists of $n + S = 846$ agents, in which we pick $S = 180$ agents as stubborn agents whose degrees are closest to the median degree of the graph. The social network has $|E| = 13,269$ edges and a mean degree of 19.92. The stubborn-normal network has a mean degree of 25.07. We simulate the randomized opinion dynamics on the above graph using a randomized broadcast gossiping protocol [47]. As the opinions converge to a random variable, we consider the estimator described in Eq. (22.36) where we set $|\mathcal{T}_k| = 5 \times 10^5$ and the sampling instances are taken uniformly from $[10^5, 5 \times 10^7]$. Fig. 22.4 compares

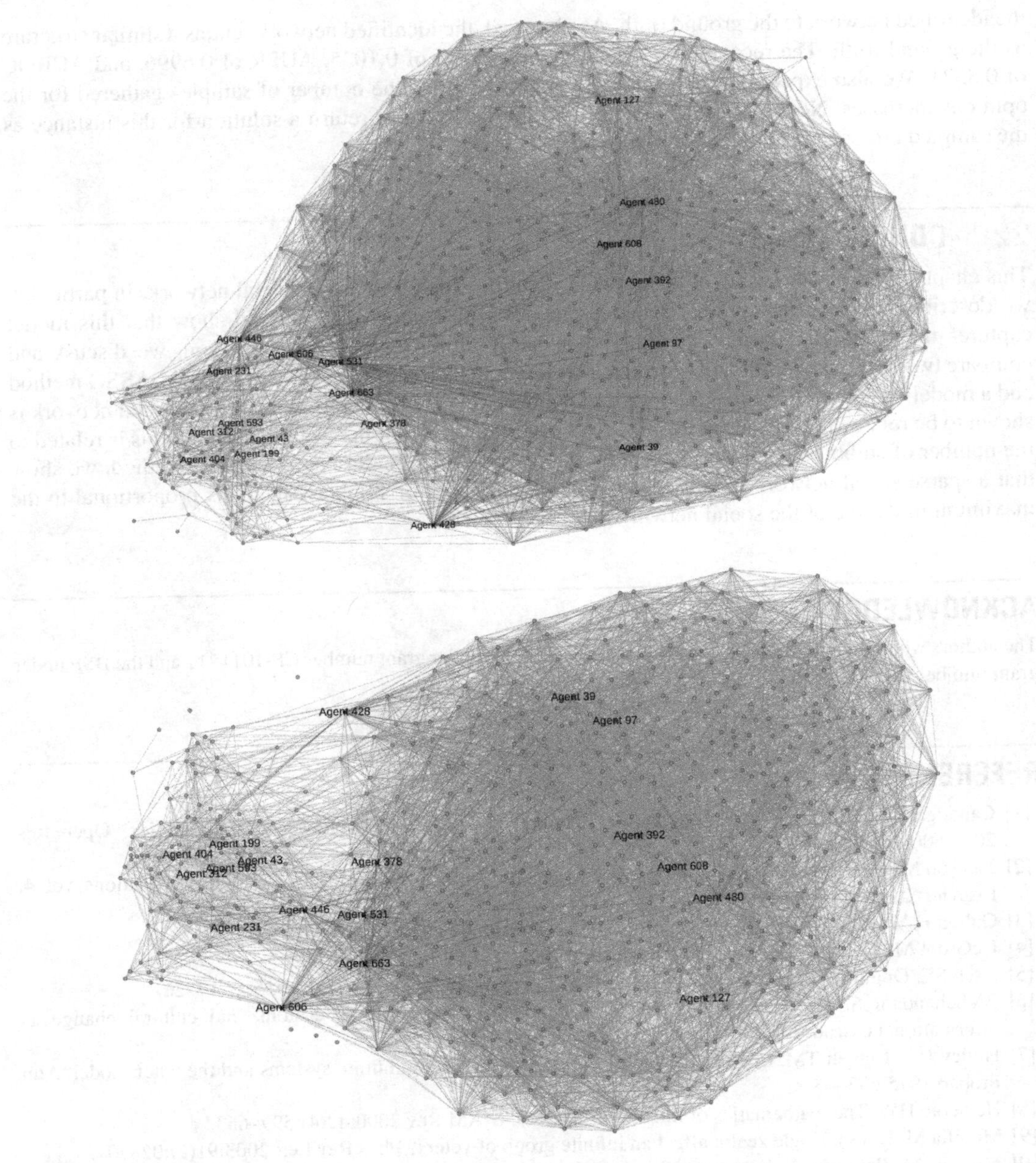

FIG. 22.4

Identifying social networks based on a real network. Examining the model-based learning approach on a more realistic setting. We simulate the randomized opinion dynamics on a real network from the Facebook [46]. The networks are illustrated using **Gephi** with the Force Atlas 2 layout with the edge weights taken into account. (Left) Ground Truth. (Right) Identified network via solving Eq. (P1) with $g^1(\cdot)$.

the identified network to the ground truth. As observed, the identified network retains a similar structure as the ground truth. The recovered network has an NMSE of 0.1035, AUPR of 0.6996, and AUROC of 0.8571. We also expect the performance to improve when the number of samples gathered for the opinions increases. Note that the graphical LASSO has failed to return a solution for this instance as the sampled covariance matrix is of low rank.

22.5 CONCLUSIONS

This chapter gives an overview of the network identification problem for social network. In particular, we describe the DeGroot opinion dynamics model with stubborn agents and show that this model captures the "low-rank" behavior observed in many opinion data obtained. Then, we discuss and compare two network identification strategies in depth, which are the popular graphical LASSO method and a model-based learning method proposed by the authors. The identifiability of the social network is shown to be related to the rank of the observed data. In the case of the DeGroot model, this is related to the number of stubborn agents needed. Importantly, using the model-based learning method, we show that a sparse social network is identifiable when the number of stubborn agents is proportional to the maximum in-degree of the social network.

ACKNOWLEDGMENTS

The authors would like to thank for the support from the NSF under grant number CF-1011811 and the BSF under grant number ISF 903/13.

REFERENCES

[1] Candogan O, Bimpikis K, Ozdaglar A. Optimal pricing in networks with externalities. Oper Res 2012;60(4):883–905.

[2] Jackson MO, Zenou Y. Games on networks. In: Handbook of game theory with economic applications, vol. 4, Elsevier; 2015. pp. 95–163.

[3] Galton F. Vox populi (the wisdom of crowds). Nature 1907;75(7):450–1.

[4] DeGroot M. Reaching a consensus. J Am Stat Assoc 1974;69:118–21.

[5] Asch SE. Opinions and social pressure. In: Readings about the social animal; 1955. p. 17–26.

[6] Bikhchandani S, Hirshleifer D, Welch I. A theory of fads, fashion, custom, and cultural change as informational cascades. J Polit Econ 1992;100(5):992–1026.

[7] Holley RA, Liggett TM. Ergodic theorems for weakly interacting infinite systems and the voter model. Ann Probab 1975:643–63.

[8] Hethcote HW. The mathematics of infectious diseases. SIAM Rev 2000;42(4):599–653.

[9] Mobilia M. Does a single zealot affect an infinite group of voters? Phys Rev Lett 2003;91(2):028701.

[10] Mobilia M, Petersen A, Redner S. On the role of zealotry in the voter model. J Stat Mech: Theory Exp 2007;2007(08):P08029.

[11] Acemoglu D, Como G, Fagnani F, Ozdaglar A. Opinion fluctuations and disagreement in social networks. Math Oper Res 2013;38(1):1–27. https://doi.org/10.1287/moor.1120.0570.

[12] Hegselmann R, Krause U. Opinion dynamics and bounded confidence models, analysis, and simulation. J Artif Soc Soc Simul 2002;5(3).

[13] Acemoglu D, Ozdaglar A, ParandehGheibi A. Spread of (mis) information in social networks. Games Econ Behav 2010;70(2):194–227.

[14] Yildiz ME, Scaglione A. Computing along routes via gossiping. IEEE Trans Signal Process 2010;58(6):3313–27.

[15] Yildiz E, Acemoglu D, Ozdaglar AE, Saberi A, Scaglione A. Discrete Opinion Dynamics with Stubborn Agents. Available at SSRN: https://ssrn.com/abstract=1744113, 2011.

[16] Yildiz ME, Ozdaglar A, Acemoglu D, Saberi A, Scaglione A. Binary opinion dynamics with stubborn agents. ACM Trans Econ Comput 2013;1(4):19.

[17] Waagen A, Verma G, Chan K, Swami A, D'Souza R. Effect of zealotry in high-dimensional opinion dynamics models. Phys Rev E 2015;91(2):022811.

[18] Das A, Gollapudi S, Munagala K. Modeling opinion dynamics in social networks. In: Proceedings of WSDM; 2014. p. 403–12.

[19] Moussaïd M, Kämmer JE, Analytis PP, Neth H. Social influence and the collective dynamics of opinion formation. PLOS One 2013;8(11):e78433.

[20] Friedman J, Hastie T, Tibshirani R. Sparse inverse covariance estimation with the graphical lasso. Biostatistics 2008;9(3):432–41.

[21] Segarra S, Marques AG, Mateos G, Ribeiro A. Network topology inference from spectral templates. IEEE Trans Signal Inf Process Netw 2017;3(3):467–83.

[22] Tang J, Lou T, Kleinberg J. Inferring social ties across heterogeneous networks. In: WSDM'12; 2012. p. 743–52.

[23] Tang J, Gao H, Liu H, Das Sarma A. etrust: understanding trust evolution in an online world. In: Proceedings of the 18th ACM SIGKDD international conference on knowledge discovery and data mining. KDD'12. New York, NY: ACM; 2012. p. 253–61. ISBN 978-1-4503-1462-6. https://doi.org/10.1145/2339530.2339574.

[24] Bresler G. Efficiently learning Ising models on arbitrary graphs. In: Proceedings of the forty-seventh annual ACM on symposium on theory of computing. STOC'15. New York, NY: ACM; 2015. p. 771–82. ISBN 978-1-4503-3536-2. https://doi.org/10.1145/2746539.2746631.

[25] Etesami J, Kiyavash N, Zhang K, Singhal K. Learning network of multivariate Hawkes processes: a time series approach; 2016. arXiv preprint arXiv:160304319.

[26] He X, Rekatsinas T, Foulds J, Getoor L, Liu Y. Hawkestopic: a joint model for network inference and topic modeling from text-based cascades. In: International conference on machine learning; 2015.

[27] Pouget-Abadie J, Horel T. Inferring graphs from cascades: a sparse recovery framework. In: Proceedings of the 32nd international conference on machine learning (ICML 2015); 2015. http://econcs.seas.harvard.edu/files/econcs/files/pouget_icml15.pdf.

[28] Ronen M, Rosenberg R, Shraiman BI, Alon U. Assigning numbers to the arrows: parameterizing a gene regulation network by using accurate expression kinetics. Proc Natl Acad Sci U S A 2002;99(16):10555–60.

[29] Timme M. Revealing network connectivity from response dynamics. Phys Rev Lett 2007;98(22):1–4. https://doi.org/10.1103/PhysRevLett.98.224101.

[30] Shen Z, Wang WX, Fan Y, Di Z, Lai YC. Reconstructing propagation networks with natural diversity and identifying hidden sources. Nat Commun 2014;5.

[31] Wang WX, Lai YC, Grebogi C, Ye J. Network reconstruction based on evolutionary-game data via compressive sensing. Phys Rev X 2011;1(2):1–7. https://doi.org/10.1103/PhysRevX.1.021021.

[32] Han X, Shen Z, Wang WX, Di Z. Robust reconstruction of complex networks from sparse data. Phys Rev Lett 2015;114:028701. https://doi.org/10.1103/PhysRevLett.114.028701.

[33] Bazerque JA, Baingana B, Giannakis GB. Identifiability of sparse structural equation models for directed and cyclic networks. In: 2013 IEEE global conference on signal and information processing (GlobalSIP). IEEE; 2013. p. 839–42.

[34] Ching ES, Lai PY, Leung C. Reconstructing weighted networks from dynamics. Phys Rev E 2015;91(3):030801.

[35] Sontag ED. Network reconstruction based on steady-state data. Essays Biochem 2008;45:161–76.
[36] Friedkin NE, Johnsen EC. Social influence network theory: a sociological examination of small group dynamics. Cambridge University Press; 2011.
[37] Blondel VD, Hendrickx JM, Olshevsky A, Tsitsiklis JN. Convergence in multiagent coordination, consensus, and flocking. In: Proceedings of CDC-ECC'05, vol. 2005; 2005. p. 2996–3000. ISBN 0780395689. https://doi.org/10.1109/CDC.2005.1582620.
[38] Khan UA, Kar S, Moura JMF. Distributed sensor localization in random environments using minimal number of anchor nodes. IEEE Trans Signal Process 2009;57(5):2000–16.
[39] Wai HT, Scaglione A, Leshem A. Active sensing of social networks. IEEE Trans Signal Inf Process Netw 2016;2(3):406–19.
[40] Wainwright MJ, Jordan MI, et al. Graphical models, exponential families, and variational inference. Found Trends Mach Learn 2008;1(1–2):1–305.
[41] Banerjee O, Ghaoui LE, d'Aspremont A. Model selection through sparse maximum likelihood estimation for multivariate Gaussian or binary data. J Mach Learn Res 2008;9(March):485–516.
[42] Eldar YC. Sampling theory: beyond bandlimited systems. New York, NY: Cambridge University Press; 2014. ISBN 0511762321, 9780511762321.
[43] Khajehnejad MA, Dimakis AG, Xu W, Hassibi B. Sparse recovery of nonnegative signals with minimal expansion. IEEE Trans Signal Process 2011;59(1):196–208. https://doi.org/10.1109/TSP.2010.2082536.
[44] Frieze A, Karoński M. Introduction to random graphs. Cambridge University Press; 2015.
[45] Alon N, Spencer JH. The probabilistic method. John Wiley & Sons; 2004.
[46] Traud AL, Mucha PJ, Porter MA. Social structure of Facebook networks. Phys A: Stat Mech Appl 2012;391(16):4165–80. https://doi.org/10.1016/j.physa.2011.12.021, 1102.2166.
[47] Aysal TC, Yildiz ME, Sarwate AD, Scaglione A. Broadcast gossip algorithms for consensus. IEEE Trans Signal Process 2009;57(7):2748–61.

CHAPTER

DYNAMIC SOCIAL NETWORKS: SEARCH AND DATA ROUTING

23

Hazer Inaltekin[a]*, **H. Vincent Poor**[†]

*Department of Electrical and Electronic Engineering, The University of Melbourne, Parkville, VIC, Australia**
Department of Electrical Engineering, Princeton University, Princeton, NJ, United States[†]

23.1 INTRODUCTION

Social networks exhibit complex and often enigmatic dynamical properties both in structure and process dimensions [1–8]. From the structure point of view, two common models for describing the connections among individuals are *scale-free* [9,10] and *small-world* [11–13] networks. The origin of scale-free networks lies in the interplay between dynamic *growth* and *preferential attachment* mechanisms, whose discovery in the specific context of citation statistics can be traced back to the 1960s [14]. It is known as the Barabási-Albert model today in the more general context of network science [9].[1] For a given network $\mathcal{G} = (V, E)$ with vertex set V and edge set E, the growth and preferential attachment mechanisms can be described as in Algorithm 23.1. It is a well-known result that the fraction of vertices ξ_m with degree m in a network $\mathcal{G} = (V, E)$ with large numbers of vertices generated according to the growth and preferential attachment dynamics scales according to

$$\xi_m \propto m^{-\gamma} \tag{23.1}$$

with degree exponent $\gamma = 3$ [16,17].[2] Many real-world networks such as citation, science collaboration, and movie actor networks exhibit a scale-free property with a degree exponent ranging from 2 to 4 [10] (Table 23.1).

[a]This work was prepared under the support of the U.S. Army Research Office under Grant W911NF-16-1-0448.
[1]Generation of the classical random graph models, on the other hand, does not obey these mechanisms, and hence they do not exhibit large variations around a typical connectivity degree [15].
[2]Any γ value greater than 2 can be obtained by changing the functional form generating the preferential attachment probability in Algorithm 23.1 [17].

Cooperative and Graph Signal Processing. https://doi.org/10.1016/B978-0-12-813677-5.00023-7

Table 23.1 Notation and Symbols Used in the Chapter

Symbol	Definition/Explanation
$\mathcal{G} = (V, E)$	graph with vertex set V and edge set E
V_u	set of vertices to which vertex $u \in V$ is connected
E_u	set of edges belonging to vertex $u \in V$
e_{uv}	an edge in E connecting the vertices $u, v \in V$
$l_e(x)$	latency function of edge $e \in E$
$\mathcal{S}$	social space in which the locations of vertices in V lie
$\rho(u, v)$	social distance between vertices $u, v \in V$ measured with respect to the geometry of $\mathcal{S}$
r	radius of local friendship circle
α	clustering parameter in Kleinberg's model and the Watts-Dodds-Newman Model
T	delay of social search
N	random number of long-range contacts per vertex in the Inaltekin-Chiang-Poor Models
Q	probability distribution of N
$\varphi(x)$	probability generating function of Q
$g_1(n) \propto g_2(n)$	$\lim_{n\to\infty} \frac{g_1(n)}{g_2(n)} = c > 0$
$g_1(n) = \Omega(g_2(n))$	$\liminf_{n\to\infty} \frac{g_1(n)}{g_2(n)} \geq c > 0$
$g_1(n) = O(g_2(n))$	$\limsup_{n\to\infty} \frac{g_1(n)}{g_2(n)} \leq c > 0$
$\mathcal{P}$	set of paths available in a given graph $\mathcal{G} = (V, E)$
M	number of source-target pairs in the process of altruistic routing
$\mathbf{A} = [a_{i,j}]_{i,j=1}^{M}$	social relationship matrix with $a_{i,j} \in [0, 1]$ and $a_{i,i} = 1$ for all $i, j = 1, \ldots, M$
f_i	routing policy of player $i \in \{1, \ldots, M\}$
Γ_i	data rate of player $i \in \{1, \ldots, M\}$
$\boldsymbol{f}$	routing policy profile
f	total network flow
$\|A\|$	cardinality of a set A
$\Pr\{A\}$	probability of an event A
$\mathrm{E}[X]$	expectation of a random variable X
$1_{\{\cdot\}}$	indicator function

Algorithm 23.1 GROWTH AND PREFERENTIAL ATTACHMENT

Input: $\mathcal{G} = (V, E)$ and a new vertex u with m_u edges to be added.

Output: Updated network $\mathcal{G} = \left(V \bigcup \{u\}, E \bigcup E_u\right)$ with E_u being the added m_u edges for u.

Preferential Attachment Part: $E_u = \texttt{P-ATTACH}(\mathcal{G}, m_u)$

Initialization: $V_u = \emptyset$ and $E_u = \emptyset$, where V_u is the set of vertices to which u is connected.
for $k = 1$ to m_u **do**
  choose a $v \in V \setminus V_u$ randomly from distribution $\{p_v\}_{v \in V \setminus V_u}$ with $p_v = \frac{m_v}{\sum_{w \in V \setminus V_u} m_w}$
  update E_u and V_u as $E_u \leftarrow E_u \bigcup \{e_{uv}\}$ and $V_u \leftarrow V_u \bigcup \{v\}$

end for
return E_u

Growth Part: $\mathcal{G} = \text{GROWTH}(\mathcal{G}, u, E_u)$
update $\mathcal{G}$ as $\mathcal{G} \leftarrow (V \bigcup \{u\}, E \bigcup E_u)$
return $\mathcal{G}$

Small-world networks, on the other hand, describe networks whose properties are intermediate to *order* and *disorder* [11–13,18]. The forces in favor of order result in clustering and local structure in a small-world network. On the other hand, the forces in favor of randomness exhibit themselves as random long-range connections in such networks. The most striking characteristic of small-world networks is their small graphical diameter even though they are highly clustered [11,19]. This property is a consequence of the phenomenon that random admixtures of long-range connections to a regular substrate graph (e.g., ring lattice, rectangular lattice, etc.) result in highly nonlinear changes in global dynamical properties of a network with, at most, a linear effect on the local clustering behavior. Algorithm 23.2 describes one possible way of constructing a small-world network. The process of generating a small-world network starting from an example regular ring lattice is illustrated in Fig. 23.1.

Algorithm 23.2 SMALL-WORLD NETWORK GENERATION

Input: A network $\mathcal{G} = (V, E)$ to which long-range connections will be attached.
Output: Updated network $\mathcal{G} = (V, E_{\text{new}})$ with E_{new} being the new edge set containing long-range connections.

Initialization: Select a rewiring probability $P > 0$. $E_{\text{new}} = E$.
for Each $u \in V$ **do**
  for Each $e \in E_u$ **do**
    determine whether or not to rewire e with probability P
    if decide to re-wire, choose a vertex $v \in V \setminus \{u\}$ with probability $\frac{1}{|V|-1}$. Update E_{new} as $E_{\text{new}} \leftarrow E_{\text{new}} \setminus \{e\} \bigcup \{e_{uv}\}$
  end for
end for
update $\mathcal{G}$ as $\mathcal{G} \leftarrow (V, E_{\text{new}})$
return $\mathcal{G}$

There are two important disjunctures between scale-free and small-world networks. To start with, small-world networks exhibit a confluence of homophily and small geodesic paths connecting any two network vertices even for very large networks, whereas the clustering coefficient measuring the cliquishness of a scale-free network approaches zero according to $\frac{(\log|V|)^2}{|V|}$ as the network size $|V|$, measured in terms of the number of vertices in a network $\mathcal{G} = (V, E)$, grows large [20]. On this point, we want to preserve a nonvanishing clustering property for social networks. Second, the probability distribution for the degree of a vertex in a small-world network decays to zero exponentially whereas the probability of finding a vertex with degree m scales according to a power law in scale-free networks as given by Eq. (23.1). On this point, we want to have a network model that can mimic the scale-free property and the existence of hubs in a social network. Hence, for the purpose of consolidating the important features of scale-free and small-world networks, we will focus on a more recent network model, which will be called the *Inaltekin-Chiang-Poor Models* in this chapter and were introduced by

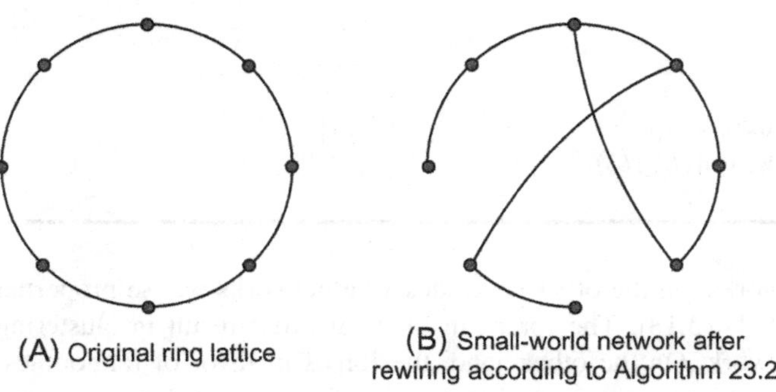

(A) Original ring lattice
(B) Small-world network after rewiring according to Algorithm 23.2

FIG. 23.1

Generation of a small-world network according to Algorithm 23.2. In this figure, we start with a regular ring lattice containing 8 vertices with each vertex connected to its immediate right and left neighbors, as depicted in (A). The small-world network constructed according to Algorithm 23.2 is illustrated in (B). The small-world networks constructed starting with larger regular lattices having different local connectivity structure are qualitatively the same. $P = 0.25$.

Inaltekin et al. [21,22] for the process of targeted social search. We do not assume any specific network topology for the process of altruistic data routing in the current chapter. The details of the Inaltekin-Chiang-Poor Models are provided in Section 23.3.

Process dimension of social networks is multifaceted and critically depends on the underlying substrate social network topology. Examples include the search for resources [2,23,24], mobilization and organization of communities for common goals [2,23], spread of epidemics [25–27], global cascades and complex contagions [1,4,6,28,29], adoption of innovations [30], emergence of social norms [31], and spread of rumors [32]. Further, the network topology can be considered to play *active* and *passive* roles in determining the dynamical behavior of these processes [12], meaning that the network can be manipulated by individuals to maximize their utilities (i.e., see [23,33] for finding a job and growing social capital) in the active sense, and that the existing network connections critically determine the dynamics of processes unfolding over time based on these connections but without room for modifying them over the time span of interest (i.e., see [1,4,27] for the spread of epidemics and global cascades) in the passive sense. Considering such a wide spectrum of processes and the role of network structure in specifying their dynamical behavior, it usually requires a substantive amount of research to interpret model parameters for each particular process of interest and to understand related dynamics in each field. For example, search agents may not exist in all applications, and diffusion processes become more appropriate to model process dynamics in cases such as the spread of viruses and diseases. Therefore, our focus in this chapter will be limited to the targeted search and altruistic data routing processes in the passive sense.

The rest of the chapter is organized as follows. In Section 23.2, we will introduce the algorithmic perspectives for the targeted social search processes in social networks and review the selected results in the literature. In Section 23.3, we explain the details of the Inaltekin-Chiang-Poor Models and present the analytical solutions for the average value of the delay of social search as well as its probability distribution. This section will also provide a discussion on the implications of the derived delay equations as well as a comparison between analytical solutions and empirical results. In Section 23.4, we will introduce the problem of altruistic data routing for general topologies of social and physical connections. Finally, Section 23.5 concludes the chapter.

23.2 TARGETED SOCIAL SEARCH PROCESS

The targeted social search process of interest to us in this chapter is motivated by the well-known letter-referral experiments, which are also known as small-world experiments, conducted by the social psychologist Stanley Milgram and his colleagues [34–36]. The purpose of these studies was to estimate the average number of intermediaries needed to connect two individuals in the United States. They found this number to be around *six*, which forms the first empirical evidence for the common notion that any two individuals are separated by *six degrees of separation*, or for the so-called *small-world phenomenon*. A striking aspect of this result is that individuals are able to navigate over the set of social connections to find these short paths by only using their local information to decide about the next-hop message holders.

After its first appearance, the letter-referral technique and its modifications attracted substantial attention and were repeated in various social contexts to discover the social network structure [37–40]. At the global scale, it was also repeated by Dodds et al. through email messages, confirming the small-world phenomenon [41]. In spite of some objections such as sample selection bias and low chain completion rates raised against these findings as well as some recent results indicating that the average length of referral chains is higher if the missing data is accounted for correctly [42,43], it is commonly agreed that people around the world are connected to each other through a small number of acquaintances [44], for which we seek an analytical confirmation in this chapter.

23.2.1 MILGRAM SOCIAL SEARCH EXPERIMENT

A typical realization of the Milgram letter-referral experiment is as follows. Two individuals are selected randomly according to a sample selection process such as mail lists, newspaper advertisements [34–36], and emails [41], which involves stochastic dynamics during recruitment. One of them is assigned as the message originator (i.e., source individual), and the other is assigned as the message recipient (i.e., target individual). Depending on the social context, source and target individuals are located in the same country but in different states [34,35], in different countries [41], in the same institution but have different professional ranks [45,46], or are intentionally chosen to belong to different racial, religious, or cultural groups [36,40].

The source individual is provided with some basic information about the target such as address and occupation, and she is allowed to send the message only to others whom she knows on a first-name basis. Therefore, the source is not allowed to send the message to the target directly unless she knows the target on a first-name basis. Intermediate message holders repeat the same step until the message reaches the target. Once the message reaches the target, the targeted social search process terminates. In the process of searching for the target, individuals traced by the message form a referral chain, and the number of links connecting the source and target individuals through such a referral chain is called the *delay of social search*.

As is clear from this discussion, this search process is inherently algorithmic, and can be described as in Algorithm 23.3. There are three important stylized assumptions we make while writing Algorithm 23.3 in order to formalize Milgram's social search experiments. The first one is on the use of a social distance metric $\rho(\cdot,\cdot)$ in order to measure the similarity between target and intermediate message holders. The second one is on the ability of individuals to select their best contacts closest to the target individual with respect to ρ. Finally, the third one is on the condition that if the selected next-hop message holder is one of the previous ones, then the targeted social search process is terminated to signal that the current message holder does not have any neighbors offering a better choice than the previously chosen ones in order to move the message forward closer to the target individual. When we use the term "targeted social search process," it will be the description provided by Algorithm 23.3 to which we refer in the rest of this chapter.

Algorithm 23.3 TARGETED SOCIAL SEARCH

Input: A network $\mathcal{G} = (V, E)$, referral chain $\mathcal{R}$ containing current message holders $v \in V$ as well as their social locations X_v, source individual $s \in V$ with social location X_s, target individual $t \in V$ with social location X_t and a social similarity measure $\rho(\cdot, \cdot)$.

Output: Delay of social search T and the updated referral chain $\mathcal{R}$.

Initialization: $T = \infty$ and $\mathcal{R} = \{s\}$.

Search Process: $(T, \mathcal{R}) = \text{SEARCH}(s, X_s, t, X_t, \rho, \mathcal{R}, \mathcal{G})$

```
if t ∈ V_s then
    R ← R ⋃ {t}
    T ← |R| − 1
    return (T, R)
else
    find v = arg min_{v∈V_s} ρ(X_v, X_t)
    if v ∈ R \ {s} then
        T ← ∞ and R ← ∅
        return (T, R)
    else
        R ← R ⋃ {v}
        (T, R) = SEARCH(v, X_v, t, X_t, ρ, R, G)
        return (T, R)
    end if
end if
```

23.2.2 KLEINBERG'S SMALL-WORLD SEARCH MODEL

Kleinberg used a rectangular grid with random long-range connections as a proxy for the social space while seeking a mathematical explanation for the small-world phenomenon from an algorithmic point of view [47,48]. This model is illustrated in Fig. 23.2 for an 8×8 square lattice. In general, the social space $\mathcal{S}$ consisting of the locations of vertices in an $R \times R$ square lattice can be represented by the collection of points

$$\mathcal{S} = \{(i,j) : i \in \{1, \ldots, R\}, j \in \{1, \ldots, R\}\} \tag{23.2}$$

and the social similarity between two vertices $u, v \in V$ with locations $X_u = (i,j) \in \mathcal{S}$ and $X_v = (k,l) \in \mathcal{S}$ is given by

$$\rho(u, v) = |i - k| + |j - l|. \tag{23.3}$$

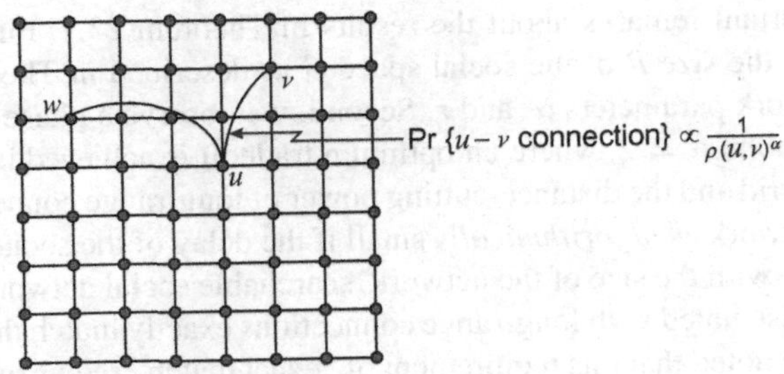

FIG. 23.2

Kleinberg's model for the targeted social search. An illustration for Kleinberg's model to obtain an analytical confirmation for the small-world phenomenon over the rectangular grid with random long-range connections. Each vertex is connected to its immediate one-hop neighbors for this particular example and long-range connections are added with probabilities proportional to an inverse power law of social similarity, i.e., $\Pr\{u - v \text{ connection}\} \propto \frac{1}{\rho(u,v)^\alpha}$.

The local friendship circle $V_{u,\text{short-range}}$ of a vertex u is then defined as the set of all vertices with social similarity less than a threshold value r, i.e.,

$$V_{u,\text{ short-range}} = \{v \in V : \rho(u,v) \leq r\}. \tag{23.4}$$

Unlike the original small-world network model [11], the probability of a long-range connection between two individuals in Kleinberg's model decreases with the social distance between them as measured on the rectangular grid according to

$$\Pr\{u - v \text{ connection}\} \propto \frac{1}{\rho(u,v)^\alpha} \tag{23.5}$$

for $v \notin V_{u,\text{short-range}}$, where $\alpha \geq 0$ determines the dynamic range spanned by the long-range connections. For small values of α, long-range connections span longer distances and the individuals in this hypothetical social space have a stronger tendency to connect with others having dissimilar characteristics. On the other hand, most of the connections are short-range if α is large. In a sense, the value of α determines the level of clustering in this model. Adding n such long-range connections, we obtain the set of long-range contacts $V_{u,\text{long-range}}$ for each $u \in V$. The resulting social network is then equal to $\mathcal{G} = (V, E)$ with $E = \bigcup_{u \in V} \{e_{uv} : v \in V_u\}$ and $V_u = V_{u,\text{short-range}} \bigcup V_{u,\text{long-range}}$. The main result in [47,48] is summarized below.

Theorem 23.1. *The targeted social search process, when it runs over the small-world network $\mathcal{G} = (V, E)$ constructed as above, produces the following bounds on the average value of the delay of social search T.*

1. $\alpha < 2 : \mathrm{E}[T] = \Omega\left(R^{\frac{2-\alpha}{3}}\right)$
2. $\alpha = 2 : \mathrm{E}[T] = O\left((\log R)^2\right)$
3. $\alpha > 2 : \mathrm{E}[T] = \Omega\left(R^{\frac{\alpha-2}{\alpha-1}}\right)$

Proof. See [48]. ■

There are three important remarks about the results in Theorem 23.1. First, the scaling behavior of $\mathrm{E}[T]$ as a function of the size R of the social space $\mathcal{S}$ as described in Theorem 23.1 is correct for any given values of network parameters α, and r. Second, we observe a phase transition in the scaling behavior of $\mathrm{E}[T]$ at the point $\alpha = 2$, where an optimum tradeoff is achieved between the navigational cues embedded over the grid and the distance-cutting power of long-range connections. To put it another way, defining a social network as *algorithmically* small if the delay of the social search grows no faster than poly-logarithmically with the size of the network, searchable social networks exist when the decay rate of the probabilities associated with long-range connections exactly match the dimension of the grid. On this point, it should be noted that this requirement of "exact match" seems quite brittle, and previous empirical studies *only* found that chain lengths connecting two individuals are small [35,36,41]. There is no empirical evidence showing that lengths of small-world search chains grow logarithmically with the network size, even though such evidence exists for the lengths of shortest paths [11,44,49]. Last but not least, these bounds, although helpful to reason about when a social network becomes searchable, do not reveal any detailed information about the functional form of $\mathrm{E}[T]$ as a function of social separation between source and target individuals participating in the targeted social search process.

23.2.3 WATTS-DODDS-NEWMAN SMALL-WORLD SEARCH MODEL

Starting from a sociologically more realistic premise than the low-dimensional network model proposed in [47,48], Watts et al. [50] showed that the space of searchable networks is, in fact, larger than what is predicted by Kleinberg's model. For their analysis, they use an ultrametric that measures the similarity between individuals along multiple social dimensions such as geography, race, profession, religion, and education. They claim that it is this multidimensioned nature of social identity that makes social networks searchable [50,51]. Their model has four important characteristics.

1. **Grouping:** Individuals are identified through their participation in social groups with some well-defined characteristics such as academic departments in universities.
2. **Social Similarity Through Hierarchy:** The set of individuals V is divided into hierarchical layers, with the top layer being the entire collection of individuals and the successively deeper layers in the hierarchy containing more but increasingly specific groups. The social distance $\rho(u, v)$ between $u, v \in V$ is defined to be 1 if they belong to the same social group, and to be the height of their lowest common ancestor in the resulting hierarchy otherwise. An example illustration for this hierarchical view of the social space is given in Fig. 23.3.
3. **Connection Probability:** The probability of any two individuals $u, v \in V$ being connected in this hierarchical world is chosen to be inversely proportional with their social similarity according to

$$\Pr\{u - v \text{ connection}\} \propto \exp(-\alpha\rho(u, v)), \tag{23.6}$$

where $\alpha > 0$ is a tunable parameter determining the level of clustering of the resulting social network. The social links E_u for vertex $u \in V$ are added according to the connection probability in Eq. (23.6) until u has a predetermined connection degree. We note that a similar clustering parameter also exists in Kleinberg's model but determines the power law decay exponent for

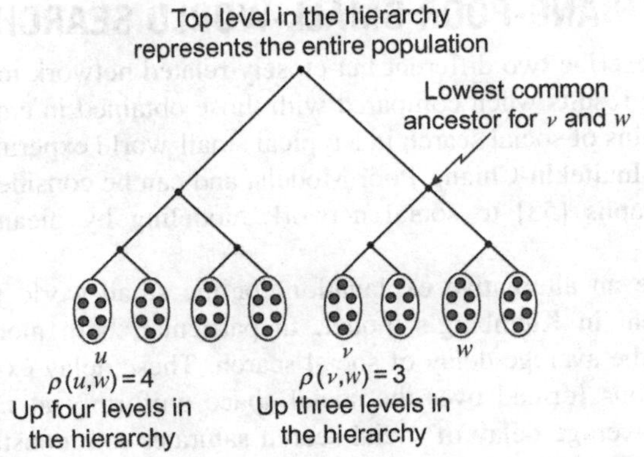

FIG. 23.3

The Watts-Dodds-Newman Model for the targeted social search. An illustration for the Watts-Dodds-Newman Model to explain the small-world phenomenon through a hierarchical model of society. The entire population is successively refined into subpopulations with a branching factor of 2 in this particular example. Social similarity between individuals *u* and *w*, represented through their group memberships, is equal to 4 because the lowest common ancestor is the top-level node with height of 4 from their respective groups. On the other hand, $\rho(v, w) = 3$ because the lowest common ancestor is up three levels from their respective groups.

long-range connection probabilities rather than specifying the exponential decay rate for the probability of existence of any social link as in the current model.

4. **Multidimensioned Nature and Ultrametric:** The described hierarchical structure appears in multiple social dimensions such as geography, race, profession, religion, and education. Hence, the perceived social similarity between any two individuals $u, v \in V$ is measured by the ultrametric as the minimum of the social distance over multiple social dimensions.

Some of the important results from [50], obtained through numerical simulations, are as follows. First, it is shown that the space of searchable social networks, as parametrized by α and the number of social dimensions, is quite larger than the one predicted by Kleinberg's model that only contains a single network type in which the dimension of the social space is matched to the power law decay rate of the long-range connection probability. Second, it is observed that the optimum number of social dimensions to use for the targeted social search process ranges from two to three, which is in line with the empirical observation reported in [52]. The reason behind this observation is that an increase in the number of social dimensions to use in targeted social search processes thins the existing social ties in each dimension, which leads to a decrease in the navigational power of the social ties of individuals because they are formed independently over each social dimension. Finally, the histogram representing the probability distribution for the delay of the social search obtained over 10^6 random chains initiated over the social network $\mathcal{G} = (V, E)$ resulting from the described hierarchical social space exhibits statistical similarity with the empirical data in [35]. In particular, after cleaning the data in [35] to include only the referral chains initiated in Nebraska, the Watts-Dodds-Newman Model produces an average delay value around 6.7 (i.e., compare this with 6.5 reported in [35]) and the standard chi-square test results in a *P*-value of 0.49, which supports the hypothesis that the empirical distribution from [35] and the one obtained through the Watts-Dodds-Newman Model are statistically similar.

23.3 INALTEKIN-CHIANG-POOR SMALL-WORLD SEARCH MODELS

In this section, we will describe two different but closely related network models, both of which lead to statistically convincing results when compared with those obtained in empirical studies, to explain why we observe short chains of social search in a typical small-world experiment [21,22]. They will be generically referred to as Inaltekin-Chiang-Poor Models, and can be considered to be a generalization of random geometric graphs [53] to social network modeling by means of random long-range connections.

These models provide an alternative explanation for the small-world phenomenon without the "exact match" requirement in Kleinberg's model. In particular, both models produce analytically tractable expressions for the average delay of social search. These delay expressions show that, even with long-range connections formed over the social space uniformly at random as in the original small-world model [11], average delay of social search saturates to a constant as the social distance between source and target individuals increases. This constant depends on the model parameters and can be made small enough, as observed in empirical studies, for reasonable choices of the number of long-range contacts per individual. One of these models, named the Inaltekin-Chiang-Poor Model 2, focuses on the targeted search processes progressing on a nested sequence of small worlds with long-range connections generated uniformly at random. This model exhibits logarithmically growing average social search delays. These results imply that the original small-world model, in which long-range connections are formed uniformly at random without depending on any particular notion of social distance, is in fact adequate to explain the existence of searchable social networks from an algorithmic point of view.

23.3.1 THE INALTEKIN-CHIANG-POOR MODEL 1

In this model, a measure-metric space is considered as a proxy for the social space $\mathcal{S}$ in order to measure the similarity between individuals. We note that this space was a rectangular grid in Kleinberg's model and the multidimensional hierarchical social grouping structure in the Watts-Dodds-Newman Model.[3] For the sake of simplicity of exposition, we will provide the results for three different social spaces in the remainder of this section: (i) rectangular, (ii) circular, and (iii) spherical. The results for general measure-metric spaces can be found in [21]. The circular social space is similar to the original small-world network model above. Different meanings can be attributed to these network models in other areas of science in which small-world networks emerge. For example, the spherical network space can find direct applications when modeling connections among nerve cells in the human brain due to geometrical similarities [54,55].

The side lengths of the rectangular social space and the radius of spherical and circular ones are R distance units. The common radius of local friendship circles of vertices is r distance units. It should be noted that we use the term "distance" here to mean social similarity rather than physical distance. The location of a vertex $u \in V$ in this model is represented by $X_u \in \mathcal{S}$, and X_u's for $u \in V$ are independent and identically distributed from a uniform distribution over the social space of interest, as

[3]All metric properties do not hold in the Watts-Dodds-Newman Model, and hence, to be more precise, the measure of social similarity must be considered as a pseudometric in this model.

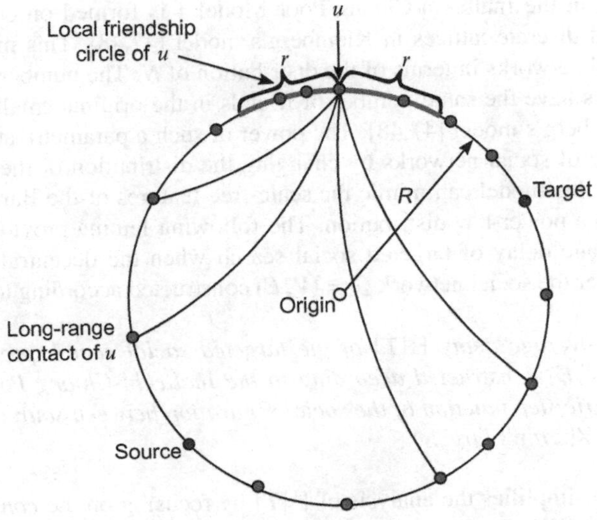

FIG. 23.4

The Inaltekin-Chiang-Poor Model 1 for the targeted social search. An illustration for the Inaltekin-Chaing-Poor Model 1 to obtain an analytical confirmation for the small-world phenomenon over a circular social space with radius R. Vertices are distributed over the social space uniformly at random. Each vertex is connected to its immediate neighbors within its local friendship circle with radius r. The random long-range contacts for vertex $u \in V$ are added uniformly over vertices that do not lie in its local friendship circle.

illustrated in Fig. 23.4 for the circular space. The local friendship circle $V_{u,\text{short-range}}$ of a vertex u is defined as the set of all vertices with social similarity less than a threshold value r, i.e.,

$$V_{u,\text{short-range}} = \{v \in V : \rho(u, v) \leq r\}, \tag{23.7}$$

where $\rho(u, v)$ measures the social distance between vertices $u, v \in V$ having locations $X_u \in \mathcal{S}$ and $X_v \in \mathcal{S}$ with respect to the particular geometry of the social space. For example, $\rho(u, v)$ is the regular Euclidean distance between X_u and X_v for the rectangular social space whereas it is equal to the length of the smaller arc connecting them for circular and spherical social spaces.

Each vertex $u \in V$ has a random number, N, of long-range contacts, who are selected uniformly at random over others that do not lie in the local friendship circle of u. For example, the number of long-range contacts of u is equal to 4 in Fig. 23.4. In general, N can be drawn from any given discrete probability distribution $Q(n)$, such as power law, Poisson, geometric, or uniform. We are able to determine the average delay of targeted social search for any given distribution of N. Adding N such long-range connections, we obtain the set of long-range contacts $V_{u,\text{long-range}}$ for each $u \in V$. The resulting social network is then equal to $\mathcal{G} = (V, E)$ with $E = \bigcup_{u \in V} \{e_{uv} : v \in V_u\}$ and $V_u = V_{u,\text{short-range}} \bigcup V_{u,\text{long-range}}$.

The Inaltekin-Chiang-Poor Model 1 captures the order-disorder properties of the small-world network model proposed by Watts and Strogatz [11] due to the threshold rule for forming local contacts.

The substrate network in the Inaltekin-Chiang-Poor Model 1 is formed on continuum social spaces whereas it consisted of discrete lattices in Kleinberg's model [47,48]. This model also parametrizes the generation of social networks in terms of the distribution of N. The number of long-range contacts is fixed, and all vertices have the same number of friends in the original small-world network model [11] as well as in Kleinberg's model [47,48]. The power of such a parametrization is that we can now generate a wide variety of social networks by changing the distribution of the number of long-range contacts. For example, this model can mimic the scale-free features of the Barabási-Albert model [9] when N is drawn from a power-law distribution. The following lemma provides important structural properties for the average delay of targeted social search when the decentralized search process in Algorithm 23.3 runs over the social network $\mathcal{G} = (V, E)$ constructed according to the Inaltekin-Chiang-Poor Model 1.

Lemma 23.1. *The average delay* $\mathrm{E}[T]$ *of the targeted social search process running over the social network* $\mathcal{G} = (V, E)$ *constructed according to the Inaltekin-Chiang-Poor Model 1 converges to a spherically symmetric step function of the social separation between source and target vertices as the number of vertices* $|V|$ *grows large.*

Proof. See [21]. ■

Lemma 23.1 greatly simplifies the analysis of $\mathrm{E}[T]$ by focusing on the *continuum* social network limit obtained by sending $|V|$ to infinity. In particular, for fixed but arbitrary locations of source s and target t vertices over the social space, we can define

$$\overline{T}_0 = 0 \text{ for } \rho(s,t) = 0,$$
$$\overline{T}_1 = 1 \text{ for } 0 < \rho(s,t) < r,$$
$$\overline{T}_k = \mathrm{E}[T] \text{ for } (k-1)r \le \rho(s,t) < kr \text{ and } k \ge 2$$

in the continuum limit, where the condition $\overline{T}_0 = 0$ indicates that there is no need to initiate the search process because source and target vertices are the same, and the condition $\overline{T}_1 = 1$ indicates that the source can reach the target in one hop because they are local friends of each other. The following theorem establishes the functional form for $\overline{T}_k$'s.

Theorem 23.2. *The average delay of the targeted social search process, when it runs over the social network* $\mathcal{G} = (V, E)$ *constructed according to the Inaltekin-Chiang-Poor Model 1, converges to*

$$\overline{T}_k = 1 + \sum_{j=1}^{k-1} \prod_{i=1}^{j} \varphi(\beta_i) \tag{23.8}$$

when $(k-1)r \le \rho(s,t) < kr$ *for* $k \ge 2$ *in the continuum limit, where* $\varphi(x)$, $x > 0$, *is the probability generating function of the distribution* Q *defined as* $\varphi(x) = \mathrm{E}[x^N]$ *and* $\beta_i, i \ge 1$, *for rectangular, spherical, and circular social spaces is given according to*

1. *Rectangular:* $\beta_i = 1 - \frac{\pi r^2 (i-1)^2}{R^2 - \pi r^2}$,
2. *Spherical:* $\beta_i = \frac{\cos(\frac{r}{R}) + \cos((i-1)\frac{r}{R})}{1 + \cos(\frac{r}{R})}$, *and*
3. *Circular:* $\beta_i = \frac{\pi - i\frac{r}{R}}{\pi - \frac{r}{R}}$.

Proof. See [21]. ■

A comparison of the analytical results in Theorem 23.2 with the simulation results for a 250r-by-250r rectangular social space is given in Fig. 23.5. The convergence behavior of social search as observed in simulations to analytical results looks essentially the same for all cases considered in [21]. For a given source-target separation, many different realizations of the social search process are considered, and the delay is averaged over all realizations for various numbers of vertices contained in the social space and all source-target separations.

The results are promising. As the number of vertices increases, the deviations between analytical and simulation results become negligible, as expected. In particular, what is more surprising regarding these simulation results is that when the average number of short-range contacts per vertex is between Dunbar's number 150 [56] and Killworth's estimate 290 [57] for the average size of personal networks, the average delay of the targeted social search estimated by the analytical formula deviates only 2.7% from simulation results, i.e., see the curve corresponding to 160 short-range contacts in Fig. 23.5. Furthermore, the gains obtained by increasing the number of vertices become quite marginal when the number of short-range contacts per vertex is larger than 80, i.e., compare the curves corresponding to 80, 160, and 320 short-range contacts per vertex in Fig. 23.5. In summary, the analytical formula derived in the continuum limit of the Inaltekin-Chiang-Poor Model 1 approximates the expected social delay curve well when the average number of short-range contacts per vertex is around the commonly accepted average sizes for personal networks.

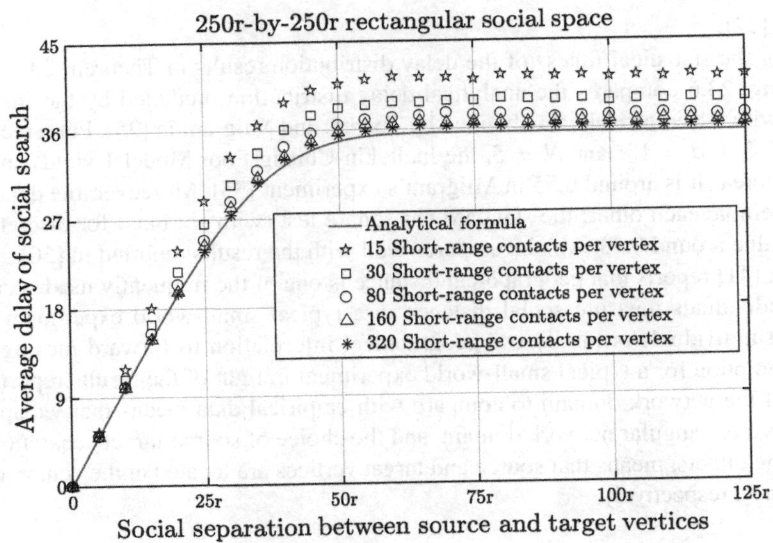

FIG. 23.5

Average delay of the targeted social search process for the Inaltekin-Chiang-Poor Model 1. Comparison of the analytical formula in Theorem 23.2 with simulation results for a 250r-by-250r rectangular social space with various numbers of short-range contacts. The number of long-range contacts per vertex is set to 1. The target vertex is placed at the center of the rectangular social space to avoid edge effects.

Beyond the average delay of social search [34,35,41,47,48], the Inaltekin-Chaing-Poor Model 1 can also generate the entire probability distribution of the delay of social search. To this end, $p_k(j)$ is defined to be the probability that the delay of social search is equal to j in the continuum limit when $\rho(s,t) \in [(k-1)r, kr)$ for $k \geq 2$. $p_1(j)$ is defined similarly for $0 < \rho(s,t) < r$. The next theorem provides the desired recursive expressions for $p_k(j)$.

Theorem 23.3. *The probability distribution of the delay of the targeted social search process, when it runs over the social network $\mathcal{G} = (V, E)$ constructed according to the Inaltekin-Chiang-Poor Model 1, converges to*

$$p_1(j) = 1_{\{j=1\}}, \text{ and}$$

$$p_k(j) = p_{k-1}(j-1)\varphi(\beta_{k-1}) + \sum_{i=1}^{k-2} p_i(j-1)\left(\varphi(\beta_i) - \varphi(\beta_{i+1})\right) \text{ for } k \geq 2$$

in the continuum limit, where $1_{\{\cdot\}}$ is the indicator function, $\varphi(x)$, $x > 0$, is the probability generating function of the distribution Q defined as $\varphi(x) = \mathrm{E}\left[x^N\right]$ and β_i, $i \geq 1$, for rectangular, spherical, and circular social spaces is given according to

1. *Rectangular:* $\beta_i = 1 - \frac{\pi r^2 (i-1)^2}{R^2 - \pi r^2}$,
2. *Spherical:* $\beta_i = \frac{\cos(\frac{r}{R}) + \cos((i-1)\frac{r}{R})}{1+\cos(\frac{r}{R})}$, *and*
3. *Circular:* $\beta_i = \frac{\pi - i\frac{r}{R}}{\pi - \frac{r}{R}}$.

Proof. See [22]. ∎

As a test for the statistical fitness of the delay distribution results in Theorem 23.3 to small-world experiments, Fig. 23.6 compares the analytical delay distribution predicted by the Inaltekin-Chiang-Poor Model 1 with the empirical data obtained by Travers and Milgram in [35]. For a rectangular social space with $R = 32r$, $d = 15r$ and $N = 5$, the Inaltekin-Chiang-Poor Model 1 yields an average delay around 6.14 whereas it is around 6.55 in Milgram's experiment [35]. Moreover, the delay distributions statistically resemble each other: the standard chi-square test (with six bins) for discrete distributions produces a P value around 0.53, which compares well with the results reported in [50].

Dodds et al. [41] reports that geographical distance is one of the frequently used social dimensions along which individuals measure social distance in a typical small-world experiment. Therefore, if we assume that individuals primarily use geographical information to forward messages, which is a reasonable assumption for a typical small-world experiment in light of the results reported in [41], the above choice of the network domain to compare with empirical data means that we approximate the United States as a rectangular network domain, and the choice of source-target separation, by ignoring the possible edge effects, means that source and target vertices are located at the center and at the edge of the social space, respectively.

23.3.2 THE INALTEKIN-CHIANG-POOR MODEL 2

Mainly motivated by Kleinberg's results [47,48], an algorithmically small social network is considered in the computer science community to be a network in which the delay of social search grows no faster than poly-logarithmically with the size of the social network. The Inaltekin-Chiang-Poor Model 2

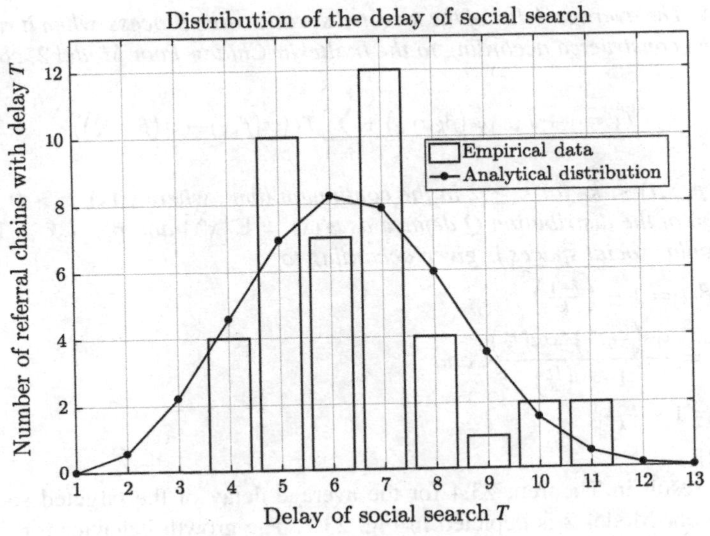

FIG. 23.6

Delay distribution of the targeted social search process for the Inaltekin-Chiang-Poor Model 1. Comparison of the analytical formula in Theorem 23.3 with empirical data in [35] for the 42 referral chains that originated in Nebraska. The 24 completed referral chains originating in Boston were excluded to keep the source-target separation in the model constant. Parameters $R = 32r$, $d = 15r$, and $N = 5$ for a rectangular social space.

provides an alternative explanation for the small-world phenomenon from an algorithmic point of view. In this model, the social space is viewed as a *nested* sequence of small worlds $\mathcal{SW}_1 \supset \mathcal{SW}_2 \supset \cdots \supset \mathcal{SW}_K$ in which the aim of each message holder, by using her long-range and short-range contacts, is to advance a message from a greater small world spanning longer social distances to a smaller one spanning shorter social distances to deliver the message to a target located in $\mathcal{SW}_K$. Long-range connections in each small world are formed uniformly at random within this world as in the original small-world model [11]. Therefore, unlike Kleinberg's model, formation of long-range connections within a small world does not depend on an individual's particular perception of distance, yet we can still find long-range contacts at all scales of social distance as the social search progresses from one small world to another one.

To make these arguments mathematically precise, we only require a vertex with the message to have N, a random number with distribution $Q(n)$, long-range contacts formed uniformly at random over all vertices lying in the disk centered around the target vertex with radius kr if the social distance between the message holder and the target vertex is in between $(k-1)r$ and kr. A message holder uses these N long-range contacts while searching for the target lying in the small world spanning the social space between them. Other than this new long-range contact formation rule, all the interpretations and assumptions given for the Inaltekin-Chiang-Poor Model 1 hold for this new model. Defining $\overline{T}_k$'s as above, the next theorem provides a recursive expression for $\mathrm{E}[T]$ in the continuum limit.

Theorem 23.4. *The average delay of the targeted social search process, when it runs over the social network $\mathcal{G} = (V, E)$ constructed according to the Inaltekin-Chiang-Poor Model 2, converges to*

$$\overline{T}_k = 1 + \overline{T}_{k-1}\varphi\left(\beta_{k-1,k}\right) + \sum_{i=1}^{k-2} \overline{T}_i \left(\varphi\left(\beta_{i,k}\right) - \varphi\left(\beta_{i+1,k}\right)\right) \tag{23.9}$$

when $(k-1)r \leq \rho(s,t) < kr$ for $k \geq 2$ in the continuum limit, where $\varphi(x)$, $x > 0$, is the probability generating function of the distribution Q defined as $\varphi(x) = \mathrm{E}\left[x^N\right]$ and $\beta_{i,k}$, $i, k \geq 1$, for rectangular, spherical, and circular social spaces is given according to

1. *Rectangular:* $\beta_{i,k} = 1 - \left(\frac{i-1}{k}\right)^2$,
2. *Spherical:* $\beta_{i,k} = \frac{\cos\left(\frac{(i-1)r}{R}\right) - \cos\left(\frac{kr}{R}\right)}{1 - \cos\left(\frac{kr}{R}\right)}$, *and*
3. *Circular:* $\beta_{i,k} = 1 - \frac{i-1}{k}$.

Proof. See [22]. ∎

The analytical result in Theorem 23.4 for the average delay of the targeted social search in the Inaltekin-Chiang-Poor Model 2 is depicted in Fig. 23.7. The growth behavior for the average social search delay for different sizes of the social space should be intuition confirming for researchers familiar with logarithmically growing social search delays. The side length of the rectangular social space is increased 10 times from $100r$ to $1000r$, but the maximum average delay of social search grows only 1.5 times from 9.17 steps to 13.82 steps. Note also that the average delay of social search under the Inaltekin-Chiang-Poor Model 2 seems to exhibit self-similarity: the two graphs look similar to each other except for a change of scale in the social distance and delay dimensions.

The next theorem provides the probability distribution for the delay of social search in the Inaltekin-Chiang-Poor Model 2.

Theorem 23.5. *The probability distribution of the delay of the targeted social search process, when it runs over the social network $\mathcal{G} = (V, E)$ constructed according to the Inaltekin-Chiang-Poor Model 2, converges to*

$$p_1(j) = 1_{\{j=1\}}, \text{ and}$$

$$p_k(j) = p_{k-1}(j-1)\varphi\left(\beta_{k-1,k}\right) + \sum_{i=1}^{k-2} p_i(j-1)\left(\varphi\left(\beta_{i,k}\right) - \varphi\left(\beta_{i+1,k}\right)\right) \text{ for } k \geq 2$$

in the continuum limit, where $\varphi(x)$, $x > 0$, is the probability generating function of the distribution Q defined as $\varphi(x) = \mathrm{E}\left[x^N\right]$ and $\beta_{i,k}$, $i, k \geq 1$, for rectangular, spherical, and circular social spaces is given according to

1. *Rectangular:* $\beta_{i,k} = 1 - \left(\frac{i-1}{k}\right)^2$,
2. *Spherical:* $\beta_{i,k} = \frac{\cos\left(\frac{(i-1)r}{R}\right) - \cos\left(\frac{kr}{R}\right)}{1 - \cos\left(\frac{kr}{R}\right)}$, *and*
3. *Circular:* $\beta_{i,k} = 1 - \frac{i-1}{k}$.

Proof. See [22]. ∎

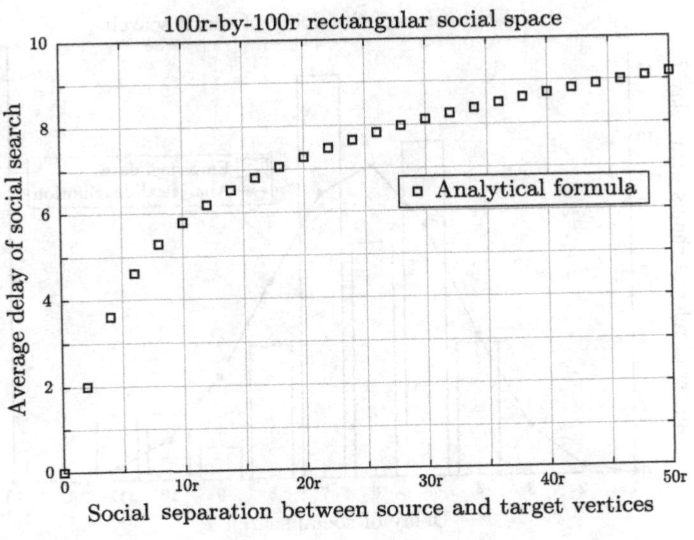

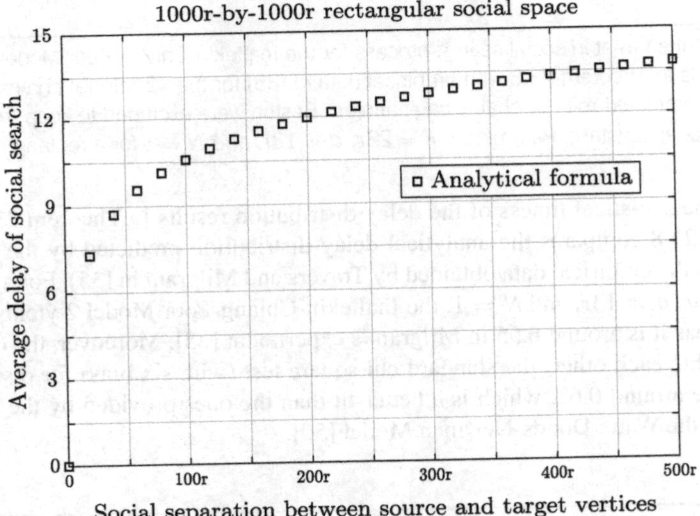

FIG. 23.7

Average delay of the targeted social search process for the Inaltekin-Chiang-Poor Model 2. Demonstration of the growth of the average delay of the targeted social search process for the Inaltekin-Chiang-Poor Model 2 in Theorem 23.4. Side lengths of a rectangular social space are increased 10 times, but the maximum delay increases only 1.5 times. $N = 1$.

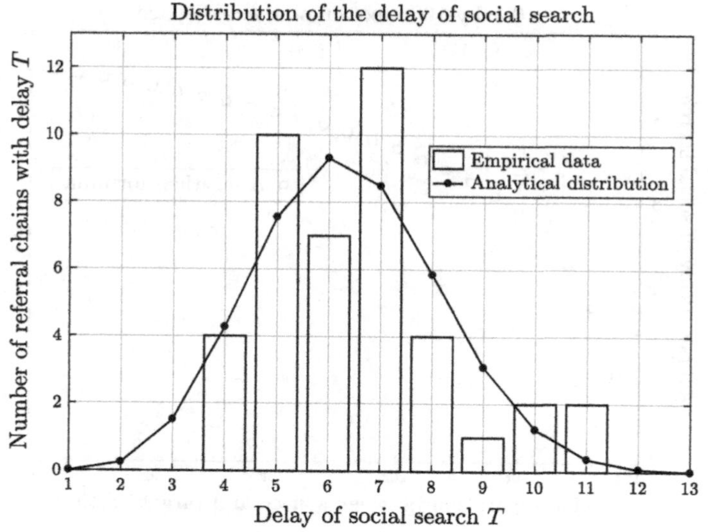

FIG. 23.8

Delay distribution of the targeted social search process for the Inaltekin-Chiang-Poor Model 2. Comparison of the analytical formula in Theorem 23.5 with empirical data in [35] for the 42 referral chains that originated in Nebraska. The 24 completed referral chains originating in Boston were excluded to keep the source-target separation in the model constant. Parameters $R = 28r$, $d = 13r$, and $N = 1$ for a rectangular social space.

As a test for the statistical fitness of the delay distribution results in Theorem 23.5 to small-world experiments, Fig. 23.8 compares the analytical delay distribution predicted by the Inaltekin-Chiang-Poor Model 2 with the empirical data obtained by Travers and Milgram in [35]. For a rectangular social space with $R = 28r$, $d = 13r$, and $N = 1$, the Inaltekin-Chiang-Poor Model 2 yields an average delay around 6.36 whereas it is around 6.55 in Milgram's experiment [35]. Moreover, the delay distributions statistically resemble each other: the standard chi-square test (with six bins) for discrete distributions produces a P value around 0.63, which is a better fit than the ones provided by the Inaltekin-Chiang-Poor Model 1 and the Watts-Dodds-Newman Model [50].

23.4 ALTRUISTIC DATA ROUTING

In this section, we will switch the focus from the targeted social search process to the altruistic data routing process for general topologies of social and physical connections. One important distinction between the two cases is that the process of altruistic data routing requires *global* knowledge of social and physical connections whereas individuals operate based on their local knowledge of social connections in the targeted social search process in order to reach a target individual. Our network model to investigate the properties of the altruistic data routing will be similar to the classical selfish routing setup [58,59]. Below, we will first present the details of the model and then characterize the global system behavior through a Nash equilibrium analysis.

23.4.1 MODEL FOR ALTRUISTIC DATA ROUTING

Network details

We consider a *physical* communication graph $\mathcal{G} = (V, E)$, where V is the set of vertices representing communication devices and E is the set of edges representing the communication links connecting these devices. There are socially connected M different source-target pairs, which will be called *players*, using $\mathcal{G}$ to communicate at rate Γ_i (measured in terms of bits per second) for $i = 1, \ldots, M$. Let $\mathcal{P}_i$ be the set of available paths connecting the ith source-target pair over $\mathcal{G}$.[4] Further, let $\mathcal{P}$ be the set of all paths connecting all M source-target pairs. That is,

$$\mathcal{P} = \bigcup_{i=1}^{M} \mathcal{P}_i.$$

The ith player can distribute her incoming data traffic rate over the paths in $\mathcal{P}$, which leads to the following definition of routing policy.

Definition 23.1. A routing policy of player i, $i \in \{1, \ldots, M\}$, is a rate distribution function $f_i : \mathcal{P} \mapsto [0, \infty)$ over the set of paths available in $\mathcal{G} = (V, E)$. We call it a *feasible* routing policy if it satisfies

$$f_i(P) \geq 0 \text{ for all } P \in \mathcal{P}_i, \sum_{P \in \mathcal{P}} f_i(P) = \Gamma_i \text{ and } f_i(P) = 0 \text{ for all } P \in \mathcal{P} \setminus \mathcal{P}_i.$$

Further, a routing policy profile $\boldsymbol{f} = (f_1, \ldots, f_M)$ is the vector of individual routing policies. ■

We will focus on feasible routing policies in the remainder of this section, even though the identifier "feasible" is not always made explicit in the text. Consider a routing policy profile $\boldsymbol{f} = (f_1, \ldots, f_M)$. $\boldsymbol{f}$ induces a total network flow function on the set of paths in $\mathcal{G}$, which is given by

$$f(P) = \sum_{i=1}^{M} f_i(P), \text{ for all } P \in \mathcal{P}.$$

We note that $\boldsymbol{f}$, in turn, determines how much traffic is routed through each edge. To see this, let f_e be the total traffic on edge $e \in E$. Then,

$$f_e = \sum_{P \in \mathcal{P}} f(P) \, 1_{\{e \in P\}}, \text{ for all } e \in E.$$

By its definition, f_e is composed of individual traffic originated from each player and using edge $e \in E$. One useful decomposition for f_e is

$$f_e = f_{e,i} + f_{e,-i},$$

where $f_{e,i} = \sum_{P \in \mathcal{P}} f_i(P) \, 1_{\{e \in P\}}$ and $f_{e,-i} = \sum_{j \neq i} f_{e,j}$. f_e determines the amount of data that flows through edge $e \in E$ per second, which is the main measure of congestion level for e. To formalize this idea further, we assign a congestion related latency function $l_e(x)$, $x \geq 0$, to each $e \in E$. We assume that

[4]A path $P \in \mathcal{P}_i$ is a collection of edges connecting the ith source-target pair. We allow the possibility that different source-target pairs can share the same paths due to virtualization of network services.

$l_e(x)$ is a nonnegative, nondecreasing, and continuous function of its parameter. Then, for an induced network flow f, the latency experienced by data traffic over the edge $e \in E$ is $l_e(f_e)$.

It is important to remember that f_e is determined by the routing policy profile $\boldsymbol{f} = (f_1, \ldots, f_M)$, and hence delay over an edge can also be viewed as a function of $\boldsymbol{f}$. Latency of a path $P \in \mathcal{P}$ is equal to the additive latencies of edges on P, which leads to[5]

$$l_P(\boldsymbol{f}) = \sum_{e \in E} l_e(f_e)\, 1_{\{e \in P\}}.$$

Optimum routing problem

The goal of each player is to minimize total latency of her data traffic by respecting the delay of other players in an altruistic way.[6] More formally, we first write the total latency of the traffic from player i as

$$L_i(\boldsymbol{f}) = \sum_{P \in \mathcal{P}} l_P(\boldsymbol{f}) f_i(P).$$

We next consider some altruistic behavior characterized by the social relationship matrix $\boldsymbol{A} = [a_{i,j}]_{i,j=1}^{M}$. The entries of $\boldsymbol{A}$ satisfy $a_{i,j} \in [0,1]$ and $a_{i,i} = 1$ for all $i, j = 1, \ldots, M$. Finally, the cost of a routing policy profile $\boldsymbol{f} = (f_1, \ldots, f_M)$ for player $i \in \{1, \ldots, M\}$ is written as

$$C_i(\boldsymbol{f}) = \sum_{j=1}^{M} \alpha_{i,j} L_j(\boldsymbol{f}).$$

As is standard, we will write $\boldsymbol{f}_{-i}$ to refer to routing policies chosen by all players except the routing policy chosen by the ith one. Then, the local optimization problem solved by each player $i \in \{1, \ldots, M\}$ is given by

$$\begin{aligned} \text{minimize} \quad & C_i(f_i, \boldsymbol{f}_{-i}) \\ \text{subject to} \quad & \textstyle\sum_{P \in \mathcal{P}} f_i(P) = \Gamma_i \\ & f_i(P) \geq 0 \text{ for all } P \in \mathcal{P} \\ & f_i(P) = 0 \text{ for all } P \in \mathcal{P} \setminus \mathcal{P}_i. \end{aligned} \tag{23.10}$$

We note that the set of paths $\mathcal{P}$ is always finite for the physical communication graphs considered in this section, and hence the decision variables for player i to minimize her cost are the *finite* collection of nonnegative real variables given by $\{f_i(P)\}_{P \in \mathcal{P}}$ subject to the above constraints. As a result, the standard techniques from the theory of convex optimization [60] under appropriate assumptions on the link latency functions can be utilized to solve primal optimization problems faced by individual players. To this end, two quantities will frequently appear in our calculations below, which deserve standalone definitions.

[5]When considering individual latency over an edge $e \in E$, it will be more convenient to represent it as a function of total flow over e, i.e., as $l_e(f_e)$. On the other hand, when considering latency of a path $P \in \mathcal{P}$, it will be more insightful to represent it as a function of routing policy profile, i.e., as $l_P(\boldsymbol{f})$.

[6]The reason behind considering the total latency as the performance metric is the assumption that each received bit contributes to the communication and signal processing quality.

Definition 23.2. A pseudoflow $\Lambda_i(P)$ seen by player $i \in \{1, \ldots, M\}$ over a path $P \in \mathcal{P}$ is equal to

$$\Lambda_i(P) = \sum_{j=1}^{M} \alpha_{i,j} f_j(P).$$

Similarly, a pseudoflow $\lambda_{e,i}$ seen by player $i \in \{1, \ldots, M\}$ over an edge $e \in E$ is equal to

$$\lambda_{e,i} = \sum_{j=1}^{M} \alpha_{i,j} f_{e,j}.$$

■

Using $\Lambda_i(P)$, an alternative way to express $C_i(f)$ is $C_i(f) = \sum_{P \in \mathcal{P}} \Lambda_i(P)\, l_P(f)$. We note that $\Lambda_i(P)$ or $\lambda_{e,i}$ is not an actual flow realized over a path or an edge. Rather, these are the flow rates conceived to occur by player i due to her altruistic behavior. As a result, they will not change the actual latencies over paths or edges, but they will rather behave as quantities capturing the importance ranking of player i for others' traffic. As is done for f_e, one useful decomposition for $\lambda_{e,i}$ is

$$\lambda_{e,i} = f_{e,i} + \lambda_{e,-i},$$

where $\lambda_{e,-i} = \sum_{j \neq i} \alpha_{i,j} f_{e,j}$. $\lambda_{e,i}$ is always smaller than or equal to f_e, and hence $\lambda_{e,-i} \leq f_{e,-i}$.

Solution concept: Group Nash equilibrium

We will use the solution concept introduced in [61] to obtain the equilibrium points for the problem of altruistic data routing defined above. The formal definition that we will follow throughout this section is given below.

Definition 23.3. A routing policy profile $f^\star = (f_1^\star, \ldots, f_M^\star)$ is called a *group Nash equilibrium* solution for the problem of altruistic data routing over $\mathcal{G} = (V, E)$ if $f_i^\star$ is a solution for Eq. (23.10) for all $i \in \{1, \ldots, M\}$ given the routing policies $f_{-i}^\star$ of all other players. ■

Note that the totally selfish case corresponds to the case $\alpha_{i,j} = 0$ for all nondiagonal entries of $\boldsymbol{A}$. In this case,

$$C_i(f) = L_i(f) \quad \text{for all } i \in \{1, \ldots, M\},$$

and hence all players care only about their own latency, without any regard to the latencies experienced by others. On the other hand, the case $\alpha_{i,j} = 1$ for all $i, j \in \{1, \ldots, M\}$ corresponds to the case of complete altruism. In this case,

$$C_i(f) = \sum_{j=1}^{M} L_j(f) \quad \text{for all } i \in \{1, \ldots, M\},$$

and hence all players care about latencies of other players as much as they care about their own, which leads to the minimization of total network latency through local adjustment of their individual routing policies. For other selections of the social relationship matrix $\boldsymbol{A} = [\alpha_{i,j}]_{i,j=1}^{M}$, the concept of group Nash equilibrium traces a continuum range of solutions in the spectrum from totally selfish behavior to complete altruism.

23.4.2 GROUP NASH EQUILIBRIA FOR THE PROCESS OF ALTRUISTIC DATA ROUTING

In this part, we establish necessary and sufficient conditions for a routing policy profile f to be a group Nash equilibrium. At a group Nash equilibrium point, all players are required to solve the primal optimization problems given by Eq. (23.10) simultaneously. Hence, as a first step for characterizing an equilibrium solution, we require that Eq. (23.10) can be solved efficiently. To this end, one condition that will enable us to obtain a solution for Eq. (23.10) is the *semiconvexity* of the edge latency functions, a notion that we define formally as below.

Definition 23.4. A function $l(x)$ is said to be semiconvex if $l(x)(x+d)$ is convex for all $d \in \mathbb{R}$. ■

For example, all linear latency functions, i.e., $l_e(x) = ax + b$ with positive constants a and b, are semiconvex. Linear latency functions are used extensively in the papers on selfish routing [58,59] to obtain the price-of-anarchy (PoA) as the efficiency metric for equilibrium solutions. The next lemma establishes the convexity of the optimization problem in Eq. (23.10).

Lemma 23.2. *If all edge latency functions are semiconvex, then the primal optimization problem faced by player i in Eq. (23.10) is a convex optimization problem for all $i \in \{1, \dots, M\}$.*

Proof. We fix $i \in \{1, \dots, M\}$. Then, the cost of a routing policy profile $f = (f_1, \dots, f_M)$ for player i can be written as

$$\begin{aligned} C_i(f) &= \sum_{P \in \mathcal{P}} \Lambda_i(P)\, l_P(f) \\ &= \sum_{j=1}^{M} \sum_{P \in \mathcal{P}} \sum_{e \in E} \alpha_{i,j} f_j(P)\, l_e(f_e)\, 1_{\{e \in P\}} \\ &= \sum_{e \in E} l_e(f_e) \sum_{j=1}^{M} \alpha_{i,j} f_{e,j} \\ &= \sum_{e \in E} l_e(f_e)\, \lambda_{e,i}. \end{aligned}$$

Hence, using the decompositions $f_e = f_{e,i} + f_{e,-i}$ and $\lambda_{e,i} = f_{e,i} + \lambda_{e,-i}$ for f_e and $\lambda_{e,i}$, respectively, we have

$$C_i(f) = \sum_{e \in E} l_e\left(f_{e,i} + f_{e,-i}\right)\left(f_{e,i} + \lambda_{e,-i}\right).$$

Let $g_e(x) = l_e\left(x + f_{e,-i}\right)\left(x + \lambda_{e,-i}\right)$ and $h_e\left(\{f_i(P)\}_{P \in \mathcal{P}}\right) = \sum_{P \in \mathcal{P}} f_i(P)\, 1_{\{e \in P\}}$. $g_e(x)$ is a convex nondecreasing function of x due to semiconvexity of $l_e(x)$ and $h_e\left(\{f_i(P)\}_{P \in \mathcal{P}}\right)$ is a linear function of $\{f_i(P)\}_{P \in \mathcal{P}}$, which shows that

$$l_e\left(f_{e,i} + f_{e,-i}\right)\left(f_{e,i} + \lambda_{e,-i}\right) = g_e \circ h_e\left(\{f_i(P)\}_{P \in \mathcal{P}}\right)$$

is convex for all $e \in E$, as a function of decision variables $\{f_i(P)\}_{P \in \mathcal{P}}$. This proves that $C_i\left(\{f_i(P)\}_{P \in \mathcal{P}}, f_{-i}\right)$ is a convex function of decision variables $\{f_i(P)\}_{P \in \mathcal{P}}$ if $l_e(x)$ is a semiconvex function. Hence, the primal optimization problem faced by player i is a convex optimization problem. ■

The next lemma provides necessary and sufficient conditions for a collection of decision variables $\{f_i(P)\}_{P\in\mathcal{P}}$ to be a solution for the ith player's individual optimization problem, which is the key result to characterize the group Nash equilibria for the problem of altruistic data routing.

Lemma 23.3. *Assume all edge latency functions are semiconvex. Then, a feasible routing policy $f_i : \mathcal{P} \mapsto [0, \infty)$ is a solution for the primal optimization problem (23.10) faced by player i if and only if*

$$l_H(f) + \sum_{P\in\mathcal{P}} \Lambda_i(P)\, c_{P\cap H}(f) \le l_G(f) + \sum_{P\in\mathcal{P}} \Lambda_i(P)\, c_{P\cap G}(f) \tag{23.11}$$

for all $H, G \in \mathcal{P}_i$ with $f_i(H) > 0$, where $c_S(f)$ is defined as $c_S(f) = \sum_{e\in S} l'_e(f_e)$ for all $S \subseteq E$ and $l'_e(f_e) = \frac{dl_e(x)}{dx}\big|_{x=f_e}$.

Proof. The proof follows from an application of the standard Karush-Kuhn-Tucker (KKT) conditions to Eq. (23.10) [60]. ■

The next theorem is the main result of this section. It puts forward necessary and sufficient conditions for a routing policy profile $f^\star = (f_1^\star, \ldots, f_M^\star)$ to be a group Nash equilibrium point for the process of altruistic data routing. It is a direct consequence of the lemmas provided above.

Theorem 23.6. *Assume all edge latency functions are semiconvex. Then, a feasible routing policy profile $f^\star = (f_1^\star, \ldots, f_M^\star)$ is a group Nash equilibrium for the altruistic data routing problem in the sense of Definition 23.3 if and only if it satisfies the following conditions*

$$l_H(f^\star) + \sum_{P\in\mathcal{P}} \Lambda_i(P)\, c_{P\cap H}(f^\star) \le l_G(f^\star) + \sum_{P\in\mathcal{P}} \Lambda_i(P)\, c_{P\cap G}(f^\star) \tag{23.12}$$

for all $i \in \{1, \ldots, M\}$ and $H, G \in \mathcal{P}_i$ with $f_i(H) > 0$.

Proof. $f^\star = (f_1^\star, \ldots, f_M^\star)$ is a group Nash equilibrium for the altruistic data routing problem if and only if $f_i^\star$ is a solution for the primal optimization problem in Eq. (23.10) for each player $i \in \{1, \ldots, M\}$, given the routing policies $f_{-i}^\star$ of the rest. Fix player $i \in \{1, \ldots, M\}$. By Lemma 23.2, the primal optimization problem faced by player i is convex because the edge latency functions are semiconvex. By Lemma 23.3, $f_i^\star$ is a solution for Eq. (23.10) if and only if the condition in Eq. (23.11) is satisfied for all $H, G \in \mathcal{P}_i$ with $f_i(H) > 0$. Repeating the same arguments for all players, we conclude that $f^\star = (f_1^\star, \ldots, f_M^\star)$ is a group Nash equilibrium if and only if the condition in Eq. (23.12) is satisfied for all $i \in \{1, \ldots, M\}$ and $H, G \in \mathcal{P}_i$ with $f_i(H) > 0$. ■

An existence result of at least one such group Nash equilibrium point is provided in the next theorem.

Theorem 23.7. *Assume all edge latency functions are semiconvex. Then, there exists a feasible routing policy profile $f^\star = (f_1^\star, \ldots, f_M^\star)$ satisfying Eq. (23.12) for all $i \in \{1, \ldots, M\}$ simultaneously.*

Proof. The optimization problems faced by players $i \in \{1, \ldots, M\}$ are all convex optimization problems due to the semiconvexity of edge latency functions. Hence, the existence of a Nash equilibrium point in pure strategies follows from the standard result in [62]. ■

23.5 CONCLUSIONS

In this chapter, we have first reviewed the commonly used graph theoretical models for social network analysis. Then, we have focused on the *decentralized* targeted social search process, which evolves based on the existing social connections and local network information, by studying three models: (i) Kleinberg's model [47,48], (ii) the Watts-Dodds-Newman Model [50], and (iii) the Inaltekin-Chiang-Poor Models [21,22]. Kleinberg's model provides an analytical justification for the small-world phenomenon when the social network is constructed on a rectangular grid through the addition of distance-dependent random long-range connections. In particular, the average delay of social search under this model grows poly-logarithmically when the decay rate of probability of finding a long-range contact at a certain social distance matches the dimension of the substrate grid topology.

The Watts-Dodds-Newman Model, on the other hand, provides a numerical justification for the small-world phenomenon when the social network is constructed based on multiple social dimensions. Through this model, it has been illustrated that the multidimensioned nature of social identity leads to a class of searchable social networks much larger than what is predicted by Kleinberg's model. The Watts-Dodds-Newman Model also exhibits statistical similarity with the empirical data collected in small-world experiments [35] after adjusting model parameters appropriately. Lastly, the Inaltekin-Chiang-Poor Models provide alternative explanations for why we observe short referral chains of social search in small-world experiments without the "exact match" requirement in Kleinberg's model. These models (i) capture the order-disorder properties of the small-world network model [11] due to a threshold rule for forming short-range contacts, (ii) are constructed on continuum measure-metric spaces as a proxy for the social space, and (iii) can mimic the scale-free features of the Barabási-Albert model [9] through a parametrization of the generation of social networks in terms of the distribution of the number of long-range contacts. An important feature of the Inaltekin-Chiang-Poor Models is that they lead to analytical expressions for the average delay of social search as well as its probability distribution. The derived analytical results provide good statistical matches with the empirical data [35], having a better chi-square fit than the one in [50].

In the final part of this chapter, we have studied the process of *centralized* altruistic data routing, which evolves based on the physical connections among devices as well as the social connections among individuals using these devices. In this case, the social ties determine the extent to which individuals care about the latency performance of others simultaneously accessing the same communication resources described through physical device connections. Using a game theoretic approach and defining the strategy set of a player as a rate distribution function over the set of paths connecting it to the target vertex, we have provided necessary and sufficient conditions for a strategy profile to be a Nash equilibrium point. We have also established the existence of a Nash equilibrium point in pure strategies satisfying these necessary and sufficient conditions.

REFERENCES

[1] Granovetter M. Threshold models of collective behavior. Am J Sociol 1978;83(6):1420–43. https://doi.org/10.1086/226707.

[2] Granovetter M. The strength of weak ties: a network theory revisited. Sociol Theory 1983;1:201–33. https://doi.org/10.2307/202051.

[3] Feld SL. Why your friends have more friends than you do. Am J Sociol 1991;96(6):1464–77. https://doi.org/10.1086/229693.

[4] Watts DJ. A simple model of global cascades on random networks. Proc Natl Acad Sci U S A 2002;99(9):5766–71. https://doi.org/10.1073/pnas.082090499.
[5] Kossinets G, Watts DJ. Empirical analysis of an evolving social network. Science 2006;311(5757):88–90. https://doi.org/10.1126/science.1116869.
[6] Centola D, Macy M. Complex contagions and the weakness of long ties. Am J Sociol 2007;113(3):702–34. https://doi.org/10.1086/521848.
[7] Schnettler S. A small world on feet of clay? A comparison of empirical small-world studies against best-practice criteria. Soc Netw 2009;31(3):179–89. https://doi.org/10.1016/j.socnet.2008.12.005.
[8] Barabási AL. Luck or reason. Nature 2012;489(7417):507–8. https://doi.org/10.1038/nature11486.
[9] Barabási AL, Albert R. Emergence of scaling in random networks. Science 1999;286(5439):509–12. https://doi.org/10.1126/science.286.5439.509.
[10] Barabási AL. Network science. Cambridge, United Kingdom: Cambridge University Press; 2016.
[11] Watts DJ, Strogatz SH. Collective dynamics of 'small-world' networks. Nature 1998;393(6684):440–2. https://doi.org/10.1038/30918.
[12] Watts DJ. Networks, dynamics, and the small-world phenomenon. Am J Sociol 1999;105(2):493–527. https://doi.org/10.1086/210318.
[13] Strogatz SH. Exploring complex networks. Nature 2001;410(6825):268–76. https://doi.org/10.1038/35065725.
[14] Price DJS. Networks of scientific papers. Science 1965;149(3683):510–5. https://doi.org/10.1126/science.149.3683.510.
[15] Bollobás B. Random graphs. 2nd ed. Cambridge Studies in Advanced Mathematics. Cambridge, United Kingdom: Cambridge University Press; 2001.
[16] Barabási AL, Albert R, Jeong H. Mean-field theory for scale-free random networks. Physica A 1999; 271(1–2):173–87. https://doi.org/10.1016/S0378-4371(99)00291-5.
[17] Krapivsky PL, Redner S, Leyvraz F. Connectivity of growing random networks. Phys Rev Lett 2000; 85(21):4629–32. https://doi.org/10.1103/PhysRevLett.85.4629.
[18] Watts DJ. The "new" science of networks. Ann Rev Sociol 2004;30(1):243–70. https://doi.org/10.1146/annurev.soc.30.020404.104342.
[19] Newman MEJ, Moore C, Watts DJ. Mean-field solution of the small-world network model. Phys Rev Lett 2000;84(14):3201–4. https://doi.org/10.1103/PhysRevLett.84.3201.
[20] Klemm K, Eguiluz VM. Growing scale-free networks with small-world behavior. Phys Rev E 2010; 65(5):057102. https://doi.org/10.1103/physreve.65.057102.
[21] Inaltekin H, Chiang M, Poor HV. Average message delivery time for small-world networks in the continuum limit. IEEE Trans Inf Theory 2010;56(9):4447–70. https://doi.org/10.1109/TIT.2010.2054490.
[22] Inaltekin H, Chiang M, Poor HV. Delay of social search on small-world graphs. J Math Sociol 2014;38(1):1–46. https://doi.org/10.1080/0022250X.2011.629062.
[23] Granovetter M. The strength of weak ties. Am J Sociol 1973;78(6):1360–80. https://doi.org/10.1086/225469.
[24] Lee NH. The search for an abortionist. 1st ed. Chicago, IL, USA: University of Chicago Press; 1969.
[25] Klovdahl AS. Social networks and the spread of infectious diseases: the AIDS example. Soc Sci Med 1985;21(11):1203–16. https://doi.org/10.1016/0277-9536(85)90269-2.
[26] Bearman PS, Moody J, Stovel K. Chains of affection: the structure of adolescent romantic and social networks. Am J Sociol 2004;110(1):44–91. https://doi.org/10.2307/3568252.
[27] Ganesh A, Massoulie L, Towsley D. The effect of network topology on the spread of epidemics. In: Proceedings of the 24th annual joint conference of the IEEE computer and communications societies, Miami, FL, March 13–17. IEEE; 2005. p. 1455–66. https://doi.org/INFCOM.2005.1498374.
[28] Centola D, Eguiluz VM, Macy MW. Cascade dynamics of complex propagation. Phys A: Stat Mech Appl 2007;374(1):449–56. https://doi.org/10.1016/j.physa.2006.06.018.
[29] Centola D. Failure in complex social networks. J Math Sociol 2009;33(1):64–8. https://doi.org/10.1080/00222500802536988.

[30] Coleman J, Katz E, Menzel H. The diffusion of an innovation among physicians. Sociometry 1957;110(4):253–70. https://doi.org/10.2307/2785979.
[31] Centola D, Willer R, Macy M. The emperor's dilemma: a computational model of self-enforcing norms. Am J Sociol 2005;110(4):1009–40. https://doi.org/10.1086/427321.
[32] Donovan P. How idle is idle talk? One hundred years of rumor research. Diogenes 2007;54(1):59–82. https://doi.org/10.1177/0392192107073434.
[33] Burt RS. Structural holes: the social structure of competition. Cambridge, MA, USA: Harvard University Press; 1992.
[34] Milgram S. The small world problem. Psychol Today 1967;1:61–7.
[35] Travers J, Milgram S. An experimental study of the small world problem. Sociometry 1969;32(4):425–43. https://doi.org/10.2307/2786545.
[36] Korte C, Milgram S. Acquaintance networks between racial groups: application of the small world method. J Pers Soc Psychol 1970;15(2):101–8. https://doi.org/10.1037/h0029198.
[37] Erickson BH, Kringas PR. Small world of politics, or seeking elites from bottom up. Can Rev Sociol 1975;12(4):585–93. https://doi.org/10.1111/j.1755-618X.1975.tb00562.x.
[38] Guiot JM. A modification of Milgram's small world method. Eur J Soc Psychol 1976;6(4):503–7. https://doi.org/10.1002/ejsp.2420060409.
[39] Lin N, Dayton PW, Greenwald P. The urban communication network and social stratification: a "small world" experiment. In: Ruben BD, editor. Communication yearbook, New Brunswick, NJ, USA; 1977. p. 107–19.
[40] Weimann G. The not-so-small world—ethnicity and acquaintance networks in Israel. Soc Netw 1983;5(3):289–302. https://doi.org/10.1016/0378-8733(83)90029-1.
[41] Dodds PS, Muhamad R, Watts DJ. An experimental study of search in global social networks. Science 2003;301(5634):827–9. https://doi.org/10.1126/science.1081058.
[42] Kleinfeld J. The small world problem. Society 2002;39(2):61–6. https://doi.org/10.1007/bf02717530.
[43] Goel S, Muhamad R, Watss DJ. Social search in "small-world" experiments. In: Proceedings of the 18th international conference on world wide web. WWW'09. New York, NY: ACM; 2009. p. 701–10. https://doi.org/10.1145/1526709.1526804.
[44] Schnettler S. A structured overview of 50 years of small-world research. Soc Netw 2009;31(3):165–78. https://doi.org/10.1016/j.socnet.2008.12.004.
[45] Shotland RL. University communication networks: the small world method. New York, NY: John Wiley & Sons; 1976.
[46] Stevenson WB, Davidson B, Manev I, Walsh K. The small world of the university: a classroom exercise in the study of networks. Connections 1997;20(2):23–33.
[47] Kleinberg JM. Navigation in a small world. Nature 2000;406(6798):845. https://doi.org/10.1038/35022643.
[48] Kleinberg JM. The small-world phenomenon: an algorithmic perspective. In: Proceedings of the 32nd annual ACM symposium on theory of computing. STOC'00. New York, NY: ACM; 2000. p. 163–70. https://doi.org/10.1145/335305.335325.
[49] Newman MEJ. The structure and function of complex networks. SIAM Rev 2003;45(2):167–256. https://doi.org/10.1137/S003614450342480.
[50] Watts DJ, Dodds PS, Newman MEJ. Identity and search in social networks. Science 2002;296(5571):1302–5. https://doi.org/10.1126/science.1070120.
[51] Watts DJ. Six degrees: the science of a connected age. New York, NY: Norton & Company; 2003.
[52] Bernard HR, Killworth PD, Evans MJ, McCarty C, Shelley GA. Studying social relations cross-culturally. Ethnology 1988;27(2):155–79. https://doi.org/10.2307/3773626.
[53] Penrose M. Random geometric graphs. New York, NY: Oxford University Press; 2003.
[54] Bassett DS, Bullmore E. Small-world brain networks. Neuroscientist 2006;12(6):512–23. https://doi.org/10.1177/1073858406293182.

[55] Bassett DS, Gazzaniga MS. Understanding complexity in the human brain. Trends Cogn Sci 2011;15(5):200–9. https://doi.org/10.1016/j.tics.2011.03.006.
[56] Hill RA, Dunbar RIM. Social network size in humans. Hum Nat 2003;14(1):53–72. https://doi.org/10.1007/s12110-003-1016-y.
[57] McCarty C, Killworth PD, Bernard HR, Johnsen E, Shelley GA. Comparing two methods for estimating network size. Hum Organ 2001;60(1):28–39. https://doi.org/10.17730/humo.60.1.efx5t9gjtgmga73y.
[58] Roughgarden T, Tardos E. How bad is selfish routing? J ACM 2002;49(2):236–59. https://doi.org/10.1145/506147.506153.
[59] Roughgarden T, Tardos E. Bounding the inefficiency of equilibria in nonatomic congestion games. Games Econ Behav 2004;47(2):389–403. https://doi.org/10.1016/j.geb.2003.06.004.
[60] Boyd S, Vandenberghe L. Convex optimization. New York, NY: Cambridge University Press; 2004.
[61] Chen X, Gong X, Yang L, Zhang J. Exploiting social tie structure for cooperative wireless networking: a social group utility maximization framework. IEEE/ACM Trans Netw 2016;24(6):3593–606. https://doi.org/10.1109/TNET.2016.2530070.
[62] Rosen JB. Existence and uniqueness of equilibrium points for concave *n*-person games. Econometrica 1965;33(3):520–34. https://doi.org/10.2307/1911749.

CHAPTER

INFORMATION DIFFUSION AND RUMOR SPREADING

24

Argyris Kalogeratos*, **Kevin Scaman***,[†], **Luca Corinzia***,[‡], **Nicolas Vayatis***

*CMLA, ENS Cachan, CNRS, University of Paris-Saclay, Cachan, France**

MSR, Inria Joint Center, Palaiseau, France[†]

ETH Zurich, Zurich, Switzerland[‡]

24.1 INTRODUCTION

Modern societies understand the world, manifest different viewpoints, and test their objectiveness by exchanging information through direct communication or, in more recent years, through online social networks. On a larger scale, this process may also create consensus and mitigate social friction through public debate, two essential aspects of a healthy democracy. Information diffusion is often represented by *pieces of information* (e.g., news, scientific, or historical facts) that spread through a network. As for the network, that consists of interacting entities such as individuals, institutions (e.g., governments, authorities, or other organizations), and private entities (e.g., media, marketing agencies).

The Internet era has offered new means to produce and share information through large-scale online social networks. The disposition of large amounts of data coming from diffusion traces has helped scientific research improve our understanding of diffusion processes arising in various disciplines, including sociology, epidemiology, marketing, and computer system security. However, the *democratization* of content creation and sharing has not been adequately coupled with effective (self-, collective, or automatic) moderation, correction, and filtering mechanisms. Consequently, the explosive volume of the available content brings forward huge challenges regarding the human capacity to process that fast-paced and gigantic information stream as well as regarding the technical aspects of data management.

Our daily information diet tends to promote the variety in the content we consume to the expense of its precision and detail. During moments of crisis, the scarcity of trustworthy information and lack of time to analyze it lead to the proliferation of false rumors. There are also various psychological factors that impact the way we participate in this exchange. For instance, people get influenced by others, but also tend to search and recall information and facts that align with their already formed belief system (*confirmation bias*).

Furthermore, users interact preferably with people of similar profiles and opinions (*homophily*), a tendency that greatly reduces the heterogeneity of the user's perceived public debate. In addition, members of any online group receive social pressure to conform to a group's beliefs; that tends to

Cooperative and Graph Signal Processing. https://doi.org/10.1016/B978-0-12-813677-5.00024-9

radicalize opinions and allow questionable ideas to gain momentum (*echo chambers*). Then, the relative isolation of small online communities may lead them to believe in false rumors, even create a false consensus against what is considered as verifiable by the majority of society. The situation may get considerably aggravated in periods of political tension where polarization and partisanship grow in well-segregated groups that reduce significantly their exposure to counterarguments.

Rumor spreading and control. There are many types of misinformation: bad or "yellow" journalism, fake news, rumors and unverified information, hoaxes, and others (for a discussion on the taxonomy see [1]). Despite the fact that many studies hardly distinguish these types, there are still notable differences with regards to the *actors* propagating unverified or untrue information (e.g., individuals, media, politicians, or authorities), their *motives* (e.g., ignorance, desire to be part of a movement, gaining visibility and revenues, or as part of a speculative communication campaign), and the way people interact with a new piece of information in each of those cases, especially during its verification process. As has been pointed out, terms such as "fake news" *are just new names for very old problems*. The particular recent concern of public opinion on fake news is, however, due to the fact that the cascading effects of misinformation gain magnitude and speed in online social networks, and thus their short-term negative impact is boosted. These effects have been recorded in numerous major events, such as terrorist attacks, social demonstrations, elections, natural disasters, and wars.

In this chapter we mainly refer to *untrue rumors*[1] that represent false information and may have malicious motives. Such rumors are usually proven false shortly after their appearance. However, the debunking may not propagate fast enough in the social network to prevent a rumor from pursuing its diffusion (this is also the case, for example, of long-lasting rumors such as conspiracy theories) and that is exactly the point where computational tools can be beneficial.

There have been many developments in recent decades concerning both information dissemination and viral epidemics on networks. Despite the particular properties of rumor spreading, it is still a type of information diffusion for which many generic models and results are therefore relevant. Early models originated from the Susceptible-Infected-Removed (SIR) epidemic model [2,3] and a detailed related work is provided in the next section. Worth mentioning though, is the modern family of Information Cascade Models (ICMs) [4], which considers heterogeneous node-to-node transmission probabilities. ICM fits well to problems related to information diffusion on social networks and, among others, finds straightforward applications in digital marketing [5]. Indeed, ICMs were used to fit real information cascade data and observed node "infection" times in the MemeTracker dataset [6]. In another work, the aim was to infer the edges of a diffusion network and estimate the transmission rates of each edge that best fit the observed data [7].

Theoretical studies have given valuable insights on diffusion processes by defining quantities tightly related with the systemic behavior (e.g., epidemic threshold, extinction time) and describing how a diffusion unfolds from an initial set of contagious nodes. Most notably, a number of studies highlighted the crucial role that the network structure plays in how the diffusion process unfolds, which is also the

[1]According to the Oxford English Dictionary, a rumor is "*a currently circulating story or report of uncertain or doubtful truth.*" Thus, a rumor is by definition uncertain and may eventually be true or false. However, what will always be problematic is the fact that rumors gain disproportional circulation speeds to their level of certainty.

subject on which this chapter is largely devoted. The relation between the network structure and the behavior of SIR epidemics has been shown in [8]. Follow-up works verified this relation and broadened the discussion to other types of diffusion models [9,10]. Similar theoretical results have then been given for ICM as well [11,12].

The quantification of systemic properties can help on the direction of risk assessment (e.g., economic, health, social risks) and, furthermore, enable *diffusion process engineering* whose aim could be either to suppress or enhance a spreading. Under ICM, this engineering task is also named in literature as *influence optimization* or *activity shaping*, whereas the maximization has received a lot of attention for its direct marketing applications. In recent years, the suppression of information diffusion processes has also become a hot topic because it is related to various security hazards, e.g., due to cascades of misinformation such as harmful rumors and fake news. Suppressive scenarios of the latter type are also possible in the ICM modeling context; the optimization problem would be the minimization of the spread of a piece of information in the network, e.g., by decreasing the probability for certain users to share the false content to their contacts. To the best of our knowledge there is no prior work on this direction and part of the contribution of this chapter is exactly on covering this gap by developing computational approaches that are able to reduce an undesired spread under the ICM.

Contribution and summary. The rest of the chapter keeps its focus on information diffusion and is structured as follows. We commence with the detailed related work (Section 24.2), the technical background regarding diffusion models (Section 24.3), and their dynamics as stochastic processes (Section 24.4). The reader may find helpful the Table 24.1 which lists the main notations we use in this chapter. Then, we discuss one of the interesting tasks arising in diffusion networks: the offline influence optimization through local intervention actions that affect the information spread (Section 24.5). The purpose can be either to minimize or maximize the influence by means of suppressive or enhancive actions, respectively. An efficient strategy should decide where on the network to perform a number of available actions (limited by a budget of resources) in order to better serve one of those two opposing aims.

To this end we extend the discussion with the novel approach that first appeared in [13], which frames this task as a generalized optimization problem under the ICM and enjoys a convex continuous relaxation. In particular, we present a class of algorithms based on the optimization of the spectral radius of the Hazard matrix using a projected subgradient method (Section 24.6). For these algorithms, which can address both the maximization and the minimization problem, we provide theoretical analysis. The suppressive case is, however, more interesting in the context of this chapter as it is straightforwardly related to the control of undesired diffusion processes such as the spread of rumors. Hence, we investigate two standard case-studies of the minimization problem (Section 24.7): the *quarantine* (e.g., see [10,14]) and the *node immunization* problem (see [15]).

Notably, among the major strengths of this framework is the fact that it can describe complex strategies that are able to use several immunization options by deploying simultaneously resources of different types (partial or total immunization of edges and nodes, etc.). We also discuss how such strategies could find practical application to rumor control scenarios. In a section with experimental results (Section 24.8), the main presented control algorithm, called *NetShape*, is compared to standard baselines and state-of-the-art competitors in synthetic and benchmark network datasets. In the last section (Section 24.9), we give our conclusions and directions of future research.

24.2 RELATED WORK

Modeling information and rumor spreading. Phenomena such as rumors are part of an old story that is adapted to the current technological context. Scientists started studying rumors and stories related to the two World Wars. Knapp [2], and soon after Allport and Postman [3,16], were among the first to analyze rumors and pose the question of their control. In the work of the latter two, it was pointed out that, loosely speaking, the spread of rumors is somewhat proportional to the general interest of the story and the ambiguity of the related evidence. The similarities between rumor and disease spreading were also noted in later literature, though Daley and Kendal were the first to connect epidemics and rumors in mathematical terms [17,18]. However, they noted that their dynamics may be strikingly different due to the particularly complex rumor-spreading mechanism. Specifically, they introduced a variant of the Susceptible-Infected-Removed (SIR) epidemic model, where stochastic recoveries are triggered either when (a) an infected node interacts with an already recovered one, or (b) two infected nodes interact and both may then recover. A slight modification was proposed in [19] concerning case (b) where only the infected node that initiates the interaction may recover.

These alterations to the basic epidemic model try to incorporate mechanisms where a person is probable to lose motivation in continuing to spread a rumor when he realizes that it is no longer novel and interesting, or has already been debunked. Interesting to note, though, there is no assumed self-recovery process and the recovery is rather brought about by crowdsourcing. This is in accordance to follow-up and recent data-driven studies on rumor spreading on `twitter`, which from one side observed self-correction to be very weak and slow to take effect while from the other side they observed an almost 1:1 ratio of users promoting important false rumors and users trying to debunk them [20,21].

Over the course of years, more refined SIR-like epidemic models were proposed for information diffusion, including rumors, that still have a permanent recovered state (for a survey on compartmental models see [22,23]). One example is the SEIR model that introduces the (E)xposed state in which the individual is infected but incubating before getting to (I) and become infectious to others. Another example is SEI[R]Z [24,25] that introduces competition among adopters at state (I) and those at state (Z) who, after infection, have become skeptics. Both adopters and skeptics recruit from the susceptible population; nodes can "exit" the system and change the population size over time. However, the state (S) also recruits from a general population that is out of the system, and one could assume that previously departed individuals may later become susceptible again.

Evidently, the most popular epidemic modeling choice for information cascades, including rumors, are the *monotonically increasing stochastic models* such as SIR that allow node transitions only toward more critical states and eventually lead to permanent recovery or removal (i.e., as if the node dies out). Indeed, such modeling fits to what is observed in high-frequency information circulation with a short life, a setting that covers the majority of the content reaching users from social networks, news broadcasts, the entertainment industry, and advertising. Nevertheless, for an information spread that spans longer time periods and may come and go in current affairs (e.g., political issues, ideas, competing products, long-lasting rumors), models that allow reinfection are definitely more relevant. In this sense, the Susceptible-Infected-Susceptible (SIS), or the more information-oriented SEI[R]Z [24,25], could be fit better and also enable dynamic approaches for suppressing a diffusion, e.g., the *priority-planning* [26] or the greedy approach of [27].

More recently, Information Cascade Models (ICMs) were introduced that have higher detail and can take advantage of the wealth of available social interaction data to fine-tune their parameters.

First, independent cascades have emerged as a relevant model for viral diffusion of ideas and opinions [7,28–30]. Similarly to SIR, independent cascades are also increasing stochastic processes. However, contrary to epidemic models, they capture the precise temporal dependencies between infection events of neighboring nodes but require larger training datasets to infer them properly. Second, multivariate Hawkes processes are self-exciting point processes that are considered the gold standard to deal with sequences of correlated events in many scientific fields, e.g., for earthquake prediction [31] as well as in biological [32], financial [33,34], and social interactions studies [35]. They were thus naturally adapted to information diffusion in social networks with the main advantage of allowing multiple events on a single node (e.g., posts, likes, or shares in the case of a social network) [36,37]. Finally, Linear Threshold Models were developed to account for more complex diffusion dynamics in which users may require more than one concordant piece of information to accept it [28].

Influence optimization. The first attempts to put forward computational approaches for assessing the influence of users in social networks were those in [38,39]. The *influence maximization* problem under the ICM was first formulated in [5]. It was proved that it is an NP-hard problem and remains NP-hard to approximate it within a factor $1 - 1/e$. It was also proven that the influence is a submodular function of the set of initially contagious nodes (referred to as *influencers*) and the authors proposed a greedy Monte Carlo-based algorithm as an approximation. A number of subsequent studies were focused on improving that technique [40,41]. Notably, today's state-of-the-art techniques on influence control under the ICM are still based on Monte Carlo simulations and a greedy mechanism to select the actions sequentially.

Besides the popularity of influence maximization, various questions regarding how one could apply suppressive interventions have also become a hot topic in recent years. However, to the best of our knowledge, there is no existing work under the ICM and, as mentioned in the introduction, the methodological contribution of this chapter is on the development of computational approaches under the ICM that are able to efficiently reduce an undesired spread (see Section 24.5).

Network structure, information spread, and control approaches. Recent theoretical studies have highlighted how crucial the structure of the underlying network is for the behavior of a diffusion process. Specifically, they have studied the way structural characteristics of the network do appear in quantities that are tightly related with the process behavior, such as the epidemic threshold and the extinction time.

An early work that drew a line between epidemic spreading and the structural properties of the underlying network is that in [8]. Under a mean field approximation of an SIR epidemic model on a graph, they found that the epidemic threshold is proportional to the *spectral radius* of the adjacency matrix. Follow-up works verified this relation and broadened the discussion to more types of diffusion and related models. In [9] the $S^*I^2V^*$ model was presented as a generalization of numerous virus propagation models of the literature. It was also made possible to generalize the result of [8] to that of generic virus models. Based on these works, several research studies have been presented on the epidemic control on networks, mainly focusing on developing *immunization* strategies (elimination of nodes) and *quarantine* strategies (elimination of edges). The eigenvalue perturbation theory was among the main analytical tools used; see for example [10,14,15].

Similar theoretical results to those discussed above have been given for ICM as well. Under discrete- or continuous-time ICM, it has been shown that the epidemic threshold depends on the *spectral radius* of a matrix built upon the edge transmission probabilities, termed as the *Hazard matrix* [11,12].

Table 24.1 Index of Main Notations

Symbol	Description
$\mathbb{1}\{$<condition>$\}$	indicator function
$\mathbf{1}$	vector with all values equal to one
$\|X\|_\ell$	ℓ-norm for a given vector X: e.g., $\|X\|_1 = \sum_{ij} X_{ij}$, or generally $\|X\|_\ell = (\sum_{ij} X_{ij}^\ell)^{1/\ell}$
$M \odot M'$	the Hadamard product between matrices (i.e., coordinate-wise multiplication)
$\mu_{\pi(1)} \geq \mu_{\pi(2)} \dots$	ordered values of vector μ using the order-to-index bijective mapping π
$\mathcal{G}, \mathcal{V}, n, \mathcal{E}, E$	network $\mathcal{G} = \{\mathcal{V}, \mathcal{E}\}$ of $n = \|\mathcal{V}\|$ nodes and $E = \|\mathcal{E}\|$ edges
(i,j)	edge $(i,j) \in \mathcal{E}$ of the graph between nodes i and j
A	network's adjacency matrix $A \in \{0,1\}^{n \times n}$
$\mathcal{S}$	state space. Example states: (S)usceptible, (I)nfeted, (R)ecovered
S_0, n_0	subset $S_0 \subset \mathcal{V}$ of $n_0 = \|S_0\|$ influencer nodes from which the IC initiates
$\mathcal{F}$	$n \times n$ *Hazard matrix* $[\mathcal{F}_{ij}]_{ij}$ of nonnegative integrable *Hazard functions* over time
$\mathbb{F}$	set of feasible Hazard matrices $\mathbb{F} \subset \mathbb{R}_+ \to \mathbb{R}_+^{n \times n}$, where $\mathcal{F}$ is one of its elements
Δ	matrix of the integrated difference of two Hazard matrices: $\Delta = \int_0^{+\infty} (\hat{\mathcal{F}}(t) - \mathcal{F}(t))dt$
τ_i	time $\tau_i \in \mathbb{R}_+ \cup \{+\infty\}$ at which the information reached node i during the process
$\sigma(S_0)$	*influence*: the final number of contagious nodes when diffusion starts from the set S
$\rho_{\mathcal{H}}(\mathcal{F})$	the largest eigenvalue of the symmetrized and integrated Hazard matrix $\mathcal{F}$
$\hat{p}(s)$	Laplace transform of the function $p(t)$
X	control actions matrix $X \in [0,1]^{n \times n}$ with the amount of action taken on each edge
x	control actions vector $x \in [0,1]^n$ with the amount of action taken on each node
k	budget of control actions $k \in (0, E)$ for actions on edges, or $k \in (0, n)$ for nodes

Related applications. Dealing with information diffusion and rumors gives rise to a series of computational and inference problems, including: credibility assessment of posts and users [42]; sentimental analysis on how individuals receive a piece of information; stance/role identification of users toward it; detection of rumors and their spreaders in content streams [43–45]; identification of influential users that could maximize the reach of a campaign by examining structural properties of the network alone or in combination with historical data (interaction traces) [5,46,47]; and finally, the development of countermeasures to suppress a rumor or information cascade [13,48], which is discussed in the technical part of the chapter.

24.3 MODELS OF INFORMATION CASCADES

Information cascades describe the dynamics of communication between individuals of a social network by capturing the way messages are shared and propagated among users. In all generality, an information cascade on a graph $\mathcal{G} = (\mathcal{V}, \mathcal{E})$ is a multivariate stochastic process $\{X_i(t) : i \in \mathcal{V}, t \geq 0\}$ where $X_i(t) \in \mathcal{S}$ denotes the state of user i at time t, and $\mathcal{S}$ is a state space that may be finite, countable, or uncountable. Depending on the specific model, the state of a user may refer to a binary quantity (e.g., $\mathcal{S} = \{Unaware, Informed\}$), to the number of messages received during $[0,t]$ (in which case $\mathcal{S} = \mathbb{N}$), or something more detailed regarding the message spread (e.g., $\mathcal{S} = \mathbb{R}^d$ a low-dimensional representation of the content of the message). In all the models, we consider that users that did not participate at all in the cascade are in a default state $0 \in \mathcal{S}$. As a rumor propagates through the network, the number of individuals participating in the cascade, called *influence*, will grow and eventually reach a saturation point. We use this quantity as our main quality metric:

Definition 24.1 (Influence $\sigma(S_0, t)$). Let $S_0 = \{i \in \mathcal{V} : X_i(0) \neq 0\} \subset \mathcal{V}$ be the set of *influencers*, i.e., users that are initially contagious. The influence of the set S_0 at time t is defined as the total number of messages received by users of the social network before time t:

$$\sigma(S_0, t) = \mathbb{E}\left[\sum_{i \in V} \mathbb{1}\{X_i(t) \neq 0\}\right]. \tag{24.1}$$

■

In the following, we denote as $n = |\mathcal{V}|$ the size of the social network, $E = |\mathcal{E}|$ the number of connections, $n_0 = |S_0|$ the number of initial influencers and the adjacency matrix of $\mathcal{G}$ as $A \in \{0,1\}^{n \times n}$ s.t. $A_{ij} = 1 \Leftrightarrow (i,j) \in \mathcal{E}$. Moreover, we denote as long-term influence the total number of received messages after the diffusion $\sigma(S_0) = \lim_{t \to +\infty} \sigma(S_0, t)$.

24.3.1 EARLY MODELS: VIRUSES SPREADING THROUGH SOCIAL NETWORKS

Epidemics are usually modeled using *Markov processes* [49], i.e., *memoryless* stochastic processes entirely defined by their transition matrix. This transition matrix defines the probability for each node to change state during an infinitesimal time window $[t, t + dt]$ (the simultaneous change of more than one node's state is considered improbable). In the following, we thus use the notation:

$$X_i(t) : \mathrm{Y} \to \mathrm{Z} \text{ at rate } C_i(t) \tag{24.2}$$

to denote the stochastic transition rate $C_i(t) \geq 0$ of node $i \in \{1, \ldots, n\}$ at time $t \geq 0$ from state Y to state Z, with Y, Z $\in \mathcal{S}$.

Due to similarities between spreading phenomena, virus models have also been used to describe information cascades on social networks. We focus on two standard such models: the SI and SIR models, and we refer the reader to the recent review in [50] for more information on the vast epidemiology literature.

Susceptible-Infected model

The Susceptible-Infected (SI) model is the simplest epidemic model, in which nodes can be either (S)usceptible or (I)nfected. An infected node transmits the disease to one of its susceptible neighbor at a rate β, and once infected a node remains infected and thus contagious.

Model 24.1 (SI model). Let $\mathcal{G}$ be a (possibly weighted) graph of n nodes and adjacency matrix A. The SI model is a continuous-time Markov process $X(t) \in \{\mathrm{S}, \mathrm{I}\}^n$ with the following transition rate:

$$X_i(t) : \mathrm{S} \to \mathrm{I} \text{ at rate } \beta \sum_j A_{ji} X_j(t), \tag{24.3}$$

where β is the transmission rate of the epidemic. ■

Because the nodes remain infected, a connected network will be totally infected at the end of the diffusion, and hence any set S_0 has influence $\sigma(S_0) = n$.

Susceptible-Infected-Removed model

The Susceptible-Infected-Removed (SIR) model [51] is a widely used epidemic model designed for scenarios in which patients present immunity to the disease after their infection and recovery. A recovered person will not transmit the disease further nor will it be subject to reinfections. An additional state is thus added to the SI model and each node of the network is either (S)usceptible, (I)infected, or (R)emoved. At $t = 0$, a subset S_0 of n_0 nodes is infected. Then, each infected node will transmit the disease to its neighbors at rate β, and recover at rate δ.

Model 24.2 (SIR model). Let $\mathcal{G}$ be a (possibly weighted) graph of n nodes and adjacency matrix A. The SIR model is a continuous-time Markov process $X(t) \in \{\mathrm{S}, \mathrm{I}, \mathrm{R}\}^n$ with the following transition rates:

$$\begin{aligned} X_i(t) &: \mathrm{S} \to \mathrm{I} \text{ at rate } \beta \textstyle\sum_j A_{ji} X_j(t) \\ X_i(t) &: \mathrm{I} \to \mathrm{R} \text{ at rate } \delta, \end{aligned} \tag{24.4}$$

where β is the transmission rate of the epidemic and δ is the recovery rate of nodes. ■

Usually, the graph is undirected and all edges have the same rate. More complex scenarios can be modeled using the *inhomogeneous SIR* model, in which each edge has its own transmission rate β_{ij} and each node its own recovery rate δ_i.

An alternative definition for this model is possible using *infection times*. One may see that each node gets infected at most once and recovers at most once as well. We can thus define, for each node i, the time τ_i^{I} at which it gets infected and the time τ_i^{R} at which it recovers, with $\tau_i^{\mathrm{I}}, \tau_i^{\mathrm{R}} \in \mathbb{R}_+ \cup \{+\infty\}$. Then, $\tau_i^{\mathrm{I}} = 0$ would indicate that user i is an influencer while $\tau_i^{\mathrm{I}} = +\infty$ would indicate that node i never got infected throughout the whole epidemic.

Proposition 24.1. *For an SIR epidemic, the infection times τ_i^{I} of not initially infected nodes verify the following equality:*

$$\forall i \notin S_0,\ \tau_i^{\mathrm{I}} = \min_{\{j \in \{1,\dots,n\} : T_{ji} < D_j\}} (\tau_j^{\mathrm{I}} + T_{ji}), \tag{24.5}$$

where T_{ji} and D_j are independent exponential random variables of expected value $1/\beta$ and $1/\delta$, respectively, and $\tau_i^{I} = +\infty$ if the set $\{j \in \{1,\dots,n\} : T_{ji} < D_j\}$ is empty. Furthermore, the recovery time of each node i is:

$$\tau_i^{\mathrm{R}} = \tau_i^{\mathrm{I}} + D_i. \tag{24.6}$$

Proof. This result relies on the fact that a node is infected as soon as at least one of its infected neighbors transmits the infection to him. Because these events are independent, the times T_{ij} required for infection along the edges of the network are also independent. For more precisions, see e.g., [11]. ■

24.3.2 INDEPENDENT CASCADES

Independent cascades were initially introduced as discrete-time diffusion processes [28], and later refined to more flexible continuous-time processes [7].

Model 24.3 (Discrete-time independent cascades $\mathcal{DTIC}(\mathcal{P})$). At time $t = 0$, only a set S_0 of influencers is infected. Given a matrix $\mathcal{P} = (p_{ij})_{ij} \in [0, 1]^{n \times n}$, each node i that receives the contagion at time t may transmit it at time $t+1$ along its outgoing edge $(i,j) \in \mathcal{E}$ with probability p_{ij}. Node i cannot infect its neighbors in subsequent rounds $t' > t+1$. The process terminates when no more infections are possible. ■

The continuous version of independent cascades requires the definition of *Hazard functions* to describe the varying transmission rates along each edge of the network.

Definition 24.2 (Hazard function $\mathcal{F}_{ij}(t)$). For every edge $(i,j) \in \mathcal{E}$ of the graph, $\mathcal{F}_{ij}$ is a nonnegative integrable function that describes the time-dependent stochastic transmission rate from node i to node j, after i's infection. ■

Model 24.4 (Continuous-time independent cascades $\mathcal{CTIC}(\mathcal{F})$). The $\mathcal{CTIC}(\mathcal{F})$ model is a stochastic diffusion process defined as follows: at time $s = 0$, only the influencer nodes in S_0 are infected. Then, each node i that receives the contagion at time τ_i may transmit it at time $s \geq \tau_i$ along an outgoing edge $(i,j) \in \mathcal{E}$ with stochastic rate of occurrence $\mathcal{F}_{ij}(s - \tau_i)$. ■

The rest of this chapter will mainly focus on the analysis and control of such information cascades. For notational purposes, we denote as $\mathcal{F} = [\mathcal{F}_{ij}]_{ij}$ the $n \times n$ *Hazard matrix* containing as elements the individual Hazard functions and, respectively, as $\mathcal{F}(t) = [\mathcal{F}_{ij}(t)]_{ij}$ the evaluation of all functions at a relative time-point t after each infection time τ_i. Essentially, network edges imply nonzero Hazard functions:

$$(i,j) \in \mathcal{E} \;\Leftrightarrow\; \exists t \geq 0 \text{ s.t. } \mathcal{F}_{ij}(t) \neq 0. \tag{24.7}$$

Note that each Hazard function $\mathcal{F}_{ij}$ is always evaluated at a *relative time-point* initialized at the infection time τ_i of the source node i.

Similarly to SIR, independent cascades are monotonically increasing stochastic processes, and each node can only be infected once. We can thus define, for each node i, the time τ_i of its first infection, which may be infinite if the node never gets infected during the contagion. Unlike SIR, no epidemic states are explicitly mentioned in the notations of $\mathcal{CTIC}$ (the reader may compare Eqs. (24.5) and (24.8)).

Proposition 24.2. *For a Continuous-Time Independent Cascade $\mathcal{CTIC}(\mathcal{F}, T)$, the infection times τ_i of noninfluencer nodes verify the following equality:*

$$\forall i \notin S_0,\; \tau_i = \min_{j \in \{1,\dots,n\}} (\tau_j + T_{ji}), \tag{24.8}$$

where $T_{ij} \in \mathbb{R}_+ \cup \{+\infty\}$ are independent random variables of subprobability density

$$p_{ij}(t) = \mathcal{F}_{ij}(t) \exp\left(-\int_0^t \mathcal{F}_{ij}(s)ds\right). \tag{24.9}$$

Proof. This result is similar to Proposition 24.1 and relies on the same observation: a node is active as soon as at least one of its active neighbors activated him. Because these events are independent (hence the name of the model), the times T_{ij} required for activation along the edges of the network are also independent. For more precisions, see for example [52]. ■

In general, $p_{ij}(t)$ is *not* a probability density over $\mathbb{R}_+$ as it does not integrate to one, and $\mathbb{P}(T_{ij} = +\infty) = 1 - \int_0^{+\infty} p_{ij}(t)dt = \exp(-\int_0^{+\infty} \mathcal{F}_{ij}(t)dt)$. Proposition 24.2 provides a simple mechanism for simulating $\mathcal{CTIC}$, as one can first draw one independent random variable T_{ij} per edge, and then use a shortest-path algorithm to compute the infection times τ_i for each node of the network.

In what follows, we focus on this model due to its expressiveness and broad use in modern social network studies. However, the large-scale dynamics of all diffusion models are relatively similar and exhibit the same threshold behavior.

24.4 LARGE-SCALE DYNAMICS OF INDEPENDENT CASCADES

At the scale of the network, the emergent behavior of information cascades displays several typical characteristics that are common in most diffusion processes, including epidemics and computer viruses. For instance, Fig. 24.1 shows the number of identified cases of Ebola during a recent crisis, the number of queries for "Pokemon go" when the game became viral, and the simulation of an independent cascade (see Model 24.4 in Section 24.3). All these diffusion processes exhibit similar behavior:

1. **Explosive start:** The cascade starts with an exponential increase and quickly reaches a nonnegligible amount.
2. **Saturation point:** After a sharp increase during the early phase of the diffusion, the process reaches a saturation point and comes to a halt. Note that, for information cascades, a residual activity may produce a linear slope after the end of the diffusion. However, we ignore this aspect in our study.

As a consequence, we focus on four main characteristics of interest to describe the large-scale dynamics of information cascades:

1. **Existence:** Is the cascade powerful enough to enter the explosive phase?
2. **Saturation point:** What is the final reach of the cascade?
3. **Time for action:** When is the explosion taking place?
4. **Explosive rate:** How fast is the initial exponential increase of the cascade?

These four characteristics are summarized in a simulated toy example on Fig. 24.1C. In the following sections, we provide estimates of these quantities depending on the diffusive properties of the process as well as the structure of the social network.

24.4.1 EXISTENCE OF A SUPERCRITICAL CASCADE

Intuitively, an information cascade may only sustain itself if, on average, people that receive the message share it to more than one of their neighbors. When the network connectivity is too low, the cascade cannot reach a large audience before dying out. This is highlighted by the following upper bound relating a measure of network connectivity introduced in [12], the *Hazard radius*, to the long-term influence.

Definition 24.3 (Hazard radius $\rho_{\mathcal{H}}(\mathcal{F})$). For a diffusion process $\mathcal{CTIC}(\mathcal{F})$, $\rho_{\mathcal{H}}(\mathcal{F})$ is the largest eigenvalue of the symmetrized and integrated Hazard matrix:

$$\rho_{\mathcal{H}}(\mathcal{F}) = \rho\left(\int_0^{+\infty} \frac{\mathcal{F}(t) + \mathcal{F}(t)^\top}{2} dt\right), \tag{24.10}$$

where $\rho(\cdot) = \max_i |\lambda_i|$ and λ_i are the eigenvalues of the input matrix. ■

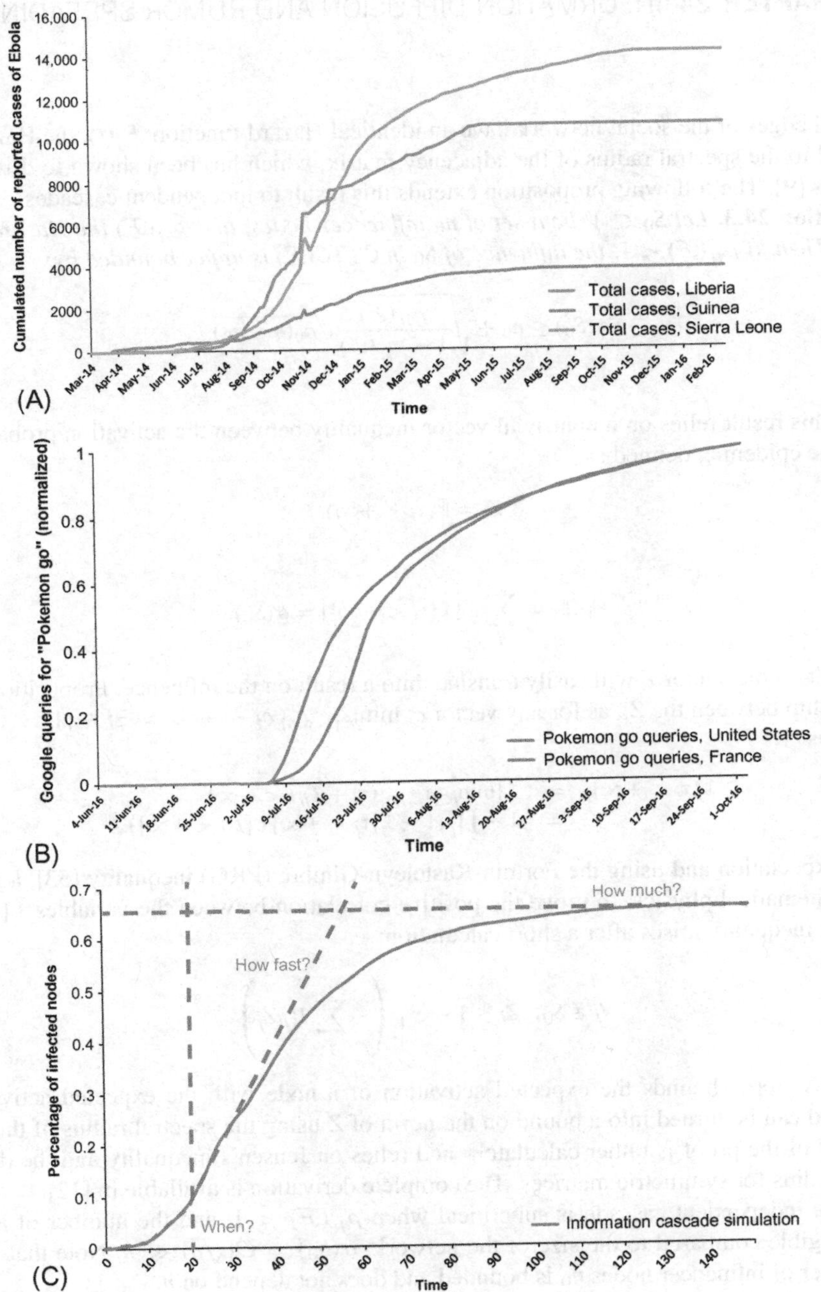

FIG. 24.1

Main large-scale characteristics of diffusion processes appearing in real and simulated cascades. (A) Number of Ebola cases in Ginea, Liberia, and Sierra Leone (source: World Health Organization). (B) Number of searches for the query "Pokemon go" on the Google search engine (source: Google Trend). (C) Simulation of a Continuous-Time Independent Cascade (see Model 24.4). The main large-scale characteristics highlighted in our analysis are also summarized: existence of outbreak, time before the explosion, explosive rate, and saturation point.

When all edges of the social network have an identical Hazard function $\mathcal{F}_{ij}(t)$, the Hazard radius is proportional to the spectral radius of the adjacency matrix, which has been shown to drive the spread of epidemics [9]. The following proposition extends this result to independent cascades.

Proposition 24.3. *Let $S_0 \subset \mathcal{V}$ be a set of n_0 influencer nodes, and $\rho_{\mathcal{H}}(\mathcal{F})$ the Hazard radius of a $\mathcal{CTIC}(\mathcal{F})$. Then, if $\rho_{\mathcal{H}}(\mathcal{F}) < 1$, the influence of S_0 in $\mathcal{CTIC}(\mathcal{F})$ is upper bounded by:*

$$\sigma(S_0) \leq n_0 + \sqrt{\frac{\rho_{\mathcal{H}}(\mathcal{F})}{1-\rho_{\mathcal{H}}(\mathcal{F})}}\sqrt{n_0(n-n_0)}. \tag{24.11}$$

■

Proof. This result relies on a nontrivial vector inequality between the activation probabilities Z_i at the end of the epidemic, defined as:

$$Z_i = \mathbb{P}(\tau_i < +\infty). \tag{24.12}$$

Note that

$$\|Z\|_1 = \sum_i \mathbb{E}[\mathbb{1}\{\tau_i < +\infty\}] = \sigma(S_0), \tag{24.13}$$

and any result on the vector Z will easily translate into a result on the influence. Proposition 24.2 leads to a relationship between the Z_i, as for any vector c, $\min_{j\in\{1,\ldots,n\}} c_j < +\infty \Leftrightarrow \exists j \in \{1,\ldots,n\}$ s.t. $c_j < +\infty$, and thus

$$\begin{aligned} \mathbb{1}\{\tau_i < +\infty\} &= \mathbb{1}\{\min_{j\in\{1,\ldots,n\}}(\tau_j + T_{ji}) < +\infty\} \\ &= 1 - \textstyle\prod_j \left(1 - \mathbb{1}\{\tau_j < +\infty\}\mathbb{1}\{T_{ji} < +\infty\}\right). \end{aligned} \tag{24.14}$$

Taking the expectation and using the Fortuin-Kasteleyn-Ginibre (FKG) inequality [53], a well-known result of mathematical physics, to prove the positive correlation between the variables $\mathbb{1}\{\tau_i < +\infty\}$, the following inequality arises after a short calculation:

$$\forall i \notin S_0,\ \ Z_i \leq 1 - \exp\left(-\sum_j \mathcal{H}_{ji} Z_j\right). \tag{24.15}$$

This inequality upper bounds the expected activation of a node with the expected activation of its neighbors, and can be turned into a bound on the norm of Z using the spectral radius of the matrix $\mathcal{H}$. The final step of the proof is rather calculatory and relies on Jensen's inequality and the definition of the spectral radius for symmetric matrices. The complete derivation is available in [12]. ■

Hence, the independent cascade is subcritical when $\rho_{\mathcal{H}}(\mathcal{F}) < 1$, and the number of active users remains negligible compared to the size of the network: $\sigma(S_0) = \mathcal{O}(\sqrt{n}) \ll n$. Note that we assume that the number of influencer nodes n_0 is bounded and does not depend on n.

24.4.2 LONG-TERM BEHAVIOR OF INDEPENDENT CASCADES

When the cascade is efficient enough to propagate to a large proportion of the network, it displays a sharp increase before saturating to a limit value. Although the precise value of this limit influence is hard to evaluate, several upper bounds have been provided and proven in the literature [12,54]. We now provide such a result relating the long-term influence to the Hazard radius of the cascade.

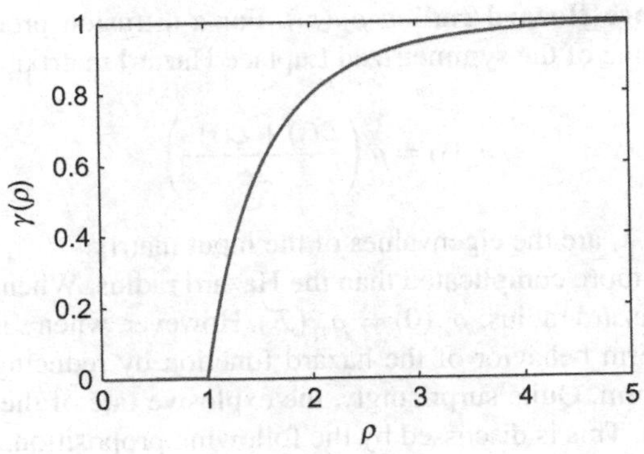

FIG. 24.2

Upper bound on the saturation point. Function γ defined in Eq. 24.17. When $\rho_{\mathcal{H}}(\mathcal{F}) < 1$, the function is equal to 0, then increases and saturates to $\gamma = 1$ as $\rho_{\mathcal{H}}(\mathcal{F})$ tends to infinity.

Proposition 24.4. *Let $S_0 \subset \mathcal{V}$ be a set of n_0 influencer nodes, and $\rho_{\mathcal{H}}(\mathcal{F})$ the Hazard radius of a $CTIC(\mathcal{F})$. Then, if $\rho_{\mathcal{H}}(\mathcal{F}) > 1$, the long-term influence of S_0 in $CTIC(\mathcal{F})$ is upper bounded by:*

$$\sigma(S_0) \leq n_0 + \gamma(n - n_0) + c_n\sqrt{n_0(n - n_0)}, \tag{24.16}$$

where $c_n = \sqrt{\frac{\eta}{1-\eta}}$, $\eta = (1 - \gamma)\rho_{\mathcal{H}}(\mathcal{F})$ and $\gamma \in [0, 1]$ is the unique positive solution of the equation:

$$\gamma = 1 - \exp\left(-\rho_{\mathcal{H}}(\mathcal{F})\gamma\right). \tag{24.17}$$

Proof. This result is also a consequence of Eq. (24.15) relating the expected activations Z_i. See [12]. ■

In essence, the proportion of active nodes after the cascade is negligible when $\rho_{\mathcal{H}}(\mathcal{F}) < 1$, and at most γ when $\rho_{\mathcal{H}}(\mathcal{F}) > 1$, where γ is defined by the implicit equation $\gamma = 1 - \exp\left(-\rho_{\mathcal{H}}(\mathcal{F})\gamma\right)$. Fig. 24.2 shows the proportion γ of Proposition 24.4 with respect to the Hazard radius $\rho_{\mathcal{H}}(\mathcal{F})$.

24.4.3 EXPLOSIVE DYNAMICS IN THE SUPERCRITICAL REGIME

Finally, the intermediate regime when the cascade grows exponentially can be analyzed using a modified version of the Hazard radius, known as the *Laplace Hazard radius.*

Definition 24.4 (Laplace Hazard matrix $\mathcal{L}(s)$). Let p_{ij} be the edge transmission probabilities defined in Eq. (24.9). For $s \geq 0$, let $\mathcal{L}(s)$ be the $n \times n$ matrix, called *Laplace Hazard matrix*, whose coefficients are:

$$\mathcal{L}_{ij}(s) = \begin{cases} -\hat{p}_{ij}(s)\left(\int_0^{+\infty} p_{ij}(t)dt\right)^{-1} \ln\left(1 - \int_0^{+\infty} p_{ij}(t)dt\right) & \text{if } (i,j) \in \mathcal{E}, \\ 0 & \text{otherwise,} \end{cases} \tag{24.18}$$

where $\hat{p}_{ij}(s)$ denotes the Laplace transform of p_{ij} defined for every $s \geq 0$ by $\hat{p}_{ij}(s) = \int_0^{+\infty} p_{ij}(t)e^{-st}dt$. ■

Definition 24.5 (Laplace Hazard radius $\rho_{\mathcal{L}}(s)$). For a diffusion process $\mathcal{CTIC}(\mathcal{F})$ and $s \geq 0$, $\rho_{\mathcal{L}}(s)$ is the largest eigenvalue of the symmetrized Laplace Hazard matrix:

$$\rho_{\mathcal{L}}(s) = \rho\left(\frac{\mathcal{L}(s) + \mathcal{L}(s)^{\mathrm{T}}}{2}\right), \tag{24.19}$$

where $\rho(\cdot) = \max_i |\lambda_i|$ and λ_i are the eigenvalues of the input matrix. ■

This concept is slightly more complicated than the Hazard radius. When $s = 0$, the Laplace Hazard radius coincides with the Hazard radius: $\rho_{\mathcal{L}}(0) = \rho_{\mathcal{H}}(\mathcal{F})$. However, when s is large, the Laplace Hazard radius captures the short-term behavior of the hazard function by reducing the impact of long times through the Laplace transform. Quite surprisingly, the explosive rate of the cascade is upper bounded by the inverse value $\rho_{\mathcal{L}}^{-1}(1)$. This is discussed by the following proposition.

Proposition 24.5. *Let $t \geq 0$, $S_0 \subset \mathcal{V}$ be a set of n_0 influencer nodes, and $\rho_{\mathcal{L}}$ the Laplace Hazard radius. Then, the short-term influence of S_0 in $\mathcal{CTIC}(\mathcal{F})$ at time t is upper bounded by:*

$$\sigma(S_0, t) \leq n_0 + (2n_0)^{1/3}(n - n_0)^{2/3} \exp\left(\rho_{\mathcal{L}}^{-1}(1)t\right). \tag{24.20}$$

Proof. This result relies on a similar equation to Eq. (24.15) describing the dynamics of the cascade instead of its long-term stable regime. More specifically, Proposition 24.2 shows that, for any $t \geq 0$, the variables $\mathbb{1}\{\tau_i < t\}$ are related according to:

$$\mathbb{1}\{\tau_i < t\} = 1 - \prod_j \left(1 - \mathbb{1}\{\tau_j + T_{ji} < t\}\right). \tag{24.21}$$

Now, denoting as $Z_i(t) = \mathbb{P}(\tau_i < t)$ the probability that node i is active at time t, one may show the following vectorial inequality relating the variables $Z_i(t)$:

$$Z_i(t) \leq 1 - \exp\left(-\sum_j (\mathcal{F}_{ji} * Z_j)(t)\right), \tag{24.22}$$

where $(f * g)(t) = \int_{\mathbb{R}} f(s)g(t - s)ds$ is the convolution product. From this inequality, one may prove an upper bound on the Laplace transform of the influence $\hat{\sigma}(s) = \int_0^{+\infty} \sigma(S_0, t)e^{-st}dt$, directly translating into an upper bound on the exponential increase of the influence. Again, the complete derivation is available in [11]. ■

This result has two implications (for more precise results see [11]):

- First, the influence is at most increasing at an exponential rate of $\rho_{\mathcal{L}}^{-1}(1)$.
- Second, this also provides a characteristic time under which the cascade is still in its early phase. More precisely, before the critical time

$$t \leq \frac{\ln n}{3\rho_{\mathcal{L}}^{-1}(1)}, \tag{24.23}$$

the cascade is *subcritical* and the influence is negligible: $\sigma(S_0, t) = \mathcal{O}(n^{2/3})$.

24.5 MONITORING INFORMATION CASCADES

Having presented the fundamental theoretical properties of diffusion processes related to information propagation over networks, we now discuss an efficient approach to the generic problem of optimizing influence (maximizing or minimizing) using actions that can shape, i.e., modify, the activity of single users. For instance, a marketing campaign may have a certain advertisement budget that can be used on targeted users of a social network. While these targeted resources are usually represented as new influencer nodes that will spread the piece of information, we rather consider the more refined and general case in which each resource will essentially alter the Hazard functions $\mathcal{F}_{ij}$ associated to a target node i, thus increasing, or decreasing, the probability for i to propagate by sharing the information with its neighbors.

Our generic framework assumes that a *set of feasible Hazard matrices* $\mathbb{F} \subset \mathbb{R}_+ \to \mathbb{R}_+^{n\times n}$ is available to the administrator. This set virtually contains all admissible policies that one could apply to the network. Then, the concern is to find the Hazard matrix $\mathcal{F} \in \mathbb{F}$ that minimizes, or maximizes depending on the task of interest, the influence. In Section 24.7 we show that two problems that have been a major focus of the literature so far, namely the edge-deletion problem [14] and the node-immunization problem [15] are particular instances of this framework. Note that this framework is generic enough to describe complex strategies that may use several immunization options by deploying simultaneously resources of different types (removal of edges, nodes, partial immunization, etc.).

Problem 24.1 (Determining the optimal feasible policy). Given a graph $\mathcal{G}$, a number of influencers n_0, and a set of admissible policies $\mathbb{F}$, find the optimal policy:

$$\mathcal{F}^* = \underset{\mathcal{F}\in\mathbb{F}}{\operatorname{argmin}}\ \sigma^*_{n_0}(\mathcal{F}), \tag{24.24}$$

where $\sigma^*_{n_0}(\mathcal{F}) = \max\{\sigma(S_0) : S_0 \subset \mathcal{V} \text{ and } |S_0| = n_0\}$ is the optimal influence (according to Eq. (24.24) this is the minimum) over any possible set of n_0 influencer nodes. ∎

Problem 24.1 cannot be solved exactly in polynomial time. The exact computation of the maximum influence $\sigma^*_{n_0}(\mathcal{F})$ is already a hard problem on its own, and minimizing this quantity adds an additional layer of complexity due to the nonconvexity of the maximum influence w.r.t. the Hazard matrix (note: $\mathcal{F} \mapsto \sigma^*_{n_0}(\mathcal{F})$ is positive, upper bounded by n and not constant).

Proposition 24.6. *For any size of the set of influencers n_0, the computation of $\sigma^*_{n_0}(\mathcal{F})$ is #P-hard.*

Proof. We prove the theorem by reduction from a known #P-hard function: the computation of the influence $\sigma(S_0)$ given a set of influencers S_0 of size n_0 (see Theorem 1 of [55]). Indeed, let $\mathcal{CTIC}(\mathcal{F})$ be an independent cascade defined on $\mathcal{G} = (\mathcal{V}, \mathcal{E})$. We can construct a new graph $\mathcal{G}' = (\mathcal{V}', \mathcal{E}')$ as follows: for each influencer node $i \in S_0$, add a directed chain of n nodes $\{v_{i,1}, \ldots, v_{i,n}\} \subset \mathcal{V}'$ and connect $v_{i,n}$ to i by letting the transmission probabilities along the edges be all equal to one. Then, the maximum influence $\sigma^*_{n_0}$ is achieved with the nodes $S'_0 = \{v_{i,1} : i \in S_0\}$ as influencer, and $\sigma^*_{n_0} = n\,n_0 + \sigma(S_0)$. The result follows from the #P-hardness of computing $\sigma(S_0)$ given S_0. ∎

The standard way to approximate the maximum influence is to employ incremental methods where the quality of each potential influencer is assessed using a Monte Carlo approach. In the following, we assume that the feasible set $\mathbb{F}$ is convex and included in a ball of radius R. Also, the requirement of Eq. (24.7), that network edges correspond to nonzero Hazard functions, holds for every feasible policy $\mathcal{F} \in \mathbb{F}$. Therefore, the number of edges E upper bounds the number of nonzero Hazard functions for any $\mathcal{F} \in \mathbb{F}$.

Remark 24.1. Although Problem 24.1 focuses on the minimization of the maximum influence, the algorithm presented in this paper is also applicable to the opposite task of influence maximization. Having a common ground for solving these opposite problems can be useful for applications where both opposing aims can interest different actors, e.g., in market competition. For the maximization, our algorithm would use a gradient ascent instead of a gradient descent optimization scheme. While the performance of the algorithm in that case may be competitive to state-of-the-art influence maximization algorithms, the nonconvexity of this problem prevents us from providing any theoretical guarantees regarding the quality of the final solution.

24.6 AN ALGORITHM FOR REDUCING INFORMATION CASCADES

As has been mentioned, solving exactly the influence optimization problem is computational intractable. Here, we propose to exploit the upper bound given in Proposition 24.4 as a heuristic for approximating the maximum influence. This approach can be seen as a *convex relaxation* of the original NP-Hard problem, and allows the use of convex optimization algorithms for this particular problem. The relaxed optimization problem thus becomes:

$$\mathcal{F}^* = \operatorname*{argmin}_{\mathcal{F}\in\mathbb{F}} \rho_{\mathcal{H}}(\mathcal{F}). \tag{24.25}$$

When the feasible set $\mathbb{F}$ is convex, this optimization problem is also convex and our proposed method called *NetShape* uses a simple *projected subgradient descent* (see e.g. [56]) in order to find its minimum and make sure that the solution lays in $\mathbb{F}$. However, special care should be taken to perform the gradient step because although the objective function $\rho_{\mathcal{H}}(\mathcal{F})$ admits a derivative w.r.t. the norm

$$\|\mathcal{F}\| = \sqrt{\sum_{i,j}\left(\int_0^{+\infty} \left|\mathcal{F}_{ij}(t)\right| dt\right)^2}, \tag{24.26}$$

the space of matrix functions equipped with this norm is only a Banach space in the sense that the norm $\|\mathcal{F}\|$ cannot be derived from a well-chosen scalar product. Because gradients only exist in Hilbert spaces, gradient-based optimization methods are not directly applicable.

In the NetShape algorithm, the gradient and projection steps are performed on the *integral* of the Hazard functions $\int_0^{+\infty} \mathcal{F}_{ij}(t)dt$ by solving the optimization problem bellow:

$$\mathcal{F}^* = \operatorname*{argmin}_{\hat{\mathcal{F}}\in\mathbb{F}} \left\| \int_0^{+\infty} \left(\hat{\mathcal{F}}(t) - \mathcal{F}(t)\right) dt + \eta\, u_{\mathcal{F}} u_{\mathcal{F}}^{\top} \right\|_2, \tag{24.27}$$

where $\eta > 0$ is a positive gradient step, $u_{\mathcal{F}}$ is the eigenvector associated to the largest eigenvalue of the matrix $\int_0^{+\infty} \frac{\mathcal{F}(t)+\mathcal{F}(t)^{\top}}{2} dt$, and $u_{\mathcal{F}} u_{\mathcal{F}}^{\top}$ is a subgradient of the objective function, as provided by the following proposition.

Proposition 24.7. *A subgradient of the objective function $f(M) = \rho(\frac{M+M^{\mathsf{T}}}{2})$ in the space of integrated Hazard functions, where M is a matrix, is given by the matrix:*

$$\nabla f(M) = u_M u_M^{\mathsf{T}}, \tag{24.28}$$

where u_M is the eigenvector associated to the largest eigenvalue of the matrix $\frac{M+M^{\mathsf{T}}}{2}$.

Proof. For any matrix M, let $f(M) = \rho(\frac{M+M^{\mathsf{T}}}{2}) = \max_{x: \|x\|_2=1} x^{\mathsf{T}} M x$, and u_M be such an optimal vector. Then, we have $f(M+\varepsilon) = u_{M+\varepsilon}^{\mathsf{T}}(M+\varepsilon)u_{M+\varepsilon} \geq u_M^{\mathsf{T}}(M+\varepsilon)u_M = f(M) + u_M^{\mathsf{T}} \varepsilon u_M$, and, because $u_M^{\mathsf{T}} \varepsilon\, u_M = \langle u_M u_M^{\mathsf{T}}, \varepsilon \rangle$, $u_M u_M^{\mathsf{T}}$ is indeed a subgradient for $f(M)$. ■

Algorithm 24.1 NETSHAPE METAALGORITHM

Input: feasible set $\mathbb{F} \subset \mathbb{R}_+ \to \mathbb{R}_+^{n \times n}$, radius $R > 0$ of $\mathbb{F}$, initial Hazard matrix $\mathcal{F} \in \mathbb{F}$, approx. parameter $\epsilon > 0$
Output: Hazard matrix $\mathcal{F}^* \in \mathbb{F}$

1: $\mathcal{F}^* \leftarrow \mathcal{F}$
2: $T \leftarrow \lceil \frac{R^2}{\epsilon^2} \rceil$
3: **for** $i = 1$ to $T - 1$ **do**
4: $\quad u_{\mathcal{F}} \leftarrow$ compute the eigenvector associated to the spectral radius $\rho_{\mathcal{H}}(\mathcal{F})$
5: $\quad \eta \leftarrow \frac{R}{\sqrt{i}}$
6: $\quad \mathcal{F} \leftarrow \operatorname{argmin}_{\hat{\mathcal{F}} \in \mathbb{F}} \left\| \int_0^{+\infty} \left(\hat{\mathcal{F}}(t) - \mathcal{F}(t)\right) dt + \eta\, u_{\mathcal{F}} u_{\mathcal{F}}^{\mathsf{T}} \right\|_2$
7: $\quad \mathcal{F}^* \leftarrow \mathcal{F}^* + \mathcal{F}$
8: **end for**
9: **return** $\frac{1}{T}\mathcal{F}^*$

The projection step of line 6 in Algorithm 24.1 is an optimization problem on its own, and the NetShape algorithm is practical if and only if this optimization problem is simple enough to be solved. In the next sections we will see that, in many cases, this optimization problem can be solved in near linear time w.r.t. the number of edges of the network (i.e., $\mathcal{O}(E \ln E)$), and is equivalent to a projection on a simplex.

24.6.1 CONVERGENCE AND SCALABILITY

Due to the convexity of the optimization problem in Eq. (24.25), NetShape finds the global minimum of the objective function and, as such, may be a good candidate to solve Problem 24.1. The complexity of the NetShape algorithm depends on the complexity of the projection step in Eq. (24.27). Each step of the gradient descent requires the computation of the first eigenvector of an $n \times n$ matrix, which can be computed in $\mathcal{O}(E \ln E)$, where E is the number of edges of the underlying graph. In most real applications, the underlying graph on which the information is diffusing is *sparse*, in the sense that its number of edges E is small compared to n^2.

Proposition 24.8. *Assume that $\mathbb{F}$ is a convex set of Hazard matrices included in a ball of radius $R > 0$ w.r.t. the norm in Eq. (24.26), and that the projection step in Eq. (24.27) has complexity at most $\mathcal{O}(E \ln E)$. Then, the NetShape algorithm described in Algorithm 24.1 converges to the minimum of Eq. (24.25). Moreover, the complexity of the algorithm is $\mathcal{O}(\frac{R^2}{\epsilon^2} E \ln E)$.*

Proof. This is a direct application of the projected subgradient descent to the problem:

$$\mathcal{H}^* = \underset{\mathcal{H}\in\mathbb{H}}{\operatorname{argmin}}\, \rho\left(\frac{\mathcal{H}+\mathcal{H}^\top}{2}\right), \tag{24.29}$$

where $\mathbb{H} = \left\{\int_0^{+\infty} \mathcal{F}(t)dt \in \mathbb{R}^{n\times n} : \mathcal{F} \in \mathbb{F}\right\}$ is the set of feasible Hazard matrices. The convergence rate of such an algorithm can be found in [56]. ■

Remark 24.2. The corresponding maximization problem is not convex anymore and only convergence to a local maximum can be expected. However, when the changes in the Hazard functions are relatively small (e.g., inefficient control actions, or only a limited number of treatments available to distribute), then NetShape achieves fairly good performance.

24.7 CASE STUDIES

In this section, we illustrate the generality of our framework by reframing well-known diffusion suppression problems that can find application in rumor control that has been discussed extensively in this chapter. Using Problem 24.1 we derive the corresponding variants of the NetShape algorithm.

For simplicity, we denote as $M \odot M'$ the Hadamard product between the two matrices (i.e., coordinate-wise multiplication), as $\Delta = \int_0^{+\infty}\left(\hat{\mathcal{F}}(t) - \mathcal{F}(t)\right)dt$ the matrix with the integrated coordinate-wise difference of two Hazard matrices in time, and as $\mathbf{1} \in \mathbb{R}^n$ the all-one vector (see notations in Table 24.1).

24.7.1 PARTIAL QUARANTINE

The *quarantine* approach aims to remove a small number of edges in order to minimize the spread of the contagion. This strategy is highly interventional in the sense that it totally removes edges, but in order to be practical it has to remain at low scale and affect a small amount of edges. This is the reason why it is mostly appropriate for dealing with the initial very few infections. The *partial quarantine* setting is a relaxation where one is interested in decreasing the transmission probability along a set of targeted edges by using local and expensive actions.

Definition 24.6 (Partial quarantine). Consider that a marketing campaign has k control actions to distribute in a network $\mathcal{G} = (\mathcal{V},\mathcal{E})$. For each edge $(i,j) \in \mathcal{E}$, let $\mathcal{F}_{ij}$ and $\hat{\mathcal{F}}_{ij}$ be the Hazard matrices *before* and *after* applying control actions, respectively. If $X \in [0,1]^{n\times n}$ is the control actions matrix and X_{ij} represents the amount of suppressive action taken on edge (i,j), then the set of feasible policies can be expressed as:

$$\mathbb{F} = \left\{(1-X)\odot\mathcal{F} + X\odot\hat{\mathcal{F}} : X\in[0,1]^{n\times n},\ \|X\|_1\leq k\right\}. \tag{24.30}$$

■

Example. For a nonnegative scalar $\epsilon \geq 0$, we may consider $\hat{\mathcal{F}} = (1-\epsilon)\mathcal{F}$ in order to model the suppression of selected transmission rates; formally:

$$\mathbb{F} = \left\{(1-\epsilon X) \odot \mathcal{F} : X \in [0,1]^{n\times n},\ \|X\|_1 \leq k\right\}. \tag{24.31}$$

Importantly, for the special case where $\epsilon = 1$, this problem becomes equivalent to the setting discussed in [10,14].

A straightforward adaptation of Algorithm 24.1 to this setting leads to the NetShape algorithm for partial quarantine described in Algorithm 24.2. The projection step is performed by Algorithm 24.3 on the flattened versions $x', \delta, y \in \mathbb{R}^E$ of the matrices X', Δ, and Y, and the parameter R is chosen to upper bound $\max_{\mathcal{F}' \in \mathbb{F}} \|\mathcal{F}' - \mathcal{F}\|_2 = \max_{X\in[0,1]^{n\times n}, \|X\|_1 \leq k} \|X \odot \Delta\|_2$. ■

Lemma 24.1. *The projection step of Algorithm 24.1 for the partial quarantine setting of Definition 24.6 is:*

$$X^* = \arg\min_{x' \in [0,1]^E,\ \|x'\|_1 \leq k} \left\|x' \odot \delta - y\right\|_2, \tag{24.32}$$

where δ and y are flattened version of, respectively, Δ and $Y = X \odot \Delta - \eta u_{\mathcal{F}} u_{\mathcal{F}}^{\mathsf{T}}$. Moreover, this problem can be solved in time $\mathcal{O}(E \ln E)$ with Algorithm 24.3, where E is the number of edges of the network.

Proof. Eq. (24.32) directly follows from Eq. (24.27) and the definition of $\mathbb{F}$. Algorithm 24.3 is an extended version of the L_1-ball projection algorithm of [57]. Karush-Kuhn-Tucker (KKT) conditions for the optimization problem of Eq. (24.32) imply that $\exists z > 0$ s.t. $\forall i,\ x_i' = \max\{0, \min\{\frac{2\delta_i y_i - z}{2\delta_i^2}, 1\}\}$. The algorithm is a simple linear search for this value. Finally, the sorting step (Algorithm 24.3, line 5) has the highest complexity $\mathcal{O}(E \ln E)$, and the loops perform at most $2E$ iterations, hence an overall complexity $\mathcal{O}(E \ln E)$. ■

Algorithm 24.2 NETSHAPE PARTIAL QUARANTINE PROBLEM

Input: graph $\mathcal{G} = (\mathcal{V}, \mathcal{E})$, matrices of Hazard functions *before* and *after* treatment $\mathcal{F}, \hat{\mathcal{F}} \in \mathbb{F}$, approximation parameter $\epsilon > 0$, number of treatments k

Output: matrix of Hazard functions $\mathcal{F}^* \in \mathbb{F}$

$X \leftarrow 0, X^* \leftarrow 0$

$F \leftarrow \int_0^{+\infty} \mathcal{F}(t)dt$

$\Delta \leftarrow \int_0^{+\infty} (\hat{\mathcal{F}}(t)dt - \mathcal{F}(t))dt$

$R \leftarrow \sqrt{k} \max_{ij} \Delta_{ij}$

$T \leftarrow \lceil \frac{R^2}{\epsilon^2} \rceil$

for $i = 1$ to $T-1$ **do**

  $M \leftarrow F + X \odot \Delta$

  $u \leftarrow$ the largest eigenvector of $\frac{1}{2}(M + M^{\mathsf{T}})$

  $Y \leftarrow X \odot \Delta - \frac{R}{\sqrt{i}} uu^{\mathsf{T}}$

  $X \leftarrow \operatorname{argmin}_{X' \in [0,1]^{n\times n}, \|X'\|_1 \leq k} \left\|X' \odot \Delta - Y\right\|_2$ // projection step (Algorithm 24.3)

  $X^* \leftarrow X^* + X$

end for

return $\mathcal{F}^* = (1 - \frac{1}{T}X^*) \odot \mathcal{F} + \frac{1}{T}X^* \odot \hat{\mathcal{F}}$

Algorithm 24.3 PROJECTION STEP FOR THE PARTIAL QUARANTINE PROBLEM

Input: $\delta, y \in \mathbb{R}^E$, budget $k \in (0, E)$
Output: control actions vector x'

for $i = 1$ to E **do**
    $\mu_i \leftarrow 2\delta_i y_i$
    $\mu_{E+i} \leftarrow 2\delta_i(y_i - \delta_i)$
end for
sort μ into $\mu_{\pi(1)} \geq \mu_{\pi(2)} \geq \cdots \geq \mu_{\pi(2E)}$
$d \leftarrow 0$
$s \leftarrow 0$
$i \leftarrow 1$
while $s < k$ and $\mu_{\pi(i)} \geq 0$ **do**
    $d \leftarrow d + \mathbb{1}\{\pi(i) \leq E\}\frac{1}{2\delta^2_{\pi(i)}} - \mathbb{1}\{\pi(i) > E\}\frac{1}{2\delta^2_{\sigma(i)-E}}$
    $s \leftarrow s + d(\mu_{\pi(i)} - \mu_{\pi(i+1)})$
    $i \leftarrow i + 1$
end while
$z \leftarrow \max\{0, \mu_{\sigma(i)} + \frac{s-k}{d}\}$
return x' s.t. $x'_i = \max\{0, \min\{\frac{2\delta_i y_i - z}{2\delta_i^2}, 1\}\}$

24.7.2 PARTIAL NODE IMMUNIZATION

More often, control actions can only be performed on the nodes rather than the network edges that was the case of the previous section. For example, imagine advertising campaigns that aim to enhance the diffusion of a product or, more relevant to the suppressive scenario, decision-makers that debunk false information targeting specific influencer nodes. In that case, the effect of the control actions must be aggregated over nodes in the following way.

Definition 24.7 (Partial node immunization). Consider that a control campaign has k control actions to distribute in a network $\mathcal{G} = (\mathcal{V}, \mathcal{E})$. For each edge $(i,j) \in \mathcal{E}$, let $\mathcal{F}_{ij}$ and $\hat{\mathcal{F}}_{ij}$ be the Hazard matrices *before* and *after* applying control actions, respectively. If $x \in [0,1]^n$ is the control actions vector and x_i represents the amount of suppressive action taken on node i, then we express the set of feasible policies as:

$$\mathbb{F} = \left\{(1 - x\mathbf{1}^\top) \odot \mathcal{F} + x\mathbf{1}^\top \odot \hat{\mathcal{F}} : x \in [0,1]^n, \|x\|_1 \leq k\right\}. \tag{24.33}$$

■

This setting corresponds to partial quarantine in which all outgoing edges of a node are impacted by a single control action. When $\hat{\mathcal{F}} = 0$, this problem corresponds to the node removal problem (or vaccination), that consists in removing k nodes from the graph in advance in order to minimize a future contagion (see [15]).

Given a vector x, the projection problem to solve is:

$$\begin{aligned} x^* &= \underset{x' \in [0,1]^n, \|x'\|_1 \leq k}{\operatorname{argmin}} \left\|(x'\mathbf{1}^\top) \odot \Delta - Y\right\|_2 \\ &= \underset{x' \in [0,1]^n, \|x'\|_1 \leq k}{\operatorname{argmin}} \sum_i x_i^2\Big(\sum_j \Delta_{ij}^2\Big) - 2x_i\Big(\sum_j \Delta_{ij} Y_{ij}\Big) \\ &= \underset{x' \in [0,1]^n, \|x'\|_1 \leq k}{\operatorname{argmin}} \|x' \odot \delta' - y'\|_2, \end{aligned} \tag{24.34}$$

where $\delta'_i = \sqrt{\sum_j \Delta_{ij}^2}$ and $y'_i = \frac{\sum_j \Delta_{ij} Y_{ij}}{\sqrt{\sum_j \Delta_{ij}^2}}$. Hence we can apply the projection step of Algorithm 24.3 for the partial node immunization problem using δ' and y', and its complexity is $\mathcal{O}(n \ln n)$.

Remark 24.3. Because the upper bound of Proposition 24.4 holds as well for SIR epidemics [51] (see also [11]), this setting may also be used to reduce the spread of a disease using, for example, medical treatments or vaccines. More specifically, the Hazard matrix for an SIR epidemic is the following:

$$\mathcal{H} = \ln\left(1 + \frac{\beta}{\delta}\right) A, \tag{24.35}$$

where δ is the recovery (or removal) rate and β is the transmission rate along edges of the network, and A the adjacency matrix. Then, a medical treatment may increase the recovery rate δ for targeted nodes, thus decreasing all Hazard functions on its outgoing edges, and the partial node immunization setting is applicable.

24.8 EXPERIMENTS

24.8.1 EXPERIMENTAL SETUP AND EVALUATION

In this section, we provide empirical evidence for the discussion of this chapter on controlling independent cascades under the ICM. We set the focus of this empirical evaluation in the offline partial node immunization problem under the ICM, as described in Section 24.7.2, and we are interested to see in practice the performance gains of the *NetShape* algorithm when compared to other baseline and state-of-the-art alternative policies.

Compared policies. We provide comparative experimental results against several strategies, namely:

(i) *Rand*: random selection of nodes;
(ii) *Degree*: selection of k nodes with highest out-degree;
(iii) *WeightedDegree*: selection of k nodes with highest sum of outgoing edge weight $w_{ij} = \int_0^{+\infty} \mathcal{F}_{ij}(t)dt$. This strategy can also be seen as the optimization of the first influence lower bound LB_1 of [54].
(iv) *NetShield* algorithm [15]. Given the adjacency matrix of a graph, this outputs the best k-nodes to totally immunize so as to decrease the vulnerability of the graph. This is done by assigning to each node a *shield value* that is high for nodes with high eigenscore and no edges connecting them. Note that, despite the fact that *NetShield* is tailored for immunization on unweighted graphs, it is not general enough to account for weighted edges and partial immunization as in our experimental setting.

Network datasets. The evaluation is performed on three benchmark real datasets (see Table 24.2) and the results are presented in subfigures of Fig. 24.4:

(a) a network of "friends lists" from `Facebook` [58];
(b) the `Gnutella` peer-to-peer file sharing network [58],
(c) the who-trust-whom online review site `Epinions.com`;

Table 24.2 Datasets: Details of the Benchmark Real Networks

Network	Nodes	Edges	Nodes in Largest SCC
SBD10ER	500	2704	500 :: 100.0%
Facebook	4039	88,234	4039 :: 100.0%
Gnutella	62,586	147,892	14,149 :: 22.6%
Epinions	75,879	508,837	32,223 :: 42.5%

The last column is the size of the strongly connected component.

(d) a synthetic random network of $n = 500$ nodes forming group structure (stochastic block-diagonal) that has been generated as follows. First, 10 equally-sized Erdös-Rényi clusters were independently formed with intracluster edge creation probability $p_{\text{inter}} = 0.1$. Then, their adjacency matrices were used to compose a block-diagonal structure with uniform intercluster rewiring probability $p_{\text{intra}} = 0.001$. Fig. 24.3A shows the structure of the final adjacency matrix (as having binary edge weights).

Note that the above networks only provide an unweighted adjacency matrix, thus only the existence, or not, of an edge between a pair of nodes is known. NetShape and the analysis of Section 24.5 is generally covering time-variable propagation functions between nodes. However, without loss of generality and for the sake of simplifying the experimental setup, we decided to use a simple class of propagation functions. For the generation of the matrix of edge-transmission probability rates $\{p_{ij}\}$ we use a *trivalency model*, according to which, the p_{ij} values are drawn chosen uniformly at random from a small set of constants. In our case that is set to $\{p_{\text{low}}, p_{\text{med}}, p_{\text{high}}\}$ and the specific used values are mentioned explicitly for each dataset at the figures' captions.

Each treatment unit of the budget can be assigned to a single node and, here, we assume that it can cause a fixed decrease to the node's transmission probability rates along all its outgoing edges (70% for the SBD10ER and 50% for the real networks).

In the experiments we evaluate the efficiency of the immunization policies on the basis of two measures, for both of which lower values are better:

- *Spectral radius decrease*. We examine the extend of the decrease of the spectral radius of the Hazard matrix $\mathcal{F}$ and, hence, the decrease of the bound of the max influence as described in Proposition 24.4.
- *Expected influence decrease*. We compare the performance of policies in terms of Problem 24.1. To this end, for each Hazard matrix $\mathcal{F}$, the influence is computed as the *average number of infected nodes* at the end of more than 1000 runs of the independent cascade $\mathcal{CTIC}$ while applying that specific Hazard matrix $\mathcal{F}$. Each time a single initial influencer is selected by the influence maximization algorithm Pruned Monte Carlo [40] by generating 1000 vertex-weighted directed acyclic graphs (DAGs).

In the empirical study, we focus on the scenario where the spectral radius of the original network is approximately one, which is the setting in which decreasing the spectral radius has the most impact

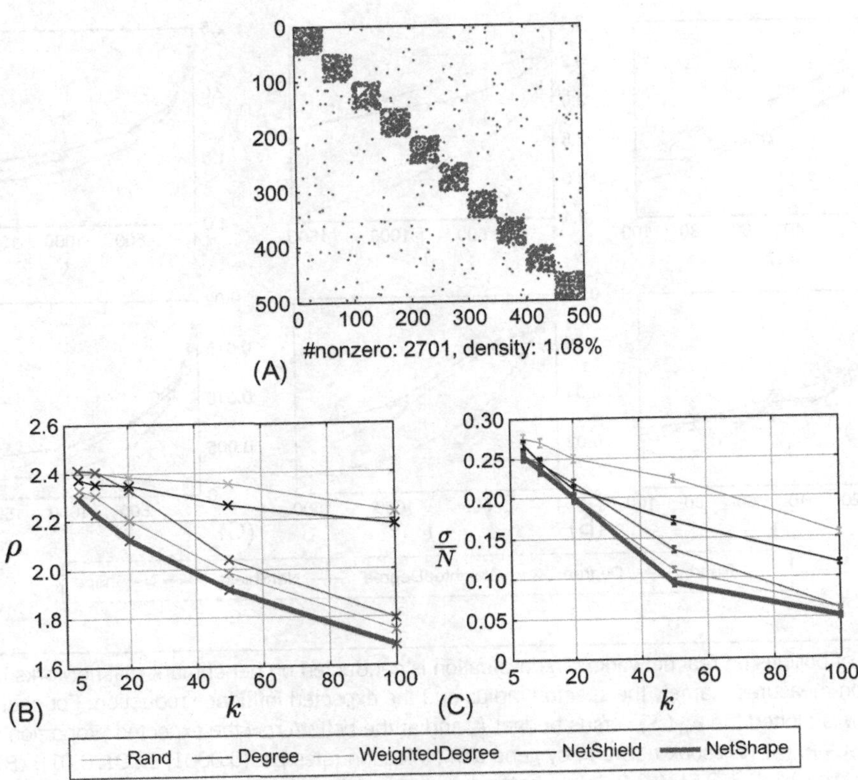

FIG. 24.3

Comparison of policies on a synthetic network. Comparison of NetShape's performance against competitors on the synthetic network **SBD10ER**, which is a composition of 10 Erdős-Rényi clusters (see details in Section 24.8.1). The values used for the trivalency model to generate edge weights are $p \in \{0.1, 0.2, 0.5\}$. The tested budget values are $k \in \{5, 10, 20, 50, 100\}$. (A) The structure of the generated nonsymmetric, block-diagonal adjacency matrix (here plotted as a binary matrix); (B) Spectral radius $\rho_{\mathcal{H}}(\mathcal{F})$ versus budget k; (C) Influence: the expected proportion of infected nodes $\frac{\sigma}{n}$ versus k. Lower values are better.

on the upper bounds in Proposition 24.4 and [12]. We believe that this intermediate regime is the most meaningful and interesting in order to test the different algorithms.

24.8.2 RESULTS

The results on the synthetic network are shown in Fig. 24.3 and those on the three real network datasets in subfigures of Fig. 24.4. The subfigures correspond to the two evaluation measures that we use for a wide range of budget size k in proportion to the number of nodes of that network.

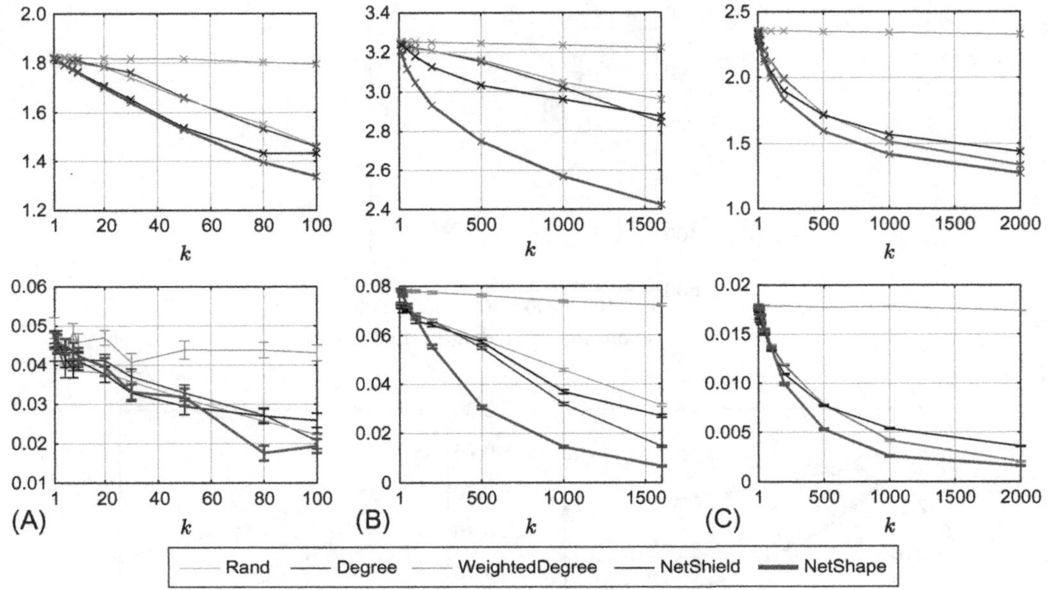

FIG. 24.4

Comparison of policies on real networks. The evaluation is conducted on benchmark real networks in terms of two evaluation measures, namely the spectral radius and the expected influence reduction. For each network, at the top row is plotted the $\rho_{\mathcal{H}}(\mathcal{F})$ versus budget k, and at the bottom row the expected proportion of infected nodes $\frac{\sigma}{n}$ versus k. (A) Facebook network, by generating infection rates $p \in \{0.0001, 0.001, 0.01\}$; (B) Gnutella network with $p \in \{0.1, 0.3, 0.6\}$; (C) Epinions network with $p \in \{0.005, 0.005, 0.05\}$. Lower values are better. (A) Facebook; (B) Gnutella; (C) Epinions.

First, we should note that the influence and spectral radius measures correlate generally well across all reported experiments; they present similar decrease w.r.t. budget increase and hence "agree" in the *order of effectiveness* of each policy when examined individually. As expected, all policies perform more comparably when very few or too many resources are available. In the former case, the very "central" nodes are highly prioritized by all methods while in the latter the significance of node selection diminishes. Even simple approaches perform well in all but `Gnutella` network where we get the most interesting results. NetShape achieves a sharp drop of the spectral radius early (i.e., for small budget k) in `Gnutella` and `Epinions` networks, which drives a large influence reduction. With regards to influence minimization, the difference to competitors is bigger, though, in `Gnutella` which is the most sparse and has the smallest strongly connected component (see Table 24.2). In `Facebook`, the reduction of the spectral radius is slower and seems less closely related with the influence, in the sense that the upper bound that we optimize is probably less tight to the behavior of the process.

Overall, the performance of the proposed NetShape algorithm is mostly as good or superior to that of the competitors, achieving up to a 50% decrease of the influence on the `Gnutella` network compared to its best competitor. Similar findings can be claimed for the experiments on the synthetic network `SBD10ER`.

24.9 CONCLUSION

The future of the diffusion networks field is full of interesting problems and potential applications. It will continue to enrich our understanding of diffusive phenomena and, at a second level, is expected to also change how information is circulated in online social networks.

The subject of this chapter was first to analyze the way information diffusion takes place in modern large-scale online social networks and the challenges regarding the control of certain types of undesired diffusion such as rumors, fake news, and others. We have presented an overview of the complex context in which these information-related diffusive phenomena appear and how individuals participate in the process acting as users of online social platforms.

To present the background of related problems, we went through various approaches for modeling information cascades, including the early used virus models and the more recent independent cascades model. Specifically for the latter model, we spoke about its large-scale dynamics and how that relates to the network properties, the existence of a threshold value that defines the point of transition between subcritical and supercritical behavior, and the connection of that threshold value to the spectral radius of the Hazard matrix of the network.

Subsequently, we discussed a framework that we proposed recently for *spectral activity shaping* under the Continuous-Time Independent Cascades Model [13] that allows the administrator for local control actions by allocating targeted resources, which can alter locally the spread of the process. The activity shaping is achieved via the optimization of the *spectral radius of the Hazard matrix*, which enjoys a simple convex relaxation when used to minimize the influence of the cascade. In addition, by reframing a number of use cases, we explained that the proposed framework is general and includes tasks such as partial quarantine that acts on edges and partial node immunization that acts on nodes. Notably, this generic framework can describe complex strategies that may use several immunization options by deploying simultaneously resources of different types (removal of edges, nodes, partial immunization, etc.). Specifically for the influence minimization that is the one directly related to rumor spreading control, we presented the *NetShape* method, which was compared favorably to baseline and a state-of-the-art method on real benchmark network datasets.

Among the interesting and challenging future work directions, on the same line to the presented framework, there can be the introduction of an "aging" feature to each piece of information that would model its loss of relevance and attraction through time, and the theoretical study and experimental validation of the maximization counterpart of *Netshape* method.

REFERENCES

[1] Zubiaga A, Aker A, Bontcheva K, Liakata M, Procter R. Detection and resolution of rumours in social media: a survey. ACM Comput Surv 2017;51(2):32:1–32:36.

[2] Knapp RH. A psychology of rumor. Public Opin Q 1944;8(1):22–37.

[3] Allport GW, Postman L. An analysis of rumor. Public Opin Q 1946;10(4):501–17.

[4] Chen W, Lakshmanan LVS, Castillo C. Information and influence propagation in social networks. Synth Lect Data Manag 2013;5(4):1–177.

[5] Kempe D, Kleinberg J, Tardos É. Maximizing the spread of influence through a social network. In: Proceedings of the ACM SIGKDD international conference on knowledge discovery and data mining; 2003. p. 137–46.

[6] Leskovec J, Backstrom L, Kleinberg J. Meme-tracking and the dynamics of the news cycle. In: Proceedings of the ACM SIGKDD international conference on knowledge discovery and data mining; 2009. p. 497–506.
[7] Gomez-Rodriguez M, Balduzzi D, Schölkopf B. Uncovering the temporal dynamics of diffusion networks. In: Proceedings of the international conference on machine learning; 2011. p. 561–8.
[8] Wang Y, Chakrabarti D, Wang C, Faloutsos C. Epidemic spreading in real networks: an eigenvalue viewpoint. In: Proceedings of the IEEE international symposium on reliable distributed systems; 2003. p. 25–34.
[9] Prakash BA, Chakrabarti D, Valler NC, Faloutsos M, Faloutsos C. Threshold conditions for arbitrary cascade models on arbitrary networks. Knowl Inf Syst 2012;33(3):549–75.
[10] Tong H, Prakash BA, Eliassi-Rad T, Faloutsos M, Faloutsos C. Gelling, and melting, large graphs by edge manipulation. In: Proceedings of the ACM international conference on information and knowledge management; 2012. p. 245–54.
[11] Scaman K, Lemonnier R, Vayatis N. Anytime influence bounds and the explosive behavior of continuous-time diffusion networks. In: Advances in neural information processing systems; 2015. p. 2017–25.
[12] Lemonnier R, Scaman K, Vayatis N. Tight bounds for influence in diffusion networks and application to bond percolation and epidemiology. In: Advances in neural information processing systems; 2014. p. 846–54.
[13] Scaman K, Kalogeratos A, Corinzia L, Vayatis N. A spectral method for activity shaping in continuous-time information cascades; 2017. ArXiv e-prints, abs/1709.05231.
[14] Van Mieghem P, Stevanović D, Kuipers F, Li C, Van De Bovenkamp R, Liu D, et al. Decreasing the spectral radius of a graph by link removals. Phys Rev E 2011;84(1):016101.
[15] Tong H, Prakash BA, Tsourakakis C, Eliassi-Rad T, Faloutsos C, Chau DH. On the vulnerability of large graphs. In: Proceedings of the IEEE international conference on data mining; 2010. p. 1091–6.
[16] Allport GW, Postman L. The psychology of rumor. J Clin Psychol 1947;3(4):402.
[17] Daley DJ, Kendall DG. Epidemics and rumours. Nature 1964;204(8):1118.
[18] Daley DJ, Kendall DG. Stochastic rumours. IMA J Appl Math 1965;1(1):42–55.
[19] Maki DP, Thompson M. Mathematical models and applications: with emphasis on the social, life, and management sciences. Englewood Cliffs, NJ: Prentice-Hall; 1973.
[20] Castillo C, Mendoza M, Poblete B. Predicting information credibility in time-sensitive social media. Internet Res 2013;23(5):560–88.
[21] Procter R, Vis F, Voss A. Reading the riots on twitter: methodological innovation for the analysis of big data. Int J Soc Res Methodol 2013;16(3):197–214.
[22] Hethcote HW. The mathematics of infectious diseases. SIAM Rev 2000;42(4):599–653.
[23] Newman M. Networks: an introduction. New York, NY: Oxford University Press; 2010.
[24] Bettencourt LMA, Cintrórn-Arias A, Kaiser DI, Castillo-Chávez C. The power of a good idea: quantitative modeling of the spread of ideas from epidemiological models. Phys A: Stat Mech Appl 2006;364(Supplement C):513–36.
[25] Jin F, Dougherty E, Saraf P, Cao Y, Ramakrishnan N. Epidemiological modeling of news and rumors on twitter. In: Proceedings of the workshop on social network mining and analysis. New York, NY: ACM; 2013. p. 1–9.
[26] Scaman K, Kalogeratos A, Vayatis N. Suppressing epidemics in networks using priority planning. IEEE Trans Netw Sci Eng 2016;3(4):271–85.
[27] Scaman K, Kalogeratos A, Vayatis N. A greedy approach for dynamic control of diffusion processes in networks. In: IEEE international conference on tools with artificial intelligence; 2015. p. 652–9.
[28] Kempe D, Kleinberg J, Tardos E. Maximizing the spread of influence through a social network. In: Proceedings of the ACM SIGKDD international conference on knowledge discovery and data mining; 2003. p. 137–46.
[29] Chen W, Wang Y, Yang S. Efficient influence maximization in social networks. In: Proceedings of the ACM SIGKDD international conference on knowledge discovery and data mining; 2009. p. 199–208.

[30] Gomez-Rodriguez M, Schölkopf B. Influence maximization in continuous time diffusion networks. In: Proceedings of the international conference on machine learning; 2012, p. 313–20.
[31] Vere-Jones D. Earthquake prediction—a statistician's view. J Phys Earth 1978;26(2):129–46.
[32] Reynaud-Bouret P, Rivoirard V, Grammont F, Tuleau-Malot C. Goodness-of-fit tests and nonparametric adaptive estimation for spike train analysis. J Math Neurosci 2014;4(1):1–41.
[33] Bauwens L, Hautsch N. Modelling financial high frequency data using point processes. Springer; 2009.
[34] Alfonsi A, Blanc P. Dynamic optimal execution in a mixed-market-impact Hawkes price model. Finance and Stochast 2015;20(1):183–218.
[35] Crane R, Sornette D. Robust dynamic classes revealed by measuring the response function of a social system. Proc Natl Acad Sci U S A 2008;105(41):15649–53.
[36] Farajtabar M, Du N, Gomez-Rodriguez M, Valera I, Zha H, Song L. Shaping social activity by incentivizing users. In: Advances in neural information processing systems; 2014. p. 2474–82.
[37] Lemonnier R, Scaman K, Kalogeratos A. Multivariate Hawkes processes for large-scale inference. In: Proceedings of the AAAI conference on artificial intelligence; 2017. p. 2168–74.
[38] Domingos P, Richardson M. Mining the network value of customers. In: Proceedings of the ACM SIGKDD international conference on knowledge discovery and data mining. San Francisco, CA: ACM; 2001. p. 57–66.
[39] Richardson M, Domingos P. Mining knowledge-sharing sites for viral marketing. In: Proceedings of the ACM SIGKDD international conference on knowledge discovery and data mining. New York, NY, USA: ACM; 2002. p. 61–70.
[40] Ohsaka N, Akiba T, Yoshida Y, Kawarabayashi K. Fast and accurate influence maximization on large networks with pruned Monte-Carlo simulations. In: Proceedings of the AAAI conference on artificial intelligence; 2014. p. 138–44.
[41] Leskovec J, Krause A, Guestrin C, Faloutsos C, VanBriesen J, Glance N. Cost-effective outbreak detection in networks. In: Proceedings of the ACM SIGKDD international conference on knowledge discovery and data mining; 2007. p. 420–9.
[42] Gupta A, Kumaraguru P, Castillo C, Meier P. Tweetcred: Real-time credibility assessment of content on twitter. In: Proceedings international conference of social informatics. Cham: Springer; 2014. p. 228–43.
[43] Shah D, Zaman T. Rumors in a network: who's the culprit? IEEE Trans Inf Theory 2011;57(8):5163–81.
[44] Seo E, Mohapatra P, Abdelzaher T. Identifying rumors and their sources in social networks. In: Proceedings of the ground/air multisensor interoperability, integration, and networking for persistent ISR III, vol. 8389; 2012. p. 83891I.
[45] Ma J, Gao W, Wei Z, Lu Y, Wong KF. Detect rumors using time series of social context information on microblogging websites. In: Proceedings of the ACM international conference on information and knowledge management. New York, NY: ACM; 2015. p. 1751–4.
[46] Kitsak M, Gallos LK, Havlin S, Liljeros F, Muchnik L, Stanley HE, Makse HA. Identification of influential spreaders in complex networks. Nature 2010;6(11):888–93.
[47] Goyal A, Bonchi F, Lakshmanan LVS. A data-based approach to social influence maximization. Proc VLDB Endowment 2011;5(1):73–84.
[48] He Z, Cai Z, Wang X. Modeling propagation dynamics and developing optimized countermeasures for rumor spreading in online social networks. In: Proceedings of the IEEE international conference on distributed computing systems; 2015. p. 205–14.
[49] Van Mieghem P, Omic J, Kooij R. Virus spread in networks. IEEE/ACM Trans Netw 2009;17(1):1–14.
[50] Pastor-Satorras R, Castellano C, Van Mieghem P, Vespignani A. Epidemic processes in complex networks. Rev Mod Phys 2015;87:925–79.
[51] Kermack WO, McKendrick AG. Contributions to the mathematical theory of epidemics. II. The problem of endemicity. Proc R Soc Lond Ser A 1932;138(834):55–83.
[52] Du N, Song L, Gomez-Rodriguez M, Zha H. Scalable influence estimation in continuous-time diffusion networks. In: Advances in neural information processing systems; 2013. p. 3147–55.

[53] Fortuin CM, Kasteleyn PW, Ginibre J. Correlation inequalities on some partially ordered sets. Commun Math Phys 1971;22(2):89–103.
[54] Khim JT, Jog V, Loh PL. Computing and maximizing influence in linear threshold and triggering models. In: Advances in neural information processing systems 29; 2016. p. 4538–46.
[55] Wang C, Chen W, Wang Y. Scalable influence maximization for independent cascade model in large-scale social networks. Data Min Knowl Disc 2012;25(3):545–76.
[56] Bubeck S. Convex optimization: algorithms and complexity. Found Trends Mach Learn 2015;8(3-4):231–357.
[57] Duchi J, Shalev-Shwartz S, Singer Y, Chandra T. Efficient projections onto the L1-ball for learning in high dimensions. In: Proceedings of the international conference on machine learning. ACM; 2008. p. 272–9.
[58] Leskovec J, Krevl A. SNAP datasets: Stanford large network dataset collection; 2014.

CHAPTER

MULTILAYER SOCIAL NETWORKS

25

Brandon Oselio*, Sijia Liu*, Alfred Hero*
*EECS Department, University of Michigan, Ann Arbor, MI, United States**

25.1 INTRODUCTION

Social media's prevalence in daily life has led to a massive increase in data for modeling human behaviors. Often, this data is structured in a way that can be thought of as a connection of links between agents, i.e., a network of social media users. This precipitates a necessity for specialized algorithms that can handle multilayer structured network data for inference, evaluation, and prediction.

We often find heterogeneous structure in social media data—there may exist more than one type of relationship between agents, and these relationships may impose different topological characteristics. For instance, people may be connected by more than one social platform. Alternatively, we may observe explicit links between agents but also infer implicit affinities based on agent features.

Another example of this heterogeneous structure arises when relationships between agents appear and disappear over time; agents begin talking to each other at one time and end at another time, possibly signifying a change in relation. Both of the above examples can be explained by a multilayer network framework.

A multilayer network is a network where a set of elementary units is connected by intralayer and interlayer relationships ("edges"). This structure is a generalization of single-layer networks where there are only intralayer relationships. These layers represent heterogeneity in the structure or labeling of the data; a layer might correspond to a type of connection or a discrete timestep. The interlayer structure represents ties among nodes in the different layers; this structure may be observed, assumed, or estimated depending on the application. The interlayer structure in a social network often corresponds to the labels of each node, so that each node in a single layer is connected to its counterparts in the other layers. If the layers represent timesteps, each entity might be connected to its counterpart in layers before and after the present layer, which represents the localization of that layer's characteristics in time.

As the multilayer structure is more complicated than its single-layer counterpart, methods for single-layer analysis must be modified to accommodate accordingly, and new methods are developed specifically for the multilayer case. This chapter will review some of the approaches for modeling multilayer networks and some of the methods that are specific to this structure.

Cooperative and Graph Signal Processing. https://doi.org/10.1016/B978-0-12-813677-5.00025-0

Table 25.1 List of Notations

Symbol	Description
$\mathcal{G} = (\mathcal{V}, \mathcal{E})$	a graph with vertex set $\mathcal{V}$ and edge set $\mathcal{E}$
$\mathcal{M}, \mathcal{G}_M, \mathbf{M}$	a multilayer network $\mathcal{M}$ with supragraph $\mathcal{G}_M$ and tensor form $\mathbf{M}$
$\mathbf{A}_M, \mathbf{L}_M$	supra-adjacency matrix $\mathbf{A}_M$ and supra-Laplacian matrix $\mathbf{L}_M$
$\mathbf{A}^{(\alpha)}, \mathbf{L}^{(\alpha)}$	adjacency matrix and Laplacian matrix for network at layer α
$[L]$	an integer set $\{1, 2, \ldots, L\}$
$\circ, \otimes$	outer/tensor product, and Kronecker product
$\mathbf{X}^{(t)}, \mathbf{Y}^{(t)}$	features from time $1, \ldots, t$

The rest of this chapter will proceed as follows: Section 25.2 will discuss the mathematical formulation of multilayer networks. Section 25.3 will cover some examples of multilayer node centralities. Section 25.4 will review some types of multilayer community detection methods. Section 25.5 will pivot and discuss the estimation of multilayer interaction networks for social media data. Section 25.6 will utilize some of the techniques discussed in the chapter on two application datasets. Finally, Section 25.7 will provide some concluding remarks. We list our notations used in this chapter in Table 25.1.

25.2 MATHEMATICAL FORMULATION OF MULTILAYER NETWORKS

In this section, we focus on the mathematical formulation of multilayer networks. Different from single-layer networks, they allow multiple types of interactions between each pair of nodes. In what follows, we introduce two network representations: supra-adjacency representation and tensor representation, each of which generalizes the notation of a single-layer network. We next show some real-life examples that involve the multilayer network structure.

25.2.1 MODELING AND REPRESENTATION

A single-layer network (also called a monoplex network) can be represented by a graph [1]. A graph is a tuple $\mathcal{G} = (\mathcal{V}, \mathcal{E})$, where $\mathcal{V}$ is the set of nodes and $\mathcal{E} \subseteq \mathcal{V} \times \mathcal{V}$ is the set of edges that connects pairs of nodes. A multilayer network generalizes the notion of a single-layer network by incorporating the interlayer connections; see Fig. 25.1A for an illustrative example. More formally, a multilayer network is a pair $\mathcal{M} = (\mathcal{T}, \mathcal{C})$ [2], where $\mathcal{T} = \{\mathcal{G}_\alpha, \alpha \in [L]\}$ is a family of graphs $\mathcal{G}_\alpha = (\mathcal{V}_\alpha, \mathcal{E}_\alpha)$ with $\mathcal{V}_\alpha \subseteq \mathcal{V}$, $[L] := \{1, 2, \ldots, L\}$, and $\mathcal{C} = \{\mathcal{E}_{\alpha\beta} \subseteq \mathcal{V}_\alpha \times \mathcal{V}_\beta, \alpha, \beta \in [L]\}$ denotes the set of interlayer connections ($\alpha \neq \beta$). Here α is the layer index, and by convention, $\mathcal{E}_{\alpha\alpha} = \mathcal{E}_\alpha$. When $L = 1$, the multilayer network $\mathcal{M}$ simplifies to a single-layer network. In the rest of the chapter, unless specified otherwise, we assume that each layer contains the same set of nodes with $|\mathcal{V}_\alpha| = |\mathcal{V}| = N$ for $\alpha \in [L]$, where $|\mathcal{V}|$ denotes the cardinality of the node set $\mathcal{V}$.

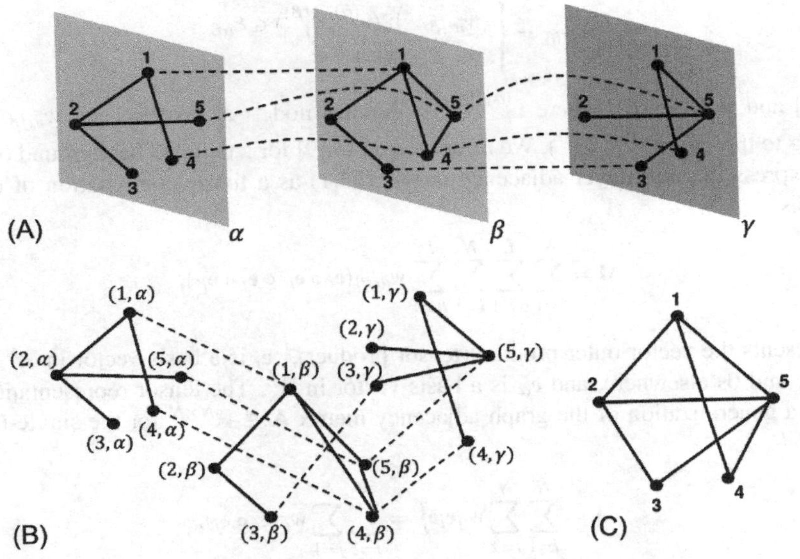

FIG. 25.1

Schematic illustration of multilayer network. Example with five nodes and three layers labeled α, β, and γ. (A) Multilayer network where *solid line* represents intralayer connection, and *dashed line* represents interlayer connection. (B) Supragraph representation. (C) Aggregated network.

Supragraph representation

Let $\mathcal{V}_M \subseteq \mathcal{V} \times [L]$ denote a set of node-layer combinations corresponding to $\mathcal{M}$, where $(v, \alpha) \in \mathcal{V}_M$ signifies that the node $v \in \mathcal{V}$ is present in layer $\alpha \in [L]$. Let $\mathcal{E}_M \subseteq \mathcal{V}_M \times \mathcal{V}_M$ be the set of edges between node-layer tuples. The multilayer network $\mathcal{M}$ can then be described by a graph $\mathcal{G}_M = (\mathcal{V}_M, \mathcal{E}_M)$, known as a supragraph, leading to a supra-adjacency matrix $\mathbf{A}_M$ and/or a supra-Laplacian matrix $\mathbf{L}_M$ [3]. Fig. 25.1B shows the supragraph representation of the multilayer network in Fig. 25.1A. Based on such a representation, many methods for single-layer networks, e.g., centrality-based network diagnostics and community detection methods, can be extended to multilayer networks [4,5].

In contrast to the supragraph, network aggregation provides the simplest representation for a multilayer network, where connections between nodes are aggregated in all layers to a single layer. The resulting graph is given by $\mathcal{G}_a = (\mathcal{V}_a, \mathcal{E}_a)$, where $\mathcal{V}_a = \cup_{\alpha=1}^{L} \mathcal{V}_\alpha$ and $\mathcal{E}_a = \cup_{\alpha=1}^{L} \mathcal{E}_\alpha$. Often the aggregated network can be cast as a convex combination (e.g., linear combination) of graph adjacency matrices across all layers [6,7]. Although such an aggregation may cause loss of information about the interlayer network structure [3], it becomes useful when modeling across networks that have very similar interlayer connectivity. Fig. 25.1C shows the aggregated representation of the multilayer network in Fig. 25.1A.

Tensor representation

A multilayer network can be represented in a tensor form [2,8,9]. Let $\mathbf{M} \in \mathbb{R}^{N\times L\times N\times L}$ denote the fourth-order adjacency tensor of the L-layer network $\mathcal{M}$. Each element of $\mathbf{M}$ is defined by

$$M_{i\alpha j\beta} = \begin{cases} w_{i\alpha j\beta} & \text{if } (v_i^{(\alpha)}, v_j^{(\beta)}) \in \mathcal{E}_{\alpha\beta} \\ 0 & \text{otherwise,} \end{cases} \tag{25.1}$$

for $i,j \in [N]$ and $\alpha, \beta \in [L]$, where $v_i^{(\alpha)} \in \mathcal{V}_\alpha$ denotes node i at layer α, and $w_{i\alpha j\beta}$ is the weight corresponding to the edge $(v_i^{(\alpha)}, v_j^{(\beta)})$. We refer readers to [9] for a detailed background on tensors.

We can express the multilayer adjacency tensor (25.1) as a linear combination of tensors in the canonical basis

$$\mathbf{M} = \sum_{i=1}^{N}\sum_{\alpha=1}^{L}\sum_{j=1}^{N}\sum_{\beta=1}^{L} w_{i\alpha j\beta}(\mathbf{e}_i \circ \mathbf{e}_\alpha \circ \mathbf{e}_j \circ \mathbf{e}_\beta), \tag{25.2}$$

where $\circ$ represents the vector outer product (tensor product),[1] $\mathbf{e}_i$ is a basis vector in $\mathbb{R}^N$ with 1 at the ith coordinate and 0s elsewhere, and $\mathbf{e}_\alpha$ is a basis vector in $\mathbb{R}^L$. The tensor representation (25.2) can be viewed as a generalization of the graph adjacency matrix $\mathbf{A} \in \mathbb{R}^{N\times N}$ for the single-layer network $\mathcal{G} = (\mathcal{V}, \mathcal{E})$,

$$\mathbf{A} = \sum_{i=1}^{N}\sum_{j=1}^{N} w_{ij}\mathbf{e}_i\mathbf{e}_j^T = \sum_{i=1}^{N}\sum_{j=1}^{N} w_{ij}(\mathbf{e}_i \circ \mathbf{e}_j), \tag{25.3}$$

where w_{ij} is the weight associated with edge $(v_i, v_j) \in \mathcal{E}$.

In addition to the fourth-order tensor representation (25.2), a multilayer network is also modeled by a third-order tensor in [10], where each slice corresponds to the network at one layer, e.g., a dynamic network at one snapshot. In contrast with the third-order tensor, the fourth-order tensor encodes detailed information on the interlayer connection between any two nodes at different layers, namely, Eq. (25.1). Also, the fourth-order tensor $\mathbf{M}$ can be flattened out to the supra-adjacency matrix $\mathbf{A}_M$ of dimension $NL \times NL$. Therefore, the fourth-order tensor is a natural representation of multilayer networks, and many techniques on tensor algebra [8] can be used for network analysis.

25.2.2 EXAMPLES OF MULTILAYER NETWORKS

We next introduce three important classes of multilayer networks: node-colored networks, edge-colored networks, and temporal networks [3].

Node-colored networks are graphs in which each node is labeled by one color. Considering each color as designating a layer, node-colored graphs can be represented as multilayer networks. They are often used to model heterogeneous networks that contain nodes of different types.

Example 1. Bibliographic information networks contain information about researchers (authors) and publications they produce (documents). Links exist between papers and/or authors by the authorship, colleagueship, published venues, or topics [11].

[1] If $\mathbf{X} = \mathbf{a}_1 \circ \mathbf{a}_2 \circ \cdots \circ \mathbf{a}_n$, then each element of the tensor $\mathbf{X}$ is given by $X_{i_1 i_2 \ldots i_n} = [\mathbf{a}_1]_{i_1}[\mathbf{a}_2]_{i_2}\cdots[\mathbf{a}_n]_{i_n}$, where $[\mathbf{x}]_i$ denotes the ith entry of $\mathbf{x}$.

Example 2. Internet of Things (IoT) denotes the internetworking of smart phones, computers, vehicles, buildings, and other devices embedded with electronics, sensors, and actuators [12]. IoT allows autonomous exchange of useful information between "heterogeneous nodes."
Edge-colored networks are graphs with multiple types of edges, where similar to node-colored networks, color distinguishes between layers. Edge-colored graphs can be represented by multilayer networks, where nodes in each layer are fixed and linked by edges with a unique color. They can be used to model multirelational networks where nodes have relations of different types [13].
Example 3. Public social networks link social entities by several types of relationships, including friendship, vicinity, kinship, and membership in the same cultural society [14].
Example 4. Urban transportation networks describe the urban ecosystem, where nodes represent spatial locations (e.g., restaurants, shopping malls, schools, parks, and other places of interest), and edges represent vehicles of different types, e.g., taxis, buses, and subways, that are used to travel between two locations [15].

A temporal network is given by an ordered sequence of graphs. It can be interpreted as a special case of an edge-colored multigraph, where the set of time instants provides the set of edge colors, and the interlayer edges are between nodes and their counterparts across all time steps. The chromatin contact map over a time course of cell growth/development is an example of a temporal network in biology [16].

25.3 DIAGNOSTICS FOR MULTILAYER NETWORKS: CENTRALITY ANALYSIS

The study of centrality, i.e., evaluating the degree of nodal importance to the network structure, is often used to identify and rank essential nodes in complex networks. A number of centrality measures are commonly used, such as degree, eigenvector, clustering coefficient, closeness, betweenness, hubness, and authority, differing in what type of influence is to be emphasized [17]. For example, degree centrality measures the total number of connections a node has while eigenvector centrality measures the importance of a node by the importance of its neighbors [18]. Most centrality methods are only directly applicable to single-layer networks. Here we generalize some important single-layer centrality methods to multilayer networks.

25.3.1 OVERLAPPING DEGREE AND MULTIPLEX PARTICIPATION COEFFICIENT

Nodal degree is the simplest feature in network diagnostics. There exist several ways to define multilayered degree centrality. The simplest way is to use network aggregation, where two nodes are considered to be adjacent if and only if the number of edges that connect them in a multilayer network is larger than a threshold [19,20]. However, this measure does not fully consider the interlayer effect.

In a multilayer network, it is essential to study how the nodal degree is distributed across different layers. We recall from Section 25.2 that $\mathcal{M}$ denotes a multilayer network with N nodes and L layers, the degree of node i on layer α becomes

$$k_i^{(\alpha)} = \sum_{j=1}^{N} A_{ij}^{(\alpha)}, \tag{25.4}$$

where $A_{ij}^{(\alpha)}$ is the (i,j)th entry of the adjacency matrix associated with graph $\mathcal{G}_\alpha$ on layer α. The degree of node i in a multilayer network is a vector quantity

$$\mathbf{k}_i = \left[k_i^{(1)}, k_i^{(2)}, \ldots, k_i^{(L)}\right], \quad i \in [N]. \tag{25.5}$$

The overlapping degree of node i across all layers is defined as [6]

$$o_i = \sum_{\alpha=1}^{L} k_i^{(\alpha)} = \mathbf{1}^T \mathbf{k}_i, \quad i \in [N], \tag{25.6}$$

where $\mathbf{1}$ is the $L \times 1$ vector of all ones. The overlapping degree (25.6) can be used to identify hubs, nodes with high degree in the network. However, a node that is a hub in one layer may only have a few connections in another layer. Thus a more suitable multilayer hub definition is the multiplex participation coefficient [6,21],

$$P_i = \frac{L}{L-1}\left[1 - \sum_{\alpha=1}^{L}\left(\frac{k_i^{(\alpha)}}{o_i}\right)^2\right]. \tag{25.7}$$

Here P_i takes values in $[0,1]$ and measures the degree to which the degree of node i is uniformly distributed among the L layers. If $P_i = 1$, then node i has exactly the same number of edges on each layer, namely, $k_i^{(\alpha)} = o_i/L$. If $P_i = 0$, all the edges of node i are concentrated in just one layer. The multiplex participation coefficient thus captures the heterogeneity of nodal degrees across layers in multilayer networks.

25.3.2 EIGENVECTOR CENTRALITY IN SUPRAGRAPH

Eigenvector centrality describes the impact of a node on the network's global structure, and is defined by the dominant eigenvector of the graph adjacency matrix. Eigenvector centrality is widely used in many applications. For example, it is closely related to hubness and authority centrality used in the hyperlink-induced topic search (HITS) algorithm [22]. Because computing the dominant eigenvalue and eigenvector can be computed in a distributed setting, eigenvector centrality is often preferable to other types of global centralities such as betweenness [23,24].

The simplest way to generalize the concept of eigenvector centrality for multilayer networks is to use network aggregation and apply single-layer based methods [25]. However, as shown in Fig. 25.1, network aggregation oversimplifies the multilayer network. Therefore, we consider the supragraph representation $\mathcal{G}_M$ of a multilayer network with L layers and N nodes. The supra-adjacency matrix

$\mathbf{A}_M \in \mathbb{R}^{NL \times NL}$ of $\mathcal{G}_M$ can be separated into two parts: the intralayer component $\mathbf{A}_M^L$ and the interlayer component $\mathbf{A}_M^I$. That is,

$$\mathbf{A}_M = \mathbf{A}_M^L + \mathbf{A}_M^I, \quad \mathbf{A}_M^L = \mathrm{diag}(\{\mathbf{A}^{(\alpha)}\}_{\alpha=1}^L), \tag{25.8}$$

where $\mathrm{diag}(\{\mathbf{A}^{(\alpha)}\}_{\alpha=1}^L)$ denotes a block-diagonal matrix with diagonal elements $\mathbf{A}^{(\alpha)}$ for $\alpha \in [L]$, and recall that $\mathbf{A}^{(\alpha)}$ is the graph adjacency matrix on layer α. The interlayer supra-adjacency matrix $\mathbf{A}_M^I$ defines the interlayer connectivity between every two layers. If the interlayer connectivity is identical for all nodes [5], then $\mathbf{A}_M^I = \mathbf{A}^I \otimes \mathbf{I}_N$, where $\mathbf{A}^I \in \mathbb{R}^{L \times L}$ is an interlayer adjacency matrix whose elements represent the strength of the connection between every pair of layers. For example, in a temporal network, if layers are connected at consecutive time steps, then the interlayer supra-adjacency matrix becomes

$$\mathbf{A}_M^I = \mathbf{A}^I \otimes \mathbf{I}_N, \quad \mathbf{A}^I = \begin{bmatrix} 0 & 1 & 0 & \cdots \\ 1 & 0 & 1 & \ddots \\ 0 & 1 & 0 & \ddots \\ \vdots & \ddots & \ddots & \ddots \end{bmatrix}. \tag{25.9}$$

Here $\mathbf{A}^I$ models an undirected chain network in which each node is adjacent to its nearest neighbors. It is worth mentioning that the decomposition of the supra-adjacency matrix in Eq. (25.8) facilitates exploring the spectral properties of multilayer networks [5].

The eigenvector centrality $\mathbf{v}_M \in \mathbb{R}^{NL}$ of the supra-adjacency matrix $\mathbf{A}_M$ can then be defined as the solution to the following eigenvalue problem

$$\mathbf{A}_M \mathbf{v}_M = \lambda_{\max} \mathbf{v}_M, \tag{25.10}$$

where $\lambda_{\max}$ denotes the largest positive eigenvalue of $\mathbf{A}_M$. The entries of $\mathbf{v}_M$ give the centralities of each node-layer pair. It is convenient to map the eigenvector centralities $\mathbf{v}_M$ to an $N \times L$ matrix

$$V_{i\alpha} = v_{N(\alpha-1)+i}, \quad i \in [N], \alpha \in [L], \tag{25.11}$$

where $V_{i\alpha}$ corresponds to the joint centrality of the node-layer pair (i, α). Based on Eq. (25.11), we introduce the marginal node centrality $\hat{v}_i$ and the marginal layer centrality $\tilde{v}_\alpha$ [24]

$$\hat{v}_i = \sum_{\alpha=1}^{L} V_{i\alpha}, \quad i \in [N], \quad \tilde{v}_\alpha = \sum_{i=1}^{N} V_{i\alpha}, \quad \alpha \in [L]. \tag{25.12}$$

Similar to the supra-adjacency matrix in Eq. (25.8), we can define the decomposition of the supra-Laplacian matrix, $\mathbf{L}_M = \mathbf{L}_M^L + \mathbf{L}_M^I$, where $\mathbf{L}_M^L = \mathrm{diag}(\mathbf{A}_M^L \mathbf{1}) - \mathbf{A}_M^L$, and $\mathbf{L}_M^I = \mathrm{diag}(\mathbf{A}_M^I \mathbf{1}) - \mathbf{A}_M^I$. The decomposition of the supra-Laplacian matrix corresponds to a diffusion process over nodes of the network. Specifically, the nodal dynamics follows the differential equation [5]

$$\dot{x}_{i\alpha} = \sum_{j=1}^{N} w_{ij}^{(\alpha)} (x_{j\alpha} - x_{i\alpha}) + \sum_{\beta=1}^{L} u_{\alpha\beta} (x_{i\beta} - x_{i\alpha}) \tag{25.13}$$

for any $i \in [N]$ and $\alpha \in [L]$, where $x_{i\alpha}$ denotes the state of node i at layer α, $w_{ij}^{(\alpha)}$ is the (i,j)th entry of the graph adjacency matrix $\mathbf{A}^{(\alpha)}$ at layer α, and $u_{\alpha\beta}$ is the interlayer coupling constant, namely, the (α, β)th entry of the interlayer adjacency matrix $\mathbf{A}^I$. The discretized matrix form of the diffusion equation (25.13) yields

$$\dot{\mathbf{x}} = -(\mathbf{L}_M^L + \mathbf{L}_M^I)\mathbf{x} = -\mathbf{L}_M\mathbf{x}. \tag{25.14}$$

Here the second smallest eigenvalue of $\mathbf{L}_M$ (also known as algebraic connectivity [1]) governs the convergence properties of the diffusion process.

25.3.3 NODAL CENTRALITY VIA TENSOR DECOMPOSITION

A fourth-order tensor was introduced in Section 25.2 to represent a multilayer network. Tensor decomposition is an effective tool for multiarray data analysis, and mono-layer centrality measures can be extended in order to identify key nodes in multilayer networks. It has been shown in [2] that the principal singular vectors obtained from the CANDECOMP/PARAFAC tensor decomposition [9] can provide hub and authority scores of all nodes in a multilayer network.

The fourth-order adjacency tensor $\mathbf{M} \in \mathbb{R}^{N\times L\times N\times L}$ of a multilayer network can be decomposed into a sum of rank-one tensors [9],

$$\mathbf{M} = \sum_{i=1}^{R} \sigma_i \mathbf{a}_i \circ \mathbf{b}_i \circ \mathbf{c}_i \circ \mathbf{d}_i, \tag{25.15}$$

where $\{\sigma_i\}_{i=1}^R$ are singular values of $\mathbf{M}$ sorted in a descending order, $\mathbf{a}_i \in \mathbb{R}^N$, $\mathbf{b}_i \in \mathbb{R}^L$, $\mathbf{c}_i \in \mathbb{R}^N$, and $\mathbf{d}_i \in \mathbb{R}^L$ are singular vectors corresponding to the singular value σ_i, and R is the rank of $\mathbf{M}$. Considering the principal quadruplet $\{\mathbf{a}_1, \mathbf{b}_1, \mathbf{c}_1, \mathbf{d}_1\}$ in Eq. (25.15), the entries of $\mathbf{a}_1$ and $\mathbf{c}_1$ correspond to hub and authority scores of all nodes while the entries of $\mathbf{b}_1$ and $\mathbf{d}_1$ give hub and authority scores of all layers. Note that if $L = 1$, then the four-order adjacency tensor reduces to the second-order adjacency matrix, and the entries of $\mathbf{a}_1$ and $\mathbf{c}_1$ give the conventional hub and authority scores of nodes in a single-layer network. Given the hub and authority scores [2], one can further generalize HITS of multilayer networks [26].

Based on Eq. (25.15), the importance of a node-layer pair (i, α) can be evaluated as

$$H_{i,\alpha} = |a_{1,i} b_{2,\alpha}| + |c_{1,i} d_{1,\alpha}|, \tag{25.16}$$

where $a_{1,i}$ denotes the ith entry of the vector $\mathbf{a}_1$. The nodal importance measure defined in Eq. (25.16) is called EDCPTD (Essential nodes Determining based on CP Tensor Decomposition) centrality [2]. Given the joint centrality of the node-layer pair in Eq. (25.16), we can then define the marginal node centrality and the marginal layer centrality following Eq. (25.12).

In addition to the hubness and authority centrality, other generalized centrality measures such as clustering coefficient, modularity, and random walk centrality can also be defined using the tensor representation; see [8] for details.

25.4 CLUSTERING AND COMMUNITY DETECTION IN MULTILAYER NETWORKS

Discovering mesoscale structures such as communities in complex networks is a wide field of study [27]. These communities are generally described as a subset of nodes in the graph that are more densely connected than other nodes in the network. This is sometimes referred to in sociology as homophily [28]. Detecting these communities in a single-layer network has and continues to be an active research field. Furthermore, research in community detection for the more general multilayer case has become increasingly prevalent in the past decade.

Community detection in social networks facilitates the interpretation of the overall structure of the network. Generally, we expect to see communities in social networks that strongly relate different agents to one another, such as common activities, interests, or memberships to organizations. For instance, students that attend the same university, play the same sport, or like the same music are more likely to be connected in a particular link type. The concept of communities becomes more complex when multiple layers are introduced (see Fig. 25.2); communities that develop in one type of interaction may not be present in another or may be subsumed by a larger, more prevalent supercommunity. It is also possible that the community structure in each layer exhibits different homophilic clusters that

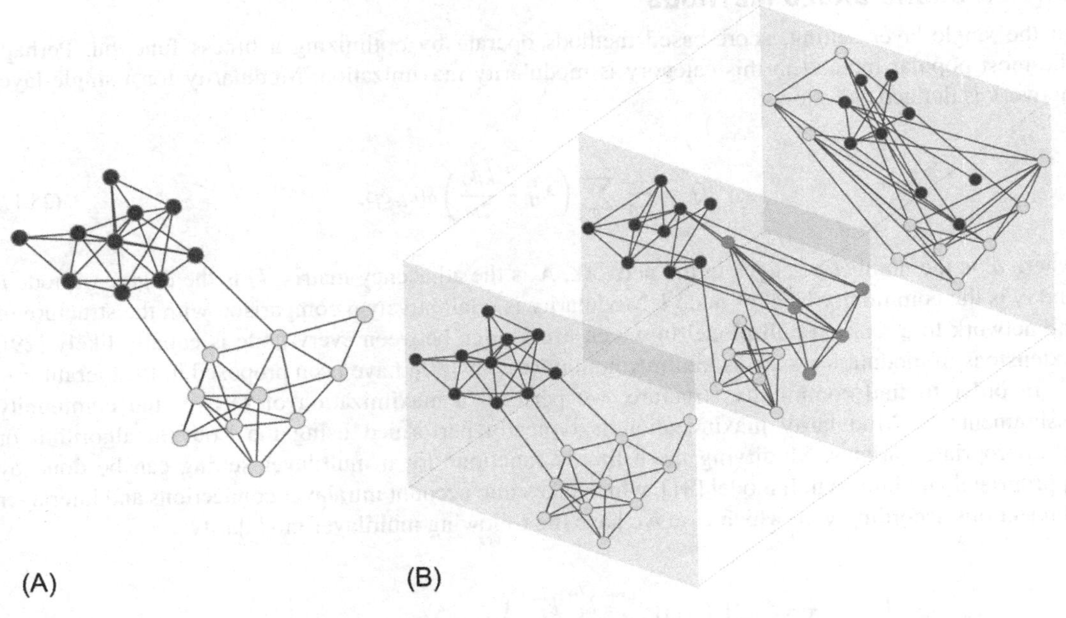

FIG. 25.2

Illustration of community detection. Examples of community detection with 20 nodes. (A) Single-layer community detection, where the community structure captures homophily among the nodes. (B) Community detection in a multilayer setting, where more complex situations can occur. The middle layer has a subcommunity in one of the larger communities displayed in the front layer while the back layer has a different latent structure altogether.

do not correlate across layers. Depending on the application, the main goal of analysis might be to utilize multiple layers to find communities that may not have been obvious in a single-layer slice of the network. In other applications, we may be interested in the similarities and dissimilarities of the community structure for each layer, which necessitates different approaches. Community detection in temporal networks deserves its own special treatment, as we often make temporal locality assumptions that allow for a more focused analysis.

We will briefly cover three types of methods for multilayer community detection: score-based methods, model-based methods, and aggregation methods. This list is by no means exhaustive nor are the types of methods meant to be canonical. Rather, we find them to be useful descriptors that tend to have reasonably well understood advantages and disadvantages in the multilayer setting. The main goal of aggregation methods is to find shared community structure by combining each slice of the multilayer graph into a single-layer network. Score-based methods rely on maximizing fitness functions based on an appropriate null model in order to detect communities. Finally, model-based methods rely on statistical models and formal inference to discover latent structure. These three types of methods are not necessarily disjointed from one another; for instance, many in the statistics community study models that are very similar to the null models that are used in score-based methods.

25.4.1 SCORE-BASED METHODS

In the single-layer setting, score-based methods operate by optimizing a fitness function. Perhaps the most popular method in this category is modularity maximization. Modularity for a single-layer network is defined as follows:

$$Q = \frac{1}{2m} \sum_{i,j \in \mathcal{E}} \left(\mathbf{A}_{ij} - \frac{k_i k_j}{2m} \right) \delta(c_i, c_j), \tag{25.17}$$

where m is the number of edges in the network, $\mathbf{A}$ is the adjacency matrix, k_i is the degree of node i, and c_i is the community label of node i. Modularity is qualitatively a comparison with the structure of the network to a random null model in which every edge between every node is equally likely [29]. Extensions of modularity, such as multiresolution variants [30], have been proposed in the literature.

In order to find community structure, we perform a maximization of Q over the community assignments c_i. Modularity maximization is typically performed using the Louvain algorithm or its appropriate variants. Modifying these fitness functions for a multilayer setting can be done by appropriately defining a null model [31], which takes into account intralayer connections and interlayer connections accordingly, in which case we have the following multilayer modularity:

$$\frac{1}{2\mu} \sum_{i,j \in \mathcal{E}_M, \alpha, \beta \in [L]} \left[\left(\mathbf{A}_{ij}^{(\alpha)} - \gamma_\alpha \frac{k_i^{(\alpha)} k_j^{(\alpha)}}{2m_\alpha} \right) \delta(\alpha, \beta) + \delta(i,j) C_{j\alpha\beta} \right] \delta(c_{i\alpha}, c_{j\beta}). \tag{25.18}$$

This model only takes into account interlayer connections between the same nodes, and their strengths are represented by $C_{j\alpha\beta}$. Further, each node in each layer has a different community label $c_{i\alpha}$, and μ is an appropriate normalization term; see [31] for details.

Another score-based method that allows for extensions to any single-layer fitness function involves Pareto optimality [32]. In this case, we assume that each node has one community label for every layer, so that $c_i = c_{i\alpha} = c_{i\beta}$. In this method, we define a fitness function for each layer, $f_1(\mathbf{c}), f_2(\mathbf{c}), \ldots, f_L(\mathbf{c})$, that we wish to jointly minimize. We could, for instance, choose (negative) modularity on each layer for our cost function. Alternatively, we could choose a similar cost function that arises when attempting to reduce the intercommunity connections—this is called spectral clustering [33]. Once we define these functions, we attempt to solve the multiobjective optimization problem:

$$\hat{\mathbf{c}} = \arg\min_{\mathbf{c}} [f_1(\mathbf{c}), f_2(\mathbf{c}), \ldots, f_L(\mathbf{c})]. \tag{25.19}$$

The objective is to find the Pareto optimal solution or solutions. A non-Pareto optimal solution $\mathbf{c}$ is a solution such that there exists at least one other solution $\mathbf{d}$ such that, for all $\alpha \in [L]$, $f_\alpha(\mathbf{d}) \leq f_\alpha(\mathbf{c})$, and $f_\beta(\mathbf{d}) < f_\beta(\mathbf{c})$ for at least one layer $\beta \in [L]$. The set of Pareto points is the set of solutions for which the above is not true. The special case of using the spectral clustering score function has been explored in [32]. Other methods for finding approximate Pareto optimal points include evolutionary algorithms, and Pareto methods have been used in anomaly detection [34] and image retrieval [35].

25.4.2 MODEL-BASED METHODS

Model-based methods assume a specified statistical model for the network, and then use statistical methods for inference in order to discover the latent community structure. These models are often variants of a ubiquitous single-layer model called the stochastic block model (SBM) [36]. This model assumes that given the community structure, each edge is drawn independently as a Bernoulli random variable according to a parameter p_{ij}, where i and j are the communities of the nodes for the edge that is being drawn. Reference [37] generalize the SBM to have discrete types of layers and communities in each type. References [38–40] explore different extensions of the single-layer SBM. The inference for these models can be quite difficult from both a computational and statistical perspective. Work to find provably computationally and statistically efficient algorithms in various model cases continues to be an active field of research in the multilayer setting.

25.4.3 AGGREGATION-BASED METHODS

Aggregation-based methods attempt to find a single-layer network that holds information about the communities in the multilayer network, and then utilize single-layer community detection methods. Examples include [41–43]. A recent paper [7] utilizes spectral clustering and convex layer aggregation to perform community detection. Specifically, given a layer weight vector $w \in \mathcal{W}_L$, where $\mathcal{W}_L = \{w : w_\alpha \geq 0, \sum_{\alpha=1}^{L} w_\alpha = 1\}$, and a supra-adjacency matrix as defined in Section 25.2, we define the weighted adjacency layer matrix and associated Laplacian as:

$$\mathbf{A}_w = \sum_{\alpha=1}^{L} w_\alpha \mathbf{A}^{(\alpha)}, \quad \mathbf{L}_w = \sum_{\alpha=1}^{L} w_\alpha \mathbf{L}^{(\alpha)}. \tag{25.20}$$

The authors in [7] discuss theoretical guarantees and limits of this method under different models, and also provide a framework for model selection.

25.5 ESTIMATION OF DYNAMIC SOCIAL INTERACTION NETWORKS

With social media data, we are often faced with the situation where given some features over time, we would like to infer an interaction graph among the agents. The network structure we consider is a dynamic graph, which as discussed above can be thought of as a multilayer graph, with each layer corresponding to a discrete time point or a series of discrete time points.

In the stationary case, under Gaussian assumptions, this problem is related to structural estimation in Gaussian graphical models (GGM) [44–46]. The extension of GGM models to the time-varying setting is an active research area [47,48]. It can be especially valuable for understanding the interaction of multiple agents over time.

In order to capture these time-varying interactions on the network, we turn to using information theoretic measures. Specifically, directed information (DI) will be used as a measure of influence among these nodes and edges. DI was originally introduced as an extension of mutual information for a channel that exhibits feedback. Consider $\mathbf{X}^{(T)} = [X_1, X_2, \ldots, X_T]$ and $\mathbf{Y}^{(T)} = [Y_1, Y_2, \ldots, Y_T]$ to be features at two respective nodes at time periods $1, \ldots, T$. Then, the DI between $\mathbf{X}^{(T)}$ and $\mathbf{Y}^{(T)}$ is defined as:

$$I(\mathbf{X}^{(T)} \to \mathbf{Y}^{(T)}) = \sum_{i=1}^{T} I(\mathbf{X}^{(i)}; Y_i | \mathbf{Y}^{(i-1)}) \tag{25.21}$$

$$= H(\mathbf{Y}^{(T)}) - H(\mathbf{Y}^{(T)} || \mathbf{X}^{(T)}), \tag{25.22}$$

where $H(\mathbf{Y}^{(T)} || \mathbf{X}^{(T)})$ is defined as the causally conditioned entropy:

$$H(\mathbf{Y}^{(T)} || \mathbf{X}^{(T)}) = \sum_{i=1}^{T} = H(Y_i | \mathbf{X}^{(i)}, \mathbf{Y}^{(i-1)}). \tag{25.23}$$

This measure has been utilized extensively in information theory [49] as well as fMRI and EEG interaction studies [50–52], financial time-series analysis [53], and video indexing and retrieval [54]. It has proven robust in detecting nonlinear relationships among objects of interest over time.

The disadvantage of DI, as with most information theoretic measures, is the lack of ability to track changes over time. Most of these estimators assume stationary processes, and estimate (either parametrically or nonparametrically) the distributions and subsequently the functionals of the distributions. With this in mind, we introduce adaptive measures for directed information that allow for changes in the distributions of interactions over time. The adaptivity is a two-stage process; we first must allow for the parameters to be adaptable, and then also allow for the functional to adapt over time. In [55], we use an exponentially weighted filter applied to the DI:

$$(ADI_{X^t \to Y^t}) = \sum_{i=1}^{t} g(t,i) I(\mathbf{X}^{(i)}; Y_i | \mathbf{Y}^{(i-1)}), \tag{25.24}$$

where $g(t,i) = e^{(t-i)\lambda} c_t$. Under (time-varying) Gaussian assumptions, we can simplify ADI in terms of the varying covariance matrices:

$$(ADI_{X^t \to Y^t}) = \frac{1}{2} \sum_{i=1}^{t} g(t,i) \log \frac{|\Sigma_{Y_i | Y_{i-1}}|}{|\Sigma_{Y_i | \mathbf{Y}^{(i-1)}, \mathbf{X}^{(i-1)}}|}. \tag{25.25}$$

In this case, we must estimate these covariances, for instance using a dynamic covariance model [56]. In other cases, such as the discrete case, other methods can be used to parametrically estimate the appropriate distributions. After the estimation of ADI, we can define our dynamic interaction graph at each time step with a directed weighted edge between each two nodes with nonzero ADI; functional p-value transformations for thresholding can also be used [55,57].

25.6 APPLICATIONS

In the following sections, we will demonstrate the utility of multilayer network methods on real data. First, we will examine a biological multilayer network to uncover topological roles in gene contact networks. We will also describe a Twitter dataset, and use the dynamic interaction graph estimation technique discussed in Section 25.5 to uncover novel interactions between US senators.

25.6.1 IDENTIFYING GENES ENCODING ALLELIC DIFFERENCES IN GENE CONTACT NETWORKS

Allelic differences between two homologous chromosomes (corresponding to paternal and maternal alleles) can affect the propensity of inheritance in humans [58]. Therefore, it is important to discriminate the contribution of the paternal (Pat) and maternal (Mat) genomes to the functional diploid human nucleome. In what follows, we perform multilayer network analysis to understand allelic differences at the gene level.

Genome technologies such as genome-wide chromosome conformation capture (Hi-C) can be used to measure the genomic structure [59–61]. Here, Hi-C evaluates long-range interactions between pairs of segments delimited by specific cutting sites using spatially constrained ligation [59]. As a result, we obtain a fragment read table, each row of which indicates a ligated pair of fragments from the genome with the coordinates of both fragments. Based on that, we can construct two-dimensional Hi-C contact maps at gene resolution [16,62]. We refer the reader to [62] for more details on data generation and preprocessing. From the network point of view, this leads to a sequence of intergene interaction networks over time (namely, cell cycle phases G1, S, and G2/M) under both Pat and Mat alleles. That is, we obtain an allele-specific multilayer network where each cell cycle stage corresponds to a layer. Our goal is to identify genes that yield significant contact differences between the Pat and Mat alleles.

We adopt the overlapping degree centrality and the multiplex participation coefficient to distinguish Pat allele from Mat allele. We recall from Section 25.3.1 that the overlapping degree centrality allows us to identify hubs from a network, and the multiplex participation coefficient can quantify the participation of a gene to different cell cycle phases. In Fig. 25.3B, we present z-scores of genes' overlapping degrees versus genes' participation coefficients. As we can see, due to allelic differences, there exist genes that play different topological roles on Pat and Mat alleles. Let z_i denote the z-score of the overlapping degree for gene i, and P_i denote its multiplex participation coefficient. We distinguish hubs (interacting with many genes) from regular nodes if $z_i \geq 2$. Motivated by [6], we call genes focused if contacts associated with them were concentrated on a single cell cycle phase, corresponding to $P_1 < 1/3$, and multiplex if their connected edges were homogeneously distributed across different cell cycle phases, corresponding to $P_1 > 2/3$. In the considered experiment, genes LEPREL1 and CTSS are hubs at Pat allele while they become regular nodes at Mat allele. And gene KBTBD2 is a

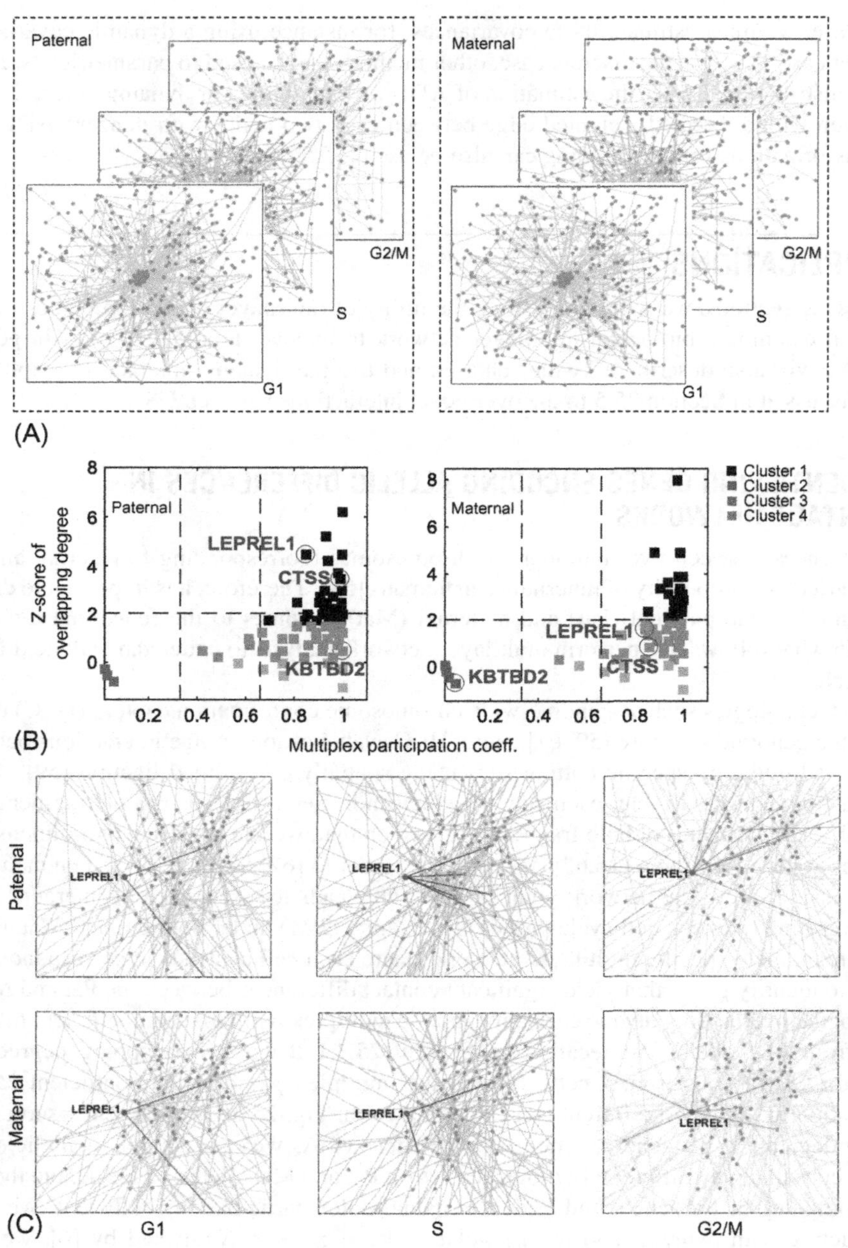

FIG. 25.3

Allele-specific intergene contact networks over different cell cycle phases. (A) Temporal network with implicit interlayer connections between genes at one cell cycle phase and their counterparts at other cell cycle phases. (B) Overlapping degree versus multiplex participation coefficient: genes are divided into 4 clusters via K-means. (C) Representative gene LEPREL1 with allelic differences in the topological structure.

multiplex node at Pat allele, but it becomes a focused node at Mat allele. We show the allelic differences in terms of contact differences of genes, e.g., LEPREL1 in Fig. 25.3C.

25.6.2 INTERACTION NETWORKS IN PRESIDENTIAL AND SENATORIAL DATASETS

In what follows, we demonstrate the method of interaction graph estimation on Twitter datasets, using the measures described in Section 25.5. The dataset is of the members of the US Senate, from October 1, 2015 to January 13, 2016. In total, the dataset consists of 96,090 tweets. In [55], adaptive directed information (ADI) was used to show time-varying interaction structure between the Twitter accounts. This is an extension of that work. Fig. 25.4 displays updated versions of ADI graphs at consecutive timesteps. Some senators are not displayed as they have no significant edges. We notice that there are nodes of high activity such as RB (Rob Bishop) and MK (Marcy Kaptur). Further, we see significant evolution in the network, with nodes adapting their behavior; this shows the method's ability to estimate changes in influence.

In Fig. 25.5 we plot the total degrees over time of ADI for a subset of senators. Total degree for a particular node is defined as the out-degree (sum of outgoing ADI) minus the in-degree (sum of incoming ADI). These senators were chosen to show examples of nodes that have high average influence (large positive total degree), senators that receive influence approximately equal to the amount they influence (small total degree), and senators that are recipients of influence on average

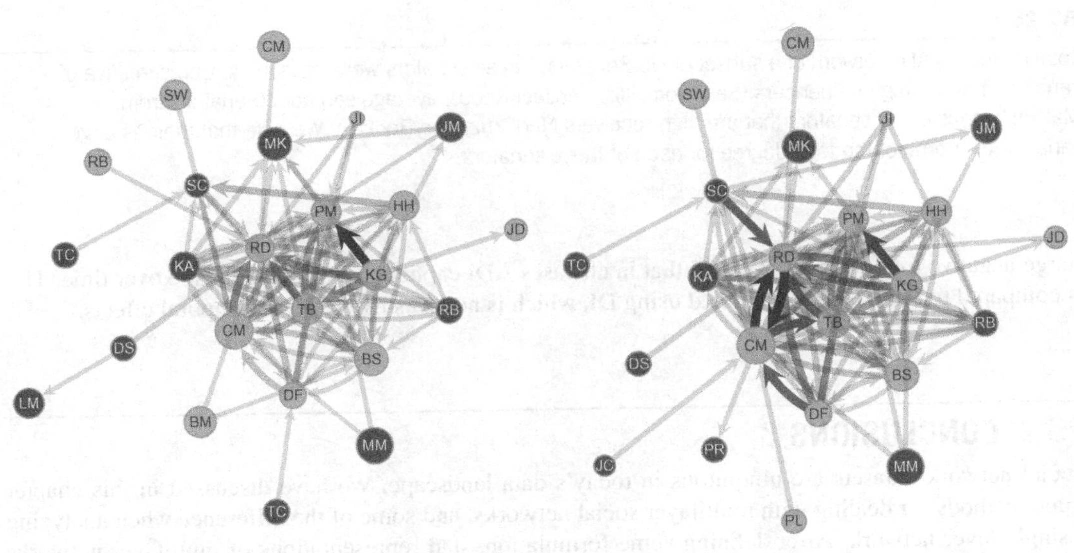

FIG. 25.4

ADI network of US Senators over two consecutive time periods from left to right. The width of the directed edge as well as the shade is related to the magnitude of the DI, and the size of each node represents the volume of tweets. We see a large connected component exhibiting mutual interaction, and significant evolution in the network, with nodes adapting their behavior.

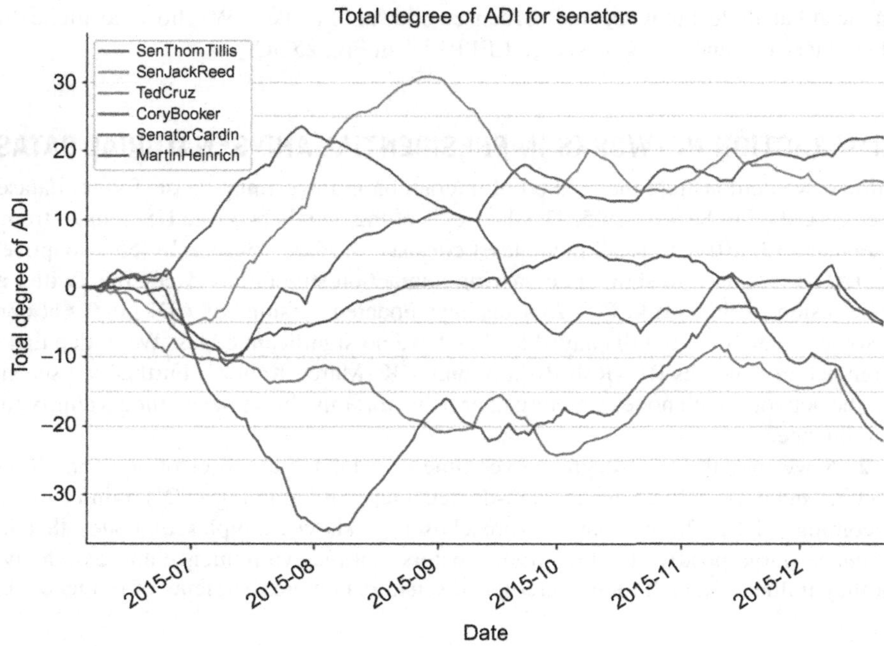

FIG. 25.5

Total degree of ADI network of a subset of US Senators. These senators were chosen as representative of senators that are high influencers (SenThomTillis, SenJackReed), average senators (SenatorCardin, MartinHeinrich), and senators that are high receivers (TedCruz, CoryBooker). We note that there is large variation over time of the total degree for each of these senators.

(large negative total degree). We note that in all cases ADI captures variation in degree over time. This is compared to total degree computed using DI, which is not sensitive to these temporal effects.

25.7 CONCLUSIONS

Social network datasets are ubiquitous in today's data landscape. We have discussed in this chapter some methods for dealing with multilayer social networks, and some of the difference when analyzing a single-layer network. After defining some formulations and representations of multilayer networks and some common examples that one might encounter, we covered some measures of multilayer network centrality. We also discussed a few types of methods for multilayer community detection, including briefly discussing some benefits and drawbacks of each type. We finally covered the problem of multilayer interaction graph estimation, with special focus on dynamic graphs. We then applied a few of the techniques to two datasets, a biological dataset and finally a social network dataset.

As the field of multilayer networks continue to grow, we expect that the methods that we have summarized here will continue to evolve and improve, and that the framework of multilayer graphs will become even more useful to the field of social network analysis in the years to come.

ACKNOWLEDGMENTS

This work was partially supported by the following grants: USAF FA8650-15-D-1845 and by ARO W911NF-15-1-0479. We also would like to thank Prof. Indika Rajapakse at Department of Computational Medicine and Bioinformatics, University of Michigan, who provided data on allele-specific Hi-C contact maps.

REFERENCES

[1] Chung FRK. Spectral graph theory, No. 92. American Mathematical Society; 1997.
[2] Wang D, Wang H, Zou X. Identifying key nodes in multilayer networks based on tensor decomposition. Chaos: Interdisciplinary J Nonlinear Sci 2017;27(6):063108.
[3] Kivelä M, Arenas A, Barthelemy M, Gleeson JP, Moreno Y, Porter MA. Multilayer networks. J Complex Netw 2014;2(3):203–71.
[4] Cozzo E, Kivelä M, De Domenico M, Solé-Ribalta A, Arenas A, Gómez S, et al. Structure of triadic relations in multiplex networks. New J Phys 2015;17(7):073029.
[5] Sole-Ribalta A, De Domenico M, Kouvaris NE, Díaz-Guilera A, Gómez S, Arenas A. Spectral properties of the Laplacian of multiplex networks. Phys Rev E 2013;88(3):032807.
[6] Battiston F, Nicosia V, Latora V. Structural measures for multiplex networks. Phys Rev E 2014;89(3):032804.
[7] Chen PY, Hero AO. Multilayer spectral graph clustering via convex layer aggregation: theory and algorithms. IEEE Trans Signal Inf Process Netw 2017;3(3):553–67.
[8] De Domenico M, Solé-Ribalta A, Cozzo E, Kivelä M, Moreno Y, Porter MA, et al. Mathematical formulation of multilayer networks. Phys Rev X 2013;3(4):041022.
[9] Kolda TG, Bader BW. Tensor decompositions and applications. SIAM Rev 2009;51(3):455–500.
[10] Gauvin L, Panisson A, Cattuto C. Detecting the community structure and activity patterns of temporal networks: a non-negative tensor factorization approach. PLOS One 2014;9(1):e86028.
[11] Sun Y, Yu Y, Han J. Ranking-based clustering of heterogeneous information networks with star network schema. In: Proceedings of the 15th ACM SIGKDD international conference on knowledge discovery and data mining. ACM; 2009. p. 797–806.
[12] Gubbi J, Buyya R, Marusic S, Palaniswami M. Internet of things (IoT): a vision, architectural elements, and future directions. Futur Gener Comput Syst 2013;29(7):1645–60.
[13] Cai D, Shao Z, He X, Yan X, Han J. Community mining from multi-relational networks. In: European conference on principles of data mining and knowledge discovery. Springer; 2005. p. 445–452.
[14] Yuan Z, Zhao C, Wang WX, Di Z, Lai YC. Exact controllability of multiplex networks. New J Phys 2014;16(10):103036.
[15] Aleta A, Meloni S, Moreno Y. A multilayer perspective for the analysis of urban transportation systems. Sci Rep 2017;7:44359.
[16] Liu S, Chen H, Ronquist S, Seaman L, Ceglia N, Meixner W, et al. Genome architecture leads a bifurcation in cell identity. bioRxiv; 2017. p. 151555.
[17] Newman M. Networks: an introduction. Oxford University Press; 2010.
[18] Bertrand A, Moonen M. Seeing the bigger picture: how nodes can learn their place within a complex ad hoc network topology. IEEE Signal Process Mag 2013;30(3):71–82.

[19] Bródka P, Kazienko P, Musiał K, Skibicki K. Analysis of neighbourhoods in multi-layered dynamic social networks. Int J Comput Intell Syst 2012;5(3):582–96.
[20] Bródka P, Skibicki K, Kazienko P, Musiał K. A degree centrality in multi-layered social network. In: International conference on computational aspects of social networks (CASoN). IEEE; 2011. p. 237–42.
[21] Guimera R, Amaral LAN. Functional cartography of complex metabolic networks. Nature 2005;433(7028): 895.
[22] Kleinberg JM. Authoritative sources in a hyperlinked environment. J ACM (JACM) 1999;46(5):604–32.
[23] Gleich DF. Pagerank beyond the web. SIAM Rev 2015;57(3):321–63.
[24] Taylor D, Myers SA, Clauset A, Porter MA, Mucha PJ. Eigenvector-based centrality measures for temporal networks. Multiscale Model Simul 2017;15(1):537–74.
[25] Solá L, Romance M, Criado R, Flores J, García del Amo A, Boccaletti S. Eigenvector centrality of nodes in multiplex networks. Chaos: Interdisciplinary J Nonlinear Sci 2013;23(3):033131.
[26] Kolda TG, Bader BW, Kenny JP. Higher-order web link analysis using multilinear algebra. In: Fifth IEEE international conference on data mining; 2005. p. 8.
[27] Fortunato S, Hric D. Community detection in networks: a user guide. Phys Rep 2016;659:1–44. https://doi.org/10.1016/j.physrep.2016.09.002.
[28] Mcpherson M, Smith-Lovin L, Cook J. Birds of a feather: homophily in social networks. Ann Rev Sociol 2001;27:415–44.
[29] Newman MEJ. Modularity and community structure in networks Proc Natl Acad Sci U S A 2006;103(23): 8577–82. https://doi.org/10.1073/pnas.0601602103.
[30] Xiang J, Hu XG, Zhang XY, Fan JF, Zeng XL, Fu GY, et al. Multi-resolution modularity methods and their limitations in community detection. Eur Phys J B 2012;85(10). https://doi.org/10.1140/epjb/e2012-30301-2.
[31] Mucha PJ, Richardson T, Macon K, Porter MA, Onnela JP. Community structure in time-dependent, multiscale, and multiplex networks. Science 2010;328(5980):876–8. https://doi.org/10.1126/science.1184819.
[32] Oselio B, Kulesza A, Hero AO. Multi-layer graph analysis for dynamic social networks. IEEE J Sel Top Signal Process 2014;8(4):514–23.
[33] Shi J, Malik J. Normalized cuts and image segmentation. IEEE Trans Pattern Anal Mach Intell 2000;22(8): 888–905. https://doi.org/10.1109/34.868688.
[34] Hsiao KJ, Xu KS, Calder J, Hero AO. Multicriteria similarity-based anomaly detection using Pareto depth analysis. IEEE Trans Neural Netw Learn Syst 2016;27(6):1307–21. https://doi.org/10.1109/TNNLS.2015.2466686.
[35] Hsiao KJ, Calder J, Hero AO. Pareto-depth for multiple-query image retrieval. IEEE Trans Image Process 2015;24(2):583–94. https://doi.org/10.1109/TIP.2014.2378057.
[36] Holland P, Laskey K, Leinhardt S. Stochastic blockmodels: first steps. Soc Networks 1983;5:109–37.
[37] Stanley N, Shai S, Taylor D, Mucha PJ. Clustering network layers with the strata multilayer stochastic block model. IEEE Trans Network Sci Eng 2016;3(2):95–105. https://doi.org/10.1109/TNSE.2016.2537545.
[38] Peixoto TP. Inferring the mesoscale structure of layered, edge-valued, and time-varying networks. Phys Rev E 2015;92:042807. https://doi.org/10.1103/PhysRevE.92.042807.
[39] De Bacco C, Power EA, Larremore DB, Moore C. Community detection, link prediction, and layer interdependence in multilayer networks. Phys Rev E 2017;95:042317. https://doi.org/10.1103/PhysRevE.95.042317.
[40] Han Q, Xu KS, Airoldi EM. Consistent estimation of dynamic and multi-layer block models. In: Proceedings of the 32Nd international conference on international conference on machine learning. ICML'15, vol. 37. JMLR.org; 2015. p. 1511–20. http://dl.acm.org/citation.cfm?id=3045118.3045279.
[41] De Domenico M, Nicosia V, Arenas A, Latora V. Structural reducibility of multilayer networks Nat Commun 2015;6:6864.
[42] Taylor D, Shai S, Stanley N, Mucha PJ. Enhanced detectability of community structure in multilayer networks through layer aggregation. Phys Rev Lett 2016; 116:228301. https://doi.org/10.1103/PhysRevLett.116.228301.

[43] Berlingerio M, Coscia M, Giannotti F. Finding redundant and complementary communities in multidimensional networks. In: Proceedings of the 20th ACM international conference on information and knowledge management. CIKM '11. New York, NY: ACM; 2011. p. 2181–4. https://doi.org/10.1145/2063576.2063921.

[44] Hero A, Rajaratnam B. Hub discovery in partial correlation graphs. IEEE Trans Inf Theory 2012;58(9):6064–78. https://doi.org/10.1109/TIT.2012.2200825.

[45] Meng Z, Eriksson B, Hero A. Learning latent variable Gaussian graphical models. In: Proceedings of the 31st international conference on machine learning (ICML-14); 2014. p. 1269–77.

[46] Zhou S, Rütimann P, Xu M, Bühlmann P. High-dimensional covariance estimation based on Gaussian graphical models. J Mach Learn Res 2011;12:2975–3026. http://dl.acm.org/citation.cfm?id=1953048.2078201.

[47] Kolar M, Song L, Ahmed A, Xing EP. Estimating time-varying networks. Ann Appl Stat 2010:94–123.

[48] Ahmed A, Xing EP. Recovering time-varying networks of dependencies in social and biological studies. Proc Natl Acad Sci U S A 2009;106(29):11878–83.

[49] Jiao J, Permuter HH, Zhao L, Kim YH, Weissman T. Universal estimation of directed information. IEEE Trans Inf Theory 2013;59(10):6220–42. https://doi.org/10.1109/TIT.2013.2267934.

[50] Mehta K, Kliewer J. Directed information measures for assessing perceived audio quality using EEG. In: Conference record—Asilomar conference on signals, systems and computers; 2016. p. 123–7. https://doi.org/10.1109/ACSSC.2015.7421096.

[51] Chen X, Syed Z, Hero A. EEG spatial decoding with shrinkage optimized directed information assessment. In: ICASSP 2012 proceedings; 2012. p. 577–80. https://doi.org/10.1109/ICASSP.2012.6287945.

[52] Quinn C, Coleman TP, Kiyavash N, Hatsopoulos NG. Estimating the directed information to infer causal relationships in ensemble neural spike train recordings. J Comput Neurosci 2011;30(1):17–44. https://doi.org/10.1007/s10827-010-0247-2.

[53] Permuter HH, Kim YH, Weissman T. Interpretations of directed information in portfolio theory, data compression, and hypothesis testing. IEEE Trans Inf Theory 2011;57(6):3248–59.

[54] Chen X, Hero AO, Savarese S. Multimodal video indexing and retrieval using directed information. IEEE Trans Multimedia 2012;14(1):3–16. https://doi.org/10.1109/Tmm.2011.2167223.

[55] Oselio B, Hero A. Dynamic reconstruction of influence graphs with adaptive directed information. In: Proceedings of IEEE international conference on acoustics, speech and signal processing. ICASSP; 2017. p. 5935–9.

[56] Chen Z, Leng C. Dynamic covariance models. J Am Stat Assoc 2016;111(515):1196–207. https://doi.org/10.1080/01621459.2015.1077712.

[57] Rao A, Hero AO, States DJ, Engel JD. Using directed information to build biologically relevant influence networks. J Bioinform Comput Biol 2008;06(03):493–519. https://doi.org/10.1142/S0219720008003515.

[58] Leung D, Jung I, Rajagopal N, Schmitt A, Selvaraj S, Lee AY, et al. Integrative analysis of haplotype-resolved epigenomes across human tissues. Nature 2015;518(7539):350.

[59] Lieberman-Aiden E, et al. Comprehensive mapping of long-range interactions reveals folding principles of the human genome. Science 2009;326(5950):289–93. https://doi.org/10.1126/science.1181369.

[60] Chen H, Chen J, Muir LA, Ronquist S, Meixner W, Ljungman M, et al. Functional organization of the human 4D Nucleome. Proc Natl Acad Sci U S A 2015;112(26):8002–7.

[61] Boulos RE, Tremblay N, Arneodo A, Borgnat P, Audit B. Multi-scale structural community organisation of the human genome. BMC Bioinformatics 2017;18(1):209.

[62] Chen H, Liu S, Seaman L, Najarian C, Wu W, Ljungman M, et al. Parental allele-specific genome architecture and transcription during the cell cycle. bioRxiv; 2017. p. 201715.

CHAPTER

MULTIAGENT SYSTEMS: LEARNING, STRATEGIC BEHAVIOR, COOPERATION, AND NETWORK FORMATION

26

Cem Tekin*, **Simpson Zhang**†, **Jie Xu**‡, **Mihaela van der Schaar**§

*Electrical and Electronics Engineering Department, Bilkent University, Ankara, Turkey** *Economics Department, University of California, Los Angeles, Los Angeles, CA, United States*† *Electrical and Computer Engineering Department, University of Miami, Coral Gables, FL, United States*‡ *Electrical Engineering Department, University of California, Los Angeles, Los Angeles, CA, United States*§

26.1 LEARNING IN MULTIAGENT SYSTEMS: EXAMPLES AND CHALLENGES

26.1.1 CONTENT AGGREGATION

Proliferation of web-based multimedia content sources and servers led to a tremendous growth in the volume and diversity of multimedia content that is consumed by a diverse set of users. This diversity results in users exhibiting a vast range of preferences over the content, which often depends on the context in which they consume the content. Such demand led to the emergence of *multimedia content aggregators* (MCAs) [1,2] that gather and fuse content from numerous multimedia sources to provide a ubiquitous content delivery experience for their users. It is thus essential for these systems to learn the context-specific content preferences of their users using past feedback of their users on the content that they provide. The context of a user includes information that is related to its content preferences, including but not limited to the location information, search query, gender, age, and the type of the device that the user is using (e.g., mobile phone, tablet, PC) to access the content [3]. Thus, the goal of the MCA is to match its users with the most appropriate content by learning how users with different contexts react to different contents. Such learning is necessary for continuous satisfaction of a user's request for content, which dynamically evolves over time depending on how the user's context evolves. This problem can be formulated as an online learning problem where the MCA learns the best content for its users through exploring how its users react to different content.

In order to maximize the user satisfaction, an MCA needs to connect with other MCAs that have access to other multimedia sources to find the right content for its users. This requires cooperation between the MCAs: In addition to serving its own users, an MCA should also serve content to the users of other MCAs when a request is made. Thus, each MCA has two types of users: direct users, who visit the website of the MCA to search for content, and indirect users, who are users of another MCA

Cooperative and Graph Signal Processing. https://doi.org/10.1016/B978-0-12-813677-5.00026-2

that requests content from the MCA. The objective of an MCA is to maximize the number of positive feedbacks (e.g., likes) given by its users (both direct and indirect). This objective can be achieved simultaneously by all MCAs through a distributed learning method, where all MCAs learn online how to match their current users with the right content (either by serving content from a directly connected multimedia source or by requesting content from another MCA) [4]. Learning the right matching using the past information collected from the users is not a trivial task because both the user and content characteristics are dynamically changing over time, and the most suitable content for the user might be located in the network of another MCA.

26.1.2 DISCOVERING EXPERTISE

An important multiagent learning problem is to assign arriving tasks to experts in a way that maximizes the performance (e.g., quality, accuracy, timeliness) of task execution. In this problem, each agent experiences inflows of tasks that needs to be completed. To achieve this, an agent needs to assign its tasks to appropriate experts. However, the most suitable expert for a task might be located out of the agent's own expert network. In such a case, the agent needs to request help from the other agents to match its task with the most suitable expert. In addition, the performance of an expert for a particular task depends on the context of the task, and how well an expert executes a particular task is not known in advance. Hence, the expertise of each expert in the system needs to be learned in a context-driven and decentralized way.

An interesting example of the multiagent learning problem mentioned above is the medical expertise discovery problem for diagnosis of patients with complex diseases. Developing new learning methods turns out to be essential for this problem due to the fact that the standard clinical methods often fail to provide an accurate diagnosis for these complex cases [5]. Fortunately, the widespread adoption of *electronic health records* (EHRs) allows for the aggregation of patients' records for use by an automated *clinical decision support system* (CDSS) that supports matching patients with the right medical expert. The decision-making process to match the patient with an expert of appropriate expertise should be guided by the available information about the patient, which may include the patient's past health condition, past hospital visits, drug administrations, and genetic and social data. This vast amount of information, also called the context, can greatly affect the diagnosis accuracy of the experts. For instance, it is possible that the experts residing in the same institution have experience confined to a specific population of patients that is admitted to the institution. In such a case, these experts will be unable to form an accurate diagnosis for a patient that has a context that is very different than the contexts observed from the other patients treated within the same institution. Therefore, it is essential to discover the level of expertise (diagnosis accuracy) of the experts for each patient context in order to match the patient with the right expert. Moreover, in order to further improve the diagnosis accuracy, the patient's institution may seek help from other institutions that might be better at handling this particular patient context. For instance, a small clinic may only have a primary care physician and may not have any specialists that are able to identify a particularly complex disease. Such a clinic may be able to request information from an established hospital that has numerous specialists and CDSSs that can be helpful in diagnosing the patient. In such a case, the control of the small clinic over the established hospital is limited. While the clinic can request information, it cannot select the expert that the hospital will assign to the case. Therefore, a learning mechanism that enables these institutions to cooperate by preserving their autonomy and privacy is necessary.

Numerous challenges need to be addressed in designing such a learning mechanism: Each agent should learn (i) whether it should rely on its own experts or request another agent to execute the task, and (ii) if it relies on its own experts, which expert it should assign to maximize the performance of the task execution. For instance, for the medical expertise discovery problem, the performance can be defined as a function of the diagnosis accuracy and the cost of diagnosis (time, money, etc.). In addition, learning should be continuous, i.e., the expert selection strategy needs to be updated every time after the performance of the task execution is observed to get the most benefit from the newly acquired information. Moreover, the contextual specializations of the agents and experts should be learned from past data. Because the number of different contexts is vast, intelligent mechanisms that aggregate information from similar past contexts should be developed.

26.1.3 POPULARITY FORECASTING

Social media has recently been used to provide situational awareness and inform predictions and decisions in a variety of application domains, ranging from live or on-demand event broadcasting to security and surveillance [6], health communication [7], disaster management [8], and economic forecasting [9]. In all these applications, forecasting the popularity of the content shared in a social network is vital due to a variety of reasons. For network and cloud service providers, accurate forecasting facilitates prompt and adequate reservation of computation, storage, and bandwidth resources [10], thereby ensuring smooth and robust content delivery at low cost. For advertisers, accurate and timely popularity predictions provide a good revenue indicator, thereby enabling targeted ads to be composed for specific videos and viewer demographics. For content producers and contributors, attracting a high number of views is paramount for attracting potential revenue through micropayment mechanisms.

While popularity prediction is a long-lasting research topic [11–14], understanding how social networks affect the popularity of the media content, then using this understanding to make better forecasts poses significant new challenges. Conventional prediction tools have mostly relied on the history of the past view counts, which worked well when the popularity solely depended on the inherent attractiveness of the content and the recipients were generally passive. In contrast, social media users are proactive in terms of the content they watch and are heavily influenced by their social media interactions; for instance, the recipient of a certain media content may further forward it or not, depending on not only its attractiveness but also the situational and contextual conditions in which this content was generated and propagated through social media [15]. For example, the latest measurement on Twitter's Vine, a highly popular short mobile video sharing service, has suggested that the popularity of a short video indeed depends less on the content itself, but more on the contributor's position in the social network [16]. Hence, being situation-aware, e.g., considering the content initiator's information and the friendship network of the sharers, can clearly improve the accuracy of the popularity forecasts. However, critical new questions need to be answered: which situational information extracted from social media should be used, how to deal with dynamically changing and evolving situational information, and how to use this information efficiently to improve the forecasts?

As social media becomes increasingly more ubiquitous and influential, the video propagation patterns and users' sharing behavior dynamically change and evolve as well. Offline prediction tools [11,17–19] depend on specific training datasets, which may be biased or outdated, and hence may not accurately capture the real-world propagation patterns promoted by social media. Moreover, popularity forecasting is a multistage rather than a single-stage task because each video may be propagated through a cascaded social network for a relatively long time and thus, the forecast can be made at any time while the video is being propagated. A fast prediction has important economic and technological benefits; however, too early a prediction may lead to a low accuracy that is less useful or even damaging (e.g., investment in videos that will not actually become popular). The timeliness of the prediction has yet to be considered in existing works that solely focus on maximizing the accuracy. Hence, developing a systematic methodology for accurate and timely popularity forecasting is essential.

26.1.4 OVERVIEW OF THE CHAPTER

A multiagent learning method that promotes cooperation and enables formation of cooperation networks between the agents is described in Section 26.2. How decentralized agents learn in a system that provides global or group feedback about the set of actions taken by the agents is discussed in Section 26.3. Learning and strategic network formation in networks with incomplete information is reviewed in Section 26.4. Concluding remarks are given in Section 26.5.

26.1.5 NOTATION

Unless otherwise stated, sets are denoted using calligraphic letters and vectors are denoted using boldface letters. $\mathcal{N}$ denotes the set of agents, $\mathcal{N}_i$ denotes the set of agents excluding agent i, $\mathcal{F}_i$ denotes the set of actions of agent i, $\mathcal{F}$ denotes the set of all actions, and $\mathcal{X}$ denotes the d-dimensional context set.

26.2 COOPERATIVE CONTEXTUAL BANDITS

In this section, we consider a multiagent decision-making and learning problem in which a set of cooperative and decentralized agents $\mathcal{N} = \{1, 2, \ldots, N\}$ help each other by selecting actions on behalf of each other to maximize their long-term rewards. In this model, each agent sequentially observes over time side information (context) about the environment that determines the expected rewards of the actions. Then, the agent decides to take an action or asks another agent to take an action on its behalf. If the agent chooses the latter option, then it is called the *requesting agent* and the agent that it asks to is called the *serving agent*. After this interaction both the requesting agent and the serving agent observe the random reward of the selected action, but the requesting agent may not observe the action selected by the serving agent. Such a situation happens in many real-world applications including real-time stream mining, where the agents may not be willing to reveal the type of classifiers they use to make predictions; network security, where agents may not be willing to reveal the number and type of the security protocols they use; and multimedia content aggregation, where an MCA does not have direct access to the content that is outside its content network. Both the requesting and serving agents benefit from this cooperation in the following way: the requesting agent makes use of the actions of the serving agent to obtain a high reward while the serving agent gets to know more about the rewards of its actions by taking an action for the requesting agent. For instance, in medical expertise discovery, the requesting institution benefits from the expertise of the expert of the serving institution by getting a more accurate diagnosis for its patient while the serving institution benefits from the experience gained through serving the other institution's patient, which may help it to match its patients with better experts in the future.

26.2.1 MODELING MULTIAGENT LEARNING USING COOPERATIVE CONTEXTUAL BANDITS

The set of serving agents of agent i is $\mathcal{N}_i := \mathcal{N} - \{i\}$, the set of actions of agent i is $\mathcal{F}_i$, and the set of all actions is $\mathcal{F} := \cup_{i \in \mathcal{N}} \mathcal{F}_i$. The set of choices of agent i is given as $\mathcal{K}_i := \mathcal{N}_i \cup \mathcal{F}_i$, which includes both its actions and its serving agents. Cardinalities of the sets $\mathcal{N}_i$, $\mathcal{F}_i$, and $\mathcal{K}_i$ are denoted by N_i, F_i, and K_i, respectively. An agent's action set is its private information. All agents know an upper bound on the number of actions that each agent has, which is denoted by $F_{\max}$. The system operates in discrete time where the following sequence of events take place at time t for all agents $i \in \mathcal{N}$:

- A d-dimensional context $x_i(t)$ arrives to agent i.
- Agent i selects a choice $a_i(t) \in \mathcal{K}_i$.
- If $a_i(t) \in \mathcal{F}_i$ agent i takes action $a_i(t)$.
- Else if $a_i(t) \in \mathcal{N}_i$, it sends $x_i(t)$ to agent $a_i(t)$, agent $a_i(t)$ takes an action in $\mathcal{F}_{a_i(t)}$ for agent i, and observes its reward.
- Agent i observes and collects the reward of the taken action.

The context set, denoted by $\mathcal{X}$, is assumed to be a bounded d-dimensional subset of the d-dimensional Euclidean space, which without loss of generality can be taken as $[0, 1]^d$. The expected reward of action f for context x is $\pi_f(x)$. Neither $\pi_f(x)$ nor its distribution is known by any of the agents. We impose a natural assumption that relates the expected rewards of an action under different contexts, which is called the *similarity assumption.*

Similarity Assumption: $\exists L > 0, \alpha > 0$ such that $\forall f \in \mathcal{F}, \forall x, x' \in \mathcal{X}$

$$|\pi_f(x) - \pi_f(x')| \le L||x - x'||^\alpha, \tag{26.1}$$

where L and α determine the strength of similarity between the expected rewards of an action under two different contexts. It is assumed that this similarity information is known by the agents. For instance, in real-time stream mining the classification accuracy of a classifier may be similar for data streams with similar metadata, or in the medical expert assignment problem, the accuracy of the diagnosis of an expert may be similar for patients with similar features. While variants of the similarity measure defined in Eq. (26.1) are commonly used in the online learning literature [20–22], there are many other similarity measures that can be used, including similarity measures defined for categorical or mixed contexts [23,24].

In this setup, the goal of each agent is to maximize its total expected reward. Thus, each agent needs to learn the best choices for its contexts. However, the agents will not be able to achieve the maximum possible total expected reward due to the fact that they do not know the reward distributions of the actions beforehand. This loss due to learning is captured using the notion of *regret*, which measures how bad an agent performs with respect to an oracle that knows the set of all actions and their expected rewards. The set of the best actions of agent i for context x is $f_i^*(x) := \arg\max_{f \in \mathcal{F}_i} \pi_f(x)$. Hence, the maximum expected reward of agent i for context x is $\pi_i(x) = \pi_{f_i^*(x)}(x)$.[1,2] Based on this, the set of the

[1] With an abuse of notation, for a set g for which $\pi_h(x) = \pi_l(x)$ for all h and l in g, we use $\pi_g(x)$ to denote the expected reward of an element of g.

[2] We also use $k_i^*(x)$, $f_i^*(x)$, and $f^*(x)$ to refer to an element of the corresponding set when the corresponding set contains more than one element.

best choices of agent i for context x is given as $k_i^*(x) := \arg\max_{k \in \mathcal{K}_i} \pi_k(x)$. On the other hand, the set of the best actions for context x is $f^*(x) := \arg\max_{f \in \mathcal{F}} \pi_f(x)$. Note that the expected reward of the best choice of agent i and the best action can be different when $k_i^*(x) \in \mathcal{N}_i$. This is due to the fact that the expected reward of $k_i^*(x)$ depends on the action chosen by agent $k_i^*(x)$. However, the maximum expected reward of agent $k_i^*(x)$ is equal to the expected reward of $f^*(x)$ in this case. Based on the definitions given above, the regret of agent i by time T is given as

$$R_i(T) := \sum_{t=1}^{T} \pi_{k_i^*(x_i(t))}(x_i(t)) - \mathrm{E}\left[\sum_{t=1}^{T} r^i_{a_i(t)}(x_i(t), t)\right], \tag{26.2}$$

where $r^i_{a_i(t)}(x_i(t), t)$ denotes the random reward of choice $a_i(t)$ of agent i at time t. Note that $r^i_{a_i(t)}(x_i(t), t)$ is a sample obtained from the reward distribution of action $a_i(t)$ when $a_i(t) \in \mathcal{F}_i$, and a sample obtained from the reward distribution of the action selected by agent $a_i(t)$ when $a_i(t) \in \mathcal{N}_i$.

The regret has two important properties: (i) it is nondecreasing in T, (ii) $R_i(T)/T$ gives the difference between the average rewards collected by the oracle and agent i. Therefore, when $R_i(T) = O(T^\gamma)$ for $\gamma < 1$ (i.e., the regret is sublinear), agent i is average reward optimal.

26.2.2 A DISTRIBUTED LEARNING ALGORITHM FOR COOPERATIVE CONTEXTUAL BANDITS

The online learning method used in cooperative contextual bandits consists of three essential components: a component that enables learning the expected rewards of the choices together for similar contexts, a component that enables agent i to learn the expected rewards of its actions and the maximum expected rewards of the set of serving agents accurately, and a component that enables agent i to maximize its total reward using what it had learned about its choices.

Because there are many contexts, estimating the expected reward of a choice for each context is in general infeasible. As a simple example, consider the case when agent i observes $x_i(t)$ for the first time, but has observed contexts that lie within the ϵ neighborhood of $x_i(t)$ many times. Without using the observations from the ϵ neighborhood of $x_i(t)$, the agent cannot do better than random guessing. On the other hand, the agent can get sufficiently accurate estimates of the expected rewards of its choices for context $x_i(t)$ by using the sample mean rewards of the choices calculated using the contexts within the ϵ neighborhood of $x_i(t)$. Thus, it is essential to form a partition of the context set $\mathcal{X}$ that is formed by sets that include similar contexts, and estimate the expected rewards of the choices for each context using the past observations of actions and rewards from the set to which the context belongs.

Two types of error must be taken into account when forming such a partition. The first type, called *dissimilarity error*, is due to the variation of the expected rewards of the contexts that lie in each set of the partition. For each set, this variation can be bounded using the diameter, i.e., the maximum distance between two contexts, of the set and the similarity assumption. The second type, called *sample size error*, is due to the limited number of random reward observations that are used to estimate the expected choice rewards. These two errors conflict with each other. For instance, decreasing the diameter of a set to reduce its dissimilarity error generally increases its sample size error because it corresponds to decreasing the number of past observations within the set. On the other hand, increasing the diameter of a set to reduce its sample size error generally increases the dissimilarity error. Thus, a balance between these two types of errors must be achieved in order to minimize the regret.

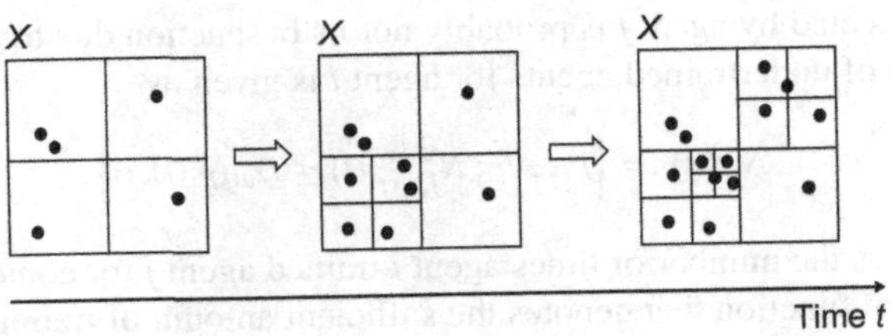

FIG. 26.1

An illustration of the contextual zooming method.

A natural way to achieve this balance is to use a technique called *contextual zooming* [21]. The idea is to adaptively form the partition of the context set based on the contexts that have arrived so far in a way that balances the dissimilarity error and the sample size error in each set of the partition. This implies that the partition will include many small sets concentrated around the region of the context set with high number context arrivals. An illustration of the contextual zooming method is shown in Fig. 26.1. Contextual zooming can be achieved by the following mechanism: Initially, agent i starts with partition $\mathcal{P}_i(1) = \{\mathcal{X}\}$, which only contains a single set. Let $\mathcal{P}_i(t)$ denote agent i's partition at time t, which consists of d-dimensional hypercubes with edge lengths coming from the set $\{2^0, 2^{-1}, 2^{-2}, \ldots\}$. A hypercube with edge length 2^{-l} is called a level-l hypercube, and for $p \in \mathcal{P}_i(t)$, $l(p)$ denotes its level. Also let $T_p^i(t)$ denote the number of contexts arrivals to $p \in \mathcal{P}_i(t)$ by time t, and $p_i(t)$ denote the hypercube in $\mathcal{P}_i(t)$ that contains context $x_i(t)$. $\mathcal{P}_i(t+1)$ is formed as follows:

- If $T^i_{p_i(t)}(t) \geq 2^{\rho l(p_i(t))}$, where $\rho > 0$ is a scale parameter, divide $p_i(t)$ into 2^d level $l(p_i(t)) + 1$ hypercubes. $\mathcal{P}_i(t+1)$ will contain all the sets in $\mathcal{P}_i(t)$ except $p_i(t)$ and will also contain the newly created level $l(p_i(t)) + 1$ hypercubes.
- Else $\mathcal{P}_i(t+1)$ will be the same as $\mathcal{P}_i(t)$.

The above procedure allows the *lifetime* of a hypercube to be an exponential function of its level. This bounds the number of hypercubes crated by time T as a sublinear function of T, and allows the agent to incur a small amount of regret in each hypercube such that when summed up over all hypercubes it still remains sublinear in T.

Next, we discuss how an agent learns the expected rewards of its choices accurately and uses them to maximize its total reward. We would like to emphasize that the agents cooperate with each other and obey the rules of the learning algorithm to choose actions on behalf of the other agents. Hence, cooperative contextual bandits do not consider strategic interactions between selfish agents. The agent maximizes its total reward by a three-phased mechanism that consists of *training*, *exploration*, and *exploitation* phases. For each context that arrives to agent i at time t, it can only be in one of these phases. Each phase is described below:

- **Training phase:** In this phase agent i trains its serving agents in $\mathcal{N}_i$. Training is done by choosing an *undertrained* agent in $\mathcal{N}_i$ as the serving agent. This agent j will receive $x_i(t)$, select an action $b_{j,i}(t) \in \mathcal{F}_j$, and observe its reward. As a result, it will update the estimated expected reward for its action $b_{j,i}(t)$. However, agent i is not going to update the estimated maximum reward for agent j

because the action selected by agent j is probably not its best action due to the fact that agent j is undertrained. The set of undertrained agents for agent i is given as

$$\mathcal{N}_i^{\mathrm{ut}}(t) := \left\{ j \in \mathcal{N}_i : N_{j,p_i(t)}^{\mathrm{tr},i}(t) \le D_{\mathrm{ut}}(p_i(t), t) \right\},$$

where $N_{j,p_i(t)}^{\mathrm{tr},i}(t)$ denotes the number of times agent i trained agent j for contexts in $p_i(t)$, and $D_{\mathrm{ut}}(p_i(t), t)$ is a control function that denotes the sufficient amount of trainings by time t that will enable agent j to form accurate estimates of the expected rewards of its own actions for $p_i(t)$.

- **Exploration phase:** In this phase agent i selects an *underexplored* choice in $\mathcal{K}_i$, and uses the reward it obtains from its selection to update the estimated expected reward of its choice. Contrary to the training phase, in the exploration phase the action chosen by the serving agent is its best action with a high probability. The set of underexplored choices of agent i is given as

$$\mathcal{N}_i^{\mathrm{ue}}(t) := \left\{ j \in \mathcal{K}_i : N_{j,p_i(t)}^{\mathrm{ue},i}(t) \le D_{\mathrm{ue}}(p_i(t), t) \right\},$$

where $N_{j,p_i(t)}^{\mathrm{ue},i}(t)$ denotes the number of times choice j is selected by agent i except selections in the training phases of agent i for contexts in $p_i(t)$, by time t, and $D_{\mathrm{ue}}(p_i(t), t)$ is a control function that denotes the sufficient amount of explorations by time t that will enable agent i to form accurate estimates of the expected rewards of its choices for $p_i(t)$.

- **Exploitation phase:** In this phase agent i selects the choice with the highest estimated expected reward in order to maximize its long-term reward.

When agent i is selecting a choice for its own context at time t, it can be in three of the above phases. If there are any underexplored actions of agent i, it will be in the exploration phase. Else if there are any undertrained agents of agent i, it will be in the training phase. Else if there are any underexplored agents of agent i, it will be in the exploration phase. Else, it will be in the exploitation phase. On the other hand, as a serving agent, agent i can either be in the exploration or the exploitation phase because it selects an action from $\mathcal{F}_i$. Due to the heterogeneity of the context arrivals, usually an agent cannot learn the expected rewards of its actions well for all possible contexts. However, cooperation enables the agents to learn the expected rewards of their actions for contexts that are observed frequently by the other agents. An agent can benefit from this learning in the future if the future contexts are not among contexts that are frequently observed by the agent thus far. On the other hand, the requesting agent gets an immediate benefit in terms of the reward through using the actions of the serving agents.

Next, we provide a result on the performance of the distributed learning algorithm discussed in this section [20].

Theorem 26.1. *Consider the distributed learning algorithm for the cooperative contextual bandit problem. Assuming that the rewards of the actions lie in* $[0, 1]$, *there exists a set of parameters* ρ, $D_{ut}(p, t)$ *and* $D_{ue}(p, t)$ *for any* p *such that*

$$R_i(T) = \tilde{O}(T^{\frac{2\alpha+d}{3\alpha+d}})$$

for any sequence of context arrivals $x_i(1), x_i(2), \ldots, x_i(T)$ *and for all* $i \in \mathcal{N}$. ■

This theorem shows that the regret of agent i is sublinear in time. The actions selected in training and exploration phases can be suboptimal. Therefore, the number of training and exploration phases must be sublinear in order to achieve sublinear regret. This implies that the agent exploits for the majority of the time when T is large. The choices selected by the agent when it exploits form its cooperation network.

The ideal cooperation network of agent i can be computed if $\pi_k(x)$ is perfectly known $\forall x \in \mathcal{X}$ and $\forall k \in \mathcal{K}_i$. The region of optimality of a choice $k \in \mathcal{K}_i$ can be defined as $\mathcal{X}^*_{i,k} := \{x \in \mathcal{X} : k \in k^*_i(x)\}$. $\mathcal{X}^*_{i,k} = \emptyset$, implies that choice k is never optimal for agent i. The ideal cooperation network of agent i consists of other agents that are optimal choices for agent i for at least one context in $\mathcal{X}$. Hence, the ideal cooperation network of agent i is given as $\mathcal{N}^*_i := \{k \in \mathcal{N}_i : \mathcal{X}^*_{i,k} \neq \emptyset\}$. Although an agent might be in the ideal cooperation network of agent i, it may not be optimal for a particular realization $\boldsymbol{x}_i := \{x_i(1), \ldots, x_i(T)\}$ of agent i's contexts. We can define the realization-specific ideal cooperation network of agent i as the set of other agents who are optimal choices for agent i for at least one context in $x_i(1), \ldots, x_i(T)$, i.e., $\mathcal{R}^*_i(\boldsymbol{x}_i) := \{k \in \mathcal{N}_i : k \in k^*_i(x) \text{ for some } x \in \{x_i(1), \ldots, x_i(T)\}\}$. Obviously, we have $\mathcal{R}^*_i(\boldsymbol{x}_i) \subseteq \mathcal{N}^*_i$.

On the other hand, the real cooperation network of agent i at time t consists of other agents that it is willing to cooperate with based on their estimated expected rewards. Let $\hat{\pi}^i_{k,p}(t)$ denote the estimated expected reward of choice k for contexts in $p \in \mathcal{P}_i(t)$ formed by agent i at time t. This can be taken as the sample mean of the rewards observed for choice k from contexts in p excluding the observations from the training phases of agent i. Based on this, the set of estimated optimal choices of agent i at time t for $p \in \mathcal{P}_i(t)$ is given as $\hat{k}^*_{i,p}(t) := \arg\max_{k \in \mathcal{K}_i} \hat{\pi}^i_{k,p}(t)$. Note that $a_i(t) \in \hat{k}^*_{i,p}(t)$ when agent i exploits at time t. Hence, the real cooperation network of agent i at time t is given as $\hat{\mathcal{N}}^*_i(t) := \{k \in \mathcal{N}_i : k \in \hat{k}^*_{i,p}(t) \text{ for some } p \in \mathcal{P}_i(t)\}$. The real cooperation network is expected to converge to the ideal cooperation network as the estimated expected rewards of the choices for all the contexts converge to the true expected rewards. Moreover, the realized ideal cooperation network is expected to converge to the ideal cooperation network when for all $k \in \mathcal{N}^*_i$, there exists at least one $x_i(t)$ in $\{x_i(1), \ldots, x_i(T)\}$ such that $x_i(t) \in \mathcal{X}^*_{i,k}$.

26.2.3 LEARNING TO COOPERATE WHEN COOPERATION IS COSTLY

Thus far we assumed that agents can cooperate with each other freely without incurring any costs. An interesting extension is to force the requesting agent i to incur a penalty, denoted by $d^i_k > 0$ every time it chooses a serving agent k. For instance, in a network security application the agents will correspond to *autonomous systems* (ASs) that collaborate with each other to detect cyberattacks. This requires exchanging network traffic information between the ASs, which can incur both communication cost and cost related to using the resources of the serving AS to detect the cyberattack. Another interesting example is expert selection for medical diagnosis described in Section 26.1.2. In this example, actions correspond to medical experts that will diagnose a patient and agents correspond to clinics. It is possible that a clinic refers the patient to another clinic, and the serving clinic charges the requesting clinic for the service. A similar situation may also arise in the multimedia content aggregation application described in Section 26.1.1.

There are numerous ways to tackle cooperation costs. A straightforward way is to define the expected payoff of each choice $k \in \mathcal{N}_i$ for context x as $\pi_k(x) - \gamma d^i_k$, where $\gamma > 0$ is a weight that defines the tradeoff between the reward and the cost. Then, agent i can run the distributed learning algorithm described in Section 26.2.2 by using estimated expected payoffs for the choices $k \in \mathcal{N}_i$, i.e., $\hat{\mu}^i_{k,p}(t) := \hat{\pi}^i_{k,p}(t) - \gamma d^i_k$. Because this modification also changes the oracle, the definition of the regret in Eq. (26.2) and the regret bound in Theorem 26.1 remains unchanged. However, the ideal and real cooperation networks of the agents will be affected by the introduction of cooperation costs. When the cooperation costs are high, agents will learn to cooperate with the other agents only if the expected payoff from the cooperation exceeds the expected reward that the agent can obtain from its own actions. This implies that the region of optimality for a choice $k \in \mathcal{N}_i$ is expected to shrink from $\mathcal{X}^*_{i,k}$ to $\emptyset$ as the cooperation cost d^i_k increases from 0 to ∞.

The cooperation networks that are formed by the agents can vary based on how the tradeoff between the reward and cooperation cost is captured. Another way to capture this tradeoff is to consider cooperative contextual bandits as a multiobjective online learning problem. Then, the reward can be taken as one objective and the cost can be taken as the other objective.

In the multiobjective online learning problem, the distributed learning algorithm can be modified to learn the Pareto front. In this case, it is natural to define the regret as the sum of the distances of the actions chosen by the agents to the Pareto front [25]. In addition, since the Pareto front depends on the context, the contextual zooming method should be adapted to learn the choices that are in the Pareto front [26].

Another alternative is to consider specific subsets of the Pareto front. One interesting case is when one objective dominates the other objective. For instance, the agents may want to select choices that maximize their reward, and if there are multiple such choices, they may also want to select choices that minimize the cost among all choices that maximize the reward. In this case, the reward dominates the cost, and hence, the cooperative contextual bandit problem can be modeled as a multiobjective contextual bandit problem with a dominant objective [27].

26.3 DISTRIBUTED LEARNING WITH GLOBAL OR GROUP FEEDBACK

In this section, we consider a multiagent decision-making and learning problem in which a set of distributed agents $\mathcal{N} = \{1, 2, \ldots, N\}$ select actions from their own action sets $\mathcal{A}_n$, with the cardinality of the action set denoted by $K_n = |\mathcal{A}_n|, \forall n \in \mathcal{N}$ in order to maximize the overall system reward $r(\boldsymbol{a})$. The system reward depends on the joint action $\boldsymbol{a} = (a_1, \ldots, a_N)$ of all agents but agents do not know *a priori* how their actions influence the overall system reward, or how their influence may change dynamically over time. Each time, agents can only observe/measure the *overall* system performance and hence, they only obtain global feedback that depends on the joint actions of all agents, yet the observations of the global feedback may be subject to individual errors. That is, each agent n privately observes $r^n_t = r_t + \epsilon^n_t$, equal to the global reward r_t plus a random noise term ϵ^n_t. The fact that individualized feedback is missing, communication is not possible, and the global feedback is noisy makes the development of efficient learning algorithms that maximize the joint reward very challenging. Fig. 26.2 portrays distributed cooperative learning with individualized feedback and global feedback with individual noises.

The considered problem has many application scenarios. For instance, in a stream-mining system that uses multiple classifiers for event detection in video streams, individual classifiers select operating points to classify specific objects or actions of interest, the results of which are synthesized to derive an event classification result (see Fig. 26.3). If the global feedback is only about whether the event classification is correct, individualized feedback about individual contribution is not available. For another instance, in a cooperative communication system, a set of wireless nodes forwards signals of

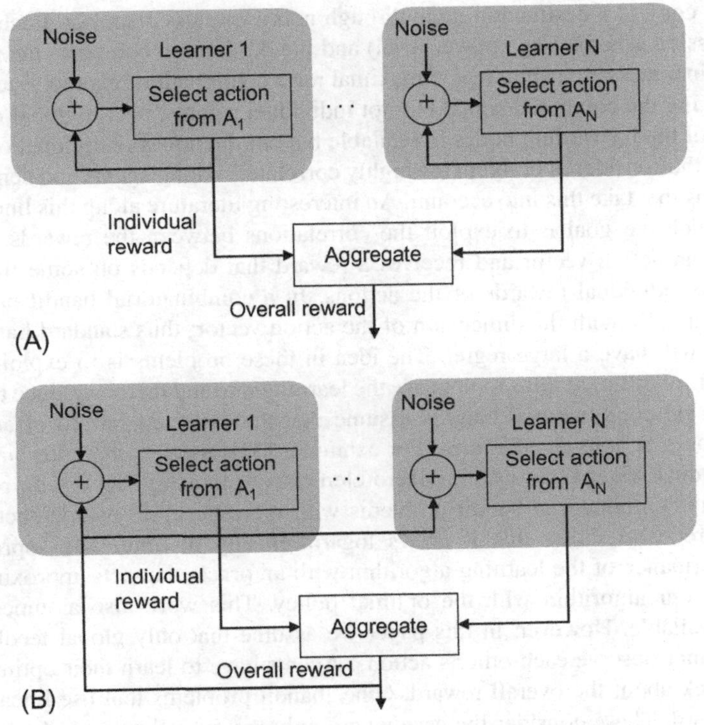

FIG. 26.2

(A) Individualized feedback; (B) Global feedback.

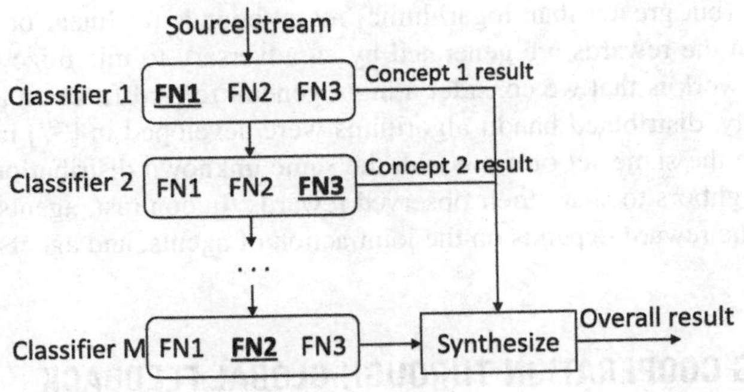

FIG. 26.3

Classifier chain for real-time stream mining.

the same message copy to a destination node through noisy wireless channels. Each forwarding node selects its transmission scheme (e.g., power level) and the destination combines the forwarded signals to decode the original message using, e.g., a maximal ratio combination scheme. Because the message is only decoded using the combined signal but not individual signals, only a global reward depending on the joint effort of the forwarding nodes is available but not the nodes' individual contributions.

The rewards in the considered problem are highly correlated among agents and hence, it is important to design algorithms that take this into account. An interesting literature along this line is combinatorial bandit [28], in which the goal is to exploit the correlations between the rewards. In this problem, the agent chooses an action vector and receives a reward that depends on some linear or nonlinear combination of the individual rewards of the actions. In a combinatorial bandit problem the set of arms grows exponentially with the dimension of the action vector; thus standard bandit policies such as the one in [29] will have a large regret. The idea in these problems is to exploit the correlations between the rewards of different arms to improve the learning rate and thereby reduce the regret [30,31]. Most of the works on combinatorial bandits assume that the expected reward of an arm is a linear function of the chosen actions for that arm. For example, [32] assumes that after an action vector is selected, the individual rewards for each nonzero element of the action vector are revealed. Another work [33] considers combinatorial bandit problems with more general reward functions, defines the approximation regret, and shows that it grows logarithmically in time. The approximation regret compares the performance of the learning algorithm with an oracle that acts approximately optimally while we compare our algorithm with the optimal policy. This work also assumes that individual observations are available. However, in this paper we assume that only global feedback is available and individuals cannot observe each other's actions. Agents have to learn their optimal actions based only on the feedback about the overall reward. Other bandit problems that use linear reward models are studied in [34–36]. These consider the case where only the overall reward of the action profile is revealed but not the individual rewards of each action. However, our analysis is not restricted to linear reward models, but instead are much more general.

Another line of work considers online optimization problems where the goal is to minimize the loss due to learning the optimal vector of actions that maximizes the expected reward function. These works show sublinear (but greater than logarithmic) regret bounds for linear or submodular expected reward functions when the rewards are generated by an adversary to minimize the gain of the agent. The difference of our work is that we consider a more general reward function and prove logarithmic regret bounds. Recently, distributed bandit algorithms were developed in [37] in network settings. In that work, agents have the same set of arms with the same unknown distributions and are allowed to communicate with neighbors to share their observed rewards. In contrast, agents in our problem have distinct sets of arms, the reward depends on the joint action of agents, and agents do not communicate at run-time.

26.3.1 ACHIEVING COOPERATION THROUGH GLOBAL FEEDBACK

We propose multiagent learning algorithms that enable the various agents to learn how to make decisions to maximize the overall system reward without exchanging information with other agents. In order to quantify the loss due to learning and operating in an unknown environment, we define the regret of an online learning algorithm for the set of agents as the difference between the expected reward of the best joint action of all agents and the expected reward of the algorithm used by the agents. We prove that, if the global feedback is received without errors by the agents, then all deterministic algorithms can be implemented in a distributed manner without message exchanges. This implies that the distributed nature of the system does not introduce any performance loss compared with a centralized system because there exist deterministic algorithms that are optimal. Subsequently, we show that if agents receive the global feedback *with* different (individual) errors, existing deterministic algorithms may break down and hence, there is a need for novel distributed cooperative algorithms that are robust to such errors. For this, we develop a class of algorithms that achieves a logarithmic upper bound on the regret, implying that the average reward converges to the optimal average reward. The upper bound on regret also gives a lower bound on the convergence rate to the optimal average reward.

We start with a basic *distributed cooperative learning* (DisCo) algorithm without making any additional assumptions on the reward structure. The DisCo algorithm is divided into phases: exploration and exploitation. Each agent using DisCo will alternate between these two phases in a way that at any time t, either all agents are exploring or all are exploiting. In the exploration phase, each agent selects an arm only to learn about the effects on the expected reward, without considering reward maximization, and updates the reward estimates of the arm it selected. In the exploitation phase, each agent exploits the best (estimated) arm to maximize the overall reward. There is a deterministic control function $\zeta(t)$ of the form $\zeta(t) = A \ln t$ commonly known by all agents. This function will be designed and determined before run-time, and thus is an input of the algorithm. Each exploration phase has a fixed length of $L_1 = \prod_{n=1}^{N} K_n$ slots, equal to the total number of joint arms. Each agent maintains two counters. The first counter $\gamma(t)$ records the number of exploration phases that they have experienced by time slot t. The second counter $E(t) \in \{0, 1, \ldots, L_1\}$ represents whether the current slot is an exploration slot and, if yes, which relative position it is at. Specifically, $E(t) = 0$ means that the current slot is an exploitation slot; $E(t) > 0$ means that the current slot is the $E(t)$th slot in the current exploration phase. Both counters are initialized to zero: $\gamma(0) = 0, E(0) = 0$. Each agent n maintains L_1 sample mean reward estimates $\bar{r}^n(l)\ \forall l \in \{1, \ldots, L_1\}$, one for each relative slot position in an exploration phase. Let b_l^n denote the arm selected by agent n in the lth position in an exploration phase. These reward estimates are initialized to be $\bar{r}^n(l) = 0$ and will be updated over time using the realized rewards. Whether a new slot t is an exploration slot or an exploitation slot will be determined by the values of $\zeta(t)$, $\gamma(t)$ and $E(t)$. At the beginning of each slot t, the agents first check the counter $E(t)$ to see whether they are still in the exploration phase: if $E(t) > 0$, then the slot is an exploration slot; if $E(t) = 0$, whether the slot is an exploration slot or an exploitation slot will then be determined by $\gamma(t)$ and $\zeta(t)$. If $\gamma(t) \leq \zeta(t)$, then the agents start a new exploration phase, and at this point $E(t)$ is set to be $E(t) = 1$. If $\gamma(t) > \zeta(t)$, then the slot is an exploitation slot. At the end of each exploration slot, counter $E(t+1)$ for the next slot is updated to be $E(t+1) \leftarrow mod(E(t)+1, L_1+1)$. When $E(t+1) = 0$, the current exploration phase ends, and hence the counter $\gamma(t+1)$ for the next slot is updated to be $\gamma(t+1) \leftarrow \gamma(t)+1$. Fig. 26.4 provides the flowchart of the phase transition for the algorithm. The algorithm prescribes different actions for agents in different slots and in different phases.

(i) *Exploration phase*: As is clear from above, an exploration phase consists of L_1 slots. In each phase, the agents select their own arms in such a way that every joint arm is selected exactly once. This is possible without communication if agents agree on a selection order for the joint arms before run-time. At the end of each exploration slot (the lth slot), $\bar{r}^n(l)$ is updated to

$$\bar{r}^n(l) \leftarrow \frac{(\gamma(t)-1)\bar{r}^n(l) + r_t^n}{\gamma(t)}. \tag{26.3}$$

Note that the observed reward realization r_t^n at time t may be different for different agents due to errors.

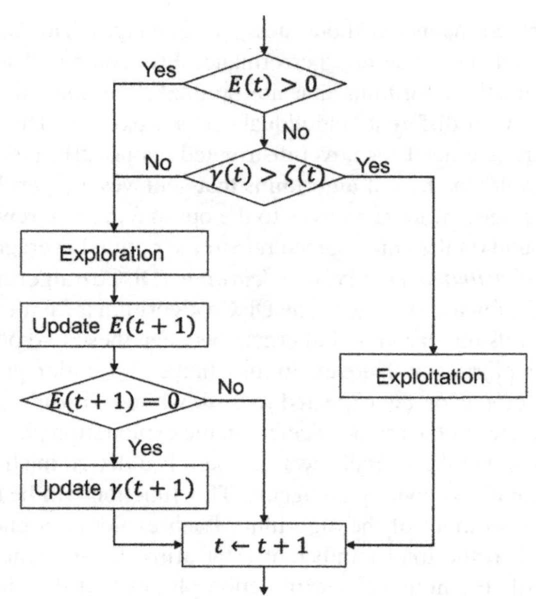

FIG. 26.4

Flowchart of the phase transition.

(ii) *Exploitation phase*: Each exploitation phase has a variable length that depends on the control function $\zeta(t)$ and counter $\gamma(t)$. At each exploitation slot t, each agent n selects $a_n = \{b^n_{l^*} : l^* = \arg\max_l \bar{r}^n(l)\}$. That is, each agent n selects the arm with the best reward estimate among $\bar{r}^n(l)$, $\forall l \in \{1, \ldots, L_1\}$. Note that in the exploitation slots, an agent n does not need to know other agents' selected arms. Because agents have individual observation noises, it is also possible that l^* is different for different agents.

Regret of the DisCo algorithm is bounded in the following theorem [38].

Theorem 26.2. *If $\zeta(t) = A \ln t$ with $A > 2\left(\frac{D}{\Delta_{\min}}\right)^2$, then the expected regret of the DisCo algorithm after any number T periods is bounded by*

$$R(T) \leq AL_1\Delta^{\max} \ln T + B_1, \tag{26.4}$$

where $B_1 = L_1\Delta^{\max} + \sum_{t=1}^{\infty} 2NL_1\Delta^{\max} t^{-\frac{A}{2}\left(\frac{\Delta^{\min}}{D}\right)^2}$ is a constant. Here $\Delta^{\max}$ represents the difference between the expected rewards of the best joint arm and the second-best joint arm, and $\Delta^{\min}$ represents the maximum expected loss due to selecting any suboptimal joint arm. ■

26.3.2 ACCELERATING LEARNING THROUGH REWARD INFORMATIVENESS

Although the time order of the regret of the DisCo algorithm is logarithmic, due to its linear dependence on the cardinality of the joint action space, which increases exponentially with the number of agents, the regret is large and the convergence rate is very slow with many agents. In many application scenarios, even if we do not know exactly how the actions of agents determine the expected overall rewards, some structural properties of the overall reward function may be known. For example, in the classification problem that uses multiple classifiers [39], the overall classification accuracy is increasing in each individual classifier's accuracy, even though each individual's optimal action is unknown a priori. Thus, some overall reward functions may provide higher levels of informativeness about the optimality of individual actions. Therefore, we also develop learning algorithms that achieve faster learning speed by exploiting such information.

Definition 26.1. An expected overall reward function $\mu(\boldsymbol{a})$ is said to be informative with respect to agent n if there exists a unique arm $a_n^* \in \mathcal{A}_n$ such that $\forall \boldsymbol{a}_{-n}$, $a_n^* = \arg\max_{a_n} \mu(a_n, \boldsymbol{a}_{-n})$, where $\mu(\boldsymbol{a})$ represents the expected reward of a joint arm $\boldsymbol{a}$. An expected overall reward function $\mu(\boldsymbol{a})$ is said to be fully informative if it is informative with respect to all agents. ■

If the overall reward function is fully informative, then the agents only need to record the relative overall reward estimates instead of the exact overall reward estimates. Therefore, the key design problem of the learning algorithm is, for each agent n, to ensure that the fraction of time selecting $\boldsymbol{a}_{-n}$ (called weight $\theta_{\boldsymbol{a}_{-n}}$) is the same for the relative reward estimates for any given arm a_n so that it is sufficient for agent n to learn the optimal arm using only these relative reward estimates. We emphasize the importance of the weights $\theta_{\boldsymbol{a}_{-n}}, \forall \boldsymbol{a}_{-n}$ being the same for all $a_n \in \mathcal{A}_n$ of each agent n even though agent n does not need to know these weights exactly. If the weights are different for different a_n, then it is possible that $\bar{r}^n(a_n') > \bar{r}^n(a_n^*)$ merely because other agents are using their good arms when agent n is selecting a suboptimal arm a_n while other agents are using their bad arms when agent n is selecting the optimal arm a_n^*. Hence, simply relying on the relative reward estimates does not guarantee obtaining the correct information needed to find the optimal arm.

Next, we describe an improved learning algorithm. We call this new algorithm the DisCo-FI algorithm where "FI" stands for "Fully Informative." The key difference from the basic DisCo algorithm is that, in DisCo-FI, the agents will maintain relative reward estimates instead of the exact reward estimates. Agents know a common deterministic function $\zeta(t)$ and maintain two counters $\gamma(t)$ and $E(t)$. Now each exploration phase has a fixed length of $L_2 = \sum_{n=1}^{N} K_n$ slots and hence, $E(t) \in \{0, 1, \ldots, L_2\}$ with $E(t) = 0$ representing that the slot is an exploitation slot and $E(t) > 0$ representing that it is the $E(t)$th relative slot in the current exploration phase. As before, both counters are initialized to be $\gamma(0) = 0, E(0) = 0$. Each agent n maintains K_n sample mean (relative) reward estimates $\bar{r}^n(a_n), \forall a_n \in \mathcal{A}_n$, one for each one of its own arms. These (relative) reward estimates are initialized to be $\bar{r}^n(a_n) = 0$ and will be updated over time using the realized rewards. The transition between exploration phases and exploitation phases is almost identical to that in the DisCo algorithm. The only difference is that at the end of each exploration slot, the counter $E(t+1)$ for the next slot is updated to be $E(t+1) \leftarrow mod(E(t)+1, L_2+1)$. Hence, we ensure that each exploration phase has only L_2 slots. The algorithm prescribes different actions for agents in different slots and in different phases.

(i) Exploration phase: As is clear from the phase transition, an exploration phase consists of L_2 slots. These slots are further divided into N subphases and the length of the nth subphase is K_n. In the nth subphase, agents take actions as follows:

1. Agent n selects each of its arms $a_n \in \mathcal{A}_n$ in turn, each arm for one slot. At the end of each slot in this subphase, it updates its reward estimate using the realized reward in this slot as follows,

$$\bar{r}^n(a_n) \leftarrow \frac{\gamma(t)\bar{r}^n(a_n) + r_t^n}{\gamma(t) + 1}. \tag{26.5}$$

2. Agent $i \neq n$ selects the arm with the highest reward estimate for every slot in this subphase, i.e., $a_i(t) = \arg\max_{a_i \in \mathcal{A}_i} \bar{r}^i(a_i)$.

(ii) Exploitation phase: Each exploitation phase has a variable length that depends on the control function $\zeta(t)$ and counter $\gamma(t)$. In each exploitation slot t, each agent n selects $a_n(t) = \arg\max_{a \in \mathcal{A}_n} \bar{r}^n(a)$.

Theorem 26.3 ([38]). *Suppose $\mu(\boldsymbol{a})$ is fully informative. If $\zeta(t) = A \ln T$ with $A \geq 2\left(\frac{D}{\Delta_{FI}^{\min}}\right)^2$, then the expected regret of the DisCo-FI algorithm after any number T slots is bounded by*

$$R(T) < AL_2\Delta^{\max} \ln T + B_2, \tag{26.6}$$

where $B_2 = L_2\Delta^{\max} + 2L_2\Delta^{\max}\sum_{t=1}^{\infty} t^{-\frac{A}{2}\left(\frac{\Delta_{FI}^{\min}}{D}\right)^2}$ is a constant number. Here $\Delta_{FI}^{\min} = \min_n \Delta_n^{\min}$ where $\Delta_n^{\min}$ is the expected reward difference between agent n's best arm and its second-best arm. ■

DisCo-FI can be extended to the more general case where the full informativeness constraint is relaxed.

Definition 26.2. An expected overall reward function $\mu(\boldsymbol{a})$ is said to be informative with respect to a group of agents g_m if there exists a unique group-joint arm $\boldsymbol{a}_m^* \in \times_{n \in g_m} \mathcal{A}_n$ such that $\forall \boldsymbol{a}_{-m}, \boldsymbol{a}_m^* = \arg\max_{\boldsymbol{a}_m \in \times_{i \in g_m} \mathcal{A}_i} \mu(\boldsymbol{a}_m, \boldsymbol{a}_{-m})$. An expected overall reward function $\mu(\boldsymbol{a})$ is said to be partially informative with respect to a group partition $\mathcal{G} = \{g_1, \ldots, g_M\}$ if it is informative with respect to all groups in $\mathcal{G}$. ■

The new algorithm is called the DisCo-PI algorithm where "PI" stands for "Partially Informative." It is a combination of the basic DisCo and DisCo-FI. In particular, each agent group is treated as a single agent in the DisCo-FI algorithm and hence, each agent group learns the relative rewards of its group-joint arms. Within each group, agents follow the idea of the basic DisCo algorithm to learn their optimal arms.

Theorem 26.4. *[38] Suppose $\mu(\boldsymbol{a})$ is partially informative with respect to a group partition $\mathcal{G}$. If $\zeta(t) = A \ln T$ with $A \geq 2\left(\frac{D}{\Delta_{PI}^{\min}}\right)^2$, then the expected regret of the DisCo-PI algorithm after any number T slots is bounded by*

$$R(T) < AL_3\Delta^{\max} \ln T + B_3, \tag{26.7}$$

where $L_3 = \sum_{m=1}^{M} \prod_{n \in g_m} K_n$ and

$$B_3 = L_3\Delta^{\max} + 2\sum_{m=1}^{M} N_m \prod_{n \in g_m} K_n \Delta^{\max} \sum_{t=1}^{\infty} t^{-\frac{A}{2}\left(\frac{\Delta_{PI}^{\min}}{D}\right)^2} \tag{26.8}$$

is a constant number. Here $\Delta_{PI}^{\min} = \min_m \Delta_m^{\min}$ where $\Delta_m^{\min}$ is the expected reward difference between the best group-joint arm of g_m and the second-best group-joint arm of g_m. ■

We illustrate the performance of the proposed learning algorithms via simulation results for the real-time stream-mining problem using multiple classifiers in Fig. 26.5. The proposed algorithms are compared against several benchmarks: random, safe experimentation, UCB1 [29], and optimal. Safe experimentation (SE) is a method used in [40] when there is no uncertainty about the accuracy of the classifiers. In each period t, each classifier selects its baseline action with probability $1 - \epsilon_t$ or selects a new random action with probability ϵ_t. When the realized reward is higher than the baseline reward, the classifiers update their baseline actions to the new action. As can been seen, SE works almost as poorly as the random benchmark in terms of event detection accuracy. Due to the uncertainty in the

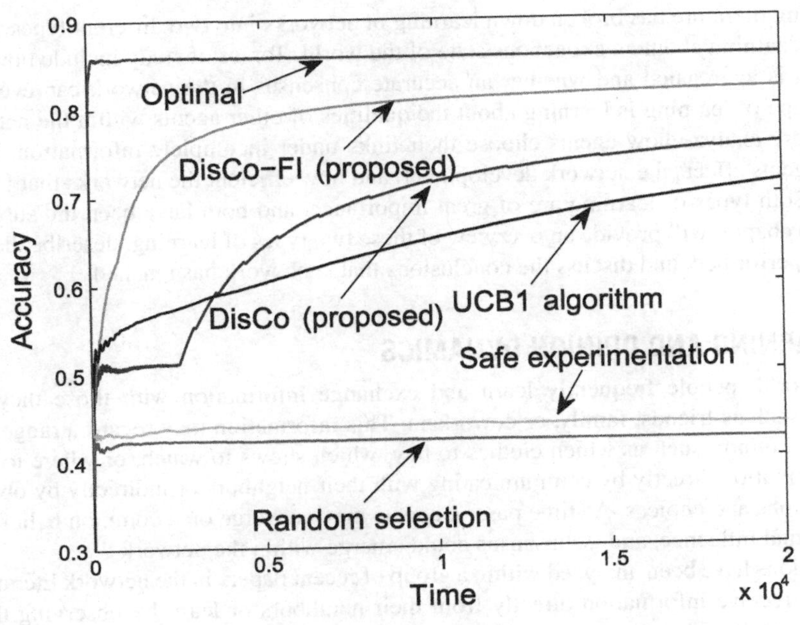

FIG. 26.5

Performance comparison [38].

detection results, updating the baseline action to a new action with a higher realized reward does not necessarily lead to selecting a better baseline action. Hence, SE is not able to learn the optimal operating points of the classifiers. UCB1 achieves a much higher accuracy than random and SE algorithms and is able to learn the optimal joint operating points over time. However, the learning speed is slow because the joint arm space is large. The proposed DisCo algorithm can also learn the optimal joint action. However, because the joint arm space is large, the classifiers have to stay in the exploration phases for a relatively long time in the initial periods to gain sufficiently high confidence in reward estimates while the exploitation phases are rare and short. Thus, the classification accuracy is low initially. After the initial exploration phases, the classifiers begin to exploit and hence the average accuracy increases rapidly. Because the reward structure satisfies the fully informative condition, DisCo-FI rapidly learns the optimal joint action and performs the best among all schemes.

26.4 LEARNING IN NETWORKS WITH INCOMPLETE INFORMATION

Learning plays a major role in the analysis of networks and network formation. Many significant real-world interactions take place over a variety of networks, such as social networks, financial networks, and trade networks. Learning and incomplete information can strongly affect the development and dynamics of such networks, and so it is important to properly model and study these forces in performing network analysis.

The existing literature has broken down learning in networks into two different types. The first type of learning is learning about an exogenous state of the world. Topics of study include how information among agents is aggregated and whether an accurate consensus in the network can eventually form. The second type of learning is learning about the qualities of other agents within the network. Papers in this literature analyze how agents choose their links under incomplete information, how learning about other agents affects the network development, and how efficient the networks that form from this process are. Both types of learning are of great importance, and both have been the subject of active research. This chapter will provide an overview of these two types of learning, describe the recent work that has been performed, and discuss the conclusions that such work has reached.

26.4.1 LEARNING AND OPINION DYNAMICS

In the real world, people frequently learn and exchange information with those they are closely connected to, such as friends, family, or coworkers. This information may regard a range of decisions both major and minor, such as which clothes to buy, which shows to watch, or where to live. Agents can learn information directly by communicating with their neighbors or indirectly by observing their neighbors' actions and choices. As time passes, agents may converge on a common belief through this process of mutual influence, and a consensus could emerge within the network.

Such situations have been analyzed within a group of recent papers in the network literature. In these papers, agents receive information directly from their neighbors or learn by observing the decisions their neighbors make. They then use this to update their own beliefs and make their own decisions. A key question that is raised is the issue of learning efficiency and the convergence of beliefs to the truth. If information is initially diffused across a network, is it possible for the information to be aggregated through a learning process such that the network arrives at the truth? Can this property be guaranteed if the network is sufficiently large or the agents sufficiently numerous? This question is important for understanding how learning occurs within very large networks, such as societies.

The existing literature has shown that the configuration of the network has a strong impact on the information that agents learn and the rate at which they can reach a consensus. Golub and Jackson [41] consider such a problem when agents utilize a form of learning known as DeGroot learning. This type of learning assumes that each period, agents update their belief based on the average beliefs of their neighbors, and can be interpreted, for instance, as agents communicating with and influencing each other during each period. The paper shows that the network converges to the correct belief if and only if no agent has too large an influence. If there is a prominent group of agents that receive too much attention, then learning will not converge to the truth.

Acemoglu et al. [42] consider a different type of opinion dynamics where the agents do not directly communicate their beliefs with others but merely observe other agents' actions. For instance, an agent may observe the car that another agent bought, and thus update its own belief more favorably for that car manufacturer. This is known in the general economic theory literature as social learning, which has its own extensive literature but was only recently applied to network theory. The paper shows that learning will be correct asymptotically as long as the observation network is large enough. This is satisfied in settings where agents are able to draw from enough observations.

While these other papers consider observations that occur in exogenously given networks, some other papers have allowed for endogenous network formation. Acemoglu et al. [43] consider a model in which agents form communication links with each other by paying a cost. Through these links agents can transmit information to each other. The authors show that learning converges to the truth if the formed network feature "information hubs" that can distribute information to the agents. Thus learning must be aggregated within a few agents. Furthermore, they show that if the network has too many social cliques that contain the spread of information, then learning may not converge to the truth.

Song [44] also allows for endogenous observations. In this model, agents make observations sequentially as in Acemoglu et al. [42], but they are able to choose which other agents to observe. Agents have a fixed capacity of observations to make, and thus they must be strategic in choosing which other agents to observe. Whether the eventual consensus in the network is accurate will depend on the specific links that agents choose to form. This model introduces important aspects of network formation, discussed further in the following section, into the social learning problem. The paper shows that under certain situations, there will indeed be equilibrium in which agents are able to correctly learn. In such equilibrium, agents observe a large enough group of other agents so that the agent signals can be aggregated correctly.

26.4.2 LEARNING AND STRATEGIC NETWORK FORMATION

The other major type of learning that has been analyzed within the network literature is learning about other agents through the process of strategic network formation. Whereas opinion dynamics models analyze situations in which agents do not directly derive benefits from their neighbors but instead learn information about an exogenous state of the world, strategic network formation models assume that agents directly derive benefits from their links with others. In forming a link, an agent weighs the benefits provided through that link versus the costs of maintaining the link. The beliefs an agent has about the benefits each neighbor can provide will strongly influence which neighbors the agent chooses to link with, and learning in turn will strongly affect the network structure and dynamics. In this section, we provide an overview of such models and summarize the main conclusions that have been reached.

Strategic network formation itself has developed over the past several decades into a large literature. However, many papers in this field have not considered the presence of incomplete information in their models. Seminal strategic network formation papers, such as Jackson and Wolinsky [45], Bala and Goyal [46], and Jackson and Watts [47], all assume that agents have complete information about others, and thus agents know exactly how much benefit they will receive from each link. Because information is perfect, agent beliefs are always accurate and never update over time.

Incorporating incomplete information into strategic network formation models represents an important step toward more accurately modeling real-world interactions. In many real-world networks, incomplete information often has a profound effect. For instance, in a social network agents who are meeting and making friends will begin with only limited information about each other. It takes mutual experiences to learn about others, and as such learning occurs the network structure will evolve.

A recent strand of network papers has developed to analyze the impact of incomplete information on the network formation process and the resulting network structures. These models operate by assuming agents have different types. Agents of different types may provide different payoffs to others, receive different payoffs from others, or both. Importantly, each agent's type is unknown to others at the beginning of the model. Instead, agents have prior beliefs about other agent types. As links are formed, information is learned about a neighbor's type through the benefits that are received, and this allows an agent to update its prior beliefs using the Bayes rule. The prior beliefs can have a strong impact on the eventual stable network, however, as they will affect the initial links that are formed, which influences subsequent learning. Thus, the initial beliefs can have lasting consequences for the network.

These models demonstrate that incomplete information can have a strong impact on the network formation process. First, incomplete information will affect the initial links that are formed among the agents. Even if an agent has a high true quality, if others believe that the agent's quality is low, then they will never form a link with that agent. Thus, it is possible for some high-quality agents to be excluded immediately from the network. This presents a source of inefficiency from the incomplete information.

Incomplete information may provide social benefits as well. If links have externalities, then it may be socially beneficial for agents to form links to bad agents. For instance, a common payoff model within the network literature is the connections model, in which agents receive benefits from direct neighbors as well as indirect benefits from neighbors of direct neighbors. Even if a direct neighbor's quality is low, if that direct neighbor has many indirect neighbors it may be beneficial to form a link to it. There can be multiple stable networks with varying levels of indirect benefits provided, and the socially optimal network may not be reachable along certain formation paths. Incomplete information can increase the amount of formation paths, and give a chance for the optimal network to develop. The paper by Song and van der Schaar [48] highlights theses effects via a discrete-time network formation model. They show that a large increase in potential networks arises under incomplete information, and they give examples where incomplete information increases social welfare.

Song and van der Schaar [49] extend previous work by considering the impact of foresight on network formation. Most network models assume that agents make myopic decisions. As a result, agents link with other agents if the payoff, or their beliefs about the payoff under incomplete information, of forming each link is positive. However, this paper extends the previous models by allowing agents to be foresighted and consider future payoffs in making links. The model is a discrete-time model like that in Song and van der Schaar [48]. It considers the limit as agents become very patient. They derive a "folk theorem" like result that shows that many networks are now possible. The idea is that when agents are very patient, it is possible to enforce a wide range of equilibrium strategies through imposing sufficiently long punishments when agents deviate from equilibrium play. Given that agents are very patient, they value the future greatly and thus will not wish to be punished. This paper shows that the range of possible networks can thus be even larger when foresight is introduced.

Zhang and van der Schaar [50] consider learning that occurs gradually, in contrast to Song and van der Schaar [48] and Song and van der Schaar [49] in which the learning is instant. Learning happens in continuous time, and the rate at which learning occurs will impact the resulting social welfare. In their model, agents who provide high benefits develop high reputations and remain in the network while agents who provide low benefits drop in reputation and become ostracized. The information to which agents have access and the speed at which they learn and act both have a tremendous impact on the resulting network dynamics. Importantly, the authors show that learning can have a negative impact on social welfare because it causes agents to become ostracized from the network more quickly. When an agent is ostracized from the network due to the learning process, although the ostracized agent may be of marginal quality it is still receiving a large benefit from the rest of the network. Thus, having the ostracized agent leave the network reduces social welfare on average. Faster learning may be undesirable as it leads to faster ostracization times. Due to the potential negative consequences of ostracism, it may be necessary to place agents with lower initial reputations at less central positions within the network. For instance, core-periphery networks with high-reputation agents in the core and low-reputation agents in the periphery can be optimal.

Zhang and van der Schaar [51] consider an alternative learning model in which agents exhibit homophilic preferences. Homophily is the preference to link with others who are similar to themselves. Agents have a limited capacity for links and thus maintain links with others learned to be similar to themselves and cut links to those learned to be dissimilar to themselves. Due to noise in the learning process, the agents may not find out which other agents are the closest in type, and thus the incomplete information network can exhibit vast differences from the complete information network.

The paper shows that higher levels of homophily decrease the (average) number of links that agents form. Mutually beneficial links may be dropped before learning is completed, thereby resulting in sparser networks and less clustering than under complete information. Homophily also exhibits an interesting interaction with the presence of incomplete information: initially, greater levels of homophily increase the difference between the complete and incomplete information networks, but sufficiently high levels of homophily eventually decrease the difference. Complete and incomplete information networks differ most when the degree of homophily is intermediate. The paper also extends the analysis to multiple stages of life/technology. Under such circumstances, the paper shows that the effects of incomplete information are large initially but fade somewhat over time.

26.5 CONCLUSION

In the modern world, applications that require multiple decentralized agents to act in harmony are ubiquitous. It is thus essential to understand what types of interactions enable them to cooperate and exchange information in order to reach their goals. This chapter presents learning strategies and algorithms for multiagent systems. It investigates how cooperative and strategic learning will lead to different types of networks between the agents, and how efficient these networks are in making the agents achieve their goals.

When the agents are cooperative, they are able to acquire knowledge about their environment by helping the other agents receive higher rewards, which also allows them to select better actions in the future. In addition, they are able to receive higher rewards by using the actions of the other agents. However, the willingness to cooperate depends on the cost of cooperation. If this cost is too high, agents will prefer to rely only on their own actions. It is also possible that agents' actions and rewards are correlated with each other. In such a case, the agents are required to follow distributed coordination strategies that will let them learn joint actions that maximize a global reward. With proper coordination, it is possible for agents to settle down to a joint action that maximizes the global reward by only observing noisy versions of it even without knowing the actions taken by the other agents.

On the other hand, strategic learning leads to many different kinds of networks between the agents, some of which can be efficient while the others can be highly inefficient. All the methods covered in this chapter shed light into how real-world networks are formed and how agents can reach their goals in these networks.

REFERENCES

[1] Schranz M, Dustdar S, Platzer C. Building an integrated pan-European news distribution network. In: Collaborative networks and their breeding environments, vol. 186; 2005. p. 587–96.

[2] Boutet A, Kloudas K, Kermarrec AM. FStream: a decentralized and social music streamer. In: Networked systems. Springer; 2013. p. 253–7.
[3] Mohan R, Smith JR, Li CS. Adapting multimedia internet content for universal access. IEEE Trans Multimedia 1999;1(1):104–14.
[4] Tekin C, van der Schaar M. Contextual online learning for multimedia content aggregation. IEEE Trans Multimedia 2015;17(4):549–61.
[5] Elmore JG, Miglioretti DL, Reisch LM, Barton MB, Kreuter W, Christiansen CL, et al. Screening mammograms by community radiologists: variability in false-positive rates. J Natl Cancer Inst 2002;94(18):1373–80.
[6] Trottier D. Social media as surveillance. Farnham: Ashgate; 2012.
[7] Chou WYS, Hunt YM, Beckjord EB, Moser RP, Hesse BW. Social media use in the United States: implications for health communication. J Med Internet Res 2009;11(4).
[8] Sakaki T, Okazaki M, Matsuo Y. Earthquake shakes Twitter users: real-time event detection by social sensors. In: Proceedings of the 19th international conference on world wide web. ACM; 2010. p. 851–60.
[9] Choi H, Varian H. Predicting the present with Google Trends. Econ Rec 2012;88(s1):2–9.
[10] Liu HH, Wang Y, Yang YR, Wang H, Tian C. Optimizing cost and performance for content multihoming. In: Proceeding of the ACM SIGCOMM 2012 conference on applications, technologies, architectures, and protocols for computer communication. ACM; 2012. p. 371–82.
[11] Szabo G, Huberman BA. Predicting the popularity of online content. Commun ACM 2010;53(8):80–8.
[12] Cha M, Kwak H, Rodriguez P, Ahn YY, Moon S. I tube, you tube, everybody tubes: analyzing the world's largest user generated content video system. In: Proceedings of the 7th ACM SIGCOMM conference on internet measurement. ACM; 2007. p. 1–14.
[13] Wu T, Timmers M, De Vleeschauwer D, Van Leekwijck W. On the use of reservoir computing in popularity prediction. In: Proceedings of the 2nd international conference on evolving internet. IEEE; 2010. p. 19–24.
[14] Pinto H, Almeida JM, Gonçalves MA. Using early view patterns to predict the popularity of YouTube videos. In: Proceedings of the 6th ACM international conference on web search and data mining. ACM; 2013. p. 365–74.
[15] Li H, Ma X, Wang F, Liu J, Xu K. On popularity prediction of videos shared in online social networks. In: Proceedings of the 22nd ACM international conference on information & knowledge management. ACM; 2013. p. 169–78.
[16] Zhang L, Wang F, Liu J. Understand instant video clip sharing on mobile platforms: Twitter's Vine as a case study. In: Proceedings of the network and operating system support on digital audio and video workshop. ACM; 2014. p. 85.
[17] Galuba W, Aberer K, Chakraborty D, Despotovic Z, Kellerer W. Outtweeting the Twitterers-predicting information cascades in microblogs. WOSN 2010;10:3–11.
[18] Hong L, Dan O, Davison BD. Predicting popular messages in Twitter. In: Proceedings of the 20th international conference on companion on world wide web. ACM; 2011. p. 57–8.
[19] Lerman K, Hogg T. Using a model of social dynamics to predict popularity of news. In: Proceedings of the 19th international conference on world wide web. ACM; 2010. p. 621–30.
[20] Tekin C, van der Schaar M. Distributed online learning via cooperative contextual bandits. IEEE Trans Signal Process 2015;63(14):3700–14.
[21] Slivkins A. Contextual bandits with similarity information. J Mach Learn Res 2014;15(1):2533–68.
[22] Lu T, Pál D, Pál M. Contextual multi-armed bandits. In: Proceedings of the AISTATS; 2010. p. 485–92.
[23] Choi SS, Cha SH, Tappert CC. A survey of binary similarity and distance measures. J Syst Cybern Inf 2010;8(1):43–8.
[24] Cha SH. Comprehensive survey on distance/similarity measures between probability density functions. City 2007;1(2):1.
[25] Drugan MM, Nowe A. Designing multi-objective multi-armed bandits algorithms: a study. In: Proceedings of the international joint conference on neural networks (IJCNN); 2013. p. 1–8.

[26] Turgay E, Oner D, Tekin C. Multi-objective contextual bandit problem with similarity information. In: Proceedings of the AISTATS; 2018. p. 1673–81.
[27] Tekin C, Turgay E. Multi-objective contextual bandits with a dominant objective. In: Proceedings of the 27th IEEE international workshop on machine learning for signal processing; 2017.
[28] Cesa-Bianchi N, Lugosi G. Combinatorial bandits. J Comput Syst Sci 2012;78(5):1404–22.
[29] Auer P, Cesa-Bianchi N, Fischer P. Finite-time analysis of the multiarmed bandit problem. Mach Learn 2002;47(2-3):235–56.
[30] Anantharam V, Varaiya P, Walrand J. Asymptotically efficient allocation rules for the multiarmed bandit problem with multiple plays—Part I: IID rewards. IEEE Trans Autom Control 1987;32(11):968–76.
[31] Anandkumar A, Michael N, Tang AK, Swami A. Distributed algorithms for learning and cognitive medium access with logarithmic regret. IEEE J Sel Areas Commun 2011;29(4):731–45.
[32] Gai Y, Krishnamachari B, Jain R. Combinatorial network optimization with unknown variables: multi-armed bandits with linear rewards and individual observations. IEEE/ACM Trans Netw 2012;20(5):1466–78.
[33] Chen W, Wang Y, Yuan Y. Combinatorial multi-armed bandit: general framework and applications. In: Proceedings of the international conference on machine learning (ICML); 2013. p. 151–9.
[34] Rusmevichientong P, Tsitsiklis JN. Linearly parameterized bandits. Math Oper Res 2010;35(2):395–411.
[35] Auer P. Using confidence bounds for exploitation-exploration trade-offs. J Mach Learn Res 2002;3:397–422.
[36] Dani V, Hayes TP, Kakade SM. Stochastic linear optimization under bandit feedback. In: COLT; 2008. p. 355–66.
[37] Szorenyi B, Busa-Fekete R, Hegedus I, Ormándi R, Jelasity M, Kégl B. Gossip-based distributed stochastic bandit algorithms. In: Proceedings of the international conference on machine learning (ICML); 2013. p. 19–27.
[38] Xu J, Tekin C, Zhang S, van der Schaar M. Distributed multi-agent online learning based on global feedback. IEEE Trans Signal Process 2015;63(9):2225–38.
[39] Foo B, van der Schaar M. A distributed approach for optimizing cascaded classifier topologies in real-time stream mining systems. IEEE Trans Image Process 2010;19(11):3035–48.
[40] Foo B, van der Schaar M. A rules-based approach for configuring chains of classifiers in real-time stream mining systems. EURASIP J Adv Signal Process 2009;2009:40.
[41] Golub B, Jackson MO. Naive learning in social networks and the wisdom of crowds. Am Econ J: Microecon 2010;2(1):112–49.
[42] Acemoglu D, Dahleh MA, Lobel I, Ozdaglar A. Bayesian learning in social networks. Rev Econ Stud 2011;78(4):1201–36.
[43] Acemoglu D, Bimpikis K, Ozdaglar A. Dynamics of information exchange in endogenous social networks. Theor Econ 2014;9(1):41–97.
[44] Song Y. Social learning with endogenous network formation; 2015. arXiv preprint arXiv:150405222.
[45] Jackson MO, Wolinsky A. A strategic model of social and economic networks. J Econ Theory 1996;71(1):44–74.
[46] Bala V, Goyal S. A noncooperative model of network formation. Econometrica 2000;68(5):1181–229.
[47] Jackson MO, Watts A. The evolution of social and economic networks. J Econ Theory 2002;106(2):265–95.
[48] Song Y, van der Schaar M. Dynamic network formation with incomplete information. Econ Theory 2015;59(2):301–31.
[49] Song Y, van der Schaar M. Dynamic network formation with foresighted agents; 2015. arXiv preprint arXiv:150900126.
[50] Zhang S, van der Schaar M. Reputational dynamics in financial networks during a crisis. Office of Financial Research, US Department of the Treasury, Working Paper: No. 18-03; 2018.
[51] Zhang S, van der Schaar M. From acquaintances to friends: homophily and learning in networks. IEEE J Sel Areas Commun 2017;35(3):680–90.

PART

5

APPLICATIONS

CHAPTER

GENOMICS AND SYSTEMS BIOLOGY

27

Daifeng Wang*, Chao Cheng†

*Department of Biomedical Informatics, Stony Brook University, Stony Brook, NY, United States**

Dartmouth-Hitchcock Medical Center, Lebanon, NH, United States†

27.1 INTRODUCTION

The Biological systems, though very complex, have evolved and been organized based on certain design and engineering principles. A variety of genomic elements and molecules work together as a coordinated *system* to maintain the basic and normal functions while also achieving biological diversity at the same time. If the coordination gets disrupted, a biological system highly likely leads the abnormal functions such as in diseases. To understand complex biological systems, advanced technologies have been developed in recent decades to generate a great number of high-throughput datasets, measuring the biological activities at the system level. These datasets provide novel resources to systematically study the biological functions, especially on a genomic scale. For example, the next-generation sequencing technologies such as RNA-seq and ChIP-seq can simultaneously measure gene transcription factor genomic occupation or other genome-wide events under a variety of biological conditions.

In addition, network modeling has become a common formalism for understanding complex systems, including gene regulatory networks. Modeling and analyzing the gene regulatory networks play a crucial role in understanding the interactions among various genes and genomic elements in biological systems [1]. To better understand biological networks, the signal processing and graph models have been widely applied to process, model, and analyze large-scale, high-throughput biological data. In this chapter, we will review the recent signal processing applications to the high-throughput data, especially the genomic data from the next generation sequencing analyses. We will then introduce how to use these data to construct, model, and analyze the gene regulatory networks—a key type of molecular network controlling the biological systems. In particular, we will focus on introducing the gene regulatory network structures that play key roles in controlling the genomic activities, such as gene expression at the transcriptional level and driving the phenotypes. The approaches and models for

Cooperative and Graph Signal Processing. https://doi.org/10.1016/B978-0-12-813677-5.00027-4

gene regulatory networks we introduce here are general purpose and thus can be used to study other types of biological networks, including protein-protein interaction networks, metabolic networks, and signaling networks.

27.2 GENE REGULATION

Gene expression is an important process to develop various biological functions and drive the phenotypes [2]. Following the molecular central dogma, a gene—a piece of DNA on the chromosome—is first transcribed to RNA (transcription). It is then translated to protein (translation), which is normally the functional product of the gene. The process from a gene to its functional product is called gene expression. The gene expression is actually used and evolutionarily conserved for all organisms from viruses and bacteria to mammals. Therefore, it is strictly controlled to make correct biological functions. In general, the mechanism controlling gene expression is called gene regulation.

Gene regulation is very complex, involving a variety of different genomic elements. It varies at different conditions such as in healthy versus cancerous cells. Gene regulation occurs at all stages of gene expression from pretranscription to posttranslation. In particular, at the transcriptional stage when genes are primarily regulated, gene regulation reveals how a piece of DNA (a gene) converts to RNA, and involves the following major proteins and genomic elements [3]:

Transcriptional factors—a group of proteins to activate or repress the gene's transcription. They bind to the regulatory regions to affect gene expression.
Promoter—a short DNA region before the transcriptional start site of a gene. The transcription factor(s) bind to the promoter region to initiate the gene transcription.
Enhancer—a DNA region distant to the gene to positively affect the gene transcription. The enhancers can be found by a group of transcriptional factors.
Silencer—a DNA region distant to the gene to negatively affect the gene transcription. The silencer can be found by a group of transcriptional factors.
Coactivators—a group of proteins that does not directly bind to the regulatory regions (e.g., promoters, enhancers) but still affects the gene transcription via the coordination with transcription factors.
Regulatory variants—the DNA variants associated with gene expression changes. They can break the transcription factor binding sites to affect gene expression [4].
microRNAs—the small noncoding RNA molecules that silence or repress gene expression [5].

The transcription factors (TFs) and coactivators are transcribed and translated from the genes as well, and thus their expressions are also regulated by each other. Moreover, after transcriptional regulation, the gene expression can be still regulated by other mechanisms such as alternative splicing [6]. Even proteins can be regulated at posttranslation stages such as phosphorylation [7]. Based on the brief introduction above, the genes are not independent. Instead, they interact with other each, such as controlling gene expression by gene regulation, and thus form a gene regulatory network to achieve biological functions. Any activities disrupting the gene regulatory network potentially drive abnormal behaviors, such as those that lead to human diseases. However, understanding the gene regulatory networks, especially their changes across various conditions including diseases, is still very challenging. Therefore, next-generation sequencing technologies have been developed in recent years and widely applied to detect the genome-wide expression and regulatory activities.

27.3 NEXT-GENERATION SEQUENCING TECHNIQUES

Next-generation sequencing (NGS) techniques have been developed and widely applied to systematically measure the gene expression and regulatory activities in recent years [8] They can sequence in parallel millions or billions of short reads from DNA or RNA molecules, providing deep and accurate measurements genome-wide. We highlight the key NGS techniques on gene expression and regulation as follows:

DNA-seq—an NGS method to identify the DNA bases and individual variants by comparing to a reference genome [9]. The DNA variants can affect gene expression, and mainly consist of four types: single nucleotide polymorphisms (SNPs), insertions, deletions, and structural variants.

RNA-seq—an NGS method to sequence both protein-coding and noncoding RNAs and measure their transcriptional activities, including gene expression [8].

ChIP-seq—an NGS method, Chromatin Immunoprecipitation Sequencing is used to detect the interactions between proteins, and DNA regions and identify the genomic binding sites of DNA-binding proteins such as transcription factors [8].

A large number of computational approaches and software has also been developed to process these NGS data. For example, using these approaches, we can first align the raw data of RNA-seq—e.g., millions of short reads to the reference genome—and then quantify the gene expression levels [10]. Moreover, the advanced signal processing methods have been used to help uniformly process the NGS data, such as optimal filtering [11], and calculate the gene expression and regulatory activities, such as multiscale modeling [12].

27.4 GENE REGULATORY NETWORK PREDICTION

Increasing next-generation sequencing datasets provides great resources to understand the gene expression and regulatory activities, particularly genome-wide. However, integrating these large-scale datasets is still very computationally challenging [13]. Thus, a variety of computational approaches and bioinformatics tools have been recently developed to predict the gene regulatory relationships and the gene regulatory networks that control gene expression. We highlight some recent and representative work as follows, especially using next-generation sequencing data.

The gene coexpression analysis has long been widely used to predict the association networks among genes [14]. The coexpressed genes in the network share similar expression patterns, and thus are likely coregulated by similar mechanisms, which also suggests that they may be involved in the same functions. Recent work has integrated the RNA-seq data and predicted the gene coexpression networks, including noncoding RNAs that have strong functional associations in disease [15]. In addition, gene expression is generally activated by the transcription factors that bind to the regulatory regions. Thus, the transcription factors that have enriched binding sites among coexpressed genes are potentially regulating the genes [16–18].

The activities of regulatory regions such as enhancers and promoters affect gene expression. Recent work has revealed the regulatory linkages between enhancers and genes for primary human cells and tissues using ChIP-seq data [19]. Moreover, the regulatory regions are also impacted by histone modifications [55]. Thus, several previous works applied the statistical models to reveal the predictive relationships between the RNA-seq gene expression levels and the ChIP-seq signals for transcription factors and histone modifications [20–22]. In addition, a probabilistic model-based method, TIP, was developed to integrate the ChIP-seq data on TF binding, histone modifications, and the RNA-seq data on gene expression to establish regulatory relationships [23]. In addition, several machine-learning approaches were developed to identify and construct the gene regulatory networks linking microRNAs and their target genes for humans and model organisms [20,22,24]. Also, the sparse Gaussian graphical model has been used to integrate the eQTLs and gene expression data to learn the regulatory networks [25].

27.5 GENE REGULATORY NETWORK ANALYSIS

Like engineering systems, gene regulatory networks, though very complex and large scale, have been found to be organized based on certain structures (e.g., modules, pathways, circuits), rather than randomly organized to maintain the basic conserved functions and to achieve tremendous biological diversity (e.g., across species). These network structures have also been reported to associate with various functions. For example, a network motif consisting of three genes was found to be a popular structure in gene regulatory network [26]. The different types of this motif structure can represent different regulatory circuits. Therefore, using the approaches from graph theory and network science, especially recent machine learning methods, we are able to systematically analyze the gene regulatory networks structures, compare them across multiple conditions, and find out how the structures are associated with various functions and phenotypes. In particular, the following topological features of gene regulatory networks have been extensively investigated. We also summarize the approaches to discover the structures.

Centrality—the nodes in gene regulatory networks are typically genes and gene regulatory factors controlling gene expression. For regulatory factors, they typically control specific target genes instead of random genes. One of the key properties of gene regulatory networks is that there exists a group of "hub" genes, playing central roles in the network. Network statistics such as "eccentricity" and "betweenness" are very useful to identify such hub genes and explain their network connectivity and behaviors [27,28]. For example, the genes with high degree centrality in gene coexpression networks have a variety of neighbors, suggesting that they are highly likely to be coregulated by similar mechanisms. The genes with high betweenness centrality mean that many regulatory pathways pass through them, implying that they play bridge roles in gene regulation such as indirect binding TFs [22]. These hub genes also make gene regulatory networks have scale free topologies; i.e., their nodes' degree of distribution follows powerlaw [29,30].

Module—the gene regulatory networks can also be subdivided into modules, which are a sub-network with enriched interactions. The genes in the same module have significantly more intra connections than interconnections between modules. The regulatory modules suggest that the modular genes share similar regulatory mechanisms (e.g., coregulation) [31]. A number of approaches have been developed to detect the gene modules in the regulatory networks; i.e., clustering [32]. For example, the widely used gene coexpression network clustering method, WGCNA, clusters the gene coexpression networks based on a nodes' topological similarity using hierarchical clustering [14]. In addition, we further developed a novel framework consisting of a cross-species multilayer network (OrthoClust) to analyze gene coexpression networks in an integrated fashion using orthologous genes across species, and found the cross-species conserved and species-specific modules [33].

Pathway—multiple genes can be regulated in a cascade fashion forming gene regulatory pathways. The pathway structures in gene regulatory networks have also been found to highly associate with biological functions. For example, recent studies reveal the distinct gene regulatory pathways for different cell types in human immunological systems [34]. Thus, advanced computational methods have been developed to identify such functional regulatory pathways, such as using gradient descent optimization [35].

Hierarchy—The directed networks have hierarchical structures representing the overall direction of the underlying information flow. Particularly, the gene regulatory networks are by nature directed and organized in a hierarchical manner to carry out biological functions. In particular, it was found that the hierarchy better reflects the importance of gene regulatory factors rather than node connectivity [20,22]. Also, comparing the degree of collaboration among the gene regulatory factors across different hierarchical levels, it has been found that in E. coli, yeast, and humans the highest degree of collaboration is between gene regulatory factors at the middle level, which is analogous to the important roles middle level managers play in managing a company [36]. Thus, we recently developed a computational approach based on simulated annealing to find the maximum hierarchy of gene regulatory networks [37].

27.6 GENE REGULATORY NETWORK MODELING

Gene regulatory networks have provided the wiring relationships among various molecules and genomic elements. For example, the specific network structures such as motifs and modules have been found to associate with the phenotypes. However, the network "wires" may activate or not at different conditions, just like switching on or off the lights. To further understand the gene regulatory mechanisms underlying the "wires," we need the sophisticated models for gene regulatory networks to reveal how they control the gene expression changes, i.e., model parameters [38]. In this section, we summarize a few typical models for gene regulatory networks as follows.

Logical modeling—Gene expression at a high level has two states: on for a high expression level and off for a low expression level. Thus, the gene regulatory networks controlling gene expression can be analogous to an electronic circuit because they both have inputs and outputs related by certain logic rules. For example, a variety of transcription factors (TFs) were found to follow logic gate operations to regulate target genes during embryonic development [39,40]. A logic gate is a discrete, high-level model that describes the relationship between Boolean input and output elements of the model. Mangan et al. [41] applied logic functions to study TF interactions in E. coli and S. cerevisiae, and found that the logic gate, though straightforward, is a very useful modeling for understanding gene regulatory cooperation. Compared to the continuous models, this model may not be able to capture the complex regulatory activities, but it is computationally efficient, especially when applied to large-scale gene regulatory networks. Therefore, using the regulatory combinatorics, a key design principle in electronics, we developed a computational method, Loregic, to characterize the gene regulatory cooperative logic, using logic gate models [42]. In particular, we validated this method with known yeast transcription factor knockout experiments, and applied it to discover how the gene regulatory cooperative logic behaves differently in human cancer.

Dynamic modeling—Gene expression is fundamentally a continuous dynamic process. Thus, the dynamic models such as differential equations have been used to understand the higher order of gene regulatory activities, in additional to binary states. For example, the ordinary differential equations were used to model the time-series gene expression dynamics to reveal the gene regulatory network [43]. The state-space model has been used to analyze gene expression dynamics [44,45]. It models the gene expression output of the gene regulatory network as a function of both the current internal system state and the external control signal. We also developed a novel computational method, DRIESS (Decomposition of gene Regulatory network into External and Internal components based on State Space models), which integrates a state space model and dimensionality reduction analysis for a given subsystem to identify the effects of other regulatory subsystems on its gene expression [46]. In particular, we applied this method to the developmental time-series gene expression datasets to show the capabilities of DREISS for studying the effects of different evolutionary gene regulatory subsystems on gene expression during the embryogenesis across distant species.

Machine learning modeling—Increasing the amount of gene expression and regulation datasets enables the machine learning modeling for gene regulatory networks [47]. In recent years, the advanced machine learning approaches have been developed to analyze large-scale genomic data and to predict the gene regulatory mechanisms for various phenotypes. For example, the deep learning approach was used to predict large-scale gene expression profiles in cancer [48]. The convolutional neural networks (CNNs) have been used to model the regulatory mechanism of genomic variants to reveal how they affect gene expression [49].

27.7 CONCLUSIONS AND FUTURE DIRECTIONS

In this chapter, we introduced the basic concepts of gene regulation and reviewed the signal processing and network science approaches applied to model, predict, and analyze the gene regulatory networks in this chapter. The specific structures of gene regulatory networks have been found to associate with various phenotypes, especially human diseases. In addition, the gene regulatory networks can change across both temporal and spatial dimensions. For example, during embryonic development, the different components of the gene regulatory network control developing various tissues, and only take effect at the specific time points and the locations where tissues need to develop. However, due to the costs, the experimental datasets remain very incomplete and noisy to systematically capture gene expression and regulation activities, especially at the spatial dimension (e.g., gene expression patterns leading to develop various tissues). Therefore, single cell technologies have been recently developed to detect the transcriptomic events at the individual cell level [50], attempting to find the spatial changes of gene regulation networks. The gene regulatory networks inferred using the single cell gene expression reveal the heterogeneity of gene regulation across different cell types, even from the same tissue [51,52]. In addition, the advanced technologies such as optogenetics also can capture in situ spatial transcriptomic activities, which provides the location information on gene expression [53]. Thus, one future direction for systems biology is how to integrate these novel data types and develop computational approaches to infer the accurate and complete gene regulatory networks that control the gene expression dynamic changes across both temporal and spatial domains.

In this chapter, we have also reviewed several typical methods to model the gene regulatory networks to reveal the regulatory mechanisms, such as logical circuit models. These models can be also used to impute the missing data, simulate the outcome of permutations to gene regulation, and eventually guide the engineering and design of gene regulatory circuits. For example, the synthetic biology approaches apply the engineering principles into gene regulatory networks and engineer the networks to change the biological functions. However, synthetic biology currently only focuses on the small-scale regulatory circuits [54]. Thus, another possible future direction is how to efficiently model and predict the controllable components in gene regulatory networks for synthetic biologists, aiming to change abnormal regulatory circuits such as in diseases.

REFERENCES

[1] Ideker T, Nussinov R. Network approaches and applications in biology. PLoS Comput Biol 2017;13: e1005771.

[2] Perdew GH, Vanden Heuvel JP, Peters JM. Regulation of gene expression: molecular mechanisms. Totowa, NJ: Humana Press; 2006.

[3] Maston GA, Evans SK, Green MR. Transcriptional regulatory elements in the human genome. Annu Rev Genomics Hum Genet 2006;7:29–59.

[4] Lee D, Gorkin DU, Baker M, Strober BJ, Asoni AL, McCallion AS, et al. A method to predict the impact of regulatory variants from DNA sequence. Nat Genet 2015;47:955–61.

[5] Cannell IG, Kong YW, Bushell M. How do microRNAs regulate gene expression? Biochem Soc Trans 2008;36:1224–31.

[6] Sultan M, Schulz MH, Richard H, Magen A, Klingenhoff A, Scherf M, et al. A global view of gene activity and alternative splicing by deep sequencing of the human transcriptome. Science 2008;321:956–60.

[7] Johnson LN. The regulation of protein phosphorylation. Biochem Soc Trans 2009;37:627–41.

[8] Goodwin S, McPherson JD, McCombie WR. Coming of age: ten years of next-generation sequencing technologies. Nat Rev Genet 2016;17:333–51.

[9] Heather JM, Chain B. The sequence of sequencers: the history of sequencing DNA. Genomics 2016;107:1–8.

[10] Conesa A, Madrigal P, Tarazona S, Gomez-Cabrero D, Cervera A, McPherson A, et al. A survey of best practices for RNA-seq data analysis. Genome Biol 2016;17:13.

[11] Kumar V, Muratani M, Rayan NA, Kraus P, Lufkin T, Ng HH, et al. Uniform, optimal signal processing of mapped deep-sequencing data. Nat Biotechnol 2013;31:615–22.

[12] Knijnenburg TA, Ramsey SA, Berman BP, Kennedy KA, Smit AF, Wessels LF, et al. Multiscale representation of genomic signals. Nat Methods 2014;11:689–94.

[13] Muir P, Li S, Lou S, Wang D, Spakowicz DJ, Salichos L, et al. The real cost of sequencing: scaling computation to keep pace with data generation. Genome Biol 2016;17:53.

[14] Langfelder P, Horvath S. WGCNA: an R package for weighted correlation network analysis. BMC Bioinformatics 2008;9:559.

[15] van Dam S, Vosa U, van der Graaf A, Franke L, de Magalhaes JP. Gene co-expression analysis for functional classification and gene-disease predictions. Brief Bioinform 2017. https://doi.org/10.1093/bib/bbw139.

[16] Marco A, Konikoff C, Karr TL, Kumar S. Relationship between gene co-expression and sharing of transcription factor binding sites in Drosophila melanogaster. Bioinformatics 2009;25:2473–7.

[17] Bar-Joseph Z, Gerber GK, Lee TI, Rinaldi NJ, Yoo JY, Robert F, et al. Computational discovery of gene modules and regulatory networks. Nat Biotechnol 2003;21:1337–42.
[18] McLean CY, Bristor D, Hiller M, Clarke SL, Schaar BT, Lowe CB, et al. GREAT improves functional interpretation of cis-regulatory regions. Nat Biotechnol 2010;28:495–501.
[19] Cao Q, Anyansi C, Hu X, Xu L, Xiong L, Tang W, et al. Reconstruction of enhancer-target networks in 935 samples of human primary cells, tissues and cell lines. Nat Genet 2017;49:1428–36.
[20] Gerstein MB, Lu ZJ, Van Nostrand EL, Cheng C, Arshinoff BI, Liu T, et al. Integrative analysis of the Caenorhabditis elegans genome by the modENCODE project. Science 2010;330:1775–87.
[21] Cheng C, Alexander R, Min R, Leng J, Yip KY, Rozowsky J, et al. Understanding transcriptional regulation by integrative analysis of transcription factor binding data. Genome Res 2012;22:1658–67.
[22] Gerstein MB, Kundaje A, Hariharan M, Landt SG, Yan K-K, Cheng C, Mu XJ, et al. Architecture of the human regulatory network derived from ENCODE data. Nature 2012;489:91–100.
[23] Cheng C, Min R, Gerstein M. TIP: a probabilistic method for identifying transcription factor target genes from chip-seq binding profiles. Bioinformatics 2011;27:3221–7.
[24] Negre N, Brown CD, Ma L, Bristow CA, Miller SW, Wagner U, et al. A cis-regulatory map of the Drosophila genome. Nature 2011;471:527–31.
[25] Zhang L, Kim S. Learning gene networks under SNP perturbations using eQTL datasets. PLoS Comput Biol 2014;10:e1003420.
[26] Milo R, Shen-Orr S, Itzkovitz S, Kashtan N, Chklovskii D, Alon U. Network motifs: simple building blocks of complex networks. Science 2002;298:824–7.
[27] Yu H, Jansen R, Stolovitzky G, Gerstein M. Total ancestry measure: quantifying the similarity in tree-like classification, with genomic applications. Bioinformatics 2007;23:2163–73.
[28] Yu H, Kim PM, Sprecher E, Trifonov V, Gerstein M. The importance of bottlenecks in protein networks: correlation with gene essentiality and expression dynamics. PLoS Comput Biol 2007;3:e59.
[29] Nicolau M, Schoenauer M. On the evolution of scale-free topologies with a gene regulatory network model. Biosystems 2009;98:137–48.
[30] Albert R. Scale-free networks in cell biology. J Cell Sci 2005;118:4947–57.
[31] Segal E, Shapira M, Regev A, Pe'er D, Botstein D, Koller D, et al. Module networks: identifying regulatory modules and their condition-specific regulators from gene expression data. Nat Genet 2003;34:166–76.
[32] Wiwie C, Baumbach J, Rottger R. Comparing the performance of biomedical clustering methods. Nat Methods 2015;12:1033–8.
[33] Yan KK, Wang D, Rozowsky J, Zheng H, Cheng C, Gerstein M. OrthoClust: an orthology-based network framework for clustering data across multiple species. Genome Biol 2014;15:R100.
[34] Koues OI, Collins PL, Cella M, Robinette ML, Porter SI, Pyfrom SC, et al. Distinct gene regulatory pathways for human innate versus adaptive lymphoid cells. Cell 2016;165:1134–46.
[35] Das M, Mukhopadhyay S, De RK. Gradient descent optimization in gene regulatory pathways. PLoS One 2010;5:e12475.
[36] Yan KK, Fang G, Bhardwaj N, Alexander RP, Gerstein M. Comparing genomes to computer operating systems in terms of the topology and evolution of their regulatory control networks. Proc Natl Acad Sci U S A 2010;107:9186–91.
[37] Cheng C, Andrews E, Yan KK, Ung M, Wang D, Gerstein M. An approach for determining and measuring network hierarchy applied to comparing the phosphorylome and the regulome. Genome Biol 2015;16:63.
[38] Karlebach G, Shamir R. Modelling and analysis of gene regulatory networks. Nat Rev Mol Cell Biol 2008;9:770–80.
[39] Peter IS, Faure E, Davidson EH. Predictive computation of genomic logic processing functions in embryonic development. Proc Natl Acad Sci U S A 2012;109:16434–42.
[40] Peter IS, Davidson EH. Evolution of gene regulatory networks controlling body plan development. Cell 2011;144:970–85.

[41] Mangan S, Alon U. Structure and function of the feed-forward loop network motif. Proc Natl Acad Sci U S A 2003; 100:11980–5.
[42] Wang D, Yan KK, Sisu C, Cheng C, Rozowsky J, Meyerson W, et al. Loregic: a method to characterize the cooperative logic of regulatory factors. PLoS Comput Biol 2015;11:e1004132.
[43] Du P, Gong H, Wurtele ES, Dickerson JA. Modeling gene expression networks using fuzzy logic. IEEE Trans Syst Man Cybern B: Cybern 2005;35:1351–9.
[44] Bansal M, Della Gatta G, di Bernardo D. Inference of gene regulatory networks and compound mode of action from time course gene expression profiles. Bioinformatics 2006;22:815–22.
[45] Huang S, Ingber DE. A non-genetic basis for cancer progression and metastasis: self-organizing attractors in cell regulatory networks. Breast Dis 2006;26:27–54.
[46] Wang D, He F, Maslov S, Gerstein M. DREISS: using state-space models to infer the dynamics of gene expression driven by external and internal regulatory networks. PLoS Comput Biol 2016;12:e1005146.
[47] Angermueller C, Parnamaa T, Parts L, Stegle O. Deep learning for computational biology. Mol Syst Biol 2016;12:878.
[48] Chen Y, Li Y, Narayan R, Subramanian A, Xie X. Gene expression inference with deep learning. Bioinformatics 2016; 32:1832–9.
[49] Kelley DR, Snoek J, Rinn JL. Basset: learning the regulatory code of the accessible genome with deep convolutional neural networks. Genome Res 2016;26:990–9.
[50] Stegle O, Teichmann SA, Marioni JC. Computational and analytical challenges in single-cell transcriptomics. Nat Rev Genet 2015;16:133–45.
[51] Chan TE, Stumpf MPH, Babtie AC. Gene regulatory network inference from single-cell data using multivariate information measures. Cell Syst 2017;5:251–67, e253.
[52] Aibar S, Gonzalez-Blas CB, Moerman T, Huynh-Thu VA, Imrichova H, Hulselmans G, et al. SCENIC: single-cell regulatory network inference and clustering. Nat Methods 2017;14:1083–6.
[53] Marx V. Neurobiology: gene expression captured on-site. Nat Methods 2017;14:1037–40.
[54] Andrianantoandro E, Basu S, Karig DK, Weiss R. Synthetic biology: new engineering rules for an emerging discipline. Mol Syst Biol 2006;2:2006.0028.
[55] Bannister AJ, Kouzarides T. Regulation of chromatin by histone modifications. Cell Res 2011;21(3):381–95.

CHAPTER 28

DIFFUSION AUGMENTED COMPLEX EXTENDED KALMAN FILTERING FOR ADAPTIVE FREQUENCY ESTIMATION IN DISTRIBUTED POWER NETWORKS

Yili Xia*, **Sithan Kanna**†, **Danilo P. Mandic**†

*School of Information Science and Engineering, Southeast University, Nanjing, PR China** *Department of Electrical and Electronic Engineering, Imperial College London, London, United Kingdom*†

28.1 INTRODUCTION

Modern electricity grids are undergoing unprecedented changes to meet several ambitious goals which include the need to reduce our reliance on fossil fuels and better management of the current aging infrastructure while creating economic benefits for society [1]. It is widely expected that these challenges are to be met by the next generation power network, commonly known as the "smart grid." The smart grid aims to make a paradigm shift from the top-down mode of operation of the current grid by incorporating distributed generation, typically from renewable sources (wind, solar), distributed storage from plug-in-vehicles, and demand-response technologies [1]. The backbone of this future grid is an intelligent, real-time, wide-area monitoring system capable of estimating key indicators of the stability of the grid [2].

One of the most closely monitored indicators is the system frequency as its deviation from the nominal frequency of 50 Hz or 60 Hz indicates a mismatch between the supply and demand in electricity or a fault. Because electricity is a nonstorable commodity, its generation and demand have to be balanced in real time. If demand is greater than generation, the frequency drops, whereas if generation is greater than demand, the frequency rises. Therefore, accurate tracking of the system frequency is a prerequisite for the task of balancing the demand and generation of electricity [3].

In three-phase power systems, none of the single phases can faithfully characterize the whole system and its properties. Therefore, a robust frequency estimator should take into account the information of all three phases, to enable a unified estimation of system frequency as a whole and provide enhanced robustness. To this end, Clarke's $\alpha\beta$ transformation has been introduced to construct a complex-valued signal with the information provided by all the three-phase voltages in a simultaneous

Cooperative and Graph Signal Processing. https://doi.org/10.1016/B978-0-12-813677-5.00028-6

way. This has equipped the classical single-phase methods with more robustness in characterizing system frequency [4]. Currently, used frequency estimation techniques along this direction include: (i) Fourier transform approaches [5,6], (ii) gradient descent and least mean square adaptive estimation [7,8], and (iii) state space methods and Kalman filters [9]. These explicitly or implicitly assume balanced conditions (equal voltage amplitudes and equally spaced phases), and would be adequate for the smart grid if the dynamics of the grid remain within the status quo [3].

However, the key challenge for frequency estimation in the future grid is the large penetration of intermittent renewable resources (wind, solar), which is expected to shorten the frequency regulation time scales [10] and degrade the quality of the voltage signals used by conventional control and estimation algorithms [11]. First, the frequency regulation time scales are expected to be shortened in future grids because replacing a large fraction of conventional generators by renewable energy sources is expected to reduce the overall inertia available in the grid [12], thus resulting in more extreme frequency excursions, necessitating a rapid response from transmission system operators [12]. The second challenge is concerned with the degradation of the signal quality (voltage and current measurements) used in estimation algorithms. This is attributed to the diverse profile of users who are plugging into and out of the grid at a more rapid pace [13]. In addition, the increased use of electronic inverter-based equipment, e.g., consumer electronics, introduces switching noise and uneven distribution of single-phase loads [11], adverse effects which are not well represented by algorithms derived in standard linear estimation theory [14,15].

Our earlier work showed that the complex-valued $\alpha\beta$ voltage admits a widely linear autoregressive (WLAR) model [16–19] under both balanced and unbalanced system conditions. It has been justified that standard, strictly linear, complex-valued estimators applied to this $\alpha\beta$ voltage introduce biased and oscillatory frequency estimates for unbalanced system conditions. Indeed, widely linear estimators (also known as "augmented" estimators) are able to provide optimal and consistent estimates of the system frequency over a range of operating conditions [15,20,21]. Therefore, the aim of this work is to extend the single-node, widely linear, state space frequency estimators in [21–24] to the distributed scenario to suit the requirements of the smart grid. Distributed estimation has already found various applications in both military and civilian scenarios [25–28], as cooperation between the nodes (sensors) provides more accurate and robust estimation over independent nodes while approaching the performance of centralized systems at much reduced communication overhead. Recent distributed approaches include diffusion least-mean-square estimation [29,30] and Kalman filtering [26,28,31]; however, these consider noisy measurements without cross-nodal correlations, which is not realistic in real-world power systems.

To this end, we propose the diffusion augmented complex extended Kalman filter (D-ACEKF) by adapting the diffusion scheme in [32] and the Kalman filtering models in [21,33] to suit a class of real-world problems where the system frequency is identical over a certain geographical area at the distribution level while the voltage imbalances and cross-nodal correlation can be different. In particular, extending the widely linear frequency estimators to the distributed case is nontrivial as the states that are to be estimated are not identical in the power network, as required by the classical diffusion scheme [32].

Notation: Lowercase letters are used to denote scalars, a, boldface letters for column vectors, $\mathbf{a}$, and boldface uppercase letters for matrices, $\mathbf{A}$. The symbols $(\cdot)^T$, $(\cdot)^*$ and $(\cdot)^H$ are respectively the transpose, complex conjugate, and Hermitian transpose operators. The symbol $\mathbb{E}\{\cdot\}$ denotes the statistical expectation operator while $\mathrm{Re}\{\cdot\}$ and $\mathrm{Im}\{\cdot\}$ are, respectively, the real and imaginary parts of a complex variable, and $j = \sqrt{-1}$.

28.2 PROBLEM FORMULATION

In order to illuminate the challenges that future low inertia grids impose on frequency estimation and tracking, we first consider a general electricity grid represented by a undirected graph $\mathcal{G} = (\mathcal{N}, \mathcal{E})$. The N buses in the network are represented by a node set, $\mathcal{N} = \{1, 2, \ldots, N\}$, while the power lines between the buses (connections) are represented by an edge set, $\mathcal{E}$. The neighborhood of a node i, denoted by $\mathcal{N}_i$, comprises all the nodes connected to node i including itself, that is $\mathcal{N}_i = \{j \,|\, (i,j) \in \mathcal{E}\}$ [34], see Fig. 28.1. Each node (e.g., node i) has access to sampled three-phase voltage measurements, at the discrete time instant k, to give the state vector

$$\mathbf{s}_{i,k} = \begin{bmatrix} v_{a,i,k} \\ v_{b,i,k} \\ v_{c,i,k} \end{bmatrix} = \begin{bmatrix} V_{a,i}\cos(\omega k + \phi_{a,i}) \\ V_{b,i}\cos\left(\omega k + \phi_{b,i} - \frac{2\pi}{3}\right) \\ V_{c,i}\cos\left(\omega k + \phi_{c,i} + \frac{2\pi}{3}\right) \end{bmatrix}, \tag{28.1}$$

where the amplitudes of the phase voltages $v_{a,i,k}$, $v_{b,i,k}$, $v_{c,i,k}$, are $V_{a,i}$, $V_{b,i}$, $V_{c,i}$ while the corresponding phase values are denoted by $\phi_{a,i}$, $\phi_{b,i}$, $\phi_{c,i}$. The angular frequency is $\omega = 2\pi f T$, with f the fundamental power system frequency, which is identical throughout the network, and T is the sampling interval.

The three-phase representation of the $\mathbf{s}_{i,k}$ in Eq. (28.1) is overparametrized and can be compactly represented as "two-phase" Clarke voltages, $v_{\alpha,i,k}$ and $v_{\beta,i,k}$, via a projection onto a new orthogonal basis using the so-called "Clarke transform," given by [4]

$$\begin{bmatrix} v_{\alpha,i,k} \\ v_{\beta,i,k} \end{bmatrix} \stackrel{\text{def}}{=} \underbrace{\sqrt{\frac{2}{3}}\begin{bmatrix} 1 & -\frac{1}{2} & -\frac{1}{2} \\ 0 & \frac{\sqrt{3}}{2} & -\frac{\sqrt{3}}{2} \end{bmatrix}}_{\text{Clarke matrix}} \begin{bmatrix} v_{a,i,k} \\ v_{b,i,k} \\ v_{c,i,k} \end{bmatrix}. \tag{28.2}$$

Moreover, the Clarke transform enables, $v_{\alpha,i,k}$ and $v_{\beta,i,k}$, to be conveniently represented jointly as a complex-valued scalar,

$$s_{i,k} \stackrel{\text{def}}{=} v_{\alpha,i,k} + j v_{\beta,i,k}. \tag{28.3}$$

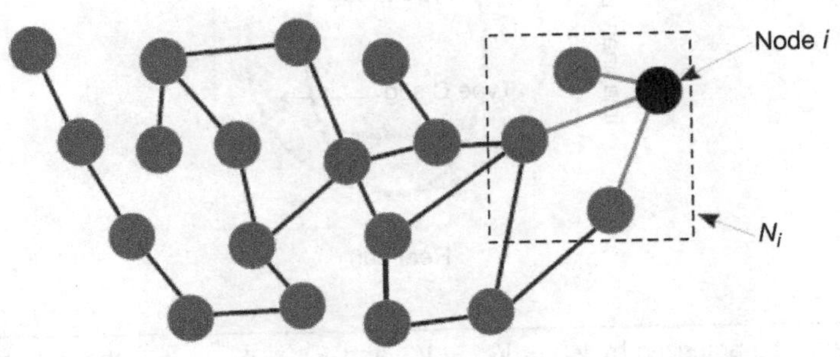

FIG. 28.1

A distributed network with $N = 20$ nodes.

It can be shown that the complex-valued Clarke voltage (also referred to as the $\alpha\beta$ voltage) in Eq. (28.3) takes the form

$$s_{i,k} = A_i e^{j\omega k} + B_i e^{-j\omega k}, \tag{28.4}$$

where the positive and negative sequence phasors A_i and B_i are given by [20,35]

$$A_i = \frac{\sqrt{6}}{6}\left[V_{a,i}e^{j\phi_{a,i}} + V_{b,i}e^{j\phi_{b,i}} + V_{c,i}e^{j\phi_{c,i}}\right], \tag{28.5a}$$

$$B_i = \frac{\sqrt{6}}{6}\left[V_{a,i}e^{-j\phi_{a,i}} + V_{b,i}e^{-j\left(\phi_{b,i}+\frac{2\pi}{3}\right)} + V_{c,i}e^{-j\left(\phi_{c,i}-\frac{2\pi}{3}\right)}\right]. \tag{28.5b}$$

Our task is to estimate the system frequency, ω, given noisy observations of the Clarke voltage $s_{i,k}$ in Eq. (28.4). In particular, the noisy observations of $s_{i,k}$ can be expressed in the form

$$y_{i,k} = s_{i,k} + \eta_{i,k}, \tag{28.6}$$

where $\eta_{i,k}$, is a zero-mean, complex-valued, white Gaussian noise process with variance $\sigma^2_{\eta_i} = \mathbb{E}\left\{|\eta_{i,k}|^2\right\}$. Notice from the noise-free signal in Eq. (28.5b), that the negative sequence phasor vanishes, that is $B_i = 0$, for a balanced system (equal amplitudes, $V_{a,i} = V_{b,i} = V_{c,i}$, and uniform phase separation, $\phi_{a,i} = \phi_{b,i} = \phi_{c,i}$), thus yielding the balanced Clarke voltage

$$s_{i,k} = A_i e^{j\omega k}. \tag{28.7}$$

Fig. 28.2 depicts the real and imaginary parts of the balanced signal in Eq. (28.7), which exhibits a circular trajectory. However, for unbalanced systems, $B_i \neq 0$, and therefore $s_{i,k}$ does not follow a uniform circular trajectory. During a fault in the voltage lines, the currents and voltages across the three phases fail to maintain uniformity and can either experience a sag (undervoltage) or a swell (overvoltage). These sags and swells are classified into distinct categories denoted by alphabet symbols

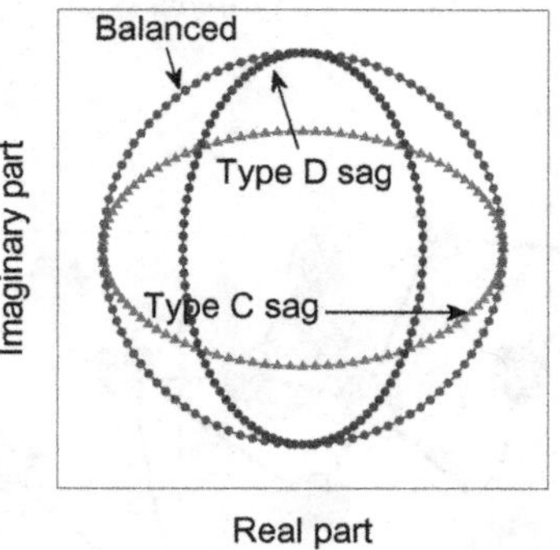

FIG. 28.2

For a balanced system, characterized by $V_{a,i} = V_{b,i} = V_{c,i}$ and $\phi_{a,i} = \phi_{b,i} = \phi_{c,i}$, the trajectory of Clarke's voltage v_k is circular. For unbalanced systems, e.g., with Type C and Type D voltage sags, the voltage trajectories are noncircular.

Table 28.1 Voltage Sags and Their Phasor Representations

Voltage Sag	$\bar{V}_a$	$\bar{V}_b$	$\bar{V}_c$
Type C	1	$-\frac{1}{2}-j\frac{\sqrt{3}\gamma}{2}$	$-\frac{1}{2}+j\frac{\sqrt{3}\gamma}{2}$
Type D	γ	$-\frac{\gamma}{2}-j\frac{\sqrt{3}}{2}$	$-\frac{\gamma}{2}+j\frac{\sqrt{3}}{2}$

from A to G [36]. For example, a power system with Type C and Type D voltage sags, whose three-phase voltage characteristics are provided in Table 28.1, exhibits noncircular Clarke voltage trajectories as explained by Eq. (28.4).

An unbalanced system condition therefore introduces a problem, as conventional complex-valued linear estimation theory does not cater for noncircular signals. In fact, it was recently shown that the standard strictly linear model for Eq. (28.4) is inadequate for unbalanced systems and a widely linear model is required [16,20,35]. Furthermore, the multisensor nature of the estimation problem in Eq. (28.6) calls for the extension of standard single-node frequency estimators to a distributed setting. In the particular application of frequency tracking in low-inertia grids, the distributed estimation algorithms are crucial to exploit spatially diverse measurements across the grid so as to compensate for the fewer temporal measurements available in the first few hundred milliseconds after a contingency. Although spatially diverse measurements, $y_{i,k}$, from Eq. (28.6) can be collected and processed in a fusion center, given the growth in the number of measurements from phasor measurements units (PMU) and smart meters, this makes the network vulnerable to a single point of failure [37,38].

28.3 WIDELY LINEAR FREQUENCY ESTIMATORS AND THEIR NONLINEAR STATE SPACE MODELS

Because the $\alpha\beta$ voltage in Eq. (28.4) can be interpreted as a sum of two phasors, one rotating clockwise (positive sequence) and the other rotating counter clockwise (negative sequence), at the same frequency, it is only natural and intuitive to consider an estimate of $s_{i,k}$, that is, $\hat{s}_{i,k}$, based on the previous value $s_{i,k-1}$ and its conjugate $s^*_{i,k-1}$ (where the conjugate represents the phasor rotating in the opposite direction) within a widely linear autoregressive (WLAR) model [16–19], given by

$$\begin{aligned}\hat{s}_{i,k} &= h^*_{i,k-1}s_{i,k-1} + g^*_{i,k-1}s^*_{i,k-1}\\ &= h^*_{i,k-1}\left(A_i e^{j\omega(k-1)} + B_i e^{-j\omega(k-1)}\right) + g^*_{i,k-1}\left(A^*_i e^{-j\omega(k-1)} + B^*_k e^{j\omega(k-1)}\right)\\ &= \left(h^*_{i,k-1}A_i + g^*_{i,k-1}B^*_i\right)e^{j\omega(k-1)} + \left(h^*_{i,k-1}B_i + g^*_{i,k-1}A^*_i\right)e^{-j\omega(k-1)},\end{aligned} \tag{28.8}$$

where $h_{i,k-1}$ and $g_{i,k-1}$ are the weight coefficients. A comparison of Eq. (28.8) to the state signal $s_{i,k}$ in Eq. (28.4), which is given by

$$s_{i,k} = \left(e^{j\omega}A_i\right)e^{j\omega(k-1)} + \left(e^{-j\omega}B_i\right)e^{-j\omega(k-1)}, \tag{28.9}$$

yields

$$e^{j\omega}A_i = h^*_{i,k-1}A_i + g^*_{i,k-1}B^*_i \quad \text{and} \quad e^{-j\omega}B_i = h^*_{i,k-1}B_i + g^*_{i,k-1}A^*_i. \tag{28.10}$$

Solving the simultaneous equations in Eq. (28.10) gives a widely linear estimate of the system frequency in the form [15,20,21]

$$\hat{\omega}_{i,k} = \tan^{-1}\left[\frac{\sqrt{\text{Im}^2\{h_{i,k-1}\} - |g_{i,k-1}|^2}}{\text{Re}\{h_{i,k-1}\}}\right]. \tag{28.11}$$

The signal model in Eq. (28.8) and its corresponding frequency estimate in Eq. (28.11) shall be referred to as the widely linear AR-I (WLAR-I) model. Note that when the system is balanced ($B_i = 0$), the coefficient $g_{i,k-1} = 0$ and the widely linear frequency estimate in Eq. (28.11) is identical to its strictly linear counterpart in [8].

Another similar widely linear frequency estimator was proposed independently in [22–24], which considers the fact that the positive and negative sequence phasors rotate in opposite directions but at an identical angular velocity ω, so that

$$s_{i,k} = e^{j\omega}\underbrace{\left(A_i e^{j\omega(k-1)}\right)}_{\overset{\text{def}}{=}\nu^+_{i,k}} + e^{-j\omega}\underbrace{\left(B_i e^{-j\omega(k-1)}\right)}_{\overset{\text{def}}{=}\nu^-_{i,k}}. \tag{28.12}$$

Both phasors, $\nu^+_{i,k}$ and $\nu^-_{i,k}$, independently obey a linear evolution process where their phases are incremented by ω at every time instant, that is,

$$\nu^+_{i,k} = e^{j\omega}\nu^+_{i,k-1}, \quad \nu^-_{i,k} = e^{-j\omega}\nu^-_{i,k-1}. \tag{28.13}$$

In this way, a weight coefficient $h_{i,k-1}$ can be used to estimate $s_{i,k}$ and the frequency ω, to yield

$$\hat{\nu}^+_{i,k} = h_{i,k-1}\hat{\nu}^+_{i,k-1}, \quad \hat{\nu}^-_{i,k} = h^*_{i,k-1}\hat{\nu}^-_{i,k-1}, \quad \hat{s}_{i,k} = h_{i,k-1}\hat{\nu}^+_{i,k} + h^*_{i,k-1}\hat{\nu}^-_{i,k}, \tag{28.14}$$

$$\hat{\omega}_{i,k} = \tan^{-1}\left[\frac{\text{Im}\{h_{i,k-1}\}}{\text{Re}\{h_{i,k-1}\}}\right]. \tag{28.15}$$

We refer to the model in Eq. (28.14) as the widely linear AR-model-II (WLAR-II), due to its resemblance to the intuition of the widely linear model in Eq. (28.8). However, it offers a simpler framework as compared with the WLAR-I in Eq. (28.8) because a single weight coefficient $h_{i,k-1}$ is involved. However, the WLAR-I provides the advantage in terms of physical interpretability, as the coefficient $g_{i,k-1}$ represents the negative sequence, which characterizes the imbalance of the complex $\alpha\beta$ voltage [15,20].

Nonlinear state space models have been investigated for frequency tracking tasks mainly through nonlinear Kalman filters [39]. This is due to the fact that nonlinear models provide greater modeling flexibility and may have near optimal performance under noisy scenarios. For generality, at each node i, consider a nonlinear state space evolution given by

$$\begin{aligned} \boldsymbol{w}_{i,k} &= \boldsymbol{f}_{i,k-1}(\boldsymbol{w}_{i,k-1}) + \boldsymbol{q}_{i,k}, \\ y_{i,k} &= \varphi_{i,k}(\boldsymbol{w}_{i,k}) + \eta_{i,k}, \end{aligned} \tag{28.16}$$

where $y_{i,k} \in \mathbb{C}$ is the observation (or measurement) of the state vector $\boldsymbol{w}_{i,k} \in \mathbb{C}^{M\times 1}$ through a known nonlinear function $\varphi_{i,k}(\cdot)$. The state vector $\boldsymbol{w}_{i,k}$ is also time varying with a known state transition function $\boldsymbol{f}_{i,k}(\cdot)$. The zero-mean white Gaussian observation noise $\eta_{i,k}$ is independent of the white Gaussian state noise vector $\boldsymbol{q}_{i,k}$. The nonlinear function $\varphi_{i,k}(\cdot)$ caters for a wide-range of models considered in frequency estimation problems, and the state $\boldsymbol{w}_{i,k}$ can be estimated/tracked using extended Kalman filters. To achieve this, the first step is to unify the considered widely linear AR models under the umbrella of the nonlinear state space formulation in Eq. (28.16). Let us consider the WLAR-I model in Eq. (28.8), whereby the signal $s_{i,k}$ and the weight coefficients $h^*_{i,k}$ and $g^*_{i,k}$ are represented as elements in the state vector $\boldsymbol{w}_{i,k} = [h^*_{i,k}, g^*_{i,k}, s_{i,k}]^T$. This admits the nonlinear state transition model given by [21]

$$\boldsymbol{w}_{i,k} = \begin{bmatrix} h^*_{i,k} \\ g^*_{i,k} \\ s_{i,k} \end{bmatrix} = \underbrace{\begin{bmatrix} 1 & 0 & 0 \\ 0 & 1 & 0 \\ s_{i,k-1} & s^*_{i,k-1} & 0 \end{bmatrix} \begin{bmatrix} h^*_{i,k-1} \\ g^*_{i,k-1} \\ s_{i,k-1} \end{bmatrix}}_{\boldsymbol{f}_{i,k-1}(\boldsymbol{w}_{i,k-1})} + \boldsymbol{q}_{i,k}, \quad y_{i,k} = \underbrace{\begin{bmatrix} 0 & 0 & 1 \end{bmatrix} \begin{bmatrix} h^*_{i,k} \\ g^*_{i,k} \\ s_{i,k} \end{bmatrix}}_{\varphi_{i,k}(\boldsymbol{w}_{i,k})} + \eta_{i,k}. \tag{28.17}$$

In a similar fashion, the process in Eq. (28.14) can be configured within a nonlinear state space model whereby the state vector contains the WLAR-II coefficients, i.e., $\boldsymbol{w}_{i,k} = [h_{i,k}, \nu^+_{i,k}, \nu^-_{i,k}]^T$, in the form

$$\boldsymbol{w}_{i,k} = \begin{bmatrix} h_{i,k} \\ \nu^+_{i,k} \\ \nu^-_{i,k} \end{bmatrix} = \underbrace{\begin{bmatrix} 1 & 0 & 0 \\ 0 & h_{i,k-1} & 0 \\ 0 & 0 & h^*_{i,k-1} \end{bmatrix} \begin{bmatrix} h_{i,k-1} \\ \nu^+_{i,k-1} \\ \nu^-_{i,k-1} \end{bmatrix}}_{\boldsymbol{f}_{i,k-1}(\boldsymbol{w}_{i,k-1})} + \boldsymbol{q}_{i,k}, \quad y_{i,k} = \underbrace{\begin{bmatrix} 0 & 1 & 1 \end{bmatrix} \begin{bmatrix} h_{i,k} \\ \nu^+_{i,k} \\ \nu^-_{i,k} \end{bmatrix}}_{\varphi_{i,k}(\boldsymbol{w}_{i,k})} + \eta_{i,k}. \tag{28.18}$$

It is important to observe that the nonlinear state space models in Eqs. (28.17) and (28.18), which in essence employ the linear prediction idea, do not use the signal $y_{i,k}$ and its past sample $y_{i,k-1}$ to predict the frequency. Instead the signal $y_{i,k}$ is only used in the observation function while the past sample $y_{i,k-1}$ is also estimated as a nuisance state.

28.4 DIFFUSION AUGMENTED COMPLEX EXTENDED KALMAN FILTERS

Early work in the field of distributed Kalman filtering focused on decentralizing the Kalman filtering operations using individual agents that communicate to a fusion center [40]. This method can be regarded as the centralized Kalman filter because a central fusion center has access to all the information in the network. Because communicating to a single fusion center makes the network vulnerable to a single point of failure, more decentralized solutions were proposed. Fully distributed Kalman filters then began to emerge where each node was required to share all its information with every other node in the network [41,42], i.e., effectively replicating the operation of the centralized Kalman filter at each node in the network [41].

More general distributed Kalman filters were proposed in the consensus estimation framework, where the constraint that nodes communicate with every other node in the network was relaxed [28,43]. To compensate for the fact that nodes only access the measurements in their neighborhood, consensus Kalman filters include a consensus step for their state estimates whereby the individual nodes exchange and average their intermediate state estimates with their neighbors several times before the next measurement is obtained [44]. Therefore, the consensus filters operated at two time scales, a longer time scale for the measurement updates and a shorter time scale for the consensus update. The consensus protocol is therefore unsuitable for problems where measurements are taken in a similar time scale as the communication protocol.

The diffusion Kalman filter, proposed in [26], is based on a wider class of diffusion adaptive algorithms [32], and enables both the measurement update and information fusion throughout the network to be applied in a single time scale. Furthermore, it was shown that diffusion strategies enable information to diffuse more thoroughly in the network compared to consensus strategies [45]. A fundamental feature in the diffusion strategy is that only the state estimates, together with observation variables, are shared in the network.

However, traditional diffusion strategies only account for strictly linear models with circular noise processes. To cater for widely linear models, the diffusion augmented Kalman filter (D-ACKF) was proposed in [46]. In this section, we extend the D-ACKF for nonlinear models using the theory of extended Kalman filtering and refer to the proposed algorithm as the diffusion augmented complex extended Kalman filter (D-ACEKF).

28.4.1 SINGLE-NODE COMPLEX KALMAN FILTERS

Consider again Fig. 28.1, where, at each time instant, k, the node i is tasked to estimate a parameter (or state) vector $\boldsymbol{w}_k \in \mathbb{C}^{M\times 1}$, which is assumed to be identical throughout the network but observed locally through measurements $\boldsymbol{y}_{i,k} \in \mathbb{C}^{L\times 1}$. The measurements and state are coupled via a state space model given by

$$\begin{aligned} \boldsymbol{w}_k &= \mathbf{F}_k \boldsymbol{w}_{k-1} + \boldsymbol{q}_k, \\ \boldsymbol{y}_{i,k} &= \mathbf{H}_{i,k}\boldsymbol{w}_k + \boldsymbol{\eta}_{i,k}, \end{aligned} \tag{28.19}$$

where $\mathbf{F}_k \in \mathbb{C}^{M\times M}$ is the state transition matrix and $\mathbf{H}_{i,k} \in \mathbb{C}^{L\times M}$ is the observation matrix. In standard complex-valued Kalman filtering literature, the state noise $\boldsymbol{q}_k \in \mathbb{C}^{M\times 1}$ and observation noise $\boldsymbol{\eta}_{i,k} \in \mathbb{C}^{L\times 1}$ are temporally uncorrelated and spatially independent zero-mean white Gaussian noise processes with a joint covariance and pseudocovariance matrices are defined as [26]

$$\mathbf{C}_k \overset{\text{def}}{=} \mathbb{E}\left\{\boldsymbol{q}_k\boldsymbol{q}_k^{\mathrm{H}}\right\}, \quad \tilde{\mathbf{C}}_k \overset{\text{def}}{=} \mathbb{E}\left\{\boldsymbol{q}_k\boldsymbol{q}_k^{\mathrm{T}}\right\}, \quad \boldsymbol{\Sigma}_{i,k} \overset{\text{def}}{=} \mathbb{E}\left\{\boldsymbol{\eta}_{i,k}\boldsymbol{\eta}_{i,k}^{\mathrm{H}}\right\}, \quad \tilde{\boldsymbol{\Sigma}}_{i,k} \overset{\text{def}}{=} \mathbb{E}\left\{\boldsymbol{\eta}_{i,k}\boldsymbol{\eta}_{i,k}^{\mathrm{T}}\right\}, \tag{28.20a}$$

$$\mathbb{E}\left\{\begin{bmatrix}\boldsymbol{q}_k\\ \boldsymbol{\eta}_{i,k}\end{bmatrix}\begin{bmatrix}\boldsymbol{q}_n^{\mathrm{H}} & \boldsymbol{\eta}_{\ell,n}^{\mathrm{H}}\end{bmatrix}\right\} = \begin{bmatrix}\mathbf{C}_k & \mathbf{0}\\ \mathbf{0} & \boldsymbol{\Sigma}_{i,k}\delta_{i,\ell}\end{bmatrix}\delta_{k,n}, \tag{28.20b}$$

$$\mathbb{E}\left\{\begin{bmatrix}\boldsymbol{q}_k\\ \boldsymbol{\eta}_{i,k}\end{bmatrix}\begin{bmatrix}\boldsymbol{q}_n^{\mathrm{T}} & \boldsymbol{\eta}_{\ell,n}^{\mathrm{T}}\end{bmatrix}\right\} = \begin{bmatrix}\tilde{\mathbf{C}}_k & \mathbf{0}\\ \mathbf{0} & \tilde{\boldsymbol{\Sigma}}_{i,k}\delta_{i,\ell}\end{bmatrix}\delta_{k,n}, \tag{28.20c}$$

where $\delta_{k,n}$ and $\delta_{i,\ell}$ are the Kronecker delta functions that satisfy

$$\delta_{a,b} = \begin{cases} 1, & \text{if } a = b, \\ 0, & \text{otherwise.} \end{cases}$$

If no collaboration is allowed, each node is able to perform its own estimation scheme with the Kalman filter given by [33]

$$\textit{Predict}: \begin{cases} \hat{\mathbf{w}}_{i,k|k-1} &= \mathbf{F}_k \hat{\mathbf{w}}_{i,k-1|k-1} \\ \mathbf{M}_{i,k|k-1} &= \mathbf{F}_k \mathbf{M}_{i,k-1|k-1} \mathbf{F}_k^{\mathrm{H}} + \mathbf{C}_{i,k} \end{cases} \tag{28.21a}$$

$$\textit{Update}: \begin{cases} \mathbf{M}_{i,k|k}^{-1} &= \mathbf{M}_{i,k|k-1}^{-1} + \mathbf{H}_{i,k}^{\mathrm{H}} \boldsymbol{\Sigma}_{i,k}^{-1} \mathbf{H}_{i,k} \\ \hat{\mathbf{w}}_{i,k|k} &= \hat{\mathbf{w}}_{i,k|k-1} + \mathbf{M}_{i,k|k} \mathbf{H}_{i,k}^{\mathrm{H}} \boldsymbol{\Sigma}_{i,k}^{-1} \left[\mathbf{y}_{i,k} - \mathbf{H}_{i,k} \hat{\mathbf{w}}_{i,k|k-1} \right], \end{cases} \tag{28.21b}$$

where the state estimate $\hat{\mathbf{w}}_{i,k|k}$ and state error covariance matrix $\mathbf{M}_{i,k|k}$ at each node are initialized as

$$\hat{\mathbf{w}}_{i,0|0} = \mathbb{E}\{\mathbf{w}_0\}, \quad \mathbf{M}_{i,0|0} = \mathbb{E}\left\{ (\mathbf{w}_0 - \mathbb{E}\{\mathbf{w}_0\})(\mathbf{w}_0 - \mathbb{E}\{\mathbf{w}_0\})^{\mathrm{H}} \right\}. \tag{28.22}$$

Note that the initialization procedure in Eq. (28.22) is technically based on some a priori information about the mean and covariance of the initial state, $\mathbf{w}_0$. However, linear Kalman filters will converge (in the mean square sense) even when initialized arbitrarily. Therefore, it is common practice to set $\hat{\mathbf{w}}_{i,0|0} = \mathbf{0}$ and $\mathbf{M}_{i,0|0} = c\mathbf{I}$, where c is a real positive constant, when no prior information is available [47].

Extension to nonlinear models

The linear state space model in Eq. (28.19) can be extended to a nonlinear state space formulation with the state and measurement evolutions given by

$$\begin{aligned} \mathbf{w}_k &= \boldsymbol{f}_k(\mathbf{w}_{k-1}) + \boldsymbol{q}_k, \\ \mathbf{y}_{i,k} &= \boldsymbol{\varphi}_{i,k}(\mathbf{w}_k) + \boldsymbol{\eta}_{i,k}, \end{aligned} \tag{28.23}$$

where the nonlinear function $\boldsymbol{f}_k(\cdot)$ is the state-transition function while $\boldsymbol{\varphi}_{i,k}(\cdot)$ is a general observation function. The statistics of the process noise, $\boldsymbol{q}_k$, and measurement noise, $\boldsymbol{\eta}_{i,k}$, remain as Eqs. (28.20b)–(28.20c).

The original Kalman filter was derived for linear Gaussian systems and to cater for nonlinear models in Eq. (28.23), the complex extended Kalman filter (CEKF) linearizes the nonlinear state and observation functions by their first order Taylor series expansions (TSE) about the state estimates $\hat{\mathbf{w}}_{i,k|k}$ and $\hat{\mathbf{w}}_{i,k|k-1}$ for each node i, so that [33]

$$\begin{aligned} \mathbf{w}_k &\approx \mathbf{F}_{i,k} \mathbf{w}_{k-1} + \tilde{\mathbf{F}}_{i,k} \mathbf{w}_{k-1}^* + \boldsymbol{u}_{i,k} + \boldsymbol{q}_k, \\ \mathbf{y}_{i,k} &\approx \mathbf{H}_{i,k} \mathbf{w}_k + \tilde{\mathbf{H}}_{i,k} \mathbf{w}_k^* + \boldsymbol{\epsilon}_{i,k} + \boldsymbol{\eta}_{i,k}, \end{aligned} \tag{28.24}$$

where the Jacobians of functions $f(\cdot)$ and $\varphi_i(\cdot)$ are defined as

$$\begin{aligned}
\mathbf{F}_{i,k} &= \left.\frac{\partial f_k(w)}{\partial w^{\mathrm{T}}}\right|_{w=\hat{w}_{i,k-1|k-1}}, \quad \tilde{\mathbf{F}}_{i,k} = \left.\frac{\partial f_k(w)}{\partial w^{\mathrm{H}}}\right|_{w^*=\hat{w}^*_{i,k-1|k-1}}, \\
\mathbf{H}_{i,k} &= \left.\frac{\partial \varphi_{i,k}(w)}{\partial w^{\mathrm{T}}}\right|_{w=\hat{w}_{i,k|k-1}}, \quad \tilde{\mathbf{H}}_{i,k} = \left.\frac{\partial \varphi_{i,k}(w)}{\partial w^{\mathrm{H}}}\right|_{w^*=\hat{w}^*_{i,k|k-1}}.
\end{aligned} \tag{28.25}$$

while the variables $u_{i,k}$ and $\epsilon_{i,k}$ in Eq. (28.24) represent the Taylor series expansion errors and can be treated as deterministic inputs to the state equations, given by

$$\begin{aligned}
u_{i,k} &= f(\hat{w}_{i,k-1|k-1}) - \mathbf{F}_{i,k-1}\hat{w}_{i,k-1|k-1} - \tilde{\mathbf{F}}_{i,k-1}\hat{w}^*_{i,k-1|k-1}, \\
\epsilon_{i,k} &= \varphi_{i,k}(\hat{w}_{i,k|k-1}) - \mathbf{H}_{i,k}\hat{w}_{i,k|k-1} - \tilde{\mathbf{H}}_{i,k}\hat{w}^*_{i,k|k-1}.
\end{aligned}$$

Notice that in Eq. (28.24), the linearized equations for a general state space admit a widely linear form by involving both the state w_k and its conjugate[1] w_k^*. For this reason, it shall be useful to always consider the augmented state space to account for the widely linear nature of the linearization procedure in Eq. (28.24). This is accomplished using an augmented version of the state, $\bar{w}_k = \left[w_k^{\mathrm{T}}, \;\; w_k^{\mathrm{H}}\right]^{\mathrm{T}}$, and observation vectors, $\bar{y}_{i,k} = \left[y_{i,k}^{\mathrm{T}}, \;\; y_{i,k}^{\mathrm{H}}\right]^{\mathrm{T}}$, so that the linearized state space in Eq. (28.24) admits an augmented form

$$\begin{aligned}
\bar{w}_k &= \bar{\mathbf{F}}_{i,k}\bar{w}_{k-1} + \bar{u}_{i,k-1} + \bar{q}_k, \\
\bar{y}_{i,k} &= \bar{\mathbf{H}}_{i,k}\bar{w}_k + \bar{\epsilon}_{i,k} + \bar{\eta}_{i,k},
\end{aligned} \tag{28.26}$$

where the augmented state transition matrix, $\bar{\mathbf{F}}_{i,k}$, the observation matrix, $\bar{\mathbf{H}}_{i,k}$, and the Taylor series errors (deterministic inputs), $\bar{u}_{i,k}$, $\bar{\epsilon}_{i,k}$, are given by

$$\bar{\mathbf{F}}_{i,k} = \begin{bmatrix} \mathbf{F}_{i,k} & \tilde{\mathbf{F}}_{i,k} \\ \tilde{\mathbf{F}}^*_{i,k} & \mathbf{F}^*_{i,k} \end{bmatrix}, \quad \bar{\mathbf{H}}_{i,k} = \begin{bmatrix} \mathbf{H}_{i,k} & \tilde{\mathbf{H}}_{i,k} \\ \tilde{\mathbf{H}}^*_{i,k} & \mathbf{H}^*_{i,k} \end{bmatrix}, \quad \bar{u}_{i,k} = \begin{bmatrix} u_{i,k} \\ u^*_{i,k} \end{bmatrix}, \quad \bar{\epsilon}_{i,k} = \begin{bmatrix} \epsilon_{i,k} \\ \epsilon^*_{i,k} \end{bmatrix}.$$

Single-node ACEKF

The complex Kalman filter methodology in Eqs. (28.21a)–(28.21b) can now be applied to the linearized state space equation in Eq. (28.26), to give the augmented complex extended Kalman filter (ACEKF) algorithm as [33]

$$\textit{Predict}: \begin{cases} \bar{\mathbf{F}}_{i,k} & = \;\; \text{Jacobian}\left(\bar{f}_k, \; \hat{\bar{w}}_{i,k-1|k-1}\right) \\ \hat{\bar{w}}_{i,k|k-1} & = \;\; \bar{f}_k(\hat{\bar{w}}_{i,k-1|k-1}) \\ \mathbf{M}_{i,k|k-1} & = \;\; \bar{\mathbf{F}}_{i,k}\mathbf{M}_{i,k-1|k-1}\bar{\mathbf{F}}^{\mathrm{H}}_{i,k} + \bar{\mathbf{C}}_k \end{cases} \tag{28.27a}$$

[1]Note that the widely linear model degenerates into a strictly linear one when the state vector is real (i.e., $w_k^* = w_k$) or if the observation and state transition functions are analytic (i.e., $\partial\varphi_i(\cdot)/\partial w_k^{\mathrm{H}} = \mathbf{0}$, or $\partial f(\cdot)/\partial w_k^{\mathrm{H}} = \mathbf{0}$).

$$\textit{Update}: \begin{cases} \bar{\mathbf{H}}_{i,k} &= \text{Jacobian}\left(\bar{\varphi}_{i,k}, \hat{w}_{i,k|k-1}\right) \\ \mathbf{M}_{i,k|k}^{-1} &= \mathbf{M}_{i,k|k-1}^{-1} + \bar{\mathbf{H}}_{i,k}^{\mathrm{H}} \bar{\boldsymbol{\Sigma}}_{i,k}^{-1} \bar{\mathbf{H}}_{i,k} \\ \hat{w}_{i,k|k} &= \hat{w}_{i,k|k-1} + \mathbf{M}_{i,k|k} \bar{\mathbf{H}}_{i,k}^{\mathrm{H}} \bar{\boldsymbol{\Sigma}}_{i,k}^{-1} \left[\bar{y}_{i,k} - \bar{\varphi}_{i,k}(\hat{w}_{i,k|k-1}) \right], \end{cases} \tag{28.27b}$$

where the augmented nonlinear functions are $\bar{f}_k = \left[f_k^{\mathrm{T}}, f_k^{\mathrm{H}}\right]^{\mathrm{T}}$ and $\bar{\varphi}_{i,k} = \left[\varphi_{i,k}^{\mathrm{T}}, \varphi_{i,k}^{\mathrm{H}}\right]^{\mathrm{T}}$ and the Jacobian($\cdot$) operator designates the computation of the Jacobian matrices based on Eq. (28.25). The ACEKF is initialized with the augmented state vector and state error covariance matrix as

$$\hat{w}_{i,0|0} = \mathbb{E}\left\{\bar{w}_0\right\}, \quad \mathbf{M}_{i,0|0} = \mathbb{E}\left\{(\bar{w}_0 - \mathbb{E}\{\bar{w}_0\})(\bar{w}_0 - \mathbb{E}\{\bar{w}_0\})^{\mathrm{H}}\right\}. \tag{28.28}$$

28.4.2 COLLABORATION SCHEMES

Centralized ACEKF

The most straightforward distributed multisensor Kalman filter is the centralized scheme where all nodes transmit their measurements to a fusion center that assumes the following centralized augmented state space formulation

$$\bar{w}_k = \bar{f}_k(\bar{w}_{k-1}) + \bar{q}_k, \tag{28.29a}$$
$$\bar{y}_{\text{cen},k} = \bar{\varphi}_{\text{cen},k}(\bar{w}_k) + \bar{\eta}_{\text{cen},k}, \tag{28.29b}$$

where the network observation vector, $\bar{y}_{\text{cen},k} \in \mathbb{C}^{NL\times 1}$, in Eq. (28.29b) represents all the observations in the network, with the collective measurement and measurement noise vectors given by

$$\bar{y}_{\text{cen},k} = \begin{bmatrix} \bar{y}_{1,k} \\ \bar{y}_{2,k} \\ \vdots \\ \bar{y}_{N,k} \end{bmatrix}, \quad \bar{\varphi}_{\text{cen},k}(\cdot) = \begin{bmatrix} \bar{\varphi}_{1,k}(\cdot) \\ \bar{\varphi}_{2,k}(\cdot) \\ \vdots \\ \bar{\varphi}_{N,k}(\cdot) \end{bmatrix}, \quad \bar{\eta}_{\text{cen},k} = \begin{bmatrix} \bar{\eta}_{1,k} \\ \bar{\eta}_{2,k} \\ \vdots \\ \bar{\eta}_{N,k} \end{bmatrix}, \tag{28.30}$$

while the augmented covariance matrices of the measurement and state noise are respectively $\bar{\boldsymbol{\Sigma}}_{\text{cen},k} \overset{\text{def}}{=} \mathbb{E}\left\{\eta_{\text{cen},k}\eta_{\text{cen},k}^{\mathrm{H}}\right\}$ and $\bar{\mathbf{C}}_k \overset{\text{def}}{=} \mathbb{E}\left\{\bar{q}_k\bar{q}_k^{\mathrm{H}}\right\}$. Following the augmented state space formulation, which leads to the ACEKF in Eqs. (28.27a)–(28.27b), the centralized ACEKF can be derived as

$$\textit{Predict}: \begin{cases} \bar{\mathbf{F}}_k &= \text{Jacobian}\left(\bar{f}_k, \hat{w}_{\text{cen},k-1|k-1}\right) \\ \hat{w}_{\text{cen},k|k-1} &= \bar{f}_k(\hat{w}_{\text{cen},k-1|k-1}) \\ \mathbf{M}_{k|k-1} &= \bar{\mathbf{F}}_k \mathbf{M}_{k-1|k-1} \bar{\mathbf{F}}_k^{\mathrm{H}} + \bar{\mathbf{C}}_k \end{cases} \tag{28.31a}$$

$$\text{Update}: \begin{cases} \bar{\mathbf{H}}_{\text{cen},k} &= \text{Jacobian}\left(\bar{\varphi}_{\text{cen},k}, \hat{w}_{\text{cen},k|k-1}\right) \\ \mathbf{M}_{k|k}^{-1} &= \mathbf{M}_{k|k-1}^{-1} + \bar{\mathbf{H}}_{\text{cen},k}^{\mathrm{H}} \bar{\mathbf{\Sigma}}_{\text{cen},k}^{-1} \bar{\mathbf{H}}_{\text{cen},k} \\ \hat{w}_{\text{cen},k|k} &= \hat{w}_{\text{cen},k|k-1} + \mathbf{M}_{k|k} \bar{\mathbf{H}}_{\text{cen},k}^{\mathrm{H}} \bar{\mathbf{\Sigma}}_{\text{cen},k}^{-1} \left[\bar{y}_{\text{cen},k} - \bar{\varphi}_{\text{cen},k}(\hat{w}_{\text{cen},k|k-1})\right]. \end{cases} \tag{28.31b}$$

Although the centralized ACEKF in Eqs. (28.31a)–(28.31b) is the optimal state-estimation algorithm that is given access to all the measurements in the network, it requires excessive communications to a single fusion center. This causes bottlenecks and imposes the risk of the distributed estimation task to a single point of failure. To avoid these issues, a distributed formulation of Eqs. (28.31a)–(28.31b) is accomplished by restricting the nodes in the network to only communicate within their neighborhood.

Local ACEKF

To remove the dependence on a fusion center, the so-called "local" ACEKF replicates the centralized ACEKF at each node using only the observations from its neighborhood. Specifically, the node i observes $\bar{y}_{i_{\text{col}},k} \in \mathbb{C}^{LN_i \times 1}$, where $N_i = |\mathcal{N}_i|$, denotes the number of nodes in neighborhood of node i, such that the local state equations are

$$\bar{w}_k = \bar{f}_k(w_{k-1}) + \bar{q}_k, \tag{28.32a}$$

$$\bar{y}_{i_{\text{col}},k} = \bar{\varphi}_{i_{\text{col}},k}(w_k) + \bar{\eta}_{i_{\text{col}},k}, \tag{28.32b}$$

with the collection of observation variables from the neighborhood of the node i

$$\bar{y}_{i_{\text{col}},k} = \begin{bmatrix} y_{i_1,k} \\ y_{i_2,k} \\ \vdots \\ y_{i_{N_i},k} \end{bmatrix}, \quad \bar{\varphi}_{i_{\text{col}},k}(\cdot) = \begin{bmatrix} \bar{\varphi}_{i_1,k}(\cdot) \\ \bar{\varphi}_{i_2,k}(\cdot) \\ \vdots \\ \bar{\varphi}_{i_{N_i},k}(\cdot) \end{bmatrix}, \quad \bar{\eta}_{i_{\text{col}},k} = \begin{bmatrix} \bar{\eta}_{i_1,k} \\ \bar{\eta}_{i_2,k} \\ \vdots \\ \bar{\eta}_{i_{N_i},k} \end{bmatrix}. \tag{28.33}$$

The augmented covariance matrix of the collective observation noise vector is then given by

$$\bar{\mathbf{\Sigma}}_{i_{\text{col}},k} = \mathbb{E}\left\{\bar{\eta}_{i_{\text{col}},k} \bar{\eta}_{i_{\text{col}},k}^{\mathrm{H}}\right\} = \begin{bmatrix} \bar{\mathbf{\Sigma}}_{i_1} & \bar{\mathbf{\Sigma}}_{i_1 i_2} & \cdots & \mathbf{\Sigma}_{i_1 i_{|\mathcal{N}_i|}} \\ \bar{\mathbf{\Sigma}}_{i_2 i_1} & \bar{\mathbf{\Sigma}}_{i_2} & \cdots & \bar{\mathbf{\Sigma}}_{i_2 i_{N_i}} \\ \vdots & \vdots & \ddots & \vdots \\ \bar{\mathbf{\Sigma}}_{i_{N_i} i_1} & \bar{\mathbf{\Sigma}}_{i_{N_i} i_2} & \cdots & \mathbf{\Sigma}_{i_{N_i}} \end{bmatrix}. \tag{28.34}$$

The local ACEKF therefore takes the form

$$\text{Predict}: \begin{cases} \bar{\mathbf{F}}_{i,k} &= \text{Jacobian}\left(\bar{f}_k, \hat{w}_{i,k-1|k-1}\right) \\ \hat{w}_{i,k|k-1} &= \bar{f}_k(\hat{w}_{i,k-1|k-1}) \\ \mathbf{M}_{i,k|k-1} &= \bar{\mathbf{F}}_{i,k} \mathbf{M}_{i,k-1|k-1} \bar{\mathbf{F}}_{i,k}^{\mathrm{H}} + \bar{\mathbf{C}}_k \end{cases} \tag{28.35a}$$

$$\textit{Update}: \begin{cases} \bar{\mathbf{H}}_{i_{\mathrm{col}},k} & = \text{Jacobian}\left(\bar{\boldsymbol{\varphi}}_{i_{\mathrm{col}},k}, \hat{\mathbf{w}}_{i,k|k-1}\right) \\ \mathbf{M}_{i,k|k}^{-1} & = \mathbf{M}_{i,k|k-1}^{-1} + \bar{\mathbf{H}}_{i_{\mathrm{col}},k}^{\mathrm{H}} \bar{\boldsymbol{\Sigma}}_{i_{\mathrm{col}},k}^{-1} \bar{\mathbf{H}}_{i_{\mathrm{col}},k} \\ \hat{\mathbf{w}}_{i,k|k} & = \hat{\mathbf{w}}_{i,k|k-1} + \mathbf{M}_{i,k|k} \bar{\mathbf{H}}_{i_{\mathrm{col}},k}^{\mathrm{H}} \bar{\boldsymbol{\Sigma}}_{i_{\mathrm{col}},k}^{-1} \left[\bar{\mathbf{y}}_{i_{\mathrm{col}},k} - \bar{\boldsymbol{\varphi}}_{i_{\mathrm{col}},k}(\hat{\mathbf{w}}_{i,k|k-1})\right]. \end{cases} \tag{28.35b}$$

One of the most widely used assumptions in the distributed Kalman filtering literature is that the observation noise vectors $\bar{\boldsymbol{\eta}}_{i,k}$ are spatially independent, resulting in a block-diagonal local observation noise covariance matrix in Eq. (28.34) in the form

$$\bar{\boldsymbol{\Sigma}}_{i_{\mathrm{col}},k} = \begin{bmatrix} \bar{\boldsymbol{\Sigma}}_{i_1} & \mathbf{0} & \cdots & \mathbf{0} \\ \mathbf{0} & \bar{\boldsymbol{\Sigma}}_{i_2} & \cdots & \mathbf{0} \\ \vdots & \vdots & \ddots & \vdots \\ \mathbf{0} & \mathbf{0} & \cdots & \bar{\boldsymbol{\Sigma}}_{i_{N_i}} \end{bmatrix}. \tag{28.35c}$$

This enables the local ACEKF update equations in Eq. (28.35b) to be expressed as sums of the individual measurement matrices as

$$\textit{Update}: \begin{cases} \bar{\mathbf{H}}_{\ell,k} & = \text{Jacobian}\left(\bar{\boldsymbol{\varphi}}_{\ell,k}, \hat{\mathbf{w}}_{i,k|k-1}\right), \quad \ell = i_1, \ldots, i_{N_i} \\ \mathbf{M}_{i,k|k}^{-1} & = \mathbf{M}_{i,k|k-1}^{-1} + \sum_{\ell \in \mathcal{N}_i} \bar{\mathbf{H}}_{\ell,k}^{\mathrm{H}} \bar{\boldsymbol{\Sigma}}_{\ell,k}^{-1} \bar{\mathbf{H}}_{\ell,k} \\ \hat{\mathbf{w}}_{i,k|k} & = \hat{\mathbf{w}}_{i,k|k-1} + \mathbf{M}_{i,k|k} \sum_{\ell \in \mathcal{N}_i} \bar{\mathbf{H}}_{\ell,k}^{\mathrm{H}} \bar{\boldsymbol{\Sigma}}_{\ell,k}^{-1} \left[\bar{\mathbf{y}}_{\ell,k} - \bar{\boldsymbol{\varphi}}_{\ell,k}(\hat{\mathbf{w}}_{i,k|k-1})\right]. \end{cases} \tag{28.35d}$$

28.4.3 PROPOSED DIFFUSION AUGMENTED COMPLEX EXTENDED KALMAN FILTER (D-ACEKF)

We next illustrate how to achieve a computationally efficient diffusion strategy for a distributed network consisting of local ACEKFs by using a weighted averaging scheme. Observe that the centralized state estimate, $\hat{\mathbf{w}}_{\mathrm{cen},k|k}$ in Eq. (28.31b), and the local EKF state estimates $\hat{\mathbf{w}}_{i,k|k}$ in Eq. (28.27b) are related through their respective weight error covariance matrices, $\mathbf{M}_{i,k|k}$, as

$$\mathbf{M}_{k|k}^{-1} \hat{\mathbf{w}}_{\mathrm{cen},k|k} = \sum_{\ell=1}^{N} \mathbf{M}_{\ell,k|k}^{-1} \hat{\mathbf{w}}_{\ell,k|k}. \tag{28.36}$$

Premultiplying both sides of Eq. (28.36) with the centralized state error covariance matrix $\mathbf{M}_{k|k}$ yields

$$\hat{\mathbf{w}}_{\mathrm{cen},k|k} = \sum_{\ell=1}^{N} \mathbf{M}_{k|k} \mathbf{M}_{\ell,k|k}^{-1} \hat{\mathbf{w}}_{\ell,k|k}, \tag{28.37}$$

which implies that the centralized solution $\hat{\boldsymbol{w}}_{\text{cen},k|k}$ in Eq. (28.31b) can be obtained by a weighted averaging operation of the individual state estimates $\hat{\boldsymbol{w}}_{\ell,k|k}$ with weighting matrices $\mathbf{M}_{k|k}\mathbf{M}_{\ell,k|k}^{-1}$. Because the centralized state error covariance matrix, $\mathbf{M}_{k|k}$, is related to its single-node counterparts $\mathbf{M}_{\ell,k|k}$ as

$$\mathbf{M}_{k|k}^{-1} = \sum_{\ell=1}^{N} \mathbf{M}_{\ell,k|k}^{-1}, \tag{28.38}$$

the weighting matrices satisfy the power-invariance condition, such that

$$\sum_{\ell=1}^{N} \mathbf{M}_{k|k}\mathbf{M}_{\ell,k|k}^{-1} = \mathbf{I}.$$

To obtain the effect of the fusion rule in Eq. (28.37) without requiring the communication (sharing) of the state error covariance matrices $\mathbf{M}_{\ell,k|k}$, we shall replace the weighting matrices $\mathbf{M}_{k|k}\mathbf{M}_{\ell,k|k}^{-1}$ with combination coefficients $a_{\ell,\text{cen}}$ such that the fusion rule in Eq. (28.37) is approximately given by

$$\hat{\boldsymbol{w}}_{\text{cen},k|k} \approx \sum_{\ell=1}^{N} a_{\ell,\text{cen}}\hat{\boldsymbol{w}}_{\ell,k|k}, \tag{28.39}$$

where the weighting coefficients $a_{\ell,\text{cen}}$ are only restricted to satisfy the power-invariance condition $\sum_{\ell} a_{\ell,\text{cen}} = 1$. The step in Eq. (28.39) resembles the diffusion step introduced in [26], and can now be incorporated with the local ACEKF in Eqs. (28.35b)–(28.35a), to give the diffusion ACEKF (D-ACEKF) in the form

$$\bar{\mathbf{F}}_{i,k} = \text{Jacobian}\left(\bar{\boldsymbol{f}}_k, \hat{\boldsymbol{w}}_{i,k-1|k-1}\right), \tag{28.40a}$$

$$\hat{\boldsymbol{w}}_{i,k|k-1} = \bar{\boldsymbol{f}}_k(\hat{\boldsymbol{w}}_{i,k-1|k-1}), \tag{28.40b}$$

$$\mathbf{M}_{i,k|k-1} = \bar{\mathbf{F}}_{i,k}\mathbf{M}_{i,k-1|k-1}\bar{\mathbf{F}}_{i,k}^{\mathrm{H}} + \bar{\mathbf{C}}_k, \tag{28.40c}$$

$$\bar{\mathbf{H}}_{i_{\text{col}},k} = \text{Jacobian}\left(\bar{\boldsymbol{\varphi}}_{i_{\text{col}},k}, \hat{\boldsymbol{w}}_{i,k|k-1}\right), \tag{28.40d}$$

$$\mathbf{M}_{i,k|k}^{-1} = \mathbf{M}_{i,k|k-1}^{-1} + \bar{\mathbf{H}}_{i_{\text{col}},k}^{\mathrm{H}}\bar{\boldsymbol{\Sigma}}_{i_{\text{col}},k}^{-1}\bar{\mathbf{H}}_{i_{\text{col}},k}, \tag{28.40e}$$

$$\boldsymbol{\psi}_{i,k|k} = \hat{\boldsymbol{w}}_{i,k|k-1} + \mathbf{M}_{i,k|k}\bar{\mathbf{H}}_{i_{\text{col}},k}^{\mathrm{H}}\bar{\boldsymbol{\Sigma}}_{i_{\text{col}},k}^{-1}\left[\bar{\boldsymbol{y}}_{i_{\text{col}},k} - \bar{\boldsymbol{\varphi}}_{i_{\text{col}},k}(\hat{\boldsymbol{w}}_{i,k|k-1})\right], \tag{28.40f}$$

$$\hat{\boldsymbol{w}}_{i,k|k} = \sum_{\ell\in\mathcal{N}_i} a_{\ell i}\boldsymbol{\psi}_{\ell,k|k} \text{ (Diffusion step).} \tag{28.40g}$$

The distinguishing feature of the proposed distributed Kalman filters in Eqs. (28.40a)–(28.40g) is that we generalize the diffusion strategy in [26] by equipping the model with state and noise models that do not impose any restrictions on: (i) the correlation properties of the cross-nodal observation noises, (ii) the signal and noise circularity at different nodes, and (iii) the widely linear nature of the underpinning system. This also allows distributed Kalman filtering algorithms proposed in [26,28,48] to be used in wider application scenarios. Also note that unlike the linear Kalman filter, the state vector estimate for the EKF has to be initialized to a value that is "close" to the true state $\bar{\boldsymbol{w}}_0$. This can be accomplished with a priori knowledge about the system. For example, in frequency estimation tasks, the state (frequency) can be initialized to the nominal frequency of the grid of $\hat{\boldsymbol{w}}_{i,0} = \boldsymbol{w}_0 = 50\,\text{Hz}$, because the grid frequency is known to stay within a small interval of 49 Hz and 51 Hz. The state error covariance matrix can be initialized as $\mathbf{M}_{i,0|0} = c\mathbf{I}$, where c is the positive constant [47].

28.5 SIMULATIONS

To verify the suitability of the proposed D-ACEKF for distributed frequency estimation, simulations through case studies were performed based on a network of six buses (nodes) where each substation had access to three-phase voltage measurements via transformers with metering capabilities. The number of connections in the network was chosen to be nine (each node is connected to less than two other nodes) as it reflected the topology of substations in a distribution network. The power system under consideration had a nominal frequency of 50 Hz, and was sampled at a rate of 1 kHz while the signal to noise ratio (SNR) was determined by the metering accuracy class of the potential transformer. The BS EN 61869-1:2009 standard for the metering accuracy of potential transformers considers six separate classes for metering requirements, which translates to an SNR range of 30 dB to 60 dB. To illuminate the robustness of our proposed augmented diffusion Kalman filters, we chose an SNR level of 35 dB in all our simulations, unless stated otherwise.

Case Study # 1: Voltage sags

First, the performances of the proposed algorithms were evaluated for an initially balanced system that became unbalanced after undergoing a Type D voltage sag starting at 0.1 s, followed by a balanced condition starting at 0.3 s. Fig. 28.3 shows that, conforming with the analysis, the widely linear algorithm, D-ACEKF, was able to converge to the correct system frequency for both balanced and unbalanced operating conditions while the strictly linear algorithm, D-CEKF, was unable to accurately estimate the frequency during the voltage sag due to undermodeling of the system (not accounting for its widely linear nature), and exhibited obvious estimation oscillations. As expected, the widely linear and strictly linear algorithms had similar performances under balanced conditions, as exemplified in the time intervals 0–0.1 s and 0.3–0.5 s.

Furthermore, it is important to note that although the standard diffusion scheme is formulated to share all the states in the network, in the case of frequency tracking in the electricity grid, it is only the frequency that is common throughout the network while the imbalance levels, amplitudes, and phase angles are not necessarily the same. Therefore, diffusing other states besides the frequency results in biased estimates. Fig. 28.4 shows the profile of the voltages at different nodes in the network. Each

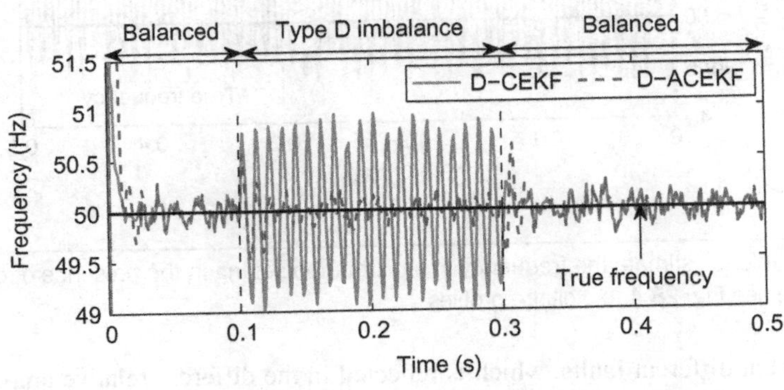

FIG. 28.3

Frequency estimation performance of the distributed algorithms (D-CEKF and D-ACEKF) for a system at 35 dB SNR. The system was balanced up to 0.1 s, it then underwent a Type D voltage imbalance followed by a balanced condition at 0.3 s.

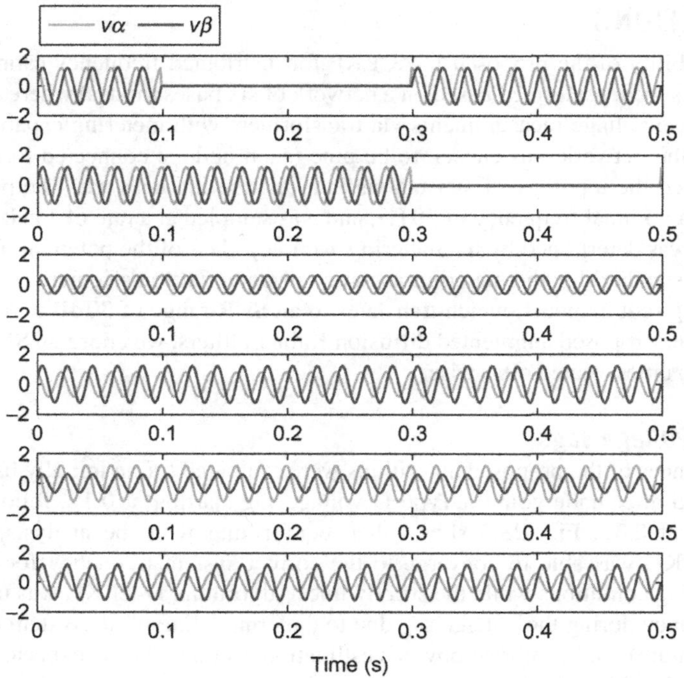

FIG. 28.4

Each substation (node) had different $\alpha\beta$ voltages, including cases where the voltage dropped to zero (*line cut*).

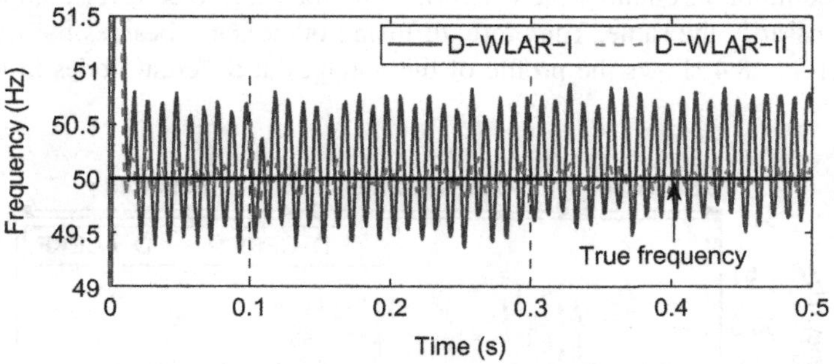

FIG. 28.5

The D-ACEKF was able to estimate the frequency in a distributed setting in the presence of different types of faults at each node; see Fig. 28.4 for voltage profiles.

substation underwent different faults, which is reflected in the different relative amplitudes and phase shifts of the $\alpha\beta$ voltages. In addition, Substations 1 and 2 underwent total line failures from 0.3 s to 0.5 s and 0.1 s to 0.3 s, respectively. Fig. 28.5 shows distributed implementations of two widely linear state space models described in Section 28.3, namely the WLAR-I and WLAR-II models. Notice that both models are capable of estimating the system frequency under both balanced and unbalanced conditions. However, within the WLAR-I model, the frequency is embedded within the coefficients h and g, which also contain information about the level of imbalance in the system. Diffusing the widely linear coefficients, will therefore lead to biased estimates because the level of imbalance across the network is not necessarily the same, as illustrated in Fig. 28.4. On the other hand, the WLAR-II model contains the frequency as an isolated state, which can be diffused across the network. Indeed, Fig. 28.5 shows that the diffusion-based WLAR-II algorithm was able to estimate the frequency of the network even with different levels of imbalance at each node while the diffusion-based WLAR-I model produced biased estimates.

Case Study #2: Frequency variations

Fig. 28.6 illustrates the performance of the D-ACEKF when a power network measurement was contaminated with white noise at 35 dB and 60 dB SNR while the system simultaneously underwent a gradual drop and increase in frequency from 0.1 s to 0.3 s. This is a typical scenario when generation does not match the load and system inertia keeps the frequency from changing too quickly. From 0.3 s to 0.5 s, the system underwent a step change followed by a linear ramp in frequency. The D-ACEKF

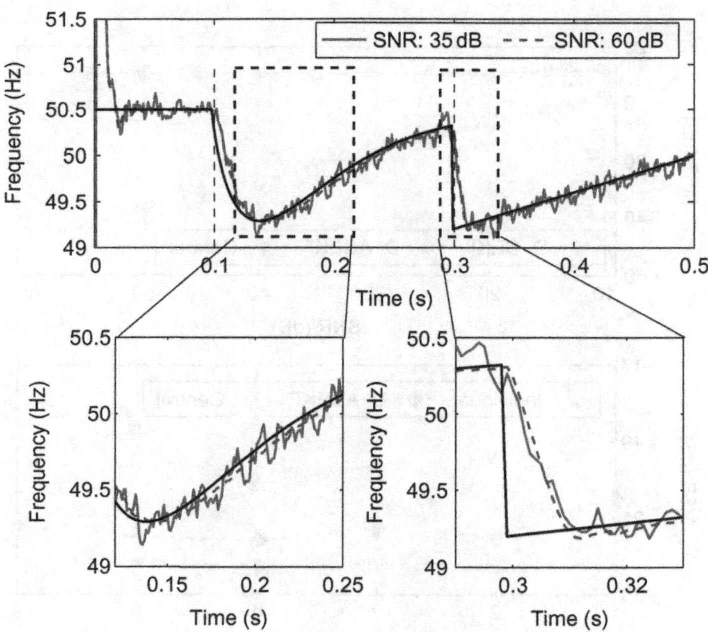

FIG. 28.6

Frequency tracking performance of D-ACEKF for system measurements at 35 dB and 60 dB SNR, which experienced a gradual change in frequency from 0.1 s to 0.3 s, a step change in system frequency to 49.2 Hz at 0.3 s and a linear ramp from 0.3 s to 0.5 s. The *solid black line* shows the true instantaneous frequency of the voltage. The proposed D-ACEKF was able to track both slow and rapid changes in frequency.

was able to track the frequency in both cases, illustrating its suitability for both the current electricity grid and future smart grids.

Case Study #3: Steady-state mean square error

Fig. 28.7 illustrates the mean square error (MSE) for the proposed distributed frequency estimators. The steady state frequency estimate at a node i for the trial m is denoted by $\hat{f}_{i,ss}[m]$. The MSE of the frequency estimators was calculated over 200 independent trials, as

$$\text{MSE} = \frac{1}{200 \cdot 6} \sum_{m=1}^{200} \sum_{i=1}^{6} \left(\hat{f}_{i,ss}[m] - f_0\right)^2, \tag{28.41}$$

where $f_0 = 50\,\text{Hz}$ is the fundamental frequency.

The algorithms were evaluated at different SNR levels for an unbalanced system undergoing a Type D voltage sag. Observe that the distributed estimation algorithm outperformed its noncooperative counterpart while the only consistent distributed estimator was the proposed D-ACEKF.

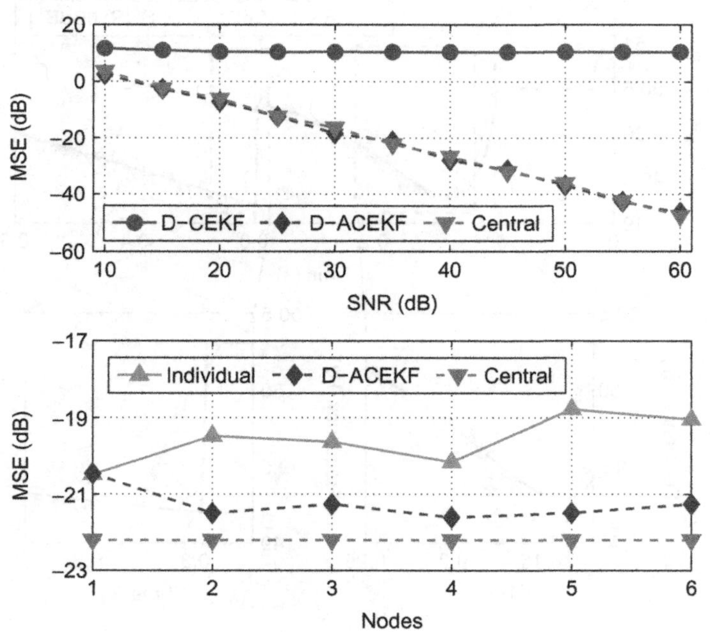

FIG. 28.7

Top panel: The average MSE of a frequency estimate by D-ACEKF is lower than that of the D-CEKF under a Type D sag. *Bottom panel*: MSE for each node in the network with and without cooperation at SNR = 35 dB shows that the diffusion strategy reduces the steady state error for all the nodes.

28.6 CONCLUSION

We have introduced a novel diffusion augmented complex extended Kalman filter (D-ACEKF) for cooperative frequency estimation in power networks. It has been shown to provide sequential state estimation of the generality of the complex $\alpha\beta$ voltage signals, both circular and noncircular, within a general and unifying framework that also caters to correlated nodal observation noises. This novel widely linear framework has been applied for distributed state space-based frequency estimation in the context of three-phase power systems, and has been shown to be optimal for both balanced and unbalanced operating conditions. Simulations over a range of balanced and unbalanced power system conditions have illustrated that the proposed D-ACEKF is a consistent estimator, offering accurate and fast frequency estimation in power networks.

ACKNOWLEDGMENTS

The work by S. Kanna and D.P. Mandic was supported by the EPSRC Pathways to Impact under Grant P96871. The work of Y. Xia was supported by the National Natural Science Foundation of China under Grants 61401094 and 61771124, the Natural Science Foundation of Jiangsu Province under Grant BK20140645, and the Fundamental Research Funds for the Central Universities under Grant 2242016K41050.

REFERENCES

[1] Ipakchi A, Albuyeh F. Grid of the future. IEEE Power Energy Mag 2009;7(2):52–62.

[2] Zhang Y, Markham P, Xia T, Chen L, Ye Y, Wu Z, et al. Wide-area frequency monitoring network (FNET) architecture and applications. IEEE Trans Smart Grid 2010;1(2):159–67.

[3] Phadke AG, Thorp JS, Adamiak MG. A new measurement technique for tracking voltage phasors, local system frequency, and rate of change of frequency. IEEE Trans Power Apparatus Syst 1983;PAS-102(5):1025–38.

[4] Clarke E. Circuit analysis of A.C. power systems. New York, NY: Wiley; 1943.

[5] Lobos T, Rezmer J. Real-time determination of power system frequency. IEEE Trans Instrum Meas 1997;46(4):877–81.

[6] Yang R, Xue H. A novel algorithm for accurate frequency measurement using transformed consecutive points of DFT. IEEE Trans Power Syst 2008;23(3):1057–62.

[7] Jeon HJ, Chang TG. Iterative frequency estimation based on MVDR spectrum. IEEE Trans Power Delivery 2010;25(2):621–30.

[8] Pradhan AK, Routray A, Basak A. Power system frequency estimation using least mean square technique. IEEE Trans Power Delivery 2005;20(3):1812–6.

[9] Dash PK, Pradhan AK, Panda G. Frequency estimation of distorted power system signals using extended complex Kalman filter. IEEE Trans Power Delivery 1999;14(3):761–6.

[10] Seneviratne C, Ozansoy C. Frequency response due to a large generator loss with the increasing penetration of wind/PV generation: a literature review. Renew Sust Energy Rev 2016;57:659–68.

[11] Bollen MHJ. Understanding power quality problems: voltage sags and interruptions. New York, NY: Wiley-IEEE; 2000.

[12] Tielens P, Van Hertem D. The relevance of inertia in power systems. Renew Sust Energy Rev 2016;55:999–1009.

[13] von Jouanne A, Banerjee B. Assessment of voltage unbalance. IEEE Trans Power Delivery 2001;26(4):782–90.

[14] Bollen MHJ, Gu IYH, Santoso S, McGranaghan MF, Crossley PA, Ribeiro MV, et al. Bridging the gap between signal and power. IEEE Signal Process Mag 2009;26(4):11–31.

[15] Xia Y, Mandic DP. Widely linear adaptive frequency estimation of unbalanced three-phase power systems. IEEE Trans Instrum Meas 2012;61(1):74–83.

[16] Picinbono B, Chevailer P. Widely linear estimation with complex data. IEEE Trans Signal Process 1995;43(8):2030–3.

[17] Mandic DP, Goh SL. Complex valued nonlinear adaptive filters: noncircularity, widely linear and neural models. New York, NY: Wiley; 2009.

[18] Schreier PJ, Scharf LL. Statistical signal processing of complex-valued data: the theory of improper and noncircular signals. Cambridge, UK: Cambridge University Press; 2010.

[19] Xia Y, Mandic DP. Complementary mean square analysis of augmented CLMS for second-order noncircular Gaussian signals. IEEE Signal Process Lett 2017;24(9):1413–7.

[20] Xia Y, Douglas SC, Mandic DP. Adaptive frequency estimation in smart grid applications: exploiting noncircularity and widely linear adaptive estimators. IEEE Signal Process Mag 2012;29(5):44–54.

[21] Dini DH, Mandic DP. Widely linear modeling for frequency estimation in unbalanced three-phase power systems. IEEE Trans Instrum Meas 2013;62(2):353–63.

[22] Dash PK, Jena RK, Panda G, Routray A. An extended complex Kalman filter for frequency measurement of distorted signals. IEEE Trans Instrum Meas 2000;49(4):746–53.

[23] Huang CH, Lee CH, Shih KJ, Wang YJ. Frequency estimation of distorted power system signals using a robust algorithm. IEEE Trans Power Delivery 2008;23(1):41–51.

[24] Talebi SP, Kanna S, Mandic DP. A non-linear state space frequency estimator for three-phase power systems. In: Proceedings of the international joint conference on neural networks, Killarney, Ireland; 2015. p. 1–7.

[25] Stadter PA, Chacos AA, Heins RJ, Moore GT, Olsen EA, Asher MS, et al. Confluence of navigation, communication, and control in distributed spacecraft systems. IEEE Aerosp Electron Syst Mag 2002;17(5):26–32.

[26] Cattivelli FS, Sayed AH. Diffusion strategies for distributed Kalman filtering and smoothing. IEEE Trans Autom Control 2010;55(9):2069–84.

[27] Olfati-Saber R. Flocking for multi-agent dynamic systems: algorithms and theory. IEEE Trans Autom Control 2006;51(3):401–20.

[28] Olfati-Saber R. Distributed Kalman filtering for sensor networks. In: Proceedings of the 46th IEEE conference on decision and control, New Orleans, USA; 2007. p. 5492–8.

[29] Lopes CG, Sayed AH. Diffusion least-mean squares over adaptive networks: formulation and performance analysis. IEEE Trans Signal Process 2018;56(7):3122–36.

[30] Xia Y, Mandic DP, Sayed AH. An adaptive diffusion augmented CLMS algorithm for distributed filtering of noncircular complex signals. IEEE Signal Process Lett 2011;18(11):659–62.

[31] Khan UA, Moura JMF. An adaptive diffusion augmented CLMS algorithm for distributed filtering of noncircular complex signals. IEEE Trans Signal Process 2008;56(10):4919–35.

[32] Sayed AH. Adaptive networks. Proc IEEE 2014;102(4):460–97.

[33] Dini DH, Mandic DP. A class of widely linear complex Kalman filters. IEEE Trans Neural Netw Learn Syst 2012;23(5):775–86.

[34] Xiao L, Boyd S. Fast linear iterations for distributed averaging. Syst Control Lett 2004;53(1):65–78.

[35] Kanna S, Mandic DP. Self-stabilising adaptive three-phase transforms via widely linear modelling. Electron Lett 2017;53(13):875–7.

[36] Bollen MHJ, Zhang L. Different methods for classification of three-phase unbalanced voltage dips due to faults. Electr Power Syst Res 2003;66(1):59–69.

[37] Amin SM, Wollenberg BF. Toward a smart grid: power delivery for the 21st century. IEEE Power Energy Mag 2005;3(5):34–41.

[38] Huang YF, Werner S, Huang J, Kashyap N, Gupta V. State estimation in electric power grids: meeting new challenges presented by the requirements of the future grid. IEEE Signal Process Mag 2012;29(5):33–43.

[39] Kelly C, Gupta S. Discrete-time demodulation of continuous-time signals. IEEE Trans Inf Theory 1972;18(4):488–93.

[40] Hashemipour HR, Roy S, Laub AJ. Decentralized structures for parallel Kalman filtering. IEEE Trans Autom Control 1988;33(1):88–94.

[41] Speyer J. Computation and transmission requirements for a decentralized linear-quadratic-Gaussian control problem. IEEE Trans Autom Control 1979;24(2):266–9.

[42] Rao BS, Durrant-Whyte HF. Fully decentralized algorithm for multisensor Kalman filtering. IEE Proc D Control Theory Appl 1991;138(5):413–20.

[43] Olfati-Saber R. Kalman-consensus filter: optimality, stability, and performance. In: Proceedings of the 48th IEEE conference on decision and control, Shanghai, China, December 15–18; 2009. p. 7036–42.

[44] Hidayat Z, Babuska R, Schutter BD, Nunez A. Decentralized Kalman filter comparison for distributed-parameter systems: a case study for a 1D heat conduction process. In: Proceedings of the 16th IEEE conference on emerging technologies factory automation; 2011.

[45] Tu SY, Sayed AH. Diffusion strategies outperform consensus strategies for distributed estimation over adaptive networks. IEEE Trans Signal Process 2012;60(12):6217–34.

[46] Dini DH, Mandic DP. Cooperative adaptive estimation of distributed noncircular complex signals. In: Conference record of the forty sixth Asilomar conference on signals, systems and computers, Pacific Grove, USA; 2012. p. 1518–22.

[47] Ljung L, Sayed AH. Asymptotic behavior of the extended Kalman filter as a parameter estimator for linear systems. IEEE Trans Autom Control 1979;24(1):36–50.

[48] Kar S, Moura J. Gossip and distributed Kalman filtering: weak consensus under weak detectability. IEEE Trans Signal Process 2011;59(4):1766–84.

CHAPTER 29

BEACONS AND THE CITY: SMART INTERNET OF THINGS

Petros Spachos*, **Konstantinos Plataniotis**†

*School of Engineering, University of Guelph, Guelph, ON, Canada** *Electrical and Computer Engineering, University of Toronto, Toronto, ON, Canada*†

29.1 INTRODUCTION

Every day more computer-based devices are connected to the Internet. Most of these devices have at least one wireless communication unit, creating opportunities for more direct integration between the physical world and computer-based systems. This is the idea behind the Internet of Things (IoT), a development of the Internet in which everyday objects have network connectivity, allowing them to send and receive data. With the current advancements in wireless communication many indoor and outdoor environments such as streets, parks, shopping malls, museums, hospitals, buildings, offices, and homes are not just structures but a new platform for smart cities. In this platform, devices and occupants can share data and information to improve all things ranging from customer experience to individual quality of life and satisfaction [1,2].

The efficient use of the data that is collected and exchanged between wireless devices is the main challenge for smart cities [3,4]. An important parameter of wireless data transmission is the communication infrastructure [5]. Various wireless communication technologies with different characteristics including Bluetooth [6], ZigBee [7], RF identification (RFID) [8,9], wireless local area networks (WLANs) [10–12], wireless cellular networks, and the fifth-generation mobile networks [13], are considered for different smart city applications. Experimental results of some of these technologies are shown in Table 29.1. Recently, more and more people are adopting Bluetooth low energy (BLE) beacon hardware as a feasible and scalable solution for many applications such as indoor positioning system (IPS) solutions [14–18]. Restaurants, hotels, airports, and museums use BLE beacons to improve their services and enhance visitor experiences. BLE beacons are small and cost-effective wireless transmitters that can convert a hospital or an airport into a smart infrastructure, without interfering with other wireless technologies. Due to their high availability, low cost, low power consumption, and ease of deployment, BLE beacons are an ideal solution in providing indoor location-based [19,20] and proximity-based services [21,22]. Smart cities could deploy a mesh of cost efficient BLE beacons to collect data and improve citizen engagement.

In this chapter, the use of BLE beacons in applications where they can improve user experience and collect useful analytics for smart cities are discussed. Their characteristics along with the available

Cooperative and Graph Signal Processing. https://doi.org/10.1016/B978-0-12-813677-5.00029-8

Table 29.1 Experimental Results of Wireless Technologies and Their Characteristics

Technology	Range	Power	Advantages	Disadvantages
IEEE 802.11	Up to 100 m	Moderate	Availability	Prone to interference
RFID	3–12 m	Low	Energy	Low accuracy
BLE	Up to 100 m	Low	Energy	Prone to interference
ZigBee	10–100 m	Low	Low power	Low security

communication protocols and their challenges and limitations are discussed, toward the usage of beacons in large-scale smart city applications. Three BLE beacon applications are described: in a hospital for asset monitoring, in a shopping mall for proximity marketing, and in a parking lot for automated check in. To further examine the performance of BLE beacons in terms of accuracy for proximity services, experiments are also conducted. BLE beacons from three different hardware vendors were used in the experiment while data smoothing and Kalman filters were used to examine their effect on beacon accuracy. The purpose is not to find which of the examined beacons can be claimed to be better than the others. Instead, the objective is to provide insights on beacon accuracy and eventually guidelines for future beacon application developers.

The rest of this chapter is organized as follows: In Section 29.2, an introduction to BLE beacons as well as their main characteristics and protocols are provided, followed by Section 29.3 with a discussion on their limitations and challenges. In Section 29.4 three use cases of BLE beacons in smart cities are presented. In Section 29.5, proximity estimation through BLE beacons is discussed and in Section 29.6 experiments with BLE beacons that examine their performance are presented. Lastly, the conclusion is in Section 29.7.

29.2 BLE BEACON CHARACTERISTICS AND PROTOCOLS

In this section, BLE beacons are described along with their characteristics and their most commonly used protocols for data transmission.

29.2.1 BLE BEACONS

BLE beacons, usually referred to as beacons, are small wireless transmitters that broadcast their identifier to nearby electronic devices using BLE. BLE, as part of Bluetooth 4.0 [23], is a wireless technology that is used for applications that do not need to exchange large amounts of data and intend to provide considerably reduced power consumption and cost while maintaining a similar communication range with classic Bluetooth.

Beacons broadcast signals at a certain interval and within a certain transmission range. An analogy of the way beacons work is with the operation of a lighthouse [24]. The lighthouse represents a known location that can be uniquely identified by its light. All the ships that can see the light know about the existence of the lighthouse. On the other hand, the lighthouse neither communicates with the ships nor does it know how many ships see its light or how many other lighthouses are in the area. Similarly, every beacon is sending out a radio signal to inform all the radio-enabled devices in its range that the

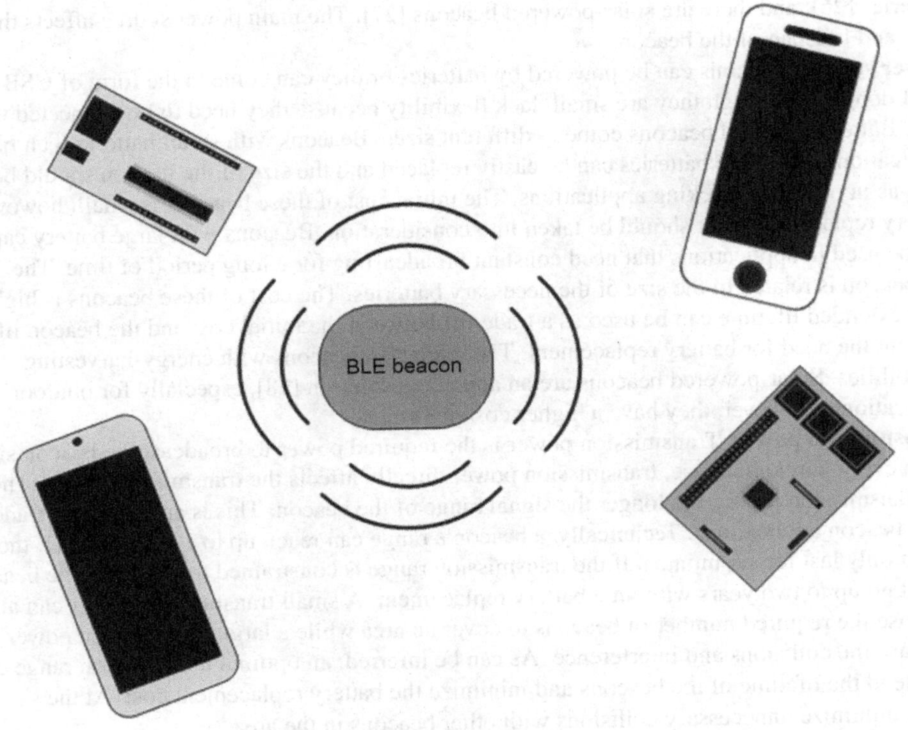

FIG. 29.1

A BLE beacon broadcasting a signal to nearby devices. Each device can receive the signal and take action in response.

beacon is there. It does not know how many beacons or receiving devices are in the area and it does not connect with them. An example of beacon operation is shown in Fig. 29.1.

A beacon broadcasts a signal to all nearby devices that can receive the Bluetooth signal, i.e., the devices that have a Bluetooth receiver and the receiver is on. In order to collect the signal from the beacon, it is necessary to have a device with a BLE receiver. This can be a smartphone or a single-board computer such as Raspberry Pi. Applications or functions can be implemented based on the signal from the beacons. However, these applications are running on the hosting device, i.e., a smartphone or a Raspberry Pi, and not on the beacon.

29.2.2 BLE BEACON CHARACTERISTICS

BLE beacons have configuration parameters and a set of values that can determine their performance and their utility for different applications.

- **Size design.** As the market for the beacons increases, so do the different design approaches. There are small beacons that work with one single coin cell battery [25], there are beacons with two AA batteries [26], and there are solar-powered beacons [27]. The main power source affects the size, cost, and lifespan of the beacon.

- **Power source.** Beacons can be powered by batteries or they can come in the form of USB dongles. USB dongles, although they are small, lack flexibility because they need to be connected to a USB port. Battery-powered beacons come in different sizes. Beacons with small batteries can be used in applications where the batteries can be easily replaced and the size of the beacon should be small, such as in occupant-tracking applications. The initial cost of these beacons is small; however the battery replacement cost should be taken into consideration. Beacons with large battery capacity can be used in applications that need constant broadcasting for a long period of time. The size of the beacon is related to the size of the necessary batteries. The cost of these beacons is higher but their extended lifetime can be used as a trade-off between the initial cost and the beacon lifetime without the need for battery replacement. There are also beacons with energy-harvesting capabilities. Solar-powered beacons are an appealing solution [28], especially for outdoor applications. However, they have a higher cost per unit.
- **Transmission power.** Transmission power is the required power to broadcast the beacon signal. As in every wireless device, transmission power directly affects the transmission range. The higher the transmission power, the longer the signal range of the beacon. This is an important trade-off for most beacon applications. Technically, a beacon's range can reach up to 70 m; however, the battery might only last for six months. If the transmission range is constrained to 2 m, then the beacon might go up to two years without a battery replacement. A small transmission power can also increase the required number of beacons to cover an area while a large transmission power can increase the collisions and interference. As can be inferred, an optimal transmission range can help to extend the lifetime of the beacons and minimize the battery replacement cost. At the same time, it can minimize unnecessary collisions with other beacons in the area.
- **Advertising interval.** Advertising interval is another characteristic that affects the overall performance of beacons. It describes the time between consecutive transmissions. Applications that need to notify or detect the users that are moving in the area need a short advertising interval while applications where the users are moving less frequently might improve their performance with a longer advertising interval. Similar to the transmission power, the advertising interval affects beacon performance. The shorter the interval, the more stable the signal from the beacon. At the same time, the shorter the interval, the higher the power consumption. Once again, there is a trade-off between beacon performance and power consumption.
- **Measured power.** Most beacons come with a factory-calibrated measured power, which represents the expected received signal strength indicator (RSSI) at a distance of 1 m from the beacon. Receiving devices can use this value to calibrate and eventually calculate the distance to the transmitting beacon. However, this value changes in different environments; hence, calibration is necessary to improve beacon performance.
- **Passive mode.** Beacons are broadcasters that do not do anything else besides sending a piece of information. The logic behind each signal is done by the supporting device, such as smartphones. Beacon signals are used by applications to trigger events and call actions, allowing the users to interact with physical things. All the implementation is done on the device while the beacons just broadcast the signal.
- **No need for Internet.** Beacons are not necessarily Internet-connected and there is no need for Internet connection. There are cases where beacons can get updates and configurations through a

smartphone that works as a gateway between the beacons and a cloud server. However, this extra functionality is not required. Beacons can be placed in remote locations and operate without Internet connectivity.

- **Platform-independent.** Beacons can be used with iOS and Android devices. Each platform requires different protocols that have different packet layouts but most platforms are able the listen to the different protocols.

29.2.3 BLE BEACON PROTOCOLS

Beacon protocols are standards of BLE communication. Each protocol describes the structure of a data packet that beacons broadcast. Beacons can use different protocols, the most popular of which are the following:

- **iBeacon**. Apple's iBeacon was the first BLE beacon technology to come out [29]. iBeacon is a proprietary, closed standard. It broadcasts four pieces of information:
  1. A universally unique identifier (UUID) that identifies the beacon.
  2. A major number identifying a subset of beacons within a large group.
  3. A minor number identifying a specific beacon within the subset.
  4. A transmission power level in 2's complement, indicating the signal strength one meter from the device. This number must be calibrated for each device by the user or manufacturer.

  iBeacon has a simple implementation and large documentation but it has fewer features in comparison with the following protocols. iBeacon works with iOS and Android, but is native to iOS.
- **Eddystone**. Eddystone, announced from Google, is another protocol that defines a BLE message format for proximity beacon messages [30]. The Eddystone protocol is able to transmit four different frame-types:
  1. UID, which is used to identify the individual beacon.
  2. URL, which can be a website link that redirects to a website that is secured using SSL, eliminating the need for a mobile app.
  3. TLM, which includes sensor and administrative data from the beacon through telemetry. Examples include the beacon's battery level and its temperature.
  4. EID, which is an encrypted ephemeral identifier that changes periodically at a rate determined during the initial registration with a web service. This frame type is intended for use in security and privacy-enhanced devices.

  Eddystone also works with both iOS and Android.
- **AltBeacon**. It is an open-source beacon protocol [31] that was designed by Radius Networks. It has the same functionality as an iBeacon but is not company-specific. This makes AltBeacon compatible with any mobile operating platform and more flexible because it has customizable source code.
- **GeoBeacon**. It is another open-source beacon protocol, designed for usage in geocaching applications [32]. It has a very compact type of data storage. GeoBeacon can provide high-resolution coordinates and is also compatible with different mobile operating platforms.

29.3 LIMITATIONS AND CHALLENGES

In this section, the limitations of beacons will be discussed, followed by the challenges and some security and privacy concerns that they pose.

29.3.1 LIMITATIONS

Beacons have a great potential for a plethora of applications. However, they do have some limitations.

First of all, there is a need for an application running on the receiving device in order to take some action when the beacon signal is received. The passive mode of the beacons makes them easy to be deployed and used, but at the same time the complexity is moved toward the development of the necessary application. If there is no application running, beacon signals are useless. Hence, a successful use case of beacons depends on the application development.

Another important limitation is the need for Bluetooth. Not only must the receiving device have a Bluetooth receiver, but the Bluetooth must be turned on. Especially when it comes to smartphone devices or prototyping boards, this can lead to high energy consumption [33,34] while also posing significant security and privacy concerns. At the same time, devices that do not have Bluetooth capabilities are not able to interact with the beacons. Hence, application developers should take into consideration users and devices with the Bluetooth turned off or no Bluetooth capabilities at all.

Beacons are used for proximity and navigation applications; however their accuracy is limited. The use of RSSI is a good indicator to estimate the distance from the beacons but is not accurate enough for every application. Beacon performance is prone to interference, especially as the number of nearby BLE devices increases. Their performance needs to be improved through the application on the receiver or further processing on a server. Hence, although an Internet connection is not required, to improve their performance, beacon data might need to be forwarded to a cloud server.

There is also a latency on the initial beacon scanning and the received signal from the beacons. This can be improved with techniques running on the received application or with a smaller advertising interval, but these solutions come with increased complexity in application design or higher power consumption. Latency should be considered when designing the application using beacons, with the user's experience and application requirements in mind.

29.3.2 SECURITY AND PRIVACY CHALLENGES

In order for a device with a Bluetooth receiver to collect the beacon signal, it is necessary to have the Bluetooth turned on. Because the Bluetooth is on and the beacon is among the trusted devices of the receiver, this poses significant security challenges.

An attacker can physically remove the trusted beacon and replace it with another device with the same packet format and communicate with the receiving device. This is an important security concern in beacons, known as spoofing. Beacons do not come with advanced security mechanisms and most of the time they broadcast their ID. An attacker can create a clone of the trusted beacons and forward malicious data to the receiving device.

Another security concern is known as piggybacking. In this case, an attacker listens for the UUID as well as the major and minor values of a beacon and adds them to a different application. The attacker can even clone the original application and try to capture crucial information from the receiver.

There are also many privacy concerns with beacons. Because the ID of the beacons is static, it is easy to mimic a trusted beacon and perform a number of attacks on the receiving device. Additionally, in many localization applications, the receiving device is a smartphone that tries to find its location based on nearby beacons. In this way, the users reveal their location as well. By trusting any beacon in the area and sharing location information, a user may also share behavioral patterns and location information to unauthorized personnel.

29.4 BLE BEACONS IN SMART CITIES APPLICATIONS

Beacons can be used in a plethora of applications in smart cities. They can be placed in many environments such as offices and museums, therefore convert, them into smart environments by providing interaction with the users. In this section, three beacon use cases are described, along with the advantages the beacons offer but also the challenges they pose.

29.4.1 ASSET TRACKING AND MONITORING

Beacons can be used to keep track of assets. For instance, a hospital can integrate beacon technology in order to maximize information exchange, as shown in Fig. 29.2. Beacons can be placed on critical medical assets and devices. The beacons will report the location of the assets in real time in a general

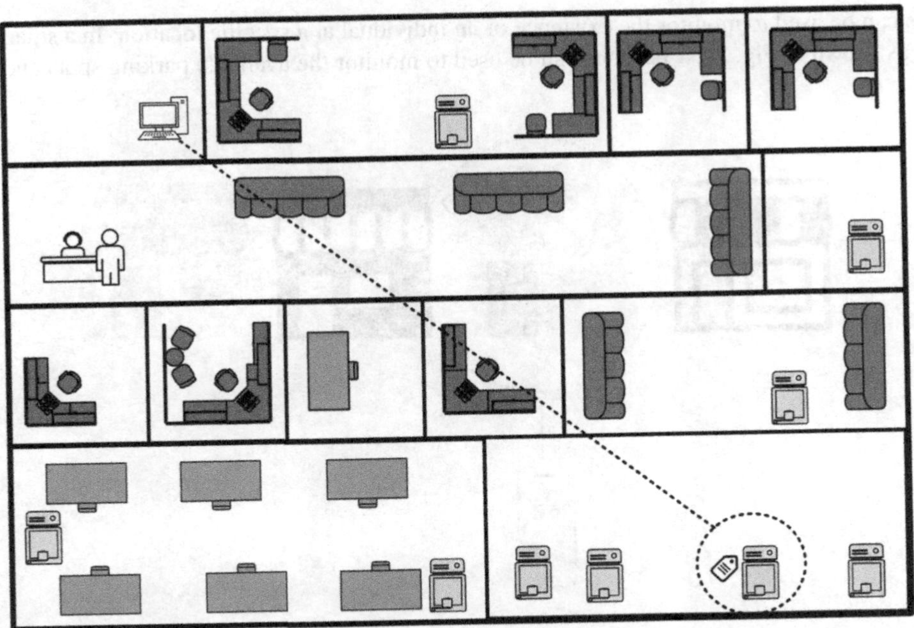

FIG. 29.2

Asset monitoring in a complex indoor environment with BLE beacons. Valuable assets have a beacon that can send location information to a general asset-management platform.

management system. When needed, the exact location of each device in a large hospital can be found. With further processing, beacons on nearby devices can also navigate the user to the required device. This can also be implemented for tracking cars, luggage, and even occupants. Mobile applications can be implemented in order to collect beacon signals and convert them into directions toward a specific asset.

The cost and ease of deployment are among the advantages of using beacons for such applications as long as there is no interference with other wireless infrastructures in the area. At the same time, there are security concerns that are related to the broadcasting nature of beacon signals. Attackers may find the location of valuable assets through beacon signals, if security mechanisms are not implemented.

29.4.2 PROXIMITY MARKETING AND ENHANCED INTERACTIVITY

Content can be delivered based on the receiver's location for proximity marketing or for enhanced interactivity. For example, in a shopping mall, offers or reviews can be provided to the users when they are about to enter a store or a restaurant, as shown in Fig. 29.3. The user only needs to turn on Bluetooth and use the application that listens to the beacons. There are many more opportunities to deliver context and enhance interactivity at the right time and place with the use of beacons. When beacons are used in such applications, they can provide useful analytics. However, the users might end up getting too many notifications or information that is not useful.

29.4.3 AUTOMATED CHECK-IN

Beacons can be used to monitor the existence of an individual at a specific location. In a smart parking system, as shown in Fig. 29.4, beacons can be used to monitor the available parking spots and forward

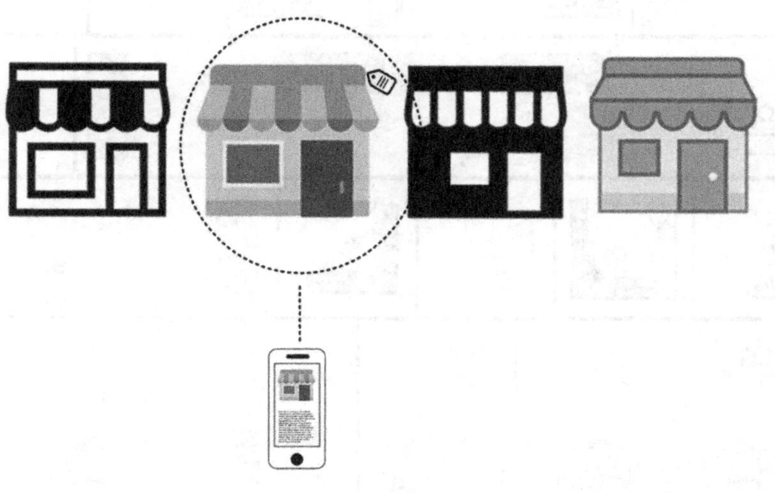

FIG. 29.3

BLE beacons in different stores in a shopping mall. As the visitors are close to one of the stores, they can get notifications about store sales and other relevant information.

FIG. 29.4

A smart parking system with BLE beacons. When the driver parks the car in front of the beacon, a message is sent from the driver's smartphone to a server to mark the specific parking spot as occupied.

the information through the user's smartphone application to a server that keeps track of all the available parking spots as well as some other useful analytics. In this way, real-time parking availability is offered to drivers through a smartphone application. However, the placement of the beacons would be a challenge, due to the fact that the beacon's performance decreases when interference increases. Hence, beacons cannot be placed too close to each other.

29.5 PROXIMITY ESTIMATION THROUGH BLE BEACONS

Proximity estimation is the key feature in all three applications described in the previous section. With BLE beacons, the distance calculation and eventually the proximity estimation are based on the RF signals. Unfortunately, RF signals are prone to interference [35,36]. In this section, the main principles for distance calculation based on RSSI values are described, followed by data filtering techniques that can improve distance estimation.

29.5.1 SIGNAL PROPAGATION AND DISTANCE CALCULATION

For distance estimation through beacons, an application should be developed on the receiver. The RSSI values can be used for the distance calculations. In an ideal propagation medium, RSSI values drops inversely proportional to the square of the distance between the transmitter and receiver. Hence, the calculation of the distance is easy and accurate.

A way to calculate the distance is with the use of the Friis transmission equation. It makes a comparison between the transmitted and received RSSI to estimate the signal attenuation over distance. With the Friis transmission equation, the received power can be written as:

$$P_r = \frac{P_t G_t G_r \lambda^2}{(4\pi R)^2}, \tag{29.1}$$

where P_r and P_t is the received and transmit ted power, respectively, G_r and G_t is the antenna gain of the receiver and the transmitter, respectively, and λ is the wavelength and R is the distance. This model can be applied only to calculations made in free space, hence in an environment without obstacles that may interfere with the signals.

Unfortunately, in a real environment RF signals are affected by interference, noise, and other channel implements that contribute to the RSSI values [37]. As the signal propagates between the beacon and the receiver, it also attenuates. This attenuation, also known as path loss, increases with the distance. Path loss relates to dissipation of signal power over distance [38]. At the same time, reflection, scattering, and absorption caused by obstacles in the environment also affect the signal, known as shadowing. The parameters on the received signal can be modeled statistically. If a log-normal distribution is assumed for the ratio of transmit-to-receive power, the combined effect of the path loss and shadowing can be expressed by the following model [38]:

$$P_r\,(\text{dB}) = P_t\,(\text{dB}) + 10\log_{10} K - 10\gamma \log_{10}\left(\frac{d}{d_0}\right) - \psi\,(\text{dB}), \tag{29.2}$$

where P_r and P_t is the received and transmit ted power, respectively, K is a constant related to the antenna and channel characteristics, γ is the path loss exponent, ψ reflects the effects of log-normal shadowing in the model, and d_0 is a reference distance for antenna far-field.

A simpler propagation model that works better for Bluetooth and the available information from the beacon packet is the following [39]:

$$RSSI = P_t + G + 20\log\left(\frac{c}{4\pi f}\right) - 10\log d, \tag{29.3}$$

where P_t is the transmit power, G is the combined antenna gain of the transmitter and the receiver, c is the speed of light, f is the frequency, d is the distance, and n is the attenuation exponent, which can be calculated as follows:

$$n = -\left(\frac{RSSI - A}{10\log_{10} d}\right), \tag{29.4}$$

where A is the RSS at 1 m distance—similar to the measured powered of the beacon signals. The antenna gain and the exact frequencies can not be acquired easily, because they are not part of the beacon packet; Eq. (29.3) can be further simplified as follows:

$$RSSI = A - 10n\log d. \tag{29.5}$$

29.5.2 KALMAN FILTER

To further improve the distance calculation, a Kalman filter can be applied to the received RSSI values [40]. Kalman filters contain statistical noise and other inaccuracies, and produce estimates of unknown variables that tend to be more accurate than those based on a single measurement alone by estimating a joint probability distribution over the variables for each time frame.

The Kalman filter is a recursive estimator. Only the estimated state from the previous time step and the current measurement are needed to compute the estimate for the current state. The state of the filter is represented by two variables:

- $\hat{\mathbf{x}}_{k|k}$, the a posteriori state estimate at time k given observations up to and including at time k;
- $\mathbf{P}_{k|k}$, the a posteriori error covariance matrix–a measure of the estimated accuracy of the state estimate.

The filter has two phases: prediction and update (correction).

i. Prediction

Prediction phase uses the state estimate from the previous time step to produce an estimate of the state at the current time step. Prediction step uses a linear-Gaussian system model.

The predicted state estimate is:

$$\hat{\mathbf{x}}(k \mid k-1) = \mathbf{F}\hat{\mathbf{x}}(k-1|k-1). \tag{29.6}$$

The predicted estimate covariance is:

$$\mathbf{P}(k \mid k-1) = \mathbf{FP}(k-1 \mid k-1)\mathbf{F}^T + Q. \tag{29.7}$$

ii. Update

Given the prediction density computed above, the update step must be carried out. To do this, the Kalman filter assumes a linear Gaussian measurement model as follows:

$$\mathbf{r}(k) = \mathbf{Hx}(k) + \mathbf{v}(k). \tag{29.8}$$

The update state estimate is:

$$\hat{\mathbf{x}}(k \mid k) = \hat{\mathbf{x}}(k \mid k-1) + \mathbf{K}(r(k) - \mathbf{H}\hat{\mathbf{x}}(k \mid k-1)). \tag{29.9}$$

The update estimate covariance is:

$$\mathbf{P}(k \mid k) = (1 - \mathbf{KH})\mathbf{P}(k \mid k-1)(1 - \mathbf{KH})^T + \mathbf{KP_rK^T}. \tag{29.10}$$

The value K is known as the Kalman gain and is computed as:

$$\mathbf{K} = \mathbf{P}(k \mid k-1)\mathbf{H}[\mathbf{HP}(k \mid k-1)\mathbf{H}^T + \mathbf{P_r}]^{-1}. \tag{29.11}$$

29.5.3 KALMAN FILTER ON BLE BEACON SIGNALS

Kalman filters can be implemented on the data from BLE beacons [41]. State x_i consists of the RSSI value while the rate of change of RSSI at time i is a function of the state at time $i-1$ and the process noise, which is equal to:

$$x_i = f(x_{i-1}, v_{i-1}). \tag{29.12}$$

The received RSSI measurements z_i at instant i from the beacon is a function of the state $i-1$ and the measured noise w_i as follows:

$$z_i = h(x_{i-1}, w_i). \tag{29.13}$$

Assuming that the process and the measurement noise are Gaussian, then Eqs. (29.12) and (29.13) are linear and can be written as:

$$x_i = Fx_{i-1} + v_i, \tag{29.14}$$

where $w_i \sim N(0, Q)$ and:

$$z_i = Hx_i + w_i, \tag{29.15}$$

where $v_i \sim N(0, R)$

Then, the prediction stage of the Kalman filter is:

$$\hat{x}_i^- = F\hat{x}_i, \tag{29.16}$$

$$P_i^- = FP_{i-1}F^T + Q, \tag{29.17}$$

and the update state is:

$$K_i = P_i^- H^T (HP_i^- H^T + R)^{-1}, \tag{29.18}$$

$$\hat{x}_i = \hat{x}_i^- + K_i(z_i - H\hat{x}_i^-), \tag{29.19}$$

$$P_i = (I - K_i H)P_i^-. \tag{29.20}$$

As can be inferred, the higher the Kalman gain, the higher the influence of the measurements on the state. The prediction and update are part of a recursive process, as shown in Fig. 29.5.

The state vector x_i consists of the RSSI value y_i and the rate of change Δy_{i-1} as follows:

$$x_i = \begin{bmatrix} y_i \\ \Delta y_i \end{bmatrix}. \tag{29.21}$$

The higher the noise, the higher the fluctuation of Δy_i. The value of the current RSSI is assumed to be the value of the previous RSSI plus the rate of change. Hence, Eq. (29.14) can be written as:

$$\begin{bmatrix} y_i \\ \Delta y_i \end{bmatrix} = \begin{bmatrix} 1 & \delta t \\ 0 & 1 \end{bmatrix} \begin{bmatrix} y_{i-1} \\ \Delta y_{i-1} \end{bmatrix} + \begin{bmatrix} v_i^y \\ v_i^{\Delta y} \end{bmatrix} \tag{29.22}$$

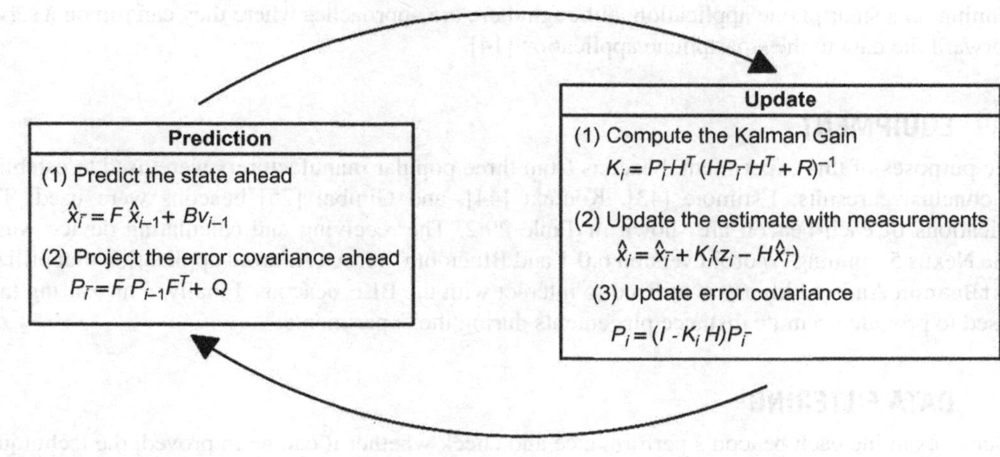

FIG. 29.5

Prediction and update steps in Kalman filter.

and the state transition matrix F is given by:

$$F = \begin{bmatrix} 1 & \delta t \\ 0 & 1 \end{bmatrix} \tag{29.23}$$

Similarly, Eq. (29.15) can be written as:

$$[z_i] = \begin{bmatrix} 1 & 0 \end{bmatrix} \begin{bmatrix} y_i \\ \Delta y_i \end{bmatrix} + [w_i^y]. \tag{29.24}$$

Once calibrated, the Kalman filter smooths the RSSI values. The smoothed RSSI value is then used in the path loss model to calculate the distance between the receiver and the beacon.

29.6 EXPERIMENTS WITH BEACONS

In this section, experiments to examine the accuracy of beacons are described. The purpose of the experiments is to measure the accuracy of beacons in terms of distance estimation, an important parameter for the smart city applications that were described in previous sections. Beacons from three popular hardware vendors were used. As discussed earlier, none of the examined beacons can be claimed to be better than the others in the parameters that were measured. The purpose of the experiment was not to find a winner, but to provide insights on each beacon's accuracy and eventually guidelines for future beacon application developers.

Because beacon signals are prone to interference, a data smoothing algorithm is implemented while the efficiency of the Kalman filter is also examined [42]. Both data smoothing and the Kalman filter are running on a smartphone application, although there are approaches where they can run on a server and forward the data to the smartphone application [14].

29.6.1 EQUIPMENT

For the purposes of this experiment, beacons from three popular manufacturers were used to establish more conclusive results: Estimote [43], Kontakt [44], and Gimbal [25] beacons were used. The specifications of each beacon are shown in Table 29.2. The receiving and calculating device was a Google Nexus 5, running Android version 6.0.1 and Bluetooth 4.0. An Android application that utilizes the `AltBeacon` Android library is utilized to interact with the BLE beacons. Finally, a measuring tape was used to provide accurate distance placements during the experiments.

29.6.2 DATA FILTERING

In order to examine each beacon's performance and check whether it can be improved, the techniques below are followed:

- **Raw data collection.** The raw data is the data collected straight from the beacon without further processing. All the data is kept and used for distance estimation.
- **Data smoothing.** During data smoothing, an average of the received data is used in the distance calculation on the receiving device.
- **Kalman filter.** The Kalman filter, as described in the previous section, is an algorithm that uses a series of measurements observed over time, containing statistical noise and other inaccuracies, and produces estimates [16,42].

29.6.3 EXPERIMENTAL ENVIRONMENT

To capture the behavior and performance of raw data, smoothed data, and Kalman filtering under different environmental conditions, the experiments were conducted in two rooms of different sizes. The first was a large lecture hall with dimensions 11 m × 9 m, and the second was a smaller meeting room with dimensions 6 m × 4 m. Both rooms were laid out with a set of tables and chairs. These rooms

Table 29.2 Beacon Specifications

	Beacon Device		
	Estimote [43]	**Kontakt [44]**	**Gimbal [25]**
Power supply	4× CR2477, 3.0 V lithium primary cell battery	2× CR2477, 1000 mAh, 3.0 V	1× CR2032, 3.0 V lithium ion battery
Radio	Bluetooth 4.2 LE	Bluetooth 4.0 LE	Bluetooth 4.0 LE
Size	Length: 62.7 mm Width: 41.2 mm Height: 23.6 mm Weight: 67 g	Length: 55 mm Width: 56 mm Height: 15 mm Weight: 23 g	Length: 40 mm Width: 28 mm Height: 5.5 mm Weight: 6.52 g

were chosen as they are a common size for many indoor environments. Furthermore, the furniture layout defines a common set of physical objects found in many indoor environments. These two rooms capture sufficient characteristics and qualities found in many indoor environments.

The variation in the two room sizes acts as differing environments to compare and contrast experimental results. Control over environmental settings during the experiments was crucial as small changes in the environment may have a large impact on beacon performance. To ensure environmental conditions remained constant in each room, between iterations of the experiment, no objects were added or removed from each room (i.e., chairs, tables, etc.) nor was anybody granted access to the rooms for the duration of all experiments.

29.6.4 EXPERIMENTAL PROCEDURE

A three-step procedure was developed to get a suitable set of results for each beacon. The procedure was conducted for each of the three types of data streams: no filter, smoothed, and Kalman-filtered.

i. The first part of the experiment obtains the raw RSSI values. The smartphone was placed at different distances from the beacon and the raw RSSI values were captured.
ii. The second part of the experiment applies a smoothing algorithm to the incoming RSSI values, eliminating the top and bottom 10% values. The top and bottom 10% are eliminated in an attempt to remove extreme outliers. The processing takes place on the smartphone. The beacon and smartphone are placed in the same locations as per the previous step.
iii. The third component of the experiment implements a Kalman filter on the smartphone. The beacon and smartphone are placed in the same locations as per the previous steps.

Within each iteration, 14 measurements are taken, leading to a total of 56 data measurements per beacon. The beacon remains stationary on a table while the smartphone moves along the table to the following set of displacements: 0.1, 0.2, 0.3, 0.4, 0.5, 0.6, 0.7, 0.8, 0.9, 1.0, 1.5, 2.0, 2.5, and 3.0 m. At each distance, the RSSI and corresponding distance is recorded. The received values are compared with the expected values, which are similar to the measured values that were described before and include the factory-calibrated RSSI measurement at 1 m.

29.6.5 RESULTS AND ANALYSIS

The performance of each beacon in the large room is shown in Fig. 29.6 and the standard deviation for the same room is shown in Table 29.3. It is clear that any filtering improves the accuracy of all the beacons. It is also clear that the performance increases when the Kalman filter is used.

When only the raw data is considered, Estimote beacons have better performance than the other two beacons. The use of the smoothing algorithm improves the performance for all three beacons but the Gimbal beacons have the best performance overall with that technique. With the use of the Kalman filter, all three beacons have further improvement in performance. In this case the Gimbal and Estimote beacons have similar performance.

The performance of each beacon in the small room is shown in Fig. 29.7 and the standard deviation for the same room is shown in Table 29.4. Similar to the previous experiment, any filtering improves the accuracy of all the beacons and the best performance is observed when the Kalman filter is used.

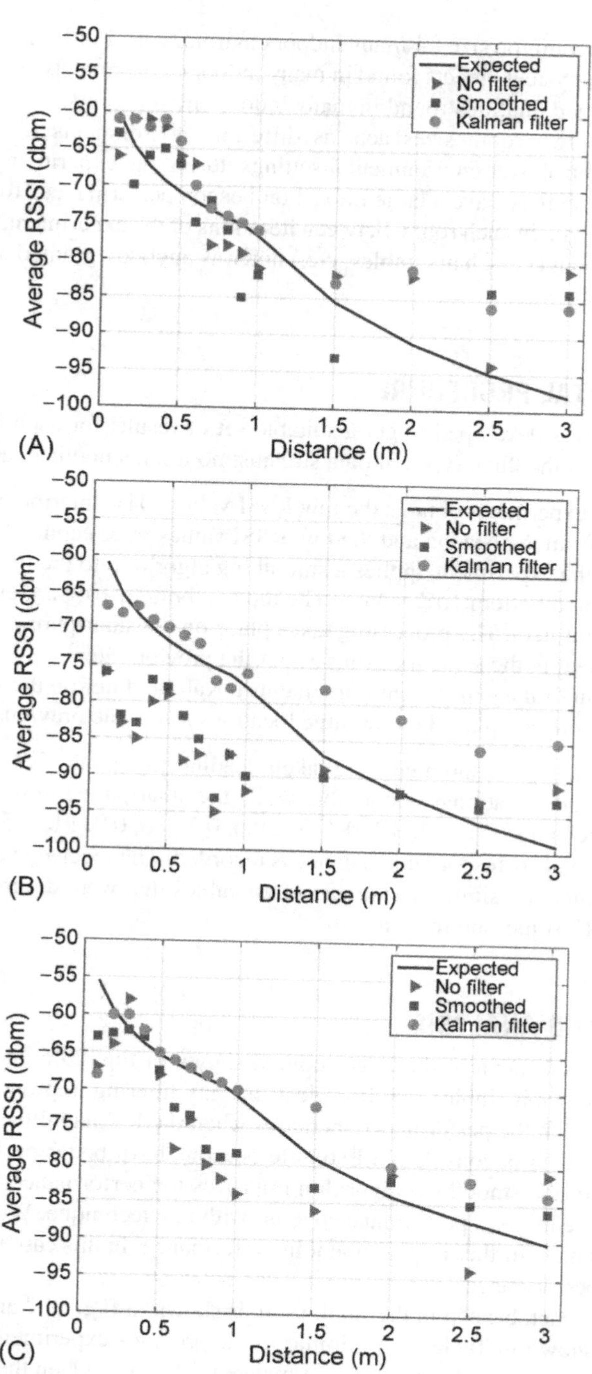

FIG. 29.6

Large room (11 m × 9 m): average RSSI value of the beacons with the use of different filtering techniques. (A) Estimote. (B) Kontakt. (C) Gimbal.

Table 29.3 Large Room (11 m x 9 m): Standard Deviation in dBm

	No Filter	Smoothed	Kalman Filter
Estimote	6.51	7.19	4.23
Kontakt	8.45	7.25	5.69
Gimbal	7.14	4.62	4.60

Table 29.4 Small Room (6 m x 4 m): Standard Deviation in dBm

	No Filter	Smoothed	Kalman Filter
Estimote	8.07	7.29	7.10
Kontakt	7.59	5.79	5.10
Gimbal	5.24	3.91	2.71

In contrast with the previous experiment, Gimbal beacons have the best performance when the raw data are examined. This is an indicator on how sensitive the beacons are to environmental changes. Another insight is the drastic performance improvement of the Gimbal and Kontakt beacons when the Kalman filter is used.

It can be inferred that the raw data from beacons does not provide high accuracy in every environment. Furthermore, the environment effects the performance of the beacons, hence beacons that perform well in one room might not have similar performance in another room of the same building. However, the implementation of simple filtering techniques can improve their performance. There are also many techniques in the literature that can take into consideration the dynamic changes of the environment and provide more efficient filtering.

29.7 CONCLUSIONS

In this chapter, an introduction to BLE beacons and their capabilities for smart city applications was presented. BLE beacons are an attractive solution for many smart city applications. The characteristics and protocols of the beacons were also described. Three use cases were discussed along with their advantages and challenges. BLE beacons have limitations and pose security and privacy challenges as well.

Experiments were conducted to examine the performance of different beacons in different rooms. According to experimental results, the environment has great effect on the performance of the beacon. Their performance can be improved through data filtering such as with the use of a Kalman filter.

BLE beacons can be used in many smart city applications. However, the selection of the proper beacon and the optimal configuration of the beacon with the application environmental parameters are important factors that should always be considered.

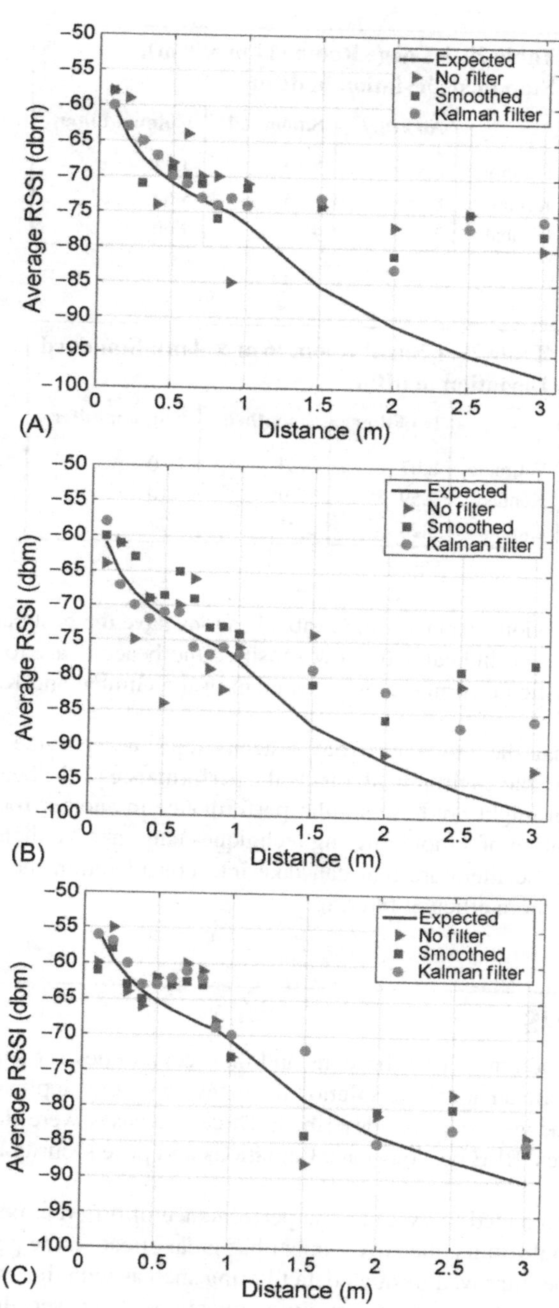

FIG. 29.7

Small room (6 m × 4 m): average RSSI value of the beacons with the use of different filtering techniques. (A) Estimote. (B) Kontakt. (C) Gimbal.

REFERENCES

[1] Li X, Lu R, Liang X, Shen X, Chen J, Lin X. Smart community: an internet of things application. IEEE Commun Mag 2011;49(11):68–75. https://doi.org/10.1109/MCOM.2011.6069711.

[2] Atzori L, Iera A, Morabito G. From "smart objects" to "social objects": the next evolutionary step of the internet of things. IEEE Commun Mag 2014;52(1):97–105. https://doi.org/10.1109/MCOM.2014.6710070.

[3] Zanella A, Bui N, Castellani A, Vangelista L, Zorzi M. Internet of things for smart cities. IEEE Internet Things J 2014;1(1):22–32. https://doi.org/10.1109/JIOT.2014.2306328.

[4] Vlacheas P, Giaffreda R, Stavroulaki V, Kelaidonis D, Foteinos V, Poulios G, et al. Enabling smart cities through a cognitive management framework for the internet of things. IEEE Commun Mag 2013;51(6):102–11. https://doi.org/10.1109/MCOM.2013.6525602.

[5] Djahel S, Doolan R, Muntean GM, Murphy J. A communications-oriented perspective on traffic management systems for smart cities: challenges and innovative approaches. IEEE Commun Surv Tutorials 2015;17(1):125–51. https://doi.org/10.1109/COMST.2014.2339817.

[6] Yoshimura Y, Krebs A, Ratti C. Noninvasive Bluetooth monitoring of visitors' length of stay at the Louvre. IEEE Pervasive Comput 2017;16(2):26–34. https://doi.org/10.1109/MPRV.2017.33.

[7] Daely PT, Reda HT, Satrya GB, Kim JW, Shin SY. Design of smart led streetlight system for smart city with web-based management system. IEEE Sens J 2017;17(18):6100–10. https://doi.org/10.1109/JSEN.2017.2734101.

[8] Saab SS, Nakad ZS. A standalone RFID indoor positioning system using passive tags. IEEE Trans Ind Electron 2011;58(5):1961–70. https://doi.org/10.1109/TIE.2010.2055774.

[9] Lau PY, Yung KKO, Yung EKN. A low-cost printed CP patch antenna for RFID smart bookshelf in library. IEEE Trans Ind Electron 2010;57(5):1583–9. https://doi.org/10.1109/TIE.2009.2035992.

[10] Aguirre E, Lopez-Iturri P, Azpilicueta L, Redondo A, Astrain JJ, Villadangos J, et al. Design and implementation of context aware applications with wireless sensor network support in urban train transportation environments. IEEE Sens J 2017;17(1):169–78. https://doi.org/10.1109/JSEN.2016.2624739.

[11] Collins K, Mangold S, Muntean GM. Supporting mobile devices with wireless LAN/MAN in large controlled environments. IEEE Commun Mag 2010;48(12):36–43. https://doi.org/10.1109/MCOM.2010.5673070.

[12] Hossain AKMM, Van HN, Jin Y, Soh WS. Indoor localization using multiple wireless technologies. In: 2007 IEEE international conference on mobile ad hoc and sensor systems; 2007. p. 1–8. https://doi.org/10.1109/MOBHOC.2007.4428622.

[13] Han T, Ge X, Wang L, Kwak KS, Han Y, Liu X. 5G converged cell-less communications in smart cities. IEEE Commun Mag 2017;55(3):44–50. https://doi.org/10.1109/MCOM.2017.1600256CM.

[14] Zafari F, Papapanagiotou I. Enhancing iBeacon based micro-location with particle filtering. In: 2015 IEEE global communications conference (GLOBECOM); 2015. p. 1–7. https://doi.org/10.1109/GLOCOM.2015.7417504.

[15] Ozer A, John E. Improving the accuracy of Bluetooth low energy indoor positioning system using Kalman filtering. In: 2016 international conference on computational science and computational intelligence (CSCI); 2016. p. 180–5. https://doi.org/10.1109/CSCI.2016.0041.

[16] Zhang K, Zhang Y, Wan S. Research of RSSI indoor ranging algorithm based on Gaussian–Kalman linear filtering. In: 2016 IEEE advanced information management, communicates, electronic and automation control conference (IMCEC); 2016. p. 1628–32. https://doi.org/10.1109/IMCEC.2016.7867493.

[17] Takahashi C, Kondo K. Accuracy evaluation of an indoor positioning method using iBeacons. In: 2016 IEEE 5th global conference on consumer electronics; 2016. p. 1–2. https://doi.org/10.1109/GCCE.2016.7800465.

[18] Yang L, Wang Q, Wang G. Positioning in an indoor environment based on iBeacons. In: 2016 IEEE international conference on information and automation (ICIA); 2016. p. 894–9. https://doi.org/10.1109/ICInfA.2016.7831945.

[19] Faragher R, Harle R. Location fingerprinting with Bluetooth low energy beacons. IEEE J Sel Areas Commun 2015;33(11):2418–28. https://doi.org/10.1109/JSAC.2015.2430281.

[20] He W, Ho PH, Tapolcai J. Beacon deployment for unambiguous positioning. IEEE Internet Things J 2017;4(5):1370–9. https://doi.org/10.1109/JIOT.2017.2708719.
[21] Kim DY, Kim SH, Choi D, Jin SH. Accurate indoor proximity zone detection based on time window and frequency with Bluetooth low energy. Proc Comput Sci 2015;56(Supplement C):88–95. The 10th international conference on future networks and communications (FNC 2015)/The 12th international conference on mobile systems and pervasive computing (MobiSPC 2015) affiliated workshops, http://www.sciencedirect.com/science/article/pii/S1877050915016804. https://doi.org/https://doi.org/10.1016/j.procs.2015.07.199.
[22] Ng PC, She J, Park S. High resolution beacon-based proximity detection for dense deployment. IEEE Trans Mobile Comput 2017. https://doi.org/10.1109/TMC.2017.2759734.
[23] Bluetooth Special Interest Group. Bluetooth 4.0 core specification. https://www.bluetooth.com/specifications/bluetooth-core-specification.
[24] 10 Things about Bluetooth beacons you need to know. http://academy.pulsatehq.com/bluetooth-beacons.
[25] Gimbal. https://gimbal.com/.
[26] Bluecats. https://www.bluecats.com/.
[27] Cyalkit-e02. http://www.cypress.com/documentation/development-kitsboards/cyalkit-e02-solar-powered-ble-sensor-beacon-reference-design.
[28] Spachos P, Mackey A. Energy efficiency and accuracy of solar powered BLE beacons. Comput Commun 2018;119:94–100.
[29] Apple. Getting started with iBeacon; 2014. https://developer.apple.com/ibeacon/Getting-Started-with-iBeacon.pdf.
[30] Eddystone. https://github.com/google/eddystone.
[31] Apple. Specifications on AltBeacon; 2016. https://github.com/AltBeacon/spec.
[32] Geobeacon. https://github.com/Tecno-World/GeoBeacon.
[33] Spachos P, James M, Gregori S. Power tradeoffs in mobile video transmission for smartphones. Comput Commun 2018;118:163–170. http://www.sciencedirect.com/science/article/pii/S0140366417305625. https://doi.org/https://doi.org/10.1016/j.comcom.2017.10.017.
[34] Spachos P, Song L, Gregori S. Power consumption of prototyping boards for smart room temperature monitoring. In: 2017 IEEE 22nd international workshop on computer aided modeling and design of communication links and networks (CAMAD); 2017. p. 1–6. https://doi.org/10.1109/CAMAD.2017.8031633.
[35] Mazuelas S, Bahillo A, Lorenzo RM, Fernandez P, Lago FA, Garcia E, et al. Robust indoor positioning provided by real-time RSSI values in unmodified WLAN networks. IEEE J Sel Top Signal Process 2009;3(5):821–31. https://doi.org/10.1109/JSTSP.2009.2029191.
[36] Wu K, Xiao J, Yi Y, Chen D, Luo X, Ni LM. CSI-based indoor localization. IEEE Trans Parallel Distrib Syst 2013;24(7):1300–9. https://doi.org/10.1109/TPDS.2012.214.
[37] Tse D, Viswanath P. Fundamentals of wireless communication. Cambridge University Press; 2004.
[38] Goldsmith A. Wireless communications. Cambridge University Press; 2005.
[39] Wang Y, Yang X, Zhao Y, Liu Y, Cuthbert L. Bluetooth positioning using RSSI and triangulation methods. In: 2013 IEEE 10th consumer communications and networking conference (CCNC); 2013. p. 837–42. https://doi.org/10.1109/CCNC.2013.6488558.
[40] Paul AS, Wan EA. RSSI-based indoor localization and tracking using sigma-point Kalman smoothers. IEEE J Sel Top Signal Process 2009;3(5):860–73. https://doi.org/10.1109/JSTSP.2009.2032309.
[41] Guvenc I. Enhancements to RSS based indoor tracking systems using Kalman filters. In: GSPx & international signal processing conference; 2003.
[42] Mackey A, Spachos P. Performance evaluation of beacons for indoor localization in smart buildings. In: 2017 IEEE global conference on signal and information processing (GlobalSIP); 2017.
[43] Estimote. https://estimote.com/.
[44] Kontakt. https://kontakt.io/.

CHAPTER

BIG DATA

30

Morteza Mardani*, Gonzalo Mateos†, Georgios B. Giannakis‡

Department of Electrical Engineering, Stanford University, Stanford, CA, United States Department of Electrical and Computer Engineering, University of Rochester, Rochester, NY, United States† Department of Electrical and Computer Engineering, University of Minnesota, Minneapolis, MN, United States‡*

30.1 LEARNING FROM BIG DATA: OPPORTUNITIES AND CHALLENGES

We live in an era of "data deluge." Pervasive sensors collect massive amounts of information on every bit of our lives, churning out enormous streams of raw data in a wide variety of formats. To get a sense of scale, users of the Facebook social network happily feed 10 billion messages per day, click the "like" button 4.5 billion times, and upload 350 million new pictures each and every day. Consumer data are collected every time we browse or purchase products online, as business models aim to provide services that are increasingly personalized. Automated sensors capture essentially every snapshot of complex phenomena of interest through high-resolution measurements. Mining information from these large volumes of data is expected to bring significant science and engineering advances along with consequent improvements in quality of life.

While big data may bring "big blessings," there are formidable challenges in dealing with large-scale datasets [1]. The *sheer volume* of data makes it often impossible to run analytics using central processors and storage units. Ubiquitous network data are also *geographically spread*, and collecting the data might be infeasible due to communication costs or privacy concerns. Consequently, datasets are often *incomplete* and thus a sizable portion of entries are missing. Moreover, large-scale data are prone to contain *corrupted measurements*, communication errors, and even suffer from *anomalies* such as cyberattacks. Furthermore, as many sources continuously generate data in *real time*, analytics must often be performed online as well as without an opportunity to revisit past data.

Conventional statistical inference tools cope with the notorious curse of dimensionality as well as with corruptions and anomalies by exploiting latent structure (of low intrinsic dimensionality) in the data. Such structure typically emerges due to dependencies present in real-world signals. In large-scale networks, complex interactions spanning the social, temporal, and spatial dimensions render such graph-indexed data highly correlated. For instance, the origin-to-destination (OD) traffic that flows in the backbone of Internet Protocol (IP) networks exhibits dependencies mainly due to traffic generation patterns [2], which can facilitate network monitoring tasks such as identifying traffic volume anomalies resulting from cyberattacks [3].

Cooperative and Graph Signal Processing. https://doi.org/10.1016/B978-0-12-813677-5.00030-4

Table 30.1 Notation

Notation	Description
$\mathbf{x}$	vector
$\mathbf{X}$	matrix
$\mathcal{X}$	set
$(\cdot)^\top$	matrix transpose
$\text{tr}\{\cdot\}$	matrix trace
σ_i	ith singular value of a matrix
$\lVert\mathbf{X}\rVert_F^2 := \text{tr}\{\mathbf{X}^\top\mathbf{X}\}$	matrix Frobenius norm
$\lVert\mathbf{X}\rVert_* := \sum_i \sigma_i$	matrix nuclear norm
$\lVert\mathbf{X}\rVert := \max_i \sigma_i$	matrix spectral norm
$\lVert\mathbf{x}\rVert_2$	vector ℓ_2-norm
$\otimes$	Kronecker product
$\text{vec}(\mathbf{X})$	concatenates columns of $\mathbf{X}$ on top of each other
$\text{unvec}(\mathbf{x})$	unfolds $\mathbf{x}$ to a matrix
$\mathcal{O}(\cdot)$	order of operation count
$\lambda_{\min}$	minimum eigen value
∇f	gradient of f
$\nabla^2 f$	Hessian of f

In this context, the goal of this chapter is to develop a framework for scalable decentralized and streaming analytics that facilitates machine learning for big data. For simplicity of exposition, the presentation focuses on the matrix completion problem with applications to IP network health monitoring. However, the scope of the ensuing framework can be broadened to accommodate other fundamental low-rank recovery tasks such as robust PCA and low-rank plus compressed-sparse recovery. Modern datasets are often indexed by several variables or dimensions giving rise to a multiway array (or tensor), in general. This chapter focuses on two-way arrays, or matrices, but extension to tensors is possible. Readers interested in delving into these generalizations are referred to [4–8]. The notations used througout the chapter are listed in Table 30.1.

30.2 MATRIX COMPLETION AND NUCLEAR NORM

Let $\mathbf{X} := [x_{l,t}] \in \mathbb{R}^{L\times T}$ be a *low-rank* matrix [rank($\mathbf{X}$) $\ll \min(L,T)$] and a set $\Omega \subseteq \{1,\ldots,L\} \times \{1,\ldots,T\}$ of index pairs (l,t) that define a sampling of the entries of $\mathbf{X}$. Given a number of (possibly) noise corrupted measurements

$$y_{l,t} = x_{l,t} + v_{l,t}, \quad (l,t) \in \Omega \tag{30.1}$$

the goal is to estimate low-rank $\mathbf{X}$ by denoising the observed entries and imputing the missing ones. Introducing the sampling operator $\mathcal{P}_\Omega(\cdot)$, which sets the entries of its matrix argument not in Ω to zero and leaves the rest unchanged, the data model can be compactly written in matrix form as

$$\mathcal{P}_\Omega(\mathbf{Y}) = \mathcal{P}_\Omega(\mathbf{X}+\mathbf{V}). \tag{30.2}$$

A natural estimator accounting for the low rank of $\mathbf{X}$ will be sought to fit the data $\mathcal{P}_\Omega(\mathbf{Y})$ in the least-squares (LS) error sense as well as minimize the rank of $\mathbf{X}$; see e.g., [9]. However, minimizing the nonconvex matrix rank demands combinatorial complexity, and it is NP-complete [10,11]. Adopting the nuclear norm $\|\mathbf{X}\|_* := \sum_i \sigma_i(\mathbf{X})$ (σ_i signifies ith singular value) as a convex surrogate of rank [12,13], one is then motivated to solve

$$\text{(P1)} \quad \min_{\mathbf{X}} \frac{1}{2}\|\mathcal{P}_\Omega(\mathbf{Y}-\mathbf{X})\|_F^2 + \lambda\|\mathbf{X}\|_*, \tag{30.3}$$

where $\lambda \geq 0$ is the rank-controlling parameter. Being convex (P1) is appealing, and it offers well-documented guarantees for stable and exact recovery in numerous tasks such as matrix completion [9,14] and low-rank plus (compressed) sparse matrix decomposition [15–17]. For the matrix completion setting, typical results establish that if the energy of $\mathbf{X}$ is sufficiently *spread out*, which can be fulfilled for instance when the singular vectors are nonspiky, then with only $\mathcal{O}(Pr\log^2(P))$ randomly chosen matrix entries ($r := \text{rank}(\mathbf{X})$ and $P := \max(L,T)$), one can accurately recover the missing entries in the complement of Ω [9,14].

However, nuclear-norm regularization lacks separability across rows and columns of $\mathbf{X}$ because singular values $\sigma_i(\mathbf{X})$ depend on all entries of the matrix. This coupling challenges streaming and decentralized data analytics, where columns of $\mathbf{X}$ are acquired sequentially in time, or rows of $\mathbf{X}$ are geographically dispersed throughout a network, respectively. The following section introduces an alternative characterization of the nuclear-norm that proves instrumental to developing scalable decentralized and online algorithms to tackle (P1).

Separable rank regularization. Being low-rank matrix $\mathbf{X}$ admits a bilinear factorization $\mathbf{X} = \mathbf{L}\mathbf{Q}^\top$, with factor matrices $\mathbf{L} \in \mathbb{R}^{L\times\rho}$ and $\mathbf{Q} \in \mathbb{R}^{T\times\rho}$. The value of $\rho \geq r$ is chosen sufficiently large to overestimate rank($\mathbf{X}$). Interestingly, the nuclear norm of $\mathbf{X}$ can be alternatively written as the solution of the following nonconvex problem [18,19]

$$\|\mathbf{X}\|_* := \min_{\{\mathbf{L},\mathbf{Q}\}} \frac{1}{2}\left\{\|\mathbf{L}\|_F^2 + \|\mathbf{Q}\|_F^2\right\}, \quad \text{s.t.} \quad \mathbf{X} = \mathbf{L}\mathbf{Q}^\top. \tag{30.4}$$

The optimization (30.4) is over all possible bilinear factorizations of $\mathbf{X}$, so that the number of columns of $\mathbf{L}$ and $\mathbf{Q}$ is also a variable. For an arbitrary matrix $\mathbf{X}$ with SVD $\mathbf{X} = \mathbf{U}_X\boldsymbol{\Sigma}_X\mathbf{V}_X^\top$, the minimum in Eq. (30.4) is attained for $\mathbf{L} = \mathbf{U}_X\boldsymbol{\Sigma}_X^{1/2}$ and $\mathbf{Q} = \mathbf{V}_X\boldsymbol{\Sigma}_X^{1/2}$. Establishing the uniqueness of such a solution requires semidefinite programming (SDP) arguments [18].

Adopting this characterization is useful because the Frobenius norm cost in Eq. (30.4) is separable across the entries of the factor matrices, but it comes at the price of nonconvexity for the corresponding recovery task. However, as argued later under certain conditions, this choice comes with no loss of optimality. Next, we leverage (30.4) to obtain a separable cost equivalent to that in (P1) that can be minimized in a decentralized fashion via the alternating-direction method of multipliers [20–22].

30.3 DECENTRALIZED ANALYTICS

The matrix completion task in (P1) assumes that the samples $\mathcal{P}_\Omega(\mathbf{Y})$ are entirely available at a central processing unit, and they can be jointly processed to infer $\mathbf{X}$. Collecting the entire data is challenging in various applications, or it can be even impossible, e.g., in wireless sensor networks (WSNs) operating under stringent power budget constraints. In other cases such as the Internet or collaborative marketing studies, agents providing private data for, e.g., fitting a low-rank preference model, may not be willing to share their training data but only the learning results. Performing the optimization in a centralized fashion raises robustness concerns as well because the central processor represents an isolated point of failure.

Several customized iterative algorithms have been proposed to solve instances of (P1), and have been shown effective in tackling low- to medium-size problems; see e.g., [4,14,18]. However, most algorithms require computation of singular values per iteration and become prohibitively expensive when dealing with high-dimensional data, as argued in [23]. In a similar vein, stochastic gradient algorithms were recently developed for large-scale problems entailing regularization with the nuclear norm [6,23]. Even though iterations in [23] are highly parallelizable, they are not applicable to networks of arbitrary topology. The aforementioned reasons motivate well developing reduced-complexity *decentralized* algorithms for nuclear-norm minimization.

Network data model. Consider N networked agents capable of performing some local computations as well as exchanging messages among directly connected neighbors. An agent should be understood as an abstract entity, e.g., a sensor in a WSN or a router monitoring Internet traffic. The network is modeled as an undirected graph $G(\mathcal{N}, \mathcal{L})$, where the set of nodes $\mathcal{N} := \{1, \ldots, N\}$ corresponds to the network agents, and the edges (links) in $\mathcal{L} := \{1, \ldots, L\}$ represent pairs of agents that can communicate. Agent $n \in \mathcal{N}$ communicates with its single-hop neighboring peers in $\mathcal{J}_n$, and the size of the neighborhood will be henceforth denoted by $|\mathcal{J}_n|$. To ensure that the data from an arbitrary agent can eventually percolate through the entire network, it is assumed that graph G is connected; i.e., there exists a (possibly) multihop path connecting any two agents.

Decentralized matrix completion. With reference to the matrix completion task in (P1), in the network setting envisioned here each agent $n \in \mathcal{N}$ acquires a few incomplete and noise-corrupted rows of matrix $\mathbf{Y} \in \mathbb{R}^{L \times T}$. Specifically, the local data available to agent n is matrix $\mathcal{P}_{\Omega_n}(\mathbf{Y}_n)$, where $\mathbf{Y}_n \in \mathbb{R}^{L_n \times T}$, $\sum_{n=1}^{N} L_n = L$, and $\mathbf{Y} := \left[\mathbf{Y}_1^\top, \ldots, \mathbf{Y}_N^\top\right]^\top = \mathbf{X} + \mathbf{V}$. The index pairs in Ω_n are those in Ω for which the row index matches the rows of $\mathbf{Y}$ observed by agent n. With regards to the decision variables, partition also $\mathbf{X} := \left[\mathbf{X}_1^\top, \ldots, \mathbf{X}_N^\top\right]^\top \in \mathbb{R}^{L \times T}$ similar to $\mathbf{Y}$, where $\mathbf{X}_n \in \mathbb{R}^{L_n \times T}$, $n = 1, \ldots, N$. Agents collaborate to form the wanted estimator (P1) in a decentralized fashion, which can be equivalently rewritten as

$$\text{(P1)} \qquad \min_{\mathbf{X}} \sum_{n=1}^{N} \left[\frac{1}{2} \|\mathcal{P}_{\Omega_n}(\mathbf{Y}_n - \mathbf{X}_n)\|_F^2 + \frac{\lambda}{N} \|\mathbf{X}\|_* \right].$$

Our objective is to develop a decentralized matrix-completion (DMC) algorithm based on in-network processing of the locally available data. The described setup naturally suggests three features that the algorithm should exhibit: (f1) agent $n \in \mathcal{N}$ should obtain an estimate of $\mathbf{X}_n$, which coincides with the corresponding solution of the centralized estimator (P1) that uses the entire data $\mathcal{P}_\Omega(\mathbf{Y})$; (f2) processing per agent should be kept as simple as possible; and (f3) the overhead for interagent communications should be affordable and confined to single-hop neighborhoods.

To facilitate reducing the computational complexity and memory storage requirements of the decentralized algorithm sought, it is henceforth assumed that the decision variable $\mathbf{X}$ in (P1) has rank at most ρ. For instance, empirical analysis of real Internet traffic data has revealed that OD flow traffic matrices typically have rank$[\mathbf{X}] \in [5, 8]$; hence, one can safely choose $\rho = 10$ [24]. In addition, recall that the rank of the solution $\hat{\mathbf{X}}$ in (P1) is controlled by the choice of λ, and can be made small enough for sufficiently large λ. As argued next, the smaller the value of ρ, the more efficient the algorithm becomes.

Because rank$(\hat{\mathbf{X}}) \leq \rho$, (P1)'s search space is effectively reduced and one can factorize the decision variable as $\mathbf{X} = \mathbf{L}\mathbf{Q}^\top$, where $\mathbf{L}$ and $\mathbf{Q}$ are $L \times \rho$ and $T \times \rho$ matrices, respectively. Adopting this reparametrization of $\mathbf{X}$ in (P1) one obtains the following equivalent optimization problem

$$\text{(P2)} \qquad \min_{\{\mathbf{L},\mathbf{Q}\}} \sum_{n=1}^{N} \left[\frac{1}{2} \|\mathcal{P}_{\Omega_n}(\mathbf{Y}_n - \mathbf{L}_n\mathbf{Q}^\top)\|_F^2 + \frac{\lambda}{N} \|\mathbf{L}\mathbf{Q}^\top\|_* \right],$$

which is nonconvex due to the bilinear terms $\mathbf{L}_n\mathbf{Q}^\top$, and $\mathbf{L} := \left[\mathbf{L}_1^\top, \ldots, \mathbf{L}_N^\top\right]^\top$. The number of variables is reduced from LT in (P1), to $\rho(L+T)$ in (P2). The savings can be significant when ρ is in the order of a few dozen, and both L and T are large. Problem (P2) is still not amenable to decentralized implementation due to: (i) the nonseparable nuclear norm present in the cost function; and (ii) the global variable $\mathbf{Q}$ coupling the per-agent summands.

Leveraging (30.4), the following reformulation of (P2) provides an important first step toward obtaining a decentralized estimator:

$$\text{(P3)} \qquad \min_{\{\mathbf{L},\mathbf{Q}\}} \sum_{n=1}^{N} \left[\frac{1}{2} \|\mathcal{P}_{\Omega_n}(\mathbf{Y}_n - \mathbf{L}_n\mathbf{Q}^\top)\|_F^2 + \frac{\lambda}{2N} \left\{ N\|\mathbf{L}_n\|_F^2 + \|\mathbf{Q}\|_F^2 \right\} \right].$$

Building on Eq. (30.4) and because rank$(\hat{\mathbf{X}}) \leq \rho$, it readily follows that the separable Frobenius-norm regularization in (P3) comes with no loss of optimality, meaning that (P1) and (P3) admit identical solutions. This equivalence ensures that by finding the global minimum of (P3) [which can have significantly fewer variables than (P1)], one can recover the optimal solution of (P1). However, because (P3) is nonconvex, it may have stationary points that need not be globally optimal. Interestingly, the next proposition offers a global optimality certificate for the stationary points of (P3). For the detailed proof, see [7, Appendix A].

Proposition 30.1. *Let $\{\bar{\mathbf{L}}, \bar{\mathbf{Q}}\}$ be a stationary point of (P3). If $\|\mathcal{P}_\Omega(\mathbf{Y} - \bar{\mathbf{L}}\bar{\mathbf{Q}}^\top)\| \leq \lambda$ (no subscript in $\|.\|$ signifies spectral norm), then $\hat{\mathbf{X}} = \bar{\mathbf{L}}\bar{\mathbf{Q}}^\top$ is the globally optimal solution of (P1).* ■

To decompose the cost function in (P3), in which summands are coupled through the global variables $\mathbf{Q}$, introduce auxiliary variables $\{\mathbf{Q}_n\}_{n=1}^N$ representing local estimates of $\mathbf{Q}$ per agent n. These local estimates are utilized to form the separable *constrained* minimization problem

$$\text{(P4)} \qquad \min_{\{\mathbf{L}_n,\mathbf{Q}_n\}} \sum_{n=1}^{N} \left[\frac{1}{2} \|\mathcal{P}_{\Omega_n}(\mathbf{Y}_n - \mathbf{L}_n\mathbf{Q}_n^\top)\|_F^2 + \frac{\lambda}{2} \|\mathbf{L}_n\|_F^2 + \frac{\lambda}{2N} \|\mathbf{Q}_n\|_F^2 \right]$$
$$\text{s.t.} \quad \mathbf{Q}_n = \mathbf{Q}_m, \quad m \in \mathcal{J}_n, n \in \mathcal{N}.$$

Clearly, (P3) and (P4) are equivalent optimization problems because the graph G is assumed to be connected. The equivalence should be understood in the sense that $\hat{\mathbf{Q}}_1 = \hat{\mathbf{Q}}_2 = \cdots = \hat{\mathbf{Q}}_N = \hat{\mathbf{Q}}$, where

$\{\hat{\mathbf{Q}}_n\}_{n\in\mathcal{N}}$ and $\hat{\mathbf{Q}}$ are the optimal solutions of (P4) and (P3), respectively. Of course, the corresponding estimates of $\mathbf{L}$ will coincide as well. Even though consensus is a fortiori imposed within neighborhoods, it extends to the whole (connected) network and local estimates agree on the global solution of (P3). To arrive at the desired decentralized algorithm, it is convenient to reparameterize the consensus constraints in (P4) as

$$\mathbf{Q}_n = \bar{\mathbf{F}}_n^m, \quad \mathbf{Q}_m = \tilde{\mathbf{F}}_n^m, \quad \text{and} \quad \bar{\mathbf{F}}_n^m = \tilde{\mathbf{F}}_n^m, \; m \in \mathcal{J}_n, \; n \in \mathcal{N}, \tag{30.5}$$

where $\{\bar{\mathbf{F}}_n^m, \tilde{\mathbf{F}}_n^m\}_{n\in\mathcal{N}}^{m\in\mathcal{J}_n}$ are auxiliary optimization variables that will be eventually eliminated.

Alternating-direction method of multipliers. To tackle the constrained minimization problem (P4), associate dual variables $\bar{\mathbf{D}}_n^m$ and $\tilde{\mathbf{D}}_n^m$ with the consensus constraints in Eq. (30.5). Next introduce the quadratically *augmented* Lagrangian function

$$\begin{aligned}\mathcal{L}_c(\mathcal{V}_1, \mathcal{V}_2, \mathcal{V}_3, \mathcal{M}) = &\sum_{n=1}^{N}\left[\frac{1}{2}\|\mathcal{P}_{\Omega_n}(\mathbf{Y}_n - \mathbf{L}_n\mathbf{Q}_n^\top)\|_F^2 + \frac{\lambda}{2N}\{N\|\mathbf{L}_n\|_F^2 + \|\mathbf{Q}_n\|_F^2\}\right] \\ &+ \sum_{n=1}^{N}\sum_{m\in\mathcal{J}_n}\left\{\langle\bar{\mathbf{C}}_n^m, \mathbf{Q}_n - \bar{\mathbf{F}}_n^m\rangle + \langle\tilde{\mathbf{C}}_n^m, \mathbf{Q}_m - \tilde{\mathbf{F}}_n^m\rangle\right\} \\ &+ \frac{c}{2}\sum_{n=1}^{N}\sum_{m\in\mathcal{J}_n}\left\{\|\mathbf{Q}_n - \bar{\mathbf{F}}_n^m\|_F^2 + \|\mathbf{Q}_m - \tilde{\mathbf{F}}_n^m\|_F^2\right\},\end{aligned} \tag{30.6}$$

where c is a positive penalty coefficient, and the primal variables are split into three groups $\mathcal{V}_1 := \{\mathbf{Q}_n\}_{n=1}^N$, $\mathcal{V}_2 := \{\mathbf{L}_n\}_{n=1}^N$, and $\mathcal{V}_3 := \{\bar{\mathbf{F}}_n^m, \tilde{\mathbf{F}}_n^m\}_{n\in\mathcal{N}}^{m\in\mathcal{J}_n}$. For notational convenience, collect all multipliers in $\mathcal{M} := \{\bar{\mathbf{C}}_n^m, \tilde{\mathbf{C}}_n^m,\}_{n\in\mathcal{N}}^{m\in\mathcal{J}_n}$. The remaining constraints in Eq. (30.5), namely $C_{\mathcal{V}} := \{\bar{\mathbf{F}}_n^m = \tilde{\mathbf{F}}_n^m,\ m \in \mathcal{J}_n,\ n \in \mathcal{N}\}$, have not been dualized.

To minimize (P4) in a decentralized fashion, a variation of the alternating-direction method of multipliers (ADMM) will be adopted here. The ADMM is an iterative augmented Lagrangian method especially well suited for parallel processing [20,21], which has been proven successful to tackle the optimization tasks encountered, e.g., with decentralized estimation problems [22,25,26]. The proposed solver entails an iterative procedure comprising four steps per iteration $k = 1, 2, \ldots$

[S1] Update dual variables for all $n \in \mathcal{N}$, $m \in \mathcal{J}_n$:

$$\bar{\mathbf{C}}_n^m[k] = \bar{\mathbf{C}}_n^m[k-1] + \mu(\mathbf{Q}_n[k] - \bar{\mathbf{F}}_n^m[k]), \tag{30.7}$$
$$\tilde{\mathbf{C}}_n^m[k] = \tilde{\mathbf{C}}_n^m[k-1] + \mu(\mathbf{Q}_m[k] - \tilde{\mathbf{F}}_n^m[k]). \tag{30.8}$$

[S2] Update the first group of primal variables:

$$\mathcal{V}_1[k+1] = \arg\min_{\mathcal{V}_1} \mathcal{L}_c(\mathcal{V}_1, \mathcal{V}_2[k], \mathcal{V}_3[k], \mathcal{M}[k]). \tag{30.9}$$

[S3] Update the second group of primal variables:

$$\mathcal{V}_2[k+1] = \arg\min_{\mathcal{V}_2} \mathcal{L}_c(\mathcal{V}_1[k+1], \mathcal{V}_2, \mathcal{V}_3[k], \mathcal{M}[k]). \tag{30.10}$$

[S4] Update the auxiliary primal variables:

$$\mathcal{V}_3[k+1] = \arg\min_{\mathcal{V}_3 \in C_V} \mathcal{L}_c(\mathcal{V}_1[k+1], \mathcal{V}_2[k+1], \mathcal{V}_3, \mathcal{M}[k]). \tag{30.11}$$

This four-step procedure implements a block-coordinate descent method with dual variable updates. At each step of minimizing the augmented Lagrangian, the variables not being updated are treated as fixed and are substituted with their most up-to-date values. Different from ADMM, the alternating-minimization step here generally cycles over three groups of primal variables $\mathcal{V}_1$–$\mathcal{V}_3$ (cf. two groups in ADMM [20]). In [S1], $\mu > 0$ is the step size of the subgradient ascent iterations on the dual problem. While it is common in ADMM implementations to select $\mu = c$, a distinction between the step size and the penalty parameter is made explicit here in the interest of generality.

Reformulating the estimator (P1) to its equivalent form (P4) renders the augmented Lagrangian in Eq. (30.6) highly decomposable. The separability comes in two flavors, both with respect to the variable groups $\mathcal{V}_1$-$\mathcal{V}_3$, as well as across the network agents $n \in \mathcal{N}$. This in turn leads to highly parallelized, simplified recursions corresponding to the aforementioned four steps. Specifically, it is shown in [7, Appendix B] that if the multipliers are initialized to zero, [S1]–[S4] constitute the DMC algorithm tabulated under Algorithm 30.1. Careful inspection of Algorithm 30.1 reveals that the inherently redundant auxiliary variables and multipliers $\{\mathbf{F}_n^m, \bar{\mathbf{F}}_n^m, \mathbf{C}_n^m\}$ have been eliminated. Agent n does not need to *separately* keep track of all its nonredundant multipliers $\{\bar{\mathbf{C}}_n^m\}_{m \in \mathcal{J}_n}$, but only to update their respective (scaled) sums $\mathbf{O}_n[k] := 2\sum_{m \in \mathcal{J}_n} \bar{\mathbf{C}}_n^m[k]$. To derive Algorithm 30.1 it is useful to recognize that linearity of $\mathcal{P}_{\Omega_n}$ implies that $\text{vec}(\mathcal{P}_{\Omega_n}(\mathbf{Z})) = \boldsymbol{\Omega}_n \text{vec}(\mathbf{Z})$, where $\boldsymbol{\Omega}_n \in \{0,1\}^{L_nT \times L_nT}$ is a diagonal matrix.

Algorithm 30.1 DMC ALGORITHM PER AGENT $N \in \mathcal{N}$

Input $\mathbf{Y}_n, \Omega_n, \lambda, c, \mu$

Initialize $\mathbf{O}[0] = \mathbf{0}_{T \times \rho}$, and $\mathbf{L}_n[1]$, $\mathbf{Q}_n[1]$ at random
for $k = 1, 2, \ldots$ **do**
Receive $\{\mathbf{Q}_m[k]\}$ from neighbors $m \in \mathcal{J}_n$
[S1] Update local dual variables:

$\mathbf{O}_n[k] = \mathbf{O}_n[k-1] + \mu \sum_{m \in \mathcal{J}_n} (\mathbf{Q}_n[k] - \mathbf{Q}_m[k])$

[S2] Update first group of local primal variables:
$\mathbf{E}_n[k+1] = \left\{(\mathbf{I}_T \otimes \mathbf{L}_n^\top[k])\boldsymbol{\Omega}_n(\mathbf{I}_T \otimes \mathbf{L}_n[k]) + (\lambda/N + 2c|\mathcal{J}_n|)\mathbf{I}_{\rho T}\right\}^{-1}$
$\mathbf{G}_n[k+1] := (\mathbf{I}_T \otimes \mathbf{L}_n^\top[k])\boldsymbol{\Omega}_n \text{vec}(\mathbf{Y}_n) - \text{vec}(\mathbf{O}_n^\top[k]) + c\,\text{vec}(\sum_{m \in \mathcal{J}_n} (\mathbf{Q}_n^\top[k] + \mathbf{Q}_m^\top[k]))$
$\mathbf{Q}_n^\top[k+1] = \text{unvec}(\mathbf{E}_n[k+1]\mathbf{G}_n[k+1])$
[S3] Update second group of local primal variables:
$\mathbf{D}_n[k+1] := \left\{(\mathbf{Q}_n^\top[k+1] \otimes \mathbf{I}_{L_n})\boldsymbol{\Omega}_n(\mathbf{Q}_n[k+1] \otimes \mathbf{I}_{L_n}) + \lambda \mathbf{I}_{\rho L_n}\right\}^{-1}$
$\mathbf{L}_n[k+1] = \text{unvec}\left(\mathbf{D}_n[k+1]\,(\mathbf{Q}_n^\top[k+1] \otimes \mathbf{I}_{L_n})\boldsymbol{\Omega}_n \text{vec}(\mathbf{Y}_n)\right)$
Broadcast $\{\mathbf{Q}_n[k+1]\}$ to neighbors $m \in \mathcal{J}_n$
end for
Return $\mathbf{Q}_n, \mathbf{L}_n$

Computational and communication cost. The per-agent computational complexity of the DMC algorithm is dominated by repeated inversions of $\rho \times \rho$ and $\rho L_n \times \rho L_n$ matrices to obtain $\mathbf{E}_n[k+1]$ and $\mathbf{D}_n[k+1]$, respectively, and matrix multiplications to update $\mathbf{Q}_n[k+1]$ and $\mathbf{L}_n[k+1]$. Notice that $\mathbf{E}_n[k+1] \in \mathbb{R}^{\rho T \times \rho T}$ has block-diagonal structure with blocks of size $\rho \times \rho$. Overall, the per-iteration complexity across the network is upper bounded by $\mathcal{O}(\rho^3 NT)$, which grows linearly with the network size. This is affordable because in practice ρ is typically small for a number of applications of interest (cf. the low-rank assumption). In addition, L_n, the number of row vectors acquired per agent, and T, the number of time instants for data collection, can be controlled by the designer to accommodate a prescribed maximum computational complexity. One can also benefit from the decomposability of Eqs. (30.9) and (30.10) across rows of $\mathbf{L}$ and $\mathbf{Q}$, respectively, and parallelize the row updates. This way, one only needs to invert $\rho \times \rho$ matrices.

On a per-iteration basis, network agents communicate their updated local estimates $\mathbf{Q}_n[k]$ only with their neighbors in order to carry out the updates of primal and dual variables during the next iteration. In terms of communication cost, $\mathbf{Q}_n[k]$ is a $T \times \rho$ matrix and its transmission does not incur significant overhead for small values of ρ. Observe that the dual variables $\mathbf{O}_n[k]$ need not be exchanged, and the overall communication cost does not depend on the network size N.

Convergence and optimality. When employed to solve nonconvex problems such as (P4), ADMM (or its variant used here) offers no convergence guarantees. However, there is ample experimental evidence in the literature that supports empirical convergence of ADMM, especially when the nonconvex problem at hand exhibits "favorable" structure. For instance, (P4) is biconvex and gives rise to the strictly convex optimization subproblems (30.9)–(30.11), which admit unique closed-form solutions per iteration. This observation and the linearity of the constraints endow Algorithm 30.1 with good convergence properties—extensive numerical tests in [7] demonstrate that this is indeed the case. The following proposition proved in [7, Appendix C] asserts that upon convergence, Algorithm 30.1 attains consensus and global optimality thus yielding the performance of the centralized estimator (P1).

Proposition 30.2. *If the sequence of iterates $\{\mathbf{Q}_n[k], \mathbf{L}_n[k]\}_{n\in\mathcal{N}}$ generated by Algorithm 30.1 converge to $\{\bar{\mathbf{Q}}_n, \bar{\mathbf{L}}_n\}_{n\in\mathcal{N}}$, and G is connected, then: i) $\bar{\mathbf{Q}}_n = \bar{\mathbf{Q}}_m$ $n, m \in \mathcal{N}$; and ii) if $\|\mathcal{P}_\Omega(\mathbf{Y}-\bar{\mathbf{L}}\bar{\mathbf{Q}}_1^\top)\| \leq \lambda$, then $\hat{\mathbf{X}} = \bar{\mathbf{L}}\bar{\mathbf{Q}}_1^\top$, where $\hat{\mathbf{X}}$ is the global optimum of (P1).* ■

30.3.1 INTERNET DELAY CARTOGRAPHY

End-to-end network latency information is critical toward enforcing quality-of-service constraints in many Internet applications. However, probing all pairwise delays becomes infeasible in large-scale networks. If one collects the end-to-end latencies of source-sink pairs (i,j) in a delay matrix $\mathbf{X} := [x_{i,j}] \in \mathbb{R}^{N\times N}$, strong dependencies among path delays render $\mathbf{X}$ low rank [27]. This is mainly because the paths with nearby end nodes often overlap and share common bottleneck links. This property of $\mathbf{X}$ along with the decentralized processing requirements of large-scale networks motivates well the adoption of the DMC algorithm for network-wide path latency prediction. Given that the nth row of $\mathbf{X}$ is partially available to agent n, the goal is to impute the missing delays through agent collaboration.

End-to-end flow latencies are collected from the operation of the Internet-2 backbone network during August 18–22, 2011 [28]. The Internet-2 network (Fig. 30.1 (left)) comprises $N = 9$ agents, $L = 26$ links, and $F = 81$ flows. Spectral analysis of the delay matrix reveals that the first four singular

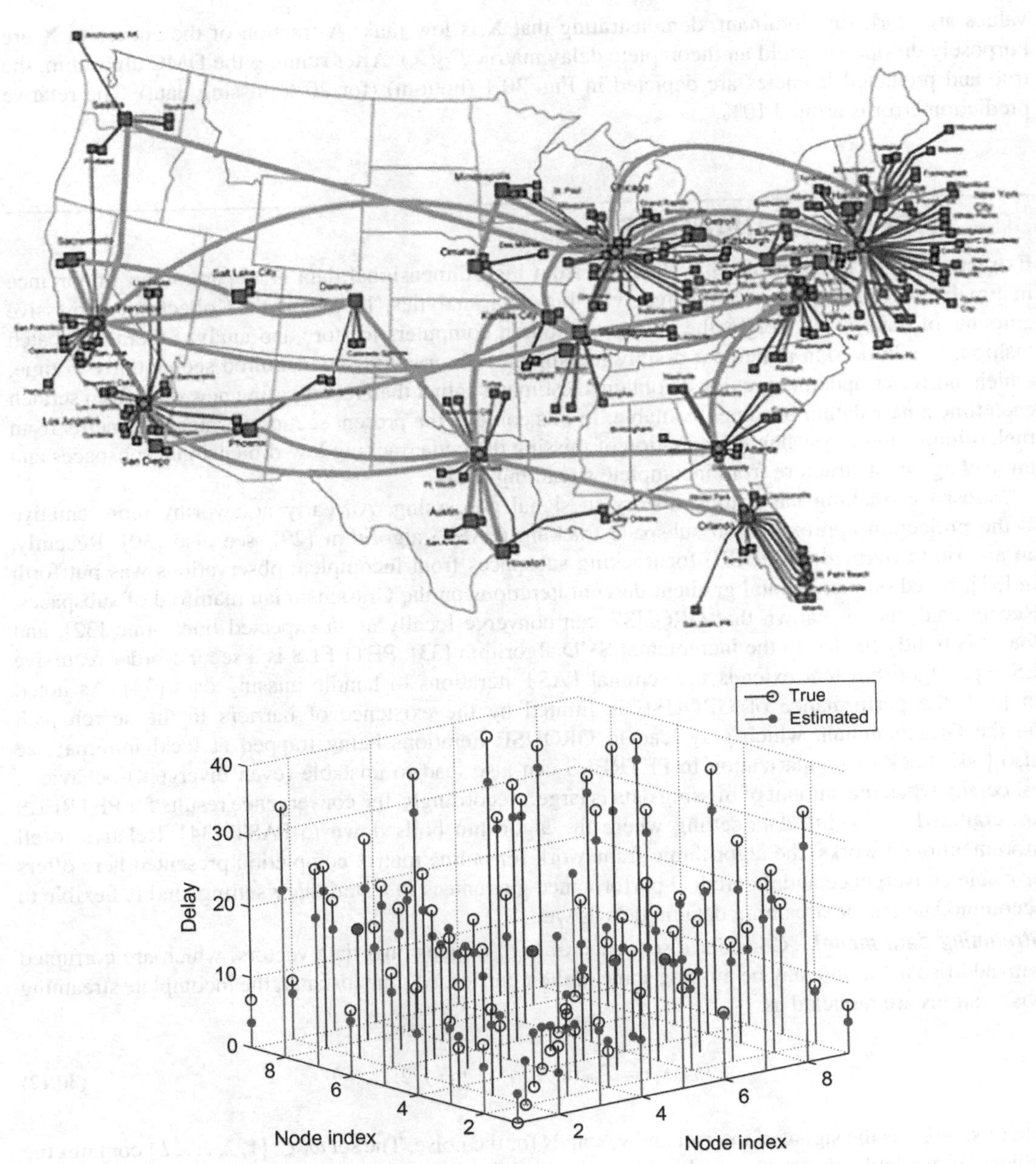

FIG. 30.1

Internet-2 end-to-end delay prediction. (top) Topology of the Internet-2 backbone network. (bottom) Predicted and true end-to-end delays of the Internet-2 network when only 20% of the paths are randomly sampled.

values are markedly dominant, demonstrating that $\mathbf{X}$ is low rank. A fraction of the entries in $\mathbf{X}$ are purposely dropped to yield an incomplete delay matrix $\mathcal{P}_{\Omega}(\mathbf{X})$. After running the DMC algorithm, the true and predicted latencies are depicted in Fig. 30.1 (bottom) (for 20% missing data). The relative prediction error is around 10%.

30.4 STREAMING ANALYTICS

Extracting latent low-dimensional structure from high-dimensional data is of paramount importance in timely inference tasks encountered with big data analytics. However, the collection of massive amounts of data far outweighs the ability of modern computers to store and analyze them in a batch fashion. In addition, in practice (possibly incomplete) observations are acquired sequentially in time, which motivates updating previously obtained estimates rather than recomputing new ones from scratch each time a new datum becomes available. In this context, the present section permeates benefits from rank minimization to scalable imputation of missing data, via tracking low-dimensional subspaces and unraveling latent structure from incomplete streaming data.

Subspace tracking has a long history in signal processing. An early noteworthy representative is the projection approximation subspace tracking (PAST) algorithm [29]; see also [30]. Recently, an algorithm (termed GROUSE) for tracking subspaces from incomplete observations was put forth in [31], based on incremental gradient descent iterations on the Grassmannian manifold of subspaces. Recent analysis has shown that GROUSE can converge locally at an expected linear rate [32], and that it is tightly related to the incremental SVD algorithm [33]. PETRELS is a second-order recursive LS type algorithm that extends the seminal PAST iterations to handle missing data [34]. As noted in [35], the performance of GROUSE is limited by the existence of barriers in the search path on the Grassmannian, which may lead to GROUSE iterations being trapped at local minima; see also [34]. Lack of regularization in PETRELS can also lead to unstable (even divergent) behaviors, especially when the amount of missing data is large. Accordingly, the convergence results for PETRELS are confined to the full-data setting where the algorithm boils down to PAST [34]. Relative to all aforementioned works, the algorithmic framework for online matrix completion presented here offers provable convergence and theoretical performance guarantees in a stationary setting, and is flexible to accommodate tensor streaming data models as well.

Streaming data model. Consider a sequence of high-dimensional data vectors, which are corrupted with additive noise and may be missing some of their entries. At time instant t, the incomplete streaming observations are modeled as

$$\mathcal{P}_{\omega_t}(\mathbf{y}_t) = \mathcal{P}_{\omega_t}(\mathbf{x}_t + \mathbf{v}_t), \quad t = 1, 2, \ldots, \tag{30.12}$$

where $\mathbf{x}_t \in \mathbb{R}^L$ is the signal of interest and $\mathbf{v}_t$ stands for the noise. The set $\omega_t \subset \{1, 2, \ldots, L\}$ contains the indices of available observations while the corresponding sampling operator $\mathcal{P}_{\omega_t}(\cdot)$ sets the entries of its vector argument not in ω_t to zero, and keeps the rest unchanged; note that $\mathcal{P}_{\omega_t}(\mathbf{y}_t) \in \mathbb{R}^L$. Depending on the application, these acquired vectors could, e.g., correspond to (vectorized) images, link traffic measurements collected across physical links of a computer network, or movie ratings provided by

Netflix users. Suppose that the sequence $\{\mathbf{x}_t\}_{t=1}^{\infty}$ lives in a *low-dimensional* ($\ll L$) linear subspace $\mathcal{L}_t$, which is allowed to change slowly over time. Given the incomplete observations $\{\mathcal{P}_{\omega_\tau}(\mathbf{y}_\tau)\}_{\tau=1}^{t}$, ensuing sections deal with online (adaptive) estimation of $\mathcal{L}_t$, and reconstruction of $\mathbf{x}_t$ as a byproduct. The reconstruction here involves imputing the missing elements, and denoising the observed ones.

Online matrix completion. Collect the indices of available observations up to time t in the set $\Omega_t := \cup_{\tau=1}^{t}\omega_\tau$, and the actual batch of observations in the matrix $\mathcal{P}_{\Omega_t}(\mathbf{Y}_t) := [\mathcal{P}_{\omega_1}(\mathbf{y}_1), \ldots, \mathcal{P}_{\omega_t}(\mathbf{y}_t)] \in \mathbb{R}^{L\times t}$. Likewise, introduce matrix $\mathbf{X}_t$ containing the signal of interest. Because $\mathbf{x}_t$ lies in a low-dimensional subspace, $\mathbf{X}_t$ is (approximately) a *low-rank* matrix. A natural estimator leveraging the low rank property of $\mathbf{X}_t$ attempts to fit the incomplete data $\mathcal{P}_{\Omega_t}(\mathbf{Y}_t)$ to $\mathbf{X}_t$ in the LS sense, and minimize the rank of $\mathbf{X}_t$. This motivates recovering $\mathbf{X}_t$ by solving (P1).

Scalable imputation algorithms for streaming observations should effectively overcome the following challenges: (c1) the problem size can easily become quite large, because the number of optimization variables Lt grows with time; (c2) existing batch iterative solvers for (P1) typically rely on costly SVD computations per iteration; see e.g., [14]; and (c3) (columnwise) nonseparability of the nuclear-norm challenges online processing when new columns $\{\mathcal{P}_{\omega_t}(\mathbf{y}_t)\}$ arrive sequentially in time. To limit the computational complexity and memory storage requirements of the algorithm sought, it is henceforth assumed that the dimensionality of the underlying time-varying subspace $\mathcal{L}_t$ is bounded by a known quantity ρ. Accordingly, it is natural to require $\text{rank}(\hat{\mathbf{X}}_t) \le \rho$. Because $\text{rank}(\hat{\mathbf{X}}_t) \le \rho$, one can factorize the matrix decision variable as $\mathbf{X} = \mathbf{L}\mathbf{Q}^\top$, where $\mathbf{L}$ and $\mathbf{Q}$ are $L \times \rho$ and $t \times \rho$ matrices, respectively. Such a bilinear decomposition suggests $\mathcal{L}_t$ is spanned by the columns of the tall matrix $\mathbf{L}$ while the rows of $\mathbf{Q}$ are the projections of $\{\mathbf{x}_t\}$ onto $\mathcal{L}_t$.

Leveraging once more the separable nuclear-norm regularization in Eq. (30.4), a possible adaptive counterpart to (P1) is the exponentially weighted LS (EWLS) estimator found by minimizing the empirical cost

$$\text{(P5)}\qquad \min_{\{\mathbf{L},\mathbf{Q}\}} \sum_{\tau=1}^{t} \theta^{t-\tau}\left[\frac{1}{2}\|\mathcal{P}_{\omega_\tau}(\mathbf{y}_\tau - \mathbf{L}\mathbf{q}_\tau)\|_2^2 + \frac{\bar{\lambda}_t}{2}\|\mathbf{L}\|_F^2 + \frac{\lambda_t}{2}\|\mathbf{q}_\tau\|_2^2\right],$$

where $\mathbf{Q} := [\mathbf{q}_1, \ldots, \mathbf{q}_t]$, $\bar{\lambda}_t := \lambda_t / \sum_{\tau=1}^{t}\theta^{t-\tau}$, and $0 < \theta \le 1$ is the so-termed forgetting factor. When $\theta < 1$, data in the distant past are exponentially downweighted, which facilitates tracking in nonstationary environments. In the case of infinite memory ($\theta = 1$) and for $\lambda_t = \lambda$, the formulation (P5) coincides with the batch estimator (P1). This is the reason for the time-varying factor $\bar{\lambda}_t$ weighting $\|\mathbf{L}\|_F^2$.

Toward deriving a real-time, computationally efficient, and recursive solver of (P5), an alternating-minimization (AM) method is adopted in which iterations coincide with the time-scale t of data acquisition. Per time instant t, a new datum $\{\mathcal{P}_{\omega_t}(\mathbf{y}_t)\}$ is drawn and $\mathbf{q}_t$ is estimated via

$$\mathbf{q}[t] = \arg\min_{\mathbf{q}}\left[\frac{1}{2}\|\mathcal{P}_{\omega_t}(\mathbf{y}_t - \mathbf{L}[t-1]\mathbf{q})\|_2^2 + \frac{\lambda_t}{2}\|\mathbf{q}\|_2^2\right], \tag{30.13}$$

which is an ℓ_2-norm regularized LS (ridge-regression) problem. It admits the closed-form solution

$$\mathbf{q}[t] = \left(\lambda_t \mathbf{I}_\rho + \mathbf{L}^\top[t-1]\boldsymbol{\Omega}_t\mathbf{L}[t-1]\right)^{-1} \mathbf{L}^\top[t-1]\mathcal{P}_{\omega_t}(\mathbf{y}_t), \tag{30.14}$$

where diagonal matrix $\boldsymbol{\Omega}_t \in \{0,1\}^{L\times L}$ is such that $[\boldsymbol{\Omega}_t]_{l,l} = 1$ if $l \in \omega_t$, and is zero elsewhere. In the second step of the AM scheme, the updated subspace matrix $\mathbf{L}[t]$ is obtained by minimizing (P5) with respect to $\mathbf{L}$, while the optimization variables $\{\mathbf{q}_\tau\}_{\tau=1}^t$ are fixed and take the values $\{\mathbf{q}[\tau]\}_{\tau=1}^t$, namely

$$\mathbf{L}[t] = \arg\min_{\mathbf{L}} \left[\frac{\lambda_t}{2}\|\mathbf{L}\|_F^2 + \sum_{\tau=1}^{t} \theta^{t-\tau}\frac{1}{2}\|\mathcal{P}_{\omega_\tau}(\mathbf{y}_\tau - \mathbf{L}\mathbf{q}[\tau])\|_2^2\right]. \tag{30.15}$$

Notice that Eq. (30.15) decouples over the rows of $\mathbf{L}$ which are obtained in parallel via

$$\mathbf{l}_l[t] = \arg\min_{\mathbf{l}} \left[\frac{\lambda_t}{2}\|\mathbf{l}\|_2^2 + \sum_{\tau=1}^{t} \theta^{t-\tau}\omega_{l,\tau}(y_{l,\tau} - \mathbf{l}^\top\mathbf{q}[\tau])^2\right] \tag{30.16}$$

for $l = 1,\ldots,L$, where $\omega_{l,\tau}$ denotes the lth diagonal entry of $\boldsymbol{\Omega}_\tau$. For $\theta = 1$ and fixed $\lambda_t = \lambda$, $\forall t$, subproblems (30.16) can be efficiently solved via recursive LS (RLS) [36]. Upon defining $\mathbf{s}_l[t] := \sum_{\tau=1}^t \theta^{t-\tau}\omega_{l,\tau}y_{l,\tau}\mathbf{q}[\tau]$, $\mathbf{H}_l[t] := \sum_{\tau=1}^t \theta^{t-\tau}\omega_{l,\tau}\mathbf{q}[\tau]\mathbf{q}^\top[\tau] + \lambda_t\mathbf{I}_\rho$, and $\mathbf{M}_l[t] := \mathbf{H}_l^{-1}[t]$, one updates

$$\mathbf{s}_l[t] = \mathbf{s}_l[t-1] + \omega_{l,t}y_{l,t}\mathbf{q}[t]$$

$$\mathbf{M}_l[t] = \mathbf{M}_l[t-1] - \omega_{l,t}\frac{\mathbf{M}_l[t-1]\mathbf{q}[t]\mathbf{q}^\top[t]\mathbf{M}_l[t-1]}{1+\mathbf{q}^\top[t]\mathbf{M}_l[t-1]\mathbf{q}[t]}$$

and forms $\mathbf{l}_l[t] = \mathbf{M}_l[t]\mathbf{s}_l[t]$, for $l = 1,\ldots,L$.

However, for $0 < \theta < 1$ the regularization term $(\lambda_t/2)\|\mathbf{l}\|_2^2$ in Eq. (30.16) makes it impossible to express $\mathbf{H}_l[t]$ in terms of $\mathbf{H}_l[t-1]$ plus a rank-one correction. Hence, one cannot resort to the matrix inversion lemma and update $\mathbf{M}_l[t]$ with quadratic complexity only. Based on direct inversion of each $\mathbf{H}_l[t]$, the alternating LS algorithm for subspace tracking from incomplete data is tabulated under Algorithm 30.2.

Algorithm 30.2 ALTERNATING LS FOR SUBSPACE TRACKING FROM INCOMPLETE OBSERVATIONS

input $\{\mathcal{P}_{\omega_\tau}(\mathbf{y}_\tau), \omega_\tau\}_{\tau=1}^\infty$, $\{\lambda_\tau\}_{\tau=1}^\infty$, and θ.
initialize $\mathbf{G}_l[0] = \mathbf{0}_{\rho\times\rho}$, $\mathbf{s}_l[0] = \mathbf{0}_\rho$, $l = 1,\ldots,L$, and $\mathbf{L}[0]$ at random.
for $t = 1,2,\ldots$ **do**
  $\mathbf{D}[t] = \left(\lambda_t\mathbf{I}_\rho + \mathbf{L}^\top[t-1]\boldsymbol{\Omega}_t\mathbf{L}[t-1]\right)^{-1}\mathbf{L}^\top[t-1]$.
  $\mathbf{q}[t] = \mathbf{D}[t]\mathcal{P}_{\omega_t}(\mathbf{y}_t)$.
  $\mathbf{G}_l[t] = \theta\mathbf{G}_l[t-1] + \omega_{l,t}\mathbf{q}[t]\mathbf{q}[t]^\top$, $l = 1,\ldots,L$.
  $\mathbf{s}_l[t] = \theta\mathbf{s}_l[t-1] + \omega_{l,t}y_{l,t}\mathbf{q}[t]$, $l = 1,\ldots,L$.
  $\mathbf{l}_l[t] = \left(\mathbf{G}_l[t] + \lambda_t\mathbf{I}_\rho\right)^{-1}\mathbf{s}_l[t]$, $l = 1,\ldots,L$.
  return $\hat{\mathbf{x}}_t := \mathbf{L}[t]\mathbf{q}[t]$.
end for

Before moving on to reduced-complexity subspace trackers it is worth commenting that the basic idea of performing online rank-minimization leveraging the separable nuclear-norm regularization was first introduced in [6], in the context of unveiling network traffic anomalies. Since then, the approach has gained popularity in real-time nonnegative matrix factorization for singing voice separation from its music accompaniment [37] and online robust PCA [38], to name a few examples.

Low-complexity stochastic-gradient subspace updates. To further reduce Algorithm 30.2's computational complexity in updating the subspace $\mathbf{L}[t]$, here we develop lightweight algorithms that better suit big data applications. To this end, the basic AM framework is retained and the update for $\mathbf{q}[t]$ will be identical [cf. Eq. (30.14)]. However, instead of exactly solving an unconstrained quadratic program per iteration to obtain $\mathbf{L}[t]$ [cf. Eq. (30.15)], the subspace estimates will be obtained via stochastic-gradient descent (SGD) iterations. As shown later on, these updates can be traced to inexact solutions of a certain quadratic program related to Eq. (30.15).

For $\theta = 1$, it is shown in Section 30.4.1 that Algorithm 30.2's subspace estimate $\mathbf{L}[t]$ is obtained by minimizing the empirical cost function $\hat{C}_t(\mathbf{L}) = (1/t)\sum_{\tau=1}^{t} f_\tau(\mathbf{L})$, where

$$f_t(\mathbf{L}) := \frac{1}{2}\|\mathcal{P}_{\omega_t}(\mathbf{y}_t - \mathbf{L}\mathbf{q}[t])\|_2^2 + \frac{\lambda}{2t}\|\mathbf{L}\|_F^2 + \frac{\lambda}{2}\|\mathbf{q}[t]\|_2^2, \quad t = 1, 2, \ldots. \tag{30.17}$$

By the law of large numbers, if $\{\mathcal{P}_{\omega_t}(\mathbf{y}_t)\}_{t=1}^{\infty}$ are stationary, solving $\min_{\mathbf{L}} \lim_{t\to\infty} \hat{C}_t(\mathbf{L})$ yields the desired minimizer of the *expected* cost $\mathbb{E}[C_t(\mathbf{L})]$, where the expectation is taken with respect to the unknown probability distribution of the data. A standard approach to achieve this same goal—typically with reduced computational complexity—is to drop the expectation (or the sample averaging operator for that matter), and update the subspace via SGD; see e.g., [36]

$$\mathbf{L}[t] = \mathbf{L}[t-1] - (\mu[t])^{-1}\nabla f_t(\mathbf{L}[t-1]), \tag{30.18}$$

where $(\mu[t])^{-1}$ is the step size, and $\nabla f_t(\mathbf{L}) = -\mathcal{P}_{\omega_t}(\mathbf{y}_t - \mathbf{L}\mathbf{q}[t])\mathbf{q}^\top[t] + (\lambda/t)\mathbf{L}$. The subspace update $\mathbf{L}[t]$ is nothing but the minimizer of a second-order approximation $Q_{\mu[t],t}(\mathbf{L}, \mathbf{L}[t-1])$ of $f_t(\mathbf{L})$ around the previous subspace estimate $\mathbf{L}[t-1]$, where

$$Q_{\mu,t}(\mathbf{L}_1, \mathbf{L}_2) := f_t(\mathbf{L}_2) + \langle \mathbf{L}_1 - \mathbf{L}_2, \nabla f_t(\mathbf{L}_2)\rangle + \frac{\mu}{2}\|\mathbf{L}_1 - \mathbf{L}_2\|_F^2.$$

To tune the step size, the backtracking rule is adopted, whereby the nonincreasing step-size sequence $\{(\mu[t])^{-1}\}$ decreases geometrically at certain iterations to guarantee the quadratic function $Q_{\mu[t],t}(\mathbf{L}, \mathbf{L}[t-1])$ majorizes $f_t(\mathbf{L})$ at the new update $\mathbf{L}[t]$. Other choices of the step size are discussed in Section 30.4.1. Different from Algorithm 30.2, no matrix inversions are involved in the update of the subspace $\mathbf{L}[t]$. In the context of adaptive filtering, first-order SGD algorithms such as Eq. (30.17) are known to converge slower than RLS. This is expected because RLS can be shown to be an instance of Newton's (second-order) optimization method [36, Ch. 4].

Algorithm 30.3 ONLINE SGD FOR SUBSPACE TRACKING FROM INCOMPLETE OBSERVATIONS

input $\{\mathcal{P}_{\omega_\tau}(\mathbf{y}_\tau), \omega_\tau\}_{\tau=1}^{\infty}, \rho, \lambda, \eta > 1$.
initialize $\mathbf{L}[0]$ at random, $\mu[0] > 0$, $\tilde{\mathbf{L}}[1] := \mathbf{L}[0]$, and $k[1] := 1$.
for $t = 1, 2, \ldots$ **do**

$\mathbf{D}[t] = \left(\lambda \mathbf{I}_\rho + \mathbf{L}^\top[t-1]\mathbf{\Omega}_t\mathbf{L}[t-1]\right)^{-1}\mathbf{L}^\top[t-1]$

$\mathbf{q}[t] = \mathbf{D}[t]\mathcal{P}_{\omega_t}(\mathbf{y}_t)$

Find the smallest nonnegative integer $i[t]$ such that with $\bar{\mu} := \eta^{i[t]}\mu[t-1]$

$$f_t(\tilde{\mathbf{L}}[t] - (1/\bar{\mu})\nabla f_t(\tilde{\mathbf{L}}[t])) \le \mathcal{Q}_{\bar{\mu},t}(\tilde{\mathbf{L}}[t] - (1/\bar{\mu})\nabla f_t(\tilde{\mathbf{L}}[t]), \tilde{\mathbf{L}}[t])$$

holds, and set $\mu[t] = \eta^{i[t]}\mu[t-1]$.

$\mathbf{L}[t] = \tilde{\mathbf{L}}[t] - (1/\mu[t])\nabla f_t(\tilde{\mathbf{L}}[t])$.

$k[t+1] = \frac{1+\sqrt{1+4k^2[t]}}{2}$.

$\tilde{\mathbf{L}}[t+1] = \mathbf{L}[t] + \left(\frac{k[t]-1}{k[t+1]}\right)(\mathbf{L}[t] - \mathbf{L}[t-1])$.

end for
return $\hat{\mathbf{x}}[t] := \mathbf{L}[t]\mathbf{q}[t]$.

Building on the increasingly popular *accelerated* gradient methods for batch smooth optimization [39,40], the idea here is to speed-up the learning rate of the estimated subspace (30.18) without paying a penalty in terms of computational complexity per iteration. The critical difference between standard gradient algorithms and the so-called Nesterov's variant is that the accelerated updates take the form $\mathbf{L}[t] = \tilde{\mathbf{L}}[t] - (\mu[t])^{-1}\nabla f_t(\tilde{\mathbf{L}}[t])$, which relies on a judicious linear combination $\tilde{\mathbf{L}}[t-1]$ of the previous pair of iterates $\{\mathbf{L}[t-1], \mathbf{L}[t-2]\}$. Specifically, the choice $\tilde{\mathbf{L}}[t] = \mathbf{L}[t-1] + \frac{k[t-1]-1}{k[t]}(\mathbf{L}[t-1] - \mathbf{L}[t-2])$, where $k[t] = \left[1 + \sqrt{4k^2[t-1]+1}\right]/2$, has been shown to significantly accelerate batch gradient algorithms resulting in a convergence rate no worse than $\mathcal{O}(1/k^2)$; see e.g., [40] and references therein. Using this acceleration technique in conjunction with a backtracking stepsize rule [41], a fast online SGD algorithm for imputing missing entries is tabulated under Algorithm 30.3. Clearly, a standard (nonaccelerated) SGD algorithm with backtracking step size rule is subsumed as a special case, when $k[t] = 1$, $t = 1, 2, \ldots$. In this case, complexity is $\mathcal{O}(|\omega_t|\rho^2)$ mainly due to the update of $\mathbf{q}_t$ while the accelerated algorithm incurs an additional cost $O(P\rho)$ for the subspace extrapolation step.

Computational cost. Careful inspection of Algorithm 30.2 reveals that the main computational burden stems from $\rho \times \rho$ inversions to update the subspace matrix $\mathbf{L}[t]$. The per iteration complexity for performing the inversions is $\mathcal{O}(|\omega_t|\rho^3)$ (which could be further reduced if one leverages also the symmetry of $\mathbf{G}_l[t]$), while the cost for the rest of operations is $\mathcal{O}(|\omega_t|\rho^2)$. The overall cost of the algorithm per iteration can thus be safely estimated as $\mathcal{O}(|\omega_t|\rho^3)$, which can be affordable because ρ is typically small (cf. the low-rank assumption). In addition, for the infinite memory case $\theta = 1$ where the RLS update is employed, the overall cost is further reduced to $\mathcal{O}(|\omega_t|\rho^2)$. The first-order Algorithm 30.3 reduces this cost by order ρ, that is in the same order as GROUSE and PETRELS, which incur costs of $\mathcal{O}(P\rho + |\omega_t|\rho^2)$ and $\mathcal{O}(|\omega_t|\rho^2)$, respectively.

30.4.1 PERFORMANCE GUARANTEES

This section studies the performance of the proposed first- and second-order online algorithms for the infinite memory special case; that is $\theta = 1$. In the sequel, to make the analysis tractable the following assumptions are adopted:

(a1) Processes $\{\omega_t, \mathcal{P}_{\omega_t}(\mathbf{y}_t)\}_{t=1}^{\infty}$ are independent and identically distributed (i.i.d.);
(a2) Sequence $\{\mathcal{P}_{\omega_t}(\mathbf{y}_t)\}_{t=1}^{\infty}$ is uniformly bounded; and
(a3) Iterates $\{\mathbf{L}[t]\}_{t=1}^{\infty}$ lie in a compact set.

To clearly delineate the scope of the analysis, it is worth commenting on (a1)–(a3) and the factors that influence their satisfaction. Regarding (a1), the acquired data is assumed statistically independent across time as is customary when studying the stability and performance of online (adaptive) algorithms [36]. While independence is required for tractability, (a1) may be grossly violated because the observations $\{\mathcal{P}_{\omega_t}(\mathbf{y}_t)\}$ are correlated across time (cf. the fact that $\{\mathbf{x}_t\}$ lies in a low-dimensional subspace). Still, in accordance with the adaptive filtering folklore, e.g., [36], as $\theta \to 1$ or $(\mu[t])^{-1} \to 0$ the upshot of the analysis based on i.i.d. data extends accurately to the pragmatic setting whereby the observations are correlated. Uniform boundedness of $\mathcal{P}_{\omega_t}(\mathbf{y}_t)$ [cf. (a2)] is natural in practice as it is imposed by the data acquisition process. The bounded subspace requirement in (a3) is a technical assumption that simplifies the analysis, and has been corroborated via extensive computer simulations [5].

Convergence of the second-order algorithm. Convergence of the iterates generated by Algorithm 30.2 (with $\theta = 1$) is established first. Upon defining

$$g_t(\mathbf{L}, \mathbf{q}) := \frac{1}{2}\|\mathcal{P}_{\omega_t}(\mathbf{y}_t - \mathbf{L}\mathbf{q})\|_2^2 + \frac{\lambda_t}{2}\|\mathbf{q}\|_2^2$$

in addition to $\ell_t(\mathbf{L}) := \min_{\mathbf{q}} g_t(\mathbf{L}, \mathbf{q})$, Algorithm 30.2 aims at minimizing the following *average* cost function at time t

$$C_t(\mathbf{L}) := \frac{1}{t}\sum_{\tau=1}^{t} \ell_\tau(\mathbf{L}) + \frac{\lambda_t}{2t}\|\mathbf{L}\|_F^2. \quad (30.19)$$

Normalization (by t) ensures that the cost function does not grow unbounded as time evolves. For any finite t, Eq. (30.19) is essentially identical to the batch estimator in (P3) up to a scaling, which does not affect the value of the minimizer. Note that as time evolves, minimization of C_t becomes increasingly complex computationally. Hence, at time t the subspace estimate $\mathbf{L}[t]$ is obtained by minimizing the *approximate* cost function

$$\hat{C}_t(\mathbf{L}) = \frac{1}{t}\sum_{\tau=1}^{t} g_\tau(\mathbf{L}, \mathbf{q}[\tau]) + \frac{\lambda_t}{2t}\|\mathbf{L}\|_F^2 \quad (30.20)$$

in which $\mathbf{q}[t]$ is obtained based on the prior subspace estimate $\mathbf{L}[t-1]$ after solving $\mathbf{q}[t] = \arg\min_{\mathbf{q}} g_t(\mathbf{L}[t-1], \mathbf{q})$ [cf. Eq. (30.13)]. Obtaining $\mathbf{q}[t]$ this way resembles the projection approximation adopted in [29]. Because $\hat{C}_t(\mathbf{L})$ is a smooth convex quadratic function, the minimizer $\mathbf{L}[t] = \arg\min_{\mathbf{L}} \hat{C}_t(\mathbf{L})$ is the solution of the linear equation $\nabla \hat{C}_t(\mathbf{L}[t]) = \mathbf{0}_{L\times\rho}$.

So far, it is apparent that because $g_t(\mathbf{L}, \mathbf{q}[t]) \geq \min_{\mathbf{q}} g_t(\mathbf{L}, \mathbf{q}) = \ell_t(\mathbf{L})$, the approximate cost function $\hat{C}_t(\mathbf{L}[t])$ overestimates the target cost $C_t(\mathbf{L}[t])$, for $t = 1, 2, \ldots$. However, it is not clear whether the subspace iterates $\{\mathbf{L}[t]\}_{t=1}^{\infty}$ converge, and most importantly, how well can they optimize the target cost function C_t. The good news is that $\hat{C}_t(\mathbf{L}[t])$ asymptotically approaches $C_t(\mathbf{L}[t])$, and the subspace iterates null $\nabla C_t(\mathbf{L}[t])$ as well, both as $t \to \infty$. This result is summarized in the next proposition.

Proposition 30.3. *Under (a1)–(a3) and $\theta = 1$ in Algorithm 30.2, if $\lambda_t = \lambda$ and $\lambda_{\min}[\nabla^2 \hat{C}_t(\mathbf{L})] \geq c$ for some $c > 0$, then $\lim_{t\to\infty} \nabla C_t(\mathbf{L}[t]) = \mathbf{0}_{L\times\rho}$ almost surely (a.s.), i.e., the subspace iterates $\{\mathbf{L}[t]\}_{t=1}^{\infty}$ asymptotically fall into the stationary point set of the batch problem (P3).* ■

It is worth noting that the pattern and the amount of missing data, summarized in the sampling sets $\{\omega_t\}$, play a key role toward satisfying the Hessian's positive semidefiniteness condition. In fact, random misses are desirable because the Hessian $\nabla^2 \hat{C}_t(\mathbf{L}) = \frac{\lambda}{t}\mathbf{I}_{L\rho} + \frac{1}{t}\sum_{\tau=1}^{t}(\mathbf{q}[\tau]\mathbf{q}^{\top}[\tau]) \otimes \mathbf{\Omega}_\tau$ is more likely to satisfy $\nabla^2 \hat{C}_t(\mathbf{L}) \succeq c\mathbf{I}_{L\rho}$, for some $c > 0$.

The proof of Proposition 30.3 is inspired by [42], which establishes convergence of an online dictionary learning algorithm using the theory of martingale sequences. Details can be found in [5], and in a nutshell the proof procedure proceeds in the following two main steps:

(S1) Establish that the approximate cost sequence $\{\hat{C}_t(\mathbf{L}[t])\}$ asymptotically converges to the target cost sequence $\{C_t(\mathbf{L}[t])\}$. To this end, it is first proved that $\{\hat{C}_t(\mathbf{L}[t])\}_{t=1}^{\infty}$ is a quasimartingale sequence, and hence convergent a.s. This relies on the fact that $g_t(\mathbf{L}, \mathbf{q}[t])$ is a *tight* upper bound approximation of $\ell_t(\mathbf{L})$ at the previous update $\mathbf{L}[t-1]$, namely, $g_t(\mathbf{L}, \mathbf{q}[t]) \geq \ell_t(\mathbf{L})$, $\forall \mathbf{L} \in \mathbb{R}^{L\times\rho}$, and $g_t(\mathbf{L}[t-1], \mathbf{q}[t]) = \ell_t(\mathbf{L}[t-1])$.

(S2) Under certain regularity assumptions on g_t, establish that convergence of the cost sequence $\{\hat{C}_t(\mathbf{L}[t]) - C_t(\mathbf{L}[t])\} \to 0$ yields convergence of the gradients $\{\nabla\hat{C}_t(\mathbf{L}[t]) - \nabla C_t(\mathbf{L}[t])\} \to 0$, which subsequently results in $\lim_{t\to\infty} \nabla C_t(\mathbf{L}[t]) = \mathbf{0}$.

Optimality. Beyond convergence to stationary points of (P3), one may ponder whether the online estimator offers performance guarantees of the batch nuclear-norm regularized estimator (P1), for which stable/exact recovery results are well documented, e.g., in [9,14]. Specifically, given the learned subspace $\bar{\mathbf{L}}[t]$ and the corresponding $\bar{\mathbf{Q}}[t]$ [obtained via Eq. (30.13)] over a time window of size t, is $\{\hat{\mathbf{X}}[t] := \bar{\mathbf{L}}[t]\bar{\mathbf{Q}}^{\top}[t]\}$ an optimal solution of (P1) as $t \to \infty$? This in turn requires asymptotic analysis of the optimality conditions for (P1) and (P3), and a positive answer is established in the next proposition whose proof is available in [5].

Proposition 30.4. *Consider the subspace iterates $\{\mathbf{L}[t]\}$ generated by either Algorithm 30.2 (with $\theta = 1$), or Algorithm 30.3. If there exists a subsequence $\{\mathbf{L}[t_k], \mathbf{Q}[t_k]\}$ for which (c1) $\lim_{k\to\infty} \nabla C_{t_k}(\mathbf{L}[t_k]) = \mathbf{0}_{L\times\rho}$ a.s., and (c2) $\frac{1}{\sqrt{t_k}}\sigma_{\max}[\mathcal{P}_{\Omega_{t_k}}(\mathbf{Y}_{t_k} - \mathbf{L}[t_k]\mathbf{Q}^{\top}[t_k])] \leq \frac{\lambda_{t_k}}{\sqrt{t_k}}$ hold, then the sequence $\{\mathbf{X}[k] = \mathbf{L}[t_k]\mathbf{Q}^{\top}[t_k]\}$ satisfies the optimality conditions for (P1) [normalized by t_k] as $k \to \infty$ a.s.* ■

Regarding condition (c1), even though it holds for a time-invariant rank-controlling parameter λ as per Proposition 30.3, numerical tests indicate that it still holds true for the time-varying case; see [5, Remark 2] for guidelines on the choice of λ_t. Under (a2) and (a3) one has $\sigma_{\max}[\mathcal{P}_{\Omega_t}(\mathbf{Y}_t - \mathbf{L}[t]\mathbf{Q}^{\top}[t])] \approx \mathcal{O}(\sqrt{t})$, which implies that the quantity on the left side of (c2) cannot grow unbounded. Moreover, upon choosing $\lambda_t \approx \mathcal{O}(\sqrt{t})$ the term in the right side of (c2) will not vanish, which suggests that the qualification condition can indeed be satisfied [5]. Effective heuristic rules are devised in [5,9] for tunning λ.

30.4.2 REAL-TIME NETWORK TRAFFIC MONITORING

Accurate estimation of OD flow traffic in the backbone of large-scale IP networks is of paramount importance for proactive network security and management tasks [43]. Several experimental studies have demonstrated that OD flow traffic exhibits low rank, mainly due to common temporal patterns across OD flows, and periodic trends across time [2]. However, due to the massive number of OD pairs and the high volume of traffic, measuring the traffic of all possible OD flows is impossible for all practical purposes [2,43]. Only the traffic level for a small fraction of OD flows can be measured via the NetFlow protocol [2].

Aggregate OD-flow traffic is collected from operation of the Internet-2 during December 8–28, 2003, containing 121 OD pairs [28]. The measured OD flows contain spikes (anomalies), which are discarded to end up with an anomaly-free data stream $\{\mathbf{y}_t\} \in \mathbb{R}^{121}$. The detailed description of the considered dataset can be found in [6]. A fraction π of the entries of $\mathbf{y}_t$ are then randomly sampled to yield the input of Algorithm 30.2. Evolution of the running-average traffic estimation error is depicted in Fig. 30.2(left) for different subspace trackers and π values. Evidently, Algorithm 30.2 outperforms the competing alternatives when λ_t is adaptively tuned as in [5]. When only 25% of the total OD flows are sampled by NetFlow, Fig. 30.2(right) depicts how Algorithm 30.2 accurately tracks representative OD flows.

30.4.3 LARGE-SCALE MACHINE LEARNING

The advocated subspace learning framework identifies latent low-dimensional structure in streaming data, and can facilitate large-scale machine learning tasks beyond matrix completion. The scope could be, for instance, broadened to accommodate large-scale dimensionality reduction and feature extraction

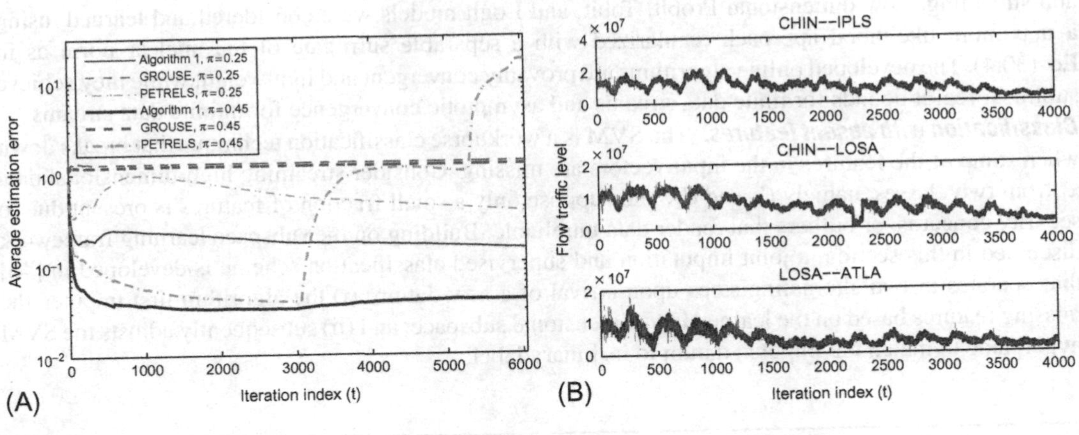

FIG. 30.2

Internet-2 flow traffic estimation. Traffic estimation performance for Internet-2 data when $\rho = 10$ and $\theta = 0.95$, and a variable fraction π of data is available. (Left) Average estimation error for $\pi = 0.25$ (*thin solid*) and $\pi = 0.45$ (*thick solid*). (Right) Algorithm 30.2's estimated (*thin dash*) versus true (*thick solid*) OD flow traffic for 75% missing data ($\pi = 0.25$).

from multiway tensors [44] as well as categorical and finite-alphabet datasets [45]. In addition, the developed subspace trackers can be adopted in conjunction with support vector machines (SVMs) to classify incomplete data online [46]. These generalizations are briefly summarized next.

Multilinear decomposition and dimensionality reduction. Many applications involve data indexed by three, or more variables giving rise to a tensor, instead of just two variables as in the matrix settings dealt with so far. It is not uncommon that one of these variables indexes time [5,47], and that sizable portions of the data are missing [48,49]. Examples of time-indexed, incomplete tensor data include: (i) dynamic social networks represented through a temporal sequence of adjacency matrices while it may be the case that not all pairwise interactions among nodes can be sampled; and (ii) multidimensional nuclear magnetic resonance (NMR) analysis, where missing data are encountered when sparse sampling is used in order to reduce the experimental time. Various data analytic tasks aim at unveiling underlying latent structures, which calls for high-order tensor factorizations even in the presence of missing data [48, 49]. With this objective in mind, [5] puts forth for the first time an online (adaptive) algorithm for decomposing low-rank tensors with missing entries; see also [44] for an adaptive algorithm to obtain parallel factor analysis (PARAFAC) decompositions—a natural extension of the bilinear model in (P5) to the multilinear case. The proposed online algorithm offers a viable approach to solving large-scale tensor decomposition (and completion) problems, even if the data is not actually streamed but they are so massive that do not fit in the main memory.

Sketching of categorical data. With the scale of data growing every day, reducing the dimensionality (aka sketching) of high-dimensional data has emerged as a task of paramount importance. Critical challenges arise with increasingly ubiquitous datasets comprising incomplete categorical samples. For instance, in movie recommender systems, observation $\mathbf{y}_t$ represents the users' categorical ratings (e.g., like/dislike or in a 1–5 integer-valued scale) for the tth movie. Because each user rates only a small fraction of movies, ratings for a sizable portion of movies will be missing. In this context, effective sketching tools were developed in [45] for large-scale categorical data that are incomplete and streaming. Low-dimensional Probit, Tobit, and Logit models were considered and learned, using a maximum likelihood approach regularized with a separable surrogate of the nuclear norm as in Eq. (30.4). The developed online algorithms are provably convergent and lightweight while they achieve sublinear regret bounds for finite data streams and asymptotic convergence for infinite data streams.

Classification with absent features. The SVM is a workhorse classification technique that breaks down when some of the features in the input vectors are missing. Consider streaming high-dimensional data $\mathbf{x}_t$ from two classes, namely C_1 and C_2, and suppose only a small fraction of features is present due to security concerns, or outliers that render data unreliable. Building on the subspace learning framework discussed in this section, a joint imputation and supervised classification scheme is developed in [46] that operates in two alternating steps upon arrival of a new datum: (i) the algorithm first imputes the missing features based on the learned low-dimensional subspace; and (ii) subsequently adjusts the SVM hyperplane to match the imputed datum to its binary label.

30.5 CONCLUDING SUMMARY

Nowadays machine learning tasks deal with *sheer volumes* of data of possibly *incomplete*, *decentralized*, and *streaming* nature that demand on-the-fly processing for real-time decision-making. Conventional inference analytics mine such big data by leveraging their intrinsic parsimony, e.g., via models that include rank and sparsity regularization or priors. Convex nuclear and ℓ_1-norm surrogates are typically adopted and offer well-documented guarantees in recovering informative low-dimensional structure from high-dimensional data. However, the computational complexity of the resulting algorithms tends to scale poorly due to the nuclear norm's entangled structure, which also impedes streaming and decentralized analytics. To mitigate this computational hurdle, this chapter discussed a framework that leverages a bilinear characterization of the nuclear norm to bring separability at the expense of nonconvexity. This challenge notwithstanding, under mild conditions stationary points of the nonconvex program provably coincide with the optimum of the convex counterpart. Using this idea along with the theory of alternating minimization, lightweight algorithms are developed with low communication overhead for in-network processing. Provably convergent online subspace trackers that are suitable for streaming analytics are developed as well. Remarkably, even under the constraints imposed by decentralized computing and sequential data acquisition, one can still attain the performance offered by the prohibitively-complex batch analytics. While the ideas were presented for a matrix completion problem, the scope of the presented framework can be broadened to be jointly adopted with downstream machine learning tasks such as dimensionality reduction, clustering, classification, multidimensional scaling, and anomaly detection.

ACKNOWLEDGMENTS

The work was supported in part by NSF grants NSF 1500713, 1509040, 1514056, 1711471, and ARO W911NF-15-1-0492.

REFERENCES

[1] Slavakis K, Giannakis GB, Mateos G. Modeling and optimization for Big Data analytics. IEEE Signal Process Mag 2014;31(5):18–31.

[2] Lakhina A, Papagiannaki K, Crovella M, Diot C, Kolaczyk ED, Taft N. Structural analysis of network traffic flows. In: Proceedings of ACM SIGMETRICS, New York, NY; 2004.

[3] Mateos G, Rajawat K. Dynamic network cartography: advances in network health monitoring. IEEE Signal Process Mag 2013;30(3):129–43.

[4] Mardani M, Mateos G, Giannakis GB. Recovery of low-rank plus compressed sparse matrices with application to unveiling traffic anomalies. IEEE Trans Inf Theory 2013;59:5186–205.

[5] Mardani M, Mateos G, Giannakis GB. Subspace learning and imputation for streaming big data matrices and tensors. IEEE Trans Signal Process 2015;63:2663–77.

[6] Mardani M, Mateos G, Giannakis GB. Dynamic anomalography: tracking network anomalies via sparsity and low rank. IEEE J Sel Top Signal Process 2013;7:50–66.

[7] Mardani M, Mateos G, Giannakis GB. Decentralized sparsity regularized rank minimization: applications and algorithms. IEEE Trans Signal Process 2013;61:5374–88.

[8] Wright J, Ganesh A, Min K, Ma Y. Compressive principal component pursuit. In: Proceeding of international symposium on information theory, Cambridge, MA; 2012. p. 1276–80.

[9] Candès EJ, Plan Y. Matrix completion with noise. Proc IEEE 2009;98:925–36.

[10] Chistov A, Grigorev D. Complexity of quantifier elimination in the theory of algebraically closed fields. In: Mathematical foundations of computer science. Lecture notes in computer science, vol. 176. Berlin/Heidelberg: Springer; 1984. p. 17–31.

[11] Natarajan BK. Sparse approximate solutions to linear systems. SIAM J Comput 1995;24:227–34.
[12] Fazel M, Hindi H, Boyd SP. A rank minimization heuristic with application to minimum order system approximation. In: Proceedings of American control conference, vol. 6; 2001. p. 4734–9.
[13] Candes EJ, Tao T. Decoding by linear programming. IEEE Trans Inf Theory 2005;51(12):4203–15.
[14] Candes EJ, Recht B. Exact matrix completion via convex optimization. Found Comput Math 2009;9(6):717–22.
[15] Candes EJ, Li X, Ma Y, Wright J. Robust principal component analysis? J ACM 2011;58(1):1–37.
[16] Chandrasekaran V, Sanghavi S, Parrilo PR, Willsky AS. Rank-sparsity incoherence for matrix decomposition. SIAM J Optim 2011;21(2):572–96.
[17] Mardani M, Mateos G, Giannakis GB. Exact recovery of low-rank plus compressed sparse matrices. In: 2012 IEEE statistical signal processing workshop (SSP). IEEE; 2012. p. 49–52.
[18] Recht B, Fazel M, Parrilo PA. Guaranteed minimum-rank solutions of linear matrix equations via nuclear norm minimization. SIAM Rev 2010;52(3):471–501.
[19] Srebro N, Shraibman A. Rank, trace-norm and max-norm. In: Proceedings of learning theory; 2005. p. 545–60.
[20] Bertsekas DP, Tsitsiklis JN. Parallel and distributed computation: numerical methods. 2nd ed. Athena-Scientific; 1999.
[21] Boyd S, Parikh N, Chu E, Peleato B, Eckstein J. Distributed optimization and statistical learning via the alternating direction method of multipliers. Found Trends Mach Learn 2010;3:1–122.
[22] Giannakis GB, Ling Q, Mateos G, Schizas ID, Zhu H. Decentralized learning for wireless communications and networking. In: Glowinski R, Osher S, Yin W, editors. Splitting methods in communication and imaging, science and engineering, scientific computation. Springer New York; 2016. p. 461–97.
[23] Recht B, Ré C. Parallel stochastic gradient algorithms for large-scale matrix completion. Math Program Comput 2013;5(2):201–26.
[24] Lakhina A, Crovella M, Diot C. Diagnosing network-wide traffic anomalies. In: Proceeding of ACM SIGCOMM, Portland, OR; 2004.
[25] Schizas ID, Giannakis GB, Luo ZQ. Distributed estimation using reduced-dimensionality sensor observations. IEEE Trans Signal Process 2007;55:4284–99.
[26] Mateos G, Bazerque JA, Giannakis GB. Distributed sparse linear regression. IEEE Trans Signal Process 2010;58(10):5262–76.
[27] Liao Y, Du W, Geurts P, Leduc G. DMFSGD: a decentralized matrix factorization algorithm for network distance prediction. IEEE/ACM Trans Network 2011;21(5):1511–24. See also arXiv:1201.1174v1 [cs.NI].
[28] The Internet Observatory Data Collection; 2012. http://internet2.edu/observatory/archive/data-collections.html.
[29] Yang B. Projection approximation subspace tracking. IEEE Trans Signal Process 1995;43(1):95–107.
[30] Yang JF, Kaveh M. Adaptive eigensubspace algorithms for direction or frequency estimation and tracking. IEEE Trans Acoust Speech Signal Process 1988;36(2):241–51.
[31] Balzano L, Nowak R, Recht B. Online identification and tracking of subspaces from highly incomplete information. In: Proceeding of Allerton conference on communication, control, and computing, Monticello, USA; 2010.
[32] Balzano L, Wright SJ. Local convergence of an algorithm for subspace identification from partial data. Found Comput Math 2015;15(5):1279–314.
[33] Balzano L, Wright SJ. On GROUSE and incremental SVD. In: IEEE 5th international workshop on computational advances in multi-sensor adaptive processing (CAMSAP). IEEE; 2013. p. 1–4.
[34] Chi Y, Eldar YC, Calderbank R. Petrels: parallel subspace estimation and tracking by recursive least squares from partial observations. IEEE Trans Signal Process 2013;61(23):5947–59.
[35] Dai W, Milenkovic O, Kerman E. Subspace evolution and transfer (SET) for low-rank matrix completion. IEEE Trans Signal Process 2011;59(7):3120–32.

[36] Solo V, Kong X. Adaptive signal processing algorithms: stability and performance. Prentice Hall; 1995.
[37] Sprechmann P, Bronstein AM, Sapiro G. Real-time online singing voice separation from monaural recordings using robust low-rank modeling. In: Proceedings of the annual conference of the international society for music information retrieval, Porto, Portugal; 2012.
[38] Feng J, Xu H, Yan S. Online robust PCA via stochastic optimization. In: Proceedings of the advances in neural information processing systems, Lake Tahoe, NV; 2013.
[39] Nesterov Y. A method of solving a convex programming problem with convergence rate $o(1/k^2)$. Sov Math Doklady 1983;27:372–6.
[40] Beck A, Teboulle M. A fast iterative shrinkage-thresholding algorithm for linear inverse problems. SIAM J Imag Sci 2009;2:183–202.
[41] Bertsekas DP. Nonlinear programming. 2nd ed. Athena-Scientific; 1999.
[42] Mairal J, Bach F, Ponce J, Sapiro G. Online learning for matrix factorization and sparse coding. J Mach Learn Res 2010;11:19–60.
[43] Kolaczyk ED. Statistical analysis of network data: methods and models. Springer; 2009.
[44] Mardani M, Giannakis GB, Ugurbil K. Tracking tensor subspaces with informative random sampling for real-time MR imaging; 2016. arXiv preprint arXiv:160904104.
[45] Shen Y, Mardani M, Giannakis GB. Online categorical subspace learning for sketching big data with misses. IEEE Trans Signal Process 65(15):4004–18.
[46] Sheikholeslami F, Mardani M, Giannakis GB. Streaming support vector classification of big data with misses. In: Proceedings of Asilomar conference on control, signal and systems; 2014. p. 516–20.
[47] Nion D, Sidiropoulos ND. Adaptive algorithms to track the PARAFAC decomposition of a third-order tensor. IEEE Trans Signal Process 2009;57(6):2299–310.
[48] Acar E, Dunlavy DM, Kolda TG, Mrup M. Scalable tensor factorizations for incomplete data. Chemom Intell Lab Syst 2011;106(1):41–56.
[49] Bazerque JA, Mateos G, Giannakis GB. Rank regularization and Bayesian inference for tensor completion and extrapolation. IEEE Trans Signal Process 2013;61(22):5689–703.

CHAPTER

GRAPH SIGNAL PROCESSING ON NEURONAL NETWORKS

31

Selin Aviyente*, Marisel Villafañe-Delgado*
*Department of Electrical and Computer Engineering, Michigan State University, East Lansing, MI United States**

31.1 INTRODUCTION

The study of the human brain has attracted a lot of research over the centuries. Many noninvasive imaging modalities such as topographical techniques based on the electromagnetic field potential, e.g., EEG and magnetoencephalography (MEG) and tomography approaches including positron emission tomography (PET) and magnetic resonance imaging (MRI), have been used widely to study the neural basis of cognition, perception, and emotion. Early analysis of neuroimaging data was based on correlational evidence of brain regions being related to specific functions, known as functional segregation [1]. However, over the past 20 years it has become obvious that the brain operates as a global complex system with many interactions, also known as functional integration [2]. This revelation has increased the interest in the study of statistical dependencies between anatomically distinct brain regions, also known as functional connectivity (FC).

In the last decade, the use of advanced tools from statistical physics, graph theory, statistics, and signal processing has led to the representation of the functional connectivity of the brain as a complex network and revolutionized the analysis of brain connectivity patterns estimated from neuroimaging data [3–5]. The construction of these networks consists of identifying the vertices that correspond to specific brain regions defined by the neuroimaging modality, estimating the edges of the network by measuring the FC between the activity of brain vertices, and encoding this information in a connectivity matrix. Once the network has been mapped, it can be analyzed with respect to either its connectivity or topology [6,7]. Connectivity analysis concentrates on variations in the type and strength of connectivity between brain regions. Topological analysis, based on graph theory, is concerned with understanding how connections are arranged with respect to each other and provides insight into key organizational principles of the connectome. Topological analysis has relied on different graph theoretic measures at the local, intermediate, and global scales [5]. At the local scale, the degree distribution and centrality of vertices have been used to identify hubs in the network. At the intermediate scale, the focus has been on community detection through network clustering methods and modularity [8]. At the global scale, the efficiency, clustering coefficient, path length, and small-worldness of the network have been computed to quantify the interplay between functional integration and segregation [9,10]. Selecting the appropriate topological metric depends mostly on the research question at hand. The different graph

Cooperative and Graph Signal Processing. https://doi.org/10.1016/B978-0-12-813677-5.00031-6

theoretic measures have been used as features to classify and characterize changes in brain dynamics due to pathology, or cognitive state change. This network-centric perspective has provided fundamental insights in terms of the organization of the healthy and diseased brain, how information processing is distributed across the network, and the resilience of the network architecture to trauma [10,11].

While this network-centric view of the brain has been successfully applied to the study of different neurological diseases, it does not fully consider the multivariate signals that live on the vertices of the brain graph. Graph signal processing (GSP) offers a new direction for analyzing multivariate neuroimaging data by taking into account both the connectivity features of the brain graph, such as structural and functional connectivity, as well as the time series that live on the vertices of this graph. Unlike classical multivariate signal processing, which treats the data as vectors in a high-dimensional Euclidean space, GSP can take advantage of the nonlinearities in the space where the data lives. In particular, GSP can extract information from complex neuroimaging data using the underlying graph structure as the non-Euclidean space or the manifold that the signal is defined over. With the underlying network informing the multivariate signal analysis, GSP can improve the extraction of low-dimensional representations where the subspaces are defined by different levels of spatial variability across the brain. In this chapter, we will illustrate different GSP methodologies developed for brain graphs, including dimensionality reduction and classification, graph-based filtering (GBF), graph learning, and spatial filtering.

31.2 BASIC CONCEPTS OF GRAPH SIGNAL PROCESSING IN NEUROSCIENCE

In this section, we will review the basic concepts and notations that will be utilized in this chapter, with the complete list of notations given in Table 31.1. The application of GSP methods on brain connectivity networks starts with the definition of a weighted graph and the associated matrices that describe its structure. We can characterize the structure of brain imaging data by a weighted graph $G(\mathcal{V}, \mathcal{E}, \mathbf{W})$, consisting of a finite set $\mathcal{V}(|\mathcal{V}| = N)$ of vertices corresponding to brain regions, a finite set of edges $\mathcal{E} \subseteq \mathcal{V} \times \mathcal{V}$ and $\mathbf{W}$ with $W_{ij} \geq 0$ quantifying the similarity between vertices i and j.

Table 31.1 List of Commonly Used Notations

Symbol	Definition
$\mathbf{W}$	weighted adjacency matrix
$\mathbf{D}$	weighted degree matrix
$\mathbf{L}$	graph Laplacian
$\mathbf{U}$	eigenvectors of the Laplacian
$\mathbf{u}_i$	ith eigenvector
λ_i	ith eigenvalue
$\mathbf{x}$	graph signal
$\tilde{\mathbf{x}}$	graph Fourier transform (GFT)

In general, the mapping between the brain space and the vertices is based on the particular neuroimaging modality. Voxel-based modalities, such as fMRI, define vertices in the measurement space whereas sensor-based modalities, such as EEG and MEG, offer a choice between assigning vertices directly to sensors or to reconstructed sources. In voxel-based modalities, there are several approaches to define brain vertices including single voxels, anatomically defined regions of interests (ROIs), or data-driven methods such as ICA [12]. In sensor-based modalities, brain vertices are commonly assigned directly to sensors or electrodes [11,13], even though the brain vertices may suffer from a biased nonneural dependence. Once the vertices of the graph are uniquely defined, a time series that corresponds to each vertex is defined. In the case of fMRI, either a temporal average time series of all the voxels within a region or components from spatial ICA are used to define the time series.

After defining the brain vertices, assigning links or edges between them is the subsequent crucial modeling step. Different data-driven and model-based methods have been used to construct the adjacency matrix, **W** [14]. Some common choices for **W** in the context of GSP include the physical distance in the case of structural imaging modalities and data-driven metrics such as the absolute value of the correlation or coherence in the case of functional imaging modalities. In particular, structural graphs that model the geometric structure of the brain can be computed as follows:

$$W_{i,j} = \exp\left(\frac{d(v_i, v_j)^2}{2\sigma^2}\right), \tag{31.1}$$

where $d(v_i, v_j)$ is the Euclidean distance between the barycenters of the different brain areas and σ is an empirically chosen spread parameter. This is known as the full geometric graph as it is fully connected [15]. An alternative to this is proposed through the geometric graph, which only connects close brain areas by applying a threshold to $d(v_i, v_j)$ resulting in

$$W_{i,j} = \begin{cases} \exp\left(-\frac{d(v_i,v_j)^2}{2\sigma^2}\right), & \text{if } d(v_i, v_j) < \alpha, \\ 0, & \text{otherwise,} \end{cases} \tag{31.2}$$

where α is an empirically determined threshold.

For functional connectivity graphs, given the time series at two vertices of the graph, $x_i(t)$ and $x_j(t)$, different connectivity measures such as absolute correlation, frequency domain coherence, and phase synchrony can be used. Correlation, $\rho(x_i, x_j)$, is a basic estimation of statistical dependency for functional connectivity, especially for fMRI data. Coherence connectivity is the frequency domain analog of the cross-correlation coefficient and is commonly computed through the Fourier transform or Wavelet transform of the time series data with

$$W_{i,j} = \left| \frac{\frac{1}{T}\sum_{t=1}^{T} A_{x_i}(t)A_{x_j}(t)e^{j[\phi_{x_i}(t)-\phi_{x_j}(t)]}}{\sqrt{\frac{1}{T}\sum_{t=1}^{T} A_{x_j}^2(t)}\sqrt{\frac{1}{T}\sum_{t=1}^{T} A_{x_j}^2(t)}} \right|, \tag{31.3}$$

where $A_{x_i}(t)$ and $\phi_{x_i}(t)$ refer to the amplitude and phase of the analytic signal, respectively. Similarly, phase synchrony or phase locking value (PLV) is commonly used to quantify connectivity for neurophysiological recordings such as the EEG and MEG. PLV separates the amplitude effects from the phase and quantifies the consistency of the phase differences. The PLV connectivity between two regions i and j is defined as [16]:

$$W_{i,j} = \frac{1}{T}\left|\sum_{t=1}^{T} e^{j[\phi_{x_i}(t)-\phi_{x_j}(t)]}\right|. \tag{31.4}$$

Finally, mixed graphs that combine structural and functional connectivity can be used such as:

$$W_{i,j} = \exp\left(\frac{-(1-\rho(x_i,x_j))^2}{2\sigma_1^2}\right)\exp\left(-\frac{d(v_i,v_j)}{2\sigma_2^2}\right), \tag{31.5}$$

where $\rho(x_i, x_j)$ is the correlation coefficient and $d(v_i, v_j)$ is the normalized Euclidean distance between two brain regions or sensors.

As GSP deals with the spectral properties of the brain graph, in addition to the adjacency matrix, we introduce the graph Laplacian as another important graph-associated matrix. Denoting $\mathbf{D}$ as the diagonal degree matrix with $D_{ii} = \sum_{j=1}^{N} W_{ij}$, the graph Laplacian is defined as $\mathbf{L} = \mathbf{D} - \mathbf{W}$. In GSP applications, it is important to choose $\mathbf{W}$ with $W_{ij} \geq 0$ and $W_{ij} = W_{ji}$ to ensure the positive definitiveness and the symmetry of the resulting Laplacian matrix. A symmetric and positive definite $\mathbf{L}$ will have a complete set of orhonormal eigenvectors $\{\mathbf{u}_i\}_{i=1,\ldots,N}$ with the associated eigenvalues $\{\lambda_i\}_{i=1,\ldots,N}$ real and nonnegative with the smallest eigenvalue being zero. The eigenvalues can be increasingly sorted as $0 = \lambda_1 \leq \lambda_2 \leq \cdots \leq \lambda_N$ and the corresponding eigendecomposition of the graph Laplacian can be written as $\mathbf{L} = \mathbf{U}\boldsymbol{\Lambda}\mathbf{U}^T$, where $\boldsymbol{\Lambda}$ is a diagonal matrix with $\Lambda_{ii} = \lambda_i$ being the ith smallest eigenvalue and the ith column of $\mathbf{U}$, $\mathbf{u}_i$ is the associated eigenvector.

Once the graph-associated matrices are defined, we can define the graph signal as a vector in $\mathbb{R}^N$. This vector can be denoted by $\mathbf{x}$ where $x(i)$ or x_i refer to the value of the signal at the ith vertex. In practice, PET scans or fMRI time series at a particular time point t can be defined as graph signals.

Similar to the Fourier transform, which encodes the temporal variation in time domain signals, the graph Fourier transform (GFT) is defined to encode a notion of spatial variation in graph signals. Using the eigenvector matrix $\mathbf{U}$ of the graph Laplacian, we can define the GFT of a graph signal $\mathbf{x}$ as [17]:

$$\tilde{\mathbf{x}} = \mathbf{U}^H\mathbf{x}. \tag{31.6}$$

The inverse GFT (iGFT) of $\tilde{\mathbf{x}}$ with respect to $\mathbf{L}$ is defined as:

$$\mathbf{x} = \mathbf{U}\tilde{\mathbf{x}} = \sum_{k=1}^{N} \tilde{x}_k \mathbf{u}_k. \tag{31.7}$$

Because $\mathbf{U}\mathbf{U}^H = \mathbf{I}$, the iGFT is the inverse of GFT. As k increases, the graph Laplacian eigenvectors fluctuate more rapidly. Therefore, the eigenvalues of the graph Laplacian, λ_ks, encode the graph frequency information with small ks corresponding to low frequency and large ks corresponding to high frequency in the graph domain. Fig. 31.1 illustrates an example of a graph signal defined on a brain connectivity graph.

This variation in frequency can also directly be captured through the graph Laplacian. For a graph signal $\mathbf{x}$, the graph Laplacian behaves as a difference operator on it and can be used to define the total variation (TV) of the graph signal with respect to the network as:

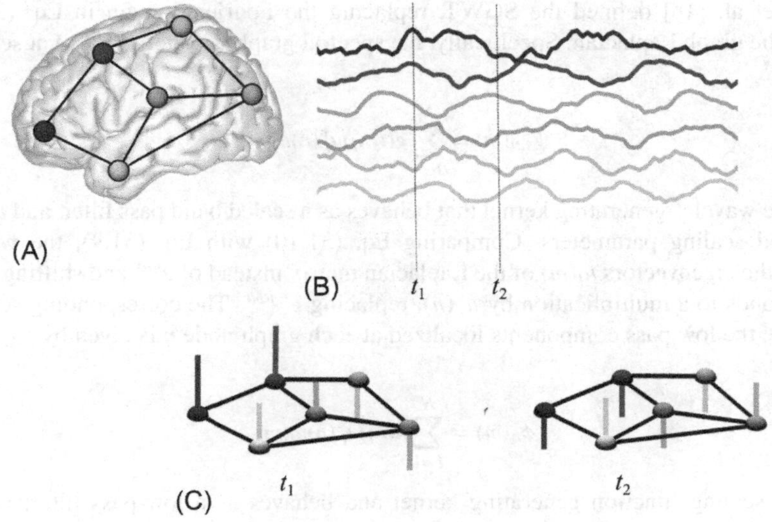

FIG. 31.1

An illustration of the construction of brain networks and the graph signals defined on these networks. (A) Illustration of a graph over the human brain. Vertices (*circles*) represent different electrodes or brain regions, and their relationship is described by edges interconnecting them (*lines*). (B) Time-series corresponding to each vertex in (A). (C) Signals defined over the graphs at two different time instances. The graph signals at each vertex are defined by the amplitude of the time series from that vertex at a specific time.

$$TV(\mathbf{x}) = \mathbf{x}^T \mathbf{L} \mathbf{x} = \sum_{i \neq j} W_{ij}(x_i - x_j)^2. \tag{31.8}$$

$TV(\mathbf{x})$ is a measure of how much the signal changes with respect to the network. When W_{ij} is large we expect the values x_i and x_j to be similar as a large weight W_{ij} signifies the similarity between brain regions i and j. The contribution of the difference $(x_i - x_j)^2$ to the total variation is amplified by the weight W_{ij}. A signal with small total variation corresponds to one that changes slowly over the graph.

Similar to GFT, other well-known signal transforms have been extended to the graph domain. One of the major spectral transforms extended to brain graphs is spectral graph wavelet transform (SGWT) [18]. The one-dimensional continuous wavelet function $\Psi_{s,a}(x) = \frac{1}{s}\psi(\frac{x-a}{s})$ at scale s and location a can be equivalently rewritten in the Fourier domain as

$$\psi_{s,a}(x) = \frac{1}{2\pi} \int_{-\infty}^{\infty} \hat{\psi}(s\omega) e^{-j\omega a} e^{j\omega x} \, d\omega, \tag{31.9}$$

where the shifting is given by the term $e^{j\omega l}$ and scaling ψ by $\frac{1}{s}$ corresponds to scaling $\hat{\psi}$ by s.

Hammond et al. [18] defined the SGWT, replacing the Fourier domain in Eq. (31.9) with the eigenspace of the graph Laplacian. Specifically, the spectral graph wavelet, $\psi_{s,n}(m)$ at scale s and node n is defined as

$$\psi_{s,n}(m) = \sum_{l=1}^{N} g(t_s\lambda_l)u_l^*(n)u_l(m), \tag{31.10}$$

where $g(\cdot)$ is the wavelet-generating kernel that behaves as a scaled band pass filter, and $t_s, s = 1, \ldots, J$ are the sampled scaling parameters. Comparing Eq. (31.10) with Eq. (31.9), the wavelet is now represented by the eigenvectors $u_l(m)$ of the Laplacian matrix instead of $e^{j\omega x}$ and shifting the wavelet to node n corresponds to a multiplication by $u_l^*(n)$, replacing $e^{-j\omega a}$. The corresponding scaling function that will capture the low pass components localized at each graph node n is given by

$$\phi_n(m) = \sum_{l=1}^{N} h(\lambda_l)u_l^*(n)u_l(m), \tag{31.11}$$

where h is the scaling function generating kernel and behaves as a low-pass filter; i.e., $h(0) > 0$, $h(\lambda) = 0$ as $\lambda \to \infty$.

Similar to the continuous wavelet transform, SGWT can be computed by the inner product of the graph signal $\mathbf{x}$ with the wavelets and the scaling function as

$$W_\psi(s,n) = \langle \mathbf{x}, \psi_{t,n}\rangle = \sum_{l=1}^{N} g(t_s\lambda_l)\tilde{x}_l u_l(n), \tag{31.12}$$

and

$$W_\phi(n) = \langle \mathbf{x}, \phi_l\rangle = \sum_{l=1}^{N} h(\lambda_l)\tilde{x}_l u_l(n), \tag{31.13}$$

where $\tilde{x}_l$ is the lth frequency component of the GFT.

31.3 DIMENSIONALITY REDUCTION AND SUPERVISED CLASSIFICATION THROUGH GSP

One common problem in the analysis of brain imaging data such as EEG, MEG, or fMRI is the large amount of high-dimensional spatiotemporal measurements that commonly require dimensionality reduction before being used for classification applications such as the ones encountered in brain-machine interface (BMI) and task-based fMRI. Traditional methods for dimensionality reduction of large amounts of neuroimaging data include principal component analysis (PCA), independent component analysis (ICA) and Laplacian eigenmaps. The novelty of graph-based approaches in comparison to these traditional methods is that they can use the underlying graph model as side information to help reduce the dimensionality in a more robust manner. By leveraging the graph structure that may contain more accurate long-term correlation information compared to the instantaneous measurements, one can learn a more accurate low-dimensional subspace. In effect, these methods model the nonlinear data manifold with data-dependent graph structures. Moreover, the proposed projector is independent of the training dataset and does not require the estimation of the statistics of the actual data, which may be noisy. Experimental results show that the proposed dimensionality reduction methods work well in classification accuracy even in the case where the size of available data for training is very small [19,20].

In GSP-based dimensionality reduction, the first r eigenvectors of the graph Laplacian, $\mathbf{L}$, computed from the adjacency matrix constructed from either the functional connectivity or the structural network, are used to form the low-dimensional subspace, $\mathbf{U}_r$. The cut-off frequency is found by sorting the eigenvalues from higher to lower and the eigenvectors whose eigenvalues are larger than the cut-off frequency are used for building the subspace. Subsequently, the noisy measurements, or signals living on the graph, are projected to this low-dimensional subspace to reduce the dimensionality of the signals, i.e., $\mathbf{U}_r^T\mathbf{X}$, where $\mathbf{X} \in \mathbb{R}^{N\times T}$ is the collection of graph signals across time. Finally, the reduced dimensional signals can be used as features in a standard classifier such as a support vector machine (SVM). A simple application of this approach has been tested on both synthetically generated brain networks and MEG and fMRI data with different types of connectivity graphs. It has been shown that graph-based dimensionality reduction with a Granger causality connectivity graph gave the best overall performance compared to correlation, coherence, and phase locking value-based graphs. Moreover, the proposed dimensionality reduction performs better than conventional approaches including PCA, LDA, and LE [19].

More recently, this approach has been extended for dimensionality reduction of sample covariance matrices [20] for decoding motor imagery (MI) in BMI applications using a common spatial pattern (CSP). The CSP is defined as a vector that maximizes the ratio between the variances of two MI tasks and can be found as the solution of a generalized eigenvalue problem for a pair of covariance matrices. In practice, these covariance matrices are estimated as empirical average of sample covariance matrices (SCMs). The limitation of this method is that they reduce SCMs to single eigenvectors and are applicable only to two-class classification problems. To address these limitations, SCMs have been modeled as points on a Riemannian manifold and each point can then be mapped to a tangent space. The computation of tangent mapping methods requires accurate estimation of SCMs. GFT-based dimensionality reduction methods can be used to obtain low-dimensional accurate estimates of SCMs. More specifically, if $\mathbf{X} \in \mathbb{R}^{N\times T}$ is an N-channel, T-time point data, the proposed approach estimates the SCMs not from $\mathbf{X}$ but from the dimensionality-reduced version. The dimensionality reduction of the column space can be obtained as $\mathbf{Y} = \mathbf{U}_r^T\mathbf{X}$, where $\mathbf{U}_r$ is a projection matrix obtained from the first r eigenvectors of the graph Laplacian.

Hu et al. [21] generalized these projection based dimensionality reduction methods by offering a more formal signal detection and classification theory for neural graph signals defined on brain graphs. The matched signal detection (MSD) theory was extended to graph signals where the subspace for the graph signal is formed by the eigenvectors of the graph Laplacian matrix, $\mathbf{L}$. Three graph signal models are considered to derive the corresponding hypothesis tests. In the first case, the graph signals are assumed to be band-limited. Band-limited graph signals are defined as signals that are only supported on lower-frequency components corresponding to the first few smaller eigenvalues. This case is equivalent to the dimensionality reduction described above [19,20]. In the second model, prior information about the signal $\mathbf{x}$ is incorporated via a constraint on its GFT coefficients. The constraint can be put in a normalized quadratic form described as:

$$C(\tilde{x}) = \frac{\sum_{i=1}^{N} \alpha_i \tilde{x}_i^2}{\sum_{i=1}^{N} \tilde{x}_i^2}, \tag{31.14}$$

where α_is are nonnegative penalty weights. When $\alpha_i = \lambda_i$, the constraint function is equivalent to the normalized TV. This constraint function assures that the graph signals have bounded variations on the graph. To satisfy the constraint of smoothness on the graph, the signal should have a higher energy concentration on the lower-frequency components compared with the higher-frequency components. In the third model, a probabilistic graph signal model is considered. In this case, the distribution of the GFT coefficients is modeled as a degenerate Gaussian distribution. A test statistic is derived under each of these three models. For bounded variation signals described in Eq. (31.14), the test reduces to a weighted energy detector. For the random signal model, the test statistic is the difference of signal variations on associated graphs. The application of MSD to brain imaging data classification of Alzheimer's disease (AD) from two different imaging modalities revealed higher accuracy and earlier detection of AD compared to standard methods such as PCA, SVM, and LDA. MSD methods on graphs detect which orthonormal subspace derived from the AD or the healthy group best captures the properties of a signals.

31.4 GRAPH FILTERING OF BRAIN NETWORKS

Dimensionality reduction described in Section 31.3 is a special case of graph-based filtering (GBF), where the filtering focuses only on the low-frequency components. Graph filters offer a generalization of this concept as they permit the decomposition of a graph signal into components that represent different modes of variability and not just the low-dimensional components corresponding to the spatially smooth variations.

Given a graph signal $\mathbf{x}$ with GFT $\tilde{\mathbf{x}}$, one can isolate the different frequency components by defining the filtered spectrum. For example, the K_L low graph frequency components $\tilde{\mathbf{x}}_L = \tilde{\mathbf{H}}_L\tilde{\mathbf{x}}$ can be defined as $\tilde{x}_{Lk} = \tilde{x}_k$ for $k < K_L$ and $\tilde{x}_{Lk} = 0$, otherwise. The filter $\tilde{\mathbf{H}}_L := diag(\tilde{\mathbf{h}}_L)$ can be written as a diagonal matrix where the diagonal elements, i.e., the vector $\tilde{\mathbf{h}}_L$, take the value 1 for frequencies smaller than K_L and are equal to zero otherwise. Utilizing the spectral decomposition of the network and the definition of GFT, filtering is equivalent to $\mathbf{x}_L = \mathbf{H}_L\mathbf{x}$ in the graph vertex domain, where the filter $\mathbf{H}_L := \mathbf{U}\tilde{\mathbf{H}}_L\mathbf{U}^{-1}$ can be written as $\mathbf{H}_L = \sum_{k=0}^{n-1} h_{Lk}\mathbf{L}^k$ in terms of Laplacian powers [17]. The coefficients h_{Lk} in this expansion are elements of the vector $\mathbf{h}_L = \mathbf{\Psi}^{-1}\tilde{\mathbf{h}}_L$ where $\mathbf{\Psi}$ is the Vandermonde matrix defined by the eigenvalues of $\mathbf{L}$. The coefficients h_{Lk} tend to be concentrated in small indexes k, and the expansion is dominated by small powers $\mathbf{L}^k$. In general, $\mathbf{L}^k\mathbf{x}$ describes interactions between k-hop neighbors.

Other types of graph filters can be defined in a similar manner. A graph band pass filter $\mathbf{H}_M$ and a graph high pass filter $\mathbf{H}_H$, whose graph frequency responses are defined as:

$$\tilde{h}_{Mk} = I[K_L \le k < K_L + K_M],$$
$$\tilde{h}_{Hk} = I[K_L + K_M \le k]. \tag{31.15}$$

Using the filtered signals, the original signal can be written as the sum $\mathbf{x} = \mathbf{x}_L + \mathbf{x}_M + \mathbf{x}_H$. This decomposition provides an analysis of spatial variability of brain activity across regions of the brain with respect to the underlying connectivity network.

In [22,23], this framework was applied to fMRI data collected across time where the subjects learned a simple motor task. The brain graph was constructed using the magnitude squared spectral coherence as the adjacency matrix, $\mathbf{W}$. The fMRI time series data was then decomposed into three frequency bands defined by this adjacency matrix. The magnitude of the decomposed signals, $\mathbf{x}_L, \mathbf{x}_M, \mathbf{x}_H$, for each brain region was averaged across all sample signals. It was shown that for $\mathbf{x}_L$, the magnitudes on adjacent brain regions tend to possess highly similar values, whereas for $\mathbf{x}_H$, neighboring signals exhibit highly dissimilar values. Moreover, the signals from the different frequency bands were correlated with the learning rate, which showed that most of the association between learning comes from the brain signals that either vary smoothly ($\mathbf{x}_L$) or rapidly ($\mathbf{x}_H$) with respect to the brain network.

The analysis of spatial variation discussed above can also be carried out at a more local scale. In particular, visual and motor modules are known to be associated with motor learning. In order to assess the fluctuations or spatial variation within a module, a modified TV measure can be defined. Given an eigenvector $\mathbf{u}_k$, its variation on any given module, i.e., a subset of vertices, V_s, can be defined as:

$$TV(\mathbf{u}_k) = \frac{\sum_{i,j \in V_s, i \neq j} W_{ij}(u_k(i) - u_k(j))^2}{\sum_{i,j, \in V_s, i \neq j} W_{ij}}. \tag{31.16}$$

This measure computes the difference for signals on this module for each unit of edge weight. This total variation can be averaged over the eigenvectors corresponding to different frequency bands, resulting in TV_s^L, TV_s^M, TV_s^H. Moreover, the variation of eigenvectors on edges across modules can be quantified in a similar manner:

$$TV_{s,t}(\mathbf{u}_k) = \frac{\sum_{i \in V_s, j \in V_t} W_{ij}(u_k(i) - u_k(j))^2}{\sum_{i \in V_s, j \in V_t} W_{ij}}. \tag{31.17}$$

This analysis revealed that as learning takes place, the motor module becomes more strongly connected and the visual module weakens. Moreover, an analysis of the temporal variation of the decomposed signals revealed that brain activities with smooth spatial variations, $\mathbf{x}_L$, exhibit the most rapid temporal variation. As the temporal variation is associated with better performance in tasks, this indicates a stronger contribution of low graph frequency components during the learning process.

Along similar lines, Medaglia et al. [24] examined the relationship between functional alignment with anatomical networks and its relationship to cognitive flexibility. In this setting, the underlying network was constructed from the white matter graph, and the graph signals are defined from the average fMRI BOLD signals over different regions of interest (ROIs). Here, it is of interest to assess the portions of the graph signals that are aligned to the anatomical network. Based on graph filtering, it was found that alignment relevant to cognitive flexibility was more present in the anterior cingulate cortex and regions related to mechanisms of cognitive flexibility. Moreover, it was shown that these results cannot be observed from unimodal studies alone, demonstrating the advantage of multimodal analyses based on GSP.

31.5 GRAPH LEARNING

Understanding network features of brain pathology is essential for determining the underlying causes of neurodegeneration such as Alzheimer's Disease, which is not possible with analysis focused only on region-wise differences of neuroimaging data. Various statistical methods have been developed to infer functional connectivity networks from neuroimaging data. The most common technique for fMRI data is correlation analysis, which estimates the correlation matrix by computing the sample correlation of the data. Other methods include the coherence and phase synchrony metrics presented in Section 31.2. However, these methods cannot distinguish between indirect and direct connections. Partial correlation, i.e., the precision matrix, addresses this problem, but is often ill-conditioned due to a limited amount of samples [25,26]. One common way of overcoming this problem is graphical Lasso [27], which estimates the partial correlation with l_1-regularization because the sample correlation matrix is often singular due to the small sample size. However, all these methods require a lot of data for reliable estimation of the functional connectivity networks.

Recently, two novel GSP-based frameworks for learning brain connectivity from neuroimaging data have been proposed. The first one, referred to as a graph regression model (GRM), makes the assumption that the observed data are smooth signals on a deterministic graph [28]. This reduces the required number of samples to estimate the graph because the GRM does not depend on a specific probability distribution of the data. As a result, GRM produces more balanced network structures, in contrast to those from sample correlation matrices.

The smoothness, or signal variation, can be quantified through metrics such as the total variation, or its variations such as the metric $M_G(\mathbf{x}) = \frac{\mathbf{x}^T\mathbf{L}^s\mathbf{x}}{\mathbf{x}^T\mathbf{x}}$, where $s > 0$ is an adjustable parameter usually chosen in the range of 1 to 3. s controls the impact of the different graph frequencies. In a smooth signal regularization, if s is increased, there will be a greater penalty on higher frequency components of $\mathbf{x}$. Given N brain regions of interest with M samples or subjects and $\mathbf{X} \in \mathbb{R}^{N\times M}$ as the data observed on these regions, GRM can be formulated by minimizing the total variation across all samples, i.e., $\sum_{m=1}^{M} M_G(\mathbf{x}_m)$ which is equivalent to $tr(\mathbf{X}^T\mathbf{L}^s\mathbf{X})$. Minimizing the total variation of the observed signals with respect to the unknown graph Laplacian and adding a regularization term on the graph Laplacian yields:

$$\begin{aligned} &\min_{\mathbf{L}} tr(\mathbf{X}^T\mathbf{L}^s\mathbf{X}) - \beta\|\mathbf{L}\|_F^2 \\ &\text{s.t.} \quad tr(\mathbf{L}) = N, \qquad \mathbf{L1} = 0. \end{aligned} \tag{31.18}$$

The first term in the objective function finds the graph Laplacian that would yield the smallest variation across all samples while the second term controls the uniformity of the graph Laplacian. The constraints normalize the sum of all the connection weights to prevent the solution of a null graph. Because the constraints are linear, this optimization problem can be solved using the projected gradient descent method. In contrast to the noisy sample correlation, the resulting Laplacian matrix extracts cleaner and potentially more meaningful information from the data, yielding the most significant connections within each lobe.

A similar line of work was introduced to learn the graph Laplacian from brain imaging data with an added constraint that the high dimensional neuroimaging data is low rank [29]. Thus, an optimization function that minimizes the total variation of the signal on the graph and the smoothness of the Laplacian with the added terms that the underlying data is low rank has been introduced as:

$$\begin{aligned} &\min_{\mathbf{L},\mathbf{S},\mathbf{M}} \|\mathbf{M}\|_* + \delta\|\mathbf{S}\|_1 + \gamma tr(\mathbf{M}^T\mathbf{L}\mathbf{M}) + \beta\|\mathbf{L}\|_F^2 \\ &\text{s.t.} \quad \mathbf{X} = \mathbf{M} + \mathbf{S}, \end{aligned} \tag{31.19}$$

where $\mathbf{M}$ and $\mathbf{S}$ correspond to the low rank and sparse parts of the observed neuroimaging data, respectively. The first two terms correspond to the cost function of robust PCA [30] for the observed data and the last two terms correspond to the graph smoothness regularization with the constraint that the underlying graph topology captures the data correlation. This approach is an extension of GRM such that it not only estimates the underlying connectivity structure but also learns the low-dimensional structure of the observed data, thus denoising the neuroimaging data. The addition of these terms requires the use of an alternating minimization approach. In the first step, the robust PCA problem is solved for a fixed Laplacian with an additional constraint on the underlying (unknown) low-rank data $\mathbf{M}$, where $tr(\mathbf{M}^T\mathbf{L}\mathbf{M})$ forces the low-rank representations of the time series to be similar for highly correlated sensors or brain regions. In the second step, the graph Laplacian is learned given $\mathbf{S}, \mathbf{M}$ which is similar to the GRM problem. This method is suitable for learning both the low-rank component and graph simultaneously for cases where the perturbations to the neuroimaging data are sparse in nature.

31.6 SPECTRAL GRAPH WAVELETS

Over recent decades, the wavelet transform has contributed significantly to the multiscale analysis of neuronal data. From single-unit recordings to noninvasive neuroimaging modalities such as EEG, MEG, and fMRI [31], the wavelet transform has been employed in classification [32], artifact suppression and denoising [33], synchronization and computation of functional connectivity [34], and signal compression [35]. More recently, its extension to graph-signals SGWT has been found to be useful in a wide variety of neuroscience applications, including source estimation in EEG and MEG, fMRI activation mapping and denoising, and connectivity analysis [36–39].

In EEG applications, the signals recorded by the sensors in the scalp are the result of electric potentials generated by dipole current sources and volume conduction. The goal of source estimation is to estimate the current sources from the electrode's measurements. Estimating the location of these sources from the recorded EEG signals is an ill-posed problem, as there are more sources than sensors. Specifically, the signals recorded by $\phi \in \mathbb{R}^{N_c \times 1}$ are the result of the linear superposition $\phi = \mathbf{K}\,\mathbf{j}$, where each entry j_i of $\mathbf{j} \in \mathbb{R}^{N_d \times 1}$ corresponds to a source dipole, and $\mathbf{K} \in \mathbb{R}^{N_c \times N_d}$ is the lead-field matrix. Approaches to estimate such sources include minimum norm estimates and beamforming techniques and may rely on prior information such as the total number of dipoles [40]. In recent work, Hammond et al. [36] incorporated knowledge of the anatomical connectivity of the brain from diffusion tensor imaging (DTI) in the regularization for the inverse problem of source estimation for EEG. Specifically, the authors constructed a wavelet frame on the cortical connectome graph based on the assumption that cortical sources are sparse in the cortical graph wavelet frame. The cortical connectome graph is formed by constructing a hybrid adjacency matrix consisting of a tractography-based connectome and a local connectivity graph from DTI. As cortical graph wavelets are defined on this cortical connectome graph, they are good at representing signals that are coherent across graph edges and localized in the graph. An alternative formulation of the inverse problem is presented where both the objective function and the penalty term are defined in the cortical graph wavelet domain. If $\mathbf{c} \in R^{(s+1)N_d}$ is the vector of the wavelet and scaling coefficients and $\mathbf{W} \in R^{(s+1)N_d \times N_d}$ is the matrix representation of the SGWT, then $\mathbf{j} = \mathbf{W}^T\mathbf{c}$ and the inverse problem becomes $\mathbf{c}^* = \text{argmin}_{\mathbf{c}} \|\phi - \mathbf{K}\mathbf{W}^T\mathbf{c}\|_2^2 + \tau\|\mathbf{c}\|_1$. The cortical sources can then be found as $\mathbf{j}^* = \mathbf{W}^T\mathbf{c}^*$. Using data from a motor potential study where subjects were required to press a button as response to a visual task, the authors showed that the proposed method was effective at estimating the sources in the motor cortex and the resulting sources were less diffused than those obtained from the minimum norm solution.

Leonardi and Van De Ville [39] extended the formulation of SGWT to tight (Parseval) graph frames because of their important property of energy conservation. Parseval graph wavelet frames were generated by designing the wavelet and scaling generating kernels, $g(\cdot)$ and $h(\cdot)$, in the spectral graph domain and then designing Meyer-like wavelets with the spatial scales defined as $s_j = \frac{a}{\lambda_{\max}} 2^j, j = 1, \ldots, J$. These tight frame graph wavelets were applied to the problem of fMRI activation mapping. Statistical parametric mapping (SPM), widely used in the estimation of stimulus-related activity, fits a general linear model (GLM) on each voxel using regressors defined by the experimental paradigm. Recently, transform-based SPM, including wavelet SPM (WSPM), has been introduced to take advantage of the spatial localization of brain activity, which enables the sparse encoding of a cluster of active voxels, thus providing a higher sensitivity than voxel-wise testing. Behjat et al. [38] extended this framework by constructing SGWTs over structural connectivity graphs, i.e., a gray matter adapted graph, consisting of the cerebellar and the cerebral graphs and applying these to functional data for activation mapping. In this manner, the resulting wavelets are adaptive to the convolutive structure of the gray matter and further avoid the assumptions made by WSPM, which are defined in regular Euclidean spaces and are stationary and quasi shift-invariant. Moreover, tight-frame SGWT prevents linear irreversible smoothing and allows for assessment at multiple scales. Results show an increase in type-1 error control, higher detection sensitivity, and improved spatial localization when compared to the traditional SPM and WSPM.

In [39], the tight frame graph wavelets were adapted to multilayer networks such as dynamic functional connectivity networks (dFCNs) constructed from fMRI data. In this two-step approach, first "eigennetworks" of a multislice graph, where each slice corresponds to a weighted adjacency matrix $\mathbf{W}$ for a different time window (T time points), are extracted through Higher Order Singular Value Decomposition (HoSVD) of the connectivity tensor, $\mathcal{W} \in \mathbb{R}^{N \times N \times T}$.

$$\mathcal{W} = \mathcal{C} \times_1 \mathbf{B} \times_2 \mathbf{B} \times_3 \mathbf{F}, \tag{31.20}$$

where $\mathcal{C} \in \mathbb{R}^{N \times N \times T}$ is the core tensor, and $\mathbf{B} \in \mathbb{R}^{N \times N}, \mathbf{F} \in \mathbb{R}^{T \times T}$ are the factor matrices along the modes. Through this tensor decomposition, the authors construct eigennetworks that capture the variability of the network across time. In particular, the kth eigennetwork is defined as $\mathcal{C}'_{::k} = \sum_{t=1}^{T} f_{kt} \mathcal{W}_{::t}$, which is equivalent to a weighted average of the different adjacency matrices. Treating these eigennetworks as the building blocks of network connectivity, different linear combinations of $\mathbf{C'}$ are generated to define new adjacency matrices, $\mathbf{W'}$. From these new adjacency matrices, Laplacian matrices and the corresponding tight frame spectral graph wavelets are constructed. The results suggest that eigennetworks obtained from the HoSVD of a dFCN capture relevant variability across graph edges, such that each eigennetwork may correspond to different well-known resting-state networks such as the default mode network (DMN). Thus, the resulting adjacency matrices and the fMRI data decomposed with the SGWTs built from them emphasize different types of brain activity and different experimental conditions.

In [37], SGWTs were used to offer a multiresolution analysis of structural networks rather than functional ones. Unlike other applications of GSP with graph signals corresponding to functional activity over the vertices of a brain network, in this case the application domain is solely the structural connectivity network such as DTI. In order to apply SGWT to structural connectivity networks, the notion of line graphs is used. The structural connectivity networks are first transformed to line graphs such that the edges of the original network become the signal on the graph. In this manner, the edge values in the structural connectivity network become the new domain of analysis. Using wavelets, multiscale filtering of edges can be efficiently performed by removing high-frequency components tied to finer scales.

31.6.1 DYNAMIC CONNECTIVITY ANALYSIS

Most applications of GSP framework to neuroimaging data make the assumption that the functional or structural networks are stable for a window of time as the networks are constructed through long-term correlation between the neurophysiological time series. Brain activity can vary more frequently, forming multiple samples of brain signals defined on a common underlying network. However, recent research has shown that functional connectivity networks also change dynamically in short time scales and exhibit task-related patterns [41–45]. This continuous formation and destruction of functional connectivity also controls the emergence of a unified neural process [45].

The current literature on dynamic functional connectivity networks (dFCNs) relies on the computation of bivariate connectivity over sliding windows, which causes the connectivity measures to depend on the window length. As a result, the dFCNs constructed through this method do not have the same high temporal resolution of the original data, as in the case of EEG. In a recent study, Smith et al. [46] introduced GSP as an alternative framework for the assessment of connectivity dynamics for EEG signals recorded during visual short-term memory (VSTM) tasks. Specifically, the authors construct a weighted graph using connectivity analysis between signals using metrics defined in Section 31.2 and take the graph signal as the EEG epoch at each electrode. Analysis at two different time scales, short-term and long-term, and at two different spatial scales, local and global, is then proposed through Dirichlet energy (DE) and graph modular Dirichlet energy (MDE). The Dirichlet energy of a graph $\mathcal{G}$ is defined as

$$E(\mathcal{G}) = \sum_{i,j=1}^{N} W_{ij}(x_i - x_j)^2, \tag{31.21}$$

where W_{ij} is the weight between the ith and the jth electrodes, and x_i and x_j are the amplitudes of the signals at a given time on each electrode. $E(\mathcal{G})$ quantifies the agreement between the correlation between EEG signals i and j and the difference between the graph signals x_i and x_j at time t. It's important to note that the Dirichlet energy is equivalent to the total variation defined in Eq. (31.8). Four different cases can be considered when the signals are highly correlated. First, if the weights W_{ij} are positive and the difference $x_i - x_j$ is large, it means that there is a discrepancy between the connectivity and the amplitude of the signals at time t, and the resulting $E(\mathcal{G})$ is large and positive. On the other hand, if the difference $x_i - x_j$ is small it implies that both the weights and the signals' amplitude are in agreement and $E(\mathcal{G})$ is small. In the third case, if the weights W_{ij} are negative and the difference $x_i - x_j$ is large, it implies that there is a strong agreement between the weights and the signal amplitudes, and $E(\mathcal{G})$ is large and negative. Lastly, if the weights are negative and the difference $x_i - x_j$ is small, it implies that both the weights and the signals' amplitude are in disagreement and $E(\mathcal{G})$ is negative and small.

To provide quantification of the spatial smoothness at the local level, graph Modular Dirichlet Energy (MDE) is defined over a module $\mathcal{G}_x$ as:

$$MDE(\mathcal{G}_x) = \sum_{i\in\mathcal{V}_x}\sum_{j\in\mathcal{V}} W_{ij}\left(x_i - x_j\right)^2, \tag{31.22}$$

where $\mathcal{V}_x$ consists of the vertices in the module $\mathcal{G}_x$.

DE and MDE can be used to analyze the temporal dynamics of brain activity as the DE of the signal during any time period is the sum of the individual DE at each point in time. This way, one can look at both short and long intervals of time to probe the connectivity information for dynamic behavior within the epoch from which the graphs are constructed. DE and MDE analysis offer a different perspective to existing dFCN methods as a single network of general connectivity patterns over a long epoch is constructed rather than a collection of time-varying networks. Thus, the dynamic activity is encoded in the graph signal rather than in the edge weights of a time-varying graph.

In the application of this framework to EEG data from a visual short-term memory binding task, two time windows corresponding to the encoding and maintenance periods are considered. Similarly, two modules corresponding to the frontal and occipital regions of the brain are identified. Through MDE, it was possible to discern driving effects in the occipital module coinciding with the P100 visually evoked potential within the 100–140 ms interval, and a driving effect in the frontal region in the 140–180 ms interval.

31.7 AN ILLUSTRATION: ERP CLASSIFICATION ON EEG FUNCTIONAL CONNECTIVITY NETWORKS

In this section, we illustrate the application of the GSP principles presented in this chapter on the classification of event-related potentials (ERPs). The EEG data is collected from a cognitive control experiment based on a Flanker speeded reaction task [47]. During the experiment, subjects are required to identify a target letter in a five-letter string, which can be congruent (MMMMM) or incongruent (MMNMM). An event-related potential (ERP) of interest in this experiment is the error-related negativity (ERN), which is observed as a large negative amplitude occurring 0–100 ms after the subjects commit errors. The ERN potential is illustrated in Fig. 31.2A, as well as the EEG time series from correct responses.

Previous studies assessing functional connectivity during cognitive control have shown increased synchrony in the theta band (4–8 Hz) in the frontal and lateral regions from error responses compared to correct responses [47]. Fig. 31.2B shows the functional connectivity networks constructed using PLV [48], averaged over the time interval 25–75 ms and over subjects. Fig. 31.2C shows the topographic distribution of the graph signals across different frequency bands averaged over the 25–75 ms time interval. The graph signals at each time point in the 25–75 ms time window are filtered into the low- middle- and high-frequency bands determined by the eigenvalues of the graph Laplacian and averaged over time [22]. Fig. 31.2D shows topoplots of the filtered signals. It can be observed that the low-frequency component of the graph signals from error responses have a great contribution to the large negative potential in the frontal-central regions, and to the large positive amplitudes in the frontal-lateral and medial regions. On the other hand, in the case of correct responses, there is also a strong contribution from the low-frequency signals, although in a less structured manner. The middle- and high-frequency range also show some of the large negative potential corresponding to the ERN in the frontal-central regions for the error response, indicating the persistence of this activity across different levels of spatial variation.

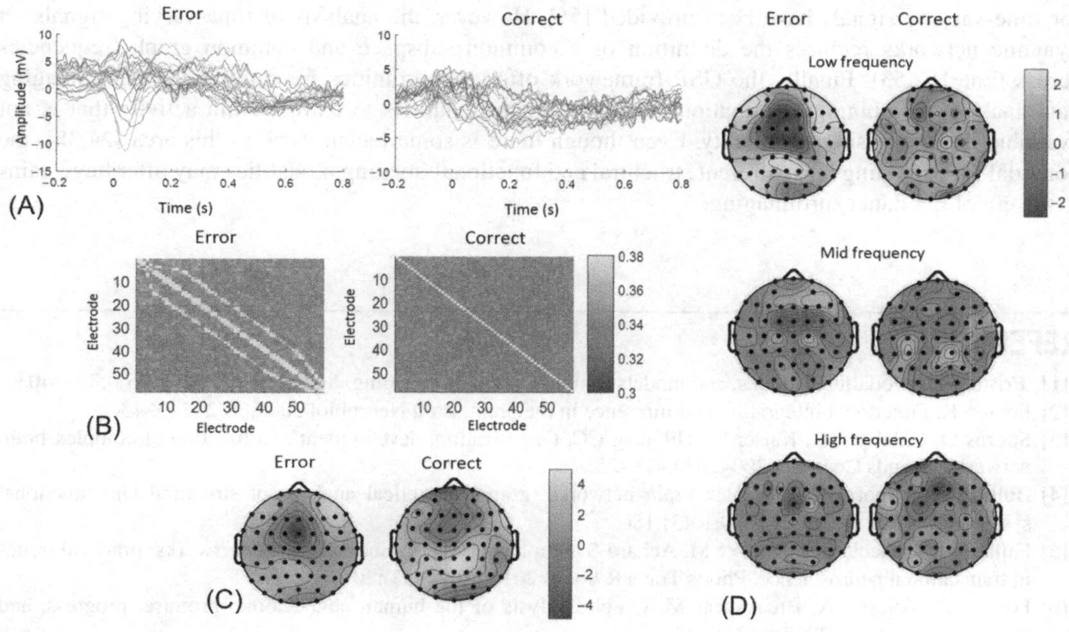

FIG. 31.2

An illustration of the construction of brain networks and the graph signals defined on these networks. (A) EEG signals from error and correct responses. The error-related negativity potential is observed as a large negative peak between 0 and 100 ms in the error signals. (B) Functional connectivity networks from PLV averaged over the 25–75 ms time interval for error and correct responses. (C) Topoplots of the EEG graph signals averaged over the 25–75 ms time interval for error and correct responses. (D) Topoplots of filtered graph signals in the low-, middle-, and high-frequency bands, from error and correct responses averaged over the 25–75 ms time interval.

31.8 CONCLUSIONS AND FUTURE DIRECTIONS

As illustrated in this chapter, the current state-of-the-art methods in GSP have made it possible to extract rich brain activity from multivariate neuroimaging data. However, there are still some remaining challenges and promising directions in the application of GSP methods to brain connectivity networks. First, all the previously discussed methods assume that the connectivity network is symmetric. It is well known that FC is directional and a variety of measures, such as Granger causality, have been defined to quantify this directional nature of the interactions [49]. This requires the development of graph spectral transforms for directed adjacency matrices. Recent developments in this area [50] can provide the tools to extract brain activity across different levels of spatial variation simultaneously with the direction of information flow between regions. Another important area of growing interest is the study of dFCNs, as discussed in Section 31.7. Current methods of dFCN analysis using GSP assume a stationary network with the graph signal changing with time. However, in reality the connectivity is quasistationary and these methods are limited to small epochs. Recently, new definitions of GFT for time-varying signals have been provided [51]. However, the analysis of time-varying signals on dynamic networks requires the definition of a common subspace and common graph frequencies across time [52,53]. Finally, the GSP framework offers opportunities for multimodal neuroimaging data analysis, combining information across different modalities to extract brain activity that is not available through a single modality. Even though there is some recent work in this area [24,38], the potential for extending it to different structural and functional imaging modalities may offer huge gains in the era of big data neuroimaging.

REFERENCES

[1] Friston KJ. Modalities, modes, and models in functional neuroimaging. Science 2009;326(5951):399–403.
[2] Friston K. Functional integration and inference in the brain. Prog Neurobiol 2002;68(2):113–43.
[3] Sporns O, Chialvo DR, Kaiser M, Hilgetag CC. Organization, development and function of complex brain networks. Trends Cogn Sci 2004;8(9):418–25.
[4] Bullmore E, Sporns O. Complex brain networks: graph theoretical analysis of structural and functional systems. Nat Rev Neurosci 2009;10(3):186.
[5] Fallani FDV, Richiardi J, Chavez M, Achard S. Graph analysis of functional brain networks: practical issues in translational neuroscience. Philos Trans R Soc B 2014;369(1653):20130521.
[6] Fornito A, Zalesky A, Breakspear M. Graph analysis of the human connectome: promise, progress, and pitfalls. Neuroimage 2013;80:426–44.
[7] Zalesky A, Fornito A, Bullmore ET. Network-based statistic: identifying differences in brain networks. Neuroimage 2010;53(4):1197–207.
[8] Meunier D, Lambiotte R, Fornito A, Ersche KD, Bullmore ET. Hierarchical modularity in human brain functional networks. Front Neuroinform 2009;3:37.
[9] Bassett DS, Bullmore E. Small-world brain networks. Neuroscientist 2006;12(6):512–23.
[10] Achard S, Salvador R, Whitcher B, Suckling J, Bullmore E. A resilient, low-frequency, small-world human brain functional network with highly connected association cortical hubs. J Neurosci 2006;26(1):63–72.
[11] Braun U, Muldoon SF, Bassett DS. On human brain networks in health and disease. eLS 2015:1–9.
[12] Jafri MJ, Pearlson GD, Stevens M, Calhoun VD. A method for functional network connectivity among spatially independent resting-state components in schizophrenia. Neuroimage 2008;39(4):1666–81.
[13] Stam CJ, Reijneveld JC. Graph theoretical analysis of complex networks in the brain. Nonlinear Biomed Phys 2007;1(1):3.
[14] Fingelkurts AA, Fingelkurts AA, Kähkönen S. Functional connectivity in the brain—is it an elusive concept? Neurosci Biobehav Rev 2005;28(8):827–36.
[15] Ménoret M, Farrugia N, Pasdeloup B, Gripon V. Evaluating graph signal processing for neuroimaging through classification and dimensionality reduction; 2017. arXiv preprint arXiv:170301842.
[16] Lachaux JP, Rodriguez E, Martinerie J, Varela FJ, et al. Measuring phase synchrony in brain signals. Hum Brain Mapp 1999;8(4):194–208.
[17] Shuman DI, Narang SK, Frossard P, Ortega A, Vandergheynst P. The emerging field of signal processing on graphs: extending high-dimensional data analysis to networks and other irregular domains. IEEE Signal Process Mag 2013;30(3):83–98.
[18] Hammond DK, Vandergheynst P, Gribonval R. Wavelets on graphs via spectral graph theory. Appl Comput Harmon Anal 2011;30(2):129–50.
[19] Rui L, Nejati H, Cheung NM. Dimensionality reduction of brain imaging data using graph signal processing. In: 2016 IEEE international conference on image processing (ICIP). IEEE; 2016. p. 1329–33.

[20] Tanaka T, Uehara T, Tanaka Y. Dimensionality reduction of sample covariance matrices by graph Fourier transform for motor imagery brain-machine interface. In: 2016 IEEE statistical signal processing workshop (SSP). IEEE; 2016. p. 1–5.

[21] Hu C, Sepulcre J, Johnson KA, Fakhri GE, Lu YM, Li Q. Matched signal detection on graphs: theory and application to brain imaging data classification. NeuroImage 2016;125:587–600.

[22] Huang W, Goldsberry L, Wymbs NF, Grafton ST, Bassett DS, Ribeiro A. Graph frequency analysis of brain signals. IEEE J Sel Top Signal Process 2016;10(7):1189–203.

[23] Goldsberry L, Huang W, Wymbs NF, Grafton ST, Bassett DS, Ribeiro A. Brain signal analytics from graph signal processing perspective. In: 2017 IEEE international conference on acoustics, speech and signal processing (ICASSP). IEEE; 2017. p. 851–5.

[24] Medaglia JD, Huang W, Karuza EA, Thompson-Schill SL, Ribeiro A, Bassett DS. Functional alignment with anatomical networks is associated with cognitive flexibility; 2016. arXiv preprint arXiv:161108751.

[25] Varoquaux G, Gramfort A, Poline JB, Thirion B. Brain covariance selection: better individual functional connectivity models using population prior. In: Advances in neural information processing systems. 2010. p. 2334–42.

[26] Marrelec G, Krainik A, Duffau H, Pélégrini-Issac M, Lehéricy S, Doyon J, et al. Partial correlation for functional brain interactivity investigation in functional MRI. Neuroimage 2006;32(1):228–37.

[27] Friedman J, Hastie T, Tibshirani R. Sparse inverse covariance estimation with the graphical lasso. Biostatistics 2008;9(3):432–41.

[28] Hu C, Cheng L, Sepulcre J, Johnson KA, Fakhri GE, Lu YM, et al. A spectral graph regression model for learning brain connectivity of Alzheimer's disease. PLOS One 2015;10(5):e0128136.

[29] Rui L, Nejati H, Safavi SH, Cheung NM. Simultaneous low-rank component and graph estimation for high-dimensional graph signals: application to brain imaging. In: 2017 IEEE international conference on acoustics, speech and signal processing (ICASSP). IEEE; 2017. p. 4134–8.

[30] Candès EJ, Li X, Ma Y, Wright J. Robust principal component analysis? J ACM (JACM) 2011;58(3):11.

[31] Hramov AE, Koronovskii AA, Makarov VA, Pavlov AN, Sitnikova E. Wavelets in neuroscience. Springer; 2015.

[32] Nazimov AI, Pavlov AN, Hramov AE, Grubov VV, Koronovskii AA, Sitnikova E. Adaptive wavelet-based recognition of oscillatory patterns on electroencephalograms. In: Proceedings of SPIE, vol. 8580; 2013. p. 85801D-1.

[33] Ahmadi M, Quiroga RQ. Automatic denoising of single-trial evoked potentials. NeuroImage 2013;66:672–80.

[34] Lynall ME, Bassett DS, Kerwin R, McKenna PJ, Kitzbichler M, Muller U, et al. Functional connectivity and brain networks in schizophrenia. J Neurosci 2010;30(28):9477–87.

[35] Srinivasan K, Dauwels J, Reddy MR. Multichannel EEG compression: wavelet-based image and volumetric coding approach. IEEE J Biomed Health Inform 2013;17(1):113–20.

[36] Hammond DK, Scherrer B, Malony A. Incorporating anatomical connectivity into EEG source estimation via sparse approximation with cortical graph wavelets. In: 2012 IEEE international conference on acoustics, speech and signal processing (ICASSP). IEEE; 2012. p. 573–6.

[37] Kim WH, Adluru N, Chung MK, Charchut S, GadElkarim JJ, Altshuler L, et al. Multi-resolutional brain network filtering and analysis via wavelets on non-Euclidean space. In: International Conference on Medical Image Computing and Computer-Assisted Intervention. Springer; 2013. p. 643–51.

[38] Behjat H, Leonardi N, Sörnmo L, Van De Ville D. Anatomically-adapted graph wavelets for improved group-level FMRI activation mapping. NeuroImage 2015;123:185–99.

[39] Leonardi N, Van De Ville D. Tight wavelet frames on multislice graphs. IEEE Trans Signal Process 2013;61(13):3357–67.

[40] Grech R, Cassar T, Muscat J, Camilleri KP, Fabri SG, Zervakis M, et al. Review on solving the inverse problem in EEG source analysis. J Neuroeng Rehabil 2008;5(1):25.

[41] Prabhakaran R, Blumstein SE, Myers EB, Hutchison E, Britton B. An event-related FMRI investigation of phonological-lexical competition. Neuropsychologia 2006;44(12):2209–21.
[42] Goebel R, Esposito F, Formisano E. Analysis of functional image analysis contest (FIAC) data with brain-voyager QX: from single-subject to cortically aligned group general linear model analysis and self-organizing group independent component analysis. Hum Brain Mapp 2006;27(5):392–401.
[43] Boveroux P, Vanhaudenhuyse A, Bruno MA, Noirhomme Q, Lauwick S, Luxen A, et al. Breakdown of within-and between-network resting state functional magnetic resonance imaging connectivity during propofol-induced loss of consciousness. Anesthesiol: J Am Soc Anesthesiol 2010;113(5):1038–53.
[44] Schrouff J, Perlbarg V, Boly M, Marrelec G, Boveroux P, Vanhaudenhuyse A, et al. Brain functional integration decreases during propofol-induced loss of consciousness. Neuroimage 2011;57(1):198–205.
[45] Chang C, Glover GH. Time-frequency dynamics of resting-state brain connectivity measured with FMRI. Neuroimage 2010;50(1):81–98.
[46] Smith K, Ricaud B, Shahid N, Rhodes S, Starr JM, Ibá nez A, et al. Locating temporal functional dynamics of visual short-term memory binding using graph modular Dirichlet energy. Sci Rep 2017;7:42013.
[47] Moran TP, Bernat EM, Aviyente S, Schroder HS, Moser JS. Sending mixed signals: worry is associated with enhanced initial error processing but reduced call for subsequent cognitive control. Soc Cogn Affect Neurosci 2015;10(11):1548–56.
[48] Aviyente S, Bernat EM, Evans WS, Sponheim SR. A phase synchrony measure for quantifying dynamic functional integration in the brain. Hum Brain Mapp 2011;32(1):80–93.
[49] Babiloni F, Cincotti F, Babiloni C, Carducci F, Mattia D, Astolfi L, et al. Estimation of the cortical functional connectivity with the multimodal integration of high-resolution EEG and FMRI data by directed transfer function. Neuroimage 2005;24(1):118–31.
[50] Sardellitti S, Barbarossa S, Di Lorenzo P. On the graph Fourier transform for directed graphs. IEEE J Sel Top Signal Process 2017;11(6):796–811.
[51] Loukas A, Foucard D. Frequency analysis of temporal graph signals; 2016. arXiv preprint arXiv:160204434.
[52] Mahyari AG, Aviyente S. Fourier transform for signals on dynamic graphs. In: 2014 48th Asilomar conference on signals, systems and computers. IEEE; 2014. p. 2001–4.
[53] Villafa ne-Delgado M, Aviyente S. Dynamic graph Fourier transform on temporal functional connectivity networks. In: 2017 IEEE international conference on acoustics, speech and signal processing (ICASSP). IEEE; 2017. p. 949–53.

Index

Note: Page numbers followed by *f* indicate figures, *t* indicate tables, and *b* indicate boxes.

B

D

G

H

I

T

U

V

W

Y

Z

Printed in the United States
By Bookmasters

Printed in the United States
By Bookmasters